TABLE C.1 Cumulative probabilities and percentiles of the standard normal distribution

(a) Cumulative probabilities

Entry is area a under the standard normal curve from $-\infty$ to $z(a)$.

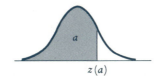

z	.00	.01	.02	.03	.04	.05	.06	.07	.08	.09
.0	.5000	.5040	.5080	.5120	.5160	.5199	.5239	.5279	.5319	.5359
.1	.5398	.5438	.5478	.5517	.5557	.5596	.5636	.5675	.5714	.5753
.2	.5793	.5832	.5871	.5910	.5948	.5987	.6026	.6064	.6103	.6141
.3	.6179	.6217	.6255	.6293	.6331	.6368	.6406	.6443	.6480	.6517
.4	.6554	.6591	.6628	.6664	.6700	.6736	.6772	.6808	.6844	.6879
.5	.6915	.6950	.6985	.7019	.7054	.7088	.7123	.7157	.7190	.7224
.6	.7257	.7291	.7324	.7357	.7389	.7422	.7454	.7486	.7517	.7549
.7	.7580	.7611	.7642	.7673	.7704	.7734	.7764	.7794	.7823	.7852
.8	.7881	.7910	.7939	.7967	.7995	.8023	.8051	.8078	.8106	.8133
.9	.8159	.8186	.8212	.8238	.8264	.8289	.8315	.8340	.8365	.8389
1.0	.8413	.8438	.8461	.8485	.8508	.8531	.8554	.8577	.8599	.8621
1.1	.8643	.8665	.8686	.8708	.8729	.8749	.8770	.8790	.8810	.8830
1.2	.8849	.8869	.8888	.8907	.8925	.8944	.8962	.8980	.8997	.9015
1.3	.9032	.9049	.9066	.9082	.9099	.9115	.9131	.9147	.9162	.9177
1.4	.9192	.9207	.9222	.9236	.9251	.9265	.9279	.9292	.9306	.9319
1.5	.9332	.9345	.9357	.9370	.9382	.9394	.9406	.9418	.9429	.9441
1.6	.9452	.9463	.9474	.9484	.9495	.9505	.9515	.9525	.9535	.9545
1.7	.9554	.9564	.9573	.9582	.9591	.9599	.9608	.9616	.9625	.9633
1.8	.9641	.9649	.9656	.9664	.9671	.9678	.9686	.9693	.9699	.9706
1.9	.9713	.9719	.9726	.9732	.9738	.9744	.9750	.9756	.9761	.9767
2.0	.9772	.9778	.9783	.9788	.9793	.9798	.9803	.9808	.9812	.9817
2.1	.9821	.9826	.9830	.9834	.9838	.9842	.9846	.9850	.9854	.9857
2.2	.9861	.9864	.9868	.9871	.9875	.9878	.9881	.9884	.9887	.9890
2.3	.9893	.9896	.9898	.9901	.9904	.9906	.9909	.9911	.9913	.9916
2.4	.9918	.9920	.9922	.9925	.9927	.9929	.9931	.9932	.9934	.9936

TABLE C.1 Cumulative probabilities and percentiles of the standard normal distribution

(a) Cumulative probabilities

Entry is area a under the standard normal curve from $-\infty$ to $z(a)$.

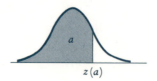

$z(a)$

z	.00	.01	.02	.03	.04	.05	.06	.07	.08	.09
2.5	.9938	.9940	.9941	.9943	.9945	.9946	.9948	.9949	.9951	.9952
2.6	.9953	.9955	.9956	.9957	.9959	.9960	.9961	.9962	.9963	.9964
2.7	.9965	.9966	.9967	.9968	.9969	.9970	.9971	.9972	.9973	.9974
2.8	.9974	.9975	.9976	.9977	.9977	.9978	.9979	.9979	.9980	.9981
2.9	.9981	.9982	.9982	.9983	.9984	.9984	.9985	.9985	.9986	.9986
3.0	.9987	.9987	.9987	.9988	.9988	.9989	.9989	.9989	.9990	.9990
3.1	.9990	.9991	.9991	.9991	.9992	.9992	.9992	.9992	.9993	.9993
3.2	.9993	.9993	.9994	.9994	.9994	.9994	.9994	.9995	.9995	.9995
3.3	.9995	.9995	.9995	.9996	.9996	.9996	.9996	.9996	.9996	.9997
3.4	.9997	.9997	.9997	.9997	.9997	.9997	.9997	.9997	.9997	.9998

(b) Selected percentiles

Entry is $z(a)$ where $P[Z \le z(a)] = a$.

a:	.10	.05	.025	.02	.01	.005	.001
z(a):	−1.282	−1.645	−1.960	−2.054	−2.326	−2.576	−3.090
a:	.90	.95	.975	.98	.99	.995	.999
z(a):	1.282	1.645	1.960	2.054	2.326	2.576	3.090

EXAMPLE: $P(Z \le 1.96) = 0.9750$ so $z(0.9750) = 1.96$.
TEXT REFERENCE: Use of this table is discussed on pp. 215–220.

FOURTH EDITION

Applied Statistics

JOHN NETER   University of Georgia
WILLIAM WASSERMAN   Syracuse University
G. A. WHITMORE   McGill University

ALLYN AND BACON

Boston   London   Sydney   Toronto

Editor-in-chief, Business: Rich Wohl
Editorial assistant: Dominique Vachon
Editorial-production service: Technical Texts, Inc.
Text designer: Sylvia Dovner
Cover designer: Susan Paradise
Cover administrator: Linda Dickinson
Composition and manufacturing buyer: Louise Richardson

Library of Congress Cataloging-in-Publication Data
Neter, John.
    Applied statistics / John Neter, William Wasserman, G.A. Whitmore.
  — 4th ed.
      p.   cm.
    Includes bibliographical references and index.
    ISBN 0-205-13478-5
    1. Social sciences—Statistical methods. 2. Management—
Statistical methods. 3. Economics—Statistical methods.
4. Statistics.   I. Wasserman, William.   II. Whitmore, G. A.
III. Title.
HA29.N437   1992   92-17067
519.5—dc20   CIP

ISBN 0-205-13478-5

Printed in the United States of America
10 9 8 7 6 5 4 3 2 1   97 96 95 94 93 92

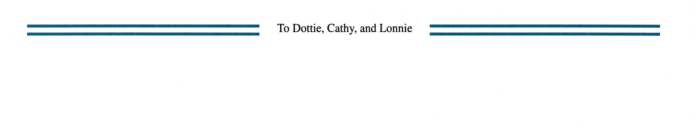

To Dottie, Cathy, and Lonnie

QUANTITATIVE METHODS AND APPLIED STATISTICS SERIES

ALLYN AND BACON

Barry Render, Consulting Editor

Roy E. Crummer Graduate School of Business, Rollins College

Readings in Production and Operations Management: A Productivity Approach
Ahmadian, Afifi, and Chandler

Introduction to Operations Research Techniques, Second Edition
Daellenbach, George, and McNickle

Managing Production: The Adventure
Fulmer

POMS: Production and Operations Management Software
Gupta, Zanakis, and Mandakovic

Business Forecasting, Fourth Edition
Hanke and Reitsch

Production and Operations Management: Strategies and Tactics, Second Edition
Heizer and Render

Statistical Analysis for Decision Making
Jarrett and Kraft

Management Science, Third Edition
Lee, Moore, and Taylor

Micro Management Science: Microcomputer Applications of Management Science, Second Edition
Lee and Shim

Micro Manager 2.0 Software—IBM
Lee and Shim

Service Operations Management
Murdick, Render, and Russell

Applied Statistics, Fourth Edition
Neter, Wasserman, and Whitmore

Microcomputer Software for Management Science and Operations Management, Second Edition
Render, Stair, Attaran, and Foeller

Introduction to Management Science
Render and Stair

Quantitative Analysis for Management, Fourth Edition
Render and Stair

Cases and Readings in Management Science, Second Edition
Render, Stair, and Greenberg

Topics in Just-In-Time Management
Schniederjans

Statistical Reasoning, Third Edition
Smith

Production and Operations Management: A Self-Correcting Approach, Third Edition
Stair and Render

Introduction to Management Science, Third Edition
Taylor

Brief Business Statistics
Watson, Billingsley, Croft, and Huntsberger

Statistics for Management and Economics, Fourth Edition
Watson, Billingsley, Huntsberger, and Croft

PC–POM: Software for Production and Operations Management
Weiss

Production and Operations Management, Second Edition
Weiss and Gershon

The Application of Regression Analysis
Wittink

Contents

UNIT FOUR

Estimation and Testing II   433

15 NONPARAMETRIC PROCEDURES   435

16 GOODNESS OF FIT   477

17 MULTINOMIAL POPULATIONS   503

UNIT FIVE

Linear Statistical Models   529

UNIT SEVEN

Time Series Analysis and Index Numbers   751

UNIT EIGHT

Bayesian Decision Making   851

28 BAYESIAN DECISION MAKING WITH SAMPLE INFORMATION   883

Preface

To Our Readers:

Applied Statistics, Fourth Edition, is written for students in basic statistics courses in business, economics, and other social sciences, as well as for people already engaged in these fields who desire an introduction to statistical methods and their application. Our aim is to offer you a balanced presentation of fundamental statistical concepts and methods, along with practical advice on their effective application to real-world problems. The conceptual foundation of each subject is developed carefully up to that level needed for prudent and beneficial use of statistical methods in practice.

Features of Applied Statistics

Clear Explanations.   The book is written so that the explanations of important principles and concepts are clear and always contain illustrations by one or more examples drawn from real life. In addition, many case applications are presented so that you can understand statistical concepts and methods in the context of actual use.

Broad Coverage.   The topical coverage of the text is broad, but it is in no sense a miscellany of statistical tools. Topics have been selected on the basis of two criteria: (1) their importance in actual applications of statistics, and (2) their contribution to the development and understanding of material presented subsequently in the book.

Clear Structure.   Units, chapters, and sections have been organized and sequenced in a way that always keeps the main track of the subject clear to you. Technical notes and secondary observations are presented in *Comments* sections. Topics that are not essential to the main development of statistical ideas are presented in *Optional Topic* sections at the ends of chapters and can be omitted without loss of continuity. Important definitions and formulas are set out in a distinctive manner to aid you in learning and to facilitate ready reference.

Wide Use of Computer Output and Plots.   Statistical packages play a major supporting role in applications of statistics and will be encountered by everyone, not just by statistical specialists. We have therefore integrated statistical computer output and plots into the text presentation in every subject area in a routine and natural manner, employing them just as they would be used in practice.

No Advanced Mathematics.   We believe that fundamental statistical ideas can be conveyed with only a modest use of mathematics and that this approach in an introductory text leads to a fuller appreciation of statistics than a more mathematical approach. Use of this book therefore requires only knowledge of college-entrance algebra but not calculus. Mathematical demonstrations are included where they make a significant contribution to your understanding of the subject, but all such demonstrations are segregated so that they do not interfere with the main presentation. Occasionally, a mathematical demonstration that requires calculus is presented in an optional section. These sections are marked "calculus needed" and may be omitted without loss of continuity.

Numerous Problems, Exercises, and Studies.   Large numbers of questions and problems are given at the ends of chapters to assist your understanding of concepts and to enable you to obtain experience in applying statistical techniques in practical situations and in interpreting results of statistical investigations. The *Problems* sections contain basic problems and drill questions. The *Exercises* sections present questions dealing with more technical concepts, as well as extensions of ideas developed in the chapters. Finally the *Studies* sections contain major comprehensive problems and case studies. Numerical answers for selected problems (identified by an asterisk in the margin) are given at the end of the book to facilitate immediate checking by you.

A computer will facilitate the working out of many of the problems, exercises, and studies. Those problems for which use of a computer is essential are marked by the logo of a computer. Most of the larger data sets in problems, tables, and figures are contained in files on a computer disk that is available with the text. These data sets are identified by the logo of a floppy disk when first introduced in problems, tables, and figures.

Student Supplements

The following student supplements, specifically keyed to the Fourth Edition of *Applied Statistics,* are available to you.

1. *Study Guide* (prepared by William Sanders, Clarion University). The *Study Guide* is a self-learning study manual to help you in understanding the concepts and techniques presented in the text. It incorporates principles of programmed learning so that correct understanding is immediately reinforced as you work through self-contained units at your own speed. The *Study Guide* emphasizes understanding and calculations are kept to a minimum. Full verbal answers and step-by-step numerical answers are given to enable you to better understand concepts and techniques and to identify causes of any mistakes. The *Study Guide* also contains detailed answers for the starred problems in the the text.

2. *Software Workbooks for Statistical Packages.* Software workbooks have been prepared for the Minitab and MYSTAT statistical software systems for use with this text. Each manual introduces you to the software and assists you in using the system for statistical computations and analysis with this text. The two manuals are:

> *Minitab for **Applied Statistics*** (prepared by Warren Dixon, State University of New York at Plattsburg)

> *Business MYSTAT with QCTOOLS for **Applied Statistics*** (prepared by Kaye A. de Ruiz, Air Force Academy)

Appendix E of this textbook contains a brief technical manual for Business MYSTAT and QCTOOLS that is a handy reference for these systems.

3. *Data Disk.* Each software workbook contains a data disk for your use. Files of the larger data sets in the tables, figures, and problems, as well as the four large data sets in Appendix D, have been assembled in ASCII format on the disk. You can import these data files readily into Minitab, MYSTAT, and most other statistical packages.

A Final Word

You are about to embark on a journey that will show you how statistical thinking plays an important role in everyday life and how statistical concepts and methods are used in business, economics, and the other social sciences. Read the Introduction first and see how the CEO Compensation case and the Space Shuttle Challenger case illustrate the importance of statistical thinking. Then proceed into the chapters that follow. We are confident that you will enjoy and benefit from the journey.

Acknowledgments

We are greatly indebted to many individuals and organizations who have helped us in the preparation of this book. Our sincere thanks go to all who have provided us with case materials and illustrations that demonstrate the usefulness of statistical methods in business, economics, and the other social sciences. Many persons, including colleagues and reviewers, have made helpful suggestions and otherwise assisted us in this and previous editions, for which we are most grateful. We particularly wish to thank for their help Robert F. Berner, Murray J. Cote, R. V. Erickson, John E. Floyd, Diwakar Gupta, Derek Hart, Edgar Hickman, Shu-Ping Hodgson, Burt S. Holland, Oswald Honkalehto, H. K. Hsieh, Allan Humphrey, Raj Jaganathan, William Meeker, Robert Norland, Jr., Eddy Patuwo, Thomas Pray, Thomas Rothrock, Barbara Rufle, J. Michael Ryan, Al Schainblatt, Kenneth C. Schneider, Randolph Shen, Brian Smith, Ehsan S. Soofi, Erland Sorensen, Albert Teitlebaum, Stephen Vardeman, Dean Wichern, and Morty Yalovsky.

We are also indebted to our respective universities for providing an environment in which the undertaking could be brought to fruition. A number of people were most helpful in working with us in preparing and checking the manuscript. We wish to express our appreciation to Karine Benzacar, Sandra June-Hatfield, Linda Keith, Karen Robertson, and Erika Whitmore for their valuable assistance. The staffs of Technical Texts and Allyn and Bacon helped us in many ways. Of course, we thank our families for their patience and encouragement while we completed this book.

Introduction

Statistical data and statistical concepts exercise a profound influence in almost every field of human activity. Statistical methods are used to improve agricultural products, to design space equipment, to plan traffic control, to forecast epidemics, and to attain better management in business and in government.

The need for sound statistical thinking is illustrated by the following two examples.

EXAMPLE □ **CEO Compensation**

Are chief executive officers of U.S. corporations paid too much? An article in *The Economist* (Ref. 1; see end-of-chapter references) addressed this question using the chart in Figure 1. In brief, the chart shows that during the decade from 1980 to 1989, the pay of CEOs rose by 160 percent, the wages of production workers rose by 50 percent, and corporate profits barely changed. The article states, ". . . the pay of American bosses has grown faster than that of other Americans, and faster than the profits

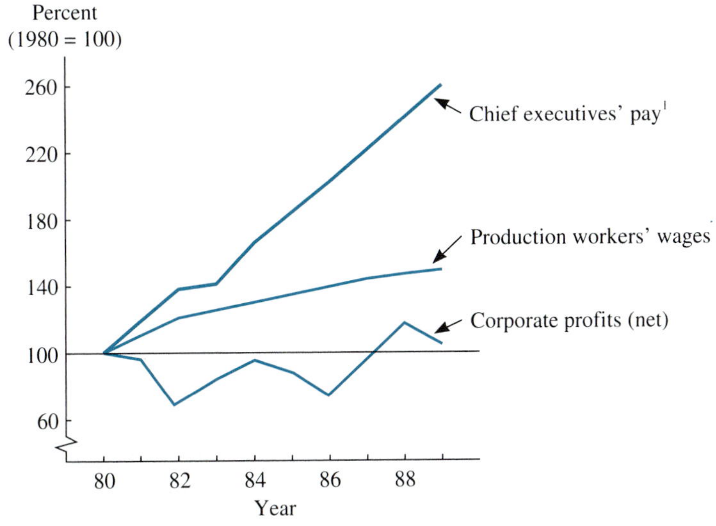

FIGURE I
Chart comparing changes between 1980 and 1989 in pay of chief executives, wages of production workers, and profits of corporations— CEO compensation example

NOTE: 1. Salary and annual incentive

Source: Data from U.S. Department of Commerce; Sibson & Company. Redrawn and reproduced, by permission, from *The Economist,* May 5, 1990, Vol. 315, No. 7653, p. 10.

of American companies. If shareholders thought and acted like owners, they would long since have rebelled against the trend in that chart.''

The chart seems to make a significant point. Yet, does it tell the full story? Several questions come to mind in studying the graph and the quoted statement. Which corporations were covered by the surveys on which the chart is based? Is it meaningful to compare the three statistical series that are presented in the graph? Is 1980 an appropriate year from which to measure changes in these statistics? Should adjustments have been made in the series for benefits received, taxes paid, risk-bearing, and so forth? If so, have these adjustments been made? None of these questions is answered by the article.

The study of statistics and the mode of thinking that it engenders prepare a person to evaluate objectively and effectively whether the information that charts such as Figure 1 communicate is relevant and adequate. Of course, interpretation of charts such as Figure 1 is not only a matter of statistics because an understanding of business and economics is also involved. □

Space Shuttle Challenger □ EXAMPLE

The investigation of the fatal explosion of the NASA space shuttle Challenger in 1986 provides a powerful, but tragic, demonstration of the need for sound statistical thinking (Ref. 2).

The primary responsibility for the explosion was attributed to the failure of O-ring seals in one section of a main booster rocket. The O-rings were designed to flatten under pressure from ignited fuel and thereby prevent the escape of potentially destructive hot gases.

The evening before the scheduled launch of Challenger, extensive deliberations were held by engineers and managers about whether or not to proceed with the launch, because the temperature at launch time was anticipated to be 29°F, a much colder temperature than at any previous launch. These deliberations focused on the possible relation between the temperature of the rings and their ability to function as designed.

Recovered sections of booster rockets from previous shuttle flights provided data on the extent of damage to O-rings from escaping gases. O-rings had shown measurable damage on four earlier flights. Figure 2a presents a plot of the damage to O-rings and the launch temperatures for these four flights. The damage is measured by the total depth of erosion in the seals (in thousandths of an inch). Two features of the plot are noteworthy. First, the Challenger launch temperature was expected to be much lower than any experienced in the four previous flights plotted on the graph. Second, the four plotted points do not reveal any definite relation between damage and temperature, although the data are too sparse to have been confident about this conclusion. The apparent absence of a relationship in these data had a strong influence on the decision to proceed with the flight.

Figure 2a is incomplete, however, because it does not contain information about 18 other shuttle flights in which no erosion was detected. Figure 2b shows a plot of damage against launch temperature for all 22 flights, including the 18 that experienced no measurable damage. This relationship was not considered in the pre-flight deliberations. The plot is very revealing. Most of the 18 flights with zero damage were launched at relatively high temperatures. Figure 2b shows a definite tendency for damage to increase with lower launch temperature. An extrapolation of the apparent relation suggests that substantial damage might occur at Challenger's launch temperature

FIGURE 2

Plots of O-ring damage and launch temperature—Space shuttle Challenger example. Damage refers to total depth of erosion in O-ring seals.

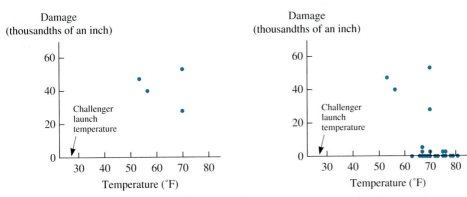

(a) Plot for Flights with O-Ring Damage

(b) Plot for All Flights

Source of Data: Lighthall, Frederick F., "Engineering Management, Engineering Reasoning, and Engineering Education: Lessons from the Space Shuttle Challenger," in *Proceedings of the 1990 International Engineering Management Conference,* Santa Clara, Calif., 1990, pp. 369–377.

of 29°F. The omission of the zero-damage flights from the set of data considered by the engineers and managers can be seen, in retrospect, to have been disastrous.

Had a plot like that in Figure 2b been seen by the engineers and managers the evening before the launch, it is very likely that the flight of Challenger would have been postponed. This conclusion was essentially the finding of the Presidential Commission that was charged with investigating the disaster.

This example points out the importance of clear statistical reasoning and of utilizing complete and pertinent data. Most encounters with statistical problems are not as dramatic as this one, but the lesson to be drawn is the same in every case. Even if a person does not intend to specialize in statistics, basic training in statistical reasoning has the potential to provide a better understanding of issues encountered in the real world and the means for coping with them. ☐

WHAT IS STATISTICS?

In common usage, the word *statistics* refers to numerical data. Vital statistics, for example, are numerical data on births, deaths, marriages, divorces, and communicable diseases; business and economic statistics are numerical data on employment, production, prices, and sales; social statistics are numerical data on housing, crime, education, and social assistance. The word *statistics* also refers to the set of methodologies for the collection, presentation, and analysis of data. Unless data are accurate, complete, effectively presented, and correctly analyzed, they may be dangerously misleading, as we have seen by the two examples. Everyone is a consumer of statistics, so it is important for all of us, not only professional statisticians, to acquire some knowledge of statistical methodology.

The word *statistician* also has several meanings. It can refer to a person who performs routine operations with statistical data, to an analyst who is highly trained in statistical methodology and uses this methodology in the collection and interpretation of data, or to an applied mathematician who utilizes advanced mathematics in the development of new statistical methods. Statisticians are needed in all these capacities in order to use statistical data effectively.

ROLE OF STATISTICS

Statistical data have been used for many centuries by governments as an aid in administration. In antiquity, statistics were compiled to ascertain the number of citizens liable for military service and taxation. After the Middle Ages, governments in Western Europe were interested in vital statistics because of the widespread fear of devastating epidemics and the belief that population size could affect political and military power. As a result, data were compiled from registrations of christenings, marriages, and burials. In the sixteenth through eighteenth centuries, when mercantilistic aspirations set nation-states in search of economic power for political purposes, data began to be collected on economic subjects, such as foreign trade, manufacturing, and food supply.

Today, data are collected, transmitted, stored, and retrieved in diverse and comprehensive information systems that supply individuals and organizations with the statistical intelligence required to carry out their activities. The expansion in computerized information systems has been accompanied by a rapid development of statistical methodology and data analysis. Not only are more comprehensive networks of data available today to serve as a basis for drawing valid conclusions and making decisions, but the purpose in assembling the data has shifted from record keeping to evaluation and action, based on timely information. Statistical analysis, in turn, has become chiefly concerned with the present and the future, rather than the past.

Some knowledge of statistics is essential today for people pursuing careers in almost every area of industry, government, public service, or the professions because evaluations and decisions are increasingly being made on the basis of data rather than hunches. Modern organizations are dependent on statistical data to obtain factual information about their internal operations and their social, economic, and ecological environments. Statistical data are concise, specific, capable of being analyzed objectively with powerful formal procedures, and well suited for making comparisons. Hence, they are especially useful for choosing among alternatives, setting goals, evaluating performance, measuring progress, and locating weaknesses.

The use of statistics is greatly facilitated by the extensive array of statistical computer packages that are available. These packages not only make the application of statistical methods and the handling of data more cost effective, but they also permit the use of powerful and sophisticated statistical techniques. Statistical procedures and models are integral parts of many management computer systems, such as decision support systems and expert systems. This expanded computerization of statistical methods increases the need for users of statistics to have an adequate understanding of statistical concepts and methods. Today, the subject of statistics is an essential component of professional education in business, economics, and administration.

STATISTICAL COMPUTER PACKAGES

Many commercial software packages are available to support the practical application of statistical data analysis. Some of these packages have been especially designed or adapted for use in the educational environment. All of the more widely used packages contain the principal analytical routines that we require for this text, although they vary in their design, computational capability, and range of specialized routines. We have chosen to use the following three statistical packages for illustrative purposes in this book:

— MINITAB, Minitab, Inc., State College, Pa. (Ref. 3)
— SAS, SAS Institute, Inc., Cary, N.C. (Ref. 4)
— SYSTAT, SYSTAT, Inc., Evanston, Ill. (Ref. 5)

SYSTAT includes a graphics module called SYGRAPH (Ref. 6). The business educational version of SYSTAT is called Business MYSTAT (Ref. 7). For quality control, we have used the SYSTAT module called QCSTAT (Ref. 8), as well as MINITAB.

The computer output for basic statistical analyses is quite similar for the three featured packages, as well as for most other statistical packages. When the output contains additional results that are not required in the text presentation, we show only those portions of the output that are required. Terms used in headings and labels occasionally differ from those used in the book. We explain the appropriate correspondences where there is any risk of confusion. Our aim is to expose the reader to the typical array of computer output that is likely to be encountered in practice, and to do so in a way that is clear and instructive.

CITED REFERENCES

1. *The Economist,* May 5, 1990, Volume 315, No. 7653, p. 10.
2. Lighthall, Frederick F. "Engineering Management, Engineering Reasoning, and Engineering Education: Lessons from the Space Shuttle Challenger," in *Proceedings of the 1990 International Engineering Management Conference.* Santa Clara, Calif., 1990, pp. 369–377.
3. *MINITAB Reference Manual, Release 7.* State College, Pa.: Minitab, Inc., 1989.
4. SAS Institute, Inc. *SAS/STAT User's Guide, Version 6, Fourth Edition.* Cary, N.C.: SAS Institute, Inc., 1989.
5. Wilkinson, Leland. *SYSTAT: The System for Statistics.* Evanston, Ill.: SYSTAT, Inc., 1990.
6. Wilkinson, Leland. *SYGRAPH: The System for Graphics.* Evanston, Ill.: SYSTAT, Inc., 1990.
7. *Business MYSTAT, Version 1.2.* Evanston, Ill.: SYSTAT, Inc., 1990.
8. Stenson, Herb. *QCSTAT: A Supplementary Module for SYSTAT.* Evanston, Ill.: SYSTAT, Inc., 1990.

Data

Data Acquisition and Management

Statistical analysis requires that the facts of interest in an investigation be assembled and organized in a useful manner. We refer to such facts as *data*. If the data are not properly assembled and organized, misleading or erroneous conclusions may be drawn from them.

EXAMPLE ☐

A TV program on campus crime (Ref. 1.1) ranked universities and colleges according to their campus crime rates. Unfortunately, the data on campus crime rates had many faults. They were obtained by asking schools to send in their crime statistics. Some of the schools did not reply. Those who did reply used different definitions. Some reported only crimes against undergraduates even though they had large graduate enrollments. Some universities with medical schools included crimes in the university hospital, while others did not. Crimes were not classified by severity, so a felony crime counted the same as a misdemeanor. No distinction was made between institutions in large urban areas and those in small college towns.

After protests by some schools, the rank listing (which had been sent to TV stations in advance of the TV program for publicity) was withdrawn from the program. The TV reviewer concluded: "The most dangerous lesson in this special was not intended by the producers: how wretched methodology can lead to deceptive statistics." ☐

This example illustrates the importance of proper acquisition and management of data, the topic of this chapter.

We begin with the concept of a data set.

DATA SETS   1.1

> **(1.1)**
> A **data set** is a collection of facts assembled for a particular purpose.

Data sets are found all around us. The financial section of our daily paper contains price data for securities and commodities; an economic report shows inflation rates for different countries; a computer file contains the names and addresses of members of a professional association.

FIGURE 1.1
Data set for a study of physical fitness profiles of six police officers

Case	Name	Age	Gender	Systolic Blood Pressure	Diastolic Blood Pressure	Triceps Skinfold Thickness[1] (cm)	Number of Sit-Ups	Fitness Rank
1	Anders	32	M	120	80	1.50	100	1
2	Colm	28	F	118	75	1.96	35	3
3	Greene	46	M	138	90	1.79	45	4
4	Keene	23	F	121	75	2.30	29	5
5	Osman	36	M	141	95	3.05	18	6
6	Waldorn	22	M	123	75	1.91	75	2

Variable → Age

Observation → 123

Case → 3

[1]Index of body fat

☐ **EXAMPLE**

Figure 1.1 shows a data set for a study of physical fitness profiles of six police officers. Data on name, age, gender, and various physical and fitness characteristics for each officer were obtained in this study.   ☐

Characteristics of Data Sets

We now consider the key characteristics of data sets.

Element.   A data set provides data about a collection of elements and contains, for each element, information about one or more characteristics of interest. In Figure 1.1, an element of the data set is a particular police officer.

Variable.   A variable is a characteristic of interest about an element. In Figure 1.1, one characteristic of interest is the age of a police officer. This characteristic takes on different values for different officers; hence, age is called a variable. Age is a quantitative variable because its outcomes are numerical in nature. On the other hand, the variable gender in Figure 1.1 is a qualitative variable because its outcomes are nonnumerical. Of the eight variables in the data set in Figure 1.1, six are quantitative and two (name and gender) are qualitative. The following are formal definitions.

> **(1.2)**
> A characteristic that can take on different possible outcomes is called a **variable.** The variable is said to be **quantitative** if the outcomes are numbers and **qualitative** if the outcomes are nonnumerical qualities or attributes.

Case.   The information on all variables for one element in the data set is called a *case* or a *record;* occasionally, it is called an *observation vector.* Thus, the information on the eight variables for Officer Colm constitutes a case. In Figure 1.1, each row of data represents a case, so there are six cases in the data set.

Observation.   The information about a single variable for an element of the data set is called an *observation,* a *measurement,* a *reading,* or an *outcome.* Thus, 123 is the observation on the systolic blood pressure variable for Officer Waldorn.

Comments

1. Sometimes, a quantitative variable is converted into a qualitative one by grouping the possible numerical outcomes into nonnumerical categories. This conversion is done to facilitate reporting, interpreting, or analyzing the data. Thus, with eggs for retail sale, the quantitative variable weight (measured in grams) is converted into a qualitative variable by dividing all possible weights into categories such as extra large and large. At other times, the reverse is done to convert a qualitative variable into a quantitative one by assigning an arbitrary numerical value or code to each nonnumerical category. For instance, the variable gender in Figure 1.1 might be quantified by using the number 0 for male and the number 1 for female.
2. Data sets may be distinguished on the basis of the number of variables they contain. A *univariate data set* contains one variable; a *bivariate data set,* two variables; and a *multivariate data set,* three or more variables. Thus, the data set in Figure 1.1 is multivariate.

Types of Data

Statistical data are of several different types. Data that represent measurements of amounts, capacities, or similar characteristics are called *measurement data.* For example, the data on blood pressure in Figure 1.1 are measurement data obtained by employing a medical instrument. Likewise, data on the total area of cultivated land in each U.S. state obtained from satellite imagery are measurement data.

Data that are counts or frequencies and, hence, are necessarily whole numbers, are called *count data.* For example, the data on numbers of sit-ups in Figure 1.1 are obtained by counting. Likewise, the numbers of pilots hired by commercial airlines last year are count data.

Data obtained by ranking or ordering elements are called *rank data.* For example, the data on fitness in Figure 1.1 were obtained by ranking the officers, with the most fit officer assigned rank 1 and the least fit officer, rank 6. Likewise, a published list gives rank data for the world's longest rivers, with the Nile shown as the longest (rank 1), the Amazon as the second longest (rank 2), and so on.

Finally, data where classes or categories are set up and each element is assigned to its appropriate category are called *classification data.* For example, the data on gender in Figure 1.1 are classification data because each police officer is assigned to one of the two categories, male or female. Likewise, the assignments of persons in the labor force to occupational categories such as secretary, computer operator, salesperson, and the like constitute classification data.

DATA SOURCES   1.2

Statistics is concerned not only with organizing and analyzing data once they are assembled but also with the sources of data and how data are collected for a study. The first stage of any investigation involves a specification or definition of the problem to be studied. The problems are usually concerned with the effects of one or more variables, called *factors,* on the variable of interest.

1. What are the effects of age and income on the amount of food expenditures? Here, age and income are the factors and food expenditures is the variable of interest.

2. What are the effects of make of car and number of cylinders in the engine on miles per gallon? Here, make of car and number of cylinders are the factors and miles per gallon is the variable of interest.   □

The specification of the problem leads to an identified need for particular types of data to deal with the problem. The question then is where or how to obtain the necessary data.

Internal and External Sources

Some data are available from an *internal source,* such as an organization's operating and accounting records. Most organizations keep such routine data in computer data files for efficient entry, storage, and retrieval of information. These data files, together with the associated computer programs, constitute the organization's internal data base.

Other data are obtainable from an *external source.* Such external data may be published in a reference book or statistical periodical. Often, external data are available in a computerized form, such as disks or tapes, or are accessible on-line from an external computer data bank. The external-source organization may be a government agency, a trade association, or a private service company.

1. A market researcher wished to study the effect of customers' locations on their propensity to purchase a particular company product. The researcher found much of the required data in an internal source—the accounting records of the company.

2. A store location analyst for a retail outlet chain was concerned with the effects of city size, average income, and similar factors on crime rates. The analyst found the necessary data in external sources—government publications containing criminal, judicial, and population statistics, as well as in computerized police records.   □

Cautions in the Use of Data Sources

When using any data source, the user should be thoroughly acquainted with the nature and limitations of the data. Limitations may include imperfect or improper methods of data collection, recording, and classification, as well as errors of omission or commission when data are transferred from one stage to another. The user also needs to determine whether the definitions employed in compiling the data are appropriate for the purpose. For example, a user of family income data will need to know if income includes investment income or not. It is also important to check whether changes in concepts, definitions, and data collection methods have occurred over the time period of interest and, if so, to determine the effect of these changes on the data. The user should have access to pertinent information and explanatory comments about the manner of compilation, accuracy, and proper interpretation of the data. The need for caution in the use of data applies to all data sources, whether they be internal or external, manual or computerized.

When the data needed for an investigation are not available in existing sources, some method for obtaining them directly must be considered. Two major methods of data collection are experimental studies and observational studies. We now illustrate the distinction between these two types of studies.

EXAMPLE ☐

An insurance company hired a large number of new programmers when it moved its headquarters to another city. The director of training for the company set up a voluntary training program for new programmers in which about one-half of the new programmers participated. Later, the director compared data on the job progress of those new programmers who took the training program and those who did not. On the whole, it was found that the programmers with the training had made more progress on the job than the others.

In this study, the variable of interest is progress on the job and the factor whose effect on job progress is to be investigated is the training program. However, the design of this study does not permit an unambiguous assessment of the effect of the training program on job progress because the programmers volunteered for the training. It is likely that the programmers who volunteered for the training were ones who were better motivated and more achievement-oriented. Such individuals might have made more job progress than the others even without the training program.

The fact that the training program was voluntary, so that no control was exercised over any factors that might affect progress on the job, makes this a nonexperimental or *observational study.*

Some years later, the director of training had the opportunity to assess the effect of the training program more definitively. The company expanded its data-processing staff substantially and hired 50 new programmers. Twenty-five of these were selected at random and assigned to the training program, and the other 25 programmers were immediately assigned to operations. Follow-up studies showed that the job progress was about the same for both groups of programmers.

The second study is an *experimental study* because control was exercised over the factor under study (training program) and randomization was employed to balance out all other, uncontrolled factors that might affect job progress, such as motivation, age, and experience of the programmers. Thus, randomization led to two groups of programmers that were approximately balanced by motivation, age, experience, and any other factors that might affect job progress. Hence, if any difference in job progress had been observed between the two groups, it could have been attributed confidently to the training program because of the experimental control that was exercised in the study.   ☐

We now consider experimental and observational studies in more detail.

Experimental Studies

The use of experiments for collecting data in the fields of business, economics, and the social sciences continues to expand.

(1.3)

In an **experimental study,** one or more factors are controlled so as to obtain information about their influence on the variable of interest, and randomization is employed to balance out the influence of any uncontrolled factors that might affect the variable of interest.

□ **EXAMPLES**

1. Thirty stores were randomly assigned one of three in-store promotional displays to study the effect of the displays on store sales of a product.

2. Four hundred consumers were randomly assigned one of four brands of all-purpose flour for routine home use over a period of one month to study consumer reactions to the brands.   □

Observational Studies

In spite of the increasing use of experimental studies, researchers and analysts in business, economics, and the social sciences must often rely on observational studies.

(1.4)

In an **observational study,** no experimental controls are exercised over factors influencing the variable of interest.

□ **EXAMPLES**

1. A survey of 1000 residents of a metropolitan area was conducted to obtain information about frequency of attending concerts and plays (the variable of interest) and about various possible explanatory factors, including income, age, and education of the resident.

2. A study was undertaken by a large corporation with five plants to obtain information about plant productivity (the variable of interest) and about several possible explanatory factors, including type and age of machinery used in the plant, education and age characteristics of the production employees in the plant, and type of wage-incentive program in the plant.   □

Choice Between Experimental and Observational Studies

Both experimental and observational studies can be extremely useful for investigating the effects of one or more factors on the variable of interest. As noted earlier, however, experimental studies provide stronger evidence of these effects than observational studies. Experiments are especially advantageous in investigating cause-and-effect patterns, such as determining whether or not subliminal advertising leads to increased sales of the advertised product.

Despite the advantages of experimental studies, much of statistical analysis in business, economics, and the social sciences is based on observational studies. One reason is that most available data, such as internal data on company operations and external data on the economy and consumer behavior, are observational data. Another reason is that it is often not feasible or may not be desirable to exercise the experimental controls required in experimental studies. For example, an economist interested in the effect of family size on the proportion of income saved cannot select a group of newlyweds and tell each couple to have a family of certain size. Observational studies of existing families, on the other hand, can provide information on family size, income, and expenditures, from which the relation between size of family and proportion of income saved can be studied.

A variety of procedures for acquiring data are employed in experimental and observational studies. Three commonly used ones are observation, interview, and self-enumeration.

DATA ACQUISITION 1.4

Data Acquisition Procedures

Observation. Data acquisition by observation entails direct examination and recording of an ongoing activity.

EXAMPLES ☐

1. In a study of family decision making, a researcher observed and recorded interactions between husband and wife as they decided on which make of home computer to buy.
2. In an engineering study, data about the internal temperature of a kiln were obtained by reading an instrument inserted in the kiln.
3. In a financial study, an analyst employed the observed daily closing prices of several publicly traded common stocks. ☐

The observation procedure has certain advantages:

1. The directness of the procedure avoids problems such as incomplete or distorted recall.
2. Data can be gathered more or less continuously over an extended time period.

Limitations of the method include the following:

1. The observer (or the instrument) must be able to accurately record the events of interest. Human observers usually require thorough training so that they will record precisely what they observe and so that different observers will record the same events in the same manner.
2. Individuals who are under observation and aware of this fact may alter their behavior; as a result, observations of their behavior may be distorted or biased.

Interview. In an interview, an interviewer asks questions from a questionnaire and records the respondent's answers. Interviews may be conducted in person or over the telephone.

☐ **EXAMPLES**

1. A household member was interviewed at home about purchases of toothpastes and mouthwashes and about family characteristics such as income and family size.

2. A household member was interviewed over the telephone about television viewing, including viewing at the moment of call, the channel viewed, and the number of persons viewing.

3. A company's financial executive was interviewed at the office by a representative of a trade association about the firm's plans for capital expenditures in the coming year.   ☐

Both the advantages and the limitations of securing data through interviews arise from the direct contact between the respondent and the interviewer. Advantages include the following:

1. Persons will tend to respond when they are approached in person or by telephone; hence, the interview procedure usually yields a high proportion of usable returns from those persons who are contacted.

2. The direct contact generally enables the interviewer to clear up misinterpretations of questions by the respondent, to observe the respondent's reactions to particular questions, and to collect relevant supplementary information.

There are several limitations of the interview method:

1. The interviewer may not follow directions for selecting respondents. For instance, the interviewer may select members of the family other than the ones designated, and this may introduce a bias into the results.

2. The interviewer may influence the respondent by the manner in which the questions are asked or by other actions. A slight inadvertent gesture of surprise at an answer, for example, can exert subtle, undetected pressures on the respondent.

3. The interviewer may make errors in recording the respondent's answers.

Self-Enumeration.   With self-enumeration, the respondent answers questions printed on a questionnaire or displayed on a computer monitor.

☐ **EXAMPLES**

1. A recent high school graduate received a self-enumeration questionnaire through the mail that requested information about educational activities following graduation.

2. A shopper entering a department store was invited to answer questions displayed on a portable computer monitor concerning the shopper's intended purchases in the store, age, gender, and occupation.

3. A person completed a registration certificate for a motor vehicle, supplying information on make, model, and year of manufacture.

4. A purchaser of a toaster filled out the warranty card, giving information on family characteristics and on the primary method by which attention was directed to this appliance (for example, word of mouth, television commercial).   ☐

Both the advantages and the limitations of the self-enumeration procedure arise from the elimination of interviewers. The types of interviewer errors discussed earlier are thus avoided. On the other hand, the absence of interviewers creates two serious problems:

1. When a questionnaire is sent to a household or an organization, there is no control over which person answers the questions.
2. The absence of interviewers can lead to low response rates.

A low response rate can be a source of serious distortion or bias in the survey results, because the persons who do answer the questionnaires often are not representative of the entire group contacted. For example, in an opinion survey conducted by mail on the proposed rezoning of a residential property site to a commercial designation, persons who feel strongly about the issue will be more likely to reply than will other residents of the community. Therefore, the survey results could be misleading about the feelings of the residents of the whole community.

Where there is a material rate of nonresponse, some follow-up of nonrespondents will be valuable. In a well-conducted mail survey, for example, some or all of the nonrespondents will be contacted as a routine procedure. Nonrespondents may be contacted by means of "reminder" letters, telephone calls, or special personal interviews.

It is important that survey results be accompanied by information about the response rate because it is a key factor for assessing the magnitude of potential bias in the survey results. The integrity and usefulness of survey results are much greater if the response rate is 95 percent than if it is 20 percent.

Questionnaire Design

Since questionnaires are a major vehicle in data collection, we now examine briefly some important considerations in designing and using a questionnaire so that accurate data are obtained.

Types of Questions. The two basic types of questions employed in questionnaires are multiple-choice and free-answer, or open-end, questions.

The *multiple-choice question* presents the respondent with a choice from among two or more prespecified answers. The following question directed at students is of this type:

How many months of full-time study are required to complete the degree or certificate for which you are enrolled?

6 or fewer	☐	25 to 36	☐
7 to 12	☐	37 to 48	☐
13 to 24	☐	More than 48	☐

The alternatives in a multiple-choice question should be clear-cut; mutually exclusive (not overlapping); and when dealing with an issue, more or less evenly distributed on both sides of the issue. Ideally, the choices should cover all answers likely to arise in response to the question. If too many alternatives are presented, however, they may not be clear-cut enough for the respondent to make a meaningful selection. One disadvantage of multiple-choice questions is that they tend to suggest an answer to the respondent in terms of the alternatives stated when actually the question might have been answered differently if no choices had been suggested. Another disadvantage is that the respondent's answer can be influenced by the order in which the question alternatives are presented.

These disadvantages are avoided by the other type of question, the *free-answer* or *open-end question,* which elicits answers in the respondent's own words. The following question is of the free-answer type:

> In your opinion, what things are good or bad about the Blank Company as a place to work?

In requesting answers in the respondent's own words, the free-answer question yields a multitude of replies that must be classified into a limited number of categories before statistical analysis can be undertaken. This classification becomes a difficult task when thousands of replies to a free-answer question have been received, because the wording of each reply must be interpreted carefully. Hence, free-answer questions are usually employed only in small-scale studies or in the exploratory stages of large-scale ones to determine the patterns into which the responses to a question tend to fall. The information obtained through the use of exploratory free-answer questions in the early stages of a large-scale study is then used to construct the multiple-choice questions utilized later for the bulk of the data collection.

Multiple-choice and free-answer formats sometimes are combined in constructing a question. The following question directed at hotel guests is of this type:

> What is the principal reason that you chose our hotel for this stay?
>
> ☐ Recommended by friend or business associate
> ☐ Recommended by travel agent
> ☐ Selected from a travel directory
> ☐ Stayed at hotel on a previous occasion
> ☐ Other—specify _____

The use of the category titled "other" allows the respondent to provide an answer other than the ones given.

Order of Questions.   The questions in a questionnaire need to be assembled in a proper order. The initial questions should establish rapport with the respondent and should be simple to answer. Where several topics are involved, it is usually best to complete the questions for one topic before going to another topic, rather than to jump back and forth. The order of questions often affects the answers given by the respondent, because questions draw the respondent's attention to a particular complex of thoughts or feelings, in the context of which later questions are answered. In market-research surveys, for example, questions mentioning a specific product or firm tend to bias answers to questions that follow; consequently, such identifying questions should be placed toward the end of the questionnaire whenever possible.

Need for Clarity in Questions.   A question must have approximately the same meaning for all respondents if the data obtained from the replies are to be meaningful. Vaguely defined terms should be avoided in framing questions, and questions should be stated simply.

Need to Avoid Bias in Question Framing.   The replies of respondents are frequently influenced in a one-sided manner by the way in which questions are phrased. One type of question that tends to produce biased replies is a *leading question*—that is, one that suggests a particular answer. The question "Wouldn't you agree that the extra quality obtained in this brand is well worth the few extra dollars in cost?" so obviously suggests a particular answer that few persons will take the replies seriously. In many instances, though, the suggestive nature of a question is less easily recognized.

Another type of question that may lead to bias is a *prestige question,* in which respondents are questioned directly about their motives, accomplishments, or other subjects involving their prestige or self-esteem. In these situations, many respondents are prone to rationalize or to exaggerate their replies. Questions need to be carefully worded so as not to arouse such sentiments. For example, instead of asking, "Have you ever served your country through military service?" interviewers might be instructed to ask, "Have you ever served in the military?" The latter avoids reference to service to one's country so the respondent will not be made to feel unpatriotic or lacking in public spirit if he or she has not served in the military.

Value of Pretesting. Once a questionnaire has been designed, it is often pretested in a preliminary small-scale study. Pretesting usually involves between a few dozen and several hundred respondents and brings out unforeseen difficulties in the layout of the questionnaire, in the arrangement and wording of the questions, and in the clarity of instructions to interviewers when interviewing is used. The difficulties can then be corrected before the full-scale study is begun.

Computer-Assisted Use of Questionnaires. Computers are frequently used in interviews and self-enumeration. In both of these uses, computers facilitate the administration of the questionnaire and the recording of the responses. In *computer-assisted interviewing* (CAI), the computer guides the interviewer through the series of questions, skipping automatically over those that do not apply to a respondent based on the answers to previous questions. The same guidance occurs when computers are utilized in self-enumeration. The computer can also check immediately for inconsistencies in responses to related questions. A disadvantage of the use of computers in self-enumeration is that some respondents may be intimidated by having to respond to questions in a computer.

DEFICIENCIES IN DATA SETS 1.5

Data sets are subject to two kinds of deficiencies—data may be erroneous and data sets may be incomplete. In this section, we discuss these deficiencies and show examples of how they arise.

Errors in Data

Errors in data may arise during the acquisition stage or at later stages.

Data Acquisition Errors

(1.5)

An **error in data acquisition** is any discrepancy between the actual result obtained and the correct result that would be provided by an ideal procedure.

EXAMPLES □

1. Some Canadians completing a travel questionnaire overlooked trips to the United States in answering a question on the number of foreign trips taken in the past year. These errors led to incorrect counts of total number of foreign trips.

2. In answering a question relating to occupation, some respondents who are "retired" mistakenly recorded themselves as "unemployed," possibly because the definitions of the categories were misunderstood. These errors led to incorrect counts for the different categories.

3. Wind velocity and direction measured on an offshore drilling rig were distorted as a result of the air turbulence created by the rig itself; consequently, incorrect measurements were recorded.

4. An industrial scale showed incorrect weights for loaded trucks because it was not calibrated properly and thereby provided incorrect measurements.   ☐

As the examples show, errors in acquiring data may be caused by imperfect recall by respondents, inaccurate measurements by an instrument, misinterpretations of a question, misunderstandings of a definition—the list is long. The procedure chosen for acquiring data should keep important errors at tolerable levels so that the data set obtained will be useful. Thus, a personal or telephone interview may be preferred over a mailed self-enumeration questionnaire in a study involving many technical definitions because the interviewers can help respondents avoid misunderstandings.

Other Data Errors.   In addition to errors in data acquisition, data may be erroneous for other reasons; for instance, because of typographical errors or computer programming mistakes.

Incomplete Data

Data sets may be incomplete for many reasons. Sometimes entire cases may be missing.

☐ **EXAMPLES**

1. In a study of weather conditions at different locations, the data for one location were missing as a result of instrument malfunction. Consequently, the data set on weather conditions is incomplete.

2. In a survey of families residing in a community to obtain information about labor skills, families away on vacation at the time were not included. As a result, the labor skills data set has some missing cases.   ☐

Data sets may also be incomplete because some individual observations are missing or are only partially or inexactly known.

☐ **EXAMPLES**

1. In a survey on consumer expenditures, a few respondents refused to provide data on family income although they gave information on expenditures and family characteristics. As a result, the data set contains missing observations and is incomplete.

2. Ten identical car door latches were placed on test machines that repeatedly opened and closed the latch, to determine how many cycles each latch would function before failing. Two of the 10 latches had not failed after 10,000 cycles when the test was terminated. The observations of the number of cycles to failure for these

two latches are incomplete; the numbers are known to be larger than 10,000 but their exact values are not known. These incomplete data are called *censored data*. □

Statistical methods to help deal with missing observations and observations that are only partially known are described in more advanced books. Although these methods are helpful, they are never as good as obtaining the actual data in the first place. Hence, it is good practice in any study to keep incomplete data to a minimum by careful design, planning, and implementation.

COMPUTERIZED DATA MANAGEMENT 1.6

Most data sets that are employed in statistical investigations are created and then used in the form of computer data files within some statistical or data management computer software system. Although data-handling procedures vary greatly from one system to another, there are some important features that are common to most systems. We now discuss two key features of computerized data management systems—data entry and data handling.

Data Entry

Data are entered in a variety of ways in different systems. The common aim, however, is to have data organized so that they can be displayed in a cases-by-variables rectangular array, such as that shown in Figure 1.1. One method of data entry involves entering the observations for all variables for each case, one case after the other. For example, in Figure 1.1, the observations for case 1 would be entered first in the order Anders, 32, M, 120, . . . , 1. Then the observations for the second case would be entered in the order Colm, 28, F, 118, . . . , 3; and so on until all six cases have been entered.

A second method of data entry involves entering the observations for all cases for each variable, one variable after the other. For example, in Figure 1.1, the names of all six officers would be entered first in the order Anders, Colm, . . . , Waldorn. Next, the observations of age would be entered for all six cases in the order 32, 28, . . . , 22; and so on until all eight variables have been entered. Although the systems may store data files in different physical forms, as far as the user is concerned the data files may be thought of as having the form illustrated by Figure 1.1.

Data entry usually requires that the user give names to the variables that are being created and also possibly identify in advance how many variables and cases are in the data set. Conventions for naming variables vary widely. A common requirement is to distinguish qualitative variables from quantitative variables. Observations for qualitative variables may be permitted to contain numerical, alphabetical, and perhaps other characters. Observations for quantitative variables are usually restricted to numerical values. Generally, arithmetic operations on data are restricted to quantitative variables. Data management systems also usually have some procedure for identifying missing observations. A common procedure is to have a missing observation represented by a designated character such as a dot ($\cdot$) or an asterisk ($*$).

In addition to data entry itself, most data management systems allow the user to edit data files. The editing may involve correcting entry errors, altering observations, deleting or adding cases and variables, and reordering cases and variables.

Data Handling

Most data management systems have facilities for transferring a data file from a different computer system into the user's system, ready for use. Likewise, data management systems usually have the capability of transferring a data file from the user's system to another system or to store it on a disk or tape.

In statistical investigations, it is often necessary to create new variables from the original ones in the data set by means of mathematical or logical operations. These operations are referred to as *data transformations*. We will encounter data transformations in later chapters. For the present, we shall illustrate their general nature by two examples.

☐ **EXAMPLES**

1. A data set for a study of the characteristics of company employees contained a current age variable (CURRENT) for each employee and a variable that represents the age of the employee when the employee started working for the company (START). The investigator wanted to create a variable that would represent the years of service of each employee (SERVICE). The new variable was created from the other two variables by means of the following transformation:

   ```
   SERVICE = CURRENT - START
   ```

   The investigator's data management system automatically calculated the arithmetic differences required in this transformation and created the new variable SERVICE.

   Figure 1.2 illustrates how the MYSTAT statistical package handles the data transformation in this example. Figure 1.2a shows a data set of four cases with observations for the variables CURRENT and START. A column for the SERVICE variable has been created but not filled (the missing values are represented by dots). Figure 1.2b shows the MYSTAT data transformation command that calculates the SERVICE observations. Figure 1.2c shows the SERVICE observations that were calculated by the transformation command.

2. A data set for a study of consumer behavior contained variables for each consumer's age (AGE) and years of education (EDUC). The researcher wished to create a new variable (SEG) representing the market segment to which each consumer belonged (1, 2, 3) based on the following definitions:

Market Segment	Age	Years of Education
1	Under 25	Under 16
2	Under 25	16 or more
3	25 or more	Any number

   The researcher used the following sequence of logical operations to create the new variable:

   ```
   IF AGE<25 AND EDUC<16 THEN SEG=1
   IF AGE<25 AND EDUC=>16 THEN SEG=2
   IF AGE=>25 THEN SEG=3
   ```

   The researcher's data management system automatically performed the logical operations required in this transformation and created the new variable SEG. ☐

Calculation of a new variable using a data transformation command in MYSTAT

FIGURE 1.2

(a) Data Set Before Transformation **(b) Data Transformation Command** **(c) Data Set After Transformation**

		CURRENT	START	SERVICE
CASE	1	43.000	21.000	.
CASE	2	58.000	18.000	.
CASE	3	31.000	25.000	.
CASE	4	48.000	30.000	.

`>let SERVICE=CURRENT-START`

		CURRENT	START	SERVICE
CASE	1	43.000	21.000	22.000
CASE	2	58.000	18.000	40.000
CASE	3	31.000	25.000	6.000
CASE	4	48.000	30.000	18.000

CITED REFERENCE

1.1 Phil Kloer. "It's a Crime How Sloppy 'Dangerous Lessons' Is with Campus Statistics." *The Atlanta Journal and Constitution,* Saturday, December 1, 1990, p. E-4.

PROBLEMS

1.1 The following data set, compiled by a firm's senior management committee, relates to candidates for promotion to product manager:

Candidate's Name	Age	Gender	Number of Training Programs Attended	Committee's Order of Preference
Andrews, Milton	46	M	2	2
Morris, John	31	M	0	3
Parker, Kim	42	F	1	1
Stanton, Roger	38	M	1	5
Williams, Ellen	35	F	0	4

a. What constitute the elements of this data set? How many elements are there?
b. Identify the variables in the data set. For each variable, indicate whether it involves measurement data, count data, rank data, classification data, or some other type of data.
c. What constitutes the case for John Morris? What is the observation for Morris on the age variable?

1.2 Municipal Debt. The following data set was compiled by a financial analyst who was examining the public debt burden of several municipalities:

Municipality	Population (thousand)	Debt per Capita ($ thousand)	Ratio of Debt to Property Value (percent)
Carullo	457	4.8	6.9
Edwards	1322	5.9	7.7
Girvan	181	3.7	8.3
Marois	665	5.5	8.0
Morency	690	6.1	7.3
Pratt	234	4.9	10.1

a. What constitute the elements of this data set? How many elements are there?
b. Identify the variables in the data set. For each variable, indicate whether it is quantitative or qualitative. Is the data set univariate or multivariate?
c. What constitute the observations on the debt per capita variable?

1.3 **House Listing.** The following data set pertains to six houses that are listed for sale:

Listing Number	Subdivision	Asking Price ($ thousand)	Number of Bedrooms	Selling Agency
80344	Maxwell	199	4	Peabody
15671	Pine	176	4	Peabody
14006	Maxwell	195	5	Realax
73317	Bayview	205	4	Realax
28180	Bedford	185	3	Realax
36641	Pine	160	3	Peabody

a. Identify the variables in the data set. For each variable, indicate whether it involves measurement data, count data, rank data, classification data, or some other type of data.
b. What constitute the observations on the selling agency variable?
c. What constitutes the case for the house with listing number 28180?

1.4 For each of the following variables, indicate whether it is quantitative or qualitative and whether the observations on the variable are obtained by measurement, counting, ranking, or classification.

a. Embassy posting of a diplomat.
b. Position of a company among the top 500 companies, ordered by sales revenue.
c. Number of students enrolled in a university.
d. Dividend income reported by a taxpayer in the current year's tax return.

1.5 For each of the following variables, indicate whether it is quantitative or qualitative and whether the observations on the variable are obtained by measurement, counting, ranking, or classification.

a. League standing of a baseball team.
b. Country of registry of an oil tanker.
c. Store price of a can of peaches.
d. Number of company employees who retired last month.

1.6 A public utility company has distributed its annual financial statement to each of its industrial customers. Is this statement an internal or external source of data for one of its industrial customers? For the public utility company itself? Explain.

1.7 A multinational corporation operates in 27 countries. Would the *Monthly Bulletin of Statistics* published by the United Nations be an internal or external source of data for this corporation? Explain.

1.8 A company's computer data base contains production data for the various divisions of the company. Is this data base an internal or external source of data for the company? Explain.

1.9 A company has been selling one of its popular beverage products in bottles in supermarkets and in cans in other outlets. An analyst is studying sales data for the beverage to see which type of container is preferred by customers.

a. Is this study experimental or observational? Why?
b. Why does the distinction in part a matter to the analyst?

1.10 Thirty sales trainees were grouped into 15 pairs, the trainees in each pair being similar in age, sales experience, and educational background. One trainee in each pair was selected at random to attend a sales training program based on role playing. The other trainee was assigned to a sales training program based on case studies. At the end of the programs, the trainees were tested for sales skills and knowledge. The test scores are now being evaluated to determine whether one type of program produces better scores than the other.

a. Is this study experimental or observational? Why?
b. An observer asks why the trainees in each pair could not simply have decided between themselves who would attend which program. Answer the observer's question.

1.11 A manufacturer is designing a new energy-efficient metal door. Three designs are under consideration. An experiment was conducted in which four doors of each design were fabricated and one of these 12 doors was installed in each of 12 new houses. The 12 doors were assigned to the houses at random. Measurements of energy loss were made over a 12-month period for each installed door.

a. Which variable is the factor that was controlled in this study? Which is the variable of interest?
b. What was the purpose of the randomization that was employed in this study?
c. What are the advantages of measuring energy loss for 12 months rather than for one month?

1.12 Refer to the **Power Cells** data set (Appendix D.2). These data are from an experimental study.

a. Which variables represent factors that were controlled in this study?
b. Which are the variables of interest in this study?
c. Would operating the same number of power cells until they fail without attempting to control any of the factors identified in part a yield the same information as the experiment? Discuss.

1.13 Each purchaser of a new personal computer receives an information card that is to be completed by the purchaser and returned to the equipment manufacturer.

a. Which method of data acquisition (observation, interview, or self-enumeration) is involved here?
b. Would your answer change if the information card were completed by the salesperson on the basis of questions put to the purchaser?
c. Is the response rate likely to be affected by whether or not the postage for the information card is prepaid by the manufacturer? Explain.

1.14 For each of the following, state which method of data acquisition (observation, interview, or self-enumeration) is most appropriate for obtaining the information, and justify your choice.

a. Quarterly data from employers on the number of employees who worked at any time during the quarter.
b. Data from technicians of a large research organization on hazardous aspects of their jobs.
c. Data on blood cholesterol levels of Canadians.

1.15 For each of the following, state which method of data acquisition (observation, interview, or self-enumeration) is most appropriate for obtaining the information, and justify your choice.

a. Data from persons in a national register of scientists concerning field of specialization, nature of present employment, and academic training.
b. Data on sequences in which shoppers purchase items in supermarkets.
c. Data from manufacturers on sales, inventory levels, and new orders.

1.16 In a mail survey of 1500 practicing physicians conducted by a pharmaceutical firm, a series of questions related to prescriptions for an antidepressant drug code-named ADV. The proportion of responding physicians who stated that they had prescribed the drug at least once in the previous month was 0.60.

a. Suppose that 90 percent of the questionnaires had been completed and returned. By how much might the proportion 0.60 be changed if every questionnaire had been completed and returned?
b. Answer part a on the supposition that the response rate was 30 percent.
c. What are the implications of your answers in parts a and b for interpreting data from studies with low response rates?
d. What factors might cause a physician not to respond in this type of mail survey? Might any of these factors be related to whether the physician prescribes this particular drug? For the purposes of the survey, does it matter whether or not such a relation exists? Discuss.

1.17 A questionnaire on wine consumption asked the following question:

What is the typical quantity of wine that you personally drink with a restaurant meal? __ None; __ one glass; __ half bottle; __ whole bottle; __ more than a whole bottle.

a. Is this question multiple choice or free answer?
b. What difficulties, if any, might a respondent have in replying to this question?

1.18 Identify any major faults in each of the following questions, and reword them to eliminate those faults. Indicate any assumptions you made when rewording the questions.

a. How many calls for telephone directory assistance did you make in the past 24 months?

b. What is your family income?

c. Which one of the following reasons describes why you are not employed currently? __ Temporary layoff; __ illness; __ unable to find employment.

1.19 Identify any major faults in each of the following questions, and reword them to eliminate those faults. Indicate any assumptions you made when rewording the questions.

a. Does Sony or some other brand come to mind when a videocassette recorder is mentioned?

b. Do you suffer from severe depression?

c. Which affects the elderly more seriously, price inflation or high interest rates?

1.20 Consider the following two alternative questions for ascertaining consumer plans for purchasing a new car: (1) Do you plan to buy a new car within the next year? (2) Do you plan to buy a new car within the next year, and how certain are you of this plan on a scale from 0 (highly uncertain) to 10 (highly certain)?

a. Which question do you think will yield more useful information? Explain.

b. Would the clarity of the question be improved if "within the next year" were replaced by "within the next 12 months"? Explain.

1.21 A class on opinion research was divided at random into two groups. One group was asked the following question: "Should university officials take primary responsibility for reducing rowdyism at basketball games?" The second group was asked the same question, except that "student leaders" was substituted for "university officials." About 80 percent of the students in each group answered affirmatively.

a. What fault in question design is illustrated by this experiment?

b. Design a better question to obtain student opinions on who should take primary responsibility to reduce rowdyism at basketball games. Explain how your approach avoids the fault identified in part a.

1.22 Explain whether each of the following entails an error from defects in the data acquisition procedure or some other type of error.

a. A schoolchild, when asked to indicate the number of brothers or sisters at home who have not yet started school, overlooked a new baby brother and wrote "1" instead of "2" on the questionnaire.

b. An incorrect entry of data to a hospital computer file caused a baby's weight to be shown as 8400 grams when the actual weight on the hospital birth form was 4800 grams.

c. In filling out a questionnaire sent to owner-operated restaurants, an owner misinterpreted the definition of "employee" in answering a question on number of full-time employees and counted himself as one of the employees.

1.23 In a consumer mail survey, respondents were asked to name their regular brand of toothpaste. In follow-up personal interviews conducted with 60 of the initial respondents, each respondent was again asked to name his or her regular brand of toothpaste. Twelve of the 60 respondents gave different answers to this question in the interview than in the mail questionnaire. Do the different answers represent errors in data acquisition, other types of errors, or something else? Comment.

1.24 A question on family income in a questionnaire provides the following response categories: __ Under $30,000; __ $30,000–under $50,000; __ $50,000 or more. In what sense does an answer to this question provide incomplete data about family income?

1.25 Refer to **Municipal Debt** Problem 1.2. Create a new variable representing total municipal debt by multiplying the population variable by the debt per capita variable. Give the outcomes of the new variable for all cases. In what units is the total municipal debt variable expressed?

1.26 Refer to **House Listing** Problem 1.3. Create a new variable representing a numerical code for the selling agency. Let the numerical code equal 1 if the agency is Peabody and 2 if it is Realax. Give the outcomes of the new variable for all cases.

1.27 For each of the following, ascertain a library source where data for the most recent period can be obtained:

a. U.S. annual birthrate
b. U.S. monthly unemployment rate
c. U.S. monthly consumer price index for all urban consumers

1.28 Answer Problem 1.27 for data pertaining to Canada.

1.29 For each published data source assigned from the following list, state who publishes the source, how often it is published, and give three specific items of information from the most recent issue of that source:

— *Business Conditions Digest*
— *Federal Reserve Bulletin*
— *International Financial Statistics*
— *Monthly Labor Review*
— *Moody's Industrial Manual*
— *Survey of Current Business*

1.30 For each published data source assigned from the following list, state who publishes the source, how often it is published, and give three specific items of information from the most recent issue of that source:

— *Bank of Canada Review*
— *Canadian Life and Health Insurance Facts*
— *Canadian Economic Observer*
— *Direction of Trade Statistics*
— *The Conference Board Statistical Bulletin*
— *Toronto Stock Exchange Review*

1.31 A study indicated that high school students who took a voluntary driver training course tended later to have better driving records than students who did not take the course. However, the study team noted that this difference could be due simply to differences in personal qualities of the students in the two groups. Among other things, relatively more women than men took the course, and young women overall tend to have better driving records than young men. The team also noted that students taking the course might have been better motivated to be good drivers than students not taking the course.

a. Why was this study not an experimental study?
b. Develop a design for an experimental study in a high school to evaluate the impact of the driver training course. Your design should control the effects of gender and motivation of students.

1.32 A professional association composed of economists and statisticians surveyed its members to study how they liked the association's journal. A high proportion of the responses indicated an unfavorable opinion. Subsequently, it was learned that the response rate was much higher for economists than for statisticians. Construct a numerical example to show how the differential response rates could have biased the survey results.

1.33 It is frequently stated that errors in responses are not too serious because they will tend to balance out. One researcher conducted a survey of 1643 library users in a community to study the nature of response errors. The researcher compared responses for the number of books borrowed from the library in the preceding three months with the library's computerized records. The results were as follows:

Respondent Statement	Percent of Respondents	Average Number of Books Borrowed	
		Library Record	Respondent's Recollection
Accurate	23	6.1	6.1
Overstated	43	2.8	5.6
Understated	34	6.4	5.4

a. By what percentage did overstaters overstate their borrowings? Did understaters understate their borrowings to the same extent?
b. What is the overall percent error in the reported average number of books borrowed? Did the response errors largely balance out?
c. What are some factors that might account for the errors in response here?

Data Patterns

2

Once the data needed for a study are in hand, the statistical investigation can begin. It usually starts with a study of the underlying pattern of the data. First, however, data sets must be organized and reduced to manageable proportions and displayed in an effective fashion. The human mind simply is not capable of assimilating and interpreting a large number of facts and figures in raw form. In this chapter, we present a number of methods for organizing and displaying data sets so that the patterns in the data can be analyzed, interpreted, and reported.

We discuss first some simple methods for displaying data for quantitative variables, that is, variables having numerical outcomes.

SIMPLE DISPLAYS 2.1 OF QUANTITATIVE DATA

Arrays

A useful first step for discerning a pattern in quantitative data—when the number of observations is not too large—is to list the observations in increasing or decreasing order of magnitude. Such an ordering, called an *array*, can greatly facilitate inspection of the data. Most data management software systems have a sorting routine that provides an ordering of quantitative data.

EXAMPLE ☐ **Semiconductor Failure**

An industrial engineer examined failure data for the bond between a wire and a semiconductor wafer. An array of the breaking stresses (in milligrams) of 17 bonds, arranged in increasing order of magnitude, follows:

1   4   31   43   47   50   51   51   58   59   62   63   66   71   75   88   113

This array quickly tells us that there is a large range of breaking stresses, from 1 to 113 milligrams.

☐

Stem-and-Leaf Displays

A *stem-and-leaf display* provides more information about the pattern of data than an array does. It assists in the discovery of (1) concentrations or clusters of particular values, (2) outlying or extreme observations, and (3) the extent of symmetry or lack thereof in the distribution of observations.

Concert Subscribers   ☐ **EXAMPLE**

An analyst who was commissioned to study characteristics of concert subscribers in a metropolitan area pretested a preliminary version of the questionnaire on a sample of 70 subscribers. One question asked for the subscriber's age (in years). An array of the responses to this question follows (only part of the data are shown):

20   25   25   25   29   30   ...   76   77

Figure 2.1a shows a stem-and-leaf display for the reported ages. The digits to the left of the vertical line are the *stems* and the digits to the right are the *leaves.* The digit 2 constitutes the first stem. All age responses from 20 through 29 are recorded on this stem in ascending order of magnitude. Only the second digit in each age response is written because the first digit is given by the stem. We see that the leaves on the first stem consist of the digits 0, 5, 5, 5, 9, indicating that five respondents reported ages in the interval 20–29, with the reported ages being 20, 25, 25, 25, and 29.

The analyst, viewing the display, concluded that it fit the usual age pattern for concert subscribers. The age pattern is compact and symmetrical and is centered on an age of about 50 years. The ages in the data set range from 20 to 77 years. The analyst also noted the disproportionately large number of 0's and 5's among the second digits (leaves) in the age responses. This result was unanticipated since each final digit from 0 to 9 was expected to be roughly equally represented. We shall consider this finding again in a later section.   ☐

☐ **EXAMPLE**

Figure 2.1b shows a MINITAB stem-and-leaf display for the 17 breaking stresses in the semiconductor failure example. The tens' and hundreds' digits of the observations form the stems of the display, and the units' digits form the leaves. A single space separates the stems from the leaves.

The engineer could see immediately from the display that two of the semiconductor bonds had broken under negligible stress (1 and 4 milligrams), possibly because of

---

FIGURE 2.1
Stem-and-leaf displays—
Concert subscribers and semi-
conductor failure examples

(a) Subscribers' Ages

```
2 | 0 5 5 5 9
3 | 0 0 1 2 3 3 4 5 5 6 7 8
4 | 0 0 0 0 1 2 2 3 4 4 5 5 6 7 7 8 9 9
5 | 0 0 0 0 0 1 1 2 3 3 4 5 5 5 6 7 8 8 9
6 | 0 0 2 3 4 5 5 6 7 9
7 | 0 1 5 5 6 7
```

(b) Breaking Stresses (MINITAB output)

```
Stem-and-leaf of STRESS   N = 17
Leaf Unit = 1.0

    0 14
    1
    2
    3 1
    4 37
    5 01189
    6 236
    7 15
    8 8
    9
   10
   11 3
```

defective manufacture. The engineer also noted that the bulk of the observations was concentrated between 31 and 88 milligrams. Finally, one bond had withstood considerable stress before breaking (113 milligrams). This bond was subsequently subjected to special study to determine the reasons for its superior performance. ☐

When two data sets are to be compared, the two stem-and-leaf displays can be combined by means of a *back-to-back stem-and-leaf display*.

EXAMPLE ☐ Cake Ingredient

A food scientist studied the effects of two different dairy ingredients (lean, rich) on the volume of baked cakes (measured in milliliters per 100 grams). Twenty cakes were baked in total, with 10 of the cakes using each ingredient. Identical recipes and procedures were followed in baking except for the dairy ingredient. The experimental results are shown in Figure 2.2 by a back-to-back stem-and-leaf display. Note the use of common stems to facilitate the comparison. The display readily indicates that cake volumes tend to be larger for the rich ingredient than for the lean ingredient. The display also shows that the distribution patterns of cake volumes for the two ingredients are similar, although the average levels differ. ☐

Dot Plots

When a data set is not too large, a *dot plot* is another effective display for viewing the data pattern. A dot plot is constructed by plotting a dot for each observation above its value on a line scale. Two data sets can be readily compared by means of dot plots that are aligned on a common scale.

EXAMPLES ☐

1. Figure 2.3a presents a MINITAB dot plot of the data for the semiconductor failure example. Note how readily the outlying observations stand out, as well as the concentration of many observations in the middle.

2. Figure 2.3b presents two MINITAB dot plots that are aligned on the same scale for the data in the cake ingredient example. The dot plots show easily that the cake volumes tend to be larger for the rich ingredient than for the lean ingredient. ☐

Lean		Rich
8 1	35	
9	36	
4 0	37	
9 6 1	38	0
8	39	8 9
0	40	2 9
	41	4 5 6
	42	7
	43	4

FIGURE 2.2
Back-to-back stem-and-leaf display of cake volumes for two dairy ingredients—Cake ingredient example

FIGURE 2.3
MINITAB dot plots—Semi-conductor failure and cake ingredient examples

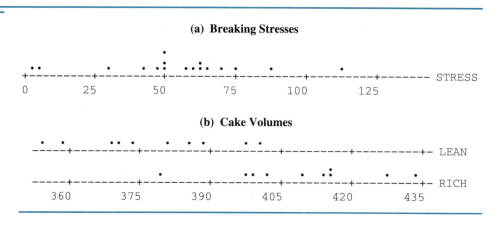

(a) Breaking Stresses

(b) Cake Volumes

Cumulative Distributions

The cumulative distribution of the observations for a quantitative variable describes the number of observations above or below given values. It is particularly useful for studying the distribution pattern of the observations. The cumulative distribution is constructed by assigning ranks to the arrayed observations, as we shall now illustrate by an example.

☐ EXAMPLE

The array of breaking stresses in the semiconductor failure example and their assigned ranks are as follows (only some of the observations are shown):

Observation:	1	4	. . .	50	51	51	58	59	. . .	88	113
Rank:	1	2	. . .	6	7	8	9	10	. . .	16	17

The ranks are simply the case numbers of the arrayed observations. The smallest observation (1) is assigned rank 1, the second smallest observation (4) is assigned rank 2, and so on for the whole data array. If there are several observations with the same value, such as 51, they receive individual consecutive ranks. This array, with the corresponding ranks, is called the cumulative distribution.

We can generally see from the rank of an observation in the cumulative distribution how many observations in the data set are smaller or larger. Thus, since value 58 is 9th in the array, there are eight breaking stresses in the data set that are smaller than 58 milligrams, and there are eight that are larger. When several observations have the same value, as for the value 51, the rank of each such observation cannot be interpreted in this direct fashion.

A plot of the cumulative distribution provides a useful graphic display. Figure 2.4 shows a SYGRAPH plot of the cumulative distribution of the breaking stresses. The smallest observation has breaking stress 1 and rank 1, and it is plotted at point (1, 1) on the graph. The second smallest observation has breaking stress 4 and rank 2, and it is plotted at point (4, 2). The other observations are similarly plotted. We see readily from Figure 2.4 that several observations are outlying and that most of the observations are concentrated in the range from 40 to 80 milligrams.   ☐

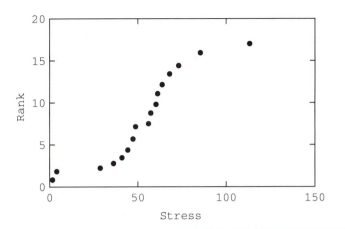

FIGURE 2.4
SYGRAPH plot of cumulative distribution of breaking stresses—Semiconductor failure example

As was demonstrated by this example, a cumulative distribution provides helpful information about the pattern in a data set. A formal definition of a cumulative distribution follows.

(2.1)
A **cumulative distribution** is the rank-ordered array of observations on a quantitative variable in a data set.

The type of cumulative distribution we have considered, in which the data are arrayed in ascending order and rank 1 is given to the smallest observation, is called a *less-than cumulative distribution.* Occasionally, the data are arrayed in descending order and rank 1 is assigned to the largest observation. Such distributions are called *more-than cumulative distributions.* Unless stated otherwise, we shall use the term cumulative distribution to refer to the less-than type.

Sometimes interest lies in the *percentage* of observations in the data array that are above or below given values. For that purpose, we express the rank as a percent of the total number of observations. In the semiconductor failure example, ranks 1, 2, . . . , 17 would be replaced by the percentages $100(1/17) = 5.9$, $100(2/17) = 11.8, . . . , 100(17/17) = 100$, respectively. We shall refer to the cumulative distribution in this case as a *cumulative percent distribution.*

When the number of observations on a quantitative variable is large, classifying the observations into a tabular distribution is helpful for studying the data pattern.

**FREQUENCY
DISTRIBUTIONS**   **2.2**

Savings Accounts   ☐ **EXAMPLE**

A bank has 30,794 savings accounts. The balances of these accounts are to be studied to assist management in revising the service charge schedule for savings account transactions. The analyst making the study concluded that an array of the balance amounts would be too long to be examined effectively and that neither a stem-and-leaf display nor a dot plot would be effective here because of the large number of observations. The analyst decided instead to prepare a tabular presentation showing the account balances classified into the five categories in column 1 of Table 2.1.

A routine in a statistical package was employed to read the balance amount for each account from the data file, assign it to the appropriate class, and count the number of accounts in each class. The counts for the five classes are shown in column 2 of Table 2.1. The total of the counts is the total number of accounts in the data file, namely, 30,794. The statistical routine also computed the percentage of all accounts that are in each class, as shown in column 3 of Table 2.1. For example, the first class (0–under 5000) contains 10,196 accounts that represent $100(10{,}196/30{,}794) = 33.1$ percent of all accounts. Thus, about one-third of all savings accounts in the bank had balances under $5000.   ☐

The tabular presentation in Table 2.1 is called a frequency distribution.

> **(2.2)**
> A **frequency distribution** is the classification of the elements of a data set by a quantitative variable.

Note that the frequency distribution in Table 2.1 has five categories or *classes.* The number of elements in a class is called the *frequency* of that class. Thus, the frequency of the class 0–under 5000 is 10,196. When the frequency of each class is expressed as a percentage of the total number of elements in the data set, it is called a *percent frequency* or *relative frequency,* and the resulting distribution is called a *percent frequency distribution* or *relative frequency distribution.*

A percent frequency distribution should show the total number of elements in the data set, as in Table 2.1 in which the total number of accounts (30,794) is shown in parentheses under the 100.0 percent total. Otherwise, the user does not know whether the percent fre-

---

TABLE 2.1
Frequency distribution of savings accounts by balance amount—Savings accounts example

(1) Balance Amount (dollars)	(2) Number of Accounts	(3) Percent of Accounts
0–under 5,000	10,196	33.1
5,000–under 10,000	15,335	49.8
10,000–under 15,000	1,812	5.9
15,000–under 20,000	1,798	5.8
20,000–under 25,000	1,653	5.4
Total	30,794	100.0 (30,794)

quencies are derived from a small or a large number of cases. The following percent frequency distribution on family shopping in supermarkets appears to provide authoritative data:

Number of Shopping Trips During Week	Percent of Families
1	21
2	64
3 or more	15
Total	100

However, the table does not disclose that it is based on a survey of only 14 families. Without information on the total number of elements in the data set, a percent frequency distribution can be easily misinterpreted.

Construction of Frequency Distributions

The classes of a frequency distribution must meet the requirements of any classification system.

> **(2.3)**
> The classes in any system of classification must be mutually exclusive and exhaustive.

In other words, every element in the data set must fall into one and only one class of the system.

The classification system used in Table 2.1 satisfies the requirements set out in (2.3). Each balance amount is assigned to *one* class among the five classes (so the five classes are exhaustive), and each balance amount qualifies for *only one* of the five classes (so the five classes are mutually exclusive, that is, not overlapping).

The determination of the classes to be utilized in constructing a frequency distribution requires care. The objective is to select a set of classes that loses as little essential information as possible while still effectively summarizing the data. Often, several systems of classification need to be tried before a final selection is made. Nevertheless, some general considerations are helpful for constructing a frequency distribution. We take these up now.

Number of Classes.   The larger the number of classes in a frequency distribution, the more detail is shown. If the number of classes is too large, though, the table loses its effectiveness for summarizing the data. Too few classes, on the other hand, condense the information so much as to leave little insight into the pattern of the distribution. The best number of classes in a frequency distribution often needs to be determined by experimentation. Usually, an effective number of classes is somewhere between 4 and 20.

EXAMPLE

In a study of the capability of untrained subjects to discriminate between different food preparations, 40 subjects were each given two food samples and asked for a rating of each sample. Unknown to the subjects, both food samples were identical in every respect. Each subject tasted both food samples and gave each sample a rating between 0 and 10. The difference between a subject's ratings for the first and second samples was taken as a measure of preference for the first sample over the second. Tables 2.2a, 2.2b, and 2.2c show the subjects' rating differences classified into 3, 7, and 21 classes, respectively.

At one extreme, the frequency distribution in Table 2.2c has so many classes that it is difficult to discern a meaningful pattern in the rating differences. At the other extreme, the frequency distribution in Table 2.2a has so few classes that any pattern present is almost lost. Finally, in Table 2.2b we have a frequency distribution that effectively reveals the pattern of variation in the rating differences. We note first that the rating differences are approximately symmetrically distributed about zero, which is not unexpected because both food samples were identical. An even more interesting observation is that the frequency of the class -1 to $+1$ is much lower than the frequencies of the classes immediately above and below it. It would therefore appear that the subjects tended to give different ratings to the two food samples even though they were identical. The subjects apparently believed the two samples must be different (for what other reason would the study be undertaken?) and consequently gave different ratings to them. These important characteristics of the rating differences, which are easily discernible from Table 2.2b, are not readily apparent from the other two frequency distributions.   ☐

TABLE 2.2

Three frequency distributions of subjects' rating differences for two food samples

(a) 3 Classes		(b) 7 Classes		(c) 21 Classes	
Rating Difference	Number of Subjects	Rating Difference	Number of Subjects	Rating Difference	Number of Subjects
-10 to -4	13	-10 to -8	1	-10	0
-3 to $+3$	19	-7 to -5	7	-9	1
$+4$ to $+10$	8	-4 to -2	12	-8	0
Total	40	-1 to $+1$	4	-7	2
		$+2$ to $+4$	11	-6	3
		$+5$ to $+7$	3	-5	2
		$+8$ to $+10$	2	-4	5
		Total	40	-3	3
				-2	4
				-1	2
				0	0
				$+1$	2
				$+2$	3
				$+3$	5
				$+4$	3
				$+5$	1
				$+6$	2
				$+7$	0
				$+8$	1
				$+9$	0
				$+10$	1
				Total	40

The very clear lesson provided by this example is that consideration of alternative numbers of classes is helpful for finding the frequency distribution that best yields insights into the data set.

Width of Classes. The choice of the class width or class interval is related to the determination of the number of classes. It is generally best if all the classes have the same width. If the classes are not equally wide, it is often difficult to tell whether differences in class frequencies result mainly from differences in the concentration of items or from differences in the class widths.

Sometimes, one must use unequal class intervals. Consider the construction of a frequency distribution for the annual salaries of a company's employees, where the salaries range from $20,000 to $360,000. Suppose that about six classes are to be used in the distribution. If equal class intervals were used, each would have a width of about $60,000, the first class perhaps being $0–under $60,000. This grouping provides equal class intervals but destroys the usefulness of the frequency distribution. Most of the employees earn less than $60,000 and hence would be included in the first class. Thus, no information would be provided about the distribution of salaries of the majority of the employees except that they earned less than $60,000. Subsequent classes would not reveal much information either, since relatively few employees earn over $60,000 in this company, yet five of the six classes would be devoted to a classification of their salaries.

In cases of this nature, unequal class intervals are generally used. For instance, equal class intervals of, say, $20,000 in width might be used for the range wherein most of the salaries fall, after which the interval might increase to, say, $60,000. In fact, when all but a few of the salaries have been classified, an open-end class might be used to account for the remainder. An *open-end class* is one that has only one limit, either upper or lower. Thus, if only three officials in the company earned over $200,000, the last class might be $200,000 and more. In this way, only one class is needed to account for these top three salaries. Open-end classes may be necessary at the upper end of the distribution, as just illustrated, sometimes at the lower end of the distribution, and occasionally at both ends.

Class Limits. Still another issue that must be considered in constructing a frequency distribution is the choice of class limits. Calculations from a frequency distribution often use the midpoint of each class to represent all the observations in the class. The *midpoint* of a class is the value halfway between the two class limits. It is usually suggested as good statistical practice that the class limits be chosen so that the midpoint of each class is approximately equal to the arithmetic average of the observations falling in that class. Often, this objective can be accomplished without special concern about the class limits. If, however, the observations tend to be bunched at certain periodic points throughout the range of the data—for example, rents are often multiples of $50 or $100—one may have to experiment in deciding upon class limits so that each midpoint will approximately equal the arithmetic average of the observations in that class.

In stating class limits, one should be careful to be unambiguous. For instance, the limits $300–$400, $400–$500 are not clear because one cannot be sure in which class $400 is included. Stating limits as $300–$399, $400–$499 is clear when the data are expressed in dollars. When they are, the midpoint of the first class is $(300 + 399)/2 = \$349.50$, or for most practical purposes, $350.

The limits $300–under $400, $400–under $500 are clear. However, without additional information, it may not be possible to determine the midpoints accurately. If no additional information is provided, we shall follow the convention of considering the midpoint of a class to be the arithmetic average of the two limits. Thus, the midpoint of the class \$300–under \$400 would be taken to be $(300 + 400)/2 = \$350$.

In some frequency distributions, each class corresponds to a single value. Table 2.2c illustrates this. As another example, in a frequency distribution of number of automobiles owned by a family, the classes might be the numbers 0, 1, 2, and so on. In each of these cases, the individual values correspond to the class midpoints.

Graphic Presentation of Frequency Distributions

Graphic displays of frequency distributions facilitate presentation and analysis. Two common methods of graphic display are the histogram and the frequency polygon.

> **(2.4)**
> A **histogram** is a rectangular graph of a frequency distribution. A **frequency polygon** is a line graph of a frequency distribution.

The method of constructing these graphic displays depends upon whether equal or unequal class intervals are present in the frequency distribution. We shall consider each case in turn.

Equal Class Intervals.   The following example illustrates both graphic methods when the frequency distribution has equal class intervals.

Farm Operators   ☐ **EXAMPLE**

Figure 2.5a presents a percent frequency distribution of ages of 27,303 farm operators classified in equal, 10-year intervals. The age data were obtained from a recent survey of farm operators conducted by a farm equipment manufacturer. Figures 2.5b and 2.5c show a histogram and frequency polygon, respectively, of this frequency distribution. In the histogram, adjoining rectangles are drawn with a width spanning the class interval and a height corresponding to the frequency (or percent frequency) of that class. In the frequency polygon, each class frequency (or percent frequency) is plotted corresponding to the midpoint of that class. The plotted points are then connected by straight lines. Both the histogram and the frequency polygon show clearly that the ages of the farm operators tend to be concentrated in the range from 35 to 75 years.   ☐

In both graphic presentations, the magnitudes of the class frequencies are conveyed visually by areas—in the case of the histogram, by the areas of the rectangles; and for the frequency polygon, by the area under the line graph. These areas are proportional to the frequencies and hence provide visual information about the pattern of the data.

Unequal Class Intervals.   When the class intervals of the frequency distribution are not equal, the heights of the histogram rectangles or the polygon points must be adjusted to make the *areas* proportional to the frequencies (or percent frequencies) of the classes. We make this adjustment by choosing a unit width and expressing all frequencies (or percent frequencies) in terms of this unit width. The procedure is illustrated in the following example.

(a) Percent Frequency Distribution

Age (years)	Percent of Farm Operators
15–under 25	2
25–under 35	10
35–under 45	19
45–under 55	27
55–under 65	25
65–under 75	16
75–under 85	1
Total	100 (27,303)

FIGURE 2.5

Graphic presentation of a frequency distribution with equal class intervals—Farm operators example

(b) Histogram

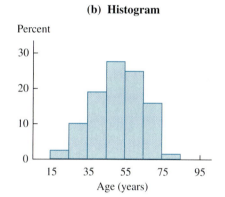

(c) Frequency Polygon

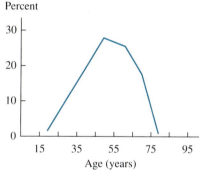

EXAMPLE ☐ **Legislative Committee**

A state legislative committee is reviewing income tax exemptions for taxpayers with annual incomes under $200,000. The percent frequency distribution of these taxpayers' incomes is shown in Figure 2.6a. Note that the frequency distribution has unequal class intervals. A graphic presentation of this frequency distribution is to be made to the committee.

   An interval of $10 thousand has been chosen arbitrarily as the unit width. We first determine the number of unit widths making up each class interval. Column 2 in Figure 2.6a shows these numbers. For example, the classes 0–under 10, 10–under 20, and 20–under 30 each have one unit width because each has a class width of $10 thousand. Class 30–under 50 has two unit widths because its class width is $20 thousand, which is equivalent to two $10 thousand units. Similar calculations are made for the remaining classes. We next calculate the percent frequency per unit width for each class by dividing the percent frequency in column 1 by the corresponding number of unit widths in column 2. The resulting adjusted percent frequencies are given in column 3. For instance, the adjusted percent frequency for the class 30–under 50 is 29.3/2 = 14.65. This adjusted frequency tells us that within the class 30–under 50, the average relative frequency for a $10 thousand subinterval (e.g., 30–under 40) is 14.65 percent.

FIGURE 2.6   **Graphic presentation of a frequency distribution with unequal class intervals—Legislative committee example**

(a) Percent Frequency Distribution

Income ($ thousand)	(1) Percent of Taxpayers	(2) Number of $10 thousand Widths	(3) Percent of Taxpayers per $10 thousand Width
0–under 10	0.6	1	0.6/1 = 0.60
10–under 20	10.2	1	10.2/1 = 10.20
20–under 30	13.4	1	13.4/1 = 13.40
30–under 50	29.3	2	29.3/2 = 14.65
50–under 100	36.4	5	36.4/5 = 7.28
100–under 200	10.1	10	10.1/10 = 1.01
Total	100.0 (1,049,916)		

(b) Histogram

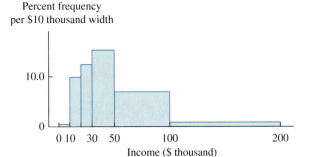

(c) Frequency Polygon

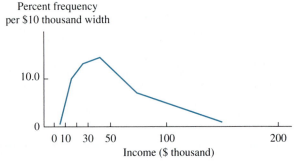

Once the adjusted percent frequencies are obtained, they are used to plot a histogram or frequency polygon in the usual fashion, as illustrated in Figures 2.6b and 2.6c.   ☐

Comparison of Frequency Polygons.   Two or more frequency polygons can be compared readily if (1) they have the same class intervals and (2) they have the same total frequency or are both expressed in percentage form.

☐ **EXAMPLE**

Figure 2.7 shows the age distribution of farm operators (taken from Figure 2.5c) as well as the age distribution of the nonfarm civilian labor force (as reported in a recent economic study), each expressed in percent frequencies. It is easy to compare the two distributions from Figure 2.7. We readily see that the farm operators are relatively much older than the nonfarm civilian labor force.   ☐

In contrast to frequency polygons, it is quite difficult to obtain an effective comparison of two or more frequency distributions by superimposing their histograms.

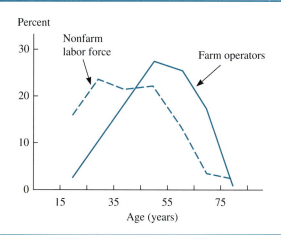

FIGURE 2.7
Comparison of age distributions of farm operators and
nonfarm civilian labor
force—Farm operators
example

Cumulative Frequency Distributions

When data are solely available in the form of a frequency distribution, the cumulative distribution of the original data set can be approximated by expressing the frequency distribution in the form of a cumulative frequency distribution. Such a distribution provides an effective summary of the number or percentage of the observations that are above or below given values.

(2.5)
The cumulative form of a frequency distribution is called a **cumulative frequency distribution.**

EXAMPLE ☐

Figure 2.8a repeats the percent frequency distribution of the ages of farm operators from Figure 2.5a, and Figure 2.8b shows the cumulative percent frequency distribution. For instance, the cumulative percent frequency distribution in Figure 2.8b indicates that 2 percent of farm operators are less than 25 years of age and 12 percent are less than 35 years of age.

Figure 2.8c presents a plot of the cumulative percent frequency distribution. Here the cumulative frequency or percentage is plotted against the corresponding class limit. Thus, the points plotted begin with (15, 0) and (25, 2) and end with (85, 100). The points are then connected by straight lines to permit interpolation. Thus, one can estimate from Figure 2.8c (see the dashed line) that 50 percent of farm operators are less than 52 years of age. ☐

The use of straight lines to connect the plotted points of the cumulative frequency distribution assumes that the observations within each class are uniformly distributed throughout the class. Since this condition is usually not met exactly, the resulting graph is

FIGURE 2.8
Cumulative percent frequency distribution of ages of farm operators—Farm operators example

(a) **Percent Frequency Distribution**

Age (years)	Percent of Farm Operators
15–under 25	2
25–under 35	10
35–under 45	19
45–under 55	27
55–under 65	25
65–under 75	16
75–under 85	1
Total	100

(b) **Cumulative Percent Frequency Distribution**

Less than This Age (years)	Cumulative Percent of Farm Operators
15	0
25	2
35	12
45	31
55	58
65	83
75	99
85	100

(c) **Plot of Cumulative Percent Frequency Distribution**

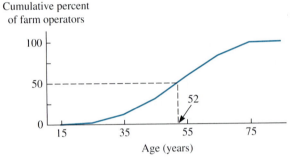

only an approximation to the actual cumulative distribution of the observations. The straight-line approximation is adequate in many cases.

No special procedures are required for plotting cumulative frequency distributions with unequal class intervals. (Recall that unequal class intervals entail special procedures for constructing a histogram or a frequency polygon.) Moreover, plots of different cumulative frequency distributions can be compared readily on the same graph even if the distributions have different class intervals, as long as the total frequencies are the same or percent frequencies are used.

When each class of a frequency distribution consists of a single value, the plot of its cumulative frequency distribution should be in the form of a step function.

Recreation Center **EXAMPLE**

Figure 2.9a shows the cumulative percent frequency distribution of family size for the 315 families belonging to a community recreation center. Figure 2.9b shows the plot of the cumulative percent frequency distribution. The plot has a step at each whole number because no intervening values can occur. Note that the height of a step equals the percentage of all families of that size. Also note that the plot in Figure 2.9b is exact, not an approximation, because each class consists of a single value. □

A more-than cumulative frequency distribution can be constructed in the same manner as a less-than cumulative frequency distribution.

Cumulative percent frequency distribution of family sizes—Recreation center example. The plot of the cumulative frequency distribution is a step function when each class consists of a single value.

FIGURE 2.9

(a) Cumulative Percent Frequency Distribution

This Number or Less	Cumulative Percent of Families
1	0
2	33
3	57
4	76
5	87
6	100

Total number of families = 315

(b) Plot of Cumulative Percent Frequency Distribution

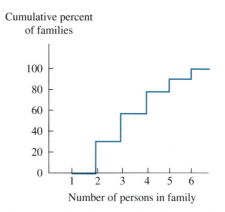

EXAMPLE ☐

The more-than cumulative percent frequency distribution of the ages of farm operators, based on the data in Figure 2.8a, is as follows:

This Age or More (years)	Cumulative Percent of Farm Operators
15	100
25	98
35	88
45	69
55	42
65	17
75	1
85	0

☐

DISPLAYS OF QUALITATIVE DATA 2.3

We now turn our attention to data for qualitative variables (that is, variables having non-numerical outcomes). As in the case of quantitative variables, patterns in large data sets are most readily studied if the data are classified into a distribution.

(2.6)

A **qualitative distribution** is the classification of the elements of a data set by a qualitative variable.

Construction of Qualitative Distributions

The general principles of classification discussed for frequency distributions apply equally to qualitative distributions. Thus, the classes of a qualitative distribution must be mutually exclusive and exhaustive. Also, the number of classes utilized should be small enough to provide an effective summary, yet large enough to avoid losing essential information.

Occupational Profile

 EXAMPLE

An analyst for a large retailer wished to study the occupational profile of the company's credit card customers and to compare it with the occupational profile of the total labor force as reported in a recent government publication. The analyst expected that the comparison would provide insights useful for credit policy, advertising, and billing.

The analyst had available a computer data file containing occupational data obtained from a recent survey of the company's customers. Altogether, information was available for 131,845 customers who were in the labor force. To effectively compare the occupational profiles of customers and the total labor force, the analyst decided to classify the occupations of the credit card customers into the seven categories utilized for the government labor force study shown in Table 2.3. A routine in a statistical package was used to read each customer's occupation from the data file (carpenter, dietitian, civil engineer, and so on), assign it to the appropriate occupational category, and count the number of customers in each category. The results are shown in Table 2.3, column 1. The routine also was used to compute the percentage of customers who are in each occupational category, as shown in column 2 of Table 2.3, so that the company profile could be readily compared to that of the total labor force.

In column 3, the analyst added the corresponding percentages for the total labor force. In comparing the two distributions, the analyst discovered, among other things, that over half of the customers (53.2 percent) fall into the managerial and professional and technical categories, whereas these categories contain only 19.1 percent of the total labor force. This one fact alone had significant implications for the company. □

TABLE 2.3
Qualitative distribution of customers by occupation— Occupational profile example

Occupation	(1) Number of Customers	(2) Percent of Customers	(3) Percent of Total Labor Force
Managerial	38,835	29.5	9.2
Professional and technical	31,262	23.7	9.9
Service and recreation	14,011	10.6	10.9
Clerical and sales	12,797	9.7	20.7
Crafts and production workers	11,090	8.4	24.2
Laborers and unskilled workers	1,577	1.2	5.0
Others	22,273	16.9	20.1
Total	131,845	100.0 (131,845)	100.0

Graphic Presentation of Qualitative Distributions

A useful display of a qualitative distribution is by means of a *bar chart.* Figure 2.10a presents a *simple bar chart* for the qualitative distribution of customers' occupations in Table 2.3. Each bar corresponds to one class of the qualitative distribution, with the length of the bar denoting the number (or percentage) of customers in the class.

   Several aspects of the arrangement of the bars in Figure 2.10a should be noted: (1) The bars differ only in length, not in width; (2) a space has been left between each bar to make it easier to identify each by its label; (3) the bars have been ranked by order of magnitude to facilitate analysis. The order may be a decreasing one, as here, or it may be an increasing one. If an "others" or "miscellaneous" category is present, it is usually shown as the lowest bar regardless of its magnitude, because this type of category most often is a collection of relatively unimportant classes.

   When a pictorial comparison of two or more qualitative distributions with the same classes is desired, a *combined bar chart* can be utilized. Figure 2.10b illustrates this type of chart, presenting the occupational profiles of both the company's customers and the total labor force. Note that both distributions are expressed in percentage terms to provide a meaningful comparison.

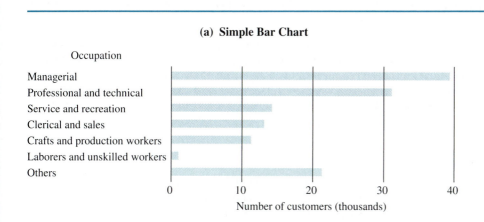

(a) Simple Bar Chart

FIGURE 2.10
Bar charts of qualitative distributions—Occupational profile example

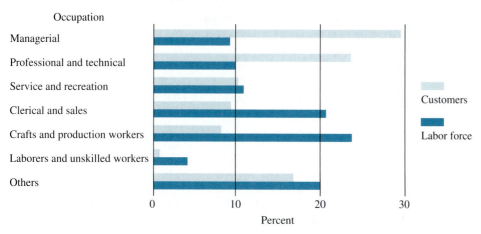

(b) Combined Bar Chart

Modifications can be made in the bar charts in Figure 2.10 to show additional details or to highlight particular features of the data. For example, the bars in Figure 2.10a could be segmented to show the numbers of male and female customers in each occupational category.

Graphics packages enable users to readily generate various types of bar charts for display and analysis of qualitative distributions.

2.4   DISPLAYS OF BIVARIATE DATA

Like the analysis of univariate data, the analysis of bivariate data usually begins with a study of the underlying pattern of the data. One of the major objectives for analyzing bivariate data is to gain insights into the nature of the relationship between the two variables. Bivariate data sets may be based on two quantitative variables, two qualitative variables, or one variable of each type. We shall begin with the case of two quantitative variables.

Quantitative Bivariate Data

Both graphic and tabular displays may be used to study the pattern of relationship between two quantitative variables.

Scatter plots.   A simple graphic display that is very helpful in studying the relationship between two quantitative variables is a *scatter plot*. In a scatter plot, the observations for the two variables for each element in the data set are plotted in a two-dimensional graph. The pattern of points indicates whether a relationship exists between the two variables and, if so, the nature of the relationship.

Delivery Fleet   □ **EXAMPLE**

A company's accountant examined bivariate data on the annual cost of maintaining and operating each of the 12 vehicles in the company's delivery fleet and the distance that each vehicle traveled during the year. To examine the relationship between the two variables, the accountant prepared the SYGRAPH scatter plot shown in Figure 2.11. The coordinates of the point for a vehicle are the distance traveled (in thousand miles) and the annual cost (in thousand dollars). The pattern of points suggests that annual cost tends to increase with the distance traveled, as would be expected.   □

Time Series Plots.   Frequently, one of the two variables in a bivariate data set is time. For instance, the bivariate data set may consist of company sales for the past 10 years. A *time series plot,* also called a *time series line graph,* is an effective graphic display for such bivariate data.

Capital Expenditures   □ **EXAMPLE**

In preparing for a meeting of the company's board of directors to discuss capital expenditures plans for the next year, the president of a major oil company asked for a graphic display of the company's annual capital expenditures during 1981–1990. Figure 2.12 presents a SYGRAPH time series plot of the capital expenditures. The points in the

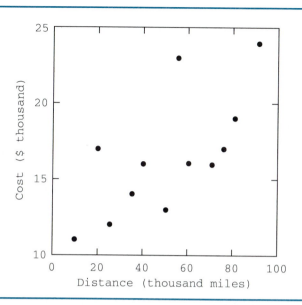

FIGURE 2.11
SYGRAPH scatter plot of
annual cost and distance
traveled—Delivery fleet
example

plot, corresponding to consecutive years, have been joined by straight lines to show more clearly the time pattern of the annual capital expenditures. The graph shows that capital expenditures experienced a generally upward trend over the decade.   □

Bivariate Frequency Distributions.   When the data set is large, a tabular distribution can be useful for revealing the nature of the relationship between two quantitative variables, as well as for effectively summarizing the data. The preparation of the tabular distribution requires that the elements of the bivariate data set be classified into a number of classes that are defined simultaneously in terms of both variables. This classification process is called *cross-classification* or *cross-tabulation*. The resulting cross-classified distribution is called a *bivariate frequency distribution*. The principles discussed earlier for classifying univariate data apply equally to bivariate data.

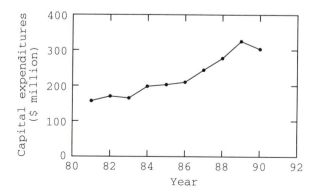

FIGURE 2.12
SYGRAPH time series plot of
capital expenditures—Capital expenditures example

European Tours   ☐ **EXAMPLE**

A government tourist bureau compiled a list of 322 escorted European tours of eight days or less that are offered by private agencies. Two characteristics of interest for each tour are its duration in days and its cost in U.S. dollars. To summarize the data set and to study the relation between the two variables, the bureau used the classes shown in Table 2.4. Each tour from the data set is counted in the cell of the table that corresponds to its duration and cost. Note that the marginal totals of the columns form the univariate frequency distribution of the tour durations, and the marginal totals of the rows form the univariate frequency distribution of the tour costs.

The bivariate frequency distribution in Table 2.4 brings several interesting facts to light. One is that an escorted tour of three days, costing up to $600, is the most commonly offered type of tour. Another is that the cost of an escorted tour tends to vary directly with its duration—a very reasonable relationship.   ☐

Quantitative and Qualitative Bivariate Data

When the data set consists of a quantitative variable and a qualitative variable, both graphic and tabular displays can be used to summarize the data and examine the relationship between the two variables. We have already illustrated the use of graphic displays for this case, but without drawing attention to this fact.

☐ **EXAMPLES**

1. The back-to-back stem-and-leaf display for the cake ingredient example in Figure 2.2 effectively shows the relation between the quantitative variable cake volume and the qualitative variable dairy ingredient. Note that this display can only be used if the qualitative variable has two classes.

2. Figure 2.3b displays aligned dot plots for the cake ingredient example to study the relation between cake volume and dairy ingredient. Note that because more than two dot plots can be aligned in the same display, this graphic method is applicable for any number of classes that the qualitative variable may have.   ☐

When the bivariate data set is large, the relationship between a quantitative variable and a qualitative variable can often be usefully examined by a tabular display showing the percent frequency distribution for the quantitative variable for each class of the qualitative variable. If the number of classes of the qualitative variable is not too large, the frequency polygons for these distributions can be plotted on the same graph for effective comparison.

TABLE 2.4
Bivariate frequency distribution of escorted tours, cross-classified by cost and duration — European tours example

Cost (U.S. $)	Duration (days)						Total
	3	4	5	6	7	8	
1– 600	53	43	9	–	–	–	105
601–1200	20	45	37	12	26	1	141
1201–1800	4	9	11	6	17	12	59
1801–2400	–	–	1	1	5	10	17
Total	77	97	58	19	48	23	322

EXAMPLE ☐ Parkinson's Disease

Figure 2.13a shows a bivariate distribution for 214 patients suffering from Parkinson's disease, cross-classified by age at onset of the disease and patient's gender. To study the relationship between the two variables, we construct percent frequency distributions for each gender, shown in Figure 2.13b. Figure 2.13c presents the frequency polygons of the percent frequency distributions for men and women on the same graph. The data clearly reveal that the patterns of age at onset differ for men and women, the age at onset tending to come later for women than for men. ☐

Qualitative Bivariate Data

We consider, finally, tabular and graphic methods for studying the relationship between two qualitative variables in a bivariate data set. We shall illustrate these with an example.

EXAMPLE ☐ Oil Wells

In a recent year, 2504 oil and gas wells were drilled in the four districts of a region. Each well was one of three possible types: new-field wildcat well, other exploratory well, or development well. Table 2.5a shows the *bivariate qualitative distribution* for these data by type of well and district. Each of the 2504 wells is counted in the cell that corresponds to its type and district.

Observe in Table 2.5a that the marginal totals of the columns form the univariate

---

FIGURE 2.13
Parkinson's disease patients cross-classified by age at onset of disease and by gender—Parkinson's disease example

(a) Bivariate Distribution

Age at Onset of Disease

Gender	30–39	40–49	50–59	60–69	70–79	80–89	Total
Male	10	28	43	37	7	1	126
Female	6	12	26	29	10	5	88
Total	16	40	69	66	17	6	214

(b) Percent Frequency Distributions

Age at Onset (years)	Male Patients	Female Patients
30–39	7.9	6.8
40–49	22.2	13.6
50–59	34.1	29.5
60–69	29.4	33.0
70–79	5.6	11.4
80–89	0.8	5.7
Total	100.	100.
	(126)	(88)

(c) Frequency Polygons

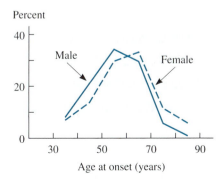

TABLE 2.5
Bivariate qualitative distribution of oil and gas wells drilled, cross-classified by type of well and district —Oil wells example

(a) Bivariate Distribution

Type	District				Total
	I	II	III	IV	
New-field wildcat	338	60	113	10	521
Other exploratory	223	11	28	3	265
Development	1131	72	461	54	1718
Total	1692	143	602	67	2504

(b) Percent Distribution for Each District

Type	District			
	I	II	III	IV
New-field wildcat	20.0	42.0	18.8	14.9
Other exploratory	13.2	7.7	4.7	4.5
Development	66.8	50.3	76.6	80.6
Total	100.	100.	100.	100.
	(1692)	(143)	(602)	(67)

Note: Percentages may not add to 100 because of rounding.

qualitative distribution of the district variable. Similarly, the marginal totals of the rows form the univariate qualitative distribution of the type-of-well variable.

It is not too easy to examine the nature of the relationship between type of well and district from Table 2.5a. Use of a tabular display based on percent distributions will bring out the nature of the relationship between the two variables more effectively. Table 2.5b presents the percent distribution of type of well for each district. It can be seen by comparing these distributions that district II differs from the other three districts, with relatively many new-field wildcat wells and relatively fewer development wells.

FIGURE 2.14

Component-part bar chart showing percent distribution of oil and gas wells drilled for each district—Oil wells example

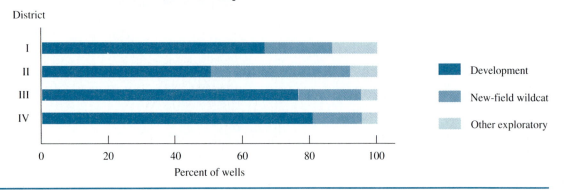

An effective graphic display of the percent distributions in Table 2.5b is provided by the *component-part bar chart* in Figure 2.14. This chart contains a bar for each district that is segmented into components representing the three types of wells. Each component has a length equal to the percentage of wells that are of this type. Different shadings or colors are generally used to help the reader distinguish the components and make comparisons. Here we have used different shades of color for this purpose, with a legend giving the correspondence between the shades and the types of wells. Again note that district II differs markedly from the other three districts in terms of the types of wells drilled.   □

DISPLAYS OF   2.5
MULTIVARIATE DATA

Multivariate data sets can be analyzed for patterns and relationships along the same lines as bivariate data sets, but the analysis is more difficult when the number of variables is large. Therefore, powerful multivariate statistical methods have been developed, such as multiple regression analysis (taken up in Chapter 20). We now illustrate the use of a simple graphic display, called a *symbolic scatter plot,* which is useful when there are three or four variables in the data set.

EXAMPLE □

Figure 2.15a contains, for the delivery fleet example, the annual cost of maintaining and operating a vehicle, the distance traveled by the vehicle during the year, and

Data set and SYGRAPH symbolic scatter plot—Delivery fleet example   **FIGURE 2.15**

(a) Data Set

Vehicle	Distance (thousand miles)	Cost ($ thousand)	Type of Vehicle
1	10	11	Van
2	25	12	Van
.	.	.	.
.	.	.	.
.	.	.	.
11	80	19	Truck
12	90	24	Truck

(b) Symbolic Scatter Plot

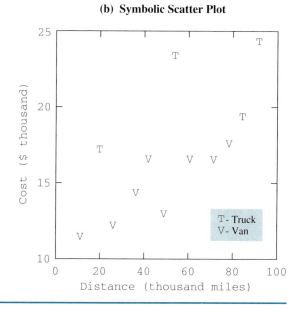

whether the vehicle is a van or a truck, for some of the 12 vehicles in the fleet. The accountant wished to examine graphically the effects of both distance traveled and type of vehicle on annual cost. The accountant therefore prepared the SYGRAPH symbolic scatter plot in Figure 2.15b. This plot is the same as the scatter plot in Figure 2.11, but uses different symbols (V and T) for vans and trucks. Note that vehicle 1 is plotted at the position (10, 11) corresponding to its distance (10) and cost (11), as in Figure 2.11, but now the plotting symbol is a V because the vehicle is a van. It is not only readily apparent from the plot in Figure 2.15b that cost tended to increase with distance, but also that cost tended to be higher for trucks than for vans, as would be expected.   □

2.6   DATA MODELS AND RESIDUAL ANALYSIS

We have presented a number of tabular and graphic methods that are effective in revealing patterns or relationships in data. Often, we wish to compare these observed patterns or relationships with ones that are anticipated on the basis of a theoretical model or from previous experience. Residual analysis is an effective technique for making such comparisons.

A *residual* is the difference between an *observed value* and the corresponding *anticipated value*. A graphic presentation of residuals, called a *residual plot,* is useful for highlighting major departures between the observed and the anticipated patterns or relationships in a data set. We shall illustrate residual analysis and residual plots by two examples.

□ EXAMPLE

In the concert subscribers example, we mentioned earlier that the analyst noted that respondents frequently reported their ages with second digits of 0 or 5. Figure 2.16a repeats the stem-and-leaf display from Figure 2.1a. Figure 2.16b shows a residual plot for the analyst's anticipation that the final digits in age would occur with equal frequency. Since the data set contains 70 readings and 10 digits are involved, each final digit should appear about 7 times under the expectation of equal frequency. The final digit 0 appears 15 times in the leaves, so the residual is $15 - 7 = 8$ for this digit. This

FIGURE 2.16
Stem-and-leaf display and residual plot—Concert subscribers example

(a) **Stem-and-Leaf Display**

Subscribers' ages (years)

```
2 | 0 5 5 5 9
3 | 0 0 1 2 3 3 4 5 5 6 7 8
4 | 0 0 0 0 1 2 2 3 4 4 5 5 6 7 7 8 9 9
5 | 0 0 0 0 0 1 1 2 3 3 4 5 5 5 6 7 8 8 9
6 | 0 0 2 3 4 5 5 6 7 9
7 | 0 1 5 5 6 7
```

(b) **Residual Plot**

residual is represented by the leftmost dot in the residual plot. The final digit 1 appears 5 times; hence, its residual is $5 - 7 = -2$. This residual is represented by the second dot from the left. The residuals for the other digits are calculated and plotted in the same manner. Note how readily the residual plot shows the excessive numbers of 0 and 5 final digits and the correspondingly lower-than-anticipated frequencies for the other digits.

The analyst concluded that the most likely explanation for this behavior of the residuals is end-digit preferences by some respondents who tend to round their reported ages to multiples of 5 or 10. The analyst therefore decided to ask for year of birth on the questionnaire for the large-scale study and then obtain each subscriber's age from the year of birth as a data transformation.   □

EXAMPLE □   **Motor Rod**

(Adapted from Ref. 2.1, pp. 280–281.) A company that manufactures small motors learned that some of the motors produced recently were breaking down. One possible cause was a rod in the motor being too loose in the bearing within which it rotates. Other possible causes included defects in the bearing. A consultant who was asked to locate the cause of the breakdowns learned that the diameter of each rod is inspected by the manufacturing department and that any rod with a diameter less than the lower specification limit of 1.000 centimeter is to be scrapped. A histogram of the diameters of rods recently inspected, based on records kept by the inspectors, is shown in Figure 2.17a. Also shown there is a smooth curve that the consultant drew as the expected model for the data. This model reflects the pattern of variability found by the consultant in similar situations in other companies.

Note that the histogram and the model are in close accord except in the interval around the lower specification limit, which is marked by an arrow in the figure. The residuals for each class were obtained and are plotted in Figure 2.17b. The consultant was interested in the size of the large negative residual just below the specification limit and the large positive residual in the next-higher class. Some systematic factor appeared to be operating here that tended to create large residuals in opposite directions in the two adjacent classes. Upon further investigation, the consultant found that

(a) Histogram and Model   **(b) Residual Plot**

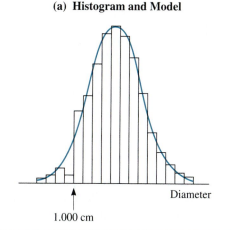

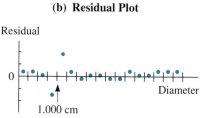

FIGURE 2.17
Histogram and residual plot—Motor rod example

the inspectors were under pressure to keep the scrappage rate low and had been mis-classifying borderline-defective rods by recording their diameters as borderline-acceptable. This misclassification led to too few rods recorded just below 1.000 centimeter and too many rods recorded just above the specification limit. Further investigation showed that the misclassified rods were in most cases the cause of the motors breaking down.   □

These examples illustrate how effective residual plots are for discovering and analyzing major differences or confirming close agreement between the observed patterns or relationships in a data set and anticipated ones based on theory or related experience. We shall make further use of residual plots in later chapters.

2.7   MISLEADING GRAPHIC PRESENTATIONS

Graphic presentations of data are widely encountered in daily life—for example, in newspapers, magazines, and annual reports. Often the graphic presentations are appropriate, sometimes they are misleading. The misrepresentations may be intentional at times, but often are the result of inadequate understanding of proper graphic presentation. We shall briefly consider misuses of pictorial charts and line graphs.

Pictorial Charts

Pictorial charts use symbols to convey information about magnitudes. They are widely used in popular presentations. When symbols of the same size are employed in a bar chart, the graph presents a proper picture. When the symbols vary in size, however, the graphic presentation may be misleading because one reader may compare the heights or widths of the symbols while another may compare the areas. In any case, it is usually difficult to obtain a proper impression of magnitudes from symbols of different sizes. We shall illustrate this point by an example.

Moscow's Top Attractions   □   **EXAMPLE**

Figure 2.18a illustrates a pictorial bar chart that compares the average daily number of visitors to the Lenin mausoleum and to McDonald's restaurant in Moscow (Ref. 2.2). The symbols are all of the same size and it is easy to compare the lengths of the bars.

Figure 2.18b presents the same data by means of two symbols of different sizes. Note how difficult it is to compare the magnitudes represented by the two symbols.   □

Line Graphs

Line graphs are also widely used for popular presentations. Misleading impressions may result from the origin and scale of the Y axis, from the proportions of the X and Y scales, and from the perspective of the plot. We shall illustrate proper and improper presentations by line graphs through the following example.

Illustrations of pictorial charts—Moscow's top attractions example

FIGURE 2.18

(a) Pictorial Bar Chart

Lining Up for Moscow's Top Attractions

1000 visitors
per day

(b) Pictorial Symbols

Lenin mausoleum   McDonald's

Source: (a) Redrawn and reproduced with permission from *Fortune,* "McDonald's Beats Lenin 3 to 1," December 17, 1990, p. 13.

EXAMPLE ☐

The four panels in Figure 2.19 show different plots of the same time series on annual sales revenues (in $ million) for a consumer goods company during the period 1983–1992. Observe that each panel gives a different visual impression of the pattern of sales over the period. Figure 2.19a is in a more or less conventional graphic format. In this figure, the sales fluctuations from year to year appear to be moderate, with an overall upward trend during the latter half of the period. The dimensions of the graph are balanced. The vertical scale has a natural origin corresponding to sales of $0, and the scale ranges from $0 to $40 million—the latter value being somewhat bigger than the largest value in the time series ($30 million). Figure 2.19b gives the impression of significant year-to-year fluctuations in sales, with a sharp increase from 1987 to 1992. This effect results because the vertical scale begins arbitrarily at about $24 million and is elongated relative to the horizontal scale. Figure 2.19c gives the impression that sales scarcely fluctuated from year to year and were quite level during the period. This result is due to the fact that the figure has substantially different proportions for the two scales. Moreover, the vertical scale ranges from $0 to $60 million—the latter value being twice as big as the largest value in the time series. Figure 2.19d is a perspective representation of the plot in Figure 2.19a. The perspective may give the careless reader

FIGURE 2.19
Four plots of the same time series that give different visual impressions of the time series pattern

(a) Conventional Format

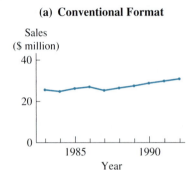

(c) Elongated _X_ Scale

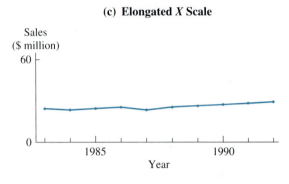

(b) Broken and Elongated _Y_ Scale

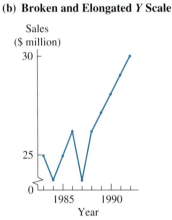

(d) Perspective Representation

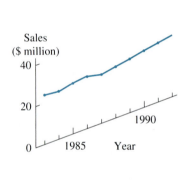

the impression that sales have increased rapidly over the period because the time series seems to rise quickly from left to right.   ☐

These examples demonstrate the need for care in preparing graphic charts and in interpreting them.

CITED REFERENCES

2.1   Deming, W. Edwards, "Making Things Right." Pages 279–286 in _Statistics: A Guide to the Unknown,_ 2nd ed. Edited by J. W. Tanur. San Francisco: Holden-Day, 1978.

2.2   "McDonald's Beats Lenin 3 to 1," _Fortune,_ December 17, 1990, p. 13.

PROBLEMS

* **2.1 Data Transmission.** A communications engineer is studying error bursts in the transmission of digital data. The following are the numbers of erroneous data bits transmitted in each hour during the past 24 hours over one transmission link:

5	24	36	2	1	3	19	55	1	2	4	20
4	1	24	30	0	1	8	2	27	0	2	22

a. Array the observations by increasing order of magnitude.
b. Present the observations in a stem-and-leaf display.
c. Present the observations in a dot plot.
d. In what ranges do the observations tend to cluster? Do any observations appear to be outlying?

2.2 The annual long-distance telephone expenditures (in $ thousand) last year of 18 sales representatives were as follows:

6.2	5.4	6.7	5.8	4.5	4.7	7.1	6.3	6.5
3.4	5.6	3.1	6.4	5.0	6.1	6.1	5.6	5.1

a. Array the observations by increasing order of magnitude.
b. Present the observations in a stem-and-leaf display.
c. Present the observations in a dot plot.
d. How many sales representatives had expenditures of less than $4 thousand for the year? Do these expenditures appear unusual relative to the others? Comment.

2.3 Hotel Chain. The numbers of rooms in the 16 hotels of the Cathay chain are as follows:

200	252	182	352	364	180	226	164
192	477	315	600	296	249	110	117

a. Array the observations by increasing order of magnitude.
b. Present the observations in a stem-and-leaf display, using only the first two digits of each number. For example, record 200 as 20, 252 as 25, and so on.
c. Present the observations in a dot plot.
d. (1) How many rooms does the largest hotel in the chain have? The smallest? (2) Would it be reasonable to claim that most hotels in this chain have between 100 and 400 rooms? Comment.

2.4 Refer to **Hotel Chain** Problem 2.3. The Pacific Rim chain is a competitor of the Cathay chain. The numbers of rooms in the 10 hotels of the Pacific Rim chain are as follows:

507	308	714	265	426	290	248	342	335	311

a. Present the data sets for the two chains in a back-to-back stem-and-leaf display, using only the first two digits of each number. For example, record 507 as 50, 308 as 30, and so on.
b. Present the two data sets in dot plots aligned on the same scale.
c. (1) Which chain tends to have the larger hotels as measured by number of rooms? (2) For each chain, what is the range of rooms in which the majority of hotels are concentrated? Are these ranges consistent with your conclusion in part c(1)? Comment.

2.5 Repair Costs. An experiment was conducted to compare the repair costs of two types of pumps (A, B). Repair cost records were kept for 16 pumps of each type for one year of pump operation. The annual repair costs of the pumps (in $ thousand) were as follows:

Type A:	0.4	1.8	4.0	1.8	2.3	1.9	0.8	1.4
	3.4	1.0	2.0	0.5	1.6	3.9	1.6	1.8
Type B:	1.2	0.8	2.7	0.8	0.5	1.1	0.7	1.3
	0.9	3.9	1.4	2.6	0.6	0.6	1.5	2.0

a. Present the two data sets in a back-to-back stem-and-leaf display.
b. Present the two data sets in dot plots aligned on the same scale.
c. Does one type of pump seem to have a worse repair cost record than the other? Comment.

* **2.6** Refer to **Data Transmission** Problem 2.1.

a. Plot the less-than cumulative distribution for the data set, as in Figure 2.4.
b. (1) What observation has rank 1? Rank 12? (2) How many observations are 19 or smaller?

2.7 Refer to **Hotel Chain** Problem 2.3.

a. Plot the less-than cumulative distribution for the data set, as in Figure 2.4.

b. (1) How many hotels in the chain have 364 or fewer rooms? (2) The smallest eight hotels in the chain have how many rooms or less?

2.8 Refer to the data set in Figure 2.2 for the lean dairy ingredient.

a. Plot the less-than cumulative percent distribution for this data set.

b. From your plot in part a, determine the percentage of cakes baked with the lean ingredient that have volumes of 370 or less.

2.9 For each of the following systems of classification, indicate whether or not it meets the formal requirements in (2.3).

a. Classes $5.00–$10.00, $10.00–$15.00, $15.00 or more, for hourly wages rates.

b. Classes 3 or less, 4 or more, for number of children in families.

c. Classes 0°–under 90°, 100°–under 190°, 200° and higher, for the temperature of ovens.

2.10 A frequency distribution of monthly house rents is to be constructed. The rents tend to be in multiples of $100 and range from $1000 to $1900. Class intervals of $200 are to be employed. What class limits would be most appropriate? Why?

∗ **2.11 Response Times.** The distribution of response times for alarms answered last year by the fire company in a township follows:

Response Time (minutes)	Number of Alarms
2.5–under 7.5	5
7.5–under 12.5	18
12.5–under 17.5	15
17.5–under 22.5	9
22.5–under 27.5	3
Total	50

a. What is the width of the first class?

b. What is the midpoint of the first class?

c. Are any of the classes open-end?

2.12 Horse Trials. The age distribution of horses competing in the Winston County Trials during a recent year follows:

Age (years at last birthday)	Number of Horses	Age (years at last birthday)	Number of Horses
5	26	10	11
6	60	11	8
7	51	12	3
8	44	13	0
9	23	14	1
		Total	227

a. What is the width of age class 5?

b. What is the midpoint of age class 5? What would be your answer if age were measured in years to nearest birthday?

2.13 Refer to the frequency distribution in Table 2.1.

a. Do the classes have equal widths?

b. What is the midpoint of the second class?

c. Is there an open-end class?

∗ **2.14** Refer to **Data Transmission** Problem 2.1.

a. Construct a frequency distribution for these data, using equal class intervals 0–9, 10–19, and so on.

b. In separate graphs, plot the frequency distribution in part a as a frequency polygon and as a histogram.

c. Does the histogram in part b show that the observations tend to cluster in several distinct ranges? Comment.

2.15 Refer to **Hotel Chain** Problem 2.3.

a. Construct a percent frequency distribution for these data, using equal class intervals 50–under 150, 150–under 250, and so on.

b. In separate graphs, plot the percent frequency distribution in part a as a frequency polygon and as a histogram.

c. At what number of rooms does the frequency polygon achieve its greatest frequency? Are the observations in the data set more concentrated about this number than any other? Comment.

2.16 **Universal Health Care.** Thirty-six persons were asked to rate their approval of universal health care on an eight-point scale ranging from 1 (strong disapproval) to 8 (strong approval). The ratings follow:

5	2	7	6	3	7	8	3	2	6	3	6
3	7	5	3	6	7	3	7	6	4	3	5
8	6	5	4	3	6	6	5	7	8	4	3

a. Construct a frequency distribution of ratings, using classes 1, 2, . . . , 8.

b. Plot the frequency distribution as a histogram.

c. Does the pattern of the histogram in part b suggest that the 36 persons are concentrated in two groups with differing points of view on universal health care? Comment.

2.17 Refer to **Response Times** Problem 2.11.

a. Convert the frequency distribution into a percent distribution and plot it as a frequency polygon and as a histogram in separate graphs.

b. Which of the two graphs in part a would be more effective for comparing the distribution of response times with that for a fire company in a nearby township? Explain.

2.18 Refer to **Horse Trials** Problem 2.12.

a. Convert the frequency distribution into a percent distribution and plot it as a histogram.

b. Does it appear from the histogram that horses under 5 years of age were not permitted to compete? Horses over 14 years of age? Explain.

c. Suppose the frequencies for the age classes 5 and 6 were 60 horses aged 5 and 26 horses aged 6. Would this new pattern suggest that an unusual factor is present that affects the ages of entering horses? Explain.

* **2.19** **Stock Holdings.** The distribution of a company's stock according to the number of shares held (including fractional shares) follows:

Number of Shares Held	Percent of Stockholders
0–under 50	25
50–under 100	19
100–under 500	41
500–under 1000	12
1000–under 5000	3
Total	100

a. Plot a histogram for the distribution, using 100 shares as the unit width.

b. Which class of the histogram has the largest adjusted percent frequency? Does the adjusted percent frequency decline consistently as the number of shares held increases? Comment.

c. Would a pair of histograms or a pair of frequency polygons be more effective for comparing this company's distribution of stock with that of another company? Explain.

2.20 Product Sales. An industrial chemical company sells 102 different products. The distribution of the annual sales amounts last year for these products follows:

Annual Sales ($ thousand)	Number of Products
0–under 10	19
10–under 50	44
50–under 100	22
100–under 200	13
200–under 500	4
Total	102

a. Upon studying this distribution, an executive of the company stated that products with annual sales between $10 thousand and $50 thousand are the most common. Do you agree or disagree? Explain.

b. Construct a frequency polygon for the distribution, using $50 thousand as the unit width. Does your graph support your answer in part a? Explain.

2.21 The age distribution of employees who were absent more than 10 days last year follows:

Age (years to nearest birthday)	Percent of Employees	Age (years to nearest birthday)	Percent of Employees
15–19	7	40–49	7
20–24	18	50–59	4
25–29	30	60–69	1
30–39	33	Total	100

a. Construct a frequency polygon for the distribution, using five years as the unit width (e.g., 15–19). Which class of the frequency polygon has the largest adjusted percent frequency?

b. Upon studying this frequency polygon, an observer stated that employees in the 25–29 age class had the worst absentee record of all the employees. Do you agree or disagree? Explain.

2.22 Refer to **Hotel Chain** Problems 2.3 and 2.4.

a. For each chain, construct a percent frequency distribution of numbers of rooms, using equal class intervals 100–under 200, 200–under 300, and so on.

b. Plot the percent frequency distributions as frequency polygons on the same graph, as in Figure 2.7.

c. What does a comparison of the frequency polygons tell us about which chain tends to have larger hotels?

2.23 Refer to **Repair Costs** Problem 2.5.

a. For each type of pump, construct a frequency distribution of repair costs, using equal class intervals 0–under 1, 1–under 2, and so on.

b. Plot the frequency distributions as frequency polygons on the same graph, as in Figure 2.7. Why was it not necessary to convert the frequencies to percent frequencies before comparing the frequency polygons?

c. What does a comparison of the frequency polygons tell us about which type of pump tends to have higher repair costs for the year?

✱ 2.24 Refer to **Response Times** Problem 2.11.

a. Construct a less-than cumulative percent frequency distribution of the response times. Use classes less than 2.5, less than 7.5, and so on.

b. Plot the cumulative percent frequency distribution in part a. Reading from the plot, what percentage of calls had a response time of less than 20 minutes?

2.25 Refer to **Product Sales** Problem 2.20.

a. Construct a less-than cumulative frequency distribution of the annual sales amounts. Use classes less than 0, less than 10, and so on.

b. Plot the cumulative frequency distribution in part a. Reading from the plot, how many products had annual sales of less than $80 thousand last year?

c. The plotted points in part b are connected by straight lines. What does this use of straight lines imply about the distribution of annual sales amounts within each class of the frequency distribution?

2.26 Refer to **Stock Holdings** Problem 2.19.

a. Construct a more-than cumulative percent frequency distribution. Use classes 0 or more, 50 or more, and so on.

b. Plot the cumulative percent frequency distribution in part a. Reading from the plot: (1) What percentage of the stockholders held 75 or more shares each? (2) The 20 percent of the stockholders with the most shares each held how many shares or more?

2.27 Refer to **Universal Health Care** Problem 2.16.

a. Construct a less-than cumulative frequency distribution for the ratings. Use classes 1 or less, 2 or less, and so on.

b. Plot the cumulative frequency distribution in part a as a step function, like that in Figure 2.9b. Reading from the plot, how many persons expressed some degree of disapproval with universal health care, that is, gave a rating of 1 through 4?

* 2.28 Sixty individuals with eye troubles were examined by an ophthalmologist. The examination outcomes—coded as astigmatism (A), cataract (C), glaucoma (G), injury (I), strabismus (S), and other (O)—were as follows:

A I G A A A A I A S A C A C C A A A A A A O I A A A A A O A C
O A A C A A I A A A S I A I A O A A A A A A O I A A S G A A A

a. Construct a qualitative distribution of individuals by eye examination outcome. Also give the corresponding percent distribution.

b. Construct a simple bar chart of the percent distribution in part a.

c. (1) What was the most common examination outcome for this group of individuals? (2) What percentage of nonastigmatic outcomes was accounted for by eye injuries?

2.29 Refer to the **Financial Characteristics** data set (Appendix D.1). The qualitative distribution of firms classified by industry follows:

Industry		Number
Code	Name	of Firms
1	Crude oil producers	60
2	Textile products	52
3	Textile apparel manufacturers	75
4	Paper	24
5	Electronic computer equipment	37
6	Electronics	21
7	Electronic components	57
8	Auto parts and accessories	57
	Total	383

a. In which industry class is firm 261 counted?

b. Construct the percentage form of this qualitative distribution, and display it as a simple bar chart.

c. Which industry has the largest number of firms in the data set? If the three electronics industry classes were combined to form a single class, would your answer change? If so, what is the implication for the construction and interpretation of qualitative distributions?

2.30 **Rider Complaints.** A bus company in a Northeastern city has categorized complaints from riders for last May and June as follows:

	Number of Complaints	
Type of Complaint	May	June
Buses too hot	14	31
Fare too high	11	12
Buses sometimes don't stop at bus stops	17	16
Bus stops have no shelters	10	21
Bus travel too expensive	12	13
Buses should be air-conditioned	14	41
Drivers are rude	13	11
Bus schedules make transfers difficult	14	14
Route maps not available	25	21
Total	130	180

a. For each month, construct a qualitative distribution using the following complaint categories: comfort, service, and cost.

b. Would other persons necessarily assign the complaints to the same complaint classes in part a as you have? If not, what is the implication for interpreting qualitative distributions for free-answer data classified by others?

c. Construct a combined bar chart patterned after Figure 2.10b to compare the distributions for the two months. Briefly report your findings.

2.31 Refer to **Rider Complaints** Problem 2.30.

a. For each month, construct a qualitative distribution using the following complaint categories: bus schedules, bus drivers, equipment, and other.

b. Construct a combined bar chart patterned after Figure 2.10b to compare the distributions for the two months. Briefly report your findings.

2.32 Refer to **Municipal Debt** Problem 1.2.

a. Construct a scatter plot of debt per capita and population for the six municipalities. Plot population on the horizontal axis.

b. Describe any relationship between the two variables revealed by the scatter plot in part a.

2.33 Refer to **House Listing** Problem 1.3.

a. Construct a scatter plot of asking price and number of bedrooms for the six houses. Plot number of bedrooms on the horizontal axis.

b. Describe any relationship between the two variables revealed by the scatter plot in part a.

2.34 **Sales Growth.** A new product was expected to show a 10 percent sales increase each quarter for the first three years, starting with a sales volume of 40 thousand units. Actual and expected sales (in thousands) during this period were as follows:

Quarter:	1	2	3	4	5	6
Actual:	37.9	46.2	51.1	50.4	55.7	67.0
Expected:	40.0	44.0	48.4	53.2	58.6	64.4

Quarter:	7	8	9	10	11	12
Actual:	74.4	74.8	84.0	96.5	105.7	111.4
Expected:	70.9	77.9	85.7	94.3	103.7	114.1

a. Construct time series plots of actual sales and expected sales for the 12 quarters on the same graph.

b. Construct a scatter plot of actual sales and expected sales for the 12 quarters. Plot expected sales on the horizontal axis.

c. Comment on whether actual sales were in line with expected sales over the 12 quarters, making specific reference to the plots in parts a and b.

2.35 Grain Production. Data on sales revenue (in $ thousand) and volume (in metric tons) for wheat and barley grown on a farm during the last seven crop years follow:

Year:	1	2	3	4	5	6	7
Wheat Revenue:	12.7	20.7	11.8	9.8	8.9	9.6	10.9
Wheat Volume:	64	119	76	63	50	72	93
Barley Revenue:	13.6	5.7	20.8	15.6	16.3	11.0	4.4
Barley Volume:	103	51	138	109	136	139	68

a. Construct a scatter plot of wheat revenue and wheat volume for the seven years. Plot wheat volume on the horizontal axis. Describe any relationship between the two variables that is revealed by the plot.

b. Calculate the ratio of wheat revenue to wheat volume for each year. What does this ratio measure?

c. Construct a time series plot of the ratio calculated in part b. Describe any trend in the ratio revealed by the plot.

2.36 Refer to **Grain Production** Problem 2.35.

a. Construct a scatter plot of barley revenue and barley volume for the seven years. Plot barley volume on the horizontal axis. Describe any relationship between the two variables that is revealed by the plot.

b. Calculate the ratio of barley revenue to barley volume for each year. What does this ratio measure?

c. Construct a time series plot of the ratio calculated in part b. Describe any trend in the ratio revealed by the plot.

* **2.37** Refer to Table 2.4.

a. What is an element of this data set? What are the variables in the data set?

b. (1) How many tours cost $1200 or less? (2) How many tours have a duration of three days? (3) How many tours cost $1200 or less and have a duration of three days? (4) What is the second most commonly offered type of tour?

2.38 Refer to Figure 2.13a.

a. (1) What is the most common age interval for onset of the disease for male patients? (2) What percentage of patients who experienced onset of the disease at 70 years or more are female? (3) What percentage of all patients experienced onset of the disease before 60 years of age?

b. Present the univariate frequency distribution of all patients by age at onset.

2.39 Refer to Table 2.5a.

a. What is an element of this data set? How many elements are there?

b. Identify the variables in this data set. Present the univariate qualitative distribution for each variable.

c. If you had *only* the univariate qualitative distributions in part b, could you construct Table 2.5a? Explain.

* **2.40** The following table shows a bivariate frequency distribution of 125 unemployed construction workers by age and duration of the current spell of unemployment:

Duration	Age (years)	
(days)	Under 35	35 or more
1–7	36	21
8–30	30	5
More than 30	14	19

a. What percentage of workers who have been unemployed 30 days or less are under 35 years of age?

b. Obtain the univariate frequency distribution of the durations of unemployment for all workers.

c. For each age class, construct a percent frequency distribution of unemployment durations. Describe the relationship between age and duration of unemployment for these construction workers.

2.41 Failure Times. An aluminum reduction cell has two modes of failure, metal contamination (C) or structural distortion (D). The failure times (in days) for 25 cells, arrayed by increasing order of magnitude within each mode of failure, follow:

Mode C:	909	1293	1601	1616	2012	2016	2180	2201	2442
Mode D:	824	1082	1135	1308	1359	1372	1401	1412
	1601	1638	1641	1674	1709	1805	1947	2208

a. Construct the bivariate distribution of modes of failure (C, D) and failure times (using equal class intervals 500–under 1000, 1000–under 1500, and so on) for these data.

b. For each mode of failure, construct the percent frequency distribution of failure times. Describe the relationship between mode of failure and failure time for these data.

2.42 Flight Simulation. The coded observations that follow pertain to errors made in handling emergencies in training sessions on a flight simulator. The observations are cross-classified by phase of flight—takeoff (T), cruising (C), or landing (L)—and by cause of error—misinterpretation of instrument readings (M) or other cause (O). For instance, the coded observation TM involves a take-off error caused by misinterpretation.

```
TM TO LM LO CO TM CM LO TM CO
LO CM LO TM LO CM TM CO LO LM
TM TO TO LM TM LO LO TM TM LO
LO CO LO LM TM LM TM LM TM CM
TM TM TO LO TO
```

a. Construct the bivariate qualitative distribution for these observations.

b. (1) What percentage of the M-type errors occurred in the landing phase? (2) What percentage of the landing phase errors were of the O type?

c. For each phase of flight, construct the percent distribution of causes of error. Describe the relationship between cause of error and phase of flight for these observations.

d. Construct a component-part bar chart corresponding to the percent distributions in part c, as in Figure 2.14.

2.43 Refer to Figure 2.13a.

a. Present this bivariate distribution as a component-part bar chart. For each age class, represent the percent frequency distribution of gender by a bar that is segmented into components representing male and female patients.

b. Does the graph in part a make it evident that age at onset of the disease comes later for women than for men? Comment.

2.44 Two hundred business executives were asked independently to evaluate nine major firms, four being foreign-owned (F) and five domestically owned (D). Each executive was asked to rate each firm on the economic competitiveness and the labor–management relations of the firm on a scale from -3 (very negative) to 3 (very positive). The average ratings of the 200 executives were as follows:

Firm:	1	2	3	4	5	6	7	8	9
Competitiveness:	-0.6	1.2	0.8	2.1	-0.3	0.6	1.4	1.5	-0.9
Labor Relations:	-1.2	0.3	-0.1	1.0	0.1	0.4	1.0	0.5	0.0
Ownership:	D	D	D	F	D	F	F	D	F

a. Construct a symbolic scatter plot of labor relations rating and competitiveness rating for the nine firms, using ownership (F, D) as a plotting symbol. Plot competitiveness rating on the horizontal axis.

b. What does the plot in part a indicate about the relationship between labor relations rating and competitiveness rating for domestic firms? For foreign firms? In what major respect do domestic and foreign firms differ in their rating outcomes?

2.45 Refer to **Grain Production** Problem 2.35.

a. Construct a symbolic scatter plot of revenue and volume for both crops for the seven years, using W (wheat) and B (barley) as plotting symbols. Plot volume on the horizontal axis.

b. What does the plot in part a indicate about the relationship between revenue and volume for each grain? In what major respect does this relationship differ for wheat and barley?

* **2.46** Refer to **Sales Growth** Problem 2.34. For each of the 12 quarters, obtain the residual as the difference between actual and expected sales. Display the residuals in a time series plot. What does the plot show?

2.47 Technicians in a health survey were instructed to take and record blood pressure readings for each subject to the nearest millimeter. The distribution of the end digits in 500 of these blood pressure readings follows:

End Digit:	0	1	2	3	4	5	6	7	8	9
Number of Readings:	60	0	80	0	112	40	88	0	120	0

a. What are the expected frequencies for the end digits if all were equally likely?

b. Construct a residual plot to assess whether each end digit actually was equally likely. What does the plot show?

2.48 The observed percent frequency distribution of the numbers of defects per strand detected in 90 optical-fiber strands and the expected percent frequency distribution if the process was in control are as follows:

	Percent of Strands	
Defects per Strand	Observed	Expected
0	78	90
1	6	9
2	1	1
3	4	0
4 or more	11	0
Total	100	100
	(90)	

Obtain the residuals for the percent frequencies. Does it appear from the residuals that the process was in control? Comment.

2.49 Obtain an example of a pictorial chart from a newspaper, magazine, or other publication. Discuss the strong and weak points of the graphic presentation in this chart.

2.50 Obtain an example of a time series plot from a newspaper, magazine, or other publication. Discuss the strong and weak points of the graphic presentation in this chart.

EXERCISE

2.51 In a large training experiment for a new welding technology, trainees were classified by previous welding experience—no previous experience (A_1) or previous experience (A_2)—and half of each group was randomly assigned to a concentrated program (B_1) while the other half was assigned to an extended program (B_2). The following four tables show different possible outcome patterns for the percentages of trainees in each experience-training group who later used the new technology successfully.

	Pattern 1			Pattern 2			Pattern 3			Pattern 4	
	B_1	B_2		B_1	B_2		B_1	B_2		B_1	B_2
A_1	60%	60%	A_1	60%	90%	A_1	60%	60%	A_1	60%	80%
A_2	60%	60%	A_2	60%	90%	A_2	90%	90%	A_2	70%	90%

Interpret each of the four possible outcome patterns as to the effects of previous experience and type of training on the likelihood of successful use of the new technology.

STUDIES

2.52 Plant Wages. Weekly wages (in dollars) of 60 wage earners in a plant during the week of January 5 were as follows:

609	655	586	595	610	591	627	575	575	607
601	582	625	598	602	621	579	646	609	635
592	583	610	589	627	603	601	587	631	586
604	610	598	621	600	597	610	572	631	637
569	582	608	605	599	605	578	618	653	609
625	589	600	650	576	565	615	645	615	585

a. Array the observations by increasing order of magnitude.
b. Present the observations in a stem-and-leaf display.
c. (1) What were the weekly wages of the lowest wage earner? The highest wage earner? (2) Which stem of the display in part b has the most observations? What dollar interval of wage amounts does this stem potentially span?
d. Construct a frequency distribution for these data, using equal class intervals 560–under 580, 580–under 600, and so on.
e. Plot the frequency distribution in part d as a histogram. Is there much variability in the weekly wages? Explain.

2.53 Hospital Costs. The average daily costs (in dollars) of hospital care last year in 50 hospitals were as follows:

257	315	327	392	318	305	342	308	384	309
274	313	267	312	272	254	257	245	276	286
319	368	318	265	235	271	252	252	341	268
282	306	326	249	241	287	282	318	289	335
253	230	255	276	309	258	267	331	249	278

a. Array the observations by increasing order of magnitude.
b. Present the observations in a stem-and-leaf display.

c. (1) Which hospital cost observations, if any, appear to be unusually large? Unusually small? (2) In what ranges do the observations appear to cluster?

d. An analyst wishes to investigate whether the frequency distribution of costs has one or two peaks and whether any of the hospitals have relatively extreme costs. To determine the most effective number of classes, construct frequency distributions with 3, 5, 7, 9, and 11 classes, each using equal class intervals. Plot the histograms for these frequency distributions on different graphs. Which frequency distribution appears to you to be most effective here? Explain.

2.54 The percent frequency distributions of numbers of days of bed care for recent burn cases in two treatment centers follow.

	Percent of Cases	
Number of Days	Center A	Center B
0–under 5	16	11
5–under 10	20	24
10–under 15	34	42
15–under 25	14	7
25–under 50	16	16
Total	100	100
	(1110)	(498)

a. Plot frequency polygons of the two distributions on the same graph. Employ a unit width of five days.

b. Compare the frequency polygons plotted in part a and state your findings.

c. Plot less-than cumulative percent frequency distributions for the two centers on the same graph. Explain how these plots also indicate your findings in part b.

d. Answer the following questions based on the plots in part c. (1) What percentage of the cases entailed a bed stay of less than eight days in center A? In center B? (2) The longest 25 percent of the bed stays in center A entailed a stay of how many days or more? What is the comparable figure for center B?

2.55 Refer to the **Financial Characteristics** data set (Appendix D.1). Consider only the data for textile apparel manufacturers in year 2.

a. Use a data transformation to calculate the rate of return on investment for each firm. Rate of return is the ratio of net income to net assets, expressed in percent.

b. Construct a scatter plot of rate of return on investment and net assets for the textile apparel firms. Plot net assets on the horizontal axis. Describe any relationship between the variables that is revealed by the plot.

2.56 Refer to the **Power Cells** data set (Appendix D.2).

a. Array the observations of the number of cycles before failure in ascending order. Plot the less-than cumulative percent distribution for these data. What percent of the cells survived 150 cycles or fewer before failure?

b. Use a data transformation to calculate the logarithm (to base 10) of the number of cycles before failure (log-cycles).

c. Construct a symbolic scatter plot of log-cycles before failure and ambient temperature, using mode of failure (0, 1) as a plotting symbol. Plot ambient temperature on the horizontal axis.

d. Briefly describe what the plot reveals about the relationship between log-cycles before failure and ambient temperature for each mode of failure.

Data Summary Measures

3

In the preceding chapter, we examined methods for studying patterns in quantitative and qualitative data. In this chapter, we focus on quantitative data and consider measures that summarize important characteristics of a data set. Summary measures are useful for exploratory data analysis, comparisons of data sets, and reporting final results of a study.

Summary measures for a data set are often referred to as *descriptive statistics*. Most statistical packages have routines for producing descriptive statistics, as the following example illustrates.

DESCRIPTIVE STATISTICS   **3.1**

EXAMPLE ☐ **Oil Spills**

In 1979, a government agency was established to monitor and control pollution in an inland waterway, including oil spills. An agency analyst now wishes to compare the time intervals between oil spills after the agency was established with comparable data available for a limited period prior to the agency's establishment. The intention is to determine whether or not the agency's controls have lengthened the intervals between oil spills.

Data for 40 time intervals between oil spills (in days) before pollution controls and 109 time intervals after pollution controls are given in arrayed form in Table 3.1. Figure 3.1a shows aligned dot plots of the two data sets. The plots show clear evidence that the time intervals between oil spills have become longer, which suggests that the controls have had a positive effect.

To compare the before-controls and after-controls data sets by means of several summary measures, the analyst used the descriptive statistics routine in the MYSTAT statistics package. Selected results are shown in Figure 3.1b. These summary measures will be explained in detail later. We note here, however, that the output for each data set gives the number of observations in the data set (N OF CASES), the smallest and largest observations in the data set (MINIMUM and MAXIMUM), and the sum of the observations in the data set (SUM). The before-controls data set, for example, has 40 observations. The smallest interval between oil spills is 1 day and the largest is 73 days. The sum of the 40 observations is 835. ☐

Summary measures fall into three broad categories: measures of position, variability, and skewness. We now consider each of these categories in turn.

TABLE 3.1
Arrays of time intervals
between oil spills (in days)—
Oil spills example

(a) Before Pollution Controls			(b) After Pollution Controls					
1	19	1	25	33	38	51	65
1	19	2	26	33	39	51	66
1	20	5	27	34	40	52	67
2	20	7	28	34	41	53	69
2	21	11	29	34	41	53	71
3	22	15	30	35	42	54	73
3	24	16	30	35	43	55	74
4	28	18	30	35	44	56	74
5	30	19	31	35	45	57	76
5	31	20	31	35	45	57	77
6	31	21	31	36	46	58	79
7	33	22	32	36	47	59	81
8	36	22	32	36	47	60	83
11	37	23	32	36	48	61	84
14	38	23	32	37	48	61	87
15	41	23	32	37	49	63	90
16	45	24	32	37	50	63	101
16	48	24	32	37	51	64	115
17	64	25	33	37			
18	73						

FIGURE 3.1
Dot plots and descriptive statistics for the before-controls
and after-controls data sets—
Oil spills example

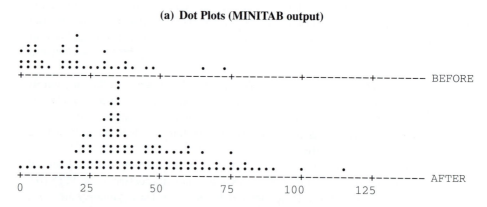

(a) Dot Plots (MINITAB output)

(b) Descriptive Statistics (MYSTAT output)

Before-Controls Data Set		After-Controls Data Set	
N OF CASES	40	N OF CASES	109
MINIMUM	1.000	MINIMUM	1.000
MAXIMUM	73.000	MAXIMUM	115.000
RANGE	72.000	RANGE	114.000
MEAN	20.875	MEAN	43.697
STANDARD DEV	17.352	STANDARD DEV	21.344
SKEWNESS(G1)	1.046	SKEWNESS(G1)	0.677
SUM	835.000	SUM	4763.000

A measure of position for a data set describes where the observations are concentrated. A comparison of such a measure for two data sets can indicate in direct fashion a shift in the location of the data, such as the one noted in the oil spills example.

We now discuss several useful measures of position.

Mean

One measure of position is the arithmetic average, or mean.

> **(3.1)**
> The **mean** of a set of observations $X_1, X_2, \ldots, X_n$, denoted by $\overline{X}$ (read "X bar"), is the sum of the observations divided by the number of observations; that is:
>
> $$\overline{X} = \frac{\sum\limits_{i=1}^{n} X_i}{n}$$

(Note that Appendix A, Section A.1, contains a review of the summation symbol Σ.)

EXAMPLE ☐

As we noted earlier from the MYSTAT output in Figure 3.1b, the before-controls data set in the oil spill example consists of 40 time intervals whose sum is 835. Therefore, the mean time interval between oil spills before the agency was created is $\overline{X} = 835/40 = 20.9$ days. By a similar calculation, it is found that the mean of the 109 intervals after pollution controls is $4763/109 = 43.7$ days. The MYSTAT output in Figure 3.1b gives these same values for the means of the two data sets, except that more digits are retained there. The increase in the mean interval from 20.9 to 43.7 days summarizes quickly and easily the lengthening of the time intervals between oil spills after pollution controls. ☐

Properties of Mean.   The mean has a number of unique properties.

1. The sum of the observations in a data set is equal to the mean multiplied by the number of observations.

> **(3.2)** $\sum\limits_{i=1}^{n} X_i = n\overline{X}$

This result is obtained by multiplying both sides of (3.1) by n.

EXAMPLE ☐

A restaurant owner is informed that the mean expenditure per customer was $9.800 for the past week when 650 customers visited the restaurant. Total receipts for the week therefore were $650(9.800) = \$6370$. ☐

2. The sum of the deviations of the X_i observations from their mean $\overline{X}$ is zero.

$$(3.3) \qquad \sum_{i=1}^{n}(X_i - \overline{X}) = 0$$

□ EXAMPLE

Consider the set of three observations 2, 6, and 13. Their mean is $\overline{X} = 7$, and their deviations sum to zero:

$$(2 - 7) + (6 - 7) + (13 - 7) = -5 - 1 + 6 = 0$$

Figure 3.2 illustrates these deviations $X_i - \overline{X}$ and shows that the X_i observations deviate from the mean $\overline{X}$ in a balanced fashion. Thus, in this sense, $\overline{X}$ is centered among the X_i observations.   □

To derive (3.3), we utilize (3.2) as follows:

$$\sum_{i=1}^{n}(X_i - \overline{X}) = \sum_{i=1}^{n}X_i - \sum_{i=1}^{n}\overline{X} = \sum_{i=1}^{n}X_i - n\overline{X} = 0$$

3. In the next section, we shall be concerned with the variability of the observations in a data set. One measure of variability that we shall consider depends on an expression of the following form, where A denotes a fixed value:

$$\sum_{i=1}^{n}(X_i - A)^2$$

Note that this expression represents the sum of the squared deviations of the X_i observations from the value A. An important property of the mean is that the sum of these squared deviations is a minimum when $A = \overline{X}$.

$$(3.4) \qquad \sum_{i=1}^{n}(X_i - A)^2 \text{ is a minimum when } A = \overline{X}.$$

FIGURE 3.2
Illustration of the balancing of observations about their mean

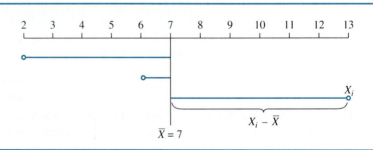

Consider again the three observations 2, 6, and 13. For $A = \bar{X} = 7$, we have:

$$\sum (X_i - 7)^2 = (2 - 7)^2 + (6 - 7)^2 + (13 - 7)^2 = 62$$

while for any other value of A this sum is larger. For $A = 0$, for instance:

$$\sum (X_i - 0)^2 = (2 - 0)^2 + (6 - 0)^2 + (13 - 0)^2 = 209$$

☐

4. When a data set is divided into two or more subsets, the overall mean is a weighted average of the means of the subsets.

(3.5) $$\bar{X} = \frac{f_1\bar{X}_1 + f_2\bar{X}_2 + \cdots + f_k\bar{X}_k}{n} = \frac{\sum_{i=1}^{k} f_i\bar{X}_i}{n}$$

where: k is the number of subsets
$\bar{X}_i$ is the mean of the ith subset
f_i is the number of observations in the ith subset
$n = \sum_{i=1}^{k} f_i$ is the total number of observations

A trade union has 1500 male members whose mean age is 41 years and 300 female members whose mean age is 25 years. The mean age of the entire trade union membership is:

$$\bar{X} = \frac{1500(41) + 300(25)}{1500 + 300} = 38.3 \text{ years}$$

☐

Trimmed Mean.   A major disadvantage of the mean as a measure of position is that it is greatly influenced by extreme or outlying observations in a data set. Consider the following array of 12 observations:

| 1 | 1 | 2 | 4 | 5 | 7 | 8 | 9 | 12 | 12 | 13 | 130 |

for which the mean is $\bar{X} = 17$. Note that 11 of the 12 observations in the data set fall below the mean. The mean lies away from the main concentration of data here because of the influence of the outlying observation, 130.

To lessen the effect of outlying observations on the mean, a modified mean called the trimmed mean is employed at times.

(3.6)

The **50 percent trimmed mean** is the mean of the central 50 percent of the arrayed observations of the data set.

Thus, to calculate a 50 percent trimmed mean, we simply eliminate 25 percent of the observations at each end of the array and calculate the mean of the remaining middle observations.

☐ **EXAMPLE**

Consider the earlier data set with the outlying observation. To obtain the 50 percent trimmed mean, we need to eliminate 25 percent of the 12 observations at each end. Thus, we eliminate the smallest three and largest three observations in the array.

Trimmed									Trimmed		
1	1	2	4	5	7	8	9	12	12	13	130

The 50 percent trimmed mean therefore is:

$$\frac{4 + 5 + 7 + 8 + 9 + 12}{6} = 7.5$$

Note that the 50 percent trimmed mean falls near the middle of the array.   ☐

A trimmed mean can be based on percentages other than the central 50 percent. For example, to calculate the 80 percent trimmed mean of a data set consisting of 50 observations, we trim the five smallest and five largest observations from the set, and the trimmed mean is the mean of the remaining central 40 observations.

Median

In data sets where the distribution pattern is not symmetrical, such as in the oil spills data, the mean tends to be located somewhat away from the concentration of the observations. For example, the mean time interval of 20.9 days for the before-controls data set is only exceeded in 16 of the 40 cases, as may be confirmed from the arrayed data in Table 3.1a. Similarly, the mean of 43.7 days for the after-controls data set is only exceeded in 47 of the 109 cases. Also, when a data set contains one or more outlying observations, as in the trimmed-mean example, the mean can be unduly influenced by the outlying observations.

A measure of position that is more central in the distribution, in the sense that it divides the arrayed observations into two equal parts, is the median.

> **(3.7)**
> The **median** of a data set, denoted by Md, is the value of the middle observation in an array of the data set. For a data array with n observations, Md is the value of the observation with rank $(n + 1)/2$.

Convenience Stores   ☐ **EXAMPLE**

Changes in annual sales between 1990 and 1991 for a small chain of 5 convenience stores operated by a single owner were as follows (data in thousands of dollars): 14, 17, −13, 41, 12. Arraying the observations in ascending order and giving them ranks, we obtain:

Observation:	− 13	12	14	17	41
Rank:	1	2	3	4	5

Rank $(n + 1)/2 = (5 + 1)/2 = 3$ corresponds to the value 14. Hence, $Md = \$14$ thousand. Note that the median divides the array of observations equally, with two observations being to one side and two to the other of the median. ☐

When the data set contains an even number of observations, rank $(n + 1)/2$ is not an integer. Then the median is taken to be the average of the two adjacent observations.

EXAMPLE ☐

In the convenience stores example, suppose that the owner operated 6 stores, with the following annual sales changes (in thousands of dollars): 14, 17, − 13, 41, 12, 66. Arraying the observations in ascending order and giving them ranks, we obtain:

Observation:	− 13	12	14	17	41	66
Rank:	1	2	3	4	5	6

Here, $(n + 1)/2 = (6 + 1)/2 = 3.5$. Hence, the two adjacent observations are the ones with ranks 3 and 4; these observations have values 14 and 17. Hence, $Md = (14 + 17)/2 = \$15.5$ thousand. Again, we find that the median divides the array of observations equally, three being to one side and three to the other side of the median. ☐

Medians, like means, are useful for comparing several data sets.

EXAMPLE ☐

In the oil spills example, the before-controls data set contains $n = 40$ observations. The middle rank is therefore $(n + 1)/2 = (40 + 1)/2 = 20.5$. The array in Table 3.1a shows that ranks 20 and 21 correspond to observations 18 and 19 so that the median time interval between oil spills is $(18 + 19)/2 = 18.5$ days. The after-controls data set contains $n = 109$ observations, so the middle rank is $(109 + 1)/2 = 55$. From the array in Table 3.1b, we find the corresponding observation is 37 days. Hence, the median interval between oil spills increased from 18.5 to 37 days with the pollution controls. Comparing the two medians therefore shows readily that the time intervals between oil spills have tended to be substantially longer after the pollution controls were instituted. ☐

Properties of Median. The following are some important properties of the median.

1. In large data sets where observations are not repeated extensively, about one-half of the observations will be smaller than the median and about one-half will be larger. Thus, if the median family income in a community is $30,000, then about one-half the families have incomes under $30,000 and about one-half have incomes over $30,000.

When a data set contains repeated observations, this interpretation may not be precise. For example, in a community of 100 families, 20 families consist of two persons, 40 consist of three persons, and the other 40 consist of four persons. The arrayed data set is as follows:

Observation:	2	...	2	3	...	3	3	...	3	4	...	4
Rank:	1	...	20	21	...	50	51	...	60	61	...	100

The median of the data set is $Md = 3$, the average of the two observations with ranks 50 and 51. However, it is not accurate to say here that about one-half the family sizes are less than 3. In fact, only 20 percent of family sizes are smaller.

2. Consider the sum of the absolute deviations of the observations X_i in a data set from a fixed value A, denoted by:

$$\sum_{i=1}^{n} |X_i - A|$$

(The symbol $|\ |$ denotes absolute value.) An important property of the median is that the sum of the absolute deviations is a minimum when $A = Md$.

(3.8)	$\sum_{i=1}^{n} |X_i - A|$ is a minimum when $A = Md$.

☐ **EXAMPLE**

Consider the observations 14, 17, -13, 41, 12. For $A = Md = 14$, we have:

$$\sum |X_i - 14| = |14 - 14| + |17 - 14| + |-13 - 14| + |41 - 14|$$
$$+ |12 - 14| = 59$$

For any other value of A, this sum will be larger. For instance, if A is set equal to the mean of these observations ($\overline{X} = 14.2$), the sum is:

$$\sum |X_i - 14.2| = |14 - 14.2| + |17 - 14.2| + |-13 - 14.2|$$
$$+ |41 - 14.2| + |12 - 14.2| = 59.2$$

☐

3. The median is not affected by outlying observations in a data set. For this reason, it is often preferred to the mean as a measure of position for data sets containing a few outlying observations. Income data sets are a case in point because they often contain some extremely large observations. The median in this situation tends to be more indicative of typical income levels than the mean.

4. The median can be viewed as a special case of a trimmed mean, where all observations except the middle one or two are trimmed away or excluded. Thus, the 50 percent trimmed mean is a compromise between the mean, which is based on all observations in the data set, and the median, which excludes all but the middle one or two observations.

Mode

The mode is still another measure of position for a data set. The *mode* is that value around which the greatest number of observations are concentrated. We shall discuss the mode only in connection with data sets that have been classified in a frequency distribution with equal class intervals.

(3.9)

The **modal class** in a frequency distribution with equal class intervals is the class with the largest frequency.

  If the frequency polygon of a distribution has only a single peak, the distribution is said to be **unimodal**. If the frequency polygon has two peaks, the distribution is said to be **bimodal**.

EXAMPLE ☐

The age distribution of farm operators in Figure 2.5a shows that ages in the class 45–under 55 are the most frequent. The class 45–under 55 is therefore the modal age class.

                       ☐

  Consideration of whether a frequency polygon is unimodal or bimodal can provide insights into the nature of the underlying data set, as illustrated by the following example.

EXAMPLE ☐   **Melting Points**

A metal alloy manufacturer was concerned with customer complaints about the lack of uniformity in the melting points of one of the firm's alloy filaments. Sixty filaments were selected from the production process and their melting points determined. The resulting frequency polygon is shown in Figure 3.3a. Note its bimodal nature. Closer examination of the data revealed that 26 of the filaments were produced by the first shift and the other 34 by the second shift. The frequency polygons for each shift separately are shown in Figure 3.3b. It is now clear that the bimodal nature of the original frequency distribution is the result of a difference in the positions of the distributions for the two shifts. An investigation revealed that the second shift was using the wrong alloy mixture specification. Correction of the mixture resulted in production of filaments with more uniform melting points.

                       ☐

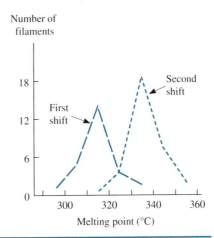

(a) Entire Data Set

(b) Separation by Shift

FIGURE 3.3
Bimodal frequency polygon of melting points—Melting points example

Percentiles

Percentiles are important and widely used summary measures for the position of a data set. The median is actually a percentile. It is the 50th percentile, the value that divides the data array into two equal parts. Other widely used percentiles are the 25th and 75th percentiles and the 10th and 90th percentiles. In general, a percentile is a value such that a given percentage of a data set is at or below this value. Percentiles can be easily found from a plot of the cumulative percent distribution. For example, in Figure 2.8c the 50th percentile or median age is shown to be 52 years. However, it is not necessary to plot the cumulative percent distribution to ascertain a percentile, as we shall now illustrate.

☐ **EXAMPLE**

In the convenience stores example, the owner of the 5 stores wanted to ascertain the 25th and 75th percentiles of the data set to compare the annual sales changes for the 5 stores with industry data. For the industry as a whole, the 25th and 75th percentiles for annual sales changes of convenience stores were $10 thousand and $15 thousand, respectively.

The array of annual sales changes for the 5 stores is repeated in Figure 3.4a; also given there are the ranks and the ranks expressed in percent form. Recall from Section 2.1 that ranks are expressed in percent form by dividing each rank by the number of observations in the data set and then multiplying by 100. Thus, since there are 5 stores in the data set, rank 1 in percent form is $100(1/5) = 20$ percent. The other percents are obtained in similar fashion.

To find any percentile, we simply identify the observation associated with the percent interval in which the percentile falls. To show clearly what occurs, we utilize the *line diagram* in Figure 3.4b. Here is shown the percent scale from 0 to 100 percent. On the scale are marked the cumulative percents associated with the ranks. In each interval is shown the corresponding observation from the data array. For example, observation -13 corresponds to the percent interval from 0 to 20, observation 12 corresponds to the percent interval from 20 to 40, and so on.

We see at once that the 25th percentile is $12 thousand, since the 25th percentile falls in the interval from 20 to 40. Similarly, we find that the 75th percentile is $17 thousand. Thus, the data for the small chain and the industry compare as follows:

Percentile	Small Chain ($ thousand)	Industry ($ thousand)
25	12	10
75	17	15

FIGURE 3.4
Determination of percentiles from the cumulative percent distribution—Convenience stores example

(a) Cumulative Percent Distribution

Observation:	-13	12	14	17	41
Rank:	1	2	3	4	5
Cumulative Percent:	20	40	60	80	100

(b) Line Diagram

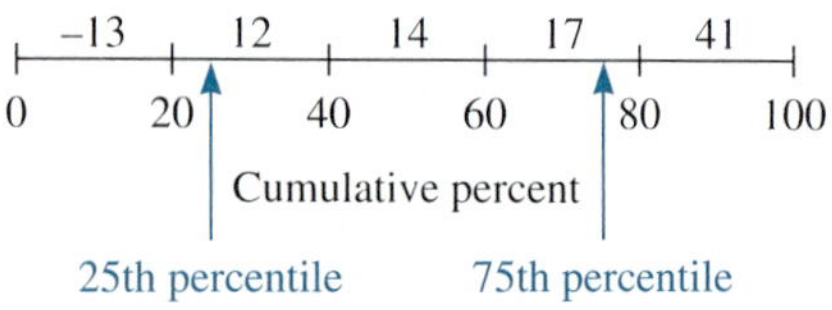

These results indicate that the 5 convenience stores tended to experience greater increases in sales than the industry as a whole.

Ordinarily, we do not need to prepare a line diagram such as in Figure 3.4b to find a percentile. The cumulative percent distribution in Figure 3.4a is sufficient to identify any percentile. For example, we see from Figure 3.4a that the 90th percentile is $41 thousand, since 90 percent falls in the percent interval associated with the fifth observation in the array.

Occasionally, the desired percentile falls on the boundary between two intervals. In that case, we shall follow the same convention employed for the median, namely, we take the average of the two adjacent observations. For example, the 40th percentile of the data set in Figure 3.4a is $(12 + 14)/2 = 13 thousand.   □

In large data sets where observations are not repeated extensively, about 25 percent of the observations will be smaller than the 25th percentile and about 75 percent will be larger. Other percentiles are interpreted correspondingly. When a data set contains repeated observations, this interpretation may no longer be appropriate, as we illustrated for the median.

A single observation can correspond to more than one percentile. For instance, the observation 12 in Figure 3.4a is not only the 25th percentile but is also every percentile between 20 and 40.

Many percentiles are known by other names. Percentiles that are multiples of 25 are called *quartiles, hinges,* or *fourths.* Thus, the 25th percentile is also called the *first quartile* and the 75th percentile the *third quartile.* Similarly, percentiles that are multiples of 10 are called *deciles.* For instance, the 70th percentile is also called the 7th decile.

Statistical packages determine percentiles by a variety of methods, usually involving some form of interpolation. The method presented here is a simple one. In any case, the various methods tend to lead to about the same percentile values for large data sets.

MEASURES OF VARIABILITY   3.3

The variability of the observations in a data set is often another important feature of interest when data sets are summarized. We now consider several summary measures of variability, sometimes also called measures of *dispersion* or *spread.*

Range

Consider a data set that consists of the amounts of active ingredient in the 300 tablets in a shipment from a pharmaceutical manufacturer. One factor that is important in assessing the quality of these tablets is the maximum extent of variation in the amounts of active ingredient per tablet.

> **(3.10)**
> The **range** is the difference between the largest and smallest observations in a data set.

EXAMPLES □

1. The largest observation in the data set for the 300 tablets is 1.38 grams, and the smallest observation is 0.96 gram, The range then is $1.38 - 0.96 = 0.42$ gram, indicating a relatively large extent of variability in the amounts of active ingredient.

2. In the oil spills example, it can be ascertained from Table 3.1 that the range for the before-controls data set is $73 - 1 = 72$ days, while the range for the after-controls data set is $115 - 1 = 114$ days. Thus, the after-controls data set has a substantially larger spread. The MYSTAT output in Figure 3.1b gives these same values for the ranges of the two data sets. □

A limitation of the range as a measure of the variability of a data set is that it depends only on the largest and smallest observations and does not consider the other observations. Thus, a data set might contain observations quite close to each other with the exception of one outlying observation. Despite the concentration of almost all the observations, the range would be large because of the one outlying observation.

Another limitation of the range as a measure of variability is that it is affected by the number of observations in the data set. The larger the number of observations, the larger the range tends to be. In the oil spills example, this effect may partly explain why the after-controls data set has a larger range than the before-controls data set. The after-controls data set contains 109 observations, while the before-controls data set contains only 40.

Sometimes, the range of a data set is indicated by presenting the smallest and largest observations in the data set. Thus, stock market summaries in newspapers often provide the high and low prices for the day. This form of presentation not only provides information about the variability of a stock's price during the day (because the range can be easily calculated) but also provides information about the location of the stock's price distribution. Another advantage is that the daily highs and lows can be used to determine the price range for a longer period, such as a week.

Interquartile Range

Because the range depends only on the smallest and largest observations in a data set, a modified range is sometimes used that reflects the variability of the middle 50 percent of the observations in the array. This modified range is called the interquartile range.

> **(3.11)**
> The **interquartile range** is the difference between the third and first quartiles of the data set.

□ **EXAMPLES**

1. An economist, studying the variation in family incomes in a community, found that the first quartile income is $22,400 and the third quartile income is $29,100. Thus, the middle 50 percent of families have incomes that vary from $22,400 to $29,100, a range of $6700. The range of $6700 is the interquartile range, reflecting the extent of variation among the middle 50 percent of family incomes.

2. In the oil spills example, the first and third quartiles and the interquartile ranges for the two data sets are as follows:

	Before Controls	After Controls
Third Quartile:	31	57
First Quartile:	5.5	31
Interquartile Range:	25.5	26

The upward shift of the quartiles reflects the lengthening of the intervals between oil spills with the institution of pollution controls. The constancy in the interquartile range indicates that the middle 50 percent of intervals in the after-controls data set are concentrated to the same extent as those in the before-controls data set. □

The interquartile range may be considered to be approximately the range for a trimmed data set in which the smallest 25 percent and the largest 25 percent of observations have been removed.

Variance

The most commonly used measure of variability in statistical analysis is called the variance. It is a measure that takes into account all the observations in a data set.

EXAMPLE □

In the melting points example of Figure 3.3, customer complaints about increased variability in the melting points of alloy filaments had led the manufacturer to investigate the cause of the problem. One of the customers who had complained had selected a sample of 15 filaments from a shipment received in April and another sample of 15 filaments from a shipment received in May. The data on the melting points for these two samples are given in Figure 3.5a.

Data sets and deviations from the mean—Melting points example   ▣ **FIGURE 3.5**

(a) Data Sets (melting points of filaments in °C)   **(b) Deviations from Mean**

April Shipment		
330	334	321
358	318	337
373	325	328
346	295	348
343	288	318

May Shipment		
302	365	343
348	318	317
374	378	385
279	294	304
364	357	362

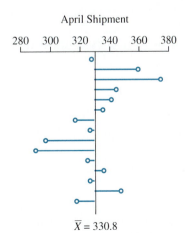

$\bar{X} = 330.8$

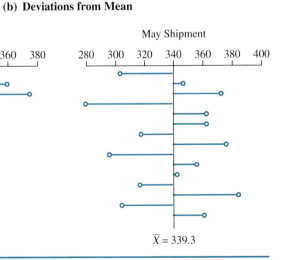

$\bar{X} = 339.3$

To compare the variability of melting points for the two data sets, the customer considered the deviations of the observations in each data set from their respective means. The means of the two data sets are 330.8 and 339.3, respectively. The two sets of deviations are shown graphically in Figure 3.5b. This figure shows readily that there was greater variability for the May filaments than for the April filaments.

The variance is a measure that provides quantified information about the variability in a data set. The first step in calculating the variance is to square the deviations from the mean. Then, to obtain an overall measure of variability per observation, we take the average of the squared deviations. This average squared deviation is called the variance. Because the data being considered here are sample data, a divisor of $n - 1$ rather than n is used in obtaining the average of the squared deviations. We shall explain later the distinction between sample and other types of data and why a divisor of $n - 1$ is used in calculating the average squared deviation.

The calculation of the variance for each of the two data sets in Figure 3.5a proceeds as follows:

$$\text{April: } \quad \frac{(330 - 330.8)^2 + \cdots + (318 - 330.8)^2}{15 - 1} = 487.5$$

$$\text{May: } \quad \frac{(302 - 339.3)^2 + \cdots + (362 - 339.3)^2}{15 - 1} = 1161.1$$

The variance for the May filaments is more than twice that for the April filaments. Based on these findings, the customer undertook additional studies and then complained about the increased variability of the melting points of the alloy filaments. ☐

We now define the variance formally.

(3.12)

The **variance** of a set of observations $X_1, X_2, \ldots, X_n$, denoted by s^2, is defined:

$$s^2 = \frac{\displaystyle\sum_{i=1}^{n} (X_i - \bar{X})^2}{n - 1}$$

where $\bar{X}$ is the mean of the X_i observations.

☐ **EXAMPLE**

Consider the data set 5, 17, 12, 10, whose mean is $\bar{X} = 11$. We calculate s^2 by (3.12):

$$s^2 = \frac{(5 - 11)^2 + (17 - 11)^2 + (12 - 11)^2 + (10 - 11)^2}{4 - 1} = 24.67$$

☐

The greater the variability of the observations in a data set, the greater the variance. If there is no variability of the observations—that is, if all are equal and hence all are equal to the mean—then $s^2 = 0$.

Comments

1. Many calculators are programmed to calculate the variance of a data set. Some calculator models employ a different definition of the variance than the one in (3.12), using n as the divisor instead of $n - 1$. A variance with n in the denominator can be converted to one with $n - 1$ by multiplying by $n/(n - 1)$.
2. An equivalent formula for calculating the variance s^2 that is generally easier for manual calculations is the following.

$$(3.12a) \quad s^2 = \frac{\sum_{i=1}^{n} X_i^2 - \frac{\left(\sum_{i=1}^{n} X_i\right)^2}{n}}{n - 1}$$

Standard Deviation

The variance s^2 is expressed in units that are the square of the units of measure of the variable under study. For instance, each variance in the melting points example is expressed in Celsius degrees squared. Often, it is desirable to return to the original units of measure. We obtain the original units by taking the positive square root of the variance. The resulting value is called the standard deviation and is also used as a measure of variability.

(3.13)

The positive square root of the variance is called the **standard deviation** and is denoted by s:

$$s = \sqrt{s^2}$$

EXAMPLES ☐

1. For the melting points example, the standard deviations are $s = \sqrt{487.5} = 22.1°C$ for the April filaments and $s = \sqrt{1161.1} = 34.1°C$ for the May filaments. The standard deviations again show that there was greater variability of melting points in the May filaments than in the April filaments.

2. For the oil spills example, the MYSTAT output in Figure 3.1b gives the standard deviation of each data set. The after-controls data set can be seen to have greater variability ($s = 21.344$) than the before-controls data set ($s = 17.352$). The difference in variability also can be seen visually in the dot plots in Figure 3.1a. ☐

Coefficient of Variation

The standard deviation is a measure of *absolute variability* in a data set. Sometimes, a measure of *relative variability* is more relevant. The most commonly used measure of relative variability is the coefficient of variation.

(3.14)

The **coefficient of variation,** denoted by C, is the ratio of the standard deviation to the mean expressed as a percentage:

$$C = 100\left(\frac{s}{\overline{X}}\right)$$

☐ **EXAMPLES**

1. A sales analyst studied the variability of a company's sales revenues for one-week and four-week time periods. The analyst determined that the mean weekly revenues during the past year were $1.36 million, and that the standard deviation of the 52 weekly revenues was $0.28 million. The mean revenues for the 13 four-week periods during the same year were $5.44 million, and the standard deviation of the four-week revenues was $0.50 million. The analyst felt that a direct comparison of the two standard deviations would not be appropriate here because the absolute variability in revenues is affected by the fact that revenues for four-week periods, on the average, are four times as large as weekly revenues. Instead, the analyst computed the coefficient of variation for each case:

Weekly Revenues	Four-Week Revenues
$C = 100\left(\dfrac{0.28}{1.36}\right) = 20.6\%$	$C = 100\left(\dfrac{0.50}{5.44}\right) = 9.2\%$

From these results, the analyst concluded that during the past year revenues were relatively less variable for four-week periods ($C = 9.2\%$) than for weekly periods ($C = 20.6\%$). The analyst further noted that measuring revenues for four-week periods cut the relative variability of revenues approximately in half as compared to weekly revenues.

2. A traffic analyst for an air carrier needed to compare the variability in volumes and weights of containerized air-cargo shipments. The analyst found for the volumes of the shipments that the mean volume was 0.467 cubic meter and the standard deviation was 0.0778 cubic meter, while the corresponding mean and standard deviation for the shipment weights were 29.1 kilograms and 8.32 kilograms, respectively. The standard deviations cannot be compared here because volume is measured in cubic meters and weight in kilograms. The coefficients of variation, in contrast, are dimensionless, because the units of measurement cancel when the standard deviation is divided by the mean. The coefficients of variation here are as follows:

Volume	Weight
$C = 100\left(\dfrac{0.0778}{0.467}\right) = 16.7\%$	$C = 100\left(\dfrac{8.32}{29.1}\right) = 28.6\%$

Hence, the relative variability of the shipment volumes is smaller than that of the shipment weights.   ☐

The coefficient of variation is usually employed only when all observations are positive, such as for sales, volume, and weight data. When the values can be both positive and negative, as for profit-and-loss data, the coefficient of variation is not an appropriate measure of relative variability. For such data, the mean can be zero or negative, in which case the coefficient of variation is either undefined or meaningless.

MEASURES **3.4**
OF SKEWNESS

Measures of position are concerned with the location around which the observations are concentrated, and measures of variability consider the extent to which the observations vary. Measures of skewness are still another kind of measure; they summarize the extent to which the observations are symmetrically distributed.

Nature of Skewness

A data set with observations that are not symmetrically distributed is said to be *skewed*. The skewness of a data set is readily studied if the data are presented as a frequency distribution or a stem-and-leaf display. For example, the stem-and-leaf display for subscribers' ages in Figure 2.1a for the concert subscribers example shows some asymmetry or skewness, with a slightly sharper drop-off from the modal class toward older subscribers than toward younger subscribers. In contrast, the frequency polygon of taxpayer incomes in Figure 2.6c shows pronounced skewness. Most taxpayer incomes are in the lower portion, but the distribution is spread far to the right, with a few taxpayers having high incomes.

Another way of studying the skewness of a data set is to compare the values of the mode, the median, and the mean. The mode is the position on the scale that has the greatest concentration of observations. For the median, we know that generally about half the observations will lie below it and about half above it. In contrast, the mean tends to be pulled in the direction of outlying observations.

The three frequency polygons in Figure 3.6 show the comparative positions of the mean, the median, and the mode in unimodal frequency distributions with differing degrees and directions of asymmetry. Figure 3.6a is an example of a symmetrical unimodal frequency distribution. The values of the mean, median, and mode in any such distribution

Examples of symmetrical and skewed unimodal frequency distributions.   In skewed unimodal distributions, the mean is typically closest to the tail, and the median falls between the mean and the mode.

FIGURE 3.6

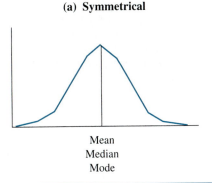

(a) Symmetrical

Mean
Median
Mode

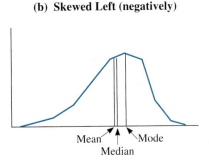

(b) Skewed Left (negatively)

Mean   Mode
Median

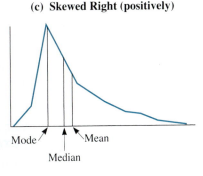

(c) Skewed Right (positively)

Mode   Mean
Median

are identical. Figure 3.6b shows an example of a unimodal frequency distribution that is skewed to the left (negatively), and Figure 3.6c shows one skewed to the right (positively). These figures indicate the relationships typically existing among the mean, the median, and the mode in unimodal frequency distributions that are moderately skewed: The mean is closest to the tail of the distribution, and the median is between the mean and the mode.

Thus, when the mean differs substantially from the median and the mode, we have evidence of skewness in the data set.

☐ **EXAMPLE**

In the semiconductor failure example (Figure 2.1b), the mean breaking stress is 55 milligrams and the median is 58 milligrams. This relatively small difference indicates that the breaking stresses are fairly symmetrically distributed.   ☐

Standardized Skewness Measure

In addition to informal analysis of the symmetry or lack of symmetry of a data set, we can also utilize some direct measures. One such measure is related to the variance, defined in (3.12). We have seen that the variance s^2 is the average of the squared deviations about the mean, $(X_i - \overline{X})^2$. The variance sometimes is called the *second moment about the mean* and denoted by m_2. A corresponding measure of skewness for a data set is obtained by averaging the cubed deviations about the mean, $(X_i - \overline{X})^3$. The resulting measure is called the third moment about the mean and is denoted by m_3. As with the variance s^2, the average cubed deviation for sample data is computed by using $n - 1$ in the denominator.

(3.15)
The **third moment about the mean** of a set of observations $X_1, X_2, \ldots, X_n$, denoted by m_3, is defined:

$$m_3 = \frac{\sum_{i=1}^{n}(X_i - \overline{X})^3}{n - 1}$$

where $\overline{X}$ is the mean of the X_i observations.

Because of the cubing operation, large deviations $X_i - \overline{X}$ tend to dominate the sum in the numerator of m_3. If the large deviations are predominantly positive, m_3 will be positive because $(X_i - \overline{X})^3$ has the same sign as $X_i - \overline{X}$. On the other hand, if the large deviations are predominantly negative, m_3 will be negative. Since large deviations are associated with the long tail of a distribution, we can see from Figure 3.6 that m_3 will be positive or negative according to whether the direction of skewness is positive (right) or negative (left). If the observations in the data set are symmetrically distributed about the mean, the third moment m_3 will be zero.

☐ **EXAMPLE**

For the data set $X_1 = -3$, $X_2 = 5$, $X_3 = 40$, we have $\overline{X} = 14$. Hence:

$$m_3 = \frac{(-3 - 14)^3 + (5 - 14)^3 + (40 - 14)^3}{3 - 1} = 5967$$

Since m_3 is positive, the third moment indicates that the data set is skewed to the right. $\square$

The magnitude of the third moment about the mean reflects the units of the observations in the data set. For example, if the observations have units of dollars, the third moment has units of dollars-cubed. A standardized skewness measure is generally more useful because its magnitude does not depend on the units of measure. It is obtained by dividing m_3 by the cube of the standard deviation. The resulting quantity is called the standardized skewness measure.

(3.16)
The **standardized skewness measure** for a data set, denoted by m_3', is defined:

$$m_3' = \frac{m_3}{s^3}$$

where: m_3 is given by (3.15)
         s is given by (3.13)

A data set is skewed positively or negatively according to whether m_3' is positive or negative. If the values in the data set are symmetric about the mean, $m_3' = 0$.

EXAMPLES $\square$

1. For the data set $X_1 = -3$, $X_2 = 5$, $X_3 = 40$, we found earlier that $m_3 = 5967$. Also, $s = 22.87$ for this data set. Hence, $m_3' = 5967/(22.87)^3 = 0.499$, indicating some positive skewness in the data set.

2. For the before-controls data set in the oil spills example, $m_3 = 5397.9$ (calculation not shown) and $s = 17.352$ (see Figure 3.1b). Hence, $m_3' = 5397.9/(17.352)^3 = 1.033$, so the data set is positively skewed. The MYSTAT output in Figure 3.1b gives a standardized skewness measure, called SKEWNESS(G1), that is defined using a divisor of n for s and m_3. It therefore takes a slightly different value than m_3'. $\square$

The standardized skewness measure m_3' can be unreliable when an unusual data set is encountered, such as one containing outlying values.

Comment
The fourth moment about the mean of a data set, denoted by m_4, is sometimes computed as a measure of the *peakedness* or *kurtosis* of the data set. A standardized form of this measure is often found in output from statistical packages.

(3.17)

The **standardized kurtosis measure** for a data set, denoted by m_4', is defined:

$$m_4' = \frac{m_4}{s^4}$$

where: $m_4 = \dfrac{\sum\limits_{i=1}^{n}(X_i - \overline{X})^4}{n - 1}$

s is given by (3.13)

3.5 STANDARDIZED OBSERVATIONS

It is often helpful to examine the distribution pattern of observations in a data set in relation to their mean and standard deviation. This approach is also helpful for judging whether some observations are outlying. An analytical procedure useful for both of these purposes is called *standardizing* the observations.

(3.18)

The **standardized observations** $Y_1, \ldots, Y_n$ corresponding to a set of observations $X_1, \ldots, X_n$ are defined:

$$Y_i = \frac{X_i - \overline{X}}{s} \qquad i = 1, 2, \ldots, n$$

where: $\overline{X}$ is the mean of the observations, as given by (3.1)
s is the standard deviation of the observations, as given by (3.13)

We see that a standardized observation Y_i simply measures the distance that the observation X_i lies from the mean $\overline{X}$ in units of the standard deviation s.

☐ **EXAMPLE**

Figure 3.7a repeats the April shipment data for the melting points example from Figure 3.5a and also shows the standardized observations for this data set calculated by using (3.18). For example, the calculations for Y_1 and Y_2 are as follows ($\overline{X} = 330.8$ and $s = 22.1$):

$$Y_1 = \frac{X_1 - \overline{X}}{s} = \frac{330 - 330.8}{22.1} = -0.04$$
$$Y_2 = \frac{X_2 - \overline{X}}{s} = \frac{358 - 330.8}{22.1} = 1.23$$

The standardized value $Y_1 = -0.04$ indicates that $X_1 = 330$ lies 0.04 standard deviation below $\overline{X} = 330.8$, while $Y_2 = 1.23$ indicates that $X_2 = 358$ lies 1.23 standard deviations above $\overline{X} = 330.8$.

Figure 3.7b shows a histogram of the April shipment data. The horizontal scale of

(a) Observations and Standardized Observations $(\bar{X} = 330.8,\ s = 22.1)$

i	Observation X_i	Standardized Observation Y_i	i	Observation X_i	Standardized Observation Y_i	i	Observation X_i	Standardized Observation Y_i
1	330	−0.04	6	334	0.14	11	321	−0.44
2	358	1.23	7	318	−0.58	12	337	0.28
3	373	1.91	8	325	−0.26	13	328	−0.13
4	346	0.69	9	295	−1.62	14	348	0.78
5	343	0.55	10	288	−1.94	15	318	−0.58

(b) Histogram of Data Set and Correspondence of Original and Standardized Scales

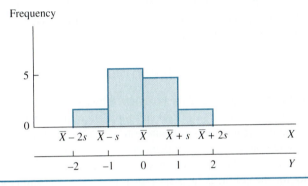

the histogram is given both in terms of the original units (X) and, equivalently, in terms of the standardized units (Y). The histogram shows that all but four of the 15 observations lie within one standard deviation of $\bar{X}$ $(-1 \le Y_i \le 1)$, and none lies beyond two standard deviations from $\bar{X}$ $(Y_i < -2$ or $Y_i > 2)$. ☐

An observation whose standardized value is large in absolute value (say, larger than 3 or 4) is often identified as an *outlier* of the data set.

EXAMPLE ☐

In the melting points example in Figure 3.7, the data set for the April shipment does not contain any outliers because no standardized observation exceeds 2 in absolute value. ☐

Comments

1. A standardized observation Y_i is a unitless quantity because the numerator and the denominator are expressed in the same units of measure (dollars, degrees Celsius, etc.) and hence cancel out. For any data set, the mean of the standardized observations is always zero and the variance is always one.
2. Sometimes data analysis is carried out in terms of the deviations about the mean, $X_i - \bar{X}$. These deviations are referred to as *centered observations*. Centering and standardizing observations in a data set are transformation procedures that often are available in statistical packages.

3.6   BOX PLOTS

A variety of graphic displays can be used to give an effective visual summary of key descriptive statistics for a data set. One of these is a *box plot*, which is also called a *box-and-whisker plot*. As will be seen, the names are quite descriptive of the plot. A box plot shows the smallest and largest values of a data set, the first and third quartiles, and the median. By presenting these key descriptive statistics, the box plot also shows the range and the interquartile range.

Figure 3.8 shows a prototype box plot. The plot is displayed horizontally here, but it could also be displayed vertically. The extremities of the plot (the ends of the whiskers) correspond to the smallest and largest observations in the data set; the distance between these represents the range. The ends of the box are positioned at the first and third quartiles of the data set; hence the length of the box represents the interquartile range. The partitioning line within the box is drawn at the median of the data set. The following example shows how effective a box plot is in giving a snapshot of the distribution pattern of a data set and in facilitating comparisons of several data sets.

Work Force   ☐ EXAMPLE

A regional economic report presented the box plots shown in Figure 3.9a to summarize data on the number of employees in firms in four related industries: leather, textile, knitting, and clothing. The summary measures on which the plots are based are given in Figure 3.9b. Note that the upper whiskers have been broken here to facilitate the plotting of the maximum values because they are so large.

The box plot for each of the industries gives a concise visual summary of the work force sizes for that industry. For the leather industry, for instance, we see from the upper end of the box (the third quartile) that about 75 percent of firms in this industry group employ fewer than 74 employees. We also see that the distribution of number of employees in this industry is heavily right-skewed (in the positive direction). This feature is evident from the length of the whisker in the positive direction and from the median being so much closer to the first quartile than to the third quartile. When the median lies closer to the third quartile than to the first, the distribution is left-skewed,

FIGURE 3.8
Main elements of a box plot

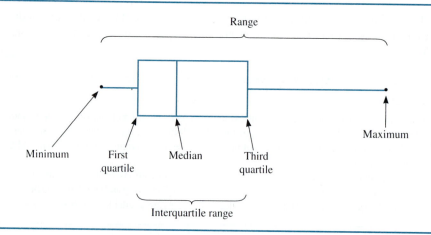

and when the median is centered between the first and third quartiles, the distribution is symmetrical in the central range.

The box plots for the four industries provide an effective means of comparing the work force sizes in the four industries. We see that the median number of employees is largest for the knitting industry and that the 75th percentile is much larger for this industry than for the other three industries. However, the sizes of the firms vary considerably in the knitting industry, the interquartile range being the largest among the four industries (though the total range is the smallest). The textile and clothing industries tend to have smaller firms, their distributions being quite similar with respect to position and variability. The distributions of firm sizes are skewed to the right for all four industry groups. □

Statistical packages prepare box plots in a variety of ways, although the main characteristics usually adhere to the structure illustrated in Figure 3.8.

EXAMPLE □

Figure 3.10 contains MINITAB box plots for the before-controls data set (period 1) and the after-controls data set (period 2) in the oil spills example. The quartiles of each

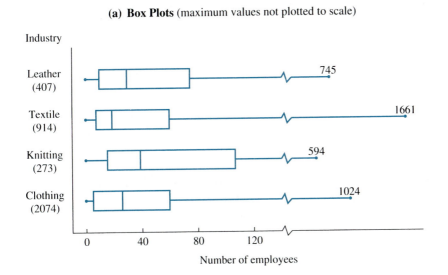

(a) Box Plots (maximum values not plotted to scale)

FIGURE 3.9
Use of box plots in a comparative study—Work force example

(b) Summary Measures

Industry Group	Number of Employees					Number of Firms
	Minimum	1st Quartile	Median	3rd Quartile	Maximum	
Leather	1	9	30	74	745	407
Textile	1	7	18	59	1661	914
Knitting	2	16	39	107	594	273
Clothing	1	4	26	59	1024	2074

FIGURE 3.10
MINITAB box plots for the before-controls and after-controls data sets—Oil spills example

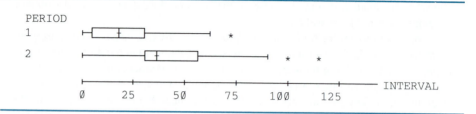

data set are computed slightly differently than we have described. The box plot also displays observations that are considered to be outlying. These are marked by an asterisk (*).

□

3.7   OPTIONAL TOPIC—CALCULATION OF SUMMARY MEASURES FROM FREQUENCY DISTRIBUTIONS

Whenever there is a choice, summary measures of a data set should be calculated from the individual observations rather than from a frequency distribution. Occasionally, however, the data are presented only in the form of a frequency distribution. In this situation, approximations of the summary measures for the data set can be calculated from the frequency distribution. In general, the more detailed the frequency distribution (i.e., the smaller the class widths), the better will be the approximations. We now present procedures for calculating the mean, percentiles, and the variance from a frequency distribution.

Mean

An approximation of the mean $\overline{X}$ of a data set can be calculated from the frequency distribution as follows.

(3.19) $$\overline{X} \simeq \frac{\sum_{i=1}^{k} f_i M_i}{n}$$

where: k is the number of classes in the frequency distribution
M_i is the midpoint of the ith class
f_i is the frequency of the ith class
$$n = \sum_{i=1}^{k} f_i$$

(The symbol $\simeq$ is used throughout the text to denote approximate equality.)

□ EXAMPLE

Table 3.2a contains a frequency distribution of the time intervals between oil spills before pollution controls, constructed from the array in Table 3.1a for the oil spills example. The table illustrates the calculation of the mean from a frequency distribution. Note that the calculated mean $\overline{X} = 23.6$ days is simply a weighted average of the midpoints M_i, with the class frequencies f_i used as weights.

□

TABLE 3.2
Calculation of summary measures from a frequency distribution—Oil spills example

(a) Mean

Time Interval (days)	Number of Cases f_i	Class Midpoint M_i	$f_i M_i$
0–under 15	15	7.5	112.5
15–under 30	13	22.5	292.5
30–under 45	8	37.5	300.0
45–under 60	2	52.5	105.0
60–under 75	2	67.5	135.0
Total	40		945.0

$$\overline{X} = \frac{945.0}{40} = 23.625$$

(b) Median

Time Interval (days)	Number of Cases	Cumulative Number of Cases	
0–under 15	15	15	
15–under 30	13	28	← Class containing 20
30–under 45	8	36	
45–under 60	2	38	
60–under 75	2	40	
Total	40		

$$Md = 15 + \frac{5}{13}(15) = 20.77$$

where: $\frac{n}{2} = \frac{40}{2} = 20, L = 15, n_1 = 5, n_2 = 13, I = 30 - 15 = 15$

(c) Variance

Time Interval (days)	Number of Cases f_i	Class Midpoint M_i	Deviation $M_i - \overline{X}$	Deviation Squared $(M_i - \overline{X})^2$	$f_i(M_i - \overline{X})^2$
0–under 15	15	7.5	−16.125	260.0156	3,900.234
15–under 30	13	22.5	−1.125	1.2656	16.453
30–under 45	8	37.5	13.875	192.5156	1,540.125
45–under 60	2	52.5	28.875	833.7656	1,667.531
60–under 75	2	67.5	43.875	1925.0156	3,850.031
Total	40				10,974.374

$$\overline{X} \approx 23.625 \quad \text{(from part a)} \qquad s^2 \approx \frac{10,974.374}{40 - 1} = 281.39$$

Formula (3.19) assumes that the midpoint of each class is approximately equal to the mean value of the observations included in that class. This assumption is adequate in many cases, but not always. In Table 3.2a, for instance, the approximation of $\overline{X}$ computed from the frequency distribution (23.6 days) differs substantially from the exact mean of the 40 time intervals, namely, 20.9 days. When the midpoints of the classes in the frequency distribution coincide exactly with the means of the observations in the respective classes, then formula (3.19) becomes identical to formula (3.5) and the weighted average of the midpoints gives the exact mean of the data set.

Unequal class widths in a frequency distribution do not affect the computation of the mean by formula (3.19). On the other hand, if the distribution has an open-end class, it is not possible to calculate the mean by (3.19) without additional information because an open-end class has no defined midpoint.

The mean of a percent frequency distribution can be computed by rewriting formula (3.19) as follows.

$$(3.20) \qquad \bar{X} \simeq \sum_{i=1}^{k} \left(\frac{f_i}{n} \right) M_i$$

Here, f_i/n is the percent frequency of the ith class expressed in decimal form.

Median and Other Percentiles

Median.   An approximation of the median Md of a data set can be calculated from the frequency distribution as follows.

$$(3.21) \qquad Md \simeq L + \left(\frac{n_1}{n_2} \right) I$$

where: L is the lower limit of the median class (the class containing the median)
n_1 is the number of observations that must be covered in the median class to reach the median
n_2 is the frequency of the median class
I is the width of the median class

☐   **EXAMPLE**

The calculation of Md for the before-controls frequency distribution for the oil spills example is illustrated in Table 3.2b. Note that $n/2$ is used to locate the median of the frequency distribution [rather than using the rank $(n + 1)/2$ as we did for locating the median in the original data array]. In Table 3.2b, the median of the frequency distribution corresponds to a cumulative frequency of $n/2 = 40/2 = 20$, which lies in the second class.

☐

Formula (3.21) assumes that the observations in the median class are spread evenly throughout that class. This assumption is adequate in many cases, but not always. For the oil spills example, the approximation is only fair, the median for the original data set being 18.5, while the approximate value calculated from the frequency distribution is 20.8.

Other Percentiles.   An approximation of the pth percentile can be obtained in the same manner as for the median, using the following formula.

$$(3.22) \qquad p\text{th percentile} \simeq L + \left(\frac{n_1}{n_2}\right) I$$

where: L is the lower limit of the pth percentile class
(the class containing the pth percentile)
n_1 is the number of observations that must be covered in the
pth percentile class to reach the pth percentile
n_2 is the frequency of the pth percentile class
I is the width of the pth percentile class

As for the median, the formula assumes that the observations are spread evenly throughout the pth percentile class.

EXAMPLE

The 80th percentile for the frequency distribution in Table 3.2b corresponds to cumulative frequency $0.80(40) = 32$. To reach this cumulative frequency, note that the first two classes include 28 observations and hence 4 observations must be covered in the interval 30–under 45. This interval contains 8 observations altogether, so:

$$80\text{th percentile} \simeq 30 + \frac{4}{8}(15) = 37.5$$

Thus, the estimate of the 80th percentile based on the frequency distribution is 37.5 days.

Variance

An approximation of the variance of a data set can be calculated from the frequency distribution as follows.

$$(3.23) \qquad s^2 \simeq \frac{\sum_{i=1}^{k} f_i(M_i - \overline{X})^2}{n - 1}$$

where: k is the number of classes in the frequency distribution
f_i is the frequency of the ith class
M_i is the midpoint of the ith class
$\overline{X}$ is the mean of the frequency distribution
as approximated by (3.19)

$$n = \sum_{i=1}^{k} f_i$$

□ **EXAMPLE**

Table 3.2c illustrates the calculation of the variance for the before-controls frequency distribution for the oil spills example. Note that the approximate value of $\overline{X}$ as computed from the frequency distribution in Table 3.2a is used in the calculation of s^2. □

Formula (3.23) assumes that each observation in a class has a value equal to the midpoint of that class. This assumption is adequate in many cases, but not always.

Since the mean cannot be computed from an open-end frequency distribution without further information, neither can the variance. Also, as for the mean, the calculation of the variance is not affected by unequal class intervals.

3.8 OPTIONAL TOPIC—GEOMETRIC AND HARMONIC MEANS

Geometric Mean

When a data set contains ratios, a geometric mean may be a useful measure of position. For instance, the beginning-of-year prices of a product in four consecutive years were $30.00, $33.00, $33.66, and $41.74. Consider the ratio of the price at the beginning of one year to the price at the beginning of the previous year. The first price ratio is 33.00/30.00 = 1.10, and the other two are 1.02 and 1.24. A single measure is now desired to describe the typical value of the three price ratios. The geometric mean may be appropriate here.

> **(3.24)**
> The **geometric mean** of a set of observations $X_1, X_2, \ldots, X_n$, denoted by G, is the antilogarithm of the arithmetic mean of the logarithms of the observations:
>
> $$\log G = \frac{\log X_1 + \log X_2 + \cdots + \log X_n}{n}$$

The geometric mean is defined only for data sets that contain positive observations. The logarithms in (3.24) are to the base 10, although they could be to any base. (Appendix A, Section A.2, contains a review of logarithms.)

□ **EXAMPLE**

The geometric mean price ratio in the example where $X_1 = 1.10$, $X_2 = 1.02$, $X_3 = 1.24$ is:

$$\log G = \frac{\log 1.10 + \log 1.02 + \log 1.24}{3}$$
$$= \frac{0.04139 + 0.00860 + 0.09342}{3} = 0.04780$$
$$G = \text{antilog } 0.04780 = 1.11635$$

The geometric mean here is an average ratio showing that the annual price increase, on the average, was 11.635 percent. It is a property of the geometric mean that the price in the final period can be obtained from the price in the initial period and the average year-to-year price ratio, as follows:

$$30.00(1.11635)^3 = \$41.74 \qquad \square$$

By taking the antilogarithm of both sides of formula (3.24), we obtain the following equivalent formula for the geometric mean.

(3.24a) $\quad G = (X_1 X_2 \cdots X_n)^{1/n}$

Harmonic Mean

When a data set contains values that represent rates of change, the harmonic mean may be a useful measure of position. For instance, in a study of traffic flow, it was noted that a vehicle traveled at 50 miles per hour over one 10-mile stretch of highway and at 30 miles per hour over a second 10-mile stretch. It is now desired to obtain a measure of the average of the two speeds, reflecting the average speed traveled over the entire 20-mile stretch. The harmonic mean may be appropriate here.

(3.25)

The **harmonic mean** of a set of observations $X_1, X_2, \ldots, X_n$, denoted by H, is the reciprocal of the arithmetic mean of the reciprocals of the observations:

$$\frac{1}{H} = \frac{\dfrac{1}{X_1} + \dfrac{1}{X_2} + \cdots + \dfrac{1}{X_n}}{n}$$

The harmonic mean is defined only for data sets that contain positive observations.

EXAMPLE $\square$

In the example where $X_1 = 50$, $X_2 = 30$, we have:

$$\frac{1}{H} = \frac{\dfrac{1}{50} + \dfrac{1}{30}}{2} = 0.0267$$

so the harmonic mean speed is:

$$H = \frac{1}{0.0267} = 37.5 \text{ miles per hour}$$

The harmonic mean is an appropriate measure here because the vehicle took 0.533 hour to cover the entire 20 miles of highway, and this represents an average speed of $20/0.533 = 37.5$ miles per hour. $\qquad \square$

By taking the reciprocal of both sides of the equation in (3.25), we obtain the following equivalent formula for the harmonic mean.

$$(3.25a) \quad H = \frac{n}{\dfrac{1}{X_1} + \dfrac{1}{X_2} + \cdots + \dfrac{1}{X_n}}$$

PROBLEMS

* **3.1 Travel Expenditures.** Last year's travel expenditures (in dollars) by the 12 members of a university's physics department were as follows:

| 0 | 0 | 173 | 378 | 441 | 733 | 759 | 857 | 958 | 985 | 1434 | 2063 |

Calculate the mean travel expenditures per member. Also calculate the mean travel expenditures of the 10 members who traveled.

3.2 Emergency Admissions. The daily numbers of admissions to the emergency ward of a hospital during the evening shift for the past 11 evenings were as follows:

| 2 | 3 | 2 | 9 | 0 | 3 | 5 | 3 | 1 | 4 | 2 |

a. Calculate the mean number of admissions per evening shift.
b. Verify that identity (3.3) holds for these data.

3.3 Lot Defectives. The numbers of defective items in 12 recent production lots of 100 items each were as follows:

| 3 | 1 | 0 | 0 | 2 | 21 | 4 | 1 | 1 | 0 | 2 | 5 |

a. Calculate the mean number of defective items per lot.
b. The mean number of defective items per lot in a group of 16 lots from another plant was 3.5. What was the total number of defective items in these 16 lots?

3.4 Enrollment Changes. The changes in enrollment between this year and last year for seven colleges are:

| −614 | −103 | 41 | 258 | 313 | 387 | 490 |

a. Calculate the mean enrollment change per college.
b. Does the fact that the mean calculated in part a is positive imply that the combined enrollment for all seven colleges is larger this year than last year? Comment.

3.5 A shipment of 500 turkeys has a mean weight of 6.3 kilograms per turkey.

a. What is the total weight of the shipment?
b. If the cost of the turkeys in the shipment is $3 per kilogram plus $1 per turkey as a handling charge, what is the total cost of the turkeys in this shipment, including the handling charge? What is the mean cost per turkey, including the handling charge?

3.6 A police department has 125 officers whose mean length of service is 8.5 years and 23 civilian employees whose mean length of service is 6.1 years. Calculate the mean length of service of all 148 members of the department.

3.7 Twenty-five thousand shares of a common stock were traded on a particular day at a mean price of $45.50 per share, while 11,500 shares of the same stock were traded on the next day at a mean price of $43.00 per share. Calculate the mean price of all shares traded during the two-day period.

* **3.8** Refer to **Travel Expenditures** Problem 3.1.

a. Calculate the 50 percent trimmed mean for the data set.

b. Compare the values of the trimmed mean and the untrimmed mean. Is the difference between them relatively large here? Will it always be? Discuss.

3.9 Refer to **Lot Defectives** Problem 3.3.

a. Array the observations in ascending order. Calculate the 50 percent trimmed mean.

b. Why can the total number of defective items in the 12 lots not be calculated from the trimmed mean in part a?

3.10 Refer to the array of 40 observations in the before-controls data set in Table 3.1a.

a. Calculate (1) the 50 percent trimmed mean and (2) the 80 percent trimmed mean of this data set. Are the two trimmed means substantially different from each other? Comment.

b. The untrimmed mean for this data set is 20.875. Are the two trimmed means in part a substantially different from the untrimmed mean? Comment.

* **3.11** Refer to **Travel Expenditures** Problem 3.1.

a. Obtain the median travel expenditures per department member.

b. Did exactly one-half of the department's members spend more than this median amount on travel? Does the median always have this property? Explain.

3.12 Refer to **Emergency Admissions** Problem 3.2.

a. Array the observations in ascending order. Obtain the median number of admissions per evening shift.

b. Is the median equal to an actual observation in the data set here? Must it be for every data set? Explain.

3.13 Refer to **Lot Defectives** Problem 3.3.

a. Array the observations in ascending order. Obtain the median of this data set.

b. Why does the median differ substantially from the mean here?

3.14 Refer to **Enrollment Changes** Problem 3.4.

a. Obtain the median enrollment change for the seven colleges.

b. Can the median in part a be multiplied by 7 to obtain the total enrollment change for all seven colleges combined? Explain.

3.15 Refer to **Response Times** Problem 2.11.

a. What is the modal class of the frequency distribution? Is the distribution unimodal?

b. Would the modal class in part a be a reasonable choice to predict the response time for an individual alarm answered by this fire company? Would the use of a combination of two classes be preferable? Explain.

3.16 Hourly Earnings. The percent frequency distribution of hourly earnings for the 2100 plant employees of the Calder Corporation follows:

Hourly Earnings (dollars)	Percent of Employees
14–under 16	6
16–under 18	14
18–under 20	44
20–under 22	31
22–under 24	5
Total	100
	(2100)

a. What is the modal class of this frequency distribution? Is the distribution bimodal?

b. Does the modal class contain the median of the data set? Must this be the case for every frequency distribution?

3.17 On learning that the modal number of children in U.S. families is zero, a commentator stated, "The birthrate has fallen so low that the majority of U.S. families have no children at all!" Comment.

3.18 A service worker stated, "The Internal Revenue Service has calculated that my average tip is $3.00. I have to admit they are correct. But I'll tell you, most of my tips are below the average!" What type of average must the Internal Revenue Service have used here? Would some other average be more appropriate for the agency's purposes? Discuss.

3.19 A state information brochure reported, "The average household in our state owns 1.79 automobiles." What type of average must have been used? Would some other type of average have been more meaningful here? Discuss.

3.20 An analyst stated, "When a data set has one or several outlying observations that might have arisen from errors in data acquisition, I use the median or a trimmed mean in preference to the untrimmed mean." Why might the analyst take this approach?

* **3.21** Refer to **Travel Expenditures** Problem 3.1. Obtain the 80th percentile for this data set by using a line diagram, as in Figure 3.4b. Interpret the meaning of the 80th percentile here.

3.22 A newspaper reported that exchange rates for seven major currencies against the domestic currency changed by the following percentages during the past six weeks:

$$-3.5 \qquad 1.8 \qquad 2.1 \qquad 2.1 \qquad 5.1 \qquad 6.6 \qquad 9.2$$

a. Obtain the third quartile of this data set by using a line diagram, as in Figure 3.4b.
b. What interval of percentiles corresponds to the observation 5.1?

3.23 **Utility Stocks.**   Rates of return (in percent) earned on 10 utility stocks during the past year were as follows:

18.7	21.5	22.7	29.2	21.8	20.5	24.2	20.8	28.4	20.9

a. Array these observations in ascending order. Obtain the 25th, 50th, and 75th percentiles by using a line diagram, as in Figure 3.4b.
b. Does the 50th percentile in part a lie halfway between the 25th and 75th percentiles? Would the 50th percentile lie halfway between the 25th and 75th percentiles if the observations were symmetrically distributed about the median of the data set? Explain.

3.24 Refer to **Emergency Admissions** Problem 3.2.

a. Array the observations in ascending order. Obtain the 20th percentile and the 6th decile using the cumulative percent distribution, as in Figure 3.4a.
b. Do the repeated values in the data set need to be taken into account in interpreting the measures obtained in part a? Discuss.

* **3.25** Refer to **Travel Expenditures** Problem 3.1.

a. Obtain the range and the interquartile range for the data set.
b. Must the interquartile range always be smaller than the range for a data set? Explain.

3.26 Refer to **Lot Defectives** Problem 3.3.

a. Array the observations in ascending order. Obtain the range and the interquartile range for the data set.
b. If the observation with value 21 happened to be an incorrect count of defective items in that lot, would the range be affected by this error? Would the interquartile range be affected?

3.27 Refer to **Enrollment Changes** Problem 3.4.

a. Obtain the range and the interquartile range for the data set.
b. If the enrollment change for an eighth college were added to the data set, might your answers in part a be different? Explain.

3.28 Swimming team A has nine divers whose scores had a range of 2.3 in recent competition. Swimming team B has five divers whose scores had a range of 1.6 in the same competition. A sports reporter commented that team B has more uniform diving talent because its scores had a smaller range. Do you agree with the reporter? Discuss.

* **3.29** Refer to **Travel Expenditures** Problem 3.1. Calculate the variance and the standard deviation of the data set. In what units is each measure expressed?

3.30 Refer to **Lot Defectives** Problem 3.3.

a. Calculate the variance and the standard deviation of the data set.
b. Which observation in the data set makes the largest contribution to the magnitude of the variance through the sum of squared deviations? Which observation makes the smallest contribution? What general conclusions are implied by these findings?

3.31 Refer to **Enrollment Changes** Problem 3.4.

a. Calculate the standard deviation of this data set. In what units is this measure expressed?
b. Does the standard deviation calculated in part a convey any information about the value of the mean of this data set? Comment.

3.32 Ionization Energy. A scientist made four measurements of the ionization energy of helium (in electron volts), obtaining 24.547, 24.623, 24.590, and 24.571. Calculate the standard deviation of these measurements.

* **3.33** Refer to **Travel Expenditures** Problem 3.1.

a. Calculate the coefficient of variation of the data set.
b. In the university's biology department, the mean travel expenditures per member last year were about the same as in the physics department, but the coefficient of variation was substantially smaller. What does this comparison tell about the travel expenditures patterns in the two departments?

3.34 Refer to **Grain Production** Problem 2.35. The means and standard deviations of the revenue and volume data for the two grains are as follows:

	Revenue		Volume	
Grain	Mean	Standard Deviation	Mean	Standard Deviation
Wheat	12.06	4.03	76.71	22.87
Barley	12.49	5.89	106.29	35.35

a. For each grain, calculate the coefficients of variation for the revenue and volume data.
b. Which grain has the greater absolute variability in volume? The greater relative variability in volume? Are the same results true for revenue?
c. Is revenue or volume relatively more variable for wheat? Is the same result true for barley?

3.35 A survey of college students engaged in full-time work last summer gave the following statistics for hours worked and total earnings:

Statistic	Hours Worked	Earnings ($)
Mean	607	5768
Standard deviation	111	1916

a. Calculate the coefficients of variation for hours worked and earnings. In what units is each coefficient expressed?
b. Which exhibits greater variability, hours worked or earnings? Be specific and explain your choice of the measure of variability used in the comparison.

* **3.36** Refer to **Travel Expenditures** Problem 3.1.

a. What does a comparison of the mean and the median indicate about the direction or absence of skewness in this data set?
b. Calculate the third moment about the mean and the standardized skewness measure for this data set. Are your observations about skewness in part a consistent with the signs of these two measures? Explain.

3.37 Refer to **Lot Defectives** Problem 3.3.

a. What does a comparison of the mean and the median indicate about the direction or absence of skewness in this data set?

b. Calculate the third moment about the mean. Which of the 12 observations makes the largest contribution to the magnitude of this measure?

c. Calculate the standardized skewness measure for this data set. In what units is this measure expressed?

3.38 Refer to **Utility Stocks** Problem 3.23.

a. Calculate the third moment about the mean and the standardized skewness measure for this data set. In what units are these measures expressed?

b. What direction of skewness is indicated by the sign of the standardized skewness measure in part a? Are the relative positions of the mean and the median consistent with the sign of the standardized skewness measure? Explain.

c. Present the observations in a dot plot. Is the pattern of asymmetry seen in the plot consistent with your conclusion in part b? Comment.

3.39 Refer to **Horse Trials** Problem 2.12. Identify the modal class of this frequency distribution. Identify the class that contains the median. What does a comparison of these two classes indicate about the symmetry or lack of symmetry of this frequency distribution?

* **3.40** Refer to **Travel Expenditures** Problem 3.1. Which of the 12 observations lies farthest from the mean? What is its standardized value? Is this observation an outlier? Discuss.

3.41 The standardized values of the grades of nine students in a graduate course were:

$$-1.92 \qquad -0.68 \qquad -0.34 \qquad -0.11 \qquad -0.11 \qquad 0.11 \qquad 0.45 \qquad 1.01 \qquad 1.58$$

a. Verify that the mean and the variance of this set of standardized observations are 0 and 1, respectively, except for rounding effects.

b. How many standard deviations apart are the lowest and highest grades in this course? Are there any outlying grades here?

c. The mean and the standard deviation of the nine grades are 73.0 and 8.9, respectively. What was the highest grade in the course?

3.42 Refer to **Utility Stocks** Problem 3.23.

a. Calculate the standardized observations for this data set. In what units are the standardized observations expressed?

b. A financial analyst uses standardized values of stock returns as indicators of their relative performance. He calls a stock with a standardized value exceeding 2 in absolute value an "extreme performer." Are any of the 10 utility stocks extreme performers?

3.43 Refer to **Ionization Energy** Problem 3.32.

a. Calculate the standardized observations for this data set. In what units are the standardized observations expressed?

b. What is the sum of the standardized observations? Must the sum of any set of standardized observations be this value? Explain.

* **3.44** Refer to **Travel Expenditures** Problem 3.1.

a. Construct a box plot for this data set.

b. Does the box plot reveal whether the data set is skewed? Explain.

3.45 Refer to **Lot Defectives** Problem 3.3.

a. Construct a box plot for this data set.

b. In the box plot in part a, the median lies closer to the first quartile than to the third quartile. What does this fact imply about the symmetry of the data set?

3.46 Refer to **Utility Stocks** Problem 3.23. The following summary measures were obtained for the rates of return (in percent) earned on nine manufacturing stocks during the past year: Minimum—4.2; first quartile—10.1; median—13.4; third quartile—17.1; maximum—22.9.

a. Construct box plots for the data sets of the two types of stocks on the same graph.

b. What does a comparison of the two box plots in part a reveal about differences in the rate-of-return distributions for the two types of stocks?

3.47 Refer to **Failure Times** Problem 2.41.

a. Construct box plots for the data sets of the two modes of failure on the same graph.

b. What does a comparison of the two box plots in part a reveal about differences in the failure-times distributions for the two modes of failure?

* **3.48** Refer to **Response Times** Problem 2.11.

a. Calculate the following summary measures from this frequency distribution: (1) mean, (2) median, (3) first quartile, (4) standard deviation. In what units is each measure expressed?

b. Why are the measures calculated in part a only approximations to the measures that would be obtained from the actual data set of 50 response times?

3.49 Refer to **Hourly Earnings** Problem 3.16.

a. Convert the percent frequencies to actual frequencies and then calculate the following summary measures from this frequency distribution: (1) mean, (2) median, (3) standard deviation.

b. The lowest 10 percent of hourly earnings were how many dollars or less?

c. If every employee received a 10 percent increase in hourly earnings, what would be the mean hourly earnings after the increase? The median hourly earnings after the increase? Explain.

3.50 As part of a program to improve profits, the Stokes Company studied customer order sizes. The distribution of order sizes for the past fiscal year follows:

Order Size (dollars)	Number of Orders	Order Size (dollars)	Number of Orders
5–under 25	1852	100–under 250	354
25–under 50	1120	250–under 500	471
50–under 100	693	500–under 1000	530
		Total	5020

a. Calculate the following summary measures from this frequency distribution: (1) mean, (2) median, (3) 80th percentile, (4) standard deviation. In what units is each of these measures expressed?

b. Under what condition would the mean calculated in part a equal the mean of the original data set for the 5020 orders?

c. What total dollar volume is represented by these 5020 orders? What percentage of orders were for less than $250? What percentage of the total dollar volume do these orders represent?

3.51 The charges per day for a semiprivate room in a hospital were as follows in four consecutive years:

Year:	1	2	3	4
Charge ($):	250	260	284	305

a. Obtain the ratio of the charge in one year to that in the preceding year for years 2, 3, and 4.

b. Obtain the geometric mean of the three ratios in part a. Show how the charge in year 4 can be obtained from knowledge of the charge in year 1 and the geometric mean.

3.52 The monthly rates of return on a stock for the past four months follow. The rates of return are based solely on price changes for the stock and ignore dividend payouts.

Month:	1	2	3	4
Rate of Return (%):	1.8	−0.6	1.3	2.6

a. The rate of return for the first month, 1.8 percent, indicates that the ratio of the closing stock price for month 1 to the closing stock price for the preceding month (month 0) was 1.018. Determine the corresponding price ratios for the other months (that is, the ratio of each month's closing price to the preceding month's closing price).

b. Calculate the geometric mean of the four price ratios determined in part a. On the basis of the geometric mean, what has been the average monthly rate of return on this stock over the past four months?

c. The closing stock price for month 0 was $50. Use this price and the geometric mean in part b to calculate the closing stock price for month 4.

3.53 The accident record for a plant follows. The record shows the annual numbers of hours worked per accident during a four-year period in which 4 million hours were worked in the plant each year.

Year:	1	2	3	4
Hours per Accident:	12,535	10,810	11,691	14,735

a. Calculate the harmonic mean number of hours per accident for the four-year period.

b. Is the harmonic mean an appropriate measure of position here? (*Hint:* How many accidents occurred during the four-year period in which 16 million hours were worked?)

c. Would your answer in part b be affected if the annual number of hours worked was not constant during the four-year period? Explain.

3.54 A pulp mill has five pumps of the same type that each operated for approximately the same total time last year. The numbers of hours of operation per repair for the five pumps during the year were as follows.

Pump:	1	2	3	4	5
Hours per Repair:	617	866	703	452	620

a. Calculate the harmonic mean number of hours of operation per repair for the five pumps. In what units is this harmonic mean expressed?

b. If pump 1 operated for 8640 hours last year, how many repairs did it have during the year?

c. If the total number of hours of operation for all five pumps last year were divided by the total number of repairs of the five pumps last year, would the harmonic mean in part a be obtained? Explain.

EXERCISES

3.55 Prove property (3.4). [*Hint:* Write the sum as $\Sigma(X_i - \bar{X} + \bar{X} - A)^2$.]

3.56 Use (3.8) to demonstrate that the mean distance (irrespective of direction) of observations in a data set from their median is smaller than the mean distance from any other value.

3.57 Rearrange (3.12a) to show that $\Sigma X_i^2 = n\bar{X}^2 + (n - 1)s^2$.

3.58 Consider the weights (in grams) of three batteries: $X_1 = 10.98$, $X_2 = 11.01$, $X_3 = 10.97$.

a. Calculate the mean and the standard deviation of the three observations. Now express each observation as a deviation from the target weight 11.00 (that is, make the transformation $Y_i = X_i - 11.00$) and calculate the mean and the standard deviation of the transformed observations. How is the mean of the transformed observations related to the mean of the original observations? What is the relationship for the two standard deviations?

b. Express the three battery weights in units of kilograms (that is, make the transformation $Y_i = X_i/1000$). Calculate the mean and the standard deviation of the transformed observations. How is the mean of the transformed observations related to the mean of the original observations? What is the relationship for the two standard deviations?

c. Consider the three temperature readings (in °C): $X_1 = 15.3$, $X_2 = 21.3$, $X_3 = 17.4$. The mean and the standard deviation of the three observations are $\bar{X} = 18.000$ and $s = 3.045$. Based on the findings in parts a and b, what must be the mean and the standard deviation of the three temperature readings expressed in degrees Fahrenheit? [*Hint:* $Y_i = 32 + (9/5)X_i$ converts the data in degrees Celsius to degrees Fahrenheit.]

3.59 Show that the third moment of the standardized observations of a data set equals the standardized skewness measure (3.16) for the data set.

3.60 Prove that the mean and the variance of the standardized observations of a data set are 0 and 1, respectively.

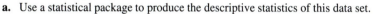

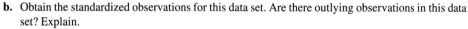

3.61 Refer to **Plant Wages** Problem 2.52.

STUDIES

a. Use a statistical package to produce the descriptive statistics of this data set.
b. Obtain the standardized observations for this data set. Are there outlying observations in this data set? Explain.
c. Construct a box plot for the data set. Is there evidence that the distribution of weekly wages is skewed; if so, in what direction? Is your conclusion about skewness consistent with the relative positions of the mean and median of the data set? Comment.

3.62 Refer to **Hospital Costs** Problem 2.53.

a. Use a statistical package to produce the descriptive statistics of this data set.
b. Obtain the standardized values for the smallest and largest observations. Are either of these extreme observations outlying? Comment.
c. Array the observations. Obtain the 10th percentile of the data set. Why might a hospital administrator be interested in studying the 10 percent of hospitals with the lowest costs?
d. Construct a box plot for the data set. State the main points shown by the box plot.

3.63 Refer to **Failure Times** Problem 2.41.

a. Use a statistical package to produce the descriptive statistics of each of the two data sets.
b. Obtain the standardized observations for each data set. Are there outlying observations in either data set? Explain.
c. Construct box plots for the two data sets on the same graph. What are the effects of mode of failure on the failure times of the cells? Discuss.

3.64 Refer to the **Financial Characteristics** data set (Appendix D.1). Consider only the data on net sales of firms in the paper and electronics industries (industries 4 and 6) in year 1.

a. Use a statistical package to produce the descriptive statistics for the sales data of each industry.
b. Construct box plots for the sales data of each industry on the same graph.
c. Use the results in parts a and b to compare net sales of firms in the two industries and briefly describe noteworthy differences and similarities.

3.65 The price movements of common stocks traded on an exchange such as the New York Stock Exchange are of importance to government, business, academicians, and the public. Effective summarization of the mass of price data generated each trading day requires careful selection of appropriate statistical measures.

a. Describe the principal statistical measures used by newspapers to report the general trend of common stock prices on a daily basis and to report the price movements of any single stock on a daily, weekly, monthly, and yearly basis.
b. How appropriate are the measures used? What are their advantages and disadvantages compared with other measures that might be used for summarizing the price data?

3.66 In a sample survey about family health insurance, the following data on family income, health insurance coverage, and age of family head were obtained:

Family Income (dollars)	Percent of All Families	Percent of Families with Health Insurance	Mean Age of Family Head
Under 25,000	18	24	28
25,000–under 50,000	43	55	39
50,000–under 75,000	27	83	51
75,000 and over	12	91	57

a. What percentage of all families had health insurance?

b. What is the mean age of the family head for all families?

c. How is it possible that the mean age can be calculated in part b, but the mean income of all families cannot because of the open-end class?

d. How does the median income of families with health insurance compare with the median income of all families? Make specific reference to the data to support your answer.

Probability

Basic Probability Concepts

4

Data acquisition and the study of data patterns and summary measures, discussed in the preceding chapters, are important first steps in using data for analysis and decision making. In this chapter we take up basic probability concepts. These concepts are important in their own right. They are used by managers and analysts in assessing odds, in constructing probability models, and in selecting courses of action where risk must be taken into account. In addition, probability concepts are a major building block in the logical foundation of statistical inference, which is the central topic in later units of this book.

Random Trials

Activities that have uncertain or chance outcomes are common.

RANDOM TRIALS, 4.1
SAMPLE SPACES, AND
EVENTS

EXAMPLES

1. An auditor selects a voucher and examines it. Whether or not the voucher contains an error is uncertain in advance of the examination. If it does contain an error, the type of error is uncertain in advance, since a number of different types of errors may occur.

2. A farmer plants and eventually harvests a crop. The crop yield is uncertain in advance because of the chance influences of weather and other natural factors.

3. A laboratory centrifuge experiences operational breakdowns from time to time. The cause of the next breakdown is uncertain, there being several possible causes, including different kinds of mechanical and electrical failures.

Each of the foregoing examples involves a random trial.

> **(4.1)**
> A **random trial** is an activity in which there are two or more different possible outcomes, and uncertainty exists in advance as to which outcome will prevail.

In Example 1, the random trial is the auditor's examination of a voucher. The possible outcomes are no error, or one or more errors of particular types. In Example 2, the random trial is the planting and harvesting of a crop, and the possible outcomes are different crop yields. In Example 3, the random trial is the operation of the centrifuge leading to a break-down, and the possible outcomes are the different causes of breakdowns. In each example, the outcome is uncertain until the trial has been conducted.

Sample Spaces and Basic Outcomes

The outcomes of a random trial are usually defined according to the purpose of the study.

Delivery Service □ **EXAMPLE**

An office supply store is studying its delivery service. The delivery of a customer's order is the random trial. If interest lies in whether or not the delivery is made on the day the customer's order is placed, the outcomes of interest are the two shown in Figure 4.1a. Here, delivery on day 1 signifies delivery on the same day, and delayed delivery signifies delivery after day 1. One of these two outcomes will result from the random trial.

If the purpose of the study is to investigate in detail the nature of delivery delays, interest might be in knowing whether delayed delivery occurred on the second day, on the third day, or after the third day. This more detailed array of outcomes is shown in Figure 4.1b.   □

We refer to the set of outcomes of interest for a random trial as the sample space.

> **(4.2)**
> The different possible outcomes of a random trial are called the **basic outcomes,** and the set of all basic outcomes is called the **sample space.**

Figures 4.1a and 4.1b contain two different sample spaces for the delivery service random trial. The first has two basic outcomes; the second is more detailed and has four basic outcomes.

To facilitate subsequent discussion, we shall denote the sample space of a random trial by S, and the basic outcomes contained in it by $o_1, o_2, \ldots, o_k$. The basic outcomes must be mutually exclusive and exhaustive so that the outcome of a random trial will always be one and only one of the basic outcomes in the sample space. In Figure 4.1b, for example, the timing of delivery of a customer's order must be one and only one of the four basic outcomes shown. A sample space and its basic outcomes are therefore equivalent in structure to a system of classification and the classes of the system, respectively.

> **Comment**
>
> The definition of a sample space in (4.2) applies to random trials that have only a finite number of basic outcomes. This case, which is frequently encountered, is convenient for the presentation of basic concepts. There are, however, many other settings in which an infinity of possible outcomes is conceivable. For example, the weight of the wheat crop in North America next year can be any of an infinity of values in an interval; thus, the sample space for the weight of the crop has an infinity of basic outcomes. With only a few qualifications, the concepts for finite sample spaces carry over directly to infinite sample spaces.

FIGURE 4.1

Two univariate sample spaces for the random trial, delivery of customer's order—Delivery service example

(a) Two Basic Outcomes

Delivery on day 1	Delayed delivery

(b) Four Basic Outcomes

Delivery on day 1	Delivery on day 2	Delivery on day 3	Delivery after day 3

FIGURE 4.1

Two univariate sample spaces for the random trial, delivery of customer's order—Delivery service example

Univariate, Bivariate, and Multivariate Sample Spaces. The sample spaces in Figures 4.1a and 4.1b are called *univariate sample spaces* because the basic outcomes refer to a single characteristic—in this case, the timing of delivery. For many random trials, each basic outcome refers to two, or more than two, characteristics. The corresponding sample spaces are then called *bivariate* and *multivariate sample spaces,* respectively.

EXAMPLE □

In the delivery service example, interest may be not only in the timing of the delivery but also in whether the correct order is delivered to the customer. The resulting bivariate sample space can be displayed in a tabular format, such as in Figure 4.2a. Note that the sample space here contains eight basic outcomes and that they have been arranged in the same manner as in a bivariate cross-classification table.   □

A *tree diagram* is a useful graphic device for visualizing bivariate or multivariate sample spaces. It is especially helpful when there are more than two characteristics, since multivariate sample spaces are difficult to portray in a tabular arrangement.

EXAMPLE □

Figure 4.2b shows a tree diagram for the bivariate sample space in Figure 4.2a. Note that the first branching (on the left) shows the two possible outcomes for the correctness status of the order. Each of these outcomes has a second branching corresponding to the four possible outcomes for the timing of delivery. Thus, the tree ends on the

Bivariate sample space for the random trial, delivery of customer's order, presented in tabular format and in tree diagram—Delivery service example

FIGURE 4.2

(a) Tabular Format

		Timing of Delivery			
		Day 1	Day 2	Day 3	After Day 3
Correctness Status	Correct	o_1	o_2	o_3	o_4
	Incorrect	o_5	o_6	o_7	o_8

S ← Sample Space

(b) Tree Diagram

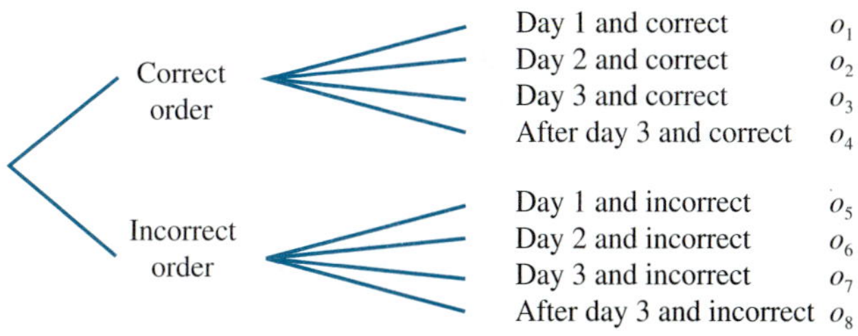

Correct order:
- Day 1 and correct o_1
- Day 2 and correct o_2
- Day 3 and correct o_3
- After day 3 and correct o_4

Incorrect order:
- Day 1 and incorrect o_5
- Day 2 and incorrect o_6
- Day 3 and incorrect o_7
- After day 3 and incorrect o_8

right with eight branches corresponding to the eight basic outcomes of the bivariate sample space. Of course, the tree diagram could have been drawn with the first branching for the timing of delivery and the second for the correctness status of the order and it would still represent the same sample space. ☐

Events

Interest frequently centers on certain basic outcomes of a sample space.

☐ **EXAMPLES**

1. The sample space of a student's grade in a course consists of the basic outcomes corresponding to the letter grades A, B, C, D, and F. When there is interest in whether the student's grade is passing, interest is in the grades A, B, C, and D, which are a subset of the sample space.

2. In the delivery service example, the office supply store is very much concerned with poor delivery service. The store defines poor delivery service as delivery of an incorrect order, or delivery after two days, or both. Here, interest is in the subset of the sample space shown in Figure 4.3. We see that six basic outcomes constitute poor delivery service. ☐

The subset of outcomes of interest is called an event. We say that an event occurs if any one of the basic outcomes in the subset occurs. For instance, in the delivery service example, a poor-service event occurs when any one of the six basic outcomes $o_3, o_4, \ldots,$ o_8 occurs.

> **(4.3)**
> An **event** is a subset of the basic outcomes of the sample space. An event is said to *occur* if any one of its basic outcomes is realized in the random trial.

We shall denote events by capital letters such as E and F, or E_1 and E_2. The poor-service event shown in Figure 4.3 is denoted by E.

Many events can be defined on the same sample space. In the delivery service example, there may also be interest in whether delivery occurs within two days. Denoting the

---

FIGURE 4.3
Poor-service event for the random trial, delivery of customer's order—Delivery service example

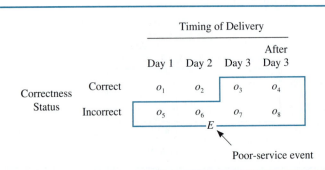

event "delivery within two days" by F, it can be seen from the sample space in Figure 4.2a that event F consists of the basic outcomes o_1, o_2, o_5, o_6.

Complementary Event. Occasionally, interest centers on whether an event of interest does not occur. The nonoccurrence is itself an event consisting of all the basic outcomes of the sample space that are *not* contained in the event of interest.

(4.4)
The set of all basic outcomes *not* contained in an event E is called the **complementary event to E** and is denoted by $E*$.

EXAMPLE

Figures 4.4a and 4.4b repeat the displays of the sample space S and the poor-service event E for the delivery service example. Figure 4.4c shows the basic outcomes in the complement to the poor-service event E. This complementary event $E*$ is a good-

(a) Sample Space, S

Timing of Delivery

		Day 1	Day 2	Day 3	After Day 3
Correctness Status	Correct	o_1	o_2	o_3	o_4
	Incorrect	o_5	o_6	o_7	o_5

S

(b) Event, E

o_1 o_2 o_3 o_4
o_5 o_6 o_7 o_8 E

(c) Complementary Event, $E*$

$E*$ o_1 o_2 o_3 o_4
o_5 o_6 o_7 o_8

(d) Occurrence of F and G, $F \cap G$

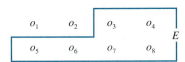

(e) Occurrence of F or G, $F \cup G$

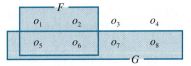

(f) Mutually Exclusive Events, H and J

H J
o_1 o_2 o_3 o_4
o_5 o_6 o_7 o_8

FIGURE 4.4
Various types of delivery service events—Delivery service example

service event that consists of all basic outcomes constituting good delivery service—namely, delivery of correct order on day 1 (o_1) or on day 2 (o_2). □

Joint Occurrence of Two Events.   Often we are interested in the joint occurrence of two events.

□ **EXAMPLE**

For the delivery service example, Figure 4.4d shows two events. Event F represents delivery within two days, and event G represents delivery of an incorrect order. The joint occurrence of both events F and G represents delivery of an incorrect order *and* delivery within two days—in other words, delivery of an incorrect order within two days. Thus, the joint occurrence of events F and G may be viewed as another event that consists of those basic outcomes that are common to the events F and G. As Figure 4.4d shows, the basic outcomes common to the events F and G are o_5 and o_6, as indicated in the shaded region. □

> **(4.5)**
> The joint occurrence of two events E_1 *and* E_2 is another event, to be denoted by $E_1 \cap E_2$, that consists of the basic outcomes common to E_1 and E_2.

The joint event $E_1 \cap E_2$ may be denoted equivalently by $E_2 \cap E_1$ since either notation indicates that both of the events must occur jointly.

The concept of joint occurrence of two events readily generalizes to more than two events.

Comment

In set terminology, the event representing the joint occurrence of two events consists of the basic outcomes in the *intersection* of the two events.

Occurrence of One Event or Another.   We also are often interested in whether one or the other of two events occurs.

□ **EXAMPLE**

For the delivery service example, Figure 4.4e shows the same two events F and G as before. The occurrence of event F or event G represents delivery within two days *or* delivery of an incorrect order. This occurrence of one event or another is itself an event that consists of all basic outcomes contained in either event F or event G. The shaded region in Figure 4.4e shows the six basic outcomes in this event. Included among these six outcomes are the two basic outcomes found in *both* events F and G. □

> **(4.6)**
> The occurrence of event E_1 *or* event E_2 is another event, to be denoted by $E_1 \cup E_2$, that consists of all basic outcomes contained in either E_1 or E_2 or in both E_1 and E_2.

The event $E_1 \cup E_2$ may be denoted equivalently by $E_2 \cup E_1$ since either notation indicates that the event contains the same basic outcomes.

The concept of the occurrence of one event or another readily generalizes to more than two events.

Comment

In set terminology, the event representing the occurrence of one event or another consists of the basic outcomes in the *union* of the two events.

Mutually Exclusive Events.   Some events of a sample space have no basic outcomes in common. They are said to be mutually exclusive events.

(4.7)
If events E_1 and E_2 have no basic outcomes in common, they are said to be **mutually exclusive events.**

EXAMPLE ☐

In Figure 4.4f, events H and J are mutually exclusive. Here, H represents delivery on day 1, and J represents delivery of a correct order on day 3 or later.   ☐

Any event E and its complement E^* are mutually exclusive. Thus, in the delivery service example, the poor- and good-service events E and E^* shown in Figures 4.4b and 4.4c are mutually exclusive.

Probability Measure

PROBABILITY   4.2

We have seen that the outcome of a random trial will be one of the basic outcomes in its sample space. Since it is not known with certainty which particular basic outcome will be realized in advance of the random trial, probability analysis associates with each basic outcome o_i a number called its *probability*. This number is a measure of how likely it is that o_i will be the realized outcome. We shall denote the probability of o_i by $P(o_i)$. In like manner, we shall let $P(E)$ denote the probability that event E will occur; that is, the probability that one of the basic outcomes of event E will be the realized outcome. We temporarily postpone discussion about the meaning of the probability measure and how probability values are obtained.

EXAMPLES ☐

1. If the probability of basic outcome o_1 in Figure 4.4a is 0.57, we write $P(o_1) = 0.57$. Thus, the probability of a correct order delivered on day 1 is 0.57.

2. If the probability of event E in Figure 4.4b is 0.25, we write $P(E) = 0.25$. Thus, the probability that the basic outcome is a poor-service outcome—o_3, o_4, o_5, o_6, o_7, or o_8—is 0.25.   □

Probability Postulates

Probability analysis is based on the following simple postulates.

(4.8)

Postulate 1.   The probability for any basic outcome o_i is a number between 0 and 1; that is, $0 \leq P(o_i) \leq 1$.

Postulate 2.   The probability for any event E is the sum of the probabilities for the basic outcomes constituting the event; that is:

$$P(E) = \sum_E P(o_i)$$

where the summation is over the basic outcomes contained in E.

Postulate 3.   The probability that some basic outcome in the sample space will occur is 1 and the probability that none will occur is 0.

If the occurrence of basic outcome o_i is impossible, then $P(o_i) = 0$. If basic outcome o_i is certain to occur, then $P(o_i) = 1$. The closer the probability $P(o_i)$ is to 0, the less likely it is that o_i will occur; and the closer $P(o_i)$ is to 1, the more likely it is that o_i will occur.

Some Consequences of the Postulates.   A number of consequences follow directly from the probability postulates. Some are given now. Others are presented later as probability theorems.

1. The sum of the probabilities of all basic outcomes in the sample space is 1.

(4.9) $$\sum_{i=1}^{k} P(o_i) = 1$$

---

2. Every event has a probability value between 0 and 1.

(4.10) $0 \leq P(E) \leq 1$ for any event E

---

3. If E_1 and E_2 are two mutually exclusive events, then the probability of both occurring is zero.

(4.11) $P(E_1 \cap E_2) = 0$ for mutually exclusive events E_1 and E_2

Meaning of Probability

We have noted already that the probability for any basic outcome and for any event is a number between 0 and 1. For instance, suppose that event E denotes the occurrence "It rains in Seattle on August 17" and that the probability of this event is $P(E) = 0.3$. How is this probability value interpreted, and how does one determine such a value in practice? We now present two interpretations of probability, both of which provide useful points of view. We follow this presentation with a brief discussion of how probability values are assessed.

Objective Interpretation.   In the *objective* interpretation of probability, the probability value of an event is equated with the relative frequency of occurrence of the event in the long run under constant causal conditions. Let us again consider the event E, "It rains in Seattle on August 17." With the objective interpretation, the probability statement $P(E) = 0.3$ is understood to mean that in an indefinitely large number of August 17's in Seattle, with basic climatic conditions remaining the same as at the present time, there will be rain on about 30 percent of these days.

The objective interpretation of probability applies only to repeatable events and not to unique events. For instance, from this point of view, one would not talk about the probability that John William Jones, aged 40, will die during the next year; he either will or won't. John William Jones's 41st year is not a repeatable event. But it does make sense, according to this interpretation of probability, to say that the probability of a male of age 40 dying during the next year is 0.002. This probability is interpreted to mean that among many males of age 40, about 0.2 percent of them will die during their 41st year. Thus, probability is interpreted from the objective point of view with reference to a large number of random trials under the same causal conditions, not with reference to a specific random trial.

Subjective Interpretation.   Since many events of interest are not repeatable under constant causal conditions, there are many situations where the objective interpretation of probability cannot be applied. The *subjective* interpretation of probability (also called the *personal* interpretation) relates probability to degree of personal belief. It is not restricted to repeatable events but applies also to unique events. Thus, under this approach, one would be willing to consider the probability that John William Jones, aged 40, will die during this coming year, or the probability that the marketing of a new breakfast cereal will be successful. The subjective interpretation even allows one to assign a probability number to an event for a random trial that has occurred in the past. For instance, a police detective investigating the burglary of a store last week might assign a probability of $P(A) = 0.9$ to suspect A having committed this burglary. For all of these events, it would not be possible to interpret probabilities as relative frequencies in the long run since these events are inherently nonrepeatable.

Personal probability is intimately related to the person making the probability evaluation, reflecting subjective feelings and the information available at the time. Thus, two insurance examiners who have identical information about John William Jones might still arrive at different personal probabilities that he will die during his 41st year. Indeed, each insurance examiner might revise his or her probability assessment upon more reflection. Even with reference to a repeatable event, two individuals might subjectively assess the probabilities differently, depending on the information available to each and how this information is evaluated. On the other hand, with the objective probability interpretation, the probabilities of repeatable events are considered to be determined by the causal conditions that are operating and therefore do not depend on personal factors.

Probability Expressed as Odds.   Sometimes a probability value is expressed as *odds.*

☐ **EXAMPLES**

1. If the probability of candidate A winning an election is 0.6, then the odds in favor of candidate A winning are 0.6 to 0.4, or 3 to 2.
2. If the probability of dying in the next year is 0.002, then the odds of death are 0.002 to 0.998, or 1 to 499.   ☐

Assessment of Probability Values

Probability values can be assessed in a variety of ways. When the objective interpretation of probability is applicable, the observed relative frequency of an event in a large number of trials under approximately constant causal conditions may be used as an estimate of the probability of occurrence. For instance, if a production process has produced 30 defective items in the last 5000 produced and if conditions for this production run have been stable and are likely to remain unchanged, then $30/5000 = 0.006$ is an estimate of the probability that the next item produced will be defective.

The use of observed relative frequencies is not without difficulties. It requires repeated experience with the process and constant causal conditions for the repeated trials.

When a subjective probability for an event is required, the most straightforward way to obtain it is to ask a *direct question* about it. A direct probability assessment question can be phrased in various ways. We illustrate three questions for a product manager about the likelihood that the sales of a new product will exceed the sales target.

1. What is the probability that sales of this new product will exceed the sales target?
2. How many chances out of 10 does this new product have of achieving sales greater than the sales target?
3. What are the odds in favor of this new product's sales exceeding the sales target?

Like the use of observed relative frequencies, the direct-question approach to probability assessment has some inherent difficulties. For example, the respondent may find it difficult to express probabilistic beliefs in quantitative terms, or the answer may vary depending on when the question is asked and how much thought is given in answering it.

A host of methods have been developed for eliciting subjective probabilities by *indirect means* to overcome the limitations of the direct approach. Even so, these indirect methods have their own limitations. Much research continues into the best means of obtaining probability assessments in view of the importance of risk assessments in daily life, ranging from assessing investment risks to risks of nuclear accidents.

Theoretical probability models can also provide probability numbers. For instance, in the motor rod example, probabilities can be developed from the probability model represented by the smooth frequency curve in Figure 2.17a, which describes the variation in motor rod diameters that would be anticipated under normal manufacturing conditions. Like the other assessment methods, the use of probability models has limitations, such as the difficulty in being able to select the correct probability model.

Probability assessments must provide a probability number for each basic outcome in the sample space. These probabilities must be numbers between 0 and 1, and their sum must be 1 since one of the basic outcomes in the sample space is certain to occur. A probability distribution shows how the total probability of 1 is allocated by the probability assessments across the basic outcomes in the sample space. It thus indicates the relative chance of occurrence of each basic outcome in the sample space.

PROBABILITY DISTRIBUTIONS

4.3

EXAMPLE ☐

Table 4.1a contains the probability distribution for the sample space in Figure 4.1b relating to the delivery service example. The probabilities were assigned on the basis of relative frequencies in the recent past. Notice that none of the probability values is less than 0 and that the sum of all the probabilities equals 1, in accordance with (4.8) and (4.9). We see from Table 4.1a that delivery is most likely to occur on day 1, and is least likely on day 3 or thereafter. ☐

A formal definition of a probability distribution follows.

TABLE 4.1
Univariate and bivariate probability distributions—Delivery service example

(a) Univariate Probability Distribution

Basic Outcome	Timing of Delivery	Probability
B_1	Day 1	0.60
B_2	Day 2	0.20
B_3	Day 3	0.10
B_4	After day 3	0.10
	Total	1.00

(b) Bivariate Probability Distribution

Correctness Status	Day 1 B_1	Day 2 B_2	Day 3 B_3	After Day 3 B_4	Total
A_1: Correct	0.57	0.18	0.08	0.07	0.90
A_2: Incorrect	0.03	0.02	0.02	0.03	0.10
Total	0.60	0.20	0.10	0.10	1.00

(c) Bivariate Probability Distribution in Symbolic Form

	B_1	B_2	B_3	B_4	Total
A_1:	$P(A_1 \cap B_1)$	$P(A_1 \cap B_2)$	$P(A_1 \cap B_3)$	$P(A_1 \cap B_4)$	$P(A_1)$
A_2:	$P(A_2 \cap B_1)$	$P(A_2 \cap B_2)$	$P(A_2 \cap B_3)$	$P(A_2 \cap B_4)$	$P(A_2)$
Total	$P(B_1)$	$P(B_2)$	$P(B_3)$	$P(B_4)$	1.00

> **(4.12)**
> An assignment of probabilities to each of the basic outcomes in the sample space is called a **probability distribution**.

Univariate, Bivariate, and Multivariate Probability Distributions

The probability distribution in Table 4.1a is called a *univariate probability distribution* because it is based on a univariate sample space. Probability distributions for bivariate and multivariate sample spaces can be constructed as well. They are referred to as *bivariate* and *multivariate probability distributions,* respectively.

☐ **EXAMPLE**

Table 4.1b contains a bivariate probability distribution for the bivariate sample space given earlier in Figure 4.2a. The two variables forming the bivariate sample space are correctness status of order and timing of delivery. Note that delivery of correct order on day 1 is by far the most likely basic outcome.   ☐

Joint, Marginal, and Conditional Probability Distributions

Joint Probabilities.   As discussed earlier, in bivariate and multivariate sample spaces, each basic outcome refers to two or more characteristics. Thus, in Table 4.1b, the basic outcome of delivery of a correct order on day 1 refers to the two characteristics (1) correct order (A_1) and (2) delivery on day 1 (B_1). Each of these characteristics can, in our terminology, be called an event of the bivariate sample space. Thus, the basic outcome of delivery of a correct order on day 1 can be considered a *joint outcome* of the two events A_1 and B_1.

☐ **EXAMPLES**

1. The basic outcome of delivery of a correct order on day 1 is denoted symbolically by $A_1 \cap B_1$ since it is the joint occurrence of events A_1 and B_1. The probability of this joint occurrence is denoted by $P(A_1 \cap B_1)$ in Table 4.1c. From Table 4.1b, we see that this probability is $P(A_1 \cap B_1) = 0.57$.
2. The basic outcome of an incorrect order delivered on day 3 is the joint occurrence of the events incorrect order (A_2) and delivery on day 3 (B_3). This joint occurrence is denoted by $A_2 \cap B_3$, and the corresponding probability according to Table 4.1b is $P(A_2 \cap B_3) = 0.02$.   ☐

The probability of a joint outcome is called a *joint probability*. Thus, $P(A_1 \cap B_1) = 0.57$ is a joint probability.

A bivariate or multivariate probability distribution, such as that in Table 4.1b, is also called a *joint probability distribution* when it is desired to emphasize the fact that the basic outcomes are determined jointly by two or more characteristics. Table 4.1c shows the joint probability distribution for the delivery service example in symbolic form.

Marginal Probabilities.   With a bivariate probability distribution, we are frequently interested in the probability distributions of the individual variables considered separately. For instance, management may wish to know the probability distribution for the correctness of the order, irrespective of the timing of delivery, or to know the probability distribution for the timing of delivery, irrespective of whether or not the order is correct. The univariate probability distribution for each of the variables can be obtained by summing the joint probabilities across the columns or rows, as the case may be.

EXAMPLES

1. The univariate probability distribution in Table 4.1a corresponding to the variable timing of delivery may be obtained by summing each column in Table 4.1b. The resulting probabilities are shown in the bottom row of Table 4.1b and are boxed for emphasis. For example, we have $P(B_1) = 0.60$.

2. The univariate probability distribution for the variable correctness status is obtained by summing each row in Table 4.1b. The resulting probabilities are shown in the rightmost column and are boxed for emphasis. For example, we have $P(A_2) = 0.10$.

We will explain the rationale for summing joint probabilities to obtain univariate probabilities with reference to the probability of delivery on day 1, $P(B_1)$. Since event B_1, if it occurs, must occur jointly with either correct order (A_1) or incorrect order (A_2) but not both, it follows that the joint outcomes $A_1 \cap B_1$ and $A_2 \cap B_1$ are mutually exclusive and exhaustive of the possible outcomes leading to the occurrence of B_1. Therefore:

$$P(B_1) = P(A_1 \cap B_1) + P(A_2 \cap B_1) = 0.57 + 0.03 = 0.60$$

Since the probabilities obtained by summing across either one of the classifications are shown in the margins of Table 4.1b, they are often called *marginal probabilities,* and the resulting probability distributions are then called *marginal probability distributions.*

EXAMPLE ☐

Table 4.1a contains the marginal probability distribution for timing of delivery for the bivariate probability distribution in Table 4.1b.   ☐

In general, the marginal probabilities are found from a bivariate probability distribution as follows.

(4.13) $$P(A_i) = \sum_j P(A_i \cap B_j)$$

$$P(B_j) = \sum_i P(A_i \cap B_j)$$

where the summations are over all events B_j and A_i, respectively.

Table 4.1c illustrates the appropriate summations to be performed for obtaining the marginal probabilities.

Conditional Probabilities.   Often, we need to know the probability of an event occurring, given that a second event occurs. For instance, referring to Table 4.1b, we may wish to know the probability that an order is incorrect (A_2), given that the order is delivered on the third day (B_3). We shall denote this probability by $P(A_2 \mid B_3)$. The vertical rule in the notation is read as "given." In other words, $P(A_2 \mid B_3)$ means "probability of A_2 occurring, given that B_3 occurs." This type of probability is called a conditional probability.

> **(4.14)**
>
> If E_1 and E_2 are any two events and $P(E_2)$ is not equal to zero, the **conditional probability of E_1, given E_2,** is denoted by $P(E_1 \mid E_2)$ and defined:
>
> $$P(E_1 \mid E_2) = \frac{P(E_1 \cap E_2)}{P(E_2)}$$

We illustrate the rationale of the definition of conditional probability by an example.

Job Placement   □   **EXAMPLE**

Table 4.2 shows, for job placement at a college, the joint probability distribution for the two variables (1) location of a graduate's first job (C_i) and (2) duration of time the graduate remains in this job (D_j). We wish to obtain the conditional probability that a graduate remains in the first job for less than two years, given that the job is in the private sector; that is, we wish to obtain $P(D_1 \mid C_1)$. In accordance with the relative frequency interpretation of the probability $P(C_1) = 0.60$ in Table 4.2, we know that, of every 100 placements in first jobs, on the average 60 are in the private sector. Furthermore, we can state, on the basis of the probability $P(C_1 \cap D_1) = 0.45$, that of every 100 first-job placements, on the average 45 are in the private sector *and* have a duration of less than two years. Thus, of every 60 first-job placements in the private sector, on the average 45 entail a stay of less than two years. The proportion $45/60 = 0.75$ is the conditional probability that a graduate remains in the first job for less than two years, given that the job is in the private sector.

Formula (4.14) provides this same result formally, as follows:

$$P(D_1 \mid C_1) = \frac{P(C_1 \cap D_1)}{P(C_1)} = \frac{0.45}{0.60} = 0.75$$

□

TABLE 4.2
Joint probability distribution of location and duration—Job placement example

Location (C_i)	Duration (D_j)		Total
	Less than Two Years (D_1)	Two Years or Longer (D_2)	
C_1: Private sector	0.45	0.15	0.60
C_2: Public sector	0.35	0.05	0.40
Total	0.80	0.20	1.00

The direct computation of conditional probabilities by (4.14) is illustrated by two additional examples.

1. For the delivery service example in Table 4.1b, we wish to obtain $P(A_2 \mid B_3)$, that is, the conditional probability of the order being incorrect, given that the order is delivered on day 3. Since $P(B_3) = 0.10$ and $P(A_2 \cap B_3) = 0.02$, we obtain:

$$P(A_2 \mid B_3) = \frac{P(A_2 \cap B_3)}{P(B_3)} = \frac{0.02}{0.10} = 0.20$$

2. For the delivery service example in Table 4.1b, we wish to find the conditional probability of delivery on day 1, given that the order is correct. We obtain:

$$P(B_1 \mid A_1) = \frac{P(A_1 \cap B_1)}{P(A_1)} = \frac{0.57}{0.90} = 0.63$$

Conditional Probability Distributions. A conditional probability distribution describes how the total probability of 1 is allocated across the different basic outcomes contained in the conditioning event.

EXAMPLES

1. For the job placement example in Table 4.2, we wish to obtain the conditional probability distribution for duration of first job (D_j), given that the job is in the private sector (C_1). We require the set of conditional probabilities $P(D_j \mid C_1)$. From previous calculation, we know that $P(D_1 \mid C_1) = 0.75$. We calculate, from Table 4.2, that $P(D_2 \mid C_1) = P(D_2 \cap C_1)/P(C_1) = 0.15/0.60 = 0.25$. Hence, the conditional probability distribution for duration of first job, given that the job is in the private sector, is:

Duration	Conditional Probability, Given C_1
D_1: Less than two years	0.75
D_2: Two years or longer	0.25
Total	1.00

Note that the conditional probability distribution has the attributes of any probability distribution: It involves outcome probabilities between 0 and 1, and these probabilities sum to 1.

2. For the delivery service example in Table 4.1b, we wish to obtain two conditional probability distributions: (1) for correctness status, given that delivery is on day 3, and (2) for timing of delivery, given that the order is correct. Table 4.3 presents these two conditional distributions and shows how they were obtained.

TABLE 4.3
Conditional probability distributions of correctness status and timing of delivery—Delivery service example

(a) Probability Distribution of Correctness Status, Given Delivery on Day 3

Correctness Status	Conditional Probability, Given B_3
A_1: Correct	$P(A_1 \mid B_3) = 0.08/0.10 = 0.80$
A_2: Incorrect	$P(A_2 \mid B_3) = 0.02/0.10 = \underline{0.20}$
	Total 1.00

(b) Probability Distribution of Timing of Delivery, Given Correct Order

Timing of Delivery	Conditional Probability, Given A_1
B_1: Day 1	$P(B_1 \mid A_1) = 0.57/0.90 = 0.63$
B_2: Day 2	$P(B_2 \mid A_1) = 0.18/0.90 = 0.20$
B_3: Day 3	$P(B_3 \mid A_1) = 0.08/0.90 = 0.09$
B_4: After day 3	$P(B_4 \mid A_1) = 0.07/0.90 = \underline{0.08}$
	Total 1.00

4.4 BASIC PROBABILITY THEOREMS

Several useful theorems follow from the probability concepts presented previously.

Addition Theorem

We often wish to obtain the probability for the occurrence of one event or another.

□ EXAMPLES

1. In the delivery service example in Table 4.1b, what is the probability that the order is correct or the delivery is on day 1, or both; that is, what is $P(A_1 \cup B_1)$?

2. In the job placement example in Table 4.2, what is the probability that the first job is in the private sector or the graduate remains in the job for less than two years, or both; that is, what is $P(C_1 \cup D_1)$?   □

The addition theorem may be used to find the probability that one or the other, or both, of two events occur.

(4.15)
Addition Theorem. For any two events E_1 and E_2:

$$P(E_1 \cup E_2) = P(E_1) + P(E_2) - P(E_1 \cap E_2)$$

EXAMPLES ☐

1. Refer to Table 4.1b. We wish to find the probability that the order is correct or the delivery is on day 1, or both; that is, we wish to find $P(A_1 \cup B_1)$. We have $P(A_1) = 0.90$, $P(B_1) = 0.60$, and $P(A_1 \cap B_1) = 0.57$. Hence, by (4.15):

$$P(A_1 \cup B_1) = 0.90 + 0.60 - 0.57 = 0.93$$

2. Refer to Table 4.2. We wish to obtain the probability that the job placement is in the private sector or the duration is less than two years, or both; that is, we wish to find $P(C_1 \cup D_1)$. We have $P(C_1) = 0.60$, $P(D_1) = 0.80$, and $P(C_1 \cap D_1) = 0.45$. Hence, by (4.15):

$$P(C_1 \cup D_1) = 0.60 + 0.80 - 0.45 = 0.95$$

3. Refer to the sample space in Figure 4.5a. The sample space has six basic outcomes, and the probability values for these outcomes are shown. Two events are defined on this sample space: Event E contains four basic outcomes, and event F contains three. Note that two outcomes are common to the two events. For these two events, we have:

$$P(E) = 0.15 + 0.34 + 0.10 + 0.07 \qquad = 0.66$$
$$P(F) = \qquad\qquad 0.10 + 0.07 + 0.22 = 0.39$$
$$P(E \cap F) = \qquad\qquad 0.10 + 0.07 \qquad = 0.17$$

Substituting these values into addition theorem (4.15), we obtain:

$$P(E \cup F) = 0.66 + 0.39 - 0.17 = 0.88$$

By adding the probabilities for the basic outcomes in the event $E \cup F$ directly from Figure 4.5a, we obtain:

$$P(E \cup F) = 0.15 + 0.34 + 0.10 + 0.07 + 0.22 = 0.88$$

which is the result provided by the addition theorem. ☐

(a) Addition Theorem

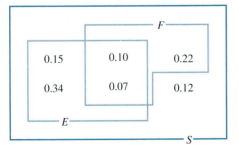

(b) Complementation Theorem

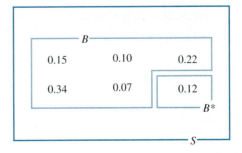

FIGURE 4.5

Illustrations of addition and complementation theorems

From the third example, it is clear why $P(E \cap F)$ must be subtracted from $P(E) + P(F)$ to obtain $P(E \cup F)$. Note that when $P(E)$ and $P(F)$ are added, the probabilities of the two basic outcomes these events have in common (0.10 and 0.07) are included twice. Since $E \cap F$ is the set of basic outcomes the two events have in common, $P(E \cap F)$ is subtracted to correct for the double counting.

Mutually Exclusive Events.   The addition theorem takes on a special form when events E_1 and E_2 are mutually exclusive. From (4.11), we know that $P(E_1 \cap E_2) = 0$ in this case. Therefore, the addition theorem simplifies as follows.

(4.16)
For any two mutually exclusive events E_1 and E_2:

$$P(E_1 \cup E_2) = P(E_1) + P(E_2)$$

☐ **EXAMPLE**

In Table 4.1b, the events delivery on day 1 (H) and correct delivery after day 2 (J) are mutually exclusive. Since $P(H) = 0.57 + 0.03 = 0.60$ and $P(J) = 0.08 + 0.07 = 0.15$, we have by (4.16) that $P(H \cup J) = 0.60 + 0.15 = 0.75$.   ☐

Formula (4.16) extends in a direct fashion for more than two mutually exclusive events.

(4.17)
For n mutually exclusive events $E_1, E_2, \ldots, E_n$:

$$P(E_1 \cup E_2 \cup \cdots \cup E_n) = P(E_1) + P(E_2) + \cdots + P(E_n)$$

Complementation Theorem

Sometimes it is more difficult to find the probability of an event than it is to find the probability of its complement. The probability of the event can then be obtained by the complementation theorem.

(4.18)
Complementation Theorem. For any event E:

$$P(E) = 1 - P(E^*)$$

where E^* is the complementary event to E.

☐ **EXAMPLES**

1. For the delivery service example in Table 4.1b, we wish to find the probability that service is not perfect, that is, that the correct order is not delivered on day 1. This

event E consists of all outcomes other than $A_1 \cap B_1$. Hence, we consider the complementary event $E^* = A_1 \cap B_1$. We have $P(E^*) = 0.57$, and by the complementation theorem, $P(E) = 1 - 0.57 = 0.43$.

2. In Figure 4.5b, $P(B) = 1 - P(B^*) = 1 - 0.12 = 0.88$. □

Multiplication Theorem

Frequently, we wish to obtain the probability of the joint occurrence of two events.

EXAMPLES □

1. Sally McQue has applied to two law schools. What is the probability that she will be accepted by both schools?

2. At a police road checkpoint, each vehicle is checked for having a current safety inspection sticker, and also the driver is checked for having a current driver's license. What is the probability that a vehicle will have a current inspection sticker and the driver a current driver's license? □

The probability of two events occurring jointly is a joint probability that may be obtained by rearranging the definition of conditional probability in (4.14). The result is the multiplication theorem.

(4.19)

Multiplication Theorem. For any two events E_1 and E_2:

$$P(E_1 \cap E_2) = P(E_1)P(E_2 \mid E_1) = P(E_2)P(E_1 \mid E_2)$$

EXAMPLES □

1. In the law school admissions example, let L_1 and L_2 denote acceptance by schools 1 and 2, respectively. The probability that Sally McQue is accepted by law school 1 is known to be $P(L_1) = 0.6$. Given acceptance by law school 1, the conditional probability of acceptance by law school 2 is known to be $P(L_2 \mid L_1) = 0.9$. We wish to find the probability of acceptance by both schools. Using (4.19), we obtain:

$$P(L_1 \cap L_2) = P(L_1)P(L_2 \mid L_1) = 0.6(0.9) = 0.54$$

2. For the police checkpoint example, let event S denote current safety inspection sticker and let event D denote current driver's license. It is known that 90 percent of vehicles have current safety inspection stickers and that 95 percent of drivers of vehicles with current inspection stickers have current driver's licenses. We wish to find the probability that both driver and vehicle will be found to have current documents. We have $P(S) = 0.90$ and $P(D \mid S) = 0.95$. Hence, we obtain by (4.19):

$$P(S \cap D) = P(S)P(D \mid S) = 0.90(0.95) = 0.855$$ □

When the multiplication theorem is used to find several related joint probabilities, a *probability tree* is often helpful to portray the situation.

Computer Chip Inspection

☐ **EXAMPLE**

A computer manufacturer inspects each memory chip before it enters assembly operations because of recent difficulties in the manufacturing process. Let D denote that a chip is defective and $D*$ that it is acceptable, that is, not defective. Also, let A denote that a chip is approved for assembly by the inspector, and $A*$ that it is not approved. From past experience, it is known that the probability of a defective chip is $P(D) = 0.10$. Also, it is known that the probability of an inspector passing a chip, given it is defective, is $P(A \mid D) = 0.005$, while the corresponding probability, given the chip is acceptable, is $P(A \mid D*) = 0.999$. The tree diagram in Figure 4.6 is called a probability tree and summarizes the situation.

We use multiplication theorem (4.19) to find the joint probability that a chip is defective and is approved for assembly:

$$P(D \cap A) = P(D)P(A \mid D) = 0.10(0.005) = 0.0005$$

Similarly, the joint probability that a chip is acceptable and is approved for assembly is:

$$P(D* \cap A) = P(D*)P(A \mid D*) = 0.90(0.999) = 0.8991$$

The other joint probabilities in Figure 4.6 have been obtained in similar fashion.

Note that since a chip is either acceptable or defective, we can find the probability that a chip is approved for assembly as follows:

$$P(A) = P(D \cap A) + P(D* \cap A) = 0.0005 + 0.8991 = 0.8996 \qquad ☐$$

FIGURE 4.6

Probability tree for chip inspections—Computer chip inspection example

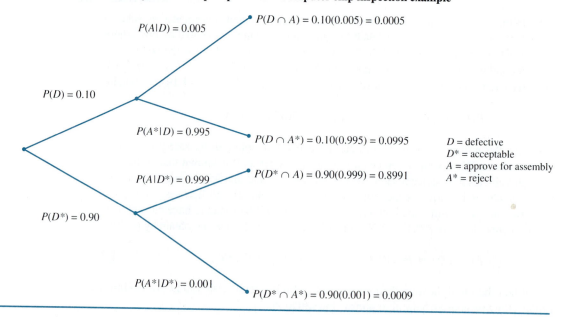

$P(A|D) = 0.005$

$P(D \cap A) = 0.10(0.005) = 0.0005$

$P(D) = 0.10$

$P(A*|D) = 0.995$

$P(D \cap A*) = 0.10(0.995) = 0.0995$

$P(A|D*) = 0.999$

$P(D* \cap A) = 0.90(0.999) = 0.8991$

$P(D*) = 0.90$

$P(A*|D*) = 0.001$

$P(D* \cap A*) = 0.90(0.001) = 0.0009$

D = defective
$D*$ = acceptable
A = approve for assembly
$A*$ = reject

Multiplication theorem (4.19) can be extended to more than two events. For three events, the multiplication theorem is as follows.

(4.20)
For any three events E_1, E_2, and E_3:

$$P(E_1 \cap E_2 \cap E_3) = P(E_1)P(E_2 \mid E_1)P(E_3 \mid E_1 \cap E_2)$$

RELATIONSHIPS BETWEEN VARIABLES   4.5

The concept of relationships between variables is very important in statistical analysis. We shall illustrate the presence and absence of relationships between two variables by the following examples.

EXAMPLE ☐   **Computer Chip Inspection—continued**

Table 4.4 contains two conditional probability distributions for the computer chip inspection example: the conditional probability distribution of inspector's action, given that the chip is defective, and the conditional probability distribution of inspector's action, given that the chip is acceptable. Note that the two conditional probability distributions differ. For instance, the conditional probability of the inspector approving the chip is very low if the chip is defective, and it is very high if the chip is acceptable. The differences in the conditional probabilities imply that the inspector's action is related to the quality of the chip. One would hope, of course, that such a relation exists.

☐

EXAMPLE ☐   **Hospital Stay**

A hospital administrator is studying the relationship between length of stay in hospital and whether or not the patient has hospital insurance, for patients with a certain illness. The joint probability distribution is shown in Table 4.5a. The conditional probability distributions for length of stay, given insurance status, are shown in Table 4.5b. The conditional probabilities are obtained in the usual manner. For example, $P(A_1 \mid B_1) = 0.42/0.70 = 0.60$. Note that the two conditional probability distributions are the same.

TABLE 4.4

Conditional probability distributions for the inspector's action, given the quality of the chip—Computer chip inspection example.   The differences between the two conditional probability distributions indicate that quality of chip and inspector's action are dependent variables.

		Conditional on	
Action		Chip Defective D	Chip Acceptable D^*
Approve (A)		0.005	0.999
Reject (A^*)		0.995	0.001
	Total	1.000	1.000

TABLE 4.5   **Joint and conditional probability distributions of length of stay—Hospital stay example.**   The identity of the conditional probability distributions indicates that length of stay and insurance status are independent variables.

(a) Joint Probability Distribution

Length of Stay (days)	Insurance Status Insured B_1	Uninsured B_2	Total
A_1: Less than 5	0.42	0.18	0.60
A_2: 5–10	0.21	0.09	0.30
A_3: Over 10	0.07	0.03	0.10
Total	0.70	0.30	1.00

(b) Conditional Probability Distributions of Length of Stay, Given Insurance Status

Length of Stay (days)	Conditional on Insured B_1	Uninsured B_2	Marginal Probability Distribution
A_1: Less than 5	0.60	0.60	0.60
A_2: 5–10	0.30	0.30	0.30
A_3: Over 10	0.10	0.10	0.10
Total	1.00	1.00	1.00

Also note that for each length of stay, the conditional probabilities are equal to the marginal probability, that is, $P(A_i \mid B_j) = P(A_i)$ for all i and j. For convenience, the marginal probability distribution from Table 4.5a is repeated in Table 4.5b. The equality of the conditional and marginal probability distributions in Table 4.5b indicates that insurance status of the patient is unrelated to length of stay. A similar result is found if we look at the conditional probability distributions of insurance status, given the length of stay of the patient.   □

When no relationship exists as in the hospital stay example, we say that *independence* is present. The terms *statistical independence* and *probabilistic independence* are also used at times. When a relationship does exist as in the computer chip inspection example, we say that *dependence* is present. We shall now take up in more detail the concept of independence, first as it relates to events and then as it relates to variables. Following this discussion, we shall consider the concept of dependence.

Independence

Independent Events.   Two events are unrelated if the conditional probability that one event occurs is unaffected by whether or not the other event occurs. When two events are unrelated in this sense, we say that they are *independent events*.

Symbolically, the independence of two events E_1 and E_2 means that $P(E_2 \mid E_1) = P(E_2)$ or, equivalently, that $P(E_1 \mid E_2) = P(E_1)$. Substituting either of these relations into multiplication theorem (4.19) yields the following definition.

(4.21)

Two **events** E_1 and E_2 are **independent** if:

$$P(E_1 \cap E_2) = P(E_1)P(E_2)$$

Thus, two events are independent if their joint probability equals the product of their marginal probabilities. Equivalently, we can define independence of two events in terms of conditional probabilities.

(4.21a)

Two **events** E_1 and E_2 are **independent** if:

$$P(E_1 \mid E_2) = P(E_1) \quad \text{or} \quad P(E_2 \mid E_1) = P(E_2)$$

EXAMPLE ☐

For the hospital stay example in Table 4.5a, A_1 and B_1 are independent events since $P(A_1 \cap B_1) = 0.42$ and $P(A_1)P(B_1) = 0.60(0.70) = 0.42$. In other words, a length of stay of less than five days is independent of a patient being insured.

☐

Formula (4.21) may be extended as follows.

(4.22)

For n independent events $E_1, E_2, \ldots, E_n$:

$$P(E_1 \cap E_2 \cap \cdots \cap E_n) = P(E_1)P(E_2) \cdots P(E_n)$$

EXAMPLE ☐ **Security Check**

In a new security system being tested at a military base, visitors wishing to enter a secure area must pass three security checks in succession—a voice-pattern check, a fingerprint check, and a handwriting check. Studies indicate that an unauthorized visitor will pass any one check with a probability of 0.02 and that the three checks are independent. We wish to find the probability that an unauthorized visitor passes all three checks.

Let V, F, and H denote the events that an unauthorized visitor passes the voice, fingerprint, and handwriting checks, respectively. We have $P(V) = P(F) = P(H) = 0.02$. Hence, the probability that an unauthorized visitor passes all three checks is represented by $P(V \cap F \cap H)$. We have by (4.22):

$$P(V \cap F \cap H) = P(V)P(F)P(H) = 0.02(0.02)(0.02) = 0.000008$$

Thus, the probability of an unauthorized visitor passing all three checks is quite small.

☐

Independent Variables. The concept of two independent events extends to the concept of two independent variables. We denote the classes of the first variable by A_1, A_2, and so on, and the classes of the second variable by B_1, B_2, and so on.

(4.23)

Two **variables** A and B are **independent** if:

$$P(A_i \cap B_j) = P(A_i)P(B_j) \qquad \text{for all } A_i \text{ and } B_j$$

Note that *all* joint probabilities must equal the product of their corresponding marginal probabilities for the two variables to be independent.

☐ **EXAMPLE**

In Table 4.5a for the hospital stay example, length of stay and insurance status are independent variables since the joint probability of any two events A_i and B_j is equal to the product of the respective marginal probabilities. For example, $P(A_1 \cap B_2) = 0.18$, and $P(A_1)P(B_2) = 0.60(0.30) = 0.18$.

Independence of the two variables length of stay and insurance status implies that the conditional distribution of length of stay is the same for insured patients as for uninsured patients. Independence further implies that both conditional distributions are equal to the marginal distribution of length of stay, as we noted earlier in Table 4.5b. ☐

Dependence

Dependent Events. If two events are not independent, they are related and said to be *dependent events.*

☐ **EXAMPLE**

In the job placement example of Table 4.2, the events C_1 (private sector placement) and D_1 (duration of job less than two years) are dependent events because $P(C_1 \cap D_1) = 0.45$ while $P(C_1)P(D_1) = 0.60(0.80) = 0.48$. ☐

Dependent Variables. If two variables are not independent, they are related and said to be *dependent variables.* Dependence is present in many bivariate probability distributions.

☐ **EXAMPLE**

In the delivery service example in Table 4.1b, correctness status and timing of delivery are dependent variables. This dependence can be seen quickly by multiplying corresponding pairs of marginal probabilities. For instance, $P(A_2)P(B_1) = 0.10(0.60) = 0.06$, which does not equal $P(A_2 \cap B_1) = 0.03$. To be sure, some of the joint probabilities in Table 4.1b are equal to the product of the corresponding marginal probabilities. However, since the equality does not hold for *all* joint probabilities, there is dependence between the two variables. ☐

TABLE 4.6

Conditional probability distributions of timing of delivery—Delivery service example. The distributions show the nature of the relationship between timing of delivery and correctness status of the order.

Timing of Delivery	Conditional on	
	Correct Order A_1	Incorrect Order A_2
B_1: Day 1	0.63	0.30
B_2: Day 2	0.20	0.20
B_3: Day 3	0.09	0.20
B_4: After day 3	0.08	0.30
Total	1.00	1.00

Nature of Dependence. The nature of the relationship between two dependent variables can be studied readily by comparing the conditional probability distributions for one variable, conditional on each possible outcome of the other variable. This method of analysis is analogous to the one discussed in Section 2.4 for examining the nature of the relationship between two qualitative variables in a bivariate data set.

EXAMPLES ☐

1. For the delivery service example, the conditional probability distribution for the timing of delivery, given that the order is correct, was derived in Table 4.3b and is repeated in Table 4.6. The conditional probability distribution, given that the order is incorrect, is also presented in Table 4.6. A comparison of these two conditional probability distributions facilitates the study of the relationship between timing of delivery and correctness status of the order. We see that incorrect orders are much less likely to be delivered on day 1 than correct orders, and they are much more likely to be delivered on day 3 or thereafter.

2. For the computer chip inspection example in Table 4.4, we noted earlier that the relationship between the inspector's action and the quality of a chip is consistent with an effective quality control system—namely, a high approval probability for good chips and a high rejection probability for defective chips. ☐

PROBLEMS

4.1 Explain why each of the following situations involves a random trial and list two possible outcomes of the trial.

a. A cake is baked at 200°C and taken from the oven. The cake's internal temperature 10 minutes later is of interest.

b. A pilot is about to start the three engines of the aircraft. The number of engines that start without trouble is of interest.

c. A woman shopper is approached in a store and asked which of four perfume scents she finds most pleasant. Her response is of interest.

4.2 Explain why each of the following situations involves a random trial and list two possible outcomes of the trial.

a. A police squad is about to begin its night patrol. The number of incidents the squad will investigate that night is of interest.

b. A student turns on a calculator before an examination to check it. Its working order is of interest.

c. A person mails a letter. How many days later the letter reaches its destination is of interest.

4.3 **Food Inspection.** A food product is inspected at the processing plant by an inspector and given a quality grade of A, B, C, or D. Major food stores are the only stores that sell the grade A product. Grade B and grade C products are sold only through discount outlets. Grade D products are not fit for human consumption and are sold to processors of animal foods.

a. Describe the sample space for an inspected food product for: (1) different quality outcomes, (2) different channels of market distribution, and (3) whether or not the product is fit for human consumption.

b. Are the sample spaces in part a univariate or bivariate? Explain.

* 4.4 Refer to **Food Inspection** Problem 4.3. Consider the sample space for different quality outcomes. Denote the basic outcomes by A, B, C, and D.

a. Let E_1 denote the event that the product is sold through discount outlets. What basic outcomes constitute E_1? What basic outcomes constitute E_1^*?

b. Let E_2 denote the event that the product is fit for human consumption. What basic outcomes constitute E_2?

c. Are E_1 and E_2 mutually exclusive events? Complementary events? Explain.

4.5 **Manuscript Review.** Two reviewers (1, 2) for a publishing house independently screen manuscripts that arrive unsolicited in the mail. Each reviewer gives a grade of good (G), fair (F), or poor (P) to a manuscript.

a. Describe the sample space for the outcomes of the joint review of a manuscript.

b. Is the sample space in part a univariate or bivariate? Explain.

4.6 Refer to **Manuscript Review** Problem 4.5. Denote the basic outcome in which reviewers 1 and 2 give respective grades of good and poor to a manuscript by (G, P), grades of fair and good by (F, G), and similarly for the other basic outcomes.

a. Let E_1 denote the event that at least one reviewer assigns a grade of good. What basic outcomes constitute E_1? What basic outcomes constitute E_1^*?

b. Let E_2 denote the event that the reviewers give different grades. What basic outcomes constitute E_2?

c. Are E_1 and E_2 mutually exclusive events? Complementary events? Explain.

4.7 **Canal Lock.** A canal has a single lock. Ships may be waiting upstream or downstream. Assume that the number waiting on a given side never exceeds three.

a. Describe the sample space for both the number of ships waiting upstream and the number waiting downstream.

b. How many basic outcomes does the sample space in part a contain? Is the sample space univariate or bivariate? Explain.

c. Describe the sample space for the total number of ships waiting on both sides of the lock. Is this sample space univariate or bivariate?

4.8 Refer to **Canal Lock** Problem 4.7. Consider the sample space described in part a of the problem. Denote the basic outcome in which one ship is waiting upstream and no ship is waiting downstream by (1, 0), and similarly for the other basic outcomes.

a. Let E_1 denote the event that one or more ships wait upstream and one or more ships wait downstream. What basic outcomes constitute E_1? What basic outcomes constitute the complementary event to E_1?

b. Let E_2 denote the event that no ship is waiting upstream, or no ship is waiting downstream, or both. What basic outcomes constitute E_2?

c. Are E_1 and E_2 mutually exclusive events? Complementary events? Explain.

d. Let E_3 denote the event that a total of three ships are waiting at the lock. What basic outcomes constitute E_3?

4.9 Muscle Deterioration. Two astronauts (1, 2) are given medical examinations after a long space flight. The degree of muscle deterioration from weightlessness is assessed for each astronaut as negligible (N), moderate (M), or severe (S).

a. Describe the univariate sample space for the degree of muscle deterioration for one astronaut.
b. Describe the bivariate sample space for the degrees of muscle deterioration of both astronauts. How many basic outcomes does this sample space contain?

4.10 Refer to **Muscle Deterioration** Problem 4.9. Consider the sample space described in part b of that problem. Denote the basic outcome in which astronauts 1 and 2 have negligible and moderate muscle deterioration, respectively, by (N, M), severe and moderate deterioration by (S, M), and similarly for the other basic outcomes.

a. Let E_1 denote the event that neither astronaut has a severe degree of muscle deterioration. What basic outcomes constitute E_1?
b. Let E_2 denote the event that at least one of the astronauts has moderate or severe muscle deterioration. What basic outcomes constitute E_2?
c. Are E_1 and E_2 mutually exclusive events? Complementary events? Explain.

4.11 Refer to Figure 4.2b. Construct the corresponding tree diagram where the first branching is for timing of delivery.

4.12 Refer to **Canal Lock** Problem 4.7. Construct a tree diagram representing the sample space in part a of that problem. Associate the first branching with the number of ships waiting upstream.

4.13 Refer to **Muscle Deterioration** Problem 4.9. Construct a tree diagram representing the sample space in part b of that problem. Associate the first branching with the first astronaut.

* **4.14** Refer to Figure 4.4. List the basic outcomes constituting each of the following events: (1) H^*, (2) $E \cap H$, (3) $H \cup J$, (4) $J \cap G$.

4.15 Refer to **Manuscript Review** Problems 4.5 and 4.6. List the basic outcomes constituting each of the following events: (1) $E_1 \cap E_2$, (2) $E_1 \cup E_2$, (3) $E_1^* \cap E_2$, (4) $E_1^* \cup E_2^*$.

4.16 Refer to **Canal Lock** Problems 4.7a and 4.8. List the basic outcomes constituting each of the following events: (1) E_2^*, (2) $E_1 \cap E_2$, (3) $E_1 \cup E_2$, (4) $E_1 \cup E_2^*$.

4.17 Refer to **Muscle Deterioration** Problems 4.9b and 4.10. Use notation involving the events E_1 or E_2 or both as defined in Problem 4.10 to represent the following: (1) the event that at least one of the astronauts has a moderate degree of muscle deterioration, but neither astronaut has a severe degree of muscle deterioration; (2) the event that one or both astronauts has a severe degree of muscle deterioration.

4.18 For each of the following probabilities, indicate whether it is amenable to an objective interpretation, a subjective interpretation, or both. Also, comment on how a probability assessment might be made for each.

a. Probability that a certain type of automobile fuel pump will not fail during the first 1000 kilometers of operation.
b. Probability that the common stock price of Sabine, Inc., will fall more than $5 per share during the next five trading days.

4.19 For each of the following probabilities, indicate whether it is amenable to an objective interpretation, a subjective interpretation, or both. Also comment on how a probability assessment might be made for each.

a. Probability that leukemia contracted by John Doe was caused by exposure to hazardous chemicals during a period of employment overseas 10 years ago.
b. Probability that a shipment of tires from a regular supplier will contain no defective tires.

4.20 For each of the following statements, give your probability assessment that the statement is correct: (1) Ottawa, Canada, lies farther east than Washington, D.C.; (2) Cleopatra's first child was male; (3) a thumbtack will land point up if tossed to the floor. Explain whether each probability is objective or subjective, and do not consult any outside sources in arriving at your assessments.

* **4.21** An economist states that the odds are 6 to 1 in favor of the prime interest rate being higher a year from now.

a. What is the probability that the economist is assigning to the event of a higher prime interest rate a year from now?
b. Is the economist's probability in part a an objective or subjective probability assessment? Explain.

4.22 An engineer in a construction company stated, "The odds are 1 to 3 in favor of us getting the contract." A cost analyst replied, "I'll make a small bet with you at those odds." What does this exchange imply about the cost analyst's probability assessment?

4.23 A university president stated, "There is a 20 percent chance that a graduating science student will remain at our university for graduate study."

a. Express the president's statement as (1) a probability and (2) odds.
b. Is the president's statement consistent with the statement that fewer than 1 out of 7 graduating science students remain at this university for graduate study? Explain.
c. Can you tell whether the president's statement is based on an objective or subjective interpretation of probability? Explain.

4.24 A young professor sent an article to a professional journal and stated, "There is 1 chance in 5 that this article will be accepted outright for publication and 2 chances in 5 that it will be accepted after some revision."

a. Define the sample space here and give the professor's probability for each basic outcome.
b. Since the submission of this article to the journal is not a repeatable event, what interpretation can be given to the probabilities in part a?
c. A colleague stated, "I think that the odds are even that the article will be accepted outright." Does there appear to be a discrepancy between the probability assessments of the two individuals? Explain.
d. The professor's statement is based on the fact that the journal accepts 20 percent of the submitted articles outright and 40 percent after some revision. Would you now conclude that the professor's probability assessment is more credible than the colleague's? Discuss.

* **4.25** An engineering report gives the following probability distribution for the number of errors that occur in the transmission of a given quantity of data by satellite:

Errors:	0	1	2	3 or more
Probability:	0.75	0.15	0.05	0.05

a. Is this probability distribution univariate or multivariate?
b. What is the most likely number of errors to occur in a transmission? What is the probability that two or more errors occur in a transmission?
c. What is the probability that three or more errors occur in a transmission, given that one or more errors occur?

4.26 In-Flight Service. A questionnaire about in-flight service asks passengers to rate the quality of the meal served. The following probability distribution applies to passengers' ratings:

Rating:	Poor	Fair	Good	Excellent
Probability:	0.1	0.2	0.5	0.2

a. Verify that this probability distribution satisfies condition (4.9).
b. What is the least likely rating of the meal by a passenger? What is the probability that a passenger rates the meal as either good or excellent?
c. What is the probability that a passenger rates the meal as poor, given that the passenger has not rated the meal as excellent?

4.27 Refer to **Food Inspection** Problem 4.3. The following probability distribution applies to the quality grade of inspected food product:

Grade:	A	B	C	D
Probability:	0.55	0.30	0.10	0.05

a. What is the probability that a product will not qualify for sale through a major food store?

b. What is the probability that a product will qualify for sale only through discount outlets, given that it is fit for human consumption?

* **4.28 Internal Welds.** A device for checking internal welds in metal kegs is designed to signal when the weld is defective. The probability distribution for the status of the weld and the response of the device follows:

	Response of Device		
	Signals	Does Not Signal	
Status of Weld	B_1	B_2	Total
A_1: Defective	0.2	0.0	0.2
A_2: Not defective	0.1	0.7	0.8
Total	0.3	0.7	1.0

a. Give the symbolic notation for the probability of each of the following events: (1) The weld is defective; (2) the weld is defective and the device signals; (3) the device signals, given that the weld is defective; (4) the weld is defective, given that the device signals.

b. Explain what each of the following probabilities means here: (1) $P(A_1 \cap B_2)$, (2) $P(B_1 | A_2)$, (3) $P(B_2)$.

c. Determine each of the probabilities in parts a and b.

* **4.29** Refer to **Internal Welds** Problem 4.28.

a. Obtain the marginal probability distribution for the response of the device.

b. Obtain the conditional probability distribution for the response of the device, given that the weld is not defective.

4.30 Traffic Movements. The probability distribution for directions of traffic movements through an intersection in late afternoon follows:

	Vehicle Going to				
	North	South	East	West	
Coming from	B_1	B_2	B_3	B_4	Total
A_1: North	0.00	0.28	0.06	0.07	0.41
A_2: South	0.12	0.00	0.03	0.03	0.18
A_3: East	0.08	0.04	0.00	0.05	0.17
A_4: West	0.07	0.06	0.11	0.00	0.24
Total	0.27	0.38	0.20	0.15	1.00

a. Give the symbolic notation for the probability of each of the following events: (1) The vehicle goes north; (2) the vehicle comes from the west and goes south; (3) the vehicle goes north, given that it comes from the west.

b. Explain what each of the following probabilities means here: (1) $P(A_3)$, (2) $P(A_3 \cap B_4)$, (3) $P(A_3 \cup B_4)$.

c. Determine each of the probabilities in parts a and b.

4.31 Refer to **Traffic Movements** Problem 4.30.

a. Obtain the marginal probability distribution for the direction from which a vehicle enters the intersection.

b. Obtain the conditional probability distribution for the direction from which a vehicle enters the intersection, given that it leaves to the south.

4.32 Stock Movements. The joint probability distribution for the directions of price change for two common stocks on a given trading day follows:

	Stock B			
	Increase	No Change	Decrease	
Stock A	B_1	B_2	B_3	Total
A_1: Increase	0.20	0.10	0.05	0.35
A_2: No change	0.10	0.25	0.10	0.45
A_3: Decrease	0.05	0.05	0.10	0.20
Total	0.35	0.40	0.25	1.00

a. Give the symbolic notation for the probability of each of the following events: (1) Both stock prices increase; (2) the price of stock A increases; (3) the price of stock B decreases, given that the price of stock A is unchanged.
b. Explain what each of the following probabilities means here: (1) $P(A_2 \cap B_2)$, (2) $P(A_3 \cup B_1^*)$, (3) $P(B_3 \mid A_1 \cup A_2)$.
c. Determine each of the probabilities in parts a and b.

4.33 Refer to **Stock Movements** Problem 4.32.

a. Obtain the marginal probability distribution for the direction of price change of stock A. What is the most probable outcome for this distribution?
b. Obtain the conditional probability distribution for the direction of price change of stock A, given that the price of stock B increases. What is the most probable outcome for this distribution?

4.34 Consider the following sample space for the home ownership status and income (in $ thousand) of families.

	Family Income		
Ownership	Under 30	30–under 60	60 or More
Status	B_1	B_2	B_3
A_1: Own			
A_2: Rent			

The following probabilities are known: $P(A_2) = 0.52$, $P(B_1) = 0.50$, $P(B_3) = 0.10$, $P(A_1 \cap B_1) = 0.10$, $P(A_1 \cap B_3) = 0.08$.

a. Develop the joint probability distribution of ownership status and family income.
b. Determine the probability that (1) a family owns its home; (2) a family owns its home, given that its income is under $30 thousand; (3) a family's income is under $30 thousand, given that the family owns its home.
c. Obtain the conditional probability distribution for family income, given that the family rents its home.

* **4.35** Refer to **Internal Welds** Problem 4.28. Find the following probabilities by the multiplication, complementation, or addition theorems and verify your answers by obtaining the probabilities directly from the joint probability distribution: (1) $P(A_1 \cup B_1)$, (2) $P(A_2 \cap B_1)$, (3) $P(A_2^*)$.

4.36 Refer to **Traffic Movements** Problem 4.30. Find the following probabilities by the multiplication, complementation, or addition theorems and verify your answers by obtaining the probabilities directly from the joint probability distribution: (1) $P(A_2 \cup B_1)$, (2) $P(A_2 \cap B_4)$, (3) $P(A_1^*)$.

4.37 A corporation has just launched a new product. The total sales of this product in the next five years will fall in one of the following categories: A_1: Under $1 million, A_2: $1 million–under $10 million, A_3: $10 million or more. The following probability assessments were made: $P(A_1) = 0.4$, $P(A_2) = 0.2$, $P(A_3) = 0.4$. Use the multiplication, complementation, or addition theorems to obtain the following: (1) $P(A_2^*)$, (2) $P(A_1 \mid A_2)$, (3) $P(A_1 \cup A_2)$, (4) $P(A_1 \mid A_2^*)$, (5) $P(A_1 \cup A_2^*)$.

* **4.38** An auditor is checking the payroll records of a company. The probability that a randomly chosen record has an error is 0.05. An error in a record is either large or small. Two out of 10 errors, on average, are large.

a. What is the probability that a randomly chosen record has a large error?
b. Given that a randomly chosen record does not have a large error, what is the probability that it does not have an error?

4.39 An electronic system contains two parallel circuits (A, B). The system will operate if at least one of the two circuits is operational. Each circuit has a probability of 0.85 of operating at least 1000 hours. If one circuit operates at least 1000 hours, the probability is 0.90 that the other circuit will also operate at least 1000 hours. What is the probability of the system operating at least 1000 hours?

4.40 Several leading law schools have developed an information exchange under which applicants may apply to a maximum of two schools for admission. Applicant Blanque wishes to study at one of three schools (A, B, C), all of which are participants in the information exchange program. On the basis of experiences of similar graduates from his college, Blanque makes the following probability assessments, where $P(A)$ denotes the probability of his being accepted by law school A and the other probabilities are interpreted analogously: $P(A) = 0.60$, $P(B) = 0.80$, $P(C) = 0.50$, $P(B \mid A) = 0.90$, $P(C \mid A) = 0.20$, $P(C \mid B) = 0.50$.

a. To which two of the three schools should Blanque apply to have the greatest probability of being accepted by one or the other (or both)? Show your calculations.
b. Do schools A and B appear to have similar or dissimilar criteria for admission? Schools A and C? Discuss.

4.41 Commodity Trader. A trader specializing in commodities A and B has made the following subjective probability assessments about the prices of these commodities during the next two days: (1) The probability is 0.8 that the price of A will rise; (2) the probability is 0.5 that the price of B will rise; (3) if the price of A rises, the conditional probability is 0.6 that the price of B will rise.

a. Are these subjective probability assessments mutually consistent? Explain.
b. Would the assessments be consistent if the probability in (2) were 0.7 instead of 0.5?

4.42 Refer to **Internal Welds** Problem 4.28. Display the joint probability distribution in the form of a probability tree, as in Figure 4.6. Place "response of device" in the first branching.

4.43 Refer to **Stock Movements** Problem 4.32. Display the joint probability distribution in the form of a probability tree, as in Figure 4.6. Place "price change for stock B" in the first branching.

4.44 Refer to Figure 4.6. Reconstruct the probability tree, placing "inspector's action" in the first branching.

4.45 A company claims that its commercial rocket has a probability of 0.90 of successfully deploying its payload in a single launch and a probability of 0.95 of successfully deploying at least one of the two payloads in two separate launches. Are the outcomes of the two launches assumed to be independent here? Explain.

4.46 Refer to **Commodity Trader** Problem 4.41. Are the events of price rises for commodities A and B assumed to be independent by the trader? Explain.

* **4.47** Refer to **In-Flight Service** Problem 4.26. Assume that the meal ratings by different passengers are independent.

a. What is the probability that two passengers will both rate the meal as excellent? That three passengers will all rate the meal as excellent?
b. For ratings by two passengers, what is the probability that one will rate the meal as excellent and the other as good?

4.48 A salesperson is evaluated according to whether the week's sales exceed the sales quota (A), meet the sales quota (B), or fall below it (C). The probabilities for an experienced salesperson are: $P(A) = 0.3$, $P(B) = 0.6$, $P(C) = 0.1$. Consider the performances of an experienced salesperson in two successive weeks and assume that these performances are independent.

a. Obtain the bivariate probability distribution. What is the probability that the salesperson will exceed the quota in both weeks?

b. If the performances in two successive weeks were not independent, could you obtain the bivariate probability distribution with no additional information? Explain.

4.49 In handling a customer's order, the order department will fill it incorrectly with probability 0.04. Also, the order will be delivered to the wrong address with probability 0.02. Assume that the two variables, "filling of order" and "delivery of order," are independent.

a. Obtain the bivariate probability distribution.

b. What is the probability that a customer's order is filled incorrectly and delivered to the wrong address? That it is filled correctly and delivered to the right address?

c. If the two variables were dependent, then what would be the maximum probability that a customer's order is filled incorrectly and delivered to the wrong address?

* **4.50** Refer to **Internal Welds** Problem 4.28.

a. Show that the two variables are dependent.

b. Obtain the conditional probability distributions for the response of the device, given that the weld is defective and given that it is not defective. Describe the nature of the relationship between the two variables.

4.51 Refer to **Stock Movements** Problem 4.32.

a. Show that the two variables are dependent.

b. Obtain the conditional probability distributions for the stock A price change, given that the price of stock B (1) increases, (2) does not change, (3) decreases. Describe the nature of the relationship between the two variables.

c. Why might price changes of two common stocks exhibit the type of dependence noted in part b?

4.52 The conditional probability distributions for the numbers of days absent in a month for employees from each of the two divisions of a firm follow:

		Days Absent				
Division	0	1	2	3	4	Total
A	0.50	0.35	0.10	0.05	0.00	1.00
B	0.40	0.30	0.15	0.10	0.05	1.00

a. Are the two variables, "division" and "days absent," dependent or independent? Explain. If the two variables are dependent, describe the nature of the relationship between them.

b. Suppose that 40 percent of the firm's employees work in division A. Obtain (1) the joint probability distribution for the two variables and (2) the marginal probability distribution for the numbers of days absent.

4.53 Refer to the *Security Check* example in this chapter. Suppose (1) the probability an unauthorized visitor will pass the voice-pattern check is 0.02; (2) the probability an unauthorized visitor will pass the fingerprint check if he or she passes the voice-pattern check is 0.50; and (3) the probability an unauthorized visitor will pass the handwriting check if he or she passes both the other two checks is 0.90.

a. Use (4.20) to obtain the probability that an unauthorized visitor will pass all three checks.

b. How does the probability in part a compare with the corresponding probability in the text example? What are the implications of this result for the security system?

EXERCISES

4.54 Two students will be selected at random from the top three students in a political science class to attend a mock political convention. The first student selected will be the voting delegate, and the second one selected will be the alternate. Before the selection begins, one of the three students states that she has one chance in three of being the voting delegate and one chance in two of being the alternate. Do you agree? [*Hint:* Before the selection begins, is the probability of being chosen second a marginal probability, a joint probability, or a conditional probability?]

4.55 For each of the following probability statements, state whether it is always true, always false, or neither, for any pair of events E and F of a sample space. Justify each answer.

a. $P(E) < P(E \cap F)$
b. $P(E) > P(E \cup F)$
c. $P(E \cup F) \leq P(E) + P(F)$

4.56 For each of the following probability statements, state whether it is always true, always false, or neither, for any pair of events E and F of a sample space. Justify each answer.

a. Both $P(E \mid F) < P(E)$ and $P(E \mid F^*) < P(E)$
b. $P(E^* \mid F) = 1 - P(E \mid F)$
c. $P(E^* \cap F^*) = 1 - P(E \cup F)$

4.57 Prove extension (4.20) of multiplication theorem (4.19).

4.58 **Extended Addition Theorem.** Addition theorem (4.15) extends to more than two events. For any three events, E_1, E_2, and E_3 of a sample space, use (4.15) to show that:

$$P(E_1 \cup E_2 \cup E_3) = P(E_1) + P(E_2) + P(E_3) - P(E_1 \cap E_2) - P(E_1 \cap E_3)$$
$$- P(E_2 \cap E_3) + P(E_1 \cap E_2 \cap E_3)$$

[*Hint:* Denote $E_2 \cup E_3$ by F initially and expand $P(E_1 \cup F)$.]

4.59 A mathematical statistics book states that if E and F are two events of a random trial with nonzero probabilities, then E and F cannot be both mutually exclusive and independent. Prove this statement.

STUDIES

4.60 Consider the event: "The next U.S. president will hold office for less than four years."

a. Use the direct-question method described in the text to elicit the subjective probability for this event from 10 persons not in your class. Record your results.
b. Are the 10 probability assessments highly variable? Calculate a suitable measure and interpret it.

4.61 **Bid Probabilities.** A supervisor of a large construction firm is responsible for preparing project bids. In an internal report on each bid, the supervisor gives a subjective probability that the bid will be successful. A tabulation of the supervisor's record on the last 205 bids follows:

Supervisor's Subjective Probability of Successful Bid	Number of Bids	Number of Successful Bids
0.1	3	0
0.2	10	2
0.3	29	9
0.4	48	18
0.5	37	18
0.6	35	22
0.7	21	10
0.8	11	5
0.9	7	2
1.0	4	1
Total	205	87

a. For bid probability level 0.9, what is the anticipated proportion of successful bids if the supervisor's probability assessments are accurate? What, in fact, was the observed proportion of successful bids at this bid probability level? What is the residual here?

b. Obtain the residuals for all bid probability levels, and prepare a residual plot. Is there any range of bid probability levels where the supervisor's assessment performance is particularly poor?

c. Overall, how does the total number of bids that were anticipated to be successful compare with the observed number (87)?

d. Since the firm has won 87 of the last 205 bids, it has been suggested that probability $87/205 = 0.42$ be used as the probability of winning each bid rather than relying on the supervisor's probability assessment. Evaluate this proposal.

4.62 The contents of a computer file will be lost only if all three of the following events occur: The file is erased from current memory (A), no backup electronic copy of the file exists (B), no printed copy of the file exists (C).

a. If the probability of each of these events for a file is 0.05 and they are independent events, what is the probability that a file will be lost?

b. In contrast to part a, suppose (1) the probability is 0.05 that a file is erased from current memory; (2) the probability is 0.15 that no electronic copy exists for a file erased from memory; (3) the probability is 0.60 that no printed copy exists for a file that is erased from memory and for which no electronic copy exists. Using (4.20), calculate the probability that a file will be lost.

c. For the situation in part b, what is the probability that a file will be erased from current memory and no electronic copy exists but a printed copy does exist?

4.63 The probability that a household has color television is 0.85, that it has one automobile is 0.30, that it has two or more automobiles is 0.54, that it has color television and one automobile is 0.25, and that it has color television and two or more automobiles is 0.53. Use the extended addition theorem in Exercise 4.58 to find the probability that a household has either color television or at least one automobile.

4.64 The probability distribution of the week of settlement of strikes in a particular industry follows. (Week 1 is the week when the strike begins, and so on.)

Week of Settlement:	1	2	3	4	5 or later
Probability:	0.63	0.23	0.09	0.03	0.02

a. A strike has just begun. What is the probability that it will be settled in week 1? In week 5 or later?

b. If a strike is just entering week 2, what is the probability that it will be settled this week?

c. Calculate the probability that a strike is not settled in week 1 and is settled in week 2.

d. For each of weeks 2, 3, and 4, obtain the probability that a strike will be settled in that week, given that it has not been settled before that week. Does it appear to become more or less difficult to settle strikes as the duration of the strike increases?

Random Variables

In Chapter 4, we considered probability distributions in general. In this chapter, we are concerned with probability distributions in which the outcomes are quantitative. Special interest exists in these distributions because of the importance and pervasiveness of quantitative measures in practical problems and the relative ease with which they lend themselves to statistical analysis.

Definition of Random Variable

When the outcomes of a random trial are quantitative, we associate them with a random variable. The concept of a random variable is best illustrated by an example.

EXAMPLE ☐ Tire Molding

In a manufacturing process, automobile tires are molded in pairs. The random trial consists of one mold and the quantitative characteristic of interest is the number of defective tires in the mold. Thus, there are three basic outcomes in the sample space: o_1, zero defective tire; o_2, one defective tire; and o_3, two defective tires. We shall use X to denote the number of defective tires in the pair. Clearly X can assume one and only one of the values 0, 1, and 2. Which value X will assume, however, is uncertain. For this reason, X is called a random variable. ☐

> **(5.1)**
> A **random variable** is a variable whose numerical value is determined by the outcome of a random trial.

Initially, it is useful to distinguish notationally between a random variable and the possible values it can assume. We shall use an uppercase letter, such as X, to denote the random variable and the corresponding lowercase letter, x in this case, to denote a particular value assumed by the random variable.

EXAMPLES ☐

1. In the tire molding example, random variable X denotes the number of defective tires in a pair, a number that is uncertain before the random trial occurs. The notation x designates the actual number of defective tires. Here, x is either 0, 1, or 2.

2. Let random variable X denote the number of games a football team will win in a 9-game regular season. The notation x designates the actual number of games won by the team. Here, x is either 0, 1, . . . , or 9.   □

Discrete Random Variables

A random variable that may assume only distinct values on a scale is called a *discrete random variable*. For example, the number of defective tires in a pair is a discrete random variable; it may assume one of three distinct values (0, 1, 2) but no intermediate values. Similarly, the number of games the football team wins in a season of 9 games is a discrete random variable; it may assume one of 10 distinct values (0, 1, . . . , 9) but no intermediate values.

A discrete random variable may assume a finite number of possible values, as in the tire molding example, or an infinite number of distinct values. For instance, the number of traffic violations committed in a large city during a one-year period may be treated as having possible outcomes 0, 1, 2, . . . , ad infinitum. Although there is some upper limit to the actual number of violations, it is often useful to model the number of possible outcomes as potentially infinite.

Continuous Random Variables

A random variable that may assume any value on a continuum is called a *continuous random variable*. For example, the temperature in a warehouse may be any value on the temperature continuum between, say, −40°C and 45°C. Of course, any measurement can be made only to a finite number of significant digits, and so, strictly speaking, measured temperature is a discrete random variable. However, it is often conceptually advantageous to view measured temperature as being inherently continuous. Additional examples of measured variables that are often treated as continuous are family income, IQ, and height of person.

We consider discrete random variables first, and then we take up continuous random variables in Section 5.5.

5.2   DISCRETE RANDOM VARIABLES

Probability Distribution

The probability distribution of a discrete random variable, like any probability distribution, shows how the total probability of 1 is distributed across the basic outcomes of the sample space. For a discrete random variable, the basic outcomes are quantitative. We denote the probability that random variable X will assume value x by $P(X = x)$. For instance, $P(X = 0)$ denotes the probability that X will assume the value 0.

> **(5.2)**
> The **probability distribution** for a discrete random variable X associates a probability $P(X = x)$ with each distinct outcome x.

For convenience, we will often abbreviate the notation $P(X = x)$ to $P(x)$. For instance, $P(0)$ will denote $P(X = 0)$, $P(1)$ will denote $P(X = 1)$, and so on. The probability distribution of a discrete random variable is also called the *probability function.*

EXAMPLE

In the tire molding example, the probabilities of 0, 1, and 2 defective tires in the mold are assessed to be $P(0) = 0.75$, $P(1) = 0.10$, and $P(2) = 0.15$, respectively. The probability distribution for X therefore is:

x:	0	1	2
$P(x)$:	0.75	0.10	0.15

Figure 5.1a presents this probability distribution in graphic form, and Figure 5.1b shows an alternative tabular form.

The probability distribution for any discrete random variable X has the usual properties of a probability distribution, namely, each probability is a number between 0 and 1 and the sum of the probabilities is 1.

Cumulative Probability Distribution

In some applications, we are interested in the probability that a random variable X takes a value less than or equal to some specified value x. Such a probability is called a *cumulative probability* and is denoted by $P(X \leq x)$.

EXAMPLE

In the tire molding example, a production trial is profitable if the number of defective tires X is 1 or less. This event happens if either 0 or 1 defective tire occurs in the trial.

(a) Graphic Representation

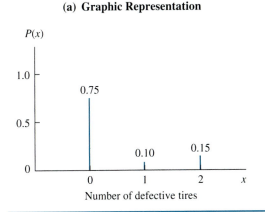

(b) Tabular Representation

x	$P(x)$
0	0.75
1	0.10
2	0.15
Total	1.00

FIGURE 5.1
Graphic and tabular representations of a probability distribution for a discrete random variable—Tire molding example

We see from Figure 5.1b that:

$$P(X \leq 1) = P(0) + P(1) = 0.75 + 0.10 = 0.85$$

Thus, the probability is 0.85 that a production trial will be profitable.   ☐

A cumulative probability $P(X \leq x)$ can be associated with every possible outcome x. The cumulative probability distribution provides all of these possible cumulative probabilities.

> **(5.3)**
> The **cumulative probability distribution** for a discrete random variable X associates a cumulative probability $P(X \leq x)$ with every possible outcome x.

A cumulative probability distribution is also called a *cumulative probability function.*

Clinic Visits   ☐   **EXAMPLE**

A health systems analyst obtained the probability distribution in Figure 5.2a for the annual number of visits by families to a clinic. The cumulative probability distribution for this random variable is developed in Figure 5.2b. For instance, we obtain $P(X \leq 0) = P(0) = 0.37$ and $P(X \leq 1) = P(0) + P(1) = 0.37 + 0.40 = 0.77$. We see that the probability is 0.97 that a family makes three or fewer visits during a year.

FIGURE 5.2   Cumulative probability distribution of number of visits—Clinic visits example

(a) Probability Distribution

Number of Visits x	$P(x)$
0	0.37
1	0.40
2	0.15
3	0.05
4	0.03
Total	1.00

(b) Cumulative Probability Distribution

Number of Visits x	$P(X \leq x)$
0	0.37
1	0.77
2	0.92
3	0.97
4	1.00

(c) Graph of Cumulative Probability Distribution

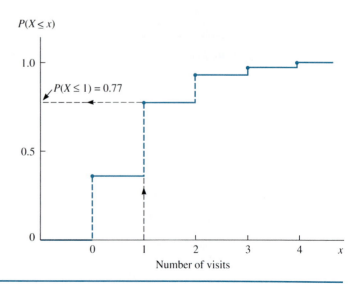

A graph of the cumulative probability distribution for clinic visits is shown in Figure 5.2c. The graph is a step function because the random variable is discrete. The graph is constructed in exactly the same way as a step function plot for a cumulative distribution. Each step of the graph occurs at an outcome x and has a size equal to the probability $P(x)$. For instance, the step size at $x = 2$ is $P(2) = 0.15$. Figure 5.2c shows how we can graphically determine the cumulative probability for any value x by reading the height of the step function at x. We see, for instance, that $P(X \leq 1) = 0.77$.   ☐

Two important properties of any cumulative probability distribution are:

1. All cumulative probabilities $P(X \leq x)$ lie between 0 and 1.
2. The cumulative probability $P(X \leq x)$ never decreases as x increases.

These two properties are illustrated in Figure 5.2c, where we see that the cumulative probability distribution starts at 0 on the bottom left and ends at 1 on the top right.

Joint, Marginal, and Conditional Probability Distributions

In Chapter 4, we defined joint, marginal, and conditional probability distributions. When the outcomes are quantitative, the joint, marginal, and conditional probability distributions play the same roles as before. We review the main ideas by means of an example.

EXAMPLE ☐ Truck Repairs

A *bivariate probability distribution* for two discrete random variables is illustrated in Table 5.1. The two variables are the number of trucks from a delivery fleet of three trucks that are in a repair depot on two consecutive nights. Variable X denotes the number of trucks in the repair depot during the first night, and Y denotes the number during the second night. The possible outcomes for X are 0, 1, 2, 3 trucks, and the same holds for Y.

The notation for joint probabilities for two discrete random variables follows that used in Chapter 4 for events. The joint probability that x trucks are in the depot the first night *and* y trucks are in the depot the second night is written $P(X = x \cap Y = y)$. We see from Table 5.1, for instance, that $P(X = 1 \cap Y = 2) = 0.03$.

The two *marginal probability distributions* are found in the margins of Table

Number of Trucks in Depot During First Night (x)	Number of Trucks in Depot During Second Night (y)				
	0	1	2	3	Total
0	0.58	0.06	0.01	0.00	0.65
1	0.06	0.10	0.03	0.01	0.20
2	0.01	0.03	0.05	0.01	0.10
3	0.00	0.01	0.01	0.03	0.05
Total	0.65	0.20	0.10	0.05	1.00

TABLE 5.1
Bivariate probability distribution for number of trucks in depot—Truck repairs example

5.1—that for X in the rightmost column and that for Y in the bottom row. The marginal probabilities are obtained by summing over rows and columns, as usual, in accordance with (4.13). Thus, we have:

$$P(X = 0) = P(X = 0 \cap Y = 0) + P(X = 0 \cap Y = 1)$$
$$+ P(X = 0 \cap Y = 2) + P(X = 0 \cap Y = 3)$$
$$= 0.58 + 0.06 + 0.01 + 0.00 = 0.65$$

We see that the probability of 0 trucks in the depot the first night is 0.65.

Conditional probabilities are obtained analogously to (4.14). Thus, the conditional probability of 0 trucks in the depot the second night, given 1 truck in the depot the first night, is:

$$P(Y = 0 \mid X = 1) = \frac{P(X = 1 \cap Y = 0)}{P(X = 1)} = \frac{0.06}{0.20} = 0.30$$

When we assemble all conditional probabilities for the number of trucks in the depot on the second night, given that one truck is in the depot during the first night, we obtain the *conditional probability distribution* of Y, given $X = 1$:

Number of Trucks in Depot During Second Night y	Conditional Probability $P(Y = y \mid X = 1)$
0	0.06/0.20 = 0.30
1	0.10/0.20 = 0.50
2	0.03/0.20 = 0.15
3	0.01/0.20 = 0.05
Total	1.00

Independence

We can apply the definition of independence in (4.23) to random variables.

(5.4)
Two **random variables** X and Y are **independent** if:

$$P(X = x \cap Y = y) = P(X = x)P(Y = y) \qquad \text{for all } x \text{ and } y$$

☐ EXAMPLE

In the truck repairs example of Table 5.1, the two random variables are not independent. For example, $P(X = 0 \cap Y = 1) = 0.06$, while $P(X = 0)P(Y = 1) = 0.65(0.20) = 0.13$. An examination of Table 5.1 indicates the nature of the relationship between X and Y, namely, that the number of trucks in the depot on two successive nights is likely to be the same.   ☐

Expected Value of Random Variable

Often, we are interested in the mean outcome of a random variable in many trials. This mean is known as the expected value of the random variable.

> **(5.5)**
> The **expected value** of a discrete random variable X is denoted by $E\{X\}$ and defined:
>
> $$E\{X\} = \sum_x xP(x)$$
>
> where the summation is over all outcomes x. The notation $E\{\ \}$ is read "expectation of."

EXAMPLE ☐

We repeat for the clinic visits example of Figure 5.2a the probability distribution of X, the annual number of family visits:

x:	0	1	2	3	4
$P(x)$:	0.37	0.40	0.15	0.05	0.03

The expected value of X is obtained by using (5.5):

$$E\{X\} = 0(0.37) + 1(0.40) + 2(0.15) + 3(0.05) + 4(0.03)$$
$$= 0.97 \text{ visit per family}$$

☐

EXAMPLE ☐ **Jewelry Loss**

A piece of jewelry worth \$10,000 has probability 0.003 of being stolen or lost during a year, and probability 0.997 of not being stolen or lost. Let X denote the amount of loss during a year. The probability distribution of X is as follows:

Loss x:	0	10,000
$P(x)$:	0.997	0.003

The expected loss during a year is, using (5.5):

$$E\{X\} = 0(0.997) + 10,000(0.003) = \$30$$

☐

Meaning of $E\{X\}$. The expected value of a random variable may be likened to a measure of position for the corresponding probability distribution. Indeed, $E\{X\}$ is simply a weighted mean of the possible outcomes, with the probability values being the weights.

For this reason, $E\{X\}$ is often called the *mean* of the probability distribution of X, or simply the mean of X.

There is another way of understanding the meaning of $E\{X\}$. Consider again the clinic visits example of Figure 5.2, where the expected value of X is $E\{X\} = 0.97$ visit per family. If the random trial associated with the annual number of family clinic visits is repeated independently many times, the relative frequency interpretation of probability suggests that about 37 percent of families will have no visit in a year, 40 percent will have one visit, and so on. The mean outcome over many independent trials would be about $E\{X\}$ $= 0.97$ visit per family.

Observe that $E\{X\}$, like any mean, may be a number that does not correspond to any of the possible outcomes. In the clinic visits example, we have $E\{X\} = 0.97$ visit per family, while all the outcomes of X are whole numbers.

Variance of Random Variable

The outcome of a random variable will vary from trial to trial, and it is useful to have a measure of the variability over many trials. A key measure is the variance of the random variable.

(5.6)

The **variance** of a discrete random variable X is denoted by $\sigma^2\{X\}$ and defined:

$$\sigma^2\{X\} = \sum_x (x - E\{X\})^2 P(x)$$

where the summation is over all outcomes x. The notation $\sigma^2\{\ \}$ is read "variance of."

☐ EXAMPLES

1. In the clinic visits example of Figure 5.2a, we found that $E\{X\} = 0.97$. Hence, the variance of X according to (5.6) is:

$$\sigma^2\{X\} = (0 - 0.97)^2(0.37) + (1 - 0.97)^2(0.40) + (2 - 0.97)^2(0.15)$$
$$+ (3 - 0.97)^2(0.05) + (4 - 0.97)^2(0.03) = 0.9891$$

2. In the jewelry loss example, we found that $E\{X\} = 30$. Hence:

$$\sigma^2\{X\} = (0 - 30)^2(0.997) + (10,000 - 30)^2(0.003) = 299,100 \qquad ☐$$

Meaning of $\sigma^2\{X\}$.   From definition (5.6) we can see that the variance of a random variable X is a weighted average of squared deviations. The deviations are the differences between the outcomes of X and the expected value or mean $E\{X\}$, and the weights are the respective probabilities of occurrence. Therefore, $\sigma^2\{X\}$ measures the extent to which the outcomes of X depart from the mean in the same way that the variance of a data set measures the variability of the observations about their mean.

Comment

As just noted, the variance of a random variable X is a weighted average of the squared deviations, $(x - E\{X\})^2$, with the respective probabilities as weights. Thus the variance may be defined as the following expected value.

(5.7) $\sigma^2\{X\} = E\{(X - E\{X\})^2\}$

An algebraically equivalent expression is given next.

(5.7a) $\sigma^2\{X\} = E\{X^2\} - (E\{X\})^2$

Standard Deviation of Random Variable

The variance $\sigma^2\{X\}$ is expressed in the squared units of X. When we take the positive square root of $\sigma^2\{X\}$, we return to the original units and obtain the standard deviation of X.

(5.8)
The positive square root of the variance of X is called the **standard deviation** of X and is denoted by $\sigma\{X\}$:

$$\sigma\{X\} = \sqrt{\sigma^2\{X\}}$$

The notation $\sigma\{\ \}$ is read "standard deviation of."

EXAMPLES ☐

1. In the clinic visits example, we found that the variance is $\sigma^2\{X\} = 0.9891$. Hence, the standard deviation of X is $\sigma\{X\} = \sqrt{0.9891} = 0.9945$ visit per family. We see that the standard deviation is about as large as the mean $E\{X\} = 0.97$ here.

2. In the jewelry loss example, we found the variance to be $\sigma^2\{X\} = 299,100$. Hence, $\sigma\{X\} = \sqrt{299,100} = \$546.90$. We see that the standard deviation is much larger than the mean $E\{X\} = \$30$ here. ☐

STANDARDIZED **5.4**
RANDOM VARIABLES

In Chapter 3, we introduced the concept of standardizing the observations in a data set. We noted that a standardized observation measures the distance between the observation and the mean in units of the standard deviation. A corresponding standardized form for random variables is frequently useful.

(5.9)

The **standardized form of a random variable** X is denoted by Y and defined by:

$$Y = \frac{X - E\{X\}}{\sigma\{X\}}$$

where: $E\{X\}$ is the expected value of random variable X

$\sigma\{X\}$ is the standard deviation of random variable X

The standardized form of a random variable is called a *standardized random variable*. Each standardized random variable has expected value 0 and variance 1. Thus, we have $E\{Y\} = 0$ and $\sigma^2\{Y\} = 1$ for any standardized random variable.

□ EXAMPLE

For the clinic visits example, we wish to obtain the probability distribution for the standardized form of the random variable X. The probability distribution of X is given in Figure 5.2a. We determined earlier that $E\{X\} = 0.97$ and $\sigma\{X\} = 0.9945$. Hence, corresponding to the outcome $x = 0$, we have the standardized value:

$$y = \frac{0 - 0.97}{0.9945} = -0.98$$

Likewise, the standardized value for the outcome $x = 1$ is:

$$y = \frac{1 - 0.97}{0.9945} = 0.03$$

Table 5.2 shows the probability distribution of the standardized random variable Y. Note that each outcome x and its associated standardized value y have the same probability, because both pertain to the same basic outcome. For instance, $P(X = 0) = P(Y = -0.98) = 0.37$. It can be verified by direct calculation, using (5.5) and (5.6), that $E\{Y\} = 0$ and $\sigma^2\{Y\} = 1$.   □

TABLE 5.2

Probability distribution of standardized random variable—Clinic visits example

Number of Visits x	Standardized Value $y = \dfrac{x - E\{X\}}{\sigma\{X\}}$	Probability $P(X = x) = P(Y = y)$
0	-0.98	0.37
1	0.03	0.40
2	1.04	0.15
3	2.04	0.05
4	3.05	0.03
	Total	1.00

$$E\{X\} = 0.97 \qquad \sigma\{X\} = 0.9945$$

Probability Density Function

As noted earlier, a continuous random variable may assume any value on a continuum. Unlike discrete random variables, it is not meaningful to associate a probability value with each possible outcome on a continuum. Instead, for continuous random variables we associate probability values with intervals on the continuum.

EXAMPLE ☐ **Crop Yield**

Consider a continuous random variable X that represents the yield of a crop in tons per acre. Suppose the yield can be any value between 0 and 2 tons. Figure 5.3 shows a continuous probability distribution with the property that, for any interval, the area between the axis and the curve corresponds to the probability that X will have a value in this interval. Note that the total area under the probability distribution is 1.0 since the distribution is rectangular with height 1/2 and width 2. The probability that the crop yield will be somewhere between 0 and 0.4 ton corresponds to the shaded area in Figure 5.3. This shaded area is 0.2 since 20 percent of the total area of 1 lies to the left of 0.4 ton. Therefore, the probability that the yield will be somewhere between 0 and 0.4 ton per acre is 0.2. Symbolically, we state $P(0 \leq X \leq 0.4) = 0.2$. The probability corresponding to any other interval is found in a similar manner, namely, by determining the area under the curve corresponding to that interval. ☐

In the crop yield example, it is simple to determine the area under the probability distribution curve for any desired interval. In general, numerical and analytical methods may be needed for calculating areas. An example of such a probability distribution follows.

EXAMPLE ☐ **Earthquake**

Figure 5.4a presents the probability distribution of the magnitude X of earthquakes in a region near a fault line. The shaded area under the curve represents the probability that the magnitude of the next earthquake is 5.0 or less on the Richter scale. That is, the shaded area corresponds to the probability $P(X \leq 5.0)$. We can see visually that the shaded area is a large proportion of the total area under the curve, so it is likely that the next earthquake will have a magnitude of 5.0 or less. However, we cannot determine the exact area by a simple geometric argument, as in the crop yield example. ☐

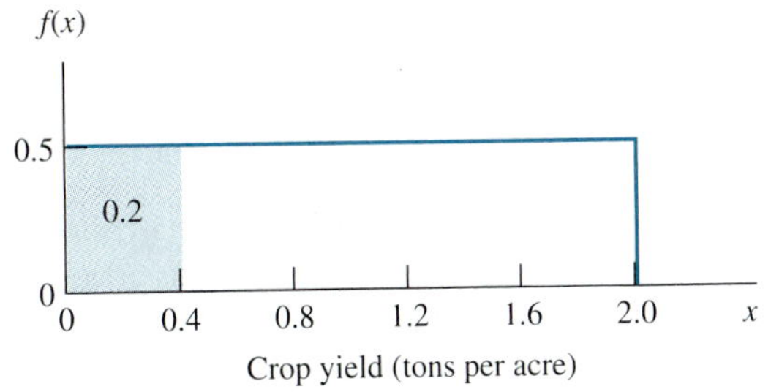

FIGURE 5.3
Probability density function
$f(x) = 1/2$ **for** $0 \leq x \leq 2$—
Crop yield example

FIGURE 5.4
Probability density function
$f(x)$ **and its cumulative prob-**
ability function $F(x)$**—Earth-**
quake example

(a) Density Function

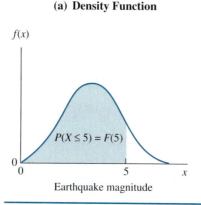

Earthquake magnitude

(b) Cumulative Probability Function

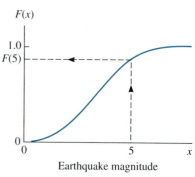

Earthquake magnitude

In later chapters, we shall discuss how areas under continuous probability distributions can be determined.

The curve of a continuous probability distribution, such as the ones in Figures 5.3 and 5.4a, is called a probability density function.

(5.10)

The **probability density function** of a continuous random variable X is a mathematical function for which the area under the curve corresponding to any interval is equal to the probability that X will take on a value in the interval. The probability density function is denoted by $f(x)$. The value $f(x)$ is called the **probability density** at x.

The probability density function of a random variable is often simply called the *probability function, density function,* or *probability distribution.*

☐ **EXAMPLE**

For the crop yield example, the probability density function shown in Figure 5.3 is as follows.

(5.11) $f(x) = \dfrac{1}{2}$ $0 \le x \le 2$

For this simple density function, the probability density at any value of x between 0 and 2 is 1/2. Thus, the height of the curve is a constant 1/2.   ☐

A probability density function can never lie below the x axis because the probability that X will lie in any interval cannot be negative. Thus, $f(x)$ is always nonnegative. Further, because X must assume some value on the axis under the curve, the total area under a probability density function is always 1.

The probability that a continuous random variable will take on any particular value on a continuum is zero. This fact follows because a single point corresponds to an interval of zero width, and hence there is zero area under the probability density function. As a result, in any probability statement of the type $P(a \leq X \leq b)$ for a continuous random variable, it is immaterial whether or not the endpoints of the interval are included; that is:

$$P(a \leq X \leq b) = P(a < X \leq b) = P(a \leq X < b) = P(a < X < b)$$

Cumulative Probability Function

Sometimes, we wish to work with cumulative probabilities for continuous random variables.

> **(5.12)**
>
> The **cumulative probability function** for a continuous random variable X associates a cumulative probability $P(X \leq x)$ with every value of x. The function is denoted by $F(x)$ so that:
>
> $$F(x) = P(X \leq x)$$

In other words, the cumulative probability function $F(x)$ indicates the probability that the outcome of X in a random trial will be less than or equal to any specified value x. Thus, $F(x)$ corresponds to the area under the probability density function to the left of x.

EXAMPLE □

The cumulative probability function for the earthquake example is shown in Figure 5.4b. Note that the cumulative probability function $F(x)$ is a smooth continuous curve when the random variable is continuous, rather than a step function as for discrete random variables. It still has the properties that $F(x)$ is always a value between 0 and 1 and never decreases as x increases.

In Figure 5.4, we illustrate the correspondence between areas under the density function and the cumulative probability function. The shaded area under the density function in Figure 5.4a corresponds to the probability that the earthquake magnitude X is less than or equal to 5, that is, $P(X \leq 5)$. By definition (5.12), this probability is also denoted by $F(5)$. The cumulative probability function is shown in Figure 5.4b. The cumulative probability $F(5)$, corresponding to $x = 5$, is identified there. This cumulative probability equals the shaded area under the density function.   □

PROBLEMS

5.1 In each of the following situations, indicate whether the random variable is discrete or continuous and describe its sample space (that is, its set of possible outcomes).

a. The number of passengers on a scheduled flight that has a capacity of 220 passengers.

b. The number of employees of a firm with 200 employees who are absent on a given day because of sickness.

c. The length of time a machine is idle during an eight-hour working day.

5.2 In each of the following situations, indicate whether the random variable is discrete or continuous and describe its sample space (that is, its set of possible outcomes).

a. The height of a plant that never grows taller than 1 meter.
b. The current volume of palm oil in a 4000-liter storage tank.
c. The number of pills left from a 20-pill prescription.

* **5.3 Job Applicant.** The probability distribution of X, the number of positions held previously by a job applicant, follows:

x:	0	1	2	3	4
P(x):	0.60	0.20	0.10	0.05	0.05

a. Interpret each of the following expressions: (1) $P(0)$, (2) $P(X \leq 1)$, (3) $P(1 \leq X \leq 3)$.
b. Obtain each of the probabilities in part a.

5.4 Helicopter Shuttle. The probability distribution of X, the number of passengers on a daily helicopter shuttle run from airport A to airport B, follows:

x:	1	2	3	4
P(x):	0.1	0.2	0.3	0.4

a. Interpret each of the following expressions: (1) $P(3)$, (2) $P(X \leq 2)$, (3) $P(2 \leq X \leq 4)$.
b. Obtain each of the probabilities in part a.

5.5 Telephone Calls. The probability distribution of X, the number of long-distance calls placed during an hour from a telephone, follows:

x:	0	1	2	3
P(x):	0.3	0.5	0.1	0.1

a. Give the notational representation for the probability that (1) one long-distance call is placed during an hour, (2) one or more long-distance calls are placed during an hour, (3) two or three long-distance calls are placed during an hour. Obtain each of these probabilities.
b. Obtain $P(0 \leq X \leq 2)$. Interpret this probability.

5.6 Ceramic Filters. The probability distribution of X, the number of damaged ceramic filters in a shipment of three, follows:

x:	0	1	2	3
P(x):	0.90	0.02	0.02	0.06

a. Give the notational representation for the probability that (1) all filters in the shipment are damaged, (2) some filters in the shipment are damaged, (3) fewer than three filters in the shipment are damaged. Obtain each of these probabilities.
b. Obtain $P(0 < X < 3)$. Interpret this probability.
c. The two most probable outcomes for X are zero and three damaged filters. What practical consideration in shipping ceramic filters might explain this fact?

* **5.7** Refer to **Job Applicant** Problem 5.3.

a. Construct a graph of the probability distribution.
b. Construct a graph of the cumulative probability distribution. From this graph, find $P(X \leq 3)$. Interpret this probability.

5.8 Refer to **Helicopter Shuttle** Problem 5.4.

a. Construct a graph of the probability distribution.
b. Construct a graph of the cumulative probability distribution. From this graph, find $P(X \leq 3)$. Interpret this probability.

5.9 Refer to **Telephone Calls** Problem 5.5.

a. Construct a graph of the probability distribution.

b. Construct a graph of the cumulative probability distribution. From this graph, find $P(X \leq 1)$. Interpret this probability.

5.10 Refer to **Ceramic Filters** Problem 5.6.

a. Construct a graph of the probability distribution.

b. Construct a graph of the cumulative probability distribution. From this graph, find the probability that two or fewer filters in the shipment are damaged.

5.11 Refer to **Job Applicant** Problem 5.3.

a. Give the notational representation for the probability that a job applicant held no position previously, given that the applicant held three or fewer positions previously. Obtain this probability.

b. Obtain $P(X \geq 2 \mid X \geq 1)$. Interpret this probability.

5.12 Refer to **Ceramic Filters** Problem 5.6.

a. Give the notational representation for the probability that three filters in a shipment are damaged, given that at least one filter in the shipment is damaged. Obtain this probability.

b. Obtain $P(1 \leq X \leq 2 \mid X \leq 2)$. Interpret this probability.

* **5.13 Computer Operations.** The bivariate probability distribution for number of computer stoppages in a day (X) and number of computer operators sick for the day (Y) follows:

Number of Stoppages	Number of Sick Operators		
	y		
x	0	1	2
0	0.40	0.15	0.02
1	0.30	0.05	0.01
2	0.04	0.03	0

From inspection of the bivariate probability distribution, obtain and interpret each of the following probabilities: (1) $P(X = 1 \cap Y = 0)$,   (2) $P(X = 2)$,   (3) $P(Y \geq 2)$,   (4) $P(X = 1 \mid Y = 0)$,   (5) $P(Y > 0 \mid X = 1)$.

5.14 Bottle Flaws. The bivariate probability distribution of the number of bubble flaws (X) and the number of solid-particle flaws (Y) in a glass bottle follows:

Bubble Flaws	Particle Flaws	
	y	
x	0	1
0	0.4	0.1
1	0.1	0.1
2	0.1	0.2

From inspection of the bivariate probability distribution, obtain and interpret each of the following probabilities: (1) $P(X = 0 \cap Y = 1)$,   (2) $P(Y = 0)$,   (3) $P(X \leq 1)$,   (4) $P(Y = 1 \mid X = 2)$,   (5) $P(X \leq 1 \mid Y = 1)$.

5.15 Refer to **Helicopter Shuttle** Problem 5.4. Let X_1 and X_2 denote the numbers of passengers on days 1 and 2, respectively. Assume that X_1 and X_2 are independent, each with the probability distribution given in Problem 5.4.

a. Construct the bivariate probability distribution of X_1 and X_2.

b. From inspection of the bivariate probability distribution, obtain and interpret each of the following probabilities: (1) $P(X_1 = 2 \cap X_2 = 1)$,   (2) $P(X_2 = 4 \mid X_1 = 2)$,   (3) $P(X_2 = 4)$,   (4) $P(X_1 \leq 1 \cap X_2 \leq 1)$.

5.16 Refer to **Ceramic Filters** Problem 5.6. Let X_1 and X_2 denote the numbers of damaged ceramic filters in shipments 1 and 2, respectively. Assume that X_1 and X_2 are independent, each with the probability distribution given in Problem 5.6.

a. Construct the bivariate probability distribution of X_1 and X_2.
b. From the bivariate probability distribution, give the notational representation for the probability that (1) one filter is damaged in shipment 1 and none is damaged in shipment 2; (2) two filters are damaged in shipment 2, given that none is damaged in shipment 1; (3) three filters are damaged in shipment 1; (4) at least one filter is damaged in each shipment. Obtain each of these probabilities.

∗ 5.17 Refer to **Computer Operations** Problem 5.13.

a. Obtain the marginal probability distribution of the number of computer stoppages.
b. Obtain the conditional probability distribution of the number of computer stoppages when no operator is sick.
c. How does the conditional distribution in part b compare with the marginal distribution in part a? Does this result imply that X and Y are dependent? Explain.

5.18 Refer to **Bottle Flaws** Problem 5.14.

a. Obtain the marginal probability distribution of the number of bubble flaws.
b. Obtain the conditional probability distribution of the number of bubble flaws when no particle flaw is present. Also obtain the conditional distribution when one particle flaw is present.
c. Does it appear from the conditional distributions in part b that the number of bubble flaws and the number of particle flaws in a bottle are related? Explain.

5.19 Refer to **Helicopter Shuttle** Problems 5.4 and 5.15.

a. From the bivariate probability distribution, obtain the conditional probability distribution of the number of passengers on day 2, given that there are two passengers on day 1.
b. Explain why the conditional distribution in part a must be the same as the marginal distribution of the number of passengers on day 2.

∗ 5.20 Refer to **Job Applicant** Problem 5.3. Calculate $E\{X\}$. Interpret its meaning here in relative frequency terms. Should one expect the measure to be a whole number here? Explain.

5.21 Refer to **Helicopter Shuttle** Problem 5.4. Calculate $E\{X\}$. Interpret its meaning here in relative frequency terms.

5.22 Refer to **Ceramic Filters** Problem 5.6.

a. Calculate the mean of the probability distribution. What notation is used in the text to denote this quantity?
b. If the random trial represented by the probability distribution here were repeated independently for a very large number of shipments, in about what percentage of the shipments would no filters be damaged? What would be the approximate mean number of damaged filters per shipment?

5.23 Underwriting Syndicate. An underwriting syndicate will insure an offshore gas production platform for one year. The syndicate's potential loss X (in \$ million) has the following probability distribution:

x:	0	20	150
$P(x)$:	0.990	0.009	0.001

a. What is the syndicate's expected loss? What is the probability that the syndicate's actual loss will be smaller than the expected loss?
b. The risk manager of the company that owns the platform has suggested that \$300,000 would be a fair premium for the syndicate to charge for assuming the potential loss under the insurance contract. Do you agree? Discuss.

∗ 5.24 Refer to **Job Applicant** Problem 5.3. Calculate the variance and the standard deviation of the probability distribution. In what units is the standard deviation expressed?

5.25 Refer to **Helicopter Shuttle** Problem 5.4. Calculate $\sigma^2\{X\}$ and $\sigma\{X\}$. In what units is $\sigma\{X\}$ expressed?

5.26 Refer to **Ceramic Filters** Problem 5.6.

a. Calculate $\sigma\{X\}$.
b. Is the standard deviation here large relative to the mean? If an improvement in the quality of packaging changed the probability distribution of X so that $E\{X\}$ became smaller, would you expect $\sigma\{X\}$ to become smaller also? Discuss.

5.27 Refer to **Underwriting Syndicate** Problem 5.23.

a. Calculate the variance and the standard deviation of the probability distribution. In what units is the standard deviation expressed?
b. Which of the three possible outcomes of X contributes the most to the magnitude of the variance in part a?

* 5.28 Refer to **Job Applicant** Problem 5.3.

a. Obtain the probability distribution of the standardized form of random variable X.
b. Verify by direct calculation that the probability distribution in part a has mean 0 and standard deviation 1.

5.29 Refer to **Underwriting Syndicate** Problem 5.23.

a. Obtain the probability distribution of the standardized form of random variable X.
b. How many standard deviations from the mean is the outcome of X that lies farthest from the mean?

5.30 The probability distribution of the standardized form Y of a random variable X follows:

y:	-0.375	0.562	8.991
$P(y)$:	0.69	0.30	0.01

a. How many standard deviations apart are the smallest and the largest possible outcomes of X?
b. The mean and the variance of X are 0.40 and 1.14, respectively. What is the value of the smallest possible outcome of X?
c. Verify by direct calculation that $E\{Y\} = 0$ and $\sigma\{Y\} = 1$ here, except for rounding differences.

* 5.31 Refer to Figure 5.4a. For this density function, $P(X \le 5.0) = 0.841$ and $P(3.5 \le X \le 5.0) = 0.341$.

a. Obtain the following: (1) $P(X > 5.0)$, (2) $P(X < 3.5)$, (3) $F(5.0)$. Interpret the probability in (1).
b. If you were asked to find $P(X \ge 5.0)$ rather than $P(X > 5.0)$, why would the answer not change? Explain.

5.32 **Suspension Cable.** When a 200-meter suspension cable breaks, it is equally likely that the break occurs at any point along its length. Let X denote the distance from one end of the cable to the break; then X has the probability density function:

$$f(x) = \frac{1}{200} \qquad 0 \le x \le 200$$

a. Construct a graph of the probability density function of X. Is the area under the density function equal to 1?
b. Use the geometric properties of $f(x)$ to obtain the following probabilities: (1) $P(X \le 100)$,(2) $P(X > 50)$, (3) $P(50 \le X \le 100)$.
c. Obtain the cumulative probability function for X and construct a graph of it.
d. From your graph in part c, obtain the values of (1) $F(80)$, (2) $F(150)$. Interpret the meaning of the latter value.

5.33 **Shelf Life.** A melon's shelf life (X, in days) is treated as a continuous random variable. The probability density function of X is the right triangular probability function:

$$f(x) = 0.005(20 - x) \qquad 0 \le x \le 20$$

a. Construct a graph of this probability density function. Is the area under the density function equal to 1?

b. Use the geometric properties of $f(x)$ to obtain the following: (1) $P(X \le 15)$, (2) $P(5 < X \le 15)$, (3) $F(5)$.

5.34 Wind Speed. The cumulative probability function for the wind speed X, in kilometers per hour, at a given place and time of day is:

$$F(x) = \frac{5x}{100 + 4x} \qquad 0 \le x \le 100$$

a. Obtain the following: (1) $F(80)$, (2) $P(10 \le X \le 80)$. Interpret the first probability.

b. Construct a graph of $F(x)$ on the interval $0 \le x \le 100$. Does $F(x)$ exhibit the properties of a cumulative probability function on this interval? Explain.

EXERCISES

5.35 Refer to **Computer Operations** Problem 5.13. Find each of the following probabilities: (1) $P(X = 2 \mid Y \le 1)$, (2) $P(X \le 1 \mid Y \ge 1)$, (3) $P(X = Y \mid Y \le 1)$.

5.36 Verify the identity (5.7a).

5.37 Refer to **Bottle Flaws** Problem 5.14. Let $E\{X \mid Y = y\}$ denote the expected value of the conditional probability distribution of X, given $Y = y$.

a. Obtain $E\{X \mid Y = 0\}$ and $E\{X \mid Y = 1\}$.

b. The two expected values in part a are different. Explain why this fact confirms that X and Y are dependent random variables. If the two expected values had been the same, why would this not necessarily imply that X and Y are independent?

c. Show that $E\{X \mid Y = y\} = E\{X\}$ for any pair of independent discrete random variables X and Y.

STUDIES

5.38 A potato farmer has a contract to sell the forthcoming harvest to a food cannery. The entire harvest will receive a grade of A, B, or C. If the harvest is graded A, the farmer will receive $7.20 per bushel. For grades B and C, the prices per bushel will be $5.80 and $5.20, respectively. The probabilities for the grade of the harvest, if the harvesting is done at the normal time, are $P(A) = 0.35$, $P(B) = 0.45$, and $P(C) = 0.20$. The yield of the harvest at the normal time is expected to be 60,000 bushels. If the farmer harvests earlier, the probabilities would be changed to $P(A) = 0.40$, $P(B) = 0.60$, and $P(C) = 0$, and the yield would be reduced to 57,000 bushels.

a. Obtain the probability distribution of total revenue under each of the following alternatives: (1) harvest earlier, (2) harvest at the normal time.

b. Which alternative in part a leads to the larger expected total revenue? The larger standard deviation of total revenue?

5.39 Let X_1 and X_2 denote the interest rates that will be paid on one-year certificates of deposit that are issued on the first day of next year (year 1) and the following year (year 2), respectively. The bivariate probability distribution of X_1 and X_2 follows:

		Year 2 x_2	
Year 1 x_1	9	10	11
9	0.1	0.1	0
10	0.1	0.3	0.2
11	0	0.1	0.1

a. Obtain $P(X_1 < X_2)$. Interpret this probability.

b. For each year, obtain the expected value and the standard deviation of the interest rate. Is the interest rate expected to be higher in year 2 than in year 1? Is there more uncertainty about the interest rate in year 2 than in year 1? Explain.

c. Obtain the probability distribution of Z, the larger of X_1 and X_2; that is, $Z = \max(X_1, X_2)$. Must Z necessarily have a larger expected value than either X_1 or X_2? Calculate the expected value of Z and comment.

d. An investment fund plans to buy \$10 million of the one-year certificates of deposit issued on the first day of year 1 and to reinvest this amount and the earned interest in the certificates issued in year 2. Assume that interest on the certificates is paid at the end of the year they are issued. Let T denote the value of this investment, including earned interest, at the end of year 2. Obtain the probability distribution of T and its expected value.

Additional Topics in Probability

6

n the preceding two chapters we have presented the basic concepts of probability and random variables. In this chapter we take up several additional concepts that are valuable for understanding probability and statistical methods.

Probability Revision Using Bayes' Theorem

BAYES' THEOREM 6.1

Bayes' theorem represents a special application of conditional probabilities. The theorem describes how initial probabilities for a variable, called *prior probabilities,* may be revised to reflect the acquisition of additional information about the variable. The revised probabilities are called *posterior probabilities.*

EXAMPLE ☐ **Martin Wilson**

Martin Wilson has submitted an application for a trainee position and is interested in knowing the probability that he will be hired. It is known that the company hires 4 percent of applicants. We shall denote the event that applicant is hired by A_1, so $P(A_1) = 0.04$. We shall denote the event that applicant is not hired by A_2, so $P(A_2) = 1 - 0.04 = 0.96$, because A_1 and A_2 are complementary events. The outcomes of events A_1 and A_2 are the outcomes of interest, and the probabilities $P(A_1)$ and $P(A_2)$ are therefore called the prior probabilities, reflecting the original information. Thus, initially, Martin Wilson knows that his probability of being hired is $P(A_1) = 0.04$.

Additional information becomes available prior to hiring since only some of the applicants are called for a formal interview. We shall denote the event "called for interview" by B_1 and the event "not called for interview" by B_2. Thus, the events B_1 and B_2 constitute additional information for Martin Wilson. It is known that among all applicants hired, 98 percent receive interviews (the other 2 percent are hired on the basis of written credentials alone). Hence, we know that $P(B_1 \mid A_1) = 0.98$. Furthermore, it is known that among all applicants not hired, only 1 percent are interviewed; hence, $P(B_1 \mid A_2) = 0.01$. Martin Wilson has been called for an interview and now wishes to know the revised probability that he will be hired; that is, he wishes to determine $P(A_1 \mid B_1)$. The desired revised probability $P(A_1 \mid B_1)$ is called a posterior probability. It is a revised probability because it is based on additional information—namely, that Martin Wilson has been called for an interview. ☐

EXAMPLE ☐ **Test Market**

A new frozen food product will ultimately turn out to be a failure in the market (A_1), a marginal success (A_2), or a major success (A_3). These are the outcomes of interest. It is known from past experience with similar products that the prior probabilities are $P(A_1) = 0.80$, $P(A_2) = 0.15$, and $P(A_3) = 0.05$.

The product was introduced in a test market to see whether it would achieve a small market share (B_1) or a large market share (B_2) in the test market. It was found that the frozen food product obtained a large market share (B_2). This test market outcome represents additional information. It is known from past experience that the conditional probabilities of a large share in the test market (B_2), given the ultimate success of the product (A_i), are as follows:

$$P(B_2 \mid A_1) = 0.02 \qquad P(B_2 \mid A_2) = 0.24 \qquad P(B_2 \mid A_3) = 0.98$$

The marketing manager now wishes to know the revised probabilities of ultimate success of the product, given the test market result. In other words, the posterior probabilities $P(A_1 \mid B_2)$, $P(A_2 \mid B_2)$, and $P(A_3 \mid B_2)$ are desired. $\square$

The desired posterior probabilities in these examples may be obtained by using Bayes' theorem.

(6.1)

Bayes' Theorem. The posterior probability $P(A_i \mid B_j)$ of outcome A_i occurring, given information outcome B_j occurs, is obtained as follows:

$$P(A_i \mid B_j) = \frac{P(A_i)P(B_j \mid A_i)}{\sum_i P(A_i)P(B_j \mid A_i)}$$

where: $P(A_i)$ are the prior probabilities for the possible outcomes A_i of the variable of interest, $i = 1, 2, \ldots$

   $P(B_j \mid A_i)$ are the conditional probabilities for the known outcome B_j of the additional information variable, given the possible outcomes A_i of the variable of interest, $i = 1, 2, \ldots$

□ **EXAMPLES**

1. In the Martin Wilson example, we know that $P(A_1) = 0.04$, $P(A_2) = 0.96$, $P(B_1 \mid A_1) = 0.98$, and $P(B_1 \mid A_2) = 0.01$. Substituting into Bayes' theorem (6.1), we obtain:

$$P(A_1 \mid B_1) = \frac{P(A_1)P(B_1 \mid A_1)}{P(A_1)P(B_1 \mid A_1) + P(A_2)P(B_1 \mid A_2)}$$

$$= \frac{0.04(0.98)}{0.04(0.98) + 0.96(0.01)} = \frac{0.0392}{0.0488} = 0.803$$

Thus, the posterior probability of Martin Wilson being hired, given that he has been called for an interview, is $P(A_1 \mid B_1) = 0.803$. Note how much larger this revised probability of being hired is than the prior probability, $P(A_1) = 0.04$. This larger probability is the result of Wilson having been called for an interview.

Since A_2 is the complementary event of A_1, the posterior probability that Martin Wilson will not be hired, given that he has been called for an interview, is:

$$P(A_2 | B_1) = 1 - P(A_1 | B_1) = 1 - 0.803 = 0.197$$

2. In the test market example, the desired posterior probabilities can best be obtained by means of systematic calculations, such as those shown in Table 6.1. Column 2 contains the prior probabilities $P(A_i)$ of ultimate success for the new product. Column 3 contains the known conditional probabilities $P(B_2 | A_i)$ of a large test market share (B_2), given the ultimate success of the product (A_i). Column 4 contains the products of columns 2 and 3, representing the joint probabilities $P(A_i \cap B_2)$. The sum of this column is the denominator of Bayes' theorem (6.1). The ratio of each entry in column 4 to the sum of column 4 is the posterior probability $P(A_i | B_2)$ of each ultimate success outcome (A_i), given a large test market share (B_2). For instance, the probability of ultimate major success, given a large test market share, is $P(A_3 | B_2) = 0.0490/0.1010 = 0.485$.

The probabilities in column 5 constitute the posterior probability distribution of ultimate market success, given a large test market share. Note how this posterior probability distribution differs from the prior probability distribution in column 2. As a result of finding a large test market share, it is much more likely now that the new product will be either a marginal success or a major success. ☐

Relation to Basic Probability Theorems

Bayes' theorem (6.1) is simply a restatement of the definition of a conditional probability in (4.14). In the current notation, this definition is expressed as follows.

$$(6.2) \qquad P(A_i | B_j) = \frac{P(A_i \cap B_j)}{P(B_j)}$$

The equivalence of the numerators of (6.1) and (6.2) follows from multiplication theorem (4.19).

$$(6.3) \qquad P(A_i \cap B_j) = P(A_i)P(B_j | A_i)$$

(1) Ultimate Market Success A_i	(2) Prior Probability $P(A_i)$	(3) Conditional Probability $P(B_2 | A_i)$	(4) Joint Probability $P(A_i \cap B_2)$ $(2) \times (3)$	(5) Posterior Probability $P(A_i | B_2)$ $(4) \div 0.1010$	
A_1: Failure	0.80	0.02	0.0160	0.158	**TABLE 6.1** Calculation of posterior probabilities using Bayes' theorem—Test market example
A_2: Marginal success	0.15	0.24	0.0360	0.356	
A_3: Major success	0.05	0.98	0.0490	0.485	
Total	1.00		0.1010	1.00	

The equivalence of the denominators of (6.1) and (6.2) follows from the definition of a marginal probability in (4.13) and the result in (6.3).

$$(6.4) \qquad P(B_j) = \sum_i P(A_i \cap B_j) = \sum_i P(A_i)P(B_j \,|\, A_i)$$

☐ EXAMPLE

We illustrate by a numerical example how the basic probability theorems are combined in Bayes' theorem. We return to the Martin Wilson example, where we wished to find $P(A_1 \,|\, B_1)$. We start with the following information about the joint probability distribution of job hiring (A_i) and interview status (B_j):

	B_1	B_2	Total
A_1	—	—	0.04
A_2	—	—	0.96
		Total	1.00

$$P(B_1 \,|\, A_1) = 0.98 \qquad P(B_1 \,|\, A_2) = 0.01$$

We use this information to complete the joint probability distribution. Multiplication theorem (4.19) enables us to obtain:

$$P(A_1 \cap B_1) = P(A_1)P(B_1 \,|\, A_1) = 0.04(0.98) = 0.0392$$
$$P(A_2 \cap B_1) = P(A_2)P(B_1 \,|\, A_2) = 0.96(0.01) = 0.0096$$

The remaining joint probabilities can then be obtained by subtraction:

$$P(A_1 \cap B_2) = 0.04 - 0.0392 = 0.0008$$
$$P(A_2 \cap B_2) = 0.96 - 0.0096 = 0.9504$$

We have therefore obtained the following joint probability distribution:

	B_1	B_2	Total
A_1	0.0392	0.0008	0.04
A_2	0.0096	0.9504	0.96
Total	0.0488	0.9512	1.00

Let us now obtain $P(A_1 \,|\, B_1)$ by formula (6.2) for an ordinary conditional probability. We see from the joint probability distribution that the numerator is $P(A_1 \cap B_1) = 0.0392$ and the denominator is $P(B_1) = 0.0488$. Hence, we obtain by formula (6.2):

$$P(A_1 \,|\, B_1) = \frac{P(A_1 \cap B_1)}{P(B_1)} = \frac{0.0392}{0.0488} = 0.803$$

This result, as expected, is the same as we obtained earlier by using the formula for Bayes' theorem in (6.1).　　　　　　　　　　　　　　　　　　　　　　　☐

Random variables that are functions of other random variables are frequently encountered. We now consider two types of functions that often arise in applications: a random variable that is a linear function of another random variable, and a random variable that is a sum or difference of other random variables.

FUNCTIONS OF   **6.2**
RANDOM VARIABLES

Linear Function of Random Variable

Often, interest centers on a random variable W that is a linear function of a random variable X; that is, $W = a + bX$, where a and b are constants.

EXAMPLE ☐   **Telephone Calls**

Let X denote the number of local calls made from a pay telephone in a day, and let W denote the daily revenue from the local calls. Each call costs $0.25, so we have $W = 0.25X$. Here, $a = \$0$ and $b = \$0.25$. If the number of calls in a day is $x = 3$, for instance, the revenue for that day is $w = 0.25(3) = \$0.75$. ☐

EXAMPLE ☐   **Production Run**

Let X denote the size of a production run in a shop, and let W denote the total cost of the production run. It is known that $W = 200 + 4X$, where $a = \$200$ is the fixed setup cost of a run and $b = \$4$ is the cost per unit produced. If the size of the production run is $x = 40$, for instance, the total cost is $w = 200 + 4(40) = \$360$. ☐

Expected Value. To find the expected value of a linear function of X, we could obtain the probability distribution of $W = a + bX$ from that of X and then use the definition of expected value in (5.5) to find $E\{W\}$.

EXAMPLE ☐

In the telephone calls example, the probability distribution of the number of local calls X is known to be as follows:

x:	0	1	2	3
$P(x)$:	0.2	0.4	0.3	0.1

Because each call costs $0.25, the probability distribution of the total revenue from these calls (W) must correspond to that of the number of calls (X) as follows:

w:	0	0.25	0.50	0.75
$P(w)$:	0.2	0.4	0.3	0.1

Using (5.5) to obtain $E\{W\}$, we find:

$$E\{W\} = 0(0.2) + 0.25(0.4) + 0.50(0.3) + 0.75(0.1) = \$0.325$$

Thus, the average daily revenue from local calls at the pay telephone is $0.325. ☐

When the expected value of X is already known, there is no need to find the probability distribution of W first. The following theorem can be used directly.

(6.5) $E\{a + bX\} = a + bE\{X\}$

Since $W = a + bX$, this theorem says that $E\{W\}$ is the same linear function of $E\{X\}$ as W is of X. There are three important special cases of (6.5).

(6.5a) $E\{a\} = a$
(6.5b) $E\{bX\} = bE\{X\}$
(6.5c) $E\{a + X\} = a + E\{X\}$

□ EXAMPLES

1. In the telephone calls example, $E\{X\} = 1.3$ (calculations not shown). Hence, using theorem (6.5b), we obtain for $W = 0.25X$:

    $E\{W\} = 0.25(1.3) = \$0.325$

    This result, as expected, is the same as that obtained when $E\{W\}$ was calculated from the probability distribution of W.

2. In the production run example, it is known that $E\{X\} = 50$. Hence, using theorem (6.5), we obtain for $W = 200 + 4X$:

    $E\{W\} = 200 + 4(50) = \$400$

    Thus, the mean cost per production run is $400.   □

Variance.   To find the variance of W, we could again use the probability distribution of W derived from that of X to make the calculation, based on the definition of the variance in (5.6). Alternatively, when the variance of X is already known, we can obtain the variance of $W = a + bX$ directly from the following theorem.

(6.6) $\sigma^2\{a + bX\} = b^2\sigma^2\{X\}$

Two important special cases of (6.6) are given next.

(6.6a) $\sigma^2\{a + X\} = \sigma^2\{X\}$
(6.6b) $\sigma^2\{bX\} = b^2\sigma^2\{X\}$

EXAMPLES ☐

1. In the telephone calls example, the variance of X is $\sigma^2\{X\} = 0.81$ (calculations not shown). Hence, using theorem (6.6b), we find for $W = 0.25X$:

$$\sigma^2\{W\} = (0.25)^2(0.81) = 0.0506$$

The standard deviation of W is $\sigma\{W\} = \sqrt{0.0506} = \$0.225$.

2. In the production run example, it is known that $\sigma^2\{X\} = 300$. Hence, using (6.6), we obtain for $W = 200 + 4X$:

$$\sigma^2\{W\} = (4)^2(300) = 4800$$

so $\sigma\{W\} = \sqrt{4800} = \$69$.

Note that the constant setup cost $a = \$200$ had no effect on the variance of the total cost W. This is intuitively reasonable since the fixed setup cost does not vary from one production run to another. The fixed setup cost only shifts the position of the probability distribution but does not affect the variability of the distribution. ☐

Comment

Any standardized random variable Y, as defined in (5.9), is a linear function of X. To see this, we rearrange the terms in definition (5.9):

$$Y = \frac{X - E\{X\}}{\sigma\{X\}} = -\frac{E\{X\}}{\sigma\{X\}} + \left(\frac{1}{\sigma\{X\}}\right)X$$

Thus, we see that Y is of the form $Y = a + bX$ where $a = -E\{X\}/\sigma\{X\}$ and $b = 1/\sigma\{X\}$. To show that any standardized random variable Y has variance 1, we have, by (6.6):

$$\sigma^2\{Y\} = b^2\sigma^2\{X\} = \left(\frac{1}{\sigma\{X\}}\right)^2 \sigma^2\{X\} = 1$$

Sum or Difference of Two Independent Random Variables

The sum or difference of two independent random variables is frequently encountered.

EXAMPLE ☐ **Sales Bonus**

Let X denote the bonus amount received by salesperson Anderson, and let Y denote the bonus amount received by salesperson Brody. Then $T = X + Y$ represents the total bonus amount received by the two salespeople. ☐

EXAMPLE ☐ **Advertisement Response**

Let X denote the number of responses to a classified advertisement in city 1 and Y the number of responses to the advertisement in city 2. Then $R = X + Y$ represents the total number of responses to the advertisement in the two cities. ☐

Gross Profit □ **EXAMPLE**

Let X denote quarterly sales revenues and Y quarterly direct costs. Then $W = X - Y$ represents quarterly gross profit. □

Probability Distributions. The probability distributions of sums or differences of two independent random variables can be derived from the probability distributions of the individual random variables.

□ **EXAMPLE**

In the sales bonus example, the probability distribution of X, the bonus amount received by Anderson, is as follows:

Bonus x:	0	500
$P(x)$:	0.6	0.4

Variable Y, the bonus amount received by Brody, follows the same probability distribution as X. Further, X and Y are independent random variables. It then follows that the joint probability distribution of X and Y is:

x	y 0	500	Total
0	0.36	0.24	0.6
500	0.24	0.16	0.4
Total	0.6	0.4	1.0

These joint probabilities are obtained from the independence definition in (5.4), which states that for independent random variables each joint probability equals the product of the corresponding marginal probabilities. For instance, we have:

$$P(X = 0 \cap Y = 0) = P(X = 0)P(Y = 0) = 0.6(0.6) = 0.36$$

The other joint probabilities are obtained in similar fashion.

We can now obtain the probability distribution of the total bonus amount, $T = X + Y$. First, we need to recognize that $T = 0$ if neither salesperson receives a bonus. Hence, $P(T = 0) = 0.36$. Similarly, $T = 1000$ if both salespeople receive a bonus, hence $P(T = 1000) = 0.16$. Finally, $P(T = 500) = 0.24 + 0.24 = 0.48$, since the total bonus amount is $500 when either salesperson (but not both) receives the bonus. Thus, the probability distribution of T is as follows:

Total Bonus t:	0	500	1000
$P(t)$:	0.36	0.48	0.16

□

Expected Values and Variances. If the probability distribution of a sum or difference of two random variables is available, the expected value and the variance of the sum or difference can be found by direct calculation in the usual fashion.

EXAMPLE □

In the sales bonus example, we find the expected total bonus amount using (5.5) as follows:

$$E\{T\} = 0(0.36) + 500(0.48) + 1000(0.16) = \$400$$

Thus, the expected total bonus amount is $400.

Similarly, we can calculate the variance using (5.6) and obtain $\sigma^2\{T\} = 120{,}000$ so that the standard deviation is $\sigma\{T\} = \sqrt{120{,}000} = \$346$. □

When the expected values and the variances of both random variables are already known, we do not need to obtain the probability distribution of the sum or difference since the expected value and the variance of the probability distribution can be obtained directly from two theorems. We give the theorem for the *sum* of two independent random variables first.

(6.7)
The expected value and the variance of the sum of two independent random variables X and Y are as follows:

(6.7a)   *Expected Value* $E\{X + Y\} = E\{X\} + E\{Y\}$

(6.7b)   *Variance* $\sigma^2\{X + Y\} = \sigma^2\{X\} + \sigma^2\{Y\}$

Note that the expected value of a sum of two independent random variables is simply the sum of the expected values of each of the two random variables, and similarly for the variance.

The expected value and the variance of a *difference* of two independent random variables are presented next.

(6.8)
The expected value and the variance of the difference of two independent random variables X and Y are as follows:

(6.8a)   *Expected Value* $E\{X - Y\} = E\{X\} - E\{Y\}$

(6.8b)   *Variance* $\sigma^2\{X - Y\} = \sigma^2\{X\} + \sigma^2\{Y\}$

Note that the expected value of a difference of two independent random variables is simply the difference of the expected values of the two random variables. Note also that the variance of a difference is the same as the variance of the sum when the two random variables are independent.

EXAMPLES □

1. In the sales bonus example, $E\{X\} = E\{Y\} = \$200$, and $\sigma^2\{X\} = \sigma^2\{Y\} = 60{,}000$ (calculations not shown). Hence, for the total bonus amount $T = X + Y$,

we obtain by (6.7):

$$E\{T\} = 200 + 200 = \$400$$
$$\sigma^2\{T\} = 60{,}000 + 60{,}000 = 120{,}000$$

These are the same results, of course, as those obtained from the probability distribution of T.

2. In the advertisement response example, it is known that $E\{X\} = 40$, $E\{Y\} = 70$, $\sigma^2\{X\} = 15$, $\sigma^2\{Y\} = 10$, and X and Y are independent. For the total number of responses $R = X + Y$, we obtain by (6.7):

$$E\{R\} = 40 + 70 = 110 \text{ responses}$$
$$\sigma^2\{R\} = 15 + 10 = 25$$

The standard deviation of R is $\sigma\{R\} = \sqrt{25} = 5$ responses.

3. In the gross profit example, it is known that $E\{X\} = \$100{,}000$, $E\{Y\} = \$70{,}000$, $\sigma^2\{X\} = 8{,}000{,}000$, $\sigma^2\{Y\} = 4{,}000{,}000$, and X and Y are independent. For the gross profit $W = X - Y$, we obtain by (6.8):

$$E\{W\} = 100{,}000 - 70{,}000 = \$30{,}000$$
$$\sigma^2\{W\} = 8{,}000{,}000 + 4{,}000{,}000 = 12{,}000{,}000$$

The standard deviation of W is $\sigma\{W\} = \sqrt{12{,}000{,}000} = \$3464$. □

Sum of Several Independent Random Variables

In statistical analysis, we are frequently concerned with the sum of several random variables. For example, suppose $X_1, X_2, \ldots, X_s$ represent the amounts an insurance company must pay out for fire losses during the year on its s policies. Then $T = X_1 + X_2 + \cdots + X_s$ represents the total disbursement for fire losses during the year on these policies.

The expected value and the variance of a sum of independent random variables are given in the next theorem.

(6.9)
The expected value and the variance of the sum $T = X_1 + X_2 + \cdots + X_s$ when the random variables $X_1, X_2, \ldots, X_s$ are independent are as follows:

(6.9a)   *Expected Value* $E\{T\} = \sum_{i=1}^{s} E\{X_i\}$

(6.9b)   *Variance* $\sigma^2\{T\} = \sum_{i=1}^{s} \sigma^2\{X_i\}$

□ **EXAMPLES**

1. The loss amounts that must be paid out during the year on three fire insurance policies (X_1, X_2, X_3) are independent random variables with the following expected values and variances:

Policy	Expected Value $E\{X_i\}$ ($ thousand)	Variance $\sigma^2\{X_i\}$
1	10	9900
2	8	3100
3	25	11,900

Let $T = X_1 + X_2 + X_3$ denote the total amount that must be paid out. By (6.9), we obtain the expected value and the variance of T:

$$E\{T\} = 10 + 8 + 25 = \$43 \text{ thousand}$$
$$\sigma^2\{T\} = 9900 + 3100 + 11,900 = 24,900$$

The standard deviation of T is $\sigma\{T\} = \sqrt{24,900} = \$158$ thousand. The standard deviation of $158 thousand is large in comparison with the expected value of $43 thousand, indicating that there is relatively large variability in the probability distribution of the total loss amount.

2. A carton contains 100 cans. The weight X_i of each can ($i = 1, 2, \ldots, 100$) has expected value $E\{X_i\} = 8$ ounces and variance $\sigma^2\{X_i\} = 0.017$. In addition, the weights X_i are independent random variables. Let $T = X_1 + X_2 + \cdots + X_{100}$ denote the total weight of the cans in the carton. We obtain by (6.9):

$$E\{T\} = 100(8) = 800 \text{ ounces}$$
$$\sigma^2\{T\} = 100(0.017) = 1.7$$

The standard deviation of T is $\sigma\{T\} = \sqrt{1.7} = 1.3$ ounces. Thus, there is relatively little variability of the actual total carton weight around the mean weight of 800 ounces. □

Effect of Dependence

Dependence of the random variables affects the variance formulas for sums and differences. We take this up in the next section. We point out here, however, that the formulas for expected values of sums and differences in (6.7a), (6.8a), and (6.9a) are valid even if the random variables are dependent.

In later chapters, we shall be concerned with the extent to which two random variables are linearly associated with each other. We now present two measures of linear association for a pair of random variables; these measures are called the *covariance* and the *coefficient of correlation*.

COVARIANCE AND CORRELATION 6.3

Covariance

We begin with the concept of concurrent variation or covariation.

EXAMPLES □

1. In a Midwestern city, the expected midday temperature on July 4 is 25°C and the expected relative humidity is 45 percent. Last July 4, the midday temperature was

30°C and the relative humidity was 70 percent. The covariation of temperature and humidity on that day is the product of the deviations from the expected values; that is, $(30 - 25)(70 - 45) = 125$. Covariation can be positive, as here, negative, or zero. The covariation is positive here because both temperature and humidity exceeded their respective expected values on that day.

2. The expected annual yield of stock A is 5 percent and that of stock B is 10 percent. The actual yields last year were 8 percent each. Therefore, the covariation equals $(8 - 5)(8 - 10) = -6$. Here, the covariation is negative, indicating that one stock performed above its expected value and the other below its expected value.   □

As has been seen in the preceding examples, *covariation* for the outcomes x and y is defined as follows.

(6.10)   Covariation $= (x - E\{X\})(y - E\{Y\})$

The expected value of the covariation of two random variables—that is, their mean covariation over repeated trials—is called their covariance.

(6.11)

The **covariance** of two discrete random variables X and Y is denoted by $\sigma\{X, Y\}$ and defined:

$$\sigma\{X, Y\} = \sum_x \sum_y (x - E\{X\})(y - E\{Y\})P(x, y)$$

where the summation is over all possible outcomes x and y, and $P(x, y)$ denotes $P(X = x \cap Y = y)$. The notation $\sigma\{\ ,\ \}$ is read "covariance of."

(Readers who are unfamiliar with double summation notation should refer to Appendix A, Section A.1.)

□ **EXAMPLE**

Random variables X and Y have the following joint probability distribution:

		y	
x	5	10	Total
10	0.3	0.2	0.5
30	0.1	0.4	0.5
Total	0.4	0.6	1.0

The calculation of the covariance of X and Y is best carried out in a systematic fashion, as shown in Table 6.2. Columns 1 and 2 contain the joint probability distribution of X and Y in a columnar form, where $P(x, y)$ in column 2 denotes $P(X = x \cap Y = y)$. Columns 3 and 4 contain the deviations $x - E\{X\}$ and $y - E\{Y\}$, respectively. The expected values of the two random variables are $E\{X\} = 20$ and $E\{Y\} = 8$ (calcu-

(1)	(2)	(3)	(4)	(5)	(6)	
			Deviation	Deviation		Weighted
			$x - E\{X\}$	$y - E\{Y\}$	Covariation	Covariation
x	y	$P(x, y)$	$= x - 20$	$= y - 8$	(3) × (4)	(5) × (2)
10	5	0.3	-10	-3	30	9.0
10	10	0.2	-10	2	-20	-4.0
30	5	0.1	10	-3	-30	-3.0
30	10	0.4	10	2	20	8.0
	Total	1.0				$\sigma\{X, Y\} = 10.0$
			$E\{X\} = 20$	$E\{Y\} = 8$		

TABLE 6.2
Calculation of covariance

lations not shown). The covariations $(x - E\{X\})(y - E\{Y\})$ appear in column 5. Finally, column 6 contains the covariations weighted by their probabilities $P(x, y)$. Thus, for the outcome $x = 10$, $y = 5$, the deviations are $10 - 20 = -10$ and $5 - 8 = -3$, and the covariation is $(-10)(-3) = 30$. This covariation is weighted by the probability $P(X = 10 \cap Y = 5) = P(10, 5) = 0.3$ to yield the weighted covariation $30(0.3) = 9.0$, as shown in column 6. The sum of column 6 is the covariance. We see that $\sigma\{X, Y\} = 10.0$. Note that the covariance is positive because the most likely outcomes are $(x, y) = (10, 5)$ and $(x, y) = (30, 10)$, for each of which both x and y are above or below their respective means together. □

Interpretation of Covariance. Because covariance is expected covariation, the interpretation of covariance parallels that of covariation for an individual joint outcome. When $\sigma\{X, Y\}$ is positive, we say that X and Y exhibit positive covariance. When $\sigma\{X, Y\}$ is negative or zero, we say that X and Y exhibit negative or zero covariance, respectively. The magnitude of the covariance measure $\sigma\{X, Y\}$ is usually not informative because it depends on the units of X and Y and generally will change when the units of X and Y are changed. Hence, the main information provided by the covariance measure about the association between X and Y is whether $\sigma\{X, Y\}$ is positive, negative, or zero.

In Table 6.3, we present three bivariate probability distributions exhibiting different amounts and direction of covariance. In all three distributions, $E\{X\} = 1$ and $E\{Y\} = 3$. Table 6.3a contains a bivariate probability distribution exhibiting negative covariance. Note that X and Y tend to vary inversely here. If X has a value greater than its mean 1, Y is likely to have a value less than its mean 3, and if X has a value less than 1, Y is likely to have a value greater than 3. The bivariate probability distribution in Table 6.3b exhibits positive covariance. Here, X and Y tend to vary in the same direction. Finally, Table 6.3c shows a bivariate probability distribution exhibiting zero covariance. Here, the covariations between X and Y tend to balance out.

An important property of the covariance measure follows.

(6.12)   When X and Y are independent, $\sigma\{X, Y\} = 0$

Thus, if X and Y are independent random variables, theorem (6.12) tells us that the covariance between them must be zero.

TABLE 6.3
Three bivariate probability distributions with different amounts of covariance and correlation

(a) Negative Covariance

	$y = 1$	$y = 3$	$y = 5$	Total
$x = 0$	0.00	0.05	0.25	0.30
$x = 1$	0.05	0.30	0.05	0.40
$x = 2$	0.25	0.05	0.00	0.30
Total	0.30	0.40	0.30	1.00

$$\sigma\{X, Y\} = -1.0 \qquad \rho\{X, Y\} = -0.833$$

(b) Positive Covariance

	$y = 1$	$y = 3$	$y = 5$	Total
$x = 0$	0.30	0.00	0.00	0.30
$x = 1$	0.00	0.40	0.00	0.40
$x = 2$	0.00	0.00	0.30	0.30
Total	0.30	0.40	0.30	1.00

$$\sigma\{X, Y\} = 1.2 \qquad \rho\{X, Y\} = 1.0$$

(c) Zero Covariance

	$y = 1$	$y = 3$	$y = 5$	Total
$x = 0$	0.10	0.10	0.10	0.30
$x = 1$	0.10	0.20	0.10	0.40
$x = 2$	0.10	0.10	0.10	0.30
Total	0.30	0.40	0.30	1.00

$$\sigma\{X, Y\} = 0 \qquad \rho\{X, Y\} = 0$$

Although independence implies zero covariance, the converse is not always true. It is possible that $\sigma\{X, Y\} = 0$, yet X and Y are dependent variables. This possibility is illustrated by Table 6.3c where $\sigma\{X, Y\} = 0$, yet X and Y are dependent variables. Note, for instance, that $P(X = 2 \cap Y = 1) = 0.10$, which differs from the product of the corresponding marginal probabilities, namely, $0.30(0.30) = 0.09$. Thus, two random variables may be associated in a fashion that the covariance measure does not reflect.

Comment

The covariance can be defined formally as expected covariation.

$$(6.13) \qquad \sigma\{X, Y\} = E\{(X - E\{X\})(Y - E\{Y\})\}$$

An algebraically equivalent expression is as follows.

$$(6.13a) \qquad \sigma\{X, Y\} = E\{XY\} - E\{X\}E\{Y\}$$

Coefficient of Correlation

As we noted earlier, the magnitude of the covariance $\sigma\{X, Y\}$ depends on the units of X and Y and thus usually cannot be directly compared for different pairs of variables. The coefficient of correlation is an adjusted form of the covariance that is unit free and therefore can be compared.

> **(6.14)**
> The **coefficient of correlation** of two random variables X and Y, denoted by $\rho\{X, Y\}$ (Greek rho), is defined:
>
> $$\rho\{X, Y\} = \frac{\sigma\{X, Y\}}{\sigma\{X\}\sigma\{Y\}}$$
>
> where: $\sigma\{X\}$ and $\sigma\{Y\}$ are the standard deviations of X and Y
> $\sigma\{X, Y\}$ is the covariance of X and Y

Note that $\rho\{X, Y\}$ and $\sigma\{X, Y\}$ must always have the same sign because the standard deviations in the denominator of $\rho\{X, Y\}$ in (6.14) are always positive.

EXAMPLES ☐

1. For the example in Table 6.2, we found that $\sigma\{X, Y\} = 10.0$. Further, $\sigma\{X\} = 10$ and $\sigma\{Y\} = 2.449$ (calculations not shown). Hence, by (6.14):

$$\rho\{X, Y\} = \frac{10.0}{10(2.449)} = 0.41$$

   Here $\rho\{X, Y\}$ is positive, just like $\sigma\{X, Y\}$.

2. In Table 6.3, we have shown the coefficients of correlation for each of the three bivariate distributions. The coefficient of correlation in Table 6.3a is negative, as is the covariance, and the coefficient of correlation in Table 6.3b is positive, as is the covariance.

☐

Interpretation of Coefficient of Correlation.   It can be shown that the coefficient of correlation $\rho\{X, Y\}$ takes on values between -1 and 1.

> **(6.15)** $-1 \le \rho\{X, Y\} \le 1$

The coefficient -1 is attained when Y is a linear function of X; that is, when $Y = a + bX$ and b is negative. Likewise, the coefficient 1 is attained when $Y = a + bX$ and b is positive. The latter case is illustrated in Table 6.3b, where all possible pairs of X and Y values satisfy the linear relationship $Y = 1 + 2X$. For instance, the pair $(X, Y) = (2, 5)$ satisfies the relation $Y = 1 + 2(2) = 5$. When the coefficient of correlation equals -1 or 1, X and Y are said to have *perfect linear association*.

Although cases of perfect linear association rarely exist in practice, values of the coefficient of correlation near -1 or near 1 do occur. The closer $\rho\{X, Y\}$ is to either -1 or 1, the stronger is the degree of linear association between the two random variables. In Table

6.3a, the coefficient of correlation $\rho\{X, Y\} = -0.833$ is moderately close to -1, indicating a moderately strong linear association between X and Y here.

A coefficient of correlation of 0 indicates that there is *no linear association* between X and Y, as is illustrated in Table 6.3c. We say that X and Y are uncorrelated then.

Variance of Sum or Difference of Dependent Random Variables

One use of the covariance is found in the variance of a sum or difference of two dependent random variables. If two random variables are dependent, the variances of the sum and difference include the covariance term.

> **(6.16)** $\sigma^2\{X + Y\} = \sigma^2\{X\} + \sigma^2\{Y\} + 2\sigma\{X, Y\}$
>
> **(6.17)** $\sigma^2\{X - Y\} = \sigma^2\{X\} + \sigma^2\{Y\} - 2\sigma\{X, Y\}$

☐ EXAMPLES

1. Let X denote the output by machine 1 and Y the output by machine 2. Suppose $E\{X\} = 200$, $E\{Y\} = 100$, $\sigma^2\{X\} = 400$, $\sigma^2\{Y\} = 80$, and $\sigma\{X, Y\} = -10$. Let $T = X + Y$ denote the total output. We then have:

$$E\{T\} = 200 + 100 = 300 \qquad \text{by (6.7a)}$$
$$\sigma^2\{T\} = 400 + 80 + 2(-10) = 460 \qquad \text{by (6.16)}$$

The standard deviation of T is $\sigma\{T\} = \sqrt{460} = 21.4$.

Note that if the covariance had been positive, $\sigma^2\{T\}$ would have been larger. For instance, if $\sigma\{X, Y\} = 10$, $\sigma^2\{T\} = 400 + 80 + 2(10) = 500$. Thus, the total output is more variable with positively related machine outputs, and less variable with negatively related machine outputs.

2. In Example 1, suppose interest is in the difference D in outputs of the two machines, where $D = X - Y$. We then have:

$$E\{D\} = 200 - 100 = 100 \qquad \text{by (6.8a)}$$
$$\sigma^2\{D\} = 400 + 80 - 2(-10) = 500 \qquad \text{by (6.17)} \qquad ☐$$

Comment

When random variables X and Y are independent, formulas (6.16) and (6.17) reduce to the expressions in (6.7b) and (6.8b) given earlier for independent variables, since $\sigma\{X, Y\} = 0$ for independent variables.

6.4   OPTIONAL TOPIC—CALCULATIONS FOR CONTINUOUS RANDOM VARIABLES (Calculus Needed)

For continuous random variables, the methods of finding probabilities, cumulative probability functions, expected values, and variances become more complex. Integral calculus often may be used for these purposes. Computerized numerical procedures also are useful at times. We shall illustrate now the use of integral calculus.

Determining Probabilities

Probabilities for a continuous random variable X correspond to areas under its probability density function $f(x)$, as described in Section 5.5. To find an area under a probability density function $f(x)$ corresponding to the interval from a to b, we determine the value of the definite integral.

$$(6.18) \qquad P(a \leq X \leq b) = \int_a^b f(x)\,dx$$

EXAMPLE ☐

Consider the following probability density function:

$$(6.19) \qquad f(x) = 12x(1 - x)^2 \qquad 0 \leq x \leq 1$$

For this probability function, we wish to find $P(0.5 \leq X \leq 0.7)$. We obtain:

$$P(0.5 \leq X \leq 0.7) = \int_{0.5}^{0.7} 12x(1 - x)^2 dx = 0.229$$

☐

Cumulative Probability Function

The cumulative probability function $F(x)$ for a continuous random variable is defined as follows.

$$(6.20) \qquad F(x) = \int_{-\infty}^{x} f(u)\,du$$

where u denotes the variable of integration

EXAMPLE ☐

The cumulative probability function corresponding to probability density function (6.19) is:

$$F(x) = \int_0^x 12u(1 - u)^2 du = 3x^4 - 8x^3 + 6x^2 \qquad 0 \leq x \leq 1$$

We see, for instance, that:

$$F(0.5) = 3(0.5)^4 - 8(0.5)^3 + 6(0.5)^2 = 0.688$$

Thus, the probability is 0.688 that random variable X will be 0.5 or less.   ☐

When the algebraic form of the cumulative distribution function $F(x)$ of a continuous random variable is given, the probability density function $f(x)$ may be obtained as the first derivative of $F(x)$, as follows.

$$(6.21) \qquad f(x) = \frac{dF(x)}{dx}$$

Expected Value and Variance

The expected value and variance of a continuous random variable are defined as follows.

$$(6.22) \qquad Expected\ Value \quad E\{X\} = \int_{-\infty}^{\infty} xf(x)dx$$

$$(6.23) \qquad Variance \quad \sigma^2\{X\} = \int_{-\infty}^{\infty} (x - E\{X\})^2 f(x)dx$$

Note that formulas (6.22) and (6.23) are analogous to formulas (5.5) and (5.6) for discrete random variables, with integration replacing summation and the probability density differential $f(x)dx$ serving in place of the probability $P(x)$.

□ **EXAMPLE**

For the probability density function (6.19), the expected value is, by (6.22):

$$E\{X\} = \int_0^1 x[12x(1-x)^2]dx = 0.4$$

The variance is, by (6.23):

$$\sigma^2\{X\} = \int_0^1 (x - 0.4)^2 [12x(1-x)^2]dx = 0.04$$

□

PROBLEMS

* **6.1** Refer to **Internal Welds** Problem 4.28.

a. Obtain the following probabilities: $P(A_1)$, $P(A_2)$, $P(B_1 \mid A_1)$, $P(B_1 \mid A_2)$. Using these values, calculate $P(A_1 \mid B_1)$ and $P(A_2 \mid B_1)$ by means of Bayes' theorem (6.1).

b. Which probabilities in part a are the prior probabilities? Which are the posterior probabilities? Interpret the prior and posterior probabilities for this application.

c. Verify the probability value for $P(A_1 \mid B_1)$ obtained in part a by using (4.14) directly.

6.2 Sixty percent of the graduates of a driver-training school pass the official driver's test on the first attempt, and the other 40 percent fail. The school gives a pretest to graduates before they take the official test. Of the graduates who pass the official test on the first attempt, 80 percent passed the pretest. Of the graduates who fail the official test on the first attempt, 10 percent passed the pretest.

Let A_1 denote that a graduate passes the official test on the first attempt, A_2 that the graduate fails, and B_1 that the graduate passed the pretest.

a. Give the values of the following probabilities: $P(A_1)$, $P(A_2)$, $P(B_1 \mid A_1)$, $P(B_1 \mid A_2)$.

b. A graduate has passed the pretest. Use Bayes' theorem (6.1) to obtain the posterior probabilities $P(A_1 \mid B_1)$ and $P(A_2 \mid B_1)$. Interpret these probabilities for this application. Has the information provided by the pretest led to a substantial modification of the prior probabilities? Discuss.

6.3 The probabilities that an offshore tract has no gas (A_1), a minor gas deposit (A_2), or a major gas deposit (A_3) are 0.60, 0.30, and 0.10, respectively. A test well drilled in the tract will yield no gas (B_1) if none is present and will yield gas (B_2) with probability 0.3 if a minor deposit is present and with probability 0.9 if a major deposit is present. Using a tabular format such as that in Table 6.1, calculate the posterior probability distribution for the size of the gas deposit, given that a test well drilled in the tract yielded gas. Did the information provided by the test well lead to a substantial modification of the prior probabilities? Discuss.

6.4 An electronic instrument can fail in one of two ways (T_1, T_2) with equal probability. The two types of failures occur, however, in different locations (L_1, L_2, L_3) with different probabilities. The following are the conditional probability distributions for failure location, conditional on the type of failure:

Location	Conditional on	
	T_1	T_2
L_1	0.1	0.6
L_2	0.1	0.3
L_3	0.8	0.1
Total	1.0	1.0

a. Using a tabular format such as that in Table 6.1, calculate the posterior probability distribution for the type of failure if the failure is known to have occurred in the location: (1) L_1, (2) L_2, (3) L_3.

b. From the results in part a, does it appear that knowledge of failure location provides substantial information about the type of failure that has occurred? Comment.

* 6.5 The probability distribution of X, the number of persons in a restaurant party, follows:

x:	1	2	3	4	5	6
$P(x)$:	0.05	0.15	0.25	0.40	0.10	0.05

The restaurant serves a buffet lunch costing $9.50 per person. Let Y denote the total cost of the buffet lunch for a party.

a. Express Y in terms of X.

b. Calculate $E\{X\}$ and $\sigma\{X\}$. Use these results to obtain $E\{Y\}$ and $\sigma\{Y\}$ by (6.5) and (6.6).

c. Verify the results in part b by obtaining the probability distribution of Y and calculating $E\{Y\}$ and $\sigma\{Y\}$ directly from this probability distribution.

6.6 Refer to **Helicopter Shuttle** Problem 5.4. The fare for the trip is $50. Let Y denote the total fare revenue for the run.

a. Express Y in terms of X.

b. Calculate $E\{Y\}$ and $\sigma\{Y\}$ from $E\{X\}$ and $\sigma\{X\}$, using (6.5) and (6.6).

c. Verify the results in part b by obtaining the probability distribution of Y and calculating $E\{Y\}$ and $\sigma\{Y\}$ directly from this probability distribution.

6.7 Refer to **Ceramic Filters** Problem 5.6. Let Y denote the total cost of a shipment, including the shipping charge and the cost of damaged filters. The shipping charge is $40, and the cost per damaged filter is $100.

a. Express Y in terms of X.
b. Calculate $E\{Y\}$ and $\sigma\{Y\}$ from $E\{X\}$ and $\sigma\{X\}$, using (6.5) and (6.6).
c. Which would increase the expected total cost of a shipment more, an increase in the shipping charge to $50 or an increase in the cost per damaged filter to $150?
d. Is the standard deviation of the total cost of a shipment affected by the shipping charge? By the cost per damaged filter? Explain.

6.8 A construction company has successfully bid for a job that will pay it $800 thousand. The total cost of completing the job (in $ thousand) is a random variable X, with $E\{X\} = 600$ and $\sigma\{X\} = 150$. The company's gross income Y from the job is therefore $Y = 800 - X$.

a. Obtain $E\{Y\}$ and $\sigma\{Y\}$. In what units is each of these measures expressed?
b. Compare the coefficients of variation (that is, the ratio of standard deviation to mean, expressed as percentage) of X and Y. Which is relatively more variable, the total cost of the job or the gross income from the job?
c. If the total cost of the job were 10 percent higher because of poor cost control, thereby changing the company's gross income from the job to $Y' = 800 - 1.10X$, what would be the resulting values of the mean and the standard deviation of the gross income?

* **6.9 Sports Match.** The probability distributions of the number of goals scored by the home team (X) and the visiting team (Y) in a sports match follow:

x:	0	1	2		y:	0	1	2
$P(x)$:	0.4	0.3	0.3		$P(y)$:	0.5	0.3	0.2

Here X and Y are independent random variables. Let $T = X + Y$ denote the total number of goals scored in the match.

a. Calculate $E\{T\}$ and $\sigma\{T\}$, using (6.7). Interpret each measure in terms of relative frequency.
b. Obtain the probability distribution of T. Verify the results in part a by direct calculations from this probability distribution.

* **6.10** Refer to **Sports Match** Problem 6.9.

a. What does the difference $D = X - Y$ represent here? Calculate $E\{D\}$ and $\sigma\{D\}$, using (6.8).
b. Obtain the probability distribution of D. What is the probability of a tie game? Of the home team winning?

6.11 Refer to **Helicopter Shuttle** Problems 5.4 and 5.15. Let $T = X_1 + X_2$ denote the total number of passengers on the two days' runs.

a. Calculate $E\{T\}$ and $\sigma\{T\}$, using (6.7). Does either of these results depend on the fact that X_1 and X_2 are independent?
b. Obtain the probability distribution of T. Verify the results in part a by direct calculations from this probability distribution.

6.12 Refer to **Helicopter Shuttle** Problems 5.4 and 5.15.

a. What does the difference $D = X_2 - X_1$ represent here?
b. Calculate $E\{D\}$ and $\sigma\{D\}$, using (6.8). Interpret each measure in terms of relative frequency.

6.13 The percentage grades that two students will obtain in a history course this semester are denoted by X_1 and X_2. Assume X_1 and X_2 are independent, with the following means and standard deviations:

$$E\{X_1\} = 85 \qquad \sigma\{X_1\} = 5$$
$$E\{X_2\} = 75 \qquad \sigma\{X_2\} = 10$$

a. Obtain the mean and the standard deviation of $D = X_1 - X_2$. Does either of these results depend on the fact that X_1 and X_2 are independent?
b. Obtain the mean and the standard deviation of $T = X_1 + X_2$.
c. Use the results in part b and theorems (6.5) and (6.6) to obtain the mean and the standard deviation of the average grade of the two students—that is, of $T/2$.

* **6.14** The daily output X in a bottling plant is a random variable, with $E\{X\} = 10{,}000$ bottles and $\sigma\{X\} = 500$ bottles. Assume that the outputs on successive days are independent.

a. What is the expected total output for 10 successive days? For 30 successive days?
b. Obtain the standard deviation of total output for 10 days. Also obtain the standard deviation of total output for 30 days. How do these two standard deviations compare relative to their respective means?

6.15 Refer to **Helicopter Shuttle** Problem 5.4. Assume that the numbers of passengers on different runs are independent, each with the probability distribution given in Problem 5.4. Let T denote the total number of passengers on thirty successive runs.

a. Obtain $E\{T\}$ and $\sigma^2\{T\}$.
b. Is the standard deviation of T thirty times as large as the standard deviation of X? Discuss.

6.16 Let X_1, X_2, and X_3 denote the service lives (in operating hours) of three drive belts installed in a machine. The total service provided by all three belts is $T = X_1 + X_2 + X_3$. Assume that X_1, X_2, and X_3 are independent, with the following means and standard deviations:

	X_1	X_2	X_3
Mean:	410	530	590
Standard Deviation:	60	75	80

a. Obtain $E\{T\}$ and $\sigma\{T\}$. Does either of these results depend on the fact that X_1, X_2, and X_3 are independent?
b. Suppose that the first belt has an actual service life of $X_1 = 473$ hours. Obtain the mean and the standard deviation of $X_2 + X_3$, the total service provided by the second and third belts, given that $X_1 = 473$.

* **6.17** **Product Demand.** The bivariate probability distribution of the price received (P) and the quantity sold (Q) of a product follows:

	q (cartons)	
p ($ per carton)	30	40
6	0.2	0.4
8	0.3	0.1

a. Calculate $E\{P\}$ and $E\{Q\}$.
b. Obtain the covariation of $p = 6$ and $q = 40$. What information is provided by the sign of the measure of covariation?
c. Calculate $\sigma\{P, Q\}$. What does its sign indicate about the nature of the linear association between P and Q?

6.18 Refer to **Computer Operations** Problem 5.13.

a. Calculate $E\{X\}$ and $E\{Y\}$.
b. Obtain the covariation of $x = 1$ and $y = 2$. In what units is the covariation expressed?
c. Calculate $\sigma\{X, Y\}$. What does its sign tell us about the nature of the linear association between X and Y?

6.19 Refer to **Bottle Flaws** Problem 5.14.

a. Calculate $E\{X\}$, $E\{Y\}$, and $\sigma\{X, Y\}$.
b. Are the two random variables independent? Discuss.

* **6.20** Refer to **Product Demand** Problem 6.17.

a. Calculate the coefficient of correlation between price received and quantity sold. In general, what range of values can this coefficient take on?

b. Using your results in part a, describe the direction and the strength of the linear association between price and quantity.

6.21 Refer to **Computer Operations** Problem 5.13.

a. Calculate $\rho\{X, Y\}$. Does $\rho\{Y, X\}$ necessarily have the same sign and magnitude as $\rho\{X, Y\}$? Explain.

b. Using your result in part a, describe the direction and the strength of the linear association between X and Y.

6.22 Refer to **Bottle Flaws** Problem 5.14.

a. Calculate $\rho\{X, Y\}$. In what units is this measure expressed?

b. Does the positive value of $\rho\{X, Y\}$ here guarantee that a bottle with more particle flaws than another must also have more bubble flaws? Discuss.

* 6.23 Refer to Table 5.1. Here $E\{X\} = E\{Y\} = 0.55$, $\sigma^2\{X\} = \sigma^2\{Y\} = 0.7475$, and $\sigma\{X, Y\} = 0.5675$.

a. Calculate $E\{X + Y\}$, $E\{X - Y\}$, $\sigma\{X + Y\}$, and $\sigma\{X - Y\}$. Interpret each of these numbers.

b. What would be the values calculated in part a if X and Y had the marginal probability distributions shown in Table 5.1 but were independent?

6.24 Refer to **Bottle Flaws** Problem 5.14. Let $T = X + Y$ denote the total number of flaws in a bottle.

a. Calculate $E\{X\}$, $E\{Y\}$, $\sigma^2\{X\}$, $\sigma^2\{Y\}$, and $\sigma\{X, Y\}$. Then obtain $E\{T\}$ and $\sigma\{T\}$, using (6.7a) and (6.16).

b. Verify the results for $E\{T\}$ and $\sigma\{T\}$ in part a by obtaining the probability distribution of T and calculating the two measures directly from this distribution.

6.25 Let X_1 and X_2 denote the earnings (in \$ million) of a firm in two consecutive years. Assume that X_1 and X_2 have the same mean 15 and the same standard deviation 3.

a. Calculate the mean and the standard deviation of the total earnings $X_1 + X_2$ in the two years, assuming (1) $\rho\{X_1, X_2\} = -0.5$, (2) $\rho\{X_1, X_2\} = 0.5$. For which of these two correlations will the firm's total earnings in the two years be less variable? [*Hint:* Use (6.14) to obtain the covariance.]

b. Calculate the mean and the standard deviation of the difference in earnings $X_1 - X_2$, assuming (1) $\rho\{X_1, X_2\} = -0.5$, (2) $\rho\{X_1, X_2\} = 0.5$. For which of these two correlations will the difference in earnings for the two years be less variable?

EXERCISES

6.26 Refer to Bayes' theorem (6.1). If information outcome B_j yields a posterior probability for outcome A_i that is identical to the prior probability for A_i, what is the nature of the relationship between events B_j and A_i?

6.27 Verify the identities (6.5) and (6.6).

6.28 Verify property (6.12).

6.29 Verify the identity (6.13a).

6.30 Let Y_1 and Y_2 denote the standardized forms of random variables X_1 and X_2, respectively. Show that $\rho\{X_1, X_2\} = \sigma\{Y_1, Y_2\}$. [*Hint:* Use (6.13a).]

6.31 *(Calculus needed.)* Refer to **Suspension Cable** Problem 5.32. Obtain the following: (1) $F(x)$, (2) $E\{X\}$, (3) $\sigma^2\{X\}$.

6.32 *(Calculus needed.)* Refer to **Shelf Life** Problem 5.33. Obtain the following: (1) $F(x)$, (2) $E\{X\}$, (3) $\sigma^2\{X\}$.

6.33 *(Calculus needed.)* Refer to **Wind Speed** Problem 5.34.

a. Obtain the probability density function using (6.21) and construct a graph of it.
b. Obtain $E\{X\}$. Interpret its meaning in this setting.
c. Obtain the median wind speed. [*Hint:* $F(Md) = 0.5$.]

6.34 On the basis of a physical examination and symptoms, a physician assesses the probabilities that the patient has no tumor, a benign tumor, or a malignant tumor as 0.70, 0.20, and 0.10, respectively. A thermographic test is subsequently given to the patient. This test gives a negative result with probability 0.90 if there is no tumor, with probability 0.80 if there is a benign tumor, and with probability 0.20 if there is a malignant tumor.

a. What is the probability that a thermographic test will give a negative result for this patient?
b. Obtain the posterior probability distribution for the state of this patient when the test result is negative. Interpret this probability distribution. What is the most likely state for the patient? Did the test result lead to substantial changes in the prior probabilities? Discuss.
c. Obtain the posterior probability distribution for the state of this patient when the test result is positive. How does this posterior distribution differ from the one in part b? Discuss.

6.35 A national chain operates 400 similar restaurants. The probability distribution of loss by fire in any one of these restaurants in a year (X, in dollars) follows:

x:	0	150,000
$P(x)$:	0.998	0.002

Assume that the losses in the different restaurants ($X_1, X_2, \ldots, X_{400}$) are independent. Let $T = \Sigma X_i$ denote the total loss by fire for the whole chain in a year.

a. Calculate $E\{X_i\}$ and $\sigma\{X_i\}$ for any one restaurant.
b. Obtain $E\{T\}$ and $\sigma\{T\}$. Is random variable T relatively more predictable than the random variable X_i for any one restaurant? Comment.
c. What is the probability that the total loss by fire for the chain will be (1) \$0, (2) \$150,000 or less? [*Hint:* What is the probability that none of the 400 restaurants experiences a loss? That one restaurant experiences a loss?]
d. The chain self-insures its 400 restaurants against fire loss because the expected loss by fire for the chain is substantially less than the cost of fire insurance for a year. If a middle-income person were to own and operate a single one of these restaurants, why might this person not be willing to self-insure against fire loss?

6.36 A small investor is considering a \$200 investment in growth stocks for two years. Let X denote the gain in buying and holding for two years one \$100 share of stock 1, and let Y be defined similarly for one \$100 share of stock 2. Assume that $E\{X\} = 50$, $E\{Y\} = 50$, $\sigma^2\{X\} = 64$, $\sigma^2\{Y\} = 81$, and $\sigma\{X, Y\} = -60$. Consider the following options: (1) buy two shares of stock 1, (2) buy two shares of stock 2, (3) buy one share each of stocks 1 and 2.

a. Is any one of these options better than the others in terms of expected gain? Explain.
b. Calculate $\sigma^2\{2X\}$, $\sigma^2\{2Y\}$, and $\sigma^2\{X + Y\}$ and explain what each quantity represents. The investor states that option 3 is "less risky" than options 1 and 2. How is the investor interpreting risk here?
c. Repeat the calculations in part b for $\sigma\{X, Y\} = 60$. Which option now has the smallest variance? What does this finding suggest about the covariance conditions that will "spread the risk" when one is diversifying with several different securities?

6.37 Refer to **Product Demand** Problem 6.17.

a. The dollar cost of producing quantity Q is given by $C = 100 + 3Q$. Obtain $E\{C\}$ and $\sigma^2\{C\}$.
b. Denote the revenue from the sale of the product by R, so $R = PQ$. Find $E\{R\}$ and interpret its meaning. [*Hint:* Use (6.13a).]
c. The profit from the sale of the product is $R - C$. Find $E\{R - C\}$.

Common Probability Distributions: Discrete Random Variables

7

A n infinite variety of probability distributions exist. However, a limited number of *families,* or types, of probability distributions are used in a wide range of applications. In this chapter, we consider four important families of distributions for discrete random variables. In the next chapter, we take up several families of distributions for continuous random variables.

A random variable that can take on any integer value within a given interval with equal probability is called a *discrete uniform random variable.* This type of random variable has several special uses, including applications in statistical sampling.

EXAMPLE ☐ **Stock Market**

The final digit in today's volume of stock market transactions will be denoted by X. This random variable can take on any of the 10 digits from 0 to 9, each with probability 0.10. Hence its probability distribution is:

x:	0	1	2	3	4	5	6	7	8	9
$P(x)$:	0.10	0.10	0.10	0.10	0.10	0.10	0.10	0.10	0.10	0.10

☐

Discrete Uniform Probability Function

As we have seen in Chapter 5, the probability distribution for a discrete random variable X associates a probability $P(x)$ with each basic outcome x. Often, the probability function $P(x)$ can be expressed by a mathematical formula so that the probability $P(x)$ for each possible outcome x can be calculated from the formula. The probability function for a discrete uniform random variable is of this type.

(7.1)
The **discrete uniform probability function** is:

$$P(x) = \frac{1}{s}$$

where: $x = a + 1, a + 2, \ldots, a + s$

 $a + 1, s$ are integers, with $s > 0$

Specific values of $a + 1$ and s identify one discrete uniform probability distribution from the family of all such probability distributions. The constants $a + 1$ and s are called the *parameters* of the discrete uniform probability distribution. Parameter $a + 1$ denotes the smallest outcome, and s denotes the number of distinct outcomes.

☐ **EXAMPLE**

In the stock market example, where the possible outcomes are the integers from 0 to 9, the two parameters of the discrete uniform distribution are $a + 1 = 0$ and $s = 10$. A graph of this probability distribution is shown in Figure 7.1. Note that the probability of each digit is $1/s = 1/10 = 0.10$.   ☐

Characteristics of Discrete Uniform Probability Distributions

The mean and the variance of a discrete uniform probability distribution depend on the parameters $a + 1$ and s.

(7.2)

The mean and the variance of a discrete uniform probability distribution are:

(7.2a) $E\{X\} = (a + 1) + \dfrac{s - 1}{2}$

(7.2b) $\sigma^2\{X\} = \dfrac{s^2 - 1}{12}$

☐ **EXAMPLE**

For the stock market example, the discrete uniform probability distribution has parameters $a + 1 = 0$ and $s = 10$. Hence:

$$E\{X\} = 0 + \frac{10 - 1}{2} = 4.5 \qquad \sigma^2\{X\} = \frac{10^2 - 1}{12} = 8.25$$

Thus, the standard deviation of the distribution is $\sigma\{X\} = \sqrt{8.25} = 2.87$.   ☐

FIGURE 7.1
Discrete uniform probability distribution with $a + 1 = 0$, $s = 10$—Stock market example

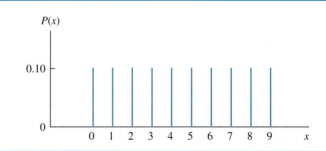

Bernoulli Random Variable

A *Bernoulli random trial* is a random trial that has two basic outcomes of a qualitative nature. To quantify these outcomes, we arbitrarily assign one outcome the value 1 and the other the value 0. We shall let p denote the probability that the outcome takes value 1. It then follows from complementation theorem (4.18) that the probability of outcome 0 is $1 - p$. We shall let B denote the random variable associated with a Bernoulli random trial. We have then that $P(B = 1) = p$ and $P(B = 0) = 1 - p$. We shall refer to random variable B as a Bernoulli random variable.

(7.3)

The probability distribution of a **Bernoulli random variable** B is called a **Bernoulli probability distribution** and has the form:

b	$P(b)$
0	$1 - p$
1	p
Total	1

where $0 < p < 1$.

EXAMPLES

1. The calculation of a payroll check may be correct or incorrect. Let $B = 1$ correspond to an incorrectly calculated check and $B = 0$ to a correctly calculated one. If the probability of an incorrectly calculated check is $p = 0.05$, then the Bernoulli probability distribution for random variable B is:

b	$P(b)$
0	0.95
1	0.05
Total	1.00

2. A consumer either recalls a television commercial ($B = 1$) or does not recall it ($B = 0$). If the probability that a consumer recalls the commercial is $p = 0.30$, then the Bernoulli probability distribution for random variable B is:

b	$P(b)$
0	0.70
1	0.30
Total	1.00

Bernoulli Process

In statistical applications, we are seldom interested in a single Bernoulli random variable. More commonly, a sequence $B_1, B_2, \ldots, B_n$ of Bernoulli random variables is under consideration. If the random variables in the sequence (1) are independent and (2) all have the same Bernoulli probability distribution with $P(B_i = 1) = p$, then the sequence is called a Bernoulli process.

> **(7.4)**
> A **Bernoulli process** is a sequence of **independent and identically distributed** Bernoulli random variables.

Spoon Manufacturing ☐ **EXAMPLE**

In a process for manufacturing spoons, the ith spoon may be either defective ($B_i = 1$) or not ($B_i = 0$). A quality control program had been instituted that brought the manufacturing process into control. As a result, the manufacturing process has the following two characteristics: (1) Whether one spoon is defective or not is independent of the outcomes for the other spoons in this process; (2) the probability of a spoon being defective is the same for all spoons in this process. Consequently, for the next n spoons produced by this process, the sequence of random variables $B_1, B_2, \ldots, B_n$ constitutes a Bernoulli process.   ☐

A Bernoulli process need not refer only to sequences over time, as the next example illustrates.

Plant Specimens ☐ **EXAMPLE**

In an agricultural experiment, n plant specimens will be treated for a particular fungus. At the end of the experiment, the ith specimen either will have fungus present ($B_i = 1$) or will be fungus free ($B_i = 0$). The experimental conditions are such that the outcome for each specimen is independent of the outcomes for the other specimens and each specimen has the same probability of having fungus present at the end of the experiment. Hence the sequence of random variables $B_1, B_2, \ldots, B_n$ for the n specimens constitutes a Bernoulli process.   ☐

Because the Bernoulli random variables in a Bernoulli process are independent, the probability of any sequence can be readily determined by extending definition (5.4) for two independent random variables to n independent variables.

☐ **EXAMPLE**

In the spoon manufacturing example, the probability of a defective spoon is $P(B_i = 1) = p = 0.005$ and the probability of an acceptable spoon is $P(B_i = 0) = 1 - p = 0.995$. The probability that only the first spoon in the next sequence of five spoons is defective is of interest. This probability can be obtained by extending definition (5.4) to more than two independent variables:

$$P(B_1 = 1)P(B_2 = 0)P(B_3 = 0)P(B_4 = 0)P(B_5 = 0)$$
$$= p(1 - p)(1 - p)(1 - p)(1 - p)$$
$$= 0.005(0.995)^4$$
$$= 0.0049 \qquad \square$$

Binomial Random Variable

In a Bernoulli process, interest often centers on how many random variables in the sequence $B_1, B_2, \ldots, B_n$ have the outcome 1.

EXAMPLES

1. In the spoon manufacturing example, the total number of defective spoons in the next production run of $n = 100$ spoons is of interest. Since $B_i = 1$ if the ith spoon is defective and $B_i = 0$ otherwise, it follows that the sum:

$$B_1 + B_2 + \cdots + B_{100}$$

contains as many 1's as there are defective spoons among the 100 and contains 0's otherwise. Thus, the sum equals the total number of defective spoons in the production run. We shall denote this sum by X.

2. In the plant specimens example, $n = 30$ specimens are employed in the experiment and the total number of specimens with fungus present at the end of the experiment is of interest. The sum:

$$X = B_1 + B_2 + \cdots + B_{30}$$

equals the total number of specimens with fungus because $B_i = 1$ if the ith specimen has fungus and $B_i = 0$ otherwise. □

(7.5)
The sum of n independent and identically distributed Bernoulli random variables is denoted by X and called a **binomial random variable:**

$$X = B_1 + B_2 + \cdots + B_n$$

Binomial Probability Function

The binomial probability function has two parameters, n and p, where n is the number of Bernoulli random variables in the sum X and p denotes the probability $P(B_i = 1)$ for each Bernoulli random variable. The binomial probability distribution is a discrete probability distribution because random variable X—the number of 1's in n independent Bernoulli trials—can take only one of the $n + 1$ values $0, 1, \ldots, n$.

(7.6)
The **binomial probability function** is:

$$P(x) = \binom{n}{x} p^x (1 - p)^{n - x}$$

where: $x = 0, 1, \ldots, n$
   $0 < p < 1$

Here, $\binom{n}{x}$ represents a *binomial coefficient* and is defined as follows.

(7.6a) $$\binom{n}{x} = \frac{n!}{x!(n-x)!}$$

where $a! = a(a-1)\cdots(2)(1)$ and $0! = 1$.

For example, we have:

$$\binom{5}{2} = \frac{5!}{2!3!} = \frac{5(4)(3)(2)(1)}{2(1)(3)(2)(1)} = 10$$

(A review of permutations and combinations can be found in Appendix A, Section A.3.)

☐ **EXAMPLE**

In the plant specimens example, a preliminary experiment involved $n = 3$ plant specimens. Suppose that the probability of any individual specimen having fungus at the end of the experiment is $p = 0.2$. The binomial probability distribution for X, the total number of specimens having fungus present, may then be computed using (7.6). We obtain for $P(0)$, the probability that none of the three specimens has fungus:

$$P(0) = \binom{3}{0}(0.2)^0(0.8)^3 = \frac{3!}{0!3!}(1.0)(0.5120) = 0.5120$$

In similar fashion, we calculate the probabilities of the other possible outcomes and obtain the binomial probability distribution:

x	$P(x)$
0	$\binom{3}{0}(0.2)^0(0.8)^3 = 0.5120$
1	$\binom{3}{1}(0.2)^1(0.8)^2 = 0.3840$
2	$\binom{3}{2}(0.2)^2(0.8)^1 = 0.0960$
3	$\binom{3}{3}(0.2)^3(0.8)^0 = 0.0080$
	Total 1.0000

We see, for instance, that the probability of all three specimens having fungus present is $P(3) = 0.0080$, so that this outcome is not likely. We also note that the probability of one or fewer specimens having fungus present is high, since $P(X \le 1) = P(0) + P(1) = 0.5120 + 0.3840 = 0.8960$.   ☐

Comment

We shall explain the derivation of the binomial probability function (7.6) in terms of the plant specimens example in which $n = 3$ and $p = 0.2$. The probability tree diagram in Figure 7.2 is helpful in this connection. It shows all possible outcome sequences for the three Bernoulli random variables B_1, B_2, and B_3.

We begin by finding $P(X = 1) = P(1)$. Figure 7.2 shows that three sequences yield exactly one specimen with fungus present, namely, S_4, S_6, and S_7. Let us obtain the probability that sequence S_4 occurs. Since the random variables B_1, B_2, and B_3 are independent, and since $P(B_i = 1) = 0.2$ and $P(B_i = 0) = 0.8$ for each, we obtain:

$$P(S_4) = P(B_1 = 1)P(B_2 = 0)P(B_3 = 0)$$
$$= 0.2(0.8)(0.8) = (0.2)^1(0.8)^2$$

Similarly, we find:

$$P(S_6) = (0.2)^1(0.8)^2$$
$$P(S_7) = (0.2)^1(0.8)^2$$

Finally, because S_4, S_6, and S_7 are mutually exclusive events, we have, by addition theorem (4.17):

$$P(X = 1) = P(1) = P(S_4) + P(S_6) + P(S_7)$$
$$= 3(0.2)^1(0.8)^2 = 0.3840$$

Similar reasoning allows us to obtain the probability of any other outcome of X. For instance:

$$P(X = 2) = P(2) = P(S_2) + P(S_3) + P(S_5)$$
$$= 3(0.2)^2(0.8)^1 = 0.0960$$

Note that these two probabilities are the same as those calculated earlier from the binomial probability function (7.6).

We now generalize the derivation.

1. For n independent Bernoulli random variables, for each of which $P(B_i = 1) = p$, the probability of obtaining x 1's and $n - x$ 0's in a *specific sequence* or *permutation* is:

$$p^x(1 - p)^{n-x}$$

2. The number of distinct sequences or permutations of x 1's and $n - x$ 0's is given by the binomial coefficient:

$$\binom{n}{x}$$

3. Combining these results, we see that $P(x)$ is simply the sum of $\binom{n}{x}$ sequence probabilities, each sequence probability being $p^x(1 - p)^{n-x}$. Thus:

$$P(x) = \binom{n}{x}p^x(1 - p)^{n-x}$$

FIGURE 7.2
Derivation of the binomial probability function for $n = 3$ and $p = 0.2$—Plant specimens example

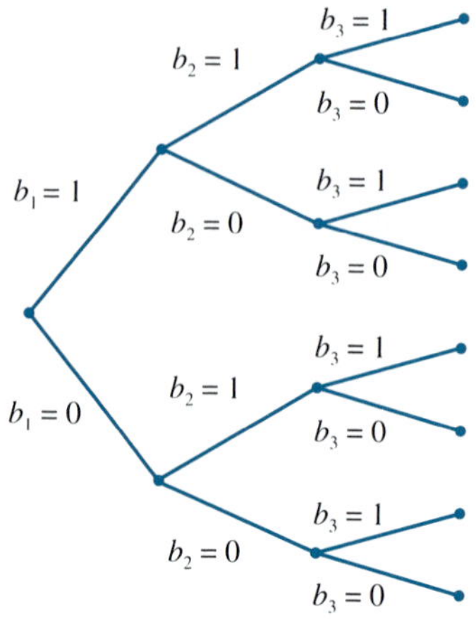

Sequence	$x = b_1 + b_2 + b_3$	Probability of Sequence
S_1	3	$(0.2)^3(0.8)^0$
S_2	2	$(0.2)^2(0.8)^1$
S_3	2	$(0.2)^2(0.8)^1$
S_4	1	$(0.2)^1(0.8)^2$
S_5	2	$(0.2)^2(0.8)^1$
S_6	1	$(0.2)^1(0.8)^2$
S_7	1	$(0.2)^1(0.8)^2$
S_8	0	$(0.2)^0(0.8)^3$

Probability Table

Binomial probabilities can be computed readily with many calculators and statistical packages. They can also be obtained from binomial probability tables. Table C.5 (see Appendix C) gives binomial probabilities for selected values of p and selected values of n up to 20. For large n, approximation methods may be used. We shall take these up in Chapter 13.

☐ EXAMPLES

1. In a psychological experiment, $n = 20$ children are each asked to solve a puzzle independently in a fixed amount of time. Interest lies in the number of children who succeed in completing the task. The experimental conditions make it reasonable to consider the number who succeed to be a binomial random variable with $p = 0.40$. We wish to find the probability of exactly $x = 6$ children succeeding. We see from Table C.5 that it is $P(6) = 0.1244$. This value is located in the column marked $p = 0.40$ at the top and in the row corresponding to $n = 20$ and $x = 6$ as labeled on the left-hand side of the table.

   If the probability of successful completion were $p = 0.70$, we would then find from the probability table that $P(6) = 0.0002$. This value is located in the column marked $p = 0.70$ at the bottom and in the row corresponding to $n = 20$ and $x = 6$ as labeled on the right-hand side of the table. Note, therefore, that the (n, x) values on the left-hand side correspond to p values of 0.5 or less, and those on the right-hand side correspond to p values of 0.5 or more.

   To obtain cumulative binomial probabilities from Table C.5, we simply sum the appropriate individual probabilities. For example, if $P(X \leq 2)$ is desired for the case where $p = 0.40$ and $n = 20$, we find:

$$P(X \leq 2) = P(0) + P(1) + P(2)$$
$$= 0.0000 + 0.0005 + 0.0031 = 0.0036$$

2. In an insurance company, $n = 10$ claims adjusters are asked, as part of a training program, to consider independently a hypothetical claim. Interest is in the number of adjusters who handle the claim correctly. It is reasonable to assume here that the number who handle the claim correctly is a binomial random variable with $p = 0.90$. We wish to find $P(X \leq 8)$. For convenience, we use complementation theorem (4.18) to recognize that $P(X \leq 8) = 1 - P(X \geq 9)$. From Table C.5, we find $P(X \geq 9) = P(9) + P(10) = 0.3874 + 0.3487 = 0.7361$. Hence, $P(X \leq 8) = 1 - 0.7361 = 0.2639$. □

Characteristics of Binomial Probability Distributions

Mean and Variance. Two important characteristics of a binomial probability distribution are its mean and variance.

> **(7.7)**
> The mean and the variance of a binomial probability distribution are:
>
> **(7.7a)** $E\{X\} = np$
>
> **(7.7b)** $\sigma^2\{X\} = np(1 - p)$

EXAMPLE □

The number of books with substandard bindings produced in a production run of $n = 3000$ books is a binomial random variable X with probability $p = 0.02$ of a book with a substandard binding. The expected number of books with substandard bindings in the production run is then $E\{X\} = 3000(0.02) = 60$ books. The variance of the number of books with substandard bindings in the production run is $\sigma^2\{X\} = 3000(0.02)(0.98) = 58.8$. Thus, the standard deviation of the number of books with substandard bindings is $\sigma\{X\} = \sqrt{58.8} = 7.67$ books. This shows that the variability in the number of defective books from one production run to another is moderately large relative to the mean number of defective books, which is 60. □

> **Comment**
>
> The mean and the variance of a binomial random variable can be readily derived by using the probability concepts discussed in earlier chapters. To show that $E\{X\} = np$, for instance, we first find $E\{B_i\}$ for each Bernoulli random variable. Applying the definition of expected value in (5.5) to the Bernoulli probability distribution in (7.3), we obtain $E\{B_i\} = 1(p) + 0(1 - p) = p$. Since $X = B_1 + B_2 + \cdots + B_n$, we have, by (6.9a), that $E\{X\} = np$.

Distribution Shape. Binomial probability distributions can take on a variety of shapes. Figure 7.3 shows three binomial probability distributions. Note that the distribution is skewed right if $p < 0.5$, is skewed left if $p > 0.5$, and is symmetrical if $p = 0.5$. The closer p is to 0 or 1 for a given n, the more pronounced is the skewness.

FIGURE 7.3   **Three binomial probability distributions.**   The binomial distribution is skewed right when $p < 0.5$, skewed left when $p > 0.5$, and symmetrical when $p = 0.5$.

(a) $n = 3$, $p = 0.2$

(b) $n = 8$, $p = 0.7$

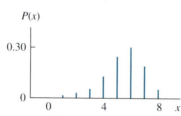

(c) $n = 15$, $p = 0.5$

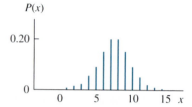

7.3

POISSON PROBABILITY DISTRIBUTIONS

The Poisson probability distribution is useful for many random phenomena dealing with the number of events occurring in a fixed time interval.

☐ EXAMPLES

1. The number of machines X in a plant that break down during a day is a Poisson random variable. Management wishes to know how likely it is that there are no breakdowns in a day.

2. The number of units X of an item sold from stock during a week is a Poisson random variable. The inventory controller wishes to know the probability that more than five units will be sold in a week.

3. The number of persons X arriving at an automatic bank teller between 9 A.M. and 10 A.M. is a Poisson random variable. A systems consultant needs to know the probability that more than 10 persons will arrive during the one-hour period.   ☐

The Poisson probability distribution also applies to random occurrences that are not related directly to time, such as the number of typographical errors on a page or the number of aphids on a plant.

Poisson Probability Function

A Poisson random variable is a discrete variable that can take on any integer value from 0 to infinity. The Poisson probability function follows.

(7.8)
The **Poisson probability function** is:

$$P(x) = \frac{\lambda^x \exp(-\lambda)}{x!}$$

where: $x = 0, 1, \ldots$
$0 < \lambda < \infty$

Here and elsewhere in the text, $\exp(a)$ represents e^a, where $e = 2.71828\ldots$ is the base of natural logarithms. Thus, $\exp(-\lambda)$ represents $e^{-\lambda}$. (A review of exponents and logarithms can be found in Appendix A, Section A.2.)

The Poisson probability distribution has only one parameter, λ (Greek lambda), which may be any positive number. Even though a Poisson random variable may assume any indefinitely large integer value, it is often useful for modeling the number of occurrences in real-world phenomena that are bounded. We shall demonstrate this by an example.

EXAMPLE  **Crimes**

The number of crimes occurring in a city district in the one-hour period between 1 A.M. and 2 A.M. is a Poisson random variable with $\lambda = 0.2$. We obtain the probability that no crime occurs during this period, $P(0)$, by substituting into (7.8):

$$P(0) = \frac{(0.2)^0 \exp(-0.2)}{0!} = \frac{(1)(0.8187)}{1} = 0.8187$$

The probabilities of other possible outcomes are obtained in similar fashion by using (7.8):

x	$P(x)$
0	$\dfrac{(0.2)^0 \exp(-0.2)}{0!} = 0.8187$
1	$\dfrac{(0.2)^1 \exp(-0.2)}{1!} = 0.1637$
2	$\dfrac{(0.2)^2 \exp(-0.2)}{2!} = 0.0164$
3	$\dfrac{(0.2)^3 \exp(-0.2)}{3!} = 0.0011$
4	$\dfrac{(0.2)^4 \exp(-0.2)}{4!} = 0.0001$

We see, for instance, that the probability of four crimes occurring during this hour is very small, $P(4) = 0.0001$. On the other hand, the probability of two or fewer crimes occurring during the hour is very large:

$$P(X \le 2) = P(0) + P(1) + P(2)$$
$$= 0.8187 + 0.1637 + 0.0164 = 0.9988$$

For outcomes of five crimes or more, the probabilities of occurrence are so small that we have not shown calculations. Here is the reason why the Poisson probability distribution is often a useful model for bounded phenomena, namely, because the probabilities of large values are negligibly small. $\square$

Probability Table

Poisson probabilities can be computed readily with many calculators and statistical packages. They can also be obtained from Poisson probability tables. Table C.6 contains Poisson probabilities for selected values of λ up to 20.

☐ EXAMPLES

1. Daily demand for a replacement part of a videocassette recorder model is Poisson distributed with $\lambda = 0.9$. We see from Table C.6 that the probability of no demand for the part on any day is $P(0) = 0.4066$. This value is found in the column headed $\lambda = 0.9$ and the row marked $x = 0$.

    For cumulative probabilities, we need only add the appropriate individual probabilities. For example, we find the probability that 4 or fewer replacement parts are demanded on any day as follows:

$$P(X \leq 4) = P(0) + P(1) + \cdots + P(4)$$
$$= 0.4066 + 0.3659 + \cdots + 0.0111 = 0.9977$$

2. The number of accidents in an office building during a four-week period is a Poisson random variable with $\lambda = 2$. Interest is in the probability that there is 0 or 1 accident in a four-week period. We find from Table C.6:

$$P(X \leq 1) = P(0) + P(1) = 0.1353 + 0.2707 = 0.4060 \qquad \square$$

Characteristics of Poisson Probability Distributions

Mean and Variance.   The mean and the variance of a Poisson probability distribution depend on the parameter λ in a simple way.

(7.9)

The mean and the variance of a Poisson probability distribution are:

(7.9a) $E\{X\} = \lambda$

(7.9b) $\sigma^2\{X\} = \lambda$

Thus, the mean and the variance of a Poisson distribution are equal.

☐ EXAMPLE

For the crimes example, the Poisson parameter is $\lambda = 0.2$, so the mean number of crimes per hour is $E\{X\} = 0.2$ crime and the variance is $\sigma^2\{X\} = 0.2$. The standard deviation therefore is $\sigma\{X\} = \sqrt{\lambda} = \sqrt{0.2} = 0.447$ crime, which is more than twice as large as the mean and shows that this distribution has relatively large variability. ☐

Distribution Shape.   Figure 7.4 contains three Poisson probability distributions, with parameter values $\lambda = 0.3$, 2.0, and 5.0. All Poisson probability distributions are skewed to the right, although they become more symmetrical as λ becomes larger.

Comment

In some applications, we are interested in the sum of two or more independent Poisson random variables. The following theorem is helpful then.

Three Poisson probability distributions.   Right-skewness decreases as λ increases.

FIGURE 7.4

(a) $\lambda = 0.3$ (b) $\lambda = 2.0$ (c) $\lambda = 5.0$

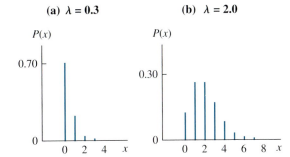

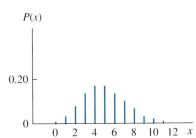

(7.10)
The sum $X_1 + X_2 + \cdots + X_n$ of n independent Poisson random variables with parameters λ_1, $\lambda_2, \ldots, \lambda_n$, respectively, is also a Poisson random variable, with parameter $\lambda = \lambda_1 + \lambda_2 + \cdots + \lambda_n$.

We illustrate this theorem by considering broadcasting interruptions at a television station that arise from equipment breakdowns and human errors. The numbers of interruptions per 1000 broadcasting hours from these two sources are independent Poisson random variables with parameters $\lambda_1 = 0.4$ and $\lambda_2 = 1.1$, respectively. By theorem (7.10), we know that the total number of interruptions from both sources is a Poisson random variable with parameter $\lambda = 0.4 + 1.1 = 1.5$. Thus, the probability of no interruptions from either source in the next 1000 hours of broadcasting is $P(0) = 0.2231$ (refer to Table C.6, column $\lambda = 1.5$, row $x = 0$).

The hypergeometric probability distribution is a discrete distribution that finds applications in statistical sampling. Like the binomial distribution, the hypergeometric distribution pertains to a random variable that is a sum of Bernoulli random variables. Unlike the binomial distribution, the constituent Bernoulli random variables are related in a special way and, hence, are not independent.

**OPTIONAL   7.4
TOPIC—
HYPERGEOMETRIC
PROBABILITY
DISTRIBUTIONS**

EXAMPLE   ☐   **Senate Committee**

A state senate consists of $N = 50$ senators, of whom $C = 10$ are women. What is the probability that a committee of $n = 5$ senators formed at random contains no women?

The total number of women on the senate committee will be denoted by X. It can be represented by the sum of Bernoulli random variables $X = B_1 + B_2 + \cdots + B_5$, where $B_i = 1$ if the ith senator assigned to the committee is female and $B_i = 0$ otherwise. The probability $P(X = 0)$ is required. The binomial probability distribution is not applicable here even though the probability of the ith senator being female is the same for each Bernoulli random variable, namely, $P(B_i = 1) = 10/50$. The reason is that the Bernoulli random variables are dependent. For example, the conditional probability of the second senator selected being female, given that the first senator selected is female, is $P(B_2 = 1 \mid B_1 = 1) = 9/49$, which differs from the marginal probability, $P(B_2 = 1) = 10/50$.

☐

Electric Generators ☐ **EXAMPLE**

Of the $N = 15$ generators in an electricity generating plant, $C = 2$ have a defective rotor shaft. If $n = 4$ different generators are selected at random for inspection, what is the probability that exactly $X = 1$ is found to have a defective shaft?

☐

Hypergeometric Probability Function

A hypergeometric random variable is a discrete random variable. Its probability function depends on the parameters N, C, and n, all of which are integers.

(7.11)

The **hypergeometric probability function** is:

$$P(x) = \frac{\binom{C}{x}\binom{N-C}{n-x}}{\binom{N}{n}}$$

where: $0 \le x \le C$
        $0 \le n - x \le N - C$

The binomial coefficient $\binom{a}{b}$ is defined in (7.6a). (A review of permutations and combinations can be found in Appendix A, Section A.3.) Note that the possible outcomes of the random variable X depend on C, N, and n.

☐ **EXAMPLES**

1. In the senate committee example, we have $N = 50$, $C = 10$, and $n = 5$. The possible outcomes of X, the number of women on the committee, are $0, 1, \ldots, 5$. We wish to find $P(X = 0)$. Substituting $x = 0$ into (7.11), we obtain:

$$P(0) = \frac{\binom{10}{0}\binom{40}{5}}{\binom{50}{5}} = \frac{\frac{10!}{0!10!}\frac{40!}{5!35!}}{\frac{50!}{5!45!}} = 0.311$$

Thus, the probability is 0.311 that a committee of five senators formed at random will not contain any women.

2. In the electric generators example, we have $N = 15$, $C = 2$, and $n = 4$. The possible outcomes of X, the number of inspected generators with a defective shaft, are 0, 1, or 2. We wish to find $P(X = 1)$. Substituting $x = 1$ into (7.11), we obtain:

$$P(1) = \frac{\binom{2}{1}\binom{13}{3}}{\binom{15}{4}} = \frac{\frac{2!}{1!1!}\frac{13!}{3!10!}}{\frac{15!}{4!11!}} = 0.419$$

Thus, the probability is 0.419 that, among four generators selected at random, exactly one will be found to have a defective shaft. □

Characteristics of Hypergeometric Probability Distributions

In what follows, we let p denote the proportion C/N. In the senate committee example, $p = 10/50 = 0.20$ is the proportion of senators who are women. In the electric generators example, $p = 2/15 = 0.133$ is the proportion of generators with a defective shaft.

(7.12)
The mean and the variance of a hypergeometric probability distribution are:

(7.12a) $E\{X\} = np$

(7.12b) $\sigma^2\{X\} = \left(\dfrac{N - n}{N - 1}\right)np(1 - p)$

where: $p = \dfrac{C}{N}$

Note the similarity of these formulas to those for the mean and variance of a binomial random variable in (7.7a) and (7.7b). The formulas here are identical except for the term $(N - n)/(N - 1)$ in formula (7.12b).

═══════════
EXAMPLES □

1. In the senate committee example, $p = 0.20$, $n = 5$, and $N = 50$. Hence:

   $E\{X\} = 5(0.20) = 1.0$

   $\sigma^2\{X\} = \left(\dfrac{50 - 5}{50 - 1}\right)(5)(0.20)(0.80) = 0.735$

2. In the electric generators example, $p = 0.133$, $n = 4$, and $N = 15$. Hence:

   $E\{X\} = 4(0.133) = 0.53$

   $\sigma^2\{X\} = \left(\dfrac{15 - 4}{15 - 1}\right)(4)(0.133)(0.867) = 0.363$ □

When N is large relative to n, the hypergeometric probability function (7.11) is approximated well by the binomial probability function (7.6) with the same values of n and p.

☐ **EXAMPLE**

In the senate committee example, $N = 50$ is large relative to $n = 5$. Hence we can approximate the hypergeometric probability function by the binomial probability function with parameters $n = 5$ and $p = 0.20$. For instance, the approximating binomial probability for $P(X = 0)$ is found from Table C.5 to be 0.3277. The exact hypergeometric probability was calculated earlier as 0.311.

☐

PROBLEMS

* **7.1** The number of students who participate on any day in discussion in Professor Bird's class is a discrete uniform random variable X with $a + 1 = 0$ and $s = 14$.

a. Plot the probability distribution of X.
b. Obtain each of the following probabilities: (1) $P(0)$, (2) $P(X > 10)$, (3) $P(1 \leq X \leq 5)$.
c. Obtain $E\{X\}$ and $\sigma\{X\}$. Interpret each of these measures.

7.2 A computer random number generator produces one of the numbers 000, 001, ... , 999 with equal probabilities.

a. What are the parameter values of the discrete uniform probability distribution for this random variable?
b. What is the probability that number 123 will be generated? That a number starting with the digit 1 will be generated?
c. What are the mean and the standard deviation of the probability distribution?

7.3 Insurance Calculation. The calculation of a home insurance premium involves seven consecutive computational steps. An auditor is reviewing a large number of premium calculations that contain an error. Let X denote the step at which the error is made. Assume that X has a discrete uniform probability distribution.

a. Plot the probability distribution of X.
b. On the average, how many steps must the auditor check in each premium calculation to find the error?
c. Obtain $\sigma\{X\}$. In what units is this measure expressed?

7.4 Refer to **Insurance Calculation** Problem 7.3.

a. If locations of the errors for different premium calculations are independent, what is the probability that two premium calculations have errors that occur at step 4?
b. Would you actually expect an error to occur with equal probability at each step? Explain.

7.5 Which of the following random trials is a Bernoulli trial: (1) the inspection classification (accept, rework, or scrap) for a component produced on an assembly line, (2) the gender of a chick in a brood bred for egg production, (3) the weight of a package of ground beef in a supermarket?

7.6 Which of the following random trials is a Bernoulli trial: (1) the absence or presence of an employee at work on a particular day, (2) the number of patients who visit a hospital eye clinic during a week, (3) the status of an indicator light (on, off) in the control panel of an oil refinery?

* **7.7** There are 9 participants in a 10-mile run for charity. Let $B_i = 1$ if the ith participant finishes the run and $B_i = 0$ otherwise.

a. Under what conditions does the sequence $B_1, \ldots, B_9$ constitute a Bernoulli process? Would you expect these conditions to be satisfied here? Comment.
b. Assuming that the sequence in part a is a Bernoulli process and that the probability of finishing the run is 0.9 for each of the nine participants, what is the probability that all participants except the 3rd and 8th finish the run?

7.8 Birth Order. A family has three children. Let $B_i = 1$ if the ith child in order of birth is a girl and $B_i = 0$ otherwise.

1 2 7 1 2 1 7

a. Under what conditions does the sequence B_1, B_2, B_3 constitute a Bernoulli process?

b. Assuming that the sequence in part a is a Bernoulli process and that the probability of a girl is 0.55 for each of the three children, what is the probability that only the middle child is a girl?

7.9 Shirt Buttons. A sample of five shirts is taken from a production line. Let $B_i = 1$ if the ith shirt has a missing button and $B_i = 0$ otherwise.

a. Under what conditions does the sequence $B_1, \ldots, B_5$ constitute a Bernoulli process?

b. Assuming that the sequence in part a is a Bernoulli process and that the probability of a missing button is 0.005 for each of the five shirts, what is the probability that (1) none of the shirts has a missing button; (2) only the fifth shirt has a missing button?

7.10 Refer to **Birth Order** Problem 7.8. What is represented by the random variable $X = B_1 + B_2 + B_3$? Under what conditions is X a binomial random variable?

7.11 Refer to **Shirt Buttons** Problem 7.9. What is represented by the random variable $X = B_1 + \cdots + B_5$? Under what conditions is X a binomial random variable?

* **7.12 Inaccurate Gauges.** The number of inaccurate gauges in a batch of four is a binomial random variable X with $n = 4$ and $p = 0.05$.

a. Describe the underlying Bernoulli trial. What different values can X take on (that is, what is its sample space)?

b. Using (7.6), obtain the following: (1) $P(0)$, (2) $P(2)$, (3) $P(X \le 2)$. Confirm your answers by using Table C.5.

7.13 Transmission Work. The number of trucks in a fleet of eight that will require major transmission work next year is a binomial random variable X with $n = 8$ and $p = 0.2$.

a. Describe the underlying Bernoulli trial. What different values can X take on (that is, what is its sample space)?

b. Using (7.6), obtain the following: (1) $P(0)$, (2) $P(3)$, (3) $P(X \le 3)$. Confirm your answers by using Table C.5.

7.14 The number of graduate students in a group of six who will incur less than $1000 of research computing charges next term is a binomial random variable X with $n = 6$ and $p = 0.65$.

a. Describe the underlying Bernoulli trial. What does the outcome $X = 5$ represent?

b. Using (7.6), obtain the following: (1) $P(5)$, (2) $P(X \le 2)$, (3) $P(X > 4)$. Confirm your answers by using Table C.5.

7.15 Use Table C.5 to obtain the following binomial probabilities: (1) $P(1)$ for $p = 0.09$, $n = 10$; (2) $P(3)$ for $p = 0.35$, $n = 6$; (3) $P(8)$ for $p = 0.95$, $n = 9$; (4) $P(X \ge 4)$ for $p = 0.8$, $n = 6$.

7.16 Use Table C.5 to obtain the following binomial probabilities: (1) $P(2)$ for $p = 0.05$, $n = 12$; (2) $P(2)$ for $p = 0.25$, $n = 8$; (3) $P(3)$ for $p = 0.90$, $n = 7$; (4) $P(X \le 2)$ for $p = 0.7$, $n = 10$.

* **7.17** Refer to **Inaccurate Gauges** Problem 7.12.

a. Obtain $E\{X\}$, $\sigma^2\{X\}$, and $\sigma\{X\}$. In what units is $\sigma\{X\}$ expressed?

b. Plot the probability distribution of X. Is the distribution skewed? If so, in which direction?

7.18 Refer to **Transmission Work** Problem 7.13.

a. Obtain the mean and the standard deviation of the probability distribution of X. In what units are these measures expressed?

b. Plot the probability distribution of X. Is the distribution skewed? If so, in which direction?

7.19 The number of persons in a university class of 250 students who will be employed one year after graduation is a binomial random variable X with $n = 250$ and $p = 0.98$.

a. What are the mean and the standard deviation of the probability distribution of X?

b. Is the number of persons in the class who will not be employed one year after graduation also a binomial random variable? What are the mean and the standard deviation of the number not employed?

7.20 In a survey of 15 manufacturing firms, the number of firms that use LIFO (a last-in first-out accounting procedure for inventory) is a binomial random variable X with $n = 15$ and $p = 0.2$.

a. What is the probability that five or fewer firms will be found to use LIFO? Is it unlikely that more than 10 firms will be found to use LIFO? Comment.

b. Plot the probability distribution of X. Is the distribution highly skewed?

c. What are the mean and the standard deviation of X? What is the coefficient of variation of X? What would be the value of this coefficient if 150 firms were surveyed rather than 15 (continue to assume that $p = 0.2$)? Describe the effect of the larger survey on the relative variability of X.

7.21 In a group of 12 patients scheduled for dental appointments tomorrow morning, the number of patients who will cancel their appointments is a binomial random variable X with $n = 12$ and $p = 0.05$.

a. What is the underlying Bernoulli trial here?

b. What is the probability that exactly 2 out of the 12 appointments will be canceled? That two or fewer will be canceled?

c. What are the mean and the variance of the number of cancellations? Are these two measures almost equal here? Does (7.7) assure this result for any binomial random variable when p is small? Explain.

7.22 Which of the following random trials is possibly a Poisson trial: (1) the thickness of a coat of paint on a metal sheet, (2) the number of trout in a lake, (3) the number of blue ink pens among the next five pens sold in a store?

7.23 Which of the following random trials is possibly a Poisson trial: (1) the number of women among six scholarship winners, (2) the length of a crack in a timber, (3) the pollen count in a cubic centimeter of air?

* **7.24 Police Calls.** The number of calls to a police dispatcher between 8:00 P.M. and 8:30 P.M. on Fridays is a Poisson random variable X with $\lambda = 3.5$.

a. Using (7.8), obtain the probability (1) of no calls during this period, (2) of two calls, (3) of two or fewer calls. Confirm your answers by using Table C.6.

b. What is the expected number of calls received by the dispatcher during this period? What is the variance of the number of calls?

c. Plot the probability distribution of X. Is the distribution skewed? If so, in which direction?

7.25 The number of irregularities in one kilometer of optical fiber is a Poisson random variable X with $\lambda = 0.8$.

a. Using (7.8), obtain the following: (1) $P(0)$, (2) $P(1)$, (3) $P(X \le 1)$. Confirm your answers by using Table C.6.

b. Obtain $E\{X\}$ and $\sigma\{X\}$. In what units are these measures expressed?

c. Plot the probability distribution of X. Is the distribution skewed? If so, in which direction?

7.26 Facsimile Unit Warranty. A new facsimile unit is fully warranted during its first year. The number of warranty service calls for a unit during its first year is a Poisson random variable X with $\lambda = 2.0$.

a. Using (7.8), obtain the probability (1) of exactly two warranty calls for a unit during the first year, (2) of four or more warranty calls. Confirm your answers by using Table C.6.

b. What are the mean and the standard deviation of the probability distribution of X? In what units are these measures expressed?

c. Plot the probability distribution of X. Is the distribution skewed? If so, in which direction?

7.27 Use Table C.6 to obtain the following: (1) $P(3)$ for $\lambda = 3.5$, (2) $P(X \le 2)$ for $\lambda = 1.5$, (3) the value of λ for which $P(0) = 0.0041$.

7.28 Use Table C.6 to obtain the following: (1) $P(4)$ for $\lambda = 2.0$, (2) $P(X > 4)$ for $\lambda = 0.8$, (3) the most probable outcome for $\lambda = 6.5$.

7.29 The numbers of prescriptions for a certain medication that will be written tomorrow by two physicians are denoted by X_1 and X_2. Assume that X_1 and X_2 are independent Poisson random variables with $\lambda_1 = 2.5$ and $\lambda_2 = 1.5$, respectively. Let $T = X_1 + X_2$.

a. What does the random variable T represent here?

b. Obtain $P(T = 2)$. Is this probability the same as the joint probability $P(X_1 = 1 \cap X_2 = 1)$? Explain.

c. Obtain the following: (1) $E\{T\}$, (2) $\sigma\{T\}$. In what units are these measures expressed?

7.30 Wallpaper Blemishes. The number of coating blemishes in 10-square-meter rolls of customized wallpaper is a Poisson random variable X_1 with $\lambda_1 = 0.3$. The number of printing blemishes in these 10-square-meter rolls of customized wallpaper is a Poisson random variable X_2 with $\lambda_2 = 0.1$. Assume that X_1 and X_2 are independent and let $T = X_1 + X_2$.

a. What does the random variable T represent here?
b. What is the most probable total number of blemishes in a roll?
c. If rolls with a total of two or more blemishes are scrapped, what is the probability that a roll will be scrapped?
d. What are the mean and the standard deviation of the probability distribution of T?

* 7.31 An accounting department has six tenured and four untenured faculty members. A committee of $n = 3$ members is to be selected at random. Let X denote the number of committee members who are tenured.

a. What different values can X take on (that is, what is its sample space)?
b. What is the probability that: (1) exactly one of the committee members is tenured; (2) the number of tenured members on the committee exceeds the number of untenured members?
c. Obtain the mean and the standard deviation of the number of committee members who are tenured.

7.32 Two of the 12 generators in a power plant have worn rotor shafts. Three of the 12 generators are selected at random for inspection. Let X denote the number of generators selected for inspection that have worn rotor shafts.

a. What different values can X take on (that is, what is its sample space)?
b. Obtain the probability distribution of X.
c. Calculate the mean and the variance of X from the probability distribution in part b. Verify that the same values are obtained from (7.12).

7.33 Fluid Trading. An investment analyst uses the term "fluid trading" to describe a certain trading condition for stocks. In a group of 20 stocks, 8 exhibit fluid trading. The analyst will select 5 stocks at random from the group of 20. Let X denote the number of selected stocks that exhibit fluid trading.

a. What different values can X take on (that is, what is its sample space)?
b. Obtain (1) $P(0)$, (2) $P(3)$, (3) $P(X \le 3)$.
c. What are the mean and the standard deviation of X?

7.34 Refer to **Fluid Trading** Problem 7.33. Approximate $P(3)$ in part b by a binomial probability. Is N large enough here for the binomial probability to be a good approximation? Comment.

7.35 Consider the hypergeometric probability function with $N = 100$, $C = 10$, and $n = 3$.

a. Calculate $P(2)$, using (7.11).
b. Approximate $P(2)$ by a binomial probability. Is the approximation reasonably close here? Comment.

EXERCISES

7.36 Let $B_i (i = 1, 2, \ldots, n)$ be independent and identically distributed Bernoulli random variables with $P(B_i = 1) = p$.

a. Show that $\sigma^2\{B_i\} = p(1 - p)$.
b. Use the result in part a to verify (7.7b).

7.37 For binomial random variable X with $n = 20$, obtain $\sigma\{X\}$ for $p = 0, 0.1, \ldots, 0.9, 1.0$. Plot these $\sigma\{X\}$ values on a graph as a function of p. What effect does the magnitude of p have on the variability of the probability distribution?

7.38 In a sports match, two teams (1, 2) play each other in consecutive games until one team has won two games more than the other. Tied games are impossible. Let $B_i = 1$ if team 1 wins game i and $B_i = 0$ if team 2 wins game i. Assume that the game outcomes B_i are independent and that $P(B_i = 1) = 0.6$ for all i. Consider a match that ends in exactly four games.

a. Is $X = B_1 + \cdots + B_4$ a binomial random variable for this match? Explain.
b. Obtain the probability distribution of X in part a.

7.39 Each use of an electric switch is an independent Bernoulli trial in which the switch either functions ($B = 0$) or fails ($B = 1$). Let $p = P(B = 1)$. Consider the associated Bernoulli process, and let Y be a random variable representing the number of trials up to and including the trial in which the switch fails for the first time. Show that the probability function of Y is:

$$P(y) = p(1 - p)^{y-1} \qquad y = 1, 2, \ldots$$

7.40 Prove (7.9a). $\left[\textit{Hint: } \exp(\lambda) = \sum_{x=0}^{\infty} \lambda^x/x!. \right]$

7.41 Prove that a Poisson probability distribution has a mode at 0 (that is, its maximum probability is at 0) only if $\lambda < 1$.

STUDIES

7.42 A glass company has received an order for two large lenses that must be specially cast. The probability that a given lens will prove to be acceptable (that is, free from flaws) when the glass has cooled is 0.7. The number of acceptable lenses X in a casting run of n lenses is a binomial random variable. Management wishes to know how many lenses to cast in the run to have a probability of at least 0.99 that a minimum of two lenses are acceptable. Find the smallest size of the casting run that will provide this assurance.

7.43 Ten persons in a taste-test experiment will be served a portion of sausage prepared according to the current recipe (recipe 1) and another portion prepared from a recipe designed to give the same taste as the current recipe but having a longer shelf life (recipe 2). The order of the two servings will be randomized. Each person must independently state after tasting the two recipes which of the two (1, 2) is preferred. Let X denote the number of persons of the 10 who prefer recipe 2.

a. Describe the underlying Bernoulli trial here. Does X necessarily follow a binomial distribution here?
b. Suppose that recipe 2 has exactly the same taste as recipe 1, so that each person will prefer recipes 1 or 2 with equal probabilities. What is the probability, then, that the proportion X/n of persons in the study who prefer recipe 2 lies between 0.4 and 0.6 inclusive?
c. How much larger would be the probability in part b if 20 persons were included in the experiment? What are the implications of this result?

7.44 The numbers of typographical errors on each of the n pages of a book are independent and identically distributed Poisson random variables, each with $\lambda = 0.2$. Let T denote the total number of errors on the n pages.

a. Obtain the mean, the standard deviation, and the coefficient of variation of T for (1) $n = 10$, (2) $n = 100$, (3) $n = 1000$. From a comparison of the coefficients of variation for these three cases, what is the effect on the relative variability of T as the number of pages n increases? Discuss.
b. What is the probability that T lies within one standard deviation of its mean value for each of the three cases in part a? Does this probability appear to be approaching a limit as n increases? Comment.

7.45 The daily number of disabling breakdowns in cabs of a taxi fleet is a Poisson random variable X with $\lambda = 0.7$. The loss of revenue to the company on the day of the breakdown of a taxi is $200 if a standby cab is not available and $0 if a standby cab is available. Disabled cabs can be repaired by the next day. Standby cabs (without drivers) can be leased for $70 per day. Assume that the probability of a breakdown in a standby cab is negligibly small.

a. Let C denote the daily revenue loss from breakdowns plus the cost of leasing standby cabs, and let m denote the number of standby cabs leased for the day. Express C as a function of X and m.

b. Evaluate $E\{C\}$ for $m = 0, 1, 2, 3$.

c. How many standby cabs should the company have on hand at the beginning of a day to minimize the expected value of C?

Common Probability Distributions: Continuous Random Variables

8

In this chapter, we take up three important families of continuous probability distributions.

The discrete uniform probability distribution discussed in Section 7.1 has a continuous analog known as the *continuous uniform probability distribution*. This continuous distribution is also sometimes referred to as the *rectangular probability distribution*. This distribution was used in the crop yield example in Figure 5.3. The continuous uniform distribution is useful in a variety of situations, including in random number generation for simulation studies.

Density Function

A *continuous uniform random variable* can take any value in an interval from a to b and has a uniform (i.e., constant) probability density over this interval. The lower endpoint a and the upper endpoint b of the interval are the parameters of the density function.

(8.1)

The **continuous uniform probability density function** is:

$$f(x) = \frac{1}{b - a}$$

where $a \leq x \leq b$.

 **EXAMPLE** □ **Population**

A geographer has found that a continuous uniform distribution is a good approximation for the distribution of population sizes for towns with populations between 5000 and 7500 persons. The density function here is:

$$f(x) = \frac{1}{2500} \qquad 5000 \leq x \leq 7500$$

Note that the two parameters here are $b = 7500$ and $a = 5000$. A graph of this density function is shown in Figure 8.1. As the graph illustrates, the use of the term *rectangular* in the name of this distribution is quite descriptive. □

Characteristics of Continuous Uniform Probability Distributions

The mean and the variance of a continuous uniform distribution depend on the parameters a and b.

(8.2)
The mean and the variance of a continuous uniform probability distribution are:

(8.2a) $\quad E\{X\} = \dfrac{b + a}{2}$

(8.2b) $\quad \sigma^2\{X\} = \dfrac{(b - a)^2}{12}$

□ **EXAMPLE**

The continuous uniform distribution for the population example in Figure 8.1 has mean and variance:

$$E\{X\} = \frac{7500 + 5000}{2} = 6250 \text{ persons}$$

$$\sigma^2\{X\} = \frac{(7500 - 5000)^2}{12} = 520{,}833$$

The standard deviation is $\sigma\{X\} = \sqrt{520{,}833} = 721.7$ persons. Note that the mean falls at the midpoint of the interval between 5000 and 7500, as is to be expected in view of the uniform density over the interval. □

Determining Probabilities and Percentiles

Recall from Chapter 5 that the probability of a continuous random variable X taking on a value in a specified interval is found by determining the corresponding area under the

FIGURE 8.1
Continuous uniform probability distribution with $a = 5000, b = 7500$—Population example

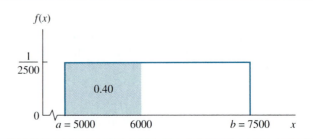

probability density function $f(x)$. Any desired probability for a continuous uniform distribution may be found very easily by using the cumulative probability function.

> **(8.3)**
> The cumulative probability function $F(x)$ for a continuous uniform random variable X is:
>
> $$F(x) = P(X \le x) = \frac{x - a}{b - a}$$
>
> where $a \le x \le b$.

Percentiles can also be readily found from this cumulative probability function.

EXAMPLE □

For the population example in Figure 8.1, the probability that a town in the 5000–7500 population group has a population of 6000 persons or less is desired. This probability is:

$$F(6000) = P(X \le 6000) = \frac{6000 - 5000}{7500 - 5000} = 0.40$$

This probability is shown as the shaded area in Figure 8.1. Thus, the 40th percentile of the distribution of population sizes for towns in this group is 6000 persons.

To find the 75th percentile of the probability distribution, we simply write:

$$F(x) = \frac{x - 5000}{7500 - 5000} = 0.75$$

and solve for x. We find that the 75th percentile of the probability distribution is 6875 persons.

□

Standard Uniform Distribution

The continuous uniform distribution for the interval from 0 to 1—that is, the distribution with parameters $a = 0$ and $b = 1$—is called the *standard uniform distribution*. Many calculators and statistical packages have routines for generating random outcomes from this distribution. A probability theorem then permits the random outcomes from this standard uniform distribution to be converted to random outcomes from any probability distribution for use in simulation studies.

The family of normal probability distributions is one of the most important in statistics. Many random phenomena involve a *normal random variable*.

NORMAL PROBABILITY DISTRIBUTIONS

8.2

☐ **EXAMPLES**

1. The temperature X at noon on August 15 in a Southeastern city is a normal random variable. A weather forecaster wishes to know the probability that the noon temperature will exceed 35° C.

2. The weight X of a metal ingot produced in a smelter is a normal random variable. An engineer needs to know the probability that the weight of an ingot will be between 500 and 540 pounds.

3. The height X of women who are 20–29 years old is a normal random variable. A clothing designer wishes to know the proportion of women in this age group who are less than 5 feet tall.   ☐

In addition to describing many random phenomena, the normal distribution is also used frequently in drawing inferences from data, as will be seen in later chapters.

Density Function

A normal random variable is a continuous random variable that can take on any value between minus infinity and plus infinity. It has the following density function.

(8.4)

The **normal probability density function** is:

$$f(x) = \frac{1}{\sqrt{2\pi}\sigma} \exp\left[-\frac{1}{2}\left(\frac{x - \mu}{\sigma}\right)^2 \right]$$

where: $-\infty < x < \infty$
$-\infty < \mu < \infty$
$\sigma > 0$
$\pi = 3.14159\ldots$

The notation $\exp(a)$ represents e^a, where $e = 2.71828\ldots$ is the base of natural logarithms. (A review of exponents and logarithms can be found in Appendix A, Section A.2.)

Even though a normal random variable is not bounded in magnitude, we shall show that it is, nevertheless, a good model for many bounded real-world phenomena.

Characteristics of Normal Probability Distributions

Distribution Shape.   The normal distribution has two parameters, μ (Greek mu) and σ (Greek sigma), with σ positive. Each different pair of (μ, σ) values corresponds to a different probability distribution of the family of normal distributions. The normal distribution is bell-shaped and symmetrical, as illustrated in Figure 8.2. The normal distribution is centered at μ, which is the mean of the distribution and determines its position on the x axis. The parameter σ is the standard deviation of the normal distribution and determines its variability—the larger is the value of σ, the greater is the spread of the distribution. The examples in Figure 8.2 illustrate these facts. The distributions in Figures 8.2a and 8.2b have the same standard deviation but different means, while the distributions in Figures 8.2b and 8.2c have the same mean but different standard deviations.

Figure 8.2 also shows that almost all of the probability in a normal distribution is located in a limited range about its mean, so the probability of an observation falling outside this range is practically zero. It is for this reason that the normal distribution is useful for modeling bounded phenomena, such as the heights of females aged 20–29 or the temperatures at noon on August 15 in a city.

Mean and Variance.   As the foregoing discussion has indicated, the mean and the variance of a normal distribution are related to its parameters as follows.

(8.5)

The mean and the variance of a normal probability distribution are:

(8.5a) $E\{X\} = \mu$

(8.5b) $\sigma^2\{X\} = \sigma^2$

Three normal probability distributions.   The mean μ determines the position and the standard deviation σ determines the variability of the distribution.

FIGURE 8.2

(a) $\mu = 56.0,\ \sigma = 2.7$

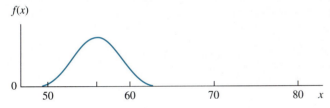

(b) $\mu = 66.5,\ \sigma = 2.7$

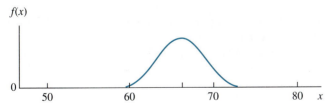

(c) $\mu = 66.5,\ \sigma = 4.1$

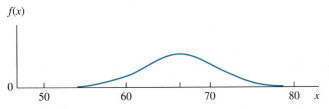

Note that we have used the same Greek symbol σ^2 in the notation for the variance of random variable X as for the variance of a normal distribution. The context and the notation will make it clear whether the symbol refers to the variance of a random variable or to the parameter of a probability distribution.

$N(\mu, \sigma^2)$ Notation.   The normal distribution is used frequently throughout the text, so we shall adopt the compact notation $N(\mu, \sigma^2)$ to denote a normal distribution with mean μ and variance σ^2. Thus, when we state that X is distributed as $N(100, 20)$, we mean that X is a normal random variable with mean $\mu = 100$ and variance $\sigma^2 = 20$.

Standard Normal Probability Distribution

The *standard normal probability distribution* is that particular member of the family of normal distributions that has mean $\mu = 0$ and variance $\sigma^2 = 1$. The normal random variable corresponding to the standard normal distribution is called the *standard normal variable*. Thus, in terms of our compact notation, the standard normal variable is distributed as $N(0, 1)$.

The significance of the standard normal distribution results from the fact that any normal random variable can be transformed into the standard normal variable by means of the standardizing transformation in (5.9).

> **(8.6)**
> The standardized form of a normal random variable X with mean μ and standard deviation σ is denoted by Z and defined:
>
> $$Z = \frac{X - \mu}{\sigma}$$

The standardized form Z of any normal random variable X is a standard normal variable for the following reasons:

1. We know from Chapter 5 that $E\{Z\} = 0$ and $\sigma^2\{Z\} = 1$ since these properties hold for any standardized variable.
2. From (8.6) it can be seen that Z is a linear function of X:

$$Z = -\frac{\mu}{\sigma} + \left(\frac{1}{\sigma}\right)X$$

3. The following theorem tells us then that Z is a normal random variable.

> **(8.7)**
> Any linear function of a normal random variable is also a normal random variable.

Thus, the standardized form of any normal random variable is distributed as $N(0, 1)$.

Determining Probabilities and Percentiles for Standard Normal Distribution

The standardizing transformation in (8.6) permits us to obtain probabilities and percentiles for any normal distribution from those for the standard normal distribution. The probabilities and percentiles for the standard normal distribution can be obtained with many types of calculators and statistical packages. They also can be obtained from a probability table for the standard normal distribution, such as Table C.1. We now explain Table C.1.

Standard Normal Probability Table.   In accordance with our earlier notation, we employ the lowercase letter z to denote a particular outcome of the standard normal random variable Z. The row and column labels in Table C.1 specify different z outcomes, the row label providing the first decimal place value and the column label the second decimal place value. Each cell entry for a given value z is the cumulative probability $P(Z \leq z)$. This cumulative probability is represented by the shaded area labeled a in the figure at the top of Table C.1. The percentile corresponding to cumulative area a is denoted by $z(a)$.

We can use Table C.1 in two ways: (1) to determine the area a that corresponds to a specified value of z, and (2) to find the percentile $z(a)$ that corresponds to a specified area a. We begin by explaining the first use.

Probabilities.   Figure 8.3 contains several examples of finding probabilities for the standard normal distribution.

EXAMPLES ☐

1. Refer to Figure 8.3a. We wish to find $P(Z \leq 0.45)$. In entering Table C.1 at the row labeled 0.4 and the column labeled 0.05, we find, for $z = 0.45$, the cumulative area $a = 0.6736$. Hence, $P(Z \leq 0.45) = 0.6736$. It is reasonable that this probability

Determining probabilities for the standard normal distribution $N(0, 1)$ **FIGURE 8.3**

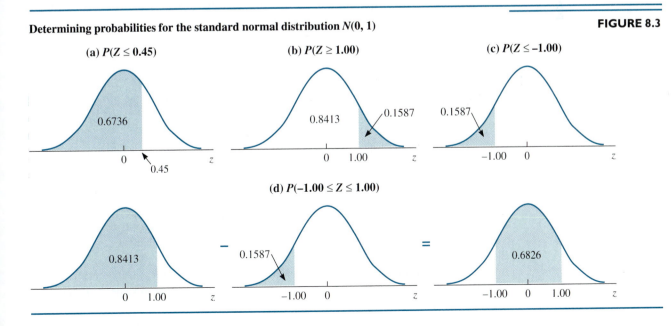

exceeds 0.50 because the standard normal distribution is symmetrical about its mean 0 so that the area to the left of 0.45 must exceed 0.50.

2. Refer to Figure 8.3b. We wish to find $P(Z \geq 1.00)$. Since Table C.1 shows areas to the left of any z value, we use complementation theorem (4.18):

$$P(Z \geq 1.00) = 1 - P(Z < 1.00)$$

Remember that $P(Z < 1.00) = P(Z \leq 1.00)$ because Z is a continuous random variable. From Table C.1, we find that $P(Z < 1.00) = 0.8413$. Therefore, we have $P(Z \geq 1.00) = 1 - 0.8413 = 0.1587$.

3. Refer to Figure 8.3c. We wish to find $P(Z \leq -1.00)$. Table C.1 does not show probabilities for negative values of z. The reason is the symmetry of the standard normal distribution about its mean 0. From this symmetry, we have:

$$P(Z \leq -1.00) = P(Z \geq 1.00)$$

Using the result from the previous example, we conclude that $P(Z \leq -1.00) = 0.1587$.

4. Refer to Figure 8.3d. We wish to find $P(-1.00 \leq Z \leq 1.00)$. We find this probability from Table C.1 by first obtaining the probability $P(Z \leq 1.00)$ and then subtracting the probability in the left tail that is not to be included; that is, $P(Z < -1.00)$. We know from Example 2 that $P(Z \leq 1.00) = 0.8413$ and from Example 3 that $P(Z < -1.00) = 0.1587$. Hence:

$$P(-1.00 \leq Z \leq 1.00) = P(Z \leq 1.00) - P(Z < -1.00)$$
$$= 0.8413 - 0.1587 = 0.6826 \qquad \square$$

Percentiles.   We stated earlier that $z(a)$ denotes the value on the z scale to the left of which the cumulative probability is a, as illustrated in Figure 8.4a. Thus, $z(a)$ is the $100a$ percentile of the standard normal distribution.

> **(8.8)**
> The $100a$ percentile of the standard normal distribution, denoted by $z(a)$, is defined by:
>
> $$P[Z \leq z(a)] = a$$

□ **EXAMPLES**

1. We wish to find the 67.36th percentile of the standard normal distribution; that is, we wish to find $z(0.6736)$. We found earlier in Figure 8.3a that for $z = 0.45$, the cumulative area to the left is 0.6736. Hence, $z(0.6736) = 0.45$, so 0.45 is the 67.36th percentile. This means that 67.36 percent of the time, a random selection from the standard normal distribution will result in an observed value for Z of 0.45 or less.

2. We wish to find $z(0.96)$, the 96th percentile. The cell entry nearest 0.96 in Table C.1 is $a = 0.9599$. The corresponding z value is 1.75 because entry 0.9599 is found

FIGURE 8.4

Determining percentiles for the standard normal distribution $N(0, 1)$

(a) $z(a)$

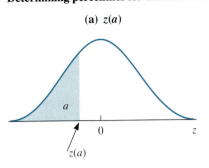

(b) $z(0.96)$

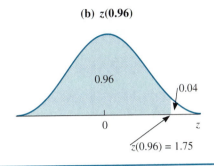

(c) $z(0.04)$

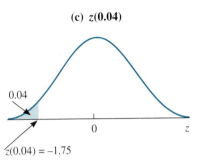

in the row labeled 1.7 and the column labeled 0.05. Thus, $z(0.96) = 1.75$. This percentile is illustrated in Figure 8.4b.

3. We wish to find $z(0.95)$, the 95th percentile. The cell entries nearest 0.95 in Table C.1 are $a = 0.9495$ and $a = 0.9505$. The corresponding z values are 1.64 and 1.65, respectively. By interpolating linearly, we obtain 1.645 for the 95th percentile; that is, $z(0.95) = 1.645$. □

In the preceding examples we found percentiles above the 50th. Note that these percentiles are all positive, because the standard normal distribution is symmetrical around its mean at 0, so that 50 percent of the area lies to the left of 0. Hence percentiles above the 50th must lie to the right of 0. For the same reason, percentiles below the 50th lie to the left of 0 and are all negative. They are related to percentiles above the 50th in the following fashion.

(8.9) $z(a) = -z(1 - a)$

EXAMPLE □

We wish to find the 4th percentile; that is, $z(0.04)$. By relation (8.9), we know that $z(0.04) = -z(0.96)$. In preceding Example 2, we found $z(0.96) = 1.75$; hence, $z(0.04) = -1.75$, as illustrated in Figure 8.4c. □

Some percentiles of the standard normal distribution are used so frequently that they are shown separately in part b of Table C.1. These percentiles can also be obtained from the cumulative probabilities in part a of Table C.1, in the manner just illustrated, but not to the accuracy shown in part b of Table C.1.

EXAMPLES □

1. We wish to find $z(0.98)$. From the table of selected percentiles, we obtain $z(0.98) = 2.054$.

2. We wish to find $z(0.001)$. From the table of selected percentiles, we obtain $z(0.001) = -3.090$. □

Determining Probabilities and Percentiles for Any Normal Distribution

Probabilities.   Probabilities for any normal distribution may be obtained from those for the standard normal distribution by means of standardizing transformation (8.6).

Ingot Weight   ☐ **EXAMPLE**

The weight X of a metal ingot produced in a smelter is normally distributed, with the mean and the standard deviation of X being $\mu = 520$ pounds and $\sigma = 11$ pounds, respectively.

1. Refer to Figure 8.5a. We wish to find the probability of an ingot weighing 525 pounds or less; that is, we wish to find $P(X \leq 525)$. Figure 8.5a illustrates the desired probability and the correspondence between the probability distributions of X and Z by the vertical alignment of the x and z axes. The value z for the standard normal distribution corresponding to $x = 525$ is obtained by the standardizing transformation (8.6):

$$z = \frac{x - \mu}{\sigma} = \frac{525 - 520}{11} = 0.45$$

The value 0.45 indicates that 525 is 0.45 standard deviation above the mean. Consequently, $P(X \leq 525) = P(Z \leq 0.45)$, as shown in Figure 8.5a. We can find this probability directly from Table C.1. Actually, we found this probability earlier in Figure 8.3a, where it was seen to equal 0.6736. Hence, the probability that an ingot will weigh 525 pounds or less is $P(X \leq 525) = 0.6736$.

2. Refer to Figure 8.5b. We wish to find $P(509 \leq X \leq 531)$. We calculate the z values corresponding to the specified values of x as follows:

$$z = \frac{x - \mu}{\sigma} = \frac{531 - 520}{11} = 1.00$$

$$z = \frac{x - \mu}{\sigma} = \frac{509 - 520}{11} = -1.00$$

FIGURE 8.5
Determining probabilities for the normal distribution with $\mu = 520$ and $\sigma = 11$—Ingot weight example

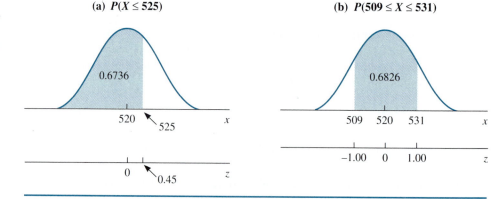

(a) $P(X \leq 525)$ (b) $P(509 \leq X \leq 531)$

Hence, $P(509 \leq X \leq 531) = P(-1.00 \leq Z \leq 1.00)$. The desired probability equals 0.6826, as was found earlier in Figure 8.3d. Hence, the probability that an ingot weighs between 509 and 531 pounds is $P(509 \leq X \leq 531) = 0.6826$. □

Percentiles. Percentiles for any normal probability distribution may be obtained from the corresponding percentiles for the standard normal distribution by again using the standardizing transformation (8.6).

EXAMPLE □

Refer to the ingot weight example in Figure 8.6a. We wish to obtain the 96th percentile of the ingot weight distribution, which we shall denote by $x(0.96)$. We begin by obtaining the 96th percentile of the standard normal distribution. We found this percentile earlier in Figure 8.4b to be $z(0.96) = 1.75$. Using the standardizing transformation (8.6), we then have:

$$z(0.96) = \frac{x(0.96) - \mu}{\sigma} = 1.75$$

which may be rewritten as:

$$x(0.96) = \mu + 1.75\sigma$$
$$= 520 + 1.75(11) = 539.3$$

Thus, the 96th percentile of the ingot weight distribution is 539.3 pounds. Hence, the probability is 0.96 that an ingot will weigh 539.3 pounds or less. Note that 539.3 is 1.75 standard deviations above the mean weight of 520. □

This example shows how the $100a$ percentile for any normal distribution can be found.

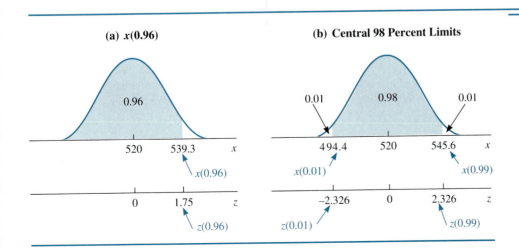

(a) x(0.96)

(b) Central 98 Percent Limits

(8.10)

The $100a$ percentile for any normal distribution, denoted by $x(a)$, can be obtained from the following relation:

$$x(a) = \mu + z(a)\sigma$$

where: μ is the mean of the normal distribution
σ is the standard deviation of the normal distribution
$z(a)$ is the $100a$ percentile of the standard normal distribution

EXAMPLE ☐

Refer to Figure 8.6b. We wish to find the central 98 percent probability limits of ingot weight. We see from Figure 8.6b that we must find the 1st and 99th percentiles of the ingot weight distribution, which we shall denote by $x(0.01)$ and $x(0.99)$, respectively. Formula (8.10) indicates that to obtain these percentiles, we require $z(0.01)$ and $z(0.99)$. From part b of Table C.1, we find $z(0.01) = -2.326$ and $z(0.99) = 2.326$. Hence, the desired percentiles are:

$$x(0.01) = 520 - 2.326(11) = 494.4 \text{ pounds}$$
$$x(0.99) = 520 + 2.326(11) = 545.6 \text{ pounds}$$

Thus, the probability is 0.98 that an ingot will weigh between 494.4 and 545.6 pounds; that is, $P(494.4 \leq X \leq 545.6) = 0.98$.   ☐

Three Important Normal Probabilities.   Three sets of central probability limits for normal distributions are used so frequently that it is worthwhile recording them. They are as follows:

1. $\mu \pm 1\sigma$ contains 68.3 percent of the area under a normal distribution.
2. $\mu \pm 2\sigma$ contains 95.4 percent of the area under a normal distribution.
3. $\mu \pm 3\sigma$ contains 99.7 percent of the area under a normal distribution.

For instance, in the ingot weight example, almost all of the ingots (99.7 percent) will weigh between $520 \pm 3(11)$, or between 487 and 553 pounds.

Cumulative Normal Probability Function

The cumulative probability function for a normal random variable has an S-shaped appearance, as illustrated in Figure 8.7 for the ingot weight example. The function is plotted by obtaining cumulative normal probabilities of the form $P(X \leq x)$ for a range of x values.

☐ **EXAMPLE**

Figure 8.7 shows the cumulative probability function for the weight X of an ingot in the ingot weight example. To illustrate how the function is plotted, consider $F(x)$ at $x = 525$. Recall from Figure 8.5a that $P(X \leq 525) = P(Z \leq 0.45) = 0.6736$. Hence, $F(525) = 0.6736$, as shown in Figure 8.7. Other points for the cumulative probability function are obtained in similar fashion.   ☐

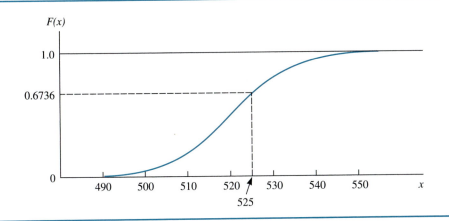

FIGURE 8.7
Cumulative probability function for the normal distribution with $\mu = 520$ and $\sigma = 11$—Ingot weight example

Sum of Independent Normal Random Variables

The following theorem is helpful when we work with the sum of two or more independent normal random variables.

> **(8.11)**
> The sum $X_1 + X_2 + \cdots + X_n$ of n independent normal random variables with means $\mu_1, \mu_2, \ldots, \mu_n$ and variances $\sigma_1^2, \sigma_2^2, \ldots, \sigma_n^2$, respectively, is also a normal random variable, with mean $\mu = \mu_1 + \mu_2 + \cdots + \mu_n$ and variance $\sigma^2 = \sigma_1^2 + \sigma_2^2 + \cdots + \sigma_n^2$.

EXAMPLE □

The cost of materials for a future construction project is a normal random variable X_1 with mean $\mu_1 = \$60$ million and standard deviation $\sigma_1 = \$4$ million. The cost of labor for this same project is an independent normal random variable X_2 with mean $\mu_2 = \$20$ million and standard deviation $\sigma_2 = \$3$ million. What is the probability that the total cost of materials and labor for this project will be \$85 million or less?

By (8.11), the total cost $X_1 + X_2$ is a normal random variable with mean $\mu = 60 + 20 = \$80$ million and variance $\sigma^2 = (4)^2 + (3)^2 = 25$; hence, the standard deviation of the total cost is $\sigma = \sqrt{25} = \$5$ million. To find the desired probability, we employ standardizing transformation (8.6) and obtain $z = (85 - 80)/5 = 1.00$. From Table C.1, we see that $P(Z \leq 1.00) = 0.8413$. Hence, the probability is 0.8413 that the total cost of materials and labor will be \$85 million or less. □

Exponential probability distributions are an important family of probability distributions that are useful in describing duration phenomena.

8.3
EXPONENTIAL PROBABILITY DISTRIBUTIONS

Police Response **EXAMPLE**

A police department is studying the lengths of time required by patrol cars to answer calls. The police commissioner wishes to know the percentage of calls that are answered within 15 minutes. The response time X, measured in minutes, is known to have an exponential distribution. The desired probability is $P(X \leq 15)$.

☐ **EXAMPLE**

The length X of long-distance telephone calls has an exponential distribution. An analyst who is studying telephone rate setting would like to know the proportion of calls completed within 3 minutes. The required probability is $P(X \leq 3)$.

Density Function

An *exponential random variable* is a continuous random variable that may take on any positive value. The density function for an exponential random variable follows.

(8.12)
The **exponential probability density function** is:

$$f(x) = \lambda \exp(-\lambda x)$$

where: $0 < x < \infty$
$\lambda > 0$

The constant λ is the only parameter of the exponential distribution. It is always positive. Each different value of λ yields a different probability distribution of the exponential family.

☐ **EXAMPLE**

In the police response example, the response time X is an exponential random variable with $\lambda = 0.2$. The corresponding exponential density function is:

$$f(x) = 0.2 \exp(-0.2x)$$

This probability distribution is plotted in Figure 8.8a. The value of the density function at $x = 5$, for instance, is:

$$f(5) = 0.2 \exp[-0.2(5)] = 0.0736$$

Characteristics of Exponential Probability Distributions

The mean and the variance of an exponential distribution depend on the parameter λ in simple ways.

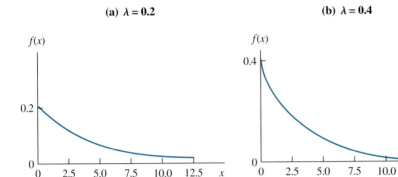

Two exponential probability distributions. The larger is λ, the smaller are the mean and the variance of the distribution.

FIGURE 8.8

(a) $\lambda = 0.2$

(b) $\lambda = 0.4$

(8.13)
The mean and the variance of an exponential probability distribution are:

(8.13a) $E\{X\} = \dfrac{1}{\lambda}$

(8.13b) $\sigma^2\{X\} = \dfrac{1}{\lambda^2}$

Note that the standard deviation for an exponential distribution is $\sigma\{X\} = 1/\lambda$, and hence is equal to the mean.

EXAMPLE □

In the police response example, where $\lambda = 0.2$, the mean response time is $E\{X\} = 1/0.2 = 5$ minutes. The standard deviation of the response time is the same—that is, $\sigma\{X\} = 5$ minutes—and hence the variance is $\sigma^2\{X\} = 5^2 = 25$. □

Figure 8.8 shows, in addition to the exponential distribution for the police response example, an exponential distribution corresponding to parameter value $\lambda = 0.4$. Note that each exponential distribution is markedly right-skewed, and that the probability density decreases steadily as x increases. Further, note that the larger is the value of λ, the less is the spread of the probability distribution and the closer is the mean to the origin.

Determining Probabilities and Percentiles

Any desired probability or percentile for an exponential distribution may be calculated readily from the cumulative probability function.

(8.14)

The cumulative probability function for an exponential random variable X is:

$$F(x) = P(X \le x) = 1 - \exp(-\lambda x)$$

where $0 < x < \infty$.

☐ EXAMPLE

For the police response example, the cumulative probability function for $\lambda = 0.2$ is shown in Figure 8.9. The value of the cumulative probability function at $x = 5$, for instance, is:

$$F(5) = 1 - \exp[-0.2(5)] = 0.6321$$

This cumulative probability is shown in Figure 8.9. Other values of the cumulative probability function are obtained in similar fashion.   ☐

We now present several examples to illustrate how probabilities and percentiles can be calculated using formula (8.14) for the cumulative probability function.

☐ EXAMPLES

1. For the police response example, the probability of a response within 15 minutes is of interest. For $\lambda = 0.2$ and $x = 15$, we substitute into (8.14) and find:

$$P(X \le 15) = F(15) = 1 - \exp[-0.2(15)] = 0.9502$$

FIGURE 8.9
Cumulative probability function for the exponential distribution with $\lambda = 0.2$—Police response example

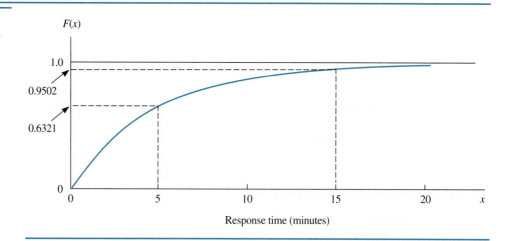

Response time (minutes)

This probability is shown in Figure 8.9.

2. The waiting time X (in minutes) in the express checkout lane of a food store is an exponential random variable with $\lambda = 0.5$. We wish to find the probability that a customer has to wait more than five minutes. Using (8.14), we first obtain:

$$P(X \le 5) = F(5) = 1 - \exp[-0.5(5)] = 0.9179$$

Then, by complementation theorem (4.18), we have:

$$P(X > 5) = 1 - P(X \le 5) = 1 - 0.9179 = 0.0821$$

3. In Example 2, we also wish to find the 95th percentile of the waiting time distribution; that is, we wish to find the value of x such that $P(X \le x) = F(x) = 0.95$. Since $\lambda = 0.5$, substitution into (8.14) gives:

$$F(x) = 1 - \exp(-0.5x) = 0.95$$

Solving for x, we obtain:

$$x = \frac{\ln(1 - 0.95)}{-0.5} = 5.99$$

where $\ln(a)$ denotes the natural logarithm of a. Hence, the 95th percentile of the waiting time distribution is 5.99 minutes. □

OPTIONAL TOPIC—RELATION BETWEEN EXPONENTIAL AND POISSON DISTRIBUTIONS

8.4

The exponential and Poisson probability distributions are closely related through an important type of probabilistic process called a *Poisson process*. We begin by describing this type of process.

Poisson Process

Many real-world phenomena involve events occurring randomly in time with no apparent pattern, such as computer breakdowns, house fires, or violent storms. If the random occurrences satisfy four conditions, or postulates, that we describe next, then the occurrences are said to be generated according to a Poisson process. In such a process, the numbers of occurrences in time intervals of fixed length follow a Poisson distribution, and the durations of the time intervals between successive occurrences follow an exponential distribution.

Poisson Process Postulates. Consider occurrences that happen randomly on the time continuum. Figure 8.10 illustrates three such occurrences, indicated by small circles. Imagine this continuum to be divided into many small nonoverlapping intervals of length Δt, as shown in Figure 8.10. The time between t and $t + \Delta t$ denotes a typical interval.

FIGURE 8.10
Illustration of a Poisson process

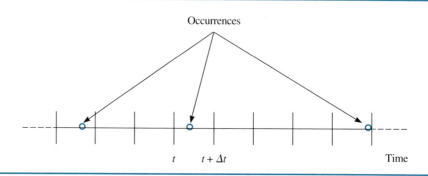

When these four postulates hold, the numbers of occurrences in unit time intervals follow a Poisson probability distribution with parameter λ. Note that postulates 1 and 2 require that the numbers of occurrences in nonoverlapping time intervals be independent and identically distributed random variables. These conditions are the same as those in (7.4) for a Bernoulli process.

(8.15)

The postulates of a Poisson process with parameter λ are as follows:

Postulate 1. The numbers of occurrences in nonoverlapping time intervals are independent.

Postulate 2. The number of occurrences in a time interval has the same probability distribution for all time intervals.

Postulate 3. The probability of one occurrence in any time interval $(t, t + \Delta t)$ is approximately proportional to Δt, the length of the interval. Specifically, the probability of one occurrence is approximately $\lambda \Delta t$.

Postulate 4. The probability of two or more occurrences in any time interval $(t, t + \Delta t)$ is negligibly small, relative to the probability of one occurrence in the interval.

☐ **EXAMPLE**

Consider the number of phone calls received at a switchboard in a one-hour period. Assume that these calls are received at a mean rate of λ per hour. Divide this one-hour period into a large number of nonoverlapping intervals of length Δt—for instance, into 3600 intervals of length $\Delta t = 1$ second. Postulate 1 requires that the numbers of phone calls received during these intervals be independent of one another. Postulate 2 requires that the number of phone calls in each interval be identically distributed for all intervals. This postulate would be violated, for instance, if some of the intervals correspond to busy periods and others to slack periods of telephone activity. Postulate 3 requires that the probability of receiving one phone call in any interval of length Δt be approximately proportional to Δt, with the constant of proportionality being λ. For instance, the probability for $\Delta t = 2$ seconds must be about twice as large as the probability for $\Delta t = 1$ second. Finally, postulate 4 requires that the probability of receiving two or more phone calls in any interval be very small. If these four postulates are satisfied by the telephone-calling process, then the number of calls received in one hour will be a Poisson random variable with parameter λ.   □

Relation Between Exponential and Poisson Distributions

Figure 8.11 illustrates the relation between the exponential and Poisson distributions in a Poisson process. The figure shows the time continuum marked off in intervals of unit length, with occurrences indicated by small circles. In the first interval, there are three occurrences; in the second, there is one; and so on. The durations between consecutive occurrences are denoted by x_1, x_2, and so forth, and are shown by braces. Statistical theory gives us the following relationship between the numbers of occurrences and the durations between successive occurrences in a Poisson process.

(8.16)
If occurrences are generated by a Poisson process with parameter λ, then (1) the numbers of occurrences in nonoverlapping unit intervals are independent random variables having the same Poisson distribution with parameter λ, and (2) the durations between successive occurrences are independent random variables having the same exponential distribution with parameter λ.

Theorem (8.16) implies that the mean number of occurrences per time interval is λ (the Poisson parameter) and the mean duration between successive occurrences is $1/\lambda$.

EXAMPLE   □

Accidents in a plant occur according to a Poisson process, with $\lambda = 1.5$ accidents per month. Therefore, by theorem (8.16), the numbers of accidents in the plant each month are independent outcomes from a Poisson distribution with mean $E\{X\} = \lambda = 1.5$ accidents per month. Moreover, the durations between successive accidents in the plant are independent outcomes from an exponential distribution, with the mean duration being $E\{X\} = 1/\lambda = 1/1.5 = 0.67$ month.   □

Relation between exponential and Poisson probability distributions.   The numbers of occurrences in the unit intervals are independent Poisson variables with parameter λ. The durations between successive occurrences are independent exponential variables with parameter λ.

FIGURE 8.11

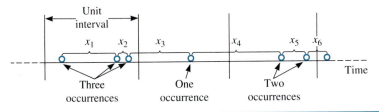

PROBLEMS

*   **8.1** A team preparing a bid on an excavation project assesses that the lowest competitive bid is a continuous uniform random variable X with $a = \$250$ thousand and $b = \$300$ thousand.

   a. State the density function of X and plot it on a graph.

   b. Obtain the mean and the variance of the probability distribution.

   c. Obtain the following: (1) the probability that the lowest competitive bid is between $250 thousand and $270 thousand, (2) the 10th percentile of the probability distribution. Interpret the measure obtained in (2).

8.2 A service station is located at one end of a four-kilometer tunnel. Any vehicular breakdown in the tunnel that requires servicing is equally likely to be at any point in the tunnel. Let X denote the position of a disabled vehicle (measured in kilometers from the service station).

   a. State the density function of X and plot it on a graph.

   b. Obtain $E\{X\}$ and $\sigma\{X\}$. In what units are these measures expressed?

   c. Obtain the following: (1) the probability that a service call to the tunnel will involve a vehicle positioned between 1 and 3 kilometers from the service station, (2) the 80th percentile position of a disabled vehicle. Interpret the measure obtained in (2).

8.3 Refer to **Suspension Cable** Problem 5.32.

   a. What are the values of the parameters a and b here?

   b. Obtain $E\{X\}$ and $\sigma\{X\}$. When the cable breaks, it breaks into two segments. Is $E\{X\}$ the same as the expected length of the shorter segment? Explain.

   c. Obtain (1) $F(40)$, (2) $P(40 < X \le 160)$, (3) the 40th percentile of the probability distribution. Interpret each of these quantities.

8.4 A calculator has a key that generates a random outcome X from the standard uniform distribution.

   a. State the density function of X and plot it on a graph.

   b. Obtain the mean and the standard deviation of X.

   c. Obtain the following: (1) the probability that X will be less than 0.834, (2) the probability that X will lie between 0.410 and 0.834, (3) the 20th percentile of the probability distribution.

   d. An observer states that because the calculator only displays a fixed number of decimal positions (for example, displays such as 0.834, 0.410, and so on are presented to three decimal positions), the random outcome X actually has a discrete uniform distribution. Do you agree? Comment.

8.5 Consider a normal probability distribution with $\mu = 6$ and $\sigma = 3$.

   a. Obtain the probability densities at $x = 0, 3, 6, 9, 12$.

   b. Plot the probability density function.

8.6 Consider a normal probability distribution with $\mu = 2$ and $\sigma = 1$.

   a. Obtain the probability densities at $x = 0, 1, 2, 3, 4$.

   b. Plot the probability density function.

8.7 Consider the family of normal probability distributions.

   a. Identify the probability distributions denoted by (1) $N(20, 9)$, (2) $N(0, 9)$, (3) $N(5, 1)$.

   b. For each of the normal random variables in part a, obtain expression (8.6) for the standardized normal variable Z.

8.8 Consider the family of normal probability distributions.

   a. Identify the probability distributions denoted by (1) $N(40, 100)$, (2) $N(0, 100)$, (3) $N(-3, 1)$.

   b. For each of the normal random variables in part a, obtain expression (8.6) for the standardized normal variable Z.

*   **8.9** Consider the standard normal variable Z and Table C.1.

   a. Find the areas a that correspond to the z values (1) 0, (2) 1.06, (3) 2.50.

   b. Find the following probabilities: (1) $P(Z \le 0)$, (2) $P(Z \ge 2.50)$, (3) $P(-1.06 \le Z \le 2.50)$.

c. Find the following percentiles from Table C.1a: (1) $z(0.5000)$, (2) $z(0.9066)$, (3) $z(0.0934)$.

d. Find the following percentiles from Table C.1b: (1) $z(0.99)$, (2) $z(0.05)$.

8.10 Consider the standard normal variable Z and Table C.1.

a. Find the areas a that correspond to the z values (1) 0.57, (2) 1.73, (3) 2.46.

b. Find the following probabilities: (1) $P(Z \leq -2.46)$, (2) $P(Z \geq 1.73)$, (3) $P(-2.46 \leq Z \leq 1.73)$.

c. Find the following percentiles from Table C.1a: (1) $z(0.5199)$, (2) $z(0.1841)$.

d. Find the following percentiles from Table C.1b: (1) $z(0.95)$, (2) $z(0.025)$.

8.11 Consider the standard normal variable Z and Table C.1.

a. Find the areas a that correspond to the z values (1) 0.22, (2) 1.46, (3) 3.10.

b. Find the following probabilities: (1) $P(Z \leq 1.46)$, (2) $P(Z \geq 0.22)$, (3) $P(-1.46 \leq Z \leq -0.22)$.

c. Find the following percentiles from Table C.1a: (1) $z(0.8980)$, (2) $z(0.0571)$.

d. Find the following percentiles from Table C.1b: (1) $z(0.005)$, (2) $z(0.90)$.

* **8.12** The weekly total long-distance telephone charges for a travel agency are a normal random variable X with $\mu = \$2800$ and $\sigma = \$150$.

a. For a randomly selected week, what is the probability the total charges will be (1) less than \$2950, (2) more than \$3000, (3) between \$2560 and \$3040?

b. Obtain the 75th percentile of the probability distribution. Interpret its meaning here.

c. Within what interval centered about μ will the weekly total charges fall with probability 0.50?

8.13 The duration of a flight between two cities is a normal random variable X with $\mu = 3.6$ hours and $\sigma = 0.15$ hour.

a. Obtain the following probabilities: (1) $P(X \leq 4.0)$, (2) $P(X > 3.7)$, (3) $P(3.2 \leq X \leq 4.0)$.

b. Obtain the 20th percentile of the probability distribution. Interpret its meaning here.

c. Within what interval centered about μ will the flight duration fall during 90 percent of the flights?

8.14 Noon Temperature. The temperature at noon on August 15 in a Southeastern city is a normal random variable X with $\mu = 30.1°C$ and $\sigma = 2.3°C$.

a. What is the probability that the noon temperature (1) will exceed 35.0°C; (2) will be between 25.0°C and 35.0°C?

b. Obtain the following: (1) $x(0.95)$, (2) $x(0.01)$. Interpret the meaning of each measure here.

8.15 Graduate Test Scores. The aptitude test scores of applicants to a university graduate program are normally distributed with mean 500 and standard deviation 60.

a. Applicants need a test score higher than 530 to be admitted into the graduate program. What proportion of applicants qualify?

b. If the university wishes to set the cutoff score for graduate admission so that only the top 10 percent of applicants qualify for admission, what is the required cutoff score?

c. What percentage of applicants have test scores within two standard deviations of the mean?

8.16 Refer to **Noon Temperature** Problem 8.14.

a. Obtain the values of $F(x)$ for $x = 26, 28, 30, 32, 34$.

b. Plot the cumulative probability function of X.

8.17 Refer to **Graduate Test Scores** Problem 8.15.

a. Obtain the values of $F(x)$ for $x = 400, 450, 500, 550, 600$.

b. Plot the cumulative probability function of X.

8.18 In restoring the exterior of a historic mansion, the required number of hours of carpentry X_1 is assessed to be $N(265, 256)$. The required number of hours of painting X_2 is assessed to be $N(208, 144)$. Here X_1 and X_2 are independent random variables. Let $T = X_1 + X_2$.

a. What does T represent here? Describe the probability distribution of T.

b. The labor cost of either type of work is \$20 per hour. A sum of \$10,000 has been budgeted for the total labor cost of carpentry and painting. What is the probability that the total labor cost will exceed the budget?

8.19 A firm has two major products. Let X_1 and X_2 denote the annual sales (in $ million) of the two products. Here X_1 and X_2 are independent random variables with probability distributions $N(38, 25)$ and $N(65, 49)$, respectively. Let $T = X_1 + X_2$.

a. What does T represent here? Describe the probability distribution of T.

b. Obtain the probability that total sales of the two products (1) will exceed $100 million, (2) will lie between $90 million and $110 million.

c. Obtain the 10th percentile of the probability distribution of T. Interpret this measure.

* 8.20 The time required to check out a customer at a supermarket is an exponential random variable X with a mean of 200 seconds.

a. What is the value of the parameter λ here?

b. Obtain $\sigma\{X\}$.

c. Obtain each of the following: (1) $P(X \leq 100)$, (2) $P(X > 400)$, (3) the 90th percentile of the probability distribution.

d. Obtain $F(x)$ for $x = 0, 100, 400, 1000$. Plot the cumulative probability function. How can you tell from this plot that the probability distribution is skewed?

8.21 The time (in hours) between consecutive arrivals of ambulances at a hospital emergency ward is an exponential random variable X with $\lambda = 0.8$.

a. Obtain $E\{X\}$ and $\sigma\{X\}$. In what units are these measures expressed?

b. Obtain the probability that an ambulance will arrive (1) within one hour after the previous arrival, (2) between 30 minutes and one hour after the previous arrival.

c. Obtain the 25th and 75th percentiles of the probability distribution. What is the interquartile range, and how is this range interpreted here?

d. Obtain $F(x)$ for $x = 0, 1, 2, 3$. Plot the cumulative probability function.

8.22 Digital Display. The failure time (in years) of a digital display unit in an instrument panel is an exponential random variable X with $\lambda = 0.4$.

a. Obtain $f(x)$ for $x = 0, 2, 4, 6$. Plot the probability density function.

b. Obtain $E\{X\}$ and $\sigma\{X\}$. In what units are these measures expressed?

c. Obtain each of the following probabilities: (1) $P(X \leq 4)$, (2) $P(X > 8)$, (3) $P(1 \leq X < 4)$.

d. Obtain the following measures of the probability distribution: (1) the 90th percentile, (2) the median. Interpret the latter measure.

8.23 The particle life (in days) of a radioactive isotope used in medical research is an exponential random variable X with mean 3.8 days.

a. Obtain the probability that a particle of this isotope will survive (1) beyond 7 days, (2) beyond 14 days.

b. What is the half-life (that is, the median life) of this isotope?

8.24 Refer to **Police Calls** Problem 7.24. An observer noted that calls to the dispatcher sometimes come in bunches relating to the same event (for example, calls from several witnesses of a road accident). Which, if any, of the postulates in (8.15) for a Poisson process are violated by this situation?

8.25 Refer to **Facsimile Unit Warranty** Problem 7.26. For each of the following situations, explain which, if any, of the postulates of a Poisson process in (8.15) are violated: (1) The warranty calls for a unit often are bunched toward the end of the warranty year; (2) a warranty call for a unit is often followed in a short while by a second warranty call for the same unit to fix the same or a related problem; (3) about half of the units have no warranty calls during the warranty year, while the other half have three or more calls.

8.26 The number of units of a water heater element demanded as replacements follows a Poisson process with $\lambda = 3$ units per year. Let Δt denote an interval of one day in a year having 365 days. According to the postulates for a Poisson process in (8.15), state the approximate probability that (1) exactly one unit will be demanded on a given day, (2) no unit will be demanded on a given day, (3) two or more units will be demanded on a given day.

* **8.27** Calls to an emergency ambulance service in a large city are generated by a Poisson process with $\lambda = 3$ calls per hour.

 a. What is the nature of the probability distribution for the lengths of time between successive calls to the ambulance service?
 b. Obtain the mean length of time between successive calls to the ambulance service.
 c. What is the probability that less than 0.25 hour will elapse between successive calls?

8.28 Large percent changes in a common stock price index on any single day are rare. Consider a change in the index (either up or down) of 10 percent or more on a single day to be an "occurrence." Assume that these occurrences follow a Poisson process with $\lambda = 0.3$ per year.

 a. What is the nature of the probability distribution for the lengths of time between successive occurrences for this index?
 b. What is the mean time between successive occurrences for this index?
 c. What is the probability that (1) there are exactly two such occurrences in one year, (2) an occurrence is followed by the next occurrence within six months?

8.29 Collisions of ships in a shipping channel follow a Poisson process with $\lambda = 0.5$ collision per year. Let Y denote the number of collisions in a year and let X denote the time interval between successive collisions.

 a. What is the nature of the probability distribution of X? Of Y?
 b. Obtain $P(X > 1)$ and $P(Y = 0)$. Why must these probabilities be the same?

EXERCISES

8.30 (*Calculus needed.*) Derive (8.2a) and (8.2b).

8.31 Let $X_1, X_2, \ldots, X_{12}$ be 12 independent standard uniform random variables and define W as follows:

$$W = X_1 + X_2 + \cdots + X_{12} - 6$$

It can be shown theoretically that W is approximately a standard normal variable. Verify that $E\{W\} = 0$ and $\sigma^2\{W\} = 1$.

8.32 Variable X is $N(10, 4)$. Find the following conditional probabilities: (1) $P(X \geq 14 \mid X \geq 12)$, (2) $P(9 \leq X \leq 11 \mid 8 \leq X \leq 14)$.

8.33 The 20th and 60th percentiles of a normal random variable are -4.3 and 15.8, respectively. Find the mean and the standard deviation of the random variable.

8.34 (*Calculus needed.*) Derive (8.13a). [*Hint:* Use integration by parts.]

8.35 (*Calculus needed.*) Derive (8.14).

8.36 Show that the following relation holds for any exponential random variable X:

$$P(X > x + x_0 \mid X > x_0) = P(X > x) \qquad \text{for any positive } x \text{ and } x_0$$

STUDIES

8.37 A cosmetics firm must purchase a bottle-filling machine and has narrowed the choice to one of two machines (1, 2). The purchased machine will be used to fill 100 milliliter (ml) bottles of cologne. Both machines can control the mean fill at any desired level, but the standard deviation of fill is 0.3 ml for machine 1 and 0.1 ml for machine 2. The fills with either machine are normally distributed.

 a. At what mean fill level should each machine be set so that 99.9 percent of filled bottles contain 100 ml or more of cologne?
 b. Consider the mean fill levels obtained in part a. The cost of cologne is $0.10 per ml. The cost of using machine 1 to fill 1000 bottles is $50. The corresponding cost for machine 2 is $80. Other pro-

duction and operating costs for the two machines are identical. Obtain for each machine the expected total cost of machine usage and of the cologne for filling 1000 bottles. What is the expected total cost per bottle for each machine? On this basis, which machine is the better choice?

8.38 A battery-powered safety light is required to operate 400 hours without failing. The light fails when its battery pack fails. Two configurations of the battery pack (1, 2) are being considered. In configuration 1, the pack consists of a single battery having a normally distributed operational life-time with a mean of 450 hours and a standard deviation of 30 hours. In configuration 2, the pack consists of two batteries, designed so that the second battery begins to operate when the first battery fails. Each of these two batteries has a normally distributed operational lifetime with a mean of 220 hours and a standard deviation of 15 hours. The operational lifetimes of the two batteries are inde-pendent. Let T_1 and T_2 denote the operational lifetimes of battery packs with configurations 1 and 2, respectively.

a. Specify the nature of the probability distribution of T_2.
b. For each configuration, what is the probability that the battery pack will have an operational lifetime of 400 hours or longer? According to these probabilities, which is the better configura-tion? Explain.
c. In a single trial of two battery packs, one with each configuration, what is the probability that the configuration 1 pack will have a longer operational lifetime than the configuration 2 pack? Assume that the lives of the two packs are independent. Which appears to be the better configu-ration based on this result? Reconcile this choice with your choice in part b. [*Hint:* Consider the difference $T_1 - T_2$.]
d. Suppose that there is a probability of 0.05 that the second battery in configuration 2 will not commence operation when the first battery fails. In this case, what will be the probability that a configuration 2 pack will have an operational lifetime of 400 hours or longer?

8.39 A large display will contain many colored light bulbs. Because the labor cost of replacing the bulbs individually when they burn out is expensive, the operators plan to replace all the bulbs simul-taneously. Two replacement schedules are under consideration: (1) replace when 15 percent of the bulbs have burned out; (2) replace when 30 percent of the bulbs have burned out.

a. For bulbs of make A, the lifetimes in hours are independent and exponentially distributed with $\lambda = 0.0004$. What is the expected number of operating hours before the bulbs are replaced under schedule 1? Under schedule 2?
b. For bulbs of make B, the lifetimes are independent and normally distributed with $\mu = 1100$ hours and $\sigma = 250$ hours. What is the expected number of operating hours before the bulbs are replaced under schedule 1? Under schedule 2?
c. Assuming that bulbs of make A cost the same as bulbs of make B, which make of bulb do you recommend that the operators use under schedule 1? Under schedule 2? Why?

8.40 Refer to **Digital Display** Problem 8.22.

a. Obtain the following probabilities: (1) $P(X > 3)$, (2) $P(X > 8 \mid X > 5)$, (3) $P(X > 10 \mid X > 7)$.
b. Obtain the following probabilities: (1) $P(X > 1)$, (2) $P(X > 6 \mid X > 5)$, (3) $P(X > 8 \mid X > 7)$.
c. Describe the property of the exponential probability distribution that is illustrated by the results in parts a and b. Does this property imply that a functioning, used digital display unit is as good as a new one? Explain.
d. When the digital display unit in the instrument panel fails, it is immediately replaced by a new one. The number of failures of the display unit during a year follows a Poisson process with $\lambda = 0.4$. Describe the probability distribution of the number of failures of this display that will be experienced during the next year. Is this probability distribution affected by how long the cur-rently installed display unit had been operating prior to the beginning of the year? Explain.

Estimation and Testing I

Statistical Sampling

9

We are now ready to apply the principles of probability to the analysis of statistical data. For this purpose, we must distinguish between data sets that are considered to be populations and data sets that are considered to be samples from populations.

In any statistical investigation, there is a collection of elements of interest, called the population. Since it is often not possible to consider all elements in the population, a subset, or sample, is frequently selected to provide information about the population.

EXAMPLE ☐

A U.S. congressional staff assistant is studying Medicaid abuses to help in drafting new legislation. The data set consisting of all Medicaid claims in the previous year is the population of interest. The assistant cannot possibly examine all Medicaid claims in the previous year because the number is much too large. Instead, a probability sample of 1000 claims will be selected. This sample data set will then be used to draw conclusions about the characteristics of abuses in the population of all Medicaid claims last year. ☐

The use of probability principles to draw sound conclusions about population characteristics from the corresponding characteristics of a sample is the subject of *statistical inference*. In this chapter, we consider some basic concepts of sampling. We define a population and a sample, discuss why sampling is so widely used, and explain what a simple random sample is and how such a sample can be selected. We close the chapter by considering the relationship between population and sample characteristics.

We present three examples of populations.

POPULATIONS 9.1

EXAMPLES ☐

1. The manager of an automobile-leasing agency is interested in the fuel economy of the passenger cars in the company's fleet. Here, the population consists of all passenger cars in the fleet. The elements of the population are the individual cars.

2. A candidate for city council is concerned about the attitudes of voters toward the metropolitan transit agency. The population here consists of all eligible voters in the city. The elements of the population are the individual voters.

3. A quality assurance manager desires information about the current quality level of the firm's manufacturing process for computer memory chips. Here the population consists of all the chips that could be produced by the current process operating under the same conditions. The elements of the population are the individual chips. ☐

A formal definition of a population follows.

> **(9.1)**
> A **population** is the total set of elements of interest for a given study.

A population is sometimes also referred to as a *universe.*

We distinguish between finite and infinite populations. Examples 1 and 2 are illustrations of finite populations, while Example 3 illustrates an infinite population.

Finite Populations

A *finite population* is one that consists of a finite number of elements, such as the set of passenger cars owned by the leasing agency or the set of eligible voters in the city. Many of the populations of interest in business, economics, and the social sciences are finite, although they are often large. The set of all college students in the United States is an example of a large finite population.

Ph.D. Students ☐ **EXAMPLE**

The population of interest consists of the five Ph.D. students currently enrolled in an economics department. The variable of interest is the current grade point average of each of these students. These grade point averages are:

$$X_1 = 3.74 \qquad X_2 = 3.89 \qquad X_3 = 4.00 \qquad X_4 = 3.68 \qquad X_5 = 3.69$$

These five values constitute the finite population of interest. ☐

Infinite Populations

An *infinite population* is one that consists of an infinite number of elements. In general, an infinite population refers to a *process,* and its elements consist of all the outcomes of the process if the process were to operate indefinitely under the same conditions. Thus, the manufacturing process for memory chips constitutes an infinite population whose elements are the chips that would be produced by the process if it were to operate indefinitely under the same conditions. The outcomes of the process for the characteristic of interest are described by a probability distribution.

☐ **EXAMPLE**

Consider the infinite population pertaining to the process of molding tires in pairs, as described in the tire molding example in Chapter 5. The number of defective tires in a pair, X, is the characteristic of interest. We cannot simply list the outcomes of the process for all of the elements in the infinite population. Instead, we describe them by a probability distribution:

x:	0	1	2
$P(x)$:	0.75	0.10	0.15

The probability $P(0) = 0.75$ indicates that 75 percent of the trials in the infinite population lead to zero defective tires in the pair. Similarly, $P(1) = 0.10$ indicates that 10 percent of the trials in the infinite population lead to one defective tire in the pair. ☐

CENSUSES AND SAMPLES   9.2

Census

Information about a finite population can at times be obtained by a census.

> **(9.2)**
> A **census** is a study of a finite population that includes every element of the population.

A census is usually employed only for small populations, for large populations that are readily accessible, and when the investigation requires complete data on the population.

EXAMPLES ☐

1. An organization with 50 employees is interested in employees' preferences for a new pension plan. The population here is small, and it is easy to reach every employee; hence, a census is conducted.

2. A bank is interested in the monthly closing balances of the savings accounts of its 100,000 customers. The balances are readily available from the bank's computerized operations data base. Hence, a census is feasible even though the population is large.

3. Grants for municipal public works are allocated by a state on the basis of counts of the number of persons in each municipality. The most recent census of the population in the state is used to provide the population counts for the municipalities.   ☐

Comment

Censuses are not necessarily successful in including every population element. Thus, the U.S. and Canadian population censuses are known to miss significant numbers of people even though great efforts are made to include everyone.

Sample

Often, information about a population is obtained from a sample. We present three examples of samples.

☐ **EXAMPLES**

1. As part of a study of Canadian fiscal policy, an economist considers data on savings plans by 1000 households in a sample selected from the population of Canada.

2. An analyst uses data on the unemployment rates for different regions of the state in setting state priorities for public works. The unemployment data were gathered in a sample survey of households from each region of the state.

3. Water samples were drawn from a river at a number of locations in a survey designed to measure the levels of several contaminants in the river's water.   ☐

In each of these examples, inferences are made about the population from the sample.

> **(9.3)**
> A **sample** is a part of the population selected so that inferences can be drawn from it about the population. The process of designing and executing a study based on a sample is called a **sample survey.**

Reasons for Sampling

When a population is infinite, information about it can be obtained only from a sample. When a population is finite, on the other hand, information can be obtained by a census or by a sample survey. Still, a sample survey often is preferred to a census for a number of reasons.

1. *Cost.* A sample survey often provides reliable and useful information at much lower cost than does a census. For example, the cost of a census of the inhabitants of a city to obtain information about educational television viewing habits would be extremely high. A sample survey of the inhabitants can provide data with sufficient reliability at a fraction of this cost.

2. *Timeliness.* A sample survey usually provides more timely information than a census, because fewer data have to be collected and processed. This feature is particularly important when information is needed quickly, such as when a legislative committee must determine how much of a proposed tax reduction might be saved by families.

3. *Accuracy.* A sample survey often provides information as accurate as, or even more accurate than, a census because data errors typically can be controlled better in a small undertaking than in a large one. For instance, staff training and supervision are generally easier and more effective in a sample survey than in a census.

4. *Detailed information.* Frequently, more time can be spent in getting more detailed information with a sample survey than with a census, as in an opinion survey in which respondents' attitudes and motivations are probed. Such probing is expensive and usually needs to be confined to small-scale surveys.

5. *Destructive testing.* When a test involves the destruction of an item, sampling must be used. For example, a manufacturer who wishes to study the life of batteries in a produc-

tion lot by subjecting batteries to a life test must use a sample if there are to be any batteries left to sell.

Sampling and Nonsampling Errors

Sample results are usually in error to some degree, simply because a sample involves a study of only part of the population. This kind of error is called a sampling error.

> **(9.4)**
>
> A **sampling error** is the difference between the result obtained from a sample and the result that would be obtained from a census conducted by using the same procedures as in the sample.

Sampling errors can be controlled by effective sample design and by use of an appropriate sample size.

Sampling errors need to be distinguished from errors in data that were described in Chapter 1, such as faulty respondent recall or instrument measurement errors. Errors in data will be present whether a sample or a census is conducted because they arise from the method of data collection and recording. Errors that are present in data irrespective of whether they are obtained from a sample or a census are called *nonsampling errors.*

Probability and Judgment Samples

We distinguish between two types of samples, probability samples and judgment samples, based on the manner in which population elements are selected for the sample. The following two examples illustrate the distinction.

EXAMPLES

1. A sample of four cities was selected from a population of 20 cities. The selected cities are to be used as test markets for a new product. The selection procedure used the fact that 4845 different combinations of four cities can be drawn from a population of 20 cities. The procedure gave each combination an equal probability (1/4845) of being chosen as the actual sample. The resulting sample is a probability sample because a probability mechanism was used for the sample selection and the probabilities of selection are known.

2. A television show was previewed in an auditorium. The director of the show selected 40 persons she judged to be representative of the audience of 435 persons for detailed interviews about their impressions of the show. The selection is a judgment sample because the director judged it to be a "representative" sample. □

Probability Sample.   Probability samples are widely used.

> **(9.5)**
>
> A sample in which the selection of elements from the population is made according to known probabilities is a **probability sample.**

The selection of population elements according to known probabilities allows discretion about the probabilities assigned to each population element but not about which particular elements in the population enter the sample. Probability selection has two major advantages:

1. The sample data can be evaluated by statistical methods to provide information about the margin of error in the results due to sampling.
2. Biases are avoided that could enter if judgment were used to select the population elements for the sample.

Throughout much of the remainder of this text, we shall be concerned with probability samples and their statistical evaluation.

Judgment Sample.   Judgment samples also are widely used.

> **(9.6)**
> A sample for which judgment is used to select representative elements from the population or to infer that it is representative of the population is a **judgment sample.**

In contrast to probability sampling, expert opinion is used in judgment sampling to select particular elements for the sample or to determine whether the sample is representative of the population. Judgment samples may at times provide useful results. Statistical methods cannot be utilized, however, to evaluate the sample results for purposes of assessing the margin of error due to sampling. Also, some judgment samples may be poor because the judgment is unsound and the sample turns out to be unrepresentative.

A common type of judgment sample used in surveys is a *quota sample*. For instance, interviewers in a consumer survey may be given target numbers or *quotas* of persons with certain characteristics, such as age, gender, and income, to be selected for the sample. The purpose of the quotas is to make the sample a cross section of the population under study. The actual selection of persons is left to the interviewers. Although interviewers are expected to use good judgment in selecting persons who meet the quotas, they may choose persons readily accessible, such as homemakers at home during the day. Also, an interviewer can simply substitute another person from the same quota for a person who is not immediately available for interview. In these and other ways, biases can enter the survey. These biases do not enter a probability sample survey that is properly carried out.

Comments

1. Some samples designed as probability samples turn out to be judgment samples upon completion. For example, a probability sample of new-car purchasers was carefully selected. Unfortunately, only 50 percent of the purchasers in the sample replied to the questionnaire about the performance of their car. After some analysis of the characteristics of the respondents, the market researcher decided that the purchasers who responded are typical of all new-car purchasers. Hence, the sample actually obtained should be viewed as a judgment sample because the market researcher made a key judgment about the representativeness of the responses obtained.
2. Some studies are based on data that do not fulfill either the definition of a probability sample or that of a judgment sample. This situation occurs when a study is based on population elements that happen to be conveniently at hand. An example is the use of students enrolled in a marketing course to study the effectiveness of different advertisements. These students

were not selected as representative of the relevant population nor were they selected according to known probabilities. Such a study group is called a *chunk* or a *convenience sample*. A chunk can be useful for limited purposes, but it cannot be relied on for credible inferences about the population.

Many types of probability samples are available for sampling a finite population. The most basic is a simple random sample, which we discuss now. Other types of probability samples are discussed in Chapter 23.

**SIMPLE RANDOM 9.3
SAMPLING FROM A
FINITE POPULATION**

Definition

(9.7)
A **simple random sample** from a finite population is a sample selected so that each possible sample combination of the specified size has equal probability of being chosen.

For brevity, a simple random sample is often called a *random sample*.

EXAMPLE ☐

Consider again the Ph.D. students example. The population consists of five economics Ph.D. students. Let us denote these students by A, B, C, D, and E. If a sample of two students is to be selected from this population, there are altogether 10 possible sample combinations, as follows:

A, B	A, D	B, C	B, E	C, E
A, C	A, E	B, D	C, D	D, E

To obtain a simple random sample of two students from the population of five, each of the 10 sample combinations must have probability $1/10 = 0.10$ of being chosen as the actual sample. ☐

Comments

1. Definition (9.7) refers to simple random sampling *without replacement,* in which a population element can enter the sample only once. Simple random sampling *with replacement,* in which the same population element may enter the sample more than once, is rarely used with finite populations.

2. Definition (9.7) implies that each population element has an equal probability of being selected. However, an equal selection probability for each population element is not a sufficient condition for a sample to be a simple random one. For example, suppose that a random sample of 50 oranges is to be selected from a crate of 100 oranges. We could divide the 100 oranges into two groups consisting of the 50 largest and the 50 smallest oranges, and then select one of the two groups with equal probability. This procedure gives each orange in the population an equal (and known) probability of 0.5 of entering the sample, and hence it qual-

ifies as probability sampling. Yet it does not qualify as simple random sampling because every combination of 50 oranges does not have equal probability of being selected. For example, a mixture of small and large oranges has no chance of being selected.

3. The number of possible sample combinations of n elements that can be selected from a finite population of N elements is given by $_NC_n$ in formula (A.23) in Appendix A. Thus, the number of possible sample combinations of $n = 2$ students from the population of $N = 5$ Ph.D. students is:

$$_5C_2 = \frac{5!}{2!3!} = 10$$

Selection of Simple Random Sample

A practical method of meeting the requirement that each possible sample combination have an equal probability of being chosen involves selecting sample elements one at a time.

(9.8)
The following procedure selects a simple random sample of n elements without replacement from a finite population of N elements:
1. Select the first sample element by giving each of the N population elements equal probability of being chosen; that is, probability $1/N$.
2. Select the second sample element by giving each of the remaining $N - 1$ population elements equal probability of being chosen; that is, probability $1/(N - 1)$.
3. Repeat this process until all n sample elements have been selected.

Frame.   Selecting a simple random sample from a finite population requires a frame.

(9.9)
A **frame** is a listing of all the elements of the finite population.

☐ **EXAMPLE**

A computer listing of students at a university is a frame for the population of students.

☐

Plywood Shipment   ☐ **EXAMPLE**

A shipment of plywood sheets consisting of a row of 40 bundles, with 50 sheets stacked in each bundle, is the population of interest. The physical arrangement of the plywood constitutes a frame of the population of sheets.   ☐

Sometimes, the frame available for sampling is not a perfect one for the population of interest.

County Lawyers   ☐ **EXAMPLE**

A random sample of lawyers is to be selected from the population of all lawyers in a county. The frame for sampling is a listing of all lawyers who were members of the county bar association as of last month. This listing does not include lawyers who joined the association since last month or lawyers in the county who are not members of the county bar association. ☐

In cases such as the county lawyers example, we distinguish between the *target population* and the *sampled population*. The target population is the population of interest (all lawyers in the county), while the sampled population consists of the elements in the frame (all lawyers in the county who were members of the county bar association as of last month). When the frame available for sampling differs materially from the target population, the sample results may have only limited relevance. Hence, great efforts are made in practice to obtain a frame that coincides closely with the target population.

Sample Selection Using Random Numbers. Random numbers can be used to select a simple random sample from a population frame following the procedure in (9.8). The random numbers required for this purpose can be produced by a calculator or computer routine or from a table of random digits, such as Table C.7. We shall discuss the use of Table C.7 first.

A table of random digits contains outcomes of independent random trials from the discrete uniform probability distribution in Figure 7.1, which has possible outcomes 0, 1, . . . , 9 with equal probability. Thus, for each position in the table, every digit from 0 to 9 has equal probability of appearing in that position, and the outcomes for the various positions in the table are independent.

Tables of random digits are generated by computer and are usually tested carefully to ensure close adherence to the required properties of equal probability and independence. These properties permit the user to form random numbers by combining random digits. For instance, consider forming pairs of random digits. Since every digit has equal probability of appearing in a position and all positions are filled independently, pairs of positions are filled with numbers from 00 to 99 with equal probability and independently. It is this capability of forming random numbers of any size, which are equally likely and independent, that enables us to use a table of random digits for selecting a simple random sample.

EXAMPLE ☐

In the county lawyers example, the frame contains 950 lawyers. For convenience, we assign numbers from 001 to 950 to the lawyers. Since the frame contains 950 elements, we shall select three-digit numbers in the range from 000 to 999. Table 9.1 contains an extract of Table C.7, consisting of the first 10 rows of columns 1–5. Suppose we use the first three digits in each row as the three-digit number and read downward. The procedure, then, is as follows:

1. The first number is 132. Lawyer 132 is therefore the first element for the sample. A number over 950 or number 000 would have been disregarded. Thus, each of the 950 lawyers has equal probability of being the first sample element.

2. The second number is 212. Lawyer 212 is therefore the second element for the sample. A number over 950 or number 000 would have been disregarded. Also, number 132 would have been disregarded because that lawyer is already in the sample. Thus, each of the remaining 949 lawyers has equal probability of being the second sample element.

TABLE 9.1
Use of Table C.7, a table of random digits—County lawyers example

Row	Columns 1–5	Lawyer
1	13284 ⟶	132
2	21224 ⟶	212
3	99052 ⟶	Disregarded
4	00199 ⟶	001
5	60578 ⟶	605
6	91240 ⟶	912
7	97458 ⟶	Disregarded
8	35249 ⟶	352
9	38980 ⟶	389
10	10750 ⟶	107

3. This procedure is repeated until the required number of lawyers for the sample has been selected.

Table 9.1 shows the disposition for the first 10 random numbers.

Numbers may be chosen from a table of random digits in any manner as long as the procedure is systematic and determined in advance. Thus, three-digit numbers in the county lawyers example might have been obtained by reading across in a row, or by taking the first digit in each of three columns, or by reading upward in a column.

Instead of using a table of random digits, a computer or calculator may be employed to generate the random numbers directly. Using a computer is especially helpful when the sample is large, since many computer packages also order the random numbers for the user. Typically, the user specifies the smallest and the largest values of the random numbers to be generated and the sample size. Thus, in the county lawyers example, the user would specify 1 and 950 as the smallest and the largest possible values and also would specify the sample size. In some computer applications, the random numbers generated by the computer are used internally to identify those elements of a computerized data base that are selected for the sample, and only the descriptions of the sample elements are printed out.

Comments

1. Sometimes the elements in a frame are prenumbered, such as invoices that have serial numbers or students who have been assigned ID numbers. These numbers can be used for sampling identification provided that there are no duplicate assignments of the same number to several population elements. Often these number assignments have gaps, such as when an invoice is voided or when a student has graduated. Such gaps cause no problem; when a selected random number refers to a blank, it is simply disregarded and the next random number is selected.

2. At times, identification numbers need not be assigned explicitly to the elements in the frame. The plywood shipment example illustrates this case. The frame consists of 2000 plywood sheets arranged in 40 bundles of 50 sheets each. Once we specify the order of counting bundles (for instance, from front to back) and the order of counting sheets within a bundle (for instance, from the top down), any plywood sheet designated by a random number can be identified by counting. For example, random number 0017 refers to the 17th sheet in the first bundle. Similarly, random number 1317 refers to the 17th sheet in the 27th bundle.

When a population is infinite, we cannot use a probability mechanism for selecting the sample. The process associated with the infinite population simply furnishes observations. For instance, a manufacturing process furnishes the number of units produced each day.

**SIMPLE RANDOM 9.4
SAMPLING FROM AN
INFINITE POPULATION**

Definition

Observations generated by a process are random variables prior to the random trials. They constitute a simple random sample from an infinite population if two conditions are met.

> **(9.10)**
> The n random variables $X_1, X_2, \ldots, X_n$ generated by a process constitute a **simple random sample from an infinite population** if (1) they are independent and (2) they come from the same probability distribution. The common probability distribution for $X_1, X_2, \ldots, X_n$ is the infinite population.

EXAMPLE ☐

Consider again the tire molding example. If the process is stable (that is, the probability distribution does not change) and the successive observations are independent, the next five observations from the process will constitute a random sample from the infinite population with which the probability distribution is associated. The observations on the number of defective tires in the next five pairs might be:

$$X_1 = 0 \qquad X_2 = 0 \qquad X_3 = 0 \qquad X_4 = 1 \qquad X_5 = 0$$

Note that most of the pairs contain no defective tires, consistent with the high probability of no defective tires in a pair, $P(0) = 0.75$. ☐

Comments

1. The two conditions given in (9.10) for data generated by a process to constitute a simple random sample can be stated equivalently as requiring the observations to be (1) independently and (2) identically distributed random variables. These are the same conditions that we encountered in the definitions of Bernoulli and Poisson processes in (7.4) and (8.15), respectively.

2. A sample from an infinite population may, for some other purpose, be considered a finite population. Thus, the memory chips produced by a stable manufacturing process during a production run represent a sample from the infinite population associated with the process. When, however, these chips are sent to a computer firm that wishes to sample them to determine whether this shipment is of acceptable quality, the chips in the shipment constitute a finite population for this purpose.

3. No sample information is required about an infinite population for which the associated probability distribution is known because the probability distribution contains all of the information pertaining to the population. Usually, we do not fully know the probability distribution associated with an infinite population and therefore require sample data to make inferences about the infinite population.

Diagnostic Procedures for Checking Randomness of Data

In general, we do not know in advance whether data generated by a process, such as the number of units produced by a manufacturing process in each of the last 40 days, can be treated as a random sample from the process. That is, we do not know whether the process has been stable so that all observations came from the same probability distribution or whether there has been some shift or other change in the probability distribution. Nor do we usually know whether the observations are independent or whether they are related cyclically or in some other way. Therefore, we must ordinarily employ various diagnostic procedures to examine if the two conditions for a simple random sample in (9.10) are met by a data set generated by a process. Some of these diagnostic procedures are informal, descriptive procedures while others involve formal statistical inferences. If these procedures reveal that the observations in the data set are dependent or that they were drawn from different probability distributions, we will not be able to treat the data set as a simple random sample from a process. We consider now some simple diagnostic procedures.

Since observations from a process often come in time order, a helpful informal diagnostic procedure is to plot the data in time sequence. A *time series plot* can show whether there is a trend in the data, which would suggest that the observations come from different probability distributions. The plot also can show whether there are cyclical swings, which would suggest either that the observations are dependent or that the process is not stable. A plot of points in time sequence that is random around the mean for the data set suggests that the assumptions of independence and constant probability distribution are met.

A *histogram* or *frequency polygon* of the data can be helpful by revealing whether or not the distribution has several modes. The finding of several modes would suggest that the probability distribution may not have been constant.

A *scatter plot* of each observation against the preceding observation is also useful. It should show no pattern if the observations are independent.

FIGURE 9.1

Time series plot and scatter plot of number of compressors assembled—Manufacturing process example.   The data set exhibits no trend or cyclical persistence and therefore may be treated as a random sample from the process.

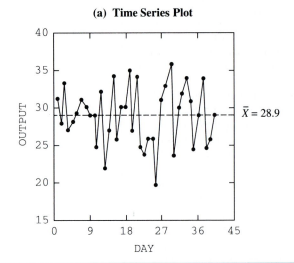

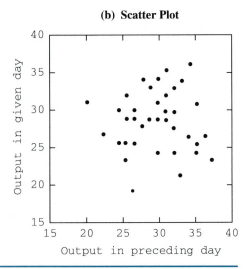

We shall illustrate the use of these informal diagnostic procedures by three examples. In Chapter 15, we shall take up a formal statistical test that is helpful for deciding whether a data set can be treated as a simple random sample.

EXAMPLE ☐ **Manufacturing Process**

Figure 9.1a shows the daily number of compressor units assembled in a manufacturing process (OUTPUT) for the most recent 40 days, plotted in time order (DAY). The mean number of units assembled per day, $\bar{X} = 28.9$, is shown also on the plot. No trend is evident in the data, nor do there appear to be any cyclical swings. The points appear to fall randomly around the central line.

Figure 9.1b shows a scatter plot of the daily number of compressors assembled, with each day's observation plotted against that of the preceding day. The scatter appears to be random, with no pattern in evidence.

These diagnostic checks suggest that the manufacturing process has been stable and that the 40 observations can be treated as a simple random sample from the infinite population associated with the process. ☐

EXAMPLE ☐ **Oil Discoveries**

Figure 9.2 shows the volume of oil (in million cubic meters) for each of the 20 oil discoveries in a field, in order of discovery. The mean volume per discovery for the data set, $\bar{X} = 30.35$ million cubic meters, is also shown in the plot. The plot shows that the volume of oil, with some variations from one discovery to the next, has tended to decline with the order of discovery. This suggests that the probability distribution has not been stable and has tended to shift downward. Therefore, one can conclude that the data set does not qualify as a simple random sample. ☐

Time series plot of volume of oil—Oil discoveries example. The data set exhibits a trend and therefore does not constitute a simple random sample.

FIGURE 9.2

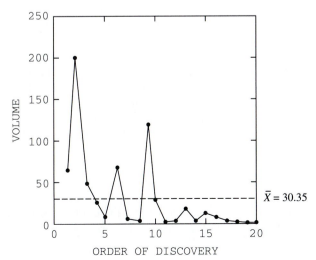

FIGURE 9.3

Scatter plot of price of eggs in given week and price in preceding week—Egg prices example.
The data set exhibits dependence and therefore does not constitute a simple random sample.

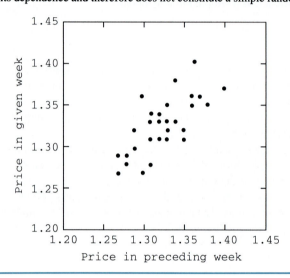

Egg Prices **EXAMPLE**

A scatter plot of data on the price of grade A eggs (in dollars per dozen) received by
Canadian producers for 30 consecutive weeks is shown in Figure 9.3, with the price in
a given week plotted against the price in the preceding week. The scatter plot shows a
distinct positive relationship—the higher the price is in a week, generally the higher
the price is in the following week. If the consecutive prices were independent, the
scatter plot would exhibit no systematic relationship. Thus, it can be concluded that
the egg price data set does not constitute a simple random sample.   □

9.5   **SAMPLE
STATISTICS AND
POPULATION
PARAMETERS**

In the introduction to this chapter we noted that statistical inference is concerned with
drawing sound conclusions about population characteristics from the corresponding characteristics of a sample. The characteristics of interest may be of many kinds, but they principally include measures of position (such as the mean or median), measures of variability
(such as the range or variance), and other measures that we discussed in Chapter 3. To
distinguish the characteristics of a population from the characteristics of a sample, we shall
use different terms for them.

(9.11)
A characteristic of a population is referred to as a **population parameter,** or **parameter** for short. A characteristic of a sample is referred to as a **sample statistic,** or
statistic for short.

Sample Statistics

When the data set $X_1, X_2, \ldots, X_n$ represents a sample of size n from a population, the various summary measures described in Chapter 3 may be computed. Henceforth, we shall use the notation for summary measures in Chapter 3 to represent sample statistics. For example, when we are dealing with a sample data set, $\overline{X}$ as defined in (3.1) will denote the *sample mean,* s^2 as defined in (3.12) will denote the *sample variance,* and so on. The number of observations in the sample data set, $n,$ will be referred to as the *sample size.*

Population Parameters

Parameters are defined for populations in a similar fashion as statistics are defined for samples. We need, however, to maintain a distinction between finite and infinite populations in these definitions. We first consider a finite population.

Finite Population.   Let $X_1, X_2, \ldots, X_N$ be the observations on the N elements of a finite population. We shall refer to N as the *finite population size.* To distinguish population parameters from sample statistics, we shall use different symbols for parameters—generally, the symbols will be Greek letters. For example, we shall use μ (Greek mu) to denote the *population mean,* σ (Greek sigma) to denote the *population standard deviation,* and σ^2 (read "sigma squared") to denote the *population variance.* The definitions of the population mean, variance, and standard deviation follow. Definitions for other population parameters will be introduced in later chapters as required.

(9.12)
For a finite population with observations $X_1, X_2, \ldots, X_N,$ the population mean, variance, and standard deviation are defined as follows:

(9.12a)   *Population Mean*

$$\mu = \frac{\sum_{i=1}^{N} X_i}{N}$$

(9.12b)   *Population Variance*

$$\sigma^2 = \frac{\sum_{i=1}^{N} (X_i - \mu)^2}{N}$$

(9.12c)   *Population Standard Deviation* $\sigma = \sqrt{\sigma^2}$

Note that the definition of a population mean μ is the same as that for a sample mean $\overline{X}$ in (3.1). Also, a population variance σ^2 is defined almost the same as a sample variance s^2 in (3.12), except that the denominator is N and not $N - 1$. When N is large, the difference in the two definitions is negligible. We shall explain the reason for this difference in definitions in a later chapter. For the moment, remember to use divisor N for the variance when the data set refers to a finite population and $n - 1$ when it refers to a sample.

EXAMPLE

Refer again to the Ph.D. students example in which the finite population consists of $N = 5$ Ph.D. students. For convenience, we repeat their current grade point averages:

$$X_1 = 3.74 \qquad X_2 = 3.89 \qquad X_3 = 4.00 \qquad X_4 = 3.68 \qquad X_5 = 3.69$$

The population mean is:

$$\mu = \frac{3.74 + 3.89 + 4.00 + 3.68 + 3.69}{5} = 3.80$$

The population variance is:

$$\sigma^2 = \frac{(3.74 - 3.80)^2 + (3.89 - 3.80)^2 + \cdots + (3.69 - 3.80)^2}{5} = 0.01564$$

Thus, the population standard deviation is $\sigma = \sqrt{0.01564} = 0.125$.   □

Infinite Population.   The parameters of an infinite population, represented by random variable X, are the characteristics of the associated probability distribution. Thus, the population mean is the expected value of X, and the population variance is the variance of X.

(9.13)
For an infinite population represented by random variable X, the population mean, variance, and standard deviation are defined as follows:

(9.13a)   *Population Mean* $\mu = E\{X\}$

(9.13b)   *Population Variance* $\sigma^2 = \sigma^2\{X\}$

(9.13c)   *Population Standard Deviation* $\sigma = \sqrt{\sigma^2} = \sigma\{X\}$

□ **EXAMPLES**

1. Consider the infinite population pertaining to the process of molding tires in pairs, as described in the tire molding example. The random variable X is the number of defective tires in the pair. The probability distribution for X is:

x:	0	1	2
$P(x)$:	0.75	0.10	0.15

Using definition (5.5) for the expected value of a discrete random variable, the population mean is:

$$\mu = E\{X\} = 0(0.75) + 1(0.10) + 2(0.15) = 0.40$$

Using definition (5.6) for the variance of a discrete random variable, the population variance is:

$$\sigma^2 = \sigma^2\{X\} = (0 - 0.40)^2(0.75) + (1 - 0.40)^2(0.10) + (2 - 0.40)^2(0.15)$$
$$= 0.54$$

Hence, the population standard deviation is $\sigma = \sigma\{X\} = \sqrt{0.54} = 0.73$.

2. The infinite population of interest pertains to the heights of 20-year-old trees of a

particular species (X; in feet). The probability distribution of X is $N(38, 9)$. Thus, the population of interest is normally distributed, and the population mean and variance are $\mu = 38$ and $\sigma^2 = 9$, respectively. □

Comment

The parameters of a finite population may be related to those of a probability distribution in the following way. Consider selecting one household from a population of N households in such a way that each household in the population has equal probability of being chosen. The probability of any household being chosen must then be $1/N$. Let X be the random variable denoting the household income observed for the household selected at random. The random variable X can then take on the values $X_1, X_2, \ldots, X_N$ (the household incomes in the population) with probability $1/N$ each. Hence, by definition of the expected value of a discrete random variable in (5.5), we have:

$$E\{X\} = \sum_{i=1}^{N} X_i \left(\frac{1}{N}\right) = \frac{\sum_{i=1}^{N} X_i}{N} = \mu$$

Similarly, by definition of the variance of a discrete random variable in (5.6), we have:

$$\sigma^2\{X\} = \sum_{i=1}^{N} (X_i - \mu)^2 \left(\frac{1}{N}\right) = \sigma^2$$

Thus, the population mean μ and variance σ^2 for a finite population correspond, respectively, to the expected value and variance of the random variable associated with the equal-probability selection of one population element.

Inferences from Sample Statistics About Population Parameters

Intuition tells us that the sample mean $\overline{X}$ provides information about the population mean μ, and that the sample standard deviation s provides information about the population standard deviation σ. In succeeding chapters, we examine how the information provided by these and other sample statistics may be used to make inferences about the population parameters of interest.

PROBLEMS

 9.1 In each of the following cases, identify the population and explain whether it is finite or infinite.

a. Tax returns for last year are sampled to estimate the mean deduction for interest expenses.

b. A company sends a large number of letters by mail to estimate the mean number of days required for delivery.

c. Outgoing orders for books are sampled prior to mailing to estimate the proportion of orders filled incorrectly by the current process.

9.2 In each of the following cases, identify the population and explain whether it is finite or infinite.

a. A poultry specialist feeds a hen flock an experimental high-mineral diet to measure its impact on the strength of eggshells.

b. The children of Westview Elementary School are sampled to estimate the proportion who have had a dental checkup in the past year.

c. An auditor studies the accuracy of the total amount of fixed assets recorded by a large company by examining a sample of the fixed assets.

* **9.3** The number of photocopying machines X in a large office that are out of order at any time has the following probability distribution:

x:	0	1	2	3	4
$P(x)$:	0.80	0.11	0.05	0.03	0.01

Describe the infinite population associated with this probability distribution. What is the proportion of the observations in this infinite population that have the value 0?

9.4 An office building contains three elevators. The probability distribution of the number of elevators X waiting on the ground floor at any time during business hours follows:

x:	0	1	2	3
$P(x)$:	0.40	0.30	0.20	0.10

Describe the infinite population associated with this probability distribution. What is the proportion of the observations in this infinite population that have the value 2 or more?

9.5 The number of telephone calls handled in an hour by a commodity trader is a Poisson random variable X with parameter value $\lambda = 5$. Describe the infinite population associated with this probability distribution. What is the proportion of the observations in this infinite population that have the value 5?

9.6 For each of the following, discuss whether a sample or a census would be preferable. Indicate any assumptions you make.

a. A survey of Canadian households to obtain information on the proportion of households that used a microwave oven at least once during the preceding week.

b. An examination of three turbines to be installed in a dam to obtain information about possible damage in shipment.

c. An examination of a large shipment of light bulbs to obtain information on the proportion of bulbs that burn out prematurely in normal use.

9.7 For each of the following, discuss whether a sample or a census would be preferable. Indicate any assumptions you make.

a. A survey of the 25 universities in a regional association to obtain current enrollment data.

b. A survey of the 3000 employees of a company to study their attitudes toward the company.

c. An examination of dwellings in a large city to obtain information about the extent of present home insulation and the cost to bring substandard dwellings up to minimum insulation standards.

9.8 An appliance manufacturer wishes to study consumers' color preferences for refrigerators.

a. Identify the population of interest in this study.

b. Would a sample or a census be preferable to obtain the desired information? Discuss.

9.9 Respond to each of the following arguments.

a. "Samples are to be preferred to censuses because samples cost less."

b. "Censuses are to be preferred to samples because censuses are more accurate."

9.10 The families living in a city constitute the population of interest. A sample survey of these families was conducted to estimate the mean family expenditures on children's footwear during the past 12 months. The mean expenditures reported in the survey were $188. The responses in this survey were based on unaided recall of expenditures by respondents and were frequently in error. Suppose, in fact, that the actual mean expenditures for the families in the survey were $253. Furthermore, suppose that a census of this population, using the same survey methodology (based on unaided recall), would have yielded mean expenditures of $175. What is the magnitude of the sampling error in the reported mean expenditures? What is the magnitude of the nonsampling error?

9.11 A shipment of 10,000 transistors is the population of interest. One key operating characteristic of a transistor is its gain. A test instrument was used to measure the gain of each transistor in a sample of 200 transistors from the shipment. The sample test results showed that 1.00 percent of the 200 transistors failed to meet the quality specification for gain. Suppose, in fact, that 0.45 percent of the 10,000 transistors would fail to meet the quality specification for gain if all were tested by this instrument.

a. If the test instrument is known to be perfectly calibrated, what is the magnitude of the sampling error in the sample percentage of transistors that failed to meet the quality specification? What is the magnitude of the nonsampling error?

b. If the test instrument is imperfectly calibrated to an unknown degree, what can be said about the magnitudes of the sampling and nonsampling errors in the sample result?

9.12 A manufacturer of veterinary surgical supplies plans to survey a sample of practicing veterinarians to obtain their opinions of a new surgical instrument. Veterinarians participating in the survey will be given two of the surgical instruments as a gift for their participation. A survey consultant points out that the gift may cause respondents to express opinions that are more favorable toward the new instrument than they would express without this gift. Is the consultant concerned here about a sampling error or a nonsampling error? Explain.

9.13 An information specialist in O.K. Financial Services, Inc., selected six branch offices in which to test a new computerized decision support system, because these offices represent the variety of informational requirements found in the company's branch offices throughout the country.

a. Do these six branch offices represent a probability sample, a judgment sample, or a convenience sample? Explain.

b. Would your answer in part a change if the offices were selected because they were within easy driving distance of the company's home office? Discuss.

9.14 The names of 200 students were selected from a computer file that contains the names of all students currently attending a university. The selection was done in a manner that gave each student in the file an equal chance to be selected. Each selected student is now to be interviewed in depth about problems of coping with examination stress.

a. Do the 200 selected students constitute a probability sample or a judgment sample? Explain.

b. If some of the selected students cannot be contacted for an interview or refuse to be interviewed, will the completed study still be based on the kind of sample indicated in your answer to part a? Explain.

9.15 An association of professional personnel managers has 7200 members of whom 830 attended the association's annual conference in Chicago. At the conference, each attending member was given a questionnaire concerning preferences for the location, format, and dates of the next conference. A total of 410 of these questionnaires were completed and returned.

a. Do the respondents constitute a probability sample, a judgment sample, or a convenience sample of the association's membership? Explain.

b. Do the respondents constitute a probability sample, a judgment sample, or a convenience sample of the members attending the conference in Chicago? Explain.

* 9.16 A simple random sample of two stores is to be selected without replacement from a population of four stores (A, B, C, D).

a. List the different possible sample combinations.

b. What probability of selection must each sample combination in part a have?

c. Must you list all the possible sample combinations to select a simple random sample? Explain.

9.17 A simple random sample of three cars is to be selected without replacement from a population of six cars (A, B, C, D, E, F).

a. List the different possible sample combinations.

b. What probability of selection must each sample combination in part a have?

c. Must you list all the possible sample combinations to select a simple random sample? Explain.

9.18 **Computer Users.** The computer users of a company have identification codes numbered consecutively from 001 to 630. Ten percent of the users are to be surveyed about their potential use of a new programming package. One of the 10 digits from 0 to 9 will be selected randomly, and every user with an identification code ending with this selected digit will be surveyed.

a. State the probability that (1) digit 8 will be the one selected, (2) the user with identification code 228 will be selected.

b. Does the set of users selected for the survey in this manner constitute a probability sample? A simple random sample? Explain.

9.19 **Retail Establishments.** The retail establishments operating in a county in March are to be sampled, using a directory published in October of the preceding year.

a. What is the frame here?

b. Describe several ways in which the sampled population may differ from the target population.

9.20 The adult population of a city is to be sampled next week. Some persons will be out of town during that time. What is the target population here? What is the sampled population?

9.21 Bulk-food stores that carry a line of exotic spices are to be sampled, using a list of all bulk-food stores. Each store selected from the list will be contacted initially to learn if it carries the line of spices. If it does, the store will be kept in the sample. Otherwise, the store will be discarded from the sample. What is the population from which the final sample is drawn here? Is it the same as the target population? Comment.

9.22 Refer to **Computer Users** Problem 9.18.

a. Explain how you would use a table of random digits to select a simple random sample of 63 computer users without replacement.

b. If the users' codes were divided into the three groups (001–210, 211–420, 421–630) and simple random samples of 21 codes each were drawn without replacement and independently from the three groups, would the combined sample of 63 users constitute a simple random sample of the population of 630 users? Explain.

9.23 Refer to **Retail Establishments** Problem 9.19. The retail establishments directory contains 100 pages. Fifty establishments are listed to a page, except on the last page, on which 31 establishments are listed.

a. Explain how you would use a table of random digits to select a simple random sample of 100 establishments without replacement from the directory.

b. If the table of random digits were used to select one establishment at random from each page of the directory, would the resultant sample be a probability sample? A simple random sample? Explain.

9.24 In each of the following cases, explain how you would select a simple random sample without replacement by means of random numbers generated by a computer or a statistical calculator.

a. Fifty school buses to be selected from the 4000 school buses registered in a state.

b. One hundred floor tiles to be selected from a warehouse floor laid with 150 rows of such tiles, 70 tiles to the row.

c. One hundred points in time (designated in minutes) to be selected during the working hours next week (9:00 A.M. to 5:00 P.M. Monday through Friday).

9.25 In each of the following cases, explain how you would select a simple random sample without replacement by means of a table of random digits.

a. Eighty members to be selected from the 760,000 members of an automobile association.

b. Forty-five time points (designated in minutes) to be selected during the five hours of a production run.

c. Five 1-cubic-centimeter cubes to be selected from a block of wood measuring (in centimeters) $15 \times 15 \times 120$.

9.26 Book Orders. The book orders leaving a publisher yesterday were checked at the end of the day to see whether each order was filled correctly ($X = 1$) or not ($X = 0$). For several hours during the day, the experienced employee filling the book orders was replaced by a substitute who tends to make more errors. Do the data on the correctness of the filled orders for the day constitute a random sample from an infinite population? Explain.

9.27 Computer Breakdowns. The time intervals X between successive breakdowns of a computer have been measured during the past month. During this period, a persistent problem existed that was fixed temporarily several times, each time being followed shortly by another breakdown, until finally the problem was properly diagnosed and corrected. Do the data on time intervals between successive breakdowns for the past month constitute a random sample from an infinite population? Explain.

9.28 Refer to **Book Orders** Problem 9.26. Suppose that for subsequent days, the correctness of the filled orders constitutes a Bernoulli process as defined in (7.4). Does a sequence of outcomes for five consecutive orders constitute a simple random sample from the infinite population defined by this process? Explain.

9.29 Refer to **Computer Breakdowns** Problem 9.27. Suppose that for subsequent months, the breakdowns of the computer occur according to a Poisson process as defined in (8.15). Four successive time intervals between breakdowns have been observed recently. Do these constitute a simple random sample from the infinite population defined by this process? Explain.

9.30 Refer to the time series plot for the **Capital Expenditures** example in Figure 2.12. Do the 10 observations on annual capital expenditures appear to constitute a simple random sample from a process? Comment.

9.31 Weekly sales of automobiles by a small dealer for a 20-week period follow:

Week:	1	2	3	4	5	6	7	8	9	10
Sales:	10	10	9	9	13	10	9	7	15	12
Week:	11	12	13	14	15	16	17	18	19	20
Sales:	10	8	7	8	10	8	12	11	11	6

a. Construct a time series plot of the weekly sales data.
b. Construct a scatter plot of sales each week against sales of the preceding week.
c. Do the plots in parts a and b show any evidence that the data do not constitute a simple random sample from a process? Comment.

9.32 The excess weights of jam (in grams) in 20 successive jars leaving a filling machine follow:

Jar:	1	2	3	4	5	6	7	8	9	10
Excess Weight:	5	6	6	6	8	7	7	6	10	9
Jar:	11	12	13	14	15	16	17	18	19	20
Excess Weight:	9	8	8	8	10	9	11	11	12	9

a. Construct a time series plot of the excess weight data.
b. Construct a scatter plot of the excess weight of each jar against the excess weight of the preceding jar.
c. Do the plots in parts a and b show any evidence that the data do not constitute a simple random sample from a process? Comment.

9.33 Refer to **Data Transmission** Problem 2.1. The successive observations are arranged from left to right, the first row pertaining to the first 12 hours of the day and the second row pertaining to the remaining hours of the day.

a. Construct a time series plot of the data.

b. Construct a scatter plot of the number of erroneous data bits transmitted in each hour against the number of erroneous data bits transmitted in the preceding hour.

c. Do the plots in parts a and b show any evidence that the data do not constitute a simple random sample from a process? Comment.

9.34 The following table classifies the occurrence of rain or no rain on each of 60 consecutive days in a tropical city according to the outcome on the preceding day:

Preceding Day	Current Day	
	Rain	No Rain
Rain	30	6
No rain	7	17

Do the tabulated data appear to constitute a simple random sample from a process? Comment.

* 9.35 A population consists of seven employees whose ages (in years) are as follows:

Employee:	1	2	3	4	5	6	7
Age:	27	34	47	44	37	29	41

a. Obtain the population mean and the population standard deviation. In what units is each measure expressed?

b. Employees 2, 3, and 6 were selected for a sample. Obtain the sample mean and the sample standard deviation. What is the magnitude of the sampling error in the sample mean as an estimate of the population mean?

9.36 A population consists of five assembly plants. The floor areas of these plants (in thousands of square meters) are as follows:

Plant:	1	2	3	4	5
Area:	10.1	6.8	12.4	7.9	8.0

a. Obtain the population mean and the population variance. Are these measures parameters or statistics?

b. Plants 2 and 5 were selected for a sample. Obtain the sample mean and the sample variance. What is the magnitude of the sampling error in the sample mean as an estimate of the population mean?

9.37 A population consists of six stores. Their annual sales last year (in $ million) were as follows:

Store:	1	2	3	4	5	6
Sales:	7.1	6.8	5.3	10.4	8.0	9.9

a. Obtain the population mean and the population standard deviation. Are these measures parameters or statistics? In what units are these measures expressed?

b. Stores 2 and 6 were selected for a sample. Obtain the sample mean and the sample standard deviation. What is the magnitude of the sampling error in the sample mean as an estimate of the population mean?

9.38 Refer to **Book Orders** Problems 9.26 and 9.28. Assume that the process currently is a Bernoulli process with $p = 0.995$.

a. What are the population mean and the population variance? Are these measures parameters or statistics?

b. A simple random sample of five observations from this process yields 1, 1, 1, 1, 1. Obtain the sample mean and the sample variance. Are these measures equal to the population mean and the population variance, respectively? Would you expect them to be? Explain.

9.39 Refer to **Computer Breakdowns** Problems 9.27 and 9.29. Assume that the process currently is a Poisson process with $\lambda = 2.5$ breakdowns per week.

a. What are the population mean and the population standard deviation of the time intervals X between successive breakdowns? In what units is each measure expressed?

b. A simple random sample of four time intervals from this process yields 0.132, 0.100, 0.344, 1.564. Obtain the sample mean and the sample standard deviation. Are these measures equal to the population mean and the population standard deviation, respectively? Would you expect them to be? Explain.

9.40 A population consists of four resistors. Their resistances (in ohms) are as follows:

Resistor:	1	2	3	4
Resistance:	108	103	104	112

One of these four resistors is chosen in a manner that gives each an equal probability of being selected. Let random variable X denote the resistance of the chosen resistor.

a. Describe the probability distribution of X.

b. Obtain $E\{X\}$ and $\sigma^2\{X\}$. Are these two measures the same as the population mean and the population variance, respectively? Explain.

EXERCISES

9.41 For a population of $N = 4$ distinct elements from which a random sample of $n = 2$ elements is to be selected, show that procedure (9.8) yields equal probabilities for each possible sample combination. Draw a probability tree representing selection process (9.8) here and identify the outcomes constituting a given sample combination.

9.42 A simple random sample of $n = 4$ elements is to be drawn without replacement from a population of $N = 11$ distinct elements.

a. How many different sample combinations can be drawn from the population?

b. What is the probability that a particular one of these sample combinations will be chosen?

c. What is the probability that a particular population element will appear in the selected random sample?

9.43 A two-digit number is to be selected from a table of random digits. What is the probability that (1) it will be 12, (2) it will be even, (3) the two digits will be the same?

9.44 In a table of random digits, what is the probability that (1) three successive digits are each 2, (2) three successive digits are the same?

9.45 A simple random sample of 200 vouchers is to be selected from a population of 3000 vouchers that are serially numbered from 0001 to 3000. The sampling plan calls for the selection of four-digit random numbers. If the first digit is 0, 1, or 2, it will be treated as 0; if it is 3, 4, or 5, it will be treated as 1; if it is 6, 7, or 8, it will be treated as 2; and if it is 9, the four-digit number will be discarded. The other three digits in the number will remain unchanged. Thus, 1245 becomes 0245; 7391 becomes 2391; and 9218 is discarded. The transformed number 0000 will correspond to voucher 3000. Is this a valid way of obtaining a simple random sample in this case? Explain.

9.46 A city planning commission, using an up-to-date listing of all addresses in the city, will select a simple random sample of 300 addresses for a survey of households. The number of households that currently live at each address may be zero, one, two, or more than two.

a. Suppose that all households living at the 300 selected addresses will be included in the survey. Does each household in the city have an equal chance of being included? Will the households at the 300 addresses constitute a simple random sample of households in the city? Explain.

b. Alternatively, suppose that when a selected address contains more than one household, one of those households is chosen at random for inclusion in the survey and the other households at that address are not included. Will the households included in the survey constitute a simple random sample of all households in the city? Explain.

9.47 Explain why simple random sampling with replacement from a finite population is equivalent to simple random sampling from an infinite population. [*Hint:* Refer to definition (9.10).]

STUDIES

9.48 Refer to the **Financial Characteristics** data set (Appendix D.1). The population of textile apparel manufacturing firms is to be sampled to obtain information about net sales in year 2. You are asked to select a simple random sample of 10 firms without replacement.

a. How many different possible sample combinations of 10 firms can be selected from this population?

b. Using random numbers, select a random sample of 10 firms. Explain how you used the random numbers.

c. Calculate the sample mean and the sample standard deviation.

d. The population mean is $\mu = \$153.43$ million, and the population standard deviation is $\sigma = \$256.00$ million. Given the large variability in the population [the coefficient of variation is $100(256.00/153.43) = 167$ percent], do you expect the sample mean and the sample standard deviation based on a sample of 10 firms to be close to the population parameters? How close are your sample statistics in part c to the corresponding population parameters?

9.49 Ten hospitals have agreed to participate in a joint research program. The research procedures are to be tested initially in a simple random sample of 3 of these 10 hospitals:

Carmel	Maple Hill	Southridge
Central Valley	Mercy	St. Francis
Community	Northmount	Winchester General
Franklin General		

a. Using random numbers, select a simple random sample of $n = 3$ hospitals without replacement. Explain how you used the random numbers.

b. To study how a different random sample of $n = 3$ hospitals might be composed, select another simple random sample of 3 hospitals without replacement from the 10 hospitals in the population, using random numbers independent of those used in part a. How much overlap is there between the hospitals in the samples in parts a and b? What is the probability of no overlap in the second sample, given the results for your first sample? What is the probability of no overlap in advance of selecting either of the two samples?

c. Continue selecting independent random samples of $n = 3$ hospitals from the population of 10 hospitals until you have selected 50 random samples of $n = 3$. For each hospital in the population, determine the number of samples in which it was selected. What is the expected number for each hospital? Are the actual frequencies close to the expected frequencies? Discuss.

9.50 Refer to the **Investment Funds** data set (Appendix D.3). Consider only the sequence of quarterly unit values for the equity fund. Denote these values by V_i ($i = 0, 1, \ldots, 44$). Note that $V_0 = 1.0780$.

a. Calculate the percent changes X_i in the quarterly unit values, using the transformation

$$X_i = 100\left(\frac{V_i - V_{i-1}}{V_{i-1}}\right) \qquad i = 1, 2, \ldots, 44$$

b. Construct a time series plot of the percent change data.

c. Construct a scatter plot of each percent change against the preceding percent change.

d. Do the plots in parts b and c show any evidence that the data do not constitute a simple random sample from a process? Comment.

Point Estimation and Sampling Distribution of $\overline{X}$

10

We noted in the previous chapter that samples are selected to provide information about populations and, in particular, that sample statistics are used to draw inferences about population parameters. In this chapter, we begin a study of the use of sample statistics to draw inferences about population parameters. We first consider inferences about the population mean because it is often the population parameter of interest, as the following examples illustrate.

EXAMPLES

1. A legislative committee needs to know whether unemployed persons tend to be unemployed for a short time while shifting to another job or for a long time because they are unable to find work. A random sample of unemployed persons will be selected to estimate the mean duration of unemployment (μ) for persons currently unemployed.

2. An environmental control agency requires information about the mean pollution emission per car (μ) for a particular make of passenger car. A random sample of cars from the production line will be selected and tested to estimate the population mean pollution emission for the manufacturing process.

3. An auditor needs information about the mean audit amount per account (μ) in a population of N accounts to estimate the total audit amount for all the accounts in the population. The total audit amount is given by $N\mu$. A random sample of accounts will be used to obtain information about the total audit amount for all accounts in the population.

10.1 POINT ESTIMATION

Information derived from a sample about a population parameter, such as the population mean μ, usually takes the form of a sample statistic that is calculated from the sample observations. We call the sample statistic an *estimate* of the population parameter.

EXAMPLE — Length of Service

An estimate of the mean length of service (μ) for a population of 3580 employees of a company was required for collective-bargaining negotiations. The personnel office selected a simple random sample of 50 employees, determined the length of service for each, and calculated the sample mean $\overline{X}$ of the 50 lengths of service. The sample mean was $\overline{X} = 6.3$ years. Thus, 6.3 years is the estimate of μ here.

(10.1)

The process of estimating a population parameter by a single number derived from the sample is called **point estimation.**

Features of Point Estimation

The length of service example has illustrated the main features of point estimation. These features are as follows.

1. *Parameter.* An unknown population parameter is to be estimated. We shall use the symbol θ (Greek theta) to represent this parameter in the general case.

2. *Estimate.* A sample of n observations, $X_1, X_2, \ldots, X_n$, is selected from the population. Some statistic, which is a function of these n sample observations, is used as an estimate of the parameter θ. We shall use the symbol S to represent this statistic in the general case.

3. *Sampling distribution.* Prior to the selection of the actual sample, the sample observations $X_1, X_2, \ldots, X_n$ are random variables, and hence, the statistic S to be calculated is also a random variable. The probability distribution of a statistic is called its sampling distribution.

(10.2)

The probability distribution of a sample statistic S is called the **sampling distribution** of S.

☐ **EXAMPLE**

In the length of service example, the parameter θ corresponds to the population mean μ. The sample statistic S is $\overline{X}$, and from the definition of $\overline{X}$, we know it is the following function of $X_1, X_2, \ldots, X_n$:

$$\overline{X} = \frac{X_1 + X_2 + \cdots + X_n}{n}$$

Prior to the selection of the sample, $\overline{X}$ is a random variable, and its probability distribution is called the sampling distribution of $\overline{X}$. ☐

Since a sample statistic is a random variable prior to sample selection but is simply a number after sample selection, statisticians use different terms to distinguish between these two situations.

(10.3)

Prior to sample selection, a sample statistic is a random variable and is called a **point estimator** of a population parameter. After sample selection, a sample statistic is a number and is called a **point estimate** of the population parameter.

For simplicity, we shall usually drop the adjective "point" and use the terms, *estimator* and *estimate*.

In the length of service example, $\bar{X}$ is used as the point estimator of μ. The number obtained for the particular sample selected in the study, $\bar{X} = 6.3$ years, is the point estimate of μ.

☐

Alternative Point Estimators

A population parameter can usually be estimated by many point estimators. In the length of service example, for instance, the personnel office could have used other estimators besides the sample mean $\bar{X}$ to estimate the population mean length of service μ. Other point estimators include the sample median Md and the sample midrange, which is defined as follows.

(10.4)

The **sample midrange** is the mean of the smallest (X_S) and the largest (X_L) observations in the sample:

$$\frac{X_S + X_L}{2}$$

How should we choose between the different possible point estimators? Essentially, the choice will be made on the basis of the sampling distribution of each point estimator. We shall choose that estimator whose sampling distribution has the most desirable characteristics. Thus, we need to consider next the nature of sampling distributions. We shall do this by examining the sampling distribution of the sample mean $\bar{X}$. This sampling distribution is extremely important because of the wide use of $\bar{X}$ as an estimator of the population mean μ.

**EXPERIMENTAL 10.2
STUDY OF** $\bar{X}$

As we have noted already, the sample mean $\bar{X}$ provides information about the population mean μ. Intuition tells us, however, that if we were to take several simple random samples of given size from a population and calculate the sample mean $\bar{X}$ for each, we would obtain a different value of $\bar{X}$ for each sample. The reason is that different samples usually will include different population elements. While, in practice, we select only a single sample, knowledge about the behavior of the sample statistic $\bar{X}$ in repeated samples from a population will enable us to assess and control the sampling error that is present in any given sample.

We shall now describe an experiment conducted to study the behavior of the sample mean $\bar{X}$ as it varies from one sample to another. After this, we shall present relevant statistical theory and show how the experimental results illustrate the theory. The experiment is based on the following example.

Accounts Receivable ☐ **EXAMPLE**

An auditor conducted an experiment involving the 8042 accounts receivable of a freight company. The auditor wished to demonstrate the use of sampling to estimate the mean audit amount per account for the population of 8042 accounts. The audit amount for an account is the amount that the auditor deems to be correct for the account. The audit amount may differ from the amount carried on the company's books for various reasons, including use of improper tariff, incorrect calculations, and dispute over damages. For purposes of this experiment, each of the 8042 accounts receivable was initially audited. Thus, the audit amount for each account receivable in the population was known. Ordinarily, the auditor would only audit a sample of these accounts and use the sample mean $\bar{X}$ as an estimate of the population mean μ.

Table 10.1 contains a frequency distribution of the audit amounts of the 8042 accounts in the population. The population mean and standard deviation of these amounts, calculated from the actual 8042 observations, are $\mu = \$30.303$ and $\sigma = \$30.334$. Figure 10.1a shows the frequency polygon of the population. The distribution is substantially skewed to the right, which is typical of many populations in business and economics.   ☐

Sampling Distribution of $\bar{X}$

Sample Size $n = 3$.   We begin the study of the behavior of the sample mean $\bar{X}$ by starting with a very small sample size, $n = 3$. A simple random sample of three accounts was selected from the population, and the audit amounts shown in the first row of Table 10.2 were obtained. The sample mean for this sample is:

$$\bar{X} = \frac{\sum X_i}{n} = \frac{30.96 + 38.20 + 22.45}{3} = 30.537$$

Additional random samples of size $n = 3$ were selected, always from the full population of 8042 accounts. Table 10.2 contains the results for the first five of these samples. Note that all five sample means differ from one another and that none of them is equal to the population mean $\mu = 30.303$. In fact some of the sample means differ substantially from the population mean. The departures of the sample means from the population mean here are the result of sampling, and represent sampling errors as explained in Chapter 9.

TABLE 10.1
Frequency distribution of the audit amounts of 8042 accounts receivable— Accounts receivable example

Audit Amount (dollars)	Number of Accounts	Audit Amount (dollars)	Number of Accounts
0–under 10	1008	70–under 80	148
10–under 20	2856	80–under 90	99
20–under 30	1926	90–under 100	116
30–under 40	780	100–under 150	244
40–under 50	326	150–under 200	121
50–under 60	258	Total	8042
60–under 70	160		

FIGURE 10.1

Frequency polygons for the sampling experiment—Accounts receivable example. The population is highly skewed to the right. As the sample size n increases, the sampling distribution of $\bar{X}$ remains centered at μ, becomes less variable and less skewed, and approaches a normal distribution as n becomes large.

(a) Population

Mean	Standard Deviation
30.303	30.334

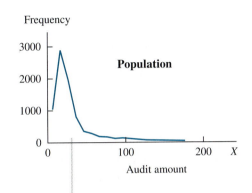

(b) 600 $\bar{X}$ Values

$n = 3$

Mean	Standard Deviation
30.68	17.60

$n = 10$

Mean	Standard Deviation
30.23	9.13

$n = 100$

Mean	Standard Deviation
30.31	3.05

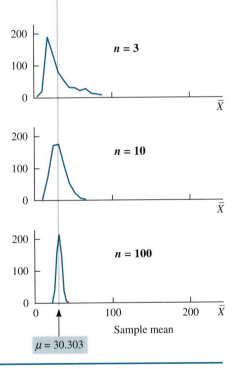

Altogether, 600 samples of size $n = 3$ were selected for this experiment. Figure 10.1b for $n = 3$ shows a frequency polygon of the 600 sample means, as well as the mean and standard deviation of the 600 $\bar{X}$ values. Three important results emerge.

1. While the 600 sample means are widely divergent, their mean of 30.68 is close to the population mean $\mu = 30.303$.
2. The standard deviation of the 600 $\bar{X}$ values, which is 17.60, shows that the variability of the sample means is substantially smaller than the variability of the audit amounts

TABLE 10.2
Sample means $\overline{X}$ for five random samples of $n = 3$—Accounts receivable example

Sample	Observation 1	2	3	$\overline{X}$
1	30.96	38.20	22.45	30.537
2	18.91	6.75	15.45	13.703
3	10.60	14.08	9.15	11.277
4	51.82	20.76	50.79	41.123
5	23.05	31.20	25.15	26.467

in the population ($\sigma = 30.334$). In fact, the standard deviation of the 600 $\overline{X}$ values is only about six-tenths as great as the population standard deviation.
3. The distribution of the 600 sample means is skewed to the right, like the population, but not as much.

Sample Size $n = 10$. Next in the experiment, 600 random samples of size $n = 10$ were selected from the population of 8042 accounts receivable. Figure 10.1b for $n = 10$ shows the frequency polygon of these 600 $\overline{X}$ values, as well as their mean and standard deviation. Four important results are evident.

1. The 600 sample means for sample size $n = 10$ have a mean of 30.23, which is close to the population mean $\mu = 30.303$. This was also true for sample size $n = 3$.
2. The standard deviation of the 600 $\overline{X}$ values, which is 9.13, is substantially smaller than the population standard deviation $\sigma = 30.334$. In fact, it is only about three-tenths as large.
3. The 600 $\overline{X}$ values for sample size $n = 10$ are less variable than the 600 $\overline{X}$ values for sample size $n = 3$. The respective standard deviations are 9.13 and 17.60.
4. The distribution of the 600 $\overline{X}$ values for sample size $n = 10$ is only moderately skewed to the right, in contrast to the marked positive skewness of the population.

Sample Size $n = 100$. The final part of the experiment was to select 600 random samples of size $n = 100$. Figure 10.1b for $n = 100$ shows the relevant results. The following conclusions are of interest.

1. The mean of the 600 sample means, 30.31, again is close to the population mean $\mu = 30.303$.
2. The standard deviation of the 600 $\overline{X}$ values is smaller than that for either sample size $n = 3$ or $n = 10$, and is only about one-tenth as great as the population standard deviation.
3. The distribution of the 600 $\overline{X}$ values is quite symmetrical and appears, when enlarged and smoothed, to be similar to a normal distribution.

Empirical Conclusions

On the basis of this experiment, the following conjectures appear to be warranted:

1. The distribution of $\overline{X}$ values for simple random sampling is centered around the population mean, regardless of sample size.
2. The standard deviation of the $\overline{X}$ values decreases with increasing sample size; that is, the distribution of $\overline{X}$ values becomes more concentrated around the population mean as the sample size gets larger.

3. The distribution of the $\bar{X}$ values becomes more symmetrical as the sample size gets larger and is approximately normal for large sample sizes.

We now turn to theoretical results about the behavior of the sample mean $\bar{X}$, which confirm the conjectures from our experiment. These theoretical results apply to simple random sampling from:

THEORETICAL 10.3 RESULTS ABOUT $\bar{X}$

1. Infinite populations
2. Finite populations whenever the sample size n is small relative to the population size N

The latter case applies to the experimental study in which $N = 8042$ and n varied from 3 to 100.

Sampling Distribution of $\bar{X}$

As we noted earlier, a sample statistic is a random variable in advance of sampling, and the associated probability distribution is called the sampling distribution of the statistic. The probability distribution associated with the sample mean $\bar{X}$ is called the *sampling distribution of $\bar{X}$* or the *sampling distribution of the mean*. In our experiment, the 600 $\bar{X}$ values obtained for a given sample size n represent 600 observations from the sampling distribution of $\bar{X}$ for that sample size.

There is a different sampling distribution of $\bar{X}$ for each population and sample size. We now consider the characteristics of sampling distributions of $\bar{X}$ and how they relate to the population sampled and the sample size.

Expected Value of $\bar{X}$

The expected value of $\bar{X}$, or equivalently, the mean of the sampling distribution of $\bar{X}$, is denoted by $E\{\bar{X}\}$. With simple random sampling, the mean of the sampling distribution of $\bar{X}$ is always equal to the population mean μ.

(10.5) $E\{\bar{X}\} = \mu$

The experimental results are consistent with theorem (10.5).

Sample Size (n)	Mean of 600 $\bar{X}$ Values
3	30.68
10	30.23
100	30.31

Population mean $\mu = 30.303$

Of course, we would not expect the mean of 600 trials to equal μ exactly, because there is always some random variation present in a limited number of trials.

Comment

Theorem (10.5) can be proven readily as follows, using earlier results on expectations.

$$E\{\overline{X}\} = E\left\{\frac{X_1 + X_2 + \cdots + X_n}{n}\right\}$$

$$= \frac{1}{n}E\{X_1 + X_2 + \cdots + X_n\} \qquad \text{by (6.5b)}$$

$$= \frac{1}{n}(E\{X_1\} + E\{X_2\} + \cdots + E\{X_n\}) \qquad \text{by (6.9a)}$$

$$= \frac{n\mu}{n} = \mu \qquad \text{because } E\{X_i\} = \mu \text{ for each } i$$

Variance of $\overline{X}$

The variance of $\overline{X}$ or, equivalently, the variance of the sampling distribution of $\overline{X}$, is denoted by $\sigma^2\{\overline{X}\}$. This variance is often also called the *variance of the mean*. The standard deviation of $\overline{X}$ is denoted by $\sigma\{\overline{X}\}$ and is often called the *standard deviation of the mean* or the *standard error of the mean*.

The variance $\sigma^2\{\overline{X}\}$ and the standard deviation $\sigma\{\overline{X}\}$ measure the variability of the possible sample means $\overline{X}$ that can be obtained with simple random sampling from a population. It turns out that these measures depend on the population variance σ^2 and sample size n in the following simple way.

$$\textbf{(10.6)} \qquad \sigma^2\{\overline{X}\} = \frac{\sigma^2}{n} \qquad\qquad \sigma\{\overline{X}\} = \frac{\sigma}{\sqrt{n}}$$

The experimental results again are consistent with this theorem. From theorem (10.6), we know that the standard deviation of $\overline{X}$ for our experiment is $\sigma\{\overline{X}\} = 30.334/\sqrt{n}$ since $\sigma = 30.334$ here.

Sample Size n	Standard Deviation of 600 $\overline{X}$ Values	$\dfrac{30.334}{\sqrt{n}}$
3	17.60	17.51
10	9.13	9.59
100	3.05	3.03

Again, we must anticipate some variation from theoretical expectations in the results of a limited number of trials.

Effect of Sample Size. Theorem (10.6) indicates that the standard deviation of $\overline{X}$ decreases in inverse proportion to the square root of the sample size. Thus, the larger the sample size, the more concentrated the sampling distribution of $\overline{X}$. This result is the counterpart to our intuition that larger samples lead to more precise results.

However, because the standard deviation of $\overline{X}$ decreases in inverse proportion to the *square root* of the sample size, it becomes increasingly difficult to reduce $\sigma\{\overline{X}\}$ by increasing n. Thus, if the standard deviation of $\overline{X}$ based on sample size $n = 100$ is to be cut in half, the sample size must be quadrupled to $n = 400$. In turn, if that standard deviation is to be cut in half, the sample size must be quadrupled to $n = 1600$.

Effect of Population Variability.   Theorem (10.6) indicates that, for any given sample size, the greater the population variability, the larger the variability of the sampling distribution of $\overline{X}$. Thus, for any given n, $\overline{X}$ tends to vary more about the population mean μ when the population is highly variable than when it is concentrated.

Comment

Theorem (10.6) can be proven readily for simple random sampling from infinite populations. From definition (9.10), we know that in this case the observations must be independent and identically distributed. Hence, we obtain:

$$\sigma^2\{\overline{X}\} = \sigma^2\left\{\frac{X_1 + X_2 + \cdots + X_n}{n}\right\}$$

$$= \frac{1}{n^2}\sigma^2\{X_1 + X_2 + \cdots + X_n\} \qquad \text{by (6.6b)}$$

$$= \frac{1}{n^2}(\sigma^2\{X_1\} + \sigma^2\{X_2\} + \cdots + \sigma^2\{X_n\}) \qquad \text{by (6.9b)}$$

$$= \frac{n\sigma^2}{n^2} = \frac{\sigma^2}{n} \qquad \text{because } \sigma^2\{X_i\} = \sigma^2 \text{ for each } i$$

Central Limit Theorem

One of the most important theorems in statistics is the central limit theorem. This theorem is important because from it we know that, for almost any population, the sampling distribution of $\overline{X}$ is approximately normal when the random sample size is reasonably large.

(10.7)

Central Limit Theorem. For almost all populations, the sampling distribution of $\overline{X}$ is approximately normal when the simple random sample size is sufficiently large.

The experimental results displayed in Figure 10.1 are consistent with theorem (10.7). Although the distribution of the 600 $\overline{X}$ values for $n = 3$ is skewed, it is less skewed for $n = 10$ and is almost symmetrical for $n = 100$. Furthermore, the appearance of the distribution of the $\overline{X}$ values in Figure 10.1b for $n = 100$ suggests normality.

To study this aspect more formally, recall that the population mean and standard deviation are $\mu = 30.303$ and $\sigma = 30.334$. Hence, we know that the sampling distribution of $\overline{X}$ for $n = 100$ has the following characteristics, according to (10.5) and (10.6):

$$E\{\overline{X}\} = \mu = 30.303 \qquad \sigma\{\overline{X}\} = \frac{\sigma}{\sqrt{n}} = \frac{30.334}{\sqrt{100}} = 3.033$$

If the sampling distribution of $\overline{X}$ for $n = 100$ is approximately normal, we can determine from Table C.1 the proportion of $\overline{X}$ values expected to fall in any interval. Consider the interval 30.30 ± 3.50, for example. The deviation 3.50 is equivalent to $3.50/\sigma\{\overline{X}\} = 3.50/3.033 = 1.15$ standard deviations from the mean. We find from Table C.1 that $P(-1.15 \leq Z \leq 1.15) = 0.750$. The actual proportion of the 600 sample means that fell in this interval was 0.752. This is certainly in close agreement with expectations based on the central limit theorem.

We have calculated similar probabilities for several other intervals and obtained the following comparisons of experimental results and theoretical expectations with $n = 100$.

	Experimental Results		Theoretical Expectations
Interval	Number of $\overline{X}$ Values in Interval	Proportion of $\overline{X}$ Values in Interval	Normal Probability Based on Central Limit Theorem
30.30 ± 3.50	451	0.752	0.750
30.30 ± 5.50	559	0.932	0.930
30.30 ± 7.50	594	0.990	0.986
30.30 ± 9.50	599	0.998	0.998

The close accord between the experimental results and the theoretical expectations based on the central limit theorem strongly supports the applicability of the central limit theorem here.

What is a sufficiently large sample size for the central limit theorem to apply? The requisite size depends on the nature of the population and on the degree of approximation to the normal distribution required. In general, for skewed populations such as the one used in the experiment, a larger random sample size is required for the sampling distribution of $\overline{X}$ to be approximately normal than is required for a population that is fairly symmetrical. In the case of a continuous uniform population, such as is illustrated in Figure 8.1, the sampling distribution of $\overline{X}$ will be approximately normal for sample sizes that are as small as 5.

Comment

The central limit theorem applies whenever the population standard deviation σ is finite, which is the case in almost all problems encountered in practice.

Sampling Finite Populations

In presenting theorems (10.5) and (10.6) for the mean and variance of the sampling distribution of $\overline{X}$ and central limit theorem (10.7), we stated that they apply when (1) the population is infinite and (2) when the population is finite but with the sample size n small relative to the population size N; that is, with fraction n/N small.

(10.8)

The ratio n/N of the sample size n to the population size N is called the **sampling fraction**.

A working rule often employed is the following.

> **(10.9)**
> Theorems (10.5) and (10.6) for the mean and variance of the sampling distribution of $\bar{X}$ and central limit theorem (10.7) apply to sampling of finite populations if the sampling fraction is 5 percent or smaller.

In practice, working rule (10.9) is usually met because the sampling fractions generally employed are small. In the audit sampling experiment, for instance, the sampling fraction for $n = 100$ was only $n/N = 100/8042 = 0.012$. Similarly, when a sample of $n = 1500$ households is selected from the population of a state containing $N = 1,000,000$ households, the sampling fraction is only $n/N = 1500/1,000,000 = 0.0015$. Even when a sample of $n = 75$ retail outlets is selected from the population of $N = 1800$ retail outlets of a chain, the sampling fraction is only 0.042.

Throughout this text, unless otherwise noted, we assume that finite population sizes are large relative to sample sizes and that working rule (10.9) is met. In Chapter 23, we take up the case when this rule is not met.

USING THE NORMAL APPROXIMATION FOR $\bar{X}$ 10.4

To use central limit theorem (10.7) for making probability statements about $\bar{X}$ for reasonably large sample sizes, we require the following standardized variable.

> **(10.10)** $$Z = \frac{\bar{X} - E\{\bar{X}\}}{\sigma\{\bar{X}\}} = \frac{\bar{X} - \mu}{\sigma/\sqrt{n}}$$

This variable corresponds to the earlier definition of a standard normal variable in (8.6), except that the variable being standardized is now $\bar{X}$ instead of X.

We illustrate the use of the central limit theorem to make probability statements about $\bar{X}$ with two examples.

EXAMPLES ☐

1. Suppose that the auditor in the accounts receivable example was actually planning to use a simple random sample of 250 accounts receivable. The auditor would like to know the probability that the sample mean will be within \$4 of the population mean audit amount; that is, between \$26.30 and \$34.30.

   To obtain the desired probability, we invoke the central limit theorem, which tells us that the sampling distribution of $\bar{X}$ is approximately normal for that sample size. Further, theorem (10.5) indicates that the mean of the sampling distribution of $\bar{X}$ is equal to the population mean $\mu = 30.303$, and theorem (10.6) states that the standard deviation of the sampling distribution of $\bar{X}$ is:

$$\sigma\{\bar{X}\} = \frac{\sigma}{\sqrt{n}} = \frac{30.334}{\sqrt{250}} = 1.92$$

FIGURE 10.2
Sampling distributions of $\bar{X}$
for $n = 250$ and $n = 100$—
Accounts receivable example

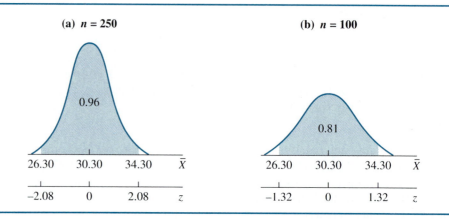

Figure 10.2a shows the sampling distribution of $\bar{X}$ as approximated by the normal distribution. The shaded area represents the desired probability. We calculate the z values for the specified $\bar{X}$ values by (10.10) in the usual fashion:

$$z = \frac{26.30 - 30.30}{1.92} = -2.08 \qquad z = \frac{34.30 - 30.30}{1.92} = 2.08$$

We find that $P(26.30 \le \bar{X} \le 34.30) = P(-2.08 \le Z \le 2.08) = 0.96$.

Thus, there is a very high probability that the auditor will obtain a sample mean that is within \$4 of the population mean. In ordinary practice, of course, the auditor takes only one random sample of 250 accounts, on the basis of which the population mean audit amount is estimated. Since the probability is so high that the sample mean is within \$4 of the population mean for a sample of 250 accounts, the auditor has, by this knowledge, a measure of the sampling error for the particular sample mean $\bar{X}$ that will be obtained.

In this way, the sampling distribution of $\bar{X}$ provides information about the sampling error for the one sample result that is actually obtained—namely, by indicating the probability that the sample mean $\bar{X}$ will be within a given proximity of the population mean μ.

2. To study the effect of sample size, let us find the same probability as in Example 1 for a random sample of 100 accounts. Again, the sampling distribution of $\bar{X}$ will be approximately normal according to the central limit theorem. The mean of the sampling distribution of $\bar{X}$ will still be $\mu = 30.303$, because it is unaffected by sample size. The standard deviation of the sampling distribution of $\bar{X}$ will, however, be larger:

$$\sigma\{\bar{X}\} = \frac{30.334}{\sqrt{100}} = 3.03$$

Figure 10.2b shows the effect of the smaller sample size on the sampling distribution of $\bar{X}$. It is more variable than the distribution in Figure 10.2a for sample size 250. Hence, the probability of $\bar{X}$ falling in a given interval around the population mean μ is smaller for $n = 100$ than for $n = 250$. Let us again find the probability that $\bar{X}$ falls within \$4 of μ. The required z values now are:

$$z = \frac{26.30 - 30.30}{3.03} = -1.32 \qquad z = \frac{34.30 - 30.30}{3.03} = 1.32$$

Hence, $P(26.30 \leq \overline{X} \leq 34.30) = P(-1.32 \leq Z \leq 1.32) = 0.81$. Recall that the corresponding probability for $n = 250$ is 0.96. The larger probability for the larger sample size corresponds to our intuition that a sample mean based on a large sample is a more precise estimator of the population mean than one based on a small sample.   □

In actual practice, the population mean and standard deviation, which are required for obtaining the mean and the standard deviation of the sampling distribution of $\overline{X}$, are usually unknown. We shall explain in the next chapter how the sample mean $\overline{X}$ and the sample standard deviation s can be used to estimate the population mean μ and the population standard deviation σ, respectively, and how these estimates in turn enable us to estimate the mean and the standard deviation of the sampling distribution of $\overline{X}$.

10.5 EXACT SAMPLING DISTRIBUTION OF $\overline{X}$

The central limit theorem tells us the approximate nature of the sampling distribution of $\overline{X}$ when the sample size is reasonably large. In certain cases, the exact sampling distribution of $\overline{X}$ can be ascertained. We shall consider two of these cases.

Normal Population

When the population is normal, the following theorem provides us with the exact sampling distribution of $\overline{X}$.

> (10.11)
> When the population is a normal probability distribution, the sampling distribution of $\overline{X}$ is exactly normal for *any* random sample size.

EXAMPLE □

Reasoning aptitude scores of students are known to be normally distributed, with mean $\mu = 500$ and standard deviation $\sigma = 100$. We wish to know the probability that the mean aptitude score for a random sample of $n = 5$ students will be less than 450. From (10.11) we know that the sampling distribution of $\overline{X}$ is normal here because the population is normal. Moreover, from (10.5) and (10.6) we know that the sampling distribution of $\overline{X}$ has mean $\mu = 500$ and standard deviation:

$$\sigma\{\overline{X}\} = \frac{100}{\sqrt{5}} = 44.72$$

The required z value is:

$$z = \frac{450 - 500}{44.72} = -1.12$$

Hence, the desired probability is $P(\overline{X} < 450) = P(Z < -1.12) = 0.13.$ □

> **Comment**
>
> Other populations for which statistical theory permits us to derive the exact sampling distribution of $\overline{X}$ include exponential and Poisson populations.

Construction by Enumeration

For simple discrete populations, the exact sampling distribution of $\overline{X}$ can be derived by enumeration of all possible samples. This procedure is tedious, however, and is rarely used. We present a very simple example to illustrate the basic principles.

☐ **EXAMPLE**

The number of microcomputers sold in week 1 is denoted by X_1 and the number sold in week 2 by X_2. The random variables X_1 and X_2 are independent and are known to have the same probability distribution, as follows:

x:	2	4
$P(x)$:	0.9	0.1

For this probability distribution, $E\{X\} = 2.2$ and $\sigma\{X\} = 0.6$ (calculations not shown). We wish to obtain the exact sampling distribution of $\overline{X}$, the mean number of microcomputers sold in the two weeks.

We list all possible sample outcomes and proceed as follows:

Sample Outcome $\begin{cases} X_1: \\ X_2: \end{cases}$	2 2	2 4	4 2	4 4
Sample Mean $\overline{X}$:	2	3	3	4
Probability:	0.81	0.09	0.09	0.01

The probabilities are obtained by theorem (5.4) because X_1 and X_2 are independent random variables. For example, $P(X_1 = 2 \cap X_2 = 2) = P(X_1 = 2)P(X_2 = 2) = 0.9(0.9) = 0.81$.

We can now obtain the exact sampling distribution of $\overline{X}$ for $n = 2$:

$\overline{X}$	2	3	4
$P(\overline{X})$:	0.81	0.18	0.01

Note that $P(\overline{X} = 3) = 0.09 + 0.09 = 0.18$ because two sample outcomes yield this value of $\overline{X}$. ☐

10.6 CRITERIA FOR CHOOSING POINT ESTIMATOR

We noted earlier that many point estimators are usually available to estimate the population parameter of interest. We also stated that the quality of a point estimator depends on its sampling distribution. Now that we have studied the sampling distribution of $\overline{X}$ in some detail, we can consider several criteria that will help us to distinguish good estimators from poor ones. These criteria are all related to the overall goal that a good estimator should provide reasonable assurance that the estimate will be close to the parameter being estimated.

Unbiasedness

The first criterion, unbiasedness, is based on the premise that the sampling distribution of a good estimator should be located near the parameter to be estimated.

(10.12)

An estimator S is **unbiased** if the mean of its sampling distribution is equal to the population parameter θ to be estimated; that is, S is an unbiased estimator of θ if:

$$E\{S\} = \theta$$

If estimator S is biased, the amount of its **bias** is:

$$\text{Bias} = E\{S\} - \theta$$

Unbiasedness in point estimators refers to the tendency of sampling errors to balance out over all possible samples. For any one sample, of course, the sample estimate will usually differ from the population parameter.

Figure 10.3 shows two estimators, S_1 being unbiased and S_2 having substantial bias. Clearly, S_2 will tend to give estimates far from θ, while estimates obtained from S_1 will tend to be nearer θ.

A biased estimator may still be a desirable estimator if the bias is not large, provided the estimator has other desirable properties.

EXAMPLES

1. Theorem (10.5) states that $E\{\overline{X}\} = \mu$. Hence, $\overline{X}$ is an unbiased estimator of μ.

2. The sample median Md is a biased estimator of μ when the sampled population is skewed; that is, $E\{Md\} \neq \mu$ then. For example, median income in a random sample

---

Sampling distributions of unbiased and biased estimators. The estimator S_2 is biased because $E\{S_2\} \neq \theta$. The magnitude of the bias is $E\{S_2\} - \theta$.

FIGURE 10.3

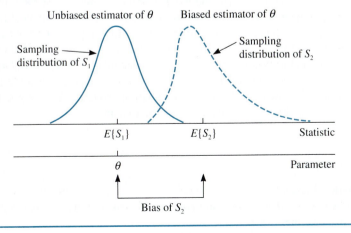

of families will, on the average, understate mean family income in the population when the income distribution is right-skewed, as it usually is.   □

In Chapter 3, we noted that the sample variance s^2, defined in (3.12), uses the denominator $n - 1$ rather than n for obtaining the average squared deviation. Now we can explain why the denominator $n - 1$ is used. The reason is that when s^2 is defined with denominator $n - 1$, it is an unbiased estimator of the population variance σ^2 for infinite populations; that is, $E\{s^2\} = \sigma^2$. Even though the sample variance is an unbiased estimator of σ^2, the sample standard deviation s, defined in (3.13), is not an unbiased estimator of the population standard deviation σ; that is, $E\{s\} \neq \sigma$. However, the bias is small for reasonably large samples.

Efficiency

The criterion of efficiency is based on the premise that among two estimators that have no bias, we prefer the one that has the smaller variability (that is, the tighter sampling distribution) because its outcomes will tend to lie closer to the population parameter. The criterion of efficiency is therefore a relative one, since it involves the comparison of two estimators.

(10.13)
The **efficiency** of an unbiased estimator is measured by the variance of its sampling distribution. If two estimators based on the same sample size are both unbiased, the one with the smaller variance is said to have greater **relative efficiency** than the other. Thus, S_1 is relatively more efficient than S_2 in estimating θ if:

$$\sigma^2\{S_1\} < \sigma^2\{S_2\} \quad \text{and} \quad E\{S_1\} = E\{S_2\} = \theta$$

□ **EXAMPLE**

The mean shelf life of packages of a new breakfast cereal is to be estimated. The distribution of shelf life is normal, and the question is whether to use the sample mean $\overline{X}$ or the sample median Md for estimating μ. It is known from statistical theory that both estimators are unbiased when sampling from a normal population. We also know from (10.6) that $\sigma^2\{\overline{X}\} = \sigma^2/n$. From statistical theory, it can be shown that $\sigma^2\{Md\} \simeq 1.57(\sigma^2/n)$ for random sampling from a normal population when n is large. Thus, $\sigma^2\{\overline{X}\} < \sigma^2\{Md\}$, and consequently $\overline{X}$ is relatively more efficient than Md as an estimator of μ here. Figure 10.4 illustrates this situation. Note that $\overline{X}$ has a greater probability than Md of falling within any specified interval about μ.   □

Consistency

A third criterion frequently employed to identify a good point estimator is consistency. It is based on the premise that a good estimator should lie nearer the population parameter as the sample size becomes larger.

FIGURE 10.4

Two unbiased estimators of μ that differ in relative efficiency when sampling a normal population.   Both estimators are unbiased, but the sampling distribution of $\overline{X}$ is tighter than that of *Md*.

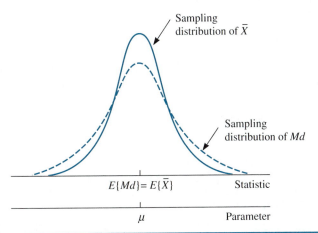

(10.14)
An estimator is a **consistent** estimator of a population parameter θ if the larger the sample size, the more likely it is that the estimate will be close to θ.

EXAMPLE

The sample mean $\overline{X}$ is a consistent estimator of μ. Figure 10.5 illustrates, for sampling from a normal population, how the sampling distribution of $\overline{X}$ tightens around μ as the sample size increases.

FIGURE 10.5

$\overline{X}$ **as a consistent estimator of μ when sampling a normal population.**   The sampling distributions of $\overline{X}$ for increasing sample sizes become more concentrated around μ.

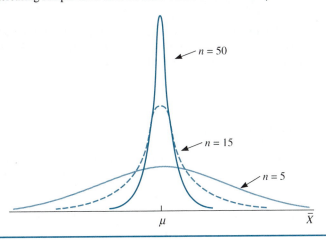

Comments

1. A consistent estimator may be biased. Consistency ensures, however, that the bias becomes smaller as the sample size becomes larger.
2. A formal definition of consistency follows.

(10.14a)

S is a **consistent** estimator of population parameter θ if for any small positive value ϵ (Greek epsilon):

$$\lim_{n \to \infty} P(|S - \theta| < \epsilon) = 1$$

The definition states that the probability is nearly 1 that a consistent estimator will have an outcome close to the parameter value (that is, within $\pm \epsilon$) when n is sufficiently large. A difficulty with this criterion is that the sample size may have to be very large before the assurance of closeness of the estimate takes effect.

Mean Squared Error

A fourth criterion, mean squared error, combines the criteria of unbiasedness and efficiency. It is useful when one or both of two estimators being compared is biased. The mean squared error of an estimator S combines the variance of the sampling distribution of S, $\sigma^2\{S\}$, and the bias of the estimator, $E\{S\} - \theta$, in the following way.

(10.15)

The **mean squared error** of an estimator S for estimating a population parameter θ is defined as follows:

$$\text{Mean squared error} = \sigma^2\{S\} + (E\{S\} - \theta)^2$$

In a comparison of two estimators, the one with the smaller mean squared error is said to have greater **relative mean squared error efficiency** than the other.

An estimator that is slightly biased but has a very concentrated sampling distribution near the population parameter θ will be preferred according to this criterion to an unbiased estimator with a sampling distribution that is highly variable.

☐ **EXAMPLE**

Consider the following two estimators:

Estimator S_i	Variance $\sigma^2\{S_i\}$	Bias $E\{S_i\} - \theta$	Mean Squared Error $\sigma^2\{S_i\} + (E\{S_i\} - \theta)^2$
S_1	20	5	$20 + 5^2 = 45$
S_2	80	0	$80 + 0^2 = 80$

Here, S_1 is the preferred estimator according to the mean-squared-error criterion. Its mean squared error is much smaller even though it has a small bias because its sampling distribution is so much less variable. ☐

Good point estimators for particular circumstances are not always intuitively obvious. Two widely used methods of finding good estimators are the *method of maximum likelihood* and the *method of least squares*. The latter method is discussed in Chapter 18 in connection with regression analysis. We now take up the method of maximum likelihood, which has wide applicability. Its implementation differs slightly for discrete and continuous populations, so we shall explain the method of maximum likelihood by two examples.

OPTIONAL   10.7
TOPIC—MAXIMUM
LIKELIHOOD
ESTIMATION

Discrete Population

EXAMPLE ☐   **Calculator Demand**

The weekly demand for a programmable calculator at a retail outlet follows a Poisson distribution with unknown parameter λ. Observations for three weeks are available; the number of calculators sold in each of these three weeks were $X_1 = 2$, $X_2 = 6$, and $X_3 = 1$. We assume that these three observations are independent observations from the same Poisson distribution and hence constitute a random sample.

In choosing an estimate of the parameter λ, we clearly would like this estimate to be consistent with the observed sample. Hence, the method of maximum likelihood requires that we calculate the probability of obtaining the observed sample for different values of λ and find that value of λ that is most consistent with the observed sample.

Recall that the Poisson probability function, given by (7.8), is:

$$P(x) = \frac{\lambda^x \exp(-\lambda)}{x!}$$

Because the three sample observations are independent, the joint probability of obtaining the sample equals the product $P(2)P(6)P(1)$. As this joint probability depends on the value of the unknown parameter λ, we shall denote it by $L(\lambda)$. Thus, we have:

$$L(\lambda) = P(2)P(6)P(1)$$

For any value of λ, the individual probabilities in this product can be computed from the formula for the probability function $P(x)$ or can be obtained from Table C.6.

We now proceed on a systematic search to find the value of λ that maximizes the function $L(\lambda)$; that is, the value of λ that yields the largest joint probability for the sample. First, we consider $\lambda = 4.5$. From Table C.6, we obtain $P(2) = 0.1125$, $P(6) = 0.1281$, and $P(1) = 0.0500$. Hence, the joint probability is:

$$L(4.5) = P(2)P(6)P(1) = 0.1125(0.1281)(0.0500) = 0.000721$$

Next, we consider $\lambda = 3.0$. We now have:

$$L(3.0) = P(2)P(6)P(1) = 0.2240(0.0504)(0.1494) = 0.001687$$

In the same manner, we can calculate $L(\lambda)$ for other values of λ. We have made these calculations and plotted the function $L(\lambda)$ in Figure 10.6. It appears from there that $L(\lambda)$ is largest at $\lambda = 3.0$. Thus, 3.0 is the estimate of λ that maximizes the joint probability of the observed sample and hence is the estimate that is most consistent with the observed sample.

It is no coincidence that $\overline{X}$ for this sample also happens to equal 3.0 because it can be shown analytically that $\overline{X}$ is the maximum likelihood estimator of λ for a random sample from a Poisson distribution. Thus, in this case, we do not need to obtain the likelihood function to find the maximum likelihood estimate. Rather, we can utilize the knowledge that $\overline{X}$ is the maximum likelihood estimator of the Poisson parameter λ and simply calculate $\overline{X} = 3.0$ as the maximum likelihood estimate.   □

The function $L(\lambda)$ in the preceding example is called the *likelihood function*. The values of $L(\lambda)$ for different values of the parameter λ are called *likelihood values*. For instance, $L(4.5) = 0.000721$ is the likelihood value for $\lambda = 4.5$. The value of the parameter λ that maximizes the likelihood function ($\lambda = 3.0$ in the example) is the desired estimate and is called the *maximum likelihood estimate* of λ.

General Approach.   We now summarize the method of maximum likelihood for a discrete population whose parameter, denoted by θ, is to be estimated.

1. We start with the population probability function $P(x)$, which depends on the unknown parameter θ that is to be estimated.
2. We then consider the joint probability of the observed sample $X_1, X_2, \ldots, X_n$, which, because of the independence of the sample observations, is $P(X_1)P(X_2) \cdots P(X_n)$. This product, when viewed as a function of θ for given $X_1, X_2, \ldots, X_n$, is the likelihood function $L(\theta)$.

FIGURE 10.6

Likelihood function when population is Poisson—Calculator demand example.   The maximum likelihood estimate is that value of λ that maximizes the likelihood function.

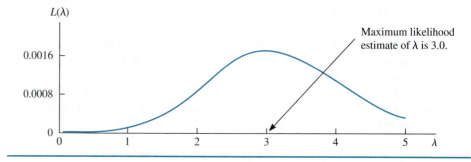

3. Finally, we maximize the likelihood function with respect to θ. The value of θ that maximizes the likelihood function depends on the sample observations and is the maximum likelihood estimator of θ.

(10.16)

The **method of maximum likelihood** for estimating a parameter θ of a discrete probability distribution selects as a point estimator that value of θ that maximizes the likelihood function:

$$L(\theta) = P(X_1)P(X_2) \cdots P(X_n)$$

where $P(x)$ denotes the population probability function.

EXAMPLE

To illustrate the general approach, we return to the calculator demand example.

1. The Poisson probability function $P(x)$ was given earlier. The population parameter to be estimated in this application is λ.

2. The joint probability of the sample observations X_1, X_2, X_3, when viewed as a function of λ, is the likelihood function $L(\lambda)$. Here, $L(\lambda)$ equals:

$$
\begin{aligned}
L(\lambda) &= P(X_1)P(X_2)P(X_3) \\
&= \frac{\lambda^{X_1}\exp(-\lambda)}{X_1!} \cdot \frac{\lambda^{X_2}\exp(-\lambda)}{X_2!} \cdot \frac{\lambda^{X_3}\exp(-\lambda)}{X_3!} \\
&= \frac{\lambda^{(X_1+X_2+X_3)}\exp(-3\lambda)}{X_1!X_2!X_3!}
\end{aligned}
$$

For the specific sample outcomes, we have:

$$L(\lambda) = \frac{\lambda^{(2+6+1)}\exp(-3\lambda)}{2!6!1!} = \frac{\lambda^9\exp(-3\lambda)}{1440}$$

It is this function that is plotted in Figure 10.6.

3. The maximum of the function $L(\lambda)$ can be shown analytically to occur at $\lambda = 3.0$. Thus, as noted before, the maximum likelihood estimate of λ is 3.0. □

Continuous Population

When the population probability distribution is continuous, the probability function $P(x)$ in (10.16) is simply replaced by the probability density function $f(x)$ of the continuous random variable. The following example illustrates the method.

Forecast Errors **EXAMPLE**

A trading company has a computerized procedure for forecasting exchange rates for major currencies. Let X denote the percentage error of this forecasting procedure for one major currency. The random variable X is assumed to be normally distributed. Since the forecasting procedure is known to be unbiased, on the average neither over-predicting nor underpredicting, the mean of the normal distribution is $\mu = 0$. Thus, only the standard deviation σ of the probability distribution of forecast errors is unknown. Hence the probability distribution associated with the population of forecast errors is assumed to be $N(0, \sigma^2)$. The probability density function is therefore given by the normal density function in (8.4), with $\mu = 0$:

$$f(x) = \frac{1}{\sqrt{2\pi}\sigma} \exp\left[-\frac{1}{2}\left(\frac{x}{\sigma}\right)^2\right]$$

It is desired to estimate the parameter σ.

The two latest forecasts yielded percentage errors of $X_1 = 8$ and $X_2 = -2$. These observations may be regarded as a random sample from the forecasting process. The joint probability density of the observed sample is therefore given by the product of the densities for the two sample observations, as follows:

$$\begin{aligned}
L(\sigma) &= f(X_1)f(X_2)\\[4pt]
&= \frac{1}{\sqrt{2\pi}\sigma} \exp\left[-\frac{1}{2}\left(\frac{8}{\sigma}\right)^2\right] \cdot \frac{1}{\sqrt{2\pi}\sigma} \exp\left[-\frac{1}{2}\left(\frac{-2}{\sigma}\right)^2\right]\\[4pt]
&= \frac{1}{2\pi\sigma^2} \exp\left(-\frac{34}{\sigma^2}\right)
\end{aligned}$$

The joint probability density of the sample observations, when viewed as a function of the unknown population standard deviation σ, is the likelihood function $L(\sigma)$. Figure 10.7 shows a plot of the likelihood function for this example. For instance, the likelihood value of $\sigma = 5.0$ is:

$$L(5.0) = \frac{1}{2\pi(5.0)^2} \exp\left[-\frac{34}{(5.0)^2}\right] = 0.001634$$

FIGURE 10.7
Likelihood function when population is $N(0, \sigma^2)$—Forecast errors example

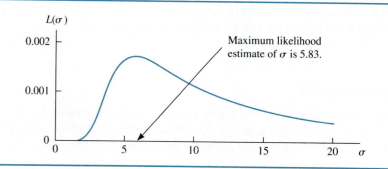

Inspection of Figure 10.7 shows that the maximum of the likelihood function occurs near $\sigma = 6$. Exact analysis of the function $L(\sigma)$ shows that the maximum likelihood estimate of σ is 5.83. $\square$

Properties of Maximum Likelihood Estimators

Under quite general conditions, maximum likelihood estimators are consistent estimators. Also, they are usually approximately normally distributed and more efficient than other estimators for reasonably large samples. These properties make maximum likelihood estimators particularly useful in statistical analysis.

For many well-known probability distributions, the maximum likelihood estimators are already known and are simple functions of the sample observations, such as the sample mean $\overline{X}$. In these cases, we need not derive the maximum likelihood estimator each time. There are occasions, however, when the maximum likelihood estimator is not a simple function of the sample observations. It is then often necessary to find the maximum likelihood estimate by efficient computerized numerical-search procedures.

PROBLEMS

* **10.1** The claims for damages filed last year with a railroad constitute the population of interest for an internal auditor. Suppose that, unknown to the auditor, the population mean claim amount is $\mu = \$3540$. A random sample of three claims drawn from this population yields the observations $\$120,890$, $\$135$, and $\$1200$.

a. The sample mean is to be used to estimate the population mean. (1) What is the value of the parameter being estimated? (2) What is the estimator here? To what does the sampling distribution of the estimator refer here? (3) What is the value of the estimate here? What is the magnitude of the sampling error in this estimate?

b. The sample median and the sample midrange are alternative estimators of the population mean. What are their values for this sample? Which estimate—mean, median, or midrange—has the smallest sampling error for this sample? Would this estimator necessarily have the smallest sampling error for another sample?

10.2 The trees in a timber tract form the population of interest to a forester. Suppose that, unknown to the forester, the mean growth of these trees during the previous decade, as measured by their last 10 annual growth rings, is $\mu = 8.38$ centimeters. A random sample of four trees from the tract yields growth measurements of 8.76, 7.93, 7.88, and 8.55 centimeters.

a. The sample mean is to be used to estimate the population mean. (1) What is the value of the parameter being estimated? (2) What is the estimator here? To what does the sampling distribution of the estimator refer here? (3) What is the value of the estimate here? What is the magnitude of the sampling error in this estimate?

b. The sample median and the sample midrange are alternative estimators of the population mean. What are their values for this sample? Which estimate—mean, median, or midrange—has the smallest sampling error for this sample? Would this estimator necessarily have the smallest sampling error for another sample?

10.3 The mean weight of chicks of a particular breed at 15 weeks of age is of interest. A specialist has just received a random sample of 100 chicks of this breed to raise. Assume that all chicks survive to 15 weeks of age.

a. Describe the parameter of interest here.

b. The sample mean weight of the 100 birds will be determined when they reach 15 weeks of age. At the present moment, is the sample mean an estimator or an estimate? To what does the sampling distribution of the sample mean refer here?

10.4 A student, on studying sampling, was puzzled about how there can be a sampling distribution of $\overline{X}$ when, in practice, only a single sample is selected. Explain to this student the concept of a sampling distribution and its significance.

10.5 A statistics teacher used a computer routine to generate 1000 independent simple random samples, each of size 30, from an infinite theoretical population. The routine also computed the sample mean for each of the 1000 samples. The teacher then constructed a histogram of the 1000 sample means and stated that the histogram is "a graphical representation of the sampling distribution of the sample mean with a sample size of 30 for this population." Is the teacher's statement correct? Explain.

* **10.6 Brick Production.** A plant produces sulfur bricks that are used for structures in high-stress industrial environments. The production process is currently molding bricks with a mean weight of $\mu = 1.740$ kilograms and a standard deviation of $\sigma = 0.030$ kilogram. What are the mean and standard deviation of the sampling distribution of $\overline{X}$ for a random sample of size (1) $n = 9$, (2) $n = 50$?

10.7 Aptitude Scores. Scholastic aptitude scores of college-bound high school seniors have a mean of $\mu = 475$ and a standard deviation of $\sigma = 80$. What are the mean and standard deviation of the sampling distribution of $\overline{X}$ for a random sample of size (1) $n = 225$, (2) $n = 36$?

10.8 Bean Growth. The mean time from germination to harvest under laboratory conditions for a certain variety of beans is $\mu = 61$ days, and the standard deviation is $\sigma = 2$ days. What are the mean and standard deviation of the sampling distribution of $\overline{X}$ for a random sample of size (1) $n = 16$, (2) $n = 100$?

10.9 Weekly soft drink purchases by adults have a mean of $\mu = \$7.50$ and a standard deviation of $\sigma = \$4.80$. What are the mean and standard deviation of the sampling distribution of $\overline{X}$ for a random sample of size (1) $n = 121$, (2) $n = 49$?

10.10 Refer to **Brick Production** Problem 10.6.
a. For which sample size ($n = 9, 50$) is it more likely that the sample mean is within ± 0.010 kilogram of μ? Why?
b. If a random sample of 9 bricks were selected from the process, and independently another random sample of 50 bricks were selected, would the sample mean based on the larger sample necessarily be closer to $\mu = 1.740$ kilograms? Discuss.

10.11 Refer to **Bean Growth** Problem 10.8.
a. How many seedlings would be required in the sample if the standard deviation of the sampling distribution of $\overline{X}$ is to be $\sigma\{\overline{X}\} = 0.25$?
b. What would be the answer in part a if the population standard deviation were $\sigma = 4$ days instead of $\sigma = 2$ days? What is the effect of larger population variability?

10.12 Refer to Figure 10.1b. The frequency polygons for $n = 3$, $n = 10$, and $n = 100$ are located in almost identical positions on the $\overline{X}$ scale, but they differ in variability. What theorems in Chapter 10 are illustrated by this comparison?

10.13 An analyst remarked, "When the sample size is large, the sample mean is as likely to lie above the population mean as below it." Do you agree? Explain.

10.14 On hearing an explanation of the central limit theorem, a listener stated, "The theorem guarantees that the n observations in a random sample will be approximately normally distributed if n is sufficiently large." Does the listener understand the central limit theorem? Explain.

10.15 A sample of $n = 3000$ persons is to be taken from a population of 1,000,000 persons in the labor force. What is the sampling fraction here? Is it large or small according to working rule (10.9)?

10.16 A sample of $n = 200$ parts is to be taken from a lot of 50,000 parts. What is the sampling fraction here? Is it large or small according to working rule (10.9)?

* **10.17** Refer to **Brick Production** Problem 10.6.
a. For a random sample of 50 bricks, find the probability that the sample mean will (1) exceed 1.745, (2) lie within ± 0.005 of μ.

b. For a random sample of 50 bricks, within what interval centered around μ will $\overline{X}$ fall with probability 0.90? What is the corresponding interval when $n = 100$? What is the effect of the larger sample size on the width of the interval?

10.18 Refer to **Aptitude Scores** Problem 10.7.

a. For a random sample of 225 seniors, find the probability that the sample mean will lie within ± 10 of the population mean.

b. For a random sample of 225 seniors, within what interval centered around μ will $\overline{X}$ fall with probability 0.95? What is the corresponding interval when $n = 450$? What is the effect of the larger sample size on the width of the interval?

c. Would the widths of the intervals in part b be twice as large if σ were 160 rather than 80? Explain.

10.19 Refer to **Bean Growth** Problem 10.8. A random sample of $n = 100$ seedlings is being grown in the laboratory.

a. What is the probability that the sample mean time from germination to harvest will lie within ± 0.5 day of the population mean?

b. What would be the answer in part a if the population standard deviation were $\sigma = 4$ days instead of $\sigma = 2$ days? What is the effect of larger population variability?

10.20 An automatic machine cuts ribbons of hot steel into bars of specified length. The variability in the lengths of the bars is $\sigma = 0.07$ meter. A random sample of 60 bars is to be selected.

a. What is the probability that the sample mean $\overline{X}$ will not differ from the process mean μ by more than ± 0.01 meter?

b. What is the probability that the sample mean will exceed the population mean?

10.21 For each of the following cases, indicate whether the functional form of the sampling distribution of $\overline{X}$ is exactly normal, approximately normal, or not approximately normal for sample sizes (1) $n = 3$, (2) $n = 300$.

a. The mean duration of a random sample of satellite transmissions. The durations of satellite transmissions follow an exponential distribution.

b. The mean weight of a random sample of rough gems. The population of gem weights is normally distributed.

10.22 For each of the following cases, indicate whether the functional form of the sampling distribution of $\overline{X}$ is exactly normal, approximately normal, or not approximately normal for sample sizes (1) $n = 2$, (2) $n = 200$.

a. The mean useful life of a random sample of drive chains. The useful lives of drive chains are normally distributed.

b. The mean number of interruptions per hour in an assembly line in a random sample of one-hour intervals. Interruptions in an assembly line follow a Poisson process with a mean of $\lambda = 0.2$ interruption per hour.

* **10.23** Refer to **Brick Production** Problem 10.6. Assume that the weights of bricks are normally distributed. For a random sample of 16 bricks, find the probability that the sample mean will lie within ± 0.005 of μ.

10.24 Refer to **Aptitude Scores** Problem 10.7. Assume that the aptitude scores of seniors are normally distributed. For a random sample of four seniors, find the probability that $\overline{X}$ will exceed 525. What is this probability for an individual score?

10.25 Refer to **Bean Growth** Problem 10.8. Assume that the times from germination to harvest for seedlings are normally distributed. For a random sample of nine seedlings, within what interval centered around μ will $\overline{X}$ lie with probability 0.90?

10.26 The number of breakdowns X of a chain drive during a week has the following probability distribution:

x:	0	1	2
$P(x)$:	0.70	0.20	0.10

a. Obtain the mean and the standard deviation of this infinite population.
b. Obtain the exact sampling distribution of $\overline{X}$ for a random sample of $n = 2$ weeks.
c. Calculate the mean and the standard deviation of the sampling distribution of $\overline{X}$ in part b. Verify that the results agree with theorems (10.5) and (10.6).

10.27 The number of rework cycles X before a machine component passes quality specifications or is scrapped has the following probability distribution:

x:	0	1	2
$P(x)$:	0.95	0.04	0.01

a. Obtain $E\{X\}$ and $\sigma\{X\}$.
b. Obtain the exact sampling distribution of $\overline{X}$ for a random sample of $n = 2$ components.
c. Calculate $E\{\overline{X}\}$ and $\sigma\{\overline{X}\}$ for the sampling distribution of $\overline{X}$ in part b. Verify that the results agree with theorems (10.5) and (10.6).

10.28 A travel agents' association commissioned a random sample survey of retired persons to ascertain their winter travel plans and preferences. One item of particular interest was the mean anticipated expenditures per retired person for travel during the coming winter. The survey was conducted during the third week in September, and the response rate was 70 percent. Interviewers, when commenting on reasons for nonresponses, often noted that the nonrespondent was away on a trip. Let $\overline{X}$ denote the mean anticipated expenditures per respondent in the sample survey. Describe the population mean μ that $\overline{X}$ is estimating here. Is $\overline{X}$ an unbiased estimator of μ here? Is $\overline{X}$ an unbiased estimator of the quantity that the travel agents' association wishes to estimate? Comment.

10.29 A sample of four tires of a new type was subjected to wear out on a life-testing machine. The first tire to wear out did so at X miles. Is X an unbiased estimator of the mean life of this type of tire? Explain.

10.30 As noted in the text, in a normal population $\overline{X}$ is relatively more efficient than Md as an estimator of μ. Does this mean that (1) in every random sample from a normal population $\overline{X}$ will lie closer to μ than Md does? (2) A smaller sample size is needed when $\overline{X}$ is used to estimate μ than when Md is used? Explain.

10.31 A sample survey of firms in an industrial district is being planned to measure their anticipated research and development expenditures for the coming year. Two unbiased estimators of the population mean expenditures μ are being considered for use in the survey. The survey aims to estimate μ within a specified margin of error. A consulting statistician has advised that, for the planned sample size, one estimator has a higher probability of being within the specified margin of error than the other. Does the statistician's statement indicate that the two estimators differ in efficiency? Comment.

10.32 If a point estimator is biased but consistent, does this imply that the bias must approach zero as the sample size gets larger? Explain.

10.33 If two different statistics computed from the same sample are both consistent estimators of the same population parameter, does this imply that there is a high probability that the two statistics will be close in value when the sample size is large? Explain.

10.34 In an econometric study of the shoe industry, two estimators (S_1 and S_2) of the marginal productivity of labor (θ) are under consideration. The variances and the biases of the two estimators are as follows:

S_i	$\sigma^2\{S_i\}$	$E\{S_i\} - \theta$
S_1	0.103	0
S_2	0.058	-0.167

a. Does estimator S_2 tend to underestimate or overestimate the marginal productivity of labor? Explain.
b. Calculate the mean squared error of each estimator. Which estimator has greater relative mean squared error efficiency?

10.35 Two estimators are being considered for estimating mean annual rainfall in a county (θ). Estimator S_1 is the sample mean annual rainfall in the county, based on the previous 10 years of rainfall data. Estimator S_2 is the sample mean annual rainfall for the whole state, based on the previous 10 years of rainfall data.

a. If both estimators are unbiased but S_2 has a smaller variance than S_1, which estimator has greater relative mean squared error efficiency? Might your answer be different if S_2 were biased? Explain.

b. Suppose that the variances and the biases of the two estimators are as follows:

S_i	$\sigma^2\{S_i\}$	$E\{S_i\} - \theta$
S_1	10.7	0
S_2	3.5	2.8

Calculate the mean squared error of each estimator. Which estimator has the greater relative mean squared error efficiency?

c. Why might S_2 be a biased estimator of θ here? Why might S_2 have a smaller variance than S_1 here?

* 10.36 The numbers of medal winners from a high school in three successive state gymnastic competitions were $X_1 = 1$, $X_2 = 0$, $X_3 = 2$. Assume that these three observations constitute a random sample from a Poisson distribution with unknown parameter value λ.

a. Obtain the values of the likelihood function $L(\lambda)$ for this sample at $\lambda = 0, 0.5, 1.0, 1.5, 2.0$.

b. Sketch a graph of $L(\lambda)$, as in Figure 10.6. Does the sample mean $\overline{X}$ appear to be the maximum likelihood estimator of λ, as the theory in the text states? Explain.

10.37 The numbers of visits made to the dentist last year by four children were $X_1 = 0$, $X_2 = 1$, $X_3 = 3$, $X_4 = 1$. Assume that these four observations constitute a random sample from a Poisson distribution with unknown parameter value λ.

a. Obtain the values of the likelihood function $L(\lambda)$ for this sample at $\lambda = 0, 0.5, 1.0, 1.5, 2.0$.

b. Sketch a graph of $L(\lambda)$, as in Figure 10.6. Does the sample mean $\overline{X}$ appear to be the maximum likelihood estimator of λ, as the theory in the text states? Explain.

10.38 An investment analyst has developed a procedure for calculating the *excess return* (in percent) earned monthly on common stock investments. The excess return is the difference (either positive or negative) between the actual return earned on a stock and the return that would be expected if the stock's price moved in unison with other stock prices in the market generally. The excess return for the particular stock now under consideration, denoted by X, is normally distributed with mean 0 and unknown variance σ^2. The excess returns earned by this stock in three consecutive months were $X_1 = 2.15$, $X_2 = -1.43$, $X_3 = 0.51$. The analyst assumes that these observations constitute a random sample from $N(0, \sigma^2)$.

a. Obtain the values of the likelihood function $L(\sigma^2)$ for this sample at $\sigma^2 = 1.00, 2.00, 3.00, 4.00$.

b. The maximum likelihood estimate of σ^2 for this sample can be shown to be 2.31. What is the likelihood value for $\sigma^2 = 2.31$?

c. Sketch a graph of $L(\sigma^2)$, similar to Figure 10.7, based on your results in parts a and b. Is there a large relative difference in the likelihood values for $\sigma^2 = 2.31$ and 2.00? For 2.31 and 3.00? What do these comparisons suggest about how precisely σ^2 is known from this sample? Explain.

EXERCISES

10.39 Let τ denote a finite population total; that is:

$$\tau = \sum_{i=1}^{N} X_i = N\mu$$

Consider the sampling distribution of random variable $T = N\overline{X}$ when the sampling fraction n/N is small. Derive $E\{T\}$ and $\sigma\{T\}$, using (10.5) and (10.6).

10.40 Consider the following probability distribution for an infinite population:

x:	0	2
$P(x)$:	0.5	0.5

a. Verify that $\sigma = 1$ for this population.

b. For a simple random sample of $n = 2$ from this population, construct (1) the sampling distribution of s^2 and (2) the sampling distribution of s.

c. As noted in the text, s^2 is an unbiased estimator of σ^2 for an infinite population, but s is a biased estimator of σ. Verify these facts for the sampling distributions in part b. What is the amount of bias in s here?

10.41 A population is normal with $\mu = 200$.

a. Assume $\sigma = 20$. Calculate $P(195 \leq \overline{X} \leq 205)$ for $n = 4, 25, 100$. Plot the probabilities against n on a graph. Describe the effect of increasing the sample size on the probability that $\overline{X}$ will be within ± 5 of the population mean μ.

b. Let $n = 25$. Calculate $P(195 \leq \overline{X} \leq 205)$ for $\sigma = 10, 20, 30$. Plot the probabilities against σ on a graph. Describe the effect of larger population variability on the probability that $\overline{X}$ will be within ± 5 of the population mean μ.

10.42 Let X_1, X_2, X_3 be the outcomes of $n = 3$ independent Bernoulli trials, each with the same parameter p. Obtain the sampling distribution of $\overline{X}$.

10.43 The mean μ of a normal population is to be estimated by either the sample midrange (10.4) or the sample mean $\overline{X}$. The random sample size will be greater than two. The sampling distribution of the sample midrange is symmetrical about μ in this case but has a larger variance than the sampling distribution of $\overline{X}$.

a. Explain why the sample midrange is an unbiased estimator of μ here.

b. (1) Which of the two estimators is relatively more efficient in this situation? (2) Which estimator would require the larger sample size to estimate μ with a given precision?

10.44 For a patient suffering from a certain chronic disease, the time until relapse has an exponential probability distribution. In a random sample of two patients, the numbers of days until relapse were $X_1 = 60, X_2 = 100$.

a. State the likelihood function $L(\lambda)$ for this sample, and evaluate it at $\lambda = 0.010, 0.0125, 0.015, 0.020$.

b. Sketch a graph of the likelihood function of λ for this sample. Confirm by inspection that $1/80$ is the maximum likelihood estimate of λ.

c. Verify that for a random sample of n patients, the likelihood function of λ in this case is $L(\lambda) = \lambda^n \exp(-\lambda n \overline{X})$, where $\overline{X}$ denotes the sample mean relapse time.

10.45 (*Calculus needed.*) Refer to Problem 10.44c. Show that $1/\overline{X}$ is the maximum likelihood estimator of λ in this case.

STUDIES

10.46 Consider the discrete uniform probability distribution (7.1) with possible outcomes $X = 0, 1, 2, 3, 4$.

a. Obtain the mean and the standard deviation of this infinite population.

b. Use enumeration to construct the exact sampling distribution of $\overline{X}$ for a random sample of size $n = 2$ from this infinite population. Calculate $E\{\overline{X}\}$ and $\sigma\{\overline{X}\}$ from the exact sampling distribution and verify that they agree with theorems (10.5) and (10.6).

c. Use a uniform random number generator to generate 100 random samples of size $n = 2$ from this infinite population. Obtain $\overline{X}$ for each sample. [*Hint:* Let $X = \text{int}(5U)$, where U is a standard uniform random number and $\text{int}(A)$ denotes the integer part of the number A.]

d. Construct a frequency distribution of the 100 sample means in part c, using the nine possible outcome values. Plot the distribution as a frequency polygon. Does the frequency polygon have approximately the same shape as the exact sampling distribution of $\overline{X}$ constructed in part b? Should it? Explain.

10.47 Consider a uniform random number generator.

a. Use the uniform random number generator to generate 200 random samples of size $n = 2$ from the continuous uniform probability distribution (8.1) with $a = 0$ and $b = 100$. Obtain $\overline{X}$ for each sample. [*Hint:* Let $X = 100U$, where U is a standard uniform random number.]

b. Construct a frequency distribution of the 200 sample means in part a, using 10 equal class intervals. Plot the distribution as a frequency polygon.

c. Repeat parts a and b for $n = 5$ and $n = 10$.

d. For which sample sizes ($n = 2, 5, 10$) does the sampling distribution of $\overline{X}$ appear to be approximately normal?

10.48 Refer to the **Financial Characteristics** data set (Appendix D.1). Consider the population of all firms.

a. Determine the population mean and the population standard deviation for net income in year 2.

b. Select 300 independent random samples, each of $n = 15$ firms. Use sampling with replacement; that is, when a firm is selected, do not remove it from the population. Calculate the sample mean $\overline{X}$ for each sample.

c. Select 300 independent random samples, each of $n = 50$ firms. Again, use sampling with replacement. Calculate the sample mean $\overline{X}$ for each sample.

d. Use your simulation results to demonstrate that the sampling distribution of $\overline{X}$ approaches the normal distribution as the sample size increases and that the mean and the standard deviation of $\overline{X}$ are given by (10.5) and (10.6), respectively.

Interval Estimation of Population Mean

Point estimates have the limitation that they do not provide information about the *precision* of the estimate—that is, about the magnitude of the sampling error. Often, such information is essential for proper interpretation of the sample result.

EXAMPLE □

A legislative committee concerned with stimulating consumer spending was informed about a sample survey of veterans conducted to ascertain how much of a proposed bonus of $1000 the veterans planned to spend. The survey found that the veterans planned to spend $540 on average of the proposed bonus of $1000 and to save the remainder. If the precision of the estimate of $540 were ± $200, so that the mean amount to be spent could be as high as $740 or as low as $340, the estimate would not be useful to the committee because of the wide margin of error. On the other hand, if the precision of the estimate were ± $20, the estimate would be quite useful since the committee would know that the mean amount of the bonus to be spent is somewhere between $520 and $560. A point estimate by itself does not supply this information about its precision.

□

In this chapter, we consider statistical procedures for estimating the population mean in terms of an interval, where the width of the interval indicates the precision of the estimate. We also take up related inference procedures for estimating a population total and predicting a new observation drawn from a population.

The inference procedures to be discussed all assume that the sample is a simple random one. When the population is finite, the probability selection procedure must therefore meet the requirements in (9.7). When interest is in a process, so that the population is infinite, the data set needs to be examined by the diagnostic procedures discussed in Section 9.4 to determine whether it can be treated as a simple random sample from the infinite population. A formal test for randomness to be discussed in Section 15.4 is also useful for determining whether the data set can be treated as a random sample.

INTERVAL ESTIMATION **11.1**

An interval estimate of a population parameter uses a pair of limits or bounds to indicate the range within which the parameter is estimated to be.

> **(11.1)**
>
> An **interval estimate** of a population parameter θ consists of two bounds L and U within which θ is estimated to lie:
>
> $$L \le \theta \le U$$
>
> where L is the **lower bound** and U is the **upper bound**.

For instance, an interval estimate for the population mean μ will have the form $L \le \mu \le U$.

☐ **EXAMPLE**

We return to the accounts receivable example of Chapter 10. The auditor has selected a random sample of 100 accounts from the 8042 accounts receivable of the freight company to estimate the mean audit amount μ of the accounts in the population. Some of the sample data are presented in Table 11.1. The sample mean is $\overline{X} = \$33.19$, and we know that this estimator has some desirable properties such as unbiasedness and consistency. Still, we also know that almost surely $\overline{X} = \$33.19$ is not equal to μ. We therefore wish to construct an interval estimate for μ.   ☐

To develop an interval estimate that reflects the precision of the estimator $\overline{X}$, we need to estimate the variability of the sampling distribution of $\overline{X}$.

Estimated Standard Deviation of $\overline{X}$

Recall from Chapter 10 that the variability of the sampling distribution of $\overline{X}$ indicates how likely it is that the sample mean $\overline{X}$ is close to the population mean μ. The smaller is the variability of the sampling distribution, the greater is the probability that $\overline{X}$ will fall within any specified interval around μ.

Even though we take only one sample from the population, we can estimate the variability of the sampling distribution of $\overline{X}$. The reason is that $\sigma^2\{\overline{X}\}$, the variance of $\overline{X}$, is a simple function of the population variance σ^2 as shown by (10.6):

$$\sigma^2\{\overline{X}\} = \frac{\sigma^2}{n}$$

To estimate $\sigma^2\{\overline{X}\}$, we therefore simply replace the population variance σ^2 in this expression by the sample variance s^2. The resulting estimator is denoted by $s^2\{\overline{X}\}$:

$$s^2\{\overline{X}\} = \frac{s^2}{n}$$

and the estimated standard deviation of $\overline{X}$ is denoted by $s\{\overline{X}\}$:

$$s\{\overline{X}\} = \frac{s}{\sqrt{n}}$$

This latter standard deviation is often called the *estimated standard error of the mean*.

i	X_i	$X_i - \overline{X}$	$(X_i - \overline{X})^2$
1	80.29	47.10	2,218.41
2	6.97	−26.22	687.49
3	4.55	−28.64	820.25
.	.	.	.
.	.	.	.
.	.	.	.
99	51.51	18.32	335.62
100	10.30	−22.89	523.95
Total	3318.73	0	117,674.67

$$n = 100 \qquad \overline{X} = \frac{3318.73}{100} = 33.19 \qquad s^2 = \frac{117,674.67}{100 - 1} = 1188.63 \qquad s = 34.48$$

TABLE 11.1
Results for random sample of 100 accounts—Accounts receivable example

(11.2)
Estimators of the variance and the standard deviation of the sampling distribution of $\overline{X}$ are, respectively:

$$s^2\{\overline{X}\} = \frac{s^2}{n} \qquad s\{\overline{X}\} = \frac{s}{\sqrt{n}}$$

These estimators are appropriate for infinite populations and also for finite populations as long as the sampling fraction n/N is 5 percent or less according to rule (10.9).

EXAMPLE ☐

In the accounts receivable example, we wish to estimate the variability of the sampling distribution of $\overline{X}$ from the sample results in Table 11.1. We have $s^2 = 1188.63$ and $n = 100$. The sampling fraction here is small, $n/N = 100/8042 = 0.012$. Hence, we can use (11.2) and obtain:

$$s^2\{\overline{X}\} = \frac{1188.63}{100} = 11.886 \qquad s\{\overline{X}\} = \sqrt{11.886} = 3.448$$

Since the sample size $n = 100$ is large, the sampling distribution of $\overline{X}$ is approximately normal by central limit theorem (10.7). Consequently, it is highly probable that the sample mean $\overline{X}$ is within, say, two standard deviations of the population mean. Thus, since $s\{\overline{X}\} = 3.448$, $\overline{X}$ is likely to be within $2(3.448) = \$6.90$ of μ.

Incidentally, we know from our experiment (p. 263) that $\sigma = 30.334$ and so $\sigma\{\overline{X}\} = 3.03$ when $n = 100$. Thus, our sample estimate $s\{\overline{X}\} = 3.448$ is reasonably close. ☐

Interval Estimate

We just found for the accounts receivable example that $s\{\overline{X}\} = 3.448$ and concluded that the sample mean $\overline{X}$ is likely to be within $2(3.448) = \$6.90$ of the population mean. If that is so, we might reason that $\overline{X} \pm 6.90$ should give us limits that are likely to include μ. Here, the limits would be:

$$L = \overline{X} - 2s\{\overline{X}\} = 33.19 - 2(3.448) = 26.29$$
$$U = \overline{X} + 2s\{\overline{X}\} = 33.19 + 2(3.448) = 40.09$$

and we would therefore have the interval estimate:

$$26.29 \leq \mu \leq 40.09$$

Thus, we would estimate that the mean audit amount per account in the population is somewhere between \$26.29 and \$40.09. Since we happen to know from our experiment that $\mu = \$30.303$, we see that this particular interval estimate is correct.

Note that the approach used involves the following steps:

Step 1. Select a random sample of size n.
Step 2. Obtain $\overline{X}$ and s for the sample.
Step 3. Estimate $\sigma\{\overline{X}\}$ by $s\{\overline{X}\} = s/\sqrt{n}$.
Step 4. Calculate the following interval.

$$(11.3) \qquad \overline{X} - 2s\{\overline{X}\} \leq \mu \leq \overline{X} + 2s\{\overline{X}\}$$

The interval in formula (11.3) represents an interval estimate of the population mean μ, and the term $2s\{\overline{X}\}$ represents the calculated precision of the estimate $\overline{X}$.

Behavior of Interval Estimate

We shall study the behavior of interval estimate (11.3) by continuing our earlier experiment. We selected 600 independent random samples of 100 accounts from the population of 8042 accounts receivable. The auditor's sample in Table 11.1 is the first of the 600. In the second sample, we obtained $\overline{X} = 31.89$ and $s = 34.94$. If the auditor had selected this second sample, we would have calculated $s\{\overline{X}\} = 34.94/\sqrt{100} = 3.494$ and obtained the calculated precision $2(3.494) = 6.99$, giving limits of 31.89 ± 6.99. Hence, we would have obtained the interval $24.90 \leq \mu \leq 38.88$. Again, this interval contains $\mu = 30.303$. Figure 11.1 shows the interval estimates for a few of the 600 samples, including the two already mentioned. Note that all but one of the interval estimates displayed contain $\mu = 30.303$. The exception is sample 39, which provides an interval estimate lying entirely above μ. In all, 559 of the 600 intervals in the experiment, or 93.2 percent, contain μ.

This percentage of correct intervals is a measure of the confidence we can have in the interval estimation procedure. Here it is quite likely that the interval estimate obtained for the one sample actually selected will contain μ, because 93.2 percent of the 600 interval estimates were correct. Clearly, the percentage of correct intervals is a function of the multiple of $s\{\overline{X}\}$ used in calculating the interval. In this illustration, we used a multiple of 2. In general, as we shall now see, we can specify the multiple to yield any desired probability of a correct interval.

Interval estimates of the form $\overline{X} \pm 2s\{\overline{X}\}$ for several of the 600 samples—Accounts receivable example. The intervals differ in location because $\overline{X}$ varies from sample to sample, and they differ in width because s varies from sample to sample.

FIGURE 11.1

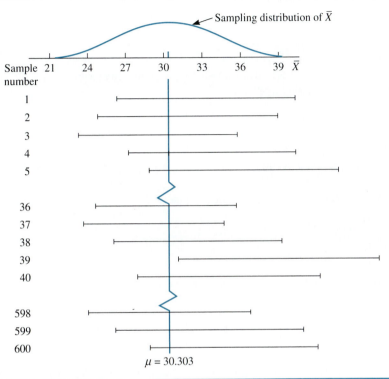

We now consider the construction of an interval estimate for any specified probability of a correct interval. The procedure is applicable *when the random sample size is large.*

CONFIDENCE INTERVAL FOR μ— LARGE SAMPLE

11.2

> **(11.4)**
> The probability that a correct interval estimate is obtained is called the **confidence coefficient** and is denoted by $1 - \alpha$. The interval:
>
> $$L \leq \mu \leq U$$
>
> is called the **confidence interval.** The limits L and U are called the **lower** and **upper confidence limits,** respectively.

The numerical confidence coefficient is often expressed as a percent. For example, confidence coefficient 0.95 is frequently expressed as 95 percent. A confidence interval that has an associated confidence coefficient of $1 - \alpha$ is therefore called either a $1 - \alpha$ confidence interval or a $100(1 - \alpha)$ percent confidence interval. For instance, a 0.95 confidence interval and a 95 percent confidence interval both describe a confidence interval with a confidence coefficient of 0.95.

(11.5)

The approximate $1 - \alpha$ confidence limits for the population mean μ, when the random sample size is reasonably large, are:

$$\overline{X} \pm zs\{\overline{X}\}$$

where: $z = z(1 - \alpha/2)$

$s\{\overline{X}\}$ is given by (11.2)

The approximate $1 - \alpha$ confidence interval for μ is:

$$\overline{X} - zs\{\overline{X}\} \leq \mu \leq \overline{X} + zs\{\overline{X}\}$$

Recall from Chapter 8 that $z(1 - \alpha/2)$ denotes the $100(1 - \alpha/2)$ percentile of the standard normal distribution.

☐ **EXAMPLES**

1. Refer to Table 11.1 for the accounts receivable example. We wish to construct a confidence interval for μ with confidence coefficient $1 - \alpha = 0.954$. Hence, $\alpha = 0.046$ and $1 - \alpha/2 = 0.977$. We find from Table C.1 that $z(0.977) = 2.0$. We previously obtained:

$$\overline{X} = 33.19 \qquad s\{\overline{X}\} = 3.448$$

Hence:

$$L = 33.19 - 2(3.448) = 26.29 \qquad U = 33.19 + 2(3.448) = 40.09$$

and:

$$26.29 \leq \mu \leq 40.09$$

We thus conclude, with 95.4 percent confidence, that the population mean audit amount per account is between \$26.29 and \$40.09.

Note that this confidence interval is the one we obtained earlier. Now we know that it has an approximate confidence coefficient of 95.4 percent. Recall that the percentage of correct intervals in the 600 trials was 93.2 percent, which is close to the 95.4 percent confidence coefficient.

2. Refer to the accounts receivable example in Table 11.1 again. Suppose we wish to use a confidence coefficient of $1 - \alpha = 0.90$. We then require $z(0.95) = 1.645$.

The confidence limits now are $33.19 \pm 1.645(3.448)$, and the 90 percent confidence interval is:

$$27.52 \le \mu \le 38.86$$

3. In a random sample of $n = 40$ bricks from a production process, the mean weight was $\overline{X} = 3.724$ pounds and the standard deviation was $s = 0.0710$ pound. We wish to estimate the process mean μ with a 99 percent confidence interval. We require:

$$s\{\overline{X}\} = \frac{0.0710}{\sqrt{40}} = 0.0112 \qquad z(0.995) = 2.576$$

and we obtain the confidence limits $3.724 \pm 2.576(0.0112)$. The 99 percent confidence interval therefore is:

$$3.70 \le \mu \le 3.75$$

We thus conclude, with 99 percent confidence, that the process mean weight of bricks is between 3.70 and 3.75 pounds. □

Many statistical packages provide confidence intervals for the population mean. Figure 11.2 shows MINITAB output for the 95.4 percent confidence interval for the accounts receivable example. The output includes the sample size (labeled N), the estimated standard deviation $s\{\overline{X}\}$ of the sampling distribution of $\overline{X}$ (labeled SE MEAN), and the 95.4 percent confidence limits (labeled C.I.).

Development of Confidence Interval

We now show the basis of confidence interval (11.5). We begin with a formal definition of a confidence interval.

(11.6)
The interval $L \le \mu \le U$ is a $1 - \alpha$ confidence interval for the population mean μ if the following probability holds prior to sampling:

$$P(L \le \mu \le U) = 1 - \alpha$$

This definition states that, in advance of selecting the random sample, the probability is $1 - \alpha$ that a $1 - \alpha$ confidence interval calculated from the sample results will include μ. Therefore, if many random samples of size n are drawn from a population, $100(1 - \alpha)$ percent of the confidence intervals calculated from the samples will include μ and consequently will be correct.

N	MEAN	STDEV	SE MEAN	95.4 PERCENT C.I.
100	33.19	34.48	3.45	(26.29, 40.09)

FIGURE 11.2
MINITAB output for confidence interval for population mean—Accounts receivable example

The validity of confidence interval (11.5) depends on the following extension of central limit theorem (10.7).

(11.7)

For almost any population, $(\overline{X} - \mu)/s\{\overline{X}\}$ follows approximately a standard normal distribution when the random sample size is sufficiently large; that is:

$$\frac{\overline{X} - \mu}{s\{\overline{X}\}} \simeq Z$$

For populations that are not highly skewed, sample sizes need not be too large for theorem (11.7) to apply. For highly skewed populations, however, the sample size may need to be quite large.

Assuming n is large enough so that $(\overline{X} - \mu)/s\{\overline{X}\}$ is distributed approximately as a standard normal variable Z, we can state:

$$P\left[z(\alpha/2) \le \frac{\overline{X} - \mu}{s\{\overline{X}\}} \le z(1 - \alpha/2) \right] \simeq 1 - \alpha$$

Figure 11.3 illustrates why this probability is $1 - \alpha$. In each tail, the area is $\alpha/2$, so the central area is $1 - \alpha$. But by (8.9), $z(\alpha/2) = -z(1 - \alpha/2)$. Hence:

$$P\left[-z(1 - \alpha/2) \le \frac{\overline{X} - \mu}{s\{\overline{X}\}} \le z(1 - \alpha/2) \right] \simeq 1 - \alpha$$

Rearranging the inequalities, we obtain:

$$P[\overline{X} - z(1 - \alpha/2)s\{\overline{X}\} \le \mu \le \overline{X} + z(1 - \alpha/2)s\{\overline{X}\}] \simeq 1 - \alpha$$

But this expression fits the definition of a confidence interval in (11.6), where the lower and upper confidence limits are:

$$L = \overline{X} - z(1 - \alpha/2)s\{\overline{X}\} \qquad U = \overline{X} + z(1 - \alpha/2)s\{\overline{X}\}$$

These are, of course, the confidence limits given in (11.5).

Figure 11.3 summarizes the reasoning that underlies the confidence interval for μ in (11.5). In effect, the risk α of an incorrect confidence interval is divided equally in the two tails of the standard normal distribution.

Comment

We can now see why confidence interval (11.5) has an approximate confidence coefficient of $1 - \alpha$. The reason is that the central limit theorem only ensures that $(\overline{X} - \mu)/s\{\overline{X}\}$ is *approximately* a standard normal variable for large n.

FIGURE 11.3

Two-sided confidence interval for μ—Large sample. Before the sample is taken, the probability is $1 - \alpha$ that the quantity $(\overline{X} - \mu)/s\{\overline{X}\}$ will fall in the shaded interval. The interval estimate $\overline{X} \pm z(1 - \alpha/2)s\{\overline{X}\}$ will be correct (that is, will contain μ) if $(\overline{X} - \mu)/s\{\overline{X}\}$ falls in the shaded interval.

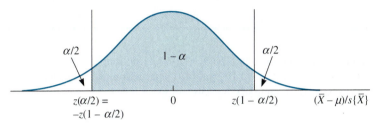

Confidence Coefficient

Meaning of Confidence Coefficient. From the definition of a confidence interval in (11.6), we know that, in advance of selecting the random sample, the probability is $1 - \alpha$ that the confidence interval we obtain will contain the population mean μ. The particular confidence interval obtained, however, may or may not contain μ. Thus, any one confidence interval result will be either correct or incorrect, and we do not know for certain which is the case. Still, we act as if the result is correct when the confidence coefficient $1 - \alpha$ is sufficiently large because the procedure gives us assurance that in $100(1 - \alpha)$ percent of the cases, the confidence interval result will be correct.

Selecting the Confidence Coefficient. In the best of all worlds, we would like the confidence interval to be very precise (that is, very narrow) and we would also like to be very confident that the interval contains μ. Unfortunately, for any fixed sample size, the confidence coefficient can only be increased by increasing the width of the confidence interval. Figure 11.4 illustrates this relationship graphically. It shows how the width of confidence interval (11.5) varies with the confidence coefficient for any given sample result. Note how rapidly the confidence interval widens as the confidence coefficient gets near 100 percent.

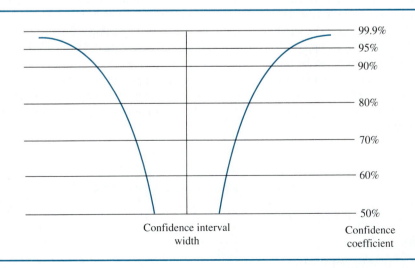

FIGURE 11.4
Relation between confidence coefficient and confidence interval width

The choice of confidence coefficient $1 - \alpha$ will vary from case to case, depending on how much risk can be taken of obtaining an incorrect interval. Confidence coefficients of 90, 95, and 99 percent are often used in practice.

11.3 CONFIDENCE INTERVAL FOR μ— SMALL SAMPLE

When the random sample size is small, $(\overline{X} - \mu)/s\{\overline{X}\}$ no longer follows the standard normal distribution as described in (11.7), and the construction of a confidence interval for μ depends on the nature of the underlying population.

Normal Population

We shall first consider the construction of confidence intervals for μ *when the population is normal.* The following theorem helps us to construct a confidence interval for μ in this case.

(11.8)

For a simple random sample of size n from a normal population:

$$\frac{\overline{X} - \mu}{s\{\overline{X}\}} = t(n - 1)$$

The notation $t(n - 1)$ in (11.8) denotes the t distribution with $n - 1$ degrees of freedom. Theorem (11.8) tells us, therefore, that the statistic $(\overline{X} - \mu)/s\{\overline{X}\}$ follows the t distribution with $n - 1$ degrees of freedom when the sampled population is normal.

The t distribution is continuous, unimodal, and symmetrical, with mean of 0. It is similar in appearance to a standard normal distribution but is more variable. We know from Chapter 10 that $(\overline{X} - \mu)/\sigma\{\overline{X}\}$ has a standard normal distribution when the sampled population is normal. The substitution of the estimate $s\{\overline{X}\}$ for $\sigma\{\overline{X}\}$ in (11.8) introduces additional variability into the statistic and the t distribution correctly takes this additional variability into account. As the sample size n becomes larger, and hence the degrees of freedom $n - 1$ become larger, the additional variability introduced by using $s\{\overline{X}\}$ becomes smaller and the t distribution approaches the standard normal distribution. (A detailed discussion of the t distribution may be found in Appendix B, Section B.2.)

In view of theorem (11.8), the confidence limits for μ when the population is normal will be of exactly the same form as those for large samples, except that an appropriate percentile of the t distribution is used instead of a percentile from the standard normal distribution.

(11.9)

The $1 - \alpha$ confidence limits for μ, when the population is normal, are:

$$\overline{X} \pm ts\{\overline{X}\}$$

where: $t = t(1 - \alpha/2; n - 1)$

$s\{\overline{X}\}$ is given by (11.2)

These confidence limits are appropriate for any sample size, small or large, as long as the sampled population is normal. The reasonableness of the assumption of a normal population can be examined from a dot plot, histogram, or frequency polygon for the sample data, and from a normal probability plot and a formal test of normality to be described in Chapter 16.

EXAMPLE □

A sample of five cans of tomatoes in puree was taken at random from a production line immediately after filling. The solid content of each can was weighed to serve as a basis for estimating the process mean drained weight per can (μ). Past experience indicates that the distribution of drained weights is normal. It is desired to construct a 99 percent confidence interval for the process mean μ.

The following are the sample results (in ounces):

i:	1	2	3	4	5
X_i:	23.0	23.5	23.5	25.0	24.5

$$n = 5 \qquad \overline{X} = 23.9 \qquad s = 0.822$$

Hence, $s\{\overline{X}\} = s/\sqrt{n} = 0.822/\sqrt{5} = 0.368$. The desired confidence coefficient is $1 - \alpha = 0.99$, so $\alpha = 0.01$ and $1 - \alpha/2 = 0.995$. The degrees of freedom for the t distribution are $n - 1 = 5 - 1 = 4$. From Table C.3, we find that $t(0.995; 4) = 4.604$. Using (11.9), we obtain the confidence limits $23.9 \pm 4.604(0.368)$, which gives the 99 percent confidence interval:

$$22.2 \leq \mu \leq 25.6$$

We can now state, with 99 percent confidence, that the mean drained weight for the current process is somewhere between 22.2 and 25.6 ounces. □

Comments

1. The confidence limits in (11.9) become the same as the large sample confidence limits in (11.5) for large sample sizes. The reason is that the t multiple in (11.9) approaches the standard normal multiple as the sample size increases.
2. Occasionally, the population standard deviation σ is known and need not be estimated by the sample standard deviation s. In this case, $\sigma\{\overline{X}\}$ is also known exactly. When the population is normal, then $(\overline{X} - \mu)/\sigma\{\overline{X}\}$ follows a standard normal distribution by (10.11) and (8.6). Consequently, the appropriate confidence limits in this case are $\overline{X} \pm z\sigma\{\overline{X}\}$, where, irrespective of sample size, $z = z(1 - \alpha/2)$.

Nonnormal Populations

Confidence limits (11.9) for normal populations are useful also for many populations that are not normal. The limits are applicable to nonnormal populations *when the population does not depart too markedly from normality and the sample size is not exceedingly small.* Under these conditions, statistical theory shows that $(\overline{X} - \mu)/s\{\overline{X}\}$ is distributed *approximately* as $t(n - 1)$.

(11.10)

The approximate $1 - \alpha$ confidence limits for the population mean μ, when the population does not depart too markedly from normality and the sample size is not exceedingly small, are:

$$\overline{X} \pm ts\{\overline{X}\}$$

where: $t = t(1 - \alpha/2; n - 1)$

$s\{\overline{X}\}$ is given by (11.2)

Populations that do not depart too markedly from normality include ones that are not highly skewed, such as the uniform population in (8.1). For instance, the confidence limits in (11.10) would be applicable for a random sample of 10 observations from a uniform distribution.

Statistical procedures that continue to be approximately valid even when their exact theoretical requirements are not met are said to be *robust*. The confidence limits in (11.9), based on the t distribution, are therefore robust with respect to nonextreme departures from normality.

☐ EXAMPLE

A random sample of $n = 15$ physicians was selected in a community to estimate the mean charge for a certain medical procedure in the community. The sample results were $\overline{X} = \$74.35$ and $s = \$4.28$. It is known that the distribution of charges by the physicians in the community is somewhat skewed, but not highly so. Hence, confidence limits (11.10) are applicable. The desired confidence coefficient is $1 - \alpha = 0.95$. We require $t(0.975; 14) = 2.145$. We also need $s\{\overline{X}\} = s/\sqrt{n} = 4.28/\sqrt{15} = 1.105$. The confidence limits therefore are $74.35 \pm 2.145(1.105)$. Hence, with 95 percent confidence, we estimate that the mean charge for the procedure in the community is between $\$72.0$ and $\$76.7$.   ☐

When a population is markedly nonnormal, a mathematical transformation of the data often can be found that makes the population close enough to normal so that the confidence limits (11.10) can be applied to the transformed data. For instance, the logarithmic transformation is often employed with income and other right-skewed data because the log X data are generally less skewed than the untransformed data X.

☐ EXAMPLE

Figure 11.5a shows a histogram of income data for persons in a community. Figure 11.5b presents the same data on a log X scale. The logarithmic transformation has produced a much less skewed distribution.   ☐

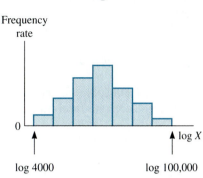

(a) **Distribution of Incomes**
X

Frequency
rate

0

4000   100,000

(b) **Distribution of Log-Incomes**
log X

Frequency
rate

0

log 4000   log 100,000

FIGURE 11.5
Logarithmic transformation
making skewed distribution
more normal

CONFIDENCE 11.4 INTERVAL FOR POPULATION TOTAL

A confidence interval for the population total of a finite population is often desired.

EXAMPLE

In the accounts receivable example, the auditor was actually interested in the total audit amount of the 8042 accounts receivable, and the concern with the mean audit amount (μ) was merely a step toward obtaining a confidence interval for the population total.

Let τ (Greek tau) denote the *population total.*

$$(11.11) \quad \tau = \sum_{i=1}^{N} X_i$$

In the accounts receivable example, τ denotes the total of all audit amounts for the $N = 8042$ accounts in the population.

Since $N\mu = \Sigma X_i$, by definition (1.12a) of the population mean, it follows that:

$$(11.12) \quad \tau = N\mu$$

Consequently, confidence limits for τ can be obtained simply by multiplying the confidence limits for μ by N.

☐ **EXAMPLE**

Refer to the accounts receivable example in Table 11.1. We wish to obtain a 95.4 percent confidence interval for the total audit amount of the accounts in the population. Earlier, we found the 95.4 percent confidence interval for μ:

$$26.29 \leq \mu \leq 40.09$$

Since $N = 8042$, the 95.4 percent confidence limits for the total audit amount of the company's accounts receivable are:

$$8042(26.29) = 211,424 \qquad 8042(40.09) = 322,404$$

So the 95.4 percent confidence interval for τ is:

$$211,424 \leq \tau \leq 322,404$$

We conclude, with 95.4 percent confidence, that the total audit amount of the 8042 accounts is between \$211.4 thousand and \$322.4 thousand.   ☐

11.5 PLANNING OF SAMPLE SIZE

Need for Planning

Users of statistical information often would like to specify in advance the precision and confidence level required in an interval estimate. This control over an interval estimate requires appropriate planning of the sample size.

☐ **EXAMPLES**

1. A marketing manager wants to know the mean income of households living in the vicinity of a shopping center within $\pm \$500$, with a 95 percent confidence coefficient. The number of households to be included in the sample needs to be determined.

2. In a merger negotiation, the mean number of days of sick leave accrued by the employees of one of the corporations must be estimated. Lawyers on both sides agree that it will be sufficient to estimate this mean within ± 2 days, with a confidence coefficient of 99 percent. The number of employee files to be included in the sample needs to be determined.   ☐

From such specifications, appropriate sample sizes can often be determined that will approximately yield the desired precision at the specified confidence level.

A systematic planning approach to determining the needed sample size is more reasonable than using an arbitrary sample size. With an arbitrary sample size, the width of the confidence interval may be tighter than needed, indicating that the sample size was too large, or it may be too wide, indicating that the sample size was inadequate. If too large a sample is taken, money will have been wasted in getting greater precision than necessary. If too small a sample is taken, it may be costly or even impossible to enlarge the sample subsequently.

Assumptions

The planning procedure we shall explain is based on the following assumptions:

1. The random sample size ultimately determined is reasonably large.
2. The population is infinite or, if finite, is large relative to the resulting sample size.

Planning Procedure

The planning procedure requires specifications of the desired precision of the estimate and the desired confidence coefficient. Furthermore, because we are planning the sample size, we do not yet have the sample standard deviation s available to compute the estimated standard deviation of the mean, $s\{\overline{X}\}$. Inasmuch as the population standard deviation σ is generally unknown, we shall also need to specify a planning value for σ to find the needed sample size.

EXAMPLE ☐ **Acreage Plans**

A farming region contains more than 40,000 farms. In the early spring, a survey based on a simple random sample of farm operators is planned to determine the acreage each operator intends to sow in summer wheat. The survey forms part of a larger study designed to estimate total agricultural production in the next 12 months. It is desired to have the wheat acreage survey produce an interval estimate of the mean wheat acreage per farm within ± 5 acres, with a confidence level of 95 percent. Ample recent historical data about farm wheat acreage in the region are available, and consultation with agricultural experts indicates that the variability of wheat acreages among farms in the coming year is not likely to differ sharply from that exhibited in the recent past. From the historical data, a planning value $\sigma = 85$ acres seems reasonable. ☐

As shown by the acreage plans example, the three specifications to be made for determining the needed sample size are as follows.

1. *The desired precision.* The desired precision will be denoted by $\pm h$ and indicates how close the confidence limits must be around the point estimate $\overline{X}$. In the acreage plans example, $h = 5$ acres. Thus, the confidence limits $\overline{X} \pm h$ should be $\overline{X} \pm 5$. The symbol h is called the desired *half-width* of the confidence interval; the total width of the confidence interval will be $2h$.
2. *The desired confidence coefficient.* In the acreage plans example, the desired confidence coefficient is $1 - \alpha = 0.95$.
3. *The planning value for the population standard deviation.* Information about σ is often available from past experience with the same or a similar study or can be obtained from a pilot study. Approximate information about the order of magnitude of σ is usually adequate for planning purposes. In the acreage plans example, the planning value is $\sigma = 85$ acres.

From (11.5), we know that the half-width h of the confidence interval is:

$$h = z\sigma\{\overline{X}\} \qquad \text{where } z = z(1 - \alpha/2)$$

Note that here we use the notation for the true standard deviation of the mean, $\sigma\{\overline{X}\}$, rather than that for the estimated standard deviation, $s\{\overline{X}\}$, because we are in the planning phase.

We have, by (10.6):

$$\sigma\{\overline{X}\} = \frac{\sigma}{\sqrt{n}}$$

Hence:

$$h = z\sigma\{\overline{X}\} = z\frac{\sigma}{\sqrt{n}}$$

Squaring both sides and solving for n, we obtain:

$$n = \frac{z^2\sigma^2}{h^2}$$

(11.13)

The needed random sample size for a specified half-width h, confidence coefficient $1 - \alpha$, and planning value σ is:

$$n = \frac{z^2\sigma^2}{h^2}$$

where: $z = z(1 - \alpha/2)$

σ is the planning value for the population standard deviation

---

Formula (11.13) must be modified when the necessary sample size turns out to be large relative to the population size. We consider this case in Chapter 23.

☐ EXAMPLE

In the acreage plans example, the specified precision is $h = 5$ acres, the planning value for the population standard deviation is $\sigma = 85$ acres, and the desired confidence coefficient is $1 - \alpha = 0.95$. Hence, $\alpha = 0.05$ and $1 - \alpha/2 = 0.975$. Thus, $z(0.975)$ is required. We find from Table C.1 that $z(0.975) = 1.960$. Substituting into (11.13) gives the needed sample size:

$$n = \frac{(1.960)^2(85)^2}{(5)^2} = 1110$$

If the planning value of σ is at all reliable, a random sample of 1110 farms should produce a 95 percent confidence interval with a half-width of about 5 acres. Note that formula (11.13) is appropriate here because the resultant n is large while n/N is small.   ☐

Once the sample size has been determined and the actual sample selected, the construction of a confidence interval proceeds in the usual manner. We now use, of course, the sample standard deviation s and not the planning value for σ in constructing the confidence interval.

In the acreage plans example, a random sample of 1110 farms was selected and yielded $\overline{X} = 75.6$ acres and $s = 80.1$ acres. We calculate $s\{\overline{X}\} = 80.1/\sqrt{1110} = 2.404$ by (11.2). Hence, the 95 percent confidence limits for the mean wheat acreage per farm by (11.5) are $75.6 \pm 1.960(2.404)$. The 95 percent confidence interval therefore is:

$$70.9 \leq \mu \leq 80.3$$

Observe that the attained half-width is 4.7 acres. This value is slightly smaller than the 5-acre value specified in the planning stage. The extra precision resulted because the sample standard deviation of 80.1 acres is a little smaller than the planning value of 85 acres.   ☐

Thus far, our confidence intervals for a population mean μ have had both upper and lower limits. Sometimes, one-sided confidence intervals are useful.

**OPTIONAL   11.6
TOPIC—ONE-SIDED
CONFIDENCE
INTERVALS FOR μ**

A bank officer contemplating making a loan to a firm on its inventory is interested in an upper limit for the mean value μ of obsolete inventory items. Thus, the officer requires an upper one-sided confidence interval of the form $\mu \leq U$, where U is the upper limit.   ☐

One-sided confidence intervals are easily constructed.

> **(11.14)**
> A **lower one-sided confidence interval** for μ has the form $\mu \geq L$, and an **upper one-sided confidence interval** for μ has the form $\mu \leq U$.
> A one-sided $1 - \alpha$ confidence interval for μ is obtained by using the relevant limit for a two-sided confidence interval for μ where the two-sided interval is constructed with confidence coefficient $1 - 2\alpha$.

Note from (11.14) that a 95 percent upper or lower confidence limit must be obtained from a 90 percent two-sided confidence interval. Similarly, a 99 percent upper or lower confidence limit must be obtained from a 98 percent two-sided confidence interval. The reason for this is that with one-sided confidence intervals, we incur risk of an incorrect interval on only one side of the sampling distribution of $\overline{X}$, as may be seen from Figure 11.3.

To implement (11.14) in an actual application, we do not need to calculate a two-sided confidence interval first; we can construct the desired limit for a one-sided interval directly.

☐ **EXAMPLE**

Refer to the accounts receivable example in Table 11.1 again. We wish to obtain a lower 95 percent confidence interval for the mean audit amount per account in the population. Hence, we must calculate the lower limit of a two-sided 90 percent confidence interval using (11.5). We know from before that $\bar{X} = 33.19$ and $s\{\bar{X}\} = 3.448$. We require $z(0.95) = 1.645$. Hence, the lower confidence limit is $33.19 - 1.645(3.448) = 27.52$. Thus, with 95 percent confidence, we can assert that $\mu \geq 27.52$. In terms of the total audit amount of the company's 8042 accounts receivable, the auditor can conclude, with 95 percent confidence, that $\tau \geq 8042(27.52) = 221,316$, or \$221.3 thousand.   ☐

If a one-sided confidence interval for μ is required for a population that is normal or for one that does not depart too markedly from normality, based on a small random sample, then the appropriate confidence limit (upper or lower) can be calculated using (11.9) or (11.10).

11.7   OPTIONAL TOPIC—PREDICTION INTERVAL FOR NEW OBSERVATION

Occasionally, we may wish to construct an interval estimate for a *new* observation using sample data. Such an interval estimate is called a *prediction interval.* One important use of prediction intervals is to judge whether or not a process is in control.

Chemical Process ☐ **EXAMPLE**

A chemical process converts a batch of raw material into final product in each operating cycle. The yields of final product in a random sample of 20 operating cycles when the process was in control are to be used to construct a prediction interval for the yield in the next operating cycle. A comparison of this new yield with the prediction interval will permit the process operator to judge whether or not the process is in control in the next cycle.   ☐

Another use of prediction intervals is to help in identifying outliers.

☐ **EXAMPLE**

A government meat inspector has measured the percentage of fat on 200 beef carcasses. The inspector wishes to obtain a prediction interval for a new carcass so the fattiness of the new carcass can be compared with the prediction interval to determine whether the new carcass is an outlier with respect to its fattiness.   ☐

Assumptions

We shall let X_{new} denote the new observation for which a prediction interval is required. The procedure we shall describe for constructing a prediction interval is based on the following assumptions.

1. The new observation X_{new} is selected independently from the same population as the random sample $X_1, X_2, \ldots, X_n$.
2. The population is normal.

These assumptions describe conditions that are encountered frequently in practice.

Population Parameters Known

To explain the construction of a prediction interval, we first consider the case where the parameters μ and σ of the normal population are known. The more common case where these parameters are unknown will be considered next.

When X_{new} is randomly drawn from a normal population with mean μ and standard deviation σ, we know from the normal probability calculations discussed in Figure 8.6b that the probability is $1 - \alpha$ that the following interval is correct:

$$\mu - z\sigma \le X_{new} \le \mu + z\sigma \qquad \text{where } z = z(1 - \alpha/2)$$

Hence, this interval is a $1 - \alpha$ prediction interval for X_{new}.

EXAMPLE ☐

In the ingot weight example of Chapter 8, a 98 percent prediction interval for the weight X_{new} of a new metal ingot is to be determined. The normal population has mean $\mu = 520$ pounds and standard deviation $\sigma = 11$ pounds. We require $z(0.99) = 2.326$. Hence, the 98 percent prediction limits are $520 \pm 2.326(11)$ and the prediction interval is:

$$494.4 \le X_{new} \le 545.6$$

We can thus state, with 98 percent confidence, that the weight of the new ingot will be between 494.4 and 545.6 pounds.   ☐

Population Parameters Unknown

In the more common case where the parameters μ and σ of the normal population are unknown, we use a random sample of size n from the normal population to provide the sample statistics $\overline{X}$ and s as estimators of the parameters μ and σ, respectively. The prediction interval uses the estimated variance $s^2\{X_{new}\}$.

$$(11.15) \qquad s^2\{X_{new}\} = s^2 + \frac{s^2}{n} = s^2\left(1 + \frac{1}{n}\right)$$

Note that this estimated variance has two components that are related to two sources of variation. The first component, s^2, measures the variability of individual observations in the population. The second component, s^2/n, reflects the variability of $\overline{X}$ as an estimator of μ.

It can be shown that prediction limits based on the t distribution are appropriate.

(11.16)

The prediction limits for a new observation X_{new} with confidence coefficient $1 - \alpha$, when the random sample and X_{new} are independently drawn from the same normal population, are:

$$\overline{X} \pm ts\{X_{new}\}$$

where:　$t = t(1 - \alpha/2; n - 1)$

　　　　$s\{X_{new}\}$ is given by (11.15)

☐ **EXAMPLE**

In the chemical process example, the yields (in kilograms) of final product for $n = 20$ operating cycles were obtained. The sample statistics are $\overline{X} = 238.0$ and $s = 4.30$. Hence, from (11.15), we calculate:

$$s^2\{X_{new}\} = (4.30)^2\left(1 + \frac{1}{20}\right) = 19.41 \qquad s\{X_{new}\} = \sqrt{19.41} = 4.406$$

A 99 percent prediction interval for the yield in the next operating cycle is desired. We require $t(1 - \alpha/2; n - 1) = t(0.995; 19) = 2.861$. Following (11.16), the prediction limits are $238.0 \pm 2.861(4.406)$, and the desired prediction interval is:

$$225 \le X_{new} \le 251$$

Thus, the process operator may predict, with 99 percent confidence, that if the process is still in control, the yield in the next operating cycle will be between 225 and 251 kilograms. Should the yield in the next cycle be 213 kilograms, for instance, then the operator would conclude that the process was not in control during that cycle.　　☐

Comments

1. The prediction limits in (11.16) are not robust to departures from normality. Their validity requires that the sampled population be normal or close to normal. When nonnormal populations are encountered, a mathematical transformation of the sample data, such as the logarithmic transformation illustrated in Figure 11.5, may be helpful in making the data approximately normal. A second alternative is to employ a prediction procedure that does not require normality of the population. Such procedures are described in specialized statistical reference books.

2. When predicting a new observation X on the basis of an earlier sample, the relevant variance is that of the deviation of the new observation from the sample mean—that is, the deviation $X - \overline{X}$. Because of the independence of the new observation and the earlier sample, we have:

$$\sigma^2\{X - \overline{X}\} = \sigma^2\{X\} + \sigma^2\{\overline{X}\}$$

Any observation from the population has variance σ^2; hence, $\sigma^2\{X\} = \sigma^2$. We further know from (10.6) that $\sigma^2\{\bar{X}\} = \sigma^2/n$. Hence, we have:

$$\sigma^2\{X - \bar{X}\} = \sigma^2 + \frac{\sigma^2}{n} = \sigma^2\left(1 + \frac{1}{n}\right)$$

Replacing σ^2 by the sample estimator s^2 yields (11.15), where for brevity the estimated variance is denoted by $s^2\{X_{new}\}$.

PROBLEMS

* **11.1 Parcel Postage.** The amounts of postage were recorded for a random sample of $n = 400$ parcels handled by the postal system on a particular day. The sample mean and standard deviation were $\bar{X} = \$5.21$ and $s = \$3.70$, where X denotes the amount of postage on a parcel (in dollars). It is desired to obtain an interval estimate of μ, the mean amount of postage on all parcels handled by the postal system that day.

a. Estimate $\sigma\{\bar{X}\}$, the standard deviation of the sampling distribution of $\bar{X}$. What notation is used for this estimate?
b. Calculate the confidence limits $\bar{X} \pm 2s\{\bar{X}\}$. What confidence coefficient is associated with these limits?
c. Obtain a 90 percent confidence interval for μ.
d. If 100 random samples of $n = 400$ parcels were selected independently and a 90 percent confidence interval for μ calculated for each of the 100 samples, what proportion of the 100 confidence intervals would be expected to contain μ?

11.2 As part of a profitability analysis, a refrigeration company wished to estimate μ, the mean number of hours expended by company personnel per service call last year. A random sample of 144 service calls yielded $\bar{X} = 1.34$ hours and $s = 1.82$ hours, where X denotes the hours expended in a service call.

a. Calculate the confidence limits $\bar{X} \pm s\{\bar{X}\}$. How confident can one be that these limits cover μ?
b. A person, commenting on the analysis, said, "Because the confidence limits always include the mean $\bar{X}$, the confidence interval corresponding to $\bar{X} \pm s\{\bar{X}\}$ should always be correct." Do you agree? Comment.
c. Obtain a 90 percent confidence interval for the population mean. Interpret the confidence interval.

11.3 Trade Association. An apparel and shoe trade association has a large number of member firms. As part of a study of the impact of new minimum-wage legislation on association members, a random sample of 225 member firms was selected to estimate μ, the mean number of hourly paid employees in member firms. An analysis of the sample results showed that $\bar{X} = 8.31$ employees and $s = 4.80$ employees, where X denotes the number of hourly paid employees in a firm.

a. Construct a 95 percent confidence interval for μ. Interpret the confidence interval.
b. Explain why the confidence coefficient of the interval estimate in part a is only approximately 95 percent.
c. If you had constructed a 99 percent confidence interval, would it have been wider or narrower than the one in part a? Which interval would involve the greater risk of an incorrect estimate?

11.4 Airline Reservations. An airline researcher studied reservation records for a random sample 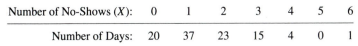 of 100 days to estimate μ, the mean number of no-shows (persons who fail to keep their reservations) on the daily 4 P.M. commuter flight to New York City. The records revealed the following:

Number of No-Shows (X):	0	1	2	3	4	5	6
Number of Days:	20	37	23	15	4	0	1

a. Verify that $\overline{X} = 1.500$ and $s = 1.185$. Obtain $s\{\overline{X}\}$. What does $s\{\overline{X}\}$ estimate here?

b. Construct a 99 percent confidence interval for μ. Interpret the confidence interval.

c. Would a different random sample of 100 days have provided the same interval estimate as the one in part b? Explain.

11.5 Data Transmissions. Data transmission errors tend to occur in clusters of 1, 2, 3, . . . characters. A communications engineer selected a random sample of 200 error clusters to estimate μ, the mean cluster size. The sample results showed $\overline{X} = 2.51$ characters and $s = 1.64$ characters, where X denotes the number of characters in the error cluster.

a. Construct a 90 percent confidence interval for μ. Interpret the confidence interval.

b. Explain why the confidence coefficient of the interval estimate in part a is only approximately 90 percent.

11.6 A research report stated that the mean return on invested capital earned by furniture manufacturers is between 6.1 and 10.8 percent per year, with a confidence coefficient of 95 percent. One person interpreted this result as meaning that 95 percent of furniture manufacturers had investment returns between 6.1 and 10.8 percent. Another person interpreted it in the sense that if many random samples were taken, 95 percent of them would have sample means between 6.1 and 10.8 percent. Why are these interpretations incorrect?

11.7 Refer to Figure 11.4.

a. In the construction of a confidence interval for a population mean based on a large random sample, what is the ratio of the width of a 90 percent confidence interval to that of an 80 percent confidence interval? What is the ratio of the width of a 99 percent confidence interval to that of a 90 percent confidence interval? What generalization is suggested by these results?

b. Can a 100 percent confidence interval for a population mean be constructed? Comment.

11.8 A work study is undertaken to estimate, with a 95 percent confidence interval, the mean time required to weld a metal seam. Assume that welding times are normally distributed and consider a random sample of n trial welds.

a. Obtain the t percentile for the confidence limits in (11.9) if n is (1) 3, (2) 15, (3) 30.

b. Does the t percentile in part a shrink by much when n increases from 3 to 15? From 15 to 30? What is the implication of this finding for the effect of the t multiple on the width of a confidence interval with a large sample size?

* 11.9 Nutrition Study. In a study of the effects of nutrition on work efficiency, a random sample of 16 workers was asked to follow a rigid dietary regimen. In one part of this study, the blood sugar level X of each sample worker was measured two hours after breakfast. The results (in milligrams of sugar per 100 cubic centimeters of blood) were $\overline{X} = 112.8$ mg and $s = 9.6$ mg. Blood sugar levels may be considered to be normally distributed.

a. Construct a 95 percent confidence interval for the mean blood sugar level under this regimen. Interpret the confidence interval.

b. If blood sugar levels are actually not normally distributed but the departure from normality is not marked, what assurance do you have about the actual confidence level of the confidence interval in part a?

11.10 Vehicle Fleet. An examination of the records for a random sample of nine motor vehicles in a large fleet shows the following operating costs (in cents per mile):

$$38.3 \quad 36.4 \quad 37.0 \quad 32.5 \quad 33.5 \quad 39.1 \quad 36.8 \quad 36.7 \quad 40.9$$

Let μ denote the mean operating cost in cents per mile for the vehicles in the fleet. The distribution of vehicle operating costs is not normal, although the departure from normality is not too marked.

a. Construct a 90 percent confidence interval for μ. Interpret the confidence interval.

b. What is the effect of the distribution of operating costs not being normal on the actual confidence level of the confidence interval in part a?

11.11 A scientist is studying the effects of different communication modes on the problem-solving capability of teams. For one of the communication modes under study, a random sample of 16 teams completed a specific task in an average of 25.9 minutes, with a standard deviation of 4.6 minutes. Completion times are approximately normally distributed.

a. Construct a 95 percent confidence interval for the mean time required to complete the task with this communication mode. Interpret the confidence interval.

b. What is the effect of the distribution of completion times not being exactly normal on the actual confidence level of the confidence interval in part a?

11.12 Computer Software Rates of Return. A computer industry publication wished to estimate the mean percentage return on equity last year for all computer software firms. A random sample of 12 firms was drawn from the many firms in the software sector. The results were as follows:

8.37	31.38	−6.52	24.87	−8.47	24.83
−3.15	7.20	15.45	0.35	11.28	21.01

Consider that returns on equity are approximately normally distributed.

a. Construct a 90 percent confidence interval for the population mean percentage return on equity. Interpret the confidence interval.

b. If percentage returns on equity are not approximately normally distributed but the departure from normality is not too marked, what assurance do you have about the actual confidence level of the interval estimate in part a?

* **11.13** Refer to **Parcel Postage** Problem 11.1. Two million parcels were handled by the postal system on the particular day. Obtain a 90 percent confidence interval for the total amount of postage on the parcels handled that day. Interpret the confidence interval.

11.14 Refer to **Trade Association** Problem 11.3. The association has 9100 member firms. Construct a 95 percent confidence interval for the total number of hourly paid employees of all member firms. Interpret the confidence interval.

11.15 Refer to **Vehicle Fleet** Problem 11.10. All of the vehicles in the fleet travel about the same number of miles during a month. Construct a 90 percent confidence interval for the expected total cost of operating the fleet in a month when a total of 5 million vehicle miles is accumulated. Interpret the confidence interval.

* **11.16** A sample survey of kindergarten children in a state is being planned to estimate, among other things, the mean number of older siblings of kindergarten children. It is desired to estimate this mean within ±0.08 sibling, with a 90 percent confidence coefficient. A reasonable planning value for the population standard deviation σ is 0.6 sibling.

a. What sample size is needed to estimate the mean number of older siblings?

b. If the desired precision were tightened to ±0.06, would the required sample size be increased substantially? Explain.

11.17 A nationwide survey of practicing physicians is to be undertaken to estimate μ, the mean number of prescriptions written during the survey day. The desired precision is ±0.7 prescription, with a 99 percent confidence coefficient. A pilot study revealed that a reasonable planning value for the population standard deviation σ is 5 prescriptions.

a. How many physicians should be included in the survey to estimate μ with the desired precision?

b. If the specified confidence coefficient were lowered to 95 percent, would the required sample size be reduced substantially? Explain.

11.18 Refer to **Airline Reservations** Problem 11.4. Sometime after this study, the airline instituted a new reservations system that discourages customers from holding simultaneous reservations on different flights to the same destination. It is now desired to estimate again the mean number of no-shows for the daily 4 P.M. flight to New York City. A 95 percent confidence interval, with a half-width of 0.2 no-show, is wanted. The standard deviation of the original sample ($s = 1.185$ no-shows) is to be taken as the planning value for σ.

a. How many days should be included in the sample to prepare the estimate?

b. If the specified confidence coefficient were increased to 99 percent (other factors being unchanged), how many days would have to be included in the sample? Is the resultant change in sample size substantial relative to the increase in the confidence coefficient?

c. What is the likely effect on the confidence interval if the planning value for σ is too large? Why can you not discuss the effect on the confidence interval for any one sample?

11.19 Refer to **Data Transmissions** Problem 11.5. Suppose that the sample size has not yet been specified. The communications engineer wishes to estimate the mean cluster size within ± 0.1 character, with a 90 percent confidence coefficient. A reasonable planning value for σ is 1.6 characters. Find the number of error clusters that must be included in the sample to provide the desired confidence interval.

* **11.20** Refer to **Parcel Postage** Problem 11.1.

a. Construct a lower 95 percent confidence interval for μ. Interpret the confidence interval.

b. Based on the interval estimate in part a, can the postal authorities confidently claim that the total amount of postage for the 2 million parcels handled that day exceeded $9 million? Explain.

11.21 Refer to **Trade Association** Problem 11.3.

a. Obtain an upper 97.5 percent confidence interval for the mean number of hourly paid employees per firm. Interpret the confidence interval.

b. The trade association is contemplating developing a mailing list containing all hourly paid employees of the 9100 member firms and needs an indication of the minimum number of hourly paid employees that will be contained in the list. Obtain the appropriate 99 percent confidence limit.

11.22 Refer to **Nutrition Study** Problem 11.9.

a. Construct a lower 97.5 percent confidence interval for the mean blood sugar level under this regimen.

b. Why can one not conclude that 97.5 percent of workers' blood sugar levels two hours after eating the regimen breakfast will exceed the confidence limit in part a?

11.23 When the soldering process for wire connections in an electric appliance meets quality assurance standards, the breaking strengths of connections have a normal distribution with $\mu = 1400$ milligrams and $\sigma = 150$ milligrams. An inspector will choose one connection at random for testing.

a. Construct a 99 percent prediction interval for the breaking strength of the connection if the process standards are being met.

b. The breaking strength of the connection turned out to be 750 milligrams. What does this imply?

* **11.24** The monthly rates of return on a common stock over the past six months were (in percent):

$$-0.2 \quad 1.3 \quad 1.0 \quad -0.6 \quad 2.1 \quad 0.8$$

Assume that these monthly rates of return constitute independent random observations from a normal population.

a. Construct a 95 percent prediction interval for next month's rate of return on the stock. Interpret the prediction interval.

b. Would a 95 percent confidence interval for the mean monthly rate of return for this stock be wider or narrower than the prediction interval in part a? Explain.

11.25 The demand elasticity of a price promotion of a product is the ratio of the proportionate change in sales volume induced by the price promotion to the proportionate change in the price of the product. In three recent price promotions of a product, the demand elasticities were -1.80, -1.93, and -1.65. Assume that these three elasticities constitute a random sample from a normal population.

a. Construct a 95 percent prediction interval for the elasticity of the next price promotion. What assumptions did you make in constructing the prediction interval?

b. If the elasticity of the next promotion happens to be -0.95, would this outcome suggest that the assumptions in part a might not be valid? Discuss.

11.26 The mean and the standard deviation of the annual snowfalls in a northern city for the past 20 years are 2.03 meters and 0.45 meter, respectively. Assume that annual snowfalls for this city are random observations from a normal population. Construct a 95 percent prediction interval for next year's snowfall. Interpret the prediction interval.

11.27 An oil company executive was furnished a 95 percent prediction interval for the flow of a new test well being drilled in a field, based on experience with four other test wells in the same field. The prediction interval was $300 \leq X_{new} \leq 1550$ barrels per day. The executive interpreted the interval as stating that the new well will have daily flows varying between 300 and 1550 barrels per day. Why is the executive's interpretation not accurate?

11.28 Given that $E\{s^2\} = \sigma^2$ for a random sample from an infinite population, show that $s^2\{\overline{X}\}$ is an unbiased estimator of $\sigma^2\{\overline{X}\}$.

11.29 With simple random sampling, what distribution does each of the following statistics follow if (1) the population is normal and n is small? (2) The population is normal and n is large? (3) The population is not normal but the departure from normality is not too marked and n is small? (4) The population is not normal and n is large?

a. $\overline{X}$ **b.** $\dfrac{\overline{X} - \mu}{\sigma\{\overline{X}\}}$ **c.** $\dfrac{\overline{X} - \mu}{s\{\overline{X}\}}$

11.30 It is desired to estimate the mean μ of a Poisson population with a relative precision of 10 percent (that is, within $\pm 0.1\mu$), using a 95 percent confidence interval. A reasonable planning value for μ is 1.5. What is the required sample size? [*Hint:* $\mu = \sigma^2$ for a Poisson population.]

11.31 For a random sample of size n from a normal population, show that the interval estimate $\mu \geq L$ is a lower $1 - \alpha$ confidence interval for μ, where $L = \overline{X} - t(1 - \alpha; n - 1)s\{\overline{X}\}$. [*Hint:* Show that $P(L \leq \mu) = 1 - \alpha$ prior to sampling.]

11.32 Generate 50 independent random samples, each of size $n = 3$, from a standard normal distribution.

a. For each sample, calculate an 80 percent confidence interval for μ of the form $\overline{X} \pm z\sigma\{\overline{X}\}$. [*Hint:* See Comment 2, p. 299.] How many of the 50 intervals are correct; that is, how many contain $\mu = 0$?

b. Repeat part a, using intervals of the form $\overline{X} \pm ts\{\overline{X}\}$.

c. Explain why, in both parts a and b, 40 out of the 50 intervals are expected to contain $\mu = 0$. How do the actual numbers compare with the expected numbers in the two cases?

11.33 Electrolytic Cells. An aluminum company is experimenting with a new design for electro- lytic cells in smelter potrooms. A major design objective is to maximize the expected service life of the cells. Thirty cells of the new design were operated under similar conditions and failed at the following ages (in days):

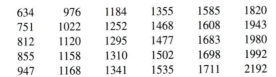

634	976	1184	1355	1585	1820
751	1022	1252	1468	1608	1943
812	1120	1295	1477	1683	1980
855	1158	1310	1502	1698	1992
947	1168	1341	1535	1711	2192

Two items of concern to management are (1) the mean service life for cells of this design and (2) the comparative performance of cells of this design with cells of the standard industry design, which are known to have a mean service life of 1300 days. Management does not want to conclude that the new design is superior to the standard one unless the evidence is fairly strong.

a. Assuming that the distribution of service lives of the cells is normal, calculate an appropriate 95 percent confidence interval for the mean service life of cells of the new design. Justify your choice of a one-sided or a two-sided confidence interval. Should management conclude that cells of the new design have a longer mean service life than cells of the standard design? Comment.

b. It has been suggested that the logarithms of service lives are more normally distributed than the original observations. Take logarithms (to base 10) of the service life data, and calculate the same type of confidence interval as in part a for the mean log-service-life of cells of the new design. Cells of the standard design are known to have a mean log-service-life of 3.095 (to base 10). Can management claim with confidence that the mean log-service-life is greater for cells of the new design than for cells of the standard design?

c. Plot histograms of the original data and of the log data in separate graphs. Does the distribution of log-service-lives appear to be more normal than the distribution of service lives as suggested? Comment.

d. A management objective is to obtain a large total service life for the cells. Is the mean service life or the mean log-service-life the more relevant measure for this purpose? Explain.

11.34 A state employment office will survey business establishments in the state about their hiring plans for college students next summer. A questionnaire is to be mailed to a random sample of the 31,800 establishments in the state to obtain information from each sample establishment about the number of summer job positions it plans to create and the total number of student weeks of employment for the summer positions.

a. The state employment office desires to estimate the total number of summer positions to be created and the total number of student weeks for all establishments by means of 95 percent confidence intervals with half-widths of at most 3000 positions and 50,000 student weeks, respectively. Planning values for the population standard deviations, based on similar surveys in previous years, are 1.1 positions and 19.3 student weeks, respectively. Assume that there will be a 100 percent response rate for purposes of planning the sample size. What is the smallest sample size that will give both interval estimates with the desired precision at the desired confidence level?

b. It was finally decided to select a random sample of 650 establishments and responses were obtained from each. The results were as follows:

Variable	Sample Mean	Sample Standard Deviation
Positions	1.040	1.212
Student weeks	14.23	17.40

Obtain the desired 95 percent confidence intervals and interpret them.

c. Suppose that the survey questionnaire had actually been mailed to 1000 establishments selected at random and only the 650 establishments referred to in part b replied. Furthermore, the responding establishments were found to be larger (as measured by the size of their work force), on average, than for the population as a whole. Would these circumstances affect the interpretation of the confidence intervals in part b? Explain.

Tests for Population Mean

ample data are frequently used in a *statistical test* to draw a conclusion about whether a population mean, or another population parameter, differs from a specified standard or has changed from its previous level. We now give three illustrations of statistical test situations involving the population mean.

1. A manufacturing concern has produced 36 units of an experimental machine component and measured their service lives. It wishes to test whether or not the mean service life of the experimental component exceeds the mean service life of the standard component.

2. A financial analyst has measured the current return offered by mutual funds in a sample of 15 funds. The analyst wishes to test whether or not mutual funds in general now offer a mean return that is lower than in earlier decades.

3. A psychologist who is concerned with job performance has conducted an experiment to measure the attention span for a task when an employee works alone and when several employees work in the same room. The psychologist wishes to test whether the mean attention span decreases when several employees work in the same room. ☐

In this chapter, we discuss statistical tests for tne population mean μ that involve two alternative conclusions. This type of test is commonly encountered in practice and provides a basis for understanding other statistical tests to be discussed subsequently. All of the inference procedures taken up in this chapter require that the sample data come from a simple random sample. It is therefore important to confirm that the requirements of simple random sampling have been met. When the population of interest is infinite, the data set needs to be examined by the diagnostic procedures discussed in Section 9.4 or subjected to the test for randomness to be described in Section 15.4 to determine whether it can be treated as a random sample. For finite populations, the requirements in (9.7) must be met.

Specifying the Alternative Conclusions

STATISTICAL TESTS　**12.1**

The first step in conducting a statistical test is to specify the two alternative conclusions, one of which is to be chosen on the basis of the sample data. To illustrate how the alternatives are specified, let us look more closely at Example 1.

Machine Component **EXAMPLE**

An ultrasonic equipment manufacturer employs a component in one of its machines that must withstand considerable stress from vibration. An all-metal component has been used for some years. Extensive past experience has shown that this component has a mean service life of 1100 hours. The research division of the supplier of the all-metal component has just developed an experimental component constructed from bonded metal and plastic. The manufacturer wishes to know whether or not the mean service life of this bonded component (μ) exceeds the mean service life of 1100 hours of the all-metal component. Thus, the manufacturer wishes to choose between the following two alternative conclusions:

1. The mean service life of the bonded component is 1100 hours or less.
2. The mean service life of the bonded component is more than 1100 hours.

We state these alternatives symbolically as follows:

$$H_0: \mu \leq 1100$$
$$H_1: \mu > 1100$$

Here, H_0 and H_1 denote the alternative conclusions. To conclude H_0 in this case is to conclude that the mean service life of the bonded component (μ) is 1100 hours or less. To conclude H_1 is to conclude the opposite of H_0, namely, that μ is more than 1100 hours. ☐

Note three important aspects of this decision problem. It has been formulated as a choice between two alternatives that are mutually exclusive with respect to μ. Thus, either H_0 is true because $\mu \leq 1100$ or H_1 is true because $\mu > 1100$, but not both. Furthermore, which of the two alternatives is true is not known. Finally, the alternatives are defined in terms of a value (1100 in this case) that is the standard against which the population mean is to be compared. We denote this standard by μ_0. In this illustration, $\mu_0 = 1100$.

We next illustrate two additional types of alternatives encountered in tests of the population mean μ.

☐ **EXAMPLE**

An engineer has observed the output of a machine in each of 20 one-hour periods selected at random. The engineer is concerned with whether or not the mean hourly output is less than $\mu_0 = 40$ units. The alternatives of interest may be stated symbolically as follows:

$$H_0: \mu \geq 40$$
$$H_1: \mu < 40$$

To conclude H_0 here is to conclude that the mean hourly output (μ) is 40 units or more. To conclude H_1 is to conclude the opposite, namely, that μ is less than 40 units. ☐

Lake Acidity ☐ **EXAMPLE**

A scientist made pH measurements of acidity on 18 water samples drawn from a lake. The scientist wishes to know whether the mean pH level for the lake differs from

$\mu_0 = 7.0$ (the neutral pH level). The alternatives of interest may be stated symbolically as follows:

$$H_0: \mu = 7.0$$
$$H_1: \mu \neq 7.0$$

To conclude H_0 here is to conclude that the mean pH level for the lake is 7.0 and, hence, that its water is neutral. To conclude H_1 is to conclude the opposite, namely, that the mean pH level of the lake is not 7.0 and, hence, that its water is not neutral. ☐

In general, we shall consider three forms of decision alternatives in this chapter.

(12.1)
The three forms of decision alternatives are as follows.

(12.1a)   **One-Sided Upper-Tail Alternatives:**

$$H_0: \mu \leq \mu_0$$
$$H_1: \mu > \mu_0$$

(12.1b)   **One-Sided Lower-Tail Alternatives:**

$$H_0: \mu \geq \mu_0$$
$$H_1: \mu < \mu_0$$

(12.1c)   **Two-Sided Alternatives:**

$$H_0: \mu = \mu_0$$
$$H_1: \mu \neq \mu_0$$

Notice that the name used for each form of alternative in (12.1) describes the range of values for μ given in alternative H_1. Thus, in (12.1c), the form is called *two-sided* because H_1 includes values of μ on both sides of μ_0. In (12.1a), the form is called *one-sided upper-tail* because H_1 includes values of μ that are larger than μ_0. The reason for the term *upper-tail* will become apparent shortly. Also notice in the three forms of decision alternatives in (12.1) that we follow the convention of labeling the alternative containing the equality sign as H_0.

In statistical testing, the alternative conclusions H_0 and H_1 are often called *hypotheses* since each conclusion may be thought of as a hypothesis about the true value of the parameter. For instance, $H_0: \mu \leq 1100$ represents the hypothesis that μ is 1100 or less. In this terminology, H_0 is referred to as the *null hypothesis* of the test.

Statistical Decision Rule

The choice between the alternative conclusions H_0 and H_1 in a statistical test is made on the basis of sample data. The specific rule that is used for deciding on the basis of the sample data which conclusion should be chosen is called a statistical decision rule.

(12.2)

A **statistical decision rule** specifies for each possible sample outcome which alternative, H_0 or H_1, should be selected.

☐ **EXAMPLES**

1. In the machine component example, it is desired to choose between H_0: $\mu \leq 1100$ and H_1: $\mu > 1100$ on the basis of the sample mean life $\overline{X}$ of 36 bonded machine components. If $\overline{X}$ is much smaller than 1100, it will be reasonable to conclude H_0; that is, the bonded component is not superior to the all-metal component with respect to mean service life. If $\overline{X}$ is much larger than 1100, the opposite conclusion (H_1) will be drawn. The statistical decision rule might be as follows:

   If $\overline{X} \leq 1150$, conclude H_0 ($\mu \leq 1100$).
   If $\overline{X} > 1150$, conclude H_1 ($\mu > 1100$).

   With this rule, H_0 is concluded if $\overline{X}$ is 1150 or less, and H_1 is concluded if $\overline{X}$ is greater than 1150.

2. For the lake acidity example, the alternative conclusions are H_0: $\mu = 7.0$ and H_1: $\mu \neq 7.0$. If the mean pH of the 18 water samples, $\overline{X}$, is close to 7.0, the scientist should conclude that the lake water is neutral, that is, conclude H_0. If $\overline{X}$ differs substantially from 7.0, then the scientist should conclude that the lake water is not neutral, that is, conclude H_1. The statistical decision rule here might be as follows:

   If $6.8 \leq \overline{X} \leq 7.2$, conclude H_0 ($\mu = 7.0$).
   If $\overline{X} < 6.8$ or $\overline{X} > 7.2$, conclude H_1 ($\mu \neq 7.0$).

   With this rule, the scientist concludes H_0 if $\overline{X}$ is in the interval from 6.8 to 7.2 inclusive and concludes H_1 if $\overline{X}$ is either smaller than 6.8 or larger than 7.2.   ☐

Before discussing how statistical decision rules are constructed, we must first consider the kinds of incorrect decisions that can be made in performing a test using a statistical decision rule because of the presence of sampling variability.

Errors of Inference

We know from our study of the sampling distribution of $\overline{X}$ in Chapter 10 that the sample mean $\overline{X}$ is subject to sampling variability and generally will differ from the population mean μ. Consequently, there are risks of drawing incorrect conclusions when $\overline{X}$ is used in a statistical decision rule to choose between the two alternative conclusions about μ.

☐ **EXAMPLE**

In the lake acidity example, suppose that the lake water is neutral (that is, $\mu = 7.0$). Yet because of sampling variability, the sample mean may be $\overline{X} = 7.4$. Using the earlier decision rule, the scientist would then mistakenly conclude H_1 ($\mu \neq 7.0$) because $\overline{X} = 7.4$ is larger than 7.2.

To consider another possibility, assume that the lake is somewhat acidic, with $\mu = 6.7$. Yet because of sampling variability, the sample mean could be $\overline{X} = 7.1$. Using the same decision rule, the scientist would then mistakenly conclude H_0 ($\mu = 7.0$) because $\overline{X} = 7.1$ lies between 6.8 and 7.2. ☐

Drawing the incorrect conclusion in a statistical test is called an *error of inference* because it entails drawing an incorrect inference or conclusion from the sample about the population parameter. There are two possible types of errors when choosing between two alternatives, of which one must be true. Table 12.1 shows the general situation.

Type I Errors. When H_0 is the true alternative, we make a correct inference if we draw conclusion H_0 from the sample and an incorrect inference if we draw conclusion H_1. This incorrect inference is called a *Type I error.*

EXAMPLE ☐

In the machine component example, the two alternatives are:

H_0: $\mu \leq 1100$
H_1: $\mu > 1100$

A Type I error occurs when we conclude that the bonded component has a longer mean service life (H_1) when, in fact, it is not longer. ☐

We shall show shortly that the probability or risk of making a Type I error can be ascertained mathematically. The risk of making a Type I error is called the α *risk* (Greek alpha).

Type II Errors. Referring again to Table 12.1, we see that when H_1 is the true alternative, we make a correct inference if we draw conclusion H_1 from the sample and an incorrect inference if we draw conclusion H_0. This incorrect inference is called a *Type II error.*

EXAMPLE ☐

In the machine component example, a Type II error occurs when we conclude that the bonded component does not have a longer mean service life (H_0) when, in fact, it is longer. ☐

As with Type I errors, the probability or risk of making a Type II error can be ascertained mathematically. The risk of making a Type II error is called the β *risk* (Greek beta).

Definitions. We now present formal definitions of the two types of inferential errors and their associated risks.

Conclusion Drawn from Sample	True Alternative	
	H_0	H_1
H_0	Correct conclusion	Type II error (β risk)
H_1	Type I error (α risk)	Correct conclusion

TABLE 12.1
Two types of errors of inference

(12.3)

When H_0 is true, the incorrect decision is to conclude H_1. This incorrect decision is called a **Type I error,** and the probability of making it is called the **α risk.**

When H_1 is true, the incorrect decision is to conclude H_0. This incorrect decision is called a **Type II error,** and the probability of making it is called the **β risk.**

Control of Risks.   When using a statistical decision rule for a test, we must accept some risks of incorrect conclusions since the conclusion is based on sample data. However, we would like to control the magnitudes of the α and β risks at reasonably low levels. Control over both kinds of risks can be achieved by appropriate planning of the sample size; we consider this subject in Section 12.7.

Often, however, the sample size is predetermined, and we shall consider this case now. When the sample size is already determined, only one of the two types of risks can be controlled at a specified level. In this case, it is reasonable to control the more serious type of risk; that is, the risk of error that would result in the more costly or undesirable outcome. We shall adopt the convention of formulating the alternative conclusions for a test so that the Type I error is the more serious type of error. Consequently, the α risk is the risk to be controlled.

☐ **EXAMPLE**

In the machine component example, the more serious error is to conclude that the bonded component has a longer mean service life than 1100 hours (the mean life for the all-metal component) when, in fact, the mean life for the bonded component is not longer. This error is particularly costly because it would lead to an expensive conversion of the production process, and then a switch back to the all-metal component when the error is confirmed by subsequent experience. To make the Type I error the more serious type of error here, we set up the alternatives involving $\mu_0 = 1100$ as follows:

$$H_0\colon \ \mu \leq 1100$$
$$H_1\colon \ \mu > 1100$$

Note that a Type I error with these alternatives entails concluding $\mu > 1100$ when in reality $\mu \leq 1100$.   ☐

Appropriate Types of Decision Rules

We consider now the appropriate type of decision rule for each of the three forms of decision alternatives in (12.1).

☐ **EXAMPLE**

In the machine component example, the alternatives are $H_0\colon \mu \leq 1100$ and $H_1\colon \mu > 1100$. Clearly, if the sample mean $\overline{X}$ is much smaller than 1100, it will be reasonable to conclude $\mu \leq 1100$; that is, to conclude H_0. Similarly, if $\overline{X}$ is much larger than 1100, it will be reasonable to conclude that $\mu > 1100$; that is, to conclude H_1. If $\overline{X}$ lies near 1100, the appropriate conclusion becomes readily apparent when we recall that the alternatives were formulated so that the α risk is the more serious of the two types of risks. Hence, we should avoid concluding H_1 unless the sample evidence clearly suggests that H_1 is true. The appropriate conclusion when $\overline{X}$ is near 1100 is therefore conclusion H_0, because this choice will reduce the risk of a Type I error (that is, reduce the probability of concluding H_1 when H_0 is true). This common-sense argument tells us that the appropriate form of the statistical decision rule in this example is as follows:

If $\overline{X} \leq A$, conclude H_0 ($\mu \leq 1100$).
If $\overline{X} > A$, conclude H_1 ($\mu > 1100$).

Here A is some number larger than 1100. ☐

This decision rule, which is appropriate for all one-sided upper-tail alternatives, is illustrated in Figure 12.1a. The sample mean $\overline{X}$ is referred to as the test statistic of the decision rule. Note that the rule provides that values of the test statistic in the neighborhood of μ_0 or lower lead us to conclude, or to accept, H_0. For this reason, this range of $\overline{X}$ values

FIGURE 12.1
General form of statistical decision rule for three types of decision alternatives concerning population mean μ

(a) One-Sided Upper-Tail Alternatives

H_0: $\mu \leq \mu_0$
H_1: $\mu > \mu_0$

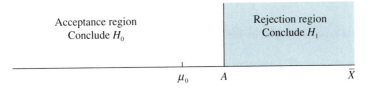

Acceptance region
Conclude H_0

Rejection region
Conclude H_1

μ_0 A $\overline{X}$

(b) One-Sided Lower-Tail Alternatives

H_0: $\mu \geq \mu_0$
H_1: $\mu < \mu_0$

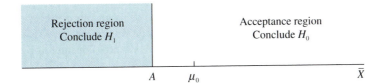

Rejection region
Conclude H_1

Acceptance region
Conclude H_0

A μ_0 $\overline{X}$

(c) Two-Sided Alternatives

H_0: $\mu = \mu_0$
H_1: $\mu \neq \mu_0$

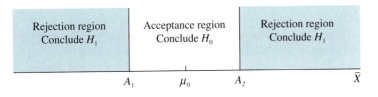

Rejection region
Conclude H_1

Acceptance region
Conclude H_0

Rejection region
Conclude H_1

A_1 μ_0 A_2 $\overline{X}$

is called the acceptance region. Correspondingly, large values of $\overline{X}$ lead us to conclude H_1, or stated equivalently, to reject H_0. Hence, this range of $\overline{X}$ values is called the rejection region. The value A that separates the acceptance and rejection regions is called the action limit of the decision rule.

> **(12.4)**
>
> The sample statistic upon which a statistical decision rule is based is called the **test statistic.**
>
> The range of values of the test statistic for which alternative H_0 is concluded is called the **acceptance region.**
>
> The range of values of the test statistic for which alternative H_1 is concluded is called the **rejection region.**
>
> The value that separates the acceptance and rejection regions is called the **action limit.**

Similar reasoning leads us to the appropriate statistical decision rules for one-sided lower-tail alternatives and for two-sided alternatives. The general form of each of these rules is illustrated in Figures 12.1b and 12.1c, respectively. Note that in the two-sided case, the lower and upper ends of the acceptance region are defined by two action limits, which we denote by A_1 and A_2, respectively.

> **Comment**
>
> The rejection region is sometimes called the *critical region* of the statistical decision rule. Correspondingly, the action limit is then called the *critical value. Concluding H_0* is often referred to as *accepting H_0.* Both of these terms stand for acting as if H_0 is correct. Similarly, *concluding H_1* is also called *rejecting H_0* or *accepting H_1*, all of which stand for acting as if H_1 is correct.

12.2 CONSTRUCTION OF STATISTICAL DECISION RULES

Figure 12.1 shows the general form of the statistical decision rule for each of the three types of decision alternatives. To construct the decision rule in any given case, we must determine the specific position(s) of the action limit(s). We now consider how this is done.

Assumptions

Our method of constructing a statistical decision rule is based on three assumptions.

1. The sample data constitute a simple random sample.
2. The simple random sample size is large.
3. The risk of a Type I error—that is, the α risk—is to be controlled at $\mu = \mu_0$.

As a result of assumptions 1 and 2, the extension of the central limit theorem in (11.7) tells us that the quantity $(\overline{X} - \mu)/s\{\overline{X}\}$ has an approximate standard normal distribution. Assumption 3 states that the decision rule will be chosen to control the α risk for the situation where the population mean μ equals the standard value μ_0.

Steps in Construction of Decision Rule

Construction and use of a statistical decision rule involve the following four steps.

Step 1. Set up the test alternatives H_0 and H_1.
Step 2. Specify the desired α risk for the test.
Step 3. Set up the statistical decision rule.
Step 4. Calculate the test statistic from the sample data and draw the appropriate conclusion from the decision rule.

One-Sided Upper-Tail Alternatives

We return to the machine component example to illustrate the steps in the construction and use of the decision rule for one-sided upper-tail alternatives.

EXAMPLE ☐

Step 1. In the machine component example, the test alternatives are defined in terms of $\mu_0 = 1100$, where the α risk is to be controlled:

$$H_0: \ \mu \leq 1100$$
$$H_1: \ \mu > 1100$$

Step 2. Management wishes to control the α risk at 0.01 when $\mu = \mu_0 = 1100$. This specification of the α risk implies that management wants only 1 chance in 100 of being led to the conclusion that the mean service life of the bonded component is longer than that of the all-metal component when in fact it is the same. Equivalently, the risk specification implies that management wants only 1 chance in 100 of having $\overline{X}$ fall in the rejection region of the decision rule when the population mean service life of the bonded component is 1100 hours.

Figure 12.2a illustrates the situation pictorially. The sampling distribution of $\overline{X}$ shown there is centered at $\mu_0 = 1100$ because the probability that $\overline{X}$ falls in the rejec-

FIGURE 12.2
Statistical decision rule and MINITAB output for one-sided upper-tail test for μ— Machine component example

(a) Decision Rule

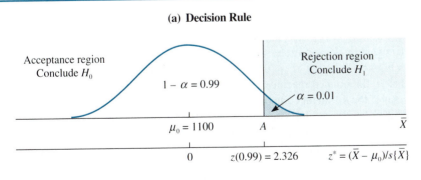

Acceptance region
Conclude H_0

$1 - \alpha = 0.99$

Rejection region
Conclude H_1

$\alpha = 0.01$

$\mu_0 = 1100$ A $\overline{X}$

0 $z(0.99) = 2.326$ $z^* = (\overline{X} - \mu_0)/s\{\overline{X}\}$

(b) MINITAB Output

```
TEST OF MU = 1100.0 VS MU G.T. 1100.0

             N      MEAN     STDEV   SE MEAN        Z   P VALUE
SERVICE     36    1121.0     222.0      37.0     0.57      0.29
```

tion region (the α risk) is to be controlled at $\mu_0 = 1100$. The tail area in the rejection region, corresponding to the α risk, must be equal to 0.01. Note that the sampling distribution of $\overline{X}$ is approximately normal because the sample size is reasonably large.

Step 3. We shall measure how far $\overline{X}$ is from the standard μ_0 in units of $s\{\overline{X}\}$, the estimated standard deviation of $\overline{X}$. Thus, we shall use the standardized variable z^*, defined as follows:

$$z^* = \frac{\overline{X} - \mu_0}{s\{\overline{X}\}}$$

Because n is large, we know from (11.7) that z^* is approximately a standard normal variable when $\mu = \mu_0$. We shall refer to z^* as the standardized test statistic. The z^* scale is also shown in Figure 12.2a.

Because the tail area of the sampling distribution of $\overline{X}$ when $\mu = 1100$ must be equal to $\alpha = 0.01$, it follows that the action limit A must correspond to the 99th percentile of the standard normal distribution, which is $z(0.99) = 2.326$. Thus, the position 2.326 on the z^* scale corresponds to the position of the action limit A on the $\overline{X}$ scale and forms the boundary between the acceptance and rejection regions. The appropriate decision rule for the test is therefore as follows:

If $z^* \leq 2.326$, conclude H_0.
If $z^* > 2.326$, conclude H_1.

Step 4. The sample results for the service lives of the random sample of $n = 36$ bonded components were $\overline{X} = 1121$ hours and $s = 222$ hours (sample data not shown). We now compute the estimated standard deviation of $\overline{X}$:

$$s\{\overline{X}\} = \frac{s}{\sqrt{n}} = \frac{222}{\sqrt{36}} = 37.0$$

We next compute the standardized test statistic z^*:

$$z^* = \frac{\overline{X} - \mu_0}{s\{\overline{X}\}} = \frac{1121 - 1100}{37.0} = 0.57$$

Referring to the decision rule, we see that $z^* = 0.57 \leq 2.326$. Equivalently, we can see in Figure 12.2a that $z^* = 0.57$ lies in the acceptance region. Hence we conclude H_0; that is, the bonded machine component has a mean service life that is not longer than that of the all-metal component.

Figure 12.2b contains a portion of the MINITAB output for this decision problem. It shows, among other items, the sample mean, $\overline{X} = 1121$, the sample standard deviation, $s = 222$, the estimated standard deviation of $\overline{X}$, $s\{\overline{X}\} = 37.0$, and the value of the standardized test statistic, $z^* = 0.57$. The entry for P-value will be explained in Section 12.3. □

Test Statistic and Decision Rule.   The standardized test statistic developed here is appropriate for all tests concerning the population mean μ when the sample size is reasonably large.

(12.5)
Tests concerning the population mean μ, when the random sample size is reasonably large, are based on the **standardized test statistic:**

$$z^* = \frac{\overline{X} - \mu_0}{s\{\overline{X}\}}$$

where: $s\{\overline{X}\} = \frac{s}{\sqrt{n}}$

When $\mu = \mu_0$, the standardized test statistic z^* follows approximately the standard normal distribution.

The decision rule for a test involving one-sided upper-tail alternatives about the population mean μ, when n is reasonably large, can be summarized as follows.

(12.6)
When the alternatives are:

$$H_0: \mu \leq \mu_0$$
$$H_1: \mu > \mu_0$$

and the random sample size is reasonably large, the appropriate decision rule to control the α risk at μ_0 is as follows:

If $z^* \leq z(1 - \alpha)$, conclude H_0.
If $z^* > z(1 - \alpha)$, conclude H_1.

where z^* is given by (12.5).

Why the test for the alternatives in (12.6) is called an upper-tail test is apparent from Figure 12.2a. Conclusion H_1 is drawn when the observed value of $\overline{X}$ falls in the upper tail of the sampling distribution.

The specified level of the α risk for a test is sometimes called the *level of significance* of the test. Thus, the level of significance in the machine component example is $\alpha = 0.01$. In this same context, statistical tests are sometimes referred to as *significance tests.*

Comments

1. In Figure 12.2a, it can be seen that action limit A lies $z(1 - \alpha)$ standard deviations to the right of μ_0, where the standard deviation is $s\{\overline{X}\}$. Thus, the value of A for a one-sided upper-tail test may be calculated as follows.

(12.7) For a one-sided upper-tail test: $A = \mu_0 + z(1 - \alpha)s\{\overline{X}\}$

For the machine component example, $A = 1100 + 2.326(37.0) = 1186$ hours.

2. Because the sampling distribution of test statistic z^* in (12.5) is only approximately standard normal for large n when $\mu = \mu_0$, the actual α risk may not be exactly at the specified level.

One-Sided Lower-Tail Alternatives

Figure 12.1b shows the general form of the decision rule for one-sided lower-tail alternatives. We construct the decision rule in this case in a similar way to that shown in Figure 12.2a for an upper-tail test, using the same test statistic (12.5), except that the rejection region is in the lower tail of the sampling distribution.

Bank Service ☐ **EXAMPLE**

A bank recently installed an upgraded computer system for its outside automatic tellers that extends the line of banking services offered on a 24-hour basis. After several months of operations of the new system, an operations analyst of the bank wishes to test whether or not the mean number of customers being served inside the main bank during regular hours (μ) has decreased from its level prior to the upgrading of the outside system. This previous level is known to have been $\mu_0 = 123$ customers per hour. The test is to be based on customer counts for a random sample of 64 hourly intervals during regular banking hours.

Step 1. The test alternatives are:

$$H_0\colon\ \mu \geq 123$$
$$H_1\colon\ \mu < 123$$

Step 2. The analyst wishes to control the α risk at 0.05 when $\mu = \mu_0 = 123$. Figure 12.3 illustrates the situation. Note that the sampling distribution of $\overline{X}$ is centered at $\mu_0 = 123$ and that the tail area in the rejection region, corresponding to the α risk, must be equal to 0.05.

Step 3. Since the sample size is large, the sampling distribution of the standardized test statistic z^* is approximately normal. As the lower-tail area of the sampling distribution in Figure 12.3 is to be 0.05, the action limit A must correspond to the 5th percentile of the standard normal distribution. This percentile is $z(0.05) = -1.645$ and is shown on the z^* scale. The appropriate decision rule for the test is therefore as follows:

If $z^* \geq -1.645$, conclude H_0.
If $z^* < -1.645$, conclude H_1.

Step 4. The customer counts for the random sample of 64 hourly intervals were obtained, and it was found that $\overline{X} = 115.8$ and $s = 24.4$ customers per hour (sample data not shown). Hence, $s\{\overline{X}\} = 24.4/\sqrt{64} = 3.05$. From (12.5), we have:

$$z^* = \frac{\overline{X} - \mu_0}{s\{\overline{X}\}} = \frac{115.8 - 123}{3.05} = -2.36$$

Referring to the decision rule, we see that $z^* = -2.36 < -1.645$. Equivalently, we see in Figure 12.3 that $z^* = -2.36$ lies in the rejection region. Hence, the analyst

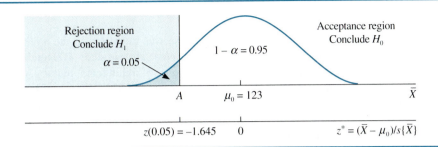

FIGURE 12.3
Statistical decision rule for
one-sided lower-tail test for
μ—**Bank service example**

should conclude H_1; that is, the upgraded outside teller system has reduced the mean number of customers per hour served inside the bank. □

Decision Rule. The decision rule for a test involving one-sided lower-tail alternatives about the population mean μ can be summarized as follows.

(12.8)
When the alternatives are:

$$H_0: \ \mu \geq \mu_0$$
$$H_1: \ \mu < \mu_0$$

and the random sample size is reasonably large, the appropriate decision rule to control the α risk at μ_0 is as follows:

If $z^* \geq z(\alpha)$, conclude H_0.
If $z^* < z(\alpha)$, conclude H_1.

where z^* is given by (12.5).

Comment
The value of the action limit A in a one-sided lower-tail test may be calculated as follows.

(12.9) For a one-sided lower-tail test: $A = \mu_0 + z(\alpha)s\{\overline{X}\}$

For the bank service example, $A = 123 - 1.645(3.05) = 118.0$.

Two-Sided Alternatives

Figure 12.1c shows the general form of the decision rule for two-sided alternatives. The decision rule is based on the same test statistic (12.5) used for one-sided tests, but the rejection region here falls in both tails of the sampling distribution.

Art Valuation ☐ **EXAMPLE**

The research department of an art-publishing house was interested in the ability of nonexpert subjects to judge the market value of art objects. In one study, 100 randomly selected subjects were asked to assess the market value of a thirteenth-century Andalusian vase that (unknown to the subjects) recently sold for $550. The test of interest is whether or not the mean dollar assessment of nonexperts (μ) differs from the actual market value of $\mu_0 = \$550$. If it were to differ, then nonexperts would be biased assessors of the market value of the vase.

Step 1. The test alternatives are:

$$H_0:\ \mu = 550$$
$$H_1:\ \mu \neq 550$$

Step 2. The research department wishes to control the α risk at 0.05 when $\mu = \mu_0 = 550$. Figure 12.4 illustrates the situation. The sampling distribution of $\overline{X}$ is centered at $\mu_0 = 550$ because the α risk is to be controlled there. Two action limits on the $\overline{X}$ axis (A_1 and A_2) define the acceptance region. These limits are placed equidistant from μ_0 in such a way that the probability of the observed $\overline{X}$ falling in either the upper or the lower rejection region when $\mu = \mu_0$ is $\alpha/2$. The total α risk is thus split equally in the two tails of the sampling distribution of $\overline{X}$.

Step 3. Since the sample size is large, the sampling distribution of the standardized test statistic z^* is approximately normal. The lower-tail area of the sampling distribution in Figure 12.4 is to be $\alpha/2 = 0.025$. Hence, the lower action limit A_1 must correspond to the 2.5th percentile of the standard normal distribution, which is $z(0.025) = -1.960$. Likewise, by symmetry, the upper action limit A_2 must correspond to $z(0.975) = 1.960$. Thus, the acceptance region includes all values of z^* between $z(0.025) = -1.960$ and $z(0.975) = 1.960$. The appropriate decision rule for the test is therefore as follows:

If $-1.960 \leq z^* \leq 1.960$, conclude H_0.
If $z^* < -1.960$ or $z^* > 1.960$, conclude H_1.

The inequality $-1.960 \leq z^* \leq 1.960$ is mathematically equivalent to $|z^*| \leq 1.960$, where $|z^*|$ denotes the absolute value of z^*. Thus, the decision rule may also be written as follows:

If $|z^*| \leq 1.960$, conclude H_0.
If $|z^*| > 1.960$, conclude H_1.

Step 4. The random sample of 100 nonexpert subjects yielded a mean judgment of the market value of the vase of $\overline{X} = \$733$ and a standard deviation of $s = \$787$. Hence, $s\{\overline{X}\} = 787/\sqrt{100} = 78.7$. From (12.5), we have:

$$z^* = \frac{\overline{X} - \mu_0}{s\{\overline{X}\}} = \frac{733 - 550}{78.7} = 2.33$$

Referring to the decision rule, we see that $|z^*| = 2.33 > 1.960$. Equivalently, we can see in Figure 12.4 that $z^* = 2.33$ lies in the upper portion of the rejection region.

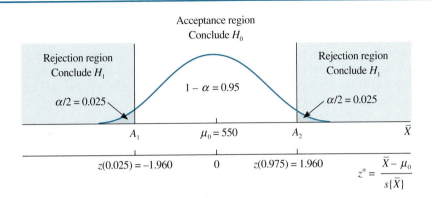

FIGURE 12.4
Statistical decision rule for two-sided test for μ—Art valuation example

Hence, we conclude H_1; that is, nonexperts are biased assessors of the market value of this Andalusian vase. □

Decision Rule. The decision rule for a test involving two-sided alternatives about the population mean μ can be summarized as follows.

(12.10)
When the alternatives are:

$$H_0: \mu = \mu_0$$
$$H_1: \mu \neq \mu_0$$

and the random sample size is reasonably large, the appropriate decision rule to control the α risk at μ_0 is as follows:

If $|z^*| \leq z(1 - \alpha/2)$, conclude H_0.
If $|z^*| > z(1 - \alpha/2)$, conclude H_1.

where z^* is given by (12.5).

Comment

The values of the action limits in a two-sided test may be calculated as follows.

(12.11) For a two-sided test: $A_1 = \mu_0 + z(\alpha/2)s\{\bar{X}\}$
$A_2 = \mu_0 + z(1 - \alpha/2)s\{\bar{X}\}$

For the art valuation example, $A_1 = 550 - 1.960(78.7) = 396$, and $A_2 = 550 + 1.960(78.7) = 704$.

12.3 P-VALUES

Nature of P-Value

The P-value of a statistical test is a probability number that measures the extent to which the sample data are consistent with conclusion H_0. It represents a compact summary of the sample findings in a statistical test. Published reports of statistical test results and the output of statistical packages frequently report the P-value.

Figure 12.5 illustrates the meaning of the P-value of a statistical test involving the population mean μ. Figure 12.5a shows the standard statistical decision rule for a one-sided upper-tail test about the population mean μ. The sampling distribution of $\overline{X}$ when $\mu = \mu_0$

FIGURE 12.5

Illustration of meaning of P-value. The P-value of a test for μ measures the consistency of the observed test statistic with μ_0 as the population mean.

(a) Decision Rule

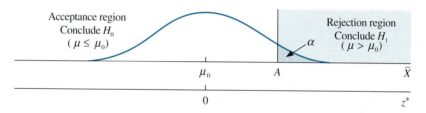

(b) P-Values for Different Sample Outcomes

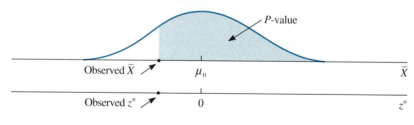

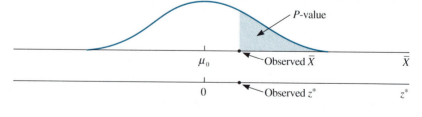

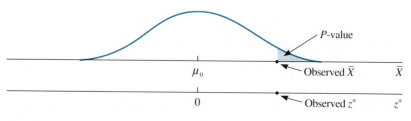

is also shown. Figure 12.5b shows for three different sample outcomes how likely it is that the values of the sample mean $\overline{X}$ and of the associated standardized test statistic z^* are greater than the observed values when in fact $\mu = \mu_0$. The P-value is the shaded area under the sampling distribution in each figure. Notice how this shaded area decreases the farther $\overline{X}$ is from μ_0 or, equivalently, the farther z^* is from 0—*in the direction of the rejection region*. Thus, the less consistent the sample evidence (in this case, $\overline{X}$ or z^*) with conclusion H_0 being true ($\mu \leq \mu_0$, in this example), the smaller the P-value.

For a one-sided test concerning the population mean, the P-value is defined as follows.

> **(12.12)**
> The **P-value** of a one-sided statistical test for μ is the probability that the standardized test statistic z^* might have been more extreme in the direction of the rejection region than was actually observed when, in fact, $\mu = \mu_0$.

Calculation of *P*-Value

In calculating the P-value of a test, the procedure differs for one-sided and two-sided test alternatives. We shall explain each case now. We assume throughout that *the sample size is reasonably large.*

One-Sided *P*-Value.   The P-values illustrated in Figure 12.5 are referred to as *one-sided P-values* because they relate to a one-sided test. We now show how one-sided P-values are calculated.

EXAMPLE ☐ **Upper-Tail Test**

For the machine component example, Figure 12.6a contains the decision rule reproduced from Figure 12.2. Recall that $\overline{X} = 1121$ and $z^* = 0.57$ were the sample outcomes. These observed values are located on their respective scales in Figure 12.6b. The one-sided P-value in this case is the probability that, if $\mu = \mu_0 = 1100$, z^* might have been farther in the direction of the rejection region than actually was observed. In other words, the P-value is the probability that, if $\mu = \mu_0 = 1100$, z^* might have been larger than its actual value of 0.57. This probability is shown by the shaded area in Figure 12.6b and can be seen from Table C.1 to equal $P(Z > z^*) = P(Z > 0.57) = 0.2843$. Thus, the one-sided P-value is 0.2843 here. The MINITAB output for the statistical test in Figure 12.2b shows this P-value as 0.29, which differs slightly from 0.2843 because of rounding.

The P-value is reasonably large here and indicates that the sample outcome is quite consistent with conclusion H_0.   ☐

EXAMPLE ☐ **Lower-Tail Test**

For the bank service example, Figure 12.7a contains the decision rule reproduced from Figure 12.3. Recall that the sample outcomes were $\overline{X} = 115.8$ and $z^* = -2.36$. The one-sided P-value here is the probability that, if $\mu = \mu_0 = 123$, z^* might have been farther in the direction of the rejection region than actually was observed. Thus, the P-value is the probability that, if $\mu = \mu_0 = 123$, z^* might have been smaller than its actual value of -2.36. This probability is shown by the shaded area in Figure 12.7b

FIGURE 12.6
Calculation of one-sided
upper-tail *P*-value—Machine
component example

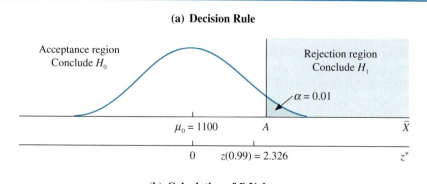

(a) Decision Rule

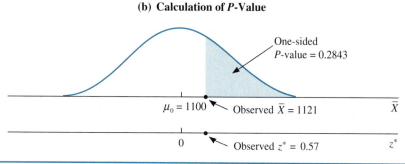

(b) Calculation of *P*-Value

FIGURE 12.7
Calculation of one-sided
lower-tail *P*-value—Bank service example

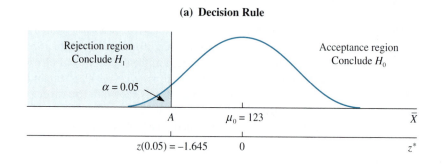

(a) Decision Rule

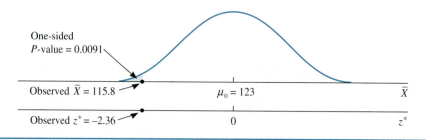

(b) Calculation of *P*-Value

and equals $P(Z < z^*) = P(Z < -2.36) = 0.0091$. Hence, the one-sided *P*-value here is 0.0091, which is so small that there is little doubt the sample outcome is inconsistent with conclusion H_0. □

Two-Sided *P*-Value. In a two-sided test, the extent to which $\overline{X}$ and z^* are consistent with μ_0 is judged by the distance toward the rejection region in either direction. Hence, in a two-sided test, the calculation of the *P*-value considers extreme departures in either direction. This calculation is done in two steps:

1. Obtain the one-sided *P*-value for the direction in which $\overline{X}$ and z^* happen to fall.
2. Double this one-sided *P*-value.

The resulting number is called a *two-sided P-value.*

EXAMPLE □

For the art valuation example, Figure 12.8a contains the decision rule reproduced from Figure 12.4. Recall that the sample outcomes were $\overline{X} = 733$ and $z^* = 2.33$. To calculate the two-sided *P*-value here, we first obtain the one-sided *P*-value corresponding to $z^* = 2.33$. This one-sided *P*-value is shown by the shaded area in Figure 12.8b and equals $P(Z > z^*) = P(Z > 2.33) = 0.0099$. Next, we double this one-sided *P*-value to obtain $2(0.0099) = 0.0198$. Thus, the two-sided *P*-value for this test is 0.0198. This number is fairly small, suggesting that the sample outcome is not consistent with conclusion H_0. □

(a) Decision Rule

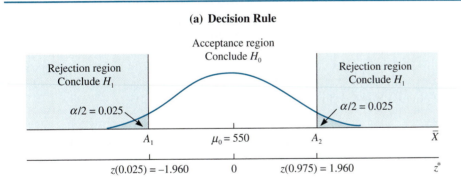

(b) Calculation of *P*-Value

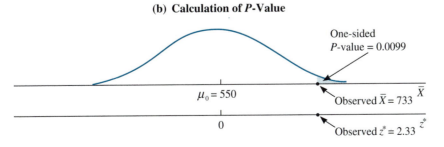

Two-sided *P*-value = 2(0.0099) = 0.0198

FIGURE 12.8
Calculation of two-sided
P-value—Art valuation
example

Definitions.   We summarize now the calculational formulas for obtaining the P-values for the three types of tests for μ.

> **(12.13)**
> Calculational formulas for obtaining P-values are as follows.
>
> **(12.13a)**   **One-Sided Upper-Tail Alternatives** (H_0: $\mu \leq \mu_0$, H_1: $\mu > \mu_0$):
>
> $$\text{One-sided } P\text{-value} = P(Z > z^*)$$
>
> **(12.13b)**   **One-Sided Lower-Tail Alternatives** (H_0: $\mu \geq \mu_0$, H_1: $\mu < \mu_0$):
>
> $$\text{One-sided } P\text{-value} = P(Z < z^*)$$
>
> **(12.13c)**   **Two-Sided Alternatives** (H_0: $\mu = \mu_0$, H_1: $\mu \neq \mu_0$):
>
> $$\text{Two-sided } P\text{-value} = \begin{cases} 2P(Z > z^*) & \text{if } z^* \geq 0 \\ 2P(Z < z^*) & \text{if } z^* < 0 \end{cases}$$

Testing with P-Values

As noted earlier, the P-value of a test for μ measures the consistency of the sample outcome with the value of μ_0 postulated in H_0. A large P-value indicates that μ_0 is plausible and, hence, that H_0 is a reasonable conclusion. A small P-value indicates that μ_0 is not plausible and, hence, that the sample evidence supports H_1.

Indeed, the P-value can be used directly to choose between H_0 and H_1, according to whether or not the P-value is larger or smaller than the specified α risk for conducting the test. The result obtained from such a test based on the P-value is mathematically equivalent to the result provided by the corresponding decision rule based on the standardized test statistic. The decision rule based on the P-value has the following form.

> **(12.14)**   If P-value $\geq \alpha$, conclude H_0.
> If P-value $< \alpha$, conclude H_1.

This decision rule applies whether the test is two-sided or one-sided, provided that a two-sided or one-sided P-value is used, as appropriate.

Upper-Tail Test ☐ **EXAMPLE**

Refer to Figure 12.6 for the machine component example, where the α risk was specified to be 0.01 and the one-sided P-value for the observed sample outcome was 0.2843. Since $0.2843 \geq 0.01$, decision rule (12.14) states that H_0 is the appropriate conclusion. This conclusion is identical to the one drawn earlier on the basis of the standardized test statistic, as it must necessarily be.   ☐

Lower-Tail Test ☐ **EXAMPLE**

Refer to Figure 12.7 for the bank service example, where $\alpha = 0.05$ and the one-sided P-value $= 0.0091$. Since $0.0091 < 0.05$, decision rule (12.14) states that H_1 is the appropriate conclusion, as was found earlier.   ☐

EXAMPLE   **Two-Sided Test**

Refer to Figure 12.8 for the art valuation example, where $\alpha = 0.05$ and the two-sided P-value $= 0.0198$. Since $0.0198 < 0.05$, decision rule (12.14) states that H_1 is the appropriate conclusion, as was found earlier.

In interpreting P-values presented in statistical reports and computer output, the user must establish whether the reported P-value is a one-sided or a two-sided P-value. Figure 12.9 (p. 337) shows MINITAB output for a test to be discussed shortly in which the reported P-value is simply labeled P VALUE. It actually is a two-sided P-value. Other labels for the P-value that are often found in computer output include P, PROBABILITY, and PROB. A P-value is also sometimes referred to as the *observed level of significance*.

In later chapters, we will show that P-values are applicable to many other statistical tests.

Comment

The equivalence of decision rule (12.14) based on the P-value and the corresponding decision rules based on the standardized test statistic z^* can be seen from Figures 12.6, 12.7, and 12.8. Note in these figures that the P-value is less than α only when the observed z^* value falls in the rejection region.

SMALL-SAMPLE TESTS   **12.4**

Construction of Decision Rule

The test procedures for μ discussed so far have assumed that the sample size is reasonably large. We turn now to the construction of decision rules when the sample size is not large. The conditions when these decision rules are applicable are exactly the same as those in Chapter 11 for establishing a confidence interval for the population mean when the sample size is small.

Assumptions

1. The population sampled is either normal or, if it is not normal, does not depart too markedly from normality.
2. The sample size is not exceedingly small.

As mentioned earlier, we shall describe in Chapter 16 a normal probability plot and a formal statistical test for examining whether the sampled population is reasonably normal.

Test Statistic.   We know from Chapter 11 that the statistic $(\overline{X} - \mu)/s\{\overline{X}\}$ is distributed exactly as $t(n - 1)$ when the population is normal, and approximately as $t(n - 1)$ when the population does not depart too markedly from a normal distribution and the sample size is not very small. This means that the same test statistic (12.5) used for large samples may also be used for small samples and that an appropriate t percentile is employed in place of the z percentile in constructing the decision rule.

We shall now use t^* to denote the earlier test statistic (12.5) to remind us that the tests for small samples are based on the t distribution.

(12.15)

Tests concerning the population mean μ, when the population is normal or the departure is not too marked and the sample size is not exceedingly small, are based on the standardized test statistic:

$$t^* = \frac{\overline{X} - \mu_0}{s\{\overline{X}\}}$$

where: $s\{\overline{X}\} = \dfrac{s}{\sqrt{n}}$

When $\mu = \mu_0$, test statistic t^* follows either exactly or approximately the t distribution with $n - 1$ degrees of freedom.

The appropriate decision rules are constructed in the same fashion as those for large samples. Table 12.2 shows the form of the decision rule for each of the three types of tests. Because the only difference between these tests and those for the large-sample case is the use of t percentiles in place of z percentiles, we proceed directly to an illustration.

☐ **EXAMPLE**

In the lake acidity example, the scientist wishes to test the mean pH level for the lake to see whether it differs from neutral, that is, from 7.0.

Step 1. The test alternatives are:

$$H_0: \ \mu = 7.0$$
$$H_1: \ \mu \neq 7.0$$

Hence, $\mu_0 = 7.0$.

Step 2. The scientist wishes to control the α risk of the test at 0.01 when $\mu = \mu_0 = 7.0$.

Step 3. Past experience indicates that pH levels in samples of lake water tend to be approximately normally distributed. The scientist has drawn a random sample of

---

TABLE 12.2

Summary of decision rules for tests of μ—Small-sample case, population normal or departure not too marked and sample size not very small

(a) Test Statistic

$$t^* = \frac{\overline{X} - \mu_0}{s\{\overline{X}\}}$$

(b) Decision Rules

Test	Alternatives	Decision Rule
(1) One-sided Upper-tail	$H_0: \mu \leq \mu_0$ $H_1: \mu > \mu_0$	If $t^* \leq t(1 - \alpha; n - 1)$, conclude H_0 If $t^* > t(1 - \alpha; n - 1)$, conclude H_1
(2) One-sided Lower-tail	$H_0: \mu \geq \mu_0$ $H_1: \mu < \mu_0$	If $t^* \geq t(\alpha; n - 1)$, conclude H_0 If $t^* < t(\alpha; n - 1)$, conclude H_1
(3) Two-sided	$H_0: \mu = \mu_0$ $H_1: \mu \neq \mu_0$	If $|t^*| \leq t(1 - \alpha/2; n - 1)$, conclude H_0 If $|t^*| > t(1 - \alpha/2; n - 1)$, conclude H_1

$n = 18$ water samples from the lake; thus, a small-sample testing procedure is required. The appropriate decision rule is the two-sided one in Table 12.2b. For $\alpha = 0.01$ and $n = 18$, we require $t(1 - \alpha/2; n - 1) = t(0.995; 17) = 2.898$. Hence, the decision rule is as follows:

If $|t^*| \leq 2.898$, conclude H_0.
If $|t^*| > 2.898$, conclude H_1.

Step 4. The sample data are given in the MINITAB output in Figure 12.9. In addition to the 18 observations, the output contains the sample statistics $n = 18$, $\overline{X} = 6.8861$, $s = 0.1048$, and $s\{\overline{X}\} = 0.1048/\sqrt{18} = 0.0247$. The standardized test statistic (12.15) is also given in the output and was calculated as follows:

$$t^* = \frac{\overline{X} - \mu_0}{s\{\overline{X}\}} = \frac{6.8861 - 7.0}{0.0247} = -4.61$$

Since $|t^*| = |-4.61| = 4.61$ and $4.61 > 2.898$, the scientist should conclude H_1; that is, the lake water is not neutral. In fact, the indication is that the lake water is acidic since the sample mean pH reading is less than 7.0. $\square$

P-Values for Small-Sample Tests

The P-value may be computed for a small-sample test based on t^* just as for a large-sample test based on z^*, except that the t distribution is now used. Because Table C.3 for the t distribution provides only a limited number of percentiles, it usually only permits us to obtain bounds for the P-value rather than the actual P-value itself. Many statistical packages and statistical calculators have built-in programs for calculating the exact P-value in these cases.

EXAMPLE $\square$

For the lake acidity example, a two-sided P-value is required. The one-sided P-value for $t^* = -4.61$ is the tail area below -4.61 in the t distribution with $n - 1 = 17$ degrees of freedom. From Table C.3, we see that the smallest percentile available is $t(0.0005; 17) = -3.965$. Since $t^* = -4.61$ is smaller than -3.965, we know that the one-sided P-value is smaller than 0.0005 and thus the two-sided P-value is smaller

---

MINITAB output containing sample data, statistics, and test results for small-sample case, population approximately normal—Lake acidity example

FIGURE 12.9

Case	pH	Case	pH
1	6.97	10	6.60
2	6.70	11	6.94
3	6.84	12	6.89
4	6.83	13	7.05
5	6.95	14	6.90
6	6.84	15	6.94
7	6.89	16	6.99
8	6.92	17	6.95
9	6.91	18	6.84

TEST OF MU = 7.0000 VS MU N.E. 7.0000

	N	MEAN	STDEV	SE MEAN	T	P VALUE
pH	18	6.8861	0.1048	0.0247	-4.61	0.0002

than $2(0.0005) = 0.001$. The MINITAB output of Figure 12.9 shows that the exact two-sided P-value is 0.0002. The small P-value indicates that the sample result is not consistent with $\mu_0 = 7.0$, neutral lake water.

As for large-sample tests, P-values for small-sample tests may be used for deciding between conclusions H_0 and H_1 by comparing the P-value with the specified α. Since the P-value in the lake acidity example is smaller than the specified α risk of 0.01, decision rule (12.14) leads us to conclude H_1, the same conclusion as that drawn earlier on the basis of the standardized test statistic t^*. □

12.5   OPTIONAL TOPIC — RELATION BETWEEN CONFIDENCE INTERVALS AND TESTS FOR μ

Statistical testing is closely connected with interval estimation. Indeed, a test of μ can be constructed on the basis of an appropriate confidence interval. A one-sided test corresponds to a one-sided confidence interval, and a two-sided test corresponds to a two-sided interval. The correspondence of tests and confidence intervals is set out formally in Table 12.3. If the confidence coefficient is $1 - \alpha$, then the corresponding risk of a Type I error is α. This correspondence holds for both large and small samples. Finally, note that a lower one-sided confidence interval corresponds to an upper-tail test and that an upper one-sided confidence interval corresponds to a lower-tail test.

□ **EXAMPLE**

In the machine component example, the alternatives were:

$$H_0: \ \mu \le 1100$$
$$H_1: \ \mu > 1100$$

and the sample results were $n = 36$, $\overline{X} = 1121$, $s = 222$, and $s\{\overline{X}\} = 37.0$. If we wish to conduct the statistical test with $\alpha = 0.01$ by means of a confidence interval, Table 12.3 indicates that we need to set up a 99 percent lower one-sided confidence interval. We therefore require $z(0.99) = 2.326$ and the lower confidence limit is:

$$L = \overline{X} - zs\{\overline{X}\} = 1121 - 2.326(37.0) = 1034.9$$

Since $\mu_0 = 1100$ exceeds $L = 1034.9$, the decision rule in Table 12.3a indicates that the appropriate conclusion is H_0, that the mean life of the bonded component does not

TABLE 12.3   Correspondence of confidence intervals and tests for μ

Test	Alternatives	Confidence Interval	Decision Rule Based on Confidence Interval
(a) One-sided Upper-tail	$H_0: \ \mu \le \mu_0$ $H_1: \ \mu > \mu_0$	Lower one-sided $\mu \ge L$	If $\mu_0 \ge L$, conclude H_0 If $\mu_0 < L$, conclude H_1
(b) One-sided Lower-tail	$H_0: \ \mu \ge \mu_0$ $H_1: \ \mu < \mu_0$	Upper one-sided $\mu \le U$	If $\mu_0 \le U$, conclude H_0 If $\mu_0 > U$, conclude H_1
(c) Two-sided	$H_0: \ \mu = \mu_0$ $H_1: \ \mu \ne \mu_0$	Two-sided $L \le \mu \le U$	If $L \le \mu_0 \le U$, conclude H_0 If $\mu_0 < L$ or $\mu_0 > U$, conclude H_1

exceed that of the all-metal component. This, of course, is the same conclusion that we reached by means of the decision rule based on the standardized test statistic z^*. □

Comment

To illustrate the mathematical correspondence between a confidence interval and the associated decision rule, consider a two-sided test based on a large sample. Table 12.3c states that H_0 is concluded if $L \leq \mu_0 \leq U$. Now, for a two-sided confidence interval based on a large sample, L and U are given in (11.5) as $\overline{X} - zs\{\overline{X}\}$ and $\overline{X} + zs\{\overline{X}\}$, respectively, where $z = z(1 - \alpha/2)$. Rearranging the inequalities as shown in the following steps, we obtain the desired result:

1. $L \leq \mu_0 \leq U$

2. $\overline{X} - zs\{\overline{X}\} \leq \mu_0 \leq \overline{X} + zs\{\overline{X}\}$

3. $-z \leq \dfrac{\overline{X} - \mu_0}{s\{\overline{X}\}} \leq z$

4. $-z(1 - \alpha/2) \leq z^* \leq z(1 - \alpha/2)$ since $z^* = (\overline{X} - \mu_0)/s\{\overline{X}\}$

5. $|z^*| \leq z(1 - \alpha/2)$

This last inequality defines the acceptance region of the decision rule for a two-sided test.

OPTIONAL TOPIC — POWER OF TEST   12.6

The decision rules for tests of μ discussed in this chapter have been set up to control the α risk of the test when $\mu = \mu_0$, since the test alternatives are always formulated so that the α risk is the more important risk to control. Still, the β risks (the risks of making a Type II error) are also of concern. Both types of risks can be controlled at desired levels by an appropriate choice of sample size, and in Section 12.7 we shall consider how to plan the sample size for a statistical test. First, however, we must consider how to evaluate the β risks of a statistical test, which we do in this section.

Assumptions

Throughout most of this section, we shall assume the following.

1. The random sample size is large.
2. The population standard deviation σ is known.

Under these assumptions we know from Chapter 10 that the sample mean $\overline{X}$ is approximately normally distributed and that the standard deviation of $\overline{X}$ is known and need not be estimated.

Rejection Probabilities

To evaluate the β risks associated with any decision rule, we first consider the probability that the rule will lead to conclusion H_1. This probability depends on the true value of the population mean μ. We denote this probability by $P(H_1; \mu)$ and call it the rejection probability.

We have already encountered one special rejection probability, namely, $P(H_1; \mu_0)$—the rejection probability at $\mu = \mu_0$. We know that this rejection probability is equal to α for each decision rule taken up because the probability of concluding H_1 when $\mu = \mu_0$ has always been controlled at α.

The rejection probability $P(H_1; \mu)$ at a value of μ for which alternative H_1 is true is often called the *power* of the decision rule at μ because it measures the probability of correctly concluding H_1. In this terminology, the rejection probability curve is called the *power curve* because power values can be read directly from this curve.

We now consider how to evaluate rejection probabilities in general and how they, in turn, enable us to assess the β risks. We shall do this by means of an example involving a one-sided upper-tail test. Subsequently, we shall present another example involving a two-sided test.

Calculation of Rejection Probabilities

Data Entry **EXAMPLE**

A marketing research company conducts regular consumer expenditures surveys, the findings of which are sold on a syndicated basis to many consumer product firms. The company has recently restructured the questionnaire used in the survey, making it longer but also more streamlined to facilitate the computer data entry of questionnaire responses. In the past, data entry has required a mean time of 23.0 minutes per questionnaire. The data entry operators are now experienced with the new questionnaire. The company wishes to test whether or not the mean data entry time for the new questionnaire is greater than $\mu_0 = 23.0$ minutes. The test will be based on the entry times for a random sample of 60 new questionnaires. The standard deviation of times for the data entry process has been $\sigma = 6.5$ minutes in the past, and the company believes it will remain the same for the new questionnaire.

The test alternatives here are:

$$H_0: \ \mu \le 23.0$$
$$H_1: \ \mu > 23.0$$

The α risk is to be controlled at $\alpha = 0.10$ when $\mu = \mu_0 = 23.0$. Hence, the decision rule has the form shown in Figure 12.10a. Note that the rejection probability at $\mu_0 = 23.0$ is $P(H_1; \mu_0 = 23.0) = 0.10$ because the α risk is controlled at 0.10 there.

In (12.7) we have shown how the value of the action limit A in a one-sided upper-tail test is calculated. To use that formula in this example, we replace $s\{\overline{X}\}$ in the formula by $\sigma\{\overline{X}\}$, because the latter is known here. Specifically, since $\sigma = 6.5$ and $n = 60$, we have $\sigma\{\overline{X}\} = \sigma/\sqrt{n} = 6.5/\sqrt{60} = 0.8391$. Furthermore, we have $z(1 - \alpha) = z(0.90) = 1.282$. Thus:

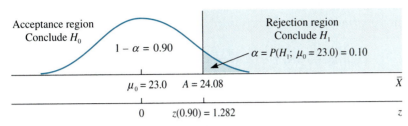

(a) Decision Rule

Acceptance region
Conclude H_0

$1 - \alpha = 0.90$

Rejection region
Conclude H_1

$\alpha = P(H_1;\ \mu_0 = 23.0) = 0.10$

$\mu_0 = 23.0$ $A = 24.08$

$\overline{X}$

0 $z(0.90) = 1.282$

z

FIGURE 12.10
Calculation of rejection probabilities for one-sided upper-tail test—Data entry example

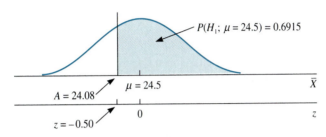

(b) Calculation of Rejection Probability at $\mu = 24.5$

$P(H_1;\ \mu = 24.5) = 0.6915$

$A = 24.08$ $\mu = 24.5$ $\overline{X}$

$z = -0.50$ 0 z

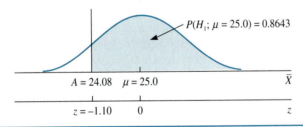

(c) Calculation of Rejection Probability at $\mu = 25.0$

$P(H_1;\ \mu = 25.0) = 0.8643$

$A = 24.08$ $\mu = 25.0$ $\overline{X}$

$z = -1.10$ 0 z

$$A = \mu_0 + z(1 - \alpha)\sigma\{\overline{X}\} = 23.0 + 1.282(0.8391) = 24.08$$

The decision rule for the test may then be written in terms of $\overline{X}$ as follows:

If $\overline{X} \leq 24.08$, conclude H_0.
If $\overline{X} > 24.08$, conclude H_1.

This form of the decision rule facilitates the evaluation of the rejection probabilities at different values of μ.

1. Consider $\mu = 24.5$. In this case, the sampling distribution of $\overline{X}$ will be centered at 24.5, as shown in Figure 12.10b. The probability of concluding H_1 when $\mu = 24.5$—that is, the rejection probability $P(H_1;\ \mu = 24.5)$—is the shaded area in the rejection region in Figure 12.10b. The action limit $A = 24.08$ is located at a distance $z = (24.08 - 24.5)/0.8391 = -0.50$ standard deviation from $\mu = 24.5$ (recall that $\sigma\{\overline{X}\} = 0.8391$). Hence, the rejection probability, which is represented by the shaded area, equals:

$$P(H_1;\ \mu = 24.5) = P(Z > -0.50) = 0.6915$$

2. Consider $\mu = 25.0$. The sampling distribution of $\overline{X}$ in this case will be centered at 25.0, as shown in Figure 12.10c. The action limit $A = 24.08$ is located $z = (24.08 - 25.0)/0.8391 = -1.10$ standard deviations from $\mu = 25.0$. Hence, the rejection probability, shown as the shaded area in Figure 12.10c, is:

$$P(H_1; \mu = 25.0) = P(Z > -1.10) = 0.8643$$ □

Rejection Probability Curve

Once a number of rejection probabilities have been calculated for different possible values of μ, the results are then summarized in a rejection probability curve.

□ **EXAMPLE**

The rejection probability curve for the decision rule in the data entry example is shown in Figure 12.11. The points previously determined:

$$P(H_1; \mu = \mu_0 = 23.0) = 0.10$$
$$P(H_1; \mu = 24.5) = 0.6915$$
$$P(H_1; \mu = 25.0) = 0.8643$$

are shown explicitly on the curve to illustrate the plotting procedure. As one would expect, the rejection probability increases steadily toward 1.0 as μ increases. Thus, the larger the mean data entry time for the new questionnaire, the more probable it is that we will be led to conclude H_1: $\mu > 23.0$ in the test. We see, for instance, that if μ is actually 27.0, the test is almost certain to lead to conclusion H_1.   □

FIGURE 12.11
Rejection probability curve
for one-sided upper-tail
test—Data entry example

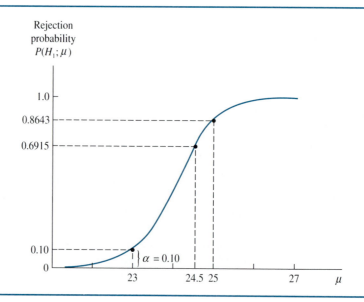

Calculation of Error Risks

The rejection probability curve gives the probability of concluding H_1 for different values of μ. The probabilities of making either a Type I error (α risk) or a Type II error (β risk) at different values of μ can be easily calculated from this curve.

EXAMPLE ☐

Refer to Figure 12.12 for the data entry example, which repeats the rejection probability curve shown in Figure 12.11. Consider any μ value below or at the past mean level $\mu_0 = 23$. The rejection probability here is the probability of concluding H_1 when, in fact, H_0 is true. Consequently, the rejection probability here is the probability of a Type I error and, hence, is an α risk.

When μ exceeds the past mean level $\mu_0 = 23$, on the other hand, H_1 is true and the rejection probability $P(H_1; \mu)$ is then the probability of a correct decision. Its complement, $1 - P(H_1; \mu)$, now equals the probability of a Type II error and, hence, is a β risk.

By this reasoning, we are able to assess the error risk at any value of μ, whether it be an α or a β risk. Referring to Figure 12.12, we see that the α risk at $\mu_0 = 23.0$ is 0.10—indeed, the decision rule was constructed so that this condition would hold. At $\mu = 24.5$, on the other hand, the error probability is a β risk and equals $1 - 0.6915 = 0.3085$. At $\mu = 25.0$, the error probability is also a β risk and equals $1 - 0.8643 = 0.1357$. This last error risk implies that if the mean data entry time for the new questionnaire is $\mu = 25.0$, then the probability is 0.1357 that the test will lead to the erroneous conclusion that the mean data entry time for the new questionnaire is not longer than it was for the old questionnaire. ☐

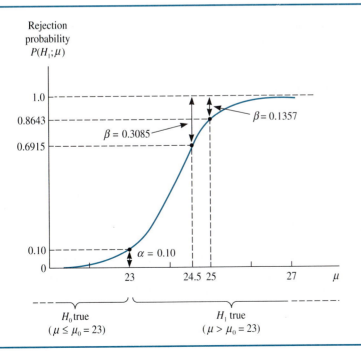

FIGURE 12.12
Determination of error risks from rejection probability curve for one-sided upper-tail test—Data entry example

Comments

1. In one-sided tests, the α risk is a maximum at $\mu = \mu_0$. This is illustrated by Figure 12.12, where $\alpha = 0.10$ at $\mu_0 = 23.0$ is the maximum α risk. Hence, controlling the α risk at $\mu = \mu_0$ guarantees that the probability of a Type I error at any other value of μ for which H_0 is true is smaller than α. The β risk in a one-sided test is a maximum when μ is close to μ_0. In Figure 12.12, for instance, the β risk approaches $1 - \alpha = 0.90$ when μ is just slightly larger than $\mu_0 = 23.0$. As a practical matter, however, β risks are only a major concern at values of μ some distance from μ_0, where a Type II error would be more costly.

2. For some applications, such as quality control, interest centers on the probability of concluding H_0 at different values of μ. Such a probability is called the *acceptance probability* or *operating characteristic* of the decision rule at μ and is denoted by $P(H_0; \mu)$. A plot of $P(H_0; \mu)$ against μ is called the *acceptance probability curve* or *operating characteristic curve*. The rejection and acceptance probabilities at a given value of μ are complementary probabilities. Hence, any acceptance probability may be obtained from the corresponding rejection probability by using complementation theorem (4.18), as follows.

    (12.17) $P(H_0; \mu) = 1 - P(H_1; \mu)$

Two-Sided Tests

The calculations of rejection probabilities and error risks for two-sided tests follow the same principles as for one-sided tests but involve slightly different procedures.

Growth Pellets   □   **EXAMPLE**

An agricultural manufacturer produces large batches of growth hormone pellets for cattle. Product specifications require the pellets to have a mean hormone content of $\mu_0 = 1280$ milligrams per pellet. A random sample of 80 pellets is selected from each batch to test whether the batch meets this specification. The process standard deviation of the hormone content per pellet is known from extensive past experience to be $\sigma = 110$ milligrams per pellet.

The test alternatives are:

$$H_0: \ \mu = 1280$$
$$H_1: \ \mu \neq 1280$$

The α risk is to be controlled at 0.05 when $\mu = \mu_0 = 1280$. Hence, the decision rule has the form shown in Figure 12.13a. Note that the rejection probability at $\mu = \mu_0 = 1280$ equals the α value for the decision rule; that is, $P(H_1; \mu_0 = 1280) = \alpha = 0.05$.

In (12.11), we have shown how the action limits A_1 and A_2 in a two-sided test are calculated. To use (12.11) here, we replace $s\{\overline{X}\}$ by $\sigma\{\overline{X}\}$ because the latter is known. Specifically, $\sigma = 110$ and $n = 80$; hence $\sigma\{\overline{X}\} = \sigma/\sqrt{n} = 110/\sqrt{80} = 12.30$. Furthermore, we have $z(\alpha/2) = z(0.025) = -1.960$ and $z(1 - \alpha/2) = z(0.975) = 1.960$. Thus:

$$A_1 = \mu_0 + z(\alpha/2)\sigma\{\overline{X}\} = 1280 - 1.960(12.30) = 1256$$
$$A_2 = \mu_0 + z(1 - \alpha/2)\sigma\{\overline{X}\} = 1280 + 1.960(12.30) = 1304$$

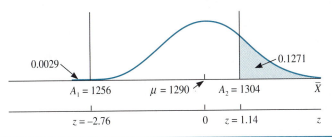

(a) Decision Rule

Acceptance region
Conclude H_0

Rejection region
Conclude H_1

$\alpha/2 = 0.025$

$1 - \alpha = 0.95$

Rejection region
Conclude H_1

$\alpha/2 = 0.025$

$A_1 = 1256$ $\mu_0 = 1280$ $A_2 = 1304$ $\overline{X}$

$z(0.025) = -1.960$ 0 $z(0.975) = 1.960$ z

$P(H_1; \mu_0 = 1280)$
$= 0.025 + 0.025$
$= 0.05$

(b) Calculation of Rejection Probability at $\mu = 1290$

0.0029

0.1271

$A_1 = 1256$ $\mu = 1290$ $A_2 = 1304$ $\overline{X}$

$z = -2.76$ 0 $z = 1.14$ z

$P(H_1; \mu = 1290)$
$= 0.0029 + 0.1271$
$= 0.1300$

FIGURE 12.13
Calculation of rejection prob-
abilities for two-sided test—
Growth pellets example

Since A_1 and A_2 define the boundaries of the acceptance and rejection regions, the decision rule for the test may be stated in terms of $\overline{X}$ as follows:

If $1256 \leq \overline{X} \leq 1304$, conclude H_0.
If $\overline{X} < 1256$ or $\overline{X} > 1304$, conclude H_1.

Using this form of the decision rule makes it convenient to evaluate the rejection probabilities of the rule at different values of μ. We shall illustrate one such calculation, at $\mu = 1290$.

If the population mean is $\mu = 1290$, then the sampling distribution of $\overline{X}$ will be centered at 1290, as shown in Figure 12.13b. The probability of concluding H_1 when $\mu = 1290$ is the sum of the two tail areas in the two parts of the rejection region. Action limit A_1 corresponds to $z = (1256 - 1290)/12.30 = -2.76$ and action limit A_2 corresponds to $z = (1304 - 1290)/12.30 = 1.14$. Hence, the two tail areas are $P(Z < -2.76) = 0.0029$ and $P(Z > 1.14) = 0.1271$, respectively. Adding the two tail probabilities, we obtain the rejection probability at $\mu = 1290$ as $P(H_1; \mu = 1290) = 0.0029 + 0.1271 = 0.1300$.

The entire rejection probability curve is shown in Figure 12.14. Note that the curve is symmetrical about $\mu_0 = 1280$ and that the rejection probability increases steadily toward 1.0 as μ departs from μ_0 in either direction.

In evaluating the error risks inherent in the decision rule, we need to recognize that an α risk exists here only at $\mu = \mu_0 = 1280$. We noted earlier that this α risk is 0.05, as may be read directly from the rejection probability curve in Figure 12.14. A β risk exists at any value of μ other than $\mu_0 = 1280$ and equals the complementary probability of the rejection probability at that value of μ. Referring to Figure 12.14,

FIGURE 12.14
Rejection probability curve for two-sided test—Growth pellets example

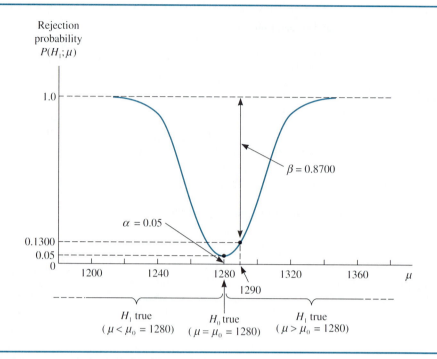

we see that the rejection probability at $\mu = 1290$ is 0.1300. Hence, the β risk when $\mu = 1290$ equals $1 - 0.1300 = 0.8700$.   □

Comment

The discussion about the evaluation of rejection probabilities and β risks was confined to the case of large samples. Rejection probabilities and β risks can also be evaluated for small samples, but the calculations are more complex.

Estimation of Rejection Probabilities and β Risks When σ Unknown

So far in our discussion of the power of a test, we have assumed that the population standard deviation σ is known. In most applications, however, σ is unknown so that we cannot calculate exact rejection probabilities. However, we can estimate rejection probabilities by using the estimated standard deviation of the mean $s\{\overline{X}\}$ instead of the true standard deviation $\sigma\{\overline{X}\}$. We shall explain the estimation of rejection probabilities when the sample size n is large. The procedure is exactly the same as in calculating exact rejection probabilities except that we use the estimated standard deviation of the mean $s\{\overline{X}\}$ instead of $\sigma\{\overline{X}\}$.

□ **EXAMPLE**

In the machine component example, the test alternatives were:

$$H_0: \ \mu \leq 1100$$
$$H_1: \ \mu > 1100$$

and the α risk was controlled at 0.01 when $\mu = \mu_0 = 1100$. We wish to estimate the rejection probability and β risk at $\mu = 1300$.

The sample results were $n = 36$, $\overline{X} = 1121$, and $s = 222$. The standard deviation of the mean $\sigma\{\overline{X}\}$ was estimated by $s\{\overline{X}\} = 222/\sqrt{36} = 37.0$.

To estimate the rejection probability at $\mu = 1300$, $P(H_1; \mu = 1300)$, we first obtain the value of the action limit from (12.7). We require $z(0.99) = 2.326$ and find:

$$A = 1100 + 2.326(37.0) = 1186.1$$

This action limit is located $z = (1186.1 - 1300)/37.0 = -3.08$ standard deviations from $\mu = 1300$. Note that we use the estimate $s\{\overline{X}\} = 37.0$ here in place of the unknown $\sigma\{\overline{X}\}$. We can now obtain the estimated rejection probability in the usual fashion:

$$P(H_1; \mu = 1300) = P(Z > -3.08) = 0.9990$$

The corresponding estimated β risk is $1 - 0.9990 = 0.0010$. Thus, if the bonded machine component has a mean service life of $\mu = 1300$ hours, a service life substantially superior to that of the all-metal component (1100 hours), there is only an estimated probability of 0.0010 that the test will fail to detect this superior performance. □

Trade-Off Between α and β Risks

We have just seen how to construct the rejection probability curve for a decision rule based on a given sample size and how to assess from that curve the error risks implicit in the rule. We can now consider an important question: What trade-off occurs between the α and β risks when the control on the α risk is changed? To answer this question, we return to the data entry example.

EXAMPLE □

Figure 12.15 contains two rejection probability curves for the data entry example. The black curve is the rejection probability curve presented initially in Figure 12.11 for the decision rule where the α risk is controlled at 0.10. The colored curve is the rejection probability curve when the α risk is controlled at 0.01. Observe that this latter curve has the same shape as the former but is shifted rightward. The effect of this shift on the error risks is to *decrease* the α risks for all values of $\mu \le \mu_0 = 23$ and to *increase* the β risks for all values of $\mu > \mu_0 = 23$. □

The preceding illustration for a one-sided upper-tail test shows that, for a given sample size, a decrease in the α risks is associated with an increase in the β risks and vice versa. This same trade-off between the two types of error risks occurs in all statistical tests when the sample size is predetermined. The trade-off reflects an important statistical principle.

(12.18)
For a given random sample size, one type of error risk can be reduced only at the expense of increasing the other type.

FIGURE 12.15
Rejection probability curves for decision rules controlling the α risk at 0.10 and 0.01—Data entry example

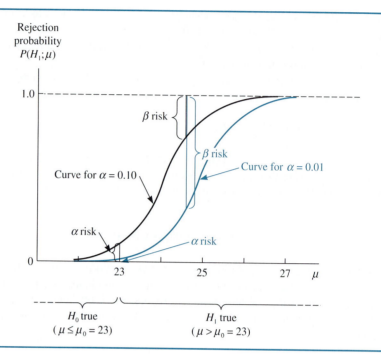

The reference to a given sample size in (12.18) is important because it is intuitively clear that if the sample size can be increased, then the larger sample should provide more precise information about the true value of μ and, hence, would offer the possibility of reducing both types of error risks simultaneously. In the next section, we consider how planning of the sample size permits control of both types of error risks.

12.7 OPTIONAL TOPIC—PLANNING OF SAMPLE SIZE

Need for Planning

Simultaneous control of both α and β risks in a test can be achieved by a planned choice of an appropriate sample size. Planning of the sample size is desirable because a sample size that is too small may entail unsatisfactorily high β risks (the α risk being fixed at the desired level), while a sample size that is too large will be uneconomical. We now discuss how to determine the sample size for a test that will keep both α and β risks at prespecified levels.

Assumptions

The planning procedure we shall use assumes the following:

1. The random sample size ultimately determined is reasonably large.
2. The population is either infinite or, if finite, is large relative to the resulting sample size.

Planning Procedure

The planning procedure requires specifications of the α and β risks to be controlled. Moreover, because the sample size is being planned, we do not yet have the sample standard deviation s available to compute $s\{\overline{X}\}$. Because the population standard deviation σ is generally unknown, we shall therefore also need to specify a planning value for σ to find the needed sample size. Recall that use of a planning value for σ is also necessary for planning the sample size for obtaining a confidence interval with specified precision.

The planning procedure involves the following four steps:

Step 1. Specify the desired α risk at $\mu = \mu_0$.

Step 2. Specify (i) the value of μ at which the β risk is to be controlled (denoted by μ_1) and (ii) the desired level of the β risk at $\mu = \mu_1$.

Step 3. Select a planning value for σ.

Step 4. Calculate the required sample size n by using formula (12.19), which will be presented shortly.

Once the required sample size is determined and the sample results are obtained, the test is conducted exactly as described in earlier sections of this chapter.

EXAMPLE □

In the machine component example, suppose that the sample size has not yet been determined.

Step 1. The manufacturer wants to control the α risk at 0.01 when $\mu = \mu_0 = 1100$, as before. Figure 12.16a shows the appropriate decision rule for controlling the α risk. The sampling distribution of $\overline{X}$ centered at $\mu_0 = 1100$ is shown on the left side of the figure. We shall now denote the z value associated with the control of the α risk by z_0. In this example, $z_0 = z(0.99) = 2.326$. The action limit A on the $\overline{X}$ scale corresponds to $z_0 = 2.326$ on the z scale.

Step 2. The manufacturer has decided that when the mean service life of the bonded component is $\mu = 1250$—that is, when it is substantially longer than that of the all-metal component—there should be only a 0.10 chance of concluding H_0 on the basis of the test. We denote this specified value of μ by μ_1. Thus, μ_1 is the specified value of μ where it is important that conclusion H_1 should be reached and where the decision maker wishes to control the β risk. The specification here is equivalent to stating that the probability of concluding H_0 when $\mu_1 = 1250$ is to be $\beta = 0.10$. The sampling distribution of $\overline{X}$ centered at $\mu_1 = 1250$ is shown on the right side of Figure 12.16a. The requirement that $\beta = 0.10$ dictates that the left-tail area in the acceptance region be equal to 0.10 and, hence, that the action limit A on the $\overline{X}$ scale correspond to $z(0.10) = -1.282$ on the z scale. We denote this z value associated with the control of the β risk by z_1. Hence, $z_1 = z(0.10) = -1.282$ here.

Step 3. The manufacturer believes that $\sigma = 250$ hours is a reasonable planning value for the population standard deviation.

Step 4. Rather than employing formula (12.19) directly to determine the required sample size, we will use Figures 12.16a and 12.16b together to explain the rationale of the formula. First, we note that the standard deviation of the sampling distribution of $\overline{X}$, given that the planning value for the population standard deviation is $\sigma = 250$, is $\sigma\{\overline{X}\} = \sigma/\sqrt{n} = 250/\sqrt{n}$. Next, we note that the required sample size n must be such that the interval on the $\overline{X}$ scale from $\mu_0 = 1100$ to the action limit A is equal to

FIGURE 12.16
Determination of the required sample size to control α and β risks—Machine component example

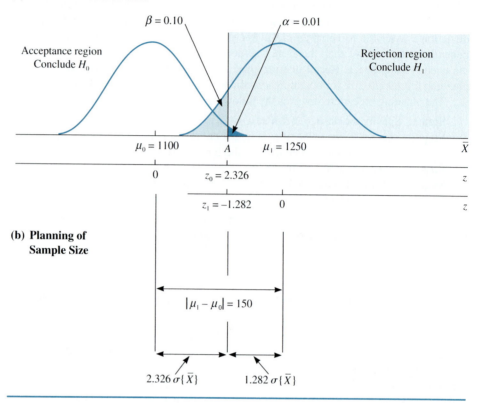

(a) Statistical Decision Rule

(b) Planning of Sample Size

$2.326\sigma\{\overline{X}\}$. Furthermore, the interval on the $\overline{X}$ scale from action limit A to $\mu_1 = 1250$ must equal $1.282\sigma\{\overline{X}\}$ (we ignore the sign of $z_1 = -1.282$ because we are concerned only with the length of the interval). Thus, the whole interval length from μ_0 to μ_1 equals:

$$2.326\sigma\{\overline{X}\} + 1.282\sigma\{\overline{X}\} = 3.608\sigma\{\overline{X}\} = 3.608\left(\frac{250}{\sqrt{n}}\right)$$

Since the interval from μ_0 to μ_1 has length $|\mu_1 - \mu_0| = |1250 - 1100| = 150$, it follows that:

$$150 = 3.608\left(\frac{250}{\sqrt{n}}\right)$$

Solving this equation for n gives $n = 36$. This sample size was, in fact, the one used earlier in the manufacturer's test.

Once the sample size is determined and the actual sample results are in hand, the test procedure is applied in the manner described earlier, using the sample standard deviation s in lieu of the planning value for σ. For the machine component example, recall that the sample results were $\overline{X} = 1121$, $s = 222$, and $s\{\overline{X}\} = 37.0$. Thus, z^* for the test was calculated as:

$$z^* = \frac{\overline{X} - \mu_0}{s\{\overline{X}\}} = \frac{1121 - 1100}{37.0} = 0.57$$

□

Comments

1. The α and β risk specifications in the machine component example can be expressed in terms of rejection probabilities as follows:

$$P(H_1; \mu_0 = 1100) = \alpha = 0.01 \qquad P(H_1; \mu_1 = 1250) = 1 - \beta = 0.90$$

These two specifications fix two points of the rejection probability curve.

2. The procedure described for planning the sample size to control both types of error risks is appropriate if the resulting sample size is reasonably large. Modifications in the procedure are needed if the required sample size turns out to be small.

3. If the planning value of σ is reasonably accurate, then the actual β risk at $\mu = \mu_1$ for the test will be close to the planned level. This is the case in the machine component example, where the planning value for σ was 250 and the sample standard deviation was $s = 222$.

General Sample Size Formula

The reasoning we have just followed for determining the required sample size in the machine component example can also be applied to lower-tail and two-sided tests. A sample size formula can be obtained that applies to all three types of tests.

(12.19)

The required random sample size to control both α and β risks, for a given planning value for σ, is:

$$n = \frac{\sigma^2(|z_1| + |z_0|)^2}{|\mu_1 - \mu_0|^2}$$

where: σ is the planning value for the population standard deviation

μ_0 and μ_1 are the μ values where the α and β risks are controlled, respectively

z_0 and z_1 are the z values associated with the specified α and β risks, respectively, and are defined as follows for each type of test:

One-Sided Upper-Tail Test $(H_0: \mu \leq \mu_0, H_1: \mu > \mu_0)$

$$z_0 = z(1 - \alpha) \qquad z_1 = z(\beta)$$

One-Sided Lower-Tail Test $(H_0: \mu \geq \mu_0, H_1: \mu < \mu_0)$

$$z_0 = z(\alpha) \qquad z_1 = z(1 - \beta)$$

Two-Sided Test $(H_0: \mu = \mu_0, H_1: \mu \neq \mu_0)$

$$z_0 = z(1 - \alpha/2) \qquad z_1 = z(\beta)$$

We illustrate the use of formula (12.19) by two examples.

One-Sided Upper-Tail Test　☐ **EXAMPLE**

For the machine component example in Figure 12.16, we require for use of formula (12.19):

$$z_0 = z(1 - \alpha) = z(0.99) = 2.326 \quad \text{so } |z_0| = 2.326$$
$$z_1 = z(\beta) = z(0.10) = -1.282 \quad \text{so } |z_1| = 1.282$$
$$|\mu_1 - \mu_0| = |1250 - 1100| = 150 \quad \text{and} \quad \sigma = 250$$

Substituting into (12.19) yields:

$$n = \frac{(250)^2(1.282 + 2.326)^2}{(150)^2} = 36$$

☐

Two-Sided Test　☐ **EXAMPLE**

In the art valuation example, suppose that the sample size had not yet been determined.

Step 1. As before, the α risk is to be controlled at 0.05 when $\mu_0 = 550$ (see Figure 12.4).

Step 2. The β risk is to be controlled at 0.10 when $\mu_1 = 850$. (By symmetry, the same sample size is obtained if the β risk is controlled at $\mu_1 = 250$, a distance of 300 below $\mu_0 = 550$.)

Step 3. A planning value of $\sigma = 925$ is deemed appropriate.

Step 4. Referring to formula (12.19) for a two-sided test, we see that:

$$z_0 = z(1 - \alpha/2) = z(0.975) = 1.960 \quad \text{so } |z_0| = 1.960$$
$$z_1 = z(\beta) = z(0.10) = -1.282 \quad \text{so } |z_1| = 1.282$$
$$|\mu_1 - \mu_0| = |850 - 550| = 300 \quad \text{and} \quad \sigma = 925$$

Substituting into (12.19), we obtain:

$$n = \frac{(925)^2(1.282 + 1.960)^2}{(300)^2} = 100$$

Thus, 100 subjects are needed in the study to achieve the desired control over the α and β risks of the test.

☐

Comments

1. Formula (12.19) describes how the required sample size n is affected by the error risk specifications and by the levels of μ at which the risks are controlled. The smaller are the desired α and β risks, the larger are $|z_0|$ and $|z_1|$, respectively, and consequently the larger is the required n. Likewise, the closer μ_1 is to μ_0, the smaller is $|\mu_1 - \mu_0|$ and the larger is the required n. Finally, the larger is the population standard deviation (as anticipated by the planning value for σ), the larger is the required sample size n to achieve the desired error control.
2. The derivation of formula (12.19) can be explained in terms of the example in Figure 12.16.

The length of the interval from μ_0 to μ_1 may be represented by $|\mu_1 - \mu_0|$. The absolute value is used here because we wish to measure the length without regard to its sign. The length of the interval from μ_0 to A equals $|z_0|\sigma\{\overline{X}\}$. Likewise, the length of the interval from A to μ_1 equals $|z_1|\sigma\{\overline{X}\}$. Again, absolute values of z_0 and z_1 are used because we wish to measure length without regard to sign. Thus, we can see from Figure 12.16b that we have the equality:

$$|\mu_1 - \mu_0| = |z_1|\sigma\{\overline{X}\} + |z_0|\sigma\{\overline{X}\} = (|z_1| + |z_0|)\sigma\{\overline{X}\}$$

Finally, we have that $\sigma\{\overline{X}\} = \sigma/\sqrt{n}$. Making this substitution in the preceding equation gives:

$$|\mu_1 - \mu_0| = (|z_1| + |z_0|)\left(\frac{\sigma}{\sqrt{n}}\right)$$

Solving this equation for n gives formula (12.19).

PROBLEMS

12.1 For each of the following test situations, (1) specify the population mean μ; (2) give the value of the standard μ_0 against which the population mean is being compared; (3) state the alternatives H_0 and H_1; (4) indicate whether the alternatives are one-sided upper-tail, one-sided lower-tail, or two-sided; and (5) describe the Type I and Type II errors that are possible.

a. The mean donation per contributor to the Community Appeal was $15.83 prior to the initiation of a new public relations campaign. A random sample of donations received while the campaign is in effect is to be used to determine whether the mean donation now exceeds $15.83. The more serious error is concluding that the campaign increases the mean donation when, in fact, it does not.

b. The mean duration of marriages that ended in divorce or annulment during the past few years in a certain state was 8.1 years. A sociologist wishes to test whether new divorce legislation has changed the mean duration, based on a random sample of divorce records accumulated since the legislation was enacted. The more serious error is concluding that a change in the mean duration has occurred when, in fact, it has not.

12.2 For each of the following test situations, (1) specify the population mean μ; (2) give the value of the standard μ_0 against which the population mean is being compared; (3) state the alternatives H_0 and H_1; (4) indicate whether the alternatives are one-sided upper-tail, one-sided lower-tail, or two-sided; and (5) describe the Type I and Type II errors that are possible.

a. An engineer has developed a leaching process for reducing the amount of sulfur in coal. Sixty-four one-kilogram samples are to be selected randomly from coal treated by this process to test whether the mean sulfur content of the treated coal is less than 15 grams per kilogram. The more serious error is concluding that the process reduces the mean sulfur content to less than 15 grams per kilogram when, in fact, it does not.

b. Bilingual children achieve a mean score of 60 on a mathematics proficiency test when mathematics is taught to them in their primary language. A random sample of 400 bilingual children has been taught the subject in their second language. An equivalent proficiency test will now be administered to see whether the mean score differs from 60. The more serious error is concluding that the mean score differs from 60 when, in fact, it does not.

12.3 A university has the option to switch to a new billing schedule for local telephone calls. This new schedule will be less costly than the present schedule if the mean duration of local calls is less than 6.0 minutes but more costly if it is greater than 6.0 minutes. A random sample of local calls will be timed to decide whether to retain the present billing schedule or switch to the new one.

a. Define the parameter μ here.

b. Specify H_0 and H_1 if the α risk is to be the risk of incurring a cost increase because of an incorrect conclusion about μ.

c. Specify H_0 and H_1 if the more serious error is to fail to obtain a cost saving when, in fact, the mean duration of calls is less than 6.0 minutes.

12.4 A firm has developed a diagnostic product for use by physicians in private practice. A decision must now be made whether or not to market the product. Marketing the product will be profitable if the mean number of units ordered per physician is greater than 3.0 and unprofitable if it is less than 3.0. The decision will be based on the sales potential shown in office demonstrations of the product with a random sample of physicians in the target market.

a. Define the parameter μ here.

b. Specify H_0 and H_1 if the more costly error is to market the product when it is not profitable.

c. Specify H_0 and H_1 if the more costly error is to fail to market the product when the mean number of units ordered per physician exceeds 3.0.

12.5 Refer to Table 12.1. In a statistical test, can a Type I error and a Type II error be made simultaneously? Explain.

12.6 A machine fills cartons with soap powder and seals the cartons. The machine is set for a mean fill of 4.500 kilograms. However, the mean fill μ can drift upward or downward from 4.500 kilograms in the course of operations. The machine operator is given the following instructions by the quality assurance engineer: "The mean fill should be 4.500 kilograms. When the machine has been operating for an hour, empty the next 10 filled cartons and weigh the soap powder in each. If the mean weight per carton for the 10 cartons is between 4.480 and 4.520 kilograms, let the machine keep operating. If the mean weight is below 4.480 or above 4.520 kilograms, stop the machine and make an appropriate adjustment."

a. Specify H_0 and H_1 here.

b. For the engineer's statistical decision rule, give (1) the test statistic, (2) the action limits, (3) the acceptance region, (4) the rejection region.

c. What constitute the Type I and Type II errors here?

12.7 A buying consortium receives shipments of fresh turkeys from a farm. A clause in the contract states that the mean dressed weight of the turkeys in a shipment should exceed 5.0 kilograms. In a test to determine whether this provision is being met in a large shipment of turkeys, a random sample of 60 turkeys is selected and the birds are weighed. Consider the decision rule for a one-sided upper-tail test here, in which the α risk is controlled at 0.01 when $\mu = 5.0$. If, in fact, $\mu = 5.0$, what is the probability (prior to selecting the sample of turkeys from the shipment) that the sample mean dressed weight will fall in (1) the rejection region of the decision rule, (2) the acceptance region?

12.8 Refer to Figure 12.1. Explain why the acceptance region of each of these decision rules includes the neighborhood of μ_0.

12.9 Refer to Figure 12.2a.

a. Using the definition of z^* in (12.5), show that $\mu_0 = 1100$ on the $\overline{X}$ axis corresponds to 0 on the z^* axis.

b. If μ were, in fact, greater than 1100, would the probability of $\overline{X}$ falling in the rejection region be larger or smaller than 0.01? Explain.

* **12.10 Commercial Loans.** The mean size of commercial loans made by a bank has been $60 thousand in the past. A recent change in the bank's credit policy allows larger amounts to be borrowed under the same terms. The credit manager now wishes to test whether the mean size of commercial loans made since the policy change is larger than $60 thousand ($H_1$) or not ($H_0$). The manager wishes to control the α risk at 0.01 when $\mu = 60$. A random sample of $n = 144$ loans made since the policy change yielded the following results (in $ thousand): $\overline{X} = 68.1$, $s = 45.0$.

a. Conduct the test. State the alternatives, the decision rule, the value of the standardized test statistic, and the conclusion.

b. If the α risk were controlled at 0.05 instead of 0.01, would the conclusion be different? Explain.

12.11 Refer to **Airline Reservations** Problem 11.4. It is known from extensive experience that the mean number of no-shows for other commuter flights of this airline is 1.310. Using the sample results in Problem 11.4, test whether the mean number of no-shows on the 4 P.M. commuter flight to New York City exceeds 1.310 (H_1) or not (H_0). Control the α risk at 0.05 when $\mu = 1.310$. State the alternatives, the decision rule, the value of the standardized test statistic, and the conclusion.

12.12 **Customs Declarations.** Travelers returning from abroad are asked to declare the value of goods they are bringing back to their country. Customs authorities are concerned about reporting errors (the difference between the declared value and the actual value) and wish to test whether the mean reporting error is negative (H_1) or not (H_0). The authorities want to control the α risk at 0.001 when $\mu = 0$. A random audit of the personal effects of $n = 100$ travelers yielded the following results on the 100 reporting errors: $\overline{X} = -\$35.41$, $s = \$45.94$.

a. Conduct the test. State the alternatives, the decision rule, the value of the standardized test statistic, and the conclusion.

b. Why is the actual α risk of the test conducted in part a only approximately at the specified level of 0.001?

12.13 Refer to **Trade Association** Problem 11.3. Shortly before enactment of the legislation, a full enumeration of all the member firms in the association showed that the mean number of hourly paid employees per firm was 9.16. It is desired to test whether the current mean number of hourly paid employees is less than 9.16 (H_1) or not (H_0), controlling the α risk at 0.05 when $\mu = 9.16$.

a. Using the sample results in Problem 11.3, conduct the test. State the alternatives, the decision rule, the value of the standardized test statistic, and the conclusion.

b. If, in fact, μ is actually 9.25, is the conclusion in part a correct? If it is not correct, has a Type I or a Type II error been made? Explain.

* **12.14** **Summer Book Loans.** In past summers in a large library system, the mean number of books borrowed per cardholder was 8.50. The library administration would like to test whether the mean number of books borrowed per cardholder this summer under modified loan arrangements differs from the level of past summers (H_1) or not (H_0). A random sample of 100 cardholders showed the following results for borrowing this summer: $\overline{X} = 9.34$ books, $s = 3.31$ books.

a. Conduct the test, controlling the α risk at 0.05 when $\mu = 8.50$. State the alternatives, the decision rule, the value of the standardized test statistic, and the conclusion.

b. Why is the actual α risk of the test conducted in part a only approximately at the specified level of 0.05?

12.15 **Follow-Up Purchases.** Financial analysts in a firm operating a large chain of retail computer stores have assumed for planning purposes that customers buying a microcomputer will spend an additional mean amount of $850 in follow-up purchases in the chain within 12 months following the computer purchase. A random sample of 81 customers showed the following results for follow-up purchases: $\overline{X} = \$746$, $s = \$380$. A test is to be conducted to determine whether the mean follow-up purchases are $850 ($H_0$) or not ($H_1$). The α risk is to be controlled at 0.01 when $\mu = 850$.

a. Conduct the test. State the alternatives, the decision rule, the value of the standardized test statistic, and the conclusion.

b. If, in fact, $\mu = 700$, is the conclusion in part a correct? If it is not correct, has a Type I or a Type II error been made? Explain.

12.16 **Auto Parts Producer.** A sales analyst in a firm producing auto parts laboriously determined, from a study of all sales invoices for the previous fiscal year, that the mean profit contribution per invoice was $16.50. For the current fiscal year, the analyst selected a random sample of 225 sales invoices to test whether the mean profit contribution this year has changed from $16.50 ($H_1$) or not ($H_0$). The sample of 225 invoices yielded the following results for the invoice profit contributions: $\overline{X} = \$17.14$, $s = \$18.80$. The α risk is to be controlled at 0.05 when $\mu = 16.50$.

a. Conduct the test. State the alternatives, the decision rule, the value of the standardized test statistic, and the conclusion.

b. What constitute Type I and Type II errors here? Given the conclusion in part a, is it possible that a Type I error has been made in this test? Is a Type II error possible here? Explain.

12.17 In a tasting session, a random sample of 100 subjects from a target consumer population tasted a food item, and each subject individually gave it a rating from 1 (very poor) to 10 (very good). It is desired to test H_0: $\mu \leq 6.0$ versus H_1: $\mu > 6.0$, where μ denotes the mean rating for the food item the target population. A computer analysis of the sample results showed that the one-sided P-value of the test is 0.0068.

  a. Does the sample mean lie above or below $\mu_0 = 6.0$?

  b. What must be the value of z^* here?

  c. The sample standard deviation is $s = 2.16$. What must be the sample mean $\overline{X}$?

  d. Does the magnitude of the P-value indicate that the sample results are inconsistent with conclusion H_0? Explain.

12.18 For each of the following large-sample test situations, (1) obtain the P-value, (2) state whether the P-value is one-sided or two-sided, and (3) comment on whether the sample data are consistent with conclusion H_0.

  a. One-sided lower-tail alternatives, $z^* = -0.32$.

  b. Two-sided alternatives, $z^* = 2.86$.

  c. One-sided upper-tail alternatives, $z^* = 3.01$.

12.19 For each of the following large-sample test situations, (1) obtain the P-value, (2) state whether the P-value is one-sided or two-sided, and (3) comment on whether the sample data are consistent with conclusion H_0.

  a. Two-sided alternatives, $z^* = 0.63$.

  b. One-sided upper-tail alternatives, $z^* = -0.86$.

  c. One-sided lower-tail alternatives, $z^* = -2.32$.

* **12.20** Refer to **Commercial Loans** Problem 12.10.

  a. Calculate the P-value of the test. Interpret its meaning here.

  b. Conduct the required test using the P-value in part a.

12.21 Refer to **Airline Reservations** Problems 11.4 and 12.11.

  a. Calculate the P-value of the test. Is it a one-sided or two-sided P-value?

  b. Conduct the required test using the P-value in part a.

  c. An assistant of the airline researcher interprets the P-value of the test calculated in part a as the probability that H_0 is true, given the sample results. Do you agree with the assistant's interpretation? Explain.

12.22 Refer to **Customs Declarations** Problem 12.12.

  a. Calculate the P-value of the test. Interpret its meaning here.

  b. Conduct the required test using the P-value in part a.

  c. Would the test conclusion in part b be different if the α risk were to be controlled at 0.0005 rather than 0.001?

12.23 Refer to **Trade Association** Problems 11.3 and 12.13.

  a. Calculate the P-value of the test. Is it a one-sided or two-sided P-value?

  b. Conduct the required test using the P-value in part a. Must the test conclusion necessarily be identical to the one drawn on the basis of the decision rule for z^*? Explain.

* **12.24** Refer to **Summer Book Loans** Problem 12.14.

  a. Calculate the P-value of the test. Is it a one-sided or two-sided P-value?

  b. Conduct the required test using the P-value in part a.

12.25 Refer to **Follow-Up Purchases** Problem 12.15.

  a. Calculate the P-value of the test. Interpret its meaning here.

  b. Conduct the required test using the P-value in part a.

  c. Would the test conclusion in part b be different if the α risk were to be controlled at 0.05 rather than 0.01?

12.26 Refer to **Auto Parts Producer** Problem 12.16.

a. Calculate the P-value of the test. Is it a one-sided or two-sided P-value?

b. Conduct the required test using the P-value in part a. Must the test conclusion necessarily be identical to the one drawn on the basis of the decision rule for z^*? Explain.

* **12.27** Refer to **Nutrition Study** Problem 11.9. The dietary regimen is intended to yield a mean blood sugar level of 110 milligrams of sugar per 100 cubic centimeters of blood. It is desired to test whether this target is being met (H_0) or not (H_1), controlling the α risk at 0.05 when $\mu = 110$.

a. Conduct the test. State the alternatives, the decision rule, the value of the standardized test statistic, and the conclusion.

b. Using Table C.3, obtain bounds for the P-value of this test. Is the P-value one-sided or two-sided? Is the P-value consistent with the test result in part a? Explain.

12.28 A random sample of 21 children of the same age participated in a special language skills program designed to build vocabulary. Measurements at the end of the program showed the children's vocabularies had a mean of $\overline{X} = 1060$ words and a standard deviation of $s = 119$ words. It is known that the mean vocabulary of children of this age ordinarily is 1000 words. An analyst wishes to test whether the mean vocabulary of children who take the special program is greater than 1000 words (H_1) or not (H_0). Assume that the vocabulary sizes of children who take the program are approximately normally distributed.

a. Conduct the test, controlling the α risk at 0.01 when $\mu = 1000$. State the alternatives, the decision rule, the value of the standardized test statistic, and the conclusion.

b. Computer output gives a one-sided P-value of 0.0158 for the test. Is this value consistent with the test result in part a? Explain.

12.29 A large bakery has installed new washing equipment for baking pans. A washing cycle with the old equipment required a mean time of 35 minutes. Eight trial cycles with the new equipment had mean cycle time $\overline{X} = 32.1$ minutes and standard deviation $s = 4.2$ minutes. Assume that the distribution of cycle times for the new equipment does not depart too markedly from a normal distribution and that the eight trials constitute a random sample from this population. It is desired to test whether the mean cycle time with the new equipment is shorter than with the old equipment (H_1) or not (H_0).

a. Conduct the test, controlling the α risk at 0.05 when $\mu = 35$. State the alternatives, the decision rule, the value of the standardized test statistic, and the conclusion.

b. Using Table C.3, obtain bounds for the P-value of this test. Is the P-value one-sided or two-sided? Is the P-value consistent with the test result in part a? Explain.

12.30 Safety regulations governing the evacuation of a high-risk area of a chemical plant call for the mean evacuation time to be less than 5 minutes. Six safety drills have produced the following evacuation times (in minutes): 3.8, 4.6, 5.3, 4.8, 4.5, 5.1. Assume that the evacuation times for the six drills constitute a random sample from a population that is approximately normal.

a. Test whether the regulation governing mean evacuation time is being met (H_1) or not (H_0). Control the α risk at 0.01 when $\mu = 5.0$. State the alternatives, the decision rule, the value of the standardized test statistic, and the conclusion.

b. Using Table C.3, obtain bounds for the P-value of this test. Is the P-value one-sided or two-sided? Is the P-value consistent with the test result in part a? Explain.

12.31 A manufacturer sells a safety light that turns on automatically when the electricity fails in an installation. The manufacturer claims that these lights, when fully charged, have a mean operating life of more than 300 minutes before recharging is required. In a test of this claim, 12 fully charged lights were selected at random, and the operating life of each was measured. The sample results (in minutes) were as follows:

290	331	329	364	332	333	346	356	352	272	316	347

Assume that operating lives of the safety light are approximately normally distributed. It is desired to test whether the mean operating life exceeds 300 minutes (H_1) or not (H_0), controlling the α risk at 0.025 when $\mu = 300$.

   a. Conduct the test. State the alternatives, the decision rule, the value of the standardized test statistic, and the conclusion.

   b. Computer output gives a one-sided P-value of 0.00117 for the test. Is this value consistent with the test result in part a? Explain.

   c. A subsequent check showed that the two sample lights with operating lives of 290 and 272 minutes were not fully charged prior to the test. Would the test conclusion in part a change if these two observations were discarded from the sample? Explain.

   d. Construct a stem-and-leaf display of the 12 sample observations. Do the two observations mentioned in part c appear to be outliers? Does the assumption of approximate normality appear to be plausible for the other 10 observations? Comment.

* 12.32 Refer to **Nutrition Study** Problem 11.9. Use the 95 percent confidence interval to test whether the mean blood sugar level is 110 (H_0) or not (H_1). State the test conclusion. What is the implied α risk of this test?

12.33 Refer to **Airline Reservations** Problem 11.4. Use the 99 percent confidence interval to test whether the mean number of no-shows is 1.0 (H_0) or not (H_1). State the test conclusion. What is the implied α risk of this test?

12.34 Refer to **Customs Declarations** Problem 12.12.

   a. What type of confidence interval will provide the relevant information for the test? What should be the confidence coefficient?

   b. Using the sample results, construct the appropriate confidence interval. Explain how this confidence interval leads to the same conclusion as the test.

12.35 Refer to **Summer Book Loans** Problem 12.14.

   a. What type of confidence interval will provide the relevant information for the test? What should be the confidence coefficient?

   b. Using the sample results, construct the appropriate confidence interval. Explain how this confidence interval leads to the same conclusion as the test.

* 12.36 New Product. A market study for a new industrial product indicates that the product should be launched by the firm only if the mean number of units purchased per customer in the first solicitation of the firm's customers is more than 2.0. Initial orders for the product are to be solicited from a random sample of 100 of the firm's many customers. The standard deviation of the numbers of units purchased in initial orders is $\sigma = 1.6$ units.

   a. The alternatives to be tested are H_0: $\mu \leq 2.0$, H_1: $\mu > 2.0$, and the α risk is to be controlled at 0.05 when $\mu = 2.0$. Verify that the following decision rule for $\overline{X}$ is appropriate for this test:

        If $\overline{X} \leq 2.263$, conclude H_0.
        If $\overline{X} > 2.263$, conclude H_1.

   b. Calculate the rejection probabilities at $\mu = 2.15, 2.45$ for the decision rule in part a, and complete the following table:

μ:	1.85	2.00	2.15	2.30	2.45
$P(H_1; \mu)$:	0.005	0.050	___	0.591	___

   c. Sketch the rejection probability curve for the decision rule in part a.

   d. What is the incorrect conclusion when $\mu = 2.45$? What is the probability that the decision rule in part a will lead to the incorrect conclusion when $\mu = 2.45$? Is this probability an α or β risk?

12.37 Color Graphics. The developer of a decision-support software package wishes to test whether users consider a color graphics enhancement to be beneficial, on balance, given its list price of $800. A random sample of 100 users of the package will be invited to try out the enhancement and rate it on a scale ranging from -5 (completely useless) to 5 (very beneficial). The test alternatives are H_0: $\mu \leq 0$, H_1: $\mu > 0$, where μ denotes the mean rating of users. The α risk of the test is to be controlled at 0.01 when $\mu = 0$. The standard deviation of users' ratings is $\sigma = 2.3$.

a. Verify that the following decision rule for $\overline{X}$ is appropriate for this test:

If $\overline{X} \leq 0.535$, conclude H_0.
If $\overline{X} > 0.535$, conclude H_1.

b. Calculate the rejection probabilities at $\mu = 0.30, 0.90$ for the decision rule in part a, and complete the following table:

μ:	0	0.30	0.60	0.90
$P(H_1; \mu)$:	0.010	____	0.610	____

c. Sketch the rejection probability curve for the decision rule in part a.
d. What is the incorrect conclusion when $\mu = 0.60$? What is the probability that the decision rule in part a will lead to the incorrect conclusion when $\mu = 0.60$? Is this probability an α or β risk?

12.38 Trade Agreement. A trade agreement governing the movement of agricultural products between two countries stipulates that the mean weight of boxes of butter must be 25.00 kilograms. A large shipment of boxes of butter is to be tested to determine whether it meets this requirement by selecting a random sample of 49 boxes and using the following decision rule:

If $24.95 \leq \overline{X} \leq 25.05$, conclude H_0 ($\mu = 25.00$).
If $\overline{X} < 24.95$ or $\overline{X} > 25.05$, conclude H_1 ($\mu \neq 25.00$).

The standard deviation of weights of boxes of butter is $\sigma = 0.15$ kilogram.

a. Calculate the rejection probabilities at $\mu = 24.90, 25.00, 25.10$ for this decision rule, and complete the following table:

μ:	24.90	24.95	25.00	25.05	25.10
$P(H_1; \mu)$:	____	0.500	____	0.500	____

b. Sketch the rejection probability curve for the decision rule.
c. What is the α risk at $\mu = 25.00$ for this decision rule? What is the error risk at $\mu = 25.10$? Does the latter relate to a Type I or a Type II error?

12.39 Foreign Applicants. An education researcher is interested in the effect of culture on the performance of applicants who take a North American graduate school admission test. North American applicants' scores are known to have a mean of 550 points. A random sample of 225 applicants from an English-speaking foreign country is scheduled to take the test. The researcher will use the sample results to test H_0: $\mu = 550$ versus H_1: $\mu \neq 550$, where μ denotes the mean test score of applicants from the English-speaking foreign country. The standard deviation of test scores for applicants from the foreign country is $\sigma = 48$ points. The following decision rule for $\overline{X}$ is to be employed:

If $544 \leq \overline{X} \leq 556$, conclude H_0.
If $\overline{X} < 544$ or $\overline{X} > 556$, conclude H_1.

a. Calculate the rejection probabilities at $\mu = 540, 544, 550, 556, 560$ for this decision rule.
b. Sketch the rejection probability curve for the decision rule.
c. What is the α risk at $\mu = 550$? Does an α risk exist at any other value of μ? Explain.
d. If applicants from this foreign country have a mean score that is 10 points above that of North American applicants, what is the probability that the decision rule will fail to detect this difference? Is this probability an α or β risk?

12.40 Refer to **New Product** Problem 12.36.

a. Construct the decision rule for $\overline{X}$ if the α risk is to be controlled at 0.01 rather than at 0.05 when $\mu = 2.0$.
b. Obtain the rejection probability at $\mu = 2.45$ for the decision rule in part a. Is the error risk at $\mu = 2.45$ smaller or larger with the decision rule in part a than with the decision rule in Problem

12.36a, where α is controlled at 0.05? Is this finding consistent with the principle in (12.18)? Explain.

12.41 Refer to **Trade Agreement** Problem 12.38.

a. Obtain the decision rule for $\bar{X}$ if the α risk is to be controlled at 0.01 when $\mu = 25.00$. Also obtain the decision rule for $\alpha = 0.10$.

b. For each decision rule in part a, obtain the β risk when $\mu = 25.10$. Are the α and β risks for the two decision rules consistent with the principle in (12.18)? Explain.

* **12.42** Refer to **Commercial Loans** Problem 12.10. Use the sample results to estimate the β risk for the test when $\mu = 70$. Interpret this risk.

12.43 Refer to **Airline Reservations** Problems 11.4 and 12.11. Use the sample results to estimate the β risk for the test when $\mu = 1.60$. Interpret this risk.

12.44 Refer to **Customs Declarations** Problem 12.12. Use the sample results to estimate the β risk for the test when $\mu = -20$. Interpret this risk.

12.45 Refer to **Trade Association** Problems 11.3 and 12.13. Use the sample results to estimate the power of the test when $\mu = 8.5$. Estimate the β risk at $\mu = 8.5$ and interpret it.

* **12.46** Refer to **Summer Book Loans** Problem 12.14. Use the sample results to estimate the β risk for the test when $\mu = 10.0$. Interpret this risk.

12.47 Refer to **Follow-Up Purchases** Problem 12.15. Use the sample results to estimate the β risk for the test when $\mu = 900$. Interpret this risk.

12.48 Refer to **Auto Parts Producer** Problem 12.16. Use the sample results to estimate the power of the test when $\mu = 20.00$. Estimate the β risk at $\mu = 20.00$ and interpret it.

* **12.49** Refer to **New Product** Problem 12.36. Suppose that the sample size for the study is still under review.

a. It is desired to control the α risk at 0.05 when $\mu = 2.0$ and the β risk at 0.10 when $\mu = 2.45$. Using $\sigma = 1.6$ as the planning value, obtain the required sample size.

b. A random sample of the size determined in part a has been selected, and the results are $\bar{X} = 2.34$ and $s = 1.71$. Conduct the required test, stating the decision rule, the value of the standardized test statistic, and the conclusion.

c. What would be the effect on the sample size in part a if the β risk were to be controlled at 0.10 when $\mu = 2.30$ instead of when $\mu = 2.45$?

12.50 Refer to **Color Graphics** Problem 12.37. Suppose that the sample size for the test is still under review.

a. It is desired to control the α risk at 0.01 when $\mu = 0$ and the β risk at 0.05 when $\mu = 0.80$. Using $\sigma = 2.3$ as the planning value, obtain the required sample size.

b. A random sample of the size determined in part a has been selected, and the results are $\bar{X} = 0.545$ and $s = 2.12$. Conduct the required test, stating the decision rule, the value of the standardized test statistic, and the conclusion.

12.51 Refer to **Trade Agreement** Problem 12.38. Suppose that the sample size for the test is still under review.

a. It is desired to control the α risk at 0.05 when $\mu = 25.00$ and the β risk at 0.05 when $\mu = 24.95$. Using $\sigma = 0.15$ as the planning value, obtain the required sample size.

b. A random sample of the size determined in part a has been selected, and the results are $\bar{X} = 25.03$ and $s = 0.183$. Conduct the required test, stating the decision rule, the value of the standardized test statistic, and the conclusion.

c. The value of the sample standard deviation s is larger than the planning value for σ used in obtaining the required sample size. What does this difference suggest about the magnitude of the actual β risk for the test at $\mu = 24.95$ in relation to the target β risk of 0.05?

12.52 Refer to **Foreign Applicants** Problem 12.39. Suppose that the sample size for the study is still under review.

a. It is desired to control the α risk at 0.05 when $\mu = 550$ and the β risk at 0.05 when $\mu = 540$. Using $\sigma = 48$ as the planning value, obtain the required sample size.

b. A random sample of the size determined in part a has been selected, and the results are $\bar{X} = 527.2$ and $s = 43.8$. Conduct the required test, stating the decision rule, the value of the standardized test statistic, and the conclusion.

c. What would be the effect on the required sample size in part a if the appropriate planning value for σ were 40 instead of 48?

EXERCISES

12.53 Demonstrate the equivalence of decision rules (12.6) and (12.14) for testing one-sided upper-tail alternatives when n is large.

12.54 Consider the standardized test statistic $(\bar{X} - \mu_0)/s\{\bar{X}\}$. What must be assumed about the sample size or the population or both for this statistic to have (1) an approximate standard normal distribution when $\mu = \mu_0$, (2) a $t(n - 1)$ distribution when $\mu = \mu_0$?

12.55 Demonstrate the equivalence of the decision rules in Tables 12.2b(1) and 12.3a for testing one-sided upper-tail alternatives when the sampled population is normal.

12.56 Consider a test of the alternatives H_0: $\mu \leq 0$ versus H_1: $\mu > 0$, controlling the α risk at 0.01 when $\mu = 0$. Assume that the population standard deviation is $\sigma = 10$.

a. Calculate the β risk at $\mu = 2.0$ for each of the sample sizes $n = 100, 200, 400$.

b. What do the results in part a imply about the effect of increasing the sample size on the β risk at a given μ for a fixed level of the α risk?

STUDIES

12.57 Refer to the **Power Cells** data set (Appendix D.2). Assume that the data on number of cycles before failure for the $n = 54$ cells constitute a random sample from a single population with mean μ.

a. Test the alternatives H_0: $\mu = 135$ versus H_1: $\mu \neq 135$, controlling the α risk at 0.01 when $\mu = 135$. State the decision rule, the value of the standardized test statistic, and the conclusion.

b. Calculate the P-value of the test in part a. Verify that it is consistent with the test result in part a.

c. Estimate the β risk for the test in part a when $\mu = 160$.

d. Why might the 54 observations on the number of cycles until failure in this experiment not be drawn from the same population? Would it be meaningful then to consider a test on μ here? Explain.

12.58 A manufacturer has developed a new drive belt for a machine. The original drive belts are known to have a mean operating life of 3500 hours. The manufacturer wants to place a random sample of the new belts on a forced-life test to establish if the mean operating life of the new belts exceeds that of the original belts (H_1) or not (H_0). The manufacturer is considering several possible sample sizes. The α risk of the test is to be controlled at 0.01 when $\mu = 3500$. A reasonable planning value for the standard deviation of the operating lives of the new belts is $\sigma = 500$ hours.

a. Evaluate the rejection probabilities at $\mu = 3500, 3600, 3700$ for each of the following sample sizes: (1) 50, (2) 100, (3) 150. Sketch the rejection probability curves for the three sample sizes on the same graph.

b. The manufacturer definitely wants to conclude that the new belts have a longer mean operating life if, in fact, their mean life is 5 percent longer than the mean life of the original belts. Based on the curves in part a, what is the power of the decision rule for each of the three sample sizes when the mean life of the new belts is 5 percent longer than that of the original belts?

c. In the end, the manufacturer placed a random sample of $n = 100$ new belts on a forced-life test and obtained these results (in hours): $\bar{X} = 3650.2$, $s = 531.4$. Calculate the P-value of the test. What conclusion should the manufacturer draw from these sample results?

12.59 A market survey is being designed to assess consumer reactions to a new household product. The main factors to be taken into account in the design are (1) the degree to which a respondent prefers the new product to its principal competitor (X_1) and (2) the respondent's subjective assessment of the dollar value of the new product (X_2). The preference variable X_1 will be measured on an 11-point scale ranging from -5 (new product much worse than that of principal competitor) to 5 (new product much better than that of principal competitor). The test alternatives of interest in connection with the preference variable are H_0: $\mu_1 \leq 0$, H_1 : $\mu_1 > 0$, where μ_1 denotes the mean of X_1. The mean subjective dollar value, denoted by μ_2, is to be estimated by a 99 percent confidence interval. In planning the sample size for the survey, it is desired that the test on the mean preference μ_1 control each of the α and β risks at 0.01 or less when $\mu_1 = 0$ and $\mu_1 = 1$, respectively. It is also desired that the half-width of the confidence interval for μ_2 not exceed \$5. Reasonable planning values for the standard deviations of X_1 and X_2 are $\sigma_1 = 2.5$ and $\sigma_2 = \$20$, respectively.

a. What number of respondents is adequately large to satisfy both the testing and the estimation requirements?

b. Consider the following two events in advance of conducting the survey: (1) The test on μ_1 leads to the incorrect conclusion; (2) the confidence interval for μ_2 is incorrect. Are these two events necessarily independent? Explain why or why not.

c. The survey was conducted using a sample size of $n = 140$, and the following sample results were obtained:

Variable	$\bar{X}$	s
Preference	1.32	2.38
Dollar value	\$183.40	\$18.10

Construct the desired 99 percent confidence interval for μ_2. Also conduct the test for μ_1, controlling the α risk at 0.01 when $\mu_1 = 0$, by obtaining the P-value and drawing the appropriate conclusion. Interpret both the confidence interval and the test conclusion.

Inferences for Population Proportion

n Chapters 11 and 12, we considered inferences about a population mean. In this chapter, we take up inferences about a population proportion.

Population Proportion

In many applications of sampling, it is desired to obtain information about a population proportion.

EXAMPLE □

A government policy maker is concerned with the outlook by U.S. households about economic conditions. The relevant population consists of all U.S. households. The population parameter of interest is the proportion p of households expecting an economic upturn within three months. □

EXAMPLE □ Insurance Applications

In the processing of insurance applications by an insurance company, the population parameter of interest is the proportion p of applications prepared by insurance agents that provide complete information. The population in this case is an infinite one because it pertains to a process. □

A formal definition of a population proportion follows.

> **(13.1)**
> A **population proportion,** denoted by p, is the proportion of the elements of a population that have the particular characteristic of interest.

Sample Proportion

When information about a population proportion is to be obtained from a simple random sample, two sample statistics are relevant. The first statistic is the number of the sample observations that have the particular characteristic of interest, to be denoted by f. The second is the proportion of the sample observations with the characteristic of interest, to be denoted by $\bar{p}$.

☐ **EXAMPLE**

In the insurance applications example, the characteristic of interest is whether an application is complete. In a random sample of $n = 50$ applications, 36 were found to be complete. Thus, the sample frequency of occurrences of complete applications here is $f = 36$ and the sample proportion of occurrences is $\bar{p} = 36/50 = 0.72$.   ☐

(13.2)

The number of sample observations with the characteristic of interest in a simple random sample of size n is called the **sample frequency** of occurrences and is denoted by f. The proportion of sample observations with the characteristic of interest is called the **sample proportion** of occurrences and is denoted by $\bar{p}$.

The sample proportion and the sample frequency are related as follows:

$$\bar{p} = \frac{f}{n}$$

Note that the sample frequency f can be $0, 1, \ldots, n$ according to how many of the n sample observations have the given characteristic. Consequently, the sample proportion $\bar{p}$ can be $0, 1/n, 2/n, \ldots, 1$.

We turn next to consider the sampling distributions of f and $\bar{p}$ to see how these sample statistics can be used to make inferences about the population proportion p.

13.2   **SAMPLING DISTRIBUTIONS OF f AND $\bar{p}$**

We already know the sampling distribution of the sample frequency f when an infinite population is sampled. It is the binomial probability distribution that we encountered in Chapter 7.

Binomial Distribution as Sampling Distribution

Let us see why the binomial probability distribution is the sampling distribution of the sample frequency f for a random sample selected from an infinite population. First, the outcome of any one sample observation in the random sample corresponds to a Bernoulli random trial. The associated Bernoulli random variable B takes the value 1 if the element has the characteristic of interest and takes the value 0 otherwise. The population proportion p is the probability that B takes the value 1; that is, $P(B = 1) = p$.

☐ **EXAMPLE**

In the insurance applications example, $B = 1$ if an application is complete and $B = 0$ otherwise. The outcome $B = 1$ occurs with probability p, where p is the population proportion of complete applications.   ☐

The second step in recognizing why the binomial probability distribution is the sampling distribution of the sample frequency f is to recall that a random sample from an infinite population, according to definition (9.10), requires independent observations from the same population. Thus, the Bernoulli observations $B_1, B_2, \ldots, B_n$ in the random sample are independent and from the same population. This implies that $P(B_i = 1) = p$ for each observation in the random sample.

But the sample frequency f for a random sample from an infinite population is simply the sum of these independent Bernoulli observations that each have the same probability $P(B_i = 1) = p$:

$$f = B_1 + B_2 + \cdots + B_n$$

Hence it follows from (7.5) that the sample frequency f is a binomial random variable.

(13.3)

When a simple random sample of size n is drawn from an infinite population with population proportion p, the sample frequency f has a sampling distribution given by the binomial probability function:

$$P(f) = \binom{n}{f} p^f (1 - p)^{n-f}$$

where: $P(f)$ denotes the probability of f elements in the random sample
having the given characteristic

$$\binom{n}{f} = \frac{n!}{f!(n - f)!}$$

$$f = 0, 1, \ldots, n$$

Since the sample proportion $\bar{p}$, as defined in (13.2), is a constant multiple of the sample frequency f, probabilities for $\bar{p}$ correspond to the binomial probabilities for f. Thus, when $n = 3$, the probability that $\bar{p} = 2/3$ is the same as the binomial probability that $f = 2$. Consequently, the binomial probabilities obtained either from (13.3) or from Table C.5 may be used to obtain the sampling distributions of both f and $\bar{p}$ for simple random sampling from an infinite population.

EXAMPLES ☐

1. Table 13.1 shows the sampling distributions of f and $\bar{p}$ when the random sample size is $n = 6$ and the population proportion is $p = 0.3$. Figure 13.1 shows these distributions graphically, the only difference between the sampling distributions of f and $\bar{p}$ being in the scaling on the horizontal axis. The probabilities were obtained from Table C.5.

2. For the preceding example, where $n = 6$ and $p = 0.3$, we wish to obtain the probability $P(\bar{p} \le 0.5) = P(\bar{p} \le 3/6)$. Using Table 13.1, we find this probability to be:

$$P(\bar{p} \le 3/6) = P(f \le 3) = 0.1176 + 0.3025 + 0.3241 + 0.1852$$
$$= 0.9294$$

Thus, the probability is 0.9294 that $\bar{p}$ is 0.5 or less.   ☐

TABLE 13.1

Sampling distributions of f and $\bar{p}$ for $n = 6$ and $p = 0.3$.   Since $\bar{p}$ is a constant multiple of f, the corresponding probabilities are the same.

Sampling Distribution of f		Sampling Distribution of $\bar{p}$	
f	$P(f)$	$\bar{p}$	$P(\bar{p})$
0	0.1176	0	0.1176
1	0.3025	1/6	0.3025
2	0.3241	2/6	0.3241
3	0.1852	3/6	0.1852
4	0.0595	4/6	0.0595
5	0.0102	5/6	0.0102
6	0.0007	1	0.0007
Total	1.000	Total	1.000

Comment

The probability function for $\bar{p}$ can be written explicitly by replacing f in (13.3) by $n\bar{p}$.

(13.4) $$P(\bar{p}) = \binom{n}{n\bar{p}} p^{n\bar{p}} (1 - p)^{n(1-\bar{p})}$$

where: $\bar{p} = 0, \dfrac{1}{n}, \dfrac{2}{n}, \ldots, 1$

Characteristics of Sampling Distributions

Mean and Variance.   The mean and variance of the sampling distribution of f correspond to those of the binomial probability distribution given in (7.7). For convenience, we repeat these now, together with the mean and variance of the sampling distribution of $\bar{p}$.

FIGURE 13.1
Sampling distributions of f and $\bar{p}$ for $n = 6$ and $p = 0.3$

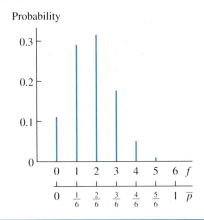

(13.5)

For simple random sampling from an infinite population:

Sampling Distribution of f	Sampling Distribution of $\bar{p}$
$E\{f\} = np$	$E\{\bar{p}\} = p$
$\sigma^2\{f\} = np(1-p)$	$\sigma^2\{\bar{p}\} = \dfrac{p(1-p)}{n}$
$\sigma\{f\} = \sqrt{np(1-p)}$	$\sigma\{\bar{p}\} = \sqrt{\dfrac{p(1-p)}{n}}$

Note that the sample proportion $\bar{p}$ is an unbiased estimator of the population proportion p.

EXAMPLE □

When $n = 6$ and $p = 0.3$:

$$E\{f\} = 6(0.3) = 1.8 \qquad\qquad E\{\bar{p}\} = 0.3$$

$$\sigma^2\{f\} = 6(0.3)(0.7) = 1.26 \qquad \sigma^2\{\bar{p}\} = \frac{0.3(0.7)}{6} = 0.035$$

$$\sigma\{f\} = \sqrt{1.26} = 1.12 \qquad\qquad \sigma\{\bar{p}\} = \sqrt{0.035} = 0.187$$

□

Comments

1. To derive the mean and variance of $\bar{p}$, we utilize the fact that $\bar{p} = f/n$ and the known results for the mean and variance of binomial random variable f in (7.7):

$$E\{\bar{p}\} = E\left\{\frac{f}{n}\right\} = \frac{1}{n}E\{f\} = \frac{np}{n} = p \qquad \text{by (6.5b)}$$

$$\sigma^2\{\bar{p}\} = \sigma^2\left\{\frac{f}{n}\right\} = \frac{1}{n^2}\sigma^2\{f\} = \frac{np(1-p)}{n^2} = \frac{p(1-p)}{n} \qquad \text{by (6.6b)}$$

2. The sample proportion $\bar{p}$ is actually a special case of the sample mean $\bar{X}$. Recall that the sample frequency f is the sum of n Bernoulli observations B_i. Here, the observations B_i happen to take on only the values 0 or 1. Hence, we have:

$$\bar{p} = \frac{f}{n} = \frac{B_1 + \cdots + B_n}{n} = \bar{B}$$

Thus, $\bar{p}$ is a sample mean of the observations B_i.

    All the earlier results for the sampling distribution of $\bar{X}$ presented in Chapter 10 carry over to the sampling distribution of $\bar{p}$, with only a change of notation. For example, the result $E\{\bar{X}\} = \mu$ in (10.5) has its counterpart here as $E\{\bar{p}\} = p$ in (13.5).

FIGURE 13.2
Sampling distributions of $\bar{p}$
for $n = 5$ and $n = 50$
when $p = 0.4$

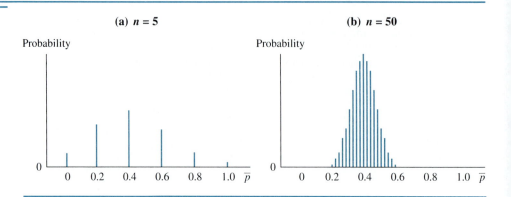

Shape of Distribution.   We know from Chapter 7 that the binomial probability distribution is skewed to the right if $p < 0.5$, is skewed to the left if $p > 0.5$, and is symmetrical if $p = 0.5$. Hence, the sampling distribution of $\bar{p}$ (or f) has these same characteristics.

Central Limit Theorem

Although the sampling distribution of $\bar{p}$ (or f) is skewed if $p \neq 0.5$, the skewness decreases as the sample size increases. Figure 13.2 illustrates this fact. Shown there are the sampling distributions of $\bar{p}$ when $p = 0.4$ for sample sizes $n = 5$ and $n = 50$. The sampling distribution of $\bar{p}$ for $n = 50$ in Figure 13.2b possesses very little skewness indeed. In fact, the outline of this sampling distribution appears to be approximately normal. This brings us to another encounter with the central limit theorem.

> **(13.6)**
> The sampling distributions of f and $\bar{p}$ are approximately normal when the simple random sample size is sufficiently large. As a working rule, the normal approximation is adequate when both $np \geq 5$ and $n(1 - p) \geq 5$.

☐ **EXAMPLES**

1. When $n = 50$ and $p = 0.4$, we have $np = 50(0.4) = 20$ and $n(1 - p) = 50(0.6) = 30$. Since both quantities exceed 5, our working rule states that the normal distribution can be used here to approximate the sampling distributions of f and $\bar{p}$.

2. When $n = 15$ and $p = 0.8$, we have $np = 15(0.8) = 12$ and $n(1 - p) = 15(0.2) = 3$. Since the latter quantity is less than 5, our working rule states that the normal distribution should not be used here to approximate the sampling distributions of f and $\bar{p}$.   ☐

Sampling Finite Populations

Up to this point, we have considered sampling of infinite populations; that is, sampling of processes. All of the results obtained for the mean and variance of the sampling distributions of f and $\bar{p}$ in (13.5) and central limit theorem (13.6) apply also to sampling of finite populations if the sampling fraction is 5 percent or less. This is the same rule as working rule (10.9) for the sampling distribution of $\bar{X}$. As we stated before, this working rule is usually met in practice. We shall take up the case when the working rule is not met in Chapter 23.

Comment

When a finite population of size N is sampled, the exact sampling distributions of f and $\bar{p}$ are given by the hypergeometric probability function (7.11), with $C = Np$ and the sample frequency f corresponding to x. The binomial probability function provides a good approximation to the hypergeometric distribution when the sampling fraction is small. Moreover, when the sample size is large, the normal distribution is a good approximation to the binomial probability distribution.

Using the Normal Approximation for f and $\bar{p}$

When using the normal approximation to the sampling distributions of f and $\bar{p}$, we employ the following standardized variables.

(13.7)
The standardized variables for approximating the sampling distributions of f and $\bar{p}$ are:

(13.7a) $\quad Z = \dfrac{f - E\{f\}}{\sigma\{f\}} = \dfrac{f - np}{\sqrt{np(1-p)}}$

(13.7b) $\quad Z = \dfrac{\bar{p} - E\{\bar{p}\}}{\sigma\{\bar{p}\}} = \dfrac{\bar{p} - p}{\sqrt{\dfrac{p(1-p)}{n}}}$

With these standardized variables, the use of the normal approximation is routine, as illustrated by the following example.

EXAMPLE ☐

In the insurance applications example, an analyst plans to select a random sample of 200 insurance applications and determine how many of them are complete. Although, ordinarily, we do not know the population proportion (the probability p that an application is complete), let us assume that we know it to be $p = 0.7$. The analyst wishes to find the probability that the sample proportion $\bar{p}$ will fall within 5 percent points of the population proportion $p = 0.7$; that is, that $\bar{p}$ will fall between 0.65 and 0.75.

Figure 13.3 presents the normal approximation to the sampling distribution of $\bar{p}$. The desired probability is shaded. The mean and standard deviation of the sampling distribution are, by (13.5):

$$E\{\bar{p}\} = 0.7 \qquad \sigma\{\bar{p}\} = \sqrt{\frac{0.7(0.3)}{200}} = 0.0324$$

FIGURE 13.3
Normal approximation to the binomial distribution for $n = 200, p = 0.7$—Insurance applications example

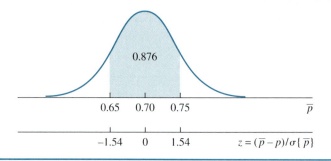

The z values therefore are:

$$z = \frac{0.65 - 0.70}{0.0324} = -1.54 \qquad z = \frac{0.75 - 0.70}{0.0324} = 1.54$$

The desired probability consequently is:

$$P(0.65 \leq \bar{p} \leq 0.75) = P(-1.54 \leq Z \leq 1.54) = 0.876$$

Comments

1. When the sample size is small, the normal approximation to the sampling distributions of f and $\bar{p}$ is improved by a *correction for continuity*. Consider again the insurance applications example. Suppose that the sample size now is only $n = 50$ and that, as before, $p = 0.7$. We wish to find the probability that 30 or more applications in the sample of 50 are complete.

    Our working rule in (13.6) indicates that the central limit theorem is applicable here because both $np = 50(0.7) = 35 \geq 5$ and $n(1 - p) = 50(0.3) = 15 \geq 5$. In using the normal distribution as an approximation to the binomial distribution for small samples, we need to recognize that we are utilizing a continuous probability distribution to approximate a discrete one. Figure 13.4 illustrates the basic problem. Figure 13.4a shows the binomial probability that $f = 30$. Since the normal distribution is continuous so that area under the normal curve corresponds to probability, we convert the discrete binomial probability to an area. We do so by treating $f = 30$ as extending from 29.5 to 30.5 and constructing a rectangle to represent the probability at $f = 30$, as shown in Figure 13.4b. Note that the area of the rectangle in Figure 13.4b is the same as the discrete probability in Figure 13.4a, since the height of the rectangle is the same and its width is 1.

    Thus, to find $P(f \geq 30)$, we will actually ascertain $P(f \geq 29.5)$ by the normal approximation, since $f = 30$ is assumed to go down to 29.5 when discrete probability is converted to area. The desired probability is shown by the shaded area in Figure 13.4c. We know from theorem (13.5) that the mean of the sampling distribution of f is $E\{f\} = np = 50(0.7) = 35$, and that the standard deviation is $\sigma\{f\} = \sqrt{np(1 - p)} = \sqrt{50(0.7)(0.3)} = 3.24$. Hence, the z value is, by (13.7a):

$$z = \frac{29.5 - 35}{3.24} = -1.70$$

We thus find that $P(f \geq 29.5) = P(Z \geq -1.70) = 0.955$. Incidentally, the true probability according to the binomial distribution is 0.952, so the normal approximation with the correction for continuity here is very close.

Normal approximation to the binomial distribution for $n = 50$, $p = 0.7$, using the correction for continuity—Insurance applications example. Here $f = 30$ is treated as extending from 29.5 to 30.5.

FIGURE 13.4

(a) Discrete Probability

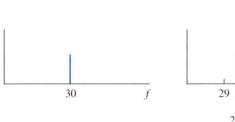

(b) Shift to Area

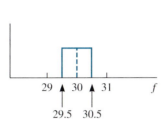

(c) Normal Approximation

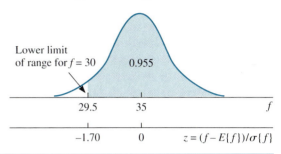

Using 29.5 rather than 30 in finding $P(f \geq 30)$ by means of the normal approximation is the correction for continuity. The correction is so named because it arises when a continuous distribution is used to approximate a discrete one. The correction for continuity becomes trivial when the sample size is large, and we shall ordinarily omit it.

2. When p is small and n is large, the binomial probability distribution can be approximated by the Poisson probability distribution (7.8), with $\lambda = np$ and the sample frequency f corresponding to x. As a working rule, the approximation is adequate when n is reasonably large and $np < 5$.

For example, suppose that we require $P(f = 0)$ for the binomial distribution with $n = 20$ and $p = 0.01$. Using the Poisson approximation, we have $\lambda = np = 20(0.01) = 0.20$. From the Poisson Table C.6 for $\lambda = 0.20$, we find $P(f = 0) = 0.8187$. The exact binomial probability from Table C.5 is $P(0) = 0.8179$, so the Poisson approximation is quite close even though n is not especially large here.

ESTIMATION OF 13.3 POPULATION PROPORTION

We frequently desire to estimate a population proportion by means of a confidence interval.

EXAMPLE ☐ **Mail Orders**

An executive of a retail firm wished to estimate the proportion p of the firm's current credit card customers who make regular use of the firm's mail-order services. A file check on a random sample of $n = 1240$ credit card customers showed that $f = 372$ of these customers used the firm's mail-order services regularly. Hence, the sample proportion is $\bar{p} = 372/1240 = 0.30$. A confidence interval for the population proportion is now desired.

☐

Confidence Interval for p—Large Sample

We consider now the construction of a confidence interval for the population proportion p. The procedure is applicable *when the sample size is large*. In that case, we know from central limit theorem (13.6) that the sampling distribution of $\bar{p}$ is approximately normal. We also know from (13.5) that the mean of the sampling distribution of $\bar{p}$ is $E\{\bar{p}\} = p$ and that the variance is $\sigma^2\{\bar{p}\} = p(1 - p)/n$.

Estimated Standard Deviation of $\bar{p}$.   To obtain confidence limits for p, we need an estimate of $\sigma\{\bar{p}\}$. This estimate can be easily obtained, since $\sigma\{\bar{p}\}$ depends only on p and n, and the sample proportion $\bar{p}$ is an unbiased point estimator of p. The estimated standard deviation of $\bar{p}$ is denoted by $s\{\bar{p}\}$ and is defined as follows.

(13.8)
The estimators of $\sigma^2\{\bar{p}\}$ and $\sigma\{\bar{p}\}$ are, respectively:

$$s^2\{\bar{p}\} = \frac{\bar{p}(1 - \bar{p})}{n - 1} \qquad s\{\bar{p}\} = \sqrt{\frac{\bar{p}(1 - \bar{p})}{n - 1}}$$

These estimators are appropriate for infinite populations, and also for finite populations as long as the sampling fraction n/N is 5 percent or less.

Development of Confidence Interval.   We utilize an extension of central limit theorem (13.6) for developing an approximate confidence interval for the population proportion p when n is large. This extension states that $(\bar{p} - p)/s\{\bar{p}\}$ follows approximately a standard normal distribution when the random sample size is sufficiently large. For finite populations, it is also required that the sampling fraction not be large.

Using the same reasoning as in finding the confidence limits for the population mean μ, we obtain analogous confidence limits for the population proportion p.

(13.9)
The approximate $1 - \alpha$ confidence limits for the population proportion p, when the random sample size is reasonably large, are:

$$\bar{p} \pm zs\{\bar{p}\}$$

where: $z = z(1 - \alpha/2)$
         $s\{\bar{p}\}$ is given by (13.8)

□ **EXAMPLE**

In the mail orders example, the sample results are $n = 1240$ and $\bar{p} = 0.30$. A 99 percent confidence interval for the population proportion p is desired. The sampling fraction is small here so that we can use (13.8):

$$s^2\{\bar{p}\} = \frac{0.30(1 - 0.30)}{1240 - 1} = 0.0001695$$

$$s\{\bar{p}\} = \sqrt{0.0001695} = 0.0130$$

For 99 percent confidence limits, we require $z(1 - \alpha/2) = z(0.995) = 2.576$. Thus, the confidence limits are $0.30 \pm 2.576(0.0130)$, and the 99 percent confidence interval is:

$$0.27 \le p \le 0.33$$

We can report, with 99 percent confidence, that between 27 and 33 percent of current credit card customers make regular use of the firm's mail-order services. ☐

Planning of Sample Size

Often the sample size for estimating a population proportion p is not predetermined and can therefore be chosen to provide a desired precision for a specified confidence coefficient.

EXAMPLE ☐ **Worker Location**

A commission is studying the spatial distribution of the labor force in a large metropolitan area. One parameter of interest is the proportion p of workers whose places of employment are within 15 miles of their residences. It is desired to select a random sample of workers of sufficient size to provide a 95 percent confidence interval for p with a half-width of $h = 0.02$. ☐

Assumptions. The planning procedure to be described assumes the following.

1. The random sample size ultimately determined is reasonably large.
2. The population is either infinite or, if finite, is large relative to the resulting sample size.

Planning Procedure. The procedure for planning the sample size to estimate the population proportion p is completely analogous to that for estimating the population mean μ. We need to specify the desired half-width of the confidence interval, denoted by h, and the specified confidence coefficient, denoted by $1 - \alpha$. We also need to specify a planning value for p. Often, this planning value can be based on past experience with the same or a similar problem.

(13.10)
The required random sample size for a specified half-width h, confidence coefficient $1 - \alpha$, and planning value p is:

$$n = \frac{z^2 p(1 - p)}{h^2}$$

where: $z = z(1 - \alpha/2)$
p is the planning value for the population proportion

EXAMPLE ☐

In the worker location example, a half-width of $h = 0.02$ and a confidence coefficient of 95 percent were specified. Hence, we require $z(1 - \alpha/2) = z(0.975) = 1.960$.

Based on other studies, it is expected that p is in the neighborhood of 0.9, so the planning value $p = 0.9$ will be used. Substituting into (13.10) yields:

$$n = \frac{(1.960)^2(0.9)(0.1)}{(0.02)^2} = 864$$

A random sample of 864 workers from the labor force should provide an interval estimate of p with approximately the desired precision for the specified confidence coefficient if the planning value for p is reasonably accurate.   ☐

Comments

1. In (13.10), the term $p(1 - p)$ takes on its largest value when $p = 0.5$, and it becomes smaller as p approaches 0 or 1. This implies that if no reliable planning value for p can be specified and the sample size must guarantee the required precision, the planning value for p should be set equal to 0.5 in computing n. This sample size may be quite conservative, however. In the worker location example, for instance, $n = 864$ is obtained when the specified planning value $p = 0.9$ is appropriate, whereas $n = 2401$ would be obtained if $p = 0.5$ were used as the planning value.

2. Formula (13.10) is obtained by noting from (13.9) that $h = z\sigma\{\bar{p}\}$. We use the notation for the true standard deviation $\sigma\{\bar{p}\}$ here because we are in the planning stage. We know from (13.5) that $\sigma^2\{\bar{p}\} = p(1 - p)/n$. Hence, it follows that:

$$h = z\sigma\{\bar{p}\} = z\sqrt{\frac{p(1 - p)}{n}}$$

Solving this equation for n yields the formula in (13.10).

13.4   TESTS FOR POPULATION PROPORTION

We now consider statistical tests concerning the population proportion p. The procedures to be described are applicable *when the simple random sample size is large.*

Tests about a population proportion p based on large samples are conducted in a similar manner as large-sample tests for the population mean μ. We shall let p_0 denote the level of p where the α risk is to be controlled. The standardized test statistic then has the usual form, except for a change in notation.

(13.11)
Tests concerning the population proportion p, when the random sample size is sufficiently large, are based on the standardized test statistic:

$$z^* = \frac{\bar{p} - p_0}{\sigma\{\bar{p}\}}$$

where: $\sigma\{\bar{p}\} = \sqrt{\frac{p_0(1 - p_0)}{n}}$

When $p = p_0$, test statistic z^* follows approximately the standard normal distribution.

Note that the denominator of z^* in (13.11) uses the standard deviation $\sigma\{\bar{p}\}$ and not its estimate $s\{\bar{p}\}$. The reason is that the standard deviation of the sampling distribution of $\bar{p}$ when $p = p_0$ is known because $\sigma\{\bar{p}\}$ depends only on n and p.

Decision rules for test statistic (13.11) are constructed in the usual fashion. We shall give an example involving a one-sided lower-tail test.

EXAMPLE ☐ **Fish Preference**

A British food company currently uses cod for fish and chips, the fish traditionally used in making the food item. Whiting, however, is a cheaper and more plentiful fish. Management wishes to switch to whiting to take advantage of the cost savings and better availability if 50 percent or more of consumers prefer whiting to cod.

Step 1. Let p denote the proportion of consumers who prefer whiting. The test alternatives are then:

$$H_0: \ p \geq 0.5$$
$$H_1: \ p < 0.5$$

Thus, the α risk is to be controlled at $p_0 = 0.5$.

Step 2. Management wishes to control the α risk at 0.05 when $p = p_0 = 0.5$. Figure 13.5 illustrates the situation. The appropriate type of decision rule is one in which small values of $\bar{p}$ lead to conclusion $H_1(p < 0.5)$, while values of $\bar{p}$ in the neighborhood of $p_0 = 0.5$ or larger lead to conclusion $H_0(p \geq 0.5)$. The sampling distribution of $\bar{p}$ is centered at $p_0 = 0.5$ because the α risk is to be controlled there. The tail area in the rejection region, corresponding to the α risk, must be equal to 0.05.

Step 3. Each subject in a random sample of $n = 265$ consumers is to be given a blind taste test of both types of fish prepared in the same fashion; $\bar{p}$ will denote the sample proportion who prefer whiting over cod. Because the sample size is large, the sampling distribution of $\bar{p}$ when $p = p_0 = 0.5$ is approximately normal, with mean $p_0 = 0.5$ and standard deviation:

$$\sigma\{\bar{p}\} = \sqrt{\frac{p_0(1 - p_0)}{n}} = \sqrt{\frac{0.5(1 - 0.5)}{265}} = 0.03071$$

Hence, the standardized test statistic in (13.11) is approximately a standard normal variable when $p = p_0$, and the action limit for the decision rule in Figure 13.5 corre-

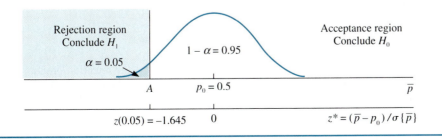

FIGURE 13.5
Statistical decision rule for large-sample, one-sided lower-tail test for p—Fish preference example

sponds to $z(\alpha) = z(0.05) = -1.645$. Thus, the appropriate decision rule for the test is as follows:

> If $z^* \geq -1.645$, conclude H_0.
> If $z^* < -1.645$, conclude H_1.

Step 4. Of the 265 consumers in the taste test, $f = 144$ preferred the whiting. Hence, $\bar{p} = 144/265 = 0.5434$. We noted in the previous step that $p_0 = 0.5$ and $\sigma\{\bar{p}\} = 0.03071$. The standardized test statistic therefore is:

$$z^* = \frac{\bar{p} - p_0}{\sigma\{\bar{p}\}} = \frac{0.5434 - 0.5}{0.03071} = 1.41$$

Since $z^* = 1.41 \geq -1.645$, the decision rule leads us to conclude H_0; that is, at least 50 percent of consumers prefer whiting to cod.

The test conclusion in this example can also be drawn from the P-value for the test. The one-sided P-value is equal to $P(Z < z^*) = P(Z < 1.41) = 0.9207$. Since $\alpha = 0.05$ and $0.9207 \geq 0.05$, we conclude H_0. This conclusion is necessarily the same as that using test statistic z^*.   ☐

13.5   OPTIONAL TOPIC—PLANNING SAMPLE SIZE FOR TEST ABOUT p

As with tests for the population mean μ, both the α and β risks of a test for the population proportion p can be controlled at desired levels by appropriate planning of the sample size. Before explaining how to plan the sample size, we first need to consider how to evaluate the β risks of tests for a population proportion.

Power of Test

To examine the exposure to β risks of a decision rule concerning the population proportion p, we need to obtain the rejection probability curve for the decision rule. Rejection probability curves for tests on p are obtained in similar fashion as those for tests on μ discussed in Section 12.6. The procedure to be explained continues to assume that *the simple random sample size is large*.

Calculation of Rejection Probabilities.   We first consider how to calculate rejection probabilities $P(H_1; p)$—that is, probabilities that the decision rule will lead to conclusion H_1 for given values of the population proportion p. There are no new principles involved, and we shall illustrate the calculation of rejection probabilities only for the one-sided lower-tail test in the fish preference example. Rejection probabilities for one-sided upper-tail and two-sided tests are calculated in an analogous fashion.

☐ **EXAMPLE**

In the fish preference example, management wishes to know the rejection probability when $p = 0.35$ for the decision rule illustrated in Figure 13.5. Figure 13.6 shows the method of calculation. Figure 13.6a reproduces the decision rule from Figure 13.5. We first need to determine the position of the action limit A that controls the α risk at 0.05

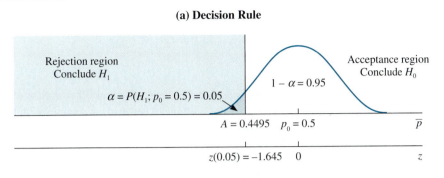

(a) Decision Rule

Rejection region
Conclude H_1

Acceptance region
Conclude H_0

$1 - \alpha = 0.95$

$\alpha = P(H_1; p_0 = 0.5) = 0.05$

$A = 0.4495 \quad p_0 = 0.5 \qquad \overline{p}$

$z(0.05) = -1.645 \quad 0 \qquad z$

FIGURE 13.6
Calculation of rejection probabilities for one-sided lower-
tail test—Fish preference
example

(b) Calculation of Rejection Probability at $p = 0.35$

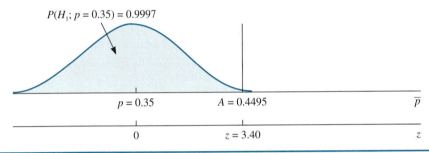

$P(H_1; p = 0.35) = 0.9997$

$p = 0.35 \qquad A = 0.4495 \qquad \overline{p}$

$0 \qquad z = 3.40 \qquad z$

when $p = p_0 = 0.5$. We found earlier that $\sigma\{\overline{p}\} = 0.03071$ when $p_0 = 0.5$. Since $z(0.05) = -1.645$, it follows that the action limit in Figure 13.6a equals:

$$A = p_0 + z(0.05)\sigma\{\overline{p}\} = 0.5 - 1.645(0.03071) = 0.4495$$

When the population proportion is $p = 0.35$, the sampling distribution of $\overline{p}$ is centered at $p = 0.35$, as shown in Figure 13.6b. The standard deviation of this distribution is:

$$\sigma\{\overline{p}\} = \sqrt{\frac{p(1 - p)}{n}} = \sqrt{\frac{0.35(1 - 0.35)}{265}} = 0.02930$$

Observe that $\sigma\{\overline{p}\}$ must be recomputed for each different value of p because $\sigma\{\overline{p}\}$ depends mathematically on p.

The area under the sampling distribution that lies in the rejection region is the desired rejection probability. This area is shaded in Figure 13.6b. We denote this area by $P(H_1; p = 0.35)$. Since the action limit $A = 0.4495$ is located $z = (0.4495 - 0.35)/0.02930 = 3.40$ standard deviations from $p = 0.35$, it follows that the rejection probability at $p = 0.35$ for the decision rule is:

$$P(H_1; p = 0.35) = P(Z < 3.40) = 0.9997$$

Thus, management is assured that the decision rule would almost certainly lead to conclusion H_1 if p were, in fact, equal to 0.35. ☐

Rejection Probability Curve.   After several rejection probabilities have been calculated, the rejection probability curve can be plotted in the usual fashion.

□ EXAMPLE

In the fish preference example, rejection probabilities for several other values of p were also calculated. In each case, $\sigma\{\bar{p}\}$ was recomputed because $\sigma\{\bar{p}\}$ varies with the value of p. These rejection probabilities are plotted as the rejection probability curve in Figure 13.7. Note that the point corresponding to $P(H_1; p = 0.35) = 0.9997$ is shown on the curve. The point at which the α risk is controlled is also shown on the curve. This corresponds to $P(H_1; p = 0.5) = 0.05$. We can see from the rejection probability curve that for any value of p less than about 0.40 the probability of concluding H_1 (and, hence, of rejecting H_0) is quite high and the β risk is quite low, indicating that the test is powerful for any value of p in this range. In other words, if only 40 percent or fewer of the consumers prefer whiting, the decision rule is almost certain to lead to the correct conclusion.   □

Comment

In a test for μ, we must have knowledge of the population standard deviation σ to construct the rejection probability curve. In contrast, in a test for p, the rejection probability curve can be constructed on the basis of the sample size n alone, as we have just illustrated, since the standard deviation $\sigma\{\bar{p}\}$ for a given value of p depends only on p and n.

FIGURE 13.7
Rejection probability curve
for one-sided lower-tail test—
Fish preference example

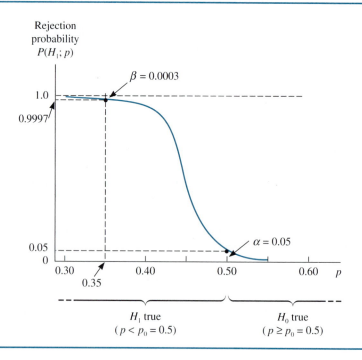

Planning of Sample Size

The procedure for determining the sample size for a test of the population proportion p that will keep both α and β risks at prespecified levels is similar to that for a test of the population mean μ.

Assumptions. The planning procedure assumes the following.

1. The random sample size ultimately determined is reasonably large.
2. The population is either infinite or, if finite, is large relative to the resulting sample size.

Planning Procedure. We shall continue to denote the value of p where the α risk is controlled by p_0. The value of p where the β risk is controlled shall be denoted by p_1. We shall denote the z values associated with the α and β risks by z_0 and z_1, respectively, as we did previously in planning the sample size for tests on μ.

The sample size formula, which is applicable to all three types of tests, follows.

(13.12)
The required random sample size to control both the α and β risks is:

$$n = \frac{\left[|z_1|\sqrt{p_1(1 - p_1)} + |z_0|\sqrt{p_0(1 - p_0)}\right]^2}{|p_1 - p_0|^2}$$

where: p_0 and p_1 are the values of p where the α and β risks are controlled, respectively

z_0 and z_1 are the z values associated with the specified α and β risks, respectively, and are defined as follows for each type of test:

One-Sided Upper-Tail Test $(H_0\colon p \leq p_0, H_1\colon p > p_0)$
$z_0 = z(1 - \alpha) \qquad z_1 = z(\beta)$

One-Sided Lower-Tail Test $(H_0\colon p \geq p_0, H_1\colon p < p_0)$
$z_0 = z(\alpha) \qquad z_1 = z(1 - \beta)$

Two-Sided Test $(H_0\colon p = p_0, H_1\colon p \neq p_0)$
$z_0 = z(1 - \alpha/2) \qquad z_1 = z(\beta)$

We illustrate the use of formula (13.12) by an example.

EXAMPLE ☐

In the fish preference example, suppose that the sample size has not yet been determined.

Step 1. We continue to specify that the α risk is to be controlled at 0.05 when $p_0 = 0.5$, as illustrated by the sampling distribution of $\bar{p}$ on the right side of Figure 13.8a.

Step 2. Management has decided that if only 40 percent of consumers prefer whiting, the probability of concluding H_0 $(p \geq 0.5)$ is to be controlled at 0.05. Thus, the β

FIGURE 13.8
Determination of the required sample size to control α and β risks—Fish preference example

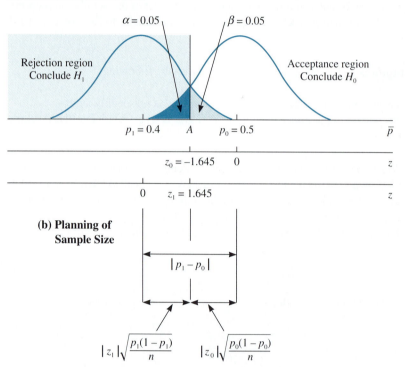

(a) Statistical Decision Rule

(b) Planning of Sample Size

risk is to be controlled at 0.05 when $p_1 = 0.4$, as illustrated by the sampling distribution of $\bar{p}$ on the left side of Figure 13.8a.

Step 3. Referring to formula (13.12) for a one-sided lower-tail test, we require:

$$z_0 = z(\alpha) = z(0.05) = -1.645 \qquad \text{so } |z_0| = 1.645$$

$$z_1 = z(1 - \beta) = z(1 - 0.05) = z(0.95) = 1.645 \qquad \text{so } |z_1| = 1.645$$

$$|p_1 - p_0| = |0.4 - 0.5| = 0.1$$

$$\sqrt{p_1(1 - p_1)} = \sqrt{0.4(0.6)} = 0.490$$

$$\sqrt{p_0(1 - p_0)} = \sqrt{0.5(0.5)} = 0.500$$

Substituting into (13.12), we obtain:

$$n = \frac{[1.645(0.490) + 1.645(0.500)]^2}{(0.1)^2} = 265$$

This is, in fact, the sample size used in the study.

Comment

The derivation of sample size formula (13.12) follows the same pattern as for tests on the population mean μ. The main difference from before is that the standard deviation $\sigma\{\bar{p}\}$ of the sampling distribution of $\bar{p}$ depends on p. Thus, the standard deviation $\sigma\{\bar{p}\}$ is as follows when $p = p_0$:

$$\sigma\{\bar{p}\} = \sqrt{\frac{p_0(1 - p_0)}{n}}$$

while it is as follows when $p = p_1$:

$$\sigma\{\bar{p}\} = \sqrt{\frac{p_1(1 - p_1)}{n}}$$

Looking at Figures 13.8a and 13.8b together, we see that the interval from p_1 to p_0 has length $|p_1 - p_0|$. This interval has two components. First, the component between p_1 and the action limit A has a length equal to:

$$|z_1|\sigma\{\bar{p}\} = |z_1|\sqrt{\frac{p_1(1 - p_1)}{n}}$$

Second, the component between the action limit A and p_0 has a length equal to:

$$|z_0|\sigma\{\bar{p}\} = |z_0|\sqrt{\frac{p_0(1 - p_0)}{n}}$$

The sum of these two components equals $|p_1 - p_0|$, as shown in Figure 13.8b. Setting these two expressions equal to each other and solving for n yields formula (13.12).

OPTIONAL TOPIC—ONE-SIDED CONFIDENCE INTERVALS FOR p 13.6

One-sided confidence intervals for the population proportion p can be readily constructed, analogous to their construction for the population mean μ. We continue to assume that *the random sample size is large.*

EXAMPLE

An auditor has selected a random sample of $n = 400$ accounts from a population of 10,000 accounts and found that $f = 51$ were overdue. The auditor wishes to obtain a 98 percent upper confidence interval for the proportion of accounts that are overdue. The sampling fraction here is small. Hence, we obtain:

$$\bar{p} = \frac{51}{400} = 0.1275 \qquad s^2\{\bar{p}\} = \frac{0.1275(0.8725)}{400 - 1} = 0.000279$$

$$z(0.98) = 2.054 \qquad s\{\bar{p}\} = 0.0167$$

The desired upper confidence limit therefore is $0.1275 + 2.054(0.0167)$, and the 98 percent upper confidence interval is $p \leq 0.162$. Hence, we can conclude with 98 percent confidence that the proportion of accounts that are overdue does not exceed 0.162.

☐

13.7   OPTIONAL TOPIC—INFERENCES FOR p WITH SMALL SAMPLE

The inference methods for a population proportion p presented in the preceding sections are valid for large random samples and cannot be used if the sample size is small. In this section we present interval estimation and test procedures for p that are valid for small samples. The procedures are exact when the population is infinite, and they can also be used for finite populations provided the sampling fraction is small.

We shall use the following example to illustrate the procedures.

Bond Issues   ☐ **EXAMPLE**

A financial analyst studied new U.S. corporate bond issues sold during the past five years. One characteristic of interest is the proportion p of the issues sold at a premium—that is, issues where the price exceeded the face value of the bond. The analyst selected a random sample of $n = 15$ new issues during the period and determined that $f = 6$ of these were sold at a premium. Hence, the sample proportion is $\bar{p} = 6/15 = 0.40$.

☐

Confidence Limits for p

The small-sample confidence limits for the population proportion p involve the sample frequency f, the sample size n, and percentiles of the F distribution. The F distribution, a continuous, unimodal distribution that is positively skewed, is related to the binomial distribution. (A detailed discussion of the F distribution may be found in Appendix B, Section B.3.)

(13.13)

The $1 - \alpha$ confidence limits L and U for the population proportion p are:

$$L = \frac{f}{f + (n - f + 1)F_1} \qquad U = \frac{(f + 1)F_2}{(f + 1)F_2 + (n - f)}$$

where: $F_1 = F[1 - \alpha/2; 2(n - f + 1), 2f]$
$F_2 = F[1 - \alpha/2; 2(f + 1), 2(n - f)]$

These confidence limits are appropriate for any sample size, small or large. When the sample size is large, however, the large-sample inference procedures discussed in previous sections are easier to utilize.

☐ **EXAMPLE**

In the bond issues example, the financial analyst wishes to construct a 90 percent confidence interval for p. The sample results are $n = 15$ and $f = 6$, and $1 - \alpha = 0.90$. Consulting Table C.4 for the F distribution, we find:

$$F_1 = F[0.95; 2(15 - 6 + 1), 2(6)] = F(0.95; 20, 12) = 2.54$$

$$F_2 = F[0.95; 2(6 + 1), 2(15 - 6)] = F(0.95; 14, 18) = 2.29$$

The latter value can be obtained by interpolation in Table C.4 or from many statistical calculators or statistical packages. Substituting into (13.13) gives:

$$L = \frac{6}{6 + (15 - 6 + 1)(2.54)} = 0.19$$

$$U = \frac{(6 + 1)(2.29)}{(6 + 1)(2.29) + (15 - 6)} = 0.64$$

Therefore, the desired confidence interval for the population proportion of bond issues sold at a premium is:

$$0.19 \leq p \leq 0.64$$

The analyst can conclude, with a confidence level of 90 percent, that the population proportion of new issues selling at a premium is between 19 and 64 percent. These limits are quite wide because the sample size is so small. Note that the point estimate $\bar{p} = 0.40$ gives no indication of this lack of precision. □

Comments

1. Unlike the large-sample confidence limits for p in (13.9), the exact confidence limits for p in (13.13) are not equally spaced about $\bar{p}$. In the bond issues example, for instance, the limits 0.19 and 0.64 are not equidistant from $\bar{p} = 0.40$.
2. The confidence coefficient for the confidence limits in (13.13) is stated to be $1 - \alpha$. In fact, $1 - \alpha$ is the guaranteed confidence level. The actual confidence level may exceed $1 - \alpha$ because of the discrete nature of the sample frequency f.

Tests for p

When the sample size is small, tests concerning the population proportion p are most easily carried out by utilizing the relations between confidence intervals and tests described in Section 12.5. Thus, to conduct a two-sided test for p when the sample size is small, we shall utilize the confidence limits in (13.13) together with the following decision rule.

(13.14)
When the alternatives are:

$$H_0: p = p_0$$
$$H_1: p \neq p_0$$

the appropriate decision rule to control the α risk at p_0 is as follows:

If $L \leq p_0 \leq U$, conclude H_0.
If $p_0 < L$ or $p_0 > U$, conclude H_1.

where L and U are given by (13.13).

☐ EXAMPLE

In the bond issues example, the analyst wishes to test if the proportion of U.S. bond issues sold at a premium is equal to the corresponding proportion in Europe for the same period, which is known from an exhaustive study by a European researcher to be 0.75. The test alternatives therefore are:

$$H_0:\ p = 0.75$$
$$H_1:\ p \neq 0.75$$

where $p_0 = 0.75$. The analyst wishes to control the α risk at 0.10 when $p = p_0 = 0.75$. Hence, 90 percent confidence limits are required. The appropriate values of L and U in (13.13) were calculated previously for this example. They were $L = 0.19$ and $U = 0.64$. Applying the decision rule in (13.14), we see that $p_0 = 0.75 > U = 0.64$. Therefore, the analyst should conclude H_1, namely, that the U.S. population proportion of bond issues sold at a premium differs from the European proportion 0.75. Indeed, since $\bar{p} = 0.40$, it appears that the U.S. population proportion is lower than 0.75.

☐

Comment

One-sided confidence intervals and one-sided tests for p can be constructed from the individual confidence limits in (13.13).

PROBLEMS

13.1 Information Systems Specialists. A random sample of $n = 9$ information systems specialists will be selected to ascertain whether they use computer virus-detection software on a regular basis. Let B_i be a Bernoulli random variable coded 1 if the ith specialist uses detection software regularly and coded 0 otherwise.

a. What does the Bernoulli parameter p denote here?
b. Let $\bar{p}$ and f be defined as in (13.2). (1) What does f denote here, and what possible values can it take on? (2) What does $\bar{p}$ denote here, and what possible values can it take on? (3) Specify the relation between f and $\bar{p}$ here.

13.2 Invoice Payments. Consider a random sample of $n = 8$ invoices. Let B_i be a Bernoulli random variable coded 1 if the ith invoice is not paid within 10 days and coded 0 otherwise.

a. What does the Bernoulli parameter p denote here?
b. Let $\bar{p}$ and f be defined as in (13.2). (1) What does f denote here, and what possible values can it take on? (2) What does $\bar{p}$ denote here, and what possible values can it take on? (3) Specify the relation between f and $\bar{p}$ here.

13.3 Customers. Consider a sample of $n = 10$ persons who enter a department store. Let f denote the number of persons in the sample who make a purchase.

a. (1) What does $\bar{p}$, as defined in (13.2), represent here? (2) What values can f take on here? (3) What values can $\bar{p}$ take on here?
b. Under what assumptions is f a binomial random variable here? Explain.

* **13.4** Refer to **Information Systems Specialists** Problem 13.1. Assume that f is a binomial random variable with $n = 9$ and $p = 0.9$.

a. Obtain the sampling distributions of f and $\bar{p}$.
b. Plot the two sampling distributions in one graph as in Figure 13.1.

c. Obtain (1) $P(f = 9)$, (2) $P(\bar{p} = 1)$, (3) $P(\bar{p} \geq 7/9)$.

d. Obtain the means and the standard deviations of the sampling distributions of f and $\bar{p}$ using (13.5).

13.5 Refer to **Invoice Payments** Problem 13.2. Assume that f is a binomial random variable with $n = 8$ and $p = 0.2$.

a. Obtain the sampling distributions of f and $\bar{p}$.

b. Plot the two sampling distributions in one graph as in Figure 13.1.

c. Obtain (1) $P(f \leq 2)$, (2) $P(\bar{p} = 1/8)$, (3) $P(\bar{p} \geq 1/8)$.

d. Obtain the means and the standard deviations of the sampling distributions of f and $\bar{p}$ using (13.5).

13.6 Refer to **Customers** Problem 13.3. Assume that f is a binomial random variable with $n = 10$ and $p = 0.3$.

a. Obtain the sampling distributions of f and $\bar{p}$.

b. Plot the two sampling distributions in one graph as in Figure 13.1.

c. Obtain (1) $P(f = 1)$, (2) $P(\bar{p} = 0.40)$, (3) $P(\bar{p} \leq 0.20)$.

d. Obtain the means and the standard deviations of the sampling distributions of f and $\bar{p}$ using (13.5).

13.7 For each of the following cases, indicate whether working rule (13.6) permits the use of a normal approximation to the binomial distribution.

a. $n = 100, p = 0.2$ **b.** $n = 30, p = 0.15$

c. $n = 30, p = 0.90$ **d.** $n = 20, p = 0.60$

13.8 A defense consultant will select 20 random moments in time to spot check whether a scrambling device for military telephone messages is in use $(B = 1)$ or not $(B = 0)$.

a. What does $\bar{p}$ represent here?

b. Will working rule (13.6) permit use of a normal approximation to the sampling distribution of $\bar{p}$ if the device is actually in use (1) 20 percent of the time, (2) 80 percent of the time, (3) 50 percent of the time? Explain.

13.9 A tire repair shop has 1000 used tires in a storage area. A random sample of 40 of the tires is selected without replacement to ascertain the proportion of the 1000 tires that have some resale value.

a. What is the sampling fraction here?

b. Do the formulas for the mean and the variance of f and $\bar{p}$ in (13.5) still apply here to a reasonable approximation, according to the working rule stated in the text? Comment.

c. If, in fact, 200 of the 1000 used tires have some resale value, will the sampling distributions of f and $\bar{p}$ be approximately normally distributed? Explain.

* **13.10** From a large population of residential water meters, a random sample of 250 meters is to be selected to estimate the proportion due for replacement within the next five years. Suppose that $p = 0.40$.

a. Obtain the mean and the standard deviation of the sampling distribution of $\bar{p}$.

b. Using a normal approximation, find the probability that (1) $\bar{p}$ is less than 0.35, (2) $\bar{p}$ lies within ± 0.05 of the population proportion p.

c. Calculate the probability in part b(2) for a sample of size 1000. What is the effect of the larger sample size?

13.11 A nationwide survey of 400 food stores selected at random is to be conducted to estimate the proportion of all food stores carrying a particular brand of curry powder. Suppose that $p = 0.30$.

a. Obtain the mean and the standard deviation of the sampling distribution of $\bar{p}$.

b. Using a normal approximation, find (1) the probability that $\bar{p}$ will exceed 0.35, (2) the interval centered about p within which $\bar{p}$ will fall with probability 0.95.

c. Calculate the interval in part b(2) for a sample of size 900. What is the effect of the larger sample size?

13.12 A public opinion survey based on a random sample of 1000 persons from the population of a large province is to be conducted to estimate the proportion of the population who believe that business conditions will worsen in the next 12 months. Using a normal approximation, find the probability that the proportion of persons in the sample who believe that business conditions will worsen is within ± 0.03 of the population proportion p if (1) $p = 0.20$, (2) $p = 0.50$, (3) $p = 0.80$. For which of these three values of p is the calculated probability smallest?

13.13 The probability that a robotized machine is in an unproductive state is $p = 0.20$. Let f denote the number of times the machine is in an unproductive state when it is observed at 50 random moments in time.

a. Verify that working rule (13.6) permits use of a normal approximation here.
b. Obtain the mean and the standard deviation of the sampling distribution of f.
c. Using a normal approximation and a correction for continuity, obtain (1) $P(f \geq 15)$, (2) $P(f < 4)$, (3) $P(f = 10)$.
d. Using the results in part c, obtain (1) $P(\bar{p} \geq 0.30)$, (2) $P(\bar{p} < 0.08)$, (3) $P(\bar{p} = 0.20)$.

13.14 The probability of a delayed shipment is $p = 0.10$. Let f denote the number of delayed shipments in a random sample of $n = 70$ shipments.

a. Verify that working rule (13.6) permits use of a normal approximation here.
b. Obtain the mean and the standard deviation of the sampling distribution of f.
c. Using a normal approximation and a correction for continuity, obtain (1) $P(f \leq 1)$, (2) $P(f > 7)$, (3) $P(f = 7)$.
d. Using the results in part c, obtain (1) $P(\bar{p} \leq 1/70)$, (2) $P(\bar{p} > 0.10)$, (3) $P(\bar{p} = 0.10)$.

13.15 A bank has many small consumer loans outstanding, of which one percent are technically in default. Using a Poisson approximation to the binomial distribution, find the probability that an auditor who examines a random sample of 300 small consumer loans will find (1) exactly one in default, (2) five or more in default.

13.16 A computer simulation model of a complex communications network is run for 500 independent trials, each trial being a simulation of the network's operation for a hypothetical 10-year period. The probability of a major network failure on any one trial is $p = 0.005$. Using a Poisson approximation to the binomial distribution, find the probability that (1) exactly 2 failures occur among the 500 trials, (2) 2 or more failures occur among the 500 trials.

* **13.17 Savings Transfers.** A savings bank noted that 180 of the 225 new accounts opened last month involved a transfer of savings from an account at another bank. It is desired to estimate the process proportion p of new accounts at this bank that involve transfers from another bank. Assume that the data for the 225 new accounts constitute a random sample from the process.

a. Calculate $\bar{p}$ and $s\{\bar{p}\}$. What does the latter estimate here?
b. Obtain a 95 percent confidence interval for the process proportion p. Interpret the confidence interval.
c. An executive of this savings bank had claimed that three-quarters of all new accounts involve a transfer from another bank. Does the confidence interval in part b suggest that the executive's claim is plausible? Comment.

13.18 Hair-Conditioning Product. In a random sample of $n = 400$ persons from the population of persons older than 40 years of age, 65 percent were found to use some hair-conditioning product regularly. It is desired to estimate the population proportion p of persons older than 40 who use some hair-conditioning product regularly.

a. Estimate the standard deviation of the sampling distribution of $\bar{p}$.
b. Construct a 95 percent confidence interval for the population proportion p. Interpret the confidence interval.
c. Why is the confidence coefficient of the confidence interval in part b only approximately 95 percent?

13.19 Chronic Illness. In a random sample of $n = 1000$ construction workers, it was found that 153 had a chronic illness. Construct a 90 percent confidence interval for the population proportion of construction workers with a chronic illness. Interpret the confidence interval.

13.20 In a shipment of 1400 ornamental glass globes received from a vendor, 28 were flawed.

a. Assume that the globes in the shipment constitute a random sample from the vendor's current production process. Construct a 95 percent confidence interval for the process proportion of flawed globes. Interpret the confidence interval.

b. An investigation showed that 700 globes in the shipment were produced in one plant, of which 25 were flawed, and the other 700 globes were produced in a second plant, of which 3 were flawed. What are the implications of this finding for the validity of the confidence interval in part a?

* **13.21** A sample survey of persons with a certain type of arthritis is to be conducted to estimate the population proportion of persons who are taking a new antiarthritic drug. A 90 percent confidence interval is desired with a half-width of 0.02. Preliminary sales data for the new drug suggest that a planning value of $p = 0.10$ is reasonable.

a. Determine the required sample size.
b. What would be the required sample size if a planning value of $p = 0.12$ were more appropriate? Is this required sample size substantially different from that determined in part a? Explain.

13.22 A graduate business school wishes to estimate the proportion of its graduates who have reached executive positions. A random sample of graduates is to be selected. The population proportion is to be estimated, using a 90 percent confidence interval with a half-width of 0.05. A reasonable planning value, based on information from other graduate schools, is $p = 0.6$.

a. How many graduates should be selected for the sample?
b. If the specified half-width were to be 0.025 instead of 0.05, what sample size would be required? Is this required sample size substantially different from that determined in part a? Explain.

13.23 A lathe operator is to be observed at random times to estimate the proportion of times the operator is in a productive state. It is desired to estimate this process proportion within ± 0.04, with a 95 percent confidence interval. It is expected that the process proportion is 0.9 or greater.

a. What sample size will be large enough to ensure that the estimate has the desired precision?
b. It was finally decided to take $n = 200$ random observations. The lathe operator was found to be in a productive state in 176 of these observations. Construct a 95 percent confidence interval for the process proportion. Interpret the confidence interval.

13.24 As part of a study of high school graduates in a city, a random sample of graduating students is to be selected to estimate the proportion who expect to enter the labor force upon graduation. It is desired to estimate the population proportion within ± 0.05, with a 90 percent confidence interval. No reliable planning value for p is available.

a. What sample size will be large enough to ensure that the estimate has the desired precision?
b. It was finally decided to include 150 students in the study. Of these, 65 stated that they expect to enter the labor force upon graduation. Construct a 90 percent confidence interval for the population proportion. Does the confidence interval provide an estimate of the proportion of graduating students who actually will enter the labor force upon graduation? Discuss.

13.25 If the planning value p used in formula (13.10) for calculating the needed sample size is incorrect, does this affect the validity of the confidence interval obtained from the sample results? Does it affect the precision of the confidence interval? Comment.

13.26 Explain why the estimated standard deviation $s\{\bar{p}\}$ in (13.8) is not used for calculating the standardized test statistic z^* in (13.11).

* **13.27 Serviceable Items.** A large stock of items has been stored under unfavorable conditions. If fewer than 70 percent of the items are serviceable, the stock should be scrapped. It is desired to test whether the proportion of items that are serviceable is less than 0.70. The α risk is to be controlled at 0.01 when $p = 0.70$. A random sample of 100 items from the stock was selected, and 65 of these items were found to be serviceable.

a. Conduct the appropriate test. State the alternatives, the decision rule, the value of the test statistic, and the conclusion.
b. Why is the α risk controlled only approximately at 0.01 when $p = 0.70$?
c. Calculate the P-value of the test. Is it a one-sided or two-sided P-value?

13.28 Refer to **Chronic Illness** Problem 13.19. In a related occupation, 13 percent of the workers have a chronic illness. It is desired to test whether the population proportion of construction workers with a chronic illness is greater than 0.13. The α risk is to be controlled at 0.05 when $p = 0.13$.

a. Conduct the appropriate test. State the alternatives, the decision rule, the value of the test statistic, and the conclusion.

b. Calculate the P-value of the test. Is this value consistent with the test result in part a? Explain.

13.29 Divorces and Annulments. Nationally, the proportion of marriages ending in divorce or annulment that involve no children is 0.40. In one state, a random sample of 1000 divorces and annulments showed that 437 involved no children. It is desired to test whether the state proportion is equal to 0.40. The α risk is to be controlled at 0.01 when $p = 0.40$.

a. Conduct the appropriate test. State the alternatives, the decision rule, the value of the test statistic, and the conclusion.

b. What is a Type II error in this test situation? Is it possible that the conclusion in part a has resulted in a Type II error? Explain.

c. Calculate the P-value of the test. Is this value consistent with the test result in part a?

13.30 Shoes. In the past, 5 percent of the shoes produced in a plant have had small defects and were classified as irregular. Management has modified the production process and now wishes to test whether the process proportion of shoes that are irregular has decreased from the past level. A random sample of 400 shoes produced under the modified process was selected, and 16 of these were classified as irregular. The α risk is to be controlled at 0.05 when $p = 0.05$.

a. Conduct the appropriate test. State the alternatives, the decision rule, the value of the test statistic, and the conclusion.

b. Calculate the P-value of the test. Is this value consistent with the test result in part a? Explain.

* **13.31** Refer to **Serviceable Items** Problem 13.27.

a. Restate the decision rule in terms of $\bar{p}$.

b. Obtain the rejection probability $P(H_1; p)$ for each of the following values of p: (1) 0.50, (2) 0.60, (3) 0.70. Why is it necessary to recalculate $\sigma\{\bar{p}\}$ for each value of p?

c. Plot the rejection probability curve for the decision rule in part a. What is the probability that this decision rule will lead to a correct conclusion if $p = 0.60$?

13.32 Refer to **Chronic Illness** Problems 13.19 and 13.28.

a. Restate the decision rule in Problem 13.28a in terms of $\bar{p}$.

b. Obtain the rejection probability $P(H_1; p)$ for each of the following values of p: (1) 0.11, (2) 0.13, (3) 0.15, (4) 0.17.

c. Plot the rejection probability curve for the decision rule in part a. What is the risk of an incorrect conclusion with this decision rule if $p = 0.15$?

13.33 Refer to **Divorces and Annulments** Problem 13.29.

a. Restate the decision rule in terms of $\bar{p}$.

b. Obtain the rejection probability $P(H_1; p)$ for each of the following values of p: (1) 0.35, (2) 0.40, (3) 0.45. Why is it necessary to recalculate $\sigma\{\bar{p}\}$ for each value of p?

c. Plot the rejection probability curve for the decision rule in part a. What is the β risk for this decision rule at $p = 0.45$?

13.34 Refer to **Shoes** Problem 13.30.

a. Restate the decision rule in terms of $\bar{p}$.

b. Obtain the rejection probability $P(H_1; p)$ for each of the following values of p: (1) 0.02, (2) 0.03, (3) 0.05.

c. Plot the rejection probability curve for the decision rule in part a. What is the probability that this decision rule will lead to a correct conclusion if $p = 0.02$?

* **13.35** An auditor will test the quality of a large population of sales invoices by examining a simple random sample of invoices. The auditor will consider the quality of the population as satisfactory

(H_0) if one percent or less of the invoices in the population have errors and unsatisfactory (H_1) otherwise. It is desired to control the α risk at 0.05 when $p = 0.01$ and the β risk at 0.05 when $p = 0.03$.

a. What random sample size should be employed?
b. A random sample of the size determined in part a was selected, and the sample proportion of invoices with errors was found to be $\bar{p} = 0.028$. What test conclusion should be reached?

13.36 A field crop was sprayed for stem rust. Two weeks later, an agronomist plans to examine a random sample of plants from the field for rust. Let p denote the proportion of plants in the field with rust. The test alternatives of interest to the agronomist are $H_0: p \geq 0.15$, $H_1: p < 0.15$. It is desired to control the α risk at 0.05 when $p = 0.15$ and the β risk at 0.05 when $p = 0.10$.

a. How many plants should be selected for the agronomist's random sample?
b. The agronomist used the sample size determined in part a and found that the proportion of plants with rust was $\bar{p} = 0.113$. What is the appropriate test conclusion?
c. Why are the α and β risks only approximately equal to 0.05 at the specified levels of p in the test performed in part b?

13.37 A political analyst is planning a random sample of eligible voters to ascertain their voting intentions in a forthcoming election between incumbent candidate A and opposition candidate B. Let p denote the population proportion of voters who would currently express an intention to vote for A. The analyst wishes to test $H_0: p = 0.50$ versus $H_1: p \neq 0.50$. It is desired to control the α risk at 0.01 when $p = 0.50$ and the β risk at 0.01 when $p = 0.45$.

a. How many voters should be selected for the random sample?
b. The analyst used the sample size determined in part a and found that the proportion of voters preferring A was $\bar{p} = 0.483$. What test conclusion should be reached?

13.38 The purchaser of a large shipment of tiles wishes to control the β risk of accepting the shipment at 0.05 when the proportion of damaged tiles is $p = 0.07$. At the same time, the vendor wishes to control the α risk of having the shipment rejected at 0.05 when the proportion of damaged tiles is $p = 0.03$. A random sample of tiles will be selected from the shipment by the purchaser, on the basis of which a decision will be made whether to accept ($H_0: p \leq 0.03$) or reject ($H_1: p > 0.03$) the shipment.

a. What random sample size is necessary to meet both the purchaser's and the vendor's specifications?
b. A random sample of the size determined in part a was selected, and the sample proportion of damaged tiles was $\bar{p} = 0.065$. What test conclusion should be reached?

* 13.39 Refer to **Savings Transfers** Problem 13.17. Construct a 95.0 percent lower confidence interval for the process proportion. Interpret the confidence interval.

13.40 Refer to **Hair-Conditioning Product** Problem 13.18. Construct a 90 percent lower confidence interval for the population proportion. Interpret the confidence interval.

13.41 Refer to **Chronic Illness** Problem 13.19. Construct a 99 percent upper confidence interval for the population proportion. Interpret the confidence interval.

* 13.42 An auditor examined a simple random sample of 15 accounts from a large population of accounts receivable and discovered that one account in the sample was in error. Estimate the error rate in the population with an exact 98 percent confidence interval. Interpret the confidence interval.

13.43 Refer to **Savings Transfers** Problem 13.17.

a. Obtain an exact 95 percent confidence interval for the process proportion p.
   [*Hint:* $F(0.975; 92, 360) = 1.36$ and $F(0.975; 362, 90) = 1.41$.]
b. Compare the exact confidence interval in part a with that in Problem 13.17b based on the normal approximation to the binomial distribution. Is the normal approximation adequate here?

13.44 A political scientist conducted in-depth interviews with a random sample of 16 lawyers and learned that 5 favor the present system of judicial appointment.

a. Construct an exact 90 percent confidence interval for the population proportion of lawyers in favor of the present system of appointment.

b. Use the confidence interval in part a to test whether the population proportion of lawyers who favor the present system of judicial appointment is 0.50. State the alternatives, the decision rule, and the conclusion. What is the implied α risk of the test?

13.45 In the recent past, 40 percent of new business ventures have failed within two years. A random sample of 12 entrepreneurs participated in a small-business management workshop prior to engaging in new business ventures. Subsequent tracking of the entrepreneurs' ventures revealed that only 1 of the 12 ventures started by the entrepreneurs failed within the first two years.

a. Using the procedure in (13.14), test whether the failure experience of new business ventures when the entrepreneurs participate in the workshop differs from that for the general population of new business ventures. Control the α risk at 0.02. State the alternatives, the decision rule, and the conclusion.

b. Is it possible that the conclusion in part a has resulted in a Type II error? What would be the practical implication of a Type II error if, in actuality, the workshop for entrepreneurs reduces the probability of failure for a new business venture?

EXERCISES

13.46 Show that $s^2\{\bar{p}\}$ as defined in (13.8) is an unbiased estimator of $\sigma^2\{\bar{p}\}$ as defined in (13.5). [*Hint:* $E\{\bar{p}^2\} = p^2 + \sigma^2\{\bar{p}\}$.]

13.47 A market researcher must determine the random sample size for a consumer survey that will be used to estimate the proportion of consumers who have tried a new dairy product. The researcher wishes to estimate the population proportion p with a relative margin of sampling error of 20 percent (that is, within $\pm 0.20p$), using a 95 percent confidence interval. A pilot study suggests that $p = 0.10$ is a reasonable planning value. What sample size is required?

13.48 A municipal airport reports that it has handled 6000 aircraft landings in the past year without a mishap. Assuming that the aircraft landings constitute a Bernoulli process, obtain a 95 percent upper confidence interval for the probability that a landing at the airport will involve a mishap. [*Hint:* Use $F(0.95; 2, \infty)$.]

13.49 A survey of $n = 15$ small manufacturers showed that $f = 3$ have increased the size of their labor force during the past 12 months. Assume that the binomial probability distribution applies.

a. In Table C.5, for which value of p is the likelihood of obtaining $f = 3$ the greatest when $n = 15$?
b. The maximum likelihood estimator of p here can be shown to be $\bar{p} = f/n$. Is the finding in part a consistent with this fact?

13.50 (*Calculus needed.*) For binomial random variable f with parameters n and p, show that the sample proportion $\bar{p} = f/n$ is the maximum likelihood estimator of p.

STUDIES

13.51 An auditor will verify all of the accounts in a population if K or more accounts in a random sample of n accounts are in error. Let p denote the population proportion of accounts in error and assume that the binomial probability distribution is applicable. The auditor wishes to choose K and n so that the probability of 100 percent verification of accounts is 0.99 if $p = 0.030$ but only 0.01 if $p = 0.005$. Search over ranges of K and n to find the combination that most nearly meets the auditor's requirements. [*Hint:* Use (13.12) to obtain a starting approximation.]

13.52 A trade association of small businesses plans to select a random sample of member firms to estimate (1) the proportion of firms with no full-time employees, (2) the proportion of firms in which the owner works more than 60 hours per week, and (3) the proportion of firms that carry no liability insurance. The random sample size should be adequate to provide 95 percent confidence intervals for each of these three population proportions with half-widths of 0.02 or smaller. Planning values for the three population proportions deemed appropriate by the research director of the trade association are 0.4, 0.7, and 0.2, respectively.

a. What sample size should the trade association employ so that the precision requirements for all three estimates can be met? Which planning value governed the choice of sample size? Explain why it determined the sample size.

b. The trade association finally decided to sample only 1000 firms to expedite the completion of the study. The final report will contain many other estimates of population proportions besides the three mentioned previously. To simplify the presentation of the precisions of the many estimates, calculate the half-widths $zs\{\bar{p}\}$ of 95 percent confidence intervals for this survey for $\bar{p} = 0.01$, 0.10, 0.20, 0.50, 0.80, 0.90, 0.99. Plot the half-widths as a function of the sample proportion $\bar{p}$. Prepare a brief statement that explains how the graph is to be used.

Comparisons of Two Populations and Other Inferences

Thus far, we have discussed statistical inferences for the mean and the proportion of a single population. In this chapter, we take up inferences for studies in which a comparison of two populations is of interest. We begin by considering the nature of comparative studies. We then describe inference methods for the difference of two population means and the difference of two population proportions. Finally, as an optional topic, we consider inference methods for the population variance.

COMPARATIVE STUDIES   14.1

The purpose of a comparative study of several populations is to establish their similarities or to detect and measure their differences. Comparative studies may be observational or experimental in nature. Recall from Chapter 1 that an experimental study involves the exercise of control over selected factors that might influence the variable of interest in an investigation, whereas an observational study does not involve any control. The effectiveness of comparative studies, whether experimental or observational, is very much influenced by their sample design. We therefore now discuss several designs for observational and experimental studies. We shall limit this discussion to comparative studies for two populations. In later chapters, we shall show how the designs presented here also apply to comparative studies for more than two populations.

Observational Studies

Two important sample designs for observational studies are the *independent samples* and the *matched samples* designs. We will illustrate each design by an example and then describe it formally.

Independent Samples Design

EXAMPLE ☐

An economist wished to compare the economic characteristics of families living in substandard dwellings in two regions of a state. For this purpose, independent random samples of families living in substandard dwellings were selected from the two regions. ☐

The economist in this observational study employed an independent samples design.

> **(14.1)**
> An **independent samples design** is an observational study design in which statistically independent samples are drawn from the populations and comparative information about the populations is derived from a comparison of the independent samples.

Matched Samples Design

Work Environment   □ **EXAMPLE**

A specialist in occupational safety wished to compare the stress of the work environment in two different plants. For this purpose, an equal number of workers from each plant were randomly selected, and all selected workers were asked to complete a questionnaire about their work environment. The questionnaire dealt with such matters as noise level, ambient air quality, and physical risks, as well as with each worker's personal characteristics, such as age, gender, job training, and work classification. The specialist then paired the workers, one from each plant, in such a way that the two workers in each pair were very similar with respect to their personal and job characteristics. Finally, each worker was asked to complete a psychological questionnaire to measure the stress of the work environment. The analysis for comparing the work stress in the two plants involved comparing the stress responses of the two workers in each pair. □

The specialist employed a matched samples design for this observational study of stress of the work environment in the two plants. The rationale of comparing the stress responses of the two workers in each pair is that any differences in the stress responses in a pair will reflect differences between the two plants, but not differences between workers' personal and job characteristics to any extent. Thus, the effects of differences between workers' personal and job characteristics are virtually eliminated from the sampling error.

> **(14.2)**
> A **matched samples design** is an observational study design in which the elements of the samples drawn from the two populations are carefully matched in pairs so that the two elements in each pair are as similar as possible with respect to characteristics that might be related to the variable of interest in the study. The observations in each pair being compared are therefore dependent by design.

Experimental Studies

Two important designs for experimental studies are the *completely randomized* and the *randomized block* designs. These designs are the counterparts of the independent samples and the matched samples designs for observational studies. We shall illustrate each design by an example and then describe it formally.

Completely Randomized Design

Marketing Instruction   □ **EXAMPLE**

An instructor who wished to compare the effectiveness of two teaching methods for an introductory marketing course randomly divided the total student enrollment into two course sections. One section was given live lectures by the instructor; the other viewed videotaped lectures prepared by the same instructor. In all other respects, the two course sections were handled in identical fashion. The instructor now wishes to compare the final grade distributions for students exposed to the two teaching methods.   ☐

The instructor in this experimental study employed a completely randomized design. In the terminology of experimental designs, the students who were randomly divided into the two course sections are considered to be the *experimental units* and the two teaching methods under investigation are considered to be the *treatments*.

(14.3)
A **completely randomized design** is an experimental study design in which the experimental units are assigned to the treatments on a completely random basis.

The correspondence of a completely randomized design for an experimental study to an independent samples design for an observational study can be seen by noting that the random assignment of the students to the two sections makes the sections equivalent to independent samples. The randomization also ensures that all factors that might affect students' grades other than the teaching method, such as a student's age and ability, tend to be balanced out in the two samples.

Randomized Block Design

EXAMPLE ☐

An industrial experiment compared the adhesion of two types of paint (P_1 and P_2) to metal alloy sheeting used in aircraft body construction. Ten rectangular sheets of the alloy were prepared for the experiment. One-half of each sheet was painted with paint P_1 and the other half with paint P_2. The half sheet for each paint was chosen at random. The adhesion of each paint was then measured and the two measures compared for each sheet.   ☐

This experiment employed a randomized block design, where the two halves of each sheet represent a *block* of two experimental units and the two paints represent the *treatments*. Within each block, the two treatments were assigned at random to the experimental units. By comparing the adhesion measures for the two paints on a given sheet, the effects of differences in the physical characteristics of different sheets are almost entirely eliminated from the sampling error.

(14.4)
A **randomized block design** is an experimental study design in which the experimental units are grouped into homogeneous **blocks** and the treatments are randomly assigned to the units within each block.

The correspondence of a randomized block design for an experimental study to a matched samples design for an observational study can be seen by noting that blocking is equivalent to matching. Comparisons within a block will be affected by the treatments, but very little by other characteristics of the experimental units that might affect the variable of interest (adhesion, in the industrial experiment) because the experimental units in a block are similar with respect to these other characteristics.

Efficiency of Designs

When a choice between study designs is available, we will want to choose the design that offers, for a given cost, the smaller sampling error. Usually, matched samples and randomized block designs are more efficient in this regard. The following simple illustration shows why.

Banana Shipment   □ **EXAMPLE**

An independent samples design and a matched samples design are under consideration for a study to obtain an estimate of the weight loss in a shipment of bananas during transit.

1. *Independent samples.* A random sample of banana bunches is selected from the lot and weighed before loading. After shipment, an independent random sample of bunches is selected and weighed during the unloading. The difference in the two sample mean weights per bunch is used as the estimate of weight loss per bunch.

2. *Matched samples.* Again, a random sample of bunches is selected and weighed before loading. After shipment, the same bunches selected before loading are weighed again, and the difference in weight for each bunch is noted. The mean of these differences is used as an estimate of the weight loss per bunch.

Measuring the same sample elements before and after some occurrence or intervention, such as the passage of time, is a special case of a matched samples design. Here, the matching is as close as possible since the same sample elements are observed before and after.

To see why the matched samples design is likely to be more effective here, consider Table 14.1, which contains data for a hypothetical population of five banana bunches, giving the weights before and after shipment. In this idealized example, each of the five bunches experiences the same weight loss of 2.0 kilograms. However, if independent samples of two bunches, say, are drawn before and after shipment, the uniform weight loss per bunch is hidden by the variation in the weights of bunches. For instance, the sample drawn before shipment might consist of bunches 1 and 5, with a mean weight of $(34 + 23)/2 = 28.5$ kilograms, while the sample drawn after shipment might contain bunches 4 and 5, with a mean weight of $(26 + 21)/2 = 23.5$ kilograms. The sample difference of 5.0 kilograms is far greater than the true difference of 2.0 kilograms. In contrast, the use of matched samples will always reveal exactly the 2.0-kilogram weight loss per bunch.   □

The example just presented is an extreme and simplified one, but it does illustrate the following general principle: If the matched observations are positively related, as are the data in Table 14.1, the matched samples design will lead to smaller sampling errors than the use of the independent samples design. The same advantage applies to the randomized block design over a completely randomized design for many experimental studies.

Bunch	Weight Before Shipment (kilograms)	Weight After Shipment (kilograms)	
1	34	32	**TABLE 14.1**
2	29	27	**Weights for a hypothetical**
3	31	29	**population of banana**
4	28	26	**bunches before and after**
5	23	21	**shipment**

Nevertheless, independent samples designs and completely randomized designs are used frequently. The cost or the physical impossibility of obtaining matched observations often necessitates the use of independent samples or completely randomized designs. For instance, in the banana shipment example, logistical difficulties may make it infeasible to tag banana bunches before shipment so that they can be identified for subsequent reweighing.

INFERENCES **14.2** ABOUT $\mu_2 - \mu_1$ —INDEPENDENT SAMPLES AND COMPLETELY RANDOMIZED DESIGNS

A comparative study of two populations often is concerned with the difference in the means of the two populations. In this section we consider interval estimates and tests for such a difference. The inference methods to be described are applicable to observational studies based on the independent samples design and to experimental studies based on a completely randomized design. First, we consider inferences when the populations are normal or do not depart too markedly from normality. Then, we take up inferences for arbitrary populations when the sample sizes are large.

Confidence Interval for $\mu_2 - \mu_1$—Normal Populations with Equal Variances

Assumptions. The procedure to be described assumes the following.

1. The two populations are normal or do not depart too markedly from normality. The population means are denoted by μ_1 and μ_2.
2. The two populations have the same variance, denoted by σ^2.
3. Independent random samples of sizes n_1 and n_2 are drawn from the two populations, respectively, or the experimental units are assigned at random to the two treatments.

The appropriateness of assumption 2, that the two populations have the same variance, can be examined by means of aligned dot plots of the two sample data sets and by the formal statistical test to be described in Section 14.5.

EXAMPLE ☐ **Admission Test Scores**

A college admissions officer wished to compare the mean admission test scores of applicants educated in rural high schools (population 1) and applicants educated in urban high schools (population 2). Independent random samples of sizes $n_1 = 15$ and $n_2 = 17$ from the two populations, respectively, were used. Analysis of the sample data and information from previous studies indicated that it is reasonable to assume here that admission test scores in both populations are normally distributed with means μ_1 and μ_2 and common variance σ^2. Interest is in estimating the difference in mean test scores, $\mu_2 - \mu_1$, by a confidence interval.   ☐

Point Estimator of $\mu_2 - \mu_1$.   A point estimator of $\mu_2 - \mu_1$ is the difference between the two sample means.

(14.5)
A point estimator of $\mu_2 - \mu_1$ is:

$$\overline{X}_2 - \overline{X}_1$$

where $\overline{X}_1$ and $\overline{X}_2$ are the sample means corresponding to populations 1 and 2, respectively.

Sampling Distribution of $\overline{X}_2 - \overline{X}_1$.   To be able to make inferences about the difference between the two population means, we need to know the properties of the sampling distribution of $\overline{X}_2 - \overline{X}_1$.

(14.6)
When two independent random samples of sizes n_1 and n_2 are selected from normal populations with means μ_1 and μ_2 and common variance σ^2, the sampling distribution of $\overline{X}_2 - \overline{X}_1$ has the following properties.

(14.6a)   **Mean:** $E\{\overline{X}_2 - \overline{X}_1\} = \mu_2 - \mu_1$

(14.6b)   **Variance:** $\sigma^2\{\overline{X}_2 - \overline{X}_1\} = \sigma^2\left(\dfrac{1}{n_2} + \dfrac{1}{n_1}\right)$

(14.6c)   **Functional form:**   Normal distribution

Note from (14.6a) that $\overline{X}_2 - \overline{X}_1$ is an unbiased estimator of the difference between the population means, $\mu_2 - \mu_1$.

Comment

The properties in (14.6) follow from earlier results.

a. Since the sampled populations are assumed to be normal, we know from (10.11) that both $\overline{X}_1$ and $\overline{X}_2$ have normal sampling distributions. The means of these sampling distributions are μ_1 and μ_2 from (10.5), and the variances are σ^2/n_1 and σ^2/n_2 from (10.6). Because the samples are independent, it follows that $\overline{X}_1$ and $\overline{X}_2$ are independent.

b. Theorem (6.8a) tells us that the expected value of $\overline{X}_2 - \overline{X}_1$ is:

$$E\{\overline{X}_2 - \overline{X}_1\} = E\{\overline{X}_2\} - E\{\overline{X}_1\} = \mu_2 - \mu_1$$

c. Theorem (6.8b) tells us that the variance of $\overline{X}_2 - \overline{X}_1$ is:

$$\sigma^2\{\overline{X}_2 - \overline{X}_1\} = \sigma^2\{\overline{X}_2\} + \sigma^2\{\overline{X}_1\} = \frac{\sigma^2}{n_2} + \frac{\sigma^2}{n_1} = \sigma^2\left(\frac{1}{n_2} + \frac{1}{n_1}\right)$$

d. The following theorem tells us that $\overline{X}_2 - \overline{X}_1$ is normally distributed.

(14.7)
Any linear combination of independent normal random variables is also normally distributed.

The difference $\bar{X}_2 - \bar{X}_1$ is such a linear combination of independent normal random variables.

Development of Confidence Interval.   To construct a confidence interval for $\mu_2 - \mu_1$, we require an estimate of the common population variance σ^2. Let s_1^2 and s_2^2 denote the sample variances of the two samples, as defined in (3.12). It can be shown that the following weighted average of s_1^2 and s_2^2 is the best unbiased estimator of σ^2.

(14.8)
$$s_c^2 = \frac{(n_1 - 1)s_1^2 + (n_2 - 1)s_2^2}{(n_1 - 1) + (n_2 - 1)}$$

The estimator s_c^2 is called a *pooled* or *combined estimator* of σ^2.

Replacing σ^2 in (14.6b) by the sample estimator s_c^2, we obtain the following estimator of $\sigma^2\{\bar{X}_2 - \bar{X}_1\}$.

(14.9)
$$s^2\{\bar{X}_2 - \bar{X}_1\} = s_c^2 \left(\frac{1}{n_2} + \frac{1}{n_1} \right)$$

where s_c^2 is given by (14.8).

A statistical theorem tells us the following.

(14.10)
For two independent random samples from normal populations with common variance:

$$\frac{(\bar{X}_2 - \bar{X}_1) - (\mu_2 - \mu_1)}{s\{\bar{X}_2 - \bar{X}_1\}} = t(n_1 + n_2 - 2)$$

where $s\{\bar{X}_2 - \bar{X}_1\}$ is given by (14.9).

Hence, the situation is parallel to that for interval estimation of μ for a normal population.

(14.11)
The $1 - \alpha$ confidence limits for $\mu_2 - \mu_1$, when the populations are normal or do not depart too markedly from normality, the populations have a common variance σ^2, and the random samples are independent, are:

$$(\bar{X}_2 - \bar{X}_1) \pm ts\{\bar{X}_2 - \bar{X}_1\}$$

continues

> where: $t = t(1 - \alpha/2; n_1 + n_2 - 2)$
> $s\{\overline{X}_2 - \overline{X}_1\}$ is given by (14.9)

The t value here is based on $n_1 + n_2 - 2$ degrees of freedom because the common variance σ^2 is estimated from two sample variances, which have a total number of degrees of freedom of $(n_1 - 1) + (n_2 - 1) = n_1 + n_2 - 2$.

☐ EXAMPLE

In the admission test scores example, the sample results were as follows:

$$n_1 = 15 \qquad \overline{X}_1 = 495 \qquad s_1 = 55$$
$$n_2 = 17 \qquad \overline{X}_2 = 545 \qquad s_2 = 50$$

A 95 percent confidence interval for $\mu_2 - \mu_1$ is desired.
   From the sample results, we obtain:

$$\overline{X}_2 - \overline{X}_1 = 545 - 495 = 50$$

$$s_c^2 = \frac{(15 - 1)(55)^2 + (17 - 1)(50)^2}{(15 - 1) + (17 - 1)} = 2745$$

$$s^2\{\overline{X}_2 - \overline{X}_1\} = 2745\left(\frac{1}{17} + \frac{1}{15}\right) = 344.47$$

$$s\{\overline{X}_2 - \overline{X}_1\} = \sqrt{344.47} = 18.56$$

$$(n_1 - 1) + (n_2 - 1) = (15 - 1) + (17 - 1) = 30$$

For a 95 percent confidence interval, we require $t(1 - \alpha/2; n_1 + n_2 - 2) = t(0.975; 30) = 2.042$. Hence, the desired confidence limits are $50 \pm 2.042(18.56)$, and the resulting 95 percent confidence interval is:

$$12.1 \leq \mu_2 - \mu_1 \leq 87.9$$

Therefore, with 95 percent confidence, it can be stated that the mean admission test score for urban applicants exceeds that for rural applicants by between 12.1 and 87.9 points. To measure this difference more precisely, larger samples would be required.   ☐

Confidence Interval for $\mu_2 - \mu_1$—Large Samples from Arbitrary Populations

We now turn to estimating $\mu_2 - \mu_1$ for arbitrary populations when the sample sizes are large.

Assumptions.   The procedure to be described assumes the following.

1. There are no restrictions on either population except that each of the sample means be approximately normally distributed for large samples.
2. Independent random samples of sizes n_1 and n_2 are selected, both of which are reasonably large, or a completely randomized design is used with large sample sizes for the two treatments.

A market researcher randomly divided a sample of 500 households into two equal groups of $n_1 = n_2 = 250$ households. Group 1 was interviewed by phone using the current manual procedure. Group 2 received the same interview, but computer-assisted interviewing was used. The researcher is interested in estimating the difference in the mean interview times for the two procedures, $\mu_2 - \mu_1$, and will use the difference between the sample means, $\bar{X}_2 - \bar{X}_1$, to obtain a confidence interval. ☐

Sampling Distribution of $\bar{X}_2 - \bar{X}_1$.   The sampling distribution of $\bar{X}_2 - \bar{X}_1$ has the following properties for this case.

(14.12)

When large independent random samples are selected, the sampling distribution of $\bar{X}_2 - \bar{X}_1$ has the following properties.

(14.12a)   **Mean:** $E\{\bar{X}_2 - \bar{X}_1\} = \mu_2 - \mu_1$

(14.12b)   **Variance:** $\sigma^2\{\bar{X}_2 - \bar{X}_1\} = \sigma^2\{\bar{X}_2\} + \sigma^2\{\bar{X}_1\}$

(14.12c)   **Functional form:**   Approximately normal distribution

---

Comment

Since both sample sizes are large, central limit theorem (10.7) guarantees that $\bar{X}_1$ and $\bar{X}_2$ are each approximately normal, and consequently, by (14.7), their difference $\bar{X}_2 - \bar{X}_1$ is also approximately normal.

Development of Confidence Interval.   To construct a confidence interval for $\mu_2 - \mu_1$, we estimate $\sigma^2\{\bar{X}_2 - \bar{X}_1\}$ with the following estimator.

(14.13) $s^2\{\bar{X}_2 - \bar{X}_1\} = s^2\{\bar{X}_2\} + s^2\{\bar{X}_1\}$

where $s^2\{\bar{X}_2\}$ and $s^2\{\bar{X}_1\}$ are given by (11.2).

Because we are dealing with large random samples, we can draw a parallel between this case and the case of interval estimation for μ based on a large random sample.

(14.14)

The approximate $1 - \alpha$ confidence limits for $\mu_2 - \mu_1$, when the random samples are independent and reasonably large, are:

$$(\bar{X}_2 - \bar{X}_1) \pm zs\{\bar{X}_2 - \bar{X}_1\}$$

where: $z = z(1 - \alpha/2)$
         $s\{\bar{X}_2 - \bar{X}_1\}$ is given by (14.13)

☐ **EXAMPLE**

In the computer-assisted interviews experiment, the sample results were as follows:

$$n_1 = 250 \qquad \bar{X}_1 = 25.3 \text{ minutes} \qquad s_1 = 2.14 \text{ minutes}$$

$$n_2 = 250 \qquad \bar{X}_2 = 18.7 \text{ minutes} \qquad s_2 = 1.69 \text{ minutes}$$

A 95 percent confidence interval for $\mu_2 - \mu_1$ is desired.

We have:

$$\bar{X}_2 - \bar{X}_1 = 18.7 - 25.3 = -6.6$$

$$s^2\{\bar{X}_2 - \bar{X}_1\} = s^2\{\bar{X}_2\} + s^2\{\bar{X}_1\}$$

$$= \frac{s_2^2}{n_2} + \frac{s_1^2}{n_1} = \frac{(1.69)^2}{250} + \frac{(2.14)^2}{250} = 0.029743$$

$$s\{\bar{X}_2 - \bar{X}_1\} = \sqrt{0.029743} = 0.1725$$

We require $z(1 - \alpha/2) = z(0.975) = 1.960$. Hence, the desired confidence limits are $-6.6 \pm 1.960(0.1725)$, and the resulting confidence interval is:

$$-6.94 \le \mu_2 - \mu_1 \le -6.26$$

Therefore, it can be concluded with 95 percent confidence that the mean interview time with computer-assisted interviewing is between 6.26 and 6.94 minutes less than that with the current manual procedure.   ☐

Statistical Tests for $\mu_2 - \mu_1$

For both the case of normal populations and that of arbitrary populations with large sample sizes, decision rules for making statistical tests concerning $\mu_2 - \mu_1$ are constructed in the same manner as for tests concerning a single population mean. We shall consider both one-sided and two-sided test alternatives:

$$H_0: \mu_2 - \mu_1 \le 0 \qquad H_0: \mu_2 - \mu_1 \ge 0 \qquad H_0: \mu_2 - \mu_1 = 0$$

$$H_1: \mu_2 - \mu_1 > 0 \qquad H_1: \mu_2 - \mu_1 < 0 \qquad H_1: \mu_2 - \mu_1 \ne 0$$

Note that $\mu_2 - \mu_1 = 0$ is equivalent to $\mu_2 = \mu_1$.

The standardized test statistics when the α risk is controlled at $\mu_2 - \mu_1 = 0$ are given next.

(14.15)

Tests concerning $\mu_2 - \mu_1$, when the two random samples are independent, are based on the following standardized test statistics.

(14.15a)   Normal Populations with Equal Variances: $t^* = \dfrac{\bar{X}_2 - \bar{X}_1}{s\{\bar{X}_2 - \bar{X}_1\}}$

where $s\{\bar{X}_2 - \bar{X}_1\}$ is given by (14.9). When $\mu_2 - \mu_1 = 0$, test statistic t^* follows the t distribution with $n_1 + n_2 - 2$ degrees of freedom.

(14.15b) Large Samples from Arbitrary Populations: $z^* = \dfrac{\overline{X}_2 - \overline{X}_1}{s\{\overline{X}_2 - \overline{X}_1\}}$

where $s\{\overline{X}_2 - \overline{X}_1\}$ is given by (14.13). When $\mu_2 - \mu_1 = 0$, test statistic z^* follows approximately the standard normal distribution.

EXAMPLE ☐

In the computer-assisted interviewing example, we are to test whether the mean interviewing time is the same for the two procedures.

Step 1. The alternatives here are as follows:

$$H_0:\ \mu_2 - \mu_1 = 0$$
$$H_1:\ \mu_2 - \mu_1 \neq 0$$

Step 2. The α risk is to be controlled at 0.05 when $\mu_2 - \mu_1 = 0$; that is, when $\mu_2 = \mu_1$.

Step 3. Figure 14.1 shows the appropriate type of decision rule. The test is two-sided, and the sampling distribution of $\overline{X}_2 - \overline{X}_1$ is centered at $\mu_2 - \mu_1 = 0$ where the α risk is to be controlled. Because the sample sizes are large, the standard normal distribution is utilized for constructing the decision rule. Since the α risk is to be controlled at 0.05, the action limits in the decision rule of Figure 14.1 correspond to $z(\alpha/2) = z(0.025) = -1.960$ and $z(1 - \alpha/2) = z(0.975) = 1.960$. Thus, the decision rule is as follows:

If $|z^*| \leq 1.960$, conclude H_0.
If $|z^*| > 1.960$, conclude H_1.

Step 4. From the sample results given earlier, we have $\overline{X}_2 - \overline{X}_1 = -6.6$ and $s\{\overline{X}_2 - \overline{X}_1\} = 0.1725$. Test statistic z^* in (14.15b) therefore is:

$$z^* = \frac{\overline{X}_2 - \overline{X}_1}{s\{\overline{X}_2 - \overline{X}_1\}} = \frac{-6.6}{0.1725} = -38.3$$

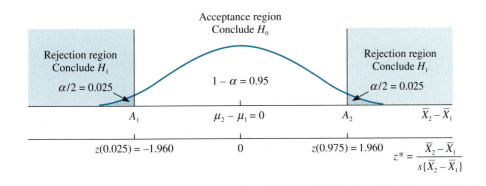

FIGURE 14.1
Statistical decision rule for large independent samples, two-sided test for $\mu_2 - \mu_1$ —Computer-assisted interviewing example

Since $|z^*| = |-38.3| = 38.3 > 1.960$, we conclude H_1; that is, the mean interviewing times are not the same. The two-sided P-value of the test is $2P(Z < z^*) = 2P(Z < -38.3) = 2(0+) = 0+$, which is very small and thus provides strong evidence that the mean interviewing times for the two procedures differ.

Recall in this connection that the 95 percent confidence interval for $\mu_2 - \mu_1$ obtained earlier for this example also indicated that the mean interviewing times are not equal. $\square$

14.3 INFERENCES ABOUT $\mu_2 - \mu_1$ —MATCHED SAMPLES AND RANDOMIZED BLOCK DESIGNS

When a matched samples design or a randomized block design is employed to make inferences about the difference of two population means, it turns out that the procedures simplify to those for a single population mean. The procedures to be discussed are applicable when an observational study is based on the matched samples design or an experimental study is based on a randomized block design.

Social Behavior $\square$ **EXAMPLE**

A social psychologist wished to compare the attitudes of people toward a particular social behavior *before* and *after* they view an informational film about the behavior, based on a random sample of $n = 10$ subjects drawn from the general adult population. For each subject, an attitude measurement was made before viewing the film (X_1), and a second measurement was made after viewing the film (X_2). Letting μ_1 and μ_2 denote the population mean values of X_1 and X_2, respectively, the psychologist wishes to estimate $\mu_2 - \mu_1$. $\square$

As we noted earlier, the before and after measurements for a subject form a matched pair of observations (X_1, X_2). To measure the change in attitude of a subject, we use the difference of the paired observations.

$$(14.16) \quad D = X_2 - X_1$$

If n subjects are studied, there are n differences that we shall denote by $D_1, D_2, \ldots, D_n$.

The differences $D_1, D_2, \ldots, D_n$ may be thought of as a random sample of observations from a population of differences D. We denote the mean and the variance of this population of differences by μ_D and σ_D^2, respectively.

$$(14.17) \quad E\{D\} = \mu_D \qquad \sigma^2\{D\} = \sigma_D^2$$

The following identity can be shown to hold.

$$(14.18) \quad \mu_D = \mu_2 - \mu_1$$

Thus, we can estimate μ_D by the earlier procedures for a single population mean, but we are really estimating $\mu_D = \mu_2 - \mu_1$, the difference between the means of the two populations.

The procedure, therefore, is to (1) calculate the sample mean of the observed differences $D_1, D_2, \ldots, D_n$, to be denoted by $\overline{D}$; (2) calculate the sample variance of the D_i, to be denoted by s_D^2 as a reminder that it is a variance of differences; and (3) calculate the estimated variance of $\overline{D}$, to be denoted by $s^2\{\overline{D}\}$.

$$(14.19) \qquad \overline{D} = \frac{\sum_{i=1}^{n} D_i}{n}$$

$$(14.20) \qquad s_D^2 = \frac{\sum_{i=1}^{n} (D_i - \overline{D})^2}{n - 1}$$

$$(14.21) \qquad s^2\{\overline{D}\} = \frac{s_D^2}{n}$$

Then we proceed as earlier, depending on whether (1) the population of differences is normal or does not depart too markedly from normality or (2) the population of differences is arbitrary and the number of differences (n) is large.

Comment

The identity (14.18) follows from (6.8a) as follows:

$$\mu_D = E\{D\} = E\{X_2 - X_1\} = E\{X_2\} - E\{X_1\} = \mu_2 - \mu_1$$

Confidence Interval for $\mu_2 - \mu_1$—Normal Population of Differences

When the population of differences D is normal or does not depart too markedly from normality, a confidence interval for $\mu_D = \mu_2 - \mu_1$ is equivalent to that in (11.9).

(14.22)
The $1 - \alpha$ confidence limits for $\mu_D = \mu_2 - \mu_1$, when the population of differences is normal or does not depart too markedly from normality, are:

$$\overline{D} \pm ts\{\overline{D}\}$$

where: $t = t(1 - \alpha/2; n - 1)$
$\overline{D}$ and $s\{\overline{D}\}$ are given by (14.19) and (14.21), respectively

☐ **EXAMPLE**

In the social behavior example, the attitude measurements of the $n = 10$ subjects are shown in Table 14.2. The greater the attitude score, the more favorable is the subject toward the social behavior. The social psychologist wishes to estimate $\mu_2 - \mu_1$ with a 90 percent confidence interval. It is reasonable here to assume that the population of differences is approximately normally distributed.

For the sample data, we have $\overline{D} = 7.640$ and $s_D = 12.575$, as shown in Table 14.2. Therefore:

$$s^2\{\overline{D}\} = \frac{s_D^2}{n} = \frac{(12.575)^2}{10} = 15.81 \qquad s\{\overline{D}\} = \sqrt{15.81} = 3.976$$

We require $t(0.95; 9) = 1.833$. The confidence limits equal $7.640 \pm 1.833(3.976)$, and the desired 90 percent confidence interval is:

$$0.4 \le \mu_2 - \mu_1 \le 14.9$$

The psychologist can assert with 90 percent confidence that the mean attitude measurement after viewing the film exceeds the mean attitude measurement before viewing by between 0.4 and 14.9 units on the measurement scale. Thus, on average, the attitude toward the social behavior becomes more favorable after viewing the film, but because of the small size of the study it is not clear whether the change is small or large.   ☐

Comment

We stated earlier that matched sampling and randomized block designs are effective when the paired observations are positively correlated. To see why, we need to consider $\sigma^2\{D\}$, the variance of the difference of a matched pair of observations. Since $D = X_2 - X_1$, we can use (6.17) to find:

$$\sigma^2\{D\} = \sigma^2\{X_2 - X_1\} = \sigma^2\{X_2\} + \sigma^2\{X_1\} - 2\sigma\{X_1, X_2\}$$

This variance was denoted by σ_D^2 earlier—that is, $\sigma^2\{D\} = \sigma_D^2$. Thus, when $\sigma\{X_1, X_2\}$, the covariance between X_1 and X_2, is positive, σ_D^2 will be smaller than when X_1 and X_2 are independent. Since $\sigma^2\{\overline{D}\} = \sigma_D^2/n$, it follows that $\sigma^2\{\overline{D}\}$ will be smaller when the X_1 and X_2 observations are positively related than when they are independent. Matched samples and randomized block designs seek to ensure a high positive correlation between X_1 and X_2 to attain this improvement in precision.

Confidence Interval for $\mu_2 - \mu_1$—Large Sample from Arbitrary Population of Differences

When the sample of differences is reasonably large, the nature of the population of differences does not matter, and the confidence interval for $\mu_D = \mu_2 - \mu_1$ is equivalent to that in (11.5).

Sample data on attitudes before and after viewing an informational film—Social behavior example

TABLE 14.2

Subject	Before X_1	After X_2	$D = X_2 - X_1$	Subject	Before X_1	After X_2	$D = X_2 - X_1$
1	41.0	46.9	5.9	6	22.5	56.8	34.3
2	60.3	64.5	4.2	7	67.5	60.7	-6.8
3	23.9	33.3	9.4	8	50.3	57.3	7.0
4	36.2	36.0	-0.2	9	50.9	65.4	14.5
5	52.7	43.5	-9.2	10	24.6	41.9	17.3

$$\overline{D} = 7.640 \qquad s_D = 12.575$$

(14.23)

The approximate $1 - \alpha$ confidence limits for $\mu_D = \mu_2 - \mu_1$, when the sample of differences is reasonably large, are:

$$\overline{D} \pm zs\{\overline{D}\}$$

where: $z = z(1 - \alpha/2)$

$\overline{D}$ and $s\{\overline{D}\}$ are given by (14.19) and (14.21), respectively

---

Statistical Tests for $\mu_2 - \mu_1$

Like the construction of confidence intervals, tests on $\mu_2 - \mu_1$ for matched samples designs and randomized block designs parallel those for a single population mean. The standardized test statistics are as follows when the α risk is controlled at $\mu_2 = \mu_1$ or, equivalently, at $\mu_D = \mu_2 - \mu_1 = 0$.

(14.24)

Tests concerning $\mu_D = \mu_2 - \mu_1$ are based on the following standardized test statistics.

(14.24a) **Normal Population of Differences:** $t^* = \dfrac{\overline{D}}{s\{\overline{D}\}}$

where $\overline{D}$ and $s\{\overline{D}\}$ are given by (14.19) and (14.21), respectively. When $\mu_D = 0$, test statistic t^* follows the t distribution with $n - 1$ degrees of freedom.

(14.24b) **Large Sample of Differences:** $z^* = \dfrac{\overline{D}}{s\{\overline{D}\}}$

where $\overline{D}$ and $s\{\overline{D}\}$ are given by (14.19) and (14.21), respectively. When $\mu_D = 0$, test statistic z^* follows approximately the standard normal distribution.

□ **EXAMPLE**

In the social behavior example, the psychologist wishes to test whether viewing the film leads to a more favorable attitude toward the social behavior—that is, to an upward shift in the mean attitude level.

Step 1. The test alternatives here are as follows:

$$H_0: \ \mu_D = \mu_2 - \mu_1 \leq 0$$
$$H_1: \ \mu_D = \mu_2 - \mu_1 > 0$$

Step 2. The α risk is to be controlled at 0.05 when $\mu_D = \mu_2 - \mu_1 = 0$.

Step 3. The population of differences can be assumed to be approximately normal, and a one-sided upper-tail test is desired. Hence, the appropriate type of decision rule is the one given in Table 12.2b(1), based on the t distribution. Since $\alpha = 0.05$ and $n = 10$, we require $t(1 - \alpha; n - 1) = t(0.95; 9) = 1.833$ and the decision rule for the test is as follows:

If $t^* \leq 1.833$, conclude H_0.
If $t^* > 1.833$, conclude H_1.

Step 4. We found earlier that $\overline{D} = 7.640$ and $s\{\overline{D}\} = 3.976$. Hence, test statistic (14.24a) here is:

$$t^* = \frac{\overline{D}}{s\{\overline{D}\}} = \frac{7.640}{3.976} = 1.922$$

Since $t^* = 1.922 > 1.833$, we conclude H_1; that is, there has been an upward shift to a more favorable attitude, on average, after viewing the film. Computer output shows that the one-sided P-value for this test is $P[t(9) > t^*] = P[t(9) > 1.922] = 0.0434$, which is less than $\alpha = 0.05$ and thus leads to the same test conclusion, as expected.

□

14.4 INFERENCES FOR $p_2 - p_1$—INDEPENDENT SAMPLES AND COMPLETELY RANDOMIZED DESIGNS

We now consider inferences about the difference of two population proportions.

Assumptions

1. The procedures are applicable to independent samples designs with observational studies and to completely randomized designs with experimental studies.
2. Throughout this section, we assume that the independent random samples of sizes n_1 and n_2 are reasonably large.

Market Survey □ **EXAMPLE**

A market survey organization carried out a product taste study in two regions. Independent random samples of $n_1 = 400$ consumers and $n_2 = 300$ consumers were selected in the two regions. Each person was given two servings of breakfast cereal and asked which one tasted better. Unknown to the subject, one serving was a new high-protein breakfast cereal and the other was an existing cereal. In the first region, the number of persons out of the 400 in the sample who preferred the new cereal was $f_1 = 220$. Hence, the sample proportion preferring the new cereal in region 1 is $\bar{p}_1 = 220/400 = 0.55$. In the second region, $f_2 = 195$ persons preferred the new cereal among the 300 persons in the sample, so that the sample proportion for the second region is $\bar{p}_2 = 195/300 = 0.65$. It is now desired to construct an interval estimate of $p_2 - p_1$, the difference in the population proportions of consumers in the two regions who prefer the new cereal. ☐

Point Estimator of $p_2 - p_1$

Estimating the difference between two population proportions is very similar to estimating the difference between two population means. A point estimator of $p_2 - p_1$ is the difference in the sample proportions.

$$(14.25) \quad \bar{p}_2 - \bar{p}_1$$

We now summarize the properties of the sampling distribution of $\bar{p}_2 - \bar{p}_1$, which parallel the properties in (14.12) for the sampling distribution of $\bar{X}_2 - \bar{X}_1$ when the sample sizes are large.

(14.26)
When large independent random samples are selected from two populations, the sampling distribution of $\bar{p}_2 - \bar{p}_1$ has the following properties.

(14.26a) **Mean:** $\quad\quad\quad\quad\quad E\{\bar{p}_2 - \bar{p}_1\} = p_2 - p_1$

(14.26b) **Variance:** $\quad\quad\quad\;\; \sigma^2\{\bar{p}_2 - \bar{p}_1\} = \sigma^2\{\bar{p}_2\} + \sigma^2\{\bar{p}_1\}$

(14.26c) **Functional form:** Approximately normal distribution

Note from (14.26a) that $\bar{p}_2 - \bar{p}_1$ is an unbiased estimator of the difference between the two population proportions.

Confidence Interval for $p_2 - p_1$

Constructing an interval estimate for $p_2 - p_1$ requires an estimator of $\sigma^2\{\bar{p}_2 - \bar{p}_1\}$. We obtain it by using $s^2\{\bar{p}_1\}$ and $s^2\{\bar{p}_2\}$, as given in (13.8), as estimators of the true variances $\sigma^2\{\bar{p}_1\}$ and $\sigma^2\{\bar{p}_2\}$. The resulting estimator is denoted by $s^2\{\bar{p}_2 - \bar{p}_1\}$.

$$(14.27) \quad s^2\{\bar{p}_2 - \bar{p}_1\} = s^2\{\bar{p}_2\} + s^2\{\bar{p}_1\}$$

where $s^2\{\bar{p}_2\}$ and $s^2\{\bar{p}_1\}$ are given by (13.8).

The confidence interval for $p_2 - p_1$ takes the usual form for the large-sample case.

(14.28)

The approximate $1 - \alpha$ confidence limits for $p_2 - p_1$, when the two random samples are independent and reasonably large, are:

$$(\bar{p}_2 - \bar{p}_1) \pm zs\{\bar{p}_2 - \bar{p}_1\}$$

where: $z = z(1 - \alpha/2)$
$s\{\bar{p}_2 - \bar{p}_1\}$ is given by (14.27)

☐ **EXAMPLE**

For the market survey example, the sample results were as follows:

$$n_1 = 400 \qquad f_1 = 220 \qquad \bar{p}_1 = 0.55$$
$$n_2 = 300 \qquad f_2 = 195 \qquad \bar{p}_2 = 0.65$$

A 90 percent confidence interval for $p_2 - p_1$ is desired. We therefore require:

$$\bar{p}_2 - \bar{p}_1 = 0.65 - 0.55 = 0.10$$

$$s^2\{\bar{p}_1\} = \frac{\bar{p}_1(1 - \bar{p}_1)}{n_1 - 1} = \frac{0.55(0.45)}{400 - 1} = 0.0006203$$

$$s^2\{\bar{p}_2\} = \frac{\bar{p}_2(1 - \bar{p}_2)}{n_2 - 1} = \frac{0.65(0.35)}{300 - 1} = 0.0007609$$

$$s^2\{\bar{p}_2 - \bar{p}_1\} = 0.0007609 + 0.0006203 = 0.0013812$$

$$s\{\bar{p}_2 - \bar{p}_1\} = \sqrt{0.0013812} = 0.0372 \qquad z(0.95) = 1.645$$

The confidence limits are $0.10 \pm 1.645(0.0372)$, and the 90 percent confidence interval is:

$$0.039 \leq p_2 - p_1 \leq 0.161$$

We conclude, with 90 percent confidence, that the proportion of consumers who prefer the new cereal in the second region exceeds the proportion for the first region by between 3.9 and 16.1 percent points.   ☐

Statistical Tests for $p_2 - p_1$

Statistical tests regarding the difference between two population proportions for the large-sample case are similar to those for the difference between two population means. We illustrate the procedure for a two-sided test and then present a general summary.

☐ **EXAMPLE**

In the market survey example, the director of marketing wishes to test whether the proportions of consumers preferring the new high-protein cereal are the same in the two regions.

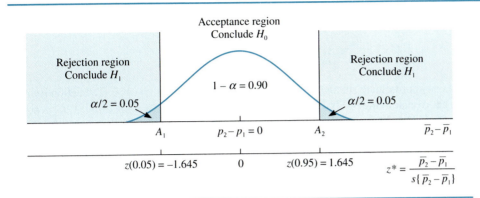

FIGURE 14.2

Statistical decision rule for large independent samples, two-sided test for $p_2 - p_1$ —Market survey example

Step 1. The test alternatives here are:

$$H_0: p_2 - p_1 = 0$$
$$H_1: p_2 - p_1 \neq 0$$

Step 2. The α risk at $p_2 - p_1 = 0$ (that is, at $p_1 = p_2$) is to be controlled at 0.10.

Step 3. Figure 14.2 shows the appropriate type of decision rule for the large-sample case. The test is two-sided, and the sampling distribution of $\bar{p}_2 - \bar{p}_1$ is centered at $p_2 - p_1 = 0$, where the α risk is controlled. To calculate the standardized test statistic z^*, we require an estimate of $\sigma\{\bar{p}_2 - \bar{p}_1\}$ for the case where $p_2 - p_1 = 0$; that is, where $p_2 = p_1$. Let p denote the common value of p_2 and p_1 when $p_2 = p_1$. To estimate p from the two samples, we combine the two samples and employ the following *pooled* estimator of p.

(14.29) $$\bar{p}' = \frac{f_1 + f_2}{n_1 + n_2} = \frac{n_1\bar{p}_1 + n_2\bar{p}_2}{n_1 + n_2}$$

Note that $\bar{p}'$ is simply a weighted average of $\bar{p}_1$ and $\bar{p}_2$, using the sample sizes as weights.

When $p_1 = p_2 = p$, we know that:

$$\sigma^2\{\bar{p}_2 - \bar{p}_1\} = \frac{p(1 - p)}{n_2} + \frac{p(1 - p)}{n_1} = p(1 - p)\left(\frac{1}{n_2} + \frac{1}{n_1}\right)$$

Hence, a sample estimator of $\sigma^2\{\bar{p}_2 - \bar{p}_1\}$ when $p_1 = p_2$, based on the pooled estimator $\bar{p}'$, is as follows.

(14.30) $$s^2\{\bar{p}_2 - \bar{p}_1\} = \bar{p}'(1 - \bar{p}')\left(\frac{1}{n_2} + \frac{1}{n_1}\right)$$

where $\bar{p}'$ is given by (14.29).

When the sample sizes are large, the standardized test statistic:

$$z^* = \frac{\bar{p}_2 - \bar{p}_1}{s\{\bar{p}_2 - \bar{p}_1\}}$$

has an approximate standard normal distribution when $p_2 - p_1 = 0$. The action limits in the decision rule of Figure 14.2 therefore correspond to $z(\alpha/2) = z(0.05) = -1.645$ and $z(1 - \alpha/2) = z(0.95) = 1.645$. Thus, the decision rule is as follows:

If $|z^*| \leq 1.645$, conclude H_0.
If $|z^*| > 1.645$, conclude H_1.

Step 4. Substituting the earlier results into (14.29) and (14.30) gives:

$$\bar{p}' = \frac{f_1 + f_2}{n_1 + n_2} = \frac{220 + 195}{400 + 300} = 0.593$$

$$s^2\{\bar{p}_2 - \bar{p}_1\} = \bar{p}'(1 - \bar{p}')\left(\frac{1}{n_2} + \frac{1}{n_1}\right)$$

$$= 0.593(0.407)\left(\frac{1}{300} + \frac{1}{400}\right) = 0.001408$$

$$s\{\bar{p}_2 - \bar{p}_1\} = \sqrt{0.001408} = 0.03752$$

Finally, we obtain the standardized test statistic:

$$z^* = \frac{\bar{p}_2 - \bar{p}_1}{s\{\bar{p}_2 - \bar{p}_1\}} = \frac{0.65 - 0.55}{0.03752} = 2.665$$

Since $|z^*| = |2.665| = 2.665 > 1.645$, we conclude H_1; that is, the proportions of consumers preferring the new cereal differ in the two regions. The two-sided P-value of the test is $2P(Z > z^*) = 2P(Z > 2.665) = 2(0.00385) = 0.0077$, which is very small and thus provides strong evidence that the proportions of consumers preferring the new cereal differ in the two regions. □

Test Statistic.   We now summarize the test procedure. The test statistic developed here is appropriate for all large-sample tests concerning $p_2 - p_1$ when the α risk is controlled at $p_2 = p_1$ or, equivalently, at $p_2 - p_1 = 0$.

(14.31)

Tests concerning $p_2 - p_1$, when the two random samples are large and independent, are based on the standardized test statistic:

$$z^* = \frac{\bar{p}_2 - \bar{p}_1}{s\{\bar{p}_2 - \bar{p}_1\}}$$

where $s\{\bar{p}_2 - \bar{p}_1\}$ is given by (14.30).
   When $p_2 - p_1 = 0$, test statistic z^* follows approximately the standard normal distribution.

The population variance σ^2 is often of interest in studies where the magnitude of the variability in a population is of concern. First, we shall consider inferences about the population variance of a single population, and then comparative inferences about the population variances of two populations.

OPTIONAL 14.5 TOPIC—INFERENCES ABOUT POPULATION VARIANCE AND RATIO OF TWO POPULATION VARIANCES

Assumptions

Throughout this section, we assume that:

1. The sampled populations are normal or nearly normal.
2. Simple random sampling is employed.

In Chapter 15, we consider inferences about the population variance when the population is not close to normal.

Confidence Interval for σ^2

EXAMPLE □ Investment Return

A financial analyst, interested in the uncertainty of an investment in a common stock, has obtained the following rates of return for the stock over the most recent eight-quarter period.

Quarter:	1	2	3	4	5	6	7	8
Rate of Return:	4.8	2.8	9.9	7.6	9.5	6.0	8.4	−5.0

The rate of return is the sum of any dividends paid and the change in the stock's price during the quarter, expressed as a percentage of the stock's opening price. The population variance σ^2 for the quarterly rates of return is to be employed as the measure of uncertainty for the common stock investment, and the analyst is therefore interested in constructing an interval estimate for σ^2. □

On previous occasions, we have used the sample variance s^2 as an unbiased point estimator of the population variance σ^2. We now require some information about the sampling distribution of s^2.

Sampling Distribution of s^2. Statistical theory provides us with the following important theorem.

(14.32)
For a random sample of size n from a normal population with variance σ^2:

$$\frac{(n-1)s^2}{\sigma^2} = \chi^2(n-1)$$

where s^2 is the sample variance defined in (3.12).

FIGURE 14.3
Sampling distribution of
$(n - 1)s^2/\sigma^2$ **for a**
normal population

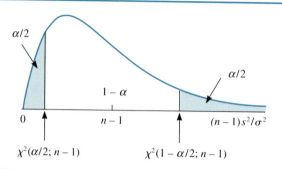

The notation $\chi^2(n - 1)$ in (14.32) denotes the χ^2 (Greek chi square) distribution with $n - 1$ degrees of freedom. Theorem (14.32) tells us, therefore, that the statistic $(n - 1)s^2/\sigma^2$ follows the χ^2 distribution with $n - 1$ degrees of freedom when the sampled population is normal. The χ^2 distribution is continuous, unimodal, and skewed to the right. (A detailed discussion of the χ^2 distribution may be found in Appendix B, Section B.1.)

Figure 14.3 contains a representative χ^2 distribution that illustrates theorem (14.32). Since a χ^2 distribution with $n - 1$ degrees of freedom has its mean at $n - 1$, the distribution in Figure 14.3 is located around $n - 1$.

Confidence Limits.   The confidence limits for σ^2 are based on theorem (14.32) and make use of the percentiles of the χ^2 distribution shown in Figure 14.3.

> **(14.33)**
> The $1 - \alpha$ confidence limits L and U for the population variance σ^2, when the population is normal or nearly normal, are:
> $$L = \frac{(n - 1)s^2}{\chi^2(1 - \alpha/2; n - 1)} \qquad U = \frac{(n - 1)s^2}{\chi^2(\alpha/2; n - 1)}$$

☐ EXAMPLE

For the investment return example, the analyst first considered the appropriateness of treating the eight quarterly rates of return as a random sample from a normal population. Previous studies and related theory supported the reasonableness of this assumption. In addition, the analyst prepared a time series plot of the observations, as well as a plot to examine the normality of the population (to be described in Chapter 16), and found that these provided no evidence to the contrary.

The analyst therefore concluded that use of the confidence limits in (14.33) is appropriate here. The sample variance of the $n = 8$ quarterly rates of return is $s^2 = 23.78$ (calculations not shown). A 95 percent confidence interval is desired; hence, we require from Table C.2:

$$\chi^2(1 - \alpha/2; n - 1) = \chi^2(0.975; 7) = 16.01$$
$$\chi^2(\alpha/2; n - 1) = \chi^2(0.025; 7) = 1.69$$

Substituting these values into (14.33) yields:

$$L = \frac{(8-1)23.78}{16.01} = 10.4 \qquad U = \frac{(8-1)23.78}{1.69} = 98.5$$

Thus, the 95 percent confidence interval for σ^2 is:

$$10.4 \le \sigma^2 \le 98.5$$

Note that the confidence limits for σ^2 are not equally spaced about s^2 because of the right-skewness of the χ^2 sampling distribution. □

A confidence interval for the population standard deviation σ is obtained by taking the square roots of the confidence limits for the population variance σ^2 in (14.33).

EXAMPLE □

For the investment return example, the 95 percent confidence interval for σ is:

$$3.2 = \sqrt{10.4} \le \sigma \le \sqrt{98.5} = 9.9$$

Hence, the standard deviation of the common stock's quarterly rates of return is between 3.2 and 9.9 percent points, with 95 percent confidence. The financial analyst concluded that investment in this common stock may involve a substantial amount of uncertainty because a standard deviation as small as 3.2 would be large relative to the mean quarterly rate of return of $\overline{X} = 5.5$. The standard deviation in fact may be as large as 9.9. □

Comments

1. One-sided confidence intervals for σ^2 can be obtained from (14.33) by selecting the appropriate limit. For example, a 95 percent upper confidence limit is obtained by calculating U in (14.33) with $\alpha/2 = 0.05$.
2. For large n, the χ^2 percentiles in (14.33) may be approximated by standard normal percentiles, using the approximation given in (B.6) of Appendix B.
3. The confidence limits in (14.33) are obtained by means of theorem (14.32), which tells us that $(n-1)s^2/\sigma^2$ follows a $\chi^2(n-1)$ distribution. It is a consequence of this theorem and the definition of percentiles for a χ^2 distribution that:

$$P\left[\chi^2(\alpha/2; n-1) \le \frac{(n-1)s^2}{\sigma^2} \le \chi^2(1-\alpha/2; n-1) \right] = 1 - \alpha$$

This probability statement is illustrated in Figure 14.3. Observe that α is divided equally between the two tails, as has been our practice in constructing two-sided confidence intervals. Rearranging the inequalities within the brackets, we obtain:

$$P\left[\frac{(n-1)s^2}{\chi^2(1-\alpha/2; n-1)} \le \sigma^2 \le \frac{(n-1)s^2}{\chi^2(\alpha/2; n-1)} \right] = 1 - \alpha$$

yielding the $1 - \alpha$ confidence limits for σ^2 in (14.33).

Statistical Tests for σ^2

Statistical tests for σ^2 may be constructed by using theorem (14.32). The value of σ^2 at which the α risk is controlled is denoted by σ_0^2 and the test statistic by X^2.

(14.34)

Tests concerning the population variance σ^2, when the population is normal or nearly normal and the α risk is controlled at $\sigma^2 = \sigma_0^2$, are based on the test statistic:

$$X^2 = \frac{(n-1)s^2}{\sigma_0^2}$$

When $\sigma^2 = \sigma_0^2$, test statistic X^2 follows the χ^2 distribution with $n - 1$ degrees of freedom.

We illustrate the use of test statistic X^2 with a two-sided test.

☐ **EXAMPLE**

In the investment return example, a theoretical financial model was developed that takes account of business risks and other related factors. This model shows that σ^2 for the common stock investment should equal 5. We wish to test whether the model is correct; that is, whether $\sigma^2 = 5$.

Step 1. The test alternatives here are:

$$H_0:\ \sigma^2 = 5$$
$$H_1:\ \sigma^2 \neq 5$$

Step 2. The α risk is to be controlled at 0.05 when $\sigma^2 = \sigma_0^2 = 5$.

Step 3. Figure 14.4 shows the appropriate type of decision rule for this test. From (14.34) we know that test statistic X^2 follows a $\chi^2(n-1)$ distribution when $\sigma^2 = \sigma_0^2$. For this example, $n - 1 = 8 - 1 = 7$. Hence, the sampling distribution of X^2 is the $\chi^2(7)$ distribution, whose mean is $n - 1 = 7$ as shown in Figure 14.4.

If s^2 is substantially smaller or larger than σ_0^2, then we should conclude H_1—that is, conclude $\sigma^2 \neq \sigma_0^2$. Correspondingly, for the test statistic $X^2 = (n-1)s^2/\sigma_0^2$, we

---

FIGURE 14.4
Statistical decision rule for normal or nearly normal population, two-sided test for σ^2—Investment return example

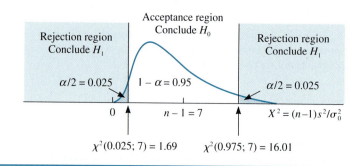

should conclude H_1 if X^2 is either much smaller or much larger than $n - 1 = 7$. The appropriate type of decision rule is therefore a two-sided one. The action limits are placed so that each tail area equals $\alpha/2$. Since $\alpha = 0.05$ here, the action limits are the following χ^2 percentiles:

$$\chi^2(\alpha/2; n - 1) = \chi^2(0.025; 7) = 1.69$$

$$\chi^2(1 - \alpha/2; n - 1) = \chi^2(0.975; 7) = 16.01$$

The decision rule for the test is therefore as follows:

If $1.69 \leq X^2 \leq 16.01$, conclude H_0.
If $X^2 < 1.69$ or $X^2 > 16.01$, conclude H_1.

Step 4. Because $s^2 = 23.78$, test statistic (14.34) equals:

$$X^2 = \frac{(n - 1)s^2}{\sigma_0^2} = \frac{(8 - 1)(23.78)}{5} = 33.3$$

Inasmuch as $X^2 = 33.3 > 16.01$, we conclude H_1. Thus, the variance of quarterly returns for the stock does not equal 5, the value derived from the model. If the model were to perform as badly for other stocks, then it should be modified or abandoned. □

Comparison of Two Population Variances

In some comparative studies, interest lies in comparing the variabilities of two populations as measured by their variances. We now discuss interval estimates and tests for the ratio of two variances for normal or near normal populations, based on independent random samples. These procedures are applicable for independent random samples designs and completely randomized designs.

EXAMPLE □ **Property Appraisal**

A behavioral scientist wished to determine whether group discussion tends to affect the closeness of judgments by property appraisers. An experimental study employing a completely randomized design was used. The scientist obtained the cooperation of 16 real estate appraisers. Six of the 16 appraisers were selected randomly and asked to make independent appraisals of a parcel of commercial property after each had participated in a group discussion of relevant appraisal factors. The other 10 appraisers were also asked to make independent appraisals of the same property but without any prior group discussion.

As a measure of the closeness of agreement among appraisers, the scientist used the variance. We shall let σ_1^2 denote the variance of appraisals made with group discussion and σ_2^2 the variance of appraisals made without group discussion. The scientist now wishes to compare the magnitudes of σ_1^2 and σ_2^2 by computing an interval estimate of the ratio σ_1^2/σ_2^2. □

Confidence Interval for σ_1^2/σ_2^2

To construct a confidence interval for σ_1^2/σ_2^2, we employ the following statistical theorem.

(14.35)

For independent random samples of sizes n_1 and n_2 from normal populations with variances σ_1^2 and σ_2^2, respectively:

$$\frac{s_1^2/\sigma_1^2}{s_2^2/\sigma_2^2} = F(n_1 - 1, n_2 - 1)$$

where s_1^2 and s_2^2 are the two sample variances.

The notation $F(n_1 - 1, n_2 - 1)$ in (14.35) denotes the F distribution with $n_1 - 1$ numerator degrees of freedom and $n_2 - 1$ denominator degrees of freedom. Theorem (14.35) tells us, therefore, that the statistic $(s_1^2/\sigma_1^2) \div (s_2^2/\sigma_2^2)$ follows this F distribution when the sampled population is normal. The F distribution is continuous, unimodal, and skewed to the right. (A detailed discussion of the F distribution may be found in Appendix B, Section B.3.)

The confidence limits for σ_1^2/σ_2^2 are based on theorem (14.35) and make use of percentiles of the F distribution.

(14.36)

The $1 - \alpha$ confidence limits L and U for σ_1^2/σ_2^2, when the populations are normal or nearly normal and the random samples are independent, are:

$$L = \frac{s_1^2}{s_2^2}\left[\frac{1}{F(1 - \alpha/2; n_1 - 1, n_2 - 1)}\right]$$

$$U = \frac{s_1^2}{s_2^2}\left[\frac{1}{F(\alpha/2; n_1 - 1, n_2 - 1)}\right]$$

☐ **EXAMPLE**

The data for the property appraisal example are presented in Table 14.3. A 90 percent confidence interval for σ_1^2/σ_2^2 is desired. The scientist has conducted analyses using methods to be described in Chapter 16 that support the reasonableness of the assumption that the two populations of appraisal values are approximately normal.

We note first that $s_1^2/s_2^2 = 9.867/194.18 = 0.05081$. For a 90 percent confidence interval, we require from Table C.4:

$$F(1 - \alpha/2; n_1 - 1, n_2 - 1) = F(0.95; 5, 9) = 3.48$$

$$F(\alpha/2; n_1 - 1, n_2 - 1) = F(0.05; 5, 9) = \frac{1}{F(0.95; 9, 5)} = \frac{1}{4.77} = 0.210$$

Therefore, by (14.36), the confidence limits are $L = 0.05081/3.48 = 0.0146$ and $U = 0.05081/0.210 = 0.242$, and the desired 90 percent confidence interval is:

$$0.0146 \le \frac{\sigma_1^2}{\sigma_2^2} \le 0.242$$

Since both confidence limits are less than 1, the scientist can conclude that group discussion does lead to a reduction in the variability of appraisal values. ☐

Experimental Group	Appraisal Values ($ thousand)						Statistics	TABLE 14.3
1. With discussion	97	101	102	95	98	103	$n_1 = 6$ $s_1^2 = 9.867$	Data on appraisal values for a commercial property based on experiment with completely randomized design— Property appraisal example
2. Without discussion	118 109 84 85 100 121 115 93 91 112						$n_2 = 10$ $s_2^2 = 194.18$	

A confidence interval for the ratio of two standard deviations, σ_1/σ_2, is obtained by taking the square roots of the confidence limits for the variance ratio in (14.36).

EXAMPLE ☐

For the property appraisal example, the 90 percent confidence interval for σ_1/σ_2 is:

$$0.12 = \sqrt{0.0146} \leq \frac{\sigma_1}{\sigma_2} \leq \sqrt{0.242} = 0.49$$

Thus, the behavioral scientist was able to conclude that group discussion reduces the variability of appraisals substantially, the standard deviation of appraisals with group discussion being only 12 percent to 49 percent as great as that without group discussion.   ☐

Comments

1. One-sided confidence intervals for σ_1^2/σ_2^2 can be obtained from (14.36) by selecting the appropriate limit. For example, a 95 percent upper confidence limit is obtained by calculating U in (14.36) with $\alpha/2 = 0.05$.
2. To obtain the confidence limits in (14.36), we use theorem (14.35) and the definition of percentiles of the F distribution to make the following probability statement:

$$P\left[F(\alpha/2; n_2 - 1, n_1 - 1) \leq \frac{s_2^2/\sigma_2^2}{s_1^2/\sigma_1^2} \leq F(1 - \alpha/2; n_2 - 1, n_1 - 1) \right] = 1 - \alpha$$

Observe that α is equally divided between the two tails, as has been our practice in constructing two-sided confidence intervals. We then rearrange the inequalities within the brackets and reexpress the F percentiles by means of (B.14) to obtain the $1 - \alpha$ confidence limits.

Statistical Tests for σ_1^2/σ_2^2

Statistical tests concerning $\sigma_1^2 = \sigma_2^2$ are constructed using theorem (14.35). We consider tests where the α risk is controlled at $\sigma_1^2 = \sigma_2^2$ or, equivalently, at $\sigma_1^2/\sigma_2^2 = 1$. The test statistic, denoted by F^*, is defined as follows.

(14.37)

Tests concerning σ_1^2/σ_2^2, when the populations are normal or nearly normal, the random samples are independent, and the α risk is controlled at $\sigma_1^2 = \sigma_2^2$, are based on the test statistic:

$$F^* = \frac{s_1^2}{s_2^2}$$

When $\sigma_1^2 = \sigma_2^2$, test statistic F^* follows the F distribution with $n_1 - 1$ numerator degrees of freedom and $n_2 - 1$ denominator degrees of freedom.

We illustrate the use of test statistic F^* with a one-sided lower-tail test.

☐ EXAMPLE

In the property appraisal example, the behavioral scientist had hypothesized that group discussion would reduce the variability of appraisals—that is, it would make σ_1^2 smaller than σ_2^2. The appropriateness of this hypothesis is to be tested.

Step 1. The test alternatives here are:

$$H_0: \frac{\sigma_1^2}{\sigma_2^2} \geq 1$$

$$H_1: \frac{\sigma_1^2}{\sigma_2^2} < 1$$

Note that alternative H_1 is equivalent to $\sigma_1^2 < \sigma_2^2$.

Step 2. The scientist wishes to control the α risk at 0.05 when $\sigma_1^2/\sigma_2^2 = 1$ or, equivalently, when $\sigma_1^2 = \sigma_2^2$.

Step 3. Figure 14.5 shows the appropriate type of decision rule for this test. From theorem (14.35), we have:

$$\frac{s_1^2/\sigma_1^2}{s_2^2/\sigma_2^2} = F(n_1 - 1, n_2 - 1)$$

Observe that when $\sigma_1^2 = \sigma_2^2$, these two terms cancel in the preceding expression, leaving the ratio s_1^2/s_2^2 on the left-hand side. Thus, when $\sigma_1^2 = \sigma_2^2$, we see that $F^* = s_1^2/s_2^2$ has an $F(n_1 - 1, n_2 - 1)$ distribution. It is this sampling distribution for F^* that is illustrated in Figure 14.5.

If test statistic $F^* = s_1^2/s_2^2$ is substantially smaller than 1, the scientist should conclude H_1—that is, conclude $\sigma_1^2/\sigma_2^2 < 1$. Thus, small values of F^* lead to conclusion H_1. The appropriate type of decision rule is therefore one-sided lower-tail, as shown in Figure 14.5. Since $\alpha = 0.05$ here, the action limit must correspond to the percentile:

$$F(\alpha; n_1 - 1, n_2 - 1) = F(0.05; 5, 9) = \frac{1}{F(0.95; 9, 5)} = \frac{1}{4.77} = 0.210$$

The decision rule for the test is therefore as follows:

If $F^* \geq 0.210$, conclude H_0.
If $F^* < 0.210$, conclude H_1.

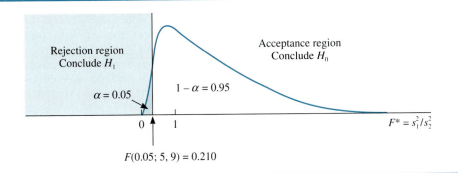

FIGURE 14.5
Statistical decision rule for independent samples, populations normal or nearly normal, lower-tail test for σ_1^2/σ_2^2—Property appraisal example

Step 4. For the two samples, we have $s_1^2 = 9.867$ and $s_2^2 = 194.18$. Hence:

$$F^* = \frac{9.867}{194.18} = 0.05081$$

Since $F^* = 0.05081 < 0.210$, the scientist should conclude H_1; that is, group discussion does reduce the variability of appraisals. ☐

Need for Normality

The inferential procedures for population variances described in this section are exact when the underlying population is normal or nearly normal, but are quite unreliable under moderate and large departures from normality. In other words, the procedures are not robust with regard to departures from normality. The remedies for this lack of robustness are exactly the same as those discussed in Section 11.3 in connection with the use of the t distribution. Transformations of the sample data, such as the logarithmic transformation, may yield data that are more normal. Another option is to employ an inferential procedure that does not require normality of the population. One such procedure is the jackknife procedure to be discussed in the next chapter.

PROBLEMS

14.1 For each of the following comparative studies, state (1) whether it is observational or experimental and (2) the design employed in the study.

a. Eighty technical trainees are randomly divided into two equal groups. One group receives field training only, while the second group receives a mixture of field and classroom training. The skill level of each of the 80 trainees is measured at the end of the training period to compare the effectiveness of the two training systems.

b. A random sample of 100 private dwellings was selected, and the owner of each dwelling was asked for the cost of heating the dwelling during the previous 12 months. Twelve months later, the survey was repeated using the same 100 dwellings. The objective was to compare heating costs for private dwellings in the population for the two 12-month periods based on the comparative heating costs for each of the 100 dwellings.

14.2 For each of the following comparative studies, state (1) whether it is observational or experimental and (2) the design employed in the study.

a. Ten television sets are randomly selected from an assembly process during hour 1 and, independently, 10 sets are randomly selected from the same assembly process during hour 2. Assembly irregularities are noted for each sample television set. The aim of the study is to compare the quality levels of television sets assembled during the two periods.

b. Two formulations of a new skin-softening lotion are to be compared as to their softening action. A random sample of 40 potential users of the lotion is selected. Each person in the sample is independently assigned at random one of the two formulations to be applied to the left arm and the other formulation to be applied to the right arm. After a lapse of eight hours, each person is asked to rate the skin-softening effect of each formulation on a 10-point scale. The ratings will be used to identify the more effective formulation based on the comparative ratings by each of the 40 persons.

14.3 A company with a large number of salespeople wished to compare the annual automobile expenses of salespeople for the current fiscal year with their expenses for the preceding fiscal year. A random sample of salespeople was selected, and the total automobile expenses for each year were determined for each of the selected salespeople.

a. Is this study observational or experimental? What design was employed here?

b. Annual automobile expenses tend to differ among salespeople because their territories differ as to size and population density. Does the study design identified in part a offer an advantage for comparing annual automobile expenses for the two years? Explain.

14.4 Refer to the study described in Problem 14.2b. The study director observed that subjects tended not to use the full 10-point range but rather gave ratings concentrated in limited ranges of the 10-point scale. For instance, some subjects tended to choose ratings for the two lotions from the interval 1 to 7, while others tended to choose ratings from the interval 4 to 10. Will the design of this study be affected favorably or unfavorably by this tendency? Explain.

* **14.5 Weather Environments.** Annual snowfalls in a Northern city in the past 21 years (period 2) averaged 284 centimeters, and their standard deviation was 66 centimeters. In the 21 years before that (period 1), the annual snowfalls averaged 264 centimeters, and their standard deviation was 65 centimeters. Assume that the two sets of observations constitute independent random samples from normal populations with the same variance.

a. Is this study observational or experimental? What study design was employed here?

b. Calculate s_c^2. What parameter does this statistic estimate?

c. Estimate $\mu_2 - \mu_1$, the difference in the mean annual snowfalls for the weather environments prevailing in the two periods, using a 90 percent confidence interval. Interpret the confidence interval.

14.6 Tread Life. The effect of tire pressure on the mean tread life for a certain make of automobile tire was studied for (1) tires inflated to the standard pressure and (2) tires inflated to a higher pressure to improve gas mileage. Thirty tires were randomly selected from the production process, and 15 of these were randomly assigned to be inflated to the standard pressure and the others were assigned to be inflated to the higher pressure. Tread life tests were conducted for all of the tires. The results, expressed in thousands of miles of tread life, follow. The size of sample 1 is shown as 14 because one of the tires assigned to the standard inflation was defective and the testing could not be completed.

Sample	Pressure Level	n	$\overline{X}$	s
1	Standard	14	48.3	1.1
2	Higher	15	46.7	1.3

Assume that both populations are normal with the same variance.

a. Identify the two populations here, taking into account the exclusion of the defective tire from the study.

b. Is this study observational or experimental? What study design was employed here?

c. Estimate $\mu_2 - \mu_1$, the difference in the mean tread lives for the two tire pressures, using a 95 percent confidence interval. Interpret the confidence interval.

14.7 Pain Relief. In a small-scale experiment to compare the pain-relieving effectiveness of two new medications for arthritis sufferers, 27 volunteers were divided at random into two groups of sizes $n_1 = 14$ and $n_2 = 13$. The volunteers in each group were given small but comparable doses of medication. In the group receiving medication 1, the mean hours of relief were $\bar{X}_1 = 6.4$ hours and the standard deviation was $s_1 = 1.5$ hours. The comparable results for the group receiving medication 2 were $\bar{X}_2 = 9.3$ hours and $s_2 = 1.4$ hours. Assume that both populations are normal with the same variance.

a. Is this study observational or experimental? What study design was employed here?
b. Estimate $\mu_2 - \mu_1$, the difference in the mean hours of relief for the two medications, using a 95 percent confidence interval. Interpret the confidence interval.

* **14.8 Electricity Use.** A simple random sample of 400 households from a large community was selected to estimate the mean electricity use per household during February of last year. Another simple random sample of 450 households was selected, independently of the first, to estimate the mean electricity use during February of this year. The sample results (expressed in kilowatt hours) were as follows:

Sample	Year	n	$\bar{X}$	s
1	Last year	400	1252	257
2	This year	450	1330	251

a. Is this study observational or experimental? What study design was employed here?
b. Construct a 99 percent confidence interval for $\mu_2 - \mu_1$. Can one conclude that the mean use per household has changed between the two years? If so, was the change small or large? Discuss.

14.9 University Salaries. The starting salaries of newly appointed assistant professors who have just graduated with doctorates from U.S. universities are to be compared for the fields of business and engineering. The following salary statistics were obtained from independent random samples of newly appointed assistant professors in each of these fields:

Sample	Field	n	$\bar{X}$	s
1	Business	170	$52,250	$6541
2	Engineering	190	$48,212	$6056

a. Is this study observational or experimental? What study design was employed here?
b. Construct a 90 percent confidence interval for $\mu_2 - \mu_1$, the difference in the mean salaries for the two fields. Interpret the confidence interval.
c. Why is the confidence coefficient for the confidence interval in part b only approximately 90 percent?

14.10 Business Directory. Two versions of a business advertising directory, differing in layouts, were distributed to a random sample of 600 households in a large community. One-half of the sample was randomly assigned version 1, and the other half was assigned version 2. Subsequently, each sample household was asked to rate the usefulness of its version of the directory on a 5-point scale from 1 (not useful) to 5 (very useful). The sample results were as follows:

Version	n	$\bar{X}$	s
1	300	3.83	1.04
2	300	3.21	1.19

a. Is this study observational or experimental? What study design was employed here?
b. Construct a 95 percent confidence interval for $\mu_2 - \mu_1$. Interpret the confidence interval.

* **14.11** Refer to **Weather Environments** Problem 14.5.

   a. Test whether the mean annual snowfall is the same in the two weather environments. The α risk is to be controlled at 0.10 when $\mu_2 - \mu_1 = 0$. State the alternatives, the decision rule, the value of the test statistic, and the conclusion.

   b. Using Table C.3, obtain bounds for the P-value of this test.

14.12 Refer to **Tread Life** Problem 14.6.

   a. Test whether the mean tread life for tires inflated to the higher pressure is less than that for tires inflated to the standard pressure. The α risk is to be controlled at 0.025 when $\mu_2 - \mu_1 = 0$. State the alternatives, the decision rule, the value of the test statistic, and the conclusion.

   b. Using Table C.3, obtain bounds for the P-value of this test.

14.13 Refer to **Pain Relief** Problem 14.7.

   a. Test whether the two medications give the same mean hours of relief. The α risk is to be controlled at 0.05 when $\mu_2 - \mu_1 = 0$. State the alternatives, the decision rule, the value of the test statistic, and the conclusion.

   b. The two-sided P-value for this test is 0.000014. Is this value consistent with the test result in part a? Would the test lead to a different conclusion if the α risk were specified as 0.001 rather than 0.05? Explain.

* **14.14** Refer to **Electricity Use** Problem 14.8.

   a. Test whether the mean household use of electricity increased from February of last year to February of this year. Control the α risk at 0.005 when $\mu_2 - \mu_1 = 0$. State the alternatives, the decision rule, the value of the test statistic, and the conclusion.

   b. What is the P-value of this test? Is this value consistent with the test result in part a?

14.15 Refer to **University Salaries** Problem 14.9.

   a. Test whether the mean salaries in the engineering and business fields are equal. Control the α risk at 0.10 when $\mu_2 - \mu_1 = 0$. State the alternatives, the decision rule, the value of the test statistic, and the conclusion.

   b. What is the P-value of this test? Is this value consistent with the test result in part a?

14.16 Refer to **Business Directory** Problem 14.10.

   a. Test whether the mean ratings for the two directories are the same, controlling the α risk at 0.05 when $\mu_2 - \mu_1 = 0$. State the alternatives, the decision rule, the value of the test statistic, and the conclusion.

   b. What is the P-value of this test?

* **14.17** **Oil Additive.** In a test involving highway cruising at 55 miles per hour, eight new cars of a certain make were randomly selected from the production line, and each was driven 1000 miles without an oil additive and 1000 miles with an oil additive. The order of the two runs was independently randomized for each car. Let X_1 and X_2 denote the miles per gallon for a car without and with the oil additive, respectively. Data for the differences $D = X_2 - X_1$ follow:

Car:	1	2	3	4	5	6	7	8
D:	1.87	1.71	2.38	2.19	1.89	1.96	1.76	2.00

Assume that the population of differences is normal.

   a. Is this study observational or experimental? What study design was employed here?

   b. Construct a 90 percent confidence interval for $\mu_2 - \mu_1$, the difference in the mean miles per gallon with and without the additive. Interpret the confidence interval.

   c. Does the confidence interval in part b indicate that the oil additive has an effect? Comment.

14.18 **Achievement Scores.** An educational psychologist studied whether two mathematics achievement tests (1, 2) lead to different achievement scores. Eighteen subjects were randomly selected, and each was given both tests. The order of the two tests was independently randomized for each subject. The test scores follow, where X_1 and X_2 denote the scores on test 1 and test 2, respectively:

Subject:	1	2	3	4	5	6	7	8	9
X_1:	93	70	79	91	83	106	94	87	119
X_2:	81⁻	103	88	85	84	100	103	102	111

Subject:	10	11	12	13	14	15	16	17	18
X_1:	111	95	93	97	73	99	108	77	98
X_2:	114	109	91	123	98	124	113	80	107

Assume that the population of test score differences $D = X_2 - X_1$ is approximately normal.

a. Is this study observational or experimental? What study design was employed here?
b. Construct a 95 percent confidence interval for $\mu_2 - \mu_1$, the difference in the mean scores for the two tests. Interpret the confidence interval.
c. Can you conclude from the interval estimate in part b that the mean scores for the two tests differ? Explain.

14.19 Diagnostic Routines. Two computer routines (1, 2) were developed to diagnose problems in the circuit logic of microprocessor chips. Twenty problems, considered to be a random sample of the types of problems of interest, were each diagnosed by both routines. The differences in the diagnosis times of the two routines ($D = X_2 - X_1$) for the 20 problems had a mean of $\overline{D} = 77$ seconds and a standard deviation of $s_D = 28$ seconds. Assume that the population of differences is normally distributed.

a. Is this study observational or experimental? What study design was employed here?
b. Construct a 99 percent confidence interval for the difference in the mean diagnosis times for the two routines. Interpret the confidence interval.
c. Use the mean and the standard deviation of the sample differences, in conjunction with the assumption of normality, to estimate the probability that routine 2 will take longer than routine 1 to diagnose a randomly selected problem.

* **14.20 Bond Yields.** A financial analyst investigated whether bond yields are affected by the presence of a certain sinking-fund provision in the bond indenture for bonds issued during the past 10 years. The analyst selected a random sample of 160 issues with the sinking-fund provision and independently selected a random sample of 160 bond issues without the sinking-fund provision. The analyst then paired the issues so that each issue in a pair was similar with respect to coupon rate, year of maturity, and investment quality but only one bond in each pair had the sinking-fund provision. Taking the difference $D = X_2 - X_1$ for each pair between the yield of the bond issue with the provision (X_2) and the yield of the bond issue without it (X_1), the analyst found that the 160 differences had a mean of $\overline{D} = -0.18$ percent point and a standard deviation of $s_D = 0.46$ percent point.

a. Is this study observational or experimental? What study design was employed here?
b. Obtain a 98 percent confidence interval for the difference in the mean yields with and without the provision. Interpret the confidence interval.
c. The analyst could have obtained the same estimate of the difference in mean yields with and without the provision by simply calculating the mean yield of the 160 issues with the provision, then calculating the mean yield of the 160 issues without the provision, and, finally, taking the difference between the two means. What is the likely benefit of using the procedure employed by the analyst?

14.21 Property Appraisals. A municipality requires that each residential property seized for non-payment of taxes be appraised independently by two licensed appraisers before it is disposed. In the past 24 months, appraisers Smith and Jones independently appraised 64 such properties. The difference in appraisals $D = X_2 - X_1$ was calculated for each sample property, where X_1 and X_2 denote Smith's and Jones's appraised values, respectively. The mean and standard deviation of the 64 differences were $\overline{D}$ = \$1.21 thousand and s_D = \$2.61 thousand, respectively. Assume that the 64 appraised properties constitute a random sample.

a. Is this study observational or experimental? What study design was employed here? What is the population about which inferences are to be made?
b. Construct a 95 percent confidence interval for the difference in the mean appraisal values for the two appraisers. Interpret the confidence interval.

14.22 Isotope Dating. An archaeologist obtained 130 artifacts and dated each by two different isotope procedures. The 130 differences between the two age datings had a mean of $\overline{D}$ = 53 years and a standard deviation of s_D = 680 years, respectively. Assume that the 130 artifacts constitute a random sample.

a. Is this study observational or experimental? What study design was employed here?
b. Construct a 90 percent confidence interval for the difference in the mean age datings for the two procedures. Interpret the confidence interval.
c. Does the confidence interval in part b indicate that the two procedures, on the average, furnish different ages? Discuss.

* **14.23** Refer to **Oil Additive** Problem 14.17.

a. Test whether the oil additive increases the mean gas mileage. Control the α risk at 0.05 when $\mu_D = \mu_2 - \mu_1 = 0$. State the alternatives, the decision rule, the value of the test statistic, and the conclusion.
b. Using Table C.3, obtain bounds for the P-value of the test.

14.24 Refer to **Achievement Scores** Problem 14.18.

a. Test whether the mean score for test 2 is higher than that for test 1. Control the α risk at 0.025 when $\mu_D = \mu_2 - \mu_1 = 0$. State the alternatives, the decision rule, the value of the test statistic, and the conclusion.
b. The one-sided P-value for this test is 0.0095. Is this value consistent with the test result in part a? Would the test conclusion be different if the α risk were specified as 0.001 rather than 0.025? Explain.
c. If all the subjects had taken test 1 first and then test 2, could one conclude that a difference in mean scores is necessarily due to differences in the two tests? Explain.

14.25 Refer to **Diagnostic Routines** Problem 14.19.

a. Test whether the mean diagnosis times for the two routines are the same. Control the α risk at 0.01 when $\mu_D = \mu_2 - \mu_1 = 0$. State the alternatives, the decision rule, the value of the test statistic, and the conclusion.
b. Using Table C.3, obtain bounds for the P-value of the test.

* **14.26** Refer to **Bond Yields** Problem 14.20. The analyst hypothesized that the sinking-fund provision reduces the mean bond yield.

a. Test whether this hypothesis is true. Control the α risk at 0.01 when $\mu_D = \mu_2 - \mu_1 = 0$. State the alternatives, the decision rule, the value of the test statistic, and the conclusion.
b. What is the P-value of the test? Is this value consistent with the test result in part a?

14.27 Refer to **Property Appraisals** Problem 14.21. An observer who has not seen the actual comparative appraisal data suspects that Jones's appraised values are higher on average than Smith's.

a. Test whether this hypothesis is true. Control the α risk at 0.025 when $\mu_D = \mu_2 - \mu_1 = 0$. State the alternatives, the decision rule, the value of the test statistic, and the conclusion.
b. What is the P-value of the test? Is this value consistent with the test result in part a?

14.28 Refer to **Isotope Dating** Problem 14.22.

a. Test whether the age datings by the two procedures, on the average, are the same. Control the α risk at 0.10 when $\mu_D = \mu_2 - \mu_1 = 0$. State the alternatives, the decision rule, the value of the test statistic, and the conclusion.
b. What is the P-value of the test? Is this value consistent with the test result in part a?

c. If $\mu_1 = \mu_2$, does this imply that the two dating procedures give the same age to an artifact? Explain.

14.29 A researcher selected a random sample of 50 families from the population of two-parent families with one or more children enrolled in schools in a large school district. The researcher asked each parent separately whether he or she favors a six-week extension in the school year to improve academic coverage. The researcher now wishes to test whether the proportions of fathers and mothers in the population who favor the extension are the same. Has the researcher obtained independent random samples of fathers and mothers from the population? Discuss.

* **14.30 Divorces and Annulments.** A random sample of $n_1 = 1000$ recent divorces and annulments in state 1 contained 437 that involved no children. In neighboring state 2, an independent random sample of $n_2 = 1200$ divorces and annulments contained 538 that involved no children. Let p_1 and p_2 denote, respectively, the proportions of divorces and annulments involving no children in the two states.

a. Calculate $\bar{p}_2 - \bar{p}_1$ and $s\{\bar{p}_2 - \bar{p}_1\}$. What does each of these statistics estimate?
b. Construct a 90 percent confidence interval for $p_2 - p_1$. Interpret the confidence interval.
c. Does it appear from the confidence interval in part b that the two state proportions are close to each other? Explain.

14.31 Employee Stability. The following data show the numbers of business school graduates hired by a firm three years ago who are still with the firm, classified by degree:

Degree	Number Hired	Number Remaining
1. Bachelor's	210	141
2. Master's	150	48

Assume that the number of graduates remaining (141 and 48) can be treated as outcomes of independent binomial random variables. Construct a 95 percent confidence interval for $p_2 - p_1$, the difference in the probabilities of Master's and Bachelor's degree graduates remaining with the firm for three years. Interpret the confidence interval.

14.32 Environmental Amendment. A public opinion research institute, taking independent random samples of $n_1 = 500$ males and $n_2 = 500$ females in a state, asked each respondent whether he or she favors the continuation of an environmental amendment in the state constitution. It was found that 335 of the males and 384 of the females were in favor of continuation.

a. Construct a 90 percent confidence interval for $p_2 - p_1$, the difference in the proportions of females and males in the state who favor continuation of the amendment.
b. Does the confidence interval in part a indicate that the proportion of females who favor continuation differs from the proportion of males? If the proportions differ, might this difference be large? Comment.

14.33 Microwave Ovens. A manufacturer obtained data on breakdowns in two makes of portable turntables designed to rotate food in microwave ovens. In a sample of $n_1 = 197$ turntables of make 1, 33 broke down within two years of the date of purchase. The comparable figure in a sample of $n_2 = 290$ units of make 2 was 25. Assume that the two samples are independent random samples. Obtain a 99 percent confidence interval for $p_2 - p_1$, the difference in the probabilities of units of the two makes breaking down within two years. Interpret the confidence interval.

* **14.34** Refer to **Divorces and Annulments** Problem 14.30. An analyst wishes to test whether the two state proportions are equal.

a. Obtain $\bar{p}'$. What does this statistic estimate here?
b. Conduct the test, controlling the α risk at 0.10 when $p_2 - p_1 = 0$. State the alternatives, the decision rule, the value of the test statistic, and the conclusion.
c. What is the P-value of the test? Is this value consistent with the test result in part b?

14.35　Refer to **Employee Stability** Problem 14.31.

a. Test whether the probability of a Bachelor's degree graduate remaining with the firm for three years or longer exceeds that of a Master's degree graduate. Control the α risk at 0.025 when $p_2 - p_1 = 0$. State the alternatives, the decision rule, the value of the test statistic, and the conclusion.

b. What is the P-value of the test? Is this value consistent with the test result in part a?

14.36　Refer to **Environmental Amendment** Problem 14.32.

a. Test whether the proportions of females and males favoring continuation of the amendment are the same. Control the α risk at 0.10 when $p_2 - p_1 = 0$. State the alternatives, the decision rule, the value of the test statistic, and the conclusion.

b. What is the P-value of the test? Is this value consistent with the test result in part a?

14.37　Refer to **Microwave Ovens** Problem 14.33.

a. Test whether the probability of a breakdown within two years is larger for a unit of make 1 than make 2. Control the α risk at 0.005 when $p_2 - p_1 = 0$. State the alternatives, the decision rule, the value of the test statistic, and the conclusion.

b. What is the P-value of the test? Is this value consistent with the test result in part a?

* **14.38　Active Ingredient.** A pharmaceutical firm produces tablets that need to contain a consistent amount of active ingredient. For a random sample of 41 tablets just taken from the production process, the standard deviation of the amounts of active ingredient was $s = 1.09$ milligrams. The amounts of active ingredient in tablets are normally distributed.

a. Construct a 95 percent confidence interval for the process variance of amounts of active ingredient in tablets.

b. Convert the confidence interval in part a into a 95 percent confidence interval for the process standard deviation. Interpret this confidence interval.

c. Why are the confidence limits in part a not equally spaced about s^2?

14.39　Fish Hatchery. The standard deviation of the lengths of 20 mature trout raised on an experimental diet in a hatchery was $s = 4.35$ centimeters. Assume that the lengths of the 20 trouts constitute a random sample from a normal population.

a. Construct a 99 percent confidence interval for the population variance.

b. Convert the confidence interval in part a into a 99 percent confidence interval for the population standard deviation. Interpret this confidence interval.

c. Why are the confidence limits in part a not equally spaced about s^2?

14.40　Refer to **Tread Life** Problem 14.6.

a. Using the sample results for tires inflated to the standard pressure, construct a 95 percent confidence interval for the population variance.

b. Convert the confidence interval in part a into a confidence interval for the population standard deviation. Interpret this confidence interval.

14.41　Heat Diffusion. A scientist studied heat diffusion in metals by measuring the temperature at one end of a metal rod 10 seconds after applying a standard source at the opposite end of the rod. The following temperature readings were obtained in five replications of the experiment:

Replication:	1	2	3	4	5
Temperature (°C):	216	221	220	218	225

The scientist wishes to obtain a confidence interval for the population standard deviation of temperature readings. Assume that the five readings constitute a random sample from a normal population.

a. Construct a 99 percent confidence interval for the population variance.

b. Convert the confidence interval in part a into a confidence interval for the population standard deviation. Interpret the confidence interval. Describe the population about which inferences are being made.

c. Calculate the ratio of the upper confidence limit to the lower confidence limit for the confidence interval in part b. Does this ratio indicate that the scientist has a precise or imprecise estimate of σ? Would the ratio be much different if the scientist had obtained 25 replications of the experiment rather than 5? Explain.

* 14.42 Refer to **Active Ingredient** Problem 14.38. Quality standards require that the process variance for the amount of active ingredient be 1.10 or less. Test the alternatives H_0: $\sigma^2 \leq 1.10$ versus H_1: $\sigma^2 > 1.10$, controlling the α risk at 0.025 when $\sigma^2 = 1.10$. State the decision rule, the value of the test statistic, and the conclusion.

14.43 Refer to **Fish Hatchery** Problem 14.39. From extensive investigations, it is known that the variance of lengths of trout raised on a standard diet is 16.32. It is desired to test whether the population variance of the lengths of trout raised on the experimental diet equals 16.32. The α risk is to be controlled at 0.01 when $\sigma^2 = 16.32$.

a. Conduct the test. State the alternatives, the decision rule, the value of the test statistic, and the conclusion.

b. Is the test result in part a consistent with the confidence interval in Problem 14.39a? Explain.

14.44 Refer to **Tread Life** Problem 14.6. Using the sample results for tires inflated to the standard pressure, test whether the population variance is more than 1.0. Control the α risk at 0.025 when $\sigma^2 = 1.0$. State the alternatives, the decision rule, the value of the test statistic, and the conclusion.

14.45 Refer to **Heat Diffusion** Problem 14.41. Theoretical calculations show that the population variance of temperature readings in this experiment should be 5.00. Test whether the theory is supported by the experimental findings. Control the α risk at 0.01 when $\sigma^2 = 5.00$. State the alternatives, the decision rule, the value of the test statistic, and the conclusion.

* 14.46 Refer to **Active Ingredient** Problem 14.38. In a second random sample of $n_2 = 31$ tablets selected independently from the production process a day later, the standard deviation of the amounts of active ingredient was $s_2 = 1.21$ milligrams. Assume that the amounts of active ingredient in tablets remain normally distributed.

a. Construct a 98 percent confidence interval for σ_1^2 / σ_2^2, the ratio of the process variances for the two days.

b. Convert the confidence interval in part a into a 98 percent confidence interval for the ratio of the process standard deviations. Interpret this confidence interval. Does this confidence interval suggest that a shift in the process variability has occurred?

14.47 Technicians. Two technicians (1, 2) have made measurements of impurity levels in a number of specimens selected from a standard solution. One technician measured 11 specimens and the other measured 9 specimens. It is desired to test whether measurements of impurity levels by the two technicians have the same variability. The technicians' observed measurements can be assumed to constitute independent random samples from normal populations. The sample results (in parts per million) are as follows:

Technician	n	s
1	11	38.6
2	9	21.7

a. Construct a 90 percent confidence interval for σ_1^2 / σ_2^2, the ratio of the variances of the technicians' measurements for specimens from this standard solution.

b. Does the confidence interval in part a indicate that measurements made by one technician differ in variability from those made by the other? Explain.

c. If technician 1 made occasional large measurement errors so that the measurements for this technician are not normally distributed, might the actual confidence level of the confidence interval in part a differ substantially from the specified level of 0.90? Explain.

14.48 The profit outcomes (in $ thousand) of two independent plays of a business game by team 1 were 68 and 96. Two independent plays of the game by team 2 yielded profit outcomes of 181 and -13 (a loss). Assume that the profit outcomes for each team are normally distributed.

a. Construct a 90 percent confidence interval for σ_1^2/σ_2^2, the ratio of the variances of this game's profit outcomes for the two teams.

b. Convert the confidence interval in part a into one for the ratio of the standard deviations.

c. Calculate the ratio of the upper confidence limit to the lower confidence limit for the confidence interval in part b. Does the confidence interval provide a precise comparison of the variability of profit outcomes for the two teams? Comment.

* **14.49** Refer to **Active Ingredient** Problems 14.38 and 14.46. Test whether the process variance has increased during the two-day period, controlling the α risk at 0.01 when $\sigma_1^2 = \sigma_2^2$. State the alternatives, the decision rule, the value of the test statistic, and the conclusion.

14.50 Refer to **Technicians** Problem 14.47. Test whether the variances of the two technicians' measurements are equal, controlling the α risk at 0.10 when $\sigma_1^2 = \sigma_2^2$. State the alternatives, the decision rule, the value of the test statistic, and the conclusion.

14.51 Refer to **Fish Hatchery** Problem 14.39. It was hypothesized that the variability in lengths of mature trout could be reduced by increasing the protein in the diet. The standard deviation of the lengths of $n_2 = 20$ mature trout raised on a higher-protein diet was found to be $s_2 = 2.76$ centimeters. Assume that the population of trout lengths here also is normal.

a. Test whether the higher-protein diet leads to a smaller variance, controlling the α risk at 0.05 when $\sigma_1^2 = \sigma_2^2$. State the alternatives, the decision rule, the value of the test statistic, and the conclusion. [*Hint: F*(0.95; 19, 19) = 2.17.]

b. Describe the nature of a Type II error here. Is it possible that the conclusion in part a is, in fact, a Type II error? Explain.

14.52 Refer to Figure 3.10. An analyst wishes to compare the variability of the intervals between oil spills for the before- and after-controls processes, using the confidence interval in (14.36). The analyst is concerned, however, by the highly skewed pattern of the data sets as indicated by the box plots in this figure. Is the analyst correct in being concerned with this feature of the data sets? Comment.

EXERCISES

14.53 A panel consisting of 300 grain farmers is surveyed annually about cropping intentions for different grains. Each panel member serves for two years, with half of the panel rotated off each year and replaced by a random sample of farmers who were not on the panel in the previous year. A key purpose of the panel is to provide information about changes in acreage intended to be planted in wheat from one year to the next.

a. Denote the change in mean acreage per farm intended to be planted in wheat by $\mu_2 - \mu_1$. Would either the interval estimation procedure in (14.14) or the one in (14.23) be appropriate in this situation if used alone? Comment.

b. Do the matched half of the panel data and the independent half of the panel data in any two consecutive years each estimate $\mu_2 - \mu_1$? Discuss. If so, why might it be advantageous to use this mixed design instead of using either an independent samples design or a matched samples design? Explain.

14.54 T_1 and T_2 are unbiased estimators of total health expenditures by U.S. families in two successive years, denoted by τ_1 and τ_2, respectively.

a. Show that $T_2 - T_1$ is an unbiased estimator of the difference $\tau_2 - \tau_1$.

b. Show that the more positive is the covariance of T_1 and T_2 (other factors being unchanged), the more efficient is $T_2 - T_1$ as an estimator of $\tau_2 - \tau_1$.

c. Explain the relevance of the results in parts a and b to the use of matched samples of families to estimate the change in total health expenditures between two years.

14.55 Demonstrate that (1) s_c^2 in (14.8) is an unbiased estimator of σ^2, (2) $(n_1 + n_2 - 2)s_c^2/\sigma^2$ has a $\chi^2(n_1 + n_2 - 2)$ distribution.

14.56 Explain why the pooled estimator $\bar{p}'$ in (14.29) is employed in tests concerning $p_2 - p_1$ using (14.31), whereas it is not employed in estimating $p_2 - p_1$ by confidence interval (14.28).

14.57 Consider the estimation of a population variance.

a. What is the smallest sample size for which the ratio of the upper and lower confidence limits in (14.33) for a 90 percent confidence interval is less than 3?

b. What is the corresponding sample size for a 95 percent confidence interval? What is the effect of increasing the confidence coefficient?

14.58 Refer to **Active Ingredient** Problem 14.38. Specifications require that at least 99 percent of tablets have an amount of active ingredient within ± 2.0 milligrams of the mean amount. Test whether the process meets this specification, controlling the α risk at 0.01. State the alternatives, the decision rule, the value of the test statistic, and the conclusion. [*Hint:* What value of σ satisfies $2/\sigma = z(0.995)$?]

STUDIES

14.59 Refer to the **Power Cells** data set (Appendix D.2). Consider power cells 1–36 only. Note that power cells 1–18 are low charge-rate cells and cells 19–36 are high charge-rate cells. Pair each of the low charge-rate cells with the corresponding cell from the high charge-rate group with the same discharge rate, depth of discharge, and ambient temperature (that is, the matched pairs are 1 and 19, 2 and 20, . . . , 18 and 36). Let X_2 and X_1 denote the numbers of cycles before failure for the low charge-rate cell and the high charge-rate cell in a pair, respectively.

a. Obtain the differences $D = X_2 - X_1$ for the 18 matched pairs of cells.

b. Assuming that the differences in part a constitute a random sample from an approximately normal distribution, obtain a 95 percent confidence interval for $\mu_2 - \mu_1$, the difference in the mean number of cycles before failure for the two charge rates.

c. Use the confidence interval in part b to test whether $\mu_D = \mu_2 - \mu_1 = 0$. What is the test conclusion?

14.60 A labor economist studied the durations of the most recent strikes in the vehicles and construction industries to see whether strikes in the two industries are equally difficult to settle. To achieve approximate normality and equal variances, the economist worked with the logarithms (to base 10) of the duration data (expressed in days), and obtained the following sample statistics:

Industry	Number of Strikes	Mean Log-Duration	Standard Deviation of Log-Durations
1. Vehicles	13	0.593	0.294
2. Construction	15	0.973	0.349

The economist believes it is reasonable to treat the data as constituting independent random samples.

a. Construct a 90 percent confidence interval for the difference in the mean log-durations of strikes in the two industries. Interpret the interval estimate.

b. What do the antilogarithms of the two confidence limits in part a represent? [*Hint:* Recall formula (3.24) and the properties of logarithms.]

c. Test whether strikes in the two industries have the same mean log-durations, controlling the α risk at 0.10. State the alternatives, the decision rule, the value of the test statistic, and the conclusion.

d. Test the economist's assumption that the log-durations of strikes in the two industries have equal variances, controlling the α risk at 0.10 when $\sigma_1^2 = \sigma_2^2$. State the alternatives, the decision rule, the value of the test statistic, and the conclusion.

14.61 **Weight Reduction.** Sixty-two persons who were seriously overweight were randomly assigned to one of two weight reduction regimens, with 31 assigned to each regimen. During the study period, one person in regimen 2 moved out of town. All other persons remained in the study. At the end of the study period, the weight losses were ascertained. The data on weight losses (in kilograms) follow:

Regimen 1

15.2	14.4	16.3	13.6	11.7	10.2
12.6	12.6	14.3	16.7	13.0	17.1
11.7	12.9	14.4	14.2	14.5	14.8
12.7	15.1	11.1	10.0	12.1	14.7
11.9	17.2	11.3	13.3	14.6	12.9
13.7					

Regimen 2

16.2	18.6	12.6	17.9	18.2	20.0
17.3	18.3	19.7	15.1	16.8	16.8
15.5	20.5	16.7	16.4	14.8	18.3
16.6	17.9	18.8	16.9	18.2	17.3
16.2	18.2	18.1	15.2	16.5	19.2

a. Construct a back-to-back stem-and-leaf display for the two data sets, using the common stems 10, 11, . . . , 20. Do the two data sets appear (1) to have the same means, (2) to have the same variability, (3) to be drawn from normal populations? Comment.

b. Assume that the two populations are normal. Obtain a point estimate of the probability of a weight loss of 16.0 kilograms or more for each of the two regimens. Does there appear to be a substantial difference between the two probabilities? Comment.

c. Continue to assume that the two populations are normal. Test whether the two population variances are equal, controlling the α risk at 0.02 when $\sigma_1^2 = \sigma_2^2$. State the alternatives, the decision rule, the value of the test statistic, and the conclusion.

d. Assume that the two populations are normal and have equal variances. Test whether the mean weight loss for regimen 2 exceeds that for regimen 1, as had been expected. Control the α risk at 0.05 when $\mu_2 - \mu_1 = 0$. State the alternatives, the decision rule, the value of the test statistic, and the conclusion.

e. Continuing to assume that the populations are normal with equal variances, estimate the difference in the mean weight losses $\mu_2 - \mu_1$, using a 90 percent confidence interval. Interpret the confidence interval.

Estimation and Testing II

Nonparametric Procedures

15

The inference procedures considered thus far have relied heavily either on the assumption that the underlying population is of a particular form (for example, normal or not a marked departure from normal) or, when little can be assumed about the nature of the population, on large sample sizes and the central limit theorem. In this chapter, we discuss procedures that require only minimal assumptions about the population and can be used for both small and large samples. These procedures are called *nonparametric* or *distribution-free* procedures. They are typically simple to implement, are generally derived from elementary probability considerations, and often employ simple statistics based on ranked or ordered sample data. Because of the minimal assumptions about the population that these nonparametric procedures require, they are generally very robust. Furthermore, they are frequently almost as efficient as procedures that make strict assumptions about the population.

We first discuss nonparametric procedures that are concerned with the median of a single population. Inferences about the population median are important in many applications for two major reasons. First, when the population is highly skewed (for instance, in studies of family incomes, store sales, and manufacturers' inventories), the population median is located more in the center of the distribution than the population mean and thus may be a more meaningful measure of position. Second, when the population is symmetrical, the population mean and the population median coincide and thus are equally meaningful measures of position.

INFERENCES 15.1 ABOUT POPULATION MEDIAN

Assumptions

The inference procedures for the population median that we consider in this section require only the following two assumptions.

1. The population is continuous.
2. The sample is a random one.

Point Estimation of Population Median

We denote the *population median* by η (Greek eta). We shall use the sample median Md as defined in (3.7) as the point estimator of η. Recall that the sample median Md is the value of the middle observation in an array of the sample data. When the sample size is even, Md is taken to be the mean of the two observations in the middle of the array.

Case Duration   **EXAMPLE**

An administrator studied how long it takes to settle claims made under a workers' compensation insurance plan for injuries received from on-the-job accidents. For one category of claims, the durations of time (in months) from the occurrence of the accident to the closing of the case were as follows for a random sample of nine settled cases:

Case:	1	2	3	4	5	6	7	8	9
Case Duration (months):	4.5	5.8	2.6	4.3	2.7	10.5	5.6	1.9	4.1

We are to estimate the median duration η for claims in this population.

Arraying the nine case durations in ascending order, we see that the sample median is $Md = 4.3$ months:

Observation:	1.9	2.6	2.7	4.1	4.3	4.5	5.6	5.8	10.5
Rank:	1	2	3	4	5	6	7	8	9

Therefore, $Md = 4.3$ months is a point estimate of the population median η for the population of cases under study here.   □

Comment

The sample median Md is an appropriate choice as a point estimator of the population median η because of the following properties of the sampling distribution of Md.

1. For a symmetrical population, it can be shown that Md is an unbiased estimator of η.
2. For any continuous population, whether symmetrical or not, the sample median has the same chance of lying above η as below it when n is odd. In other words, the median of the sampling distribution of Md equals η. The same result holds approximately when n is even. The reason why the result is approximate when n is even is our convention for calculating Md as the mean of the middle two observations in this case.
3. The sample median is a consistent estimator of η when the population is continuous (although it is not always the most efficient).

Confidence Interval for Population Median

A confidence interval for the population median η is easy to construct. Let us denote the rth smallest sample observation by L_r and the rth largest by U_r. In the case duration example, for instance, the second smallest and second largest observations are $L_2 = 2.6$ and $U_2 = 5.8$. By an appropriate choice of r, the values L_r and U_r may be used as confidence limits for the population median η.

> **(15.1)**
> A confidence interval for the population median η, when the population is continuous, is:
>
> $$L_r \leq \eta \leq U_r$$
>
> where r is suitably chosen to give the desired confidence coefficient $1 - \alpha$.

When the sample size is small, the value r is chosen from a table; for large sample sizes, it may be calculated from a formula. We shall now take up each of these cases in turn.

Small Sample.   Table C.8 provides the confidence coefficient $1 - \alpha$ for different choices of r, for sample sizes up to 15. We now illustrate the interpretation of this table.

1. If $n = 10$, choosing $r = 1$ and therefore using the confidence limits L_1 and U_1 (that is, the two extreme sample observations) involves a confidence coefficient of 0.998.
2. If $n = 11$, choosing $r = 2$ and therefore using confidence limits L_2 and U_2 (that is, the second smallest and second largest sample observations) involves a confidence coefficient of 0.988.
3. If $n = 15$, choosing $r = 4$ and therefore using the confidence limits L_4 and U_4 (that is, the fourth smallest and fourth largest sample observations) involves a confidence coefficient of 0.965.

Table C.8 allows for easy determination of r and the resulting confidence limits L_r and U_r to control the confidence coefficient near the desired level.

EXAMPLES ☐

1. In the case duration example, a confidence interval for η is desired with a confidence coefficient near 95 percent. Table C.8 shows that, for $n = 9$, the closest we can come to 95 percent is by use of $r = 2$, which involves a 96.1 percent confidence coefficient. Hence, we require L_2 and U_2. We see from the sample array presented earlier that $L_2 = 2.6$ and $U_2 = 5.8$. Hence, our confidence interval is:

$$2.6 \leq \eta \leq 5.8$$

Thus, we can conclude, with 96.1 percent confidence, that the median case duration in the population is between 2.6 and 5.8 months.
   Figure 15.1 contains the MINITAB output for this example. Shown there are:

1. In the last row, the 96.1 percent confidence interval that we just obtained.
2. In the first row, the confidence interval with the nearest confidence coefficient below the desired 95 percent level (which is 82.0 percent).
3. In the middle row, a confidence interval based on a nonlinear interpolation between intervals 1 and 2 to yield approximately a 95 percent confidence coefficient.

FIGURE 15.1   **MINITAB output for confidence interval for population median η—Case duration example**

```
                                ACHIEVED
                   N   MEDIAN   CONFIDENCE   CONFIDENCE INTERVAL   POSITION
     Duration      9   4.300     0.8203      (   2.700,    5.600)      3
                                 0.9500      (   2.623,    5.754)    NLI
                                 0.9609      (   2.600,    5.800)      2
```

The column POSITION in the output gives the value of r for each pair of confidence limits L_r and U_r. For the first row, $r = 3$; and for the last row, $r = 2$. The sample median, $Md = 4.3$, also is presented in the output.

2. A random sample of $n = 8$ households showed the following family incomes (in \$ thousand) in ascending order:

Observation:	15.9	16.4	18.3	21.2	24.9	26.3	27.6	41.7
Rank:	1	2	3	4	5	6	7	8

A confidence interval is desired for the population median family income with confidence coefficient near 90 percent. We see from Table C.8 that, for $n = 8$, use of $r = 2$ involves a 93.0 percent confidence coefficient, which is quite close to the desired 90 percent level. Hence, we use $L_2 = 16.4$ and $U_2 = 27.6$ as confidence limits, and the 93 percent confidence interval for the population median family income is:

$$16.4 \leq \eta \leq 27.6$$

The limits are wide here because of the very small sample size and the large variability in the observations. □

In general, the confidence limits L_r and U_r will not be equidistant from the sample median Md. For instance, in the case duration example, we see that $L_2 = 2.6$ and $U_2 = 5.8$ are not equidistant around $Md = 4.3$. Also note that the procedure just described does not enable us to achieve exactly a desired confidence coefficient, such as 90 percent or 95 percent. The reason is that the binomial probability distribution, which determines the confidence coefficient, is discrete. However, we can get close to the desired confidence coefficient except when the sample size is quite small.

Large Sample.   For sample sizes larger than 15, the value of r required for the confidence interval $L_r \leq \eta \leq U_r$ can be obtained by an approximation formula.

(15.2)
When the sample size n exceeds 15 and a $1 - \alpha$ confidence interval $L_r \leq \eta \leq U_r$ is desired, select the largest integer r that does not exceed:

$$0.5 \left[n + 1 - z(1 - \alpha/2)\sqrt{n} \right]$$

1. A random sample of 43 tax returns has been selected. We are to construct a 95 percent confidence interval for the population median amount of interest income per return. Since $1 - \alpha = 0.95$, we require $z(1 - \alpha/2) = z(0.975) = 1.960$, and by using (15.2) we obtain:

$$0.5(43 + 1 - 1.960\sqrt{43}) = 15.57$$

   The largest integer not exceeding 15.57 is 15. Hence, $r = 15$ should be used, and the required confidence limits for η are the 15th smallest (L_{15}) and the 15th largest (U_{15}) interest-income amounts in the sample of 43 returns.

2. A random sample of 75 vouchers has been selected. We are to construct a 90 percent confidence interval for the population median amount expended per voucher. Since $1 - \alpha = 0.90$, we require $z(0.95) = 1.645$, and we obtain:

$$0.5(75 + 1 - 1.645\sqrt{75}) = 30.88$$

   The largest integer not exceeding 30.88 is 30. Hence, $r = 30$ should be chosen, and the appropriate confidence limits are the 30th smallest (L_{30}) and the 30th largest (U_{30}) expenditure amounts.   ☐

Comments

1. The confidence coefficient for an interval estimate of η based on either Table C.8 or formula (15.2) is derived from the fact that the confidence interval $L_r \leq \eta \leq U_r$ will be correct if at least r and not more than $n - r$ out of the n sample observations fall above the median. Hence, the confidence coefficient equals the probability $P(r \leq f \leq n - r)$, where f is the frequency of the sample observations that are above the median.

(15.3)

The confidence interval $L_r \leq \eta \leq U_r$ for the population median η has a confidence coefficient given by:

$$P(r \leq f \leq n - r) = \sum_{i=r}^{n-r} \binom{n}{i} (0.5)^i (0.5)^{n-i}$$

where f is the frequency of sample observations above η.

   To see how this result is obtained, consider a sample of $n = 3$ observations and a choice of $r = 1$ so that the confidence limits L_1 and U_1 will be used. The following sample outcomes are possible:

Sample Outcome	Population Median η
1	$\underline{L_1}\ \ \underline{U_1}$
2	$\underline{L_1}\ \ \underline{U_1}$
3	$\underline{L_1}\ \ \underline{U_1}$
4	$\underline{L_1}\ \ \underline{U_1}$

Note that the confidence interval $L_1 \leq \eta \leq U_1$ will only be correct in cases 2 and 3, where at least $r = 1$ of the sample observations and not more than $n - r = 3 - 1 = 2$ fall above the median. Denoting the frequency of sample observations above the median by f, we see that the probability of a correct confidence interval here is $P(1 \leq f \leq 2)$.

The frequency f of sample observations above the median is a binomial random variable with $p = 0.50$ for the following two reasons: (1) The sample observations are independent by virtue of random sampling. (2) Each observation exceeds the population median η with probability 0.5—that is, $P(X > \eta) = 0.5$—because the median of a continuous probability distribution divides the distribution in half, as illustrated in Figure 15.2. Hence, the sample frequency f is a sum of independent Bernoulli observations, each with $p = 0.50$, and by (7.5) is therefore a binomial random variable with $p = 0.50$.

It then follows that the probability of the confidence interval $L_1 \leq \eta \leq U_1$ being correct when $n = 3$ is:

$$P(1 \leq f \leq 2) = \sum_{i=1}^{2} \binom{3}{i}(0.5)^i(0.5)^{3-i}$$

Consulting the binomial tables (Table C.5) for $n = 3$ and $p = 0.50$, we see that the probability is:

$$P(1 \leq f \leq 2) = 0.3750 + 0.3750 = 0.750$$

which is the entry for the confidence coefficient in Table C.8 for $n = 3$ and $r = 1$.

2. Formula (15.2) for r when n is not small is obtained from the normal approximation to the binomial probability in (15.3), using the correction for continuity discussed in Chapter 13.
3. The rth smallest sample value, L_r, is often referred to as the *rth order statistic* of the sample.
4. Confidence intervals for other population percentiles, such as the 25th or 75th percentiles, can be obtained by a procedure similar to the one discussed here for the median.

Sign Test for Population Median

We now describe a simple test for the population median η called a *sign test*. We consider the usual three types of test alternatives.

(15.4)	(15.4a)	(15.4b)	(15.4c)
	$H_0:\ \eta \leq \eta_0$	$H_0:\ \eta \geq \eta_0$	$H_0:\ \eta = \eta_0$
	$H_1:\ \eta > \eta_0$	$H_1:\ \eta < \eta_0$	$H_1:\ \eta \neq \eta_0$

Here, η_0 denotes the value of the population median at which the α risk is to be controlled.

FIGURE 15.2
Position of population median η in a continuous population

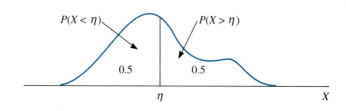

The sign test is based on the frequency of sample observations that fall above η_0. The test is actually an ordinary test concerning a population proportion p, where p here denotes the probability that a sample observation falls above η_0; that is, $p = P(X > \eta_0)$. Figure 15.2 shows the location of the population median η for a continuous population. Observe that if η_0 is below η, then $p = P(X > \eta_0) > 0.5$. Thus, the alternative H_1: $\eta > \eta_0$ corresponds to the alternative H_1: $p > 0.5$. Similarly, we see from Figure 15.2 that if η_0 is greater than η, then $p = P(X > \eta_0) < 0.5$ so that the alternatives H_1: $\eta < \eta_0$ and H_1: $p < 0.5$ are equivalent. Finally, we see from Figure 15.2 that when η_0 coincides with η (that is, when $\eta_0 = \eta$) then $p = P(X > \eta_0) = 0.5$ so that the alternatives H_1: $\eta \neq \eta_0$ and H_1: $p \neq 0.5$ are equivalent.

Hence, we can restate the alternatives concerning η in (15.4) in terms of alternatives concerning p.

(15.5)		Equivalent Alternatives for Sign Test
(15.5a)	H_0: $\eta \leq \eta_0$ H_1: $\eta > \eta_0$	H_0: $p \leq 0.5$ H_1: $p > 0.5$
(15.5b)	H_0: $\eta \geq \eta_0$ H_1: $\eta < \eta_0$	H_0: $p \geq 0.5$ H_1: $p < 0.5$
(15.5c)	H_0: $\eta = \eta_0$ H_1: $\eta \neq \eta_0$	H_0: $p = 0.5$ H_1: $p \neq 0.5$

where $p = P(X > \eta_0)$.

Thus, the sign test is simply a test about a population proportion where the α risk is controlled at $p_0 = 0.5$. We shall let f denote the frequency of sample observations larger than η_0; consequently, $\bar{p} = f/n$ denotes the proportion of sample observations larger than η_0. When $\eta = \eta_0$ (or $p = p_0 = 0.5$), then f and $\bar{p}$ follow the binomial probability distribution with $p = 0.5$.

For large n, the sign test uses test statistic z^* in (13.11) with $p_0 = 0.5$ and the normal approximation. For small n, the exact test procedure in (13.14) can be used for the sign test. We shall illustrate the sign test by an example in which n is large.

EXAMPLE ☐ **Travelers Checks**

A financial institution that issues travelers checks is concerned about the length of time (X) that a check is outstanding; that is, the length of time from issue to redemption. Management wishes to test whether the population median time outstanding for checks issued in the immediately preceding year (η) equals the median time outstanding for checks issued in earlier years, which is known to be $\eta_0 = 2.16$ months.

Step 1. The alternatives here are:

$$H_0\text{: } \eta = 2.16$$
$$H_1\text{: } \eta \neq 2.16$$

or

$$H_0\text{: } p = 0.5$$
$$H_1\text{: } p \neq 0.5$$

where $\eta_0 = 2.16$ and $p = P(X > 2.16)$.

FIGURE 15.3
Statistical decision rule and
MINITAB output for large-
sample, two-sided sign test
for η — Travelers checks
example

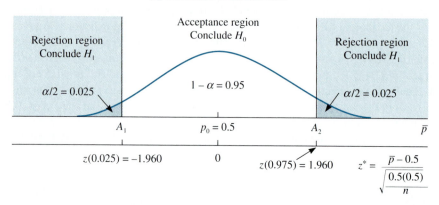

(a) Statistical Decision Rule

(b) MINITAB Output

```
SIGN TEST OF MEDIAN = 2.160 VERSUS  N.E.  2.160

            N   BELOW  EQUAL  ABOVE  P-VALUE     MEDIAN
Months     250   161     0     89    0.0000      1.400
```

Step 2. The α risk is to be controlled at 0.05 when $\eta = \eta_0 = 2.16$ or, equivalently, when $p = p_0 = 0.5$.

Step 3. A random sample of $n = 250$ travelers checks issued in the preceding year was selected, and the length of time outstanding was obtained for each check. The appropriate type of decision rule is the two-sided one shown in Figure 15.3a.

The sampling distribution of the standardized test statistic z^* in (13.11) when $p = p_0 = 0.5$ is approximately normal because of the large sample size. Hence, for $\alpha = 0.05$, we require $z(0.975) = 1.960$. Thus, the appropriate decision rule for the test is as follows:

If $|z^*| \leq 1.960$, conclude H_0.
If $|z^*| > 1.960$, conclude H_1.

Step 4. A fragment of the sample data, arrayed in *descending* order, follows.

Months Outstanding (X):	12.5	$\cdots$	2.2	2.1	$\cdots$	0.0
Rank:	1	$\cdots$	89	90	$\cdots$	250

It can be seen from the ranking that $f = 89$ observations are larger than $\eta_0 = 2.16$. Equivalently, we could have computed the differences $X - \eta_0 = X - 2.16$ and counted the number of the differences that have plus signs as follows.

Months Outstanding (X):	12.5	$\cdots$	2.2	2.1	$\cdots$	0.0
Rank:	1	$\cdots$	89	90	$\cdots$	250
$X - \eta_0$:	$+10.34$	$\cdots$	$+0.04$	-0.06	$\cdots$	-2.16
Sign:	$+$	$\cdots$	$+$	$-$	$\cdots$	$-$

We see that the number of plus signs is again $f = 89$. It is this latter procedure that gives the sign test its name.

Since $f = 89$, we have $\bar{p} = f/n = 89/250 = 0.356$. The standardized test statistic for $p_0 = 0.5$ is therefore:

$$z^* = \frac{\bar{p} - 0.5}{\sqrt{\dfrac{0.5(1 - 0.5)}{n}}} = \frac{0.356 - 0.5}{\sqrt{\dfrac{0.5(1 - 0.5)}{250}}} = -4.55$$

Since $|z^*| = 4.55 > 1.960$, we conclude H_1—that the population median time outstanding for travelers checks issued in the preceding year is no longer 2.16 months. The two-sided P-value for this test is $2P(Z < z^*) = 2P(Z < -4.55) = 0+$. This very small P-value indicates that the evidence is strong that a change in the median time outstanding has occurred.

Figure 15.3b contains the MINITAB output for the sign test. The sample frequency $f = 89$ and the P-value $= 0+$ are shown, as well as the sample median time outstanding $Md = 1.4$ months. $\square$

Comments

1. Instead of defining f as the frequency of sample observations *above* η_0, we could also have defined it as the frequency of observations *below* η_0. The rejection region for one-sided tests would then be in the opposite direction, although the conclusion would be the same, as is also the case for two-sided tests.

2. In theory, sample values should not exactly equal η_0 because the population is assumed to be continuous. In practice, sample values may occasionally equal η_0 because of rounding. It is then conventional to disregard these cases for the sign test and reduce the sample size accordingly.

15.2

INFERENCES ABOUT $\eta_2 - \eta_1$—INDEPENDENT SAMPLES AND COMPLETELY RANDOMIZED DESIGNS

A number of useful nonparametric procedures are available for comparative studies involving two populations. In this section, we describe a nonparametric procedure for comparing the positions of two population distributions that requires only that the populations be continuous and of the same shape, as illustrated in Figure 15.4. The procedure is applicable to studies based on an independent samples design or a completely randomized design. The following example involves a completely randomized design.

EXAMPLE $\square$

In the marketing instruction example that was presented in Chapter 14, an instructor wished to compare the effectiveness of two teaching methods in an introductory marketing course. The total course enrollment of 24 students was divided at random into two equal sections. Students in one section viewed videotaped lectures featuring the instructor (treatment 1), while the students in the other section received live lectures by the instructor (treatment 2). In all other respects, the two course sections were handled in identical fashion. The grade distributions for the two teaching methods are now to be compared. The instructor has concluded from previous studies that the type of

FIGURE 15.4
Two population distributions of same shape but differing in position

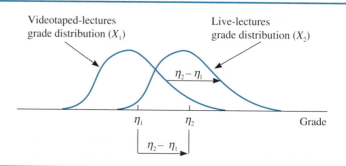

instruction should not affect the shape of the grade distribution, only its position. Figure 15.4 illustrates the situation for the case when live lectures lead to higher grades.

☐

The two grade distributions in Figure 15.4 have the same shape but differ in position. Because the distributions are of the same shape, the distance between their medians $\eta_2 - \eta_1$ and the distance between their means $\mu_2 - \mu_1$ are identical measures of the difference in position of the two distributions. Figure 15.4 illustrates the situation when the difference in positions is measured in terms of the population medians, and the following discussion will concentrate on the difference $\eta_2 - \eta_1$. If the difference $\eta_2 - \eta_1$ is positive, as in Figure 15.4, the distribution for treatment 2 lies a distance $|\eta_2 - \eta_1|$ to the right of the distribution for treatment 1. If $\eta_2 - \eta_1$ is negative, the distribution for treatment 2 lies a distance $|\eta_2 - \eta_1|$ to the left of the distribution for treatment 1. Finally, if $\eta_2 - \eta_1 = 0$, the two distributions are at the same position and hence are identical.

In comparative studies, interest usually lies first in testing whether the two population distributions differ in position and then, if they do differ, in estimating the magnitude of the difference. We turn first to a nonparametric statistical test for $\eta_2 - \eta_1$, called the *Mann–Whitney–Wilcoxon test,* or the M–W–W test for short.

Mann–Whitney–Wilcoxon Test for $\eta_2 - \eta_1$

Assumptions

1. The two population distributions are continuous and have the same shape, although they may differ in position.
2. Independent random samples of sizes n_1 and n_2 are drawn from the two populations; that is, the study is based either on an independent samples design or on a completely randomized design.

Note that assumption 1 does not specify the shape of the distributions.

Alternatives.   The M–W–W test can be used for testing the usual one-sided and two-sided alternatives concerning the difference $\eta_2 - \eta_1$.

(15.6)	(15.6a)	(15.6b)	(15.6c)
	$H_0: \eta_2 - \eta_1 \leq 0$	$H_0: \eta_2 - \eta_1 \geq 0$	$H_0: \eta_2 - \eta_1 = 0$
	$H_1: \eta_2 - \eta_1 > 0$	$H_1: \eta_2 - \eta_1 < 0$	$H_1: \eta_2 - \eta_1 \neq 0$

Test Statistic. The test statistic for the M–W–W test is simple and is calculated as follows.

1. Combine the n_1 sample observations from population 1 and the n_2 sample observations from population 2 and array the combined data in ascending order.
2. Assign ranks to the combined observations (starting with 1 for the smallest observation).
3. Sum the ranks for the n_2 sample observations from population 2 and denote this sum by S_2.

EXAMPLE ☐

Consider the following simple sample results when $n_1 = 2$ and $n_2 = 3$:

Sample from Population 1: $n_1 = 2$
Observations: 4   12
Sample from Population 2: $n_2 = 3$
Observations: 14   10   17

The array of the combined samples and the associated ranks are as follows:

Observation:	4	10	12	14	17
Rank:	1	2	3	4	5
Population:	1	2	1	2	2

The ranks shown in boxes are for the three sample observations from population 2. We see that the test statistic is:

$$S_2 = 2 + 4 + 5 = 11$$

☐

If distribution 2 lies to the right of distribution 1 (so $\eta_2 - \eta_1 > 0$, as in Figure 15.4), the observations from population 2 will tend to be larger than the observations from population 1, and, consequently, S_2 will tend to be large because it will be a sum of large ranks. Conversely, if distribution 2 is to the left of distribution 1 (so $\eta_2 - \eta_1 < 0$), the observations from population 2 will tend to be smaller than the observations from population 1, and S_2 will tend to be small because it will be a sum of small ranks. Finally, if the two distributions are identical (so $\eta_2 - \eta_1 = 0$), all possible orderings of the combined samples will be equally likely, and S_2 will tend to have an intermediate value because it will be a sum of both large and small ranks.

Thus, the appropriate type of decision rule for each set of alternatives in (15.6) is intuitively clear. For example, in a test of:

$$H_0: \eta_2 - \eta_1 = 0$$
$$H_1: \eta_2 - \eta_1 \neq 0$$

very small or very large values of S_2 will lead to H_1, and intermediate values will lead to H_0.

Sampling Distribution of S_2 When $\eta_2 - \eta_1 = 0$. To control the α risk when $\eta_2 - \eta_1 = 0$, we need to know the sampling distribution of S_2 when $\eta_2 - \eta_1 = 0$. As we mentioned earlier, all possible orderings of the combined observations are equally likely to occur when

$\eta_2 - \eta_1 = 0$. This fact allows the exact sampling distribution of S_2 to be worked out, and it is found to depend only on the sample sizes n_1 and n_2. Although exact probabilities for S_2 can be calculated, the following normal approximation is adequate for most practical applications.

(15.7)

When independent random samples of sizes n_1 and n_2 are selected from two identical populations (that is, with the same shape and $\eta_2 - \eta_1 = 0$), the sampling distribution of S_2 has mean and variance:

$$E\{S_2\} = \frac{n_2(n_1 + n_2 + 1)}{2} \qquad \sigma^2\{S_2\} = \frac{n_1 n_2(n_1 + n_2 + 1)}{12}$$

When n_1 and n_2 are sufficiently large, the sampling distribution of S_2 is approximately normal. As a working rule, the normal approximation is adequate when n_1 and n_2 are each 10 or more.

Thus, the sample sizes need not be particularly large for the normal approximation to be applicable.

Standardized Test Statistic.   Tests for the alternatives in (15.6) are based on the following standardized test statistic.

(15.8)

The M–W–W test concerning $\eta_2 - \eta_1$, when the sample sizes are sufficiently large, is based on the standardized test statistic:

$$z^* = \frac{S_2 - E\{S_2\}}{\sigma\{S_2\}}$$

where $E\{S_2\}$ and $\sigma\{S_2\}$ are given by (15.7).

When $\eta_2 - \eta_1 = 0$, test statistic z^* follows approximately a standard normal distribution.

Given the approximate normality of the sampling distribution of z^* when $\eta_2 - \eta_1 = 0$, the construction of the decision rule parallels earlier large-sample tests.

☐ **EXAMPLE**

In the marketing instruction example, the instructor wishes to test whether the method of instruction affects the position of the grade distribution.

Step 1. The alternatives here are two-sided:

$$H_0: \eta_2 - \eta_1 = 0$$
$$H_1: \eta_2 - \eta_1 \neq 0$$

Step 2. The instructor wants to control the α risk at 0.05 when $\eta_2 - \eta_1 = 0$.

Step 3. Table 15.1 contains the sample results. Note that the instructor has recorded the treatment (1 for videotaped lectures, 2 for live lectures) and the grade of each

Treatment	Grade	Rank	Treatment	Grade	Rank
2	42	1	1	75	13
1	49	2	2	78	14.5
1	52	3	2	78	14.5
1	56	4	1	80	16
2	57	5	2	84	17
1	61	6	2	85	18
2	64	7	1	87	19
2	69	8	1	88	20
2	71	9	2	91	21
1	72	10	1	94	22
2	73	11	2	97	23
1	74	12			

$$S_2 = 1 + 5 + 7 + \cdots + 21 + 23 = 149$$

Treatment 1: videotaped lectures; Treatment 2: live lectures

TABLE 15.1
Combined array and ranks for sample data—Marketing instruction example

student for both course sections combined, and has arrayed the combined grades from lowest (42) to highest (97). Also note that one student withdrew from the videotaped-lectures section, leaving 11 students in that section. Thus, $n_1 = 11$ and $n_2 = 12$. Figure 15.5a shows the appropriate decision rule. The sampling distribution of S_2 when $\eta_2 - \eta_1 = 0$ is approximately normal here because both sample sizes exceed 10. To control the α risk at 0.05, we therefore require $z(0.975) = 1.960$. Hence, the decision rule for the test is as follows:

If $|z^*| \leq 1.960$, conclude H_0.
If $|z^*| > 1.960$, conclude H_1.

Step 4. To calculate test statistic S_2, we assign ranks to the combined array of grades in Table 15.1. The ranks for students in the live-lectures section (treatment 2) are placed in boxes in Table 15.1 for emphasis. The M–W–W test procedure requires that tied observations be assigned the mean value of their individual ranks. In this example, the two grades of 78 are the only instance of tied observations. Because these two grades would normally have had ranks 14 and 15, each has been given the mean rank 14.5. Test statistic S_2 is the sum of the ranks for students who received treatment 2 (the ranks in boxes). This sum is shown at the bottom of Table 15.1 to be $S_2 = 149$.

The value of standardized test statistic (15.8) can now be calculated as follows:

$$E\{S_2\} = \frac{12(11 + 12 + 1)}{2} = 144$$

$$\sigma^2\{S_2\} = \frac{11(12)(11 + 12 + 1)}{12} = 264 \qquad \sigma\{S_2\} = \sqrt{264} = 16.25$$

$$z^* = \frac{S_2 - E\{S_2\}}{\sigma\{S_2\}} = \frac{149 - 144}{16.25} = 0.31$$

Since $|z^*| = 0.31 \leq 1.960$, we conclude H_0—that $\eta_2 - \eta_1 = 0$. Thus, this small-scale experiment suggests that the median grade for videotaped lectures does not differ from that for live lectures. The large two-sided P-value, $2P(Z > z^*) = 2P(Z > 0.31) = 2(0.3783) = 0.7566$, indicates that there is strong evidence from this experiment

FIGURE 15.5
Statistical decision rule and SAS output for large-sample, two-sided M–W–W test for $\eta_2 - \eta_1$—Marketing instruction example

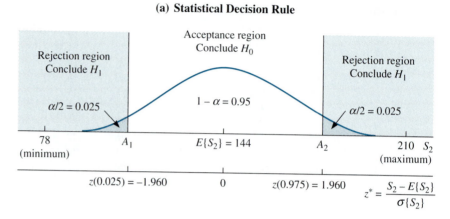

(a) Statistical Decision Rule

(b) SAS Output

LEVEL	N	SUM OF SCORES	EXPECTED UNDER HO	STD DEV UNDER HO
1	11	127.00	132.00	16.24
2	12	149.00	144.00	16.24

```
WILCOXON 2-SAMPLE TEST (NORMAL APPROXIMATION)
(WITH CONTINUITY CORRECTION OF .5)
S=  127.00      Z=-0.2770       PROB >|Z|=0.7818
```

to support concluding that the median grades for the two teaching methods do not differ. □

The M–W–W test procedure can equally well be based on the sum of the ranks of the sample from population 1, which we will denote by S_1. The expected value of S_1 when $\eta_2 - \eta_1 = 0$ is as follows.

(15.9) $E\{S_1\} = \dfrac{n_1(n_1 + n_2 + 1)}{2}$

The variance of S_1 when $\eta_2 - \eta_1 = 0$ is the same as that of S_2.

☐ **EXAMPLE**

Figure 15.5b contains a portion of the SAS output for the M–W–W test in the marketing instruction example. The test is called the WILCOXON 2-SAMPLE TEST. The output refers to ranks as SCORES. The output shows the values of the test statistics, $S_1 = 127$ and $S_2 = 149$; their expected values when $\eta_2 - \eta_1 = 0$, $E\{S_1\} = 132$ and $E\{S_2\} = 144$; and their common standard deviation, $\sigma\{S_1\} = \sigma\{S_2\} = 16.24$. SAS uses the smaller of S_1 and S_2 for the test, which happens to be $S_1 = 127$ here. The output shows the corresponding standardized statistic $z^* = -0.2770$ and that the two-sided P-value equals 0.7818. The P-value in the output differs slightly from the value we computed earlier (0.7566) because the SAS routine adjusts the standard deviation for tied ranks and also applies a correction for continuity in using the normal approximation. ☐

Comments

1. The M–W–W test accommodates tied ranks provided they are not numerous. If they are, the decision rule for the M–W–W test must be modified.

2. Although the M–W–W test uses only the ranks of the sample observations rather than the actual observations, for reasonably large samples it is almost as powerful at times as tests that depend on specific knowledge about the form of the underlying populations. For instance, the M–W–W test compares favorably with the t test described in Section 14.2 for comparing the positions of two normal distributions. That is, when both tests have the same α risk, the probability of correctly concluding H_1 when $\eta_2 - \eta_1 \neq 0$ is almost as large for the M–W–W test as for the t test.

3. The smallest possible value of statistic S_2 is the sum of the first n_2 ranks:

$$1 + 2 + \cdots + n_2 = \frac{n_2(n_2 + 1)}{2}$$

The largest possible value of S_2 is the sum of the last n_2 ranks:

$$(n_1 + 1) + (n_1 + 2) + \cdots + (n_1 + n_2) = n_1 n_2 + \frac{n_2(n_2 + 1)}{2}$$

For the marketing instruction example, for instance, where $n_1 = 11$ and $n_2 = 12$, the minimum and maximum values of S_2 are:

$$\frac{12(13)}{2} = 78 \qquad 11(12) + \frac{12(13)}{2} = 210$$

so we see that the observed result $S_2 = 149$ is not an extreme one.

   The expected value of S_2 when $\eta_1 = \eta_2$, given in (15.7), is located midway between the minimum and maximum values of S_2. This fact is illustrated in Figure 15.5a for the marketing instruction example, where $E\{S_2\} = 144$ is halfway between the minimum and maximum values of 78 and 210.

4. The statistic S_2 is also called the *Wilcoxon rank sum statistic* because it was originally developed by Wilcoxon. Mann and Whitney independently formulated an alternative test statistic called the *Mann–Whitney U statistic* that is the following simple function of S_2 and leads to identical tests.

(15.10) $$U = S_2 - \frac{n_2(n_2 + 1)}{2}$$

M–W–W Confidence Interval for $\eta_2 - \eta_1$

The M–W–W test procedure can be adapted to provide a confidence interval for $\eta_2 - \eta_1$ by generalizing the procedure so that we can test whether the difference between the two population medians is a specified amount d. The two-sided alternatives in this more general case follow.

$$(15.11) \quad H_0: \eta_2 - \eta_1 = d$$
$$\qquad\qquad H_1: \eta_2 - \eta_1 \neq d$$

Note that the alternatives in (15.6c) are a special case of those in (15.11), namely, when $d = 0$.

We carry out the general test by subtracting d from each of the observations X_2 from population 2 and denoting the new observations by X_2'.

$$(15.12) \quad X_2' = X_2 - d$$

Now when H_0 in (15.11) holds—that is, when the difference between the population medians for the X_2 and X_1 observations is d—the difference between the population medians for the new X_2' observations and the X_1 observations becomes zero because d was subtracted from each X_2 observation. Hence, we can test the generalized alternatives in (15.11) by means of the M–W–W test described earlier in the usual manner but using the X_2' observations instead of the X_2 observations.

The $1 - \alpha$ confidence interval for $\eta_2 - \eta_1$ consists of all those values of d for which the M–W–W test would lead to conclusion H_0 when testing the generalized alternatives in (15.11), with the risk of making a Type I error controlled at α.

Because the determination of the M–W–W confidence interval requires extensive calculations, the confidence limits for $\eta_2 - \eta_1$ are usually obtained from a computer routine.

☐ **EXAMPLE**

Figure 15.6 contains the MINITAB computer output for the M–W–W confidence interval for the marketing instruction example. The output shows the achieved confidence coefficient of 95.5 percent (it is not exactly equal to the desired 95 percent level because of the discrete nature of the test statistic). The output presents the confidence interval for $\eta_1 - \eta_2$, rather than the interval for $\eta_2 - \eta_1$:

$$-17 \leq \eta_1 - \eta_2 \leq 10$$

The 95.5 percent confidence interval for $\eta_2 - \eta_1$ therefore is:

$$-10 \leq \eta_2 - \eta_1 \leq 17$$

This latter interval indicates that the median grade for live lectures is between 17 points higher and 10 points lower than the median grade for videotaped lectures. Thus, the interval does not suggest that one teaching method is clearly superior to the other.

The output also provides a point estimate for $\eta_1 - \eta_2$ that is based on advanced methods not discussed here.

FIGURE 15.6
MINITAB output for M–W–W confidence interval for $\eta_1 - \eta_2$—Marketing instruction example

```
Video        N =   11      Median =          74.00
Live         N =   12      Median =          75.50
Point estimate for ETA1-ETA2 is             -3.00
95.5 pct c.i. for ETA1-ETA2 is (-17.00,10.00)
```

To illustrate how the confidence interval for $\eta_2 - \eta_1$ is obtained in principle, we reproduce the following M–W–W results for selected values of d.

d	S_2	z^*	Test Conclusion
-20	193	3.02	H_1
-10	174.5	1.88	H_0
0	149	0.31	H_0
10	127	-1.05	H_0
20	106.5	-2.31	H_1

For instance, when $d = 0$, the X_2 data are not transformed and we have the original test result in which $S_2 = 149$, $z^* = 0.31$, and test conclusion H_0 is reached. Thus, 0 is a value in the confidence interval. For $d = 20$, transformation (15.12) reduces the grades for each of the $n_2 = 12$ students in the live-lectures section by 20 points. The M–W–W test applied to the revised sample data yields $S_2 = 106.5$, $z^* = -2.31$, and test conclusion H_1 (because $|z^*| > 1.960$). Thus, 20 is a value not included in the confidence interval.

Computer programs utilize special algorithms that do not require carrying out many separate M–W–W tests and thereby can obtain the confidence limits very quickly. □

Comment

The M–W–W procedure for testing and obtaining confidence intervals has been presented here in terms of making inferences about $\eta_2 - \eta_1$. We noted in the introduction to this section that the assumption of the same shape for the two population distributions implies that $\eta_2 - \eta_1 = \mu_2 - \mu_1$. Hence, the M–W–W procedure may be used equally for making inferences about the difference between the two population means, $\mu_2 - \mu_1$.

15.3 INFERENCES ABOUT η_D —MATCHED SAMPLES AND RANDOMIZED BLOCK DESIGNS

In this section, we describe several nonparametric procedures for comparing the positions of two populations when the comparative study is based on a matched samples design or on a randomized block design. Recall from Chapter 14 that such designs involve pairing similar sample elements from the two populations. As in Chapter 14, we shall let X_1 and X_2 denote the paired sample observations from populations 1 and 2, respectively. Also similar to Chapter 14, the nonparametric procedures we now describe are based on the differences of the paired observations, $D = X_2 - X_1$.

We shall denote the median of the population of differences by η_D. The value of η_D tells us about the locations of the two distributions. When the X_2 observation is larger than the X_1 observation for the majority of differences D in the population of differences, then the median difference is positive; that is, $\eta_D > 0$. Similarly, when X_2 is smaller than X_1 for the majority of paired observations in the population of differences, then the median difference is negative; that is, $\eta_D < 0$. Finally, if the differences D in the population of differences are positive and negative with equal frequency, then $\eta_D = 0$.

Thus, η_D is a parameter that measures the difference in positions of the two population distributions. This can also be seen by considering the special case of a symmetrical distribution of differences. Here, the mean and median differences are equal; that is, $\eta_D = \mu_D$, where μ_D is the mean of the population of differences. The latter, we recall, equals the difference between the two population means; that is, $\mu_D = \mu_2 - \mu_1$. Hence, in this special case, η_D also measures the difference in locations as the difference between the two population means.

Confidence Interval for η_D

Since η_D is the median of a single population, namely, the population of differences D, the confidence interval and sign test presented in Section 15.1 for making inferences about the population median can be used here. Thus, there are no new principles involved here. We shall illustrate the inference procedures with an example for setting up a confidence interval for η_D.

Assumptions

1. The population of differences D is continuous.
2. The sample differences constitute a random sample from the population of differences.

Credit Information   **EXAMPLE**

A business researcher studied the impact of the amount of information on credit decisions by means of an experiment based on a randomized block design. Forty volunteer credit managers were grouped into $n = 20$ pairs, the managers in each pair having similar backgrounds with respect to business experience, education, and so on. Two case descriptions were prepared of a hypothetical business that is requesting trade credit. The first case description (treatment 1) provided limited information about the business (for example, product line, years in business, total assets). The second case description (treatment 2) provided detailed information, including financial and operating data for the past several years. In each pair, a random number was used to decide which credit manager received treatment 1 and which one received treatment 2. Each credit manager then studied the assigned case description and determined the maximum amount of credit he or she would extend to the business.

The sample results are presented in Table 15.2a, columns 1 and 2. The credit amounts with limited information are denoted by X_1, and those with detailed information are denoted by X_2. The researcher would like to estimate the median difference η_D of the population of differences $D = X_2 - X_1$ with a 95 percent confidence interval.

The sample differences are calculated in column 3 of Table 15.2a, and the differences are ranked in Table 15.2b. Since $n = 20$ exceeds 15, we use formula (15.2) for finding r. For $1 - \alpha = 0.95$, we require $z(1 - \alpha/2) = z(0.975) = 1.960$. Hence:

$$0.5[n + 1 - z(1 - \alpha/2)\sqrt{n}] = 0.5(20 + 1 - 1.960\sqrt{20}) = 6.12$$

or $r = 6$. We therefore choose L_6 and U_6, the 6th smallest (rank 6) and 6th largest (rank 15) sample differences, for our confidence limits. Table 15.2b shows that $L_6 = 10$ and $U_6 = 28$. Thus, the desired 95 percent confidence interval is:

$$10 \leq \eta_D \leq 28$$

(a) Calculation of Differences				TABLE 15.2
	(1)	(2)	(3)	Calculation and ranking of
	Credit Amount ($ thousand)			differences for paired sample data—Credit information
Pair of Managers	Limited Information X_1	Detailed Information X_2	Difference D	example
1	85	90	+ 5	
2	50	60	+ 10	
3	85	113	+ 28	
4	70	100	+ 30	
5	30	45	+ 15	
6	75	99	+ 24	
7	67	86	+ 19	
8	60	85	+ 25	
9	60	85	+ 25	
10	85	87	+ 2	
11	70	50	− 20	
12	85	90	+ 5	
13	70	93	+ 23	
14	77	115	+ 38	
15	70	62	− 8	
16	50	85	+ 35	
17	55	72	+ 17	
18	75	125	+ 50	
19	81	103	+ 22	
20	20	60	+ 40	

(b) Ranking of Differences

Difference:	−20	−8	...	+10	...	+28	...	+40	+50
Rank:	1	2	...	6	...	15	...	19	20

The researcher can state with 95 percent confidence that detailed information in the case description has resulted in a median increase of between $10 thousand and $28 thousand in the maximum amount of trade credit extended. These limits clearly indicate that the maximum amount of credit tends to be larger here with detailed information than with limited information. □

Wilcoxon Signed Rank Test for η_D

When we can assume that the population of differences is symmetrical, a nonparametric test called the *Wilcoxon signed rank test* is generally more powerful than the sign test for making inferences about the population median difference η_D.

The condition of a symmetrical population of differences is frequently met approximately in observational studies. In experimental studies based on a randomized block design, a symmetrical distribution of differences is to be expected when the two treatments have no differential effects, because each experimental unit has an equal chance of being assigned to either treatment.

Assumptions

1. The population of differences D is continuous and symmetrical.
2. The sample differences constitute a random sample from the population of differences.

Alternatives.   The Wilcoxon signed rank test can be employed for both two-sided and one-sided alternatives concerning η_D.

Test Statistic.   The test statistic for the Wilcoxon signed rank test is simple and is calculated as follows.

1. Obtain the absolute values of the sample differences, $|D|$. If a sample difference D should happen to equal zero, discard it and reduce the sample size accordingly.
2. Rank the absolute differences. If any absolute differences are tied, assign the mean value of the corresponding ranks, as in the M–W–W test.
3. To each rank, attach a plus or a minus sign according to whether its associated sample difference D is positive or negative.
4. Sum the signed ranks, and denote the sum by T.

□ EXAMPLE

In the credit information example, the business researcher wishes to test whether the median difference in the maximum amount of credit extended with limited and detailed information is zero. The researcher prepared a stem-and-leaf plot of the differences in Table 15.2a, column 3, which supported the reasonableness of the assumption of a symmetrical population of differences. Table 15.3 shows the calculation of the Wilcoxon test statistic T. In column 1, the sample differences D are repeated from Table 15.2a, column 3. The absolute values of the differences are shown in column 2. The ranks from smallest absolute difference (rank 1) to largest (rank 20) are shown in column 3. The signed ranks are shown in column 4. It can be seen from Table 15.3 that $T = 184$.

□

If the sample differences are mainly positive, then the positive ranks making up the sum T predominate, and T is large and positive. If the sample differences are mainly negative, then the negative ranks making up the sum T predominate, and T is large and negative. Finally, if the sample differences are positive and negative with almost equal frequency, then the positive and negative ranks making up the sum T are nearly balanced, and T is close to zero. This last situation is expected when the population of differences is symmetrically distributed about 0, so that $\eta_D = 0$. In fact, in this case, the rank associated with any absolute difference has an equal probability of being positive or negative.

Sampling Distribution of T When $\eta_D = 0$.   To control the α risk when $\eta_D = 0$, we need to know the sampling distribution of T when $\eta_D = 0$. As we just mentioned, the rank associated with any absolute difference has an equal probability of being positive or negative when $\eta_D = 0$. This fact allows the exact sampling distribution of T to be worked out. It depends only on the number of sample differences n. Although exact probabilities for T can be calculated, the following normal approximation is adequate for most practical applications.

TABLE 15.3
Wilcoxon signed rank test
calculations—Credit information example

Pair of Managers	(1) Difference D	(2) Absolute Difference $\lvert D \rvert$	(3) Rank	(4) Signed Rank
1	$+\ 5$	5	2.5	$+\ \ 2.5$
2	$+10$	10	5	$+\ \ 5$
3	$+28$	28	15	$+15$
4	$+30$	30	16	$+16$
5	$+15$	15	6	$+\ \ 6$
6	$+24$	24	12	$+12$
7	$+19$	19	8	$+\ \ 8$
8	$+25$	25	13.5	$+13.5$
9	$+25$	25	13.5	$+13.5$
10	$+\ 2$	2	1	$+\ \ 1$
11	-20	20	9	$-\ \ 9$
12	$+\ 5$	5	2.5	$+\ \ 2.5$
13	$+23$	23	11	$+11$
14	$+38$	38	18	$+18$
15	$-\ 8$	8	4	$-\ \ 4$
16	$+35$	35	17	$+17$
17	$+17$	17	7	$+\ \ 7$
18	$+50$	50	20	$+20$
19	$+22$	22	10	$+10$
20	$+40$	40	19	$+19$
				$T\ =\ \ 184$

(15.13)

When a random sample of n differences is selected from a symmetrical population of differences with $\eta_D = 0$, the sampling distribution of T has mean and variance:

$$E\{T\} = 0 \qquad \sigma^2\{T\} = \frac{n(n+1)(2n+1)}{6}$$

When n is sufficiently large, the sampling distribution of T is approximately normal. As a working rule, the normal approximation is adequate when n is 10 or more.

Thus, the sample size need not be very large for the normal approximation to apply.

Standardized Test Statistic. Tests concerning η_D are based on the following standardized test statistic.

(15.14)

The Wilcoxon signed rank test concerning η_D, when the number of sample differences is sufficiently large, is based on the standardized test statistic:

$$z^* = \frac{T}{\sigma\{T\}}$$

continues

where $\sigma\{T\}$ is given by (15.13).

When $\eta_D = 0$, test statistic z^* follows approximately a standard normal distribution.

Given the approximate normality of the sampling distribution of z^* when $\eta_D = 0$, the construction of the decision rule parallels earlier large-sample tests.

□ **EXAMPLE**

In the credit information example, the Wilcoxon signed rank test of whether the median difference η_D is zero proceeds as follows.

Step 1. The test alternatives are:

$$H_0: \eta_D = 0$$
$$H_1: \eta_D \neq 0$$

Step 2. The researcher wants to control the α risk at 0.05 when $\eta_D = 0$.

Step 3. Figure 15.7a shows the appropriate decision rule. The sampling distribution of T when $\eta_D = 0$ is approximately normal in this example because $n = 20$. Hence, to control α at 0.05, we require $z(0.975) = 1.960$, and the decision rule for the test is as follows:

If $|z^*| \leq 1.960$, conclude H_0.
If $|z^*| > 1.960$, conclude H_1.

FIGURE 15.7
Statistical decision rule and SAS output for large-sample, two-sided Wilcoxon signed rank test for η_D—Credit information example

(a) Statistical Decision Rule

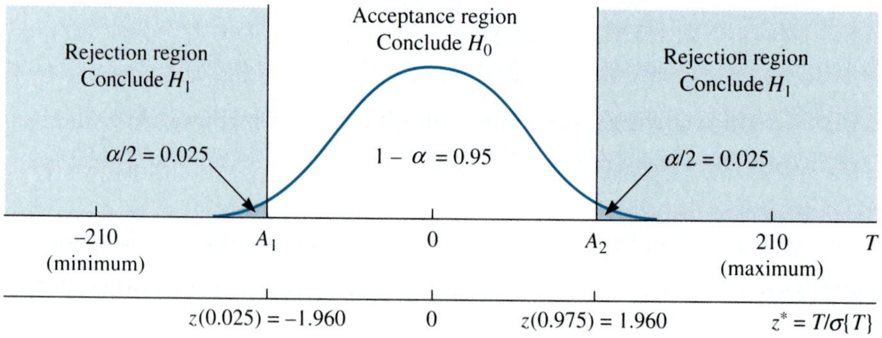

(b) SAS Output

N	20		
Mean	19.25	Sum	385
Std Dev	16.81752	Variance	282.8289
Sgn Rank	92	Prob>|S|	0.0002
Num ^= 0	20		

Step 4. The calculations in Table 15.3 gave $T = 184$. Thus, the standardized test statistic (15.14) here is:

$$\sigma^2\{T\} = \frac{20(20 + 1)[2(20) + 1]}{6} = 2870$$

$$\sigma\{T\} = \sqrt{2870} = 53.57$$

$$z^* = \frac{T}{\sigma\{T\}} = \frac{184}{53.57} = 3.43$$

Since $|z^*| = 3.43 > 1.960$, we conclude H_1—that the median difference η_D is not zero. Hence, the locations of the distributions of maximum amounts of credit extended with limited and detailed information differ. The two-sided P-value is $2P(Z > z^*) = 2P(Z > 3.43) = 0.0006$, which indicates that the evidence is quite strong that η_D does not equal zero. We saw earlier from the confidence interval for η_D that the maximum credit tends to be larger with detailed information than with limited information.

Figure 15.7b contains a portion of the SAS output for the Wilcoxon signed rank test. The upper part of the output contains some descriptive statistics for the sample differences. The lower part contains the test statistic $T/2 = 184/2 = 92$, which is labeled Sgn Rank, and the P-value $= 0.0002$, which differs slightly from our P-value of 0.0006 because the SAS routine adjusts the standard deviation $\sigma\{T\}$ for tied ranks and also applies a correction for continuity in using the normal approximation.   □

Comments

1. The Wilcoxon signed rank test accommodates tied ranks provided they are not numerous. If they are, modifications must be made in the decision rule.

2. The Wilcoxon signed rank test may also be used to test whether a single population (which is symmetrical and continuous) has a median equal to some specified value $\eta = \eta_0$. For each sample observation, the difference $D = X - \eta_0$ is computed, and the test of whether or not $\eta_D = 0$ is applied to these differences in the standard fashion. Testing whether the difference D has a zero median is equivalent to testing whether X has a median equal to η_0.

3. Like the M–W–W test, the Wilcoxon signed rank test often compares favorably with parametric tests based on stronger assumptions about the underlying population (such as the t test for a normal population of differences, discussed in Section 14.3). For this reason, the Wilcoxon test is widely used in practice.

4. The largest possible value of test statistic T occurs when all the sample differences are positive; T is then the sum of the ranks $1, 2, \ldots, n$:

$$1 + 2 + \cdots + n = \frac{n(n + 1)}{2}$$

When all the sample differences are negative, then T attains its smallest possible value, $-n(n + 1)/2$. For the credit information example, where $n = 20$, we have $n(n + 1)/2 = 210$. Figure 15.7a shows -210 and 210 as the minimum and maximum values of T.

5. The Wilcoxon signed rank test can be carried out equivalently by using only the sum of the positive ranks or the sum of the negative ranks. The output for statistical packages may not indicate which version of the Wilcoxon statistic is used.

Wilcoxon Signed Rank Confidence Interval for η_D

The Wilcoxon signed rank test procedure can be adapted to provide a confidence interval for η_D in the same fashion as we adapted the M–W–W procedure to obtain a confidence interval for the difference between two population medians, $\eta_2 - \eta_1$. We generalize the Wilcoxon procedure by considering the following alternatives.

$$(15.15) \qquad H_0:\ \eta_D = d$$
$$ H_1:\ \eta_D \neq d$$

Note that in the credit information example, the alternatives were those in (15.15) with $d = 0$.

We carry out the general Wilcoxon test procedure by subtracting d from each difference D.

$$(15.16) \qquad D' = D - d$$

Now, if H_0 in (15.15) holds and the median difference η_D equals d, then the median of the transformed differences D' will be zero since we subtracted d from each difference. Hence, we can test the alternatives in (15.15) by carrying out the usual Wilcoxon signed rank test based on the transformed differences D'.

The $1 - \alpha$ Wilcoxon confidence interval for η_D consists of all those values of d for which the Wilcoxon test would lead to conclusion H_0 when testing the alternatives in (15.15), with the risk of making a Type I error controlled at α. Computer packages are generally used to obtain the confidence limits because the calculations are extensive. These packages can obtain the Wilcoxon confidence interval quickly because they use special algorithms rather than carrying out the Wilcoxon test repeatedly for different values of d.

☐ **EXAMPLE**

For the credit information example, Figure 15.8 contains the MINITAB output for the Wilcoxon confidence interval. It shows that the achieved confidence coefficient is 95.0 percent, which happens to equal the desired 95 percent confidence level. (In some applications, the achieved confidence level may differ slightly from the desired level because of the discrete nature of the test statistic.) Figure 15.8 shows that the confidence interval for η_D is:

$$12 \leq \eta_D \leq 27.5$$

This confidence interval indicates that detailed information results in a median increase in the maximum amount of trade credit extended of between $12 thousand

---

FIGURE 15.8

MINITAB output for Wilcoxon signed rank confidence interval for η_D—Credit information example

	N	ESTIMATED MEDIAN	ACHIEVED CONFIDENCE	CONFIDENCE INTERVAL
D	20	20.2	95.0	(12.0, 27.5)

and \$27.5 thousand. The interval is very similar to the one obtained earlier using the ordered sample differences; this interval was $10 \leq \eta_D \leq 28$. This latter interval did not require the assumption of a symmetric population of differences.

The MINITAB output also provides a point estimate for η_D that is based on advanced methods not discussed here. □

OPTIONAL TOPIC—TEST FOR RANDOMNESS 15.4

In previous chapters, we have emphasized the importance of checking whether data generated by a process constitute a random sample from an infinite population. Graphic methods for checking the randomness of a data set include time series plots and scatter plots relating successive observations, as explained in Chapter 9. Now, we shall take up a formal, nonparametric statistical test for examining whether data generated by a process constitute a random sample.

Nature of Randomness

Recall from definition (9.10) that a sequence of observations generated by a process constitutes a simple random sample from an infinite population if the observations are independent and identically distributed. We shall refer to a process that generates independent and identically distributed observations as a *random process*. Thus, a random process furnishes data that constitute a simple random sample.

EXAMPLES □

1. Figure 15.9a shows a sequence of observations from a random process. Note that the sequence behaves in a random fashion, with no trend or other consistent pattern

---

Sequences of observations from random and nonrandom processes

FIGURE 15.9

(a) **Normal Random Process**

(b) **Process Involving an Upward Trend**

(c) **Process Involving Increasing Variability**

(d) **Process Involving Cyclical Behavior**

(e) **Process Involving Regular and Frequent Changes in Direction**

being present. The data in Figure 15.9a were generated by a *normal random process*. This is a random process in which the observations are drawn from the same normal distribution.

2. Bernoulli and Poisson processes, defined in (7.4) and (8.15), respectively, are examples of some other types of random processes. □

Departures from randomness can occur in many ways, some of which are illustrated in Figure 15.9. The sequence of observations in Figure 15.9b comes from a process with an upward trend, which is inconsistent with the assumption that all observations of a random process are from the same distribution. In Figure 15.9c, the sequence of observations exhibits increasing variability, which again indicates that the observations are not from the same distribution. Figures 15.9d and 15.9e contain sequences in which consecutive observations are dependent. In Figure 15.9d, the process involves a cyclical pattern, while in Figure 15.9e, the process alternates regularly.

Tests for randomness of a process are of major importance because the assumption of randomness underlies statistical inference. In addition, tests for randomness are important for time series analysis, discussed in Chapters 24 and 25.

Test of Runs Up and Down

Departures from randomness can take so many forms that no single test for randomness is best for all situations. One of the most common departures from randomness is the tendency of a sequence to persist in its direction of movement. This type of departure is found in Figure 15.9b, where the sequence contains a trend, and in Figure 15.9d, where the persistence of movement produces a pronounced cyclical pattern. We now present a nonparametric test that is particularly effective when a process contains these types of persistence. The test is also effective when a process contains excessively frequent changes in direction, as in Figure 15.9e.

Assumption.   The test assumes only that all observations come from continuous populations.

Alternatives.   The *runs-up-and-down test* may be used for detecting two different types of departures from randomness in a sequence of observations.

1. For detecting trend and other forms of persistence, the alternatives are as follows:

(15.17) H_0: Sequence generated by a random process
H_1: Sequence generated by a process containing persistence

2. For detecting departures from randomness consisting of either frequent or infrequent changes in direction, the alternatives are as follows.

(15.18) H_0: Sequence generated by a random process
H_1: Sequence generated by a process containing either persistence or frequent changes in direction

EXAMPLE ☐ **Lake Level**

Table 15.4 shows the maximum level of a lake each year for a 30-year period. We wish to test whether the 30 observations have been generated by a random process or whether the process contains a persistent trend. The presence of a trend would have significant environmental policy implications. ☐

Test Statistic. The test statistic is a simple one—namely, the total number of runs up and down. In Table 15.4, the direction of change in successive observations of the maximum lake level is shown by plus and minus signs. Thus, since $X_2 = 6.59$ is less than $X_1 = 6.63$, the change is $-$. A *run* is a succession of plus or minus signs, surrounded by the opposite sign. We see in Table 15.4 that the first run consists of two minus signs and the next run consists of one plus sign. Let R denote the total number of runs in the sequence. We find that there are $R = 12$ runs for the data in Table 15.4.

If a trend or other form of persistence is present, the number of runs R will clearly be small. The smallest possible value of R is 1 (a single run). On the other hand, if the process involves frequent changes in direction, the number of runs will be large. The largest possible value of R is $n - 1$, which occurs when the signs alternate with each observation. Thus, when testing the alternatives in (15.17), a one-sided lower-tail test is employed because small values of R are consistent with H_1. On the other hand, when testing the alternatives in (15.18), a two-sided test is employed because both small and large values of R are consistent with H_1.

Sampling Distribution of R When H_0 Holds. To control the α risk, we need to know the sampling distribution of R when the sequence is generated by a random process. Since all possible permutations of the sample observations are equally likely when the generating process is random, it is possible to derive the exact sampling distribution of R. This distribution depends only on the number of sample observations. Although exact probabilities for R can be calculated, a normal approximation is adequate for many practical applications.

Year	Level (meters)	Change		Year	Level (meters)	Change	
1	6.63			16	5.91	−	
2	6.59	− } Run 1		17	5.81	−	Run 7
3	6.46	−		18	5.64	−	
4	6.49	+ } Run 2		19	5.51	−	
5	6.45	−		20	5.31	−	
6	6.41	−		21	5.36	+ } Run 8	
7	6.38	−	Run 3	22	5.17	− } Run 9	
8	6.26	−		23	5.07	−	
9	6.09	−		24	4.97	−	
10	5.99	−		25	5.00	+ } Run 10	
11	5.92	−		26	5.01	+	
12	5.93	+ } Run 4		27	4.85	− } Run 11	
13	5.83	− } Run 5		28	4.79	−	
14	5.82	−		29	4.73	−	
15	5.95	+ } Run 6		30	4.76	+ } Run 12	

TABLE 15.4
Determination of the number of runs in the maximum lake level for 30 consecutive years—Lake level example

(15.19)

When a sequence of n observations is generated by a random process and the population distribution is continuous, the sampling distribution of R has mean and variance:

$$E\{R\} = \frac{2n-1}{3} \qquad \sigma^2\{R\} = \frac{16n-29}{90}$$

When n is sufficiently large, the sampling distribution of R is approximately normal. As a working rule, the normal approximation is adequate if n is 20 or more.

Standardized Test Statistic.   The runs-up-and-down test is based on the following standardized test statistic.

(15.20)

The runs-up-and-down test, when the number of observations is 20 or more, is based on the standardized test statistic:

$$z^* = \frac{R - E\{R\}}{\sigma\{R\}}$$

where $E\{R\}$ and $\sigma\{R\}$ are given by (15.19).

When the process is random, test statistic z^* follows approximately a standard normal distribution.

Given the approximate normality of the sampling distribution of z^* when H_0 holds, the construction of the decision rule parallels earlier large-sample tests.

□ **EXAMPLE**

In the lake level example, we wish to test whether persistence is present in the process.

Step 1. The alternatives here are those in (15.17).

Step 2. The α risk is to be controlled at 0.01 when the process is random.

Step 3. Figure 15.10 shows the appropriate type of decision rule. The test is one-sided lower-tail because, as noted earlier, evidence of persistence in a process is a small number of runs in the sequence of observations. The sampling distribution of R under H_0 is approximately normal in this example because $n = 30$. Hence, for $\alpha = 0.01$, we require $z(0.01) = -2.326$, and the decision rule for the test is as follows:

If $z^* \geq -2.326$, conclude H_0.
If $z^* < -2.326$, conclude H_1.

Step 4. Table 15.4 shows that the number of runs up and down is $R = 12$. We therefore calculate standardized test statistic (15.20) as follows:

$$E\{R\} = \frac{2(30)-1}{3} = 19.67$$

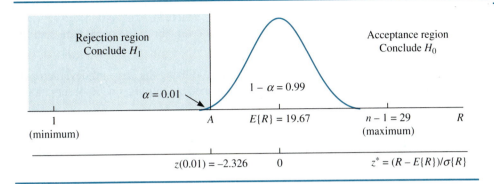

FIGURE 15.10
Statistical decision rule for
large-sample, one-sided,
lower-tail test for runs up
and down—Lake level
example

$$\sigma^2\{R\} = \frac{16(30) - 29}{90} = 5.01 \qquad \sigma\{R\} = \sqrt{5.01} = 2.24$$

$$z^* = \frac{R - E\{R\}}{\sigma\{R\}} = \frac{12 - 19.67}{2.24} = -3.42$$

Since $z^* = -3.42 < -2.326$, we conclude H_1—that the maximum lake level is under-going a persistent change. Thus, the generating process is not random and it is not appropriate to treat the 30 observations as a random sample from an infinite popula-tion. The one-sided P-value is $P(Z < z^*) = P(Z < -3.42) = 0.0003$, which indicates strongly that the data are not consistent with being generated by a random process. □

When consecutive observations are tied, we suggest that the tied observations be treated as a continuation of the existing run. For instance, tied observations following a minus sign should be treated as a minus. If ties are too numerous, the decision rule will need to be modified.

We now take up a robust estimation procedure that is applicable under a wide variety of circumstances. This procedure is called the *jackknife procedure* because it is a multipurpose tool that is reliable under quite general conditions. We shall demonstrate the procedure by applying it to the estimation of the population standard deviation.

**OPTIONAL 15.5
TOPIC—JACKKNIFE
ESTIMATION**

Jackknife Estimation of σ

In Section 14.5, we considered interval estimation of the population variance and standard deviation when the population is normal. We cautioned that the methodology is not robust and does not yield reliable results when the population distribution is not close to normal. The jackknife procedure, on the other hand, is robust and does not require the population to have any specified form. Hence, the jackknife procedure may be used where the nor-mality assumption is not met.

The jackknife procedure for estimating the population standard deviation σ is concep-tually quite simple, involving the following three steps.

Step 1. From the n sample observations, n standard deviations are calculated, each time omitting one sample observation. We shall let s_{-i} denote the sample standard deviation obtained when X_i is omitted. Clearly, each s_{-i} is an estimator of the population standard deviation σ.

Step 2. Next, n *pseudovalues* J_i are calculated, utilizing the standard deviation s for the full sample and the standard deviations s_{-i} obtained by omitting one X_i observation at a time.

$$(15.21) \qquad J_i = ns - (n-1)s_{-i} \qquad i = 1, 2, \ldots, n$$

Since both s and s_{-i} are estimators of the population standard deviation σ, each J_i is also an estimator of σ.

Step 3. Now we treat the n pseudovalues $J_1, J_2, \ldots, J_n$ as if they were a random sample from a normal population. We compute the mean and standard deviation of the pseudovalues in the usual way—here we denote them by $\bar{J}$ and s_J.

$$(15.22) \qquad \bar{J} = \frac{\sum\limits_{i=1}^{n} J_i}{n} \qquad s_J = \sqrt{\frac{\sum\limits_{i=1}^{n} (J_i - \bar{J})^2}{n-1}}$$

The statistic $\bar{J}$ is a point estimator of the population standard deviation σ because each J_i is an estimator of σ. To estimate the population standard deviation σ by means of an interval estimate, we set up a confidence interval based on the sample mean ($\bar{J}$ here). We therefore need to compute the estimated standard deviation of $\bar{J}$, denoted here by $s\{\bar{J}\}$, in the same fashion as is done for the sample mean $\bar{X}$.

$$(15.23) \qquad s\{\bar{J}\} = \frac{s_J}{\sqrt{n}}$$

Now we employ confidence interval (11.9). This interval, which utilizes the t distribution, is appropriate here because the pseudovalues J_i are treated as if they were a random sample from a normal population.

(15.24)
The approximate $1 - \alpha$ confidence limits for σ are:

$$\bar{J} \pm ts\{\bar{J}\}$$

where: $t = t(1 - \alpha/2; n - 1)$
$\bar{J}$ is given by (15.22)
$s\{\bar{J}\}$ is given by (15.23)

A random sample of seven fire insurance claims under a particular type of residential policy yielded the claim amounts (in thousands of dollars) in column 1 of Table 15.5. A 95 percent confidence interval for σ, the population standard deviation for claim amounts, is desired. Since the population of claim amounts is quite skewed, confidence limits (14.33) for a normal population are not appropriate, and the jacknife procedure is to be employed. We shall follow the three-step procedure.

Step 1. The sample mean for the $n = 7$ claims is $\overline{X} = 5.17$. Hence, the sample variance based on the full sample is:

$$s^2 = \frac{1}{7 - 1}[(6.2 - 5.17)^2 + (3.1 - 5.17)^2 + \cdots + (11.7 - 5.17)^2]$$

$$= 17.226$$

so $s = \sqrt{17.226} = 4.15$. This value is shown in the last line of Table 15.5, column 2.

Next, we calculate the standard deviations s_{-i} by omitting one of the sample observations at a time. When X_1 is omitted, for instance, the mean of the remaining six observations is 5.00, and we obtain:

$$s^2_{-1} = \frac{1}{6 - 1}[(3.1 - 5.00)^2 + (1.0 - 5.00)^2 + \cdots + (11.7 - 5.00)^2]$$

$$= 20.424$$

so $s_{-1} = \sqrt{20.424} = 4.52$. This standard deviation is shown in Table 15.5, column 2, first line. The other standard deviations s_{-i} are also shown in column 2.

Step 2. Next, we calculate the pseudovalues J_i, using (15.21). For $i = 1$, we obtain:

$$J_1 = ns - (n - 1)s_{-1} = 7(4.15) - 6(4.52) = 1.93$$

This and the other pseudovalues are shown in column 3 of Table 15.5.

	(1)	(2)	(3)
		Sample	
	Omitted	Standard	
	Observation	Deviation	Pseudovalue
i	X_i	s_{-i}	J_i
1	6.2	4.52	1.93
2	3.1	4.44	2.41
3	1.0	4.08	4.57
4	9.3	4.09	4.51
5	4.2	4.52	1.93
6	0.7	4.00	5.05
7	11.7	3.28	9.37
	None	$s = 4.15$	Total 29.77

TABLE 15.5 Calculations for a jacknife estimate of σ—Fire insurance claims example

Step 3. Now, we treat the n pseudovalues J_i in column 3 as if they were a random sample from a normal population. Using (15.22), we obtain:

$$\bar{J} = \frac{1}{7}(1.93 + 2.41 + \cdots + 9.37) = \frac{29.77}{7}$$

$$= 4.25$$

$$s_J^2 = \frac{1}{7-1}[(1.93 - 4.25)^2 + (2.41 - 4.25)^2 + \cdots + (9.37 - 4.25)^2]$$

$$= 6.862$$

so $s_J = \sqrt{6.862} = 2.62$. Finally, the estimated standard deviation of $\bar{J}$, using (15.23), is:

$$s\{\bar{J}\} = \frac{s_J}{\sqrt{n}} = \frac{2.62}{\sqrt{7}} = 0.99$$

For $1 - \alpha = 0.95$ and $n - 1 = 6$, we require $t(0.975; 6) = 2.447$. Hence, the confidence limits, using (15.24), are $4.25 \pm 2.447(0.99)$, and the 95 percent confidence interval is:

$$1.8 \leq \sigma \leq 6.7$$

We conclude with 95 percent confidence that the standard deviation of the population of claims is between $1.8 thousand and $6.7 thousand.   □

Comments

1. We noted already that jackknife estimators are robust. In addition, jackknife estimators have the desirable property that they avoid or reduce bias that may be present in the original estimator. For example, s is a biased estimator of σ, but $\bar{J}$ has almost no bias.

2. The appropriateness of confidence intervals based on the jackknife method can sometimes be improved by employing a mathematical transformation to reduce possible skewness in the distribution of the pseudovalues. In our example, for instance, we could have used $\log s_{-i}$ and $\log s$ instead of s_{-i} and s for obtaining the pseudovalues. If the population of claims were an extremely skewed distribution, this transformation would help make the pseudovalues more nearly normally distributed.

3. Because the jackknife procedure entails extensive calculations, it is usually implemented by means of a computer routine.

4. Jackknife estimation is a special type of estimation that is based on repeated sampling of the original sample. These repeated sampling methods are called *bootstrap sampling methods*. Bootstrap procedures are particularly useful for obtaining confidence intervals for parameters that must be estimated by very complicated statistics. Since this situation is common in large-scale surveys, bootstrap procedures are widely used in analyzing survey data.

15.1 A random sample of size $n = 12$ has been selected from a continuous population, and a confidence interval for the population median of the form (15.1) is to be constructed.

a. What is the value of r for which this confidence interval has a confidence coefficient closest to 0.99?

b. What is the exact confidence coefficient when $r = 2$?

15.2 A random sample of size $n = 9$ has been selected from a continuous population, and a confidence interval for the population median of the form (15.1) is to be constructed.

a. What is the value of r for which the confidence interval has a confidence coefficient closest to 0.95?

b. What is the exact confidence coefficient when $r = 3$?

* **15.3** The weights (in kilograms) of nine smallmouth bass taken from a lake were as follows:

2.83	2.96	2.47	3.00	2.08	2.54	2.84	3.04	2.31

Assume that these observations constitute a random sample from a continuous population with median η.

a. Array the observations and obtain a point estimate of the population median η.

b. Construct a confidence interval for η with a confidence coefficient near 0.95. Interpret the confidence interval.

15.4 Automobile Damage Claims. In 12 recent automobile damage suits where the claims were settled out of court, the damages agreed upon (in $ thousand) were:

5.2	5.5	3.8	12.5	8.3	2.1	1.7	20.0	4.8	6.9	7.5	10.6

Assume that these data constitute a random sample from a continuous population with median η.

a. Array the observations and obtain a point estimate of the population median η.

b. Construct a confidence interval for η with a confidence coefficient near 0.95. Interpret the confidence interval.

15.5 Chemical Yield. A chemist made eight independent measurements of the percent yield of a microchemical reaction and obtained the following:

68	43	71	65	53	62	68	49

Assume that these data constitute a random sample from a continuous population with median η.

a. Array the observations and obtain a point estimate of the population median η.

b. Construct a confidence interval for η with a confidence coefficient near 0.99. Interpret the confidence interval.

15.6 Consider a random sample of size $n = 36$ from a continuous population. Using approximation (15.2), give the value of r for which confidence interval (15.1) for the population median has a confidence coefficient of 0.99.

15.7 Consider a random sample of size $n = 25$ from a continuous population. Using approximation (15.2), give the value of r for which confidence interval (15.1) for the population median has a confidence coefficient of 0.95.

* **15.8 Failure Ages.** A group of 21 electrolytic cells of design A were installed in the potroom of an aluminum smelter. The following are the failure ages (in days) of the cells in order of failure:

518	903	1192	1477	1814	2060	2421
775	1015	1354	1604	1826	2274	2591
888	1189	1367	1708	2040	2330	2716

The 21 observations may be considered to constitute a random sample from a continuous population of failure ages.

a. Construct a 90 percent confidence interval for the population median failure age of cells of design A. Interpret the confidence interval.

b. When the distribution of failure ages is symmetrical, why does the confidence interval in part a also serve as a 90 percent confidence interval for the population mean failure age?

c. The target median failure age for cells of design A is 1800 days. Does your confidence interval in part a suggest that the target is being met? Explain.

15.9   Refer to the before-pollution-controls data array of time intervals between oil spills in Table 3.1a. Assume that the 40 observations constitute a random sample from a continuous population. Construct a 90 percent confidence interval for the population median. Interpret the confidence interval.

15.10   Refer to the stem-and-leaf display for the *Semiconductor Failure* example in Figure 2.1b. Assume that the 17 observations constitute a random sample from a continuous population.

a. Construct a 95 percent confidence interval for the population median breaking stress. Interpret the confidence interval.

b. Based on the interval in part a, would it be reasonable to claim that more than one-half of semiconductor bonds in the population have breaking stresses exceeding 40 milligrams? Comment.

* **15.11**   Refer to **Failure Ages** Problem 15.8. It is desired to use the sign test to decide whether the population median failure age of cells of design A exceeds 1800 days.

a. State the test alternatives in terms of η. Then restate the test alternatives in terms of p. Define p.

b. Conduct the sign test, controlling the α risk at 0.05 when $\eta = 1800$. State the decision rule, the value of the test statistic, and the conclusion.

c. What is the P-value of the test?

15.12   In fund-raising drives by a charitable foundation, the median contribution from regular contributors has been \$62.50 in the recent past. As an experiment, the foundation prepared an attractive brochure describing the foundation's work and distributed it to a random sample of 250 regular contributors shortly before the next fund-raising drive. A subsequent follow-up indicated that 140 of the 250 contributors gave more than \$62.50 in the drive, and 110 gave less than \$62.50. It is now desired to use the sign test to determine whether the median contribution η of regular contributors who are sent the brochure remains at \$62.50. Assume that the population of the amounts of contributions from regular contributors is continuous.

a. State the test alternatives in terms of η. Then restate the test alternatives in terms of p. Define p.

b. Conduct the sign test, controlling the α risk at 0.10 when $\eta = 62.50$. State the decision rule, the value of the test statistic, and the conclusion.

c. What is the P-value of the test?

15.13   Students enrolled in a community school system achieved a median score of 512 on a chemistry aptitude test during the past few years. Of 800 students who took the aptitude test this year, 493 had scores above 512, and 307 had scores below 512. Assume that the observations for the 800 students constitute a random sample from a continuous population. It is desired to use the sign test to decide whether the current population median score on the aptitude test equals 512.

a. State the test alternatives in terms of η. Then restate the test alternatives in terms of p. Define p.

b. Conduct the sign test, controlling the α risk at 0.01 when $\eta = 512$. State the decision rule, the value of the test statistic, and the conclusion.

c. What is the P-value of the test?

15.14   The median score on a standard psychological test designed to measure extrinsic-reward motivation is 115 in the general population. A random sample of 10 successful sales managers who took the test obtained the following scores, arrayed by magnitude:

110   116   120   121   128   131   135   138   139   145

It is desired to use the sign test to determine whether the median test score of successful sales managers is the same as that of the general population. Assume that the population of test scores of successful sales managers is continuous.

a. State the test alternatives in terms of η. Then restate the test alternatives in terms of p. Define p.

b. Use the procedure in (13.14) to conduct the sign test, controlling the α risk at 0.02 when $\eta = 115$. What is the appropriate test conclusion?

15.15 Refer to **Chemical Yield** Problem 15.5. The chemist wishes to use the sign test to determine whether the population median yield of the reaction equals the theoretically attainable yield of 60 percent.

a. State the test alternatives in terms of η. Then restate the test alternatives in terms of p. Define p.

b. Use the procedure in (13.14) to conduct the sign test, controlling the α risk at 0.10 when $\eta = 60$. What is the appropriate test conclusion?

15.16 Security System. In a random sample of $n_1 = 15$ homeowners, each was asked to state the maximum amount (in \$ thousand) he or she would pay for a home security system described in a promotional brochure. An independent random sample of $n_2 = 12$ homeowners was asked the same question but was allowed to inspect an installed system in operation, in addition to receiving the brochure. The results follow, arrayed by magnitude:

Brochure Only (X_1)					Brochure and Inspection (X_2)			
2.4	3.3	3.9	4.5	5.3	4.8	5.9	6.4	6.9
3.0	3.5	4.3	4.5	5.4	5.1	6.0	6.5	7.1
3.1	3.6	4.4	4.7	5.5	5.2	6.3	6.7	7.5

a. Construct a back-to-back stem-and-leaf display for the two samples, using 2, 3, . . . , 7 as stems. Do the plots support the applicability of the M–W–W test here? Comment.

b. Assume that the two population distributions are continuous and have the same shape. Use the M–W–W test to determine whether homeowners are willing to pay a larger amount for the system if they have inspected an installed system than if they have not. Employ test statistic S_2 and control the α risk at 0.05 when $\eta_2 - \eta_1 = 0$. State the alternatives, the decision rule, the value of the test statistic, and the conclusion.

c. What is the P-value of the test? Is it a one-sided or two-sided P-value?

d. What assumptions, if any, would be required to use test procedure (14.15a) that are not required for the M–W–W test?

15.17 Employee Productivity. The numbers of units produced by $n_1 = 25$ employees randomly selected from shift 1 and by $n_2 = 25$ employees randomly and independently selected from shift 2 are as follows, arrayed by magnitude:

Shift 1 (X_1)					Shift 2 (X_2)				
172	186	203	212	231	169	185	198	202	213
175	190	206	218	235	176	188	198	204	215
178	191	207	221	237	177	189	199	205	220
181	196	209	227	246	182	192	200	208	225
184	197	210	229	253	183	195	201	211	230

a. Construct a back-to-back stem-and-leaf display for the two samples, using the first two digits as stems. Do the plots support the applicability of the M–W–W test here? Comment.

b. Assume that the two population distributions are continuous and have the same shape. Use the M–W–W test to determine whether employee productivity in shift 2 is lower than that in shift 1. Employ test statistic S_2 and control the α risk at 0.05 when $\eta_2 - \eta_1 = 0$. State the alternatives, the decision rule, the value of the test statistic, and the conclusion.

c. What is the P-value of the test? Is this value consistent with the test result in part b? Explain.

15.18 Snow-Strip Swathing. A new method of swathing grain is under experimentation in a dry Northern region. The method leaves strips of standing grain to increase snow retention on fields during winter and thereby increase soil moisture and crop yields in the following growing season. In one experiment, 34 similar fields were divided randomly into two equal groups. Fields in one group were swathed conventionally, and fields in the other group were swathed by using the snow-strip technique. Grain yields (in bushels per acre) from the fields in the following year were as follows, arrayed by magnitude:

Conventional Swathing (X_1)				Snow-Strip Swathing (X_2)			
15.6	19.7	23.5	27.1	21.4	26.3	30.2	33.2
16.1	20.8	24.4	28.3	23.3	27.5	30.3	36.3
17.7	21.0	25.7		25.7	28.0	30.7	
18.6	22.5	26.5		25.9	28.4	31.4	
19.7	22.8	26.6		26.2	29.4	32.9	

a. Construct aligned dot plots for the two samples. Do the plots here support the applicability of the M–W–W test? Comment.

b. Assume that the two population distributions are continuous and have the same shape. Use the M–W–W test to determine whether snow-strip swathing leads to larger crop yields in the following year than does conventional swathing. Employ test statistic S_2 and control the α risk at 0.01 when $\eta_2 - \eta_1 = 0$. State the alternatives, the decision rule, the value of the test statistic, and the conclusion.

c. What is the P-value of the test?

15.19   Refer to **Security System** Problem 15.16. Computer output gives $1.4 \leq \eta_2 - \eta_1 \leq 2.9$ as a 95.2 percent M–W–W confidence interval for the difference in the two population medians. Interpret the confidence interval. Does the interval show that the difference $\eta_2 - \eta_1$ is known with considerable precision here?

15.20   Refer to **Employee Productivity** Problem 15.17. Computer output gives $-21.0 \leq \eta_2 - \eta_1 \leq 4.0$ as a 95.2 percent M–W–W confidence interval for the difference in the two population medians. Interpret the confidence interval. Is this interval also a confidence interval for the difference in the two population means? Explain.

15.21   Refer to **Snow-Strip Swathing** Problem 15.18. Computer output gives $2.70 \leq \eta_2 - \eta_1$ as a 99.5 percent M–W–W one-sided confidence interval for the difference in the two population medians. Interpret the confidence interval. Does the interval support the test conclusion in Problem 15.18b?

* 15.22   Physical Fitness.   The data that follow are indexes of the physical fitness of 24 members of a health club at the time of joining (X_1) and six months later (X_2). The index is based on six physical-fitness tests. The larger the index, the better the physical fitness. Assume that the 24 differences $D = X_2 - X_1$ constitute a random sample from a continuous population of differences.

Member:	1	2	3	4	5	6	7	8	9	10	11	12
X_1:	92	118	87	83	100	94	112	106	113	87	87	79
X_2:	100	124	101	94	113	101	117	123	116	87	97	99

Member:	13	14	15	16	17	18	19	20	21	22	23	24
X_1:	113	86	86	112	116	99	105	97	100	84	122	100
X_2:	113	98	95	109	115	95	120	122	102	82	116	124

a. Construct a 95 percent confidence interval for the median of the population of differences. Interpret the confidence interval.

b. Does it appear that club membership improves physical fitness?

15.23   Transferred Executives.   A company's personnel policy provides a transferred executive with a guaranteed price for the executive's home if it cannot be sold for a larger amount within a fixed period from the transfer notification date. The guaranteed price is the average of two appraisals (X_1, X_2) made independently by company-appointed real estate appraisers 1 and 2. The following are appraisal data (in $ thousand) for 17 houses recently appraised by this pair of appraisers. Assume that the 17 differences $D = X_2 - X_1$ constitute a random sample from a continuous population of differences.

House:	1	2	3	4	5	6	7	8	9	10
X_1:	279	249	263	247	244	272	257	242	300	237
X_2:	283	247	271	251	250	286	264	239	318	232

House:	11	12	13	14	15	16	17
X_1:	283	294	257	200	170	281	217
X_2:	282	296	266	212	171	292	230

a. Construct a 90 percent confidence interval for the median of the population of differences. Interpret the confidence interval.

b. Does it appear that the two appraisers differ systematically in their appraisals? Discuss.

15.24 A random sample of 10 boys experienced the following growths in height (in centimeters) during their sixth year of life:

$$6.0 \quad 4.7 \quad 4.2 \quad 5.0 \quad 4.6 \quad 4.0 \quad 5.4 \quad 5.3 \quad 6.5 \quad 5.5$$

a. Construct a confidence interval for the population median growth of boys in their sixth year of life. Use a confidence coefficient near 0.98. Interpret the confidence interval.

b. What assumption did you make for the estimation procedure in part a to be valid? Is the assumption satisfied here? Explain.

* **15.25** Refer to **Physical Fitness** Problem 15.22. Use the sign test to determine whether the population median change in fitness exceeds 5 points. Control the α risk at 0.025 when $\eta_D = 5$. State the alternatives, the decision rule, the value of the test statistic, and the conclusion.

15.26 Refer to **Transferred Executives** Problem 15.23. Use the sign test to determine whether the population median difference in appraisals is zero. Control the α risk at 0.10 when $\eta_D = 0$. State the alternatives, the decision rule, the value of the test statistic, and the conclusion.

15.27 Refer to **Achievement Scores** Problem 14.18.

a. Use the sign test to determine whether the population median difference in test scores is zero. Control the α risk at 0.05 when $\eta_D = 0$. State the alternatives, the decision rule, the value of the test statistic, and the conclusion.

b. Inasmuch as the population of test score differences is approximately normal, is the test in part a also a test of whether the difference $\mu_2 - \mu_1$ in the two population means is zero? Explain.

* **15.28 Age Differences.** The differences in ages between husband and wife in a random sample of 12 married couples are to be studied to determine whether the population median age difference (husband's age minus wife's age, in years) exceeds zero, using the Wilcoxon signed rank test. The sample differences follow:

Couple:	1	2	3	4	5	6	7	8	9	10	11	12
D:	8.8	-3.6	0.6	2.8	3.6	5.1	-0.4	-0.2	5.4	-0.4	1.5	4.4

a. Prepare a dot plot of the sample differences. Does the plot support the applicability of the Wilcoxon signed rank test here? Comment.

b. Assume that the population of differences in ages is continuous and symmetrical. Conduct the Wilcoxon signed rank test, controlling the α risk at 0.10 when $\eta_D = 0$. State the alternatives, the decision rule, the value of the test statistic, and the conclusion.

c. What is the P-value of the test?

15.29 Refer to **Transferred Executives** Problem 15.23. It is desired to use the Wilcoxon signed rank test to determine whether the population median difference in appraisals is zero.

a. Prepare a dot plot of the sample differences. Does the plot support the applicability of the Wilcoxon signed rank test here? Comment.

b. Assume that the population of differences is continuous and symmetrical. Conduct the Wilcoxon signed rank test, controlling the α risk at 0.10 when $\eta_D = 0$. State the alternatives, the decision rule, the value of the test statistic, and the conclusion.

c. What is the P-value of the test?

15.30   Refer to **Achievement Scores** Problem 14.18. It is desired to use the Wilcoxon signed rank test to determine whether the population median difference in test scores is zero.

a. Prepare a dot plot of the sample differences. Does the plot support the applicability of the Wilcoxon signed rank test here? Comment.

b. Assume that the population of differences is continuous and symmetrical. Conduct the Wilcoxon signed rank test, controlling the α risk at 0.05 when $\eta_D = 0$. State the alternatives, the decision rule, the value of the test statistic, and the conclusion.

c. Inasmuch as the population of test score differences is approximately normal, is use of the Wilcoxon signed rank test still appropriate here? Explain.

15.31   Refer to **Automobile Damage Claims** Problem 15.4. Assume that the population of damage claim amounts is continuous and symmetrical. A legislative staff assistant studying automobile damage settlements wished to know whether the median settlement amount exceeds $5 thousand. Conduct the Wilcoxon signed rank test, controlling the α risk at 0.025 when $\eta = 5$. State the alternatives, the decision rule, the value of the test statistic, and the conclusion.

15.32   Refer to **Failure Ages** Problem 15.8. Cells of a competing design B are known to have a median failure age of 1300 days. Assume that the population of failure ages is continuous and symmetrical and use the Wilcoxon signed rank test to determine whether the population median failure age of cells of design A exceeds 1300 days. Control the α risk at 0.05 when $\eta = 1300$. State the alternatives, the decision rule, the value of the test statistic, and the conclusion.

15.33   Refer to **Age Differences** Problem 15.28.

a. Computer output gives $0.1 \le \eta_D \le 4.4$ as a 94.5 percent Wilcoxon signed rank confidence interval for the population median difference in ages. Interpret the confidence interval.

b. Does the confidence interval in part a support a researcher's claim that, on average, husbands are two years older than their wives? Discuss.

15.34   Refer to **Transferred Executives** Problems 15.23 and 15.29.

a. Computer output gives $3.0 \le \eta_D \le 8.5$ as a 90.2 percent Wilcoxon signed rank confidence interval for the population median difference in appraisals. Interpret the confidence interval.

b. Because the population of differences is assumed to be symmetrical, explain why the confidence interval in part a is also a confidence interval for the difference $\mu_2 - \mu_1$ in the population means. Does the confidence interval in part a confirm that one appraiser gives higher appraisals on average than the other? Comment.

15.35   Refer to **Achievement Scores** Problems 14.18 and 15.30.

a. Computer output gives $1.0 \le \eta_D \le 14.5$ as a 95.0 percent Wilcoxon signed rank confidence interval for the population median difference in test scores. Interpret the confidence interval.

b. Does the confidence interval in part a support a claim that the two tests involve systematic differences in their achievement scores? Comment.

* **15.36**   **Glass Layers.** Refer to Figure 15.9a. The observations in this figure represent the deviations of the thickness of a glass layer in an optical device from the design specification, measured in micrometers, for 35 optical devices. The observations are plotted in the sequence in which the devices were manufactured and are as follows, ordered in rows:

−0.288	0.187	0.785	0.194	−0.258	1.579	1.090	0.448	−0.457
0.960	−0.491	0.219	−0.169	1.096	1.239	−0.146	−1.698	−1.041
1.620	1.047	0.032	0.151	0.290	0.873	−0.289	1.119	−0.792
0.063	0.484	1.045	0.084	−0.086	0.427	−0.528	−1.433	

Assume that the observations are drawn from continuous populations.

a. How many runs up and down are contained in the sequence of 35 observations? How many runs are expected if the process is random?

b. Conduct a two-sided runs-up-and-down test. Control the α risk at 0.10 when the sequence is generated by a random process. State the alternatives, the decision rule, the value of the test statistic, and the conclusion. The text discussion of Figure 15.9a states that the sequence is from a random process. Has the test confirmed this claim? Comment.

c. What is the P-value of the test in part b?

15.37 Interest Rates. Quarterly interest rates paid on savings accounts at a major bank, calculated as a quarterly average of the daily rates (expressed in percent), were as follows during the past five years:

Quarter	Year 1	Year 2	Year 3	Year 4	Year 5
1	6.32	7.66	8.92	10.56	7.96
2	7.18	8.65	9.57	9.39	8.68
3	7.56	8.73	9.80	9.03	8.79
4	7.47	9.25	10.69	8.16	9.17

Assume that the observations are drawn from continuous populations.

a. Prepare a line graph of the sequence of observations. Is any evidence of nonrandomness apparent in the plot? Comment.

b. Conduct a two-sided runs-up-and-down test. Control the α risk at 0.10 when the sequence is generated by a random process. State the alternatives, the decision rule, the value of the test statistic, and the conclusion.

c. Describe the types of nonrandomness that the two-sided test in part a is designed to detect. Would the test detect a quarterly seasonal pattern in interest rates for a sufficiently long data sequence? Discuss.

15.38 Marriages. The numbers of marriages in a city for each month from July of year 1 to June of year 5 follow:

Year 1		Year 2			Year 3		
J	219	J	121	J	228	J	114	J	231
A	226	F	131	A	226	F	128	A	240
S	185	M	140	S	195	M	151	S	223
O	186	A	176	O	183	A	178	O	185
N	162	M	187	N	165	M	202	N	174
D	180	J	245	D	182	J	262	D	194

Year 4			Year 5	
J	131	J	227	J	140
F	135	A	254	F	155
M	150	S	204	M	153
A	189	O	199	A	190
M	221	N	187	M	224
J	277	D	181	J	278

a. Prepare a line graph of the sequence of observations. Does your plot suggest any departure from randomness? Comment.

b. Perform the runs-up-and-down test to determine whether persistence is present. Control the α risk at 0.01 when the sequence is generated by a random process. State the alternatives, the decision rule, the value of the test statistic, and the conclusion.

c. What assumption is required for the test in part b to be valid? Is the assumption met in this application? Discuss.

* **15.39** **Rates of Return.** The rates of return (in percent) earned on the common stock of Polyphase, Inc., during each of the past five quarters are:

$$-5 \quad 10 \quad 2 \quad 19 \quad 15$$

Construct a 90 percent confidence interval for the population standard deviation of quarterly rates of return on this stock, using the jackknife procedure. Assume that the five observations constitute a random sample from the population of all quarterly rates of return. Some of the required calculational results follow:

i:	1	2	3	4	5	
X_i:	-5	10	2	19	15	
s_{-i}:	7.326	11.177	10.500	____	____	$s = 9.731$
J_i:	19.354	3.951	6.657	____	____	

15.40 In the last six sales of commercial property in a tax district, the assessed values of the properties, expressed as a percentage of market value, were:

$$38 \quad 46 \quad 22 \quad 49 \quad 27 \quad 34$$

Use the jackknife procedure to construct a 99 percent confidence interval for the population standard deviation of assessed value as a percentage of market value for commercial properties in this tax district. Assume that the six observations constitute a random sample. Some calculational results follow:

i:	1	2	3	4	5	6	
X_i:	38	46	22	49	27	34	
s_{-i}:	11.718	10.416	8.927	9.370	____	____	$s = 10.526$
J_i:	4.569	11.075	18.520	16.306	____	____	

15.41 The numbers of shirts not meeting quality specifications in the most recent six shipments of 1000 shirts each from a garment factory are:

$$5 \quad 2 \quad 7 \quad 6 \quad 4 \quad 10$$

To measure the uniformity of the quality of the shipments, a 95 percent confidence interval for the process standard deviation of the numbers of shirts in shipments not meeting specifications is to be constructed. Assume that the six observations represent a random sample from the process. Obtain the desired confidence interval, using the jackknife procedure. Some of the required calculational results follow:

i:	1	2	3	4	5	6	
X_i:	5	2	7	6	4	10	
s_{-i}:	3.033	2.302	2.966	____	____	1.924	$s = 2.733$
J_i:	1.229	4.884	1.563	____	____	6.777	

15.42 Refer to **Rates of Return** Problem 15.39.

a. Use (14.33) to construct a 90 percent confidence interval for the population standard deviation σ. What assumption is required for this confidence interval that is not required by the jackknife procedure?

b. How does the confidence interval in part a compare with that obtained in Problem 15.39 by the jackknife method?

15.43 For a random sample from a symmetrical population, explain why $E\{Md\} = \eta = \mu$.

15.44 For an odd-sized random sample from a continuous population, show that η is the median of the sampling distribution of Md. [*Hint:* Does Md have an equal probability of falling on either side of η?]

15.45 Verify (15.2). [*Hint:* Use the normal approximation to the binomial probability in (15.3), together with the correction for continuity discussed in Section 13.2.]

15.46 Use (15.3) to evaluate the confidence level of the interval $L_8 \leq \eta \leq U_8$ when $n = 20$.

15.47 Derive (15.9) from (15.7). [*Hint:* $S_1 + S_2 = (n_1 + n_2)(n_1 + n_2 + 1)/2$ for the M–W–W test statistics.]

15.48 Derive $E\{T\}$ and $\sigma^2\{T\}$ in (15.13) when $\eta_D = 0$. [*Hint:* $\sum_{i=1}^{n} i^2 = n(n + 1)(2n + 1)/6$.]

15.49 Refer to **Failure Ages** Problem 15.8. The company has also installed 13 cells of a new design C. The distribution of cell-failure ages for this design is known to be approximately normal. All 13 cells were installed on the same day. The first 10 failures among the 13 cells in the sample have occurred, and their failure ages (in days) are, in the order of failure:

808	980	1351	1543	1627	1666	1726	1993	2168	2271

The company wishes to estimate the mean failure age of cells of this design from these incomplete sample data.

a. Can any of the estimation procedures for the mean of a normal population discussed earlier in the text be employed in this case? Explain.

b. Because the mean and the median of a normal population coincide, it was decided to construct a confidence interval for the population median failure age with $r = 4$. Why do you think $r = 4$ was chosen? What is the confidence coefficient associated with $r = 4$? Obtain the confidence interval and interpret it.

c. Use the Wilcoxon signed rank test to determine whether the median failure age of cells of design C exceeds 1300 days. Control the α risk at 0.05 when $\eta = 1300$. Explain why this test procedure can be applied even though the failure ages of the last three cells are not known yet.

15.50 The annual incomes (in thousand of pounds sterling) of 10 British sales executives selected randomly are as follows:

37.0	45.7	43.1	56.8	94.1	36.9	51.4	38.3	64.1	142.9

An analyst knows that the population distribution of incomes of sales executives is continuous but also expects it to be quite skewed so that inference procedures requiring approximate normality cannot be used.

a. Estimate the population median income of British sales executives by a confidence interval, with a confidence coefficient near 95 percent. Interpret the confidence interval.

b. The analyst wishes to test whether the population median income of British sales executives exceeds 60 thousand pounds. Conduct a sign test, controlling the α risk near 0.025 when $\eta = 60$. State the alternatives, the decision rule, and the conclusion.

c. The analyst also wishes to estimate the population standard deviation σ. To improve the appropriateness of the jackknife procedure, the analyst wishes to use the logarithms of the standard deviations s and s_{-i}, as described in Comment 2, p. 466. Construct a 95 percent confidence interval for σ by first obtaining a confidence interval for $\log \sigma$, using the pseudovalues $J_i = n \log(s) - (n - 1) \log(s_{-i})$, and then converting it into a confidence interval for σ. Interpret the latter confidence interval.

15.51 Refer to the **Financial Characteristics** data set (Appendix D.1). The percent rates of return on net assets (net income as a percentage of net assets) for crude oil producers (firms 1–60) in years 1 and 2 are of interest. Assume that the 60 firms are a random sample of all crude oil producers.

a. Obtain the percent rates of return on net assets for the 60 crude oil producers for each of the years 1 and 2.

b. Obtain a 95 percent confidence interval for the population median percent rate of return of crude oil producers in year 2. Interpret the confidence interval.

c. Let X_1 and X_2 denote the percent rates of return for a firm in years 1 and 2, respectively. Obtain the differences $D = X_2 - X_1$ in the rates of return for the 60 crude oil producers. Use the sign test to determine whether the population median change in the rates of return equals zero. Control the α risk at 0.05 when $\eta_D = 0$. State the alternatives, the decision rule, the value of the test statistic, and the conclusion.

d. Construct a box plot of the changes in the percent rates of return for the 60 crude oil producers. Does the plot suggest that the test in part c could have been conducted by means of the Wilcoxon signed rank test? Comment.

Goodness of Fit

16

The functional form of a population distribution frequently is of interest. Information about the functional form may be needed (1) for confirming a theory, (2) for validating a model, or (3) for determining whether the population assumptions of a statistical inference procedure are appropriate.

EXAMPLES □

1. An engineer has formulated a theory, a consequence of which is that the life of a new electronic component should follow the exponential distribution. As part of a test of the theory, component lives will be studied to see if they are exponentially distributed.

2. An operations analyst has developed an inventory control system in which the weekly demands for the items held in inventory are assumed to follow Poisson distributions. As a check on the validity of the probability model embodied in the control system, weekly demands for selected inventory items will be studied to see if the Poisson distribution is a reasonable model for these demands.

3. A market researcher is considering the use of a *t* test for a small sample and wishes first to study whether the population sampled is normal or approximately so. □

In this chapter, we describe several graphic methods and inference procedures that are widely used for studying the functional form of a population distribution.

PROBABILITY PLOTS | **16.1**

In Chapter 2 we discussed the use of graphic methods for examining the distribution pattern of a data set, including stem-and-leaf plots, dot plots, and histograms or frequency polygons. These graphic methods can be quite effective for investigating the functional form of a population distribution. For instance, a stem-and-leaf plot of a data set can show if the data set has the bell-shaped symmetrical pattern that is indicative of a sample from a normal population.

We now supplement these graphic methods with a plotting procedure that can be used to assess whether the population distribution has a particular functional form. The plot is called a *probability plot*.

Basic Concepts

A probability plot compares each sample observation with the expected value of this observation based on the particular functional form of the population distribution under consideration. For example, when we wish to study whether the sampled population is normal by means of a probability plot, we will compare the sample observations with the corresponding expected values based on a normal distribution.

To prepare a probability plot, we first array the sample observations in ascending order. Then we obtain the expected values of these arrayed observations for the population distribution under consideration. Because the expected values are complex to obtain, we shall employ a simple approximation that makes use of percentiles. We will explain this approximation procedure by means of an example.

☐ **EXAMPLE**

Consider again the data set of changes in annual sales for the convenience stores example in Chapter 3:

$$14 \qquad 17 \qquad -13 \qquad 41 \qquad 12$$

We shall assume that these $n = 5$ observations constitute a random sample from a population, and we wish to study whether the population distribution is normal.

The approximation method we shall employ is based on Figure 16.1, which repeats the line diagram for the data from Figure 3.4b. The diagram shows the arrayed observations and their ranks from 1 to 5. It also shows the cumulative percents corresponding to the ranks. Recall that the cumulative percents 20, 40, and so on are obtained by dividing each rank by the sample size and expressing the ratio as a percent. Note in the diagram that each observation is placed in the interval on the cumulative percent scale corresponding to its rank. For instance, observation -13 has rank 1 and is placed in the interval between 0 and 20 percent on the cumulative percent scale.

The method for determining the population percentile that approximates the expected value of an observation can now be stated as follows: The middle of each interval on the cumulative percent scale is the percentile that approximates the expected value for the corresponding ranked observation. These middle positions are identified by arrows in Figure 16.1. For instance, the 10th percentile of the normal distribution (the middle of the interval from 0 to 20 on the cumulative percent scale) approximates the expected value of the observation with rank 1, which is -13 in this example. Likewise, the 30th percentile of the normal distribution approximates the expected value of the observation with rank 2, which here is 12.   ☐

We can generalize the procedure we have just described as follows.

(16.1)
When the observations in a random sample of size n are arrayed in ascending order, the approximate expected value for the ith ranked observation is the percentile corresponding to the following fraction:

$$\frac{i - 0.5}{n}$$

FIGURE 16.1

Determination of percentiles to approximate expected values of ranked sample observations—Convenience stores example

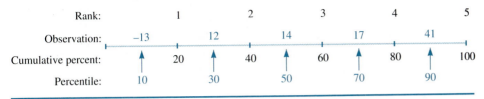

Rank:	1		2		3		4		5
Observation:	−13		12		14		17		41
Cumulative percent:		20		40		60		80		100
Percentile:	10		30		50		70		90

To see that formula (16.1) gives the required percentiles, consider again the convenience stores example, where $n = 5$. For the observation ranked 1 ($i = 1$), the formula gives:

$$\frac{i - 0.5}{n} = \frac{1 - 0.5}{5} = 0.10$$

Thus, the approximate expected value for the smallest sample observation corresponds to the 10th percentile of the normal distribution. For the third smallest observation ($i = 3$), the formula gives:

$$\frac{i - 0.5}{n} = \frac{3 - 0.5}{5} = 0.50$$

Hence, the approximate expected value for the third smallest sample observation corresponds to the 50th percentile of the normal distribution. These percentiles are exactly those shown in Figure 16.1. ☐

A probability plot is a plot of the arrayed sample observations against the corresponding expected values based on the functional form of the population under consideration (e.g., normal). If the assumed population is correct, then the observed value and the expected value for each case should be close to each other. Hence, the points plotted in the probability plot should follow approximately a straight 45° line through the origin, often called the *line of identity*. On the other hand, if the assumed population distribution is not correct, the observed and expected values will not be approximately the same, and the points in the probability plot will not follow a straight 45° line through the origin. Thus, if the probability plot of points is close to the line of identity, the plot supports the reasonableness of the assumed population distribution. On the other hand, if the plotted points deviate markedly from the line of identity, the plot provides evidence that the assumed population distribution is not the population from which the sample was drawn.

Because normal populations play such a central role in statistical methodology, we shall illustrate the construction and interpretation of probability plots with a *normal probability plot*.

Normal Probability Plot

A normal probability plot is used to examine whether the sampled population is normal. The expected values of the arrayed sample observations are therefore approximated by percentiles of the normal distribution. We obtain these approximate expected values from definition (8.10) for percentiles of a normal distribution. Because we do not know the population mean μ and the population standard deviation σ, we use the sample mean $\bar{X}$ and the sample standard deviation s as estimates of these parameters.

(16.2)

The approximate expected value of the ith ranked sample observation in a random sample of size n when the population is normal is:

$$\text{Expected value} = \bar{X} + z\left(\frac{i - 0.5}{n}\right)s \qquad i = 1, \ldots, n$$

where $z\left(\dfrac{i - 0.5}{n}\right)$ is the $100\left(\dfrac{i - 0.5}{n}\right)$ percentile of the standard normal distribution.

Milk Production □ **EXAMPLE**

A researcher for a milk producers' association obtained production data for a random sample of 40 milk cows of the same age and breed. These data are arrayed in ascending order in Table 16.1a and represent milk production per cow in thousands of pounds. The researcher wishes to estimate the variability of the outputs among cows in the population and would like to use the confidence limits in (14.33), which require the population to be approximately normal. The researcher therefore has decided to construct first a normal probability plot to check the reasonableness of assuming that the production data are drawn from an approximately normal population.

The calculation of the expected values under normality of the arrayed sample observations is illustrated in Table 16.1b. The rank of each sample observation is shown in column 1 and the arrayed sample observations are shown in column 2. The appropriate percentiles are identified in column 3, using (16.1). For instance, for the smallest sample observation (which has rank $i = 1$), formula (16.1) gives:

$$\frac{i - 0.5}{n} = \frac{1 - 0.5}{40} = 0.0125$$

In column 4, the relevant percentile for the standard normal distribution is given for each sample observation. For instance, we require $z(0.0125)$ for the smallest sample observation and find from Table C.1 that it is $z(0.0125) = -2.24$. Finally, the approximate expected values under normality are obtained in column 5, using (16.2). For instance, we obtain for the smallest sample observation ($\bar{X} = 15.96$ and $s = 2.144$, as shown in Table 16.1a):

$$\text{Expected value} = 15.96 - 2.24(2.144) = 11.16$$

Figure 16.2 contains a plot of the arrayed sample observations in Table 16.1b, column 2 (OBSERVED) against the expected values under normality in column 5 (EXPECTED). Also shown in this plot is the line of identity. Since the 40 points form

TABLE 16.1

Arrayed sample data and calculation of expected values under normality (data in thousands of pounds)—Milk production example

(a) Arrayed Sample Observations

12.39	13.25	13.76	14.20	15.79	16.16	16.58	17.54	18.08	18.79
12.43	13.29	13.97	14.62	15.99	16.32	16.75	17.81	18.23	18.79
13.00	13.32	13.98	15.04	16.08	16.40	16.93	18.04	18.36	18.97
13.20	13.63	14.12	15.25	16.12	16.56	17.32	18.05	18.74	20.55

$$\overline{X} = 15.96 \qquad s = 2.144$$

(b) Calculation of Expected Values under Normality

(1) Rank i	(2) Sample Observation	(3) $\dfrac{i - 0.5}{40}$	(4) $z\left(\dfrac{i - 0.5}{40}\right)$	(5) Expected Value Under Normality
1	12.39	0.0125	-2.24	$15.96 - 2.24(2.144) = 11.16$
2	12.43	0.0375	-1.78	$15.96 - 1.78(2.144) = 12.14$
3	13.00	0.0625	-1.53	$15.96 - 1.53(2.144) = 12.68$
⋮	⋮	⋮	⋮	⋮
38	18.79	0.9375	1.53	$15.96 + 1.53(2.144) = 19.24$
39	18.97	0.9625	1.78	$15.96 + 1.78(2.144) = 19.78$
40	20.55	0.9875	2.24	$15.96 + 2.24(2.144) = 20.76$

FIGURE 16.2

Normal probability plot—Milk production example.   OBSERVED and EXPECTED represent the sample observations and the approximate expected values under normality, respectively.

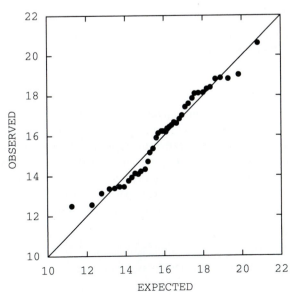

a fairly linear pattern and fall close to the line of identity, the researcher for the milk producers' association can conclude that the normal probability distribution is a reasonable model here.

☐

Many statistical software packages provide normal probability plots directly from the sample data, with no intermediate calculations required of the user. Some of these packages offer a choice of population distributions that can be selected for the probability plot, and the user simply chooses the particular population distribution (e.g., exponential) against which the sample data are to be plotted.

Comment

Several other approximations to the expected value under the assumed population distribution are widely used. Two of these identify the appropriate percentile for the ith ranked sample observation as follows.

(16.3a) $\dfrac{i}{n + 1}$

(16.3b) $\dfrac{i - 0.375}{n + 0.25}$

16.2   CHI-SQUARE TEST

A formal statistical test of the functional form of a population distribution is often required. In this and the following two sections, we describe one of the most widely used tests for this purpose, the *chi-square test*. We begin our discussion of the chi-square test by considering the nature of the test alternatives.

Test Alternatives

The following examples illustrate some of the test alternatives that are encountered in applications of the chi-square test.

Spare Part   ☐   **EXAMPLE**

An operations analyst is developing a model of replacement demand for a spare part. It is known that replacement demand for many spare parts follows a Poisson distribution. The analyst wishes to know whether the replacement demand for this particular part follows a Poisson distribution. If so, the analyst will utilize Poisson probabilities in the model to ascertain the optimal inventory level. The analyst has collected data on the number of replacement units of the spare part demanded each week for a sample of 52 weeks and now wants to use these data to test the following alternatives.

(16.4) H_0: The population distribution is Poisson.
            H_1: The population distribution is not Poisson.

☐

EXAMPLE ☐

In the milk production example, the researcher wishes to use an estimation method for the population variance that requires the population of milk production per cow to be normally distributed. The researcher therefore first wishes to test the following alternatives.

(16.5) H_0: The population distribution is normal.
         H_1: The population distribution is not normal.

☐

EXAMPLE ☐   **Hospital Stays**

Extensive experience in another country has shown that the length of hospital stay (in days) for a certain psychiatric condition has an exponential distribution with parameter value $\lambda = 0.036$. A medical researcher has collected comparable data for 10 cases in this country and wishes to test whether these data also come from an exponential distribution with $\lambda = 0.036$. Therefore, the following test alternatives are of interest.

(16.6) H_0: The population distribution is exponential with $\lambda = 0.036$.
         H_1: The population distribution is not exponential with $\lambda = 0.036$.

☐

The chi-square test is one of several statistical tests that are available for the kinds of alternatives presented in the three examples. These tests are called *goodness-of-fit tests* because whether a particular probability distribution is a good model for the sampled population will be judged by whether the probability distribution specified in H_0 is a good fit for the sample data.

The alternatives in the three examples differ in some important ways. In (16.4) the alternatives relate to a discrete probability distribution, whereas in (16.5) and (16.6) they relate to continuous probability distributions. Another difference is whether or not the parameter values are specified. The alternatives in (16.4) and (16.5) pertain to a family of probability distributions (Poisson, normal) without specifying the parameter value(s). On the other hand, H_0 in (16.6) specifies not only the family (exponential) but also the parameter value ($\lambda = 0.036$). Thus, conclusion H_1 in (16.4) or (16.5) implies that the probability distribution is not of the family type specified in H_0 (Poisson, normal). On the other hand, conclusion H_1 in (16.6) implies either that the probability distribution is not of the family type specified in H_0 (not exponential) or that it is of the family type but not with the parameter value specified in H_0 (exponential but not with $\lambda = 0.036$).

Assumptions

There are only two assumptions required for the chi-square test.

1. The sample is a simple random one from the population.
2. The sample size is reasonably large.

Nature of Test

The chi-square test is based on a comparison of the sample data with the expected outcomes if H_0 is true. We first classify the sample data into k classes. For the ith class, we denote the *sample frequency* (number of sample observations falling into that class) by f_i. Of course, the f_i must sum to the total sample size n.

$$(16.7) \qquad \sum_{i=1}^{k} f_i = n$$

Now, given the form of the distribution specified in H_0, we can determine the *expected frequency* for each class when H_0 is true. The expected frequency for the ith class when H_0 holds is denoted by F_i. As is the case for the sample frequencies, the expected frequencies F_i also must sum to the total sample size n.

$$(16.8) \qquad \sum_{i=1}^{k} F_i = n$$

The two sets of frequencies to be compared are shown symbolically in Table 16.2, columns 1 and 2.

☐ **EXAMPLE**

In the spare part example, the alternatives are as follows:

H_0: The population distribution is Poisson.
H_1: The population distribution is not Poisson.

The data on weekly demand (X) for a random sample of 52 weeks are presented in Table 16.3a. In Table 16.3b, the 52 sample observations are classified into $k = 8$ classes, and the sample frequency for each class (f_i) is shown in column 2. The reason for using these eight classes will be explained shortly.

If H_0 is true, the demand data have arisen from a Poisson distribution. We must therefore now obtain the expected frequencies when the population distribution is

TABLE 16.2
Basic structure of chi-square test for goodness of fit

Class	(1) Sample Frequency	(2) Expected Frequency Under H_0	(3) Relative Squared Residual
1	f_1	F_1	$(f_1 - F_1)^2/F_1$
2	f_2	F_2	$(f_2 - F_2)^2/F_2$
$\vdots$	$\vdots$	$\vdots$	$\vdots$
k	f_k	F_k	$(f_k - F_k)^2/F_k$
Total	n	n	$X^2 = \Sigma\,(f_i - F_i)^2/F_i$

| TABLE 16.3 |
| Chi-square test for goodness of fit of Poisson distribution—Spare part example |

(a) Weekly Replacement Demand

Weeks						Demand (X)							
1–13	5	5	2	1	3	3	3	3	8	6	5	3	6
14–26	4	6	5	3	2	6	2	5	3	2	8	2	2
27–39	6	3	6	2	5	4	6	4	5	3	4	2	2
40–52	6	6	3	7	5	0	4	4	10	1	3	4	0

$$\overline{X} = 4.0$$

(b) Frequency Distribution and Calculations for Chi-Square Test

	(1)	(2)	(3)	(4)	(5)
				Under H_0	
Class i	Number Demanded X	Sample Frequency f_i	Probability	Expected Frequency F_i	Relative Squared Residual $(f_i - F_i)^2/F_i$
1	0–1	4	0.0916	4.763	0.122
2	2	9	0.1465	7.618	0.251
3	3	11	0.1954	10.161	0.069
4	4	7	0.1954	10.161	0.983
5	5	8	0.1563	8.128	0.002
6	6	9	0.1042	5.418	2.368
7	7	1	0.0595	3.094	1.417
8	8 or more	3	0.0511	2.657	0.044
	Total	52	1.0000	52.000	$X^2 = 5.256$

Poisson. We know from (7.8) that the Poisson distribution has one parameter — its mean λ. Since H_0 does not specify the value of this parameter, we estimate it by the sample mean $\overline{X}$. For the 52 sample values, $\overline{X} = 4.0$, as shown in Table 16.3a. Now we use the Poisson distribution with $\lambda = 4.0$ to obtain the probabilities of a sample observation falling into each of the eight classes in Table 16.3b. The required probabilities may be obtained from (7.8), from Table C.6, or from a statistical package. For instance, the probability that a week's demand X will be 0 or 1 and hence fall in the first class is $P(0) + P(1) = 0.0183 + 0.0733 = 0.0916$. Similarly, the probability that X will be 2 and fall in the second class is $P(2) = 0.1465$. These probabilities are shown in column 3.

Next we obtain the expected frequencies F_i if H_0 holds by multiplying each probability by the sample size n (in this example, $n = 52$). For the first class, we obtain $F_1 = 52(0.0916) = 4.763$, and for the second class, $F_2 = 52(0.1465) = 7.618$. These expected frequencies are shown in column 4. Note that the expected frequencies need not be whole numbers.

We now have the sample frequencies f_i and the expected frequencies F_i based on the assumption that weekly demand is Poisson distributed. At this point, we are ready to calculate the test statistic. □

Test Statistic

The chi-square test statistic is based on a comparison of the sample frequencies f_i with the corresponding expected frequencies F_i under H_0. Each difference between the sample fre-

FIGURE 16.3
Statistical decision rule for
chi-square goodness-of-fit
test

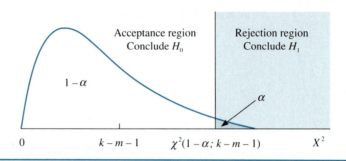

quency and the expected frequency, $f_i - F_i$, is a residual, as explained in Section 2.6. The closer the sample frequencies are to the expected frequencies, the greater is the weight of evidence in favor of H_0. The farther apart are the sample and expected frequencies, the greater is the weight of evidence in favor of H_1. The specific measure of closeness used for each class is the *relative squared residual.*

$$(16.9) \qquad \frac{(f_i - F_i)^2}{F_i}$$

Note that the measure of closeness is the ratio of the squared residual, $(f_i - F_i)^2$, to the expected frequency under H_0. These relative squared residuals are shown in Table 16.2, column 3.

The chi-square test statistic is the sum of the relative squared residuals for all classes and will be denoted by X^2.

$$(16.10) \qquad X^2 = \sum_{i=1}^{k} \frac{(f_i - F_i)^2}{F_i}$$

Note that the farther the sample frequency f_i departs in either direction from the expected frequency F_i, the larger is $(f_i - F_i)^2/F_i$ and hence the larger is X^2. On the other hand, if f_i and F_i are identical for all classes, $X^2 = 0$ because each $(f_i - F_i)^2/F_i = 0$. It follows, therefore, that large values of X^2 are indicative of H_1 being true, while small values of X^2 are indicative of H_0 being true. Thus, an upper-tail test is appropriate, as shown in Figure 16.3.

Sampling Distribution of X^2 When H_0 Holds

To control the α risk, we require the sampling distribution of X^2 when H_0 is true. The following theorem applies.

> **(16.11)**
>
> When the sampled population has the probability distribution specified in H_0 and the sample size n is reasonably large:
>
> $$X^2 \simeq \chi^2(k - m - 1)$$
>
> where: k is the number of classes
> m is the number of parameters estimated from the sample data
>
> As a working rule, n is adequately large when all of the expected frequencies F_i are 2 or more and at least 50 percent of them are 5 or more.

The notation $\chi^2(k - m - 1)$ in (16.11) denotes the χ^2 (Greek chi square) distribution with $k - m - 1$ degrees of freedom. Theorem (16.11) tells us, therefore, that the statistic X^2 follows the χ^2 distribution with $k - m - 1$ degrees of freedom when H_0 holds and none of the F_i values is too small. The χ^2 distribution is continuous, unimodal, and skewed to the right, as shown in Figure 16.3. (A detailed discussion of the χ^2 distribution may be found in Appendix B, Section B.1.)

Decision Rule

Figure 16.3 illustrates the sampling distribution of X^2 when H_0 is true. Since the risk of a Type I error is to be controlled at α, the area in the upper tail must be α. Consequently, the action limit must be the $100(1 - \alpha)$ percentile of the $\chi^2(k - m - 1)$ distribution, and hence must equal $\chi^2(1 - \alpha; k - m - 1)$. The decision rule for a chi-square test therefore has the following general form.

> **(16.12)**
>
> When the alternatives are as follows:
>
> H_0: Population distribution has specified form.
> H_1: Population distribution does not have specified form.
>
> and the sample size is sufficiently large according to working rule (16.11), the appropriate decision rule to control the α risk is as follows:
>
> If $X^2 \leq \chi^2(1 - \alpha; k - m - 1)$, conclude H_0.
> If $X^2 > \chi^2(1 - \alpha; k - m - 1)$, conclude H_1.
>
> where X^2 is given by (16.10).

The application of the chi-square goodness-of-fit test varies a little depending on whether the population distribution under consideration is discrete or continuous. We consider each of these cases in turn.

16.3

TEST INVOLVING DISCRETE POPULATION DISTRIBUTION

When a chi-square test is conducted for a discrete population distribution, an important concern is whether the expected frequencies F_i are large enough to meet the requirements of working rule (16.11). If they are not large enough, we conventionally pool or combine adjacent classes to bring the combined expected frequencies into conformance with the χ^2 approximation requirements. We shall illustrate this procedure with the spare part example.

☐ **EXAMPLE**

In the spare part example, the operations analyst wishes to test whether the probability distribution of replacement demand for the spare part is Poisson. Table 16.3a presented the data. Two of the classes in Table 16.3b were formed by pooling to meet the χ^2 approximation requirements. For example, the first class was formed as follows:

Original Classes				Pooled Class			
X	f_i	Probability Under H_0	F_i	X	f_i	Probability Under H_0	F_i
0	2	0.0183	0.952	0–1	4	0.0916	4.763
1	2	0.0733	3.812				

Note that if $X = 0$ and $X = 1$ had not been pooled, the expected frequency in the first class would have been less than 2, thereby violating working rule (16.11).

Pooling was also employed to form the class $X = 8$ or more in Table 16.3b. With the pooled classes in Table 16.3b, all of the F_i values now are at least 2, and 62.5 percent of the F_i values (5 out of 8) are at least 5. Hence, the χ^2 approximation of theorem (16.11) applies.

Step 1. The test alternatives here are as follows:

H_0: The population distribution is Poisson.
H_1: The population distribution is not Poisson.

Step 2. The α risk is to be controlled at 0.05.

Step 3. There are $k = 8$ classes here (see Table 16.3b). Also, $m = 1$ because one parameter (λ) was estimated from the sample data by $\overline{X}$. Thus, the required χ^2 distribution has $k - m - 1 = 8 - 1 - 1 = 6$ degrees of freedom. For $\alpha = 0.05$, Table C.2 shows that $\chi^2(0.95; 6) = 12.59$. The decision rule therefore is as follows:

If $X^2 \leq 12.59$, conclude H_0.
If $X^2 > 12.59$, conclude H_1.

Step 4. Table 16.3b shows the steps in calculating the test statistic X^2. We see from column 5 that $X^2 = 5.256 \leq 12.59$. Hence, we conclude H_0—that a Poisson distribution adequately fits the sample frequencies for weekly demand.

The P-value for the test equals:

$$P[\chi^2(k - m - 1) > X^2] = P[\chi^2(6) > 5.256] = 0.511$$

The high P-value confirms that the sample data are consistent with a Poisson population.   ☐

> **Comment**
>
> When H_0 is rejected in a chi-square test because of sharp departures between sample and expected frequencies, an examination of the relative squared residuals $(f_i - F_i)^2/F_i$ often shows where the population distribution hypothesized in H_0 fails to fit, and it yields clues to the true form of the population distribution.

When the population distribution specified in H_0 is continuous (for example, normal), the class intervals used to classify the sample data can be chosen in many ways. To increase the power of the test (that is, to reduce the risk of a Type II error), we employ the following procedure whenever possible.

TEST INVOLVING CONTINUOUS POPULATION DISTRIBUTION **16.4**

> **(16.13)**
> The classes of the continuous population distribution specified in H_0 should be chosen to have equal probabilities; that is, the expected frequencies for all classes should be equal.

Figure 16.4 presents this procedure graphically. If, say, $k = 10$ classes are to be used, each class should have probability $1/k = 1/10 = 0.10$. Then the percentile separating classes 1 and 2 is the $100(1/k) = 100(1/10) = 10$th percentile, the percentile separating classes 2 and 3 is the 20th percentile, and so on.

EXAMPLE □

In the milk production example, the researcher wishes to test whether or not the milk production of cows is normally distributed. The milk production data for the sample of 40 cows were presented in Table 16.1a.

Step 1. The alternatives here are as follows:

H_0: The population distribution is normal.

H_1: The population distribution is not normal.

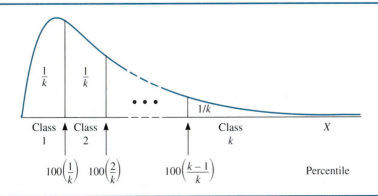

FIGURE 16.4
Partitioning a continuous probability distribution into k classes of equal probability

Step 2. The researcher wishes to control the α risk at 0.01.

Step 3. To apply the chi-square test, we first must estimate the unknown parameters μ and σ of the normal distribution because they are not specified in H_0. As before, we employ the sample estimators $\overline{X}$ and s, which were given in Table 16.1a: $\overline{X} = 15.96$, $s = 2.144$.

We shall utilize eight class intervals because $n = 40$ and so each class will then have an expected frequency $F_i = 5$. If we were to use more than eight classes, the expected frequencies F_i all would be less than 5 (remember that they are to be equal for all classes), which would violate the working rule in (16.11) that requires at least 50 percent of the expected frequencies to be 5 or more. If we were to use fewer than eight classes, we would have a less powerful test for detecting nonnormality (that is, the risk of a Type II error would be greater).

Figure 16.5 illustrates the construction of the classes. The probability of each class is to be the same; hence, it must be $1/8 = 0.125$. Therefore, the first class limit corresponds to the 12.5th percentile, the second to the 25.0th percentile, and so on. The percentiles will be obtained using the relationship in (8.10), with $\overline{X}$ and s as estimates of μ and σ, respectively. The required percentiles of the standard normal distribution can be obtained from Table C.1. We find, using $\overline{X} = 15.96$ and $s = 2.144$ from Table 16.1a:

$$X(0.125) = \overline{X} + z(0.125)s = 15.96 - 1.15(2.144) = 13.49$$
$$X(0.250) = \overline{X} + z(0.250)s = 15.96 - 0.67(2.144) = 14.52$$
$$\vdots \qquad \vdots \qquad \qquad \vdots \qquad \qquad \vdots$$
$$X(0.875) = \overline{X} + z(0.875)s = 15.96 + 1.15(2.144) = 18.43$$

We see, for example, that the 12.5th percentile of the standard normal distribution is $z(0.125) = -1.15$. Hence, the corresponding 12.5th percentile of the distribution of milk production must be 1.15 standard deviations ($s = 2.144$) below the mean ($\overline{X} = 15.96$) and thus equals 13.49. The class limits corresponding to these percentiles are shown in Table 16.4.

With this method of constructing the class limits, five sample observations are expected to lie in each class when the population is normal, as shown in Table 16.4, column 3. The 40 sample observations in Table 16.1a are now classified into these

---

FIGURE 16.5   **Partitioning a normal distribution into eight intervals of equal probability—Milk production example**

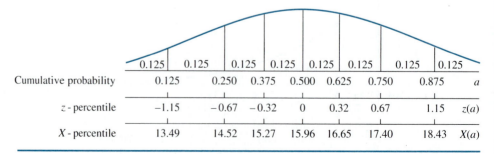

	(1)	(2)	(3)	(4)	
		Under H_0			
Milk Production X	Sample Frequency f_i	Probability	Expected Frequency F_i	Relative Squared Residual $(f_i - F_i)^2/F_i$	
Under 13.49	7	0.125	5	0.80	
13.49–14.51	6	0.125	5	0.20	
14.52–15.26	3	0.125	5	0.80	
15.27–15.95	1	0.125	5	3.20	
15.96–16.64	8	0.125	5	1.80	
16.65–17.39	3	0.125	5	0.80	
17.40–18.42	7	0.125	5	0.80	
18.43 and over	5	0.125	5	0.00	
Total	40	1.000	40	$X^2 = 8.40$	

TABLE 16.4
Chi-square test for goodness of fit of normal distribution—Milk production example

eight classes, as shown in Table 16.4, column 1. Because the data are arrayed in Table 16.1a, it is easy to count seven observations less than 13.49, six observations in the interval 13.49–14.51, and so on.

The chi-square test now proceeds in standard fashion. If H_0 holds (that is, milk production is normally distributed), then X^2 follows an approximate χ^2 distribution with $k - m - 1 = 8 - 2 - 1 = 5$ degrees of freedom. Remember that $m = 2$ parameters (μ and σ) had to be estimated here by $\bar{X}$ and s. For $\alpha = 0.01$, we require $\chi^2(0.99; 5) = 15.09$. Hence, the decision rule is as follows:

If $X^2 \leq 15.09$, conclude H_0.
If $X^2 > 15.09$, conclude H_1.

Step 4. From Table 16.4, column 4, we see that $X^2 = 8.40 \leq 15.09$. Hence, we conclude H_0—that a normal distribution is a reasonable model for the milk production population. This conclusion is exactly what we anticipated from the normal probability plot in Figure 16.2.

The P-value for this test is $P[\chi^2(5) > 8.40] = 0.136$, which is greater than $\alpha = 0.01$ and, hence, leads to the same test conclusion, as expected. $\square$

When the chi-square goodness-of-fit test is used for determining the appropriateness of an inference procedure that requires a normal population, as in the milk production example, some special considerations affect the choice of the α risk level. As we have seen in earlier chapters, some statistical procedures that assume a normal population are robust and, hence, are reliable even in the absence of exact normality. Similarly, in the development of models, exact normality is often not essential. In these cases, use of a low α risk with the chi-square test may be appropriate for the following reasons. If the departure from normality is large and hence the normal model is clearly not reasonable, the risk of a Type II error (concluding that the population is normal when it is not) will still be low. On the other hand, if the departure from normality is small and hence the normal model represents a reasonable approximation, the risk of a Type II error will be high but of no practical concern.

Comments

1. When sample data from a continuous probability distribution are already classified in a frequency distribution (for instance, in published data), it may not be possible to construct equal-probability classes as called for in (16.13). The chi-square procedure can still be used with the existing frequency classes, pooling them where needed to achieve F_i of adequate size. The resulting loss in the power of the test will, in most cases, be small.
2. A strict application of theorem (16.11) requires that any parameter estimates be based on the sample frequencies $f_1, f_2, \ldots, f_k$, rather than on the original sample observations themselves. However, when the number of classes utilized in the test is reasonably large and the expected frequencies F_i satisfy the working rule in (16.11), the customary procedure of estimating any parameters from the original sample observations is satisfactory.

16.5   OPTIONAL TOPIC—CONFIDENCE BAND AND TEST FOR CUMULATIVE PROBABILITY FUNCTION

In Chapter 5, we noted that a random variable X can be described by its cumulative probability function $F(x)$, where $F(x) = P(X \leq x)$. In this section, we discuss the Kolmogorov–Smirnov (K–S) estimation procedure, which provides a confidence band for the cumulative probability function $F(x)$. This procedure also can be used as a goodness-of-fit test.

Assumptions

The only assumption for the K–S estimation procedure is that the sample be a simple random one.

When the population is continuous, the $1 - \alpha$ confidence coefficient for the K–S estimation procedure is exact. When the population is discrete, the actual confidence level for the K–S estimation procedure will be greater than the specified $1 - \alpha$ level.

K–S Statistic

Cumulative Sample Function.   The basis of the K–S estimation procedure is the cumulative sample function.

(16.14)

The **cumulative sample function,** denoted by $S(x)$, specifies for each value of x the proportion of observations in the sample that are less than or equal to x. $S(x)$ takes the form of a step function with a vertical step of size $1/n$ at each sample observation.

☐ **EXAMPLE**

In the hospital stays example, the researcher studied the lengths of hospital stays for $n = 10$ psychiatric patients. Figure 16.6a contains the sample data for this study arrayed in ascending order. The cumulative sample function $S(x)$ for these data is plotted in Figure 16.6b, based on the following cumulative data:

x:	16	17	20	27	31	34	37	55
$S(x)$:	0.3	0.4	0.5	0.6	0.7	0.8	0.9	1.0

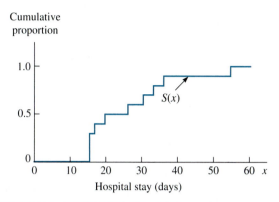

FIGURE 16.6
Hospital stays data and cumulative sample function for 10 psychiatric cases— Hospital stays example

(a) Array of Hospital Stays

Case	Stay (days)	Case	Stay (days)
TAE	16	RJV	27
MSC	16	AAM	31
AGB	16	JJT	34
KFG	17	WKR	37
JCM	20	RAM	55

(b) Cumulative Sample Function

Note that $S(x)$ is a step function with vertical steps of $1/n = 1/10 = 0.1$ at each sample observation. Because there are three observations of 16 in the sample, the vertical step at $x = 16$ is 0.3.   ☐

$D(n)$ Statistic.   The K–S estimation procedure utilizes a statistic, denoted by $D(n)$, that is based on the differences between the cumulative sample function $S(x)$ and the cumulative probability function $F(x)$ for the population.

(16.15) $$D(n) = \max_{x} |S(x) - F(x)|$$

In other words, $D(n)$ equals the largest absolute deviation of $S(x)$ from $F(x)$ when all values of x are considered.

The notation $D(n)$ is used for the K–S statistic because its sampling distribution depends on the sample size. Surprisingly, however, it does not depend on the specific form of $F(x)$ for the population. Thus, the sampling distribution of $D(n)$ can be tabulated for different values of n, irrespective of the nature of the population. Table C.9 gives percentiles of the $D(n)$ distribution for selected percentages and values of n, up to $n = 50$. For $n > 50$, the table shows formulas for computing the percentiles. We denote the $100a$ percentile of the $D(n)$ distribution by $D(a; n)$. We see from Table C.9, for instance, that $D(0.90; 10) = 0.37$. Thus the probability is 0.90 that the largest absolute deviation of $S(x)$ from $F(x)$ is 0.37 or less when $n = 10$, no matter what the nature of the population.

Confidence Band for $F(x)$

The statistic $D(n)$ provides a simple way for constructing a confidence band for the population cumulative probability function $F(x)$.

(16.16)
The $1 - \alpha$ confidence band for the population cumulative probability function $F(x)$ is of the form $L(x) \leq F(x) \leq U(x)$, where:

$$L(x) = S(x) - D(1 - \alpha; n) \qquad U(x) = S(x) + D(1 - \alpha; n)$$

☐ EXAMPLE

For the hospital stays example, a 90 percent confidence band for $F(x)$ is desired. Figure 16.7a shows the construction of the confidence band. For $n = 10$ and $1 - \alpha = 0.90$, we require from Table C.9 $D(0.90; 10) = 0.37$. Thus, $L(x)$ and $U(x)$ are obtained by constructing a pair of step functions that are identical to $S(x)$ but at vertical distances of 0.37 below and above it. We obtain the following:

x:	16	17	20	27	31	34	37	55
$S(x)$:	0.30	0.40	0.50	0.60	0.70	0.80	0.90	1.0
$L(x)$:	0	0.03	0.13	0.23	0.33	0.43	0.53	0.63
$U(x)$:	0.67	0.77	0.87	0.97	1.0	1.0	1.0	1.0

Note that there is no need to extend the confidence band above 1 or below 0 since, by definition, $F(x)$ cannot be outside these limits. In conclusion, we can state with 90 percent confidence that the cumulative probability function $F(x)$ for the lengths of hospital stay lies entirely within the confidence band in Figure 16.7a.   ☐

Comment

The K–S confidence band in (16.16) is derived from the fact that $D(n)$ is the maximum absolute deviation between $S(x)$ and $F(x)$ and is less than $D(1 - \alpha; n)$ with probability $1 - \alpha$. It follows that $S(x)$ has a probability of $1 - \alpha$ of lying within distance $D(1 - \alpha; n)$ of $F(x)$; that is:

$$P[S(x) - D(1 - \alpha; n) \leq F(x) \leq S(x) + D(1 - \alpha; n)] = 1 - \alpha$$

This probability statement provides us with the confidence band for $F(x)$ in (16.16).

Goodness-of-Fit Test for $F(x)$

The K–S estimation procedure can also be used to test whether the cumulative probability function $F(x)$ has a specified form, to be denoted by $F_0(x)$. For example, $F_0(x)$ might be the cumulative probability function for a normal population. The test alternatives may then be stated as follows.

(16.17) H_0: $F(x) = F_0(x)$ for all x
H_1: $F(x) \neq F_0(x)$ for some x

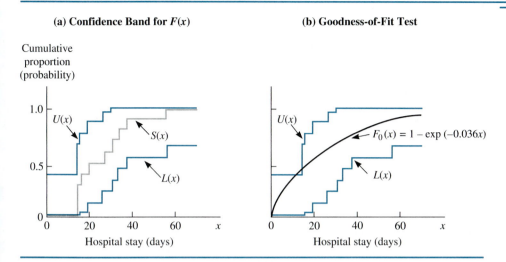

(a) Confidence Band for $F(x)$ (b) Goodness-of-Fit Test

FIGURE 16.7
K–S confidence band and goodness-of-fit test—Hospital stays example

The confidence band for $F(x)$ in (16.16) affords a convenient test of the alternatives in (16.17). We simply plot $F_0(x)$ together with the confidence band. If $F_0(x)$ falls entirely within the confidence band, then we conclude H_0; otherwise, we conclude H_1.

EXAMPLE ☐

In the hospital stays example, the researcher wishes to test whether or not hospital stays have an exponential distribution with $\lambda = 0.036$.

Step 1. We know from (8.14) that $F_0(x)$ here is $1 - \exp(-0.036x)$. Hence, the alternatives are as follows:

$$H_0: F(x) = 1 - \exp(-0.036x) \qquad \text{for all } x$$
$$H_1: F(x) \neq 1 - \exp(-0.036x) \qquad \text{for some } x$$

Step 2. The α risk is to be controlled at 0.10.

Step 3. Figure 16.7b repeats the 90 percent confidence band for $F(x)$ from Figure 16.7a and also shows the plot of $F_0(x)$. The following two examples illustrate the calculation of $F_0(x)$ for the plot:

At $x = 16$, $F_0(16) = 1 - \exp[-0.036(16)] = 0.44$.
At $x = 20$, $F_0(20) = 1 - \exp[-0.036(20)] = 0.51$.

Step 4. We note from Figure 16.7b that $F_0(x)$ does not fall entirely within the confidence band. Hence, we conclude H_1—that is, the probability distribution is not exponential with $\lambda = 0.036$. Thus, either the distribution is not exponential or it is exponential but the parameter λ does not equal 0.036.

Figure 16.7b shows that the major lack of fit is in the neighborhood of $x = 16$, where three sample observations occur. Upon further investigation, the researcher discovered that two of these observations were actually for somewhat shorter hospital stays but were incorrectly recorded at the usual minimum stay of 16 days. With the corrected data, the researcher found that an exponential distribution with $\lambda = 0.036$ is an adequate model. ☐

Comments

1. The alternatives in (16.17) can also be tested directly by calculating the maximum absolute deviation of $S(x)$ from $F_0(x)$ for all x and comparing this statistic with $D(1 - \alpha; n)$. It can be shown that the maximum absolute deviation must occur at or just below one of the sample observations.

   In the hospital stays example, the maximum absolute deviation occurs just below $x = 16$, and equals 0.4379, as may be seen from the following calculations:

x	$S(x)$	$F_0(x)$	$|S(x) - F_0(x)|$
16.0	0.30	0.4379	0.1379
15.9999	0.00	0.4379	0.4379

   Since the maximum absolute deviation 0.4379 exceeds $D(0.90; 10) = 0.37$, we conclude H_1 at the $\alpha = 0.10$ level of significance—that the population is not exponential with $\lambda = 0.036$.

   Some statistical packages show in their output for the K–S goodness-of-fit test the maximum absolute deviation and the P-value of the test.

2. For the K–S test procedure to be exact, $F_0(x)$ in (16.17) must be fully specified, including its parameter value(s). When the parameter value(s) are estimated from the sample, the test procedure will entail an actual α risk that is less than the specified one. However, the power of the test—that is, its ability to detect departures of $F(x)$ from the specified $F_0(x)$—will also be smaller.

3. When $F_0(x)$ is discrete, the actual α risk level with the K–S test will be less than the one specified.

4. In situations where both the chi-square test and the K–S test might be applied, the latter is often, but not always, more powerful in detecting departures of $F(x)$ from $F_0(x)$, where both tests have the same specified α risk.

PROBLEMS

* **16.1 Price Test.** Fifteen consumers were selected to test a new home garden tool. After using the tool, each was asked what would be a reasonable price for this tool. The replies (in dollars) were as follows:

18	23	36	33	21	26	26	16	31	30	41	27	29	40	38

Assume that the 15 observations constitute a random sample from a continuous probability distribution.

a. Present the observations in a stem-and-leaf display. Is the distribution pattern consistent with the population being a normal distribution? Comment.

b. Assuming that the population is normal, use (16.2) to obtain the approximate expected values of the smallest and 10th smallest observations in this data set. Are the observed values close to the expected values?

c. Construct a normal probability plot of the sample data. Does the plot suggest that the population is normal or approximately normal? Explain.

16.2 Real Estate Investments. The rates of return (in percent per year) earned last year by a real estate development firm on 10 short-term investment projects were:

3	18	21	6	25	12	-1	35	9	15

Assume that the 10 observations constitute a random sample from a continuous probability distribution.

a. Present the observations in a stem-and-leaf display. Is the distribution pattern consistent with the population being a normal distribution? Comment.

b. Assuming that the population is normal, use (16.2) to obtain the approximate expected values of the 4th smallest and the largest observations in this data set. Are the observed values close to the expected values?

c. Construct a normal probability plot of the sample data. Does the plot suggest that the population is normal or approximately normal? Explain.

16.3 Refer to **Chemical Yield** Problem 15.5.

a. Present the observations in a stem-and-leaf display. Is the distribution pattern consistent with the population being a normal distribution? Comment.

b. Assuming that the population is normal, use (16.2) to obtain the approximate expected values of the smallest and third smallest observations in this data set. Are the observed values close to the expected values?

c. Construct a normal probability plot of the sample data. Does the plot suggest that the population is normal or approximately normal? Explain.

16.4 Refer to **Snow-Strip Swathing** Problem 15.18. Consider the grain yields in the 17 fields receiving conventional swathing. Assume that these yields constitute a random sample from a continuous probability distribution. Prepare a normal probability plot of the sample data. Does the plot suggest that the population is normal or approximately normal? Explain.

16.5 For each of the following applications of the chi-square goodness-of-fit test, indicate (1) H_1, (2) the number of degrees of freedom associated with X^2 under H_0, (3) the value of the action limit in the decision rule.

a. H_0 specifies that the population distribution is Poisson. There are eight classes. The α risk is to be controlled at 0.01.

b. H_0 specifies that the population distribution is continuous uniform with $a = 2$ and $b = 10$. There are six classes. The α risk is to be controlled at 0.10.

c. H_0 specifies that the population distribution is normal. There are 10 classes. The α risk is to be controlled at 0.05.

16.6 For each of the following applications of the chi-square goodness-of-fit test, state (1) H_1, (2) the number of degrees of freedom associated with X^2 under H_0, (3) the value of the action limit in the decision rule.

a. H_0 specifies that the population distribution is Poisson with $\lambda = 3$. There are six classes. The α risk is to be controlled at 0.10.

b. H_0 specifies that the population distribution is binomial with $n = 4$. There are five classes. The α risk is to be controlled at 0.01.

c. H_0 specifies that the population distribution is $N(0, 1)$. There are eight classes. The α risk is to be controlled at 0.05.

16.7 Consider a chi-square goodness-of-fit test where H_0 specifies the form of the population distribution but not the parameter values. When H_0 is correct, why will the test statistic X^2 in which the parameters are estimated from the sample data tend to be smaller than if the actual (but unknown) parameter values were used in the test statistic? Use theorem (16.11) to explain how the test procedure takes this tendency into account.

* **16.8** In planning the best size for the waiting room in a medical clinic, a consultant recorded the number of persons accompanying each patient in a random sample of 300 patients. The sample results follow:

Number of Accompanying Persons:	0	1	2	3	4
Frequency:	163	104	24	8	1

The consultant wishes to determine whether the probability distribution of the number of accompanying persons is Poisson.

a. Use the chi-square procedure to conduct the test, controlling the α risk at 0.10. Pool classes as required. State the alternatives, the decision rule, the value of the test statistic, and the conclusion.
b. Examine the relative squared residuals for the test statistic in part a. Do any appear to be particularly large? Is this finding consistent with the test conclusion in part a? Explain.

16.9   Refer to **Airline Reservations** Problem 11.4. The airline researcher conjectures that the probability distribution of the number of no-shows per flight is Poisson.

a. Test the conjecture by using the chi-square procedure. Control the α risk at 0.10. Pool classes as required. State the alternatives, the decision rule, the value of the test statistic, and the conclusion.
b. What is the reason for pooling classes in conducting the test?

16.10   **False Alarms.** A city report described the results of a campaign aimed at reducing the incidence of false alarms received by city fire stations. Included in the report was the following frequency distribution of the number of false alarms received by Fire Station 27 each day during a 60-day period following the campaign:

Number of False Alarms:	0	1	2	3	4	5	6
Number of Days:	28	20	7	2	2	0	1

Assume that the data for the 60 days constitute a random sample from a process.

a. Extensive experience for this station before the campaign showed that the mean number of false alarms received each day had been 1.5. Use the chi-square procedure to test whether the probability distribution of daily number of false alarms after the campaign is Poisson with $\lambda = 1.5$. Control the α risk at 0.05. Pool classes as required. State the alternatives, the decision rule, the value of the test statistic, and the conclusion.
b. To examine further the nature of the probability distribution, use the chi-square procedure to test whether the probability distribution of daily number of false alarms after the campaign is Poisson (with no specification of the value of λ). Control the α risk at 0.05. Pool classes as required. State the alternatives, the decision rule, the value of the test statistic, and the conclusion.
c. Summarize the findings from parts a and b.

16.11   A researcher observed the final digit in the daily volume of stock transactions during the past 50 days. The frequency distribution follows:

Final Digit:	0	1	2	3	4	5	6	7	8	9
Frequency:	6	2	4	4	7	5	6	4	7	5

Assume that the data constitute a random sample from a process. Using the chi-square procedure, test whether the probability distribution of the final digits is discrete uniform for integer outcomes 0, 1, ..., 9. Control the α risk at 0.05. State the alternatives, the decision rule, the value of the test statistic, and the conclusion.

16.12   **Product Testing.** In a product-testing experiment, each of 150 subjects independently examined three patches of fabric. For each patch, a subject was asked whether the fiber is synthetic or natural. The numbers of correct identifications by the 150 subjects were as follows:

Number of Correct Identifications:	0	1	2	3
Number of Subjects:	9	40	36	65

If the subjects simply make an independent guess for each patch, the number of correct identifications by a subject will follow a binomial probability distribution with $n = 3$ and $p = 0.5$.

a. Verify that if all subjects guess, the expected numbers of subjects giving 0, 1, 2, and 3 correct identifications are 18.75, 56.25, 56.25, and 18.75, respectively.

b. Using the chi-square procedure, test whether the subjects in this experiment were guessing. Control the α risk at 0.05. State the alternatives, the decision rule, the value of the test statistic, and the conclusion.

* **16.13** Refer to **Weight Reduction** Problem 14.61. Consider the 30 persons assigned to regimen 2 and assume that they constitute a random sample. The sample mean and variance for the weight losses of these 30 persons are $\bar{X} = 17.29$ kilograms and $s = 1.69$ kilograms, respectively.

a. Using the chi-square procedure, test whether the probability distribution of weight losses of persons assigned to regimen 2 is normal. Control the α risk at 0.10. Use six equal-probability classes. State the alternatives, the decision rule, the value of the test statistic, and the conclusion.

b. Examine the relative squared residuals for the test statistic in part a. What do they suggest about the departure, if any, of the distribution of weight losses from normality?

16.14 Refer to **Employee Productivity** Problem 15.17. Consider the numbers of units produced by the 25 employees from shift 1 and assume that these data constitute a random sample. The sample mean and variance for the numbers of units produced by the 25 employees are $\bar{X} = 207.76$ units and $s = 22.95$ units, respectively.

a. Using the chi-square procedure, test whether the probability distribution of numbers of units produced by employees from shift 1 is normal. Control the α risk at 0.05. Use five equal-probability classes. State the alternatives, the decision rule, the value of the test statistic, and the conclusion.

b. Using Table C.2, obtain bounds for the P-value of the test.

16.15 Refer to **Glass Layers** Problem 15.36. Assume that the 35 observations are a random sample from a process. The sample mean and variance of these 35 observations are $\bar{X} = 0.2102$ micrometer and $s = 0.8000$ micrometer, respectively.

a. Using the chi-square procedure, test whether the probability distribution of thickness deviations of glass layers produced by this process is normal. Control the α risk at 0.05. Use seven equal-probability classes. State the alternatives, the decision rule, the value of the test statistic, and the conclusion.

b. Why would the test in part a not be meaningful if the observations did not come from a stable manufacturing process?

16.16 Hose Ruptures. An industrial machine has a 1.5-meter hydraulic hose that ruptures occasionally. The manufacturer has recorded the locations of these ruptures for 25 ruptured hoses. These locations (measured in meters from the pump end of the hose) are as follows:

1.32	1.19	1.21	1.36	0.64	1.46	0.80	1.37	0.38
1.07	0.13	1.16	0.33	1.42	1.27	0.08	0.75	1.22
1.37	1.14	1.43	0.97	1.12	0.27	1.46

Assume that the 25 rupture locations constitute a random sample from a process.

a. Using the chi-square procedure, test whether the probability distribution of rupture locations is uniform with $a = 0$ and $b = 1.5$. Control the α risk at 0.10. Use five equal-probability classes. State the alternatives, the decision rule, the value of the test statistic, and the conclusion.

b. Examine the relative squared residuals for the test statistic in part a. In which section of the hose is the rupture frequency most out of line with the expectations under H_0? What is the implication of your finding for the engineering design of this hose?

16.17 Refer to Table 3.1a and consider the 40 time intervals between oil spills before pollution controls. The corresponding descriptive statistics are provided in Figure 3.1b. Assume that the 40 observations constitute a random sample from a process. Using the chi-square procedure, test whether the probability distribution of intervals between oil spills is exponential. Control the α risk at 0.05. Use eight equal-probability classes and estimate λ by $1/\bar{X}$. State the alternatives, the decision rule, the value of the test statistic, and the conclusion.

16.18 Refer to **Response Times** Problem 2.11. The sample mean and standard deviation of the 50 response times are $\bar{X} = 13.7$ minutes and $s = 5.33$ minutes, respectively. Assume that the 50 observations constitute a random sample from a process. Using the chi-square procedure, test whether the probability distribution of response times is normal. Base the test on the class intervals and observed frequencies given in Problem 2.11. Control the α risk at 0.05. State the alternatives, the decision rule, the value of the test statistic, and the conclusion.

* **16.19** Refer to **Price Test** Problem 16.1.

   a. Prepare a plot of the 90 percent K–S confidence band for the cumulative probability function.
   b. From the confidence band in part a, determine the minimum proportion of consumers who consider a price of $20 or less to be reasonable.

16.20 Refer to **Real Estate Investments** Problem 16.2.

   a. Prepare a plot of the 95 percent K–S confidence band for the cumulative probability function.
   b. For the confidence band in part a, interpret the values of $L(15)$ and $U(15)$.

16.21 Refer to Figure 3.5a and consider the melting points data for the April shipment, which can be assumed to constitute a random sample from a process.

   a. Prepare a plot of the 95 percent K–S confidence band for the cumulative probability function.
   b. From the confidence band in part a, can you conclude that fewer than 90 percent of filaments in the population have melting points of 330°C or less? Explain.

* **16.22** Refer to **Price Test** Problems 16.1 and 16.19. Before collecting the data, the market researcher hypothesized that consumer assessments of a reasonable price follow a continuous uniform distribution with $a = \$15$ and $b = \$45$. Using the K–S confidence band for the cumulative probability function in Problem 16.19a, test the researcher's hypothesis, controlling the α risk at 0.10. State the alternatives and the conclusion.

16.23 The service lives (in thousands of operating hours) of a random sample of 10 aircraft digital-display units are as follows, arrayed by magnitude:

0.4	2.6	4.4	4.9	10.6	11.3	11.8	12.6	23.0	40.8

   a. Prepare a plot of the 95 percent K–S confidence band for the cumulative probability function.
   b. Using the confidence band in part a, test whether the probability distribution of service lives of these units is exponential with $\lambda = 0.1$ failure per thousand hours. Control the α risk at 0.05. State the alternatives and the conclusion.
   c. Would the test conclusion in part b be affected if a 90 percent K–S confidence band were employed and the α risk controlled at 0.10? Explain.

16.24 The length of time before power is restored by the electric company after a power failure is hypothesized to follow an exponential distribution with a mean of 1.0 hour. Data for the last 10 power failures (in hours) are as follows, arrayed by magnitude:

0.16	0.23	0.25	0.39	0.40	0.45	0.53	0.71	1.05	1.17

Assume that the 10 observations constitute a random sample from a process. Prepare a plot of the K–S confidence band for the cumulative probability function to test the hypothesis, controlling the α risk at 0.10. State the alternatives and the conclusion.

EXERCISES

16.25 Refer to **Hose Ruptures** Problem 16.16. A probability plot for a continuous uniform distribution can be constructed by plotting the ith ranked sample observation against its expected value under the assumption of a uniform distribution.

   a. Using (16.1), obtain an expression for the approximate expected value for the ith ranked observation in the sample of $n = 25$ when the population is a continuous uniform distribution with $a = 0$ and $b = 1.5$.
   b. Prepare a uniform probability plot for the sample data. Does the plot show any evidence that the population is not a uniform distribution? Comment.

16.26 Show that test statistic X^2 in (16.10) can be written as the following weighted sum of the residuals $f_i - F_i$:

$$X^2 = \sum_{i=1}^{k} \frac{f_i}{F_i} (f_i - F_i)$$

16.27 Consider test statistic X^2 in (16.10). Assume that H_0 fully specifies the population distribution and that, consequently, no parameters need to be estimated from the sample data.

a. Show that, for any class i, $E\{(f_i - F_i)^2/F_i\} = 1 - (F_i/n)$ when H_0 holds. [*Hint:* f_i is a binomial random variable with mean F_i and variance $F_i(n - F_i)/n$ in this case.]

b. Use the result in part a to show that $E\{X^2\} = k - 1$ when H_0 holds.

16.28 The percentile $D(1 - \alpha; n)$ approaches 0 as n increases for any α. What does this fact imply about the probability of a Type II error if the K–S confidence band for the cumulative probability function is used to test the alternatives in (16.17) when n is large?

16.29 Explain why the difference $S(x) - F(x)$ at any specified value of x is approximately normal with mean 0 and variance $F(x)[1 - F(x)]/n$ when n is large.

STUDIES

16.30 Refer to **Plant Wages** Problem 2.52. Assume that the 60 wage earners are a random sample from a large population.

a. Construct a normal probability plot of the 60 wage observations.

b. Using the chi-square procedure, test whether the population distribution of weekly wages is normal. Control the α risk at 0.05. Use 12 equal-probability classes. State the alternatives, the decision rule, the value of the test statistic, and the conclusion.

c. Does the normal probability plot in part a suggest that the population is normal or approximately normal? Is this finding consistent with the test conclusion in part b? Explain.

16.31 Refer to **Hospital Costs** Problem 2.53. Assume that the 50 hospitals are a random sample from a large population.

a. Construct a normal probability plot of the 50 average daily cost observations.

b. Using the chi-square procedure, test whether the population distribution of average daily costs is normal. Control the α risk at 0.10. Use 10 equal-probability classes. State the alternatives, the decision rule, the value of the test statistic, and the conclusion.

c. Does the normal probability plot in part a suggest that the population is normal or close to normal? Is this finding consistent with the test conclusion in part b? Explain.

d. Examine the relative squared residuals for the test statistic in part b. What do they suggest about the departure, if any, of the population distribution of average daily costs from normality?

16.32 Refer to **Product Testing** Problem 16.12. A statistician has suggested an alternative model—namely, that a proportion P of the subjects know the identification for certain and the remainder guess.

a. Using the chi-square procedure, test the aptness of this model when $P = 0.3$. Control the α risk at 0.01. State the alternatives, the decision rule, the value of the test statistic, and the conclusion.

b. How might one determine an estimate of P here utilizing the X^2 statistic?

16.33 **Capital Investment.** A financial analyst has undertaken a computer simulation study of a complex and risky capital investment proposal, the outcome of which depends on many random factors. As part of the study, the following rates of return (in percent) were obtained in 20 simulation runs:

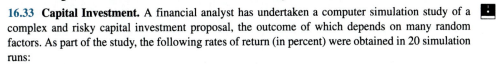

−4.7	22.7	12.3	5.1	20.4	16.7	11.1
−1.5	4.3	2.6	−11.3	1.0	13.4	23.0
12.1	10.3	−2.1	−3.4	1.1	10.1	

The 20 observations may be considered to constitute a simple random sample from the rate-of-return distribution for this investment proposal.

a. Prepare a plot of the 90 percent K–S confidence band for the cumulative probability function $F(x)$. Interpret this band.

b. From the confidence band in part a, obtain an interval estimate for $F(0)$, the probability that the investment proposal will produce a negative or zero rate of return.

c. How many simulations would have to be run in order that the probability is 0.90 that $S(x)$ will lie within a vertical distance of ± 0.04 of $F(x)$ for all values x?

d. Letting $p = F(0)$, obtain an exact 90 percent confidence interval for p, using (13.13). Why is this interval narrower than the one obtained in part b?

Multinomial Populations

17

I n many applications, the population of interest contains elements classified into several categories or classes. We now consider some important inference procedures for populations of this type.

Populations containing elements classified into several classes or categories are called multinomial populations.

(17.1)
When each element of a population is assigned to one (and only one) of two or more classes or categories, the population is called a **multinomial population.**

We encountered multinomial populations in Chapter 4. Table 4.1a provides an example. We now consider other examples.

EXAMPLES ☐

1. All of the eggs laid in a month by a hen flock constitute the population of interest. Each egg is classified according to its size, the classes being extra large, large, medium, and small.

2. All of the households in a city constitute the population of interest. Each household is classified as owning no car, one car, or two or more cars.

3. The process of producing memory chips for computers is the population of interest. A chip is classified as acceptable, having minor defects, or having major defects. The population is infinite here because it relates to a process. ☐

Parameters of Interest

We denote the number of classes into which a population element can be classified by k. Thus, in Example 1, there are $k = 4$ classes, and in Examples 2 and 3, there are $k = 3$ classes. The probability that a population element selected at random comes from the ith class is denoted by p_i ($i = 1, 2, \ldots, k$). These probabilities necessarily sum to 1.

Machine Component ☐ **EXAMPLE**

A large lot of used machine components is the population of interest. Each machine component is classified according to its condition, the classes being excellent, good, poor, and scrap. The probabilities p_i for the lot are as follows:

i	Condition of Component	p_i
1	Excellent	$p_1 = 0.4$
2	Good	$p_2 = 0.3$
3	Poor	$p_3 = 0.2$
4	Scrap	$p_4 = \underline{0.1}$
	Total	1.0

☐

The probabilities p_i for a multinomial population are the parameters of interest and are usually unknown. Sampling is then employed to make inferences about these parameters.

17.2 MULTINOMIAL PROBABILITY DISTRIBUTIONS

When a random sample of n elements is selected from a multinomial population, we denote the frequency of sample observations from the ith class by f_i. The sample frequencies f_i necessarily sum to the sample size n.

☐ **EXAMPLE**

In the machine component example, a random sample of 10 machine components yielded the following results:

i	Condition of Component	f_i
1	Excellent	$f_1 = 3$
2	Good	$f_2 = 3$
3	Poor	$f_3 = 2$
4	Scrap	$f_4 = \underline{2}$
	Total	$n = 10$

☐

When a simple random sample is selected from an infinite multinomial population, the probability of obtaining the sample frequencies $f_1, f_2, \ldots, f_k$, denoted by $P(f_1, f_2, \ldots, f_k)$, is given by the multinomial probability function.

(17.2)
The **multinomial probability function** is:

$$P(f_1, f_2, \ldots, f_k) = \frac{n!}{f_1! f_2! \cdots f_k!} p_1^{f_1} p_2^{f_2} \cdots p_k^{f_k}$$

where: $\sum_{i=1}^{k} f_i = n$

$\sum_{i=1}^{k} p_i = 1$

The multinomial probability distribution is a discrete distribution, with parameters n and $p_1, p_2, \ldots, p_k$. Each different set of parameter values defines a different probability distribution in the multinomial family.

When the multinomial population is finite, the multinomial probability function (17.2) provides approximate probabilities of the sample frequencies as long as the sampling fraction n/N is not large. Throughout this chapter, we shall assume that when finite multinomial populations are sampled, the sampling fraction is small.

EXAMPLES

1. We wish to find the probability of the sample frequencies in the machine component example. Recall that we have:

$$p_1 = 0.4 \qquad p_2 = 0.3 \qquad p_3 = 0.2 \qquad p_4 = 0.1$$
$$f_1 = 3 \qquad f_2 = 3 \qquad f_3 = 2 \qquad f_4 = 2 \qquad n = 10$$

By substituting into (17.2), we then obtain:

$$P(3, 3, 2, 2) = \frac{10!}{3!3!2!2!}(0.4)^3(0.3)^3(0.2)^2(0.1)^2 = 0.01742$$

2. For the same parameter values, we wish to find the probability of the sample frequencies $f_1 = 8$, $f_2 = 2$, $f_3 = 0$, $f_4 = 0$:

$$P(8, 2, 0, 0) = \frac{10!}{8!2!0!0!}(0.4)^8(0.3)^2(0.2)^0(0.1)^0 = 0.002654 \qquad \square$$

Binomial Distribution: A Special Case

The binomial probability distribution is a special case of the multinomial probability distribution. To see this, note that when there are only $k = 2$ classes, the two multinomial parameters p_1 and p_2 correspond to the binomial parameter p and $1 - p$, respectively. Similarly, the two multinomial sample frequencies f_1 and f_2 correspond to the binomial frequency f and $n - f$, respectively. Using these relations for $k = 2$ classes, the multinomial probability function (17.2) reduces to the binomial probability function (13.3).

We shall now take up two types of inferences concerning the parameters p_i of a multinomial population: inferences on a given parameter p_i and inferences on all of the p_i.

INFERENCES 17.3 CONCERNING THE PARAMETERS p_i

Assumptions

1. A simple random sample of size n has been selected from (1) an infinite multinomial population or (2) from a finite multinomial population with the sampling fraction small.
2. The sample size n is reasonably large.

Inferences for One p_i

When a single multinomial parameter p_i is of interest, the inference procedures are exactly the same as those of Chapter 13 for a population proportion p. To see why, consider the machine component example again. Suppose we wish to estimate p_1, the probability that a used component is in excellent condition. To estimate p_1, we do not care what the other classes are. Hence we can combine all of the other classes into one class consisting of all components that are not in excellent condition. But then we have the situation corresponding to a Bernoulli random trial in which there are two categories (excellent condition, not excellent condition), with probabilities p_1 and $1 - p_1$, respectively. Thus, all of our earlier inference procedures for a population proportion are applicable when we are interested in a single multinomial parameter p_i.

  We consider now two examples illustrating the inference procedures for a single multinomial parameter.

Confidence Interval for p_i.   The first example illustrates the construction of a confidence interval for a multinomial parameter p_i.

Bearing Failures   ☐   **EXAMPLE**

A railroad engineer, who is concerned with maintenance problems for rolling stock, studied the timing of wheel-bearing failures. Table 17.1 contains a record of 1044 bearing failures on rolling stock over the past several years, classified by quarter when the failure occurred.

  The engineer analyzed the consistency of the quarter-by-quarter patterns over the years and found them to be stable. The engineer therefore concluded that it is reasonable to treat these data as a random sample from an infinite multinomial population of bearing failures. Here, p_1 is the probability that, given a bearing failure has occurred, it happens in the first quarter, p_2 is the probability that it occurs during the second quarter, and p_3 and p_4 are similarly defined.

  To estimate the probability of a failure occurring in the first quarter, we use the sample proportion:

$$\bar{p}_1 = \frac{f_1}{n} = \frac{249}{1044} = 0.2385$$

A 90 percent confidence interval for p_1 is desired. Since $n = 1044$, we employ the large-sample confidence interval (13.9) for a population proportion. We require:

$$s\{\bar{p}_1\} = \sqrt{\frac{\bar{p}_1(1 - \bar{p}_1)}{n - 1}} = \sqrt{\frac{0.2385(1 - 0.2385)}{1044 - 1}} = 0.0132$$

$$z(1 - \alpha/2) = z(0.95) = 1.645$$

Hence, the confidence limits for p_1 are $0.2385 \pm 1.645(0.0132)$, and the 90 percent confidence interval is:

$$0.217 \leq p_1 \leq 0.260$$

Thus, the engineer can conclude with 90 percent confidence that, given a bearing failure has occurred, the probability of the failure happening during the first quarter is between 0.217 and 0.260.   ☐

Test for p_i.   The second example illustrates a test for a multinomial parameter p_i.

Quarter of Occurrence (i):	1	2	3	4	Total (n)
Number of Failures (f_i):	249	256	297	242	1044

TABLE 17.1
Number of bearing failures in rolling stock, classified by quarter of occurrence—Bearing failures example

EXAMPLE ☐

In the bearing failures example, the proportion of the railroad's annual traffic carried in the third quarter is 0.30. The engineer would like to test whether the probability of a bearing failure occurring in the third quarter is the same as the proportion of traffic in that quarter.

Step 1. The alternatives here are:

$$H_0: \ p_3 = 0.30$$
$$H_1: \ p_3 \neq 0.30$$

Step 2. The α risk is to be controlled at 0.10 when $p_3 = 0.30$. The value of the multinomial parameter where the α risk is controlled will be denoted by p_{30}. The subscript 0 on p_{30} signifies that it corresponds to the standard specified in H_0.

Step 3. For $\alpha = 0.10$, we require $z(1 - \alpha/2) = z(0.95) = 1.645$. The decision rule for the test, therefore, is as follows:

If $|z^*| \leq 1.645$, conclude H_0.
If $|z^*| > 1.645$, conclude H_1.

Step 4. To calculate the standardized test statistic z^* defined in (13.11), we note that $\bar{p}_3 = 297/1044 = 0.2845$ and that the α risk is to be controlled at $p_{30} = 0.30$. Hence, we obtain:

$$\sigma\{\bar{p}_3\} = \sqrt{\frac{0.30(0.70)}{1044}} = 0.01418$$

$$z^* = \frac{0.2845 - 0.30}{0.01418} = -1.09$$

Because $|z^*| = 1.09 \leq 1.645$, we conclude H_0—that the probability of a bearing failure occurring in the third quarter is the same as the proportion of traffic carried in that quarter. The two-sided P-value for the test is $2P(Z < -1.09) = 2(0.1379) = 0.2758$. ☐

Inferences for All p_i

In various applications, inferences about all of the parameters p_i of a multinomial population are desired.

EXAMPLE ☐

In the bearing failures example, the engineer would like to test whether bearing failures occur during the year in direct proportion to the railroad's traffic. The quarterly proportions of annual traffic are as follows:

Quarter (i):	1	2	3	4
Proportion of Annual Traffic:	0.21	0.23	0.30	0.26

Note that here a test concerning all four multinomial parameters p_i is desired.

Alternatives. In this test, the alternatives take the following form.

(17.3) H_0: $p_i = p_{i0}$ for all i
H_1: $p_i \neq p_{i0}$ for some i

Here, p_{i0} denotes the value for the parameter p_i ($i = 1, 2, \ldots, k$) specified in H_0.

☐ EXAMPLE

In the bearing failures example, the test alternatives are:

H_0: $p_1 = p_{10} = 0.21$ $p_3 = p_{30} = 0.30$
  $p_2 = p_{20} = 0.23$ $p_4 = p_{40} = 0.26$

H_1: $p_i \neq p_{i0}$ for some i

Note that H_1 states in formal terms that H_0 does not hold. Alternative H_1 does not require that all $p_i \neq p_{i0}$. For instance, alternative H_1 includes the situation where $p_1 = 0.21$ and $p_2 = 0.23$ but $p_3 = 0.40$ and $p_4 = 0.16$, even though p_1 and p_2 agree with the specified values in H_0. ☐

Test Procedure. We employ a chi-square test for the multinomial alternatives (17.3). The test uses the same test statistic as the goodness-of-fit test and is exactly of the same form. Thus, we shall use here test statistic (16.10) and decision rule (16.12).

The correspondence to the goodness-of-fit test is as follows:

1. The sample observations are classified into k classes, giving the sample frequencies f_i.
2. Alternative H_0 provides the expected frequencies F_i for the k classes when H_0 holds.
3. The test then compares the sample frequencies f_i with the expected frequencies F_i under H_0, utilizing test statistic X^2 and the χ^2 distribution.

(17.4)
The chi-square test for the multinomial alternatives (17.3) is based on the test statistic:

$$X^2 = \sum_{i=1}^{k} \frac{(f_i - F_i)^2}{F_i}$$

When H_0 holds and the sample size is reasonably large so that working rule (16.11) applies:

$$X^2 \simeq \chi^2(k - 1)$$

The appropriate decision rule to control the α risk is as follows:

If $X^2 \leq \chi^2(1 - \alpha; k - 1)$, conclude H_0.
If $X^2 > \chi^2(1 - \alpha; k - 1)$, conclude H_1.

Note that no p_i parameters need to be estimated in this test because their values are specified in H_0. Thus, $m = 0$ here and the degrees of freedom for the χ^2 distribution in (17.4) are $k - 1$.

EXAMPLE ☐

We continue with the bearing failures example.

Step 1. The alternatives for the test were given earlier.

Step 2. The α risk is to be controlled at 0.05.

Step 3. The degrees of freedom for the χ^2 distribution are, from (17.4), $k - 1 = 4 - 1 = 3$. Thus, we require $\chi^2(0.95; 3) = 7.81$. Hence, the decision rule is as follows:

If $X^2 \leq 7.81$, conclude H_0.
If $X^2 > 7.81$, conclude H_1.

Step 4. Table 17.2 contains the basic calculations for the chi-square test statistic. Column 1 repeats the sample frequencies from Table 17.1, and column 2 shows the specified probabilities p_{i0} under H_0. The expected frequencies in column 3 are obtained as usual by multiplying the probabilities p_{i0} by n; that is, $F_i = np_{i0}$. For example, $F_1 = np_{10} = 1044(0.2100) = 219.24$. Note that all of the F_i are large enough to satisfy the working rule for the χ^2 approximation in (16.11).

The test statistic (17.4) is shown in column 4 to be $X^2 = 9.121$. Since $X^2 = 9.121 > 7.81$, we conclude H_1—that bearing failures do not occur during the year in direct proportion to the railroad's traffic. The P-value for the test is $P[\chi^2(3) > 9.121] = 0.0277$.

The finding from this test is not in contradiction with our earlier conclusion that the probability of a bearing failure occurring in the third quarter is the same as the proportion of traffic carried in that quarter. Conclusion H_1 here and our earlier test conclusion together imply that the probabilities of bearing failures occurring differ from the proportions of traffic carried in some (but not all) of the quarters. In fact, the engineer noted the large relative squared residuals for the first and fourth quarters in Table 17.2. This finding, that the main discrepancies occur in the first and fourth quarters, proved to be a useful clue in the search for factors other than the traffic level that affect the occurrence of bearing failures, such as ambient temperature and company maintenance policy. ☐

	(1)	(2)	(3)	(4)
		Under H_0		Relative
	Observed		Expected	Squared
	Frequency		Frequency	Residual
Class	f_i	Probability	F_i	$(f_i - F_i)^2/F_i$
1	249	0.2100	219.24	4.040
2	256	0.2300	240.12	1.050
3	297	0.3000	313.20	0.838
4	242	0.2600	271.44	3.193
Total	1044	1.0000	1044.00	$X^2 = 9.121$

TABLE 17.2
Calculations for chi-square test concerning all parameters p_i—Bearing failures example

> **Comment**
>
> When there are only $k = 2$ classes in the multinomial population, the chi-square test for alternatives (17.3) is equivalent to the two-sided test for a population proportion in (13.11).

17.4 BIVARIATE MULTINOMIAL POPULATIONS

Often, when considering multinomial populations, we are interested in two or more characteristics of the population elements and wish to study the relationships between the variables. In this section, we focus on the analysis of bivariate multinomial populations.

> **(17.5)**
> A multinomial population classified by two variables is called a **bivariate multinomial population**.

We encountered bivariate multinomial populations in Chapter 4. Table 4.1b provides an example. We now consider several other examples.

☐ EXAMPLE

Persons in the labor force are to be classified according to age and employment status, with the following bivariate classification system:

Employment Status	Age Class			
	25 and Under	26–45	46–65	Over 65
Unemployed				
Employed part-time				
Employed full-time				

This bivariate classification system will facilitate a study of the relationship between the two variables "employment status" and "age." ☐

Instrument Failures ☐ **EXAMPLE**

Failures of an electronic instrument are to be classified by type of failure (T_1, T_2) and location of the failure in the instrument (L_1, L_2, L_3), as follows:

Type of Failure	Location of Failure		
	L_1	L_2	L_3
T_1			
T_2			

This bivariate classification system will facilitate a study of the relationship between the two variables "type of failure" and "location of failure." The nature of the relationship is important for maintenance troubleshooting and for instrument design. ☐

A bivariate multinomial population can be viewed as an ordinary multinomial population, with every cell in the bivariate classification constituting a class. For instance, the bivariate classification in Example 1 can be viewed as involving $k = 3(4) = 12$ classes, and that in Example 2 as involving $k = 2(3) = 6$ classes. The bivariate arrangement, however, facilitates the analysis of the relationship between the two variables.

Parameters of Interest

The probability parameters of a bivariate multinomial population are displayed in a convenient form in Table 17.3a. The table shows a bivariate classification system, with the row variable having r classes and the column variable having c classes; thus, in all, the multinomial population has $k = rc$ classes or cells. The symbol p_{ij} denotes the joint probability that a population element selected at random falls in the ith class of the row variable and the jth class of the column variable. The symbols $p_{i.}$ and $p_{.j}$ denote the marginal probabilities for the ith row and jth column, respectively. An example of the display in Table 17.3a with probability values is found in Table 4.1b.

Sampling a Bivariate Multinomial Population

When a random sample of size n is selected from a bivariate multinomial population, the sample frequencies will be denoted by f_{ij}, as shown in Table 17.3b. In the table, f_{ij} denotes the frequency of sample observations in the ith class of the row variable and the jth class of the column variable. The symbols $f_{i.}$ and $f_{.j}$ denote the totals of the sample frequencies for the ith row and jth column, respectively. An example of the display in Table 17.3b with sample frequencies is found in Table 17.4, which we will discuss shortly.

TABLE 17.3
Bivariate multinomial probability parameters and sample frequencies

(a) Bivariate Multinomial Probability Parameters

Class for Row Variable	Class for Column Variable				Total
	$j = 1$	$j = 2$	$\cdots$	$j = c$	
$i = 1$	p_{11}	p_{12}	$\cdots$	p_{1c}	$p_{1.}$
$i = 2$	p_{21}	p_{22}	$\cdots$	p_{2c}	$p_{2.}$
.	.	.		.	.
.	.	.		.	.
.	.	.		.	.
$i = r$	p_{r1}	p_{r2}	$\cdots$	p_{rc}	$p_{r.}$
Total	$p_{.1}$	$p_{.2}$	$\cdots$	$p_{.c}$	1.0

(b) Bivariate Multinomial Sample Frequencies

Class for Row Variable	Class for Column Variable				Total
	$j = 1$	$j = 2$	$\cdots$	$j = c$	
$i = 1$	f_{11}	f_{12}	$\cdots$	f_{1c}	$f_{1.}$
$i = 2$	f_{21}	f_{22}	$\cdots$	f_{2c}	$f_{2.}$
.	.	.		.	.
.	.	.		.	.
.	.	.		.	.
$i = r$	f_{r1}	f_{r2}	$\cdots$	f_{rc}	$f_{r.}$
Total	$f_{.1}$	$f_{.2}$	$\cdots$	$f_{.c}$	n

TABLE 17.4
Electronic instrument failures classified by type of failure and location of failure—Instrument failures example

Type of Failure	Location of Failure			Total
	L_1	L_2	L_3	
T_1	50	16	31	97
T_2	61	26	16	103
Total	111	42	47	200

Test for Independence (Contingency Table Test)

When we analyze a bivariate multinomial population, the first question of interest usually is whether the two variables are independent. If they are, we know from Chapter 4 that there is no relationship between them. If they are dependent and, hence, a relationship does exist between the two variables, the next step in the analysis is to study the nature of the relationship. We begin with the first step of the analysis, testing whether or not the two variables are independent. We shall employ a chi-square test for this purpose.

The chi-square test for independence has a structure with which we are now familiar: We compare the sample frequencies f_{ij} with the expected frequencies F_{ij} under H_0 (i.e., under the assumption that the variables are independent), and we calculate the test statistic X^2. If X^2 is small, the data support the conclusion that the variables are independent; if X^2 is large, the data support the conclusion that the variables are dependent.

A bivariate sample frequency distribution is often called a *contingency table* and the test procedure we are about to describe is therefore called a *contingency table test*.

Assumptions

1. A simple random sample of size n has been selected from (1) an infinite bivariate multinomial population or (2) from a finite bivariate multinomial population with the sampling fraction small.
2. The sample size n is reasonably large.

☐ **EXAMPLE**

For the instrument failures example, Table 17.4 displays historical data on $n = 200$ failures. An analyst examined the stability of the bivariate pattern over time and concluded that it is reasonable to treat the 200 observations as a random sample from an infinite bivariate multinomial population of failures for the electronic instrument under study. The analyst now desires to test whether or not type of failure and location of failure are independent variables.   ☐

Alternatives.   For two variables to be independent, we know from (4.23) that each joint probability in the bivariate distribution must equal the product of the corresponding row and column marginal probabilities; that is, in the notation of Table 17.3a:

$$p_{ij} = p_{i.}\, p_{.j} \qquad \text{for all } (i, j)$$

Thus, the alternatives for the test of independence are as follows.

(17.6) $H_0: p_{ij} = p_{i.} p_{.j}$ for all (i, j)
           $H_1: p_{ij} \neq p_{i.} p_{.j}$ for some (i, j)

Here H_0 represents independence and H_1 dependence. Note that dependence exists even if some (but not all) $p_{ij} = p_{i.} p_{.j}$.

Computing Expected Frequencies. Under H_0 of (17.6), $p_{ij} = p_{i.} p_{.j}$ for each cell of the bivariate multinomial population. Hence, an estimate of p_{ij} under the assumption of independence can be obtained by first estimating the marginal probabilities $p_{i.}$ and $p_{.j}$ and then forming their product. Estimates of $p_{i.}$ and $p_{.j}$, which we denote by $\bar{p}_{i.}$ and $\bar{p}_{.j}$, may be computed from the marginal sample frequencies as follows.

(17.7) $\bar{p}_{i.} = \dfrac{f_{i.}}{n}$ $\bar{p}_{.j} = \dfrac{f_{.j}}{n}$

Point estimates of the joint probabilities if H_0 holds, denoted by $\bar{p}_{ij}$, are then obtained as follows.

(17.8) $\bar{p}_{ij} = \bar{p}_{i.} \bar{p}_{.j}$

EXAMPLE ☐

For the instrument failures example, we use the sample frequencies in Table 17.4 and we find, for example:

$$\bar{p}_{1.} = \frac{97}{200} = 0.4850 \qquad \bar{p}_{.1} = \frac{111}{200} = 0.5550$$

Hence, a point estimate of the joint probability p_{11} under the assumption of independence is:

$$\bar{p}_{11} = \bar{p}_{1.} \bar{p}_{.1} = 0.4850(0.5550) = 0.2692$$

As another example, we have:

$$\bar{p}_{23} = \bar{p}_{2.} \bar{p}_{.3} = \left(\frac{103}{200}\right)\left(\frac{47}{200}\right) = 0.1210$$

         ☐

Finally, to obtain the expected frequencies F_{ij} when H_0 holds, we multiply each estimated joint probability by the sample size n.

(17.9) $F_{ij} = n\bar{p}_{ij} = n\bar{p}_{i.} \bar{p}_{.j}$

The following equivalent formula for calculating F_{ij} is obtained by substituting (17.7) into (17.9).

(17.9a) $F_{ij} = \dfrac{f_{i.}f_{.j}}{n}$

☐ EXAMPLE

For the instrument failures example, we find the expected frequency when H_0 holds for failure type T_1 in location L_1 as follows using (17.9):

$$F_{11} = n\bar{p}_{11} = 200(0.2692) = 53.84$$

or, equivalently, by using (17.9a):

$$F_{11} = \dfrac{f_{1.}f_{.1}}{n} = \dfrac{97(111)}{200} = 53.84$$

This and the other expected frequencies F_{ij} are shown in the MINITAB output in Figure 17.1. The first entry in each cell is the sample frequency f_{ij} and the second entry is the expected frequency F_{ij} under H_0. The value of F_{11} in the output (53.83) differs slightly from the value calculated here (53.84) because of rounding.   ☐

Test Procedure.   The X^2 test statistic is again utilized. It is obtained as always by summing the relative squared residuals over all classes, but a double summation is now required since each joint class (i, j) is identified in terms of the classes of the row and column variables.

(17.10) $X^2 = \displaystyle\sum_{i=1}^{r}\sum_{j=1}^{c} \dfrac{(f_{ij} - F_{ij})^2}{F_{ij}}$

FIGURE 17.1
MINITAB output for chi-square test of independence—Instrument failures example

```
Expected counts are printed below observed counts

ROWS: TYPE          COLUMNS: LOCATION
                   1         2         3      Total
       1          50        16        31         97
                53.83     20.37     22.80

       2          61        26        16        103
                57.17     21.63     24.20

    Total        111        42        47        200

ChiSq =    0.273 +   0.937 +   2.953 +
           0.257 +   0.883 +   2.781 = 8.086
    df = 2
```

(For a review of double-summation notation, refer to Appendix A, Section A.1.)

To control the α risk, we need to know the sampling distribution of X^2 when H_0 holds.

(17.11)

When the variables of a bivariate multinomial population are independent and the sample size is reasonably large according to working rule (16.11):

$$X^2 \simeq \chi^2[(r - 1)(c - 1)]$$

where X^2 is given by (17.10).

The appropriate decision rule for the alternatives in (17.6) to control the α risk is as follows:

If $X^2 \leq \chi^2[1 - \alpha; (r - 1)(c - 1)]$, conclude H_0.
If $X^2 > \chi^2[1 - \alpha; (r - 1)(c - 1)]$, conclude H_1.

EXAMPLE ☐

In the instrument failures example, the analyst wishes to test whether type of failure and location of failure are independent.

Step 1. The alternatives are given in (17.6).

Step 2. The α risk is to be controlled at 0.05.

Step 3. We see from Figure 17.1 that $r = 2$ and $c = 3$. Hence the degrees of freedom for the χ^2 distribution when H_0 holds are:

$$(r - 1)(c - 1) = (2 - 1)(3 - 1) = 2$$

Thus, we require $\chi^2(0.95; 2) = 5.99$. The appropriate decision rule is therefore as follows:

If $X^2 \leq 5.99$, conclude H_0.
If $X^2 > 5.99$, conclude H_1.

Step 4. We note first that all of the expected frequencies F_{ij} in Figure 17.1 are large enough to satisfy the working rule for the χ^2 approximation in (16.11). We then compute test statistic (17.10) from the data in Figure 17.1 as follows:

$$X^2 = \frac{(50 - 53.83)^2}{53.83} + \frac{(16 - 20.37)^2}{20.37} + \cdots + \frac{(16 - 24.20)^2}{24.20}$$

$$= 0.273 + 0.937 + \cdots + 2.781$$

$$= 8.086$$

Because $X^2 = 8.086 > 5.99$, we conclude H_1—that type of failure and location of failure are dependent. The P-value of the test is $P[\chi^2(2) > 8.086] = 0.0175$. The small P-value indicates that the sample data are not consistent with the hypothesis of independence of the two variables. The next step in the investigation therefore will be to study the nature of the relationship between type of failure and location of failure.

Note in Figure 17.1 that the MINITAB output shows the relative squared residuals for all of the cells and the value of the test statistic, $X^2 = 8.086$.   □

Comment

The degrees of freedom for the χ^2 distribution in (17.11) are obtained in the following way. There are $k = rc$ cells in the bivariate multinomial distribution. The r marginal row probabilities $p_{i.}$ and the c marginal column probabilities $p_{.j}$ must be estimated from the sample data. Since both the row and column marginal probabilities must add to 1, only $r - 1$ of the row marginal probabilities and $c - 1$ of the column marginal probabilities are independently estimated from the sample data. Thus, the total number of estimated parameters is $m = (r - 1) + (c - 1)$. The degrees of freedom for the test therefore are as follows.

(17.12) $k - m - 1 = rc - [(r - 1) + (c - 1)] - 1$
$$= (r - 1)(c - 1)$$

Examination of Nature of Dependence

When the test for independence leads to the conclusion of dependence, a simple way to examine the nature of the dependence relation is to consider the conditional probability distributions, as discussed in Chapter 4. Of course, since only sample data are available, we must work with the estimated conditional probabilities.

□ **EXAMPLE**

In the instrument failures example, the analyst was particularly interested in the effect of location of failure on type of failure. Table 17.5 shows for each of the three locations the estimated conditional probability distribution of type of failure. The estimated probabilities are computed from the sample frequencies in Table 17.4. For location L_1, for example, the estimated conditional probabilities are $50/111 = 0.45$ for failure type T_1 and $61/111 = 0.55$ for failure type T_2.

Note how the conditional probability distribution for location L_3 differs from the distributions for locations L_1 and L_2. Failure T_1 is the more likely type of failure in location L_3, while failure T_2 is the more likely type of failure in the other two locations. Design engineers can use such insights to identify potential causes of failure and to recommend design modifications aimed at improved instrument reliability.   □

TABLE 17.5
Estimated conditional probability distributions of type of instrument failure for each failure location—Instrument failures example

Type of Failure	Conditional upon Location of Failure		
	L_1	L_2	L_3
T_1	0.45	0.38	0.66
T_2	0.55	0.62	0.34
Total	1.00	1.00	1.00

Frequently, we wish to compare several multinomial populations to see if they are identical.

EXAMPLE ☐

An educator conducted a study to compare three sections of the same course, each taught by a different instructional method, to see whether the three methods produce the same letter-grade distributions. ☐

EXAMPLE ☐ Flavor Preference

A market researcher conducted an experiment to measure the proportions of persons preferring each of four different syrup flavors to see whether the preference distributions are the same for men and women. ☐

Parameters of Interest

Table 17.6 illustrates the parameters of interest. There are c multinomial populations (for example, 2 genders), each with the same r classes (for example, 4 syrup flavors). The probability that an element selected at random from population j falls into class i is denoted by p_{ij}. For each population j, the probabilities p_{ij} necessarily sum to 1.

Test of Homogeneity

A test for the identity of several multinomial populations is often called a *test of homogeneity*.

Assumptions

1. Independent random samples of sizes $n_1, n_2, \ldots, n_c$ are selected (1) from c infinite multinomial populations or (2) from c finite multinomial populations with the sampling fractions small.
2. Each of the c sample sizes $n_1, n_2, \ldots, n_c$ is reasonably large.

TABLE 17.6
Comparison of c multinomial populations, each having the same r classes

| Class | Multinomial Population | | | |
	$j = 1$	$j = 2$	$\cdots$	$j = c$
$i = 1$	p_{11}	p_{12}	$\cdots$	p_{1c}
$i = 2$	p_{21}	p_{22}	$\cdots$	p_{2c}
.	.	.		.
.	.	.		.
.	.	.		.
$i = r$	p_{r1}	p_{r2}	$\cdots$	p_{rc}
Total	1.0	1.0	$\cdots$	1.0

We denote the total number of cases in the study by n_T.

(17.13) $\quad n_T = \sum_{j=1}^{c} n_j$

Alternatives. When all of the multinomial populations are identical, the probabilities p_{ij} in any row of Table 17.6 will be the same; that is, $p_{11} = p_{12} = \cdots = p_{1c}$; also, $p_{21} = p_{22} = \cdots = p_{2c}$; and so on. Thus, a test for the identity of the c populations involves the following alternatives.

(17.14) $\quad H_0$: $p_{i1} = p_{i2} = \cdots = p_{ic}$ $\qquad$ for all i
$\qquad\qquad H_1$: Not all equalities in H_0 hold

Test Procedure. Again we shall employ a chi-square test. As before, the test statistic X^2 compares the sample frequencies with the expected frequencies under H_0. The test statistic X^2 is the same as that in (17.10) for testing independence in a bivariate multinomial population, and the procedure parallels that for earlier chi-square tests.

(17.15)
When all multinomial populations are identical and the sample sizes n_j are reasonably large according to working rule (16.11):

$$X^2 \simeq \chi^2[(r-1)(c-1)]$$

where X^2 is given by (17.10).

The appropriate decision rule for the alternatives in (17.14) to control the α risk is as follows:

$\qquad$ If $X^2 \leq \chi^2[1-\alpha;(r-1)(c-1)]$, conclude H_0.
$\qquad$ If $X^2 > \chi^2[1-\alpha;(r-1)(c-1)]$, conclude H_1.

The rationale underlying the calculation of expected frequencies for testing the identity of multinomial populations differs from that for testing independence in a bivariate multinomial population, even though the algebraic results are identical. We shall explain the rationale by means of the flavor preference example.

☐ **EXAMPLE**

In the flavor preference example, the market researcher selected a random sample of $n_1 = 250$ men and independently selected a random sample of $n_2 = 400$ women. Each person was given four syrup flavors to taste and afterward was asked for the preferred flavor. The sample results are shown in the SAS output in Figure 17.2. The first entry in each cell is the sample frequency f_{ij}, where f_{ij} denotes the sample frequency in the ith class for the jth population. Here $i = 1, 2, 3, 4$ and $j = 1, 2$. Thus, we see from

```
FLAVOR      GENDER

FREQUENCY │
EXPECTED  │      1 │        2 │  TOTAL
──────────┼────────┼──────────┼──────
       1  │     23 │      174 │   197
          │   75.8 │    121.2 │
──────────┼────────┼──────────┼──────
       2  │    169 │       34 │   203
          │   78.1 │    124.9 │
──────────┼────────┼──────────┼──────
       3  │     48 │      150 │   198
          │   76.2 │    121.8 │
──────────┼────────┼──────────┼──────
       4  │     10 │       42 │    52
          │   20.0 │     32.0 │
──────────┼────────┼──────────┼──────
   TOTAL       250        400     650


      STATISTICS  FOR  TABLE  OF  FLAVOR  BY  GENDER

STATISTIC                    DF      VALUE        PROB
─────────────────────────────────────────────────────
CHI-SQUARE                    3     256.819      0.000
```

FIGURE 17.2
SAS output for chi-square test of homogeneity—Flavor preference example

Figure 17.2 that $f_{11} = 23$, $f_{12} = 174$, and so on. We also note in Figure 17.2 that $n_T = n_1 + n_2 = 250 + 400 = 650$.

Step 1. The researcher would like to test whether the preference distributions for men and women are the same:

$$H_0: p_{i1} = p_{i2} \quad \text{for } i = 1, 2, 3, 4$$
$$H_1: \text{Not all equalities in } H_0 \text{ hold}$$

Step 2. The α risk is to be controlled at 0.01.

Step 3. There are $c = 2$ multinomial populations to be compared, each having $r = 4$ preferred flavor classes. From (17.15), it follows that the required degrees of freedom are:

$$(r - 1)(c - 1) = (4 - 1)(2 - 1) = 3$$

Hence, we need $\chi^2(0.99; 3) = 11.34$. The appropriate decision rule is therefore as follows:

If $X^2 \leq 11.34$, conclude H_0.
If $X^2 > 11.34$, conclude H_1.

Step 4. If H_0 holds, all multinomial populations are the same. Let $p_1, p_2, \ldots, p_r$ denote the parameters for the common multinomial distribution if H_0 holds. To obtain the expected frequencies when H_0 holds, we must first estimate these common parameters. An estimator for p_i is obtained by combining all samples and determining the proportion of cases falling in the ith class in the combined sample. We denote this estimator by $\bar{p}_i$ and calculate it as follows.

$$(17.16) \quad \bar{p}_i = \frac{\sum\limits_{j=1}^{c} f_{ij}}{n_T} = \frac{f_{i.}}{n_T}$$

From Figure 17.2, we see that $f_{1.} = 197$ persons of the total of $n_T = 650$ in the study preferred syrup flavor 1 (that is, are in class $i = 1$). Thus, $\bar{p}_1 = 197/650 = 0.3031$. The parameter estimates for the other classes are obtained similarly.

We shall let F_{ij} denote the frequency expected in class i for the sample from the jth population when the multinomial distributions are, in fact, identical—that is, when H_0 holds. It follows from the preceding argument that the expected frequency has the following form.

$$(17.17) \quad F_{ij} = n_j \bar{p}_i$$

For instance, in the sample of $n_1 = 250$ men, we would expect $F_{11} = n_1 \bar{p}_1 = 250(0.3031) = 75.8$ men to fall in class $i = 1$ if H_0 is true. Similarly, $F_{21} = n_1 \bar{p}_2 = 250(203/650) = 250(0.3123) = 78.1$ men would be expected to fall in class $i = 2$. The SAS output in Figure 17.2 shows the values of F_{ij} as the second entry in each cell. Note from Figure 17.2 that all F_{ij} are large enough to satisfy the working rule for the χ^2 approximation in (16.11).

Test statistic (17.10) is calculated from the data in Figure 17.2 as follows:

$$X^2 = \frac{(23 - 75.8)^2}{75.8} + \frac{(169 - 78.1)^2}{78.1} + \cdots + \frac{(42 - 32.0)^2}{32.0}$$

$$= 36.78 + 105.80 + \cdots + 3.13 = 256.8$$

Because $X^2 = 256.8 > 11.34$, we conclude H_1—that the preference distributions for men and women are different. The P-value of the test is $P[\chi^2(3) > 256.8] = 0+$. This P-value (labeled PROB), together with the relevant degrees of freedom (DF) and test statistic (VALUE), is shown at the bottom of the SAS output in Figure 17.2. The small P-value indicates that the evidence is very strong that the preference distributions for men and women are not the same.

The market researcher investigated the nature of the differences in the preference distributions for men and women by estimating the probabilities of preference for each gender from the data in Figure 17.2:

Preferred Flavor	Men	Women
1	0.092	0.435
2	0.676	0.085
3	0.192	0.375
4	0.040	0.105
Total	1.000	1.000

Clearly, men show a strong preference for flavor 2, while women have strong preferences divided between flavors 1 and 3. We could use the procedures for comparing

two population proportions when the samples are independent (Section 14.4) for analyzing further the differences in the probabilities between men and women for any of
the flavors.   □

As we noted earlier, the test statistic X^2 and the decision rule for testing the identity of
several multinomial populations are the same as those for testing independence in a bivariate multinomial population. Moreover, formulas (17.17) and (17.9a) for the expected values F_{ij} under H_0 are identical, with $n_j = f_{.j}$ and $n_T = n$. Thus, even though the theoretical
basis is not the same for the two tests, they are computationally identical. For this reason,
statistical packages usually provide only a single computational routine to do both of these
tests.

Comments

1. The degrees of freedom for the χ^2 distribution in (17.15) are obtained in the following way.
   For any one of the multinomial populations, there would be $r - 1$ degrees of freedom if H_0
   specified the common probabilities $p_1, p_2, \ldots, p_r$ completely. It can be shown that independent χ^2 random variables are additive, as are their degrees of freedom. Hence, there would be
   $c(r - 1)$ degrees of freedom for all c populations if the common probabilities were specified
   by H_0. But the r common probabilities are not specified by H_0 and have to be estimated from
   the sample, resulting in a reduction of $r - 1$ degrees of freedom. Therefore, the degrees of
   freedom for the test are:

$$c(r - 1) - (r - 1) = (r - 1)(c - 1)$$

2. When there are $c = 2$ populations and each has $r = 2$ classes, the test in (17.15) is equivalent
   to the two-sided test (14.31) for two population proportions.

PROBLEMS

* **17.1 Machine Component.** Under preventive maintenance, a machine component is replaced rou-
tinely after 2500 operating hours. At the time of preventive replacement, the component will be in
one of the following conditions: (1) excellent, (2) good, (3) poor. The respective probabilities are:
0.1, 0.2, 0.7.

a. Identify the parameters of the multinomial distribution here. Must these parameters necessarily
   sum to 1? Comment.
b. For a random sample of $n = 5$ replaced components, what is represented by $f_1 = 0, f_2 = 2, f_3 =$
   3? Obtain $P(0, 2, 3)$.
c. For a random sample of $n = 5$ replaced components, obtain the probability that three are in poor
   condition and the other two are not in poor condition. [*Hint:* Redefine the classes as poor and not
   poor, and obtain the appropriate probabilities.]

17.2 Fragile Instrument. A fragile instrument, when shipped by air freight, will require upon arrival one of the following: (1) no realignment, (2) minor realignment, (3) major realignment,
(4) replacement of parts. The respective probabilities are: 0.6, 0.2, 0.1, 0.1.

a. For $n = 4$ instruments shipped independently, obtain the probability that two will require no
   realignment, one will require minor realignment, and one will require replacement of parts.
b. Obtain $P(1, 2, 0, 1)$. What does this probability represent?
c. Construct the probability distribution of all possible outcomes for $n = 2$ instruments shipped
   independently. Which outcome is most probable?

17.3 The probability that an account will be paid within 30 days is 0.7, between 31 and 60 days is 0.2, and after 60 days is 0.1. The timings of payments for different accounts are independent.

a. For a random sample of $n = 3$ accounts, obtain the probability that (1) two will be paid within 30 days and one will be paid after 60 days, (2) all three will be paid after 60 days.

b. Construct the probability distribution of all possible payment timing outcomes in a random sample of $n = 3$ accounts. Which outcome is most probable?

c. For a random sample of $n = 4$ accounts, obtain the probability that two will be paid within 30 days and two will not be paid within 30 days. [*Hint:* Redefine the classes as paid within 30 days and not paid within 30 days, and obtain the appropriate probabilities.]

* **17.4** **Cream Preference.** In a random sample of 180 women who use a certain type of skin cream regularly, each was asked to state which of the three brands on the market (1, 2, 3) she preferred. The preference responses were as follows:

Brand Most Preferred:	1	2	3
Number of Women:	75	54	51

a. Construct a 95 percent confidence interval for p_1, the probability that a regular skin cream user prefers brand 1. Interpret the confidence interval.

b. Before the study was made, a product manager hypothesized that the market share of brand 2 among regular users is 20 percent. Test this hypothesis, controlling the α risk at 0.01 when $p_2 = 0.20$. State the alternatives, the decision rule, the value of the test statistic, and the conclusion.

17.5 **Customer Complaints.** The manager of the customer services department of a large chain store tabulated the 250 most recent complaints as follows:

Nature of Complaint	Number of Complaints
1. Quality of merchandise	89
2. Price	28
3. Service	122
4. Other	11
Total	250

Assume that the 250 complaints constitute a random sample from a process.

a. Construct a 95 percent confidence interval for p_2, the probability that a complaint deals with price. Interpret the confidence interval.

b. In the entire chain, 50 percent of complaints involve service. Test whether the probability of a complaint dealing with service at the store currently is 0.50, controlling the α risk at 0.01 when $p_3 = 0.50$. State the alternatives, the decision rule, the value of the test statistic, and the conclusion.

17.6 Refer to **Fragile Instrument** Problem 17.2. The 200 most recent shipments of this instrument have utilized a newly developed type of packaging. The shipment outcomes with this packaging were as follows:

Outcome	Number of Shipments
1. No realignment	114
2. Minor realignment	53
3. Major realignment	24
4. Replacement of parts	9
Total	200

Assume that the 200 shipments constitute a random sample from a process.

a. Construct a 90 percent confidence interval for p_1, the probability that no realignment is required.

b. Convert the confidence interval in part a into a 90 percent confidence interval for $1 - p_1$. Interpret this confidence interval.

c. A test of the alternatives H_0: $p_4 \geq 0.10$ versus H_1: $p_4 < 0.10$ is desired. Obtain the P-value for the appropriate test statistic. Which conclusion should be drawn if the α risk is to be controlled at 0.05 when $p_4 = 0.10$? Explain.

17.7 Telephone Service. As part of a quality assurance audit of customer telephone service currently provided by a government tax department, 200 telephone calls were placed to the department at randomly selected times during its open hours. The results were (1) 132 calls were answered within 10 seconds, (2) 38 calls were answered within 11 to 60 seconds, and (3) 30 calls were answered after a delay of more than 60 seconds. Quality assurance requirements for the service specify that (1) 70 percent of calls be answered within 10 seconds, (2) 20 percent of calls be answered within 11 to 60 seconds, and (3) only 10 percent of calls be answered after a delay of more than 60 seconds.

a. Based on the audit results, obtain a 95 percent confidence interval for the probability that a call is answered within 10 seconds at the current service level. Interpret the confidence interval.

b. Test whether the probability of calls being answered within 10 seconds at the current service level is 0.70, controlling the α risk at 0.05 when $p_1 = 0.70$. State the alternatives, the decision rule, the value of the test statistic, and the conclusion.

* 17.8 Refer to **Cream Preference** Problem 17.4. Current market shares for the three brands of skin cream, reflecting purchases by all users (regular users as well as nonregular users), are as follows:

Brand:	1	2	3
Market Share (percent):	35	45	20

The study director wants to determine whether the probabilities of regular users preferring brands 1, 2, and 3 are the same as the respective current market shares based on purchases by all users.

a. State the alternatives for the test. Does alternative H_1 necessarily imply that none of the three brands have the same market share among regular users as among all users? Explain.

b. Perform the appropriate test, controlling the α risk at 0.01. State the decision rule, the value of the test statistic, and the conclusion.

c. Examine the relative squared residuals for the test statistic in part b and comment on any major differences between the preference distributions for regular users and all users.

17.9 Refer to **Customer Complaints** Problem 17.5. In the entire chain, the distribution of complaints is known from extensive experience to be as follows:

Nature of Complaint	Proportion of Complaints
1. Quality of merchandise	0.25
2. Price	0.15
3. Service	0.50
4. Other	0.10
Total	1.00

a. Test whether the store's complaint pattern is the same as that for the entire chain, controlling the α risk at 0.05. State the alternatives, the decision rule, the value of the test statistic, and the conclusion.

b. Examine the relative squared residuals for the test statistic in part a and comment on any major differences between the complaint patterns for the store and for the chain.

17.10 Refer to **Fragile Instrument** Problems 17.2 and 17.6. The probabilities cited in Problem 17.2 are based on extensive experience with the original type of packaging for the instrument.

a. Test whether the shipping outcome probabilities for the new type of packaging are the same as those for the original. Control the α risk at 0.10. State the alternatives, the decision rule, the value of the test statistic, and the conclusion.

b. Does the conclusion in part a imply that the probability of an instrument requiring no realignment is different for the two types of packaging? Explain.

c. Examine the relative squared residuals for the test statistic in part a and comment on any major differences in the outcome distributions for the two types of packaging.

17.11 Refer to **Telephone Service** Problem 17.7.

a. Test whether the current service level is consistent with the three quality assurance requirements. Control the α risk at 0.10. State the alternatives, the decision rule, the value of the test statistic, and the conclusion.

b. Computer output shows that the P-value for the test in part a is 0.062. Is this value consistent with the conclusion drawn in part a? Explain.

c. Examine the relative squared residuals for the test statistic in part a. Are there any instances where the quality assurance requirements do not appear to be currently met? Comment.

* 17.12 A pharmaceutical firm has surveyed a random sample of 120 persons suffering from Parkinson's disease. Among the facts obtained is the following bivariate frequency distribution of disease duration and degree of self-reliance:

Self-Reliance	Disease Duration (years)			
	Less than 5	5–9	10–14	15 or more
Considerable	36	25	19	10
Little	4	7	9	10

a. Test whether disease duration and degree of self-reliance are independent, controlling the α risk at 0.01. State the alternatives, the decision rule, the value of the test statistic, and the conclusion.

b. For each disease duration class, obtain the estimated conditional probability distribution of the degree of self-reliance. What do these distributions show about the nature of the relationship between disease duration and self-reliance?

c. Computer output gives a P-value of 0.006 for the contingency table test in part a. Is this value consistent with the test result in part a? Explain.

17.13 An investment analyst is studying the relation between price movements of stocks in two consecutive weeks in March. A random sample of 100 stocks was selected, and the price movements of each stock during the two weeks were cross-classified as follows:

Movement in First Week	Movement in Second Week		
	Increase	No Change	Decrease
Increase	25	6	1
No change	8	32	4
Decrease	3	6	15

a. Test whether price movements in week 1 and in week 2 are independent, controlling the α risk at 0.10. State the alternatives, the decision rule, the value of the test statistic, and the conclusion.

b. Examine the relative squared residuals for the test statistic in part a. What do they suggest about how the price movements depart from independence?

c. For each price movement class in the first week, obtain the estimated conditional probability distribution of price movements in the second week. Describe the nature of the relationship between the price movements in the two weeks.

17.14 Two hundred members of a test panel were shown a preview of a film of a new musical to be released shortly. Each panel member was asked to indicate whether the amount of dancing in the film is too much, about right, or too little. The results, cross-classified by opinion and gender of viewer, were as follows:

Opinion	Gender of Viewer	
	Male	Female
Too much	18	22
About right	63	47
Too little	29	21

Assume that the 200 panel members constitute a random sample of the target audience for the film.

a. Test whether opinion and gender of viewer are independent, controlling the α risk at 0.10. State the alternatives, the decision rule, the value of the test statistic, and the conclusion.

b. Given the conclusion in part a, should the next step be to examine the relative squared residuals for the test? Explain.

17.15 A savings bank surveyed a random sample of 400 heads of households to determine how many knew the current interest rate paid on six-month certificates of deposit. The following results were obtained when the survey data were cross-classified by knowledge and occupation of head of household:

	Knowledge	
Occupation	Did Know	Did Not Know
Wage earner, clerical worker	110	100
Manager, professional	60	20
Other	40	70

The bank management would like to estimate the probability that a household head knows the current interest rate. It also wishes to test whether knowledge of the interest rate is related to occupation and, if so, to ascertain the nature of the relationship.

a. Estimate by means of a 90 percent confidence interval the probability that a household head knows the current interest rate. Interpret the confidence interval.

b. Test whether occupation and knowledge of the current interest rate are independent, controlling the α risk at 0.01. State the alternatives, the decision rule, the value of the test statistic, and the conclusion. What are the implications of the conclusion for the bank?

c. For each occupation, obtain the estimated conditional probability distribution of knowledge of the current interest rate. What do these distributions show about the nature of the relationship between occupation and knowledge of the current interest rate?

* 17.16 Refer to **Cream Preference** Problem 17.4. The study described there was preceded by another study conducted two years earlier. This earlier study used an independent random sample of 150 regular users and yielded the following preference responses:

Brand Most Preferred:	1	2	3
Number of Women:	42	55	53

a. Test whether the preference patterns in the two periods are identical, controlling the α risk at 0.10. State the alternatives, the decision rule, the value of the test statistic, and the conclusion.

b. Use (14.28) to construct a 90 percent confidence interval for the change that has occurred in the probability that a regular user prefers brand 1.

c. Computer output gives a P-value of 0.035 for the test of homogeneity in part a. Is this value consistent with the test result in part a? Explain.

17.17 Random samples of machine parts selected independently from each of three large production runs revealed the following data on the number of parts in acceptable condition:

Quality of Part	Run 1	Run 2	Run 3
Acceptable	80	90	55
Unacceptable	10	30	5
Total	90	120	60

a. Test whether the probability of an acceptable part is the same for the three runs. Control the α risk at 0.05. State the alternatives, the decision rule, the value of the test statistic, and the conclusion.

b. Runs 1 and 3 were made in the Jameston plant; run 2 was made in the Midville plant. Using the chi-square procedure, test whether the probability of a part being in acceptable condition is the same for runs 1 and 3, using $\alpha = 0.05$. State the alternatives, the decision rule, the value of the test statistic, and the conclusion. What do the results of this test and the test in part a suggest about run 2? Discuss.

17.18 A market research organization conducted three surveys at about the same time, each based on a simple random sample of families in the United States. The three samples were selected independently. The first survey—the income survey—sought information on the relation between clothing expenditures and family income. The second survey—the brand-preference survey—sought information on brand preferences among various packaged foods. The third survey—the family-planning survey—sought information on families' procreative intentions. The respective sizes of the three surveys were 1000, 1500, and 500 families. The numbers of families refusing to participate in the surveys were 140, 150, and 130, respectively. All surveys were carried out in the same manner and with the same care and supervision.

a. Test whether the subject matter of the survey affects the refusal rate, controlling the α risk at 0.01. State the alternatives, the decision rule, the value of the test statistic, and the conclusion.
b. Using the chi-square procedure, test whether the refusal rates for the income and brand-preference surveys differ, controlling the α risk at 0.01. State the alternatives, the decision rule, the value of the test statistic, and the conclusion.
c. Given the conclusion in part a, was the test in part b superfluous in the sense that the conclusion in part a logically implied the conclusion in part b? Comment.

17.19 The final grade distributions of students enrolled in the evening and day sections of Statistics 101 during the previous university term were:

Grade:	A	B	C	D	F	Total
Day Section:	50	68	55	19	9	201
Evening Section:	29	57	40	22	6	154

Assume that the students in the two sections constitute independent random samples from the respective populations of day and evening students who take this course.

a. Test whether the probability distributions of grades are the same for day and evening students. Control the α risk at 0.05. State the alternatives, the decision rule, the value of the test statistic, and the conclusion.
b. If the two sections of students took a common final examination, would this fact invalidate the test conducted in part a? Comment.

EXERCISES

17.20 Refer to **Machine Component** Problem 17.1. In a random sample of five components about to be replaced, two are in excellent condition and three are not in excellent condition. What is the conditional probability that the latter three components consist of one component in good condition and two components in poor condition?

17.21 Derive the multinomial probability function (17.2), using (A.21) in Appendix A.

17.22 Refer to multinomial probability function (17.2). Explain why $E\{f_i\} = np_i$ and $\sigma^2\{f_i\} = np_i(1 - p_i)$ for any sample frequency $f_i(i = 1, 2, \ldots, k)$.

17.23 Refer to multinomial probability function (17.2). Given that $\sigma\{f_i, f_j\} = -np_ip_j$ for $i \neq j$, derive $E\{f_if_j\}$. [*Hint:* Use (6.13a).]

17.24 Refer to **Divorces and Annulments** Problem 13.29.

a. Conduct the appropriate test by using a chi-square test procedure. State the alternatives, the decision rule, the value of the test statistic, and the conclusion.
b. Confirm that the value of the test statistic in part a is equal to the square of the value of the test statistic in Problem 13.29a.

17.25 Refer to **Environmental Amendment** Problems 14.32 and 14.36.

a. Conduct a chi-square test to decide whether the distribution of opinion (in favor, not in favor) is the same for males and females. Control the α risk at 0.10. State the alternatives, the decision rule, the value of the test statistic, and the conclusion.

b. Explain why the test alternatives here are the same as those in Problem 14.36a.

c. Confirm that the value of the test statistic in part a is equal to the square of the value of the test statistic in Problem 14.36a.

17.26 Refer to **Flight Simulation** Problem 2.42. Before collecting the data, an analyst conjectured that phase of flight and cause of error would have the following bivariate multinomial probability distribution:

| | Cause of Error | | |
Phase of Flight	M	O	Total
T	0.250	0.125	0.375
C	0.125	0.125	0.250
L	0.125	0.250	0.375
Total	0.500	0.500	1.000

Assume that the 45 observations in Problem 2.42 constitute a random sample from a process. Test the conjecture, controlling the α risk at 0.05. State the alternatives, the decision rule, the value of the test statistic, and the conclusion.

STUDIES

17.27 In a study of 1000 randomly selected automobile accidents, each accident was classified according to whether it occurred at an intersection (A_1) or elsewhere (A_2), whether it occurred during the day (B_1) or at night (B_2), and whether it involved personal injury (C_1) or no personal injury (C_2). The resulting cross-classification follows:

| | B_1 | | B_2 | |
	C_1	C_2	C_1	C_2
A_1	52	203	96	34
A_2	47	389	105	74

a. Test whether the three variables are independent, controlling the α risk at 0.01. What is your conclusion? [*Hint:* Theorem (4.23) extends directly to three variables.]

b. To analyze the nature of the relationship, obtain the marginal bivariate distributions for variables A and B, A and C, and B and C. For each bivariate distribution, test whether the two variables are independent, controlling the α risk at 0.01 in each case. Are any insights into the nature of the relationships between the three variables provided by these tests? Discuss.

c. Consider the conditional bivariate distribution of variables A and B, given C_1. Test whether variables A and B are conditionally independent here, controlling the α risk at 0.01. Repeat this test for the conditional bivariate distribution of variables A and B, given C_2. Discuss the findings.

17.28 Refer to the **Power Cells** data set (Appendix D.2). Assume that the observations for each of the three ambient temperatures constitute independent random samples.

a. Test whether the mode-of-failure distributions are the same at the three ambient temperatures (10°C, 20°C, 30°C). Control the α risk at 0.05. What is your conclusion?

b. Determine whether the test conclusion would be affected if the temperature categories for 10° and 20°C were combined (that is, pooled). Continue to control the α risk at 0.05. Explain why the test statistic here happens to have the same value as the one for the test in part a.

c. Obtain an exact 90 percent confidence interval for the probability that a cell will fail by shorting (mode 0) when operating at an ambient temperature of 30°C, using (13.13). Do the test results in parts a and b assure us that this probability differs from the corresponding probabilities for cells operating at 10° and 20°C? Explain.

17.29 Refer to the **Investment Funds** data set (Appendix D.3).

a. For each of the two funds, obtain the 44 quarter-to-quarter changes in the unit values. (Note that the unit values for January 1 of year 1 are given in the description of the data set.) Cross-classify the quarter-to-quarter changes according to whether a decrease or no change or an increase occurred, as follows:

	Change in Unit Value for Equity Fund	
Change in Unit Value for Fixed-Income Fund	Decrease	No Change or Increase
Decrease		
No change or increase		

b. Assume that the 44 pairs of quarter-to-quarter changes in part a constitute a random sample from a bivariate multinomial population. Test whether the change status of the unit value for the equity fund is independent of that for the fixed-income fund. Control the α risk at 0.01. What is your conclusion?

c. Obtain the P-value of the test in part b. Is this value consistent with the test result in part b? Explain.

Linear Statistical Models

Simple Linear Regression—I

18

In Chapter 2 we discussed the use of bivariate and multivariate frequency distributions and scatter plots for studying the relationships between two or more variables. We now consider a powerful and comprehensive methodology for analyzing relationships between quantitative variables called *regression analysis*. Regression analysis is concerned with the relation between a quantitative variable of interest, called the *dependent variable* or *response variable*, and one or more other variables called *independent, explanatory,* or *predictor variables.*

EXAMPLES ☐

1. In a recent regression study of 20 industrial customers of an oil company, the response variable was the amount of lubricating oil (in metric tons) sold to a customer during a given period, and the independent variables were the size of the customer's work force and sales by the customer during the preceding period.

2. In a regression study of 500 families, the response variable was family expenditures for food last year, and the independent variables were family size and family income last year. ☐

Regression analysis is often used to predict the response variable from knowledge of the independent variables. In the lubricating oil study, for instance, the oil company wished to predict the amount of sales of lubricating oil to each customer in the next period from knowledge of the independent variables for the current period.

At other times, regression analysis is utilized primarily for examining the nature of the relationship between the response variable and the independent variables. For example, the food expenditures study was designed to determine whether food expenditures constitute a declining proportion of family income as family income rises.

In this and the next chapter, we take up basic concepts and inference procedures for simple regression analysis in which only one predictor variable is employed. In Chapter 20, we take up multiple regression analysis in which several predictor variables are utilized.

18.1 RELATION BETWEEN DEPENDENT AND INDEPENDENT VARIABLES

We begin with the concept of a relation between a dependent variable, to be denoted by Y, and one independent variable, to be denoted by X. It is useful to distinguish between a functional relation and a statistical relation.

Functional Relation

Relations in which Y is uniquely determined from knowledge of X are called functional relations.

> **(18.1)**
> A **functional relation** between a dependent variable Y and an independent variable X is an exact relation; the value of Y is uniquely determined when the value of X is specified.

☐ EXAMPLES

1. The rental fee (Y, in dollars) for an electric motor is related to the number of hours rented (X) as follows:

$$Y = 3 + 4X$$

   Here, \$3 is a fixed handling charge and \$4 is the hourly rental charge. Thus, for any number of hours rented, there is a unique rental fee. Figure 18.1a shows the line of relationship $Y = 3 + 4X$, as well as the observations for three recent rentals of 1, 2, and 4 hours, respectively. The plotted observations are (1 hour, \$7), (2 hours, \$11), and (4 hours, \$19). Since the value of Y is uniquely determined from X, all observations fall on the line of relationship.

2. The area of a square sheet of metal (Y, in square centimeters) is related to the length of its sides (X, in centimeters) by the functional relation $Y = X^2$. Figure 18.1b shows the curve of relationship, as well as the observations for four sheets whose sides are 10, 25, 30, and 35 centimeters, respectively.   ☐

Statistical Relation

In most empirical studies, the value of a dependent variable Y is not uniquely determined when the level of the independent variable X is specified. For instance, in studying the

---

FIGURE 18.1   **Examples of functional relations.**   All observations fall exactly on the line or curve of relationship.

(a) **Linear Functional Relationship**

(b) **Curvilinear Functional Relationship**

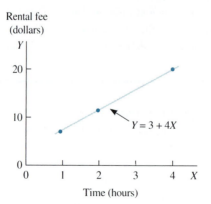

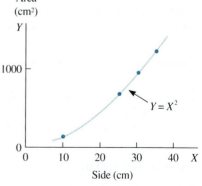

relation between family expenditures for food and family income, we are likely to find that families with the same income level will differ in their food expenditures. One main reason is that factors other than family income also play a role, such as ethnic background, family size, and living style.

Relations in which the dependent variable Y is not uniquely determined from knowledge of the independent variable X are called statistical relations.

(18.2)

A **statistical relation** between a dependent variable Y and an independent variable X is an inexact relation; the value of Y is not uniquely determined when the value of X is specified.

Some statistical relations are linear.

EXAMPLE ☐ **Bag Shipments**

A food product is shipped in bags from a company's plant. Figure 18.2a contains a scatter plot that displays the total weight Y (in pounds) and the number of bags X for 15 recent shipments from the plant. The first shipment contained $X = 40$ bags and weighed $Y = 3963$ pounds, so the point for this shipment is plotted at (40, 3963), as annotated in Figure 18.2a. The other observations are plotted similarly.

We see clearly that the weight of a shipment tends to increase with the number of bags in the shipment. To describe this tendency, we have plotted a line through the concentration of the points, as shown in Figure 18.2b. The relation is not an exact one, however, and the observations are scattered about the line of relationship. Hence, the relation is a statistical one. The statistical relation here is linear in that it follows a straight line. ☐

Other statistical relations are curvilinear.

Scatter plot of a linear statistical relation—Bag shipments example. The observations are scattered around the line of statistical relationship.

FIGURE 18.2

(a) Scatter Plot

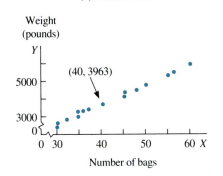

(b) Line of Statistical Relationship

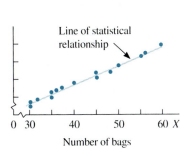

FIGURE 18.3
Scatter plot of a curvilinear statistical relation—Selling space example

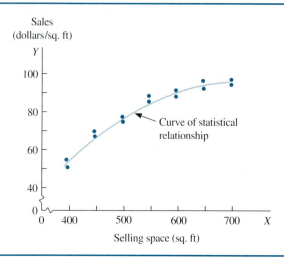

Selling Space ☐ **EXAMPLE**

In an experiment involving 14 supermarkets, the amount of selling space for a specialty foods section was controlled at seven different levels. Figure 18.3 shows a scatter plot of dollar sales of specialty foods per square foot of selling space (Y) and the number of square feet of selling space in the section (X) for the 14 supermarkets. We see from the scatter plot that the relation is a statistical one; the dollar sales per square foot vary for each pair of stores with the same amount of floor space. In contrast to Figure 18.2, the relation here is curvilinear. The growth in sales per square foot diminishes with increasing selling space. ☐

Comment

The designations "dependent variable" on the one hand and "independent variable" on the other carry no connotation that changes in X cause changes in Y. No particular cause–effect pattern between the variables is implied necessarily by a statistical relation. We discuss this point in more detail in Chapter 19.

18.2 SIMPLE LINEAR REGRESSION MODEL

Development of Model

The two examples of statistical relations just presented portray the two main features of statistical relations:

1. A tendency of the dependent variable Y to vary systematically with the independent variable X, as described by a line or curve of statistical relationship.
2. A scattering of observations around the line or curve of statistical relationship, partly because factors in addition to the independent variable X affect the dependent variable Y, and partly because of inherent variability in Y.

Regression models incorporate these features of a statistical relation by assuming the following:

1. For each level X of the independent variable, there is a probability distribution of Y.
2. The means of these probability distributions vary in a systematic fashion with X.

EXAMPLE

Consider again the linear statistical relation in the bag shipments example in Figure 18.2b. We portray the corresponding regression model in Figure 18.4. Shown there are the probability distributions of shipment weight Y when the number of bags in the shipment is $X = 40$ and $X = 50$. The regression model assumes that when a shipment contains $X = 40$ bags, the observed weight Y for that shipment will be a random selection from the probability distribution for $X = 40$, and similarly for any other number of bags in the shipment.

Note that the means of the probability distributions of Y have a systematic relation to the level of X. This systematic relation is called the *regression function*. In Figure 18.4, the regression function happens to be linear and, hence, is referred to as a regression line. The fact that the probability distributions of Y are centered on the regression line leads to sample observations that are scattered about either side of the line of statistical relationship and form a pattern from which the regression function may be inferred, as illustrated in Figures 18.2 and 18.3.

The features of a regression model that we have just illustrated can also be captured by assuming that each observation Y is the sum of the following two components:

1. A *regression component,* reflecting the line or curve of statistical relationship.
2. A *random component,* reflecting the scatter about the line or curve of statistical relationship.

Illustration of the regression model for a linear relation—Bag shipments example. The means of the probability distributions of Y are located on the regression line.

FIGURE 18.4

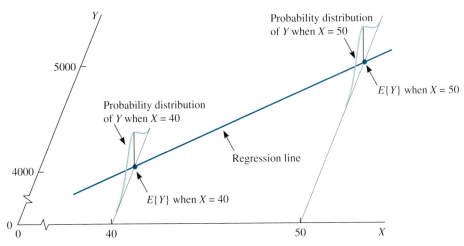

For example, if the statistical relation is linear, as in Figure 18.2b, the regression model for an observation Y takes the form:

$$Y = \beta_0 + \beta_1 X + \varepsilon$$

The linear function $\beta_0 + \beta_1 X$ represents the regression component of this model and the term ε (Greek epsilon) represents the random component. Subsequently, the random component ε will be referred to as the *error term* of the regression model.

The regression component need not be linear, as just illustrated; it also can be curvilinear. And the regression component may involve more than one independent variable.

Simple Linear Regression Model

We now consider a basic regression model utilizing one independent variable in which the statistical relation is linear. We shall denote the number of cases available for estimating the parameters of the regression model by n. Further, we shall denote the observation on the dependent variable for the ith case, called the *response,* by Y_i, and the observation on the independent variable for the ith case by X_i. The regression model can then be stated as follows.

(18.3) $Y_i = \beta_0 + \beta_1 X_i + \varepsilon_i$ $i = 1, 2, \ldots, n$

where: Y_i is the response in the ith case

X_i is the value of the independent variable in the ith case, assumed to be a known constant

β_0 and β_1 are parameters

ε_i are independent $N(0, \sigma^2)$

Model (18.3) is called the *simple linear regression model.* It has three parameters—β_0 (Greek beta), β_1, and σ^2, all of which are generally unknown. We shall now consider some important features of this regression model.

Error Terms.   Regression model (18.3) makes the following assumptions about the error terms ε_i.

1. The error terms ε_i are normally distributed random variables.
2. The expected value of each error term ε_i is 0; that is, $E\{\varepsilon_i\} = 0$.
3. The variance of each ε_i is the same; hence, the variance is constant at all levels of X and is denoted by $\sigma^2\{\varepsilon_i\} = \sigma^2$.
4. The error terms ε_i are independent random variables.

The equivalent statement of these assumptions, which we used in model statement (18.3), is that the error terms ε_i are independent $N(0, \sigma^2)$.

Comment

The normality assumption for the error terms ε_i is appropriate in many regression applications. A major reason is that there are often many factors influencing Y that are not included among the independent variables of the regression model and, hence, become incorporated in the error terms ε_i. Insofar as the effects of these factors are additive and tend to vary with a degree of mutual independence, the error terms ε_i often will tend to comply with the central limit theorem, and their distribution will be nearly normal when the number of these missing factors is large.

Regression Function. The response Y_i is a random variable because the error term component ε_i is a random variable. We wish to find the expected value of Y_i. To begin, we note that the regression component $\beta_0 + \beta_1 X_i$ is a constant because β_0 and β_1 are parameters (and, hence, constants) and X_i is assumed to be a known constant in regression model (18.3). Thus, by (6.5c), we obtain:

$$E\{Y_i\} = E\{\beta_0 + \beta_1 X_i + \varepsilon_i\} = \beta_0 + \beta_1 X_i + E\{\varepsilon_i\}$$

Since $E\{\varepsilon_i\} = 0$, we have the following result.

(18.4) $E\{Y_i\} = \beta_0 + \beta_1 X_i$

Thus, when the independent variable has the value X_i, the expected value of Y_i is $E\{Y_i\} = \beta_0 + \beta_1 X_i$.

When (18.4) is considered for any value of X, the relationship between $E\{Y\}$ and X is called the regression function.

(18.5)
The **regression function** relates $E\{Y\}$, the expected value or mean of Y, to the value of the independent variable X. For regression model (18.3), the regression function is:

$$E\{Y\} = \beta_0 + \beta_1 X$$

Thus, the regression function is the model counterpart to the line or curve of statistical relationship that we encountered when we first considered statistical relationships. The regression function is also called the *response function.*

The graph of regression function (18.5) is called the *regression line*. The parameters β_0 and β_1 are called *regression parameters* or *regression coefficients*. Parameter β_0 is the *intercept* of the regression line, and β_1 is the *slope* of the line.

EXAMPLE □ UDS

United Data Systems (UDS), a data-processing and equipment firm, leases tape drives to customers for extended periods. These drives must be tested periodically. UDS performs this testing as part of the leasing arrangement. An operations analyst for UDS wishes to study the relation between the number of minutes required on a service call to test the tape drives at the customer location (Y) and the number of tape drives tested on the call (X). The analyst considers regression model (18.3) to be appropriate for this application.

As we noted earlier, the regression parameters β_0 and β_1 are generally unknown. Suppose, however, that the parameters are known to be $\beta_0 = 15$ and $\beta_1 = 45$. The regression function then is:

$$E\{Y\} = 15 + 45X$$

Figure 18.5 illustrates the meaning of the regression parameters. The intercept, $\beta_0 = 15$, indicates that the value of the regression function at $X = 0$ is 15. The slope, $\beta_1 = 45$, signifies that for each additional tape drive tested, the expected time required to service the tape drives increases by 45 minutes.   □

Probability Distributions of Y.   We have seen that each response Y_i is a random variable with expected value $E\{Y_i\} = \beta_0 + \beta_1 X_i$. Let us next find the variance of Y_i. Since the regression component $\beta_0 + \beta_1 X_i$ is a constant, we have, by (6.6a):

$$\sigma^2\{Y_i\} = \sigma^2\{\beta_0 + \beta_1 X_i + \varepsilon_i\} = \sigma^2\{\varepsilon_i\}$$

But regression model (18.3) assumes that $\sigma^2\{\varepsilon_i\} = \sigma^2$. Hence, we have the following result.

(18.6) $\sigma^2\{Y_i\} = \sigma^2$

Thus, all Y_i have the same variability regardless of the level of the independent variable X_i.

We further know from theorem (8.7) that each Y_i is normally distributed, because Y_i is a linear function of the normal random variable ε_i. Finally, because the ε_i are assumed to be independent for the n cases, so are the Y_i.

FIGURE 18.5

Illustration of a linear regression function—UDS example.   Parameter β_0 is the intercept of the regression line, and β_1 is the slope.

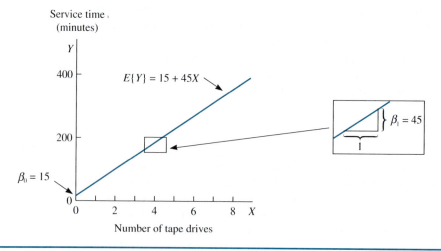

Hence, regression model (18.3) implies the following about Y_i:

1. Each Y_i is normally distributed.
2. $E\{Y_i\} = \beta_0 + \beta_1 X_i$.
3. $\sigma^2\{Y_i\} = \sigma^2$.
4. The Y_i are independent.

Therefore, an equivalent formulation of regression model (18.3) is the following.

(18.7) Y_i are independent $N(\beta_0 + \beta_1 X_i, \sigma^2)$ $i = 1, 2, \ldots, n$

In words, the Y_i are independent normal random variables with means $\beta_0 + \beta_1 X_i$ and common variance σ^2.

EXAMPLE

Figure 18.6 portrays the simple linear regression model (18.3) for the UDS example, assuming that $\beta_0 = 15$, $\beta_1 = 45$, and $\sigma = 8$. Two particular distributions of Y are displayed, at $X = 3$ and $X = 4$. Note that the means of these distributions are located on the regression line. Also note that both distributions have the same amount of variability and that both are normal.

A service call at a customer with $X_i = 4$ tape drives required $Y_i = 185$ minutes, as shown in Figure 18.6. Because the expected value of Y_i is $E\{Y_i\} = 15 + 45(4) =$

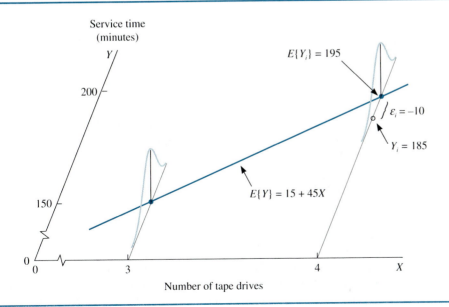

FIGURE 18.6
Illustration of simple linear regression model (18.3) with $\beta_0 = 15$, $\beta_1 = 45$, $\sigma = 8$ —UDS example

195, the error term here is $\varepsilon_i = Y_i - E\{Y_i\} = 185 - 195 = -10$. Thus, we see that the error term ε_i is simply the deviation of Y_i from its mean $E\{Y_i\}$ and hence represents the departure of the observation point from the regression line. $\quad\square$

18.3 POINT ESTIMATION OF β_0 AND β_1

The regression parameters β_0 and β_1 must usually be estimated from sample data. In this section, we take up point estimation of β_0 and β_1. Estimates of the regression coefficients often are required in their own right and also are needed when predictions for the dependent variable are desired.

Sample Data

The number of sample cases is denoted by n. Each case consists of an observation on the dependent variable, Y, and an observation on the independent variable, X. For the ith case, we shall represent these observations by (X_i, Y_i).

New Orders $\quad\square$ **EXAMPLE**

In a small-scale study based on $n = 4$ cases, the number of inquiries for a product (X) are to be related to the resulting number of orders (Y). The data are as follows.

i:	1	2	3	4
Inquiries X_i:	1	3	4	6
Orders Y_i:	1	2	1	3

With our notation, the observations for the first case are denoted by $(X_1, Y_1) = (1, 1)$, and so on for the other cases. $\quad\square$

Least Squares Estimators

Point estimates of β_0 and β_1 are usually obtained by the *method of least squares*. We explain the essential nature of this method by an example.

$\square$ **EXAMPLE**

Figure 18.7a shows the scatter plot for the new orders example. Note that a linear regression relation appears to be reasonable here. We have visually fitted the following line to the points in the scatter plot:

$$\hat{Y} = 0.3 + 0.4X$$

The symbol $\hat{Y}$ (read "Y hat") here denotes an ordinate of the fitted line. This visually fitted line is also shown in Figure 18.7a.

The vertical distances or *deviations* of the observations from the fitted line are shown by colored rules in Figure 18.7a. For instance, for the first case, in which $(X_1, Y_1) = (1, 1)$, the ordinate of the fitted line is $\hat{Y}_1 = 0.3 + 0.4(1) = 0.7$ and the vertical deviation is $Y_1 - \hat{Y}_1 = 1 - 0.7 = 0.3$. We are concerned about these vertical deviations of the observations from the fitted line because the larger they are, the poorer is

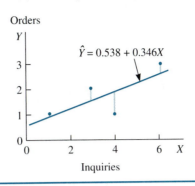

FIGURE 18.7
Finding the least squares
regression line for a scatter
plot—New orders example

the fit of the line to the data. The method of least squares measures the goodness of fit of a straight line by the sum of the squared deviations. We shall denote this sum by Q. For the visually fitted line in Figure 18.7a, the value of Q is:

$$(1 - 0.7)^2 + (2 - 1.5)^2 + (1 - 1.9)^2 + (3 - 2.7)^2 = 1.240$$

A better-fitting line is the one shown in Figure 18.7b:

$$\hat{Y} = 0.538 + 0.346X$$

For this fitted line, the sum of the squared deviations is $Q = 1.192$. In fact, it can be shown that this second line has the smallest sum of squared deviations among all fitted lines. The method of least squares considers the best-fitting line to be the one for which the sum Q is a minimum. Thus, $\hat{Y} = 0.538 + 0.346X$ is the least squares regression line; 0.538 is the least squares estimate of β_0 (the intercept of the regression line), and 0.346 is the least squares estimate of β_1 (the slope of the line). □

The *least squares line,* as we have seen, is that line for which the sum of the squared vertical distances of the Y observations from the fitted line is a minimum. We shall denote the *least squares estimators* of β_0 and β_1 by b_0 and b_1, respectively. The least squares criterion requires that b_0 and b_1 minimize Q, the sum of the squared deviations.

(18.8) $$Q = \sum_{i=1}^{n}[Y_i - (b_0 + b_1 X_i)]^2$$

It can be shown mathematically that the least squares estimators for the simple linear regression model (18.3) are obtained by solving the following two simultaneous equations.

(18.9) $$\sum Y_i = nb_0 + b_1 \sum X_i$$
$$\sum X_i Y_i = b_0 \sum X_i + b_1 \sum X_i^2$$

All summations are taken over the n sample cases ($i = 1, 2, \ldots, n$). We do not show the summation index here and throughout the regression chapters because the nature of the summation is clear.

The equations in (18.9) are called *normal equations*. When they are solved for b_0 and b_1, we obtain explicit formulas for the least squares estimators of β_0 and β_1.

(18.10)

For the simple linear regression model (18.3), the least squares estimators of β_1 and β_0 are:

(18.10a) $\qquad b_1 = \dfrac{\sum(X_i - \bar{X})(Y_i - \bar{Y})}{\sum(X_i - \bar{X})^2}$

(18.10b) $\quad b_0 = \bar{Y} - b_1\bar{X}$

where $\bar{X}$ and $\bar{Y}$ are the sample means of the X_i and Y_i observations, respectively.

Thus, the least squares estimates b_0 and b_1 are obtained by finding the deviations about the mean $Y_i - \bar{Y}$ and $X_i - \bar{X}$ for the sample data, calculating the quantities $\sum(X_i - \bar{X})(Y_i - \bar{Y})$ and $\sum(X_i - \bar{X})^2$, and substituting into the expressions in (18.10).

☐ **EXAMPLE**

In the UDS example, the analyst collected data on number of tape drives serviced and service times for 12 recent customer calls. The data are shown in columns 1 and 2 of Table 18.1, and are displayed as a scatter plot in Figure 18.8. The assumption of a linear response function appears to be reasonable here.

The basic quantities required in (18.10) for obtaining the least squares estimates

TABLE 18.1
Data and basic calculations needed to obtain least squares estimates—UDS example

Case i	(1) Number of Tape Drives X_i	(2) Service Time (minutes) Y_i	(3) $X_i - \bar{X}$	(4) $Y_i - \bar{Y}$
1	4	197	−0.58333	−20.75
2	6	272	1.41667	54.25
3	2	100	−2.58333	−117.75
4	5	228	0.41667	10.25
5	7	327	2.41667	109.25
6	6	279	1.41667	61.25
7	3	148	−1.58333	−69.75
8	8	377	3.41667	159.25
9	5	238	0.41667	20.25
10	3	142	−1.58333	−75.75
11	1	66	−3.58333	−151.75
12	5	239	0.41667	21.25
Total	$\sum X_i = 55$	$\sum Y_i = 2613$	0.0	0.0
Mean	$\bar{X} = 4.58333$	$\bar{Y} = 217.75$		

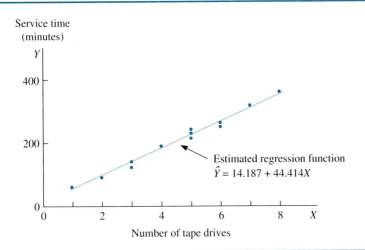

FIGURE 18.8
Scatter plot and estimated
regression function—UDS
example

of the regression parameters are calculated in Table 18.1, columns 3 and 4. We see that $\bar{X} = 4.58333$ and $\bar{Y} = 217.75$. Hence, the deviations about the mean for case 1 are:

$$X_1 - \bar{X} = 4 - 4.58333 = -0.58333$$

$$Y_1 - \bar{Y} = 197 - 217.75 = -20.75$$

The deviations for the other cases are calculated similarly and are shown in columns 3 and 4.

We can now calculate the quantities needed for the least squares formulas:

$$\sum (X_i - \bar{X})(Y_i - \bar{Y}) = (-0.58333)(-20.75)$$
$$+ (1.41667)(54.25) + \cdots + (0.41667)(21.25)$$
$$= 12.10417 + 76.85417 + \cdots + 8.85417 = 2083.7500$$

$$\sum (X_i - \bar{X})^2 = (-0.58333)^2 + (1.41667)^2 + \cdots + (0.41667)^2$$
$$= 0.34028 + 2.00694 + \cdots + 0.17361 = 46.91667$$

Now we substitute into the formulas in (18.10):

$$b_1 = \frac{2083.7500}{46.91667} = 44.41385$$
$$b_0 = 217.75 - 44.41385(4.58333) = 14.18650$$

Thus, the least squares estimates of β_0 and β_1 are $b_0 = 14.187$ and $b_1 = 44.414$.   □

Comments

1. Algebraically equivalent formulas for b_1 and b_0 suitable for hand calculations do not involve deviations about the mean.

$$(18.11) \quad b_1 = \frac{\sum X_i Y_i - \dfrac{(\sum X_i)(\sum Y_i)}{n}}{\sum X_i^2 - \dfrac{(\sum X_i)^2}{n}}$$

(continues)

$$(18.12) \quad b_0 = \frac{1}{n}\left(\sum Y_i - b_1 \sum X_i\right)$$

2. The least squares estimators b_0 and b_1 for regression model (18.3) are unbiased. In addition, it can be shown that both estimators depend linearly on the n values of Y_i. Among all unbiased estimators of β_0 and β_1 that depend linearly on the Y_i, the least squares estimators are the most efficient. These desirable properties of the least squares estimators do not depend on the normality assumption of regression model (18.3).

3. (*Calculus needed*). The least squares estimators can be derived by calculus. For the given observations (X_i, Y_i), the quantity Q to be minimized in (18.8) is a function of b_0 and b_1. To find the values of b_0 and b_1 that minimize Q, we first take partial derivatives of Q with respect to b_0 and b_1:

$$\frac{\partial Q}{\partial b_0} = \sum \frac{\partial}{\partial b_0}(Y_i - b_0 - b_1 X_i)^2 = -2\sum(Y_i - b_0 - b_1 X_i)$$

$$\frac{\partial Q}{\partial b_1} = \sum \frac{\partial}{\partial b_1}(Y_i - b_0 - b_1 X_i)^2 = -2\sum X_i(Y_i - b_0 - b_1 X_i)$$

We then set these partial derivatives equal to zero, as follows:

$$-2\sum(Y_i - b_0 - b_1 X_i) = 0$$

$$-2\sum X_i(Y_i - b_0 - b_1 X_i) = 0$$

After simplifying these equations and rearranging terms, we obtain the normal equations (18.9).

4. When the functional form of the distribution of the error terms ε_i is specified in the regression model, as in model (18.3), estimators of β_0 and β_1 can also be developed by the method of maximum likelihood described in Chapter 10. For the normal error regression model (18.3), the method of maximum likelihood yields the same normal equations as the method of least squares. Thus, the least squares estimators b_0 and b_1 for regression model (18.3) have the combined properties of least squares estimators and maximum likelihood estimators: They are unbiased, consistent, and most efficient among all unbiased estimators.

18.4

POINT ESTIMATION OF MEAN RESPONSE

Estimated Regression Function

The regression function $E\{Y\} = \beta_0 + \beta_1 X$ is estimated as follows.

$$(18.13) \quad \hat{Y} = b_0 + b_1 X$$

This equation is called the *estimated regression function*.

☐ **EXAMPLE**

We found for the UDS example that $b_0 = 14.18650$ and $b_1 = 44.41385$. Hence, the estimated regression function is:

$$\hat{Y} = 14.18650 + 44.41385X$$

This estimated regression function is also plotted in Figure 18.8. Note that the fit of the estimated regression function to the data appears to be good. ☐

Mean Response $E\{Y_h\}$

As mentioned earlier, the dependent variable Y is often called the response variable. The mean response refers to the expected value of Y for a given value of X. In the UDS example, for instance, the mean response when $X = 4$ refers to the mean of the probability distribution of service times (Y) when four tape drives are serviced.

> **(18.14)**
> The **mean response** when $X = X_h$ is denoted by $E\{Y_h\}$. For regression model (18.3), the mean response is:
>
> $$E\{Y_h\} = \beta_0 + \beta_1 X_h$$

Note that X_h denotes *any* specified level of X, in contrast to X_i, which refers to the value of the X variable for the ith sample case.

Regression analysis is often utilized to estimate mean responses. For instance, the management of UDS may be interested in the expected service time, $E\{Y_h\}$, for customers with $X_h = 4$ tape drives so that it can estimate the direct labor cost of the service calls. As another illustration, an economist may be interested in the expected amount of savings, $E\{Y_h\}$, of families whose incomes are $X_h = \$42,000$ for use in a savings model.

Point Estimator of $E\{Y_h\}$

The point estimator of the mean response $E\{Y_h\}$ is given by $\hat{Y}_h$, the value of the estimated regression function when $X = X_h$.

> **(18.15)** $\quad \hat{Y}_h = b_0 + b_1 X_h$

EXAMPLES ☐

1. For the UDS example, we wish to find the point estimate of the mean time required on calls to test $X_h = 4$ tape drives. We substitute into the estimated regression function with $X_h = 4$ and obtain:

$$\hat{Y}_h = 14.18650 + 44.41385(4) = 191.84$$

or 192 minutes.

This result signifies that in a large number of calls to test four tape drives, made under the same general conditions prevailing for the 12 calls in the sample, the mean required time on a call is in the neighborhood of 192 minutes.

2. For $X_h = 7$ tape drives, the estimated mean service time is:

$$\hat{Y}_h = 14.18650 + 44.41385(7) = 325.08 \text{ minutes}$$

☐

> **Comment**
>
> Since b_0 and b_1 are unbiased estimators of β_0 and β_1, respectively, it follows that $\hat{Y}_h$ is an unbiased estimator of $E\{Y_h\}$; that is, $E\{\hat{Y}_h\} = \beta_0 + \beta_1 X_h = E\{Y_h\}$.

18.5   RESIDUALS

For conducting inferences in regression analysis, we need to estimate the magnitude of the random component in each observation Y_i. We estimate this component by comparing Y_i, the observed value of Y for the ith case, and $\hat{Y}_i$, the value of the estimated mean response when $X = X_i$.

$$(18.16) \qquad \hat{Y}_i = b_0 + b_1 X_i$$

The value $\hat{Y}_i$ is often called the *fitted value* of Y_i. The difference between the observed value Y_i and the fitted value $\hat{Y}_i$ is called the residual for the ith observation.

(18.17)

The **residual** for the ith observation, denoted by e_i, is:

$$e_i = Y_i - \hat{Y}_i$$

The residual e_i for the ith observation is an estimate of the error term ε_i.

The residuals e_i are also needed for assessing the appropriateness of the regression model itself. We shall discuss this latter use in the next chapter.

☐ **EXAMPLE**

The residuals for the UDS example are calculated in Table 18.2. The observed and fitted values are shown in columns 2 and 3, and the residuals in column 4. To illustrate the calculations, consider the first sample case, where $X_1 = 4$ and $Y_1 = 197$. The fitted value and residual for this case are:

$$\hat{Y}_1 = 14.18650 + 44.41385(4) = 191.84$$
$$e_1 = 197 - 191.84 = 5.16$$

The observed Y values, fitted values, and residuals in Table 18.2 are portrayed in Figure 18.9.   ☐

Distinction Between e_i and ε_i

As stated in (18.17), the residual e_i is an estimate of the error term ε_i. The distinction between e_i and ε_i is an important one. The residual measures the deviation of Y_i from the

Case i	(1) Number of Tape Drives X_i	(2) Observed Value Y_i	(3) Fitted Value $\hat{Y}_i$	(4) Residual $e_i = Y_i - \hat{Y}_i$
1	4	197	191.84	5.16
2	6	272	280.67	−8.67
3	2	100	103.01	−3.01
4	5	228	236.26	−8.26
5	7	327	325.08	1.92
6	6	279	280.67	−1.67
7	3	148	147.43	0.57
8	8	377	369.50	7.50
9	5	238	236.26	1.74
10	3	142	147.43	−5.43
11	1	66	58.60	7.40
12	5	239	236.26	2.74
Total	55	2613	2613.0	0.0
Mean	4.58333	217.75	217.75	0

TABLE 18.2
Calculation of residuals— UDS example

Least squares regression line and residuals —UDS example. The observed values and residuals are not plotted to scale.

FIGURE 18.9

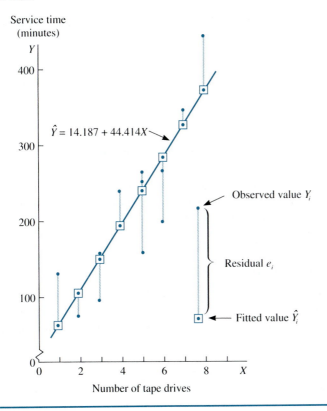

$\hat{Y} = 14.187 + 44.414X$

Observed value Y_i

Residual e_i

Fitted value $\hat{Y}_i$

Service time (minutes)

Number of tape drives

corresponding fitted value $\hat{Y}_i = b_0 + b_1 X_i$, whereas the error term measures the deviation of Y_i from the true mean $E\{Y_i\} = \beta_0 + \beta_1 X_i$. Thus, e_i and ε_i will differ to the extent that the fitted value $\hat{Y}_i$ and the true mean $E\{Y_i\}$ differ.

Properties of Least Squares Residuals

The least squares residuals (18.17) for regression model (18.3) have four important properties.

1. The residuals sum to zero.

$$(18.18) \quad \sum e_i = 0$$

   This property is evident from Table 18.2, column 4.
2. The sum of the squared residuals, $\sum e_i^2$, is a minimum. This property follows because the method of least squares minimizes Q. We see now from definition (18.8) that, for the method of least squares, Q is simply the sum of the squared residuals.
3. The sum of the weighted residuals is zero when each residual is weighted by the level of the independent variable.

$$(18.19) \quad \sum X_i e_i = 0$$

4. A consequence of properties (18.18) and (18.19) is that the sum of the weighted residuals is zero when each residual is weighted by the fitted value.

$$(18.20) \quad \sum \hat{Y}_i e_i = 0$$

18.6　ANALYSIS OF VARIANCE

Analysis of variance (ANOVA) is a highly useful and flexible mode of analysis for regression models. We introduce now some basic ANOVA concepts and procedures and use them to obtain a point estimator of the error term variance σ^2 and to measure the degree of linear association between X and Y in the sample data.

Partitioning of Total Sum of Squares

Total Sum of Squares.　Consider again the UDS example. If we wish to predict the service time Y without knowing the number of tape drives X of the customer, the uncertainty associated with a prediction is related to the variability of the Y observations around their mean $\bar{Y}$ as measured by the following deviations.

$$(18.21) \quad Y_i - \bar{Y}$$

These deviations for the UDS example are shown in Figure 18.10a. Clearly, if all service times Y_i are equal, there is no variability in the data and all deviations $Y_i - \bar{Y}$ equal zero. The greater the variability in the data, the larger the deviations $Y_i - \bar{Y}$ and the greater the uncertainty associated with a prediction of Y without utilizing knowledge of X.

Conventionally, the variability of the Y observations is measured by the sum of the squared deviations, which is denoted by *SSTO*.

(18.22) $\qquad SSTO = \sum (Y_i - \bar{Y})^2$

Here, *SSTO* stands for the *total sum of squares*. If there is no variability in the Y observations, then all deviations $Y_i - \bar{Y} = 0$ and, hence, $SSTO = 0$. The greater is the variability in the Y_i, the larger is *SSTO*.

EXAMPLE ☐

For the UDS example, the deviations $Y_i - \bar{Y}$ are shown in Table 18.1, column 4. For *SSTO* we obtain:

$$SSTO = (-20.75)^2 + (54.25)^2 + \cdots + (21.25)^2$$

$$= 92,884.2500$$

☐

Error Sum of Squares. When we use knowledge of the number of tape drives X to predict the service time Y, the uncertainty associated with a prediction is related to the variability of the Y observations around the fitted regression line as measured by the following deviations.

Partitioning of total sum of squares—UDS example. The Y values are not plotted to scale.

FIGURE 18.10

(a) Deviations in *SSTO*

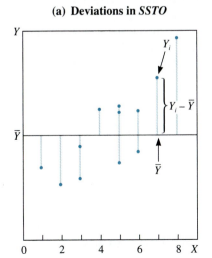

(b) Deviations in *SSE*

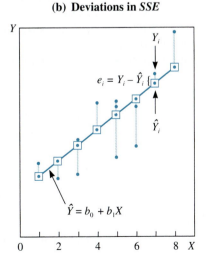

(c) Deviations in *SSR*

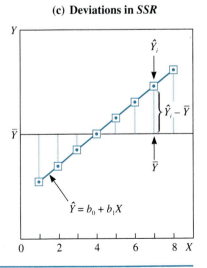

(18.23) $Y_i - \hat{Y}_i$

If all the Y observations fall on the fitted regression line, then all deviations $Y_i - \hat{Y}_i$ will equal zero. The larger the deviations $Y_i - \hat{Y}_i$, the greater the uncertainty associated with a prediction of Y utilizing knowledge of the number of tape drives X. The deviations $Y_i - \hat{Y}_i$ for the UDS example are shown in Figure 18.10b.

Again, the conventional measure of the variability around the fitted regression line is the sum of the squared deviations, which is denoted by SSE.

(18.24) $SSE = \sum (Y_i - \hat{Y}_i)^2$

Here, SSE stands for the *error sum of squares*. If all Y observations fall on the fitted regression line, $SSE = 0$. The greater are the deviations $Y_i - \hat{Y}_i$, the larger is SSE.

Note that $Y_i - \hat{Y}_i = e_i$ by (18.17); hence, SSE is simply the sum of the squared residuals.

(18.24a) $SSE = \sum e_i^2$

☐ **EXAMPLE**

We calculate SSE for the UDS example by using the residuals in column 4 of Table 18.2:

$$SSE = (5.16)^2 + (-8.67)^2 + \cdots + (2.74)^2 = 336.8810 \qquad \square$$

Regression Sum of Squares.   We see in the UDS example that the variability of the Y observations when X is utilized ($SSE = 336.8810$) is much smaller than when X is not utilized ($SSTO = 92{,}884.2500$). This reduction in the variability of the Y observations is associated with utilizing knowledge of the number of tape drives at the customer location for predicting service time. The reduction is indeed another sum of squares, denoted by SSR.

(18.25) $SSR = SSTO - SSE$

Here, SSR stands for the *regression sum of squares*.

☐ **EXAMPLE**

For the UDS example, the reduction in the variability of the Y observations is:

$$SSR = 92{,}884.2500 - 336.8810 = 92{,}547.3690 \qquad \square$$

It can be shown that *SSR* is a sum of squares involving the following deviations.

> **(18.26)** $\quad \hat{Y}_i - \bar{Y}$

Because the mean of the n fitted values $\hat{Y}_i$ equals $\bar{Y}$ (see Table 18.2, columns 2 and 3 for confirmation), each deviation $\hat{Y}_i - \bar{Y}$ represents the difference between the fitted value and the mean of the fitted values. The deviations $\hat{Y}_i - \bar{Y}$ for the UDS example are shown in Figure 18.10c. The regression sum of squares in (18.25) can therefore also be expressed as follows.

> **(18.27)** $\quad SSR = \sum(\hat{Y}_i - \bar{Y})^2$

The regression sum of squares *SSR* may be viewed as a measure of the effect of the regression relation in reducing the variability of the Y_i. If $SSR = 0$, the regression relation does not reduce the variability at all. The larger *SSR* is compared with *SSTO,* the greater is the reduction in the variability of the Y_i by utilizing knowledge of X through the regression relation.

EXAMPLE ☐

For the UDS example, the direct calculation of *SSR* by (18.27) proceeds as follows, using the data in Table 18.2, column 3:

$$SSR = (191.84 - 217.75)^2 + (280.67 - 217.75)^2 + \cdots$$
$$+ (236.26 - 217.75)^2 = 92{,}547.3690$$

This result necessarily agrees with that obtained previously by subtraction, except for possible rounding effects. ☐

Comment

An algebraically equivalent formula for *SSR* that is useful at times involves the estimated slope coefficient b_1.

> **(18.27a)** $SSR = b_1^2 \sum(X_i - \bar{X})^2$

Formal Statement of Partitioning. Since $SSR = SSTO - SSE$ by (18.25), it follows that $SSTO = SSR + SSE$. We can therefore view the process we have just completed as a partitioning of the total sum of squares *SSTO* into two additive components: (1) *SSR*—the portion of the total variability of the Y_i eliminated when the independent variable X is considered, and (2) *SSE*—the portion of the total variability of the Y_i that remains as residual or error variation when X is considered.

(18.28)

For simple linear regression, the decomposition of the total sum of squares into two additive components is:

(18.28a) $SSTO$ = SSR + SSE

(18.28b) $\sum (Y_i - \bar{Y})^2 = \sum (\hat{Y}_i - \bar{Y})^2 + \sum (Y_i - \hat{Y}_i)^2$

☐ **EXAMPLE**

In assembling the respective sums of squares for the UDS example, we obtain:

$$92{,}884.2500 = 92{,}547.3690 + 336.8810$$
$$SSTO \quad = \quad SSR \quad + \quad SSE$$

☐

Comments

1. The following formulas for the ANOVA sums of squares are algebraically equivalent to the definitional formulas in (18.22), (18.24), and (18.27) and may be more convenient for manual computation.

(18.29) $SSTO = \sum Y_i^2 - \dfrac{\left(\sum Y_i\right)^2}{n}$

(18.30) $SSR = \dfrac{\left(\sum X_i Y_i - \dfrac{\sum X_i \sum Y_i}{n}\right)^2}{\sum X_i^2 - \dfrac{\left(\sum X_i\right)^2}{n}}$

(18.31) $SSE = SSTO - SSR$

2. Formula (18.28b) can be derived by noting that the total deviation (18.21) is the sum of the error deviation (18.23) and the regression deviation (18.26):

$$Y_i - \bar{Y} = (Y_i - \hat{Y}_i) + (\hat{Y}_i - \bar{Y})$$

We now square each side and sum over the n cases:

$$\sum (Y_i - \bar{Y})^2 = \sum [(Y_i - \hat{Y}_i) + (\hat{Y}_i - \bar{Y})]^2$$

Finally, we expand the right side and obtain:

$$\sum (Y_i - \bar{Y})^2 = \sum (Y_i - \hat{Y}_i)^2 + \sum (\hat{Y}_i - \bar{Y})^2 + 2 \sum (Y_i - \hat{Y}_i)(\hat{Y}_i - \bar{Y})$$

It can be shown that the last term equals zero.

Partitioning of Degrees of Freedom

A sum of squares has an associated number of degrees of freedom. Recall that the variance s^2 in (3.12) has a denominator $n - 1$. This number is the degrees of freedom associated with the numerator sum of squares in s^2.

Corresponding to the partitioning of the total sum of squares *SSTO* into the components *SSR* and *SSE*, there is a partitioning of the degrees of freedom. *SSTO* has $n - 1$ degrees of freedom associated with it. To see why, note that there are n observations Y_i that enter into the deviations $Y_i - \bar{Y}$ for *SSTO*. These deviations have only $n - 1$ degrees of freedom, however, because they are subject to one mathematical constraint, that the sum of the deviations is zero; that is, $\Sigma(Y_i - \bar{Y}) = 0$.

SSE has $n - 2$ degrees of freedom associated with it. Again there are n observations Y_i that enter into the deviations $e_i = Y_i - \hat{Y}_i$ for *SSE*. However, the deviations are subject to the two independent mathematical constraints in (18.18) and (18.19). These two constraints arise, in essence, from the two normal equations in (18.9).

Finally, *SSR* has one degree of freedom. There are n fitted values $\hat{Y}_i$ that enter into the deviations $\hat{Y}_i - \bar{Y}$ for *SSR*. However, all of the fitted values are calculated from the same regression line. Consequently, the number of degrees of freedom associated with the n fitted values is two; these are related to the slope and intercept of the line. The two degrees of freedom are reduced to one because the deviations $\hat{Y}_i - \bar{Y}$ must satisfy the constraint $\Sigma(\hat{Y}_i - \bar{Y}) = 0$.

We thus see that the degrees of freedom (df) are additive.

$$
(18.32) \qquad \underbrace{n - 1}_{\substack{df \text{ for} \\ SSTO}} = \underbrace{1}_{\substack{df \text{ for} \\ SSR}} + \underbrace{n - 2}_{\substack{df \text{ for} \\ SSE}}
$$

EXAMPLE □

For the UDS example, where $n = 12$, *SSTO* has associated with it $12 - 1 = 11$ degrees of freedom, *SSR* has 1 degree of freedom, and *SSE* has $12 - 2 = 10$ degrees of freedom. Hence, $11 = 1 + 10$.   □

Mean Squares

A sum of squares divided by its associated degrees of freedom is called a *mean square*. The sample variance s^2 is an example of a mean square. Two mean squares of importance in regression analysis are the *regression mean square*, denoted by *MSR*, and the *error mean square*, denoted by *MSE*.

$$
(18.33) \qquad MSR = \frac{SSR}{1}
$$

$$
(18.34) \qquad MSE = \frac{SSE}{n - 2}
$$

☐ **EXAMPLE**

For the UDS example, we have $SSR = 92{,}547.3690$, $SSE = 336.8810$, and $n = 12$. Hence:

$$MSR = \frac{92{,}547.3690}{1} = 92{,}547.37 \qquad MSE = \frac{336.8810}{10} = 33.68810$$

☐

Comment

Mean squares are not additive; MSR and MSE do not sum to $SSTO/(n-1)$. For the UDS example, we have $SSTO/(n-1) = 92{,}884.2500/11 = 8444.02$, whereas:

$$MSR + MSE = 92{,}547.37 + 33.68810 = 92{,}581.06$$

ANOVA Table

It is useful to collect the sums of squares (SS), degrees of freedom (df), and mean squares (MS) in an ANOVA table. Table 18.3a shows the general format of an ANOVA table for simple linear regression, with the definitional formulas as entries. The corresponding numerical results for the UDS example are displayed in Table 18.3b.

Point Estimation of σ^2

For making statistical inferences in regression applications, we require a point estimate of σ^2, the variance of the error terms ε_i. Note that the sample variance in (3.12):

$$s^2 = \frac{\sum(X_i - \overline{X})^2}{n-1}$$

TABLE 18.3

ANOVA table for simple linear regression and results for UDS example

(a) General Format

Source of Variation	SS	df	MS
Regression	$SSR = \Sigma(\hat{Y}_i - \overline{Y})^2$	1	$MSR = \dfrac{SSR}{1}$
Error	$SSE = \Sigma(Y_i - \hat{Y}_i)^2$	$n - 2$	$MSE = \dfrac{SSE}{n - 2}$
Total	$SSTO = \Sigma(Y_i - \overline{Y})^2$	$n - 1$	

(b) UDS Example

Source of Variation	SS	df	MS
Regression	92,547.3690	1	92,547.37
Error	336.8810	10	33.68810
Total	92,884.2500	11	

is a mean square used to estimate the population variance. Similarly, we shall now use a mean square to estimate the error term variance σ^2 in the regression model.

In the regression case, the ith observation is Y_i and its estimated mean is $\hat{Y}_i$. The relevant deviations therefore are the residuals $e_i = Y_i - \hat{Y}_i$, and the relevant mean square for estimating σ^2 is the error mean square $MSE = \Sigma e_i^2/(n - 2)$.

> **(18.35)**
> The variance σ^2 of the error terms ε_i in regression model (18.3) is estimated by MSE as defined in (18.34).

It can be shown that MSE is an unbiased estimator of the error term variance σ^2; that is, $E\{MSE\} = \sigma^2$.

EXAMPLE ☐

Earlier we found for the UDS example that $MSE = 33.68810$ (Table 18.3b). Thus, the point estimate of the error term variance σ^2 is 33.68810. The corresponding point estimate of σ is $\sqrt{MSE} = \sqrt{33.68810} = 5.804$ minutes. We now have a complete description of the probability distribution of service times (Y) when the customer has any given number of tape drives. Consider customers with $X_h = 4$ tape drives. We found before that $\hat{Y}_h = 191.84$ when $X_h = 4$. Hence, the distribution of service times when four tape drives are tested has an approximate mean of 191.84 minutes and an approximate standard deviation of 5.8 minutes, and the distribution is normal according to regression model (18.3).

☐

COEFFICIENTS 18.7 OF SIMPLE DETERMINATION AND CORRELATION

We now turn to two measures that are widely used to describe the degree of linear relationship between the independent variable X and the dependent variable Y, as reflected in the simple linear regression model.

Coefficient of Simple Determination

As we saw, $SSTO$ reflects the variability of the Y observations when knowledge of X is not utilized. In contrast, SSE reflects the remaining variability of the Y observations when knowledge of X is utilized through the simple linear regression model. Finally, $SSR = SSTO - SSE$ reflects the reduction in the total variability of the Y observations associated with the use of independent variable X to predict Y. A relative measure of this reduction in variability is obtained by expressing SSR as a proportion of $SSTO$. This proportion is denoted by r^2 and is called the coefficient of simple determination.

> **(18.36)**
> The **coefficient of simple determination** r^2 is defined:
> $$r^2 = \frac{SSR}{SSTO} = \frac{SSTO - SSE}{SSTO} = 1 - \frac{SSE}{SSTO}$$

☐ **EXAMPLE**

For the UDS example, we have from Table 18.3b that $SSTO = 92,884.2500$ and $SSR = 92,547.3690$. Thus:

$$r^2 = \frac{92,547.3690}{92,884.2500} = 0.996$$

This value indicates that the variability in service times is reduced 99.6 percent when the number of tape drives is considered. This large proportionate reduction in the variability of the Y observations by the introduction of independent variable X through the simple linear regression model suggests that the regression relation may be highly useful. For example, direct labor costs might be predicted accurately on the basis of the number of tape drives to be serviced.   ☐

Limiting Values of r^2.   One limiting value of r^2 occurs when the statistical relation is perfect for the sample data; that is, when all Y observations fall on the fitted regression line so that all residuals e_i equal zero. Then, $SSE = 0$ and $SSR = SSTO - SSE = SSTO$. Hence, $r^2 = SSR/SSTO = SSTO/SSTO = 1$. This limiting case is illustrated in Figure 18.11a by the exact fit of the linear regression function to the observations.

The other limiting value of r^2 occurs when there is no linear statistical relationship between the two variables and the regression relation is of no use in reducing the variability of the Y observations. SSE is the same as $SSTO$ in this case. Hence, $SSR = SSTO - SSE = 0$, and $r^2 = SSR/SSTO = 0/SSTO = 0$. When $r^2 = 0$, the fitted regression line is horizontal; that is, $b_1 = 0$, as is evident from formula (18.27a). This condition is illustrated in Figure

FIGURE 18.11
Scatter plots showing several degrees of linear statistical relationship

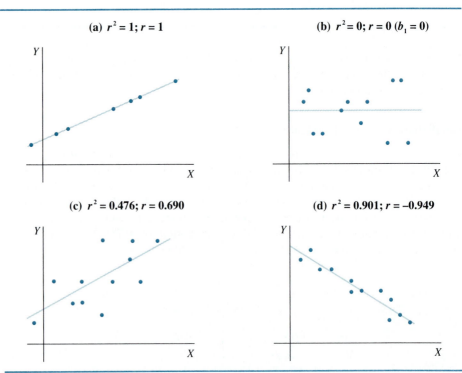

(a) $r^2 = 1; r = 1$

(b) $r^2 = 0; r = 0 \ (b_1 = 0)$

(c) $r^2 = 0.476; r = 0.690$

(d) $r^2 = 0.901; r = -0.949$

18.11b. Observe that the sample data exhibit no linear statistical relationship between the two variables X and Y.

We have, therefore, the following bounds for r^2.

$$(18.37) \quad 0 \le r^2 \le 1$$

In practice, r^2 usually falls somewhere between 0 and 1; the closer r^2 is to 1, the greater the degree of linear statistical relation in the observations. The last two panels in Figure 18.11 illustrate two scatter plots with r^2 values that lie between 0 and 1. Note how the scatter plot in Figure 18.11d, for which $r^2 = 0.901$, is rather tightly clustered about the regression line, whereas there is quite a bit of scatter in Figure 18.11c, for which $r^2 = 0.476$.

Coefficient of Simple Correlation

The square root of r^2 is another descriptive measure of the degree of linear statistical relation between X and Y. It is called the coefficient of simple correlation.

(18.38)

The **coefficient of simple correlation** r is defined:

$$r = \pm\sqrt{r^2}$$

where: the sign is that of b_1
r^2 is given by (18.36)

The sign attached to r simply shows whether the slope of the regression line is positive or negative.

Since $0 \le r^2 \le 1$ and r may be of either sign, the bounds on r are as follows.

$$(18.39) \quad -1 \le r \le 1$$

The closer is the absolute value of r to 1, the greater is the degree of linear statistical relation between X and Y in the sample cases.

The coefficient r does not have as clear-cut an operational interpretation as r^2. Even so, there has been a tendency in much applied work to use r in preference to r^2. Note that r may give the impression of a closer relation between X and Y than does r^2, since the absolute value of r is closer to 1 than is r^2 (except when r^2 equals 0 or 1).

EXAMPLES ☐

1. For the UDS example, $r^2 = 0.996$ and b_1 is positive. Hence, $r = +\sqrt{0.996} = 0.998$.

2. For the scatter plot in Figure 18.11d, $r^2 = 0.901$ and b_1 is negative. Hence, $r = -\sqrt{0.901} = -0.949$. ☐

Comment

The coefficient of simple correlation in (18.38) can be calculated directly from the following formula.

$$(18.40) \quad r = \frac{\sum (X_i - \overline{X})(Y_i - \overline{Y})}{\sqrt{\sum (X_i - \overline{X})^2 \sum (Y_i - \overline{Y})^2}}$$

18.8   COMPUTER OUTPUT

Practically all statistical packages and many types of calculators have routines for simple linear regression analysis. Figure 18.12 gives the MINITAB output for the UDS example. For convenience, we have numbered the blocks of information in the output.

FIGURE 18.12
MINITAB output for simple linear regression—UDS example

1.

	N	MEAN	STDEV
X	12	4.583	2.065
Y	12	217.7	91.9

2.   The regression equation is
    Y = 14.2 + 44.4 X

Predictor	Coef	Stdev	t-ratio	p
Constant	14.187	4.230	3.35	0.007
X	44.4138	0.8474	52.41	0.000

    s = 5.804   R-sq = 99.6%

3.   Analysis of Variance

SOURCE	DF	SS	MS	F	p
Regression	1	92547	92547	2747.18	0.000
Error	10	337	34		
Total	11	92884			

4.

ROW	Y	Fitted	Residual
1	197	191.842	5.15810
2	272	280.669	−8.66943
3	100	103.014	−3.01421
4	228	236.256	−8.25577
5	327	325.083	1.91675
6	279	280.669	−1.66943
7	148	147.428	0.57195
8	377	369.497	7.50269
9	238	236.256	1.74423
10	142	147.428	−5.42805
11	66	58.600	7.39966
12	239	236.256	2.74423

Block 1. The sample mean and the sample standard deviation for the observations on variables X and Y are given.

Block 2. The estimated regression function is given, as are the point estimates b_0 and b_1 in the column labeled Coef, on the lines for Constant and X, respectively. Other information needed for making inferences about the regression parameters β_0 and β_1 is also provided. This information will be explained in the next chapter.

The values of $\sqrt{MSE}$, labeled s, and r^2, labeled R-sq, are also given in this block.

Block 3. The entries for the ANOVA table are displayed. The degrees of freedom are shown first, followed by the sums of squares and mean squares. The columns F and p will be discussed in the next chapter.

Block 4. The observations Y_i, the fitted values $\hat{Y}_i$, and the residuals e_i are given by case number (labeled ROW).

The calculational results in Figure 18.12 are the same as those presented earlier, except for differences in the degree of rounding.

PROBLEMS

18.1 A study panel for a professional association suggested the following income guidelines for members in private practice:

Years X:	5	10	15	20	25	30
Income Y:	59.5	69.0	78.5	88.0	97.5	107.0

Here, X denotes years of experience and Y denotes target annual income in thousands of dollars.

a. Examine the relation between years of experience and income graphically.
b. Do the guidelines represent a statistical or a functional relation? Explain.

18.2 For each of the following pairs of variables, explain whether a functional or a statistical relation would most likely hold.

a. X = number of beds in hospital; Y = hospital's annual operating costs.
b. X = altitude; Y = atmospheric pressure.
c. X = company's promotional expenditures; Y = company's sales revenues.

18.3 Vending Machines. A vending machine company studied the relation between maintenance cost and dollar sales for six machines. The purpose was to identify machines whose costs are "out of line" with their sales volumes. The data (in dollars) follow:

Machine:	1	2	3	4	5	6
Cost:	95	110	80	100	125	90
Sales:	900	1250	550	850	1500	800

a. Which variable is the dependent variable here? Why?
b. Construct a scatter plot of the data. Does the maintenance cost of any machine seem "out of line"? Comment.

18.4 Spouses' Ages. The ages (in years) of the spouses of six married couples follow:

Couple i:	1	2	3	4	5	6
Wife's Age X_i:	35	25	51	25	53	42
Husband's Age Y_i:	38	25	49	31	55	44

a. Construct a scatter plot of the data.

b. Does it appear that the relation between X and Y here is statistical or functional? Linear or curvilinear? Comment.

* **18.5** Refer to Figure 18.6. For a case where $X_i = 3$, $Y_i = 146$, calculate (1) $E\{Y_i\}$, (2) ε_i. What is implied by the sign of ε_i?

18.6 The regression function relating annual agricultural exports (Y) to agricultural imports (X) for a certain country is known to be $E\{Y\} = 320 + 0.4X$, where both Y and X are expressed in millions of dollars.

a. What is the expected level of agricultural exports when agricultural imports are \$380 million?
b. For year $i = 3$, $X_3 = 600$ and $\varepsilon_3 = 35$. What is the level of agricultural exports for that year? What symbol is used to denote this quantity?

18.7 Regression model (18.3) is known to apply in a situation with $\beta_0 = 6$, $\beta_1 = 1.3$, and $\sigma^2 = 16$.

a. What is the probability that ε_i lies between -5 and 5? Explain why you can answer without knowing the value of X_i.
b. What is the value of Y_i if $X_i = 15$ and $\varepsilon_i = 4$?
c. For $X_i = 7$, obtain (1) $E\{Y_i\}$, (2) the probability that Y_i will exceed 20.

18.8 Regression model (18.3) is known to apply in a situation with $\beta_0 = 15$, $\beta_1 = 10$, and $\sigma^2 = 25$.

a. For $X_i = 8$, obtain (1) $E\{Y_i\}$, (2) the probability that ε_i lies between -10 and 10, (3) the probability that Y_i lies between 85 and 105.
b. Plot the regression line and the point for a case where $X_i = 8$, $Y_i = 92$ on the same graph. What component of regression model (18.3) is represented by the vertical deviation of the point from the regression line? What is the magnitude of this deviation?

18.9 In an expenditures study, an economist related family expenditures for clothing (Y) to family size (X) by using regression model (18.3). Critics of the study commented that numerous factors other than family size affect clothing expenditures. What are some of these other factors, and how are they accommodated by model (18.3)?

18.10 Refer to **Vending Machines** Problem 18.3.

a. Calculate the sum of squared deviations Q for the horizontal line $\hat{Y} = 100 + 0X$, where Y and X denote cost and sales, respectively.
b. Obtain the least squares line for a regression of cost on sales. Calculate the sum of squared deviations Q for the least squares line.
c. On a scatter plot of the data, show the horizontal line in part a and the least squares line in part b. Which line provides the better visual fit to the scatter plot? Is this result consistent with the values of Q obtained in parts a and b? Explain.

18.11 Refer to **Spouses' Ages** Problem 18.4.

a. Calculate the sum of squared deviations Q for the line $\hat{Y} = 2 + X$.
b. Obtain the least squares line for a regression of husband's age on wife's age. Calculate the sum of squared deviations Q for the least squares lines.
c. Compare the values of Q obtained in parts a and b. Are the results consistent with theoretical expectations? Explain.

* **18.12** **Fuel Efficiency.** The following data on automobile weight (in hundreds of pounds) and expressway gasoline mileage (in miles per gallon) were gathered in a study of the fuel efficiency of eight automobiles:

Automobile i:	1	2	3	4	5	6	7	8
Weight X_i:	21	24	23	21	22	18	20	26
Mileage Y_i:	35	27	31	38	36	40	37	28

Assume that regression model (18.3) is appropriate.

a. Construct a scatter plot of the data.

b. Obtain the estimated regression function and plot it on the graph in part a. Does a linear regression function appear to be a good fit here? Comment.

18.13 Marketing Experiment. The data that follow were obtained in an experiment to estimate the relation between the number of showings of a TV commercial in a sales territory and the territory sales of a high-technology electric blanket (in thousands of units):

Territory i:	1	2	3	4	5	6	7	8	9	10
Showings X_i:	3	1	4	0	2	4	0	3	1	2
Sales Y_i:	2.6	1.2	3.0	1.0	2.0	3.6	0.5	3.2	1.8	2.5

Each territory received a single newspaper advertisement and zero to four showings of the TV commercial. The territories were assigned at random to the numbers of TV showings. Other factors, such as TV channel used and timing of commercials, were controlled carefully so that their effects would be the same in the different territories. Assume that regression model (18.3) is appropriate.

a. Construct a scatter plot of the data.
b. Obtain the estimated regression function and plot it on the graph in part a. Does a linear regression function appear to be a good description of the statistical relation between the number of TV showings and sales? Comment.

18.14 Reconditioning Cost. Data for nine large incinerators on the cost of the most recent recon- ditioning (in $ thousand) and number of operating hours since the preceding reconditioning (in thousands) follow:

Incinerator i:	1	2	3	4	5	6	7	8	9
Hours X_i:	2.2	1.8	2.9	2.5	1.6	2.9	2.7	3.1	1.9
Cost Y_i:	5.0	4.3	6.2	5.1	3.6	5.8	5.9	6.1	4.1

Assume that regression model (18.3) is appropriate.

a. Construct a scatter plot of the data.
b. Obtain the estimated regression function and plot it on the graph in part a. Does a linear regression function appear to be a good description of the statistical relation between prior operating hours and reconditioning cost? Comment.

18.15 Bid Preparation. The following are data on the weekly number of bids prepared by Atlas Structural Corporation for the past nine weeks and the number of hours required by corporation personnel to prepare the bids:

Week i:	1	2	3	4	5	6	7	8	9
Bids X_i:	6	3	9	6	4	8	2	6	5
Hours Y_i:	170	78	232	155	107	212	56	162	131

Assume that regression model (18.3) is appropriate.

a. Construct a scatter plot of the data.
b. Obtain the estimated regression function and plot it on the graph in part a. Does a linear regression function appear to be a good fit here? Comment.

* **18.16** Refer to **Fuel Efficiency** Problem 18.12.

a. Interpret b_1.
b. Estimate the mean response when an automobile's weight is 19 hundred pounds. Interpret this estimate.
c. Is it reasonable to estimate expressway gasoline mileage for cars with a weight of 19 hundred pounds when none of the eight automobiles in the regression study was of this weight? Discuss.

18.17 Refer to **Marketing Experiment** Problem 18.13.

a. Interpret b_0 and b_1.
b. Estimate the mean sales in a territory receiving three showings of the TV commercial.

18.18 Refer to **Reconditioning Cost** Problem 18.14.

a. What is meant by "mean response" here?

b. Estimate the change in the mean response when the number of prior operating hours increases by 1.0 thousand.

c. Estimate the mean reconditioning cost when the number of prior operating hours is 2.0 thousand.

d. Are the point estimates in parts b and c subject to sampling errors? If so, do the point estimates convey any information about the magnitudes of these errors?

18.19 Refer to **Bid Preparation** Problem 18.15.

a. The numbers of hours required for bid preparation were not the same for the three weeks in which six bids were prepared. Does this represent a departure from regression model (18.3)? Explain.

b. Estimate the change in the mean response when one additional bid is prepared.

c. Estimate the mean response when five bids are prepared.

 * **18.20** Refer to **Fuel Efficiency** Problem 18.12.

a. Obtain the residuals. Which automobile has the largest positive residual? The largest negative residual? Why might an automobile buyer be interested in the former automobile?

b. Verify that (18.18) and (18.19) hold here.

18.21 Refer to **Marketing Experiment** Problem 18.13.

a. Obtain the residuals.

b. Verify that (18.18) and (18.19) hold here.

c. Refer to the residuals for territories 7 and 8. Do you expect that these residuals differ because the TV commercial was not shown in territory 7 but was shown three times in territory 8? Explain.

18.22 Refer to **Reconditioning Cost** Problem 18.14.

a. Obtain the residuals. Which incinerator has the largest positive residual? Why might a cost analyst be interested in this incinerator?

b. Verify that (18.18) and (18.19) hold here.

18.23 Refer to **Bid Preparation** Problem 18.15.

a. Obtain the residuals.

b. Verify that the residuals in part a sum to zero. Inasmuch as e_i is an estimate of ε_i when regression model (18.3) is appropriate, does it necessarily follow that $\sum_{i=1}^{n} \varepsilon_i = 0$? Explain.

 * **18.24** Refer to **Fuel Efficiency** Problem 18.12.

a. Set up the ANOVA table.

b. Which parameter of regression model (18.3) is estimated by (1) MSE, (2) $\sqrt{MSE}$?

18.25 Refer to **Marketing Experiment** Problem 18.13.

a. Set up the ANOVA table. Do $SSTO$, SSE, and SSR satisfy identity (18.28a) as expected?

b. Which parameter of regression model (18.3) is estimated by $\sqrt{MSE}$? In what units is $\sqrt{MSE}$ expressed?

18.26 Refer to **Reconditioning Cost** Problem 18.14.

a. Set up the ANOVA table.

b. Estimate σ. What information does this estimate provide about the magnitude of the residuals e_i for this regression study?

18.27 Refer to **Bid Preparation** Problem 18.15.

a. Set up the ANOVA table.

b. Obtain the estimated mean and standard deviation of the probability distribution of the number of hours required to prepare (1) six bids during a week, (2) eight bids during a week.

c. How would the entries in the ANOVA table in part a change if (1) each of the Y_i values were increased by 10 hours, (2) each Y_i value were converted from units of hours to units of minutes?

 * **18.28** Refer to **Fuel Efficiency** Problems 18.12 and 18.24.

a. Is *SSR* a large proportion of *SSTO* here? What does this fact imply about the extent to which variation in gasoline mileage among the eight automobiles is reduced when weight of automobile is considered?

b. Calculate r^2 and r. What sign did you attach to r and why?

18.29 Refer to **Marketing Experiment** Problems 18.13 and 18.25. Calculate r^2 and r. Interpret the former measure in the context of this regression application.

18.30 Refer to **Reconditioning Cost** Problems 18.14 and 18.26.

a. Calculate the coefficient of simple determination. Does knowledge of the prior operating hours appear to be helpful in reducing the variability of reconditioning costs? Comment.

b. Calculate the coefficient of simple correlation. In what units is it expressed?

18.31 Refer to **Bid Preparation** Problems 18.15 and 18.27.

a. Calculate r^2 and r. Which of the two measures has the clearer operational interpretation? Which measure is closer to 1.0 in absolute value?

b. What would be the values of r^2 and r in part a if each Y_i value were converted from units of hours to units of minutes?

18.32 A discussant stated: "Up to a point, most of the typists we hire improve in speed and accuracy as they gain experience. We have developed two tests to predict performance ratings after this improvement period is completed. A study of 100 typists has shown us that the coefficient of correlation between our speed test score and speed performance rating is 0.85, while that between the accuracy test score and accuracy performance rating is 0.82."

a. Would the predictive usefulness of the tests be reduced if the signs of the coefficients were negative instead of positive? Discuss.

b. Can we conclude from the correlation coefficients that *SSE* for speed was smaller than *SSE* for accuracy? Explain.

18.33 In a study of the effect of market share on profitability, an economist regressed rate of return on investment (Y) on market share (X) for 18 firms in an industry and obtained a coefficient of simple correlation of $r = 0.24$.

a. What does the sign of r signify here?

b. Calculate r^2. Does the market share of a firm appear to be helpful in explaining variation in the rate of return on investment for firms in this industry? Explain.

EXERCISES

18.34 Demonstrate the algebraic equivalence of the following pairs of formulas: (1) (18.10b) and (18.12), (2) (18.10a) and (18.11).

18.35 Demonstrate that the estimated regression function (18.13) passes through the point $(\bar{X}, \bar{Y})$, where $\bar{X}$ and $\bar{Y}$ denote the means of the X_i and Y_i values, respectively.

18.36 (*Calculus needed.*) Confirm that the solution to the normal equations (18.9) leads to a minimum for Q defined in (18.8).

18.37 (*Calculus needed.*) When β_0 in regression model (18.3) equals zero, the regression function (18.5) becomes $E\{Y\} = \beta_1 X$. This case is called *regression through the origin*. Obtain the least squares estimator of β_1 for this case.

18.38 Refer to Table 18.1 for the UDS example. Suppose that a 13th case were added to the study, namely, $X_{13} = 2$, $Y_{13} = 103.0142$.

a. Verify that the plotted point for this case will fall directly on the least squares line for the original 12 cases.

b. If the least squares line were calculated for all 13 cases, would this line differ from the line for the original 12 cases? Explain. [*Hint:* Would the least squares line for the 12 cases still satisfy the least squares criterion as applied to the 13 cases?]

c. How would the ANOVA table for the 13 cases differ from Table 18.3b? Be specific.

18.39 Prove properties (18.18) and (18.19) of the least squares residuals for regression model (18.3). [*Hint:* Use normal equations (18.9).]

18.40 Prove the following properties of the least squares residuals for regression model (18.3): (1) $\Sigma \hat{Y}_i e_i = 0$, (2) $\Sigma Y_i e_i = SSE$.

18.41 Demonstrate that (18.27) and (18.27a) are algebraically equivalent formulas.

18.42 In fitting regression model (18.3) to some data, we obtain $b_0 = 2.0$, $b_1 = 0$. Explain which of the following will be true and which false: (1) $SSTO = 0$, (2) $r^2 = 0$, (3) $r = 0$, (4) $SSR = SSE$, (5) $SSE = SSTO$, (6) $SSTO = SSR + SSE$.

STUDIES

18.43 Distribution Centers. Data on work loads for 10 distribution centers maintained in overseas locations by a multinational company follow:

Center i:	1	2	3	4	5	6	7	8	9	10
Units X_i:	5.0	3.5	10.0	5.0	6.5	6.0	7.1	2.5	3.0	4.2
Hours Y_i:	27.5	20.1	50.5	26.3	33.5	32.4	36.8	15.5	18.3	22.0

Here X denotes number of work units completed (in thousands) and Y denotes number of work hours required (in thousands) by the center's work force during the reporting period. An analyst calculated the ratio of work hours required to work units completed (Y/X) for each center. Center 3 appeared to the analyst to be particularly efficient because its ratio is lowest.

a. Calculate the ratio Y/X for each center, and present the ratios in the form of a dot plot. Does the ratio for center 3 appear to be particularly out of line with the ratios for the other centers?

b. Regress work hours required on work units completed using regression model (18.3). Construct a scatter plot of the data and plot the estimated regression function on the same graph.

c. Obtain the residuals and prepare a dot plot of them. Does the residual for center 3 appear to be outlying?

d. Contrast your findings with those of the analyst, paying particular attention to the matter of efficiency.

18.44 Refer to the **Financial Characteristics** data set (Appendix D.1). Consider only the data on net sales and net assets in year 2 for firms in the electronics industry (industry 6).

a. Regress net sales in year 2 on net assets in year 2 for firms in this industry using regression model (18.3). State the estimated regression function. Construct a scatter plot of the data and plot the estimated regression function on the same graph.

b. Calculate r and $\sqrt{MSE}$. Which of these two measures provides more direct information about the variability of net sales of firms with given asset size? Explain.

c. Obtain the residuals. Do any firms seem to be out of line in terms of the relation of their net sales to net assets? Is the same conclusion reached from an examination of the graph prepared in part a? Comment. What factors might explain outliers here?

d. Repeat the regression analysis in part a, but this time omitting firm 263 from the set of observations. Did this one case have a substantial effect on the values of b_0 and b_1? On $\sqrt{MSE}$? In general, why do outliers have a substantial influence on b_0 and b_1 with the method of least squares?

Simple Linear Regression—II

In Chapter 18, we discussed basic concepts of simple linear regression. In this chapter, we consider inferences about the regression parameters and predictions of new observations, as well as several other important applied regression topics.

The simple linear regression model (18.3), repeated here, will continue to be employed.

(19.1) $\quad Y_i = \beta_0 + \beta_1 X_i + \varepsilon_i \quad\quad i = 1, 2, \ldots, n$

where: Y_i is the response in the ith case

X_i is the value of the independent variable in the ith case, assumed to be a known constant

β_0 and β_1 are parameters

ε_i are independent $N(0, \sigma^2)$

19.1 CONFIDENCE INTERVAL FOR MEAN RESPONSE

As we noted in Chapter 18, regression applications frequently entail estimation of the mean of the distribution of Y at a specified level of X.

EXAMPLES

1. A loan officer needs an estimate of the mean repayment period when $5000 home improvement loans are made.
2. A university admissions officer wishes to estimate the mean grade point average of freshmen students who receive a score of 450 on a college entrance examination.

To develop a confidence interval for the mean response at a given level of X, we need to utilize information about the relevant sampling distribution.

Sampling Distribution of $\hat{Y}_h$

Recall that to estimate the mean of the probability distribution of Y when $X = X_h$, denoted by $E\{Y_h\}$, we use the point estimator in (18.15), which is denoted by $\hat{Y}_h$.

$$(19.2)\qquad \hat{Y}_h = b_0 + b_1 X_h$$

The estimate $\hat{Y}_h$ will vary in repeated samples because b_0 and b_1 will vary from sample to sample. The sampling distribution of $\hat{Y}_h$ describes the different possible values of $\hat{Y}_h$ that would be obtained if repeated samples were selected, each with the same values X_1, X_2, ..., X_n for the independent variable.

Statistical theory provides us with the following facts about the sampling distribution of $\hat{Y}_h$.

(19.3)

For the simple linear regression model (19.1), the sampling distribution of $\hat{Y}_h$ has the following properties.

(19.3a)   **Mean:** $E\{\hat{Y}_h\} = \beta_0 + \beta_1 X_h = E\{Y_h\}$

(19.3b)   **Variance:** $\sigma^2\{\hat{Y}_h\} = \sigma^2\left[\dfrac{1}{n} + \dfrac{(X_h - \bar{X})^2}{\sum(X_i - \bar{X})^2}\right]$

(19.3c)   **Functional form:** Normal distribution

As we noted in Chapter 18, $\hat{Y}_h$ is an unbiased estimator of $E\{Y_h\}$.

Estimated Variance of $\hat{Y}_h$. The variance of the sampling distribution of $\hat{Y}_h$ is usually unknown because σ^2, the variance of the error terms ε_i, is generally unknown. However, we know from Chapter 18 that *MSE*, given in (18.34), is an unbiased estimator of σ^2. Hence, we can replace σ^2 in (19.3b) by *MSE* to obtain the estimated variance of $\hat{Y}_h$, denoted by $s^2\{\hat{Y}_h\}$.

$$(19.4)\qquad s^2\{\hat{Y}_h\} = MSE\left[\frac{1}{n} + \frac{(X_h - \bar{X})^2}{\sum(X_i - \bar{X})^2}\right]$$

☐ **EXAMPLES**

1. We continue with the UDS example from Chapter 18. We wish to estimate the mean service time when $X_h = 4$ tape drives are serviced at the customer location. Hence, we need to estimate $\sigma^2\{\hat{Y}_h\}$ for $X_h = 4$ tape drives. The quantity $\Sigma(X_i - \bar{X})^2$ can be calculated from the value of STDEV for variable X given in Figure 18.12, block 1:

$$\text{STDEV} = \sqrt{\frac{\sum(X_i - \bar{X})^2}{n - 1}} = 2.065$$

Hence:

$$\sum(X_i - \bar{X})^2 = (2.065)^2(12 - 1) = 46.91$$

A direct calculation from the deviations $X_i - \overline{X}$ in column 3 of Table 18.1 yields a more precise result. We calculated this quantity in Chapter 18, and repeat it here for completeness:

$$\sum (X_i - \overline{X})^2 = (-0.58333)^2 + (1.41667)^2 + \cdots + (0.41667)^2$$
$$= 46.91667$$

The value of $\overline{X}$ was also calculated in Chapter 18 as $\overline{X} = 4.58333$.

The estimated variance of $\hat{Y}_h$ for $X_h = 4$ then is, using $MSE = 33.68810$ from Table 18.3b:

$$s^2\{\hat{Y}_h\} = 33.68810 \left[\frac{1}{12} + \frac{(4 - 4.58333)^2}{46.91667} \right] = 3.0517$$

and hence the estimated standard deviation of $\hat{Y}_h$ is:

$$s\{\hat{Y}_h\} = \sqrt{3.0517} = 1.747$$

2. We also wish to find for the UDS example the estimated variance of $\hat{Y}_h$ when $X_h = 7$. We obtain:

$$s^2\{\hat{Y}_h\} = 33.68810 \left[\frac{1}{12} + \frac{(7 - 4.58333)^2}{46.91667} \right] = 7.0009$$

The estimated standard deviation therefore is:

$$s\{\hat{Y}_h\} = \sqrt{7.0009} = 2.646$$

□

Factors Affecting Magnitude of Variance. The magnitude of the estimated variance $s^2\{\hat{Y}_h\}$ in (19.4) is affected by four factors.

1. *MSE:* The larger is the variability of the residuals e_i, measured by *MSE*, the larger $s^2\{\hat{Y}_h\}$ tends to be.
2. *Deviation of X_h from $\overline{X}$:* The further the specified level X_h is from $\overline{X}$ in either direction, the larger is $(X_h - \overline{X})^2$ and the greater $s^2\{\hat{Y}_h\}$ tends to be. This is illustrated in the UDS example where we found $s^2\{\hat{Y}_h\} = 7.0009$ when $X_h = 7$ and $s^2\{\hat{Y}_h\} = 3.0517$ when $X_h = 4$. Since $\overline{X} = 4.58333$, $X_h = 7$ is further from the mean than $X_h = 4$.
   For a given sample, $s^2\{\hat{Y}_h\}$ is smallest at $X_h = \overline{X}$.
3. *Variability of X_i:* The greater is the variability of the X_i around their mean $\overline{X}$, the larger is $\sum (X_i - \overline{X})^2$ and the smaller $s^2\{\hat{Y}_h\}$ tends to be.
4. *Sample size n:* The greater is n, the smaller is $1/n$ and the smaller $s^2\{\hat{Y}_h\}$ tends to be. In addition, the greater is n, the larger is $\sum (X_i - \overline{X})^2$ and so again the smaller $s^2\{\hat{Y}_h\}$ tends to be.

These properties can be useful when designing a regression study. For example, suppose the major purpose of a regression study is to estimate the mean response $E\{Y_h\}$ at

$X = X_h$, and the levels of the independent variable X in the sample may be chosen freely. It would then be desirable to select levels of X with mean $\overline{X} = X_h$.

Confidence Interval

A confidence interval for $E\{Y_h\}$ can be easily constructed on the basis of the following theorem.

(19.5)

For the simple linear regression model (19.1):

$$\frac{\hat{Y}_h - E\{Y_h\}}{s\{\hat{Y}_h\}} = t(n - 2)$$

The degrees of freedom for the t random variable are the $n - 2$ degrees of freedom associated with *MSE*, which is used to obtain $s^2\{\hat{Y}_h\}$.

On the basis of theorem (19.5), a confidence interval for $E\{Y_h\}$ is obtained in the usual fashion, utilizing the t distribution.

(19.6)

The $1 - \alpha$ confidence limits for $E\{Y_h\}$ for the simple linear regression model (19.1) are:

$$\hat{Y}_h \pm ts\{\hat{Y}_h\}$$

where: $t = t(1 - \alpha/2; n - 2)$
$s\{\hat{Y}_h\}$ is given by (19.4)

☐ **EXAMPLES**

1. In the UDS example, the service manager desires a 95 percent confidence interval for the mean time required for service calls involving $X_h = 4$ tape drives. We calculated in Chapter 18 that $\hat{Y}_h = 191.84$ when $X_h = 4$. Earlier in this chapter, we obtained $s\{\hat{Y}_h\} = 1.747$ when $X_h = 4$. Finally, for $n = 12$ and $1 - \alpha = 0.95$, we require $t(0.975; 10) = 2.228$. Hence, the confidence limits are $191.84 \pm 2.228(1.747)$, and the 95 percent confidence interval is:

$$188 \le E\{Y_h\} \le 196$$

Thus, with 95 percent confidence, the service manager can conclude that the mean time for servicing four tape drives is between 188 and 196 minutes.

Figure 19.1 contains the MINITAB output for estimating the mean response when $X_h = 4$. The output presents the point estimate (labeled Fit), $\hat{Y}_h = 191.84$, the estimated standard deviation of $\hat{Y}_h$ (labeled Stdev.Fit), $s\{\hat{Y}_h\} = 1.75$, and the 95 percent confidence limits (labeled 95% C.I.), 187.95 and 195.74. The limits at the right of the printout will be explained shortly.

Fit	Stdev.Fit	95% C.I.	95% P.I.
191.84	1.75	(187.95, 195.74)	(178.33, 205.35)

FIGURE 19.1
MINITAB output for estimating a mean response and predicting a new observation—UDS example

2. In the UDS example, the service manager also desires a 90 percent confidence interval for the mean service time $E\{Y_h\}$ when $X_h = 7$. In Chapter 18, we found for $X_h = 7$ that $\hat{Y}_h = 325.08$; and in this chapter, we found $s\{\hat{Y}_h\} = 2.646$. We require $t(0.95; 10) = 1.812$. Hence, the confidence limits are $325.08 \pm 1.812(2.646)$, and the 90 percent confidence interval is:

$$320 \le E\{Y_h\} \le 330 \qquad \square$$

Comments

1. The confidence coefficient for confidence interval (19.6) is interpreted in terms of repeated samples in which the set of X levels is the same from sample to sample. Thus, for the UDS example, the 95 percent confidence coefficient attached to the interval estimate of the mean response $E\{Y_h\}$ when $X_h = 4$ signifies that if many independent samples were taken where the X levels are the same as in the 12 calls in the observed sample and a 95 percent confidence interval for $E\{Y_h\}$ were constructed from each sample, about 95 percent of these intervals would be correct, that is, would contain the true value of $E\{Y_h\}$.
2. The confidence limits in (19.6) are robust. The actual confidence coefficient remains close to $1 - \alpha$ even when the error terms of the regression model are not normally distributed, provided the departure from normality is not too marked.

PREDICTION INTERVAL FOR NEW RESPONSE   19.2

In regression applications, we frequently wish to predict a new response for a given level of the independent variable, based on the fitted regression model.

(19.7)
A **new response,** denoted by $Y_{h(new)}$, is the value of Y to be observed in a new and independent observation at the level $X = X_h$ of the independent variable.

It is important to distinguish between a new response $Y_{h(new)}$ and a mean response $E\{Y_h\}$.

EXAMPLES □

1. In the UDS example, technicians will test four tape drives on a forthcoming call. The service manager wishes to predict the number of minutes required for servicing the four tape drives on this particular call. A new response $Y_{h(new)}$ is involved here. In effect, the required service time will constitute a new observation from the distribution of Y when $X_h = 4$. The prediction is not equivalent to estimating $E\{Y_h\}$,

the mean of the distribution of Y at $X_h = 4$, since the new observation in all likelihood will deviate from the mean.

2. Anne Smith has just scored 115 on the company's aptitude test. A prediction of Anne Smith's job performance rating after one year on the job, based on her test score of 115, entails a prediction of a new response $Y_{h(\text{new})}$ when $X_h = 115$. This is in contrast to an estimate of the mean performance rating after one year on the job, $E\{Y_h\}$, for all persons who score 115 on the aptitude test.

3. An analyst for a food-processing company has investigated the relation between weight loss in wheels of Swiss cheese during storage (Y) and storage temperature (X). The analyst now wishes to estimate the mean weight loss when the storage temperature is 7°C. Here, an estimate of the mean response $E\{Y_h\}$ is required and not a prediction for a single wheel of cheese.   □

Regression Parameters Known

To illustrate the basic concepts of prediction intervals in a regression setting, we assume first that the parameters β_0, β_1, and σ^2 of the simple linear regression model are known. We then extend the procedure to cover the usual case where the parameters are not known and must be estimated.

Music Publisher   □   **EXAMPLE**

The mail-order operations of a large music publisher are being studied. The time required to fill a mail order (Y, in minutes) is to be predicted from the number of different publications in the order (X). The simple linear regression model (19.1) is known to be applicable. Also, the parameters of the regression model are known to be $\beta_0 = 20$, $\beta_1 = 1.5$, and $\sigma = 6.5$.

An order for $X_h = 30$ publications is to be filled next. Figure 19.2 shows the probability distribution of Y when $X_h = 30$. The distribution is normal, with mean $E\{Y_h\} = 20 + 1.5(30) = 65$ minutes and standard deviation $\sigma = 6.5$ minutes. We wish to predict the time required to fill this next order; that is, we wish to predict a new response $Y_{h(\text{new})}$ when $X_h = 30$.

We know that the mean of the probability distribution is $E\{Y_h\} = 65$ minutes and that the new response will deviate from this mean. We therefore need to set up a pre-

FIGURE 19.2

Prediction of a new response $Y_{h(\text{new})}$ when the parameters are known—Music publisher example

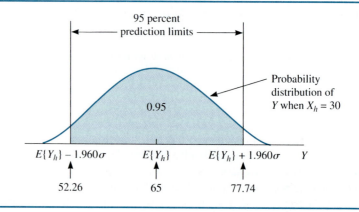

diction interval that recognizes the random deviation of $Y_{h(\text{new})}$ from $E\{Y_h\}$. Suppose we use an interval of ± 1.960 standard deviations about the mean. The prediction limits then are:

$$65 \pm 1.960(6.5)$$

and the prediction interval is:

$$52.26 \le Y_{h(\text{new})} \le 77.74$$

Since 95 percent of the area of a normal distribution is within ± 1.960 standard deviations about the mean, our procedure will give correct predictions 95 percent of the time. ☐

Regression Parameters Unknown

In the usual case when the regression parameters are unknown, β_0, β_1, and σ^2 are estimated by b_0, b_1, and MSE, respectively. To predict a new response $Y_{h(\text{new})}$ at $X = X_h$, we utilize $\hat{Y}_h = b_0 + b_1 X_h$ as the estimate of the mean of the probability distribution and MSE as the estimate of the variance of the probability distribution.

EXAMPLE ☐

In the UDS example, we wish to predict the service time on the next call, where there are four tape drives to be serviced. From Chapter 18, we know that the point estimate of the mean service time $E\{Y_h\}$ when $X_h = 4$ is:

$$\hat{Y}_h = b_0 + b_1 X_h = 14.18650 + 44.41385(4) = 191.84 \text{ minutes}$$

and that the estimate of the variance of this probability distribution is $MSE = 33.68810$. ☐

Assumptions.   There are two assumptions made in the development of the prediction interval.

1. The error term for the new response is independent of the error terms for the sample observations that are used in estimating the regression model.
2. The fitted regression model is appropriate for the new response.

Assumption 2 requires, in effect, that no change affecting the relationship of the independent and dependent variables has occurred between the time the data were collected for fitting the regression model and the time when the new observation is made.

Estimated Variance of $Y_{h(\text{new})}$.   The relevant variance in constructing the prediction interval, denoted here by $\sigma^2\{Y_{h(\text{new})}\}$, is the following.

(19.8) $\sigma^2\{Y_{h(\text{new})}\} = \sigma^2 + \sigma^2\{\hat{Y}_h\}$

Note that there are two components of the variance in (19.8), which relate to the following two sources of prediction error.

1. The inherent variability in the probability distribution of Y, represented by the error term variance σ^2.
2. The uncertainty about the mean response at $X = X_h$ because the mean response is estimated by $\hat{Y}_h$, represented by $\sigma^2\{\hat{Y}_h\}$, the variance of the sampling distribution of $\hat{Y}_h$.

The variance in (19.8) is estimated by the following statistic, denoted by $s^2\{Y_{h(\text{new})}\}$.

> **(19.9)** $s^2\{Y_{h(\text{new})}\} = MSE + s^2\{\hat{Y}_h\}$
>
> where $s^2\{\hat{Y}_h\}$ is given by (19.4).

Substituting (19.4) into (19.9) gives the following equivalent expression for the estimated variance.

> **(19.9a)** $s^2\{Y_{h(\text{new})}\} = MSE\left[1 + \dfrac{1}{n} + \dfrac{(X_h - \overline{X})^2}{\sum (X_i - \overline{X})^2}\right]$

It can be seen from (19.9a) that the same factors listed earlier that affect the magnitude of $s^2\{\hat{Y}_h\}$ also affect $s^2\{Y_{h(\text{new})}\}$. For instance, the further X_h is from $\overline{X}$, the larger is $s^2\{Y_{h(\text{new})}\}$ for any given data set used for fitting the regression model.

Prediction Interval.   The prediction interval for a new observation $Y_{h(\text{new})}$ is based on the following theorem.

> **(19.10)**
> For the simple linear regression model (19.1):
>
> $$\frac{Y_{h(\text{new})} - \hat{Y}_h}{s\{Y_{h(\text{new})}\}} = t(n - 2)$$

The prediction interval for a new observation $Y_{h(\text{new})}$ when $X = X_h$ then takes the usual form.

> **(19.11)**
> The $1 - \alpha$ prediction limits for $Y_{h(\text{new})}$ for the simple linear regression model (19.1) are:
>
> $$\hat{Y}_h \pm ts\{Y_{h(\text{new})}\}$$
>
> where: $t = t(1 - \alpha/2; n - 2)$
> $s\{Y_{h(\text{new})}\}$ is given by (19.9)

In the UDS example, the customer for the next service call has $X_h = 4$ tape drives. We are to obtain a 95 percent prediction interval for the service time $Y_{h(new)}$. From earlier work, we have for $X_h = 4$:

$$n = 12 \qquad \bar{X} = 4.58333 \qquad MSE = 33.68810$$
$$\hat{Y}_h = 191.84 \qquad s^2\{\hat{Y}_h\} = 3.0517$$

We begin by obtaining $s^2\{Y_{h(new)}\}$, using (19.9):

$$s^2\{Y_{h(new)}\} = 33.68810 + 3.0517 = 36.740$$

The estimated standard deviation therefore is $s\{Y_{h(new)}\} = \sqrt{36.740} = 6.061$. We also require $t(0.975; 10) = 2.228$. The prediction limits therefore are $191.84 \pm 2.228(6.061)$, and the desired 95 percent prediction interval is:

$$178 \le Y_{h(new)} \le 205$$

Thus, in the next call in which four tape drives are to be serviced, we predict with 95 percent confidence that the time required will be between 178 and 205 minutes.

The MINITAB output in Figure 19.1 shows these prediction limits at the right (labeled 95% P.I.).

☐

Note that the 95 percent prediction interval for $Y_{h(new)}$ in the UDS example is wider than the 95 percent confidence interval for $E\{Y_h\}$ when $X_h = 4$, obtained previously. The reason is that $s^2\{Y_{h(new)}\}$ in (19.9) contains not only $s^2\{\hat{Y}_h\}$ but also an additional component, MSE, that reflects the variability in the probability distribution of Y.

Prediction limits are useful for control purposes. Suppose in the UDS example that the service time for the next call with four tape drives was actually 230 minutes. This observation would fall outside the 95 percent prediction interval, which ranges from 178 to 205 minutes. Hence, the service manager may wish to investigate whether a change in conditions has occurred and, if so, whether remedial actions are required.

Comments

1. The relevant variance for predicting a new observation Y, at a given level $X = X_h$, can be shown to be based on the deviation of the new observation Y from $\hat{Y}_h$, the estimated mean when $X = X_h$. Because of the independence of the new observation and the earlier sample data, we have, by (6.8b) and (18.6):

$$\sigma^2\{Y - \hat{Y}_h\} = \sigma^2\{Y\} + \sigma^2\{\hat{Y}_h\} = \sigma^2 + \sigma^2\{\hat{Y}_h\}$$

In (19.8), this variance is denoted by $\sigma^2\{Y_{h(new)}\}$.

2. Unlike the confidence limits for a mean response $E\{Y_h\}$ in (19.6), the prediction limits for a new observation in (19.11) are not robust against departures from normality. Remedial procedures that may be helpful when the error terms are not normally distributed are discussed in Section 19.7.

19.3   INFERENCES CONCERNING β_1

We turn now to inferences on β_1, the slope of the regression function for regression model (19.1). Recall that β_1 indicates the change in the mean of the distribution of Y when X increases by one unit. In some regression studies, the primary objective is to estimate β_1 by means of a confidence interval.

☐ EXAMPLE

In the UDS example, the service manager wishes to ascertain how much the mean service time increases for each additional tape drive tested on a call, using a 95 percent confidence interval for β_1.   ☐

There are also occasions when a test on β_1 is appropriate.

☐ EXAMPLE

An economist wishes to ascertain whether a relation exists between the exchange rate (Y) for the currency of a country and that of its main trading partner and the difference in inflation rates (X) for the two countries by means of a test on β_1.   ☐

A test of whether a relation between X and Y exists, given that regression model (19.1) is appropriate, takes the following form.

> **(19.12)** H_0: $\beta_1 = 0$
> H_1: $\beta_1 \neq 0$

The reason is that if $\beta_1 = 0$, $E\{Y\} = \beta_0 + \beta_1 X = \beta_0$ for all levels of X. Then all distributions of Y have the same mean. Since regression model (19.1) requires all distributions of Y to be normal with equal variability, the additional condition of equal means implies that all distributions of Y are identical, regardless of the level of X. Hence, there is no relation between X and Y for regression model (19.1) when $\beta_1 = 0$.

Sampling Distribution of b_1

We require knowledge of the sampling distribution of b_1 for making inferences about β_1.

> **(19.13)**
> For the simple linear regression model (19.1), the sampling distribution of b_1 has the following properties:
>
> **(19.13a)**   **Mean:** $E\{b_1\} = \beta_1$
>
> **(19.13b)**   **Variance:** $\sigma^2\{b_1\} = \dfrac{\sigma^2}{\sum(X_i - \bar{X})^2}$
>
> **(19.13c)**   **Functional form:** Normal distribution

Estimated Variance of b_1.   The variance $\sigma^2\{b_1\}$ usually is unknown and is estimated by replacing the error term variance σ^2 by MSE. This estimated variance is denoted by $s^2\{b_1\}$.

$$\text{(19.14)} \quad s^2\{b_1\} = \frac{MSE}{\sum(X_i - \overline{X})^2}$$

EXAMPLE

In the UDS example, we know from earlier work that $MSE = 33.68810$ and $\sum(X_i - \overline{X})^2 = 46.91667$. Hence:

$$s^2\{b_1\} = \frac{33.68810}{46.91667} = 0.71804$$

and the estimated standard deviation of the sampling distribution of b_1 is:

$$s\{b_1\} = \sqrt{0.71804} = 0.84737$$

This standard deviation is shown in the MINITAB output in Figure 18.12, block 2, in the line for predictor X under the column label Stdev. $\quad\square$

Confidence Interval for β_1

A confidence interval for β_1 can be constructed readily from the following theorem.

(19.15)
For the simple linear regression model (19.1):

$$\frac{b_1 - \beta_1}{s\{b_1\}} = t(n - 2)$$

Again, the degrees of freedom for the t random variable are the $n - 2$ degrees of freedom associated with MSE.

On the basis of theorem (19.15), the confidence interval for β_1 takes the usual form.

(19.16)
The $1 - \alpha$ confidence limits for β_1 for the simple linear regression model (19.1) are:

$$b_1 \pm ts\{b_1\}$$

where: $t = t(1 - \alpha/2; n - 2)$
        $s\{b_1\}$ is given by (19.14)

EXAMPLE

We wish to obtain a 95 percent confidence interval for β_1 for the UDS example. We have $b_1 = 44.41385$ (from Chapter 18) and $s\{b_1\} = 0.84737$, and we require

$t(0.975; 10) = 2.228$. The confidence limits therefore are $44.41385 \pm 2.228(0.84737)$, and the 95 percent confidence interval is:

$$42.5 \leq \beta_1 \leq 46.3$$

Thus, with 95 percent confidence, we estimate that the mean service time increases by somewhere between 42.5 minutes and 46.3 minutes for each additional tape drive tested on a call. □

Statistical Tests for β_1

As a result of theorem (19.15), tests on β_1 are constructed in an analogous fashion to those for a population mean μ. A test of the alternatives in (19.12) uses the following standardized test statistic.

(19.17)
Tests concerning β_1 for the simple linear regression model (19.1) when the alternatives are:

$$H_0: \beta_1 = 0$$
$$H_1: \beta_1 \neq 0$$

are based on the standardized test statistic:

$$t^* = \frac{b_1}{s\{b_1\}}$$

where $s\{b_1\}$ is given by (19.14). When $\beta_1 = 0$, test statistic t^* follows the t distribution with $n - 2$ degrees of freedom.

□ **EXAMPLE**

In the UDS example, we wish to test whether a regression relation between X and Y exists.

Step 1. The alternatives here are:

$$H_0: \beta_1 = 0$$
$$H_1: \beta_1 \neq 0$$

Step 2. The α risk is to be controlled at 0.05.

Step 3. Since $\alpha = 0.05$ and $n - 2 = 10$, we require $t(0.975; 10) = 2.228$. The decision rule is as follows:

If $|t^*| \leq 2.228$, conclude H_0.
If $|t^*| > 2.228$, conclude H_1.

Step 4. We have $b_1 = 44.41385$ and $s\{b_1\} = 0.84737$. Thus, the standardized test statistic in (19.17) equals:

$$t^* = \frac{44.41385}{0.84737} = 52.41$$

Since $|t^*| = 52.41 > 2.228$, we conclude H_1—that $\beta_1 \neq 0$. This conclusion implies for regression model (19.1) that X and Y are related. The two-sided P-value for this test is $2P[t(10) > t^*] = 2P[t(10) > 52.41] = 0+$. Since this P-value is less than $\alpha = 0.05$, the same test conclusion is drawn, as expected.

The MINITAB output in Figure 18.12, block 2, shows on the line for predictor X the value of the test statistic and the two-sided P-value. The test statistic is labeled t-ratio and the P-value is labeled p. ☐

Comments

1. Occasionally, a test of whether β_1 is some value other than 0 is needed. For example, an investment analyst has regressed the daily percent change in the price of a common stock (Y) on the daily percent change in a major stock price index (X), using regression model (19.1). The analyst now wishes to test whether the slope of the regression line exceeds 1 to determine whether the common stock is more volatile (that is, tends to experience larger percent changes) than the stock index. Letting β_{10} denote the value of β_1 where the α risk is to be controlled, the test alternatives in this example are:

$$H_0: \ \beta_1 \leq \beta_{10} = 1$$
$$H_1: \ \beta_1 > \beta_{10} = 1$$

The test statistic for this test is a generalization of the standardized test statistic in (19.17).

(19.18) $$t^* = \frac{b_1 - \beta_{10}}{s\{b_1\}}$$

where $s\{b_1\}$ is given by (19.14).

When $\beta_1 = \beta_{10}$, test statistic t^* follows the t distribution with $n - 2$ degrees of freedom.

2. Both tests and confidence intervals for β_1 based on the t distribution are robust inference procedures. The actual α risk and the actual confidence coefficient remain close to their specified values even when the error terms of the regression model are not normally distributed, provided the departure from normality is not too marked.

INFERENCES CONCERNING β_0 19.4

Occasionally, inferences on the intercept β_0 are of interest. For example, when the cost of a production run (Y) is regressed on the number of units in the run (X), the intercept β_0 might represent the fixed setup cost for the production run and β_1 the variable cost per unit.

Inferences concerning β_0 are made in analogous fashion to those for β_1, relying on the following key results.

(19.19)
For the simple linear regression model (19.1):

(19.19a) $$E\{b_0\} = \beta_0$$

continues

$$(19.19b) \quad \sigma^2\{b_0\} = \sigma^2\left[\frac{1}{n} + \frac{\overline{X}^2}{\sum(X_i - \overline{X})^2}\right]$$

$$(19.19c) \quad s^2\{b_0\} = MSE\left[\frac{1}{n} + \frac{\overline{X}^2}{\sum(X_i - \overline{X})^2}\right]$$

$$(19.19d) \quad \frac{b_0 - \beta_0}{s\{b_0\}} = t(n - 2)$$

Formulas (19.19a) and (19.19b) give the mean and the variance of the sampling distribution of b_0. Since the error term variance σ^2 in (19.19b) is generally unknown, MSE is used as before as an estimator of σ^2, leading to formula (19.19c) for the estimated variance of b_0. Finally, (19.19d) states that $(b_0 - \beta_0)/s\{b_0\}$ follows the t distribution.

☐ **EXAMPLE**

We wish to construct a 95 percent confidence interval for the intercept β_0 for the UDS example. We begin by calculating the estimated variance of b_0. From earlier results, we have $b_0 = 14.18650$, $MSE = 33.68810$, $\overline{X} = 4.58333$, and $\sum(X_i - \overline{X})^2 = 46.91667$. Hence, the estimated variance of b_0, using (19.19c), is:

$$s^2\{b_0\} = 33.68810\left[\frac{1}{12} + \frac{(4.58333)^2}{46.91667}\right] = 17.8912$$

so that the estimated standard deviation of b_0 is:

$$s\{b_0\} = \sqrt{17.8912} = 4.22980$$

In view of (19.19d), the confidence limits for β_0 are of the usual form:

$$b_0 \pm ts\{b_0\}$$

where $t = t(1 - \alpha/2; n - 2)$. We require $t(0.975; 10) = 2.228$. Hence, the confidence limits are $14.18650 \pm 2.228(4.22980)$, and the desired 95 percent confidence interval is $4.8 \leq \beta_0 \leq 23.6$.

The MINITAB output in Figure 18.12, block 2, contains the estimated standard deviation of b_0 in the line corresponding to predictor Constant. Also presented there are the test statistic $t^* = b_0/s\{b_0\}$ and the two-sided P-value for a test of H_0: $\beta_0 = 0$ versus H_1: $\beta_0 \neq 0$. The small P-value here (0.007) suggests quite strongly that H_1 ($\beta_0 \neq 0$) holds. The same conclusion can be drawn from the 95 percent confidence interval for β_0.

☐

Comments

1. A confidence interval for the intercept β_0 is equivalent to a confidence interval for the mean response $E\{Y_h\}$ when $X_h = 0$. This follows because when $X_h = 0$, we have $E\{Y_h\} = \beta_0 + \beta_1 X_h = \beta_0$.

2. If it is known that $\beta_0 = 0$, regression model (19.1) simplifies to one in which the regression line goes through the origin. We consider such a model in Chapter 25 for time series data.

In Section 19.3, we discussed the t test for ascertaining whether $\beta_1 = 0$. This test, which for regression model (19.1) is equivalent to determining whether or not a relation exists between X and Y, is often utilized as a first step in the analysis of a statistical relation between two variables.

We now take up an equivalent test via the analysis of variance (ANOVA) approach. This test, called the F test, utilizes the F distribution. While the F test does not provide us with anything new here, it will be useful to understand the underlying ANOVA approach because we require this approach in Chapter 20 when we consider multiple regression, where there are two or more independent variables.

Expected Mean Squares

We stated earlier that the error mean square MSE in (18.34) is an unbiased estimator of σ^2, the variance of the error terms ε_i.

$$(19.20) \quad E\{MSE\} = \sigma^2$$

It can further be shown that the expected value of the regression mean square MSR in (18.33) is as follows for regression model (19.1).

$$(19.21) \quad E\{MSR\} = \sigma^2 + \beta_1^2 \sum (X_i - \bar{X})^2$$

Thus, when $\beta_1 = 0$, the mean of the sampling distribution of MSR is $E\{MSR\} = \sigma^2$ and then both MSR and MSE have the same expected value. This means that when $\beta_1 = 0$, MSR and MSE tend to be of the same order of magnitude. On the other hand, when $\beta_1 \neq 0$, the term $\beta_1^2 \sum (X_i - \bar{X})^2$ in (19.21) will be positive and therefore $E\{MSR\}$ will be greater than $E\{MSE\}$. Hence, if $\beta_1 \neq 0$, MSR will tend to be larger than MSE.

Statistical Test

The ANOVA test statistic, denoted by F^*, compares the two mean squares MSR and MSE, as follows.

$$(19.22) \quad F^* = \frac{MSR}{MSE}$$

If F^* is near 1 (MSR and MSE are approximately equal), this would suggest by the earlier reasoning that $\beta_1 = 0$. On the other hand, if F^* is substantially greater than 1 (MSR is much larger than MSE), this would suggest that $\beta_1 \neq 0$. Thus, an upper-tail test is appropriate.

Figure 19.3 shows the appropriate decision rule for the test, with large values of F^* leading to conclusion H_1: $\beta_1 \neq 0$. The sampling distribution of F^* when $\beta_1 = 0$, the case for which the α risk is to be controlled, is given by the following theorem.

(19.23)

For the simple linear regression model (19.1), when $\beta_1 = 0$:

$$F^* = F(1, n - 2)$$

where F^* is given by (19.22).

The notation $F(1, n - 2)$ in (19.23) denotes the F distribution with 1 numerator degree of freedom and $n - 2$ denominator degrees of freedom. Theorem (19.23) tells us, therefore, that statistic F^* follows this F distribution when regression model (19.1) is applicable and $\beta_1 = 0$. The degrees of freedom are those associated with MSR and MSE, respectively. The F distribution is continuous, unimodal, and skewed to the right, as shown in Figure 19.3. (A detailed discussion of the F distribution may be found in Appendix B, Section B.3.)

The ANOVA test procedure therefore takes the following form.

(19.24)

When the alternatives are:

$$H_0: \ \beta_1 = 0$$
$$H_1: \ \beta_1 \neq 0$$

and simple linear regression model (19.1) is applicable, the appropriate decision rule to control the α risk is as follows:

If $F^* \leq F(1 - \alpha; 1, n - 2)$, conclude H_0.
If $F^* > F(1 - \alpha; 1, n - 2)$, conclude H_1.

where F^* is given by (19.22).

The action limit of the decision rule, $F(1 - \alpha; 1, n - 2)$, is the $100(1 - \alpha)$ percentile of the F distribution with 1 degree of freedom for the numerator and $n - 2$ degrees of freedom for the denominator.

FIGURE 19.3
Statistical decision rule for F test of $\beta_1 = 0$ for simple linear regression model (19.1)

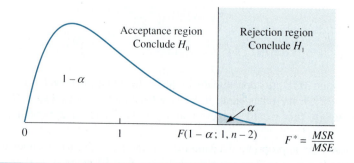

EXAMPLE ☐

For the UDS example, we wish to test whether $\beta_1 = 0$ by the ANOVA approach.

Step 1. The alternatives here are:

$$H_0: \ \beta_1 = 0$$
$$H_1: \ \beta_1 \neq 0$$

Step 2. The α risk is to be controlled at 0.05.

Step 3. For $\alpha = 0.05$ and $n - 2 = 12 - 2 = 10$, we require $F(0.95; 1, 10) = 4.96$. The decision rule therefore is as follows:

If $F^* \leq 4.96$, conclude H_0.
If $F^* > 4.96$, conclude H_1.

Step 4. We found in Table 18.3b that $MSR = 92{,}547.37$ and $MSE = 33.68810$. Hence, test statistic (19.22) here is:

$$F^* = \frac{MSR}{MSE} = \frac{92{,}547.37}{33.68810} = 2747.2$$

Since $F^* = 2747.2 > 4.96$, we conclude H_1—that $\beta_1 \neq 0$. Necessarily, this result is the same as that obtained by the t test.

The MINITAB output in Figure 18.12, block 3, contains the F^* test statistic, labeled F. Also given there is the one-sided P-value $= P[F(1, 10) > 2747.2] = 0+$. This very small P-value strongly supports the conclusion that the slope β_1 is not equal to zero. ☐

Comments

1. The equivalence of the ANOVA F test and the t test follows because the F^* statistic (19.22) is the square of the t^* statistic (19.17), in accordance with (B.19b) of Appendix B. Recall that test statistic t^* for the UDS example was $t^* = 52.41$; now $(52.41)^2 = 2747 = F^*$. Further, the action limit for the t test was 2.228, and $(2.228)^2 = 4.96$, the action limit for the ANOVA F test.
2. The F test can be used only for the two-sided alternatives $H_0: \ \beta_1 = 0$ versus $H_1: \ \beta_1 \neq 0$ and not for one-sided alternatives.

When a regression model is fitted in practice, we usually cannot be sure in advance that the model is appropriate. Consequently, the regression model needs to be checked for appropriateness or suitability, using the data themselves for guidance. Indeed, frequently several regression models need to be investigated before a final selection is made. A primary tool for studying the appropriateness of a regression model is residual analysis.

EVALUATION 19.6 OF APPROPRIATENESS OF REGRESSION MODEL

FIGURE 19.4   Residual plot and normal probability plot illustrating the appropriateness of regression model
                                   (19.1)—UDS example

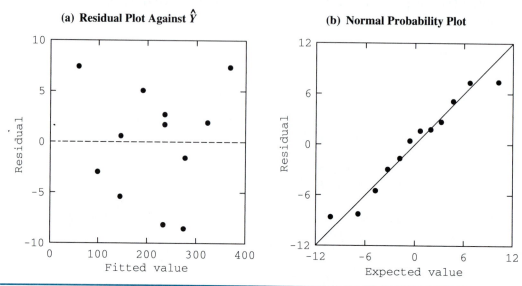

(a) Residual Plot Against $\hat{Y}$ (b) Normal Probability Plot

Residual Analysis

Residual analysis refers to a set of diagnostic methods for investigating the appropriateness
of a regression model utilizing the residuals defined in (18.17). If a regression model is
appropriate, the residuals $e_i = Y_i - \hat{Y}_i$ should reflect the properties ascribed to the model
error terms ε_i. For instance, since regression model (19.1) assumes that the ε_i are normal
random variables with constant variance, the residuals should show a pattern consistent
with these properties.

We shall describe the use of two graphic residual analysis methods here. The first
involves *residual plots,* where the residuals are plotted as a scatter plot against the corresponding fitted values. The second involves *normal probability plots* of the residuals,
where the ranked residuals are plotted against their expected values under normality. Figure
19.4 contains both a residual plot and a normal probability plot for the UDS example. As
we shall explain shortly, these plots suggest that regression model (19.1) is appropriate for
the data.

We shall examine the use of residual plots and normal probability plots for investigating the following departures from regression model (19.1).

1. The regression function is not linear.
2. The distributions of Y do not have constant variances at all levels of X; or, equivalently,
   the ε_i do not have constant variances.
3. The distributions of Y are not normal; or, equivalently, the ε_i are not normally
   distributed.
4. The error terms ε_i are not independent.

Diagnostics for Nonlinearity of Regression Function

Whether the regression function is linear or curvilinear can be readily studied from a residual plot. A systematic pattern in the plot of the residuals, such as a curvilinear pattern, often

FIGURE 19.5
Residual plot illustrating a
nonlinear regression relation—Discount card example

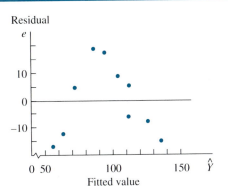

indicates a nonlinear regression relation. The residual plot in Figure 19.4a for the UDS example is based on Table 18.2. The residuals (Table 18.2, column 4) are plotted against the corresponding fitted values (Table 18.2, column 3). The absence of any systematic pattern in the points suggests that a linear regression function is appropriate here.

An illustration where a residual plot suggests that a linear regression function is not appropriate is given in the following example.

EXAMPLE ☐ **Discount Card**

Figure 19.5 contains a residual plot for a study of discount card use by students at 10 community colleges. Here, the relationship between the number of merchants in the discount card plan at a college (X) and the average discount card expenditure per student at the college (Y) was studied for the 10 colleges. A linear regression function was fitted to the data by the method of least squares, and the fitted values $\hat{Y}_i$ and the residuals e_i were obtained. These are plotted in the residual plot of Figure 19.5. Note that the residuals systematically fall below, then above, and then below the center line as the fitted value increases. This plot suggests that a linear regression function is inconsistent with the observations and that a curvilinear function is required instead. ☐

Diagnostics for Nonconstancy of Error Variances

A plot of the residuals against the fitted values also provides information as to whether or not the error terms ε_i have constant variance. If the error term variance is constant, the residual plot should show the residuals falling within a horizontal band around the center line. Figure 19.4a for the UDS example shows the residuals falling within such a horizontal band, thereby indicating that the constant variance assumption of regression model (19.1) is appropriate for the data.

A case where the residual plot suggests that the constant variance assumption is not appropriate is illustrated in the following example.

EXAMPLE ☐ **Typing**

Figure 19.6 shows a residual plot for an experiment to study the relation between typing speed (Y) and number of hours of training (X) for beginning typists using a

FIGURE 19.6
Residual plot illustrating nonconstant error variance—Typing example

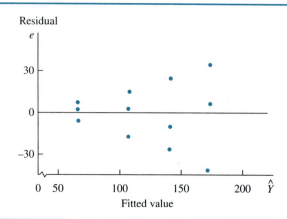

new keyboard design. Note that the vertical spread of the residuals increases as $\hat{Y}$ increases. This suggests that the distributions of Y have larger variances as $E\{Y\}$ becomes larger.

☐

Diagnostics for Lack of Normality

A diagnostic tool for examining whether the error terms are normally distributed as required by regression model (19.1) is the normal probability plot of the residuals. We discussed normal probability plots in Chapter 16. Formula (16.2) needs to be modified only slightly to take account of the fact that the observations here are the residuals. First, the sample mean $\overline{X}$ in formula (16.2) will be equal to zero here because the sum of the residuals equals zero, as noted in (18.18). Hence, the mean of the residuals, $\overline{e}$, must equal zero. Second, the population standard deviation σ here refers to the standard deviation of the error terms. As we mentioned previously, $\sqrt{MSE}$ is an estimate of this standard deviation. Hence, the sample standard deviation s in formula (16.2) is replaced by $\sqrt{MSE}$. The approximate expected value under normality for the ith ranked residual therefore is as follows.

> **(19.25)**   Expected value $= z\left(\dfrac{i - 0.5}{n}\right)\sqrt{MSE}$ $i = 1, \ldots, n$
>
> where $z\left(\dfrac{i - 0.5}{n}\right)$ is the $100\left(\dfrac{i - 0.5}{n}\right)$ percentile of the standard normal distribution.

We shall illustrate the use of a normal probability plot of the residuals by an example.

☐ **EXAMPLE**

Figure 19.4b contains a normal probability plot of the residuals for the UDS example. The calculations of the expected values under normality are illustrated in Table 19.1. Column 2 repeats the residuals, in ascending order, from Table 18.2, column 4. The

(1)	(2)	(3)	(4)	(5)	
Rank i	Residual	$\dfrac{i - 0.5}{12}$	$z\left(\dfrac{i - 0.5}{12}\right)$	Expected Value Under Normality	**TABLE 19.1** Calculation of expected values of residuals under normality—UDS example
1	−8.67	0.0417	−1.73	−10.04	
2	−8.26	0.1250	−1.15	−6.67	
⋮	⋮	⋮	⋮	⋮	
11	7.40	0.8750	1.15	6.67	
12	7.50	0.9583	1.73	10.04	

smallest residual is −8.67. The percentile needed to obtain the expected value under normality for this residual, which has rank $i = 1$, corresponds to:

$$\frac{i - 0.5}{n} = \frac{1 - 0.5}{12} = 0.0417$$

as shown in column 3. The required percentile from the standard normal distribution for the smallest residual therefore is $z(0.0417) = -1.73$, as shown in column 4. Finally, the approximate expected value under normality, using (19.25), is obtained in column 5, based on the earlier result that $MSE = 33.68810$:

$$\text{Expected value} = -1.73\sqrt{33.68810} = -10.04$$

The normal probability plot in Figure 19.4b is a plot of the ranked residuals in column 2 of Table 19.1 against the expected values under normality in column 5.

The plot of points in Figure 19.4b is reasonably linear, with the points falling fairly near the line of identity. Hence, we can conclude that the assumption of normal error terms in regression model (19.1) is reasonable for the UDS example. □

The normality assumption of regression model (19.1) can also be examined by means of a stem-and-leaf plot, a histogram, and, when the sample size is large, by the chi-square test for normality discussed in Chapter 16.

Diagnostics for Lack of Independence

In regression applications, the observations are often obtained in a time sequence, as when the dollar value of new orders received by a manufacturing corporation (Y) is related to the value of an index of business activity (X) for each of the past 48 months. In such data, there is a strong possibility that the error terms ε_i are not independent. For instance, if the error term is positive (negative) in a given month, it is often likely that the error term for the following month is also positive (negative). Such error terms are said to be *autocorrelated,* in contrast to regression model (19.1) where the ε_i are assumed to be independent.

A plot of the residuals against the time order of the observations helps to assess whether autocorrelation is present. We discuss this plot in Chapter 25, as well as a formal test for independence of the error terms based on the residuals.

19.7   REMEDIAL ACTIONS WHEN REGRESSION MODEL NOT APPROPRIATE

When residual analysis discloses that regression model (19.1) is not appropriate for the data under study, remedial action may be required. We now describe two types of remedial actions: transformations of variables and use of a different regression model.

Transformations of Variables

Transformations of variables can be employed frequently to make the data conform to the linear regression model (19.1). When the regression relation is not linear, for instance, it is often possible to linearize it by transforming the independent variable. In other cases, lack of linearity and nonconstant error variance are both found to be present. In those cases, a transformation of the dependent variable may be helpful. In still other cases, both the dependent and independent variables may need to be transformed.

Transformations that are frequently utilized include the logarithmic, square root, and reciprocal transformations.

Physical Fitness   ☐   **EXAMPLE**

Figure 19.7a presents a scatter plot of observations for 11 athletes on a physical fitness index (X) and the performance time by the athlete in a track event (Y). The regression relation is clearly not linear. Here, a reciprocal transformation on the independent variable is helpful. When the independent variable is transformed to $1/X$, the reciprocal of the fitness index, the relation becomes linear, as shown in Figure 19.7b. To illustrate the transformation, the figure shows that the original observation $(X, Y) = (10, 125)$ becomes the transformed observation $(1/X, Y) = (1/10, 125) = (0.1, 125)$. The other observations are transformed similarly.   ☐

A simple transformation of a complex regression function may sometimes yield a linear regression relation.

☐   **EXAMPLE**

The following complex regression relation is often encountered in studies of the growth of a phenomenon over time, such as sales (Y) as a function of time (X).

$$(19.26) \qquad E\{Y\} = \gamma_0 \gamma_1^X$$

When we take the logarithms of both sides, we obtain:

$$\log E\{Y\} = \log \gamma_0 + X \log \gamma_1$$

If we let $E\{Y\}' = \log E\{Y\}$, $\beta_0 = \log \gamma_0$, and $\beta_1 = \log \gamma_1$, we can write the transformed regression function as a linear regression function:

$$E\{Y\}' = \beta_0 + \beta_1 X$$

Thus, a linear regression of $\log Y$ on X may be appropriate here.   ☐

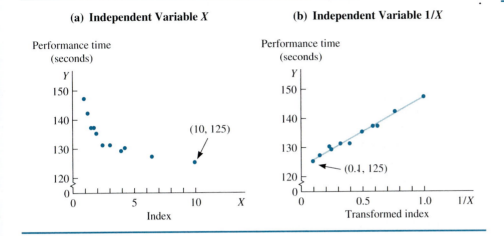

(a) Independent Variable X

(b) Independent Variable $1/X$

FIGURE 19.7
Use of reciprocal transformation on independent variable to linearize regression relation—Physical fitness example

When the distribution of the error terms is far from normal, a mathematical transformation of the Y variable may bring the distribution of the error terms closer to normality. Thus, when the distribution of the ε_i is sharply right-skewed, log Y might be used as the dependent variable instead of Y. The use of a logarithmic transformation to achieve approximate normality is illustrated in Section 11.3; in particular, see Figure 11.5 and the accompanying discussion.

Use of Different Regression Model

When the simple linear regression model (19.1) is not appropriate and transformations of variables are not helpful, a different regression model may need to be adopted. Frequently, the new model is a multiple regression model. These models are taken up in the next chapter. When the error terms are autocorrelated, a regression model that may be helpful is the autoregressive model discussed in Chapter 25.

We now discuss some important considerations in using a regression model.

PRACTICAL **19.8**
CONSIDERATIONS IN USING REGRESSION ANALYSIS

Scope of Model

Range of Observations.   Caution must be exercised in applying a regression model outside the range of the observations that were used to estimate the model's parameters and to verify the appropriateness of the model's assumptions.

EXAMPLE □

In the UDS example, we fitted regression model (19.1) using cases for which the number of tape drives serviced (X) was between 1 and 8. The fit appeared to be satisfactory for these data, but in the absence of other information, we do not know whether the model is appropriate for customers with more than eight tape drives. Thus, it may be dangerous to use the fitted model to estimate the mean service time for customers with 25 tape drives because the linear regression function might not be a good fit when extended to $X_h = 25$. A curvilinear regression function might be needed instead. □

Similarly, an estimate of the intercept β_0 in regression model (19.1) should only be used as an indication of the fixed or constant component of the relationship when there are observations near $X = 0$ or when theoretical considerations indicate that the regression function is linear for the range including $X = 0$.

Situations frequently arise in practice where inferences are required outside the range of past observations. For example, it may be desired to use disposable personal income per capita (X) as the economic indicator to predict the level of business activity (Y) for a firm. In an expanding economy, the level of X required for this prediction will frequently be outside the range of the past data. If the level of the economic indicator for which a prediction is to be made is only a little higher than in the data set, no problems usually arise. However, if the level is substantially higher, inferences obtained with the regression model must be applied with considerable judgment.

Continuation of Causal Conditions.   Whenever a regression model is utilized to make an inference for the future, the validity of the inference requires that future causal conditions be the same as during the period covered by the observed data. Thus, a prediction from the regression model in the UDS example of the service time on a customer call six months later assumes that the causal conditions affecting the servicing of tape drives have not changed.

When causal conditions have changed, the fitted model may no longer be appropriate. For example, the entry of a major new competitor may invalidate the past regression relation between a company's sales (Y) and the amount of advertising (X).

Causality

As noted earlier, the presence of a regression relation between the dependent variable Y and an independent variable X does not imply a necessary cause-and-effect relation between them. In some instances, a change in X does force a change in Y, as, for example, when an increase in drug dosage X causes a decrease in blood pressure Y. In other instances, both X and Y change in response to changes in one or more other variables without a direct causal linkage between them.

□ **EXAMPLE**

Reading ability (Y) was regressed on shoe size (X) for a sample of elementary school children and a positive regression relation was observed. This relation does not imply, of course, that larger shoes cause children to read better. The age of the child is an intervening variable in this case: Older children tend to wear bigger shoes and to read better. □

The omission of an intervening variable can sometimes also hide a relation between two variables.

Salaries   □ **EXAMPLE**

Figure 19.8a shows a scatter plot of years of education (X) and salary (Y) for 16 corporate middle managers aged 30 to 35 in a multinational firm. The regression line is horizontal, indicating no regression relation between the two variables. In Figures

Effect of intervening variable on regression relation—Salaries example. The relation between salary and education is hidden when the scatter plots for sales managers and others are combined.

FIGURE 19.8

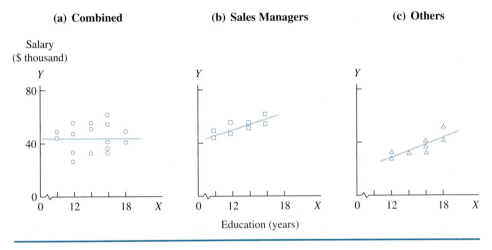

(a) Combined   **(b) Sales Managers**   **(c) Others**

Education (years)

19.8b and 19.8c, the data are divided into two groups—sales managers and others. The separate scatter plots show a persistent increase in average salary with more years of education for each group. The fact that managers who are not in sales tend to be paid less than sales managers for any given number of years of education disguises the relation between years of education and salary when all managers are combined.   □

Finally, there are instances where causation acts in the opposite direction to the regression relation in the sense that changes in Y force changes in X.

EXAMPLE □

A thermometer is being calibrated by obtaining readings at various known temperatures and regressing the actual temperature (Y) on the thermometer reading (X). Clearly, the thermometer readings X are affected by the actual temperatures Y, and not vice versa.   □

Influential Observations

In many regression applications, outlying observations may have a substantial influence on the fitted regression function, and hence, on the inferences drawn from the regression study. Figure 19.9a shows a scatter plot for 13 cases, with one observation being distinctly outlying. The solid line is the least squares regression line obtained when all cases in the data set are included in fitting the regression line. The dashed line represents the least squares regression line obtained when the outlying observation is excluded in fitting the regression line. Clearly, the fitted line is influenced greatly by the presence of the outlier.

If the outlier in this case is the result of an error in the data, such as a faulty measurement, the error needs to be corrected or the case removed from the data set before the analysis can proceed. If the outlier is a correct observation, then an explanation should be sought for the outlier. This may lead to a refined regression model that is more consistent with the data.

FIGURE 19.9
Effect of an outlying observation on the least squares fit of a regression line

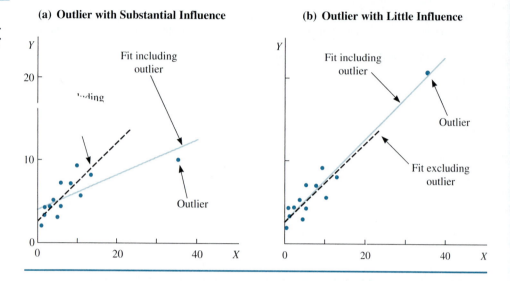

(a) **Outlier with Substantial Influence**

(b) **Outlier with Little Influence**

In some situations, an outlying observation may not be influential. Figure 19.9b shows the same scatter plot as in Figure 19.9a except that now the outlying observation has little influence on the fit of the regression line. Observe that the fitted regression line is almost the same whether the outlying observation is included in the fit or not. The accuracy of the outlying observation should still be checked, however, although no revision of the regression model may be necessary.

Independent Variable Must Be Predicted

Sometimes we must predict the value of the independent variable X that will be used to predict Y. This happens quite often in business forecasting. For example, suppose that company sales (Y) in a given quarter are regressed against an economic indicator variable (X) for the same quarter. A prediction interval is desired for company sales $Y_{h(new)}$ in the next quarter. This will require a projection X_h for the economic indicator variable for the next quarter. The sales prediction then is a conditional one, dependent on the correctness of the projected level X_h of the economic indicator variable.

Independent Variable Is Random

Sometimes, in regression applications, it is more reasonable to consider the independent variable X as random, rather than to take X as fixed in accord with regression model (19.1).

□ **EXAMPLE**

An analyst in an electric company is studying the regression relation between daily residential electricity consumption in a service area (Y) and maximum daily outdoor temperature in the area (X). Since the outdoor temperature cannot be controlled, X might be considered as random in this application.   □

When X is random, the probability distribution of Y at a given level of X is considered to be a conditional distribution with a conditional mean and a conditional variance. It can be shown that all of the results presented for regression model (19.1) where the X_i are fixed still apply when the X_i are random if the following conditions hold: (1) The conditional probability distributions of Y are normal, with conditional mean $\beta_0 + \beta_1 X$ and conditional variance σ^2; and (2) the X_i are independent random variables whose probability distribution does not depend on the parameters β_0, β_1, and σ^2.

When the independent variable is random, the interpretations of confidence coefficients and risks of errors differ from before. The interpretations now refer to repeated sampling where both the X and Y values change from one sample to the next. Thus, in the electricity consumption example, the confidence coefficient for an interval estimate for β_1 would refer to the proportion of correct interval estimates obtained if repeated random samples of n days were taken, the consumption and temperature data observed, and the confidence interval for β_1 calculated for each sample. There is now no implication that the temperatures remain the same from sample to sample.

PROBLEMS

* **19.1** Refer to **Fuel Efficiency** Problem 18.12. Obtain a 95 percent confidence interval for $E\{Y_h\}$ when $X_h = 23$. Interpret the confidence interval.

19.2 Refer to **Marketing Experiment** Problem 18.13.

a. Obtain a 90 percent confidence interval for $E\{Y_h\}$ when (1) $X_h = 0$, (2) $X_h = 2$, (3) $X_h = 4$. Interpret the last confidence interval.

b. Why are confidence intervals (1) and (3) in part a wider than interval (2)? Why do confidence intervals (1) and (3) have the same width?

19.3 Refer to **Reconditioning Cost** Problem 18.14. Construct a 95 percent confidence interval for the mean reconditioning cost of incinerators with 2.3 thousand prior operating hours. Interpret the confidence interval.

19.4 Refer to **Bid Preparation** Problem 18.15.

a. Construct a 90 percent confidence interval for the mean number of hours required for bid preparation when eight bids are prepared during the week. Interpret the confidence interval. How is the confidence coefficient interpreted here?

b. An observer comments that the number of hours required to prepare a given number of bids cannot be exactly normally distributed as required by regression model (19.1) because the number of hours is never a negative number and is always recorded as a whole number. Explain why the confidence coefficient for the confidence interval in part a will still be approximately 90 percent in spite of these departures from exact normality.

19.5 When regressing the annual contributions by company employees to the pension plan (Y) on years of job seniority (X), an analyst obtained computer output showing the 95 percent confidence interval for $E\{Y_i\}$ for each employee. The analyst expected to find that in about 95 percent of the cases, the employee contributions would be contained in the corresponding confidence intervals for $E\{Y_i\}$. Instead, the analyst discovered that this occurred in a substantially smaller percentage of cases. Explain the analyst's error in thinking.

19.6 For each of the following questions, explain whether a confidence interval for $E\{Y_h\}$ or a prediction interval for $Y_{h(new)}$ is appropriate.

a. What is the average statistics grade of students who receive 530 on the graduate admission test?

b. What will the Cramer family spend on restaurant meals next year if the family's income is $40,000?

c. What do companies with 500 full-time employees spend each year, on average, on group life insurance premiums?

19.7 For each of the following questions, explain whether a confidence interval for $E\{Y_h\}$ or a prediction interval for $Y_{h(\text{new})}$ is appropriate.

a. What will be the unemployment rate in this metropolitan area next quarter, given that the index of business activity will be 178.6?

b. What is the average score on the mathematics part of this admission test for applicants who score 600 on the verbal part?

c. How many hours of pain relief will Ms. Jones obtain from this medication when the administered dosage is 10 percent above the standard level?

* **19.8** Refer to **Fuel Efficiency** Problems 18.12 and 19.1. A 9th automobile, not included in the regression study, has a weight of 23 hundred pounds.

a. Construct a 95 percent prediction interval for the expressway gasoline mileage for this automobile. Interpret the prediction interval.

b. Why is the prediction interval in part a wider than the confidence interval in Problem 19.1?

19.9 Refer to **Marketing Experiment** Problem 18.13. Four showings of the TV commercial under the same experimental conditions are scheduled for an 11th sales territory that is similar to the territories used in the experiment.

a. Construct a 90 percent prediction interval for the sales in this 11th territory. Interpret the prediction interval.

b. If the error terms ε_i follow a highly skewed distribution here, would the confidence coefficient of the prediction interval in part a still be near 90 percent? Explain.

19.10 Refer to **Reconditioning Cost** Problem 18.14.

a. The reconditioning cost for an incinerator with 2.3 thousand prior operating hours is to be predicted. Obtain a 95 percent prediction interval. Interpret the prediction interval. Is the interval precise enough to be useful if a prediction with ± 10 percent precision is required? Discuss.

b. If the reconditioning cost for an incinerator with 4.6 thousand prior operating hours were to be predicted, would the width of the prediction interval be much wider than that in part a? Would there be other problems in this case?

19.11 Refer to **Bid Preparation** Problems 18.15 and 19.4.

a. Construct a 90 percent prediction interval for $Y_{h(\text{new})}$ when $X_h = 8$. Interpret the prediction interval.

b. Why is the prediction interval in part a wider than the confidence interval in Problem 19.4a?

c. Which interval—the prediction interval or the confidence interval—would be more useful for evaluating the actual number of hours required for bid preparation next week when eight bids are to be prepared? Explain.

* **19.12** Refer to **Fuel Efficiency** Problem 18.12. Obtain a 95 percent confidence interval for β_1. Interpret the confidence interval.

19.13 Refer to **Marketing Experiment** Problem 18.13.

a. Construct a 90 percent confidence interval for β_1.

b. The research director of the experiment has determined that to cover the cost of an additional showing of the TV commercial in a territory, the company must expect to sell more than 200 additional blankets in the territory. Obtain the lower 95 percent confidence interval for β_1 to determine whether this requirement is being met. What do you find?

19.14 Refer to **Reconditioning Cost** Problem 18.14.

a. Obtain a 95 percent confidence interval for the change in the mean response when the number of prior operating hours increases by (1) 1.0 thousand, (2) 0.5 thousand.

b. General experience with reconditioning incinerators shows that, for any given number of prior operating hours, the probability distribution of the reconditioning cost is slightly right skewed rather than exactly normal. Will this type of departure from regression model (19.1) cause the confidence level of either interval in part a to differ much from 95 percent? Explain.

19.15 Refer to **Bid Preparation** Problem 18.15.

a. Obtain a 90 percent confidence interval for β_1. Interpret the confidence interval.

b. Does the confidence interval in part a indicate that the means of the probability distributions of Y at different levels of X are not the same? Explain.

* **19.16** Refer to **Fuel Efficiency** Problem 18.12.

a. Test whether $\beta_1 = 0$. Use test statistic t^* and control the α risk at 0.05 when $\beta_1 = 0$. State the alternatives, the decision rule, the value of the test statistic, and the conclusion. What is the practical implication of the conclusion?

b. What is the P-value of the test? Is this value consistent with the test result in part a? Comment.

19.17 Refer to **Marketing Experiment** Problem 18.13.

a. Test whether $\beta_1 = 0$. Use test statistic t^* and control the α risk at 0.10 when $\beta_1 = 0$. State the alternatives, the decision rule, the value of the test statistic, and the conclusion.

b. What is the P-value of the test?

19.18 Refer to **Reconditioning Cost** Problem 18.14.

a. Test whether $\beta_1 = 0$. Use test statistic t^* and control the α risk at 0.05 when $\beta_1 = 0$. State the alternatives, the decision rule, the value of the test statistic, and the conclusion.

b. What is the P-value of the test?

c. An engineering cost study shows that the expected reconditioning cost should increase by $2.0 thousand for each additional 1 thousand prior operating hours. Test whether $\beta_1 = 2.0$, controlling the α risk at 0.05 when $\beta_1 = 2.0$. State the alternatives, the decision rule, the value of the test statistic, and the conclusion.

d. Could each of the tests in parts a and c be conducted by using a 95 percent confidence interval for β_1? Explain.

19.19 Refer to **Bid Preparation** Problem 18.15.

a. Test whether $\beta_1 = 0$. Use test statistic t^* and control the α risk at 0.10 when $\beta_1 = 0$. State the alternatives, the decision rule, the value of the test statistic, and the conclusion. What does the conclusion imply about a relationship between X and Y here? Comment.

b. What is the P-value of the test?

* **19.20** Refer to **Fuel Efficiency** Problem 18.12.

a. Construct a 95 percent confidence interval for β_0.

b. Does β_0 have an operational interpretation here? Comment.

19.21 Refer to **Marketing Experiment** Problem 18.13.

a. Construct a 90 percent confidence interval for β_0. Interpret the confidence interval.

b. Test whether $\beta_0 = 0$, controlling the α risk at 0.10 when $\beta_0 = 0$. State the alternatives, the decision rule, the value of the test statistic, and the conclusion. Is the test conclusion consistent with the confidence interval in part a? Explain.

19.22 Refer to **Reconditioning Cost** Problem 18.14.

a. Test whether $\beta_0 = 0$, controlling the α risk at 0.05 when $\beta_0 = 0$. State the alternatives, the decision rule, the value of the test statistic, and the conclusion.

b. What is the P-value of the test?

c. Does β_0 have an operational interpretation here? Comment.

19.23 Refer to **Bid Preparation** Problem 18.15.

a. Construct a 90 percent confidence interval for β_0.

b. Does it appear from the confidence interval in part a that β_0 might be 0? Explain.

c. What operational interpretation might β_0 have here?

* **19.24** Refer to **Fuel Efficiency** Problems 18.12 and 18.24.

a. Conduct an F test of whether $\beta_1 = 0$, controlling the α risk at 0.05 when $\beta_1 = 0$. State the alternatives, the decision rule, the value of the test statistic, and the conclusion.

b. What is the P-value of the test?

c. Confirm numerically that the value of test statistic F^* in part a is the square of test statistic t^* calculated from (19.17).

19.25   Refer to **Marketing Experiment** Problems 18.13 and 18.25.

a. Conduct an F test of whether $\beta_1 = 0$, controlling the α risk at 0.01 when $\beta_1 = 0$. State the alternatives, the decision rule, the value of the test statistic, and the conclusion.

b. What is the P-value of the test? Is this value consistent with the test result in part a? Comment.

19.26   Refer to **Reconditioning Cost** Problems 18.14 and 18.26.

a. Conduct an F test of whether expected reconditioning cost varies with number of prior operating hours. Control the α risk at 0.05. State the alternatives, the decision rule, the value of the test statistic, and the conclusion.

b. Confirm numerically that the value of test statistic F^* in part a is the square of test statistic t^* calculated from (19.17).

19.27   Refer to **Bid Preparation** Problems 18.15 and 18.27.

a. Conduct an F test of whether $\beta_1 = 0$, controlling the α risk at 0.01 when $\beta_1 = 0$. State the alternatives, the decision rule, the value of the test statistic, and the conclusion.

b. What is the P-value of the test? Is this value consistent with the test result in part a?

* **19.28**   Refer to **Fuel Efficiency** Problem 18.12.

a. Obtain the residuals and the fitted values.

b. Plot the residuals against the fitted values. Does the plot indicate that regression model (19.1) is appropriate here with regard to (1) linearity of the regression function, (2) constancy of the error variance? Would a larger number of cases in this regression study increase your confidence in the conclusions? Comment.

c. Construct a normal probability plot of the residuals. Does the plot support the assumption of normality of the error terms? Explain.

19.29   Refer to **Marketing Experiment** Problem 18.13.

a. Obtain the residuals and the fitted values.

b. Plot the residuals against the fitted values. Does the plot indicate that regression model (19.1) is appropriate here with regard to (1) linearity of the regression function, (2) constancy of the error variance? Comment.

c. Construct a normal probability plot of the residuals. Does the plot support the assumption of normality of the error terms? Explain.

19.30   Refer to **Reconditioning Cost** Problem 18.14.

a. Obtain the residuals and the fitted values.

b. Plot the residuals against the fitted values. Does the plot indicate that regression model (19.1) is appropriate here with regard to (1) linearity of the regression function, (2) constancy of the error variance? Would a larger number of cases in this regression study increase your confidence in the conclusions? Comment.

c. Construct a normal probability plot of the residuals. Also construct a stem-and-leaf plot of the residuals. Do these plots support the assumption of normality of the error terms? Explain.

19.31   Refer to **Bid Preparation** Problem 18.15.

a. Obtain the residuals and the fitted values.

b. Plot the residuals against the fitted values. Does the plot indicate that regression model (19.1) is appropriate here with regard to (1) linearity of the regression function, (2) constancy of the error variance? Comment.

c. The bids were prepared in the time order given. Plot the residuals against time order. What does the plot indicate about the appropriateness of regression model (19.1)?

d. Construct a normal probability plot of the residuals. Does the plot support the assumption of normality of the error terms? Explain.

19.32 Task Difficulty. A business psychologist created 16 tasks of varying difficulty, ranging from a difficulty rating of 0 (trivial) to one of 100 (extremely complex). The psychologist then organized 16 decision-making groups of similar composition and ability and randomly assigned one group to each task. The performance of the group in completing its task was rated on a scale from 0 (very low) to 100 (very high). As part of the data analysis, the psychologist regressed group performance rating (Y) on task difficulty (X) using regression model (19.1) and obtained the following fitted values and residuals:

$\hat{Y}_i$	e_i	$\hat{Y}_i$	e_i	$\hat{Y}_i$	e_i	$\hat{Y}_i$	e_i
67.4	3.6	74.5	0.5	77.1	−9.1	71.4	9.6
71.7	13.3	75.0	−10.0	70.0	13.0	67.6	7.4
64.3	−12.3	69.1	10.9	72.4	0.6	65.2	−17.2
65.5	−0.5	66.7	−3.7	72.2	6.8	75.7	−12.7

a. Plot the residuals against the fitted values. Explain why a linear regression function does not appear to be appropriate here.
b. Given the conclusion in part a, is it meaningful (1) to examine the plot in part a for constancy of the error variance? (2) To construct a normal probability plot of the residuals? Explain.

19.33 Gear Wear. A small gear that connects two large gears is made of relatively soft metal to minimize wear on the large gears. An analyst wished to study the relation between number of operating hours of the small gear accumulated to date (X) and number of remaining operating hours before replacement is required (Y). The following data (in hundreds of operating hours) were obtained for a random sample of 10 gears:

Gear i:	1	2	3	4	5	6	7	8	9	10
Accumulated Hours X_i:	38.7	5.0	9.5	15.4	30.2	17.1	40.1	14.1	25.0	42.4
Remaining Hours Y_i:	7.5	33.7	19.6	15.0	10.3	12.9	7.3	15.0	11.6	7.0

a. Construct a scatter plot of the data. Does regression model (19.1) appear to be appropriate for studying the relation between the variables here? Explain.
b. After further investigation, the analyst decided to employ the reciprocal transformation $X' = 1/X$. Construct a scatter plot of Y versus X'. Does this transformation appear to be successful in linearizing the relation between the variables here? Discuss.

19.34 Tube Extrusion. A manufacturing engineer investigated the relation between extrusion speed of plastic tubing (X) and number of blemishes in 500 feet of tubing (Y) for a new extrusion machine. Twelve trials were conducted with the machine at extrusion speeds ranging from 0.2 foot per second to 1.2 feet per second. In each trial, 500 feet of tubing were produced. The results were as follows:

Trial i:	1	2	3	4	5	6	7	8	9	10	11	12
Speed X_i:	0.2	0.2	0.4	0.4	0.6	0.6	0.8	0.8	1.0	1.0	1.2	1.2
Blemishes Y_i:	3	0	6	3	12	7	19	15	24	29	39	34

a. Construct a scatter plot of the data. Does regression model (19.1) appear to be appropriate for studying the relation between the variables here? Explain.
b. After further analysis, the engineer decided to employ the square root transformation $Y' = \sqrt{Y}$. Construct a scatter plot of Y' versus X. Does this transformation appear to be successful in linearizing the relation between the variables here? Discuss.

19.35 Refer to **Fuel Efficiency** Problem 18.12. Would it be appropriate to use the regression results in this study to estimate the mean expressway gasoline mileage of a new type of lightweight automobile weighing 10 hundred pounds? Explain.

19.36 Refer to **Marketing Experiment** Problem 18.13. A marketing executive has proposed that the regression results for the experiment be used to predict sales of the electric blanket in a territory subjected to "saturation" TV exposure. The executive's plan calls for 10 showings of the commercial in the territory. Should the executive place much reliance on the prediction obtained from the estimated regression function? Explain.

19.37 Gas Relation. A scientist is studying the relation between the volume occupied by a gas (Y, in milliliters) and temperature (X, in $°K$) while pressure is kept constant. The results of eight trials follow:

Trial i:	1	2	3	4	5	6	7	8
Temperature X_i:	200	250	300	350	200	250	300	350
Volume Y_i:	251	315	374	440	241	302	362	423

The scientist now intends to regress gas volume on gas temperature using regression model (19.1). Trials 1–4 were done by laboratory technician 1 and trials 5–8 by laboratory technician 2. Construct a symbolic scatter plot of the data, using technician (1, 2) as a plotting symbol. Should the scientist reconsider the intended use of regression model (19.1)? Explain.

19.38 An analyst has found that a statistical relation holds between thickness of silver film deposited in a process and frequency shift of a crystal used as a sensor. The analyst proposes to investigate whether regression model (19.1) can be used to estimate film thickness (taken as the dependent variable) from knowledge of the magnitude of the frequency shift (taken as the independent variable). An observer states that regression model (19.1) cannot be used in this manner because the frequency shift actually depends on the film thickness, not vice versa. Comment.

19.39 Surgical Expenditures. In a study of 10 hospitals in a buying group, the following data were gathered on the dollar expenditures per bed for a surgical item used in a hospital last year (Y) and the number of beds in the hospital at midyear (X). Regression model (19.1) is to be used to investigate the relation, if any, between the expenditures per bed and the number of beds, based on these data:

Hospital i:	1	2	3	4	5	6	7	8	9	10
Beds X_i:	196	133	678	354	216	73	454	103	247	280
Expenditures Y_i:	329	360	244	356	349	320	339	351	323	359

a. Construct a scatter plot of the expenditures per bed and the number of beds. Does hospital 3 appear to be a potentially influential observation in a regression analysis of expenditures per bed on number of beds for the 10 hospitals? Comment.

b. Obtain the estimated regression function when regressing expenditures per bed on number of beds for (1) all 10 hospitals, (2) all hospitals excluding hospital 3. Plot the two estimated regression lines on the graph prepared in part a. Describe the influence that the presence of hospital 3 has on the fitted regression line.

c. Further investigation revealed that all of the hospitals in the buying group are general hospitals except hospital 3, which is a neurological hospital with relatively fewer surgical cases per bed than the other hospitals. Might this fact explain why hospital 3 does not follow the regression pattern for the other nine hospitals? Would this fact justify dropping hospital 3 from the regression study? Discuss.

19.40 In a study of the performance of 15 firms last year, a business researcher regressed the chief executive's total pay (salary and other benefits) on the firm's net income. The researcher expected that firm 12 might not follow the pattern of the other firms because the firm was family owned and operated while the others were not. The researcher therefore also regressed pay on net income for the 14 firms excluding firm 12 and constructed a 95 percent prediction interval from this regression analysis for the chief executive's total pay in firm 12. It was found that the pay of the chief executive of firm 12 fell well outside the prediction interval. Does this finding confirm that firm 12 does not follow the regression pattern between total pay and net income for the other firms? Should the researcher remove firm 12 from the data set for the study? Discuss.

19.41 Refer to **Spouses' Ages** Problem 18.4. Consider a regression of husband's age on wife's age. Is it more appropriate to view the ages of the wives in this study as outcomes of a random variable or as fixed values? Does such a distinction matter? Comment.

EXERCISES

19.42 Refer to formula (19.8).

a. Do both components of $\sigma^2\{Y_{h(\text{new})}\}$ become smaller as the sample size increases? Explain.

b. What does the answer in part a imply about our ability to make $\sigma^2\{Y_{h(\text{new})}\}$ small by increasing n?

19.43 Derive formulas (19.19a) and (19.19b) from theorem (19.3) by letting $X_h = 0$.

19.44 A student asks why the F test for deciding between $\beta_1 = 0$ and $\beta_1 \neq 0$ is one-sided, even though the latter alternative implies that either $\beta_1 < 0$ or $\beta_1 > 0$. Answer the student, making specific reference to formulas (19.20) and (19.21).

19.45 In an application of regression model (19.1), $\sigma^2 = 100$ and $\Sigma(X_i - \bar{X})^2 = 50$. Obtain $E\{MSE\}$ and $E\{MSR\}$ when (1) $\beta_1 = 0$, (2) $\beta_1 = 10$, (3) $\beta_1 = -10$, (4) $\beta_1 = 50$. What do the results imply about the type of decision rule appropriate for the F test of a regression relation?

19.46 Prove (19.21) by using (18.27a). [*Hint:* Obtain $E\{b_1^2\}$ by using (5.7a).]

19.47 Show that the F^* test statistic in (19.22) is equal to the square of the t^* test statistic in (19.17), using (18.27a).

STUDIES

19.48 Refer to **Gear Wear** Problem 19.33.

a. Obtain the estimated regression function based on the transformed independent variable $X' = 1/X$. Also obtain the residuals and the fitted values.

b. If the distributions of Y are normal with constant variance σ^2 before the transformation $X' = 1/X$ is applied, will they continue to be so after the transformation? Explain.

c. The analyst was asked why the logarithmic transformation $Y' = \log Y$ was not used to linearize the relation between X and Y here. Regress Y' on X, and state the estimated regression function in the original Y units. Also obtain the residuals and the fitted values in the original Y units.

d. In separate graphs, plot the residuals obtained in parts a and c against their respective fitted values in the original units. Also prepare separate normal probability plots of the residuals in the original Y units. Does the logarithmic transformation of Y appear to be better than the analyst's reciprocal transformation of X? Explain.

19.49 Refer to the **Financial Characteristics** data set (Appendix D.1). Consider only the data on net income and net assets in year 2 for firms in the paper industry (industry 4). Assume that regression model (19.1) applies.

a. Regress net income for year 2 on net assets for year 2.

b. Construct (1) a 95 percent confidence interval for β_1, (2) a 95 percent confidence interval for $E\{Y_h\}$ when X_h is equal to the mean net assets for year 2 of all 24 paper firms, (3) a 95 percent prediction interval for $Y_{h(\text{new})}$ when X_h again is equal to the mean net assets for year 2 of all 24 paper firms. Interpret each interval. Is the prediction interval precise enough to be useful? Comment.

c. Obtain the residuals and the fitted values. Plot the residuals against the fitted values. Does the plot indicate that regression model (19.1) is appropriate here with regard to (1) linearity of the regression function, (2) constancy of the error variance? Comment.

d. Construct a normal probability plot of the residuals. Also construct a stem-and-leaf display of the residuals. Do the plots support the assumption of normality of the error terms? Explain.

Multiple Regression

20

In many situations, two or more independent variables are needed in a regression model to provide an adequate description of the process under study or to yield sufficiently precise inferences.

1. A regression model to control the diameter of plastic pellets produced by an extrusion process uses as independent variables the initial temperature of the plastic, the die temperature, and the extrusion rate.

2. A regression model for predicting the demands for a firm's product in each of its 25 sales territories uses as independent variables one socioeconomic variable (mean household income), two demographic variables (average family size and percentage of population over 65 years of age), and one environmental variable (mean daily temperature).

Regression models containing two or more independent variables are called *multiple regression models*. In this chapter, we extend the procedures for simple linear regression to multiple regression and also consider certain special topics that are of importance when multiple regression models are used.

20.1 MULTIPLE REGRESSION MODELS

Model in Two Independent Variables

The simple linear regression model (19.1) can be extended directly to include two independent variables, X_1 and X_2.

$$(20.1) \qquad Y_i = \beta_0 + \beta_1 X_{i1} + \beta_2 X_{i2} + \varepsilon_i \qquad i = 1, 2, \ldots, n$$

where: Y_i is the response in the ith case

X_{i1} and X_{i2} are the values of the two independent variables in the ith case, assumed to be known constants

β_0, β_1, and β_2 are parameters

ε_i are independent $N(0, \sigma^2)$

As for the simple linear regression model (19.1), we are assuming here that the error terms ε_i are independent normal random variables, with mean zero and constant variance σ^2.

Regression Function.   The regression function for model (20.1) is as follows.

$$(20.2) \qquad E\{Y\} = \beta_0 + \beta_1 X_1 + \beta_2 X_2$$

This regression function is frequently called the *response function* or the *response surface.*
The response surface in (20.2) is a plane, as illustrated in Figure 20.1.

The parameters of a multiple regression model are interpreted analogously to those in
the simple linear case. Thus, in response function (20.2):

1. β_0 is the Y intercept of the plane; it is the mean of the probability distribution of Y
   when $X_1 = 0$ and $X_2 = 0$.
2. β_1 indicates the change in the mean response $E\{Y\}$ when X_1 increases by one unit while
   X_2 remains constant.
3. β_2 indicates the change in the mean response $E\{Y\}$ when X_2 increases by one unit while
   X_1 remains constant.

☐ **EXAMPLE**

Figure 20.1 shows a response plane for regression model (20.1) where β_0 and β_1 are
positive and β_2 is negative. Just as the regression line for the simple linear regression
model (19.1) gives the mean of the distribution of Y for any level of X, the response
plane in Figure 20.1 gives the mean of the distribution of Y for any combination of
(X_1, X_2) values. Shown in Figure 20.1 is a response Y_i when $X_1 = X_{i1}$ and $X_2 = X_{i2}$.
The mean of this probability distribution, $E\{Y_i\}$, is the point on the regression plane at
(X_{i1}, X_{i2}), as indicated in the figure. The error term ε_i for this observation is represented
by the vertical distance between Y_i and $E\{Y_i\}$ on the response plane, since $\varepsilon_i = Y_i -
E\{Y_i\}$ as for simple linear regression.   ☐

FIGURE 20.1

Example of a response plane. β_0 and β_1 are positive and β_2 is negative here. Point $E\{Y_i\}$ on the
plane is the mean of the probability distribution corresponding to observation Y_i at coordinates
(X_{i1}, X_{i2}). The error term is $\varepsilon_i = Y_i - E\{Y_i\}$.

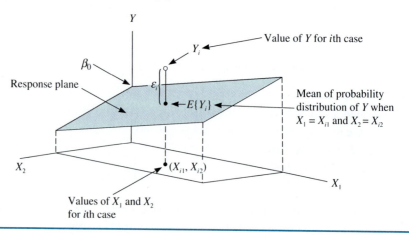

Model in $p - 1$ Independent Variables

The regression model (20.1) with two independent variables can be extended to include $p - 1$ independent variables $X_1, X_2, \ldots, X_{p-1}$.

> **(20.3)** $\quad Y_i = \beta_0 + \beta_1 X_{i1} + \beta_2 X_{i2} + \cdots + \beta_{p-1} X_{i,p-1} + \varepsilon_i \qquad i = 1, 2, \ldots, n$
>
> where: $\quad Y_i$ is the response in the ith case
> $\qquad X_{i1}, X_{i2}, \ldots, X_{i,p-1}$ are the values of the $p - 1$ independent variables in the ith case, assumed to be known constants
> $\qquad \beta_0, \beta_1, \ldots, \beta_{p-1}$ are parameters
> $\qquad \varepsilon_i$ are independent $N(0, \sigma^2)$

Note that regression model (20.3) includes regression models (19.1) and (20.1) as special cases. It becomes the simple linear regression model (19.1) when $p - 1 = 1$; and when $p - 1 = 2$, it becomes regression model (20.1), the multiple regression model with two independent variables.

The parameters in regression model (20.3) are interpreted in the standard manner.

1. β_0, the Y intercept, indicates the mean of the distribution of Y when $X_1 = X_2 = \cdots = X_{p-1} = 0$.
2. β_k ($k = 1, 2, \ldots, p - 1$) indicates the change in the mean response $E\{Y\}$ when X_k increases by one unit while all the other independent variables remain constant.
3. σ^2 is the common variance of the distributions of Y.

Most basic results for multiple regression are direct extensions of those already discussed for simple linear regression. We explain them with an illustration involving two independent variables.

SOME BASIC RESULTS FOR MULTIPLE REGRESSION MODELS 20.2

EXAMPLE ☐ Promotional Expenditures

An experiment was designed by a market researcher to study the effects of two types of promotional expenditures on sales of a line of food products in supermarkets. Sixteen localities were selected for the test. They were similar in market potential and were representative of the target market for this line of products. Sixteen combinations of media advertising expenditures levels (X_1) and point-of-sale expenditures levels (X_2) were specified for the study, and the localities were assigned at random to one of these (X_1, X_2) combinations. Table 20.1 shows the two promotional expenditures levels for each locality, together with the sales volume (Y) during the test period.

Figure 20.2 contains the MINITAB output for a regression analysis of the data in Table 20.1, using regression model (20.1). We have numbered the blocks of information in Figure 20.2 for ready identification. The format is similar to that in Figure 18.12

TABLE 20.1

Data on sales and promotional expenditures—Promotional expenditures example

Locality i	Media Expenditures ($ thousand) X_{i1}	Point-of-Sale Expenditures ($ thousand) X_{i2}	Sales Volume ($ ten thousand) Y_i
1	2	2	8.74
2	2	3	10.53
3	2	4	10.99
4	2	5	11.97
5	3	2	12.74
6	3	3	12.83
7	3	4	14.69
8	3	5	15.30
9	4	2	16.11
10	4	3	16.31
11	4	4	16.46
12	4	5	17.69
13	5	2	19.65
14	5	3	18.86
15	5	4	19.93
16	5	5	20.51

for simple linear regression. For instance, block 1 gives the mean and standard deviation for each of the variables X_1, X_2, and Y.

Correlation Matrix

The interpretation of multiple regression results often requires information about the coefficients of correlation between pairs of variables in the study. We shall denote the coefficient of simple correlation between Y and X_1, as defined in (18.38), by r_{Y1}. Similarly, r_{Y2} denotes the coefficient of simple correlation between Y and X_2. Finally, r_{12} denotes the coefficient of simple correlation between X_1 and X_2, measuring the degree of linear association between the independent variables X_1 and X_2. The simple correlation coefficients in multiple regression are frequently presented in the form of a *correlation matrix*.

☐ **EXAMPLE**

The simple correlation coefficients for the promotional expenditures example are given in block 2 of Figure 20.2 in the form of a correlation matrix. The format and the values of the coefficients for the study are as follows:

$$
\begin{array}{c}
\quad X_1 \quad X_2 \quad Y \\
\begin{array}{c} X_1 \\ X_2 \\ Y \end{array}
\begin{bmatrix}
1 & r_{12} & r_{Y1} \\
r_{12} & 1 & r_{Y2} \\
r_{Y1} & r_{Y2} & 1
\end{bmatrix}
=
\begin{bmatrix}
1 & 0 & 0.965 \\
0 & 1 & 0.225 \\
0.965 & 0.225 & 1
\end{bmatrix}
\end{array}
$$

The market researcher noted with interest that the coefficient of simple correlation between sales volume (Y) and media expenditures (X_1) is substantially greater than that between sales volume (Y) and point-of-sale expenditures (X_2)—namely, that

FIGURE 20.2
MINITAB output for multiple regression—Promotional expenditures example

1.

	N	MEAN	STDEV
X1	16	3.500	1.155
X2	16	3.500	1.155
Y	16	15.207	3.626

2.

```
1.0000000   0.0000000   0.9645865
0.0000000   1.0000000   0.2247282
0.9645865   0.2247282   1.0000000
```

3.

The regression equation is
Y = 2.13 + 3.03 X1 + 0.706 X2

Predictor	Coef	Stdev	t-ratio	p
Constant	2.1344	0.6104	3.50	0.004
X1	3.0292	0.1203	25.18	0.000
X2	0.7058	0.1203	5.87	0.000

s = 0.5379   R-sq = 98.1%   R-sq(adj) = 97.8%

4.

Analysis of Variance

SOURCE	DF	SS	MS	F	p
Regression	2	193.488	96.744	334.35	0.000
Error	13	3.762	0.289		
Total	15	197.250			

5.

Fit	Stdev.Fit	95% C.I.	95% P.I.
18.692	0.288	(18.069, 19.315)	(17.373, 20.011)

$r_{Y1} = 0.965$ while $r_{Y2} = 0.225$. Also, it was known in advance that the coefficient of simple correlation between X_1 and X_2 would equal zero—that is, $r_{12} = 0$—because this feature was designed into the experiment by setting up the (X_1, X_2) combinations in the manner shown in Table 20.1. We will discuss the reason for this design later. □

Estimated Regression Function

For regression model (20.1), the estimated regression function is as follows.

$$(20.4) \quad \hat{Y} = b_0 + b_1 X_1 + b_2 X_2$$

Here, b_0, b_1, and b_2 are the least squares estimators of the parameters β_0, β_1, and β_2.
The method of least squares leads to a system of normal equations for obtaining b_0, b_1, and b_2, as follows.

(20.5) $\quad \sum Y_i = nb_0 + b_1\sum X_{i1} + b_2\sum X_{i2}$

$\quad\quad\quad \sum X_{i1}Y_i = b_0\sum X_{i1} + b_1\sum X_{i1}^2 + b_2\sum X_{i1}X_{i2}$

$\quad\quad\quad \sum X_{i2}Y_i = b_0\sum X_{i2} + b_1\sum X_{i1}X_{i2} + b_2\sum X_{i2}^2$

The least squares estimates are obtained by solving this system of equations simultaneously. Statistical regression packages give these estimates as part of their output.

☐ **EXAMPLE**

Block 3 in Figure 20.2 presents the estimated regression function and information on the estimated regression coefficients for the promotional expenditures example. The estimated regression function is seen to be:

$$\hat{Y} = 2.1344 + 3.0292X_1 + 0.7058X_2$$

The format for the information about the estimated regression coefficients is an extension of that in Figure 18.12 for simple linear regression. As before, the printout provides for each independent variable and the constant term the estimated regression coefficient b_k, the estimated standard deviation of the regression coefficient $s\{b_k\}$, the test statistic $t^* = b_k/s\{b_k\}$ for testing whether the regression coefficient equals zero, and the two-sided P-value of the test.

☐

Analysis of Variance

Sums of Squares.　The sums of squares for the analysis of variance are defined identically in simple and multiple regression. For convenience, we repeat the definitional formulas for the total sum of squares SSTO, the error sum of squares SSE, and the regression sum of squares SSR.

(20.6) $\quad SSTO = \sum(Y_i - \bar{Y})^2$

(20.7) $\quad SSE = \sum(Y_i - \hat{Y}_i)^2 = \sum e_i^2$

(20.8) $\quad SSR = \sum(\hat{Y}_i - \bar{Y})^2$

Degrees of Freedom.　As usual, the total sum of squares SSTO has $n - 1$ degrees of freedom associated with it.

The error sum of squares SSE has $n - p$ degrees of freedom associated with it. There are n residuals $e_i = Y_i - \hat{Y}_i$ that enter into SSE. However, the residuals are subject to p independent constraints that arise because p parameters $\beta_0, \beta_1, \ldots, \beta_{p-1}$ need to be estimated to obtain the fitted values $\hat{Y}_i$.

SSR has $p - 1$ degrees of freedom associated with it. Although there are n deviations $\hat{Y}_i - \bar{Y}$ in SSR, all of the fitted values $\hat{Y}_i$ are calculated from the same regression function. Consequently, the number of degrees of freedom associated with the n fitted values is p, which is related to the p parameters $\beta_0, \beta_1, \ldots, \beta_{p-1}$ in the regression function. These p degrees of freedom are reduced by one because the deviations $\hat{Y}_i - \bar{Y}$ must meet the constraint $\Sigma(\hat{Y}_i - \bar{Y}) = 0$.

TABLE 20.2
General format of ANOVA
table for multiple regression

Source of Variation	SS	df	MS
Regression	$SSR = \Sigma(\hat{Y}_i - \bar{Y})^2$	$p - 1$	$MSR = \dfrac{SSR}{p-1}$
Error	$SSE = \Sigma(Y_i - \hat{Y}_i)^2$	$n - p$	$MSE = \dfrac{SSE}{n-p}$
Total	$SSTO = \Sigma(Y_i - \bar{Y})^2$	$n - 1$	

We thus see that the degrees of freedom are additive; that is, $n - 1 = (p - 1) + (n - p)$.

Note that for the simple linear regression case, where there are $p = 2$ parameters in the regression function, the degrees of freedom for SSE are $n - p = n - 2$ and those for SSR are $p - 1 = 1$. These, of course, are the degrees of freedom stated in Table 18.3a for simple linear regression.

Table 20.2 shows the general form of the ANOVA table for multiple regression. The mean squares MSR and MSE are defined as for simple linear regression—that is, as a sum of squares divided by the associated degrees of freedom. Block 4 in Figure 20.2 contains the ANOVA table for the promotional expenditures example.

Point Estimation of σ^2.   As in simple linear regression, the error mean square MSE is an unbiased point estimator of the error term variance σ^2. Block 4 of Figure 20.2 shows that $MSE = 0.289$ for the promotional expenditures example.

The corresponding point estimate of σ is $\sqrt{MSE} = \sqrt{0.289} = 0.538$, labeled s in block 3 of the MINITAB output.

Coefficient of Multiple Determination

The coefficient of multiple determination, denoted by R^2, is defined analogously to the coefficient of simple determination r^2.

(20.9)
The **coefficient of multiple determination R^2** is defined:

$$R^2 = \frac{SSR}{SSTO} = 1 - \frac{SSE}{SSTO}$$

Thus, R^2 measures the proportionate reduction in the total variability $SSTO$ associated with the use of all of the independent variables $X_1, X_2, \ldots, X_{p-1}$ in the multiple regression model. When $p - 1 = 1$, R^2 reduces to the coefficient of simple determination r^2. Also, as with r^2, R^2 can range from 0 to 1.

(20.10) $0 \leq R^2 \leq 1$

It can be shown that $R^2 = 0$ when $b_1 = b_2 = \cdots = b_{p-1} = 0$, and that $R^2 = 1$ when all of the Y_i observations fall directly on the estimated regression surface.

☐ **EXAMPLE**

For the promotional expenditures example, we find from the ANOVA table in Figure 20.2 that $SSR = 193.488$ and $SSTO = 197.250$. Hence:

$$R^2 = \frac{193.488}{197.250} = 0.981$$

The value of R^2 is shown as a percentage in the MINITAB output in block 3 of Figure 20.2, labeled R-sq. This value, 98.1 percent, indicates that the variability of the sales volumes for the 16 localities is reduced by 98.1 percent when media and point-of-sale promotional expenditures are considered in the regression model.   ☐

Comment

It can be shown that the coefficient of multiple determination R^2 is identical to the coefficient of simple determination obtained when the observations Y_i are regressed on the fitted values $\hat{Y}_i$ from the multiple regression model.

Coefficient of Multiple Correlation.   The positive square root of R^2 is called the coefficient of multiple correlation.

(20.11)

The **coefficient of multiple correlation R** is defined:

$$R = +\sqrt{R^2}$$

where R^2 is given by (20.9).

☐ **EXAMPLE**

For the promotional expenditures example, we have:

$$R = \sqrt{R^2} = \sqrt{0.981} = 0.990$$

  ☐

Adjusted R^2.   The coefficient of multiple determination defined in (20.9) is sometimes modified to recognize the number of independent variables in the regression model, because R^2 can generally be made larger when additional independent variables are added to the model. To see this, note that SSE tends to become smaller (it cannot increase) with each additional independent variable, while $SSTO$ remains fixed. A measure that recognizes the number of independent variables in the regression model is called the *adjusted coefficient of multiple determination* and is denoted by R_a^2.

$$(20.12) \quad R_a^2 = 1 - \left(\frac{n-1}{n-p}\right)\left(\frac{SSE}{SSTO}\right) = 1 - \frac{MSE}{\left(\frac{SSTO}{n-1}\right)}$$

When an independent variable is added to the regression model, p increases by 1 and $n - p$ decreases correspondingly. Therefore, R_a^2 can become smaller if the decrease in $n - p$ is not offset by a sufficient decrease in SSE.

EXAMPLE ☐

For the promotional expenditures example, we have:

$$R_a^2 = 1 - \left(\frac{16-1}{16-3}\right)\left(\frac{3.762}{197.250}\right) = 0.978$$

Here, the adjustment has only a small effect; note that R^2 and R_a^2 are almost equal.

The adjusted coefficient of multiple determination is shown as a percentage in the MINITAB output in Figure 20.2, block 3, labeled R-sq(adj). ☐

Coefficient of Partial Determination

The coefficient of multiple determination R^2 is a measure of the combined effects of all independent variables $X_1, X_2, \ldots, X_{p-1}$ in the regression model in reducing the total variability $SSTO$. When considering whether or not to add another independent variable to the regression model, a different measure is needed. Such a measure is the coefficient of partial determination, which shows the marginal effect of a single independent variable in further reducing the variability associated with the regression model.

EXAMPLE ☐

In the promotional expenditures example in Table 20.1, a regression of sales volume Y on point-of-sale expenditures X_2 alone yields the estimated regression function:

$$\hat{Y} = 12.73675 + 0.70575X_2$$

The error sum of squares for this simple regression model, which we shall denote by $SSE(X_2)$, is $SSE(X_2) = 187.289$. The question of concern is whether it is worth adding the other independent variable X_1, media expenditures, to the regression model. We see from block 4 of Figure 20.2 that the error sum of squares when both X_1 and X_2 are in the regression model, to be denoted now by $SSE(X_1, X_2)$, is $SSE(X_1, X_2) = 3.762$. Thus, the residual variability in sales volume when X_2 alone is in the regression model, 187.289, is reduced to 3.762 by adding X_1 to the regression model. The proportionate reduction is 98.0 percent:

$$\frac{SSE(X_2) - SSE(X_1, X_2)}{SSE(X_2)} = \frac{187.289 - 3.762}{187.289} = 0.980$$

This measure indicates the marginal effect of X_1 in reducing the variability in Y when X_2 is already in the model; it is called a coefficient of partial determination.   □

(20.13)

For regression model (20.1) with two independent variables, the **coefficient of partial determination** between Y and X_1, given that X_2 is already in the model, is denoted by $r^2_{Y1.2}$ and is defined:

$$r^2_{Y1.2} = \frac{SSE(X_2) - SSE(X_1, X_2)}{SSE(X_2)} = 1 - \frac{SSE(X_1, X_2)}{SSE(X_2)}$$

The coefficient of partial determination between Y and X_2, given that X_1 is already in the regression model, is denoted by $r^2_{Y2.1}$ and is defined in an analogous way to (20.13) by reversing the roles of X_1 and X_2.

Comment

Coefficients of partial determination may also be defined in cases where there are more than two independent variables. For instance, when X_4 enters a regression model already containing the variables X_1, X_2, and X_3, the coefficient of partial determination is given by:

$$r^2_{Y4.123} = \frac{SSE(X_1, X_2, X_3) - SSE(X_1, X_2, X_3, X_4)}{SSE(X_1, X_2, X_3)}$$

Here, $r^2_{Y4.123}$ measures the relative reduction in the error sum of squares achieved by adding X_4 into the regression model that already contains X_1, X_2, and X_3.

Coefficient of Partial Correlation.   The square root of the coefficient of partial determination is called the *coefficient of partial correlation*. It is given the same sign as the corresponding regression coefficient in the fitted regression function containing all of the independent variables.

□ **EXAMPLE**

For the promotional expenditures example, $r_{Y1.2} = +\sqrt{0.980} = 0.990$. A plus sign is attached here because b_1 in the fitted regression function is positive ($b_1 = 3.0292$ in block 3 of Figure 20.2).   □

20.3   **RESIDUAL ANALYSIS**   Before using a fitted multiple regression model, we need to investigate the appropriateness of this model. Residuals are studied for this purpose in the same manner described earlier for simple linear regression. A residual plot against the fitted values is used to examine the appropriateness of the regression function and the constancy of the error variance. A plot of the residuals against their time order is useful for examining whether the error terms are correlated, and a normal probability plot is helpful for studying the normality of the error terms.

In addition, plots of the residuals against each of the independent variables $X_1, X_2, \ldots,$ X_{p-1}, one at a time, are valuable for identifying whether the effects of any of the independent variables differ from those postulated in the regression model. Also, plots of the residuals against selected new variables not already in the regression model help to discover whether one or more important variables have been omitted from the model.

EXAMPLE ☐

Figure 20.3 contains for the promotional expenditures example a plot of the residuals against the fitted values, as well as a normal probability plot of the residuals. The residuals are obtained in the usual fashion. For instance, we have for the first case $X_{11} = 2$, $X_{12} = 2$, and $Y_1 = 8.74$. The fitted value for this case is:

$$\hat{Y}_1 = 2.1344 + 3.0292(2) + 0.7058(2) = 9.604$$

and the residual is:

$$e_1 = 8.74 - 9.604 = -0.864$$

The plot of the residuals against the fitted values in Figure 20.3a shows a random scatter about the zero line, indicating that the fitted regression function is a reasonable model for the data. Also, the points fall within a horizontal band, in accord with a constant error term variance.

The normal probability plot in Figure 20.3b shows that the points form a reasonably linear pattern and are close to the line of identity. Hence, the assumption of normally distributed error terms in regression model (20.1) appears to be reasonable here.

After studying these and other residual plots, the market researcher concluded that none of the plots challenged the appropriateness of multiple regression model (20.1). ☐

Residual plot and normal probability plot—Promotional expenditures example

FIGURE 20.3

(a) Residual Plot Against $\hat{Y}$

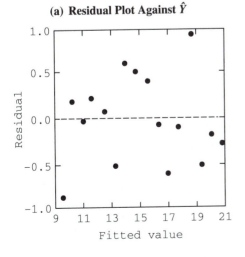

(b) Normal Probability Plot

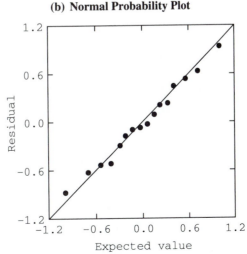

20.4   F TEST FOR REGRESSION RELATION

Frequently, the first step in the analysis of a fitted multiple regression model after its appropriateness is established is to test whether there is a relation between the dependent variable Y and the set of independent variables $X_1, X_2, \ldots, X_{p-1}$. The expectation is that such a relation exists, because the model is developed in the first place to exploit this relation.

The F test for a multiple regression relation is a direct extension of the F test in simple linear regression, where the test involves H_0: $\beta_1 = 0$. Let us first consider the case of two independent variables and regression model (20.1), for which the response function is $E\{Y\} = \beta_0 + \beta_1 X_1 + \beta_2 X_2$. We test here for a regression relation between $E\{Y\}$ and the independent variables X_1 and X_2 by testing H_0: $\beta_1 = \beta_2 = 0$. If β_1 and β_2 both equal zero, the response function reduces to $E\{Y\} = \beta_0$. Thus, the means of the distributions of Y are then the same for all (X_1, X_2) combinations, and there is no regression relation between Y and the independent variables X_1 and X_2. On the other hand, if β_1 and β_2 are not both equal to zero (i.e., if at least one of the two is nonzero), the means of Y are not constant for all (X_1, X_2) combinations, and a regression relation does exist between Y and the two independent variables as a group.

These ideas extend directly to regression model (20.3) with $p - 1$ independent variables. A test of whether Y is related to $X_1, X_2, \ldots, X_{p-1}$ involves the alternatives:

$$H_0: \beta_1 = \beta_2 = \cdots = \beta_{p-1} = 0$$
$$H_1: \text{Not all } \beta_k = 0 \qquad k = 1, 2, \ldots, p - 1$$

The test procedure is similar to that for the F test in simple linear regression. The test statistic is the same.

(20.14) $$F^* = \frac{MSR}{MSE}$$

The decision rule takes a corresponding form.

(20.15)
When the alternatives are:

$$H_0: \beta_1 = \beta_2 = \cdots = \beta_{p-1} = 0$$
$$H_1: \text{Not all } \beta_k = 0 \qquad k = 1, 2, \ldots, p - 1$$

and multiple regression model (20.3) is applicable, the appropriate decision rule to control the α risk is as follows:

If $F^* \leq F(1 - \alpha; p - 1, n - p)$, conclude H_0.
If $F^* > F(1 - \alpha; p - 1, n - p)$, conclude H_1.

Here F^* is given by (20.14).

EXAMPLE ☐

In the promotional expenditures example, the market researcher, after being satisfied from the residual analysis that regression model (20.1) is appropriate, wished next to test whether or not a regression relation between sales volume and the two types of promotional expenditures does indeed exist. If so, the researcher plans to analyze the multiple regression model in some detail.

Step 1. The test alternatives are:

$$H_0: \beta_1 = \beta_2 = 0$$

$$H_1: \text{Not both } \beta_1 = 0 \text{ and } \beta_2 = 0$$

Step 2. The researcher specified that the α risk be controlled at 0.05.

Step 3. For $\alpha = 0.05$, $n = 16$, $p - 1 = 2$, and $n - p = 13$, we require $F(0.95; 2, 13) = 3.81$. Hence, the decision rule for the test is as follows:

If $F^* \leq 3.81$, conclude H_0.
If $F^* > 3.81$, conclude H_1.

Step 4. From the MINITAB output in Figure 20.2 (block 4), we see that $MSR = 96.744$ and $MSE = 0.289$. Thus, $F^* = 96.744/0.289 = 334.75$. The MINITAB output in block 4 shows the test statistic to greater accuracy as $F^* = 334.35$. Because $F^* = 334.35 > 3.81$, we conclude H_1—that a regression relation exists between sales volume and the two kinds of promotional expenditures. The P-value of the test is also shown in block 4. Since P-value $= 0+$ is less than $\alpha = 0.05$, H_1 is the appropriate conclusion, as noted already.

☐

INFERENCES CONCERNING INDIVIDUAL REGRESSION COEFFICIENTS

20.5

Once it has been established that a regression relation exists between the dependent variable and the set of independent variables, estimation and testing of individual regression coefficients are often of interest.

Estimates and tests of individual regression coefficients β_k in the multiple regression model (20.3) are conducted in the same manner as in simple linear regression. The only difference is that the t multiple now involves $n - p$ degrees of freedom, because MSE in multiple regression has $n - p$ degrees of freedom associated with it. The confidence interval for any regression coefficient β_k takes the usual form.

(20.16)
The $1 - \alpha$ confidence limits for β_k for multiple regression model (20.3) are:

$$b_k \pm t s\{b_k\}$$

where: $t = t(1 - \alpha/2; n - p)$
$s\{b_k\}$ is the estimated standard deviation of b_k

Almost all statistical regression packages give $s\{b_k\}$, the estimated standard deviation of b_k, as part of their output.

A test of whether or not $\beta_k = 0$ is conducted in the manner explained for simple regression.

(20.17)

Tests concerning β_k for multiple regression model (20.3) when the alternatives are:

$$H_0: \ \beta_k = 0$$

$$H_1: \ \beta_k \neq 0$$

are based on the standardized test statistic:

$$t^* = \frac{b_k}{s\{b_k\}}$$

where $s\{b_k\}$ is the estimated standard deviation of b_k.

When $\beta_k = 0$, t^* follows the t distribution with $n - p$ degrees of freedom.

□ **EXAMPLES**

1. In the promotional expenditures example, the market researcher wished to estimate β_1 with a 95 percent confidence interval. From the MINITAB output in Figure 20.2 (block 3) for line X1, we obtain $b_1 = 3.0292$ and $s\{b_1\} = 0.1203$. Since $n - p = 16 - 3 = 13$, we require $t(0.975; 13) = 2.160$. The confidence limits by (20.16) are $3.0292 \pm 2.160(0.1203)$, and the desired confidence interval is:

$$2.77 \leq \beta_1 \leq 3.29$$

   The confidence interval indicates that with an increase of $1000 in media expenditures (with point-of-sale expenditures held fixed), expected sales increase by between $27,700 and $32,900. Recall from Table 20.1 that the sales volume variable is expressed in units of $ ten thousand.

2. In the promotional expenditures example, the market researcher also wished to test $H_0: \ \beta_2 = 0$ versus $H_1: \ \beta_2 \neq 0$, using an α risk of 0.05. Conclusion H_0 would imply that X_2 could be dropped from the regression model, which already contains X_1. Again, we require $t(0.975; 13) = 2.160$. Consulting the MINITAB output in Figure 20.2 (block 3) for line X2, we find $t^* = b_2/s\{b_2\} = 0.7058/0.1203 = 5.87$. Since $|t^*| = 5.87 > 2.160$, the researcher should conclude H_1—that $\beta_2 \neq 0$. This conclusion implies that there is a relation between sales volume (Y) and point-of-sale expenditures (X_2) when media expenditures (X_1) are already in the regression model. Thus, X_2 should not be dropped from the regression model.

   The two-sided P-value for the test is also shown in block 3. Since P-value $= 0+$ is much smaller than $\alpha = 0.05$, we could have directly ascertained that H_1 is the appropriate conclusion.　　　　　　　　　　　　　　　　　　　□

If the test for a regression coefficient leads to the conclusion that $\beta_k = 0$ for variable X_k, it does not necessarily follow that X_k is not related to Y. It simply means that, when the other independent variables are already in the regression model, the marginal contribution of X_k in further reducing the error sum of squares is negligible.

A key objective in many applications of multiple regression is to estimate a mean response, $E\{Y_h\}$. Our earlier discussion for simple linear regression extends readily to multiple regression.

CONFIDENCE 20.6 INTERVAL FOR E{Y_h}

We shall denote the levels of the independent variables for which the mean response is to be estimated by $X_{h1}, X_{h2}, \ldots, X_{h,p-1}$. Further, $E\{Y_h\}$ will denote the mean response when $X_1 = X_{h1}, X_2 = X_{h2}, \ldots, X_{p-1} = X_{h,p-1}$. The point estimator of $E\{Y_h\}$ is as follows.

(20.18) $\hat{Y}_h = b_0 + b_1 X_{h1} + b_2 X_{h2} + \cdots + b_{p-1} X_{h,p-1}$

The confidence interval for $E\{Y_h\}$ takes the usual form.

(20.19)
The $1 - \alpha$ confidence limits for the mean response $E\{Y_h\}$ for multiple regression model (20.3) are:

$$\hat{Y}_h \pm t s\{\hat{Y}_h\}$$

where: $t = t(1 - \alpha/2; n - p)$
$\hat{Y}_h$ is given by (20.18)
$s\{\hat{Y}_h\}$ is the estimated standard deviation of $\hat{Y}_h$

Many statistical regression packages permit the user to specify the levels of the independent variables of interest and then will calculate $\hat{Y}_h$ and $s\{\hat{Y}_h\}$.

EXAMPLE ☐

In the promotional expenditures example, the market researcher wished to estimate expected sales when media expenditures are $X_{h1} = \$5$ thousand and point-of-sale expenditures are $X_{h2} = \$2$ thousand. A 95 percent confidence interval is to be employed.

Using the results in Figure 20.2 (block 3), we obtain the point estimate of the mean sales volume in localities where $X_{h1} = 5$ and $X_{h2} = 2$:

$$\hat{Y}_h = 2.1344 + 3.0292(5) + 0.7058(2) = 18.692$$

This point estimate, $\hat{Y}_h = 18.692$, is shown in block 5 of the MINITAB output in Figure 20.2, together with the estimated standard deviation of $\hat{Y}_h$, namely, $s\{\hat{Y}_h\} = 0.288$.

We require $t(0.975; 13) = 2.160$. The confidence limits then are $18.692 \pm 2.160(0.288)$, and the 95 percent confidence interval is:

$$18.07 \le E\{Y_h\} \le 19.31$$

These confidence limits are also shown in block 5 of Figure 20.2. We therefore estimate, with 95 percent confidence, that the mean sales volume is between $181 thousand and $193 thousand when media and point-of-sale expenditures are $5 thousand and $2 thousand, respectively.   ☐

20.7   PREDICTION INTERVAL FOR $Y_{h(new)}$

Frequently, we employ a multiple regression model to predict a new response $Y_{h(new)}$ when the independent variables are at specified levels $X_{h1}, X_{h2}, \ldots, X_{h,p-1}$. The prediction interval takes the usual form.

(20.20)

The $1 - \alpha$ prediction limits for a new response $Y_{h(new)}$ for multiple regression model (20.3) are:

$$\hat{Y}_h \pm t s\{Y_{h(new)}\}$$

where: $t = t(1 - \alpha/2; n - p)$
$\hat{Y}_h$ is given by (20.18)
$s^2\{Y_{h(new)}\} = MSE + s^2\{\hat{Y}_h\}$

☐ **EXAMPLE**

In the promotional expenditures example, the market researcher wished to predict sales volume in a locality for the next period when media expenditures of $X_{h1} = \$5$ thousand and point-of-sale expenditures of $X_{h2} = \$2$ thousand will be made. A 95 percent prediction interval is desired. From the preceding example, we know that, for this case, $\hat{Y}_h = 18.692$ and $s^2\{\hat{Y}_h\} = (0.288)^2 = 0.0829$. We also know that $MSE = 0.289$. Hence, by (20.20), we calculate:

$$s^2\{Y_{h(new)}\} = 0.289 + 0.0829 = 0.3719$$

$$s\{Y_{h(new)}\} = \sqrt{0.3719} = 0.610$$

Finally, we need $t(0.975; 13) = 2.160$. The prediction limits then are $18.692 \pm 2.160(0.610)$, and the prediction interval is:

$$17.4 \le Y_{h(new)} \le 20.0$$

These prediction limits are shown at the right in block 5 of Figure 20.2. Hence, we predict with 95 percent confidence that the sales volume will be between $174 thousand and $200 thousand in the locality when media and point-of-sale expenditures are $5 thousand and $2 thousand, respectively.   ☐

20.8   INDICATOR VARIABLES

In our discussion of regression analysis up to this point, we have utilized quantitative independent variables, such as number of tape drives or dollar expenditures. Frequently, however, a qualitative independent variable is of interest. Examples of qualitative variables are job location of employee (plant, office, field), gender of respondent (male, female), and season of year (winter, spring, summer, autumn). In this section, we show how qualitative independent variables can be included in a regression model.

Representation of Qualitative Variables

We shall represent a qualitative variable in a regression model by means of 0, 1 indicator variables.

> **(20.21)**
> A **0, 1 indicator variable** is a variable that can have only the two possible values 0 or 1.

We subsequently refer to a 0, 1 indicator variable simply as an indicator variable. An indicator variable is also sometimes called a *binary variable* or a *dummy variable.*

   A qualitative variable with k classes is represented in a regression model by $k - 1$ indicator variables.

EXAMPLES

1. The independent variable gender of respondent (male, female) requires $k - 1 = 2 - 1 = 1$ indicator variable to represent it in a regression model. We denote this indicator variable by X_1 and define it as follows:

$$X_1 = \begin{array}{ll} 1 & \text{if female} \\ 0 & \text{otherwise} \end{array}$$

    Hence, the numerical values of X_1 are associated with the two classes of the variable gender as follows:

Class	X_1
Male (reference class)	0
Female	1

    The reason why the first class is called the reference class will be explained shortly.

2. The independent variable location of restaurant with three classes (highway, shopping mall, street) requires $k - 1 = 3 - 1 = 2$ indicator variables to represent it in a regression model. We denote these indicator variables by X_2 and X_3 and define them as follows:

$$X_2 = \begin{array}{ll} 1 & \text{if shopping mall location} \\ 0 & \text{otherwise} \end{array}$$

$$X_3 = \begin{array}{ll} 1 & \text{if street location} \\ 0 & \text{otherwise} \end{array}$$

    Hence, the numerical values of X_2 and X_3 are associated with the three location classes as follows:

Class	X_2	X_3
Highway (reference class)	0	0
Shopping mall	1	0
Street	0	1

The two examples show why the term *indicator* is used to describe the numerical variables. Each indicator variable takes the value 1 for one class and 0 for all other classes of the qualitative variable. Since only $k - 1$ indicator variables are employed, there is one class of the qualitative variable for which every one of the indicator variables is set to 0. This class is called the *reference class* of the system of indicator variables. We have arbitrarily selected the first class (male, highway) as the reference class in each of the two examples.

Comment

Indicator variables with numerical values other than 0 and 1 can also be used to represent a qualitative variable. In some applications, for example, the values 1 and -1 are used. In this text, we shall use only 0, 1 indicator variables.

Regression Analysis with Indicator Variables

One Qualitative Variable.   No new calculation problems are encountered when a qualitative independent variable is represented in a regression model by a set of indicator variables. The only new element involves the interpretation of the regression coefficients for the indicator variables. An example will make this interpretation clear.

Restaurant Sales **EXAMPLE**

A study for a chain of 16 restaurants examined the relationship between restaurant sales during a recent period (Y, in thousands of dollars) and number of households in the restaurant's trading area (X_1, in thousands) and location of restaurant (highway, shopping mall, street). The coding of the two indicator variables representing the location variable is that of Example 2 on page 615.

The data for the study appear in Table 20.3. We employ regression model (20.3) with three X variables.

$$(20.22) \qquad Y_i = \beta_0 + \beta_1 X_{i1} + \beta_2 X_{i2} + \beta_3 X_{i3} + \varepsilon_i$$

where: X_1 = number of households
X_2 = 1 if shopping mall location, 0 otherwise
X_3 = 1 if street location, 0 otherwise

The response function is as follows.

$$(20.23) \qquad E\{Y\} = \beta_0 + \beta_1 X_1 + \beta_2 X_2 + \beta_3 X_3$$

For highway restaurants, for which $X_2 = 0$ and $X_3 = 0$, the response function reduces to the following form.

TABLE 20.3
Data on sales volume, number of households, and restaurant location—Restaurant sales example

Restaurant i	Number of Households X_{i1} (thousand)	Location Qualitative Class	X_{i2}	X_{i3}	Sales Volume Y_i (\$ thousand)
1	155	Highway	0	0	135.27
2	93	Highway	0	0	72.74
3	128	Highway	0	0	114.95
4	114	Highway	0	0	102.93
5	158	Highway	0	0	131.77
6	183	Highway	0	0	160.91
7	178	Mall	1	0	179.86
8	215	Mall	1	0	220.14
9	172	Mall	1	0	179.64
10	197	Mall	1	0	185.92
11	207	Mall	1	0	207.82
12	95	Mall	1	0	113.51
13	224	Street	0	1	203.98
14	199	Street	0	1	174.48
15	240	Street	0	1	220.43
16	100	Street	0	1	93.19

$$(20.23a) \quad E\{Y\} = \beta_0 + \beta_1 X_1 + \beta_2(0) + \beta_3(0)$$
$$= \beta_0 + \beta_1 X_1 \qquad \text{Highway location}$$

For mall restaurants, for which $X_2 = 1$ and $X_3 = 0$, the response function becomes the following.

$$(20.23b) \quad E\{Y\} = \beta_0 + \beta_1 X_1 + \beta_2(1) + \beta_3(0)$$
$$= (\beta_0 + \beta_2) + \beta_1 X_1 \qquad \text{Mall location}$$

Finally, for street restaurants, for which $X_2 = 0$ and $X_3 = 1$, the response function becomes the following.

$$(20.23c) \quad E\{Y\} = \beta_0 + \beta_1 X_1 + \beta_2(0) + \beta_3(1)$$
$$= (\beta_0 + \beta_3) + \beta_1 X_1 \qquad \text{Street location}$$

The response functions for the three locations are portrayed in Figure 20.4. Note that each response function is linear, with the same slope β_1. The intercepts for the highway, mall, and street response functions are β_0, $\beta_0 + \beta_2$, and $\beta_0 + \beta_3$, respectively. Because of the parallel response functions, it follows that for any given number of households (X_1), the mean sales volume $E\{Y\}$ for mall restaurants differs from that

FIGURE 20.4
Meaning of indicator variable parameters—Restaurant sales example

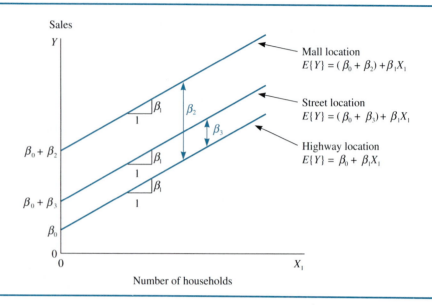

for highway restaurants by β_2; and the mean sales volume for street restaurants differs from that for highway restaurants by β_3. Note in Figure 20.4 how β_2 and β_3 reflect the differential effects of mall and street location, each with respect to the reference class highway location. It follows that the mean sales volume $E\{Y\}$ for mall restaurants differs from that for street restaurants by $\beta_2 - \beta_3$ for any given number of households (X_1). In Figure 20.4, β_0, β_1, β_2, and β_3 are portrayed as being positive, with β_2 greater than β_3.

The MYSTAT multiple regression program was used to fit regression model (20.22) to the data in Table 20.3. Figure 20.5 presents the computer output. Block 1 contains miscellaneous information including the sample size n (labeled N), R, R^2, R_a^2, and $\sqrt{MSE}$ (labeled STANDARD ERROR OF ESTIMATE). Block 2 contains the estimated regression coefficients b_k, the estimated standard deviations $s\{b_k\}$ of the regression coefficients (labeled STD ERROR), the test statistic $t^* = b_k/s\{b_k\}$ (labeled T), and the two-sided P-value for testing H_0: $\beta_k = 0$ versus H_1: $\beta_k \neq 0$. Block 3 contains the ANOVA table.

Based on the estimated regression coefficients in block 2, we see that the fitted regression function is:

$$\hat{Y} = -1.817 + 0.878X_1 + 27.298X_2 + 7.392X_3$$

The regression coefficient $b_2 = 27.298$ indicates that, for any given number of households, the mean sales volume of mall restaurants is estimated to be $27.3 thousand more than that of highway restaurants. The corresponding differential effect for street restaurants is estimated from b_3 to be $7.4 thousand. Finally, it follows that the mean sales volume of mall restaurants is estimated to be $19.9 thousand more than that of street restaurants ($b_2 - b_3 = 27.298 - 7.392 = 19.906$) for any given number of households.

A 90 percent confidence interval for β_2 is desired. It is obtained in the usual fashion from the following:

MYSTAT regression output—Restaurant sales example

FIGURE 20.5

1.

```
DEP VAR:        Y      N:   16    MULTIPLE R:   .994    SQUARED MULTIPLE R:   .988
ADJUSTED  SQUARED  MULTIPLE  R:   .985      STANDARD  ERROR  OF  ESTIMATE:      5.799
```

2.

```
      VARIABLE      COEFFICIENT      STD ERROR        T    P(2 TAIL)
    CONSTANT           -1.817          5.453      -0.333     0.745
          X1            0.878          0.035      24.752     0.000
          X2           27.298          3.620       7.540     0.000
          X3            7.392          4.177       1.770     0.102
```

3.

```
                          ANALYSIS OF VARIANCE


    SOURCE    SUM-OF-SQUARES    DF   MEAN-SQUARE      F-RATIO        P


  REGRESSION     33438.857      3    11146.286       331.403      0.000
  RESIDUAL         403.604     12       33.634
```

$n = 16$ $n - p = 12$ $t(0.95; 12) = 1.782$
$b_2 = 27.298$ $s\{b_2\} = 3.620$

Hence, the confidence limits are $27.298 \pm 1.782(3.620)$ and the 90 percent confidence interval is:

$$20.8 \le \beta_2 \le 33.7$$

Thus, with 90 percent confidence, we estimate that, for any given number of house-holds in the trading area, the expected sales in mall locations are between \$20.8 thousand and \$33.7 thousand greater than those in highway locations. □

Several Qualitative Variables. Regression models containing more than one qualitative independent variable are developed by including a set of indicator variables for each qualitative variable.

EXAMPLE □

In a regression study based on 200 plant workers, the number of absences of plant worker (Y) was regressed on age of worker (X_1), gender of worker (male, female), and work shift (first, second, third) using the following regression model:

$$Y_i = \beta_0 + \beta_1 X_{i1} + \beta_2 X_{i2} + \beta_3 X_{i3} + \beta_4 X_{i4} + \varepsilon_i$$

where: X_1 = age of worker
X_2 = 1 if female worker, 0 otherwise
X_3 = 1 if second-shift worker, 0 otherwise
X_4 = 1 if third-shift worker, 0 otherwise □

20.9 MODELING CURVILINEAR RELATIONSHIPS

The X variables in multiple regression model (20.3):

$$Y_i = \beta_0 + \beta_1 X_{i1} + \cdots + \beta_{p-1} X_{i,p-1} + \varepsilon_i$$

need not be different independent variables. In this more general form, the model is called the *general linear regression model* because it encompasses many special cases. It is called a *linear model* because it is *linear in the parameters* $\beta_0, \beta_1, \ldots, \beta_{p-1}$. Thus, no parameter appears as an exponent or is multiplied by another parameter. However, the general linear regression model is not restricted to linear relationships; it encompasses curvilinear relationships, as we shall now show.

Curvilinear relationships between the dependent variable Y and an independent variable X are frequently encountered.

☐ EXAMPLES

1. Sales (Y) may increase with advertising expenditures (X) but at a decreasing rate as saturation is approached.

2. Crop yield (Y) may increase with the amount of rain (X) up to a point and then decline as the crop becomes adversely affected by excess moisture.   ☐

Quadratic Regression

One of the more widely encountered curvilinear regression relationships is a *quadratic regression* relationship. The following is a quadratic regression model with one independent variable.

> **(20.24)** $Y_i = \beta_0 + \beta_1 X_i + \beta_2 X_i^2 + \varepsilon_i$

Here, the regression function is $E\{Y\} = \beta_0 + \beta_1 X + \beta_2 X^2$. The regression curve shown in Figure 18.3 for the selling space example is quadratic. Parameter β_1 is called the *linear effect coefficient* and β_2 the *curvature* or *quadratic effect coefficient.*

To see that regression model (20.24) is a special case of the general linear model (20.3), let $X_{i1} = X_i$ and $X_{i2} = X_i^2$. We can then write regression model (20.24) in the standard form:

$$Y_i = \beta_0 + \beta_1 X_{i1} + \beta_2 X_{i2} + \varepsilon_i$$

We now illustrate the fitting and analysis of a quadratic regression model and show how a standard multiple regression package can handle any such model.

Job Proficiency   ☐ **EXAMPLE**

An analyst was asked to investigate whether job proficiency of applicants seeking employment as assemblers could be predicted with satisfactory confidence and precision from the applicants' scores on two tests administered during the interview. Table 20.4 presents a portion of the data for a random sample of 25 applicants who were

Applicant i	(1) Manual Dexterity Score X_{i1}	(2) Depth Perception Score X_{i2}	(3) X_{i2}^2	(4) Job Proficiency Score Y_i
1	60	82	6,724	1102
2	135	163	26,569	2333
⋮	⋮	⋮	⋮	⋮
25	77	122	14,884	1513

TABLE 20.4
Partial data on job proficiency, manual dexterity, and depth perception, with a quadratic variable added—Job proficiency example

hired for purposes of the experiment irrespective of their test scores. Here, Y (column 4) denotes job proficiency score after a learning period, and X_1 (column 1) and X_2 (column 2) are the respective scores on the tests for manual dexterity and depth perception.

From earlier studies of the assembling operation, the analyst knew that manual dexterity and depth perception have linear effects on job proficiency. However, past evidence was inconclusive as to whether or not depth perception also has a curvature effect. Hence, the appropriateness of the following regression model was investigated:

$$Y_i = \beta_0 + \beta_1 X_{i1} + \beta_2 X_{i2} + \beta_3 X_{i2}^2 + \varepsilon_i$$

The purpose was to determine whether or not the quadratic effect $\beta_3 X_{i2}^2$ is really needed in the model.

After creating a data file for X_1 and X_2, the analyst employed the transformation X2SQ $= X_2^2$ to create a third X variable in the SAS statistical package, as shown in Table 20.4, column 3. Figure 20.6 presents the SAS regression output obtained by the analyst. Block 1 contains the ANOVA table. Block 2 contains miscellaneous results including R^2 and R_a^2, as well as $\sqrt{MSE}$ (labeled Root MSE). The estimated regression coefficients b_k are presented in block 3 (labeled Parameter Estimate), together with the estimated standard deviations $s\{b_k\}$ (labeled Standard Error), the test statistics $t^* = b_k/s\{b_k\}$ (labeled T), and the two-sided P-values for testing whether $\beta_k = 0$ (labeled Prob > |T|).

The analyst now wishes to test whether the quadratic effect for depth perception is needed in the regression model:

H_0: $\beta_3 = 0$ (No quadratic effect present)

H_1: $\beta_3 \neq 0$ (Quadratic effect present)

Here, H_0 implies that a quadratic effect is not present and hence can be dropped from the model, while H_1 implies that a quadratic effect is present and should be retained in the model. The analyst specified that the α risk is to be controlled at 0.10. For $\alpha = 0.10$ and $n - p = 21$, we require $t(0.95; 21) = 1.721$. As shown in block 3 of Figure 20.6, the standardized test statistic for variable X2SQ is $t^* = -1.834$. Since $|t^*| = 1.834 > 1.721$, the analyst concluded H_1—that a quadratic effect is present and should be retained in the model. The same test could have been done directly using the two-sided P-value in block 3 of Figure 20.6. Since P-value $= 0.0809 < \alpha = 0.10$, we are led to the same conclusion H_1.

FIGURE 20.6

SAS regression output—Job proficiency example. The quadratic effect is represented by variable X2SQ.

1.

Analysis of Variance

Source	DF	Sum of Squares	Mean Square	F Value	Prob>F
Model	3	3043679.8917	1014559.9639	52.279	0.0001
Error	21	407538.66830	19406.60325		
C Total	24	3451218.5600			

2.

Root MSE	139.30759	R-square	0.8819
Dep Mean	1644.24000	Adj R-sq	0.8650
C.V.	8.47246		

3.

Parameter Estimates

Variable	DF	Parameter Estimate	Standard Error	T for H0: Parameter=0	Prob > |T|
INTERCEP	1	26.916733	228.39985424	0.118	0.9073
X1	1	9.667959	1.43283327	6.747	0.0001
X2	1	9.970127	4.29145110	2.323	0.0303
X2SQ	1	-0.031218	0.01702584	-1.834	0.0809

The quadratic regression model was then examined for aptness, using various residual plots. Upon being satisfied by the examination, the analyst went on to study how well proficiency scores can be predicted from the two aptitude test scores. □

Comment

Quadratic regression is a special case of *polynomial regression.* Polynomial regression models contain one or more independent variables in powers of two or higher, such as X^2 and X^3. For example, the following is a cubic regression model for one independent variable.

$$(20.25) \quad Y_i = \beta_0 + \beta_1 X_i + \beta_2 X_i^2 + \beta_3 X_i^3 + \varepsilon_i$$

We see that this model is a special case of the general linear regression model (20.3) by letting $X_{i1} = X_i$, $X_{i2} = X_i^2$, and $X_{i3} = X_i^3$.

Regression with Transformed Variables

Another way to incorporate curvilinear effects in a regression model is to transform one or more of the independent variables, the dependent variable, or both. In Section 19.7, we discussed transforming variables for simple regression to linearize the regression relation.

Such regression models are actually curvilinear models since a linear regression relation in transformed variables can be viewed as a curvilinear relation in the original variables.

No new principles are encountered in using transformed variables in multiple regression, so we proceed directly to an example.

EXAMPLE ☐ **Ingots**

An engineer studied the cooling of ingots cast in a smelter. The curvilinear regression model that was utilized relates the rate of heat loss per unit weight of ingot (L) to the temperature differential of the ingot relative to its surroundings (D) and to the ingot's surface area (S). The regression model took the following form, with each variable expressed in logarithms:

$$\log L_i = \beta_0 + \beta_1 \log D_i + \beta_2 \log S_i + \varepsilon_i$$

To see that this model is a special case of the general linear regression model (20.3), let $Y_i = \log L_i$, $X_{i1} = \log D_i$, and $X_{i2} = \log S_i$. When this model is viewed in terms of the original variables L, D, and S, the regression relation is curvilinear. ☐

Response Surface Plot

When the regression relationship between the dependent variable and the independent variables is curvilinear, the response surface is curved rather than a plane. A computer plot of the fitted response surface is helpful for visualizing the nature of the curvilinear relationship.

EXAMPLE ☐

Figure 20.7 shows a SYGRAPH plot of the fitted response surface for the ingot example. Note how the surface is curved. It shows that L increases with D at a steady rate and decreases with S at a diminishing rate. ☐

20.10 MODELING INTERACTION EFFECTS

In some regression relationships, the effects of the independent variables on the dependent variable are not additive but rather reinforce or interfere with one another.

EXAMPLE ☐

In the promotional expenditures example, the market researcher was interested in whether the two types of promotional expenditures interfere with one another when both are at high levels so that the responsiveness of sales volume to increases in one type of promotional expenditure is reduced when the other type of promotional expenditure is high. ☐

EXAMPLE ☐ **Chemical Reaction**

A chemical engineer wanted to learn about the effects of reaction temperature and presence of a catalyst on the yield rate of a chemical reaction. However, the presence

FIGURE 20.7
**SYGRAPH plot of curvilin-
ear response surface—Ingot
example**

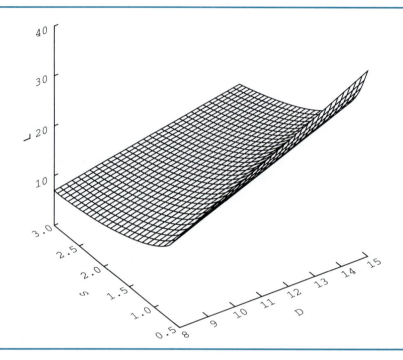

of a catalyst in the reaction is known to magnify the effect of reaction temperature on
the yield rate. Therefore, the engineer needed to include this reinforcing influence of
the catalyst in the regression model.   □

Incorporating Interaction Effects in Regression Model

The interference and reinforcement effects described in the two examples are referred to as
interaction effects. The standard way of incorporating interaction effects in a regression
model is by adding *cross-product terms* (also called *interaction terms*), such as the term
$X_{i1}X_{i2}$ in the following regression model.

$$(20.26) \quad Y_i = \beta_0 + \beta_1 X_{i1} + \beta_2 X_{i2} + \beta_3 X_{i1} X_{i2} + \varepsilon_i$$

With this model, the effect of X_1 on Y depends on the level of X_2, and similarly, the effect
of X_2 on Y depends on the level of X_1. Hence, the effects of X_1 and X_2 on Y are no longer
additive.

The fact that the effect of X_1 on Y depends on X_2 in regression model (20.26) can be
seen by writing the response function for this model in the following form.

$$(20.27) \quad E\{Y\} = \beta_0 + \beta_1 X_1 + \beta_2 X_2 + \beta_3 X_1 X_2$$
$$= (\beta_0 + \beta_2 X_2) + (\beta_1 + \beta_3 X_2)X_1$$

For a fixed value of X_2, (20.27) represents a simple linear regression function that relates Y to X_1. Note, however, that both the intercept $(\beta_0 + \beta_2 X_2)$ and the slope $(\beta_1 + \beta_3 X_2)$ depend on the fixed level of X_2. In the same way, the effect of X_2 on Y depends on the level of X_1. Whether the interaction effect is interfering or reinforcing depends on the sign of β_3.

Regression model (20.26) is still a special case of the general linear regression model (20.3). This can be seen by letting $X_{i3} = X_{i1} X_{i2}$.

To test for interaction effects in regression model (20.26), we simply test whether the regression coefficient β_3 of the interaction term equals zero, using the standard test procedure in (20.17).

EXAMPLE

In the promotional expenditures example, the market researcher employed regression model (20.26) to study whether or not an interaction effect is present. Table 20.5a contains a portion of the data. Variables X_1, X_2, and Y are repeated from Table 20.1. The interaction variable $X_1 X_2$ was created by the transformation $X_{i3} = X_{i1} X_{i2}$, as shown in column 3 of Table 20.5a. Table 20.5b presents the regression results. The estimated response function is as follows:

$$\hat{Y} = -0.8240 + 3.8745 X_1 + 1.5510 X_2 - 0.2415 X_3$$

The positive coefficients b_1 and b_2 together with the negative coefficient for the interaction term, $b_3 = -0.2415$, suggest that the two types of promotional expenditures may be interfering with one another, as the researcher suspected. To test whether this effect is indeed present, the researcher used test statistic (20.17). The test alternatives are:

$H_0: \beta_3 = 0$ (No interaction effect present)

$H_1: \beta_3 \neq 0$ (Interaction effect present)

The α risk of the test is to be controlled at 0.05. For $n - p = 12$, we require $t(0.975; 12) = 2.179$. Referring to Table 20.5b, we find $t^* = b_3/s\{b_3\} = -2.756$ for the interaction variable. Since $|t^*| = 2.756 > 2.179$, we conclude H_1—that $\beta_3 \neq 0$. Hence, the two types of promotional expenditures do interact. The two-sided P-value for the test is 0.01741, as shown in Table 20.5b. This small P-value indicates that the data support the presence of an interaction effect.

The researcher now studied the nature of the interaction effect by plotting the estimated response function relating Y to X_1 for two different levels of X_2. The levels of X_2 selected were $X_2 = 2$, when point-of-sale expenditures are at their lowest level in the experiment, and $X_2 = 5$, when these expenditures are at their highest level. The two estimated response functions are obtained by substituting $X_2 = 2$ and $X_2 = 5$, respectively, into the fitted regression model, and then simplifying the resulting expressions as follows:

$$X_2 = 2: \ \hat{Y} = -0.8240 + 3.8745 X_1 + 1.5510(2) - 0.2415 X_1(2)$$
$$= 2.2780 + 3.3915 X_1$$
$$X_2 = 5: \ \hat{Y} = -0.8240 + 3.8745 X_1 + 1.5510(5) - 0.2415 X_1(5)$$
$$= 6.9310 + 2.6670 X_1$$

These estimated response functions are plotted in Figure 20.8. The presence of the interaction effect is shown in the graph by the two estimated response functions not

TABLE 20.5
Partial data on sales and promotional expenditures, with an interaction variable added, and regression results—Promotional expenditures example

(a) Data

Locality i	(1) Media Expenditures ($ thousand) X_{i1}	(2) Point-of-Sale Expenditures ($ thousand) X_{i2}	(3) Expenditures Interaction $X_{i3} = X_{i1}X_{i2}$	(4) Sales Volume ($ ten thousand) Y_i
1	2	2	4	8.74
2	2	3	6	10.53
⋮	⋮	⋮	⋮	⋮
16	5	5	25	20.51

(b) Regression Results

Variable	b_k	$s\{b_k\}$	t^*	P-value
Constant	-0.82400	1.18296	-0.697	0.49935
X_1	3.87450	0.32196	12.034	0.00000
X_2	1.55100	0.32196	4.817	0.00042
$X_3 = X_1X_2$	-0.24150	0.08763	-2.756	0.01741

Source	SS	df	MS
Regression	194.94683	3	64.98228
Error	2.30352	12	0.19196
Total	197.25034	15	

$$F^* = 338.52 \qquad P\text{-value} = 0+$$

being parallel. The smaller slope when $X_2 = 5$ indicates that an addition to media expenditures (that is, an increase in X_1) is associated with a smaller increase in estimated mean sales volume when point-of-sale expenditures are high ($X_2 = 5$) than when they are low ($X_2 = 2$). It is in this sense that the two types of promotional expenditures interfere with one another in their relationship with mean sales volume. □

Interaction Effects Involving Indicator Variables

The variables used to form an interaction term can also be indicator variables. An example will illustrate this point.

□ **EXAMPLE**

In the chemical reaction example, the chemical engineer utilized regression model (20.26) with an interaction effect for relating the yield rate of the chemical reaction (Y, in kilograms per hour) to the reaction temperature (X_1, in degrees Celsius, °C) and whether or not a catalyst is present in the reaction ($X_2 = 1$ if catalyst present, $X_2 = 0$ otherwise). The temperature range in the experiment was 180°C to 250°C. The exper-

Estimated response functions with an interaction variable added—Promotional expenditures example. Nonparallel lines indicate the nature of the interaction between the two types of promotional expenditures.

FIGURE 20.8

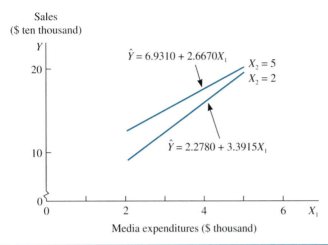

iment involved $n = 32$ trials. The estimated response function is as follows (the experimental data are not given here):

$$\hat{Y} = 112.0 + 0.103X_1 - 138.3X_2 + 0.607X_1X_2$$
$$\quad\quad\quad (0.00845)\quad (20.67)\quad\quad (0.114)$$

Notice that the estimated standard deviations of the regression coefficients are shown in parentheses under their respective coefficients. For the interaction term, we see that $b_3 = 0.607$ and $s\{b_3\} = 0.114$, so that $t^* = 0.607/0.114 = 5.32$. To test whether or not $\beta_3 = 0$ with $\alpha = 0.05$, we have $n - p = 32 - 4 = 28$ and therefore require $t(0.975; 28) = 2.048$. Since $|t^*| = 5.32 > 2.048$, we conclude that $\beta_3 \neq 0$; that is, an interaction effect is present. The positive sign of b_3 indicates that the amount of increase in mean yield with increasing temperature is larger when the catalyst is present. ☐

In some regression applications, a major objective is to measure the separate effects of the independent variables on the dependent variable. Generally, the regression coefficients are utilized for this purpose because the regression coefficient β_k in the general linear model (20.3) indicates the change in the mean of the distribution of Y when X_k increases by one unit and all the other independent variables remain constant.

Unfortunately, the separate effects of the different independent variables cannot usually be measured satisfactorily when the sample observations of the independent variables are highly correlated among themselves. Multicollinearity is then said to exist.

OPTIONAL TOPIC— MULTICOLLINEARITY

20.11

> **(20.28)**
> **Multicollinearity** is present in a regression analysis when the sample observations of the independent variables, or linear combinations of them, are highly correlated.

Effects of Multicollinearity

When multicollinearity is present, three major, related consequences are encountered.

1. The estimated regression coefficient b_k for independent variable X_k may vary substantially, depending on which other independent variables are included in the model. Consequently, the value obtained for b_k in any particular fitted model does not indicate the effect of independent variable X_k on the mean response $E\{Y\}$ in any absolute sense.
2. The estimated regression coefficients tend to have extremely large standard deviations, indicating that they vary widely in repeated samples. As a result, they will give very imprecise information about the regression parameters.
3. Fortunately, the difficulties just cited for the regression coefficients generally do not carry over to inferences on Y. As long as inferences are made within the region of sample observations on the independent variables, multicollinearity usually causes no special problems in estimating a mean response or predicting a new observation.

We shall illustrate the effects of multicollinearity by an example.

Machine Shop   **EXAMPLE**

A machine shop receives rough shafts and smooths them to specifications. When a shaft is smoothed, it is first carefully mounted on a machine; excess metal then is machined off. The total time required to process a shipment of shafts is affected by the time required to mount the shafts for smoothing, which depends on the number of shafts in the shipment. The total time also is affected by the time required to machine the shafts, which depends on the weight of the shipment since the aggregate amount of metal to be machined off varies with the total weight.

The processing time for a shipment (Y) is to be regressed on the number of shafts in the shipment (X_1) and the total weight of the shipment (X_2) for the 15 most recent shipments. Table 20.6 shows the data. The independent variables are highly correlated, as the scatter plot of X_1 versus X_2 in Figure 20.9 shows. The coefficient of simple correlation between X_1 and X_2 exceeds 0.99.

We have fitted three regression models to the data in Table 20.6 as follows:

Fitted Model	Independent Variables in Model
1	X_1
2	X_1, X_2
3	X_2

Key results are presented in Table 20.7. They illustrate the three points that we made earlier about the effects of multicollinearity:

1. The point estimate b_1 changes substantially when X_2 is added to the model. It is $b_1 = 34.85684$ when X_1 alone is in the model, but it is $b_1 = 57.75047$ when the model

Shipment i	Number of Shafts X_{i1}	Total Weight (kilograms) X_{i2}	Processing Time (minutes) Y_i
1	55	5563	1738
2	20	2041	491
3	35	3594	999
4	45	4523	1370
5	40	4082	1150
6	25	2534	684
7	55	5556	1650
8	30	3044	876
9	60	6095	1910
10	45	4561	1380
11	35	3562	995
12	25	2546	660
13	45	4576	1390
14	35	3529	1025
15	30	3056	821

TABLE 20.6
Data on processing of shafts—Machine shop example

contains both X_1 and X_2. Similarly, the value of b_2 changes when X_1 is added. In fact, the sign of the coefficient changes.

Thus, b_k does not indicate the effect of independent variable X_k on the mean response $E\{Y\}$ in any absolute sense when multicollinearity is present.

2. Table 20.7 also shows that the estimated standard deviation of b_k tends to become larger when other variables highly correlated with X_k are added to the fitted model.

Region of sample observations for independent variables—Machine shop example. The (X_1, X_2) observations fall in a narrow band here. The coefficient of correlation between X_1 and X_2 exceeds 0.99.

FIGURE 20.9

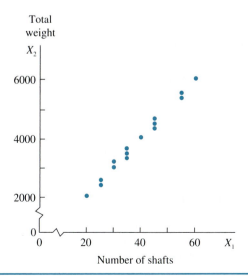

TABLE 20.7

Regression results for three fitted regression models—Machine shop example. Independent variables are highly correlated (multicollinearity is present).

Statistic	Fitted Model 1: X_1	Fitted Model 2: X_1, X_2	Fitted Model 3: X_2
b_1	34.85684	57.75047	—
$s\{b_1\}$	0.62818	39.51970	—
b_2	—	-0.22704	0.34553
$s\{b_2\}$	—	0.39187	0.00667
		$X_{h1} = 40, X_{h2} = 4060$	
$\hat{Y}_h$	1189.0758	1188.7534	1189.5460
$s\{\hat{Y}_h\}$	7.3437	7.5593	7.8622

Thus, it is $s\{b_1\} = 0.62818$ when X_1 is the only independent variable in the model, but it is $s\{b_1\} = 39.51970$ when both X_1 and X_2 are included. Similarly, $s\{b_2\}$ increases from 0.00667 to 0.39187 when X_1 is added to the model.

3. Multicollinearity causes no special problems when inferences on Y are made within the region of sample observations on the independent variables, as shown by the estimation of the mean response when $X_{h1} = 40$ and $X_{h2} = 4060$. Note from Figure 20.9 that this pair of values falls within the region of sample observations on the independent variables. As Table 20.7 shows, the estimated means $\hat{Y}_h$ and the estimated standard deviations $s\{\hat{Y}_h\}$ are practically the same for the three fitted models. This illustrates, incidentally, that in the presence of high correlation between X_1 and X_2, either variable alone contains much of the information provided by the other variable. ☐

The effects of multicollinearity also extend to tests about regression coefficients. The outcome of a test of whether or not $\beta_k = 0$ in the presence of multicollinearity depends very much on which other independent variables are included in the regression model.

☐ **EXAMPLE**

For the machine shop example, we would conclude from the fitted regression model containing both X_1 and X_2 in Table 20.7 that $\beta_1 = 0$ or that $\beta_2 = 0$ for $\alpha = 0.05$ or $\alpha = 0.01$. Yet $R^2 = 0.996$ here. The explanation for this apparent inconsistency is that, as noted previously, the test for $\beta_k = 0$ is a marginal test. It depends on the marginal effect of X_k, given that the other independent variables are already in the model. When X_1 and X_2 are highly correlated, as they are here, the marginal effect of either variable may be very small when the other is already in the model, yet there may be a strong relation between the independent variables and Y, both individually and jointly. ☐

Uncorrelated Independent Variables

When the independent variables are uncorrelated, the values of the regression coefficients and of their estimated standard deviations do not depend on which other independent variables are included in the regression model.

Regression results for three fitted regression models—Promotional expenditures example. Independent variables are uncorrelated (no multicollinearity is present).

TABLE 20.8

Statistic	Fitted Model 1: X_1	Fitted Model 2: X_1, X_2	Fitted Model 3: X_2
b_1	3.02925	3.02925	—
$s\{b_1\}$	0.22139	0.12028	—
b_2	—	0.70575	0.70575
$s\{b_2\}$	—	0.12028	0.81786

EXAMPLE ☐

In the promotional expenditures example, the two types of promotional expenditures were set up so that X_1 and X_2 are uncorrelated. The basic data were presented in Table 20.1. Table 20.8 contains pertinent results for three regression models fitted to these data, as follows:

Fitted Model	Independent Variables in Model
1	X_1
2	X_1, X_2
3	X_2

To simplify the example, we ignore the interaction effect discussed earlier.
    Table 20.8 illustrates two important points:

1. The values of b_1 and b_2 are unaffected by whether or not the other independent variable is included in the model. This is true in general for any number of independent variables that are mutually uncorrelated.

2. Neither $s\{b_1\}$ nor $s\{b_2\}$ becomes enlarged when the other independent variable is added to the model. ☐

The stability of the regression coefficients when the independent variables are mutually uncorrelated makes it desirable to use experimental control to obtain uncorrelated independent variables whenever possible. Unfortunately, experimental control is often not feasible, and correlated independent variables may then need to be employed.

Remedial Measures

Some remedial measures for the difficulties caused by multicollinearity are available. These include:

1. Obtaining additional observations that break the pattern of multicollinearity.
2. Estimating some of the regression coefficients from other data. For instance, in the machine shop example, it may be feasible to estimate β_1 by a time-and-motion study.
3. Transforming some of the independent variables to lessen the degree of multicollinearity.
4. Omitting some of the independent variables that are causing the multicollinearity.

5. Using a modified estimation procedure that reduces the impact of multicollinearity, such as ridge regression. These modified procedures are discussed in specialized texts.

In all cases, however, the interpretation of regression coefficients in the presence of multicollinearity must be done with caution.

20.12　OPTIONAL TOPIC—SELECTING THE REGRESSION MODEL

The regression models used in the examples so far have been specified in advance, and there has been little discussion of whether an alternative model with other independent variables or with a different functional relationship might have been better. Unfortunately, theoretical knowledge and conceptual understanding is often not sufficient to provide an exact specification of the most appropriate regression model in advance. Instead, numerous potential independent variables frequently exist that may be helpful in the regression model. Furthermore, in business, economic, and behavioral applications, these independent variables are typically highly intercorrelated. As we noted in Section 20.11, it is then very difficult to isolate the effect of an independent variable on the dependent variable. This makes the task of identifying the independent variables that should be included in the regression model very difficult.

Generally, we would like to include in the regression model only a small subset of all the potential independent variables—a subset that captures most of the explanatory or predictive information available in the whole set. Frequently, there will be several subsets that are potentially useful. We therefore seek to develop a few reasonable alternative subsets and to study these intensively before deciding on the final regression model to be employed.

All-Possible-Regressions Method

Several methods have been developed for identifying useful subsets of independent variables to be included in the regression model. Here, we describe one method, the all-possible-regressions selection method.

(20.29)

The **all-possible-regressions method** of selecting a regression model consists of the following three steps:

Step 1.　Initially develop a pool of potentially useful independent variables.

Step 2.　Fit a regression model for each possible combination of independent variables in the pool developed in step 1.

Step 3.　Order the regression models in step 2 by some criterion, evaluate in detail a small number of the models that are rated as best according to the criterion, and then select one of these as the regression model to be employed.

The criterion we shall employ here is that of minimizing the error mean square *MSE* for the regression model. This criterion considers regression models to be good when the variation of the observations around the fitted regression function is small. It can be shown that the criterion of minimizing *MSE* is equivalent to maximizing the adjusted coefficient of multiple determination R_a^2, as defined in (20.12).

Once the criterion of minimizing *MSE* has led to the identification of several regression models that are good according to this criterion, residual analysis and other diagnostics, together with expert judgment, are then used to identify the final regression model to be employed.

EXAMPLE ☐   **Property Selling Price**

A real estate expert was interested in developing a regression model that relates the selling price of residential properties in a large suburban development to characteristics of the properties. Data were available on 30 properties that were sold recently. The expert developed a long list of possible explanatory variables including lot size, property taxes, assessed property value, various house characteristics (such as floor area, number of rooms, and style of construction), and traffic access. After carefully sifting through this long list, the following five variables were chosen for the pool of potential independent variables:

Variable	Description
X_1	Property taxes (annual taxes, in dollars)
X_2	House size (floor area, in square feet)
X_3	Lot size (in acres)
X_4	Lot size squared $(X_4 = X_3^2)$
X_5	Attractiveness index (a composite measure of ratings for different features of the property)

The reason for including variable X_4 is that there may be a quadratic effect of lot size on selling price. The data for these X variables and for the dependent variable selling price are given in Table 20.9 for the 30 properties.

Using a statistical package that fits regression models for all combinations of X variables, the expert obtained the value of *MSE* for each combination of the five X variables in the pool. A total of $2^5 = 32$ regression models were analyzed, including the model with an intercept only. The *MSE* values for these 32 models are displayed in Table 20.10. For each number of X variables included in the model, the *MSE* values are arrayed in descending order. Figure 20.10 contains a plot of the *MSE* values against the number of X variables $(p - 1)$ in the model. The variables in the best model for each given number of X variables are identified in Figure 20.10. These minimum *MSE* points are connected by straight lines. Note, for example, that the best regression model with three X variables $(p - 1 = 3)$ contains variables X_2, X_4, and X_5. Table 20.10 shows that $MSE = 1037$ for this model.

The expert then reviewed the regression output for a few models with *MSE* values close to the minimum and made several diagnostic checks for each. Based on this study and other information, the expert chose the model with variables X_2, X_3, and X_5 for the regression model. The point corresponding to this model is shown by an arrow in Figure 20.10. Although this model does not give the smallest value of *MSE*, it is a close competitor and contains fewer independent variables than several of the other leading candidates. The three variables in the final model are house size (X_2), lot size (X_3), and attractiveness index (X_5).   ☐

TABLE 20.9

Data on selling price and potential independent variables—Property selling price example.

Property i	Property Taxes (dollars) X_{i1}	House Size (square feet) X_{i2}	Lot Size (acres) X_{i3}	Lot Size Squared X_{i4}	Attractiveness Index X_{i5}	Selling Price ($ thousand) Y_i
1	7337	3000	3.6	12.96	64	550
2	4204	2300	1.2	1.44	69	461
3	5574	3300	1.3	1.69	72	501
4	5924	2100	3.2	10.24	71	455
5	5182	3900	1.1	1.21	40	503
6	5932	3100	2.0	4.00	74	529
7	5966	3600	1.6	2.56	69	478
8	5574	2900	2.5	6.25	85	562
9	4927	2000	2.6	6.76	70	417
10	5025	3500	1.3	1.69	74	566
11	6210	3100	2.3	5.29	79	494
12	5425	3200	1.5	2.25	75	515
13	4178	2800	1.3	1.69	62	490
14	7048	3300	3.3	10.89	62	537
15	7540	3000	3.9	15.21	70	527
16	5807	3400	2.4	5.76	81	577
17	4875	2800	1.7	2.89	77	490
18	5540	2000	3.4	11.56	67	486
19	5980	2400	2.9	8.41	68	450
20	7324	3600	2.9	8.41	84	674
21	4582	2400	1.9	3.61	75	454
22	6759	3000	2.8	7.84	63	523
23	6444	2200	3.6	12.96	78	469
24	6307	3600	2.4	5.76	73	628
25	4285	2900	1.1	1.21	85	570
26	7722	3000	4.4	19.36	69	564
27	4917	3100	1.8	3.24	54	444
28	5068	2200	2.1	4.41	75	494
29	6500	2500	3.9	15.21	61	479
30	4612	2900	1.1	1.21	74	477

20.13   OPTIONAL TOPIC—F TEST FOR SEVERAL REGRESSION COEFFICIENTS

At times it is necessary to test whether a subset of two or more independent variables in a multiple regression model can be dropped from the model. In effect, we need to determine with such a test whether all of the regression coefficients for the subset of independent variables equal zero.

An F test is available for this purpose. It involves choosing between two alternative regression models, one that contains all of the independent variables (called the *full model*) and one in which the independent variables that are candidates for deletion have been dropped (called the *reduced model*).

Results for all-possible-regressions method—Property selling price example. Values of *MSE* are presented for all possible regression models involving subsets of the five *X* variables.

TABLE 20.10

Variables in Model	*MSE*	Variables in Model	*MSE*
Intercept Only	3195	X_3 X_4 X_5	3006
		X_1 X_3 X_4	2277
X_4	3262	X_1 X_4 X_5	1906
X_3	3260	X_1 X_2 X_4	1843
X_5	2877	X_1 X_2 X_3	1812
X_1	2669	X_2 X_3 X_4	1792
X_2	2013	X_1 X_3 X_5	1791
X_3 X_4	3381	X_1 X_2 X_5	1158
X_3 X_5	2928	X_2 X_3 X_5	1044
X_4 X_5	2915	X_2 X_4 X_5	1037
X_1 X_5	2297	X_1 X_2 X_3 X_4	1861
X_1 X_4	2231	X_1 X_3 X_4 X_5	1833
X_1 X_3	2193	X_2 X_3 X_4 X_5	1078
X_1 X_2	1786	X_1 X_2 X_4 X_5	1049
X_2 X_4	1785	X_1 X_2 X_3 X_5	1033
X_2 X_3	1746		
X_2 X_5	1408	X_1 X_2 X_3 X_4 X_5	1071

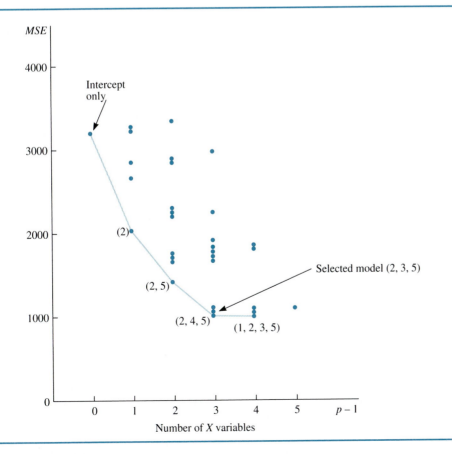

FIGURE 20.10

Plot of *MSE* against number of *X* variables for all possible regression models—Property selling price example

Arthritis Incidence □ **EXAMPLE**

A market analyst for a pharmaceutical manufacturer studied the relation in 13 districts between the incidence of arthritis (Y, in cases per thousand population) and the following independent variables:

X_1:   Percent of district population over 65 years of age
X_2:   Number of physicians per thousand population
X_3:   Mean disposable income of persons over 65 years of age (in \$ thousand)

Regression model (20.3) was assumed to be appropriate. The analyst knew from previous studies that the incidence of arthritis is related to the percent of the population over 65 but was unsure whether number of physicians and mean disposable income are also related to the dependent variable. The analyst therefore wished to choose between:

$$\text{Full model:}\quad E\{Y\} = \beta_0 + \beta_1 X_1 + \beta_2 X_2 + \beta_3 X_3$$
$$\text{Reduced model:}\quad E\{Y\} = \beta_0 + \beta_1 X_1$$

The corresponding test alternatives in terms of the regression coefficients in the full model therefore are:

$$H_0:\ \beta_2 = \beta_3 = 0$$
$$H_1:\ \text{Not both } \beta_2 \text{ and } \beta_3 \text{ equal zero}$$

□

It is inappropriate to make the decision here on the basis of separate applications of the t test for individual regression coefficients because the independent variables are correlated among themselves. Instead, we must use a special F test.

Test Procedure

To test whether a subset of independent variables can be dropped from the model, we consider whether adding this subset of variables to the regression model leads to a significant reduction of the error sum of squares. This involves fitting both the full model and the reduced model as follows:

1. Fit the full regression model to the data. Let $SSE(F)$ and df_F denote the associated error sum of squares and its degrees of freedom, respectively.
2. Fit the reduced regression model to the data. Let $SSE(R)$ and df_R denote the associated error sum of squares and its degrees of freedom, respectively.
3. Calculate the following test statistic F^*, which is a function of the amount of reduction in the error sum of squares.

$$(20.30)\qquad F^* = \left(\frac{df_F}{df_R - df_F}\right)\left[\frac{SSE(R) - SSE(F)}{SSE(F)}\right]$$

Large values of F^* are consistent with the full model because the reduction in the error sum of squares is large in that case, indicating that the addition of the independent variables in the subset to the regression model is helpful. Small values of the test statistic are consistent with the reduced model because the error sum of squares is not reduced much in that case by adding the subset of variables to the regression model. The decision rule for the test therefore takes the following form.

> **(20.31)**
> When the alternatives are:
>
> $$H_0:\ \text{All } \beta_k \text{ for the subset of variables equal zero}$$
> $$H_1:\ \text{Not all } \beta_k \text{ for the subset equal zero}$$
>
> and multiple regression model (20.3) is applicable, the appropriate decision rule to control the α risk is as follows:
>
> $$\text{If } F^* \leq F(1 - \alpha;\ df_R - df_F,\ df_F), \text{ conclude } H_0.$$
> $$\text{If } F^* > F(1 - \alpha;\ df_R - df_F,\ df_F), \text{ conclude } H_1.$$
>
> Here F^* is given by (20.30).

EXAMPLE □

The data set for the arthritis incidence example is given in Table 20.11a. The test alternatives here are:

$$H_0:\ \beta_2 = \beta_3 = 0$$
$$H_1:\ \text{Not both } \beta_2 \text{ and } \beta_3 \text{ equal zero}$$

The α risk is to be controlled at 0.01.

Selected regression output for the full and reduced models is given in Table 20.11b. Referring to the ANOVA table in each case, we find the following:

$$SSE(F) = 4.516 \qquad df_F = 9$$
$$SSE(R) = 6.962 \qquad df_R = 11$$

Thus, adding variables X_2 and X_3 to the regression model that already contains X_1 reduces the error sum of squares by:

$$SSE(R) - SSE(F) = 6.962 - 4.516 = 2.446$$

For $\alpha = 0.01$, we require $F(0.99;\ 2, 9) = 8.02$. The decision rule is as follows:

$$\text{If } F^* \leq 8.02, \text{ conclude } H_0.$$
$$\text{If } F^* > 8.02, \text{ conclude } H_1.$$

Test statistic (20.30) here is:

$$F^* = \left(\frac{9}{11 - 9}\right)\left(\frac{2.446}{4.516}\right) = 2.44$$

Since $F^* = 2.44 \leq 8.02$, we conclude H_0—that number of physicians (X_2) and mean disposable income (X_3) can be dropped from the regression model when the model already contains variable X_1, percent of the population over 65. □

TABLE 20.11
Data on arthritis incidence and independent variables, and selected regression results—Arthritis incidence example

(a) Data

District i	Percent of Population over 65 X_{i1}	Number of Physicians (per 1000) X_{i2}	Mean Disposable Income for over 65 ($ thousand) X_{i3}	Arthritis Incidence (cases per 1000) Y_i
1	10.4	1.36	16.5	8.70
2	14.8	1.49	13.8	10.44
3	12.5	1.50	13.2	9.82
4	18.0	1.91	20.6	12.38
5	12.3	1.62	15.8	8.25
6	15.6	1.25	22.6	11.04
7	10.1	1.09	19.4	7.89
8	12.5	1.99	16.1	10.08
9	12.8	1.04	15.9	9.35
10	14.9	1.72	16.2	11.57
11	11.2	1.13	13.2	7.29
12	15.6	1.34	20.7	11.49
13	11.3	1.98	14.0	10.22

(b) Analysis of Variance Tables

Full Model					*Reduced Model*			
Source	SS	df	MS		Source	SS	df	MS
Regression	23.999	3	8.000		Regression	21.553	1	21.553
Error	4.516	9	0.502		Error	6.962	11	0.633
Total	28.515	12			Total	28.515	12	

Comments

1. The test of whether all $\beta_k = 0$ $(k = 1, 2, \ldots, p - 1)$ in (20.15) is a special case of the test for a reduced model in (20.31). The equivalence can be seen by noting that when testing whether there is any regression relation, $SSE(R) = SSTO$, $SSE(F) = SSE$, $df_R = n - 1$, and $df_F = n - p$.

2. Likewise, a test of whether an individual β_k equals zero is a special case of the test for a reduced model in (20.31). In this case, t^* as defined in (20.17) and F^* as defined in (20.30) are related by $F^* = (t^*)^2$.

PROBLEMS

20.1   Consider regression model (20.1) and suppose that $E\{Y\} = 15 - X_1 + 4X_2$ and $\sigma^2 = 4$.
a. Graph the relationship between $E\{Y\}$ and X_2 when $X_1 = 10$. How is β_2 interpreted here?
b. Describe the probability distribution of Y when $X_1 = 10, X_2 = 5$.

20.2   The resale value of a certain type of road construction machine (Y, in $ thousand) is related to its age (X_1, in years) and its operating hours (X_2, in thousands) according to regression model (20.1) with the following parameter values: $\beta_0 = 230, \beta_1 = -11, \beta_2 = -6, \sigma^2 = 49$.

a. State the regression function.

b. Graph the relationship between the expected resale value and the age of machines for machines that have 5 thousand operating hours. How is β_1 interpreted here?

c. Describe the probability distribution of the resale value of machines that are 4 years old and have 5 thousand operating hours.

20.3 The earnings of firms (Y, in \$ million) are related to their sales revenues (X_1, in \$ million) and total long-term debt (X_2, in \$ million) according to regression model (20.1) with the following parameter values: $\beta_0 = -5$, $\beta_1 = 0.15$, $\beta_2 = -0.10$, $\sigma^2 = 9$.

a. State the regression function.

b. Interpret β_1 and σ here.

c. For $X_{i1} = 100$, $X_{i2} = 50$, determine (1) the value of $E\{Y_i\}$, (2) the value of Y_i when $\varepsilon_i = -8$, (3) the probability that Y_i is less than 2.

* 20.4 Fermentation Process. A chemical plant makes an organic product in a batch fermentation process that takes several days to complete. To study the factors affecting the amount of product obtained in a batch, an analyst has fitted regression model (20.3) to data from the last 23 batches. The dependent variable is amount of product (Y, in liters), and the three independent variables are the concentration of a particular saccharide in the batch material (X_1, as a percent of total saccharides), the number of hours before fermentation ceases (X_2), and the temperature range during fermentation (X_3, in °C). The data, correlation matrix, and key regression results follow:

i	X_{i1}	X_{i2}	X_{i3}	Y_i	i	X_{i1}	X_{i2}	X_{i3}	Y_i
1	50	51	2.3	48	13	38	55	2.2	47
2	36	46	2.3	57	14	34	51	2.3	51
3	40	48	2.2	66	15	53	54	2.2	57
4	41	44	1.8	70	16	36	49	2.0	66
5	28	43	1.8	89	17	33	56	2.5	79
6	49	54	2.9	36	18	29	46	1.9	88
7	42	50	2.2	46	19	33	49	2.1	60
8	45	48	2.4	54	20	55	51	2.4	49
9	52	62	2.9	26	21	29	52	2.3	77
10	29	50	2.1	77	22	44	58	2.9	52
11	29	48	2.4	89	23	43	50	2.3	60
12	43	53	2.4	67					

	X_1	X_2	X_3	Y
X_1	1.0000	0.4666	0.4977	−0.7737
X_2		1.0000	0.7945	−0.5874
X_3			1.0000	−0.6023
Y				1.0000

k:	0	1	2	3	
b_k:	162.876	−1.2103	−0.66591	−8.6130	$n = 23$
$s\{b_k\}$:	25.776	0.3015	0.82100	12.2413	$SSR = 4133.633$
					$SSE = 2011.584$

a. Which independent variable is most highly correlated with Y?

b. Give the values of (1) r_{Y2}, (2) r_{13}, (3) r_{23}. Interpret r_{23}.

20.5 Freezer Power Use. An experiment was conducted for two weeks to study the effect of freezer temperature (X_1) and freezer storage density (X_2) on freezer power use (Y, in kilowatt hours) for a certain type of commercial freezer. The independent variables are measured in terms of deviations from the levels normally used. Thus, in case 1, the temperature setting was 10°C below the normal level and the storage density was 10 percent points less than the normal density. The study results follow:

Case i:	1	2	3	4	5	6	7	8	9
Temperature X_{i1}:	-10	0	10	-10	0	10	-10	0	10
Density X_{i2}:	-10	-10	-10	0	0	0	10	10	10
Power Y_i:	228	204	142	247	190	149	248	219	180

Regression model (20.1) was employed, and the following correlation matrix and key regression results were obtained:

$$\begin{array}{c} X_1 \\ X_2 \\ Y \end{array} \begin{bmatrix} \begin{array}{ccc} X_1 & X_2 & Y \\ 1.0000 & 0 & -0.9340 \\ & 1.0000 & 0.2706 \\ & & 1.0000 \end{array} \end{bmatrix}$$

k:	0	1	2	$n = 9$
b_k:	200.778	-4.2000	1.2167	$SSR = 11{,}472.167$
$s\{b_k\}$:	3.500	0.4286	0.4286	$SSE = 661.389$

a. Which independent variable is least correlated with Y?
b. Give the value of r_{12}. Interpret this measure.
c. Does the sign of r_{Y2} indicate that power use tends to increase with higher storage density? Explain.

20.6 Hotel Equipment. A hotel equipment and supply firm has eight sales territories. A manager of the firm studied the relation of the firm's annual sales in a territory (Y, in \$ thousand) to the number of hotel rooms in the territory (X_1, in thousands) and an index of business activity for the territory (X_2). The data follow:

Territory i:	1	2	3	4	5	6	7	8
Rooms X_{i1}:	2.7	3.1	1.6	1.2	1.3	2.1	3.7	3.4
Index X_{i2}:	128	82	96	107	88	91	96	112
Sales Y_i:	88.7	77.4	71.8	73.2	71.0	75.9	85.8	87.9

The manager fitted regression model (20.1) and obtained the following correlation matrix and key regression results:

$$\begin{array}{c} X_1 \\ X_2 \\ Y \end{array} \begin{bmatrix} \begin{array}{ccc} X_1 & X_2 & Y \\ 1.0000 & 0.1307 & 0.8338 \\ & 1.0000 & 0.6386 \\ & & 1.0000 \end{array} \end{bmatrix}$$

k:	0	1	2	$n = 8$
b_k:	38.5111	5.7803	0.26651	$SSR = 373.784$
$s\{b_k\}$:	3.1953	0.4757	0.03108	$SSE = 7.3946$

a. Give the values of (1) r_{12}, (2) r_{Y1}, (3) r_{Y2}. Interpret r_{Y1}.
b. Which independent variable is most correlated with Y? What percentage reduction in the variability of Y would result if this independent variable were used alone to predict Y?

* **20.7** Refer to **Fermentation Process** Problem 20.4.

a. State the estimated regression function. Interpret b_1.
b. Obtain a point estimate of $E\{Y_h\}$ when $X_{h1} = 38$, $X_{h2} = 55$, $X_{h3} = 2.2$. Interpret this point estimate.

20.8 Refer to **Freezer Power Use** Problem 20.5.

a. State the estimated regression function.
b. Obtain point estimates of the change in expected power use (1) when freezer temperature is increased 1°C and density is held fixed, (2) when freezer temperature is increased 10°C and density is held fixed.

c. Obtain a point estimate of the expected power use when freezer temperature is 10°C below the normal level and storage density is at its normal level.

20.9 Refer to **Hotel Equipment** Problem 20.6.

a. State the estimated regression function. Interpret b_1. In what units is b_1 expressed?
b. Obtain a point estimate of mean sales in territories where the number of rooms is 1.2 thousand and the index of business activity is 107.

* 20.10 Refer to **Fermentation Process** Problem 20.4.

a. Set up the ANOVA table. What parameter in regression model (20.3) is estimated by MSE?
b. Obtain R^2 and R. Interpret R^2.
c. Obtain R_a^2 and R_a. What does the latter value take into account that R does not? Is the difference between R_a and R substantial here?

20.11 Refer to **Freezer Power Use** Problem 20.5.

a. Set up the ANOVA table.
b. Obtain $\sqrt{MSE}$. The estimated mean power use when freezer temperture is 10°C below the normal level and storage density is at its normal level is 242.8 kilowatt hours. Estimate the coefficient of variation of the probability distribution of power use in this case.
c. Obtain R^2 and R. Interpret R^2.
d. Obtain R_a^2 and R_a. Must R_a^2 necessarily be smaller than R^2, as it is here? Why?

20.12 Refer to **Hotel Equipment** Problem 20.6.

a. Set up the ANOVA table.
b. Obtain a point estimate of σ. In what units is this estimate expressed?
c. Obtain R^2 and R. Why would one expect R^2 to be larger than either r_{Y1}^2 or r_{Y2}^2, as is the case here?
d. Obtain R_a^2 and R_a.
e. When the manager added another business index (X_3) to the regression model, computer output showed $R_a^2 = 0.968$. Compare the values of R_a^2 with and without the added variable X_3. Has MSE increased or decreased with the addition of X_3 to the regression model?

* 20.13 Refer to **Fermentation Process** Problem 20.4. Note that $SSE(X_1, X_2, X_3) = 2011.584$. When Y is regressed on X_1 and X_2 only, the error sum of squares is $SSE(X_1, X_2) = 2063.998$.

a. What does the difference $SSE(X_1, X_2) - SSE(X_1, X_2, X_3)$ measure here?
b. Obtain $r_{Y3.12}^2$ and interpret its meaning here. Also obtain $r_{Y3.12}$.

20.14 Refer to **Freezer Power Use** Problem 20.5. Note that $SSE(X_1, X_2) = 661.389$. When Y is regressed on X_1 only, the error sum of squares is $SSE(X_1) = 1549.556$. Likewise, when Y is regressed on X_2 only, $SSE(X_2) = 11,245.389$.

a. Obtain (1) $r_{Y2.1}^2$, (2) $r_{Y1.2}^2$. Interpret each of these measures.
b. Obtain (1) $r_{Y2.1}$, (2) $r_{Y1.2}$.

20.15 Refer to **Hotel Equipment** Problem 20.6. Note that $SSE(X_1, X_2) = 7.3946$ and $r_{Y1} = 0.8338$.

a. Obtain $SSE(X_1)$. Explain why, in general, $SSE(X_1)$ will be larger than $SSE(X_1, X_2)$.
b. Obtain $r_{Y2.1}^2$ and interpret its meaning. Also obtain $r_{Y2.1}$.

20.16 Refer to **Fermentation Process** Problem 20.4. Figure 20.11 shows four residual plots for this regression study.

a. Figure 20.11a shows a plot of the residuals against the fitted values. Does the plot indicate that regression model (20.3) is appropriate here with regard to (1) the regression function, (2) constancy of the error variance? Comment.
b. Figure 20.11b shows a plot of the residuals in batch sequence. Do the error terms appear to be independent from batch to batch? Comment.
c. Figure 20.11c shows a plot of the residuals against independent variable X_1. Does the effect of X_1 on Y appear to differ from that postulated in the regression model? Comment.
d. Figure 20.11d shows a normal probability plot of the residuals. Does the assumption of normality of the error terms appear to be reasonable here? Comment.

FIGURE 20.11
Residual plots for Problem 20.16

(a) **Residuals Against Fitted Values**

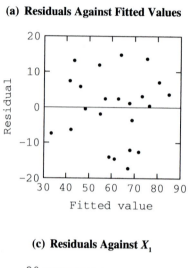

(b) **Residuals in Batch Sequence**

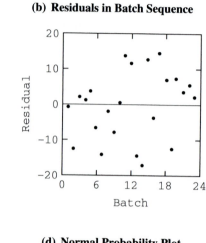

(c) **Residuals Against X_1**

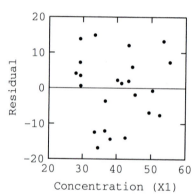

(d) **Normal Probability Plot**

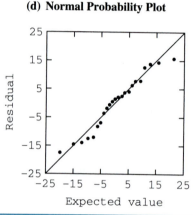

20.17 Refer to **Freezer Power Use** Problem 20.5.

a. Obtain the fitted values and the residuals. Do the residuals sum to zero as they should?

b. Plot the residuals against the fitted values. Does the plot indicate that regression model (20.1) is appropriate here with regard to (1) the regression function, (2) constancy of the error variance? Comment.

c. Construct a normal probability plot of the residuals. Does the assumption of normality of the error terms appear to be reasonable here? Comment.

20.18 Refer to **Hotel Equipment** Problem 20.6.

a. Obtain the fitted values and the residuals.

b. Plot the residuals against the fitted values. Does the plot indicate that regression model (20.1) is appropriate here with regard to (1) the regression function, (2) constancy of the error variance? Comment.

c. Plot the residuals against (1) X_1, (2) X_2. Do the effects of either independent variable on Y appear to differ from those postulated in the regression model? Comment.

d. Construct a normal probability plot of the residuals. Does the assumption of normality of the error terms appear to be reasonable here? Comment.

* **20.19** Refer to **Fermentation Process** Problem 20.4.

a. Test whether a regression relation holds, controlling the α risk at 0.01. State the alternatives, the decision rule, the value of the test statistic, and the conclusion.

b. Does conclusion H_1 imply that a regression relation holds between Y and each of the independent variables? Does conclusion H_0 imply that a regression relation does not hold between Y and any of the independent variables? Explain.

c. What is the P-value of the test in part a? Is this value consistent with the test result in part a?

20.20 Refer to **Freezer Power Use** Problem 20.5.

a. Test whether a regression relation holds, controlling the α risk at 0.01. State the alternatives, the decision rule, the value of the test statistic, and the conclusion.

b. Computer output for this study gives the upper-tail P-value for the test in part a as 0.0002. Is this value consistent with the test result in part a?

20.21 Refer to **Hotel Equipment** Problem 20.6.

a. Test whether a regression relation holds, controlling the α risk at 0.01. State the alternatives, the decision rule, the value of the test statistic, and the conclusion. Does your conclusion necessarily imply that Y is related to X_1? Explain.

b. What is the P-value of the test in part a? Is this value consistent with the test result in part a?

∗ 20.22 Refer to **Fermentation Process** Problem 20.4.

a. Construct a 95 percent confidence interval for β_2. Interpret the confidence interval.

b. Test whether $\beta_3 = 0$, controlling the α risk at 0.05 when $\beta_3 = 0$. State the alternatives, the decision rule, the value of the test statistic, and the conclusion. Does the conclusion necessarily imply that there is no regression relation between X_3 and Y? Explain.

20.23 Refer to **Freezer Power Use** Problem 20.5.

a. Construct a 95 percent confidence interval for β_2. Interpret the confidence interval.

b. Test whether $\beta_1 = 0$, controlling the α risk at 0.05 when $\beta_1 = 0$. State the alternatives, the decision rule, the value of the test statistic, and the conclusion. What is the practical interpretation of the conclusion?

c. Given the conclusion in part b, would you expect $r_{Y1.2}$ to differ substantially from 0 here? Explain.

d. Explain why β_0 is a meaningful parameter in this application and obtain a 95 percent confidence interval for it. Interpret the confidence interval.

20.24 Refer to **Hotel Equipment** Problem 20.6.

a. Construct a 99 percent confidence interval for β_2. Use the confidence interval to estimate the change in expected annual sales in territories whose index of business activity increases by 10 points, everything else remaining constant.

b. Test whether $\beta_1 = 0$, controlling the α risk at 0.05 when $\beta_1 = 0$. State the alternatives, the decision rule, the value of the test statistic, and the conclusion. What is the P-value of the test?

∗ 20.25 Refer to **Fermentation Process** Problem 20.4. Computer output gives $s\{\hat{Y}_h\} = 4.9322$ when $X_{h1} = 38, X_{h2} = 55, X_{h3} = 2.2$.

a. Construct a 95 percent confidence interval for $E\{Y_h\}$ when $X_{h1} = 38, X_{h2} = 55, X_{h3} = 2.2$. Interpret the confidence interval.

b. Construct a 95 percent prediction interval for the amount of product that will be obtained from a new batch for which the independent variables are at the levels specified in part a. Interpret the prediction interval.

20.26 Refer to **Freezer Power Use** Problem 20.5. Computer output gives $s\{\hat{Y}_h\} = 5.534$ when $X_{h1} = -10, X_{h2} = 0$.

a. Construct a 95 percent confidence interval for the mean power use when $X_{h1} = -10, X_{h2} = 0$. Interpret the confidence interval.

b. Construct a 95 percent prediction interval for the power use of a particular freezer when $X_{h1} = -10, X_{h2} = 0$. Explain why it is reasonable that this prediction interval is wider than the confidence interval in part a.

20.27 Refer to **Hotel Equipment** Problem 20.6. Computer output gives $s\{\hat{Y}_h\} = 0.7638$ when $X_{h1} = 1.2, X_{h2} = 107$.

a. Construct a 90 percent confidence interval for $E\{Y_h\}$ when $X_{h1} = 1.2, X_{h2} = 107$. Interpret the confidence interval.

b. Construct a 90 percent prediction interval for sales in a new territory for which $X_{h1} = 1.2, X_{h2} = 107$. Interpret the prediction interval.

c. Would you expect the prediction interval in part b to be much narrower if the regression function had been estimated from a very large number of territories rather than from only eight? Discuss.

20.28 Refer to regression model (20.22) for the *Restaurant Sales* example. An observer has stated that when $\beta_2 > 0$ and $\beta_3 = 0$, location has an impact on sales if the restaurant is in a shopping mall or on a highway but not if the restaurant is on a street. Comment.

* **20.29** **Bond Issues.** A financial analyst is studying the relation between bond yield (Y, in percent), bond maturity (X_1, in years), and whether the bond is a corporate or a municipal bond (X_2). Here, $X_2 = 1$ denotes a corporate bond and $X_2 = 0$ a municipal bond. The analyst has performed a regression analysis based on model (20.1) for a random sample of 300 bonds. Some regression results are: $b_0 = 8.223, b_1 = 0.1421, b_2 = 0.3381, s\{b_1\} = 0.006270, s\{b_2\} = 0.04816$.

a. Plot on the same graph $\hat{Y}$ against X_1 for corporate bonds ($X_2 = 1$) and for municipal bonds ($X_2 = 0$). Let X_1 range between 0 and 15 years in the plot. Interpret b_1 and b_2 here.

b. Construct a 95 percent confidence interval for the difference in the mean yields of corporate and municipal bonds of the same maturity. Does the confidence interval indicate that there is a difference in mean yields? Explain.

20.30 **Paperback Experiment.** Prior to general distribution of a successful hardcover novel in paperback form, an experiment was conducted in nine test markets with approximately equal sales potential. The experiment sought to assess the effects of three different price discount levels for the paperback (50, 75, 95 cents off the printed cover price) and the effects of three different cover designs (abstract, photograph, drawing) on sales of the paperback. Each of the nine combinations of price discount and cover design was assigned at random to one of the test markets. Regression model (20.3) was employed, where Y denotes thousands of copies sold in a test market during the experiment, X_1 denotes the price discount, and X_2 and X_3 are indicator variables coded as follows: $X_2 = 1$ if photograph, 0 otherwise; $X_3 = 1$ if drawing, 0 otherwise. The data and key regression results follow:

i:	1	2	3	4	5	6	7	8	9
X_{i1}:	50	50	50	75	75	75	95	95	95
X_{i2}:	0	0	1	0	0	1	0	0	1
X_{i3}:	0	1	0	0	1	0	0	1	0
Y_i:	14.89	16.34	15.02	20.21	21.55	19.31	23.41	25.42	22.13

k:	0	1	2	3	
b_k:	6.03685	0.18363	−0.68333	1.60000	$n = 9$
$s\{b_k\}$:	0.75311	0.00942	0.42468	0.42468	$SSR = 111.091$
					$SSE = 1.35266$

a. Plot on the same graph $\hat{Y}$ against X_1 for each of the three cover designs. Let X_1 range between 50 and 100 in the plot. What effect is estimated by (1) b_2, (2) b_3, (3) $b_3 - b_2$?

b. Construct a 95 percent confidence interval for (1) β_1, (2) β_3. Interpret each confidence interval.

c. Construct a 95 percent confidence interval for the mean sales in test markets when the cover design is a drawing and the price discount is 50 cents. Computer output shows that $s\{\hat{Y}_h\} = 0.37211$ for this case.

20.31 An industrial psychologist used regression model (20.3) to study the relationship between an employee's work involvement (Y, a composite behavioral rating on a 100-point scale) and the employee's age (X_1, in years), marital status ($X_2 = 1$, married; $X_2 = 0$, unmarried), and gender ($X_3 = 1$, male; $X_3 = 0$, female). Data were collected for 64 employees of the same firm, and the following regression results were obtained:

k:	0	1	2	3
b_k:	51.620	0.1514	4.9936	2.2510
$s\{b_k\}$:	11.299	0.2589	1.1627	1.6442

a. Plot on the same graph $\hat{Y}$ against X_1 for each combination of marital status and gender. Let X_1 range between 20 and 65 years in the plot. Give point estimates of the difference in the mean work involvement rating between (1) married women and unmarried women of the same age, (2) unmarried men and unmarried women of the same age.

b. Construct a 90 percent confidence interval for (1) β_1, (2) β_2. Does either interval span zero? What is the practical interpretation of this fact in each instance?

* 20.32 **Damaged Castings.** Data on the number of damaged castings in a shipment (Y) and the total number of castings in the shipment (X, in thousands) for 12 recent shipments follow:

Shipment i:	1	2	3	4	5	6	7	8	9	10	11	12
Total X_i:	5	10	4	10	7	8	8	5	10	5	12	6
Damaged Y_i:	30	51	26	52	40	43	45	31	52	30	59	36

Management wishes to know if the relation of the number of damaged castings to the total number of castings in a shipment is linear or curvilinear. An analyst has fitted quadratic regression model (20.24) to the data in an effort to obtain an answer. Some of the regression results follow:

k:	0	1	2	
b_k:	4.084	5.8478	−0.10736	$n = 12$
$s\{b_k\}$:	2.194	0.6036	0.03830	$SSR = 1273.920$
				$SSE = 4.3300$

a. Construct a scatter plot of the data. Does the plot suggest that the regression function is curvilinear? Comment.

b. State the estimated quadratic regression function. Show the data setup for computer input for the first three shipments in the format of Table 20.4.

c. Test whether the curvature effect can be dropped from the quadratic regression model, controlling the α risk at 0.05. State the alternatives, the decision rule, the value of the test statistic, and the conclusion. What is the practical interpretation of the conclusion?

20.33 A sports statistician studied the relation between the time (Y, in seconds) for a competitive swimming event and the swimmer's age (X, in years) for 20 swimmers with ages ranging from 8 to 18. The statistician employed quadratic regression model (20.24) and obtained the following results: $\hat{Y} = 147.3 - 11.11X + 0.2730X^2$, $s\{b_2\} = 0.1157$.

a. Plot the estimated regression function. Would it be reasonable to use this regression function for swimmers of age 40? Comment.

b. Construct a 95 percent confidence interval for the curvature effect coefficient. Interpret the confidence interval.

c. Test whether the curvature effect can be dropped from the quadratic regression model, controlling the α risk at 0.05. State the alternatives, the decision rule, the value of the test statistic, and the conclusion. What is the P-value of the test?

20.34 **Telephone System.** The number of telephones in service in a telephone company system (N, in millions) and the calls handled annually by the system (C, in billions) for each of the past 10 years are as follows:

i:	1	2	3	4	5	6	7	8	9	10
N_i:	1.01	1.16	1.24	1.32	1.50	1.68	1.74	1.88	1.92	1.96
C_i:	1.51	1.68	1.77	1.88	2.11	2.29	2.38	2.52	2.58	2.61

A company engineer theorizes that C and N are related by the curvilinear relation $\log C = \beta_0 + \beta_1 \log N$.

a. Construct a scatter plot of the data.

b. Make the transformations $Y = \log C$ and $X = \log N$, using common logarithms, and fit the simple linear regression model (18.3) to the transformed data. Plot the estimated regression function expressed in the original units on the scatter plot in part a. Is the curvature of the estimated regression function pronounced? Comment.

* **20.35** A company's personnel department studied salaries of employees who did not complete high school. In the study, employee's salary (Y, in \$ thousand) was regressed on number of years of work experience (X_1) and number of years of high school completed (X_2). The interaction regression model (20.26) was employed, and the estimated regression function was $\hat{Y} = 18.90 + 1.10X_1 + 3.05X_2 - 0.20X_1X_2$.

a. Plot $\hat{Y}$ against X_1 when $X_2 = 0$. On the same graph, plot $\hat{Y}$ against X_1 when $X_2 = 3$. Describe how the effect of X_1 on $\hat{Y}$ depends on the level of X_2 for this interaction model.

b. The two-sided P-value for test statistic $t^* = b_3/s\{b_3\}$ is 0.0271. Test whether the interaction term can be dropped from the regression model, controlling the α risk at 0.05. State the alternatives, the decision rule, and the conclusion.

c. Given the signs of the regression coefficients and the conclusion in part b, does it appear that X_1 and X_2 are reinforcing or interfering in their relationship with mean salary? Explain.

20.36 Refer to **Freezer Power Use** Problem 20.5. When interaction model (20.26) is fitted to these data, computer output shows that $b_3 = 0.04500$ and $s\{b_3\} = 0.05387$. Test whether the interaction term can be dropped from the regression model, controlling the α risk at 0.10. State the alternatives, the decision rule, the value of the test statistic, and the conclusion. What is the P-value of the test?

20.37 Refer to **Hotel Equipment** Problem 20.6. The manager also fitted interaction model (20.26) to the data and obtained $b_3 = 0.06855$ and $s\{b_3\} = 0.04810$. Test whether the interaction term can be dropped from the regression model, controlling the α risk at 0.05. State the alternatives, the decision rule, the value of the test statistic, and the conclusion. What is the P-value of the test?

20.38 Refer to **Reconditioning Cost** Problem 18.14. Two incinerator types (A, B) were involved in the study. Incinerators 1, 2, 3, and 7 are type A and the others are type B. Let prior operating hours be denoted by X_1, and let X_2 be an indicator variable so that $X_2 = 1$ if the incinerator is type A and $X_2 = 0$ if it is type B. Regression model (20.1) was fitted to the data, and the following results were obtained:

k:	0	1	2	
b_k:	0.86605	1.69748	0.41000	$n = 9$
$s\{b_k\}$:	0.04275	0.01712	0.01772	$SSR = 7.2314$
				$SSE = 0.0041849$

Interaction regression model (20.26) also was fitted to the data, and the following results were obtained:

k:	0	1	2	3	
b_k:	0.91561	1.67683	0.25061	0.06641	$n = 9$
$s\{b_k\}$:	0.03791	0.01536	0.06743	0.02755	$SSR = 7.2336$
					$SSE = 0.0019357$

a. For regression model (20.26), test whether $\beta_3 = 0$, controlling the α risk at 0.05 when $\beta_3 = 0$. State the alternatives, the decision rule, the value of the test statistic, and the conclusion.

b. For regression model (20.1), test whether $\beta_2 = 0$, controlling the α risk at 0.05 when $\beta_2 = 0$. State the alternatives, the decision rule, the value of the test statistic, and the conclusion.

c. Explain what the test conclusions in parts a and b imply.

20.39 In an analysis of the relation between sales (Y) and target population (X_1) and advertising expenditures (X_2) for the 10 sales territories of a large corporation, the following correlation matrix was obtained:

	X_1	X_2	Y
X_1	1.000	0.999	0.992
X_2		1.000	0.992
Y			1.000

a. What difficulties arise here in attempting to measure the separate effects of population and advertising expenditures on sales based on a multiple regression of Y on X_1 and X_2?

b. Will the difficulties cited in part a necessarily carry over to the making of inferences on Y? Explain.

c. Would simple regressions of Y on X_1 alone or of Y on X_2 alone show a high degree of linear association? Explain.

20.40 Refer to **Freezer Power Use** Problem 20.5. Results for three fitted regression models follow:

Statistic	Model 1: X_1	Model 2: X_1, X_2	Model 3: X_2
b_1	−4.20000	−4.20000	—
$s\{b_1\}$	0.60741	0.42862	—
b_2	—	1.21667	1.21667
$s\{b_2\}$	—	0.42862	1.63630

a. Construct a scatter plot of X_2 against X_1. Is multicollinearity present in these data? Can the same conclusion be drawn from the fact that $r_{12} = 0$ here? Explain.

b. Is the value of b_1 here dependent on whether or not X_2 is in the fitted model? Is the value of b_2 dependent on whether or not X_1 is in the fitted model? Are these findings consistent with the conclusion in part a? Comment.

20.41 Propeller Shafts. Data on the hardening of eight boat propeller shafts follow, where X_1 is the number of hours the shaft was in the hardening process, X_2 is the process temperature in degrees Celsius (°C), and Y is a measure of hardness of the shaft:

Shaft i:	1	2	3	4	5	6	7	8
Hours X_{i1}:	11.0	13.0	15.0	14.5	12.0	12.5	13.5	14.0
Temperature X_{i2}:	722	760	780	752	712	744	731	737
Hardness Y_i:	382	463	476	460	374	422	444	439

Results for three fitted regression models are as follows:

Statistic	Model 1: X_1	Model 2: X_1, X_2	Model 3: X_2
b_1	24.48120	13.47826	—
$s\{b_1\}$	5.62024	6.07759	—
b_2	—	0.91999	1.52585
$s\{b_2\}$	—	0.37260	0.32575
	$X_{h1} = 13.0, X_{h2} = 760$		
$\hat{Y}_h$	427.9	446.3	459.6
$s\{\hat{Y}_h\}$	7.095	9.094	8.800

a. Construct a scatter plot of X_2 against X_1. Is multicollinearity present in these data? Can the same conclusion be drawn from the fact that $r_{12} = 0.7332$ here?

b. Describe how the regression results demonstrate the problems encountered in assessing the separate effects of the independent variables in a multiple regression model when multicollinearity is present.

c. Do the difficulties cited in part b carry over to inferences on Y when such inferences are within the region of sample observations on the independent variables? Make reference to the inference results when $X_{h1} = 13.0, X_{h2} = 760$.

20.42 Refer to **Fermentation Process** Problem 20.4. The values of *MSE* for all possible regression models involving subsets of the three independent variables follow:

Variables in Model	MSE	Variables in Model	MSE
Intercept Only	297.33	X_2, X_3	185.91
X_2	191.65	X_1, X_3	104.06
X_3	186.47	X_1, X_2	103.20
X_1	117.47	X_1, X_2, X_3	105.87

a. Plot *MSE* against the number of independent variables for all possible regression models as in Figure 20.10.

b. According to the *MSE* criterion, which three models are best among all the models and should be considered for further analysis? Would these same models be identified if the criterion were to minimize *SSE*? Explain.

c. An observer suggests that the model with X_1 only, given its simplicity and predictive power, ought to be chosen to predict the amount of product from a batch. Should one reach such a conclusion without further analysis? Discuss.

20.43 Refer to the **Power Cells** data set (Appendix D.2). The *MSE* values for all possible regression models involving subsets of the four independent variables follow. Here *Y* denotes the number of discharge–charge cycles before failure, X_1 the charge rate, X_2 the discharge rate, X_3 the depth of discharge, and X_4 the ambient temperature.

Variables in Model	MSE	Variables in Model	MSE
Intercept Only	18,000	X_2, X_4	11,275
X_2	18,312	X_3, X_4	10,219
X_3	17,276	X_1, X_4	8,768
X_1	15,854	X_1, X_2, X_3	15,340
X_4	11,092	X_2, X_3, X_4	10,388
X_2, X_3	17,581	X_1, X_2, X_4	8,908
X_1, X_2	16,130	X_1, X_3, X_4	7,831
X_1, X_3	15,074	X_1, X_2, X_3, X_4	7,955

a. Plot *MSE* against the number of independent variables for all possible regression models as in Figure 20.10.

b. According to the *MSE* criterion, which three models are best among all the models and should be considered for further analysis? Would these same models be identified if the criterion were to maximize (1) R^2, (2) R_a^2? Explain.

c. What further analysis should be conducted before deciding on the regression model to be employed?

20.44 Refer to **Fermentation Process** Problem 20.4. When X_1 only is included in the regression model, the error sum of squares is $SSE(X_1) = 2466.782$.

a. Test whether both X_2 and X_3 can be dropped from the regression model containing X_1, X_2, and X_3. Control the α risk at 0.01. State the alternatives, the full and reduced regression functions, the decision rule, the value of the test statistic, and the conclusion.

b. If H_1 were concluded in part a, would this necessarily imply that both X_2 and X_3 should be retained in the model? Explain.

20.45 Refer to the *Restaurant Sales* example. The data are given in Table 20.3, and the regression results for the fitted model containing X_1, X_2, and X_3 are given in Figure 20.5. If sales volume (*Y*) is regressed on number of households (X_1) alone, the error sum of squares is $SSE(X_1) = 2566.029$.

a. Test whether X_2 and X_3 can be dropped from the regression model containing X_1, X_2, and X_3. Control the α risk at 0.05. State the alternatives, the full and reduced regression functions, the decision rule, the value of the test statistic, and the conclusion.

b. What is the implication of the conclusion in part a about the effect of location of restaurants?

20.46 (*Calculus needed.*) Derive the normal equations (20.5) for regression model (20.1).

20.47 Show that the coefficient of multiple determination R^2 is identical to the coefficient of simple determination obtained when the observations Y_i are regressed on the fitted values $\hat{Y}_i$ from the multiple regression model.

20.48

a. Show that $r_{Y1.2}^2 = (R^2 - r_{Y2}^2)/(1 - r_{Y2}^2)$, where R^2 is the coefficient of multiple determination for the regression of Y on X_1 and X_2.

b. Show for regression model (20.3) that F^* in (20.14) equals $(n - p)R^2/(p - 1)(1 - R^2)$.

20.49 Refer to **Bond Issues** Problem 20.29. What would have been the values of b_0, b_1, and b_2 if X_2 had been coded so that $X_2 = 1$ for a municipal bond and $X_2 = 0$ for a corporate bond? Would the fitted values $\hat{Y}_i$ be affected by this change in the coding scheme? Explain. [*Hint:* The original indicator variable X_2 and the new one X_2' are related by $X_2' = 1 - X_2$.]

20.50 Consider the regression function $E\{Y\} = \beta_0 X_1^{\beta_1} X_2^{\beta_2}$, where β_0, β_1, and β_2 denote regression coefficients and X_1 and X_2 denote independent variables. Show how the transformation $\log E\{Y\}$ converts this function into a linear regression function and explain the relation of the regression coefficients in the resulting linear regression function to the original regression coefficients.

20.51 A business researcher has a theory that a firm's rate of return on investment (Y) tends to increase with its accumulated research-and-development expenditures (X_1) according to a linear statistical relation. The researcher also anticipates that this linear relation between Y and X_1 may differ from one industry to another. To test the theory, data on Y and X_1 were gathered for 15 firms from two industries (coded $X_2 = 0$ and $X_2 = 1$, respectively). Variable Y was then regressed on X_1, X_2, and $X_3 = X_1 X_2$, using interaction regression model (20.26). Finally, procedure (20.31) was used to test the hypothesis H_0: $\beta_2 = \beta_3 = 0$. How should the researcher interpret the test result if (1) H_0 is concluded, (2) H_1 is concluded?

20.52 Response Time. The manager of a university computer system wishes to employ regression model (20.1) in a study of the relation between the average response time of the system (Y, in seconds) and the number of university terminal users (X_1) and commercial terminal users (X_2) signed on the system. The data for 24 randomly selected points in time follow:

X_1	X_2	Y	X_1	X_2	Y	X_1	X_2	Y	X_1	X_2	Y
10	0	0.08	48	4	0.52	23	3	0.13	69	12	1.06
36	8	0.59	21	1	0.13	66	7	0.81	58	10	0.74
75	5	0.77	66	10	0.88	10	2	0.15	26	2	0.23
16	4	0.21	30	3	0.28	70	10	1.00	14	4	0.18
35	5	0.45	55	9	0.70	44	0	0.28	62	9	0.72
50	8	0.56	49	6	0.63	42	4	0.43	63	3	0.55

a. Obtain (1) the correlation matrix, (2) the estimated regression function, (3) $s\{b_1\}$ and $s\{b_2\}$, (4) the ANOVA table, (5) the fitted values and residuals.

b. Obtain a 95 percent confidence interval to estimate the change in the mean response time of the system when (1) an additional university terminal user signs on the system and the number of commercial users is held constant, (2) an additional commercial terminal user signs on the system and the number of university users is held constant.

c. Test whether a regression relation holds, controlling the α risk at 0.05. State the alternatives, the decision rule, the value of the test statistic, and the conclusion.

d. Is multicollinearity a problem in this regression study? Comment, referring to appropriate results in part a to support your answer.

e. (1) Plot the residuals against the fitted values. (2) Plot the residuals against X_1. (3) Plot the residuals against X_2. (4) Construct a normal probability plot of the residuals. What do these plots show about the appropriateness of regression model (20.1)?

20.53 Hospital Admissions. Data on the region, number of beds (X_1), and number of admissions (Y) last year for each of 24 small acute-care hospitals follow:

Hospital i:	1	2	3	4	5	6	7	8
Region:	A	A	A	A	A	A	B	B
Beds X_{i1}:	19	120	49	100	33	22	96	48
Admissions Y_i:	460	3374	2244	3606	950	703	2958	1487

Hospital i:	9	10	11	12	13	14	15	16
Region:	B	B	B	B	B	B	B	C
Beds X_{i1}:	148	101	66	138	25	193	44	76
Admissions Y_i:	4700	3308	2696	4845	1159	5692	1576	2648

Hospital i:	17	18	19	20	21	22	23	24
Region:	C	C	C	C	C	C	C	C
Beds X_{i1}:	75	84	13	40	69	125	13	32
Admissions Y_i:	2757	2881	402	1600	1646	4825	370	987

Use X_2 and X_3 to define the regions as follows: $X_2 = 1$ if region A, 0 otherwise; $X_3 = 1$ if region B, 0 otherwise. Assume that regression model (20.3) is appropriate.

a. Obtain (1) the correlation matrix; (2) the estimated regression function; (3) $s\{b_k\}$, $k = 1, 2, 3$; (4) the ANOVA table; (5) the fitted values and residuals.

b. Interpret the meanings of b_1, b_2, and b_3. Give a point estimate of the mean number of admissions for 75-bed hospitals in region C.

c. Calculate $\sqrt{MSE}$. What does this number estimate here?

d. Construct a 95 percent prediction interval for the number of admissions to a particular 75-bed hospital in region C. Computer output shows that $s\{\hat{Y}_h\} = 128.419$ in this case.

e. Test whether a regression relation holds, controlling the α risk at 0.01. State the alternatives, the decision rule, the value of the test statistic, and the conclusion. Does the conclusion necessarily imply that mean admissions differ among the three regions for hospitals with a given number of beds? Comment.

f. Test whether X_2 and X_3 can be dropped from the regression model, controlling the α risk at 0.01. State the alternatives, the decision rule, the value of the test statistic, and the conclusion.

20.54 Refer to the *Machine Shop* example and the data in Table 20.6. A student, after studying the example, stated that the multicollinearity in the regression of Y on X_1 and X_2 would largely disappear if the total weight in excess of 100 pounds per shaft for each shipment were used in place of the total weight of shafts in the shipment, that is, if the variable $X_2' = X_2 - 100X_1$ were to replace X_2 in the regression model.

a. Calculate the coefficient of correlation between X_1 and X_2'. Do you agree with the student's statement? Explain.

b. Regress Y on X_1 and X_2', and obtain the new estimated regression coefficients b_1' and b_2' and their estimated standard deviations $s\{b_1'\}$ and $s\{b_2'\}$. How do these values compare with the corresponding values in Table 20.7 when Y is regressed on X_1 and X_2? Comment.

c. The value of R^2 for the regression of Y on X_1 and X_2 is 0.996. Obtain R^2 for the regression of Y on X_1 and X_2'. Compare the two values of R^2. Was this outcome to be expected? Discuss.

Analysis of Variance 21

Analysis of variance (ANOVA) models, like regression models, are useful for studying the statistical relation between a dependent variable and one or more independent variables. In fact, these models may be viewed as special cases of the general linear regression model. However, ANOVA models allow a study of statistical relations from a different perspective and are therefore widely used in their own right. The nature of the ANOVA approach was partly explained in the preceding chapters on regression. In this chapter, we undertake a fuller study of this approach, with particular emphasis on ANOVA models.

ANOVA MODELS 21.1

Nature of ANOVA Models

We illustrate the nature of ANOVA models by two examples.

EXAMPLE □ **Soft Drink**

A firm developing a new citrus-flavored soft drink conducted an experiment to study consumer preferences for the color of the drink. Four colors were under consideration: colorless, pink, orange, and lime green. Twenty test localities were selected that were similar in sales potential and representative of the target market for this product. Each color was then randomly assigned to five of these localities for test marketing. Aside from color, other drink characteristics that might affect sales—such as price, flavor, degree of carbonation, sweetness, and calorie content—were held fixed in all the localities.

The experimental design used here was a completely randomized design. The dependent, or response, variable was number of cases sold per 1000 population during the test period, and the single independent variable was color of the drink.

An ANOVA model was employed to study the effect of color on sales. Figure 21.1a portrays this model. For each color, there is a probability distribution of sales per 1000 population. Each of the distributions is normal, with the same variability, but the means of the distributions differ. In Figure 21.1a, the illustration shows that the green color has the largest mean and thus tends to lead to highest sales. Since all of the probability distributions are normal with constant variance, the means μ_1, μ_2, μ_3, and μ_4 of the four distributions convey the information about the effect of color on sales. Thus, Figure 21.1a illustrates the situation where sales of orange and colorless drinks tend to be substantially smaller than those of pink and green drinks. □

FIGURE 21.1
Representations of single-factor ANOVA model

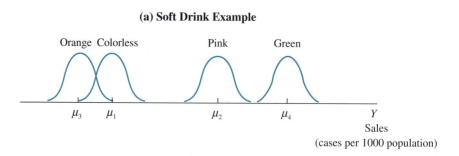

(a) Soft Drink Example

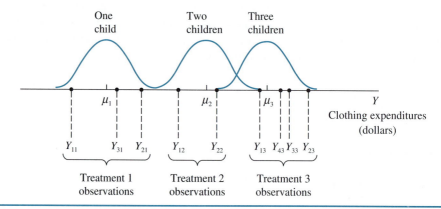

(b) Clothing Expenditures Example

Clothing Expenditures   **EXAMPLE**

Families with one, two, and three children in a middle-income school district were selected to study the effect of the number of children in the family on expenditures on children's clothing during the past year. In this observational study, the dependent, or response, variable was annual family expenditures for children's clothing (in dollars). The single independent variable was number of children in family. Three levels of the independent variable were included in the study—namely, one, two, and three children.

Figure 21.1b portrays the ANOVA model for this case. Note that there is a probability distribution of children's clothing expenditures for each level of the independent variable—that is, for each number of children in the family—and that these probability distributions are normal with the same variance. Thus, μ_1, μ_2, and μ_3 convey information about the effect of the number of children in the family on annual children's clothing expenditures. As might be expected, the positions of the means μ_1, μ_2, and μ_3 in Figure 21.1b indicate that a family's expenditures for children's clothing increase with the number of children in the family.   □

These two examples demonstrate that ANOVA models are applicable to both observational studies and experimental studies.

Factors and Treatments

The analysis of variance methodology has its own terminology. We shall now explain some of these special terms and relate them to the terminology of regression analysis.

Factors and Factor Levels.   In the special terminology of ANOVA, an independent variable is called a *factor*. Thus, in the soft drink example, the factor is color of drink. In the clothing expenditures example, the factor is number of children. A *factor level* is a particular outcome of the independent variable, such as lime green color of the soft drink or two children in the family. Observe that a factor may be a qualitative variable, such as color of drink, or a quantitative variable, such as number of children in family.

Single-Factor and Multifactor Studies.   We distinguish between *single-factor* and *multifactor* ANOVA studies, just as we did between simple and multiple regression. In multifactor ANOVA studies, there are two or more independent variables. A study to assess the effects of number of children in family (1, 2, 3) and family income (low, medium, high) on children's clothing expenditures would be a multifactor ANOVA study.

Fixed and Random Factors.   We also distinguish between *fixed* and *random* factors. A factor in an ANOVA model is fixed when our interest in the study involves only the particular levels of the factor under consideration. In the clothing expenditures example, for instance, the number of children in the family is a fixed factor because the three levels of the factor (1, 2, and 3 children) are the only levels that were of interest in the investigation. Occasionally, the factor levels included in the study are a sample of the factor levels of interest. The factor is then referred to as a random factor. For instance, consider a garment factory with 250 sewing machines. Five of the sewing machines were selected at random to study the wear characteristics of all of the factory's sewing machines. Here, the independent variable or factor is sewing machine, and there are five factor levels in the study. However, interest is not in the five machines selected but rather in all 250 sewing machines in the factory. Hence, sewing machine here is a random factor. In this chapter, we assume that the factors are fixed, which is the usual case.

Treatments.   In a single-factor ANOVA study, a factor level is also called a *treatment*. Thus, in the soft drink example, a particular color of the drink (e.g., lime green) is a treatment. In multifactor ANOVA studies, a particular combination of factor levels is called a treatment. For example, in a two-factor study of the effects of color and sweetness of a soft drink on sales, a particular color and sweetness combination (e.g., lime green and very sweet) is considered to be a treatment.

We turn now to the analysis for a single-factor ANOVA study. We begin with a description of the single-factor ANOVA model. *This model is applicable to independent samples designs with observational studies and to completely randomized designs with experimental studies.*

**SINGLE-FACTOR 21.2
ANOVA MODEL**

Description of Model

The single-factor ANOVA model, which we illustrated graphically for two examples in Figure 21.1, has the following features.

1. There are r factor levels or treatments under study.
2. For each treatment, the probability distribution of Y is normal.
3. All probability distributions of Y have constant variance, denoted by σ^2.
4. The mean of the probability distribution of Y for the jth treatment ($j = 1, 2, \ldots, r$) is denoted by μ_j and is referred to as the jth *treatment mean*. Differences in the treatment means μ_j reflect the varying effects of the treatments.

In a single-factor ANOVA study, n_j cases are selected for the jth treatment. The observation for the ith case for the jth treatment is denoted by Y_{ij}. This observation is assumed to represent a random selection from the probability distribution of Y for the jth treatment. The sampling process is illustrated in Figure 21.1b for the clothing expenditures example. Note that the sample sizes n_j need not be the same for all treatments.

As with regression models, we state the ANOVA model by indicating the components that make up an observation Y_{ij}.

(21.1)

The **single-factor ANOVA model** is:

$$Y_{ij} = \mu_j + \varepsilon_{ij} \qquad i = 1, 2, \ldots, n_j; \quad j = 1, 2, \ldots, r$$

where: Y_{ij} is the response in the ith case for the jth treatment
μ_j is the jth treatment mean
ε_{ij} are independent $N(0, \sigma^2)$

Important Features of Model

1. ANOVA model (21.1) assumes that the observed response Y_{ij} is the sum of two components: a fixed parameter μ_j and a random deviation or error term ε_{ij}. Thus, Y_{ij} is a random variable.
2. Since the error terms ε_{ij} are independent and normally distributed, so are the Y_{ij}.
3. Since $E\{\varepsilon_{ij}\} = 0$, it follows that $E\{Y_{ij}\} = \mu_j$. The parameter μ_j thus is the mean response for the jth treatment.
4. The parameter σ^2 is the common variance of the distributions of Y because $\sigma^2\{Y_{ij}\} = \sigma^2\{\varepsilon_{ij}\} = \sigma^2$ by (6.6a).
5. In view of these features, ANOVA model (21.1) can be expressed alternatively as follows.

(21.2) Y_{ij} are independent $N(\mu_j, \sigma^2)$ $i = 1, 2, \ldots, n_j; \quad j = 1, 2, \ldots, r$

Steps in Analysis

Since the probability distributions of Y corresponding to the different treatments are each normal with the same variance, differences in treatment effects are associated with differences between the treatment means μ_j. For this reason, single-factor ANOVA studies concentrate on making inferences about the treatment means μ_j and usually proceed in two stages.

1. Test whether the treatment means μ_j are equal. If they are equal, we say there is no factor effect, and no further analysis is required.
2. If the treatment means μ_j are not equal, study the nature of the treatment effects.

We now describe each of these stages of analysis.

We shall use the soft drink example to explain the test for equality of the treatment means.

**TEST FOR
EQUALITY OF
TREATMENT MEANS**

21.3

EXAMPLE ☐

Table 21.1 contains the sales data for the soft drink example. Each flavor was randomly assigned to five localities for test marketing, and the number of cases sold per 1000 population during the study period was recorded for each locality. Thus, we see from Table 21.1 that, in the first locality that was assigned to the colorless drink, $Y_{11} = 26.5$ cases per 1000 population were sold during the study period. Similarly, $Y_{12} = 31.2$.

☐

Point Estimation of Treatment Means

The least squares estimator of the treatment mean μ_j turns out to be simply the sample mean of the observations for the jth treatment, denoted by $\bar{Y}_j$.

TABLE 21.1
Data on sales—
Soft drink example
🔒

Case	Treatment j				
i	1 Colorless	2 Pink	3 Orange	4 Green	Total
1	26.5	31.2	27.9	30.8	
2	28.7	28.3	25.1	29.6	
3	25.1	30.8	28.5	32.4	
4	29.1	27.9	24.2	31.7	
5	27.2	29.6	26.5	32.8	
Total	136.6	147.8	132.2	157.3	573.9

Treatment mean: $\bar{Y}_1 = 27.32 \quad \bar{Y}_2 = 29.56 \quad \bar{Y}_3 = 26.44 \quad \bar{Y}_4 = 31.46$

Number of cases: $n_1 = 5 \quad n_2 = 5 \quad n_3 = 5 \quad n_4 = 5$

Total number of cases: $n_T = 5 + 5 + 5 + 5 = 20$

Overall mean: $\bar{\bar{Y}} = \dfrac{573.9}{20} = 28.695$

$$(21.3) \qquad \bar{Y}_j = \frac{\sum_{i=1}^{n_j} Y_{ij}}{n_j}$$

The estimated treatment mean $\bar{Y}_j$ is an unbiased estimator of μ_j.

☐ EXAMPLE

We see from Table 21.1 for the soft drink example that the point estimate of μ_1, the mean sales for the colorless drink, is $\bar{Y}_1 = 27.32$ cases per 1000 population. Note from Table 21.1 that mean sales were highest for the green color. ☐

Sums of Squares

Total Sum of Squares. As we saw from the use of ANOVA in regression analysis, ANOVA procedures utilize partitions of the total sum of squares. Recall from regression analysis that the total sum of squares, *SSTO*, is based on the deviations of the observations around their mean. We now denote the overall mean for all of the sample data by $\bar{\bar{Y}}$ and the total number of cases in the study by n_T.

$$(21.4) \qquad \bar{\bar{Y}} = \frac{\sum\sum Y_{ij}}{n_T}$$

where: $n_T = \sum n_j$

As before, we omit the indexes of summation when the nature of the summation is clear.

The deviations for the total sum of squares in the ANOVA notation then are $Y_{ij} - \bar{\bar{Y}}$, and *SSTO* is the sum of the squares of these deviations.

$$(21.5) \qquad SSTO = \sum\sum (Y_{ij} - \bar{\bar{Y}})^2$$

☐ EXAMPLE

In the soft drink example, the total sample size is $n_T = 20$ (Table 21.1), and the overall mean is:

$$\bar{\bar{Y}} = \frac{26.5 + 28.7 + \cdots + 32.8}{20} = \frac{573.9}{20} = 28.695$$

Hence, we find *SSTO* as follows:

$$SSTO = (26.5 - 28.695)^2 + (28.7 - 28.695)^2 + \cdots + (32.8 - 28.695)^2$$
$$= 115.9295$$

☐

Error Sum of Squares. The error sum of squares, *SSE*, measures the variability of the observations Y_{ij} when information about the treatment received is utilized. Hence, the deviations are taken around the estimated treatment means $\overline{Y}_j$ and are of the form $Y_{ij} - \overline{Y}_j$. *SSE* is simply the sum of the squares of these deviations, first summed within a treatment and then summed over all treatments.

$$(21.6) \qquad SSE = \sum_j \left[\sum_i (Y_{ij} - \overline{Y}_j)^2 \right]$$

Note that if all of the observations within each treatment are the same, we have $Y_{ij} \equiv \overline{Y}_j$ and hence:

$$\sum_i (Y_{ij} - \overline{Y}_j)^2 = 0$$

for each treatment. Consequently, $SSE = 0$ in this case. Also note that if all treatment means $\overline{Y}_j$ are equal, so that $\overline{Y}_j \equiv \overline{\overline{Y}}$, then a comparison of (21.5) and (21.6) shows that *SSE* equals *SSTO*. Thus, the closer the treatment means $\overline{Y}_j$ are to each other, the closer *SSE* will be to *SSTO*.

EXAMPLE ☐

For the soft drink example, we obtain *SSE* by first calculating the sums of squares within each treatment. Thus, for treatment 1, we have:

$$\sum_i (Y_{i1} - \overline{Y}_1)^2 = (26.5 - 27.32)^2 + (28.7 - 27.32)^2 + \cdots$$
$$+ (27.2 - 27.32)^2 = 10.6880$$

The corresponding calculations for the other three treatments are:

$$(31.2 - 29.56)^2 + \cdots + (29.6 - 29.56)^2 = 8.5720$$
$$(27.9 - 26.44)^2 + \cdots + (26.5 - 26.44)^2 = 13.1920$$
$$(30.8 - 31.46)^2 + \cdots + (32.8 - 31.46)^2 = 6.6320$$

SSE is then the sum of these within-treatments sums of squares:

$$SSE = 10.6880 + 8.5720 + 13.1920 + 6.6320 = 39.0840$$

☐

Treatment Sum of Squares. The difference between the total sum of squares *SSTO* and the error sum of squares *SSE* is called the *treatment sum of squares* and is denoted by *SSTR*.

$$(21.7) \qquad SSTR = SSTO - SSE$$

As we noted earlier, if the estimated treatment means $\overline{Y}_j$ are close to each other, SSE will be close to $SSTO,$ and $SSTR$ will therefore be small. If, however, the $\overline{Y}_j$ are not close to each other, $SSTR$ will tend to be larger.

The treatment sum of squares is actually a sum of squares made up of deviations of the estimated treatment means $\overline{Y}_j$ around the overall mean $\overline{\overline{Y}}$, that is, deviations of the form $\overline{Y}_j - \overline{\overline{Y}}.$ $SSTR$ is expressed in terms of these deviations as follows.

(21.8) $$SSTR = \sum n_j(\overline{Y}_j - \overline{\overline{Y}})^2$$

Thus, $SSTR$ is the sum of weighted squared deviations of the estimated treatment means $\overline{Y}_j$ around the overall mean $\overline{\overline{Y}}$, the weights being the number of cases n_j. Note from formula (21.8) that the farther apart are the estimated treatment means $\overline{Y}_j$, the larger will be $SSTR$. On the other hand, if all $\overline{Y}_j$ are equal, then $SSTR = 0$.

☐ **EXAMPLE**

In the soft drink example, by (21.7) we obtain:

$$SSTR = SSTO - SSE = 115.9295 - 39.0840 = 76.8455$$

The same result can be obtained by direct calculation of $SSTR$ using (21.8):

$$\begin{aligned} SSTR &= 5(27.32 - 28.695)^2 + 5(29.56 - 28.695)^2 \\ &\quad + 5(26.44 - 28.695)^2 + 5(31.46 - 28.695)^2 \\ &= 76.8455 \end{aligned}$$ ☐

Formal Statement of Partitioning.   As in regression analysis, we have now obtained a partitioning of the total sum of squares into two components.

(21.9)

For single-factor ANOVA model (21.1), the decomposition of the total sum of squares into two additive components is:

(21.9a) $SSTO \quad = \quad SSTR \quad + \quad SSE$

(21.9b) $\sum\sum(Y_{ij} - \overline{\overline{Y}})^2 = \sum n_j(\overline{Y}_j - \overline{\overline{Y}})^2 + \sum\sum(Y_{ij} - \overline{Y}_j)^2$

☐ **EXAMPLE**

We obtain the following decomposition for the soft drink example:

$$115.9295 = 76.8455 + 39.0840$$
$$SSTO \quad = \quad SSTR \quad + \quad SSE$$ ☐

Comments

1. Algebraically equivalent formulas for *SSTO*, *SSTR*, and *SSE* suitable for hand computation are available.

$$(21.10) \quad SSTO = \sum\sum Y_{ij}^2 - \frac{\left(\sum\sum Y_{ij}\right)^2}{n_T}$$

$$(21.11) \quad SSTR = \sum_j \frac{\left(\sum_i Y_{ij}\right)^2}{n_j} - \frac{\left(\sum\sum Y_{ij}\right)^2}{n_T}$$

$$(21.12) \quad SSE = SSTO - SSTR$$

2. *SSTR* and *SSE* are sometimes called, respectively, the *between-treatments* and *within-treatments* sums of squares.

3. Formula (21.9b) can be derived by noting that the total deviation $Y_{ij} - \overline{\overline{Y}}$ is the sum of the error deviation $Y_{ij} - \overline{Y}_j$ and the treatment deviation $\overline{Y}_j - \overline{\overline{Y}}$:

$$Y_{ij} - \overline{\overline{Y}} = (Y_{ij} - \overline{Y}_j) + (\overline{Y}_j - \overline{\overline{Y}})$$

Squaring both sides and adding over all cases yields:

$$\sum\sum(Y_{ij} - \overline{\overline{Y}})^2 = \sum\sum[(Y_{ij} - \overline{Y}_j) + (\overline{Y}_j - \overline{\overline{Y}})]^2$$

Upon expanding the right side, we obtain:

$$\sum\sum(Y_{ij} - \overline{\overline{Y}})^2 = \sum\sum(Y_{ij} - \overline{Y}_j)^2 + \sum\sum(\overline{Y}_j - \overline{\overline{Y}})^2 + 2\sum\sum(Y_{ij} - \overline{Y}_j)(\overline{Y}_j - \overline{\overline{Y}})$$

The second sum on the right can be written $\sum n_j(\overline{Y}_j - \overline{\overline{Y}})^2$ because the term $(\overline{Y}_j - \overline{\overline{Y}})^2$ is constant when summing over i and is picked up n_j times ($i = 1, 2, \ldots, n_j$). The last sum on the right can be shown to equal zero.

Degrees of Freedom

As in regression analysis, each sum of squares for the single-factor ANOVA model has associated with it a number of degrees of freedom. *SSTO* has $n_T - 1$ degrees of freedom associated with it. There are n_T deviations $Y_{ij} - \overline{\overline{Y}}$, but these are subject to one constraint, namely, $\sum\sum(Y_{ij} - \overline{\overline{Y}}) = 0$.

SSTR has $r - 1$ degrees of freedom associated with it. There are r deviations $\overline{Y}_j - \overline{\overline{Y}}$, but these are subject to one constraint, namely, $\sum n_j(\overline{Y}_j - \overline{\overline{Y}}) = 0$.

Finally, *SSE* has $n_T - r$ degrees of freedom associated with it. Each factor level contributes a component sum of squares $\sum_i (Y_{ij} - \overline{Y}_j)^2$ to *SSE*. This component sum of squares is equivalent in form to a total sum of squares and therefore has $n_j - 1$ degrees of freedom associated with it. The degrees of freedom associated with *SSE* are then the sum of the degrees of freedom for each of the r factor level components:

$$\sum_{j=1}^{r} (n_j - 1) = n_T - r$$

We see again that the degrees of freedom are additive.

(21.13) $n_T - 1 = r - 1 + n_T - r$

$\underbrace{}$ $\underbrace{}$ $\underbrace{}$
df for   *df* for   *df* for
SSTO   *SSTR*   *SSE*

☐ **EXAMPLE**

For the soft drink example, where $n_T = 20$ and $r = 4$, the degrees of freedom associated with the sums of squares are as follows:

$$
\begin{aligned}
SSTR: & \quad 4 - 1 = 3 \\
SSE: & \quad 20 - 4 = \underline{16} \\
SSTO: & \quad 20 - 1 = 19
\end{aligned}
$$

☐

Mean Squares

Our interest is in the treatment mean square, denoted by *MSTR,* and in the error mean square, denoted by *MSE.* These are obtained in the usual way by dividing the respective sums of squares by their associated degrees of freedom.

(21.14) $MSTR = \dfrac{SSTR}{r - 1}$

(21.15) $MSE = \dfrac{SSE}{n_T - r}$

☐ **EXAMPLE**

For the soft drink example, we found earlier that $SSTR = 76.8455$ and $SSE = 39.0840$. Also, $r - 1 = 3$ and $n_T - r = 16$. Thus, we calculate:

$$MSTR = \frac{76.8455}{3} = 25.61517$$

$$MSE = \frac{39.0840}{16} = 2.44275$$

☐

ANOVA Table

The ANOVA table for single-factor ANOVA model (21.1) is shown in Table 21.2. It follows the same format that we used for regression analysis.

Source of Variation	SS	df	MS	
Treatments	$SSTR = \Sigma n_j (\bar{Y}_j - \bar{\bar{Y}})^2$	$r - 1$	$MSTR = \dfrac{SSTR}{r - 1}$	**TABLE 21.2** ANOVA table for single-factor ANOVA model **(21.1)**
Error	$SSE = \Sigma\Sigma(Y_{ij} - \bar{Y}_j)^2$	$n_T - r$	$MSE = \dfrac{SSE}{n_T - r}$	
Total	$SSTO = \Sigma\Sigma(Y_{ij} - \bar{\bar{Y}})^2$	$n_T - 1$		

EXAMPLE ☐

Figure 21.2 contains MINITAB output for the analysis of variance for the soft drink example. Block 1 presents the ANOVA table. The numerical results are the same as those we obtained, except for the degree of rounding.

Block 2 contains for each treatment the sample size (labeled N), the estimated mean, and the estimated standard deviation. The factor levels are coded 1, 2, 3, and 4. Also presented in this block is $\sqrt{MSE}$, labeled POOLED STDEV. ☐

Comment

Almost all statistical packages perform single-factor ANOVA calculations.

F Test for Equality of Treatment Means

We are now in a position to carry out the first step in the strategy for the analysis of single-factor ANOVA studies—namely, to conduct a test of whether or not the treatment means μ_j are equal. The alternative conclusions for this test are the following.

(21.16) H_0: $\mu_1 = \mu_2 = \cdots = \mu_r$
H_1: Not all μ_j are equal

FIGURE 21.2
MINITAB output for single-factor ANOVA—Soft drink example

```
1.   ANALYSIS OF VARIANCE ON Y
     SOURCE      DF        SS        MS        F        p
     COLOR        3     76.85     25.62    10.49    0.000
     ERROR       16     39.08      2.44
     TOTAL       19    115.93


2.   LEVEL        N      MEAN     STDEV
       1          5    27.320     1.635
       2          5    29.560     1.464
       3          5    26.440     1.816
       4          5    31.460     1.288

     POOLED STDEV =     1.563
```

FIGURE 21.3
Statistical decision rule for F test of equality of treatment means for single-factor ANOVA model (21.1)

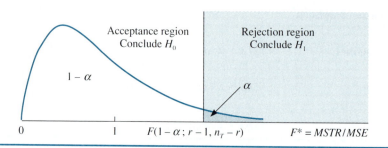

Here, H_0 implies that all distributions of Y have the same mean, and hence for ANOVA model (21.1) are identical. Alternative H_1 implies that the treatment means μ_j are not equal and hence that differential treatment effects are present. Note that H_1 is true for both examples portrayed in Figure 21.1.

The F^* test statistic is analogous to that for regression analysis.

$$\textbf{(21.17)} \qquad F^* = \frac{MSTR}{MSE}$$

Large values of F^* lead to conclusion H_1, as in testing for the presence of a regression relation. The reason is that $MSTR$ tends to be larger than MSE when H_1 holds, whereas the two mean squares tend to be of the same magnitude when H_0 holds. Figure 21.3 shows the appropriate upper-tail decision rule.

In constructing the decision rule, we make use of the following theorem, which tells us that F^* follows the F distribution when H_0 holds.

(21.18)
For single-factor ANOVA model (21.1), when $\mu_1 = \mu_2 = \cdots = \mu_r$:

$$F^* = F(r - 1, n_T - r)$$

where F^* is given by (21.17).

Figure 21.3 shows this sampling distribution of F^* when H_0 holds, the case for which the α risk is to be controlled. The figure also shows that the action limit of the decision rule is the $100(1 - \alpha)$ percentile $F(1 - \alpha; r - 1, n_T - r)$, since the risk of concluding H_1 when H_0 actually is true, represented by the shaded right-tail area, is to be controlled at α. The decision procedure may be summarized as follows.

(21.19)
When the alternatives are:

$$H_0: \mu_1 = \mu_2 = \cdots = \mu_r$$
$$H_1: \text{Not all } \mu_j \text{ are equal}$$

and single-factor ANOVA model (21.1) is applicable, the appropriate decision rule to control the α risk is as follows:

If $F^* \leq F(1 - \alpha; r - 1, n_T - r)$, conclude H_0.
If $F^* > F(1 - \alpha; r - 1, n_T - r)$, conclude H_1.

Here F^* is given by (21.17).

EXAMPLE ☐

In the soft drink example, we wish to test whether color of the drink affects mean sales.

Step 1. The alternatives here are:

$$H_0: \mu_1 = \mu_2 = \mu_3 = \mu_4$$
$$H_1: \text{ Not all } \mu_j \text{ are equal}$$

Step 2. It is desired to control the α risk at 0.05.

Step 3. Since $\alpha = 0.05$, $r - 1 = 3$, and $n_T - r = 16$, we require $F(0.95; 3, 16) = 3.24$. The decision rule is as follows:

If $F^* \leq 3.24$, conclude H_0.
If $F^* > 3.24$, conclude H_1.

Step 4. From our earlier calculations and Figure 21.2, $F^* = 25.61517/2.44275 = 10.49$. Since $F^* = 10.49 > 3.24$, we conclude H_1—that mean sales are not the same for the different colors. The small P-value of the test shown in the MINITAB output provides strong evidence that the sample data are not in accord with equal mean sales for the four colors.

The next step in the analysis will be to study the nature of the color effects. ☐

Comments

1. The expected values of the mean squares for ANOVA model (21.1) are as follows.

(21.20) $E\{MSE\} = \sigma^2$

(21.21) $E\{MSTR\} = \sigma^2 + \dfrac{\sum n_j(\mu_j - \mu)^2}{r - 1}$

where: $\mu = \dfrac{\sum n_j \mu_j}{n_T}$

Note that MSE is an unbiased estimator of the error variance σ^2, as for regression. Also note that $E\{MSTR\} = \sigma^2$ when all μ_j are equal because then $\mu_j \equiv \mu$, the weighted average of the μ_j. On the other hand, when the μ_j are not all equal, the second term on the right in (21.21) is positive and therefore $E\{MSTR\} > E\{MSE\}$.

This last result explains why $MSTR$ tends to be larger than MSE when the μ_j are not all equal and, hence, why large values of $F^* = MSTR/MSE$ lead to conclusion H_1 in test (21.19).

2. In Chapter 14, we discussed the comparison of two population means when the populations are normal with equal variance and the samples are independent. The test of $H_0: \mu_1 = \mu_2$ based on test statistic (14.15a) is a special case of the ANOVA F test here.

21.4 ANALYSIS OF TREATMENT EFFECTS

We now take up several procedures for investigating the nature of the treatment effects when the test of equality of the treatment means leads to the conclusion that the treatment means μ_j are not all equal. In particular, we shall take up interval estimation of the mean response for a given treatment and comparisons of mean responses for different treatments.

Interval Estimation of Treatment Mean

We noted earlier that the least squares estimator of the treatment mean μ_j is $\overline{Y}_j$, the sample mean of the observations for the jth treatment. The estimated variance of $\overline{Y}_j$, denoted by $s^2\{\overline{Y}_j\}$, is as follows.

$$(21.22) \qquad s^2\{\overline{Y}_j\} = \frac{MSE}{n_j}$$

Formula (21.22) has the usual form for the estimated variance of a sample mean—namely, an estimate of the population variance σ^2 (MSE in this case) divided by the sample size (n_j in this case).

A two-sided confidence interval for μ_j also takes the usual form.

(21.23)

The $1 - \alpha$ confidence limits for treatment mean μ_j for single-factor ANOVA model (21.1) are:

$$\overline{Y}_j \pm ts\{\overline{Y}_j\}$$

where: $t = t(1 - \alpha/2; n_T - r)$
$s\{\overline{Y}_j\}$ is given by (21.22)

The t multiple involves $n_T - r$ degrees of freedom because these are the degrees of freedom associated with MSE.

☐ EXAMPLE

In the soft drink example, we wish to estimate mean sales for the colorless version of the drink (treatment 1) with a 90 percent confidence interval. We know from earlier calculations and Figure 21.2 that $\overline{Y}_1 = 27.32$, $MSE = 2.44275$, $n_T - r = 16$, and $n_1 = 5$. By (21.22) we calculate:

$$s^2\{\overline{Y}_1\} = \frac{2.44275}{5} = 0.48855$$

Hence, $s\{\overline{Y}_1\} = \sqrt{0.48855} = 0.69896$. We require $t(0.95; 16) = 1.746$. The confidence limits therefore are $27.32 \pm 1.746(0.69896)$, and the 90 percent confidence interval is:

$$26.1 \le \mu_1 \le 28.5$$

Thus, it can be reported with 90 percent confidence that mean sales for the colorless version are between 26.1 and 28.5 cases per 1000 population. □

Comparison of Two Treatment Means

Frequently, we wish to compare the mean responses for two treatments. We shall denote the two treatments of interest by j and j'. The difference of interest then is as follows.

$$(21.24) \quad \mu_j - \mu_{j'}$$

Such a difference between a pair of treatment means is called a *pairwise comparison*.

An unbiased point estimator of $\mu_j - \mu_{j'}$ is the difference between the estimated treatment means.

$$(21.25) \quad \bar{Y}_j - \bar{Y}_{j'}$$

The estimated variance of this point estimator is as follows.

$$(21.26) \quad s^2\{\bar{Y}_j - \bar{Y}_{j'}\} = \frac{MSE}{n_j} + \frac{MSE}{n_{j'}}$$

The two-sided confidence interval for $\mu_j - \mu_{j'}$ is of the usual form.

(21.27)
The $1 - \alpha$ confidence limits for $\mu_j - \mu_{j'}$ for single-factor ANOVA model (21.1) are:

$$(\bar{Y}_j - \bar{Y}_{j'}) \pm t s\{\bar{Y}_j - \bar{Y}_{j'}\}$$

where: $t = t(1 - \alpha/2; n_T - r)$
$s\{\bar{Y}_j - \bar{Y}_{j'}\}$ is given by (21.26)

EXAMPLE □

In the soft drink example, the firm's advertising agency had initially recommended, on the basis of a laboratory session with a panel of consumers, that either lime green or pink be used in launching the product nationally. Hence, a major purpose of the experiment was to compare the effects of these two colors in an actual market setting.

The pairwise comparison of interest is $\mu_4 - \mu_2$, and it is to be estimated with a 90 percent confidence interval. The point estimator is $\bar{Y}_4 - \bar{Y}_2$. From Figure 21.2, we see that $\bar{Y}_4 - \bar{Y}_2 = 31.46 - 29.56 = 1.90$. We obtain $t(0.95; 16) = 1.746$ and require $s\{\bar{Y}_4 - \bar{Y}_2\}$. Since $MSE = 2.44275$, $n_4 = 5$, and $n_2 = 5$, we calculate by (21.26):

$$s^2\{\bar{Y}_4 - \bar{Y}_2\} = \frac{2.44275}{5} + \frac{2.44275}{5} = 0.97710$$

and hence $s\{\bar{Y}_4 - \bar{Y}_2\} = \sqrt{0.97710} = 0.98848$. The confidence limits therefore are $1.90 \pm 1.746(0.98848)$, and the desired 90 percent confidence interval is:

$$0.2 \leq \mu_4 - \mu_2 \leq 3.6$$

Thus, we estimate with 90 percent confidence that mean sales with green color are somewhere between 0.2 and 3.6 cases per 1000 population greater than those with pink color. □

Comment

Formula (21.26) for the estimated variance of $\bar{Y}_j - \bar{Y}_{j'}$ is based on the independence of all error terms in the single-factor ANOVA model (21.1). As a consequence, $\bar{Y}_j$ and $\bar{Y}_{j'}$ are independent and therefore, by (6.8b), the variance of $\bar{Y}_j - \bar{Y}_{j'}$ is the sum of the variances of $\bar{Y}_j$ and $\bar{Y}_{j'}$. Each of these variances is estimated by (21.22).

Simultaneous Comparisons

In single-factor ANOVA studies with three or more treatments, interest often centers on several pairwise comparisons. In the soft drink example, for instance, the sales manager might wish to compare mean sales for both green and pink colors with those for the colorless drink. The set of comparisons then would be as follows:

Comparison	Colors	Treatment Mean Difference
1	Green–colorless	$\mu_4 - \mu_1$
2	Pink–colorless	$\mu_2 - \mu_1$

Since the results of both comparisons will affect management's final marketing decision, the sales manager would like to have a known assurance that both confidence intervals in the set will be correct simultaneously. This situation calls for *simultaneous confidence intervals,* where the *joint confidence coefficient* indicates the probability that the procedure will lead to a set of confidence intervals that are all simultaneously correct. A variety of procedures are available for obtaining simultaneous confidence intervals. We shall discuss a simple procedure that is highly useful when the number of confidence intervals in the set is not too large.

Simultaneous Confidence Intervals.   When a $1 - \alpha$ confidence interval is constructed for each of m pairwise comparisons, using the usual procedure of the preceding section, the level of confidence that all m intervals are simultaneously correct will be less than $1 - \alpha$. The reason is that each confidence statement involves a risk α of being incorrect, and these risks compound when several such statements are made. Hence, if the joint confidence coefficient that all comparisons in the set are simultaneously correct is to be $1 - \alpha$, each individual confidence interval will need to be constructed with a confidence coefficient that involves a risk smaller than α.

The following theorem states how much smaller than α the risk of an incorrect confidence interval must be so that the overall risk for all m confidence intervals in the set does not exceed α.

(21.28)

If each of m confidence intervals is constructed with confidence coefficient $1 - (\alpha/m)$, the joint confidence coefficient for the set of m confidence intervals is at least $1 - \alpha$.

In other words, if the confidence interval for each pairwise comparison is constructed with confidence coefficient $1 - (\alpha/m)$, the probability that all m confidence intervals are simultaneously correct will be at least $1 - \alpha$.

Theorem (21.28) thus provides a simple procedure for guaranteeing the joint confidence level at $1 - \alpha$ when m pairwise comparisons are to be made.

(21.29)

The confidence limits for m simultaneous pairwise comparisons $\mu_j - \mu_{j'}$ with joint confidence coefficient of at least $1 - \alpha$ for single-factor ANOVA model (21.1) are:

$$(\bar{Y}_j - \bar{Y}_{j'}) \pm ts\{\bar{Y}_j - \bar{Y}_{j'}\}$$

where: $t = t(1 - \alpha/2m; n_T - r)$
$s\{\bar{Y}_j - \bar{Y}_{j'}\}$ is given by (21.26)

Note that the percentile of t that must be used is $100(1 - \alpha/2m)$. With risk $\alpha/2m$ placed in each tail, the combined risk in both tails is α/m and the confidence coefficient is therefore $1 - (\alpha/m)$ for each confidence interval.

EXAMPLE □

For the two pairwise comparisons in the soft drink example given earlier, we wish to guarantee the joint confidence coefficient at 90 percent. Hence, for $\alpha = 0.10$ and $m = 2$, each pairwise comparison requires a confidence coefficient of $1 - (\alpha/m) = 1 - (0.10/2) = 0.95$. The appropriate percentile for the t multiple therefore is the 97.5th percentile, because the risk 0.05 for a confidence interval must be divided equally in the two tails. Since $n_T - r = 16$, we obtain $t(0.975; 16) = 2.120$. The estimated standard deviation $s\{\bar{Y}_4 - \bar{Y}_2\}$ was calculated earlier from (21.26) to be 0.98848. It is applicable to all pairwise comparisons in this study because each treatment is based on the same number of cases. The remaining calculational details are as follows:

$\mu_j - \mu_{j'}$	$\bar{Y}_j - \bar{Y}_{j'}$	$\pm ts\{\bar{Y}_j - \bar{Y}_{j'}\}$
Green–colorless		
$\mu_4 - \mu_1$	$31.46 - 27.32 = 4.14$	$\pm 2.120(0.98848) = \pm 2.096$
Pink–colorless		
$\mu_2 - \mu_1$	$29.56 - 27.32 = 2.24$	$\pm 2.120(0.98848) = \pm 2.096$

Hence, the two simultaneous confidence intervals are:

$$2.04 \leq \mu_4 - \mu_1 \leq 6.24 \qquad 0.14 \leq \mu_2 - \mu_1 \leq 4.34$$

The joint confidence coefficient for the two intervals is at least 0.90. Hence, with confidence of at least 90 percent, we can conclude that both the green and pink drinks have greater mean sales than the colorless drink. □

21.5 RESIDUAL ANALYSIS

Residuals

Like all other statistical models, the appropriateness of single-factor ANOVA model (21.1) needs to be evaluated in any application. Residual plots are most helpful for this purpose. The residuals e_{ij} for single-factor ANOVA model (21.1) are defined as usual as the difference between the observed value and the fitted value. Here the fitted value $\hat{Y}_{ij}$ is equal to the estimated treatment mean $\bar{Y}_j$.

$$(21.30) \quad e_{ij} = Y_{ij} - \hat{Y}_{ij} = Y_{ij} - \bar{Y}_j$$

Thus, the residual e_{ij} represents the deviation of the observation Y_{ij} from the estimated treatment mean.

□ **EXAMPLE**

Table 21.3 contains the residuals for the soft drink example, calculated from the data in Table 21.1. For instance, the residual for the first case for color 1 is obtained as follows:

$$e_{11} = Y_{11} - \bar{Y}_1 = 26.5 - 27.32 = -0.82$$ □

Comments

1. Note from (21.6) and (21.30) that $SSE = \sum\sum e_{ij}^2$, as for regression analysis.
2. Table 21.3 shows that the residuals sum to zero for each treatment.

$$(21.31) \quad \sum_i e_{ij} = 0 \quad j = 1, 2, \ldots, r$$

Since there are r treatments, the n_T residuals are subject to r constraints of the form (21.31). This is consistent with our earlier discussion, which noted that SSE is associated with $n_T - r$ degrees of freedom.

Residual Plots

As in regression analysis, the residuals e_{ij} for ANOVA models should reflect the properties ascribed to the error terms ε_{ij} if the model is apt. ANOVA model (21.1), as we know, requires the error terms ε_{ij} to be independent and normally distributed with constant variance σ^2.

| Case | Treatment *j* | | | | |
i	1	2	3	4
1	−0.82	1.64	1.46	−0.66
2	1.38	−1.26	−1.34	−1.86
3	−2.22	1.24	2.06	0.94
4	1.78	−1.66	−2.24	0.24
5	−0.12	0.04	0.06	1.34
Total	0.00	0.00	0.00	0.00

TABLE 21.3
**Residuals for single-factor
ANOVA study—
Soft drink example**

The construction and use of residual plots for single-factor ANOVA proceeds in a manner similar to that for regression. Plots of residuals against fitted values are useful for examining the constancy of the error variance. These plots differ in appearance from corresponding plots for regression because all observations for a given treatment have the same fitted value for the ANOVA model. Dot plots of the residuals for each treatment, aligned on the same scale, also provide information about the constancy of the error variance. When the data are in time order, residual sequence plots are helpful for examining both the constancy of the error variance over time and the independence of the error terms. Normal probability plots are useful for determining whether the assumption of normality of the error terms is reasonable.

Since no new principles are involved in residual analysis for single-factor ANOVA, we proceed to three examples.

EXAMPLE □

Figure 21.4 contains two residual plots for the soft drink example. A plot of the residuals against their fitted values (i.e., against the estimated treatment means $\bar{Y}_j$) is presented in Figure 21.4a. The ranges of the scatter of the residuals are fairly similar, supporting the reasonableness of the constant variance assumption.

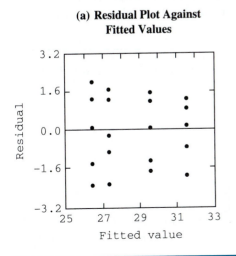

**(a) Residual Plot Against
Fitted Values**

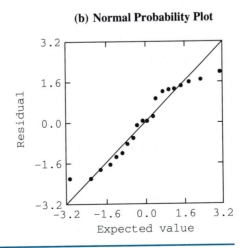

(b) Normal Probability Plot

FIGURE 21.4
**Residual plot and normal
probability plot—Soft drink
example**

FIGURE 21.5
Residual sequence plots suggesting lack of independence—Rug price example

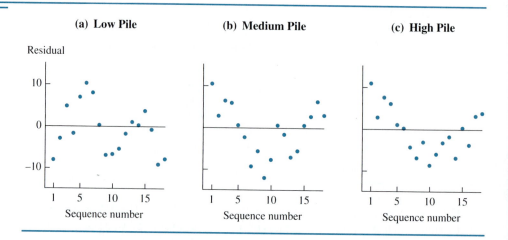

A normal probability plot of the residuals is presented in Figure 21.4b. The points form a moderately linear pattern, and most points are fairly close to the line of identity. Thus, the normality assumption for the error terms does not appear to be grossly violated here.

Rug Price ☐ **EXAMPLE**

Figure 21.5 contains residual sequence plots for a single-factor study based on ANOVA model (21.1), in which respondents were asked to estimate the price of a rug. The three treatments were different pile lengths of the rug. The residuals are plotted against observation sequence number, and the plots show tracklike patterns instead of random scatter. Subsequent investigation found that many respondents had been able to overhear the price estimate of the preceding respondent for the same rug. ☐

Risk Assessment ☐ **EXAMPLE**

Figure 21.6 contains residual sequence plots for an experiment on business risk assessments under four different incentive plans. The residuals are plotted in the sequence in

FIGURE 21.6
Residual sequence plots suggesting unequal treatment variances—Risk assessment example

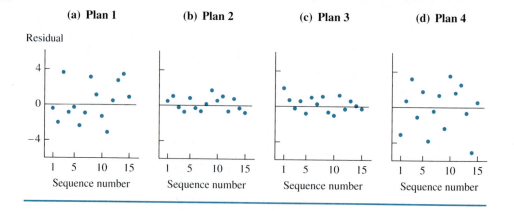

which the experiment was conducted. The plots suggest that the error variances for the four treatments are unequal, those for incentive plans 1 and 4 being greater than those for plans 2 and 3. □

Many ANOVA studies involve more than a single independent variable and are often based on matching or blocking designs. We shall briefly consider these extensions now.

OTHER TYPES 21.6 OF ANOVA STUDIES

Multifactor Studies

Just as many regression studies involve several independent variables to provide adequate explanatory or predictive power, many ANOVA studies are based on several factors.

EXAMPLE □ Power Cells

An experiment was conducted to study the effects of four factors on the life of silver–zinc power cells. The factors and the associated factor levels were as follows:

Factor	Levels
1. Charge rate	0.4, 1.0, 1.6 amperes
2. Discharge rate	1, 3, 5 amperes
3. Depth of discharge	40, 80 percent of rated ampere hours
4. Ambient temperature	10, 20, 30°C

The response variable was the life of the power cell, measured by the number of discharge–charge cycles that a power cell survives before failure. The experimental data appear in Appendix D.2.

In this experimental study, the four factors were studied at 3, 3, 2, and 3 levels, respectively. Thus, there were $3(3)(2)(3) = 54$ factor combinations or treatments in this experiment. □

Many ANOVA studies involve every possible treatment combination. The power cells experiment, for instance, included all 54 possible treatment combinations.

> (21.32)
> A multifactor study in which all combinations of the factor levels are investigated is called a **complete factorial study.**

When the number of factors or the numbers of their levels or both are large, the number of treatment combinations in a complete factorial study can be very large. For instance, with five factors, each with three levels, the number of treatment combinations is $3^5 = 243$. Complete factorial studies become unwieldy under these circumstances. The problem is usually dealt with by including only a subset of all possible treatments in the study. The particular subset chosen will depend on which individual and joint factor effects are of primary interest in the study. The subset is selected so that information is gained about these key effects. The price for using only a subset of all possible treatments in the study is that little or no information may be gained about effects that are of secondary interest.

> **(21.33)**
> A factorial study in which only a selected subset of all possible treatment combinations is investigated is called a **fractional factorial study.**

ANOVA models for multifactor studies are more complex than the single-factor ANOVA model (21.1) considered earlier. Still, the basic approach to the analyses of these studies is similar to that for single-factor studies. An ANOVA table can be constructed for a multifactor study and plays a prominent role in applying inference methods, just as it does in a single-factor study. Many statistical packages have routines for performing the analysis of variance for multifactor studies. Tests for the presence of factor effects, interval estimates for response means and for differences of response means, and other inferences can then be conducted for multifactor studies just as for single-factor studies.

Use of Matching and Blocking

The single-factor and multifactor studies described so far have been based on independent samples designs for observational studies and on completely randomized designs for experimental studies. In Chapter 14, we illustrated the potential benefits from the use of matching and blocking for comparative studies involving two populations. Matching and blocking can be used to equally good advantage with single-factor or multifactor studies. In many studies, the matching or blocking can be applied so that all treatment combinations are contained in each block. We now illustrate the use of a randomized block design for a factorial study.

Shelf Life □ **EXAMPLE**

In a study of the shelf life of a frozen food product, temperature and humidity in the freezer were each controlled at two levels (high and low), resulting in four treatments.

FIGURE 21.7
Illustration of blocking in a multifactor experiment—Shelf life example

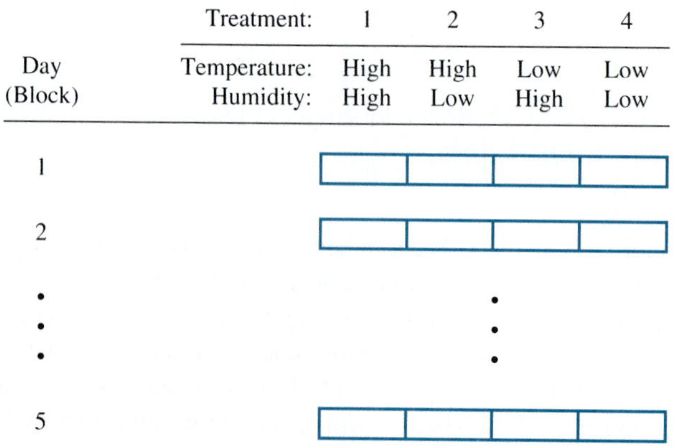

Four units of the product were selected from the production of each of five days, and the four treatments were randomly assigned to the four units for each day. The four units for each day's production constitute a block. Figure 21.7 illustrates the randomized block design for this two-factor study.

When analyzing the experimental results with this randomized block design, the treatment effects will be measured by comparing the shelf lives for the four units in each block. Thus, the effects of any day-to-day differences between the product units will be excluded from the experimental error because the four units in each block were produced on the same day. As a result, statistical tests will have greater power and estimates of treatment effects will be more precise than with a completely randomized design. □

OPTIONAL 21.7 TOPIC— CORRESPONDENCE BETWEEN ANOVA AND REGRESSION

As we noted earlier, ANOVA models may be viewed as special cases of the general linear regression model. We now illustrate the regression formulation of single-factor ANOVA model (21.1).

Development of Regression Model

In Chapter 20, we saw how indicator variables can be used in a regression model to represent a qualitative variable. Since the factor of interest in a single-factor ANOVA study (for example, color of soft drink) is equivalent to a qualitative variable, we shall use indicator variables for the factor levels in the regression function.

EXAMPLE □

For the soft drink example, the factor is color of soft drink, and the factor levels are the four colors: colorless, pink, orange, and green. The ANOVA model for the soft drink example is:

$$Y_{ij} = \mu_j + \varepsilon_{ij} \qquad j = 1, 2, 3, 4$$

To develop the equivalent regression model, we require three indicator variables, which we shall define as follows.

(21.34) $X_1 = \begin{cases} 1 & \text{if color is pink} \\ 0 & \text{otherwise} \end{cases}$

$X_2 = \begin{cases} 1 & \text{if color is orange} \\ 0 & \text{otherwise} \end{cases}$

$X_3 = \begin{cases} 1 & \text{if color is green} \\ 0 & \text{otherwise} \end{cases}$

Note that the factor level colorless has been chosen as the reference class; that is, the class for which all three indicator variables are set equal to zero.

The equivalent regression model then is as follows.

$$(21.35) \quad Y_i = \beta_0 + \beta_1 X_{i1} + \beta_2 X_{i2} + \beta_3 X_{i3} + \varepsilon_i \quad i = 1, 2, \ldots, n_T$$

where: $X_1 = 1$ if pink, 0 otherwise

$X_2 = 1$ if orange, 0 otherwise

$X_3 = 1$ if green, 0 otherwise

Here, the index i ranges over all $n_T = 20$ cases in the experiment.

Table 21.4 contains the soft drink data for the regression analysis. The sales data (Y) are repeated from Table 21.1. The indicator variables X_1, X_2, and X_3 are coded in accord with (21.34). For example, for the localities receiving reference treatment 1 (colorless), the indicator variables have the values $X_1 = 0$, $X_2 = 0$, and $X_3 = 0$. Similarly, for the localities receiving treatment 2 (pink), the indicator variables have the values $X_1 = 1$, $X_2 = 0$, and $X_3 = 0$. □

Correspondences of Parameters

We now develop the correspondences between the ANOVA parameters μ_j and the regression parameters β_k for the soft drink example. From regression model (21.35), we know that the response function for the regression model is as follows.

TABLE 21.4
Data for regression analysis of single-factor ANOVA study—Soft drink example

Case i	Treatment	X_{i1}	X_{i2}	X_{i3}	Y_i
1	Colorless	0	0	0	26.5
2	Colorless	0	0	0	28.7
3	Colorless	0	0	0	25.1
4	Colorless	0	0	0	29.1
5	Colorless	0	0	0	27.2
6	Pink	1	0	0	31.2
7	Pink	1	0	0	28.3
8	Pink	1	0	0	30.8
9	Pink	1	0	0	27.9
10	Pink	1	0	0	29.6
11	Orange	0	1	0	27.9
12	Orange	0	1	0	25.1
13	Orange	0	1	0	28.5
14	Orange	0	1	0	24.2
15	Orange	0	1	0	26.5
16	Green	0	0	1	30.8
17	Green	0	0	1	29.6
18	Green	0	0	1	32.4
19	Green	0	0	1	31.7
20	Green	0	0	1	32.8

$$(21.36) \quad E\{Y\} = \beta_0 + \beta_1 X_1 + \beta_2 X_2 + \beta_3 X_3$$

To develop the correspondences, we need the particular form of the response function for each treatment. For a case receiving reference treatment 1 (colorless), for which $X_1 = X_2 = X_3 = 0$, the response function simplifies to:

$$E\{Y\} = \beta_0 + \beta_1(0) + \beta_2(0) + \beta_3(0) = \beta_0$$

But according to the ANOVA model, the mean for treatment 1 is μ_1, so we have $\beta_0 = \mu_1$.

For a case receiving treatment 2 (pink), for which $X_1 = 1$, $X_2 = 0$, $X_3 = 0$, the response function (21.36) reduces to:

$$E\{Y\} = \beta_0 + \beta_1(1) + \beta_2(0) + \beta_3(0) = \beta_0 + \beta_1$$

Since the mean for treatment 2 in the ANOVA model is μ_2, it follows that $\beta_0 + \beta_1 = \mu_2$. However, we have just shown that $\beta_0 = \mu_1$, so it follows that $\beta_1 = \mu_2 - \mu_1$.

Continuing in this fashion, we establish the following correspondences between the parameters of the single-factor ANOVA model and the equivalent regression model.

$$(21.37a) \quad \beta_0 = \mu_1 \qquad \beta_1 = \mu_2 - \mu_1 \qquad \beta_2 = \mu_3 - \mu_1 \qquad \beta_3 = \mu_4 - \mu_1$$

$$(21.37b) \quad \mu_1 = \beta_0 \qquad \mu_2 = \beta_0 + \beta_1 \qquad \mu_3 = \beta_0 + \beta_2 \qquad \mu_4 = \beta_0 + \beta_3$$

Note from (21.37a) that β_1, β_2, and β_3 represent the differential effects of treatments 2, 3, and 4, each compared with reference treatment 1.

Inferences with Regression Model

Inferences about the treatment means in the ANOVA model can be carried out in terms of inferences about the regression coefficients of the equivalent regression model. We demonstrate two of these equivalent procedures for the soft drink example.

Estimating Treatment Means. To estimate a treatment mean μ_j from the regression model, we use the estimated regression coefficients.

EXAMPLE ☐

Table 21.5 shows the results from fitting regression model (21.35) to the soft drink data in Table 21.4 by means of a statistical package. Observe that regression coefficient b_0 is identical to the estimated mean for treatment 1 as given in Table 21.1; thus, $\bar{Y}_1 = b_0 = 27.32$. Likewise, observe that:

$$\bar{Y}_2 = b_0 + b_1 = 27.32 + 2.24 = 29.56$$

$$\bar{Y}_3 = b_0 + b_2 = 27.32 - 0.88 = 26.44$$

$$\bar{Y}_4 = b_0 + b_3 = 27.32 + 4.14 = 31.46$$

These correspondences follow from the relations between the μ_j and β_k given in (21.37b). □

Testing Equality of Treatment Means. To test the equality of treatment means with the regression model, we employ the F test for the existence of a regression relation.

□ **EXAMPLE**

The ANOVA test for the equality of the treatment means in the soft drink example involves the alternatives:

$$H_0: \ \mu_1 = \mu_2 = \mu_3 = \mu_4$$
$$H_1: \ \text{Not all } \mu_j \text{ are equal}$$

We see from (21.37a) that, when $\mu_1 = \mu_2 = \mu_3 = \mu_4$, the regression coefficients β_1, β_2, and β_3 equal zero. Hence, the equivalent test in the regression framework is:

$$H_0: \ \beta_1 = \beta_2 = \beta_3 = 0$$
$$H_1: \ \text{Not all } \beta_k = 0 \quad k = 1, 2, 3$$

These are precisely the alternatives for testing the existence of a regression relation. The appropriate test statistic, as given in (20.14), is $F^* = MSR/MSE$, and the test procedure is given in (20.15).

Referring to the regression output in Table 21.5, we see that the entries in the regression ANOVA table are the same as those in Figure 21.2 for ANOVA model (21.1), with SSR corresponding to $SSTR$. Thus, the F test for the equality of treatment means can be carried out in equivalent fashion by means of the F test for the existence of a regression relation. □

TABLE 21.5
Results for regression analysis—Soft drink example

Variable	b_k	$s\{b_k\}$	t^*	Two-Sided P-value
Constant	27.32000	0.69896	39.086	0.00000
X_1	2.24000	0.98848	2.266	0.03767
X_2	−0.88000	0.98848	−0.890	0.38652
X_3	4.14000	0.98848	4.188	0.00070

Source	SS	df	MS
Regression	76.84550	3	25.61517
Error	39.08400	16	2.44275
Total	115.92950	19	

$$F^* = 10.486 \qquad P\text{-value} = 0.00047$$

Comment

When the factor under study is qualitative, such as color of soft drink, an analyst can use the ANOVA model or the equivalent regression model with indicator variables. There is no other choice. When the factor is quantitative, however, an analyst can choose to use (1) an ANOVA model or the equivalent regression model with indicator variables, or (2) a regression model with the factor treated as a quantitative variable. Consider a study of the effect of size of crew (4, 8, 12 persons) on volume of output. If interest centers solely on the three crew sizes under study, the factor can be treated as a qualitative variable by means of the ANOVA model or the equivalent regression model with indicator variables. Neither of these two models requires any specification of a functional relation between crew size and volume of output. On the other hand, if the factor is treated as a quantitative variable in an ordinary regression model, a specification of the nature of the relation between crew size and volume of output (for instance, linear, quadratic) must be made. Making this specification is more restrictive than treating the independent variable as qualitative, but it does have the advantage of permitting inferences about crew sizes other than those in the study (for example, crews of 5 or 10 persons).

PROBLEMS

21.1 In each of the following single-factor ANOVA studies, identify (1) whether the study is experimental or observational, (2) the study design employed, (3) the factor, (4) the factor levels, (5) whether the factor is fixed or random, (6) the dependent variable.

a. A national automotive service chain operates three kinds of service centers: tire centers, brake system centers, exhaust system centers. An analyst wishes to compare quarterly profit rates for the different kinds of centers based on independent random samples of centers from the three populations of centers.

b. Three interviewers have been hired for a household survey. Households to be surveyed are randomly assigned to the interviewers in order to compare the effects of interviewers on responses about household income.

21.2 In each of the following single-factor ANOVA studies, identify (1) whether the study is experimental or observational, (2) the study design employed, (3) the factor, (4) the factor levels, (5) whether the factor is fixed or random, (6) the dependent variable.

a. In a study of the heat resistance of a certain type of wood paint, identical painted wood panels are exposed to one of three different temperatures in laboratory test rooms. Temperature levels are randomly assigned to the panels, and resistance is measured by the elapsed time until heat blistering first occurs.

b. Financial reports of companies randomly and independently selected from five different extractive industries are examined to compare the effects of a depletion tax allowance on reported annual earnings in these industries.

21.3 Television Commercials. Each of 15 persons in a consumer panel was shown one of three versions of a television commercial and was asked to rate its appeal on a scale from 0 (poor) to 10 (excellent). Five persons were assigned at random to each of the three versions. Assume that ANOVA model (21.1) is applicable and that the parameter values are known to be $\mu_1 = 6.5$, $\mu_2 = 3.3$, $\mu_3 = 5.6$, $\sigma = 1.0$.

a. Portray the model for this case graphically as in Figure 21.1a.

b. Explain the meaning of the following symbols: (1) μ_1, (2) Y_{32}, (3) ε_{41}, (4) σ.

c. Will Y_{23} necessarily be smaller than Y_{41}? Are Y_{23} and Y_{41} independent? Explain.

21.4 In a study of bonus pay earned under four bonus plans, five salespeople were assigned at random to each of these plans. Assume that ANOVA model (21.1) is applicable and that the parameter values (in dollars) are known to be $\mu_1 = 370$, $\mu_2 = 850$, $\mu_3 = 450$, $\mu_4 = 590$, $\sigma = 40$.

a. Portray the model for this case graphically as in Figure 21.1a.
b. Explain the meaning of the following symbols: (1) μ_1, (2) Y_{14}, (3) ε_{32}, (4) σ.
c. If $Y_{52} = 810$, what is ε_{52}?

21.5 For each of the following cases, state the degrees of freedom associated with (1) $SSTR$, (2) SSE, (3) $SSTO$.

a. Three treatments, seven observations for each treatment.
b. Four treatments, 15 observations for each treatment.
c. Five treatments; six observations for each of treatments 1, 2, 3, and 4, and four observations for treatment 5.

21.6 For each of the following cases, state the degrees of freedom associated with (1) $SSTR$, (2) SSE, (3) $SSTO$.

a. Three treatments, six observations for each treatment.
b. Five treatments, nine observations for each treatment.
c. Four treatments; seven observations for each of treatments 1 and 2, and eight observations for each of treatments 3 and 4.

* **21.7 Instructional Modes.** Fifteen students enrolled in a mathematics course were randomly divided into three groups of five students each. Each group was then randomly assigned to one of three instructional modes, augmenting the traditional course materials: (1) programmed text, (2) videotapes, (3) interactive computer programs. At the end of the course, each student was given the same achievement test. The test scores follow (note that the treatments are placed in rows):

Mode	Student i				
j	1	2	3	4	5
1	86	82	94	77	86
2	90	79	88	87	96
3	78	70	65	74	63

Assume that ANOVA model (21.1) is applicable.

a. State the values of (1) n_T, (2) n_1, (3) Y_{32}, (4) $\bar{Y}_2$, (5) $\bar{\bar{Y}}$.
b. For which instructional mode is the estimated treatment mean $\bar{Y}_j$ largest? Is the treatment mean μ_j for this instructional mode necessarily the largest of the three treatment means? Explain.
c. Obtain the ANOVA table.

21.8 Portable Radios. A product-rating organization tested battery life for four comparable makes of portable radios. Seven radios of each make were purchased off the shelf from local retail stores, and identical fully charged batteries were inserted on a random basis. The observations on battery playing hours at high volume follow (note that the treatments are placed in rows):

Make	Radio i						
j	1	2	3	4	5	6	7
1	5.5	5.0	5.2	5.3	4.8	4.8	5.4
2	4.7	3.9	4.3	4.5	4.1	4.3	4.0
3	6.1	5.7	5.0	5.3	5.2	6.3	5.8
4	4.5	5.1	4.3	4.1	4.5	5.1	4.2

Assume that ANOVA model (21.1) is applicable.

a. State the values of (1) n_T, (2) n_3, (3) Y_{24}, (4) $\bar{Y}_3$, (5) $\bar{\bar{Y}}$.
b. Obtain the ANOVA table. Are the tabulated results consistent with identities (21.9) and (21.13)?

21.9 Low-Energy Cookware. A kitchen utensils manufacturer selected 18 similar stores to try out three different promotional displays for a new low-energy cooking pot. The display that generates the highest sales in this study is to be used in the manufacturer's national promotion program. Each display was assigned at random to six stores. Sales (in dollars) for the stores during the two-week observation period follow (note that the treatments are placed in rows):

Display	Store i					
j	1	2	3	4	5	6
1	2161	1769	2748	1782	2830	3183
2	2379	1913	1119	1208	1962	1689
3	1479	1024	1598	963	1913	2251

Assume that ANOVA model (21.1) is applicable.

a. Obtain the three estimated treatment means. Which display generated the highest mean sales per store in the experiment? Does this display necessarily have the largest treatment mean μ_j? Explain.
b. Obtain the ANOVA table.

21.10 Dairy Ingredients. A food scientist studied the effect of the amount of a dairy ingredient on the volume of a baked cake (measured in milliliters per 100 grams). Twenty-four cakes were baked, using identical recipes and procedures except for the amount of the dairy ingredient which was varied at three levels (low: 1, medium: 2, high: 3). The 24 cakes were randomly assigned in equal numbers to each level of the ingredient. The cake volumes follow (note that the treatments are placed in rows):

Amount	Cake i							
j	1	2	3	4	5	6	7	8
1	351	369	381	386	370	358	398	375
2	390	394	406	407	415	375	374	388
3	398	409	415	399	434	427	414	420

Assume that ANOVA model (21.1) is applicable.

a. Obtain the three estimated treatment means. Do the values of these means suggest that cake volume tends to increase with the amount of the ingredient? Can one be certain of this tendency from these data? Explain.
b. Obtain the ANOVA table.

* 21.11 Refer to **Instructional Modes** Problem 21.7.

a. State the alternatives for the test of the equality of treatment means here. Specify the distribution of F^* if H_0 is correct.
b. Test for the equality of the treatment means, controlling the α risk at 0.05. State the decision rule, the value of the test statistic, and the conclusion.
c. Given the conclusion in part b, should the next step involve an examination of the differences among the individual treatment means? Comment.
d. What is the P-value of the test in part b? Is this value consistent with the test result in part b?

21.12 Refer to **Portable Radios** Problem 21.8.

a. Test for the equality of the treatment means, controlling the α risk at 0.01. State the alternatives, the decision rule, the value of the test statistic, and the conclusion.
b. Does the conclusion in part a imply that radio makes 2 and 4 have different mean playing hours at high volume? Comment.
c. If playing volume had not been controlled in the tests, would the interpretation of the results be affected? Discuss.

d. Is it likely that the differences between makes are actually due to differences in the test batteries? Explain.

e. What is the *P*-value of the test in part a? Is this value consistent with the test result in part a?

21.13 Refer to **Low-Energy Cookware** Problem 21.9.

a. Test whether the displays are equally effective in generating sales, controlling the α risk at 0.05. State the alternatives, the decision rule, the value of the test statistic, and the conclusion.

b. An observer states that the random assignment of each display to six stores makes any differences among the stores unimportant in measuring the differential sales of the displays. Respond to the observer, explaining what is gained by using 18 *similar* stores in the experiment.

c. What is the *P*-value of the test in part a? Is this value consistent with the test result in part a?

21.14 Refer to **Dairy Ingredients** Problem 21.10.

a. Test for the equality of the treatment means, controlling the α risk at 0.01. State the alternatives, the decision rule, the value of the test statistic, and the conclusion.

b. Obtain $\sqrt{MSE}$. What parameter does this statistic estimate? Interpret this parameter in the context of this experiment.

c. What is the *P*-value of the test in part a? Is this value consistent with the test result in part a?

* **21.15** Refer to **Instructional Modes** Problems 21.7 and 21.11.

a. Obtain a 95 percent confidence interval for μ_1. Interpret the confidence interval.

b. Obtain a 95 percent confidence interval for the pairwise comparison $\mu_2 - \mu_1$. Because the test for the equality of treatment means in Problem 21.11 led to conclusion H_1, what does the confidence interval obtained here suggest about the mean achievement score for instructional mode 3? Explain.

21.16 Refer to **Portable Radios** Problem 21.8.

a. Obtain a 99 percent confidence interval for the mean number of playing hours at high volume for radio make 3.

b. Does the confidence interval in part a provide appropriate limits for predicting the number of playing hours for another radio of make 3? Explain.

c. Obtain a 99 percent confidence interval for the difference in mean playing hours at high volume between radio makes 3 and 1. Interpret the confidence interval.

21.17 Refer to **Low-Energy Cookware** Problem 21.9.

a. The marketing manager had surmised in advance of the experiment that display 1 would be the most effective. Estimate the treatment mean for this display, using a 95 percent confidence interval. Interpret the confidence interval.

b. Obtain a 95 percent confidence interval for the difference in mean sales per store between displays 1 and 2. Does this confidence interval partly bear out the marketing manager's surmise described in part a? Comment.

21.18 Refer to **Dairy Ingredients** Problem 21.10. Obtain a 95 percent confidence interval for (1) μ_1, (2) $\mu_2 - \mu_1$. Interpret each of the confidence intervals.

* **21.19** Refer to **Instructional Modes** Problem 21.7. Construct simultaneous confidence intervals for the pairwise comparisons $\mu_2 - \mu_1$, $\mu_2 - \mu_3$, and $\mu_1 - \mu_3$ such that the joint confidence coefficient is at least 94 percent. Do the confidence intervals indicate that instructional mode 3 is clearly inferior to the other two modes? Explain. What other information is provided by the confidence intervals?

21.20 Refer to **Portable Radios** Problem 21.8. It is desired to construct simultaneous confidence intervals for the difference in mean playing hours for the two least expensive makes of radio (makes 2 and 4) and for the difference between the two makes with best tone (makes 1 and 4). Obtain these confidence intervals with a joint confidence coefficient of at least 90 percent. Interpret the confidence intervals.

21.21 Refer to **Low-Energy Cookware** Problem 21.9. Construct simultaneous confidence intervals for the differences in mean sales per store between displays 1 and 2 and between displays 1 and 3. Use a joint confidence coefficient of at least 0.96. Do the confidence intervals indicate that display 1 is clearly superior to both of the others? Comment.

21.22 Refer to **Dairy Ingredients** Problem 21.10. Construct simultaneous confidence intervals for $\mu_2 - \mu_1$, $\mu_3 - \mu_1$, and $\mu_3 - \mu_2$, using a joint confidence coefficient of at least 0.97. Do the confidence intervals indicate that mean cake volume increases with the amount of the ingredient? Comment.

* **21.23** Refer to **Instructional Modes** Problem 21.7.

a. Obtain the residuals. Verify that they sum to zero for each treatment.
b. Plot the residuals against the fitted values. Does the plot indicate that ANOVA model (21.1) is appropriate here with regard to the constancy of the treatment error variances? Comment.
c. Construct a normal probability plot of the residuals. Does the plot support the assumption of normality of the error terms? Explain.

21.24 Refer to **Portable Radios** Problem 21.8.

a. Obtain the residuals.
b. Construct dot plots of the residuals for each treatment aligned on the same scale. Do the plots support the reasonableness of the constant variance assumption? Comment.
c. Construct a normal probability plot of the residuals. Does the plot support the assumption of normality of the error terms? Explain.

21.25 Refer to **Low-Energy Cookware** Problem 21.9.

a. Obtain the residuals.
b. Plot the residuals against the fitted values. Does the plot support the assumption of the constancy of the treatment error variances? Comment.
c. Construct a normal probability plot of the residuals. Does the plot support the assumption of normality of the error terms? Explain.

21.26 Refer to **Dairy Ingredients** Problem 21.10.

a. Obtain the residuals.
b. For each treatment, the cakes were baked in the sequence indicated (from 1 to 8). Construct a residual sequence plot for each treatment. Do the plots support the assumptions of (1) independence of the error terms within each treatment, (2) constancy of the treatment error variances? Comment.
c. Construct a normal probability plot of the residuals. Does the plot support the assumption of normality of the error terms? Explain.

21.27 **Magazine Advertising.** In an advertising study, 48 readers were randomly divided into six groups, and the eight readers in each group were assigned one of six versions of a magazine. Each version of the magazine was identical except for one advertisement. This advertisement was varied with respect to combinations of size (half page, full page) and color (black, colored, multicolored). The six combinations were randomly assigned to the six groups. After reading the magazine, the readers were asked questions about their recall of the advertisement, including the name of the product advertised and the price of the product.

a. For this multifactor ANOVA study, identify (1) each factor and its factor levels, (2) the treatments, (3) the dependent variables, and (4) whether the study is experimental or observational.
b. Is this study a complete or a fractional factorial study?
c. Eight interviewers were employed in the experiment, each interviewing six readers receiving the six different advertisement treatments. What is the advantage of using this experimental design feature? Explain.

21.28 A random sample of 400 households was surveyed as part of a study of nonbusiness air travel to obtain information on the total expenditures on nonbusiness air travel in the preceding 12-month period. The selected households were cross-classified by (1) the number of employed persons in the household (0, 1, 2 or more), (2) whether or not the household included any children, (3) whether the household's residence was owned or rented, and (4) the total annual income of the household (under $50,000, $50,000–under $100,000, $100,000 or more).

a. For this multifactor ANOVA study, identify (1) each factor and its factor levels, (2) the treatments, (3) the dependent variable, (4) whether the study is experimental or observational.

b. Is there any guarantee in this study that every class in the cross-classification system contains the same number of households? What are the practical implications of this fact?

c. Is this study a complete factorial study? Explain.

* **21.29** Refer to **Instructional Modes** Problem 21.7. Let two indicator variables be defined as follows: $X_1 = 1$ if instructional mode is videotapes, 0 otherwise; $X_2 = 1$ if instructional mode is interactive computer programs, 0 otherwise.

a. State the regression model that is equivalent to ANOVA model (21.1) for this study.

b. State the indicator values for X_1 and X_2 for students assigned to each instructional mode.

c. The following results were obtained in fitting the regression model in part a: $b_0 = 85.0$, $s\{b_0\} = 2.769$, $b_1 = 3.0$, $s\{b_1\} = 3.916$. Construct a 95 percent confidence interval for (1) β_0, (2) β_1. What treatment characteristics are estimated by these confidence intervals?

21.30 Refer to **Portable Radios** Problems 21.8 and 21.12. Let three indicator variables be defined as follows: $X_1 = 1$ if radio is make 2, 0 otherwise; $X_2 = 1$ if radio is make 3, 0 otherwise; $X_3 = 1$ if radio is make 4, 0 otherwise.

a. State the regression model that is equivalent to ANOVA model (21.1) for this study.

b. For the regression model in part a, $SSR = 7.9129$ and $SSE = 3.3457$. Test for the equality of the treatment means using the regression approach. Control the α risk at 0.01. State the alternatives, the decision rule, the value of the test statistic, and the conclusion. Show the correspondence to the ANOVA approach in Problem 21.12a.

c. For the regression model in part a, $b_2 = 0.4857$ and $s\{b_2\} = 0.1996$. Construct a 99 percent confidence interval for β_2. What effect is estimated by this confidence interval?

21.31 Refer to **Low-Energy Cookware** Problem 21.9. Let two indicator variables be defined as follows: $X_1 = 1$ if store is assigned display 2, 0 otherwise; $X_2 = 1$ if store is assigned display 3, 0 otherwise. Selected results for the regression of store sales on X_1 and X_2 follow:

k:	0	1	2	
b_k:	_____	_____	−874.17	$SSR =$ _____
$s\{b_k\}$:	_____	_____	304.06	$SSE = 4{,}160{,}290$

a. Use the estimated treatment means and the ANOVA results from Problem 21.9 to obtain the missing regression results.

b. Test for the equality of the treatment means using the regression approach. Control the α risk at 0.05. State the alternatives, the decision rule, the value of the test statistic, and the conclusion.

c. Construct a 95 percent confidence interval for β_2. What effect is estimated by this confidence interval?

21.32 Refer to **Dairy Ingredients** Problem 21.10. Let two indicator variables be defined as follows: $X_1 = 1$ if cake has ingredient at level 2, 0 otherwise; $X_2 = 1$ if cake has ingredient at level 3, 0 otherwise. Selected results for the regression of cake volume on X_1 and X_2 follow:

k:	0	1	2	
b_k:	373.500	20.125	_____	$MSR =$ _____
$s\{b_k\}$:	_____	_____	_____	$MSE = 203.327$

a. Use the estimated treatment means and the ANOVA results from Problem 21.10 to obtain the missing regression results.

b. Test for the equality of the treatment means using the regression approach. Control the α risk at 0.01. State the alternatives, the decision rule, the value of the test statistic, and the conclusion.

c. Construct a 95 percent confidence interval for (1) β_0, (2) β_1. What treatment effects are estimated by these confidence intervals?

21.33 Refer to **Dairy Ingredients** Problems 21.10 and 21.32. What difference in assumptions is made when fitting the regression function in Problem 21.32 based on indicator variables and fitting a simple regression function in which cake volume is regressed on the actual amount of dairy ingredient in the cake? Which regression function will lead to a smaller value of SSE? Explain.

21.34 Refer to **Television Commercials** Problem 21.3. Obtain the following probabilities: (1) $P(Y_{23} < 6.0)$, (2) $P(Y_{41} - Y_{23} > 0)$.

21.35 If $\bar{Y}_j = \bar{\bar{Y}}$ for all r treatments, which of the following will be true? Which false? Why? (1) $SSTR = 0$, (2) $SSTO = SSTR + SSE$, (3) $SSE = 0$.

21.36 Refer to Comment 3 on p. 659. Show that $\Sigma \Sigma (Y_{ij} - \bar{Y}_j)(\bar{Y}_j - \bar{\bar{Y}}) = 0$.

21.37 Three treatments are investigated in an experiment, and five subjects are utilized for each treatment. Assume that ANOVA model (21.1) is applicable and that $\sigma^2 = 9$.

a. Using (21.20) and (21.21), obtain $E\{MSE\}$ and $E\{MSTR\}$ when (1) $\mu_1 = \mu_2 = \mu_3 = 10$; (2) $\mu_1 = 6$, $\mu_2 = 10$, $\mu_3 = 14$; (3) $\mu_1 = 0$, $\mu_2 = 10$, $\mu_3 = 20$.
b. What do the results in part a imply about the type of decision rule that is appropriate for the F test of the equality of treatment means?
c. Repeat part a, this time assuming that 10 subjects are utilized for each treatment. What do the results imply about the power of the test when the treatment sample sizes are increased?

21.38 Use the properties of ANOVA model (21.1) to prove (21.20).

$$\left[\text{Hint:} \quad E\left\{ \frac{\sum\limits_{i=1}^{n_j}(Y_{ij} - \bar{Y}_j)^2}{n_j - 1} \right\} = \sigma^2. \right]$$

21.39 Two confidence intervals are to be constructed from the same sample data; hence, they will not be independent. Let α_1 and α_2 denote the probabilities that the two intervals are *incorrect,* respectively. Use addition theorem (4.15) to prove that the probability of both intervals being correct is at least $1 - \alpha_1 - \alpha_2$. [*Hint:* $P(E_1 \cap E_2) = 1 - P(E_1^* \cup E_2^*)$.]

21.40 Refer to the **Financial Characteristics** data set (Appendix D.1). Consider only firms in the electronic computer equipment industry. Divide these firms into two groups: those that made a profit in year 1 and those that did not.

a. Calculate the rate of return (net income as a percentage of net assets) for each firm in year 2.
b. View the rates of return for the two groups as independent samples from normal populations with equal variances. Investigate whether the population mean rate of return in year 2 differs for firms that made a profit in year 1 and for those that did not. In the investigation, perform an F test for the equality of the population means, controlling the α risk at 0.05. Also, estimate the difference between the two population means by a 95 percent confidence interval.
c. For the test in part b, obtain test statistic t^* in (14.15a) and show numerically the equivalence of this test statistic to test statistic F^* in part b.
d. Test whether the two populations have equal variances, controlling the α risk at 0.02 when $\sigma_1^2 = \sigma_2^2$.
e. Using the assumption that the populations are normal, obtain a point estimate of the probability that a profitable firm in year 1 is unprofitable in year 2. Compare this estimate with a direct estimate of this probability based on the sample proportion. Does the direct estimate depend on any assumptions about the population distribution? Comment.

21.41 Refer to the **Power Cells** data set (Appendix D.2). It is desired to study the statistical relation between the number of cycles before failure (CYCLES) and the charge rate (CHARGE) and depth of discharge (DEPTH) for cells operating at an ambient temperature of 30°C. The study data consist of the 18 cells numbered 3, 6, 9, and so on. The six treatments under investigation are as follows:

Treatment j:	1	2	3	4	5	6
Charge Rate:	0.4	0.4	1.0	1.0	1.6	1.6
Depth of Discharge:	40	80	40	80	40	80

Assume that ANOVA model (21.1) is applicable.

a. List the three observations associated with each of the six treatments. Obtain the ANOVA table and test for the equality of the treatment means, controlling the α risk at 0.10. Interpret the test conclusion.

b. Regress the number of cycles before failure (Y) on the charge rate (X_1) and the depth of discharge (X_2), employing regression model (20.1). State the estimated regression function. Test whether a regression relation holds, controlling the α risk at 0.10. Does the test conclusion here have the same interpretation as the one in part a? Explain.

c. Which model, the ANOVA model in part a or the multiple regression model in part b, provides a better fit to the response data here as measured by the magnitude of MSE?

d. Which of the six treatments has the largest estimated mean number of cycles before failure according to the ANOVA model? Construct a 95 percent confidence interval for this mean parameter, using (1) the ANOVA model in part a, (2) the multiple regression model in part b. Are the two confidence intervals similar? Comment.

21.42 Air-Conditioning Breakdowns. A maintenance report for five aircraft in an airline's fleet contains the following data on the time intervals (in hours) between successive breakdowns in each aircraft's air-conditioning equipment. The time intervals are shown in chronological order for each aircraft:

	Aircraft				
Breakdown	1	2	3	4	5
1	17	100	54	6	13
2	31	98	15	36	270
3	179	87	33	281	91
4	45	230	68	54	38
5	27	81	132	254	603
6	198	98	67		118
7		140	60		450
8		488	229		
9		48	209		
10		89	102		

Assume that ANOVA model (21.1) is applicable when the logarithms of the time intervals are employed.

a. Test whether the mean log-time between successive breakdowns differs for the five aircraft, using the log-time data. Control the α risk at 0.01. Use logarithms to base 10.

b. Given the test conclusion in part a, would it be appropriate to combine all 38 sample observations and treat them as a random sample from the same population? Discuss.

c. Obtain the residuals for the log-time data and construct a residual sequence plot for each aircraft. Do the plots support the assumptions of (1) constant error variance and (2) independence of the error terms for each aircraft? Do these findings justify combining the residuals into one group for examining the normality of the error terms? Comment.

d. Construct (1) a normal probability plot of the residuals, (2) a stem-and-leaf display of the residuals. Is there evidence of serious lack of normality of the error terms in either plot? Explain.

Quality Improvement and Sample Surveys

Quality Control and Quality Improvement

22

Statistical thinking and statistical methods play a prominent role in quality control and quality improvement programs in business and industry. In this chapter we describe several quality-related applications of statistics that build on ideas presented in earlier chapters, including data analysis, probability, statistical inference, and the design of experiments. We begin by describing the nature of quality and the general role of statistics in achieving and maintaining quality.

Importance of Quality

QUALITY AND   **22.1**
ROLE OF STATISTICS

Quality is an important feature of any good or service, whether it is a bath towel, a physics textbook, or a telephone connection. Quality is desired in the smallest item, such as a ballpoint pen, and in the most complex system, such as a communications satellite. We shall use the term "product" to encompass any kind of item, component, good, service, or system to which the concept of quality may be applied. We shall also use the term "customer" to encompass any user of a product, whether it be an individual, a household, a firm, or some other organization.

Quality, as it relates to products, has many definitions but the following one is useful for our purposes here.

> **(22.1)**
> **Quality** is the state of a product in which it conforms to customer requirements.

This definition contains two main elements—requirements and conformance. Quality requirements for a product relate to its intended purpose or use. They are characteristics that the customer expects the product to have. For commercial products, quality requirements are determined by market needs and expectations. The following examples illustrate product quality requirements.

EXAMPLES ☐

1. A ballpoint pen is required to write for 15 hours without blotching or skipping.
2. An optical communications fiber is required to transmit light for 20 kilometers with an optical loss not to exceed 20 percent.

3. A sheet of metal alloy is required to have a thickness between 0.999 and 1.001 millimeters everywhere.

4. An airline flight is required to depart from the gate within 15 minutes of its scheduled time. ☐

A product generally is expected to meet several quality requirements. The ballpoint pen in Example 1, for instance, also has quality requirements with respect to the characteristics of its ink, clip, cap, and stem.

Note that quality requirements pertain to the needs of the customer and are not arbitrary specifications by the producer of the product. The needs of customers may vary, of course. For instance, some customers may need an automobile battery with a 36-month life, while others may need a battery with a 60-month life.

A second key element in the definition of quality in (22.1) is conformance of the product to customer requirements. A product is not a quality product unless it conforms to the customer requirements. Conformance is thus concerned with two questions: (1) Is the production process capable of creating a product that meets its quality requirements? (2) If so, is the process actually producing a quality product; in other words, is the process performing up to its capability?

For a finished product to conform to its quality requirements, all of its component parts must conform to their respective requirements. Also, each stage of the production process must adhere to its requirements. Producing a quality product therefore requires conformance throughout the whole system that creates the finished product, including conformance by purchased components, raw materials, and other productive factors that enter the production system.

Increasing worldwide competition has led to heightened attention to the quality of products. Many firms are placing strong emphasis on the quality of all processes in their organization, whether they be manufacturing, marketing, or financial ones. Leaders in the quality improvement drive, such as W. Edwards Deming and Genichi Taguchi, have emphasized the need for quality throughout the process of designing a product, designing the process to make the product, and the actual production of the product. Deming's 14 points (Ref. 22.1) stress the need to create constancy of purpose toward improvement of product and service, the need to improve constantly and always the system of production and service, the need for working as a team in solving problems, and the need for a program of education and training at all levels of the organization, including statistical education. Not only is the emphasis on quality important for its own sake, but also it frequently leads to improved productivity.

Steps to Achieving Quality

Quality is achieved (1) by careful design of the product and the process that makes the product, (2) by constant monitoring of the production process, and (3) by continuous assessments of the performance of the product in actual use by the customer. When problems are encountered at any stage, the source of the difficulty must be investigated and remedial actions taken. These may necessitate changes in the design of the product or of the production process, training of personnel, or the like. The quality improvement process is a continuous one, with feedback from each stage to earlier ones. We shall briefly describe now three important stages in the quality improvement process.

Product and Process Design.   Quality begins with the product's design. A badly designed product will not be able to conform to its quality requirements. Equally important is the design of the process that will make the product. A badly designed production process will not be capable of producing a quality product, even though the product is well designed. In fact, the designs of the product and the process must be created in tandem: A product can be said to be well designed only in relation to the design of the process that will produce it. For example, an automobile door latch may seem well designed when considered in isolation. However, if no manufacturing process is capable of producing it economically with existing technology, the latch's design is not really suitable.

Assessing Conformance to Design Requirements. Once a good product design has been achieved and a process has been developed that is capable of producing the product, the next concern is to make certain that the product conforms to the requirements of the design. Specifications and standards must be developed from the quality requirements that cover characteristics of both the process and the product. For example, specifications for an automobile door latch may require that a certain dimension of the door latch be within specified limits. The process must then be monitored to make sure that these specifications are being met. When departures from specifications occur, investigation and corrective action must be initiated. The investigation will need to track down the sources of trouble. It may be that the difficulty is connected with the management or operation of the production process, or it may be the result of the design of the product, the process, or both. Whichever the case, remedial action needs to be taken to bring the product and process back into control.

Assessing Performance. Once a product has been produced, its conformance to quality requirements must be measured in actual use. For example, a quality requirement for an automobile tire calls for it to function without failure under normal operating conditions for 100,000 kilometers. Information about conformity to this quality requirement may be obtained in various ways—for instance, from warranty claims, from customer complaints, and from customer surveys.

In some cases, actual use may be simulated under controlled conditions. For example, a ballpoint pen may be put on a test machine to see whether it can write continuously for 15 hours as required. Simulated-use tests are especially common at the product development stage, before the product reaches the customer. Such tests are not practical, however, when performance requirements involve very long time horizons under normal operating conditions. In these cases, the simulated use of a product can often be implemented under high stress conditions to speed up failure. Such tests are called *accelerated life tests*. For instance, a computer memory chip may be tested at an ambient temperature of 300°C in a laboratory environment to induce failure within a few weeks. Under normal conditions, where the ambient temperature is usually less than 40°C, failure may not be expected for several years.

The need for obtaining direct information from customers about the performance of the product exists even when simulated-use tests are employed. The reason is that these tests may not fully duplicate the actual usage conditions encountered by customers.

Role of Statistics in Quality Improvement Process

Statistical thinking and statistical methodology play a key role in the quality improvement process. The quality improvement programs of such leaders as A. V. Feigenbaum, J. M. Juran, W. E. Deming, and G. Taguchi are all based on the philosophy that decisions should be made based on facts and not hunches, that needed factual information should be obtained continuously, and that decision making must recognize the universal presence of variability.

Since decisions are made at all levels of an organization, all persons in the organization should have some training in statistics.

Many of the statistical methods useful for quality improvement are relatively simple and have already been taken up in this text. These include frequency distributions, histograms, scatter plots, bar charts, box plots, and probability plots. Several other statistical methods useful for quality improvement will be taken up in the remainder of this chapter.

Statistical methods play a role in all aspects of the quality maintenance and improvement process. The search for effective designs for new products and production processes and for improvements in existing designs is aided by the use of statistically designed experiments and the statistical analyses of the results of these experiments. Measuring the conformance of a product to requirements and ensuring that production processes are operating satisfactorily are aided by the use of statistical control procedures, including control charts and acceptance-sampling plans. Finally, the study of performance may involve statistical surveys of customers and statistical procedures related to reliability and the assessment of productive lifetimes.

In the remaining sections of this chapter, we present several of these quality applications of statistical procedures. We begin with the use of statistically designed experimental studies for product and process design.

22.2 PRODUCT AND PROCESS DESIGN

The improvement of quality by means of the design of a product and its production process is often called *off-line quality control*. Statistically designed experiments and statistical analysis play a key role in off-line quality control. We shall now briefly consider some major aspects of off-line quality control.

Pareto Chart

When experience with a product reveals that it is not conforming to its requirements, much can be learned by studying the types of nonconformities (defects) that are present and their relative frequencies of occurrence. A useful tool for this purpose is the *Pareto chart*. This chart is simply a bar chart that shows the relative frequencies of occurrence for the different types of nonconformities. The key types, which are usually few in number, will generally be readily evident from the Pareto chart. These are the ones on which off-line quality control needs to be concentrated.

Dress Shirts □ **EXAMPLE**

A manufacturer of men's dress shirts has retained a quality consultant to give advice on quality problems the firm is having with one of its product lines. As part of the consultant's investigation, 500 shirts were selected from the production process and inspected for defects to discover the types of shirt defects present and their relative frequencies of occurrence. The Pareto chart in Figure 22.1 summarizes the findings of this part of the investigation. It shows the types of defects found in the shirts, ordered by their relative frequency of occurrence. In total, 73 defects were found in the 500 shirts, with a few shirts having more than one defect. The chart clearly shows that sizing (i.e., sizes of cut pieces) and button defects were the most common nonconformities. Sizing defects alone accounted for 45 percent of the total, and button defects

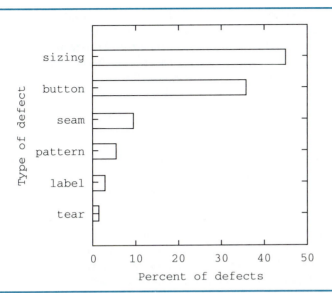

FIGURE 22.1
Pareto chart for shirt
defects—Dress shirts
example

accounted for another 36 percent. Thus, 81 percent of all defects were of these two types. ☐

As may be seen from this example, the Pareto chart gives an indication of the potential improvement in quality that can be achieved by the elimination of key types of defects. In the dress shirts example, for instance, total elimination of sizing and button defects will eliminate about 81 percent of the defects in the production process.

Comments

1. When frequency is not a good measure of the seriousness of a defect, another measure should be used in the Pareto chart, such as the cost of the defect or the time spent to remedy the defect.
2. Pareto charts sometimes show a plot of the cumulative distribution in addition to the frequency bars. This additional plot can be helpful at times in identifying the few types of defects that account for most of the defects found, but usually these are readily apparent from the bar chart alone.

Designed Experimentation

For many years, final inspection of finished product was utilized to weed out defective product. However, we have learned now that quality needs to be designed and built into the product in the first place. Inspection is too costly and unreliable and cannot transform a low-quality product design into a high-quality one. Statistically designed experiments are frequently used in studies of product and process design to identify the key variables that affect the quality characteristics and the target levels at which these variables should be controlled.

Cause-and-Effect Analysis.   Cause-and-effect analysis is helpful in identifying potential key variables that should be included in an experiment. This analysis begins with a given quality characteristic and identifies factors that affect this quality characteristic. In turn, each factor is then investigated as to other factors that affect it. The results of this cause-and-effect analysis are often presented schematically.

☐ **EXAMPLE**

In the dress shirts example, one important quality characteristic concerned the attachment of buttons. This quality characteristic is affected by the thread used, the stitching, and the like. The stitching in turn is affected by the sewing machine, the machine operator, and the like.

☐

Controllable and Noise Factors.   We distinguish between two types of factors that are identified in cause-and-effect and related analyses. Some factors represent variables that are within the control of the designer. We refer to these as *controllable factors*. For instance, the strength of a fastener or the composition of an alloy are controllable factors when they can be chosen by the designer.

Other factors affecting a quality characteristic cannot be controlled by the designer. Such uncontrollable factors are called *noise factors*. They often represent variable operating and environmental conditions to which a product may be exposed in actual use. For example, the life of an automobile battery is affected by the ambient temperatures that the battery encounters. This factor is not under the control of the product designer, because customers will use their automobiles under diverse conditions. Noise factors also include variables affecting a quality characteristic in the production process, such as temperature variations at different locations in a kiln.

Key noise factors often can be controlled in an experimental setting and should be included in quality improvement experiments, even though they cannot be controlled by the product and process designer. For example, the ambient temperature to which an automobile battery is exposed can be controlled in an experiment, even though it is a noise or uncontrollable factor as far as the product designer is concerned. The purpose of including noise factors in quality improvement experiments is to find product and process designs that make the quality characteristic insensitive to the noise factors. Such designs are called *robust designs* because they are relatively insensitive to the noise factors encountered in the production process or under actual operating conditions.

☐ **EXAMPLES**

1. An important quality characteristic of a caulking compound for bathroom tiles is resistance to discoloration. Different compounds were tested to determine the amount of discoloration when the caulking is exposed for an extended period to common bathroom chemicals, such as shampoos, conditioners, soaps, and cleaners. Here, the exposure to chemicals is a noise factor because it cannot be controlled in the product's actual use by the product designer. The objective of the experiment was to identify a compound that does not discolor when exposed to these chemicals.

2. The amount of rough handling a tomato encounters on its way to the store is a noise factor because it cannot be controlled by the product designer. A quality improvement experiment may investigate the amount of bruising of tomatoes with different

kinds of handling conditions to identify a package design that minimizes bruising under all of the handling conditions studied.   □

Target Level and Variability.   When using designed experiments to investigate the effects of controllable and noise factors on a quality characteristic, two effects of these factors are usually of importance: their effects on the level and on the variability of the quality characteristic. Consider the filling of jars of preserve, where the target amount of fill is 420 grams. Since uniformity of fill is unattainable, the quality requirement is stated in terms of a range within which the fill for each jar must fall. This specified range is 417 to 423 grams. The two limits are called the *lower specification limit (LSL)* and the *upper specification limit (USL)*, respectively.

The factors that affect the amount of fill (various control settings on the filling machine, density of the preserve, and the like) may principally affect the process level (mean fill) or the process variability (standard deviation of fills). Both the effects on process level and on process variability need to be studied because both can lead to nonconforming product. Figure 22.2a illustrates the situation when the process variability is satisfactory but the process level is unsatisfactory, and much nonconforming product is produced (i.e., the jars are frequently underfilled). Figure 22.2b illustrates the case when the process level is at the target level but the process variability is poor, again leading to much nonconforming product. Figure 22.2c illustrates the desirable case, in which both the process level and the process variability are satisfactory and conforming product is produced.

Frequently, the process level is easier to control than the process variability. A measure of the process capability to produce conforming product, when the process level can be placed at the target level and the distribution of the quality characteristic is approximately normal, is the following process capability index.

Illustrations of effects of process level and variability on conformance to specification limits   **FIGURE 22.2**

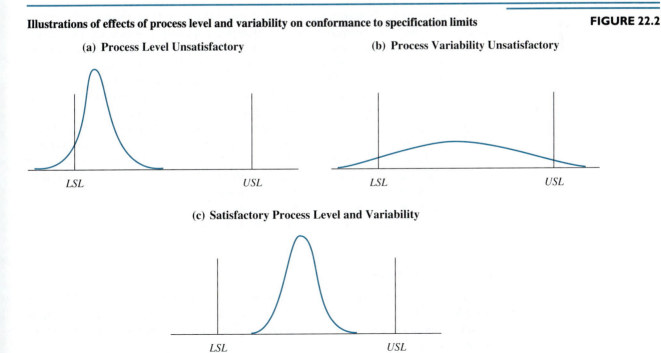

(a) Process Level Unsatisfactory

(b) Process Variability Unsatisfactory

LSL   USL   LSL   USL

(c) Satisfactory Process Level and Variability

LSL   USL

$$(22.2) \qquad \text{Process capability index} = \frac{USL - LSL}{6\sigma}$$

where: USL is the upper specification limit
LSL is the lower specification limit
σ is the process standard deviation

When this index equals 1, 99.7 percent of the product will be conforming, because the area within $\pm 3\sigma$ of the mean (i.e., a range of 6σ) in a normal distribution is 0.997. Frequently, the process capability index for a satisfactory process is required to be 1.33, in which case only 7 out of 100,000 units will be nonconforming when the distribution is normal. In cases where the quality characteristic is critical, the process capability ratio may need to be even higher.

Number of Factors and Factor Levels.   Because of the need to be concerned with both the process level and the process variability, many quality experiments have two main objectives for a given quality characteristic: (1) to identify those factors that can best bring the process level on target and (2) to identify those factors that can best minimize the process variability.

Frequently, many factors need to be included in these experiments. As a result, the number of possible treatments may be very large. This problem was noted in the previous chapter. For example, a complete factorial study with only eight factors, each at only two levels, would involve $2^8 = 256$ factor level combinations. As noted in Chapter 21, fractional factorial designs often are employed to scale down the size of these large experiments. Fractional designs, as we noted earlier, involve only a subset of the full set of possible factor combinations.

The levels of the factors in a quality experiment also need to be chosen with considerable care so that the experiment does not become too massive. For controllable factors, the selection of levels ideally should include all those that are under consideration for the product or process design. For example, if four types of oil might be used to lubricate a machine, the four types should be included in the experiment if possible. For noise factors, the selection of levels should cover the full range of variation that the production process or the product might experience from that noise factor. For example, if a water seal in a pump may experience actual pressure variations from 1200 to 2800 pounds per square inch (psi), an experiment testing such a seal should have a pressure factor that varies over this range; the selected levels might be 1000 and 3000 psi, for instance. It is often not sufficient to include only the one level of a factor that is expected to be the most adverse. For instance, in the water seal example, the most adverse condition would be the highest test pressure of 3000 psi. However, an experiment conducted only at the 3000 psi level would not reveal a leak problem if the seal does not sit tightly until it is under higher pressure.

Experimental Analysis

The analyses of data from quality experiments involve the types of statistical inference procedures that we discussed in previous chapters. Of course, the quality aims influence the questions asked and the insights drawn from the results. We illustrate these points by presenting an example.

The research division of a telecommunications company was designing a new optical fiber to be used in one of its new submarine cables to be laid across the Atlantic Ocean. One important quality characteristic was the breaking strain of the fiber, measured as the percent elongation that the fiber can tolerate without breaking. Two key factors affecting the process level of this quality characteristic were identified: the ambient temperature of the fiber and the ambient relative humidity of the fiber. Temperature was considered to be a noise factor because it depends on the conditions when the cable is laid and while resting on the ocean floor. Humidity was considered to be a controllable factor because it can be regulated during cable fabrication and by the design of the protective cable sheath.

In the experiment, three levels of temperature were studied (high, medium, low—labeled H, M, L), reflecting the range of temperatures to be encountered in the cable laying and while resting on the ocean floor. Humidity was also studied at three levels (high, medium, low—labeled H, M, L), representing three levels that could be achieved by successively tighter design and fabrication control over the humidity exposure of the fiber.

The quality requirement for the breaking strain of the fiber called for a target breaking strain of 2 percent. In this example, we shall focus on this quality requirement. In addition, there was also a quality requirement pertaining to the variability of the breaking strains of fibers. This latter requirement was also of great importance because substantial variability in the breaking strains of fibers increases the likelihood of a costly cable break.

In this experiment, all nine factor-level combinations were studied. Ninety strands of fiber of the same length were randomly assigned in equal numbers to the nine experimental treatment combinations, giving 10 strands for each treatment. Each strand of fiber was exposed to its assigned combination of temperature and humidity levels for a fixed time period. The strands were then subjected to a strain test and the breaking strain measured for each of the 90 strands. The results of the experiment are summarized in Table 22.1, which presents the estimated mean breaking strain for each treatment. For instance, the estimated mean breaking strain when temperature and humidity are at low levels was 2.08 percent. Figure 22.3 presents the experimental results graphically in the form of a *mean response plot*. For each humidity level, it shows a curve of the estimated mean breaking strain as a function of temperature. The mean response plot suggests the following conclusions.

1. The pattern of estimated mean breaking strain as a function of temperature is roughly the same at all humidity levels, although the levels of the curves differ.

Humidity Level	Temperature Level		
	Low	Medium	High
Low	2.08	2.28	1.86
Medium	1.17	0.94	0.54
High	0.82	0.83	0.43

TABLE 22.1
Estimated mean breaking strains (in percent)—Optical fiber example

FIGURE 22.3
**Mean response plot—Optical
fiber example**

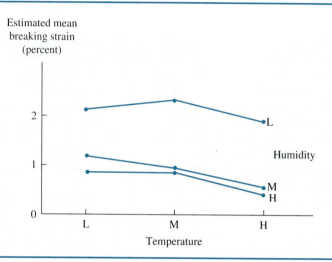

2. Humidity appears to have an adverse effect on the breaking strain. The estimated mean breaking strain decreases (i.e., the fibers are weaker) as humidity increases, irrespective of the temperature level. The decrease is especially sharp between the low and medium humidity levels.
3. The high temperature level appears to have an unfavorable effect on the breaking strain, irrespective of the humidity level, but there seems to be much less of a temperature effect between medium and low temperature levels.

Based on these findings, the company's engineers concluded that the breaking strain requirement of 2 percent could only be met if, first, the humidity level is kept at the lowest level; this could be achieved by very stringent controls on humidity during fabrication and by use of a fairly expensive cable sheath design. In addition, however, even with the lowest level for humidity, the breaking strain requirement would not be met if the highest temperature level were experienced. On considering the temperature issue, the engineers came to the conclusion that the high temperature effect could be avoided if (1) the cable were not laid in hot weather and (2) a small change were made in the sheath design that would allow greater dissipation of heat produced by electric power passing through the cable.

Several additional design options were also studied by the engineers to improve the level and uniformity of the breaking-strain characteristic of the fiber, including the use of coated fibers and the use of different glass materials that would be more impervious to humidity and less prone to breaking at high temperatures. □

22.3 ASSESSING CONFORMANCE— CONTROL CHARTS

Purpose of Control Charts

After quality designs have been achieved for a product and for the process that will make the product, the next concern is to make certain that the actual product conforms to the design requirements. Control charts are an effective tool for (1) monitoring conformance to requirements and (2) identifying possible causes of departures from requirements when these occur. A control chart is a special kind of time series plot for a quality characteristic of a product or process. It usually contains two control limits. Figure 22.4 (p. 701) illustrates a control chart. As long as the process remains within the control limits, the process is said to be *in control* and the observed variation in the process is concluded to result from random causes inherent in the process. When the observed process falls outside the control limits, an *assignable cause* (also called *special cause*) is concluded to be present and affecting the performance of the process. A search is then made for the assignable cause and remedial action taken, which frequently leads to improved process performance.

Continuous and Periodic Charting

In many applications, control charts can continuously monitor a quality characteristic.

EXAMPLES ☐

1. A batch of beer after fermentation is required to be stored in a primary aging tank for 10 days at a temperature between 4° and 8°C. Several instruments are placed in different locations within the tank to measure the temperature at each location. Control charts for the tank show continuous readouts of the temperatures of the tank's contents at the different locations. The charts are examined regularly by an operator to make certain that the correct temperature is being maintained.

2. An airline records on a control chart the difference between the actual and scheduled departure times for each of its flights. These charts are monitored by an automated administrative control system. This system brings exceptional cases to the attention of management for possible action and provides summary statistics for regular reports on operations. ☐

Continuous monitoring of a process is not feasible in many cases. Instead, conformance of a product or a process to requirements is observed by means of periodic sampling of the process.

EXAMPLE ☐   **Drug Tablets**

In the production of a drug tablet by a pharmaceutical firm, the tablet weight is an important quality characteristic. The quality requirement for the tablet weight is expressed in the form of specification limits, which are $LSL = 0.4380$ and $USL = 0.4440$ gram. It is not feasible to weigh each of the thousands of tablets produced daily. Instead, hourly samples of five drug tablets are selected and each tablet in the sample is weighed. The control chart is a plot of the hourly observations. ☐

Use of Control Charts

Statistical inference procedures are involved in the design and use of control charts because uncertainty is present in all process data. For instance, in the drug tablets example, the weight of a tablet in an hourly sample may exceed the upper specification limit. Is this one of the 7 nonconforming cases out of 100,000 that the product and process design allows for this product, or is this an indication that a change in the production process has occurred? Would the evidence of a process change be much stronger if all five tablets in the hourly sample weighed more than the upper specification limit? Statistical inference procedures are required so that management can draw useful conclusions from the control chart.

Preferably, the quality characteristic observed in a control chart is a variable, such as a dimension, weight, or breaking strain. Sometimes, however, only the conformity or non-conformity of a product to one or several quality requirements is observed, such as whether the fill of a jar of preserve is within the specification limits. Measurements for a quality characteristic are preferable to simply observing the presence or absence of conformity because they provide much more information for quality control and quality improvement.

Control charts can be used for individual observations, as well as for the mean, standard deviation, or other summary measures of periodic samples of units of the product. Control charts can also be used for observing the proportion of nonconforming units in periodic samples.

Control charts are helpful for bringing a process into control, as well as for keeping an in-control process in that state. We shall first consider the use of control charts for in-control processes, and then we shall take up briefly some special issues that arise when control charts are used to bring a process into control. Throughout this section, we shall consider control charts based on periodic sampling.

Control Chart for Process Mean

As we noted earlier, control over both the process mean and the process variability of a quality characteristic are critical so that the product will conform to the specification limits. We first consider control charts for the process mean, after which we consider control charts for the process standard deviation. We shall continue to use the drug tablets example.

☐ EXAMPLE

In the drug tablets example, the quality requirements for tablet weight call for the process mean weight to be $\mu = 0.4410$ gram, and for the process standard deviation to be $\sigma = 0.0008$ gram. Since the distribution of tablet weights is approximately normal, these levels assure management that the specification limits of 0.4380 and 0.4440 will be met almost always, the process capability ratio given in (22.2) here being $(0.4440 - 0.4380)/6(0.0008) = 1.25$.

After much effort, the production process was brought into control, with the process mean at $\mu = 0.4410$ gram and the process standard deviation at $\sigma = 0.0008$ gram. Management now wishes to use a control chart to keep the process mean at the level of $\mu = 0.4410$ gram.   ☐

Construction of Control Chart. The control chart for the process mean is based on periodic random samples of size n from the production process. A decision is made for each sample, based on the sample mean $\overline{X}$, whether or not the process mean μ has remained at the standard level μ_0. The *standard level* μ_0 for the control chart is simply the level of the process mean at which the process is to be controlled. In the drug tablets example, for instance, the standard level is $\mu_0 = 0.4410$, the process mean for the in-control process.

The control limits in the control chart correspond to the action limits of a decision rule for a two-sided test concerning the population mean, discussed in Chapter 12. The only assumption required is that *the sampling distribution of $\overline{X}$ be approximately normal.* Since the periodic sample sizes used with control charts for the process mean are frequently small, this assumption essentially requires that the distribution of the measured characteristic (e.g., the weights of drug tablets) be approximately normal or not too markedly different from a normal distribution.

The test performed with the control chart for the process mean involves the following two-sided alternatives.

(22.3) H_0: $\mu = \mu_0$ (Let process alone)
          H_1: $\mu \neq \mu_0$ (Look for cause of change)

The appropriate type of decision rule is the two-sided rule shown in Figure 12.4. For purposes of setting up the control chart, the decision rule is expressed in terms of $\bar{X}$ rather than in terms of the standardized test statistic. Also, since the process standard deviation σ is known, the standard deviation of the sampling distribution of $\bar{X}$, $\sigma\{\bar{X}\}$, is known. Recall from (10.6) that $\sigma\{\bar{X}\} = \sigma/\sqrt{n}$. The decision rule for the control chart for a specified α risk therefore is as follows.

(22.4)   If $LCL \leq \bar{X} \leq UCL$, conclude H_0 (let process alone).

          If $\bar{X} < LCL$ or $\bar{X} > UCL$, conclude H_1 (look for cause of change).

where: $LCL = \mu_0 - z\sigma\{\bar{X}\}$
         $UCL = \mu_0 + z\sigma\{\bar{X}\}$
         $z = z(1 - \alpha/2)$
         $\sigma\{\bar{X}\} = \dfrac{\sigma}{\sqrt{n}}$

Here, LCL and UCL correspond to the lower and upper action limits A_1 and A_2, respectively, in Figure 12.4. Limits LCL and UCL are called the *lower control limit* and the *upper control limit,* respectively.

In the United States and Canada, it is customary to use *three-standard-deviations control limits,* that is, to use $z = 3$. Because the sampling distribution of $\bar{X}$ is approximately normal, we know that the probability of $\bar{X}$ falling inside the three-standard-deviations control limits when the process mean has remained unchanged is very high. Specifically, this probability equals the probability of $\bar{X}$ falling within three standard deviations of μ_0, which is equal to $P(-3 \leq Z \leq 3) = 0.9974$. Hence, the α risk of $\bar{X}$ falling outside the three-standard-deviations control limits when $\mu = \mu_0$ is $1 - 0.9974 = 0.0026$. This very small α risk indicates that the decision rule is unlikely to lead one to look for a cause of change in the process mean when, in fact, no change has occurred.

EXAMPLE

For the drug tablets example, the control chart is based on hourly random samples of $n = 5$ tablets from the production process. A decision is made after each sample, on the basis of the sample mean $\bar{X}$, whether or not the process mean has remained at the standard level. The sampling distribution of $\bar{X}$ is approximately normal here, even though the sample size is small, because of the approximate normality of the population.

To calculate the control limits in (22.4), we use the standard level $\mu_0 = 0.4410$ gram and the process standard deviation $\sigma = 0.0008$ gram. The standard deviation of the sampling distribution of $\overline{X}$ therefore is:

$$\sigma\{\overline{X}\} = \frac{0.0008}{\sqrt{5}} = 0.000358$$

To obtain the three-standard-deviations control limits, we substitute into (22.4) with $z = 3$ and obtain:

$$LCL = 0.4410 - 3(0.000358) = 0.4399 \text{ gram}$$

$$UCL = 0.4410 + 3(0.000358) = 0.4421 \text{ gram}$$

Plotting Procedure.   A control chart is simply a graphic implementation of decision rule (22.4) for the sequence of samples that are drawn from the production process. A computer plot is often employed. The standard level μ_0 is plotted as a horizontal center line on the chart, and the control limits LCL and UCL are plotted on either side of the center line. The vertical axis represents the sample mean $\overline{X}$, and the horizontal axis represents the time over which the periodic samples are taken. Points representing $\overline{X}$ for each sample are plotted on the control chart and are connected by straight lines to facilitate visual interpretation. As we noted before, points that fall on or inside the control limits indicate that the process is in control, and points that fall outside the control limits indicate that the process is out of control. As long as the points do not fall outside the control limits, the process is to be let alone. If a point falls outside the control limits, however, it is concluded that the process mean has changed as the result of an assignable or special cause that was not present previously. A search for the assignable cause is then undertaken, on the basis of which appropriate remedial action is initiated.

☐ **EXAMPLE**

Figure 22.4 presents the control chart for the drug tablets example. The chart was produced by QCSTAT (Ref. 22.2) and was plotted using SYGRAPH. The sample data on which the chart is based are not presented here. The control chart covers production for 30 hours, during which time 30 quality control samples were selected hourly. Note from the chart that the plotted $\overline{X}$ points are all within the control limits, indicating that the production process has remained in control with respect to the process mean.   ☐

Information Provided by Control Charts.   One important type of information furnished by a control chart for the process mean is the approximate time when the process goes out of control. This information often provides a valuable clue to the assignable cause. For instance, a process may have gone out of control at about the same time that raw material from a new supplier was put into the process. In another instance, a point may have fallen outside the control limits shortly after an operator adjusted a machine.

Figure 22.5a illustrates a control chart when a sudden, large shift of the process mean upwards has occurred. Note that the process went out of control above the upper control limit at sample 9, and has remained out of control since then.

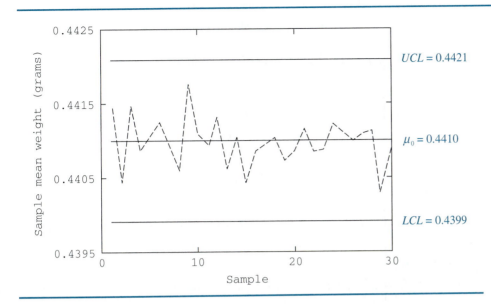

FIGURE 22.4
QCSTAT/SYGRAPH control
chart for process mean—
Drug tablets example

Frequent periodic sampling of the process assures management that the process will not operate long after it has gone out of control without this condition being detected. With infrequent sampling, the process might turn out many items that do not conform to quality requirements before the condition is discovered.

Control charts furnish other valuable clues in addition to indicating a sudden, large shift in the process mean. Through the plotting of sample results over a period of time, trends or other gradual changes in the process mean may become apparent. Figure 22.5b illustrates a control chart when the process mean is drifting downward gradually from the standard level, beginning around sample 10. A downward trend of this kind may result from toolwear, operator fatigue, and the like. A trend in the sequence of plotted points gives an early warning of an impending out-of-control situation. Early remedial action may therefore be taken before the process actually goes out of control. Statistical tests may be used to decide whether a trend pattern in plotted points indicates a real trend or is simply a chance phenomenon. The test of runs up and down described in Chapter 15 is useful for this purpose.

When an assignable cause of a process change has been discovered and appropriate remedial action taken, improvements in the process or product design may have occurred, involving changes in the process mean μ, the process standard deviation σ, or both. The control chart must then be revised to reflect the new conditions.

Control Chart for Process Standard Deviation

The process standard deviation σ is monitored by the use of a control chart similar to the one for monitoring the process mean. The sample statistic that is plotted is usually either the sample range or the sample standard deviation s. We shall take up now the control chart based on the sample standard deviation s. This control chart assumes that *the population distribution of the measured characteristic* (e.g., the drug tablet weights) *is approximately normal*.

FIGURE 22.5
MINITAB control charts showing patterns for different process mean changes

(a) Sudden, Large Shift in Process Mean

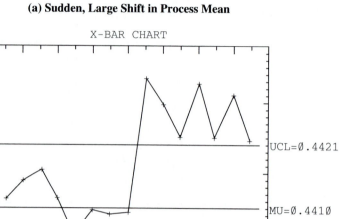

(b) Gradual Trend in Process Mean

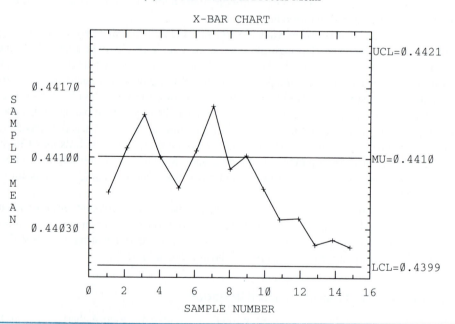

The test performed with the control chart for the process standard deviation σ involves the following two-sided alternatives.

(22.5) $H_0: \sigma = \sigma_0$ (Let process alone)

$H_1: \sigma \neq \sigma_0$ (Look for cause of change)

Here, σ_0 denotes the standard level for the control chart, that is, the level of the process standard deviation at which the process is to be controlled. The decision rule utilizes the assumption that the distribution of the measured characteristic is normal. It therefore involves percentiles of the chi-square distribution, following from theorem (14.34). For a specified α risk, the decision rule is as follows.

(22.6)   If $LCL \leq s \leq UCL$, conclude H_0 (let process alone).

If $s < LCL$ or $s > UCL$, conclude H_1 (look for cause of change).

Here: $$LCL = \sigma_0 \sqrt{\frac{\chi^2(\alpha/2; n-1)}{n-1}}$$

$$UCL = \sigma_0 \sqrt{\frac{\chi^2(1-\alpha/2; n-1)}{n-1}}$$

EXAMPLE □

Figure 22.6 shows a QCSTAT/SYGRAPH control chart for the process standard deviation σ for the drug tablets example. The control chart is for the same sample data presented in Figure 22.4 for the control chart for the process mean. The standard level for the control chart is $\sigma_0 = 0.0008$ gram, the process standard deviation for the in-control process. The control limits are based on controlling the α risk at 0.0025. The required percentiles were obtained by a statistical package; they are, for sample size $n = 5$:

$$\chi^2(0.00125; 4) = 0.1017 \qquad \chi^2(0.99875; 4) = 17.9715$$

The control limits according to (22.6) therefore are:

$$LCL = 0.0008 \sqrt{\frac{0.1017}{5-1}} \qquad UCL = 0.0008 \sqrt{\frac{17.9715}{5-1}}$$

$$= 0.00013 \qquad\qquad\qquad = 0.00170$$

These are the control limits found in Figure 22.6. Note that the control limits are not equidistant from the standard level $\sigma_0 = 0.0008$ because the chi-square distribution is not symmetric.   □

FIGURE 22.6
QCSTAT/SYGRAPH control chart for process standard deviation—Drug tablets example

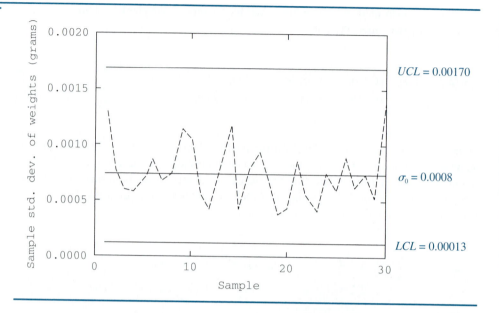

Since the control limits (22.6) are usually set so that the α risk is very small, there is only a very small probability that the sample standard deviation s will fall outside the control limits if the process standard deviation σ has not changed. Hence, one can be confident that the process standard deviation has increased from the standard level σ_0 when s falls above the upper control limit UCL. A search for an assignable cause should then ensue and remedial action should be initiated. When s falls below the lower control limit LCL, it may signal that the process is capable of producing more uniform product than originally expected. An investigation should therefore be undertaken to see if the reduced variability can be captured permanently by the process. If so, new control charts for the process mean and the process standard deviation would need to be created based on the new standard level σ_0.

◻ **EXAMPLE**

Referring to Figure 22.6 for the drug tablets example, we see that none of the 30 plotted points lies outside the control limits. Also, the plotted points show no evidence of a trend in the process standard deviation. Hence, the process standard deviation for the drug tablet weights appears to have remained unchanged.   ◻

Comments

1. The control limits in (22.6) are based on theorem (14.34), which states that for a random sample from a normal population with standard deviation σ_0, $(n - 1)s^2/\sigma_0^2$ has a χ^2 distribution with $n - 1$ degrees of freedom; that is:

$$\frac{(n - 1)s^2}{\sigma_0^2} = \chi^2(n - 1)$$

Rearranging this formula, we obtain:

$$s = \sigma_0 \sqrt{\frac{\chi^2(n-1)}{n-1}}$$

The two control limits in (22.6) involve the $100(\alpha/2)$ and $100(1 - \alpha/2)$ percentiles of the chi-square distribution.

2. Control limits for the process standard deviation are sometimes obtained by means of a large-sample approximation. Unless the sample size is very large, these limits are not as satisfactory as those in (22.6).

3. The center line in the QCSTAT control chart in Figure 22.6 is actually not exactly at the standard level $\sigma_0 = 0.0008$. Rather, it is set at 0.00075 to take into account the fact that the sample standard deviation s is a biased estimator of the population standard deviation σ. Unless the sample size is very small, this bias adjustment is not material.

Control Chart for Process Proportion Nonconforming

At times it is not possible to define a quality characteristic in terms of a measurement or it is not feasible to measure the quality characteristic. Instead, the quality characteristic is only evaluated in terms of whether it is conforming (acceptable) or nonconforming (defective).

EXAMPLES

1. The on-time performance of a flight may be evaluated simply in terms of whether or not it left the gate within 15 minutes of its scheduled time.

2. The position of the label on a can may be evaluated in terms of whether or not it is in the proper position.

3. The integrity of a glass vase may be evaluated in terms of whether or not any imperfections are present in the vase. ☐

Sometimes the conformity of a product is measured with respect to several different quality characteristics. When conformity is measured in this way, a nonconforming unit may contain several different defects.

EXAMPLE ☐

In the dress shirts example, a shirt is considered to be nonconforming if any defects (sizing, buttons, etc.) are present. ☐

The conformity of a production process may be measured by the *process proportion nonconforming p,* which is also called the *process proportion defective.* Quality demands that this proportion be as low as feasible. After a process has been brought under control at an acceptable level p_0, the objective is to keep the process proportion nonconforming at this standard level p_0.

A control chart can be used for monitoring the process proportion nonconforming. As before, samples of size n are drawn periodically from the production process. Either the sample frequency of nonconforming units f or the sample proportion nonconforming $\bar{p} = f/n$ for each sample is plotted in time order. The test performed with the control chart involves the following alternatives.

(22.7) H_0: $p = p_0$ (Let process alone)
          H_1: $p \neq p_0$ (Look for cause of change)

If the process proportion nonconforming is at its standard level p_0, then f and $\bar{p}$ follow a binomial distribution with parameters $p = p_0$ and n. Thus, the upper and lower control limits can be set by reference to appropriate percentiles of the binomial distribution. The control limits based on the sample proportion nonconforming $\bar{p}$ are as follows.

(22.8)   If $LCL \leq \bar{p} \leq UCL$, conclude H_0 (let process alone).

          If $\bar{p} < LCL$ or $\bar{p} > UCL$, conclude H_1 (look for cause of change).

Here LCL and UCL are obtained from the binomial distribution.

For a specified α risk, the control limits LCL and UCL are set so that the risk in each tail is as close as possible to, but does not exceed, $\alpha/2$. Because the binomial distribution is a discrete probability distribution, the control limits cannot be chosen to achieve the prespecified α risk exactly, but in many cases, they will yield nearly the desired risk level. Statistical packages that generate control charts usually calculate the control limits automatically.

Mail Orders   ☐ **EXAMPLE**

A mail-order house fills thousands of orders each day. Management has brought the order-filling process into control at the acceptable error rate of $p_0 = 0.004$. The process proportion nonconforming will be monitored by drawing daily random samples of $n = 150$ orders from the orders filled that day and determining the sample proportion filled incorrectly.

The α risk specified for the control chart is 0.0025. Hence, the risk in each tail should not exceed 0.00125. To find suitable control limits, we need to utilize the binomial probability distribution corresponding to $p = p_0 = 0.004$ and $n = 150$. Table 22.2 contains a portion of this probability distribution calculated from a statistics package, with column 3 showing the probabilities and column 4 the cumulative probabilities. To determine the lower control limit, we note first that if the lower control limit is $LCL = 0$, it would involve a tail risk of 0 because there can be no out-of-control points below 0. If the lower control limit were $LCL = 0.0067$, the tail risk would be 0.54815 since an out-of-control point could arise if $\bar{p} = 0$, and this probability is $P(\bar{p} = 0) = 0.54815$. This risk far exceeds the desired tail risk of 0.00125. Hence, the use of $LCL = 0$ comes closest to the desired tail risk of 0.00125 without exceeding it.

(1)	(2)	(3)	(4)	
Number of Orders Filled Incorrectly f	Proportion of Orders Filled Incorrectly $\bar{p} = f/n$	Probability $P(f) = P(\bar{p})$	Cumulative Probability	
0	0	0.54815	0.54815	
1	0.0067	0.33021	0.87836	
2	0.0133	0.09880	0.97716	
3	0.0200	0.01957	0.99674	
4	0.0267	0.00289	0.99963	
5	0.0333	0.00034	0.99996	

TABLE 22.2
Binomial probabilities for $p = p_0 = 0.004$ and $n = 150$—Mail orders example

To find the upper control limit, we note the following from column 4 in Table 22.2:

$$P(\bar{p} \leq 0.0200) = 0.99674, \text{ hence } P(\bar{p} > 0.0200) = 0.00326$$
$$P(\bar{p} \leq 0.0267) = 0.99963, \text{ hence } P(\bar{p} > 0.0267) = 0.00037$$

Thus, placing the upper control limit at 0.0200 would involve a tail risk of 0.00326, which exceeds the specified tail risk of 0.00125. On the other hand, placing the upper control limit at 0.0267 leads to a tail risk of 0.00037, which is below the required level. Hence, the appropriate control limits are $LCL = 0.0$ and $UCL = 0.0267$, which involve an α risk of $0 + 0.00037 = 0.00037$.

Figure 22.7 shows a QCSTAT/SYGRAPH control chart on which are plotted the sample proportions $\bar{p}$ for 18 consecutive days. Note that the center line is set at the standard level $p_0 = 0.004$ and that the upper and lower control limits are those that we just determined. The control chart in Figure 22.7 shows no out-of-control points during the 18-day period and no evidence of any trend in the process proportion nonconforming. Thus, the order-filling process appears to be in control. □

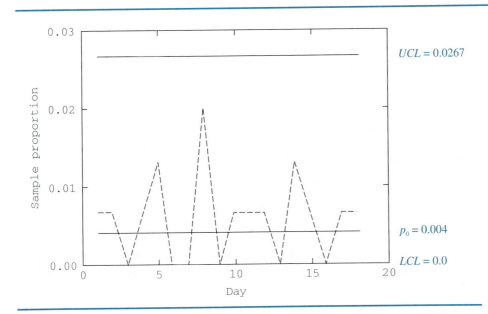

FIGURE 22.7
QCSTAT/SYGRAPH control chart for process proportion nonconforming—Mail orders example

Comments

1. When the sample size is large, the control limits in (22.8) can be calculated using the normal approximation to the sampling distribution of $\bar{p}$ presented in Chapter 13. The control limits for a specified α risk then are as follows.

$$(22.9) \qquad LCL = p_0 - z\sqrt{\frac{p_0(1 - p_0)}{n}}$$

$$UCL = p_0 + z\sqrt{\frac{p_0(1 - p_0)}{n}}$$

where $z = z(1 - \alpha/2)$.

The lower control limit in (22.9) is set equal to zero if the formula gives a negative value for *LCL*.

2. Control charts with only a one-sided upper control limit are sometimes employed in cases where there is no interest in having an out-of-control signal when the proportion nonconforming has decreased. Usually, however, two-sided control charts are more useful since out-of-control points falling below the lower control limit can indicate the presence of conditions that are favorable to permanently reducing the process proportion nonconforming.

Cusum Chart

An alternative to conventional control charts is the *cumulative sum chart*, or *cusum chart* for short. The cusum chart is especially useful for detecting a moderate departure from the standard level that persists over time. We shall now consider the cusum chart for a process mean to illustrate its use.

Let $\bar{X}_i$ be the sample mean for the *i*th sample drawn from a process; then $\bar{X}_i - \mu_0$ is a measure of the departure of the sample mean from the standard level μ_0 at that time point. The sum of these departures from the first sample to the current one is the basis of the cusum chart. This sum is denoted by C_k, where k denotes the current sample number.

$$(22.10) \qquad C_k = \sum_{i=1}^{k}(\bar{X}_i - \mu_0)$$

where: $\bar{X}_i$ is the sample mean for the *i*th sample

  μ_0 is the standard level for the process mean

Plotting the values of C_k against their time points k gives a cusum chart for the process mean. When the process mean remains stable, the consecutive values of C_k will fluctuate around 0. When there is a change in the process mean, on the other hand, the plotted points will form a trend, either upward or downward, depending on the direction of the shift in the process mean.

Sample k	Sample Mean $\overline{X}_k$	Sample Departure $\overline{X}_k - \mu_0$	Cumulative Sum $C_k = \sum_{i=1}^{k} (\overline{X}_i - \mu_0)$
1	0.4411	0.0001	0.0001
2	0.4417	0.0007	0.0008
3	0.4413	0.0003	0.0011
.	.	.	.
.	.	.	.
.	.	.	.
14	0.4413	0.0003	0.0068
15	0.4410	0.0000	0.0068

TABLE 22.3
Calculations for cusum chart for process mean when $\mu_0 = 0.4410$—Drug tablets example

EXAMPLE ☐

Table 22.3 shows a fragment of the cusum calculations for the drug tablets example based on 15 samples in a later period. Recall that the standard level here is $\mu_0 = 0.4410$ gram. The calculations for the first two cumulative sums are:

$$C_1 = 0.4411 - 0.4410 = 0.0001$$

$$C_2 = 0.0001 + (0.4417 - 0.4410) = 0.0008$$

Figure 22.8 contains the cusum chart for the 15 samples. The plotted points have been connected by straight lines to facilitate visual interpretation. The plot shows a generally steady trend upward, starting at the outset and persisting through to the end. This suggests that the process mean has shifted upward. ☐

Formal methods exist for testing when a process has gone out of control in a cusum plot. These methods are covered in specialized quality control texts.

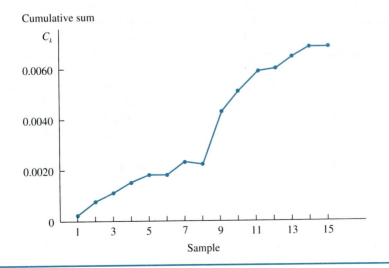

FIGURE 22.8
Cusum chart for cumulative sum of departures of sample mean from standard level $\mu_0 = 0.4410$—Drug tablets example

Other Control Charts

We have illustrated several types of control charts that are widely used in quality control applications, but many other types of charts are found in practice. Examples of other charts include charts based on individual observations, charts for number of defects per item, charts using moving averages, and charts that recognize correlations between the several quality characteristics of a product.

Use of Control Charts to Achieve Process Control

Our discussion of control charts up to this point has been oriented toward keeping an in-control process in this state. Control charts are also useful for bringing a process into control. Some special issues arise in this use of control charts. Before any control limits can be obtained, a number of samples must be selected from the production process. Based on these samples, preliminary control limits can be calculated. These are called *trial control limits* because they frequently will show some or many of the sample points outside the control limits. A search for assignable causes then follows based on clues obtained from the out-of-control points. Remedial actions are taken to correct the causes, which in turn leads to new samples and new trial control limits. Eventually, the process will be stabilized and thus brought under control.

22.4   ASSESSING CONFORMANCE— ACCEPTANCE SAMPLING

Acceptance sampling is a quality control tool that is used for determining, on the basis of a random sample from a lot of a product, whether the lot is satisfactory. Inspection of a product cannot, of course, make nonconforming product into conforming product. As we stated earlier, quality cannot be inspected into a product. Quality improvement can be achieved only by improving the product and process design and by improving the process performance. Still, acceptance sampling may be useful until adequate quality control and improvement have been achieved.

Acceptance sampling is an application of statistical testing where the choice is between *accepting* and *rejecting* a lot of items, such as an incoming shipment of material, an outgoing shipment of finished product, or a batch of clerical work. A lot constitutes a finite population and therefore, in theory, could be given 100 percent inspection (unless the testing is destructive). However, 100 percent inspection can be prohibitively expensive and usually would not detect all defects and errors anyway in view of inspection fatigue and other sources of error. Hence, sampling is frequently used to decide whether a lot should be accepted (H_0) or rejected (H_1). The latter decision may lead to 100 percent inspection of the items, a return of the lot to the vendor, complete verification of the batch of clerical work, or some other action of this nature.

We shall limit our consideration here to acceptance sampling plans based on the *lot proportion defective p.*

Single-Sampling Plans

A *single-sampling plan* for the lot proportion defective specifies the random sample size and the action limit for the sample frequency of defective items.

A firm receives machined bushings in lots of 10,000 from a supplier who does not have an adequate quality control program at present. Hence, the firm needs to rely on acceptance sampling for assurance of quality in the shipments. The alternative conclusions for any particular lot are:

$$H_0: \; p \leq p_0 = 0.04 \qquad \text{(Accept lot)}$$
$$H_1: \; p > p_0 = 0.04 \qquad \text{(Reject lot)}$$

where p denotes the proportion of defective bushings in the lot and $p_0 = 0.04$ denotes the maximum acceptable proportion of defectives in the lot. In acceptance-sampling terminology, p_0 is referred to as the *acceptable quality level*, or *AQL*. In our earlier testing terminology, p_0 is the value of p at which the α risk is to be controlled.

The firm and the supplier have reached an agreement that the following single-sampling plan be employed.

Select a random sample of 125 bushings from the lot. If 10 or fewer bushings in the sample are defective, accept the lot; if 11 or more are defective, reject the lot.

The maximum number of defectives that leads to acceptance of the lot is 10 here. This value is called the *acceptance number* of the sampling plan. The minimum number of defectives leading to rejection, here 11, is called the *rejection number*.

This single-sampling plan is described in a condensed form in Table 22.4a. It is equivalent to a decision rule for a test of a population proportion. Let f denote the sample frequency, that is, the number of defective bushings in the sample of $n = 125$. The plan then corresponds to the following decision rule:

If $f \leq 10$, conclude H_0 (accept lot).
If $f > 10$, conclude H_1 (reject lot).

TABLE 22.4
Matched single- and multiple-sampling plans— Bushings example

(a) Single-Sampling Plan

Sample Size	Acceptance Number	Rejection Number
125	10	11

(b) Multiple-Sampling Plan

Sample	Sample Size	Cumulative Sample Size	Acceptance Number	Rejection Number
First	32	32	0	5
Second	32	64	3	8
Third	32	96	6	10
Fourth	32	128	8	13
Fifth	32	160	11	15
Sixth	32	192	14	17
Seventh	32	224	18	19

Source: Military Standard 105D Tables (Ref. 22.3).

Equivalently, the plan corresponds to the following decision rule in terms of the sample proportion defective $\bar{p}$, since $\bar{p} = f/n$ and $10/125 = 0.08$:

> If $\bar{p} \leq 0.08$, conclude H_0 (accept lot).
> If $\bar{p} > 0.08$, conclude H_1 (reject lot).   □

Operating Characteristics.   In acceptance-sampling applications, a characteristic of the sampling plan that is of particular interest is the probability of accepting the lot for different values of p. These probabilities, which we denote by $P(H_0; p)$, are approximately binomial probabilities when the lot size is large relative to the sample size (as is the case in the bushings example). The plot of $P(H_0; p)$ for different values of p is called the *operating characteristic curve* of the sampling plan. The operating characteristic curve is the complement of the corresponding rejection probability curve, as can be seen from (12.17).

□ **EXAMPLE**

Figure 22.9 shows the operating characteristic curve of the sampling plan in the bushings example. The following two probabilities, corresponding to $p = 0.04$ and 0.16, were obtained from binomial probability calculations by a statistics package. Recall that H_0 (accept lot) is concluded if $f \leq 10$ for the sample of $n = 125$ bushings:

$$P(H_0; p = 0.04) = P(f \leq 10; p = 0.04) = 0.9881$$
$$P(H_0; p = 0.16) = P(f \leq 10; p = 0.16) = 0.0066$$ □

The α risk here is the probability of rejecting an acceptable shipment; for this reason, it is often called the *producer's* or *supplier's risk*. On the other hand, the β risk is the probability of accepting an unacceptable shipment and, hence, is often called the *buyer's* or *customer's risk*.

□ **EXAMPLE**

Referring again to Figure 22.9 for the bushings example, we see that the supplier's risk when $p = p_0 = 0.04$ equals:

$$\alpha = 1 - P(H_0; p = 0.04) = 1 - 0.9881 = 0.0119$$

We also see that the β risk when $p = 0.16$ equals:

$$\beta = P(H_0; p = 0.16) = 0.0066$$

Thus, the sampling plan exposes the producer to a risk of 0.0119 of having a lot rejected in which the proportion defective is 0.04, and the customer is exposed to a risk of 0.0066 of having a lot accepted in which the proportion defective is 0.16.   □

Multiple-Sampling Plans

A *multiple-sampling plan* involves sampling in two or more stages. At each stage, a decision is made whether to accept or reject the lot or to continue sampling. The use of multiple-sampling plans usually results in smaller total sample sizes than corresponding single-sampling plans with the same α and β risks.

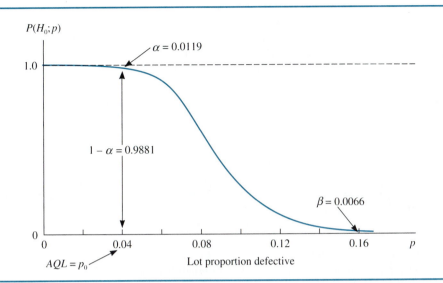

FIGURE 22.9

Operating characteristic curve of sampling plan— Bushings example

EXAMPLE ☐

Table 22.4b contains a multiple-sampling plan for the bushings example that has approximately the same operating characteristic curve as the single-sampling plan in Table 22.4a. With this multiple-sampling plan, an initial sample of 32 bushings is selected. If 0 bushings are defective (the acceptance number), the lot is accepted. If 5 (the rejection number) or more are defective, the lot is rejected. If the number of defectives is 1, 2, 3, or 4, sampling continues and a second sample of 32 bushings is selected. We now consider the *combined* sample of 64 bushings. If 3 or fewer bushings in the combined sample are defective, we accept the lot. If 8 or more are defective, we reject it. If the number of defectives is 4, 5, 6, or 7, a third sample of 32 is selected. This process continues, if necessary, until the seventh sample of 32 is selected. At this stage, the lot is either accepted (if there are 18 or fewer defectives in all seven samples combined) or rejected (if there are 19 or more defectives).

It is intuitively clear that the total sample size with this multiple-sampling plan depends on the lot proportion defective p. The plan will detect exceptionally good or poor lots at the first stage, with a resulting sample size of 32 compared with the sample size of 125 used in the single-sampling plan. Lots of intermediate quality may require two or more samples before a final decision is reached. Only infrequently, however, will the combined sample size exceed 128—that is, go beyond the fourth sample. ☐

Comments

1. Some multiple-sampling plans provide for sampling one item at each stage, so that an accept, reject, or continue decision is made with each sample observation. These plans are called *sequential-sampling plans.*
2. Extensive tables of sampling plans have been published. One set is the Military Standard 105D Tables (Ref. 22.3), from which the sampling plans in Table 22.4 have been extracted. The wide variety of plans in these tables enables the user to select readily a plan to fit the circumstances.

22.5   ASSESSING PERFORMANCE— RELIABILITY

Once a product has been produced, its conformance to quality requirements, as we noted earlier, must be assessed by its performance in actual use. One means of measuring and evaluating the performance of a product is by reliability analysis.

Meaning of Reliability

A product is judged to be reliable if it functions as required for a specified time.

> **(22.11)**
> **Reliability** is the state of a product in which it conforms to customer requirements for a specified period of time.

A comparison of this definition with the definition of quality in (22.1) shows that time is the new element introduced by the definition of reliability. Time may be measured in terms of the calendar or the clock (such as days, months, or years) or in operational terms (such as mileage or number of operating cycles).

☐ EXAMPLES

1. The manufacturer of a clock requires that it function without failure, subject only to routine replacement of its battery, for five years after it is sold. The manufacturer offers a five-year warranty to the purchaser of the clock on the basis of this quality requirement.

2. An automobile tire is required to function without failure under normal wear and road hazards for 100,000 kilometers or three years, whichever occurs first.

3. A submarine communications system is required to operate for 10 years from the time it goes into service without any failure that requires a shipboard repair. ☐

As the examples make clear, the concept of reliability requires both a definition of the event that constitutes a failure and a specified time period that the product is free from failure. In the automobile tire example, for instance, the definition of failure does not include a failure caused by a box of nails in the street or a tire injury resulting from abnormal use. The time to failure in this example is defined in terms of both calendar time (3 years) and operational time (100,000 kilometers).

Reliability Function

The time to failure and the mode of failure of an individual unit of a product are uncertain in advance and therefore are considered to be random variables. Consequently, probability concepts play a key role in reliability analysis.

The reliability of a product is often described in terms of the probability that it will function as required for longer than a specified time t. We denote the actual life of a unit of a product by T. Hence, the reliability of the product is described in terms of the probability $P(T > t)$. This probability depends on the specified time t, and may be viewed as a function of t.

(22.12)
The **reliability function** of the random variable T, describing the time to failure of a unit of a product, is denoted by $R(t)$. The reliability function $R(t)$ indicates the probability that the item will survive beyond any specified time point t; that is:

$$R(t) = P(T > t)$$

Note that the reliability function is closely related to the cumulative distribution function $F(t)$ of the random variable T since by (5.12), we have that $F(t) = P(T \le t)$. Hence, $R(t) = 1 - F(t)$.

The reliability function is often based on one of the common continuous probability distributions that we presented in Chapter 8, such as the normal distribution or the exponential distribution.

EXAMPLE

The time to failure T of a particular type of automobile tire is approximately normally distributed with mean $\mu = 150,000$ kilometers and standard deviation $\sigma = 25,000$ kilometers. A quality requirement for this tire is that it last under normal wear and road hazards for 100,000 kilometers. Thus, the probability that a tire will survive beyond 100,000 kilometers, $P(T > 100,000)$, is of interest. The standardized value corresponding to $t = 100,000$ is:

$$z = \frac{100,000 - 150,000}{25,000} = -2.00$$

We therefore have from Table C.1 for the standard normal distribution that $P(T > 100,000) = P(Z > -2.00) = 0.9772$. Thus, about 98 percent of tires of this type will conform to the reliability requirement that the tire function longer than 100,000 kilometers without failure.

The preceding calculation gives the value of the reliability function for this tire at $t = 100,000$, namely, $R(100,000) = 0.9772$. Values of the reliability function $R(t)$ for other values of t can be calculated in similar fashion. □

Analysis of Life Data

The design of a product can be used by technical experts to make a judgment about the potential reliability of the product. To substantiate such judgments, however, reliance is generally placed both on experimental life-test data and on data from actual operating experience. These data may be used to infer the functional form and the parameter values of the probability distribution of the time to failure for a product and therefore of the reliability function. Statistical methods play a central role in this aspect of quality management.

EXAMPLE □ **Ballpoint Pens**

Figure 22.10 shows a plot of the (more than) cumulative percent distribution for the times to failure (in hours) of 50 ballpoint pens of a new design. The 50 pens were

FIGURE 22.10
Cumulative percent distribution of times to failure (in
hours) of 50 pens—Ballpoint
pens example

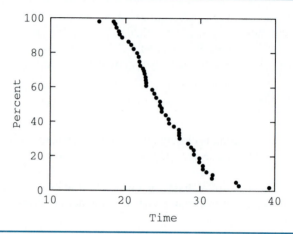

mounted on a special machine for this test. The time to failure of a pen was measured
as the time (in hours) until the pen first failed to write. The plot shows that all pens
wrote for at least 16 hours and that all had failed by 40 hours. The median time to
failure was 25 hours. The manufacturer of the pen concluded from this test that the pen
conformed to its quality requirements and made plans to produce it on a full-scale
basis.

☐

System Reliability

Some products are made up of several distinct subsystems or *components* that must function together for the product to function. The product may then be viewed as a *system* of
components. If the reliability characteristics of the individual components are known and
if it is known how the components will function jointly in the system, then probability rules
can be used to determine the reliability characteristics of the system.

Satellite ☐ **EXAMPLE**

A satellite consists of three components (1, 2, 3), each of which must function for the
satellite to function. This structure is shown schematically in Figure 22.11a. A system
that consists of such a chain of components is often referred to as a *series system*. We
shall let E_1 denote the event that component 1 survives for longer than five years without failure. E_2 and E_3 will denote corresponding events for components 2 and 3.

   The satellite is designed so that the failures of the three components are independent. It also is known from extensive experience that the probabilities of the components surviving longer than five years without failure are $P(E_1) = 0.98$, $P(E_2) = 0.96$,
and $P(E_3) = 0.90$, respectively. From multiplication rule (4.20) applied to independent
events, we find that the probability of the satellite surviving longer than five years is
given by the product:

$$P(E_1)P(E_2)P(E_3) = 0.98(0.96)(0.90) = 0.847$$

Thus, the satellite has probability 0.85 of functioning longer than five years.

FIGURE 22.11

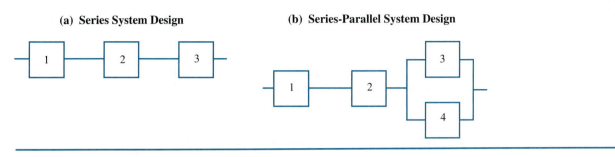

(a) Series System Design

(b) Series-Parallel System Design

The five-year survival probability for the series design in Figure 22.11a was considered unsatisfactory. The engineers noted that the low survival probability of component 3 made a major contribution to the low survival probability of the satellite. They determined that the satellite could be fitted with two identical versions of component 3 (referred to now as components 3 and 4) without compromising weight and other design constraints for the satellite. Components 3 and 4 would operate in *parallel* in such a way that the satellite would function as long as both components 1 and 2 and at least one of components 3 and 4 were functioning. Finally, it was determined that components 3 and 4 would fail independently of both the other two components and each other. Figure 22.11b illustrates the new system design.

We know from earlier reasoning that the probability of components 1 and 2 surviving longer than five years is:

$$P(E_1 \cap E_2) = P(E_1)P(E_2) = 0.98(0.96) = 0.9408$$

Now consider the probability of at least one or both of components 3 and 4 surviving longer than five years. Since components 3 and 4 are identical, we have $P(E_3) = P(E_4) = 0.90$, where E_4 denotes the event of five-year survival for component 4. Moreover, since E_3 and E_4 are independent events, we have $P(E_3 \cap E_4) = P(E_3)P(E_4)$. Finally, from addition rule (4.15), we find the probability that one or both of components 3 and 4 will function for longer than five years:

$$P(E_3 \cup E_4) = P(E_3) + P(E_4) - P(E_3 \cap E_4)$$

$$= P(E_3) + P(E_4) - P(E_3)P(E_4)$$

$$= 0.90 + 0.90 - 0.90(0.90) = 0.9900$$

Now consider the whole system. Using multiplication theorem (4.19) applied to independent events, we find that the probability of both components 1 and 2 and at least one of components 3 and 4 functioning for longer than five years is given by:

$$P(E_1 \cap E_2)P(E_3 \cup E_4) = 0.9408(0.9900) = 0.931$$

Thus, the five-year survival probability increases from 0.85 to 0.93 by using two components of type 3 in parallel rather than a single type 3 component.

The five-year survival probability was still not high enough, however, so engineers turned to other design alternatives to improve the satellite's reliability. ☐

CITED REFERENCES

22.1   Deming, W. Edwards, *Quality, Productivity, and Competitive Position*. Cambridge, Massachusetts: Massachusetts Institute of Technology, Center for Advanced Engineering Study, 1982.

22.2   Stenson, Herb, *QCSTAT: A Supplementary Module for SYSTAT*. Evanston, Illinois: SYSTAT, Inc., 1990.

22.3   *Military Standard 105D Tables: Sampling Procedures and Tables for Inspection by Attributes*. Office of the Assistant Secretary of Defense, Washington, D.C., 1963.

PROBLEMS

22.1 Specify two realistic quality requirements for each of the following products or services: (1) a computer disk, (2) a bath towel, (3) a cash withdrawal from an automatic bank machine.

22.2 Specify two realistic quality requirements for each of the following products or services: (1) a blanket, (2) a light switch, (3) a residential mail delivery.

22.3 A chemist has proposed a process for separating alcohol from water by fractionated freezing. An engineer noted that the proposal is sound in theory but cannot be implemented because of practical economic and engineering difficulties. Is the problem here one of product design, process design, or process capability? Comment.

22.4 Keyboards. A quality requirement for a computer keyboard states that 99 percent of the boards must be able to handle 1 billion key strokes without the board failing. Extensive tests with keyboards manufactured in initial production trials showed that this quality requirement was being met. The keyboard is now being shipped to customers, but there have been numerous complaints of the boards failing prematurely. Does this description suggest that fault lies with (1) the quality requirement, (2) the capability of the manufacturing process, or (3) actual production? Discuss.

22.5 Shoe Wear. A footware manufacturer tests shoes on a tread mill using two different speeds corresponding to normal and high-stress use. Why might these test results not give an accurate indication of the conformance of the shoe to durability requirements when worn by a typical wearer?

22.6 Refer to **Keyboards** Problem 22.4. Warranty records for 4109 keyboards showed the following frequencies of different keyboard defects (many keyboards had no faults and some had more than one):

Type of Defect	Frequency
Key contact	53
Jammed key	48
Jammed space bar	33
Loss of key cap	14
Space bar contact	8
Keyboard cable	2
Other	5
Total	163

a. Construct a Pareto chart for the keyboard defects as in Figure 22.1. What percent of the defects involve (1) a key contact, (2) a jammed key, (3) a jammed key or a jammed space bar?

b. Which is the most frequently occurring defect? Would your answer change if no distinction were made between a jammed key and a jammed space bar? What is the implication of this observation for interpreting Pareto charts?

c. The 163 defects occurred during the one-year warranty period. Would you expect the Pareto chart to look different if the defect records related to a two-year period? Why would the keyboard manufacturer be more concerned about defects that arise early in the life of a keyboard than late in its life?

22.7 Refer to **Shoe Wear** Problem 22.5. Tests with 120 shoes on the tread mill showed the following frequencies of different nonconformities (some shoes had more than one nonconformity):

Type of Nonconformity	Frequency
Heel separation	56
Sole penetration	28
Sole separation	14
Top stitching failure	10
Lace breakage	8
Lace eyelet failure	3
Other	12
Total	131

a. Construct a Pareto chart for the shoe nonconformities as in Figure 22.1. What percent of the nonconformities involve a heel separation?

b. If the shoe were redesigned to eliminate heel separations, would you expect the distribution of the other types of nonconformities to remain unchanged? Comment.

c. These nonconformities occurred on a tread mill. Would you expect the distribution of nonconformities in actual wearing of the shoe to match this distribution? Comment.

22.8 Refer to **Keyboards** Problems 22.4 and 22.6. For each of the following factors that affect the quality characteristics of a keyboard, state if it is a controllable or a noise factor and why. For each noise factor, suggest a way that the product and process designs might be made robust to this factor.

a. The amount of dust in the environment where the keyboard is installed.

b. The strength of the key cap.

c. The layout of the keys on the board.

d. The force with which the keys are struck by the operator.

22.9 Refer to **Shoe Wear** Problems 22.5 and 22.7. For each of the following factors that affect the quality characteristics of a shoe, state if it is a controllable or a noise factor and why. For each noise factor, suggest a way that the product and process designs might be made robust to this factor.

a. The strength of the laces.

b. The tendency for the shoe cap to scuff.

c. Uneven heel and sole wear because of poor posture.

d. The amount of exposure to moisture.

* **22.10** The lower and upper specification limits for the amount of active ingredient in a drug tablet are 20.0 and 22.4 milligrams, respectively. The amounts of active ingredient in tablets are normally distributed.

a. What is the value of the process standard deviation if the tablet production process has a process capability index of 1.3?

b. What percentage of the tablets will be nonconforming when the process mean is at the target level of 21.2 milligrams and the process capability index is 1.3?

22.11 A beverage is prepared in small batches and then bottled. Lower and upper specification limits for the bitterness level of a batch of the beverage are 10.3 and 11.5 units, respectively. The bitterness levels of batches are normally distributed.

a. Calculate the process capability index of the process if the process standard deviation is $\sigma = 0.15$ unit. What proportion of batches will be nonconforming with respect to bitterness when the process mean is at the target level of 10.9 units and the process standard deviation is at $\sigma = 0.15$ unit?

b. What proportion of batches will be nonconforming if the process mean shifts to 10.5 units and σ remains unchanged at 0.15 unit?

22.12 Refer to **Glass Layers** Problem 15.36. The lower and upper specification limits for the thickness deviation of the glass layer in the device are -2.0 and 2.0, respectively. Thickness deviations of glass layers are normally distributed.

a. Plot a histogram of the thickness deviations for the 35 devices. Show the specification limits and the target level of 0 on the horizontal scale. Do any of the thickness deviations fall outside the specification limits?

b. A process capability index of 1.3 is required for this process. What would need to be the value of the process standard deviation to obtain this index? The sample standard deviation of the 35 observations is 0.800. Does it appear that the process variability was at a satisfactory level?

22.13 An aircraft engine component must be precisely machined on a lathe. One key dimension of the component has lower and upper specification limits of -20 and 20 micrometers, respectively, where the measurement is expressed as a deviation from the target value. A manufacturing engineer has carried out an experiment to determine the effect of two lathe speeds (low, high) on the key dimension. If a high lathe speed can be used without compromising quality, components can be processed on the lathe at a faster rate with concomitant savings in manufacturing cost. The experiment involved dividing a lot of 40 components randomly into two groups of 20 each and machining one group at the low speed and the other group at the high speed. The experimental results on the key dimension are summarized below. An examination of the data, as well as engineering considerations, suggest that the deviations from the target value are normally distributed at either speed.

Experimental	Speed	
Results	Low	High
n	20	20
$\overline{X}$	0.7	-2.3
s	6.0	7.2

a. Estimate the process proportion of the components that are nonconforming at each speed. Does it appear that the lathe can be operated at the high speed without producing more nonconforming components than at the low speed? Explain.

b. If the machining operation at high speed could be changed to bring the process mean to the target level without changing the process variability, what effect on the process proportion of nonconforming components would be expected? Should a redesign of the machining operation at high speed be investigated? Comment.

* 22.14 Alloy Filaments. Quality requirements for the melting points of alloy filaments call for the process mean and the process standard deviation to be $\mu = 335°C$ and $\sigma = 21°C$, respectively. The process has recently been brought into control at these levels. As part of an ongoing quality control program, daily samples of nine alloy filaments are selected from the production process and the melting point of each filament is determined. The melting points of filaments are approximately normally distributed.

a. Construct a control chart for the process mean, using three-standard-deviations control limits. Give the values of μ_0, LCL, and UCL for the chart.

b. The sample means for the most recent 20 operating days are as follows:

Day	$\overline{X}$	Day	$\overline{X}$	Day	$\overline{X}$	Day	$\overline{X}$	Day	$\overline{X}$
1	340	5	351	9	336	13	339	17	325
2	328	6	340	10	317	14	355	18	338
3	342	7	323	11	337	15	334	19	326
4	335	8	339	12	350	16	310	20	328

Plot these $\overline{X}$ values on the control chart. Identify any out-of-control points in the plot.

c. A worker has suggested that it would be more convenient to postpone plotting the daily sample means until the end of each week and then examine the points for the entire week for out-of-control conditions. Why would such a postponement partially defeat the purpose of the control chart?

d. Under what conditions should the control limits in part a be changed?

22.15 Molded Panels. Quality requirements for the length of a molded panel used in a household appliance call for the process mean and the process standard deviation to be $\mu = 280.0$ millimeters and $\sigma = 2.0$ millimeters, respectively. The process was brought into control at these levels some months ago. As part of an ongoing quality control program, samples of four panels are taken each hour from the production process and the length of each panel is measured. The lengths of panels are approximately normally distributed.

a. Construct a control chart for the process mean, using three-standard-deviations control limits. Give the values of μ_0, *LCL*, and *UCL* for the chart.
b. The first five samples selected today yielded the following observations for the lengths (in milli-meters) of the individual panels:

| | Observation | | | |
Sample	1	2	3	4
1	279.3	280.1	279.4	281.0
2	280.2	277.5	282.4	280.5
3	277.0	278.6	280.3	278.5
4	277.6	279.2	277.3	280.4
5	275.8	283.1	277.9	282.0

Calculate and plot the sample means on the control chart. Identify any out-of-control points in the plot.
c. A production manager declared that taking samples every hour disrupts operations and suggested that it would be better to select a random sample of 32 panels from the production of each eight-hour shift at the end of the shift to check for any out-of-control situations. Evaluate the manager's proposal.

22.16 Product Weight. The weights of a product placed in containers by a machine are normally distributed with standard deviation $\sigma = 1.5$ grams. The mean fill per container depends on the setting of the machine but tends to drift during the filling operation. The standard level for the mean fill is $\mu = 375.0$ grams. A sample of nine containers is selected every hour during filling, and the sample mean weight of the product $\overline{X}$ is calculated.

a. Construct a control chart for the process mean, using three-standard-deviations control limits. Give the values of μ_0, *LCL*, and *UCL* for the chart.
b. The sample means for the most recent eight hours are as follows:

Hour:	1	2	3	4	5	6	7	8
$\overline{X}$:	374.2	375.1	374.9	376.3	376.2	376.6	376.7	377.2

Plot these sample means on the control chart. Was the process in control with respect to the mean fill during these hours? Does the plot provide any other information about the process? Explain.
c. If the sample size were increased to $n = 25$ and the control limits recalculated, would this change affect the probability of obtaining an out-of-control point when (1) the mean fill is in control at $\mu = 375.0$, (2) the mean fill is out of control at $\mu = 377.0$? Make appropriate calculations, assuming that the process standard deviation remains at $\sigma = 1.5$ grams.

22.17 The regulation of a serious medical disorder requires that the patient ingest regular minute doses of a drug. The drug is administered in the form of tablets that contain small amounts of the drug and an inert binding substance. Why would the pharmaceutical company manufacturing this tablet be concerned with controlling the process standard deviation of the amount of the drug in a tablet as well as the process mean?

* **22.18** Refer to **Alloy Filaments** Problem 22.14.

a. Construct a control chart for the process standard deviation, using control limits corresponding to an α risk of 0.0025. Give the values of σ_0, *LCL*, and *UCL* for the chart. [*Hint:* $\chi^2 (0.00125; 8) = 0.91111$, $\chi^2 (0.99875; 8) = 25.557$.]

b. The sample standard deviations for the most recent 20 operating days are as follows:

Day	s	Day	s	Day	s	Day	s	Day	s
1	8.9	5	12.4	9	20.2	13	18.8	17	27.2
2	17.8	6	29.9	10	15.0	14	21.9	18	14.5
3	12.8	7	16.5	11	15.0	15	14.1	19	45.3
4	22.7	8	28.1	12	22.9	16	15.9	20	20.9

Plot these sample standard deviations on the control chart. Identify any out-of-control points in the plot.

c. Are the control limits calculated in part a robust to departures of the distribution of melting points from normality? Comment.

22.19 Refer to **Molded Panels** Problem 22.15.

a. Construct a control chart for the process standard deviation, using control limits corresponding to an α risk of 0.01. Give the values of σ_0, LCL, and UCL for the chart.

b. Calculate the sample standard deviation s for each of the five samples in Problem 22.15b and plot them on the control chart. Identify any out-of-control points in the plot.

c. The quality assurance manager has found that when a sample point for this production process is out of control on the process mean chart, it is also often out of control on the process standard deviation chart. Explain why this joint occurrence might happen.

22.20 Refer to **Product Weight** Problem 22.16.

a. Construct a control chart for the process standard deviation, using control limits corresponding to an α risk of 0.01. Give the values of σ_0, LCL, and UCL for the chart.

b. The sample standard deviations for the most recent eight hours are as follows:

Hour:	1	2	3	4	5	6	7	8
s:	1.47	2.20	1.42	1.53	1.79	1.61	0.75	1.65

Plot these sample standard deviations on the control chart. Was the process in control with respect to the variability of the product weights? Comment.

c. A production executive asks why there is any interest in identifying cases when s falls below the lower control limit, as this occurrence does not signal a problem with product fill variability. Answer the executive's question.

* **22.21 Ceramic Figurines.** It is very difficult to produce flawless ceramic figurines. For one intricate figurine of a French violinist, the production process is considered to be satisfactory if the process proportion nonconforming is $p = 0.25$. After much effort, the process was brought into control at this level. As part of a continuing quality control program, independent random samples of $n = 20$ figurines are selected daily, and each figurine is assessed as being conforming or not.

a. Construct a control chart for the process proportion nonconforming for this figurine. Use control limits corresponding to an α risk near 0.01. Set the risk for each limit as close as possible to, but not exceeding, 0.005. Give the values of p_0, LCL, and UCL for the chart.

b. The numbers of nonconforming figurines f in the samples for the last 15 days are as follows:

Day:	1	2	3	4	5	6	7	8	9	10	11	12	13	14	15
f:	5	4	8	2	4	5	6	7	4	5	5	4	6	6	7

Calculate the sample proportion nonconforming for each day and plot these values on the control chart. Identify any out-of-control points in the plot.

c. For the control chart in part a, what is the probability that a sample point will be outside the control limits if the process proportion nonconforming is (1) in control at $p = 0.25$, (2) out of control at $p = 0.35$?

22.22 Quality requirements for the accuracy of bills sent monthly to card holders of a credit card company call for the process proportion nonconforming (that is, the proportion of bills containing one or more inaccuracies) to be $p = 0.001$. The process has been in control at this level for some time now. As part of an ongoing quality control program, independent random samples of 300 bills are selected weekly, and each is carefully checked for inaccuracies.

a. Construct a control chart for the process proportion nonconforming with only an upper control limit, corresponding to an α risk as close as possible to, but not exceeding, 0.005. Give the values of p_0 and UCL for the chart. Selected cumulative binomial probabilities for $n = 300$ and $p = 0.001$ are:

x:	0	1	2	3
$P(X \leq x)$:	0.7407	0.9631	0.9964	0.9997

b. The numbers of nonconforming bills f in the samples for the last 12 weeks are as follows:

Week:	1	2	3	4	5	6	7	8	9	10	11	12
f:	0	1	0	0	0	0	1	1	0	1	1	2

Calculate the sample proportion nonconforming for each week and plot these values on the control chart. Identify any out-of-control points in the plot.
c. Could a useful lower control limit have been calculated for this chart? Comment.

22.23 A label glued to a bottle is nonconforming if it is glued on improperly or positioned incorrectly. Quality requirements for the bottle-labeling process call for the process proportion nonconforming to be $p = 0.005$. The process has recently been brought into control at this level. As part of a continuing quality control program, samples of 144 bottles are selected from the production each hour and checked for label conformity.

a. Construct a control chart for the process proportion nonconforming with only an upper control limit, corresponding to an α risk as close as possible to, but not exceeding, 0.01. Give the values of p_0 and UCL for the chart. Selected cumulative binomial probabilities for $n = 144$ and $p = 0.005$ are:

x:	0	1	2	3	4
$P(X \leq x)$:	0.4859	0.8375	0.9638	0.9938	0.9992

b. The numbers of nonconforming labels f in the samples for hours 1 through 8 today are as follows:

Hour:	1	2	3	4	5	6	7	8
f:	0	0	1	0	3	0	2	1

Calculate the sample proportion nonconforming for each hour and plot these values on the control chart. Identify any out-of-control points in the plot.
c. Bottles wait several hours between labeling and being placed in boxes for storage and shipment. It was suggested by an employee that the labels not be inspected until just before the bottles are placed in boxes. Would this inspection delay cause any problem? Explain.

22.24 A cusum chart for a process mean initially fluctuated about the zero line of the chart and then commenced a lengthy, persistent downward drift. What does this pattern indicate about the relation of the process mean to the standard level μ_0 during the time period under consideration?

22.25 Refer to **Alloy Filaments** Problem 22.14b.

a. Calculate the daily cumulative sums C_k for the 20 days and plot these sums on a chart as in Figure 22.8.
b. Does the cusum chart provide any evidence that the process mean was out of control at any time during this period? Comment.

22.26 Refer to **Product Weight** Problem 22.16b.

a. Calculate the hourly cumulative sums C_k for the eight hours and plot these sums on a chart as in Figure 22.8.

b. Does the cusum chart give an earlier indication that the process mean may differ from $\mu_0 = 375.0$ than the control chart for the process mean in Problem 22.16b? Explain.

22.27 A new blending process is being installed in a petrochemical plant. A process engineer expressed concern when 8 of the first 25 sample points on the control chart for the process mean were found to be outside the trial control limits. How would you respond to the engineer?

22.28 A speaker stated, "Generally, acceptance-sampling plans are concerned with inferences about finite populations, whereas control charts are concerned with inferences about infinite populations." Do you agree? Discuss.

* **22.29** **Glass Containers.** Consider the following single-sampling plan for large shipments of glass containers received from a supplier.

Sample Size	Acceptance Number	Rejection Number
20	1	2

a. What action should be taken if the sample contains three defectives? If the sample contains one defective?

b. Express this single-sampling plan as a decision rule involving the sample proportion defective $\bar{p}$.

22.30 **Oil Filters.** An engine assembly plant receives large lots of oil filters from a supplier. A random sample of 100 filters from each lot is inspected as a check on incoming quality. The action for each lot is determined by the following decision rule for the sample proportion defective $\bar{p}$:

If $\bar{p} \leq 0.02$, conclude H_0 (accept lot).
If $\bar{p} > 0.02$, conclude H_1 (reject lot).

a. What are the acceptance and rejection numbers of the single-sampling plan that correspond to this decision rule and sample size?

b. What action should be taken if the sample contains no defective? If the sample contains four defectives?

22.31 **Canned Tuna.** A fish processor ships canned tuna in truckload lots to a retail food chain. The processor follows a zero-defects inspection plan for its outgoing tuna shipments whereby a random sample of 15 cans from a lot is inspected, and the lot is released for shipment only if all 15 cans conform to quality requirements.

a. Describe this inspection plan as a single-sampling plan in the format of Table 22.4a.

b. Express the zero-defects inspection plan as a decision rule involving the sample frequency f.

* **22.32** Refer to **Glass Containers** Problem 22.29. Let p denote the proportion of defective containers in a shipment. The acceptable quality level (AQL) for a shipment is $p_0 = 0.02$.

a. For the sampling plan in Problem 22.29, obtain the probability of accepting a shipment in which $p = 0.02, 0.10, 0.20$.

b. Plot the operating characteristic curve.

c. What is the probability of an incorrect conclusion with this sampling plan for a shipment in which $p = 0.02$? Is this probability the supplier's risk or the buyer's risk?

22.33 Refer to **Oil Filters** Problem 22.30. Let p denote the lot proportion defective. The acceptable quality level (AQL) for a lot is $p_0 = 0.01$.

a. For the sampling plan in Problem 22.30, obtain the points on the operating characteristic curve for $p = 0.01, 0.04, 0.07$.

b. Plot the operating characteristic curve.

c. A lot with 7 percent defectives is definitely unacceptable. What is the risk of an erroneous decision with this sampling plan when $p = 0.07$? Is this the supplier's risk or the buyer's risk?

22.34 Refer to **Canned Tuna** Problem 22.31. Let p denote the lot proportion nonconforming. The acceptable quality level (AQL) for outgoing lots is $p_0 = 0.01$.

a. Obtain the points on the operating characteristic curve of the zero-defects inspection plan in Problem 22.31 for $p = 0.01, 0.10, 0.20$.
b. Plot the operating characteristic curve.
c. What is the risk of not releasing a shipment for which $p = 0.01$? Is this the producer's risk or the customer's risk?

22.35 Consider the following two-stage sampling plan for examining incoming lots:

Sample	Sample Size	Cumulative Sample Size	Acceptance Number	Rejection Number
First	20	20	0	2
Second	20	40	1	2

a. What action should be taken if the first sample of 20 items from a lot contains (1) 0 defective, (2) 1 defective, (3) 2 defectives?
b. What action should be taken if the first sample contains 1 defective and the second contains 1 defective?

22.36 Refer to **Oil Filters** Problem 22.30. The following two-stage sampling plan has been proposed to replace the current single-sampling plan for incoming lots:

Sample	Sample Size	Cumulative Sample Size	Acceptance Number	Rejection Number
First	50	50	0	3
Second	50	100	3	4

a. What action should be taken if the first sample of 50 filters contains (1) 0 defective, (2) 1 defective, (3) 3 defectives?
b. What action should be taken if the first sample contains 1 defective and the second contains 3 defectives?

22.37 A commentator at a quality improvement conference stated, "A reliable product is one that will satisfy quality requirements for its full warranty period." Do you agree? Comment.

* **22.38** The time to failure T of an installed clock battery of a particular type is normally distributed with a mean of 18,000 hours and a standard deviation of 4000 hours.

a. Obtain the value of the reliability function for this type of battery for failure times (1) 15,000, (2) 20,000, (3) 25,000 hours.
b. The manufacturer of this battery has provided a warranty period of one year (8760 hours). If a purchaser of a new battery installs the battery immediately, what is the probability that the battery will survive beyond the end of the warranty period?

22.39 The time to failure T of an A4003 bearing in an industrial electric motor is normally distributed with a mean of 2.32 years and a standard deviation of 0.46 year.

a. Obtain $R(1.00)$ for an A4003 bearing. Interpret this measure.
b. If 1000 bearings of this type are installed and bearing failures are independent of one another, what is the expected number of these 1000 bearings that will have failed within one year? What is the probability that all but one of the 1000 bearings will survive beyond one year?

22.40 Refer to **Digital Display** Problem 8.22. Denote the failure time of a display unit by T. Obtain the values of the reliability function for $t = 0, 2, 4, 6$ years. Plot the reliability function. For what value of t is $R(t) = 0.99$?

* **22.41**   Consider the series-parallel system design for the *Satellite* example in Figure 22.11b. Assume that the four components will fail independently of one another and that the probabilities they will survive longer than five years are as follows: $P(E_1) = 0.98$, $P(E_2) = 0.99$, $P(E_3) = P(E_4) = 0.98$.

a. Obtain the five-year survival probability for the system.
b. What will be the impact on the probability in part a if the failure events for components 3 and 4 are dependent in such a way that components 3 and 4 will fail at the same time?

22.42   A lighting fixture will produce light when its switch is activated if (1) the power supply, (2) the light switch, and (3) the light bulb all function. Denote these respective events by E_1, E_2, and E_3. Assume that the three events are independent and that the respective probabilities of these events are 0.95, 0.99, and 0.98.

a. What is the probability that the lighting fixture will produce light when the switch is activated?
b. Now suppose that the fixture has two bulbs that are connected in parallel. Assume that each bulb will function with probability 0.98 and that the two bulbs, the power supply, and the light switch will fail independently of each other. What is the probability that both bulbs will light when the switch is activated? That at least one bulb will light when the switch is activated?

22.43   A repeater in a telecommunication system (1) receives a signal, (2) amplifies it, (3) rectifies it, and finally (4) sends it on to the next repeater. Consider the components that perform this series of operations. Denote the events of each of these components functioning as required for one signal by E_1, E_2, E_3, and E_4, respectively. Assume that the four events are independent and that the probability of each is 0.9999.

a. What is the probability that a repeater will function as required for one signal?
b. The telecommunication system has 50 repeaters in series, each of which will function as required for one signal with the probability in part a. Assume that the 50 repeaters function independently. What is the probability that the telecommunication system will function as required for one signal?

EXERCISES

22.44   Refer to **Molded Panels** Problem 22.15. It has been suggested that a control chart be created in which each sample observation is plotted individually.

a. Set up such a chart, using three-standard-deviations control limits. Give the values of *LCL* and *UCL* for the chart.
b. Plot the 20 observations on the control chart. Identify any out-of-control points.
c. Under the assumption of normality, what is the probability in advance of sampling that at least one of the 20 observations will fall outside the control limits when the process is in control?
d. Are the control limits in part a robust to departures of the distribution of panel lengths from normality? Comment.

22.45   Assume that a process is operating in control with mean μ_0 and variance σ^2. Let $\overline{X}_i$ ($i = 1, 2, \ldots, k$) denote the means of independent random samples of size n drawn from this process while it is operating in control.

a. Show that the cumulative sum C_k in (22.10) has expected value 0 and variance $k(\sigma^2/n)$ under these conditions.
b. Do the results in part a imply that the value of C_k is expected to lie farther from the zero line of the cusum chart as k increases when the process remains in control? Explain.

22.46   Consider the following single-sampling plan for large lots:

Sample Size	Acceptance Number	Rejection Number
500	2	3

Use the Poisson approximation to the binomial distribution discussed in Chapter 13 (Comment 2, p. 371) to calculate for this sampling plan the probability of accepting a lot containing 0.3 percent defectives.

22.47 *(Calculus needed.)*

a. Show that if the time to failure T has mean $E\{T\}$ and reliability function $R(t)$, then:

$$E\{T\} = \int_0^\infty R(t)dt$$

[*Hint:* Consider the fact that $dR(t)/dt = -f(t)$ and employ integration by parts.]

b. Confirm that the relation in part a holds when T follows the exponential distribution.

22.48 It is difficult for complex systems to be reliable. As an illustration of this truism, consider the following simplified situation. A rocket consists of N components in series, each of which has a probability R of surviving a flight. All N components must survive the flight for the rocket to survive. Assuming that the components of the rocket will fail independently, calculate the probability that the rocket will survive the flight for each of the following four cases:

Case	N	R
1	10	0.990
2	1000	0.990
3	10	0.999
4	1000	0.999

Discuss your findings.

STUDIES

22.49 Refer to the control chart for the process mean in the *Drug Tablets* example, which is shown in Figure 22.4. The decision rule for this type of chart is given in (22.4). Recall that the process standard deviation for this example is $\sigma = 0.0008$ gram and that the size of each hourly sample is $n = 5$.

a. Calculate the rejection probabilities for the control chart in Figure 22.4 at $\mu = 0.4390, 0.4400, 0.4410, 0.4420, 0.4430$. Sketch the rejection probability curve. What is the risk of failing to detect that the process is out of control with a single hourly sample if, in fact, $\mu = 0.4420$?

b. What would the required size of each hourly sample have to be if it were desired to control the β risk at 0.05 when $\mu = 0.4420$ while still employing three-standard-deviations control limits? [*Hint:* Adapt formula (12.19) to this situation.]

22.50 Refer to the two-stage sampling plan in Problem 22.35.

a. Obtain the probability that a lot will either be accepted or rejected in the first stage by this plan if the lot proportion defective is $p = 0.01, 0.02, 0.10, 0.40$.

b. Determine the expected sample size required to reach an accept–reject decision when $p = 0.01, 0.02, 0.10, 0.40$. [*Hint:* For any given lot, the total sample size will be either 20 or 40.] What is the relation between the lot proportion defective and the expected sample size required to reach a decision for this sampling plan?

22.51 A company receives a component in lots of 8000 units. A random sample of 50 units is selected from each lot, on the basis of which a decision is made to accept (H_0) or reject (H_1) the lot. A rejected lot is inspected 100 percent, all defective units are replaced with acceptable ones, and the supplier is charged for the work and the defective units. The decision rule employed has the following properties:

Lot Proportion Defective p	Acceptance Probability $P(H_0; p)$	Rejection Probability $P(H_1; p)$
0.01	0.91	0.09
0.03	0.56	0.44
0.10	0.03	0.97

a. Determine the average quality of lots after inspection when the incoming lot proportion defective is $p = 0.01, 0.03, 0.10$. Ignore defective units found in samples from accepted lots.

b. Explain why the average quality of lots after inspection is better when the lot proportion defective is 0.10 than when it is 0.03. Is it therefore desirable for the supplier to submit lots of poor quality? Explain.

c. Determine the average number of units inspected per lot when the incoming lot proportion defective is $p = 0.01, 0.03, 0.10$.

Survey Sampling

23

Sample surveys are widely employed to gather information about such diverse subjects as public opinion, consumer spending, unemployment, incidence of crimes, consumer prices, and ocean fish stocks. Generally, sample surveys are observational studies, although they occasionally have experimental components. In this chapter we take up some practical aspects of the design of sample surveys and the analysis of the data so obtained.

We first consider simple random sampling when the population is finite and the sampling fraction is sufficiently large that the finite nature of the population must be taken into account. We then describe several other probability sampling procedures that are widely used in sample surveys—namely, stratified sampling, probability-proportional-to-size sampling, cluster sampling, and systematic sampling.

Throughout the text to this point, we have assumed that when a random sample of n elements is selected from a finite population of N elements, the sampling fraction n/N is 5 percent or less, which is generally the case. There are occasions, however, when this condition is not met, as for instance, when a random sample of 5 generators is selected from a population of 15 to study the operational state of an important safety feature. In cases where the sampling fraction n/N is not small, recognition of the finite nature of the population is essential.

Estimation of Population Mean

Finite Population Correction.   When a random sample is drawn from a finite population, no matter whether the sampling fraction is small or large, the mean of the sampling distribution of $\overline{X}$ is μ; that is, $E\{\overline{X}\} = \mu$. Also, central limit theorem (10.7) continues to apply as long as the sampling fraction n/N is not close to 1, a condition that almost always holds in practice. However, the variance of $\overline{X}$ does depend on the sampling fraction. Thus, the formula for the estimated standard deviation of $\overline{X}$ in (11.2), that is, $s\{\overline{X}\} = s/\sqrt{n}$, is exact only for the case of an infinite population. When the population is finite and the sampling fraction n/N is 5 percent or smaller, formula (11.2) still applies approximately. When the sampling fraction n/N is not small, however, the estimated standard deviation $s\{\overline{X}\}$ must be modified as follows.

(23.1)

For finite populations, the estimated variance $s^2\{\bar{X}\}$ and the estimated standard deviation $s\{\bar{X}\}$ are:

$$s^2\{\bar{X}\} = \left(1 - \frac{n}{N}\right)\frac{s^2}{n} \qquad s\{\bar{X}\} = \sqrt{1 - \frac{n}{N}}\,\frac{s}{\sqrt{n}}$$

The factor $\sqrt{1 - (n/N)}$ is called the *finite population correction* because the standard deviation $s/\sqrt{n}$ for an infinite population is "corrected" by this factor to yield the proper estimate when the population is finite.

Two characteristics of the finite population correction are noteworthy.

1. For a given sample size n, the finite population correction $\sqrt{1 - (n/N)}$ is nearly equal to 1 when the sampling fraction n/N is small. For this reason, a finite population can be treated as an infinite one when the sampling fraction is small, since $s\{\bar{X}\}$ in (23.1) then equals $s/\sqrt{n}$ for all practical purposes.
2. For a given finite population, the finite population correction approaches zero as the sample size n approaches the population size N. Hence, $s\{\bar{X}\} = 0$ when $n = N$. This result is intuitively reasonable since a census of the population is taken when $n = N$. Hence $\bar{X}$ contains no sampling error and therefore is equal to the population mean μ.

Based on working rule (10.9), we shall use the finite population correction only when the sampling fraction n/N is larger than 5 percent.

Confidence Interval for μ. The estimation and testing procedures for the population mean μ when the sampling fraction is large are the same as the earlier procedures, except that (23.1) is used to obtain $s\{\bar{X}\}$. We now illustrate the construction of a confidence interval for the population mean μ when the sampling fraction is large.

□ **EXAMPLE**

A chemical company serves $N = 810$ dry-cleaning firms. It selected a simple random sample of $n = 140$ firms to estimate the mean annual use per firm (μ) of a new solvent product. The survey results gave $\bar{X} = 410$ gallons and $s = 215$ gallons. A 95 percent confidence interval for μ is desired.

The sampling fraction n/N in this case is $140/810 = 0.1728$, which is large enough to apply the finite population correction. The confidence interval for μ is of the same form as (11.5), except that $s\{\bar{X}\}$ is computed by using (23.1).

We obtain, using (23.1):

$$s\{\bar{X}\} = \sqrt{1 - \frac{140}{810}}\,\frac{215}{\sqrt{140}} = 16.526$$

For 95 percent confidence, $z(0.975) = 1.960$ is required. Hence, the confidence limits are $410 \pm 1.960(16.526)$, and the 95 percent confidence interval is $377.6 \le \mu \le 442.4$.

□

Planning of Sample Size. When planning the size of a simple random sample from a finite population, the finite nature of the population must be taken into account if the resulting sampling fraction will not be small. The appropriate planning formula for a $1 - \alpha$ confidence interval for μ is the following modification of (11.13).

$$(23.2) \qquad n = \frac{z^2 \sigma^2}{h^2 + \dfrac{z^2 \sigma^2}{N}}$$

where: $z = z(1 - \alpha/2)$
  h is the specified half-width
  σ is the planning value for the population standard deviation

Formula (23.2) still assumes that the resulting sample size will be large.

EXAMPLE ☐

A random sample from a population of $N = 900$ Canadian general hospitals is to be taken to obtain a confidence interval for the mean number of pediatric beds per hospital. The precision specifications are a half-width of $h = 5$ beds and a 95 percent confidence coefficient. Results of a previous study suggest that an appropriate planning value for the population standard deviation is $\sigma = 40$ beds. We require $z(0.975) = 1.960$ and find the requisite sample size by using (23.2):

$$n = \frac{(1.960)^2 (40)^2}{(5)^2 + \dfrac{(1.960)^2 (40)^2}{900}} = 193$$

Note that the sampling fraction will be large here; that is, $n/N = 193/900 = 0.214$. Thus, more than 21 percent of all hospitals must be sampled to obtain an estimate with the desired precision. ☐

Comment

As the population size N gets large, the second term in the denominator of (23.2) approaches zero and (23.2) simplifies to the earlier planning formula (11.13).

Estimation of Population Proportion

Confidence Interval for p. To estimate a finite population proportion p when the sampling fraction is large, the estimated standard deviation $s\{\bar{p}\}$ needs to include the finite population correction.

(23.3)

For finite populations, the estimated variance $s^2\{\bar{p}\}$ and the estimated standard deviation $s\{\bar{p}\}$ are:

$$s^2\{\bar{p}\} = \left(1 - \frac{n}{N}\right)\frac{\bar{p}(1 - \bar{p})}{n - 1} \qquad s\{\bar{p}\} = \sqrt{1 - \frac{n}{N}}\sqrt{\frac{\bar{p}(1 - \bar{p})}{n - 1}}$$

The construction of a confidence interval for the population proportion p when the sampling fraction is large proceeds as before, using (13.9), except that $s\{\bar{p}\}$ is obtained from (23.3).

☐ **EXAMPLE**

For a population of $N = 1200$ railroad tank cars, a simple random sample of $n = 200$ cars revealed that $f = 33$ had defective discharge valves. A 90 percent confidence interval for the population proportion of tank cars with defective valves is desired. We require:

$$z(0.95) = 1.645 \qquad \bar{p} = \frac{f}{n} = \frac{33}{200} = 0.165$$

$$s\{\bar{p}\} = \sqrt{1 - \frac{200}{1200}}\sqrt{\frac{0.165(1 - 0.165)}{200 - 1}} = 0.0240$$

The confidence limits therefore are $0.165 \pm 1.645(0.0240)$, and the 90 percent confidence interval is $0.13 \le p \le 0.20$.   ☐

Planning of Sampling Size.   When the required sample size for estimating a population proportion p will represent a large fraction of the population, the planning formula in (13.10) must be modified as follows.

(23.4) $$n = \frac{z^2 p(1 - p)}{h^2 + \dfrac{z^2 p(1 - p)}{N}}$$

where: $z = z(1 - \alpha/2)$
h is the specified half-width
p is the planning value for the population proportion

Formula (23.4) still assumes that the resulting sample size will be large.

☐ **EXAMPLE**

Leaders of a union local with $N = 850$ members want to survey members' opinions on a proposed change in their labor contract. It is desired to estimate the proportion p supporting the change by a confidence interval with half-width $h = 0.05$ and confidence coefficient 0.99. A planning value of $p = 0.5$ is considered to be reasonable.

We require $z(0.995) = 2.576$. The number of members needed to be included in the sample then is, using (23.4):

$$n = \frac{(2.576)^2(0.5)(1 - 0.5)}{(0.05)^2 + \dfrac{(2.576)^2(0.5)(1 - 0.5)}{850}} = 373$$

☐

Efficiency of Sampling Procedure

Simple random sampling is not always the best sampling procedure to employ. Other procedures often are more efficient. The concept of efficiency of a sampling procedure is usually defined in terms of the precision of the estimator, as measured by the width of the confidence interval for a specified confidence coefficient. Every probability sampling procedure can be evaluated in terms of the cost of attaining a given statistical precision, or, alternatively, in terms of the precision furnished for a given cost.

(23.5)

One probability sampling procedure is more **efficient** than another if it offers, at a given level of confidence, the same precision at less cost or greater precision at the same cost.

We now consider several probability sampling procedures that often are more efficient than simple random sampling and are widely used.

23.2 STRATIFIED SAMPLING

It is often more efficient to select independent random samples from subpopulations than to select a simple random sample from the entire population.

EXAMPLE ☐ **Teachers Association**

A state teachers association conducted a sample survey to study the educational qualifications of its membership, which consisted of 50,000 elementary school, high school, and college teachers. A key characteristic of interest was the mean number of years of formal education (including teacher training) of the association's teachers. The analyst conducting the study knew that the number of years of education for teachers tends to vary by level of school, being greater for teachers in higher-level educational institutions. Hence, the analyst used a sample design in which independent simple random samples of teachers were selected from among the member teachers in each of the three school levels. Column 1 of Table 23.1 shows the total number of member teachers in each school level, and column 2 shows the sample size for each level. ☐

The sampling procedure used here is called stratified random sampling.

TABLE 23.1
Stratified sampling plan—
Teachers association example

		(1)	(2)
		Number of Members	
Stratum	Level of School	In Population	In Sample
1	Elementary	25,000	150
2	High school	20,000	120
3	College	5,000	30
	Total	50,000	300

(23.6)
With **stratified random sampling,** the population is divided into a number of mutually exclusive subpopulations, or **strata,** and independent simple random samples are selected from the strata.

In the teachers association example, the 25,000 teachers in elementary schools represent one stratum of the population of 50,000 teachers.

Advantages of Stratified Sampling

Stratified sampling has three major advantages: (1) efficiency, (2) information about subpopulations, and (3) feasibility.

Efficiency.   Frequently, a stratified sample will be much more efficient than a simple random sample. We illustrate this in Table 23.2. A hypothetical population of nine teachers is shown in Table 23.2a, and the number of years of education is given for each teacher. The population variance is 2.61. This is the variability affecting the precision of the sample mean $\overline{X}$ with simple random sampling.

When we stratify the teachers by level of school, as in Table 23.2b, we encounter the strata variances shown there. Note that these variances, which measure the variability of years of education within each stratum, are much smaller than the variance for the entire population. Consequently, in this instance, a stratified random sample of given size will yield a substantially more precise estimate than a simple random sample of the same size, since much smaller variability is encountered within each stratum.

Thus, for stratified sampling to be efficient, the strata must be designed to contain relatively homogeneous elements. Homogeneity is accomplished when the basis of stratification is related to the characteristic under study. In the teachers association example, level of school in which a member teaches is highly related to number of years of education. Similarly, in a survey of firms to estimate mean sales per firm, a stratification by size of firm is likely to be highly effective in view of a strong positive relation between size of firm and sales by the firm.

While the selection of a stratified random sample may cost somewhat more than the selection of a simple random sample of equal size (for example, a separate frame is required for each stratum), this increase in cost is usually more than balanced by the substantially improved precision. Thus, a stratified sample that is designed to yield a specified precision usually is significantly smaller than the corresponding simple random sample and consequently costs less.

(a) Population

Teacher	Level of School	Years of Education	Teacher	Level of School	Years of Education
1	Elementary	15.5	6	College	20.5
2	College	19.0	7	High school	17.5
3	High school	16.5	8	College	19.5
4	Elementary	16.0	9	Elementary	16.5
5	High school	18.0			

Variance = 2.61

(b) Strata

Stratum 1: Elementary		Stratum 2: High School		Stratum 3: College	
Teacher	Years of Education	Teacher	Years of Education	Teacher	Years of Education
1	15.5	3	16.5	2	19.0
4	16.0	5	18.0	6	20.5
9	16.5	7	17.5	8	19.5
Variance = 0.17		Variance = 0.39		Variance = 0.39	

Information About Subpopulations. Stratified sampling can provide estimates of strata characteristics in addition to estimates of overall population characteristics. For example, a study by a university of the effects of tuition increases is to provide separate information for undergraduate and graduate students, as well as information for all students. Stratification by student class status will permit precise estimates for undergraduate and graduate students separately, as well as for all students.

Feasibility. At times, stratified sampling is the most feasible type of sampling. Consider the sampling of welfare claims in a state. In the one large city of the state, the claims records are in computerized files; elsewhere in the state, the records are kept locally in manual files. Administrative considerations here require separate sampling of claims in the city and elsewhere in the state.

Estimation of Population Mean

We now show how to estimate the population mean μ based on a stratified random sample.

Notation. The total number of strata into which the population is divided will be denoted by k, the number of population elements in the jth stratum by N_j, and the total population size as usual by N.

$$(23.7) \qquad N = \sum_{j=1}^{k} N_j$$

Analogously, n_j will denote the size of the sample selected from the jth stratum, and n as usual denotes the total sample size.

$$(23.8) \qquad n = \sum_{j=1}^{k} n_j$$

Finally, $\overline{X}_j$ and s_j^2 will denote the mean and the variance, respectively, of the sample drawn from the jth stratum.

☐ EXAMPLE

In the teachers association example of Table 23.1, $k = 3$ strata are employed. The stratum size and the sample size for the first stratum (elementary schools) are $N_1 = 25,000$ and $n_1 = 150$, respectively. The total population and sample sizes are $N = 50,000$ and $n = 300$.   ☐

Development of Confidence Interval.   We require a point estimator of the population mean μ.

(23.9)

The stratified sample estimator of the population mean μ, denoted by $\overline{X}_{st}$, is:

$$\overline{X}_{st} = \left(\frac{N_1}{N}\right)\overline{X}_1 + \left(\frac{N_2}{N}\right)\overline{X}_2 + \cdots + \left(\frac{N_k}{N}\right)\overline{X}_k = \sum_{j=1}^{k}\left(\frac{N_j}{N}\right)\overline{X}_j$$

Note that $\overline{X}_{st}$ is simply a weighted average of the strata sample means, each stratum weight being the proportion of the total population in the stratum. It can be shown that the estimator $\overline{X}_{st}$ is an unbiased estimator of μ.

The estimated variance of $\overline{X}_{st}$ is denoted by $s^2\{\overline{X}_{st}\}$ and is as follows.

$$(23.10) \qquad s^2\{\overline{X}_{st}\} = \left(\frac{N_1}{N}\right)^2 s^2\{\overline{X}_1\} + \left(\frac{N_2}{N}\right)^2 s^2\{\overline{X}_2\} + \cdots + \left(\frac{N_k}{N}\right)^2 s^2\{\overline{X}_k\}$$

$$= \sum_{j=1}^{k}\left(\frac{N_j}{N}\right)^2 s^2\{\overline{X}_j\}$$

where $s^2\{\overline{X}_j\}$ is given by (23.1) or (11.2), as appropriate.

In applying (23.10), we shall employ the working rule that the finite population correction is used if any stratum has a sampling fraction n_j/N_j exceeding 5 percent.

To complete the construction of the interval estimate of μ, we note that the sampling distribution of $(\overline{X}_{st} - \mu)/s\{\overline{X}_{st}\}$ is approximately normal when the total sample size n is reasonably large. Thus, the confidence limits for μ have the usual large-sample form.

TABLE 23.3
Stratified sample results—
Teachers association example

j	Level of School	N_j	n_j	$\overline{X}_j$	s_j^2	$\left(\dfrac{N_j}{N}\right)\overline{X}_j$	$\left(\dfrac{N_j}{N}\right)^2\left(\dfrac{s_j^2}{n_j}\right)$
1	Elementary	25,000	150	14.8	6.40	7.40	0.01067
2	High school	20,000	120	17.3	2.69	6.92	0.00359
3	College	5,000	30	19.3	3.57	1.93	0.00119
	Total	50,000	300			16.25	0.01545
		↑	↑			↑	↑
		N	n			$\overline{X}_{st}$	$s^2\{\overline{X}_{st}\}$

(23.11)
The approximate $1 - \alpha$ confidence limits for μ, when stratified random sampling is employed and the total sample size is reasonably large, are:

$$\overline{X}_{st} \pm zs\{\overline{X}_{st}\}$$

where: $z = z(1 - \alpha/2)$
$\overline{X}_{st}$ is given by (23.9)
$s\{\overline{X}_{st}\}$ is given by (23.10)

EXAMPLE ☐

For the teachers association example, a 95 percent confidence interval for the mean number of years of education μ is desired. Table 23.3 contains the sample results and shows the calculation of $\overline{X}_{st}$. We find that $\overline{X}_{st} = 16.25$ years. Table 23.3 also shows the calculation of $s^2\{\overline{X}_{st}\}$. Since all of the strata sampling fractions n_j/N_j are small, we have ignored the finite population corrections and used s_j^2/n_j in (11.2) for calculating $s^2\{\overline{X}_j\}$. We see in Table 23.3 that $s^2\{\overline{X}_{st}\} = 0.01545$; hence it follows that $s\{\overline{X}_{st}\} = \sqrt{0.01545} = 0.1243$ year.

For a 95 percent confidence coefficient, we require $z(0.975) = 1.960$. Hence, the confidence limits for μ are $16.25 \pm 1.960(0.1243)$, and the 95 percent confidence interval is:

$$16.0 \leq \mu \leq 16.5$$

It can be stated, with 95 percent confidence, that the mean number of years of education of the member teachers is between 16.0 and 16.5.   ☐

Comments

1. The estimated variance $s^2\{\overline{X}_{st}\}$ in (23.10) can be derived as follows. Since the estimated strata means $\overline{X}_j$ in (23.9) are based on independent random samples, we know from (6.9b) that the variances of the $(N_j/N)\overline{X}_j$ terms in $\overline{X}_{st}$ are additive. Furthermore, by (6.6b), the variance of $(N_j/N)\overline{X}_j$ is $(N_j/N)^2\sigma^2\{\overline{X}_j\}$. Hence, we obtain:

continues

$$(23.12) \quad \sigma^2\{\overline{X}_{st}\} = \left(\frac{N_1}{N}\right)^2 \sigma^2\{\overline{X}_1\} + \left(\frac{N_2}{N}\right)^2 \sigma^2\{\overline{X}_2\} + \cdots + \left(\frac{N_k}{N}\right)^2 \sigma^2\{\overline{X}_k\}$$

Finally, to estimate the variances $\sigma^2\{\overline{X}_j\}$, we employ $s^2\{\overline{X}_j\}$ in (23.1) or (11.2), as appropriate.

2. Stratified sampling can also be used to estimate a population proportion p. For example, consider an economically depressed district of a city for which the proportion of households with inadequate living space is to be estimated. The district might be stratified into groups of contiguous city blocks and a sample of households drawn from each stratum. The procedure for estimating the population proportion p from a stratified sample is analogous to that for estimating the population mean μ.

Planning Strata Sample Sizes

An important design consideration in planning a stratified random sample is the allocation of the total sample size to the individual strata.

Proportional Allocation. In Table 23.3, note that the stratum sampling fraction in the first stratum is $n_1/N_1 = 150/25{,}000 = 0.006$ or 0.6 percent, and that the sampling fractions for the other two strata are also 0.6 percent. This method of using the same sampling fraction to allocate the total sample size to the strata is called proportional allocation.

(23.13)
With **proportional allocation,** the sampling fraction is the same for all strata and, therefore, the sample size for the jth stratum is:

$$n_j = \left(\frac{N_j}{N}\right)n$$

Proportional allocation is frequently used because it is simple and often quite effective.

Optimal Allocation. Another method of allocating the total sample to the strata is to minimize the variance of $\overline{X}_{st}$, thereby providing an estimate of μ that is as precise as possible for a given total sample size. Such a method of allocation is called optimal allocation. It can be shown that the stratum sample sizes with optimal allocation are given by the following formula.

(23.14)
With **optimal allocation,** the sample size for the jth stratum is:

$$n_j = \left(\frac{N_j\sigma_j}{\sum_{j=1}^{k} N_j\sigma_j}\right)n$$

where σ_j is the standard deviation for the jth stratum.

j	Level of School	N_j	Planning Value for σ_j	$N_j \sigma_j$	$\dfrac{N_j \sigma_j}{\Sigma N_j \sigma_j}$	n_j
1	Elementary	25,000	2.5	62,500	0.6098	183
2	High school	20,000	1.5	30,000	0.2927	88
3	College	5,000	2.0	10,000	0.0976	29
	Total	50,000		102,500	1.000	300

$$\uparrow$$
$$\Sigma N_j \sigma_j$$

Note that allocation formula (23.14) involves the strata standard deviations σ_j. Since these parameters are generally unknown, planning values for the σ_j are required. Approximate information about the orders of magnitude of the σ_j is usually adequate for planning purposes.

EXAMPLE □

In the teachers association example, the analyst considered optimal allocation for the stratified sample. Table 23.4 shows the planning values for the stratum standard deviations σ_j, based on earlier surveys, together with the calculations for the optimal allocation of the total sample size to the three strata. We see that the strata sample sizes for optimal allocation differ somewhat from those for proportional allocation in Table 23.3.

In the end, the analyst did not utilize optimal allocation. The reason was that the sample was intended to provide information not only about the mean number of years of education of the members but also about other member characteristics. It turned out that the optimal allocations for estimating these other characteristics differed markedly from the one in Table 23.4, because the relative magnitudes of the σ_j for the other characteristics were not similar to those for years of education. The analyst then decided that proportional allocation represented a reasonable compromise. □

Comment

Occasionally, allocation formula (23.14) may produce a value for n_j that is larger than N_j. If this occurs, optimal allocation requires that n_j be set equal to N_j for this particular stratum (in other words, a census of the stratum needs to be undertaken) and that the remaining total sample size $n - N_j$ be allocated to the remaining strata according to formula (23.14).

When the population elements differ greatly in size, probability-proportional-to-size sampling is often efficient. For the sake of brevity, this method of sampling is called pps sampling. With pps sampling, the elements are selected for the sample with unequal probabilities, specifically with probabilities proportional to their sizes.

PROBABILITY-PROPORTIONAL-TO-SIZE SAMPLING **23.3**

> **(23.15)**
> With **probability-proportional-to-size (pps) sampling,** each element is selected for the sample with a probability that is proportional to a measure of its size.

For instance, when a population of firms is to be sampled to estimate the mean sales per firm, pps sampling requires that large firms have a greater chance of being selected than small firms. Measures of size that might be used here include firm sales during the most recent census year, number of employees, or any other variable that is closely related to current sales and that is known for all firms in the population.

☐ **EXAMPLE**

Table 23.5 illustrates a population of $N = 1000$ firms for which the annual sales during the most recent census year are known for each firm (column 1—sales in $ thousand). A probability sample of $n = 10$ firms is to be selected, using pps sampling *with replacement*. We see that the total census sales for the population are $100,000 thousand. Thus, when selecting the first sample firm with pps sampling, firm 1 should have probability 90/100,000 = 0.00090 of being selected. Similarly, the probability of selecting firm 2 should be 451/100,000 = 0.00451, and that of firm 1000 should be 304/100,000 = 0.00304. We can achieve these probabilities by first cumulating the census sales in column 2 and then developing the cumulative ranges in column 3. We now generate a discrete uniform random number between 1 and 100,000. If the generated random number is 43, we select firm 1 as the first sample firm because the random number falls in the cumulative range for firm 1 in column 3. If the random number is 99,804, we would select firm 1000 because the random number falls in the cumulative range for this firm. In this fashion, each firm has a probability of being chosen as the first sample firm that is proportional to its size measure (census sales).

Since sampling is to be done with replacement, the procedure must now be repeated for selecting sample firm 2, sample firm 3, . . . , and sample firm 10. With this sampling procedure, the possibility exists that a given firm is selected more than once, though the probability of this occurring here is not large.   ☐

A major advantage of pps sampling is that when the elements differ greatly in size and size is closely related to the characteristic of interest, pps sampling leads to much more precise estimates than simple random sampling. While stratified sampling also can be effective when the population elements differ greatly in size, pps sampling often is easier to implement.

Comment

In auditing, *dollar-unit sampling* is widely used. This sampling method is a pps sampling method. For sampling accounts receivable, for instance, the individual dollars of each account are considered to be the sampling units. For example, an account receivable with a book value of $1309 is considered to consist of 1309 dollar units. A random sample of dollar units is ‣ selected and the corresponding accounts are audited. This method of selection gives each account a selection probability that is proportional to its size, that is, its book value according to the accounting records of the company.

Firm	(1) Census Sales ($ thousand)	(2) Cumulative Census Sales ($ thousand)	(3) Cumulative Range
1	90	90	1–90
2	451	541	91–541
⋮	⋮	⋮	⋮
999	74	99,696	99,623–99,696
1000	304	100,000	99,697–100,000
Total	100,000		

TABLE 23.5
Illustration of pps sampling

CLUSTER SAMPLING 23.4

Single-Stage Cluster Sampling

Sometimes, it is efficient to sample population elements in groups rather than individually.

EXAMPLE ☐ **Inspection**

An inspector needs to examine a shipment to determine the extent of deterioration, if any, of the parts in the shipment. The shipment consists of 2000 sealed cartons, each containing 5 parts. If a simple random sample of 100 parts is to be examined, it is conceivable that as many as 100 cartons may have to be unsealed. Since unsealed cartons are difficult to store and handle, and items in unsealed cartons may be stolen or damaged, the inspector wishes to open fewer cartons. To this end, a sampling procedure will be used that involves the selection of 20 of the 2000 cartons at random and inspection of all the parts in each selected carton. ☐

The procedure just described illustrates single-stage cluster sampling. Each carton represents a group or cluster of elements, and a probability sample of these clusters is selected. All elements in the selected clusters are then included in the sample.

> **(23.16)**
> With **single-stage cluster sampling**, the population is divided into **clusters** of elements, a probability sample of clusters is selected, and all elements in the selected clusters are included in the sample.

In general, cluster sampling requires a larger number of elements to yield a specified precision than does simple random sampling, because the elements within a cluster (parts in a carton, persons in a family) tend to be more homogeneous than in the population at large. Nevertheless, cost considerations under some circumstances make cluster sampling more efficient than simple random sampling.

1. When the cost of constructing a frame of elements is high, cluster sampling may be more efficient. For example, in a survey of eligible voters in a city, a current listing of city blocks is available, but a current listing of eligible voters is not. Here, cluster sampling of city blocks has the advantage of requiring the preparation of voters' lists only for city blocks actually sampled. On the other hand, simple random sampling of eligible voters would require a voters' list for the entire city, which is expensive to compile.

2. Cluster sampling may also be efficient when it is more expensive to sample items scattered throughout the population than items that are "close" to one another. In the inspection example, it is much more costly to open 100 cartons to inspect 100 parts than to inspect the same number of parts in 20 cartons. As another example, a cluster sample of households from city blocks in a large city requires substantially less travel time by interviewers than a random sample of households scattered throughout the city.

Multistage Cluster Sampling

In many applications of cluster sampling, the selection of sample elements is done in two or more stages.

□ EXAMPLE

In a statewide survey of urban households to obtain data on food expenditures, a probability sample of cities is first selected. Within the chosen cities, probability samples of city blocks are selected, and within the selected blocks, probability samples of households are chosen.   □

The procedure just described illustrates a three-stage cluster sample. Cities are sampled in the first stage and are called primary sampling units. Blocks are sampled in the second stage and are called secondary sampling units. Households within blocks are sampled in the third stage and are called tertiary sampling units.

> **(23.17)**
> With **multistage cluster sampling,** a probability sample of **primary sampling units** is selected. From each chosen primary sampling unit, a probability sample of **secondary sampling units** is selected, and so on, until the final stage of sampling.

When the primary sampling units differ greatly in size (e.g., when cities in a state are the primary sampling units), pps sampling or stratified sampling of the primary sampling units is often desirable. Pps sampling or stratified sampling may also be used in some of the later stages of multistage cluster sampling.

Cluster sampling in more than one stage is efficient when the primary sampling units are large and heterogeneous, such as cities or counties. It is also efficient when the cost of frame construction is low for the primary sampling units and much higher for secondary and other subsequent sampling units. For instance, a frame of cities in a state is readily available, as are frames for blocks in most cities, but the cost of constructing a current frame of households in a state is very large. Also, when the elements within the clusters at any stage of sampling are homogeneous, precision will be enhanced if the budget is devoted to the selection of more clusters rather than to the selection of more elements within a cluster. The reason is that selecting more elements in a cluster will not provide much more information than is provided by a few elements, in view of the homogeneity within clusters. As an extreme example, if all persons in a household (the cluster) have the same preference among presidential candidates, a sample of one person from the household provides full information about that cluster.

Comment

When the clusters refer to geographical areas such as counties or city blocks, cluster sampling is often referred to as *area sampling*.

SYSTEMATIC SAMPLING   **23.5**

Systematic selection of sample elements is widely employed because of its ease and convenience.

EXAMPLE ☐

A market researcher wished to sample 200 sales transactions from a population of 10,000 transactions. The receipts for these transactions were serialized and stacked in order of the serial numbers. To select the sample quickly, the market researcher decided to select every $10{,}000/200 = 50$th receipt. One of the first 50 receipts was chosen at random (it happened to be random number 39) and then every 50th receipt thereafter was chosen. Thus, the sample consisted of the 39th, 89th, 139th, . . . , 9989th receipts. The actual selection was extremely easy because of the serialized ordering of the population.
☐

The sampling procedure just described is called systematic random sampling.

> **(23.18)**
> With **systematic random sampling,** every kth element in the frame is selected for the sample, with the starting point among the first k elements determined at random.

When every kth element is selected, then $100/k$ percent of the population is sampled. The resulting sample is called a *100/k percent systematic sample.* The market researcher, for instance, selected a $100/50 = 2$ percent systematic sample.

Systematic sampling has the advantage of being simple to execute and hence is economical and convenient to use. A systematic random sample frequently will also provide more precise estimates than a simple random sample of the same size. This occurs when the population frame in effect stratifies the population elements. For example, the sales receipts in the preceding example were issued consecutively. Thus, a systematic sample automatically contains receipts issued at different times. In essence, the market researcher's systematic sample was stratified by time.

> **Comments**
>
> 1. In using systematic sampling, one should avoid a sampling interval that corresponds to any natural period in the frame. For instance, in sampling closing monthly stock prices for a 30-year period, one should not sample every 12th month, or some multiple thereof. The resulting price data would always correspond to the same month of the year and might be systematically different from the data for the other months.
> 2. Systematic sampling can be used in combination with other sampling procedures. For example, a stratified sample may involve systematic selection within each stratum.

PROBLEMS

23.1

a. Calculate the finite population correction $\sqrt{1 - (n/N)}$ when (1) 3 percent of a finite population is sampled (that is, $n/N = 0.03$), (2) 30 percent is sampled.
b. By what percentage would $s\{\overline{X}\}$ be in error if it were calculated without the finite population correction in part a(1)? In part a(2)?

23.2

a. Calculate the finite population correction $\sqrt{1 - (n/N)}$ when the sample size is $n = 500$ and the size of the finite population N is (1) 1000, (2) 10,000, (3) 100,000.
b. Calculate $s\{\overline{X}\}$ when $s = 4$ for each of the three cases in part a. Is there much change in $s\{\overline{X}\}$ as N increases from 1000 to 10,000? From 10,000 to 100,000?

* **23.3** A simple random sample of 50 bank branches was selected from a population of 400 branches to estimate the mean number of certificates of deposit issued during the past month per branch. The sample results were $\overline{X} = 200$ certificates and $s = 15$ certificates. Construct a 95 percent confidence interval for the population mean. Interpret the confidence interval.

23.4 Expense Claims. An internal auditor selected a random sample of 100 expense claims from a population of 500 claims filed by company executives during the first quarter of the year and ascertained for each claim the expense amount supported by acceptable receipts. The sample results were $\overline{X} = \$418$ and $s = \$207$. Construct a 90 percent confidence interval for the population mean amount of supported expenses per claim. Interpret the confidence interval.

23.5 A random sample of 80 teachers was selected from a population of 427 teachers employed by a school system, and the number of days of accrued sick leave for each teacher in the sample was determined from personnel files. The sample results were $\overline{X} = 58$ days and $s = 12$ days. Construct a 95 percent confidence interval for the population mean number of days of accrued sick leave per teacher. Interpret the confidence interval.

* **23.6** An analyst requires a 95 percent confidence interval for the mean number of days to warranty expiration in a population of 1200 electric motor warranties. The confidence interval is to have a half-width of 2.5 days. A reasonable planning value for the population standard deviation is $\sigma = 45$ days.

a. Determine the required sample size.
b. Recompute the required sample size as though the population were infinite. Is the required sample size much larger when the population is treated as infinite? Comment.

23.7 Refer to **Expense Claims** Problem 23.4. The auditor is now planning to review the 650 expense claims filed by company executives during the second quarter of the year. What random sample size is required if the mean amount of supported expenses per claim is to be estimated by a 90 percent confidence interval with a half-width of \$25? Assume that the sample standard deviation from the first quarter's audit is a suitable planning value for σ.

23.8 A plant scientist requires a 99 percent confidence interval for the mean area per leaf (in square centimeters) for a plant, based on the surface areas of a random sample of leaves selected from the 120 leaves on the plant. The desired half-width of the confidence interval is 2 square centimeters per leaf. Previous experience indicates that an appropriate planning value for the standard deviation of the leaf areas is $\sigma = 7$ square centimeters.

a. Obtain the required sample size.
b. If the 99 percent confidence interval for the mean area per leaf were converted to a 99 percent confidence interval for the total leaf area of the plant, what would be the anticipated half-width of the latter interval?

* 23.9 In a random sample of 60 garment manufacturers selected from the population of 300 garment manufacturers operating in a state, 33 were found to pay full-time employees more than the state's minimum wage rate. Construct a 90 percent confidence interval for the proportion of garment manufacturers in the state who pay full-time employees more than the minimum wage rate. Interpret the confidence interval.

23.10 In a random sample of 250 physicians drawn from the population of 2000 physicians practicing in a sales district, a pharmaceutical company found that 40 percent write more than 20 prescriptions per working day. Construct a 95 percent confidence interval for the population proportion of physicians who write more than 20 prescriptions per working day. Interpret the confidence interval.

23.11 In a simple random sample of 100 seniors taken from the population of 427 seniors majoring in management, 32 stated that they intend to continue studying at this university to obtain a graduate degree in management. Obtain a 99 percent confidence interval for the population proportion of all management seniors who intend to obtain a graduate degree in management at this university. Interpret the confidence interval.

* 23.12 The proportion of the 3820 freshmen entering a university who own a personal computer is to be estimated.

a. What random sample size is needed to estimate this proportion by a 99 percent confidence interval with a half-width of 0.05? Assume that $p = 0.4$ is a suitable planning value.
b. Would the required sample size be much different if $p = 0.5$ were used as the planning value? Comment.

23.13 A population of 900 firms is to be surveyed to estimate the proportion of firms that have major union contracts to be renewed in the next 12 months. What random sample size is needed to estimate this proportion within ± 0.05 with 95 percent confidence? Assume that $p = 0.4$ is a suitable planning value.

23.14 A museum has acquired a collection of 1500 drawings by a sixteenth-century master that have been stored under unfavorable conditions. A random sample of the drawings is to be examined by an expert to estimate the population proportion of drawings that are significantly damaged. A 90 percent confidence interval for this proportion with a half-width of 0.03 is desired. A planning value of $p = 0.2$ is reasonable.

a. What random sample size is required?
b. If no reliable planning value for p were available and it were desired to obtain a 90 percent confidence interval that is guaranteed to have a half-width not to exceed 0.03, what planning value should be used to calculate the required sample size?

23.15 Stratified random sampling is to be employed in each of the following cases. Describe for each a basis of stratification that is likely to be useful.

a. A sample of personal checking accounts of a bank to estimate the mean number of transactions per account for May.
b. A sample of logs in a shipment to estimate the mean dollar yield of lumber per log.
c. A sample of school bus drivers from all those licensed in a state to estimate the mean number of years of driving experience per driver.

23.16 Stratified random sampling is to be employed in each of the following cases. Describe for each a basis of stratification that is likely to be useful.

a. A sample of the bales of silk in a warehouse damaged by fire to estimate the mean salvage value per bale.

b. A sample of the grain in a ship's hold to estimate the mean protein level of the grain shipment.

c. A sample of a university's faculty members who are authors of research publications listed in the university's annual report to estimate the mean annual research expenditures per faculty member for the past year.

* **23.17** A stratified random sample of 700 households was selected in a community of 28,000 households to estimate mean home-improvement expenditures last year. The strata employed were geographic areas. The strata sizes and sample results on expenditures (in dollars) follow.

Stratum (j)	N_j	n_j	$\overline{X}_j$	s_j
1	12,000	300	480	200
2	6,000	150	380	150
3	10,000	250	510	300
Total	28,000	700		

Obtain a 95 percent confidence interval for the mean home-improvement expenditures per household in the community. Ignore the finite population correction in your calculations. Interpret the confidence interval.

23.18 A dairy association with 9000 member producers selected a stratified random sample of 400 members to estimate the mean herd size per producer for the entire association. The strata were set up according to the amount of milk shipped in the previous calendar year, which was known for each member. The strata sizes and sample results on herd sizes follow:

Stratum (j)	N_j	n_j	$\overline{X}_j$	s_j
1	4500	170	15.7	8.3
2	2700	90	25.1	7.7
3	1800	140	48.2	18.0
Total	9000	400		

Obtain a 99 percent confidence interval for the population mean herd size. Use the finite population correction in your calculations for all strata. Interpret the confidence interval.

23.19 A stratified random sample of 400 stores was selected from a population of 8000 stores to estimate the mean pilferage loss per store for the past fiscal year. The strata were set up according to the floor area of the store, which was known for each store in the population. The strata sizes and sample results on pilferage losses (in dollars) follow.

Stratum (j)	N_j	n_j	$\overline{X}_j$	s_j
1	4500	125	2060	220
2	2700	75	3840	270
3	800	200	4200	350
Total	8000	400		

a. Obtain a 90 percent confidence interval for the mean pilferage loss per store in the population. Use the finite population correction in your calculations for all strata. Interpret the confidence interval.

b. Is the validity of the confidence interval in part a affected by whether or not pilferage loss in a store is related to its floor area? Does pilferage loss appear to be related to floor area here? Comment.

23.20

a. In planning a stratified random sample, what information about the population strata is required for optimal allocation that is not required for proportional allocation of the total sample size?
b. Under what special circumstances will optimal allocation and proportional allocation lead to the same allocation of a given total sample size to the strata?

23.21 Invoice Payments. A company wishes to estimate the mean number of days elapsed between the date an invoice is sent to a customer and the date payment is received. The study is to be confined to the 18,000 invoices issued in the previous fiscal quarter. Three strata, according to the amount of the invoice, are to be employed. Strata sizes and the anticipated strata standard deviations (in days) are as follows:

Stratum j:	1	2	3	Total
N_j:	9000	6000	3000	18,000
Anticipated σ_j:	8	5	3	

The budget permits a sample of 500 invoices to be examined.

a. Determine the strata sample sizes with proportional allocation.
b. Determine the strata sample sizes with optimal allocation.
c. Do the two methods of allocation lead to substantially different strata sample sizes here? Discuss.
d. If the mean delivery charge per invoice were also to be estimated, would the strata sample sizes obtained in part b also be optimal for this purpose? Discuss.

23.22 A state contains 100 large municipalities (stratum 1) and 500 small municipalities (stratum 2). It is desired to estimate the mean annual expenditures for snow-clearing last year per municipality based on a sample of 60 municipalities drawn from these two strata.

a. Determine the strata sample sizes with proportional allocation.
b. Planning values for the stratum standard deviations (in $ million) are $\sigma_1 = 3.60$ and $\sigma_2 = 1.20$, respectively. Determine the strata sample sizes with optimal allocation.
c. Do the two methods of allocation lead to substantially different strata sample sizes here? Discuss.

23.23 Probability-proportional-to-size sampling is to be employed in each of the following cases. For each, describe a variable that might be known and would serve effectively as a measure of size.

a. A sample of fields planted in canola to estimate the mean loss in yield per field caused by the infestation of a canola beetle.
b. A sample of firms to estimate the mean audit fee paid to external auditors in the previous fiscal year.
c. A sample of villages and towns to estimate the mean annual expenditures for sewage treatment.

23.24 Probability-proportional-to-size sampling is to be employed in each of the following cases. For each, describe a variable that might be known and would serve effectively as a measure of size.

a. A sample of hospitals to estimate the mean annual purchases of hospital linen and towels.
b. A sample of registered ships to estimate the mean revenue earned from cargo transport during the previous calendar year.
c. A sample of firms to estimate the mean annual expenditures for charity contributions in the previous year.

23.25 One hundred firms from a population of 3500 firms in an industrial district are to be surveyed to estimate the mean purchases of office supplies and equipment during the previous 12 months by all firms. Probability-proportional-to-size sampling with replacement will be employed. The number of employees of each firm is known and will be used as the measure of size for the pps sampling. A partial list of the firms in the population, showing the number of employees, the cumulative number of employees, and the corresponding cumulative range for each firm is given next:

Firm	Number of Employees	Cumulative Number of Employees	Cumulative Range
1	43	43	1–43
2	16	59	44–59
3	96	155	60–155
:	:	:	:
3499	11	95,022	95,012–95,022
3500	8	95,030	95,023–95,030

a. What is the probability that firm 3 will be the first firm selected for the sample? The 10th firm selected?

b. How many times more probable is it that firm 3 will be the first firm selected for the sample than firm 2? Does this number correspond to the relative sizes of these two firms as measured by their numbers of employees? Explain.

c. In drawing a discrete uniform random number between 1 and 95,030 to select the first firm, the number 21 was selected. Which firm is therefore the first to be selected for the sample? Random number 95,028 was selected next. To which firm does this random number correspond?

d. Would you expect that number of employees will be an effective measure of size for the purposes of this survey? Comment.

23.26 Thirty universities are to be surveyed from a population of 326 universities to estimate the population mean amount of research grants received from military funding sources during the previous 12 months. Probability-proportional-to-size sampling with replacement will be employed. The number of graduate students enrolled in each university is known and will be used as the measure of size for the pps sampling. A partial list of the universities in the population showing the number of graduate students, the cumulative number of graduate students, and the corresponding cumulative range for each university is given next.

University	Number of Graduate Students	Cumulative Number of Graduate Students	Cumulative Range
1	430	430	1–430
2	1239	1669	431–1669
3	196	1865	1670–1865
:	:	:	:
325	770	215,030	214,261–215,030
326	2237	217,267	215,031–217,267

a. What is the probability that university 2 will be the first university selected for the sample? The fourth university selected?

b. Is university 2 more or less likely to be selected for the sample than university 1? Explain.

c. In drawing a discrete uniform random number between 1 and 217,267 to select the first university, the number 217,133 was selected. Which university is therefore the first to be selected for the sample? Random number 1089 was selected next. To which university does this random number correspond?

d. Would you expect that number of graduate students will be an effective measure of size for the purposes of this survey? Comment.

23.27 A marketing research firm has grouped the households in a city by locational proximity, with each group consisting of 20 households. A random sample of these groups will be selected by the firm for a household survey. All households in the groups selected for the sample will be interviewed in the survey. Is such a sample a stratified random sample, in which location is used as a basis of stratification, or a single-stage cluster sample, in which the groups constitute clusters of households? Explain.

23.28 Explain how cluster sampling might be employed in each of the following situations and whether single-stage or multistage cluster sampling is preferable.

a. Sampling of peaches in a fruit orchard to estimate the percentage of peaches that suffered frost damage.

b. Sampling of fifth-grade students throughout a state to estimate the mean reading achievement score.

23.29 Explain how cluster sampling might be employed in each of the following situations and whether single-stage or multistage cluster sampling is preferable.

a. Sampling of records of patients discharged from acute-care hospitals in a state to estimate the mean length of hospital stay per patient.

b. Sampling of travelers departing on flights originating in the United States and Canada to estimate the mean number of flight segments per trip.

23.30 An auditor must check a company's inventory of raw materials and finished goods located in a large number of warehouses throughout the country. The auditor will perform the check by selecting a sample of inventory items, counting the actual inventory on hand for these items, and comparing these counts with company records. The total error in the book inventory is to be estimated.

a. Explain how the auditor might employ cluster sampling in this investigation. Does your proposal involve single-stage or multistage cluster sampling?

b. What advantages does cluster sampling offer here relative to simple random sampling? What disadvantages?

23.31 A 4 percent systematic random sample is to be selected from a computer listing of 6000 employee payroll records. With reference to definition (23.18), what is the value of k here? How many records will be selected for the sample in this case?

23.32 In a silviculture experiment, 5000 tree seedlings were planted in rows on a plot. Five years later, a 5 percent systematic sample of the young trees was selected to estimate the mean growth per tree. The successive rows of the plot were used to define the population frame.

a. How many trees were selected for the sample?

b. The southern half of the plot received more rain during this five-year period than the northern half. Does this fact imply that the selected systematic sample will reflect the benefits of stratification with respect to the effect of rainfall on growth? Explain.

23.33 A population of 5000 physicians specializing in plastic surgery is listed in order of date of certification. A systematic random sample of every 50th physician was selected from this frame to estimate the mean number of years of practice in this specialty. The analyst has evaluated the sample results as if the sample were a simple random one. Will the calculated precision so obtained likely overstate, understate, or be the same as the actual precision of the systematic random sample estimate? Explain.

EXERCISES

23.34 In the estimation of a population mean from a stratified random sample, why is the estimated variance (23.10) not applicable when the strata samples are not drawn independently?

23.35 An observer has stated that when the strata standard deviations σ_j are the same for all strata, stratified random sampling offers no greater precision than simple random sampling for the same sample size. Do you agree? Discuss. [*Hint:* If the strata standard deviations are all equal, must the population standard deviation also be the same?]

23.36 Refer to **Invoice Payments** Problem 23.21.

a. What is the anticipated standard deviation of $\bar{X}_{st}$ with proportional allocation? With optimal allocation?

b. Is optimal allocation expected to lead to a substantial gain in precision over proportional allocation here? Comment.

23.37   A pps sample of size n is to be selected with replacement from a finite population of size N.
The population total:

$$\tau = \sum_{j=1}^{N} X_j$$

is to be estimated. Show that the estimator:

$$T = \frac{1}{n} \sum_{i=1}^{n} \frac{X_i}{P_i}$$

is an unbiased estimator of τ, where P_i is the pps selection probability for the ith sample item. [*Hint:*
Show first that X_i/P_i is an unbiased estimator of τ.]

STUDIES

23.38   Refer to the **Financial Characteristics** data set (Appendix D.1). The mean net income per
firm in year 1 for all firms is of interest.

a. Select a stratified random sample of 64 firms from the population of all firms. Use the industry
groups as the strata and employ proportional allocation.

b. Estimate the population mean net income per firm in year 1 with a 95 percent confidence interval.
Use the finite population correction in your calculations. Interpret the confidence interval.

c. How would a systematic random sample of 64 firms be selected here? For purposes of analysis,
might this systematic sample be considered as approximately equivalent to the stratified sample
selected in part a? Explain.

d. Obtain the population variance σ^2 for all firms and the stratum variance σ_j^2 for each industry
group.

e. Use the results in part d and formula (23.12) to calculate $\sigma\{\overline{X}_{st}\}$ for a stratified random sample of
64 firms, allocated to the industry groups as in part a. Also, use the value of σ^2 obtained in part d
to calculate $\sigma\{\overline{X}\}$ based on a simple random sample of 64 firms. Is a stratified sample with
proportional allocation based on industry groups effective here in comparison with a simple random sample? Discuss.

23.39   Refer to **Hospital Admissions** Problem 20.53. The total number of admissions last year to
the 24 hospitals is of interest.

a. Employ pps sampling with replacement to select six hospitals from the population. Use the number of beds in each hospital as the measure of size.

b. Use the sample results in part a to estimate the total number of admissions to the population of
24 hospitals, using the formula given in Exercise 23.37.

c. Calculate the population total number of admissions τ. What is the magnitude of the sampling
error in the estimate obtained in part b? Would you expect an estimate of τ obtained from a simple
random sample of the same size to have a smaller or a larger absolute sampling error? Comment.

Time Series Analysis and Index Numbers

Time Series and Forecasting: Descriptive Methods

24

anagers and social scientists often must deal with processes that vary over time. Observations in time sequence on such a process are called a time series. Examples of time series are monthly sales of a commodity for the past 120 months and annual gross national product in the United States since 1950.

Time series are analyzed to understand, describe, control, and predict the underlying process. The analysis of a time series usually involves a study of the components of the time series, such as the trend, cyclical, and seasonal components. In this chapter, we consider basic aspects of time series analysis and discuss several descriptive methods that are widely used. In the next chapter, we consider additional methods and models for time series analysis.

We begin with a definition of a time series.

TIME SERIES   24.1

> **(24.1)**
> A **time series** is a sequence of n observations $Y_1, Y_2, \ldots, Y_t, \ldots, Y_n$ on a process at equally spaced points in time.

The equally spaced points in time may be months (as in monthly sales), years (as in annual wheat harvests), or hours (as in hourly electricity consumption).

Time series arise in two different ways.

1. Readings are taken on a variable at consecutive points in time. Examples are (a) the raw materials inventory of a manufacturer recorded at the close of each fiscal quarter and (b) the level of atmospheric pollution at a downtown location recorded at 5 P.M. each day.
2. A variable is aggregated or accumulated over an interval of time. Examples are (a) sales aggregated to obtain monthly sales data and (b) snowfalls accumulated to obtain annual data on total snowfall.

> **Comment**
>
> Time series as defined in (24.1) are *discrete* time series, because the observations pertain to separated points in time. There are also *continuous* time series, where the variable Y is measured continuously over time. An electrocardiogram of heart action is a graph of a continuous time series.

24.2   SMOOTHING

Smoothing is a statistical procedure used to average out fluctuations in a time series to obtain a *smooth component* that reflects the systematic movement of the series. We explain smoothing by an example.

Department Store   □   **EXAMPLE**

A department store extended its evening openings to five nights, Monday through Friday. The store was staffed during these extended hours primarily by new, part-time sales personnel. Shortly after the new hours went into effect, customers began to complain increasingly about incidents of rudeness and inefficiency. Management quickly initiated a remedial program. A time series plot of the number of complaints received in the evenings on Mondays through Fridays for the two weeks before the initiation of the remedial program and for four weeks thereafter is presented in Figure 24.1a.

An operations analyst perceived two important characteristics of the time series from the plot in Figure 24.1a:

1. A downward shift occurred in the underlying direction of the series coinciding with the initiation of the remedial program.
2. Marked fluctuations within each 5-day period are present both before and after the initiation of the remedial program.

To present these points effectively at a management review meeting, the analyst wished to average out the fluctuations to obtain a clearer picture of the underlying movement of the series.   □

Moving Average

A *moving average* smooths the fluctuations in a time series by taking successive averages of groups of observations. It is calculated by averaging the first k elements of the time series, then dropping the first element and adding the $(k + 1)$th element and averaging these k elements, and so on. Thus, the first sequence contains observations $Y_1, Y_2, \ldots, Y_k$; the second sequence contains observations $Y_2, Y_3, \ldots, Y_{k+1}$; and so on. For each sequence, the mean is calculated, and these means form the moving average.

The number of elements in each sequence, k, is called the *term* of the moving average. To be effective in smoothing a time series, the term k should be the same as the period of fluctuations in the series or a multiple of this period.

□   **EXAMPLE**

In the department store example, Figure 24.1a shows clearly that there is a 5-day cycle of fluctuations that is repeated week after week. Therefore, the analyst used a 5-term moving average. With five terms, each sequence—$(Y_1, Y_2, \ldots, Y_5), (Y_2, Y_3, \ldots, Y_6)$, and so on—contains an observation for each of the five days of the week that the store is open in the evening. Thus, each average contains a Wednesday, when the volume of complaints reaches a peak, and a Monday and a Friday, when the volume of complaints is lower.

Table 24.1 illustrates the procedure for calculating the 5-term moving average. Column 1 contains the number of complaints received in a portion of the evenings. Column 2 shows the totals of successive sequences of five observations. For example,

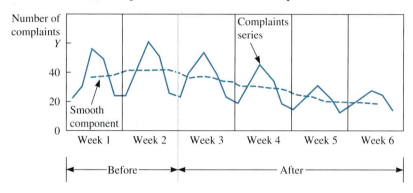

(a) Complaints Series and Smooth Component

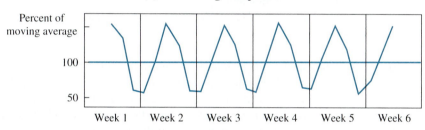

(b) Fluctuating Component

FIGURE 24.1

Complaints series and its smooth and fluctuating components—Department store example

Week and Day		t	(1) Number of Complaints Y_t	(2) 5-Term Moving Total	(3) 5-Term Moving Average	(4) Fluctuating Component $100[(1) \div (3)]$
1	M	1	22			
	T	2	30			
	W	3	57	184	36.8	155
	T	4	51	186	37.2	137
	F	5	24	197	39.4	61
2	M	6	24	203	40.6	59
	.	.	.	.	.	.
	.	.	.	.	.	.
	.	.	.	.	.	.
	F	25	10	96	19.2	52
6	M	26	13	93	18.6	70
	T	27	20	92	18.4	109
	W	28	27	92	18.4	147
	T	29	22			
	F	30	10			

TABLE 24.1

Calculation of a 5-term moving average—Department store example

the total for the first sequence is $22 + 30 + 57 + 51 + 24 = 184$. This total is placed at the middle of its sequence at $t = 3$. The total for the next sequence is $30 + 57 + 51 + 24 + 24 = 186$, which is placed at $t = 4$. The totals in column 2 are called *moving totals.*

When the totals in column 2 are divided by 5, we obtain the moving averages in column 3. For example, the moving average at $t = 3$ is $184/5 = 36.8$. Note that there are no moving averages for $t = 1$ and 2 and for $t = 29$ and 30, the first two and last two days of the time series. In general, for a k-term moving average, there will be no moving averages for the first and last $(k - 1)/2$ time periods when k is an odd number. The moving average of the complaints series, representing the smooth component, is also plotted in Figure 24.1a.

Finally, the analyst divided each original observation by its moving average and then multiplied by 100 to obtain the *fluctuating component* of the complaints series in column 4. For example, the fluctuating component for period $t = 3$ is $100(57/36.8) = 155$. The fluctuating component series, plotted in Figure 24.1b, describes the pattern of day-to-day fluctuations in complaints.

The analyst presented Figure 24.1 at the management review meeting with the following interpretation.

1. The smooth component of the complaints series, portrayed by the moving average, turned downward shortly after the remedial program was introduced. The downturn could have resulted from the remedial program, from experience gained by the salespeople, or from some combination of these or other factors. Further study is required to confirm the cause of the downturn.
2. The smooth component appears to be leveling off in the last week or so. There may be no further improvement unless the situation receives additional attention from management.
3. The fluctuating component exhibits a regular midweek peak. This peak reflects a corresponding peak in the number of transactions handled.   □

The term k of a moving average can materially affect the moving average series. As we noted from the department store example, it should be chosen to equal the period of fluctuations in the time series. Unfortunately, many time series in business, economics, and the social sciences oscillate with considerable irregularity. In these cases, the best compromise is to select the smallest value of k that will smooth the fluctuations. Many statistical packages contain routines for time series smoothing, and these make it possible to experiment with different values of k.

Centered Moving Average.   When an even number of terms is required in a moving average—for instance, in a 4-term or 12-term moving average—a special problem arises. Consider the following case where a 4-term moving average is taken:

t:	1	2	3	4	5	6	7
Y_t:	2	6	4	8	6	7	2
Moving Total ($k = 4$):			20	24	25	23	
Moving Average ($k = 4$):			5.00	6.00	6.25	5.75	

The 4-term totals and averages, placed at the centers of their sequences, fall at $t = 2.5$, $t = 3.5$, and so on. This is unsatisfactory for comparing the original observations and the corresponding moving averages. The remedy is to *center* the moving averages as follows:

t:	1	2	3	4	5	6	7
Y_t:	2	6	4	8	6	7	2
Moving Total:			20	24	25	23	
Moving Average:			5.00	6.00	6.25	5.75	
Centered M.A.:				5.50	6.13	6.00	

Note that centering simply involves taking a 2-term moving average of the 4-term moving average. For instance, the first centered average, at $t = 3$, is $(5.00 + 6.00)/2 = 5.50$. Similarly, the second centered average, at $t = 4$, is $(6.00 + 6.25)/2 = 6.13$.

With centering, we do not obtain moving averages for the first and last $k/2$ time periods. Here $k/2 = 4/2 = 2$, so there are no moving averages for the first two and last two periods.

A centered moving average may be viewed as a *weighted moving average*; that is, one in which the observations in a sequence do not receive equal weights. For example, the centered moving average in the preceding illustration is a weighted 5-term moving average where the observations in each sequence receive weights (1, 2, 2, 2, 1). Thus, the first centered moving average can be obtained directly as follows:

$$\frac{1}{8}[1(2) + 2(6) + 2(4) + 2(8) + 1(6)] = \frac{44}{8} = 5.50$$

where 8 is the sum of the weights.

Comments

1. Since a time series observation Y_t enters k successive moving averages (except at the beginning and end of the series), outlying observations can produce oscillations in the moving average series even though the original time series contains no periodicity. For this and other reasons, care must be used in interpreting oscillations in a moving average series. One means of dampening the influence of outlying observations on the moving average is to use the median, rather than the mean, as the average.
2. In the department store example, the fluctuating component was calculated as the ratio of each original observation to its associated moving average, expressed as a percentage. The fluctuating component is also calculated at times as the difference between each original observation and its associated moving average.

TIME SERIES COMPONENTS 24.3

We have seen from the department store example how a time series can be decomposed into a smooth component, reflecting the systematic movement of the series, and a fluctuating component, reflecting the shorter-term and erratic movements in the series. For many business and economic time series, however, this decomposition is inadequate and additional components need to be recognized.

Multiplicative Time Series Model

Economists have developed a model for studying economic and business time series that contains four components. These are called the trend, cyclical, seasonal, and irregular components. Two mathematical versions of the model are used. In one version, the four components are assumed to combine in a multiplicative fashion. In the other version, the components combine in an additive fashion. We consider the multiplicative model first.

> **(24.2)**
> The **multiplicative time series model** is:
>
> $$Y = T \cdot C \cdot S \cdot I$$
>
> where T, C, S, and I denote, respectively, the trend, cyclical, seasonal, and irregular components of the time series.

Fixture Shipments ☐ **EXAMPLE**

A company manufactures a large outdoor lighting fixture. A time series plot of the quarterly shipments of this fixture between 1982 and 1991 is shown in Figure 24.2a. The four components of this time series according to multiplicative time series model (24.2) are shown in panels b through e of Figure 24.2. ☐

We shall use the fixture shipments example to illustrate the nature and meaning of the four components in time series model (24.2).

Trend Component

The trend component of the fixture shipments time series is shown in Figure 24.2b. This component describes the net influence of long-term factors whose effects on the shipments tend to change gradually. Generally, these factors include (1) changes in the size, demographic characteristics, and geographic distribution of the population; (2) technological improvements; (3) economic development; and (4) gradual shifts in habits and attitudes. Since these effects tend to operate fairly gradually and in one direction over long periods of time, the trend component usually is modeled by a smooth, continuous curve spanning the entire time series. The curve employed is called the *trend curve*. The trend curve in Figure 24.2b is linear, but a different type of trend curve might be appropriate for another time series.

A major use of trend analysis is for long-term forecasting.

Cyclical Component

The cyclical component of the fixture shipments time series is shown in Figure 24.2c. Generally, business and economic activities do not grow (or decline) at a steady rate. Instead, there are alternating periods of expansion and contraction. The cyclical component of a time series describes this characteristic. Typically, the cyclical component contains cycles of expansion and contraction that are of uneven duration and amplitude. Some of the factors leading to cyclical movements in business and economic time series include buildups and depletions of inventories, shifts in rates of capital expenditures by businesses, year-to-year variations in harvests, and changes in governmental monetary and fiscal policy. The cyclical component in the shipments of outdoor lighting fixtures in Figure 24.2c reflects changes in sales arising from variations in the rate of development of new shopping

Time series components according to multiplicative model (24.2)—Fixture shipments example

FIGURE 24.2

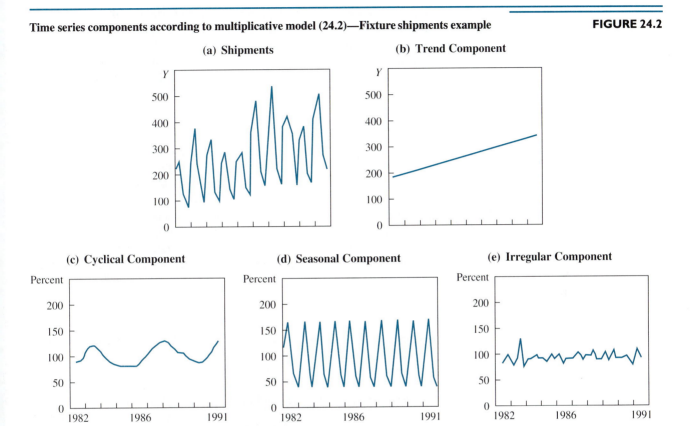

(a) Shipments

(b) Trend Component

(c) Cyclical Component

(d) Seasonal Component

(e) Irregular Component

centers and outdoor recreational facilities, which in turn are influenced by variations in underlying economic conditions.

Cyclical movements are studied for information on changes in rates of current activity. This information is useful for assessing current conditions and for making short-term forecasts.

Seasonal Component

The seasonal component of the fixture shipments time series is shown in Figure 24.2d. This component describes effects that occur regularly over a period of a year, quarter, month, week, or day. Seasonal effects generally are associated with the calendar or the clock. For example, electricity consumption in the southern United States reaches a peak in summer, year after year, because of air-conditioning demand. Department store sales reach a peak every December, because of the Christmas holidays. Subway traffic reaches a peak each weekday in early morning and late afternoon, because of passengers going to and from work.

Seasonal effects tend to recur fairly systematically. Consequently, the pattern of movement in the seasonal component tends to be more regular than the pattern in the cyclical component and therefore is more predictable, although sometimes the seasonal pattern undergoes gradual modification.

The seasonal pattern for the fixture shipments example reaches a peak in each second quarter, because the height of construction activity is in the summer. Note also that the seasonal pattern is stable over the entire period.

Seasonal movements are measured so that seasonal effects can be taken into account in evaluating past and current activity, as well as incorporated into forecasts of future activity.

Irregular Component

The irregular component describes residual movements that remain after the other components have been taken into account. Irregular movements reflect effects of unique and nonrecurring factors, such as strikes, unusual weather conditions, and international crises.

In some business and economic time series, the cyclical component is itself so irregular that any breakdown into separate cyclical and irregular components would be arbitrary. In such cases, a combined cyclical-irregular component is often developed.

Figure 24.2e contains the irregular component in the fixture shipments series. The unusual peak in the third quarter of 1983 occurred because a major competitor had a strike and could make no deliveries.

Definitions

We now define the four components in time series model (24.2) more formally.

> **(24.3)**
> The four components of a time series according to model (24.2) are as follows.
>
> **(24.3a)** The **trend component,** which describes the long-term sweep of the series and is usually modeled by a smooth curve.
> **(24.3b)** The **cyclical component,** which describes the alternating pattern of expansion and contraction in the series and consists of cycles of varying amplitude and duration.
> **(24.3c)** The **seasonal component,** which describes the short-term recurring pattern of change in the series and consists of relatively repetitious cycles of fixed amplitude and duration.
> **(24.3d)** The **irregular component,** which describes the miscellaneous erratic movements in the series and tends to have an irregular, saw-toothed pattern.

24.4   TREND COMPONENT

As we noted earlier, the trend component of a time series is usually described by a smooth curve. We now consider the use of two common trend functions to model the trend component: linear trends and exponential trends.

Linear Trend

A linear trend function is illustrated for the fixture shipments example in Figure 24.2b. It has the property of a constant amount of increase (decrease) in the trend value from one period to the next.

Fitting Linear Trend.   We shall fit a linear trend function to time series data by the method of least squares, using the formulas in (18.10) for the intercept b_0 and the slope b_1 of the fitted line. Thus, no new problems are involved in fitting a linear trend; the fitting procedure is exactly the same as in fitting a linear regression function. The only difference is that the independent variable X when fitting a linear trend function represents a sequence of time periods.

We shall follow the practice of coding the time periods by $X_1 = 1, X_2 = 2, \ldots, X_t = t, \ldots, X_n = n$. For instance, for annual data covering the 21 years from 1971 to 1991, we represent 1971 by $X_1 = 1$, 1972 by $X_2 = 2$, and 1991 by $X_{21} = 21$. Also, we shall use the subscript t to represent a time period, rather than the subscript i that we used earlier when the data did not represent a time series.

(24.4)
The fitted **linear trend function** is:

$$\hat{Y}_t = b_0 + b_1 X_t$$

where: $\hat{Y}_t$ is the trend value for period t ($t = 1, 2, \ldots, n$)
X_t is the numerical code for period t
b_0 is the intercept of the trend line
b_1 is the slope of the trend line

Almost every statistical package will fit a straight-line trend by the method of least squares.

EXAMPLE ☐ **Executive Jet**

Table 24.2, column 2, contains data obtained by an investment analyst on the annual number of executive jets sold by an aircraft manufacturer between 1982 and 1991. Figure 24.3 presents a time series plot of these data. The analyst was preparing a report on sales trend patterns for selected companies and wished to study the trend component for this company. The analyst noted from Figure 24.3 that the annual sales observations tend to fall along a straight line. This finding, together with an analysis of the factors underlying the long-term sales pattern, led the analyst to conclude that the trend of sales could be modeled appropriately by a straight line.

A statistical package was used to fit a linear regression of sales Y_t (in column 2) on the numerical codes for the years X_t (in column 1). The following fitted trend function was obtained.

(24.5) $\hat{Y}_t = 143.93333 + 8.15758X_t$

$X_t = 1$ at 1982; X in one-year units

Note that we indicate the coding scheme for X_t when we report the fitted trend function. The slope $b_1 = 8.15758$ indicates that the trend value increases by 8.2 aircraft from one year to the next.

TABLE 24.2

Fitting a linear trend—Executive jet example

Year	(1) Year, in Coded Units X_t	(2) Number of Aircraft Sold Y_t	(3) Trend Value $\hat{Y}_t$
1982	1	147	152.09
1983	2	175	160.25
1984	3	150	168.41
1985	4	191	176.56
1986	5	188	184.72
1987	6	179	192.88
1988	7	200	201.04
1989	8	220	209.19
1990	9	208	217.35
1991	10	230	225.51
Total	55	1888	

$$\hat{Y}_t = 143.93333 + 8.15758X_t$$

$X_t = 1$ at 1982; X in one-year units

The fitted trend line in (24.5) is also plotted in Figure 24.3. Trend values are calculated in the usual way. For example, the trend value for 1984 ($X_3 = 3$) is:

$$\hat{Y}_3 = 143.93333 + 8.15758(3) = 168.41$$

The trend values are presented in Table 24.2, column 3.

Projection of Trend Values.   Projection or extrapolation of trend values is performed as in regression by substituting the X_t value of interest into the trend equation.

☐ **EXAMPLE**

In the executive jet example, it was desired to project the trend to 1996 ($X_{15} = 15$). Substituting in (24.5), we obtain:

$$\hat{Y}_{15} = 143.93333 + 8.15758(15) = 266.30$$

FIGURE 24.3

Time series plot and linear trend component—Executive jet example

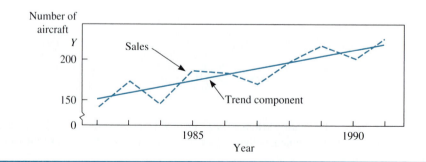

Thus, if the long-term development of jet sales continues as in the past, the trend value of sales in 1996 will be 266 aircraft. □

Exponential Trend

In an exponential trend, the trend value is changing at a constant *rate* from one period to another. This is in contrast to a linear trend, where the trend value is changing by a constant amount. Exponential trends are encountered quite often.

EXAMPLE □ **Agricultural Imports**

Figure 24.4 presents two time series plots of Canadian imports of agricultural products from the United States during 1955–1987 (expressed in millions of Canadian dollars). In Figure 24.4a, the series is plotted against an ordinary arithmetic Y scale; in Figure 24.4b, the series is plotted against a *ratio* or *logarithmic Y* scale. Whereas equal vertical distances on an arithmetic scale represent equal *amounts* of increase or decrease, equal vertical distances on a ratio or logarithmic scale represent equal *rates* or *percents* of increase or decrease. For instance, on the ratio scale in Figure 24.4b the distances from 100 to 1000 and from 1000 to 10,000 are equal because in each case the ratio of the two values is the same, namely, 1 to 10. The plotting grid in Figure 24.4b is often called a *semilogarithmic grid,* in contrast to the *arithmetic grid* in Figure 24.4a.

As may be seen from Figure 24.4a, the long-term pattern in the series bends upward, indicating that the trend component is increasing by larger increments each year. At the same time, the plot in Figure 24.4b shows a long-term pattern that is approximately a straight line, indicating that the trend component is increasing by a constant rate from year to year. An exponential trend is appropriate for such a time series. □

Plots of Canadian imports of agricultural products from the United States, 1955–1987 (millions of Canadian dollars)—Agricultural imports example

FIGURE 24.4

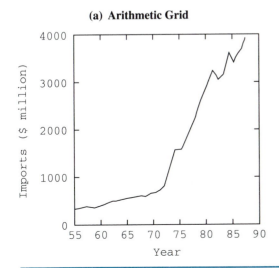

(a) Arithmetic Grid

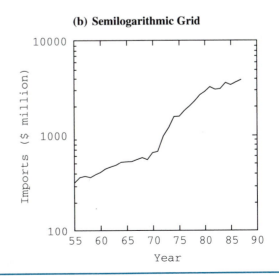

(b) Semilogarithmic Grid

Fitting Exponential Trend.   When the long-term sweep of a series is increasing (or decreasing) at a constant rate, the trend of the logarithms of the series will be linear. (A review of logarithms is found in Appendix A, Section A.2.) Hence, we fit an exponential trend by first taking logarithms of the observations Y_t, to be denoted by Y_t'. We then fit a linear trend to the Y_t' in the usual manner.

(24.6)

The fitted **exponential trend function** for the logarithms Y_t' of the observations is:

$$\hat{Y}_t' = b_0 + b_1 X_t$$

where: $\hat{Y}_t' = \log \hat{Y}_t$

$Y_t' = \log Y_t$

To obtain a trend value $\hat{Y}_t$ for the original series, we simply take the antilogarithm of the logarithmic trend value $\hat{Y}_t'$.

☐ **EXAMPLE**

Table 24.3 contains in column 2 a portion of the imports data Y_t for the agricultural imports example in Figure 24.4. Column 3 contains the logarithms $Y_t' = \log Y_t$ to base 10. The coded units X_t for the years are shown in column 1. A statistical package was used to fit a linear regression of Y_t' on X_t, yielding the following fitted exponential trend function.

(24.7) $\hat{Y}_t' = 2.37030 + 0.038251 X_t$

$X_t = 1$ at 1955; X in one-year units

TABLE 24.3
Fitting an exponential trend—Agricultural imports example

Year	(1) Year, in Coded Units X_t	(2) Agricultural Imports ($ million) Y_t	(3) $Y_t' = \log Y_t$	(4) $\hat{Y}_t'$	(5) $\hat{Y}_t$
1955	1	316	2.49969	2.40855	256.2
1956	2	362	2.55871	2.44680	279.8
1957	3	373	2.57171	2.48506	305.5
1958	4	362	2.55871	2.52331	333.7
1959	5	389	2.58995	2.56156	364.4
⋮	⋮	⋮	⋮	⋮	⋮
1985	31	3431	3.53542	3.55609	3598.3
1986	32	3638	3.56086	3.59434	3929.6
1987	33	3891	3.59006	3.63260	4291.4

$\hat{Y}_t' = 2.37030 + 0.038251 X_t$

$X_t = 1$ at 1955; X in one-year units

The statistical package also provided the logarithmic trend values $\hat{Y}'_t$ in column 4 of Table 24.3 and the trend values $\hat{Y}_t$ in the original units in column 5. The logarithmic trend values $\hat{Y}'_t$ are obtained in the standard manner. For instance, we obtain the logarithmic trend value for 1955 ($X_1 = 1$) by substituting in the fitted trend function (24.7) as follows:

$$\hat{Y}'_1 = 2.37030 + 0.038251(1) = 2.40855$$

We then take the antilogarithm to obtain the trend value $\hat{Y}_1$ in the original units:

$$\hat{Y}_1 = \text{antilog } 2.40855 = 10^{2.40855} = \$256.2 \text{ million} \qquad \square$$

Rate of Growth. The constant rate of growth implicit in a fitted exponential trend function will be denoted by g. It can be calculated as follows.

(24.8) $\qquad g = (\text{antilog } b_1) - 1 = 10^{b_1} - 1$

EXAMPLE $\square$

In the agricultural imports example, $b_1 = 0.038251$, so we have:

$$g = (\text{antilog } 0.038251) - 1 = 10^{0.038251} - 1 = 1.0921 - 1 = 0.0921$$

or 9.21 percent growth per year. $\qquad \square$

Comments

1. An exponential trend sometimes is called a *log-linear trend* because it is a linear trend for the logarithms of the observations Y_t.
2. The observations Y_t must be positive for $\log Y_t$ to be defined.

Forecasting by Trend Projections

Trend projections are used often as an aid in forecasting. This use is inherently subjective since the fitted model is extrapolated into a future that never is seen clearly. We now take up some considerations in using trend projections for forecasting.

Short-Term Versus Long-Term Forecasting. Short-term operating forecasts must be relatively precise. When the cyclical and irregular components in the activity to be forecast have been negligible and are likely to continue so in the future, it may be feasible to project the trend curve a year or two ahead to serve as a short-term forecast. In most cases, however, the cyclical and irregular components are potentially too large to permit a short-term forecast to be based on trend alone. When undertaking long-term forecasting, on the other hand, trend projections often can serve as forecasts because long-term effects are of primary importance and short-term fluctuations are of secondary consideration.

Time Period.   Ideally, the observed data to which the trend curve is fitted should be relevant to the period into which the curve will be projected. If important new conditions affecting trend have arisen in the recent past (such as a new technological breakthrough), data prior to the change of conditions may have to be discarded or given less weight. If important new conditions affecting trend are expected in the future (such as new environmental protection legislation), the trend projections may have to be modified to take account of the effects of these new conditions.

Choice of Trend Curve.   The choice of a trend curve cannot be made solely on the basis of how good the fit is to the past data. Usually, there will be several trend curves that fit the data well and yet the curves may diverge sharply when they are projected substantially beyond the range of the data. Thus, the choice of the trend curve for forecasting must be based not only on a good fit to the past data but also on expert assessment of the long-term factors determining the future course of the activity.

The choice of the trend curve is also affected by the length of the projection. An exponential trend might be reasonable for a projection of 5 years but not for a projection of 20 years.

Many statistical packages can assist in screening alternative trend functions by allowing different types of trend curves to be fitted and providing measures of fit for each. Of course, as we noted previously, the degree to which a trend curve fits the past observations should be only one of the criteria to be considered in selecting a trend curve for forecasting.

24.5   CYCLICAL COMPONENT

Cyclical movements in business and economic data are studied to assess the direction in which an economic sector is moving, to evaluate the short-term outlook for an industry or company, and to prepare intermediate and short-term forecasts for planning and controlling day-to-day operations. Unlike many time series in engineering and the physical sciences with cyclical components that are quite regular and that can be described by mathematical periodic functions, cyclical components in business and economic series tend to vary widely in both duration and amplitude from one cycle to the next. Hence, they do not lend themselves to fitting by simple periodic functions. Smoothing methods, however, can help to isolate the cyclical component for study, as we shall now illustrate.

Isolation of Cyclical Component

To obtain the cyclical component of a time series, we need to work with time series that contain no seasonal effects. Annual time series (e.g., annual sales) contain no seasonal component, so these time series are ready for the isolation of the cyclical component without any preliminary work. Many other time series do contain a seasonal component (e.g., monthly department store sales), and these time series need to have the seasonal component removed before we can isolate the cyclical component. We explain in Section 24.6 how the seasonal component of a time series can be removed.

For time series that contain no seasonal component and for time series where the seasonal component has been removed, the multiplicative time series model (24.2) simplifies as follows.

(24.9)

The multiplicative time series model with no seasonal component is:

$$Y = T \cdot C \cdot I$$

Removal of Trend Component. The first step in isolating the cyclical component of a time series containing no seasonal component is to remove the trend component T, which leaves the combined cyclical-irregular component as a remainder. We remove the trend by dividing the observations Y_t by the corresponding trend values; that is, by the fitted least squares trend values $\hat{Y}_t$. In terms of the notation for multiplicative model (24.9), we divide Y by T to give us $C \cdot I$ as follows.

(24.10) $\quad C \cdot I = \dfrac{Y}{T} = \dfrac{T \cdot C \cdot I}{T}$

The values of $C \cdot I$ are conventionally expressed as percents and are called *percents of trend*.

EXAMPLE ☐

For the agricultural imports example, column 1 of Table 24.4 reproduces a portion of the imports series from Table 24.3, and column 2 repeats the trend values based on the fitted exponential trend function (24.7). Column 3 contains the percents of trend for the imports series, obtained by dividing column 1 by column 2 and then expressing the resulting ratio as a percent. For instance, the calculation of $C \cdot I$ for 1955 is as follows:

$$\frac{Y}{T} = 100\left(\frac{316}{256.2}\right) = 123.3$$

TABLE 24.4
Calculation of percents of trend and isolation of cyclical and irregular components—Agricultural imports example

Year	(1) Agricultural Imports Y	(2) Trend Value T	(3) Percent of Trend $C \cdot I$ $100[(1) \div (2)]$	(4) Cyclical Component C M.A. of (3)	(5) Irregular Component I $100[(3) \div (4)]$
1955	316	256.2	123.3		
1956	362	279.8	129.4		
1957	373	305.5	122.1	118.0	103.4
1958	362	333.7	108.5	114.8	94.5
1959	389	364.4	106.8	110.4	96.7
⋮	⋮	⋮	⋮	⋮	⋮
1985	3431	3598.3	95.4	98.3	97.0
1986	3638	3929.6	92.6		
1987	3891	4291.4	90.7		

FIGURE 24.5
Plots of percents of trend, cyclical component, and irregular component—Agricultural imports example

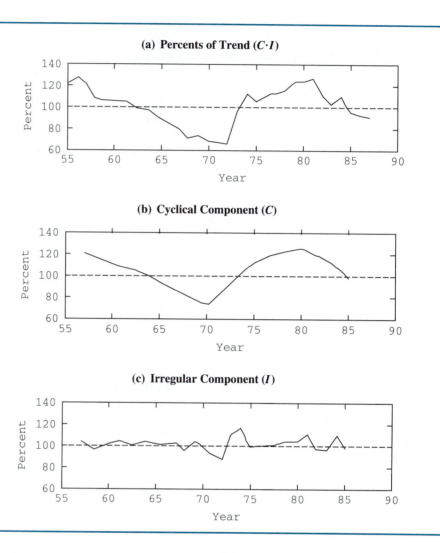

(a) **Percents of Trend (C·I)**

(b) **Cyclical Component (C)**

(c) **Irregular Component (I)**

The percents of trend are plotted in Figure 24.5a. Note that this plot shows a distinct cyclical component, together with irregular effects as reflected by the wiggles in the curve. □

Smoothing of Cyclical-Irregular Component.   The cyclical component can now be separated from the combined cyclical-irregular component by smoothing the percents of trend, using a moving average. The moving average series will represent the cyclical component C.

□ **EXAMPLE**

Column 4 of Table 24.4 contains a 5-term moving average (M.A.) of the percents of trend in column 3. The term $k = 5$ was selected after experimentation with several different terms because it was the most effective in smoothing the cyclical-irregular series. The moving average series in column 4 is calculated in the usual fashion. For example, the moving average value for 1957 is the average of the percents of trend for years 1955 through 1959. The moving average series, representing the cyclical component of the imports series, is plotted in Figure 24.5b. Observe that the cyclical component passed through a trough in 1970 and then a peak in 1979. ☐

Isolation of Irregular Component.   Occasionally, the irregular component I is of interest. In accord with multiplicative model (24.9), we obtain this component by dividing each percent of trend $(C \cdot I)$ by the corresponding cyclical component (C) and expressing the ratio as a percent.

EXAMPLE ☐

Column 5 of Table 24.4 contains the irregular component I for the agricultural imports example. It is obtained by dividing each percent of trend in column 3 by the cyclical component in column 4 and multiplying this ratio by 100. For example, the irregular component for 1957 is obtained as follows:

$$\frac{C \cdot I}{C} = 100\left(\frac{122.1}{118.0}\right) = 103.5$$

The difference in the last digit between the result here and that in Table 24.4 is because Table 24.4 contains the results of computer calculations that are more accurate.

   The irregular component of the agricultural imports series in column 5 is plotted in Figure 24.5c. As expected, the irregular component has an erratic pattern, reflecting the combined influences of many sundry factors. Note that the irregular component was especially volatile during the period 1971–1974. ☐

Cyclical Turning Points

A major objective of cyclical analysis of business and economic time series is to predict cyclical downturns and upturns in the activity of interest. Unfortunately, this is usually a difficult task. Analysts are often satisfied simply to be able to identify cyclical turning points at the time or shortly after they have occurred.

> **(24.11)**
> A **cyclical turning point** is the time point at which the cyclical component of a time series changes direction.

EXAMPLE ☐

For the agricultural imports example, we can identify from the cyclical component plot in Figure 24.5b two turning points: a trough in 1970 and a peak in 1979. ☐

   The reasons why it is so difficult to ascertain current and future cyclical developments are illustrated in Figure 24.6. Presented there is the deseasonalized quarterly series on gross private residential fixed investment (in 1982 dollars) for the United States between 1962 and 1989. We can visually trace the cyclical pattern in this series and, with some arbitrariness, identify the cyclical turning points. Note the variability of the durations and amplitudes of the cycles, which is typical of most business and economic series. In addition to

FIGURE 24.6

Example of a leading cyclical indicator series for the United States—Deseasonalized quarterly gross private residential fixed investment in 1982 dollars.   The series is expressed as an annual rate in billions of dollars and plotted on a semilogarithmic grid.

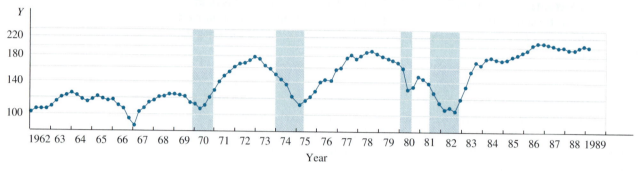

Source: U.S. Department of Commerce.

the variability in the cyclical swings, note also that substantial irregular movements are present in the series. Both of these factors make it virtually impossible to identify current or imminent cyclical behavior from inspection of the series alone.

We shall comment shortly on the meaning of the shaded and unshaded intervals in Figure 24.6.

Cyclical Indicators

Cyclical-Indicators Approach.   A useful framework for cyclical analysis of business conditions is based on the cyclical-indicators approach originally developed by the National Bureau of Economic Research (NBER). In this approach, economic time series are sought that are *leading, coinciding,* and *lagging* in comparison with cyclical movements in aggregate economic activity. A leading series, for instance, is one whose cyclical upturns and downturns tend to occur before the turns in aggregate activity.

The residential fixed investment series in Figure 24.6 is an example of a series classified as a leading series. The shaded and unshaded intervals shown in Figure 24.6 correspond to consecutive periods of cyclical contraction and expansion in *aggregate economic activity*, as identified from an assessment of many key economic time series. We can see in Figure 24.6 how the cyclical turning points in the residential fixed investment series tend to precede the general business cycle as represented by the shaded and unshaded intervals. Note, however, that the extent of the lead is variable, and in some instances, such as the downturn in the series in 1966, cyclical turning points in the series are not matched by subsequent turns in general economic activity.

Although most series are not sufficiently consistent to serve as reliable indicators, there are exceptions. The Bureau of Economic Analysis (BEA) of the U.S. Department of Commerce, working with NBER, has examined many hundreds of series and identified and listed the consistent ones. The classification process is ongoing, and the list is subject to change; it currently contains approximately 150 series. The BEA has culled a short list of cyclical indicator series from the larger list. These series represent key activities in the cyclical process, perform particularly well as indicators, and are collected and published on a timely basis. Table 24.5 presents the short list current in the early 1990s.

	TABLE 24.5
	Cyclical indicators in BEA composite indexes

(a) Leading Indicators

1. Average weekly hours of production or nonsupervisory workers, manufacturing
2. Average weekly initial claims for unemployment insurance, state programs
3. Manufacturers' new orders in 1982 dollars, consumer goods and materials industries
4. Vendor performance—slower deliveries diffusion index
5. Contracts and orders for plant and equipment in 1982 dollars
6. Index of new private housing units authorized by local building permits
7. Change in manufacturers' unfilled orders in 1982 dollars, durable goods industries
8. Change in sensitive materials prices
9. Index of stock prices, 500 common stocks
10. Money supply M2 in 1982 dollars
11. Index of consumer expectations

(b) Coincident Indicators

1. Employees on nonagricultural payrolls
2. Personal income less transfer payments in 1982 dollars
3. Index of industrial production
4. Manufacturing and trade sales in 1982 dollars

(c) Lagging Indicators

1. Average duration of unemployment in weeks
2. Ratio of manufacturing and trade inventories to sales in 1982 dollars
3. Change in index of labor cost per unit of output, manufacturing
4. Average prime rate charged by banks
5. Commercial and industrial loans outstanding in 1982 dollars
6. Ratio of consumer installment credit outstanding to personal income
7. Change in consumer price index for services

Source: U.S. Department of Commerce.

Use of Cyclical Indicators.   The leading series are studied primarily to help in anticipating cyclical turning points. The coincident and lagging series are studied mainly for assistance in recognizing a cyclical turn once it has occurred. Unfortunately, the situation is not always clear-cut. At a given time, it may be that 8 of the 11 leading indicators on the short list are moving in one direction while the others are indeterminate or are moving in the opposite direction. Similar mixed pictures can exist in the coincident and lagging series.

Various indexes have been developed to assist in the monitoring of the indicators. Each index summarizes some aspect of the movements in the group of indicators. For example, the BEA composite index of leading indicators is based on standardized versions of the 11 series in Table 24.5. Similarly, there are composite indexes of coincident and lagging indicators based on the respective groups in Table 24.5.

The cyclical-indicators approach, despite some difficulties because of inconsistent behavior among the series, has proven to be highly useful, and a broad apparatus of business-cycle analysis has developed around it. A key source that brings together the series used in the cyclical indicators approach is *Business Conditions Digest,* published monthly by the BEA, U.S. Department of Commerce. Figure 24.6 is derived from this publication.

Canadian Experience.   The cyclical indicators concept also has been evaluated for Canada. The Canadian series perform almost identically, in relation to the Canadian cycle, as their U.S. counterparts perform for the U.S. cycle.

24.6 SEASONAL COMPONENT

The seasonal component frequently is the chief source of pronounced short-term fluctuations in business and economic time series, as Figures 24.1 and 24.2 demonstrate. We now discuss how to calculate the seasonal component and how to take it into account in the analysis of past and current activity and in short-term forecasting.

Isolation of Seasonal Component

A method commonly used to measure the seasonal component for multiplicative time series model (24.2) is called the method of ratio to moving average. This method involves three main steps.

> **(24.12)**
> The three steps for measuring the seasonal component in multiplicative time series model (24.2) by the **method of ratio to moving average** are as follows.
>
> *Step 1.* Smooth the series; the smoothed values reflect the trend-cyclical component $T \cdot C$.
>
> *Step 2.* Obtain the seasonal-irregular component $S \cdot I$ by dividing the original values Y by the smoothed values $T \cdot C$:
>
> $$S \cdot I = \frac{Y}{T \cdot C} = \frac{T \cdot C \cdot S \cdot I}{T \cdot C}$$
>
> *Step 3.* Isolate the seasonal component S by averaging out the irregular effects in $S \cdot I$.

We shall illustrate this procedure with an example in which the time series is quarterly, but the method can be readily adapted to monthly, weekly, or other periodic series.

Air Taxi ☐ **EXAMPLE**

Figure 24.7 presents a time series plot of the quarterly revenue passenger miles flown between 1986 and 1991 by an air taxi service. We note a recurrent seasonal pattern marked by relatively low mileages in the first and fourth quarters and peak mileage in the third quarter. Of course, the fluctuations are not perfectly repetitive, in large part because of irregular effects present in the series. Management has requested that the seasonal component in quarterly revenue passenger miles be measured. We proceed with the three steps for the method of ratio to moving average in (24.12).

Step 1. Table 24.6, column 1, contains the quarterly revenue passenger miles flown between 1986 and 1991. Because the seasonal pattern repeats every four quarters, a centered 4-quarter moving average is appropriate here for smoothing out the seasonal fluctuations. We present the centered 4-quarter moving average in column 2. As an illustration of the calculations, the first two 4-quarter moving averages are:

$$\frac{43.20 + 90.00 + 162.00 + 64.80}{4} = 90.0000$$

$$\frac{90.00 + 162.00 + 64.80 + 42.35}{4} = 89.7875$$

Revenue
passenger miles
(thousands)

FIGURE 24.7
Revenue passenger miles
flown quarterly, 1986–1991—
Air taxi example

Year and Quarter	(1) Revenue Passenger Miles (thousands) Y	(2) Centered 4-Quarter Moving Average $T \cdot C$ M.A. of (1)	(3) Specific Seasonal Relative $S \cdot I$ $100[(1) \div (2)]$
1986 1	43.20		
2	90.00		
3	162.00	89.894	180.21
4	64.80	91.050	71.17
1987 1	42.35	93.719	45.19
2	100.10	95.688	104.61
3	173.25	96.326	179.86
4	69.30	95.078	72.89
1988 1	42.96	91.786	46.80
2	89.50	89.660	99.82
3	157.52	88.874	177.24
4	68.02	88.272	77.06
1989 1	37.95	87.582	43.33
2	89.70	86.559	103.63
3	151.80	86.525	175.44
4	65.55	87.450	74.96
1990 1	40.15	89.656	44.78
2	94.90	91.231	104.02
3	164.25	92.186	178.17
4	65.70	93.666	70.14
1991 1	47.64	95.514	49.88
2	99.25	98.034	101.24
3	174.68		
4	75.43		

TABLE 24.6
Calculation of quarterly spe-
cific seasonal relatives by the
method of ratio to moving
average—Air taxi example

and the centered average is (90.0000 + 89.7875)/2 = 89.894. This is the first entry in column 2, for quarter 3 of 1986.

The centered 4-quarter moving average in column 2 has smoothed out the seasonal and irregular components and therefore represents the combined trend-cyclical component $T \cdot C$.

Step 2. The seasonal-irregular component $S \cdot I$ can now be isolated by dividing each Y observation in column 1 by the corresponding $T \cdot C$ value in column 2 and expressing the result as a percent. For example, $S \cdot I$ for quarter 3 of 1986 is 100(162.00/89.894) = 180.21. The $S \cdot I$ component is shown in column 3 of Table 24.6. The entries are called *specific seasonal relatives* because they measure seasonal and irregular effects specific to each quarter.

Step 3. The seasonal component S for each quarter can now be isolated by averaging the specific seasonal relatives for that quarter. The idea is that irregular effects will tend to cancel one another in the average, and as a result, the average will reflect primarily the seasonal effect for that quarter. Any of several measures of position might be used as the average, including the mean, a trimmed mean, or the median. The last two measures are especially effective because they are not affected by outlying values and, hence, are immune to strong irregular effects that are often found in time series. We shall employ the median here.

The specific seasonal relatives $S \cdot I$ from column 3 of Table 24.6 are assembled by quarter in Table 24.7. We now obtain the median of the specific seasonal relatives for each quarter. The medians are shown in the second last line in Table 24.7. For instance, the median of the specific seasonal relatives for the first quarter is 45.19.

One last step remains to obtain the seasonal component. Since a seasonal pattern must balance out over a year, multiplicative model (24.2) requires that the seasonal component average out to 100 percent. The method of ratio to moving average does not contain a built-in procedure to constrain the medians to average out to 100 percent; hence, a final adjustment is usually required.

In the air taxi example, the medians sum to 399.88. They should sum to 400.00 to average out to 100. Hence, we need to simply multiply each median by the factor 400.00/399.88 = 1.00030. The adjusted medians constitute the seasonal component of the series and are referred to as the *seasonal indexes* for the four quarters; they are shown in the last line in Table 24.7.

TABLE 24.7
Specific seasonal relatives and seasonal indexes—Air taxi example

		Quarter				
Year	1	2	3	4	Total
1986			180.21	71.17	
1987	45.19	104.61	179.86	72.89	
1988	46.80	99.82	177.24	77.06	
1989	43.33	103.63	175.44	74.96	
1990	44.78	104.02	178.17	70.14	
1991	49.88	101.24			
Median:	45.19	103.63	178.17	72.89	399.88
Seasonal Index:	45.2	103.7	178.2	72.9	400.0

We see that the seasonal indexes range from a low of 45.2 in the first quarter to a high of 178.2 in the third quarter, indicating large seasonal effects. □

Comment

Some statistical packages use the mean for averaging the specific seasonal relatives and do not make the final adjustment to make the seasonal indexes average out to 100.

Examining Stability of Seasonal Component. The procedure we have just described for calculating the seasonal component assumes that this component is stable from year to year. An effective way to check this assumption is to assemble the specific seasonal relatives for each time period and plot them in time sequence.

EXAMPLE

We examine the stability of the seasonal component for the air taxi example by means of Figure 24.8, where the specific seasonal relatives for each quarter from Table 24.7 are plotted in time sequence. We study the time sequence graphs to determine whether any systematic modification is taking place in the seasonal component. For instance, if the seasonal pattern were flattening in recent years, the line graphs for the first and fourth quarters would tend upward toward 100 while that for the third quarter would tend downward toward 100. In Figure 24.8, the line graphs do not appear to have any systematic upward or downward tendencies, which is consistent with the assumption of a stable seasonal pattern. Additional analysis of the factors causing the seasonal effects here also supported the conclusion of a stable seasonal pattern. □

Plots of quarterly specific seasonal relatives in time sequence—Air taxi example. No systematic upward or downward tendencies are evident, which suggests that the seasonal pattern has been stable.

FIGURE 24.8

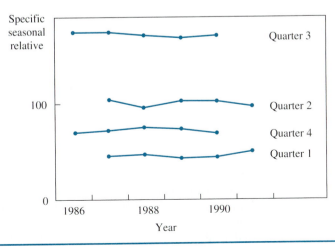

> **Comment**
>
> At times, the seasonal pattern gradually shifts because of technological developments, changes in customs or habits, or managerial actions such as promotional campaigns timed to dampen seasonal effects. In these cases, we can measure the degree of shift by fitting "trend" functions to the specific seasonal relatives for each time period (for instance, to the specific seasonal relatives for each of the four quarters). When a seasonal component for a future time period is required, these "trend" functions can be projected forward.

Applications of Seasonal Indexes

When working with time series data, questions such as the following often arise.

1. Is the recent increase in unemployment normal for this time of year?
2. Has a cyclical contraction of retail sales begun that is masked currently by the pre-Christmas buying rush?
3. A sales target of 400,000 units is reasonable for the next 12 months. How do we establish monthly sales targets that allow for seasonal effects?

Obtaining answers to these questions illustrates two important uses of seasonal indexes.

1. To adjust for seasonal effects in past and current data in order to reveal movements in the trend and cyclical time series components more clearly (questions 1 and 2).
2. To incorporate seasonal effects into forecasts (question 3).

We now consider each of these uses in turn.

Adjusting for Seasonal Effects.   When we remove the seasonal component from a time series, we call the procedure *adjusting for seasonal effects*. For multiplicative time series model (24.2), this adjustment is done simply by dividing each observation Y in the time series by the corresponding seasonal index S (in decimal form) to obtain the combined $T \cdot C \cdot I$ component.

$$(24.13) \qquad T \cdot C \cdot I = \frac{Y}{S} = \frac{T \cdot C \cdot S \cdot I}{S}$$

The value of the combined trend-cyclical-irregular component for a period is called the *seasonally adjusted value* or the *deseasonalized value* of the series for that period.

Department Store Transactions

☐ **EXAMPLE**

Table 24.8, column 1, contains the quarterly numbers of transactions of a department store for the year just ended. The seasonal indexes for the volume of transactions are given in column 2. We wish to obtain the deseasonalized number of transactions for the first quarter; the actual number is 17.36 thousand. The first-quarter seasonal index of 70 indicates that the actual number of transactions in the first quarter is 70 percent as great as if no seasonal effects were present. Thus, to find the $T \cdot C \cdot I$ component

	(1)	(2)	(3)	(4)
			Seasonally	Seasonally Adjusted
			Adjusted	Annual Rate of
	Transactions	Seasonal	Transactions	Transactions
	(thousands)	Index	$T \cdot C \cdot I$	$4(T \cdot C \cdot I)$
Quarter	Y	S		
1	17.36	70	24.80	99.20
2	21.11	85	24.84	99.36
3	21.68	92	23.57	94.28
4	32.25	153	21.08	84.32

TABLE 24.8
Calculation of seasonally adjusted quarterly and annual rates—Department store transactions example

(that is, the rate of transactions if seasonal effects were not present), we employ (24.13) as follows:

$$T \cdot C \cdot I = \frac{Y}{S} = \frac{17.36}{0.70} = 24.80 \text{ thousand}$$

The rate 24.80 thousand transactions is the deseasonalized or seasonally adjusted rate of transactions for the first quarter. It tells us that if no seasonal effects had been present, the trend-cyclical-irregular effects for the first quarter would have led to a quarterly rate of 24.80 thousand transactions.

The seasonally adjusted transactions for the other quarters are obtained in like manner. The results are shown in column 3 in Table 24.8. Note that the seasonally adjusted transactions declined in the third and fourth quarters. Although the actual volume of transactions increased during these quarters, the increases were less than would be expected on a seasonal basis alone. Hence, a decline has occurred in the trend-cyclical-irregular component that is possibly cyclical in nature and warrants attention. □

Often, seasonally adjusted data are expressed as *seasonally adjusted annual rates* to facilitate comparisons with annual data for past years. This is done by multiplying the seasonally adjusted data by the number of time periods in the year (e.g., by multiplying quarterly data by 4).

EXAMPLE □

The seasonally adjusted annual rates for the department store transactions example are given in column 4 of Table 24.8. They are obtained by multiplying the entries in column 3 by 4. The seasonally adjusted annual rate of 99.20 thousand transactions for the first quarter, for instance, indicates that if the trend-cyclical-irregular effects for the first quarter prevailed for an entire year, the annual number of transactions would be 99.20 thousand. We note again from the seasonally adjusted annual rates that the rate of nonseasonal activity declined in the third and fourth quarters. □

Incorporation of Seasonal Effects into Forecasts. In short-term forecasting, it is essential to incorporate the seasonal effects into the forecasts whenever a strong seasonal component is present in the series. The basic procedure is to start with an annual forecast. If quarterly forecasts are desired and no strong trend-cyclical effects are expected during the

TABLE 24.9
Procedure for incorporating seasonal effects into forecasts—Sales forecast example

Quarter	(1) Forecast Without Seasonal Effects	(2) Seasonal Index	(3) Forecast With Seasonal Effects $(1) \times \dfrac{(2)}{100}$
1	50	123	61.5
2	50	108	54.0
3	50	79	39.5
4	50	90	45.0
Total	200		200.0

year, the annual forecast is divided by 4 to obtain equal quarterly rates in which no seasonal effects are present. For multiplicative time series model (24.2), these quarterly rates are then multiplied by the seasonal indexes (in decimal form) to incorporate the seasonal effects into the forecasts.

Sales Forecast ☐ **EXAMPLE**

A sales manager has received an annual sales forecast of 200 thousand units for a product. The manager desires sales forecasts for each of the four quarters and expects that quarter-to-quarter changes in the trend-cyclical component will be negligible. The procedure is illustrated in Table 24.9. First, we break down the annual forecast into the quarterly rates without seasonal effects, shown in column 1. The quarterly seasonal indexes for sales of this product are given in column 2. The seasonal component now is incorporated by multiplying each entry in column 1 by the corresponding seasonal index (in decimal form). The results, entered in column 3, are the quarterly sales forecasts, which include seasonal effects.   ☐

Comment

If strong trend-cyclical effects are expected during the year, they need to be incorporated into the quarterly (or monthly) forecasts before the seasonal effects are introduced.

24.7   ADDITIVE TIME SERIES MODEL

At times, an additive version of the time series model is more appropriate than the multiplicative version in (24.2).

(24.14)
The **additive time series model** is:

$$Y = T + C + S + I$$

The isolation of the components for the additive time series model proceeds in a fashion similar to that for the multiplicative time series model, except that *differences* are used instead of ratios.

Fitting of Trend

The trend component T for additive time series model (24.14) is fitted just the same as for the multiplicative model.

Isolation of Seasonal Component

A moving average for the series is first fitted to smooth out the seasonal and irregular components, just as for the multiplicative time series model. The moving average now reflects the trend-cyclical component $T + C$. The seasonal-irregular component $S + I$ is isolated by taking the difference between each observation Y and the corresponding moving average value $T + C$.

$$(24.15) \quad S + I = Y - (T + C) = T + C + S + I - (T + C)$$

The values of $S + I$ are called *specific seasonal deviations.* The specific seasonal deviations for each time period (e.g., for each quarter) are now averaged to remove the irregular effects, and these averages are then adjusted so that they sum to zero. The adjusted averages are called *seasonal deviations* and represent the seasonal component S. As with the multiplicative model, we shall employ the median to average the specific seasonal deviations.

Deseasonalized or seasonally adjusted values are obtained by subtracting the corresponding seasonal deviation from the time series observation Y, yielding the trend-cyclical-irregular component $T + C + I$.

$$(24.16) \quad T + C + I = Y - S = (T + C + S + I) - S$$

Isolation of Cyclical Component

After the seasonal component S has been removed by means of (24.16), the cyclical-irregular component $C + I$ is isolated by subtracting the trend component T from the trend-cyclical-irregular component $T + C + I$.

$$(24.17) \quad C + I = (T + C + I) - T$$

The values $C + I$ are called *deviations from trend.*

For annual data, additive time series model (24.14) simplifies to $Y = T + C + I$. The trend component T can therefore be subtracted directly from the observations Y in this case to yield the deviations from trend $C + I$.

$$(24.17a) \quad C + I = Y - T = (T + C + I) - T \qquad \text{Annual data}$$

The cyclical component can now be isolated by taking a moving average of the deviations from trend in (24.17) or (24.17a), just as for the multiplicative time series model.

Isolation of Irregular Component

The irregular component I is isolated by subtracting the cyclical component C from the deviations from trend $C + I$.

$$(24.18) \quad I = (C + I) - C$$

☐ **EXAMPLE**

We illustrate the calculations for the additive time series model by obtaining the seasonal deviations for the air taxi example. The calculations are performed in Table 24.10. Column 1 in Table 24.10a repeats a portion of the time series data from Table 24.6, column 1. Column 2 repeats a portion of the centered 4-quarter moving average

TABLE 24.10

Calculation of seasonal component for additive time series model (24.14)—Air taxi example

(a) Calculation of Specific Seasonal Deviations

Year and Quarter	(1) Revenue Passenger Miles (thousands) Y	(2) Centered 4-Quarter Moving Average $T + C$ M.A. of (1)	(3) Specific Seasonal Deviation $S + I$ (1) − (2)
1986　1	43.20		
2	90.00		
3	162.00	89.894	72.11
4	64.80	91.050	−26.25
⋮	⋮	⋮	⋮
1991　2	99.25	98.034	1.22
3	174.68		
4	75.43		

(b) Specific Seasonal Deviations and Seasonal Deviations

Year	Quarter 1	2	3	4	Total
1986			72.11	−26.25	
1987	−51.37	4.41	76.92	−25.78	
1988	−48.83	−0.16	68.65	−20.25	
1989	−49.63	3.14	65.28	−21.90	
1990	−49.51	3.67	72.06	−27.97	
1991	−47.87	1.22			
Median:	−49.51	3.14	72.06	−25.78	−0.09
Seasonal deviation:	−49.49	3.16	72.08	−25.76	0.0

from Table 24.6, column 2. This moving average here represents the trend-cyclical component $T + C$. We now take the difference between each observation Y and the corresponding moving average value $T + C$. The differences are shown in column 3 of Table 24.10a. They are the specific seasonal deviations and represent the seasonal-irregular component $S + I$.

The specific seasonal deviations are assembled by quarter in Table 24.10b and the median deviation is calculated for each quarter. Since the medians sum to -0.09 rather than to 0, we adjust them by adding $0.09/4 = 0.0225$ to each median. This adjustment yields the seasonal deviations shown in the last row of Table 24.10b.

The seasonal deviations for the additive time series model in Table 24.10b show the same seasonal pattern as the seasonal indexes for the multiplicative model in Table 24.7, as would be expected. The lowest seasonal deviation is for the first quarter (-49.49) and the highest is for the third quarter (72.08). □

Comment

The multiplicative and additive time series models can be related mathematically by a logarithmic transformation when all observations in the time series are positive. By taking the logarithms of both sides of multiplicative model (24.2), we obtain an additive time series model in which the additive components are the logarithms of their multiplicative counterparts, as follows.

(24.19) $\log Y = \log T + \log C + \log S + \log I$

PROBLEMS

* 24.1 Pulp and Paper Production. Data on total production of pulp and paper (in millions of metric tons) in one overseas division of a multinational company during 1972–1991 follow. A 3-term moving average is also presented.

Year	Production	Moving Average	Year	Production	Moving Average
1972	3.3		1982	4.0	4.03
1973	3.7	3.63	1983	4.6	4.23
1974	3.9	3.70	1984	4.1	4.50
1975	3.5	3.43	1985	4.8	4.53
1976	2.9	3.17	1986	4.7	4.60
1977	3.1	3.17	1987	4.3	4.50
1978	3.5	3.40	1988	4.5	____
1979	3.6	3.43	1989	4.0	____
1980	3.2	3.43	1990	3.6	____
1981	3.5	3.57	1991	3.8	

a. Calculate the remaining moving average values.
b. Plot the original series and the moving average series on the same graph. What information is provided by the moving average series that is not obtained readily from the original series?

24.2 Bus Traffic. Data on daily city bus traffic (in thousands of passengers) for a recent four-week winter period follow. A 7-term moving average is also presented.

	Week 1			Week 2	
Day	Passengers	Moving Average		Passengers	Moving Average
M	211			163	136.3
T	182			143	136.4
W	199			131	142.1
T	170	158.0		173	144.7
F	206	151.1		207	151.7
S	109	145.6		149	158.0
S	29	135.9		47	166.9

	Week 3			Week 4	
Day	Passengers	Moving Average		Passengers	Moving Average
M	212	166.9		200	165.1
T	187	166.4		193	_____
W	193	166.1		196	_____
T	173	165.3		175	_____
F	204	163.6		202	
S	147	164.4		144	
S	41	164.9		46	

a. Calculate the remaining moving average values.

b. Plot the original series and the moving average series on the same graph. What information is provided by the moving average series? Is a 7-term moving average a good choice here? Explain.

24.3 Communications Link. Data on the numbers of erroneous data bits transmitted each hour over a communications link for the past 24 hours follow. A 5-term moving average is also presented.

Hour	Number	Moving Average	Hour	Number	Moving Average	Hour	Number	Moving Average
1	5		9	1	16.2	17	0	12.6
2	24		10	2	16.4	18	1	8.2
3	36	13.6	11	4	6.2	19	8	7.6
4	2	13.2	12	20	6.2	20	2	_____
5	1	12.2	13	4	10.6	21	27	_____
6	3	16.0	14	1	15.8	22	0	_____
7	19	15.8	15	24	11.8	23	3	
8	55	16.0	16	30	11.2	24	20	

a. Calculate the remaining moving average values. Why can 5-term moving averages not be computed corresponding to the first two and last two observations of the original series?

b. Plot the original series and the moving average series on the same graph. Does it appear that the 5-term moving average series adequately reflects the smooth component of the original series? Would a centered 10-term moving average yield a smoother series? Explain.

* 24.4 Refer to **Pulp and Paper Production** Problem 24.1. Calculate the first three values of a centered 6-term moving average for the production series. Would a 3-term moving average or a centered 6-term moving average smooth the series more? Explain.

24.5 Refer to **Communications Link** Problem 24.3. Calculate the first three values of a centered 10-term moving average. How many observations enter the first centered moving average value? What are the weights of these observations?

24.6 Production of a processed feed (in million pounds) was stable at the following quarterly levels during a three-year period:

Year:	1				2				3			
Quarter:	1	2	3	4	1	2	3	4	1	2	3	4
Production:	2	9	7	6	2	9	7	6	2	9	7	6

a. For the series, calculate (1) a centered 4-term moving average, (2) a 3-term moving average, (3) a 5-term moving average.
b. Why is the centered 4-term moving average series the smoothest one?

* **24.7** Refer to **Pulp and Paper Production** Problem 24.1.

a. Fit a linear trend function to the series. Set $X_t = 1$ at 1972.
b. Plot the trend line and the original series on the same graph. Does the trend line provide a good description of the trend of the series? Comment.
c. What is the year-to-year change in the trend line obtained in part a? Does the intercept of the trend line have any meaning here?
d. Obtain the projected trend value for production in 1994. Assuming that the trend line fitted in part a is appropriate through 1994 and that the best estimate of the cyclical component in 1994 is 115 percent, obtain a forecast of actual production in 1994.

24.8 Industrial Customers. Data on the number of industrial customers (in thousands, at midyear) of a gas company during the period 1974–1991 follow:

Year:	1974	1975	1976	1977	1978	1979	1980	1981	1982
Customers:	3.92	4.12	4.13	4.31	4.23	4.22	4.19	4.45	4.59

Year:	1983	1984	1985	1986	1987	1988	1989	1990	1991
Customers:	4.51	4.50	4.67	4.64	4.70	4.63	4.60	4.79	4.86

a. Fit a linear trend function to the series. Set $X_t = 1$ at 1974.
b. Plot the trend line and the original series on the same graph. Does the trend line provide a good description of the trend of the series? Comment.
c. Project the trend line to midyear 1995. Is this projection a forecast of the number of industrial customers in midyear 1995? Explain.

24.9 Consulting Contracts. Data on the total value of contracts (in $ thousand) received annually by a management consulting firm in the period 1983–1992 follow:

Year:	1983	1984	1985	1986	1987	1988	1989	1990	1991	1992
Total Value:	603	715	809	920	1008	1124	1203	1314	1412	1525

a. Fit a linear trend function to this series. Set $X_t = 1$ at 1983.
b. What is the year-to-year change in the trend line obtained in part a?
c. Obtain the projected trend value for the total value of contracts in 1995. Is this projection a forecast of the total value of contracts in 1995? Explain.

* **24.10 Sailboats.** Data on the number of sailboats sold annually by a manufacturer during 1977–1991 follow:

Year:	1977	1978	1979	1980	1981	1982	1983	1984
Number:	2089	2138	2317	2305	2473	2548	2530	2651

Year:	1985	1986	1987	1988	1989	1990	1991
Number:	2750	2906	2996	3234	3218	3366	3434

a. Fit an exponential trend function to the series. Set $X_t = 1$ at 1977.
b. Plot the original series and the trend function (expressed in the original units) on a graph with an arithmetic grid. Does the trend function provide a good description of the trend of the series? Explain.
c. Obtain the annual rate of growth in the fitted trend function.
d. Project the trend function to the year 2000 and also to 2050. Express the projected trend values in the original units. Do you believe that the trend projection for 2050 is reasonable? Discuss.

24.11 Life Insurance. Data on the amount of life insurance (in billions of Canadian dollars) under-written in Canada by U.S. companies during 1975–1989 follow:

Year:	1975	1976	1977	1978	1979	1980	1981	1982
Amount:	43.7	49.8	51.7	56.6	64.4	70.2	77.9	83.4

Year:	1983	1984	1985	1986	1987	1988	1989
Amount:	87.3	94.7	94.6	105.0	105.8	118.9	132.4

Source: Canadian Life and Health Insurance Facts, published by Canadian Life and Health Insurance Association, Inc.

a. Fit an exponential trend function to the series. Set $X_t = 1$ at 1975.
b. Plot the original series and the trend function (expressed in the original units) on a graph with an arithmetic grid.
c. Obtain the annual rate of growth in the fitted trend function.
d. Project the trend function to the year 2000. Express the projected trend value in the original units.

24.12 Production Index. The U.S. industrial production index (1987 $=$ 100) during 1977–1990 follows:

Year:	1977	1978	1979	1980	1981	1982	1983
Index:	78.2	82.6	85.7	84.1	85.7	81.9	84.9

Year:	1984	1985	1986	1987	1988	1989	1990
Index:	92.8	94.4	95.3	100.0	105.4	108.1	109.1

Source: Federal Reserve Bulletin

a. Fit an exponential trend function to the series. Set $X_t = 1$ at 1977.
b. Plot the original series and the trend function (expressed in the original units) on a graph with an arithmetic grid.
c. Obtain the annual rate of growth in the fitted trend function.
d. Project the trend function to the year 2000. Express the projected trend value in the original units. Is this projection a forecast of the production level in the year 2000? Explain.

24.13 Refer to **Life Insurance** Problem 24.11. The fitted linear trend function for this series is $T_t = 35.3238 + 5.88786X_t$ (with $X_t = 1$ at 1975). Plot this linear trend function on the graph prepared in Problem 24.11b. Do the exponential and linear trend functions provide equally good fits to the trend of the series? Do they give similar projected trend values for the year 2000? Explain.

24.14 A speaker stated: "We have developed a computer program to make short-term and long-term sales forecasts for our products. For each product, the computer will access annual sales data for the past 12 years and fit five different trend functions, including linear and exponential trends. Coeffi-cients of correlation between sales and trend will be calculated for each trend function. The trend function with the highest coefficient will be projected one year and 10 years ahead for the short-term and long-term sales forecasts for this product." Discuss the problems involved in this approach to sales forecasting.

24.15 On reading in the newspaper that the state's unemployment rate went from 5.0 percent to 6.0 percent in the previous 12 months, a coffee shop patron remarked, "If this trend continues at the same rate, the whole labor force of the state will be unemployed in 16 years." Should the patron really be concerned with trend or is a different time series component involved? Discuss.

* **24.16** Refer to **Pulp and Paper Production** Problems 24.1 and 24.7. Assume that multiplicative time series model (24.9) is appropriate for this time series.

a. Obtain (1) the percents of trend, (2) the cyclical component by using a 5-term moving average of the percents of trend, (3) the irregular component. Explain why the moving average series in (2) reflects the cyclical component of the series.

b. Plot the percents of trend, cyclical component, and irregular component in separate graphs as in Figure 24.5. Have the fluctuations of the series about the trend line been due predominantly to cyclical or irregular factors? Does the cyclical component appear to be sufficiently regular here that one can confidently project it forward? Explain.

24.17 Refer to **Industrial Customers** Problem 24.8. Assume that multiplicative time series model (24.9) is appropriate for this time series.

a. Obtain (1) the percents of trend, (2) the cyclical component by using a 5-term moving average of the percents of trend, (3) the irregular component.

b. Plot the percents of trend, cyclical component, and irregular component in separate graphs as in Figure 24.5.

c. Identify each year in which the cyclical component passes through a trough or a peak. Why can definite answers not be given for the end years?

24.18 Refer to **Consulting Contracts** Problem 24.9. Assume that multiplicative times series model (24.9) is appropriate for this time series.

a. Obtain (1) the percents of trend, (2) the cyclical component by using a 3-term moving average of the percents of trend, (3) the irregular component. Explain why the moving average series in (2) reflects the cyclical component of the series.

b. Plot the percents of trend, cyclical component, and irregular component in separate graphs as in Figure 24.5. Have the fluctuations of the series about the trend line been due predominantly to cyclical or irregular factors? Explain.

* **24.19** Refer to **Sailboats** Problem 24.10. Assume that multiplicative time series model (24.9) is appropriate for this time series.

a. Obtain (1) the percents of trend, (2) the cyclical component by using a 3-term moving average of the percents of trend, (3) the irregular component.

b. Plot the percents of trend, cyclical component, and irregular component in separate graphs as in Figure 24.5. Have the fluctuations about the trend function been due predominantly to cyclical or irregular factors? Explain.

24.20 Refer to **Life Insurance** Problem 24.11. Assume that multiplicative time series model (24.9) is appropriate for this time series.

a. Obtain (1) the percents of trend, (2) the cyclical component by using a 3-term moving average of the percents of trend, (3) the irregular component.

b. Plot the percents of trend, cyclical component, and irregular component in separate graphs as in Figure 24.5. Have the fluctuations about the trend function been due predominantly to cyclical or irregular factors? Explain.

24.21 Refer to **Production Index** Problem 24.12. Assume that multiplicative time series model (24.9) is appropriate for this time series.

a. Obtain (1) the percents of trend, (2) the cyclical component by using a 3-term moving average of the percents of trend, (3) the irregular component.

b. Plot the percents of trend, cyclical component, and irregular component in separate graphs as in Figure 24.5. Can one confidently project the irregular component forward here? Explain.

c. Identify each year in which the cyclical component passes through a trough or a peak. Why can definite answers not be given for the end years?

24.22 Data on the cyclical components of number of building permits issued for private housing units (the permits series) and value of home mortgage loans made by major lending institutions (the loans series) during a recent four-year period follow:

Year:	1				2			
Quarter:	1	2	3	4	1	2	3	4
Permits:	115	106	101	99	90	86	91	95
Loans:	105	110	103	100	95	91	90	88

Year:	3				4			
Quarter:	1	2	3	4	1	2	3	4
Permits:	99	105	103	95	92	96	105	114
Loans:	96	99	102	108	101	97	93	97

a. Plot the cyclical components of the two series on the same graph. Identify the turning points for each series.
b. Is the cyclical component of the permits series sufficiently regular that one can confidently project it forward? Explain.
c. Does the cyclical behavior of the permits series lead, lag, or coincide with that of the loans series? Is this behavior consistent? Explain.

24.23 The cyclical component of exports of coal for a recent 13-year period follows:

Year:	1	2	3	4	5	6	7	8	9	10	11	12	13
Component:	75	69	50	83	115	144	135	99	103	114	120	100	98

a. Plot the cyclical component. Identify each turning point.
b. Is the cyclical component sufficiently regular that one can confidently project it forward? Explain.

24.24 Explain why it is reasonable that average duration of unemployment is a lagging cyclical indicator while contracts and orders for plant and equipment is a leading cyclical indicator according to the BEA list in Table 24.5.

24.25 A U.S. television commentator noted that manufacturers' new orders (a leading indicator series according to the BEA list of Table 24.5) had declined during the quarter just ended. The commentator stated that a downturn in industrial production (a coincident indicator in Table 24.5) is therefore imminent. Is the commentator taking the BEA classification of the series as leading and coincident indicators too literally? Comment.

* 24.26 Electricity Sales. Data on quarterly sales revenues (in $ million) of a Southern electric company during 1986–1991 follow. Assume that multiplicative time series model (24.2) is appropriate for this time series.

Quarter	1986	1987	1988	1989	1990	1991
1	172	169	182	169	179	170
2	227	218	218	245	235	241
3	310	309	313	299	292	307
4	222	209	224	221	213	217

a. Plot the time series. Does the series have a noticeable seasonal component? A noticeable trend-cyclical component? Comment.
b. Obtain the specific seasonal relatives for the series.
c. Assume that the seasonal pattern is stable and obtain the quarterly seasonal indexes. Describe the seasonal pattern in the series.

24.27 **University Enrollment.** The enrollments during 1984–1991 at a university operating on a trimester system follow. Assume that multiplicative time series model (24.2) is appropriate for this time series.

Trimester	1984	1985	1986	1987	1988	1989	1990	1991
1	10,284	10,724	11,052	11,301	12,240	12,958	12,751	13,253
2	9,445	9,828	10,636	10,757	10,946	11,617	12,177	12,880
3	6,307	6,015	6,922	6,883	7,300	7,153	8,170	7,617

a. Plot the time series. Does the series have a noticeable seasonal component? A noticeable trend-cyclical component? Comment.
b. Obtain the specific seasonal relatives for the series. Was there any need to center the moving average in the calculations? Explain. Why are there seven specific seasonal relatives for trimesters 1 and 3 but eight for trimester 2?
c. Assume that the seasonal pattern is stable and obtain the seasonal indexes. Describe the seasonal pattern in the series.

24.28 Refer to **Interest Rates** Problem 15.37. Assume that multiplicative time series model (24.2) is appropriate for this time series.

a. Plot the time series. Does the series have a noticeable seasonal component? A noticeable trend-cyclical component? Comment.
b. Obtain the specific seasonal relatives for the series.
c. Assume that the seasonal pattern is stable and obtain the quarterly seasonal indexes.
d. The bank's chief economist expected the series to exhibit very little seasonality except possibly for slightly lower rates in the first quarter of the year. Are the results consistent with the economist's expectation? Explain.

24.29 Refer to **Electricity Sales** Problem 24.26. Plot the specific seasonal relatives in a graph as in Figure 24.8. Is the assumption of a stable seasonal pattern reasonable here? Comment.

24.30 Refer to **University Enrollment** Problem 24.27. Plot the specific seasonal relatives in a graph as in Figure 24.8. Is the assumption of a stable seasonal pattern reasonable here? Comment.

24.31 Refer to **Interest Rates** Problem 24.28. Plot the specific seasonal relatives in a graph as in Figure 24.8. Is the assumption of a stable seasonal pattern reasonable here? Comment.

24.32 If automobile manufacturers were to shift their annual retooling shutdown for new models to earlier in the year, would you expect this shift to affect the seasonal component of the quarterly series for automobile production? Comment.

* **24.33** Refer to **Electricity Sales** Problem 24.26.

a. Obtain the deseasonalized quarterly sales revenues for 1991.
b. Express the deseasonalized sales revenues for the fourth quarter of 1991 as a seasonally adjusted annual rate. Compare this annual rate with the total sales revenues for 1991 and interpret the results.
c. Obtain the irregular component of the quarterly sales revenues for 1990 by expressing each specific seasonal relative as a percentage of the corresponding seasonal index. Are the irregular movements substantial relative to the seasonal changes? Comment.

24.34 Refer to **University Enrollment** Problem 24.27.

a. Obtain the deseasonalized trimester enrollment levels for 1991.
b. Express the deseasonalized enrollment level for trimester 1 of 1991 as a seasonally adjusted annual rate. Interpret the result.
c. Obtain the irregular component of the trimester enrollment levels for 1990 by expressing each specific seasonal relative as a percentage of the corresponding seasonal index. Are the irregular movements substantial relative to the seasonal changes? Comment.

24.35 Refer to **Interest Rates** Problems 15.37 and 24.28.

a. Obtain the deseasonalized quarterly interest rates for year 5. Compare the deseasonalized interest rate for the first quarter of year 5 with the average interest rate for year 5 and interpret the result.

b. Obtain the irregular component of the quarterly interest rates for year 4 by expressing each specific seasonal relative as a percentage of the corresponding seasonal index. Are the irregular movements substantial relative to the seasonal changes? Comment.

24.36 Data on quarterly sales (in $ million) of Nettles, Ltd., during 1991 follow, together with the seasonal indexes for these sales:

Quarter:	1	2	3	4
Sales:	3.40	3.01	4.21	5.29
Seasonal Index:	83	75	102	140

a. Obtain the deseasonalized sales volume for each quarter of 1991. Should management be pleased with the relatively large sales volume experienced in the fourth quarter? Comment.

b. Suppose that only seasonal changes in sales were expected between the fourth quarter of 1991 and the first quarter of 1992. What would be the expected sales volume for the first quarter of 1992, assuming that the seasonal pattern is stable? If actual sales in the first quarter of 1992 were $3.06 million, what would this suggest?

* **24.37** Refer to **Electricity Sales** Problem 24.26.

a. In 1991, company executives forecast 1992 annual sales revenues of $930 million and expected that trend-cyclical movements during the year would be negligible. Convert this forecast into quarterly forecasts for 1992, assuming that the seasonal pattern is stable.

b. Actual sales revenues in the first quarter of 1992 were $168 million. Obtain the deseasonalized level and the seasonally adjusted annual rate of the sales revenues for this quarter and interpret each.

24.38 The seasonal indexes for the total loans made quarterly by a small credit company are 75, 91, 102, 132. Late in 1991, a forecast of total loans of $18.6 million for 1992 was made, and it was expected that trend-cyclical movements during 1992 would be very small. Assuming that the seasonal pattern is stable, convert the annual forecast into quarterly forecasts for 1992. Explain why the quarterly forecasts sum to the annual forecast of $18.6 million here.

24.39 The seasonal indexes for the numbers of visitors to a national park each month follow:

J	F	M	A	M	J	J	A	S	O	N	D
45	36	46	65	103	171	180	178	152	103	53	68

a. Total visitors to the park for 1993 were forecast to be 910,000. Assuming that trend-cyclical movements will be negligible during 1993 and that the seasonal pattern is stable, convert the annual forecast into monthly forecasts for 1993. What percent of all visitors for 1993 are expected between January and June inclusive?

b. The actual number of visitors during January 1993 was 35,214. What is the seasonally adjusted annual rate for January? What is the significance of the difference between this annual rate and the annual forecast of 910,000 visitors?

* **24.40 Bergen Corporation.** Quarterly sales (in $ million) by the Bergen Corporation for years 1986 to 1991 follow. Assume that additive time series model (24.14) is appropriate for this series.

Quarter	Year					
	1986	1987	1988	1989	1990	1991
1	38	37	37	40	41	43
2	34	36	36	38	45	42
3	40	40	38	45	46	50
4	52	55	55	53	54	56

a. Obtain (1) the trend-cyclical component and (2) the specific seasonal deviations of the series.
b. Assuming that the seasonal pattern is stable, obtain the quarterly seasonal deviations of the series. Describe the seasonal pattern in the series.
c. Obtain the deseasonalized quarterly sales for 1991. Express the deseasonalized sales for the fourth quarter of 1991 as a seasonally adjusted annual rate. Interpret the result.

24.41 Refer to **Bus Traffic** Problem 24.2. Let *season* refer here to the recurring variation in daily bus traffic within a week. Assume that additive time series model (24.14) is appropriate for this series.

a. Problem 24.2 gives the 7-term moving average series. What time series components does this series represent?
b. Obtain the specific seasonal deviations for the series.
c. Assuming that the seasonal pattern is stable from week to week, obtain the daily seasonal deviations of the series. On which day of the week is bus traffic heaviest? Interpret the value of the seasonal deviation for that day.
d. City bus traffic is expected to total 1150 thousand passengers in week 5. Assuming that trend-cyclical movements will be negligible during week 5 and that the seasonal pattern in daily traffic is stable, obtain daily traffic forecasts for week 5.

24.42 Refer to **Consulting Contracts** Problem 24.9. Assume that the additive time series model $Y = T + C + I$ is appropriate for this annual series.

a. A linear trend function was fitted in Problem 24.9a. Obtain the deviations from trend. Must these deviations necessarily add to zero when the trend line is fitted by the method of least squares? What time series components do these deviations represent?
b. Obtain (1) the cyclical component of the series by using a 3-term moving average of the deviations from trend, (2) the irregular component of the series. Interpret the values of the cyclical and irregular components for 1991.

24.43 Refer to Table 24.2 pertaining to the *Executive Jet* example. Assume that the additive time series model $Y = T + C + I$ is appropriate for this annual series.

a. A linear trend function was fitted in Table 24.2. Obtain (1) the deviations from trend, (2) the cyclical component of the series by using a 3-term moving average of the deviations from trend, (3) the irregular component of the series.
b. Plot the deviations from trend, the cyclical component, and the irregular component in three separate graphs. Have the fluctuations about the trend line been due predominantly to cyclical or irregular factors? Explain.

24.44 An analyst, before fitting a trend function to a company's monthly production volume, divided each month's production by the number of working days in that month.

a. Why did the analyst take this preliminary step before fitting a trend function? Is it always necessary to divide production data by the number of working days before fitting a trend function? Explain.
b. The analyst obtained the linear trend function $T_t = 891 + 1.23X_t$ for the adjusted monthly production data. Project the trend line to month $X_t = 40$. In this month, there will be 22 working days. What is the projected trend value of total production for this month?

24.45 The personnel manager of a large firm fitted a linear trend function to the series on annual number of job applicants handled by the personnel department and obtained $T_t = 3085 + 72X_t$, where $X_t = 1$ in year 1 and is in one-year units. The manager wishes to convert this trend function to one of the form $T_t' = b_0' + b_1'X_t'$, where T_t' is the quarterly trend value of number of job applicants and $X_t' = 1$ in the first quarter of year 1 and is in one-quarter units. Obtain the coefficients b_0' and b_1' of the converted trend function.

24.46 Construct a logarithmic scale with gradation points corresponding to 1, 2, . . . , 9, 10, 20, . . . , 90, 100. Confirm that the distances on the scale between 2 and 6 and between 30 and 90 are the same, as theoretically they should be.

24.47 Refer to **Sailboats** Problem 24.10. Plot the logarithms of the original series and the linear trend function fitted to the log-observations on a graph with an arithmetic grid. Is the linear trend function a good description of the trend of the log-series? Comment.

24.48 Refer to Figure 24.4a pertaining to the *Agricultural Imports* example.

a. An observer has suggested that the trend component for this series consists of one linear trend during the period 1955–1972 and a different linear trend during the period 1973–1987. Does the suggestion appear to have merit? Comment.

b. To follow the suggestion, multiple regression was used to obtain the following trend function for the imports series:

$$\hat{Y}_t = 283.634 + 24.8163X_{t1} - 2494.2935X_{t2} + 163.2373X_{t3}$$

Here $\hat{Y}_t$ denotes imports (in $ million) for year t; X_{t1} is a numerical code for year t with $X_{t1} = 1$ in 1955; X_{t2} is an indicator variable coded 0 if t corresponds to a year during 1955–1972 and coded 1 otherwise; and $X_{t3} = X_{t1}X_{t2}$. Plot the fitted trend function. Does the trend function consist of different linear trend segments for the two periods 1955–1972 and 1973–1987? Comment.

c. Calculate the projected trend value for the year 2000 from the fitted trend function in part b.

STUDIES

24.49 Refer to the **Investment Funds** data set (Appendix D.3). Consider the two quarterly unit value series for the equity and fixed-income funds for periods 1 to 44 inclusive. Assume that multiplicative time series model (24.2) is appropriate for each series, but ignore any seasonal component that may be present.

a. For each fund, fit an exponential trend function to the quarterly series of unit values and plot the original series and the trend function (expressed in the original units) on a graph with an arithmetic grid. Set $X_t = 1$ at period 1.

b. Evaluate the fits of the exponential trend functions to the two series. Compare the growth rates of the two series.

c. For each series, obtain (1) the percents of trend, (2) the cyclical component by using a centered 8-term moving average. Identify the cyclical turning points of each series. Do the cyclical components of the two series appear to be in phase? Discuss.

24.50 Refer to **Marriages** Problem 15.38. Assume that multiplicative time series model (24.2) is appropriate.

a. Decompose the marriages series into its trend, cyclical, seasonal, and irregular components, using the following procedures.

1. Obtain the monthly seasonal indexes.
2. Deseasonalize the series.
3. Fit a linear trend function to the deseasonalized series; set $X_t = 1$ at July of year 1, with X in one-month units.
4. Obtain the deseasonalized percents of trend.
5. Obtain the cyclical component by using a 5-term moving average of the deseasonalized percents of trend.
6. Obtain the irregular component.

b. Plot the original series and its four components on separate graphs as in Figure 24.2, and briefly describe the nature of the components.

c. Forecast the number of marriages for July of year 5. Assume that the cyclical component will be 100 in that month and that the seasonal pattern is stable.

24.51 Consult the latest issue of *Business Conditions Digest* and examine the following three cyclical indicator series: (1) contracts and orders for plant and equipment, (2) change in index of labor cost per unit of output in manufacturing, (3) manufacturing and trade sales.

a. For each of these series, indicate whether it is a leading, lagging, or coincident indicator series.
b. Find the most recent value for each of these series. Are these values seasonally adjusted?
c. What is the current cyclical phase—expansion or contraction—of each of the series? Is the phase of each series consistent with its classification as a leading, lagging, or coincident indicator? Comment.

24.52 The following specific seasonal relatives for quarterly sales of large grade A eggs by a supermarket chain were obtained for 1986–1991:

	Quarter			
Year	1	2	3	4
1986			98.9	97.1
1987	102.9	100.4	97.9	99.5
1988	101.5	99.5	99.2	107.7
1989	100.9	95.1	100.0	105.3
1990	99.5	95.1	100.4	106.7
1991	98.0	92.7		

a. Plot the specific seasonal relatives in a graph in the format of Figure 24.8. Describe the shift in the seasonal pattern of egg sales that has occurred during this period. What factors might have produced this shift?
b. Linear "trend" functions fitted to the specific seasonal relatives for the first three quarters are:

$$
\begin{array}{lll}
\text{First quarter:} & S = 104.10 - 1.18X; & X = 1 \text{ in } 1987 \\
\text{Second quarter:} & S = 102.50 - 1.98X; & X = 1 \text{ in } 1987 \\
\text{Third quarter:} & S = 97.75 + 0.510X; & X = 1 \text{ in } 1986
\end{array}
$$

Obtain the "trend" function for the fourth quarter, letting $X = 1$ in 1986.
c. Obtain seasonal indexes for 1992 by projecting each "trend" function to 1992 and adjusting the indexes so that they add to 400.0.

Time Series and Forecasting: Regression Methods and Exponential Smoothing

25

n this chapter, we present two additional methods extensively used in time series analysis and forecasting. First we discuss the use of regression models for time series data. Then we present the method of exponential smoothing.

The multiplicative and additive time series models presented in Chapter 24 are descriptive models. The irregular component for these models is not defined in probabilistic terms, and no parameters are specified about which formal statistical inferences can be made. Uncertainties are handled by judgment, not by statistical theory. Regression time series models, on the other hand, are formal models permitting statistical inferences and predictions. In this section, we take up regression models where the error terms are independent. In the next section, we consider regression models where the error terms are correlated.

Time series regression models with independent error terms are special cases of the general multiple regression model (20.3), which we repeat here for convenience.

REGRESSION MODELS WITH INDEPENDENT ERROR TERMS **25.1**

$$(25.1) \qquad Y_t = \beta_0 + \beta_1 X_{t1} + \beta_2 X_{t2} + \cdots + \beta_{p-1} X_{t,p-1} + \varepsilon_t$$

where the ε_t are independent $N(0, \sigma^2)$.

We now use the subscript t to denote the time period ($t = 1, 2, \ldots, n$). With this type of model, trend, cyclical, and seasonal effects are incorporated into the model entirely by means of independent variables. The error terms of the model represent the random or irregular component of the time series and, as already noted, are assumed to be time-independent. We shall only illustrate regression models here that combine the time series components in an additive manner.

Trend Component Model

This time series regression model is appropriate for annual time series, such as annual company sales for the past 10 years, that contain only a linear trend component and a random (irregular) component.

$$(25.2) \qquad Y_t = \beta_0 + \beta_1 X_t + \varepsilon_t$$

where $X_t = t$.

Here, X_t denotes the time period, coded as before with $X_1 = 1$, $X_2 = 2$, and so on. Thus, $\beta_1 X_t$ is the linear trend effect. Note that the trend and random components are additive effects in regression model (25.2). Also note that the assumption of independent error terms and the fact that the only independent variable relates to the trend effect together imply that no cyclical effects are present in the time series.

Trend and Seasonal Components Model

This regression model is appropriate for quarterly time series, such as quarterly shipments of television sets during the past 48 quarters, that contain a linear trend component, a seasonal component, and a random component.

$$(25.3) \qquad Y_t = \beta_0 + \beta_1 X_{t1} + \beta_2 X_{t2} + \beta_3 X_{t3} + \beta_4 X_{t4} + \varepsilon_t$$

where: $X_{t1} = t$

$$X_{t2} = \begin{matrix} 1 & \text{if period } t \text{ is second quarter} \\ 0 & \text{otherwise} \end{matrix}$$

$$X_{t3} = \begin{matrix} 1 & \text{if period } t \text{ is third quarter} \\ 0 & \text{otherwise} \end{matrix}$$

$$X_{t4} = \begin{matrix} 1 & \text{if period } t \text{ is fourth quarter} \\ 0 & \text{otherwise} \end{matrix}$$

Here $\beta_1 X_{t1}$ is the linear trend effect, and $\beta_2 X_{t2} + \beta_3 X_{t3} + \beta_4 X_{t4}$ is the seasonal effect. Variables X_2, X_3, and X_4 are three indicator variables (discussed in Section 20.8) that take on the following values for the four quarters:

Quarter	X_2	X_3	X_4
First	0	0	0
Second	1	0	0
Third	0	1	0
Fourth	0	0	1

The regression coefficient β_2 in model (25.3) represents the differential effect on the dependent variable of the second quarter as compared with the reference quarter (first quarter). The regression coefficients for the other indicator variables represent similar differential effects for the other quarters.

Note in regression model (25.3) that the trend, seasonal, and random components are additive effects. Also note that the assumption of independent error terms and the fact that the only independent variables relate to trend and seasonal effects together imply that no cyclical effects are present in the time series.

Regression model (25.3) can be readily modified for time series with monthly or other seasonal patterns by suitably modifying the seasonal indicator variables.

Regression Models with Indicator Series as Independent Variables

Time series regression models often contain indicator series as independent variables. *Indicator series* are time series for key business, economic, and social activities. The cyclical indicator series in Table 24.5 are examples of such series. Other examples include series for number of births, crime rate, and level of college enrollment.

In some applications, indicator series supplement independent variables for trend and seasonal effects. In other applications, the indicator series serve to represent the time series components, as the following examples illustrate.

EXAMPLES

1. Annual production in the shoe industry (Y) is related to the gross national product (X_1) and the consumer price index (X_2), for the past 16 years.

2. Annual kindergarten enrollment (Y) is related to number of births five years earlier (X), for the past 20 years.

3. Monthly sales of a firm (Y) are related to the firm's promotional expenditures in the previous month (X), for the past 36 months.

4. Quarterly shipments of cement (Y) are related to two independent variables for the past 20 quarters: (1) the volume of construction contracts awarded (X) during the preceding quarter and (2) the volume awarded during the quarter before that. ▢

Coincident Independent Variables.   The regression model for Example 1 involves *coincident time series*; that is, the two indicator series refer to the same time period as the dependent variable series. Assuming that the effects of the independent variables are linear and additive, the time series regression model is as follows.

$$(25.4) \qquad Y_t = \beta_0 + \beta_1 X_{t1} + \beta_2 X_{t2} + \varepsilon_t$$

Here, Y_t is industry production, X_{t1} is the gross national product, and X_{t2} is the consumer price index, each for year t.

With coincident independent variables, a forecast of the dependent variable requires forecasts of the independent variables. For instance, suppose in Example 1 that the regression model is fitted to data for the years 1976–1991 and a forecast of industry production for 1992 is desired. Forecasts for both the gross national product and the consumer price index for 1992 will then be needed. Forecasting the gross national product and the consumer price index to obtain a forecast of industry production may seem to be a circuitous procedure. If, however, accurate forecasts of the independent variables are available and there is a close relation between these variables and industry production, an accurate forecast of the dependent variable can result.

Lagged Independent Variables.   Sometimes, it is possible to construct a regression model in which some or all of the independent variables are lagged. Example 2 is an illustration. In this example, the following simple linear regression model might be useful.

$$(25.5) \qquad Y_t = \beta_0 + \beta_1 X_{t-5} + \varepsilon_t$$

Here, Y_t is kindergarten enrollment in year t and X_{t-5} is the number of births five years earlier. The variable X_{t-5} represents a *lagged time series* relative to Y_t, the lag being five years.

An advantage of a regression model with lagged independent variables is that the independent variables may not need to be predicted to forecast the dependent variable. For instance, suppose that the data for Example 2 span the years 1972–1991 and that a forecast is to be made in 1992 of kindergarten enrollment for 1995. In this case, the number of births in 1990 will be needed to forecast 1995 kindergarten enrollment. This number would be known in 1992, so there would be no need to forecast the independent variable.

A regression model may contain more than one lagged term of an independent variable. In Example 4, cement shipments in quarter t (Y_t) are related to the volume of construction contracts awarded in quarter $t - 1$ (X_{t-1}) and to the volume awarded in quarter $t - 2$ (X_{t-2}). Assuming that the effects are linear and additive, the following regression model might be appropriate.

$$(25.6) \qquad Y_t = \beta_0 + \beta_1 X_{t-1} + \beta_2 X_{t-2} + \varepsilon_t$$

This model is said to contain a *distributed time lag* because the lag is distributed over more than one period.

Indicator Series and Time Series Components.   Regression models for time series data often contain both indicator series and time series components as independent variables. An example is the following regression model for a quarterly time series.

$$(25.7) \qquad Y_t = \beta_0 + \beta_1 X_{t1} + \beta_2 X_{t2} + \beta_3 X_{t3} + \beta_4 X_{t4} + \beta_5 X_{t-1,5} + \varepsilon_t$$

where: $X_{t1} = t$

X_{t2}, X_{t3}, X_{t4} are indicator variables for quarterly seasonal effects

$X_{t-1,5}$ is a cyclical indicator series lagged one quarter

Regression model (25.7) involves a linear trend effect $\beta_1 X_{t1}$ and a seasonal effect $\beta_2 X_{t2} + \beta_3 X_{t3} + \beta_4 X_{t4}$. In addition, the indicator series effect $\beta_5 X_{t-1,5}$ will capture much of the cyclical effect that is present.

Roofing Contracts ☐ **EXAMPLE**

A consultant examined the time series for the dollar volume of contracts received annually by a firm specializing in commercial and industrial roofing. The consultant found that the series had a linear trend. The consultant also noted that the volume of contracts in any year reflected the state of general business activity in the firm's service region during the preceding year. The consultant therefore decided to tentatively forecast the dollar volume of contracts one year ahead by regressing the dollar volume of contracts against a linear trend component and a lagged index of regional business activity using the following regression model:

$$Y_t = \beta_0 + \beta_1 X_{t1} + \beta_2 X_{t-1,2} + \varepsilon_t$$

where: Y_t is the dollar volume of contracts in year t

$X_{t1} = t$ is the linear trend variable

$X_{t-1,2}$ is the index of regional business activity in year $t - 1$

Year	Linear Trend Variable	Index of Regional Business Activity (lagged one year)	Dollar Volume of Contracts ($ thousand)
t	X_{t1}	$X_{t-1,2}$	Y_t
1	1	79.7	173.7
2	2	83.1	187.0
3	3	90.9	232.1
4	4	99.1	280.0
5	5	100.0	287.0
6	6	97.5	282.8
7	7	91.3	273.9
8	8	86.6	254.7
9	9	94.4	300.8
10	10	105.5	355.7
11	11	108.7	387.2

TABLE 25.1
Data for regression with linear trend and lagged business indicator—Roofing contracts example

The data are presented in Table 25.1. MINITAB regression output is shown in Figure 25.1a. We see that the fitted regression function is:

$$\hat{Y}_t = -188.63 + 8.7723X_{t1} + 4.3507X_{t-1,2}$$

The consultant was particularly interested in whether the error terms could be assumed to be independent. A plot of the residuals against time (X_1) is presented in Figure 25.1b. The pattern of points in this residual plot appeared to be random, confirming the appropriateness of the assumption of independent error terms. A residual plot against the fitted values $\hat{Y}$ was also prepared by the consultant. It showed that the fit of the regression function is satisfactory and that the error variance is constant.

A forecast of the volume of contracts for next year, $t = 12$, by means of a 95 percent prediction interval for a new observation is desired. The trend variable for year $t = 12$ is $X_{12,1} = 12$. Also, the consultant ascertained that the index of regional business activity for the current year, $t = 11$, is $X_{11,2} = 105.8$. Hence, the fitted value for year $t = 12$ is:

$$\hat{Y}_{12} = -188.63 + 8.7723(12) + 4.3507(105.8) = 376.9$$

as shown in the MINITAB output in Figure 25.1a. For $n = 11$, $p = 3$, and $1 - \alpha = 0.95$, we require $t(0.975; 8) = 2.306$. The MINITAB output in Figure 25.1a shows that $MSE = 180/8 = 22.5$ and $s\{\hat{Y}_h\} = 3.07$. Hence, we obtain by (20.20) the estimated variance needed to construct the prediction interval:

$$s^2\{Y_{h(\text{new})}\} = MSE + s^2\{\hat{Y}_h\} = 22.5 + (3.07)^2 = 31.925$$

The estimated standard deviation therefore is:

$$s\{Y_{h(\text{new})}\} = \sqrt{31.925} = 5.65$$

The prediction limits (20.20) are $376.9 \pm 2.306(5.65)$, and the 95 percent prediction interval is:

$$364 \le Y_{h(\text{new})} \le 390$$

These prediction limits are shown in the MINITAB output in Figure 25.1a. Hence, with 95 percent confidence, the consultant was able to predict that the volume of roofing contracts in year 12 will be between \$364 thousand and \$390 thousand. ☐

FIGURE 25.1

MINITAB regression output and residual time sequence plot—Roofing contracts example

(a) Regression Output

```
The regression equation is
Y = - 189 + 8.77 X1 + 4.35 X2

Predictor        Coef       Stdev      t-ratio          p
Constant       -188.63      19.62       -9.61       0.000
X1               8.7723      0.6386     13.74       0.000
X2               4.3507      0.2343     18.57       0.000

s = 4.749        R-sq = 99.6%      R-sq(adj) = 99.4%

Analysis of Variance

SOURCE         DF           SS          MS          F          p
Regression      2         40067       20033      888.38     0.000
Error           8           180          23
Total          10         40247

     Fit   Stdev.Fit          95% C.I.           95% P.I.
  376.95        3.07    ( 369.86, 384.03)   ( 363.90, 389.99)
```

(b) Residual Time Sequence Plot

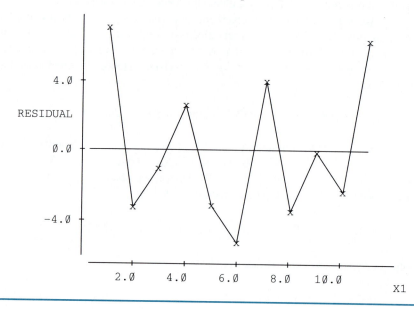

Comment

The time series regression models considered in this section are sometimes used with a transformed dependent variable, such as $Y'_t = \log Y_t$. When the dependent variable is transformed, regression models (25.1) to (25.7) will still involve additive time series effects in terms of the transformed variable, but not in terms of the original variable.

The regression models in the previous section all assume that the error terms ε_t are independent from period to period. In many business and economic applications, however, the error terms in different periods are correlated. When this is the case, the error terms are said to be *autocorrelated* or *serially correlated*.

**REGULATION 25.2
MODELS WITH
AUTOCORRELATED
ERROR TERMS**

Autocorrelated Error Terms

Autocorrelated error terms arise for a variety of reasons. One major cause is the omission of one or more key variables from the fitted regression model, such as the omission of an interest rate variable in a model predicting bank lending where interest rates vary cyclically and affect bank lending accordingly. Another important reason is that major random effects often tend to persist for several periods, as when the effects of a very poor harvest of an agricultural commodity are felt for several years thereafter.

First-Order Autoregressive Error Model. When the error terms are related over time, the first-order autoregressive error model is frequently employed.

(25.8)
The **first-order autoregressive error model** is:

$$\varepsilon_t = \rho\varepsilon_{t-1} + u_t$$

where: ρ (Greek rho) is the autocorrelation parameter, $-1 < \rho < 1$
u_t are independent $N(0, \sigma^2)$

This model assumes that the error term ε_t for period t contains two components: (1) a component resulting from the error term ε_{t-1} for the preceding period (when $\rho \neq 0$) and (2) a random disturbance term u_t that is independent of earlier time periods.

The parameter ρ is called the *autocorrelation parameter*. It is, in fact, the coefficient of correlation between ε_t and ε_{t-1} as defined in (6.14). When ρ is positive, the autocorrelation is positive; when ρ is negative, the autocorrelation is negative. In most business and economic series with time-dependent error terms, the autocorrelation is positive. For example, large sales in one month because of good cyclical conditions are likely to persist into the next month.

In the special case where $\rho = 0$, error model (25.8) simplifies to $\varepsilon_t = u_t$. The ε_t are then independent $N(0, \sigma^2)$, the case assumed for all regression models up to now.

Effects of Autocorrelation.   The following consequences arise if the usual method of least squares is employed for fitting a regression model when the error terms are autocorrelated according to (25.8).

1. The least squares regression coefficients are still unbiased estimators, but tend to be relatively inefficient.
2. The error mean square *MSE* tends to seriously underestimate the true variance of the error terms.
3. The standard procedures for confidence intervals and tests using the t and F distributions are no longer strictly applicable.

Several methods are available to avoid these difficulties caused by autocorrelated error terms. We shall discuss a method that involves a simple transformation of the variables. First, however, we present a test for the presence of autocorrelation.

Durbin–Watson Test

A widely used test for examining whether or not the error terms in a regression model are autocorrelated is the Durbin–Watson test. This test is based on the first-order autoregressive error model (25.8). The test alternatives for positive autocorrelation, the usual case for business and economic applications, are as follows.

$$(25.9) \qquad H_0: \rho \leq 0$$
$$H_1: \rho > 0$$

Here, H_0 implies that the error terms are either uncorrelated or negatively correlated, while H_1 implies that they are positively correlated.

The Durbin–Watson test statistic d is based on the differences between adjacent residuals, $e_t - e_{t-1}$, as follows.

$$(25.10) \qquad d = \frac{\sum_{t=2}^{n}(e_t - e_{t-1})^2}{\sum_{t=1}^{n}e_t^2}$$

where: e_t is the regression residual for period t
n is the number of time periods used in fitting the regression model

The Durbin–Watson test statistic d is often included in regression computer output.

When the error terms are positively autocorrelated, adjacent residuals will tend to be of similar magnitude so that the differences $e_t - e_{t-1}$ of adjacent residuals will be small, and hence the numerator of the test statistic d will be small. If the error terms are not correlated or are negatively correlated, the adjacent residuals will tend to differ more and hence the numerator of the test statistic will be larger. Thus, small values of the test statistic d are consistent with H_1, while large values are consistent with H_0.

The exact action limit for the one-sided Durbin–Watson test has a complex form. However, a lower bound d_L and an upper bound d_U for the action limit are available. When the test statistic d in (25.10) is less than the lower bound d_L, we conclude H_1, that positive autocorrelation is present. When the test statistic exceeds the upper bound d_U, we conclude H_0, that positive autocorrelation is not present. When the test statistic falls in the interval between d_L and d_U, we do not know for certain which conclusion to reach, although it is usually prudent to conclude that positive autocorrelation may be present.

The decision rule therefore has the following form.

(25.11)

When the alternatives are:

$$H_0: \rho \leq 0$$
$$H_1: \rho > 0$$

and the first-order autoregressive error model (25.8) applies, the appropriate decision rule to control the α risk at $\rho = 0$ is:

If $d > d_U$, conclude H_0.
If $d < d_L$, conclude H_1.
If $d_L \leq d \leq d_U$, the test is inconclusive.

Here d is given by (25.10).

Appendix Table C.10 contains values of the bounds d_L and d_U for various sample sizes (n), number of independent variables in the regression model ($p - 1$), and two α levels (0.05 and 0.01).

EXAMPLE ☐ **Company Sales**

A securities analyst studied the relation between sales of a company (Y_t) and industry sales (X_t). A portion of the deseasonalized data for the past 20 quarters is presented in columns 1 and 2 of Table 25.2. The data are in millions of dollars.

The analyst expected high positive autocorrelation. The first step was to fit the simple linear regression model $Y_t = \beta_0 + \beta_1 X_t + \varepsilon_t$ by means of the MYSTAT pack-

TABLE 25.2
First-differences calculations—Company sales example ⬇

Quarter t	(1) Deseasonalized Company Sales ($ million) Y_t	(2) Deseasonalized Industry Sales ($ million) X_t	(3) First Differences Y_t'	(4) First Differences X_t'	(5) $X_t' Y_t'$	(6) $(X_t')^2$
1	77.044	746.512	—	—	—	—
2	78.613	762.345	1.569	15.833	24.842	250.684
3	80.124	778.179	1.511	15.834	23.925	250.716
⋮	⋮	⋮	⋮	⋮	⋮	⋮
20	102.481	1006.882	1.745	17.592	30.698	309.478
				Total	667.250	6612.943

FIGURE 25.2   **MYSTAT regression output and residual time sequence plot—Company sales example**

(a) Regression Output

```
DEP VAR:      Y       N:   20     MULTIPLE R: .999     SQUARED MULTIPLE R:   .998
ADJUSTED SQUARED MULTIPLE R:   .998      STANDARD ERROR OF ESTIMATE:        0.3433


    VARIABLE     COEFFICIENT     STD ERROR        T      P(2 TAIL)
   CONSTANT          4.1678        0.8568     4.8646      0.0001
        X            0.0971        0.0010    98.5629      0.0000


DURBIN-WATSON D STATISTIC      .477
```

(b) Residual Time Sequence Plot

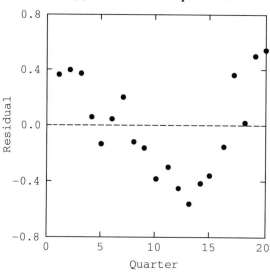

age using the ordinary least squares method. MYSTAT output is shown in Figure 25.2a. We see that the fitted regression function is:

$$\hat{Y}_t = 4.1678 + 0.0971X_t$$

The residuals were also obtained and are plotted in time order in Figure 25.2b. Note the succession of positive residuals, then negative residuals, and finally positive residuals, with generally gradual changes in the residuals from one quarter to the next. This pattern suggests high positive autocorrelation.

To test formally for positive autocorrelation, the analyst utilized the Durbin–Watson test. The alternatives are as given in (25.11). The MYSTAT output in Figure 25.2a shows the value of the test statistic, $d = 0.477$. The analyst wished to control the α risk at 0.05 when $\rho = 0$. When entering Table C.10 for $\alpha = 0.05$, $p - 1 = 1$ independent variable, and $n = 20$ cases, we find $d_L = 1.20$ and $d_U = 1.41$. Decision rule (25.11) therefore is:

If $d > 1.41$, conclude H_0 ($\rho \leq 0$).
If $d < 1.20$, conclude H_1 ($\rho > 0$).
If $1.20 \leq d \leq 1.41$, the test is inconclusive.

Since $d = 0.477 < 1.20$, we conclude H_1, that is, the error terms are positively autocorrelated.   □

Comment

The test of runs up and down discussed in Section 15.4 is not strictly applicable to regression residuals but can be used as an approximate test for independence of the error terms when the number of residuals is large.

Method of First Differences

The method of first differences is a simple method for dealing with the difficulties caused by positively autocorrelated error terms. The basic idea is to use a simple transformation of the variables that will make the error terms approximately uncorrelated so that ordinary least squares regression methods can be used to their full advantage. We shall explain the method of first differences with reference to the simple linear regression model, but the method extends readily to multiple regression models.

Development of Method.   The simple linear regression model with first-order autoregressive error terms is as follows.

$$(25.12) \quad Y_t = \beta_0 + \beta_1 X_t + \varepsilon_t$$

where: $\varepsilon_t = \rho \varepsilon_{t-1} + u_t$
u_t are independent $N(0, \sigma^2)$

The method of first differences is most helpful when the autocorrelation parameter is positive and large, which is the typical case in business and economic applications. Let us see what happens to the error term ε_t in (25.12) in the extreme case when $\rho = 1$.

$$(25.13) \quad \varepsilon_t = \varepsilon_{t-1} + u_t \quad \text{when } \rho = 1$$

The difference $Y_t - Y_{t-1}$ is referred to as a *first difference*. For the model in (25.12), we see that the first difference equals:

$$Y_t - Y_{t-1} = (\beta_0 + \beta_1 X_t + \varepsilon_t) - (\beta_0 + \beta_1 X_{t-1} + \varepsilon_{t-1})$$

Replacing ε_t by the expression in (25.13) when $\rho = 1$, we obtain for the first difference:

$$Y_t - Y_{t-1} = (\beta_0 + \beta_1 X_t + \varepsilon_{t-1} + u_t) - (\beta_0 + \beta_1 X_{t-1} + \varepsilon_{t-1})$$

which simplifies to the following expression.

$$(25.14) \quad Y_t - Y_{t-1} = \beta_1(X_t - X_{t-1}) + u_t \qquad \text{when } \rho = 1$$

Let us denote $Y_t - Y_{t-1}$ by Y_t' and $X_t - X_{t-1}$ by X_t'. We then can express (25.14) as follows.

$$(25.15) \quad Y_t' = \beta_1 X_t' + u_t \qquad \text{when } \rho = 1$$

where: u_t are independent $N(0, \sigma^2)$
$$Y_t' = Y_t - Y_{t-1}$$
$$X_t' = X_t - X_{t-1}$$

Note that model (25.15) is a simple linear regression model with *independent* error terms u_t and the intercept set to zero; that is, $\beta_0 = 0$.

The result in (25.15) suggests that when the error terms are positively correlated and the autocorrelation parameter ρ is large, the ordinary regression procedures become approximately applicable when used in the following way.

1. *First differences* $Y_t' = Y_t - Y_{t-1}$ and $X_t' = X_t - X_{t-1}$ are used for the dependent and independent variables, respectively.
2. *Regression through the origin* is employed; that is, regression with $\beta_0 = 0$.

Note that the slope parameter β_1 in the first-differences model (25.15) is the same as in the original regression model (25.12). Thus, we are still estimating the same slope parameter.

Regression Through Origin.   Procedures for regression through the origin follow those already discussed for the standard regression model. The key formulas for regression-through-the-origin model (25.15) are as follows.

(25.16)
The key formulas for simple linear regression through the origin, for the first-differences model $Y_t' = \beta_1 X_t' + u_t$, in (25.15), are:

(25.16a)   Least squares estimator of β_1: $b_1 = \dfrac{\sum X_t' Y_t'}{\sum (X_t')^2}$

(25.16b)   Error mean square: $MSE = \dfrac{\sum (Y_t' - b_1 X_t')^2}{n_d - 1}$

(25.16c)   Estimated variance of b_1: $s^2\{b_1\} = \dfrac{MSE}{\sum (X_t')^2}$

(25.16d)   Estimated variance of $\hat{Y}_h'$ at X_h': $s^2\{\hat{Y}_h'\} = \dfrac{(X_h')^2 MSE}{\sum (X_t')^2}$

(25.16e)   Estimated variance for predicting new observation $Y_{h(new)}'$ at X_h':

$$s^2\{Y_{h(new)}'\} = MSE\left[1 + \dfrac{(X_h')^2}{\sum (X_t')^2}\right]$$

where: $Y'_t = Y_t - Y_{t-1}$

$\quad\quad\quad X'_t = X_t - X_{t-1}$

$\quad\quad\quad n_a$ is the number of first differences

Observe in the formulas in (25.16) that n_a denotes the number of first differences for the time series and that the summations are over the n_a first differences. Confidence intervals and tests are set up in the usual way by means of the t distribution with $n_a - 1$ degrees of freedom, the number of degrees of freedom associated with MSE in regression through the origin.

EXAMPLE □

In the company sales example of Table 25.2, the analyst decided to use the method of first differences to overcome the difficulties associated with the presence of high positive autocorrelation in the error terms. The first differences are shown (in part) in columns 3 and 4 of Table 25.2. For example, $Y'_2 = Y_2 - Y_1 = 78.613 - 77.044 = 1.569$.

Note that the first-differences transformation yields one fewer first difference than the number of original observations (that is, $n_a = 19$ while $n = 20$). The necessary calculations for estimating the slope β_1 are shown in columns 5 and 6 of Table 25.2. Using (25.16a), we obtain the estimated slope:

$$b_1 = \frac{\sum X'_t Y'_t}{\sum (X'_t)^2} = \frac{667.250}{6612.943} = 0.10090$$

and the fitted first-differences model is:

$$\hat{Y}'_t = 0.10090 X'_t$$

Figure 25.3a shows MYSTAT regression output for the first-differences model (Y_DIFF and X_DIFF refer to Y'_t and X'_t, respectively). Note the absence of a line for CONSTANT, since a model with no intercept was fitted.

The 19 residuals for this fitted model were also obtained and are plotted in time order in Figure 25.3b. The pattern now appears reasonably random and consistent with time-independent error terms. Hence, the analyst was satisfied by the first-differences model in this regard.

The analyst now desired to forecast company sales for next quarter, $t = 21$, and obtained a reliable forecast of deseasonalized industry sales for next quarter, $X_{21} = 1015.9$. Because deseasonalized industry sales in quarter 20 were $X_{20} = 1006.9$, the projected first difference is $X'_{21} = 1015.9 - 1006.9 = 9.0$. Hence, a point estimate of the expected change in company sales, $E\{Y'_{21}\}$, is:

$$\hat{Y}'_{21} = 0.10090(9.0) = 0.9$$

This estimate indicates that deseasonalized company sales are expected to increase by $0.9 million between quarters 20 and 21.

FIGURE 25.3   **MYSTAT first-differences regression output and residual time sequence plot—Company sales example**

(a) Regression Output

```
MODEL CONTAINS NO CONSTANT


DEP VAR:  Y_DIFF      N:   19    MULTIPLE R:  .993    SQUARED MULTIPLE R:  .987
ADJUSTED SQUARED MULTIPLE R:  .987    STANDARD ERROR OF ESTIMATE:      0.22565


   VARIABLE      COEFFICIENT      STD ERROR     T     P(2 TAIL)
   X_DIFF            0.10090        0.00277   .36E+02   0.00000
```

(b) Residual Time Sequence Plot

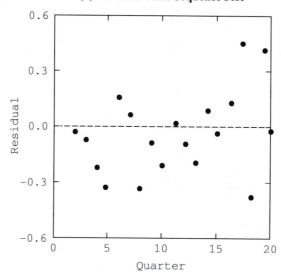

The final step in developing the forecast is to take the deseasonalized company sales in quarter 20, $Y_{20} = 102.5$, and adjust them by the expected increase:

$$\text{Forecast for quarter } 21 = 102.5 + 0.9 = 103.4$$

The analyst thus estimates that expected deseasonalized company sales in quarter 21 will be \$103.4 million. □

Comments

1. The residuals of a regression model through the origin are not constrained to sum to zero when fitted by the method of least squares, in contrast to regression models with an intercept term. This fact explains why the residual plot in Figure 25.3b is not balanced around the zero line of the graph.
2. The Durbin–Watson test in (25.11) is not appropriate for regression through the origin.

Many forecasting applications require forecasts for a large number of time series on a frequent, periodic basis.

**EXPONENTIAL 25.3
SMOOTHING FOR
STATIONARY TIME
SERIES**

EXAMPLES ☐

1. A manufacturer stocks thousands of different types of spare parts, raw materials, components, and finished products. Monthly demand forecasts for these items are required for procurement and production decisions. The forecasts are the key to effective inventory cost control. Inventory carrying costs and out-of-stock costs can be kept at low levels only if reasonably reliable forecasts of requirements are available.

2. An airline requires forecasts of weekly passenger traffic on each of the airline's several hundred regularly scheduled flights. The forecasts are used in operations management, budget planning, and financial control.

3. An investment firm that manages many investment portfolios requires the daily monitoring of rates of return for a large number of securities. ☐

The methods of time series analysis and forecasting discussed thus far require much attention by skilled professionals and large amounts of data handling and storage. Therefore, these methods would be very expensive to apply to hundreds or thousands of time series on a frequent, periodic basis. Exponential smoothing is a forecasting method that is well suited to these situations. It is readily computerized, requires a minimum of professional attention, makes modest demands on data handling and storage, and adapts effectively to structural changes in time series.

The system of exponential smoothing models to be presented here is known as the *Holt–Winters exponential smoothing system*. This system has been successfully applied in many diverse settings. We shall take up here the additive models of the exponential smoothing system. The multiplicative models involve a similar approach.

We present in this section the exponential smoothing model for time series that contain neither trend nor seasonal components. We then extend this basic model to time series with a trend component and to time series with both trend and seasonal components.

Stationary Time Series

A stationary time series is one that has an irregular component but no trend, cyclical, or seasonal components. The irregular component is assumed to be random in nature, with its probability distribution unchanging over time. We describe a stationary time series formally in the following way.

(25.17)
A **stationary time series** $Y_1, Y_2, \ldots, Y_t, \ldots, Y_n$ is one in which the consecutive observations Y_t are outcomes of random variables that have the same probability distribution.

The following are examples of stationary time series.

☐ EXAMPLES

1. The number of traffic lights failing daily in a city.
2. The monthly demand per capita for a food staple, such as rice.
3. The closing monthly balance of a family's checking account.   ☐

Smoothing Procedure

To forecast a stationary time series with the exponential smoothing model, we first smooth the time series with a moving average to isolate the systematic or smooth component of the series. We then project this smooth component into the future. The moving average employed by the exponential smoothing model is a special type of weighted moving average.

The smooth component of a stationary time series may be considered as a succession of estimates of the mean level of the underlying probability distribution of Y. We shall denote the estimate of the mean level in period t by A_t, and we shall refer to A_t as the *smoothed estimate* for period t. The exponential smoothing model calculates the current smoothed estimate A_t as a weighted average of the current observation Y_t, with weight a, and the smoothed estimate of the preceding period A_{t-1}, with weight $1 - a$. The weight a is called the *smoothing constant* and generally is a value between 0 and 1.

> **(25.18)**
>
> The **exponential smoothing procedure for stationary time series** is as follows.
>
> Update the smoothed estimate: $A_t = aY_t + (1 - a)A_{t-1}$ $t = 1, 2, \ldots$
>
> where: A_t is the smoothed estimate for the current period t
>
> A_{t-1} is the smoothed estimate for the preceding period $t - 1$
>
> Y_t is the observation for the current period t
>
> a is the smoothing constant, $0 < a < 1$

To apply formula (25.18), we require a value for the smoothing constant a and a *starting value* A_0 for the mean level of the series. We shall discuss how the smoothing constant a and the starting value A_0 are chosen in practice after the following example. In the example, the smoothing constant and starting value are prespecified.

Rice Orders   ☐ **EXAMPLE**

Every week a food store chain restocks inventories of staple items in its stores based on orders received by its central warehouse from each of the stores. Column 1 of Table 25.3a shows the time series Y_t of total weekly orders (in cases) received by the warehouse for a package of long-grain rice. The analyst has concluded that this time series is stationary, so formula (25.18) may be used to smooth and forecast the series. The weekly smoothed estimates A_t of the mean level of the series are shown in column 2. These smoothed estimates were computed using (25.18) with starting value $A_0 = 4000$

						TABLE 25.3

(a) Smoothing Procedure

	(1) Total Orders (cases)	(2) Smoothed Estimate		(1) Total Orders (cases)	(2) Smoothed Estimate
Week t	Y_t	A_t	Week t	Y_t	A_t
1	3169	3917	10	2502	3960
2	3682	3893	11	5006	4064
3	2655	3770	12	6885	4346
4	4500	3843	13	4196	4331
5	3682	3827	14	2728	4171
6	3568	3801	15	5262	4280
7	5045	3925	16	3719	4224
8	4733	4006	17	5707	4372
9	5164	4122	18	4580	4393

TABLE 25.3

Exponential smoothing without trend or seasonal components—Rice orders example

(b) Forecasting Procedure

Week $t + k$	Forecast F_{t+k}
19	4393
20	4393
21	4393

and smoothing constant $a = 0.10$. For instance, the first two smoothed estimates were obtained as follows:

$$A_1 = aY_1 + (1 - a)A_0 = 0.10(3169) + 0.90(4000) = 3917$$
$$A_2 = aY_2 + (1 - a)A_1 = 0.10(3682) + 0.90(3917) = 3893$$

Figure 25.4 presents time series plots of the original and smoothed series. Note how the smoothed series remains more or less centered in the original series and "tracks" the original series as it unfolds week by week. ☐

Forecasting Procedure

The exponential smoothing model for a stationary time series uses the current smoothed estimate of the mean level as the forecast, because the smoothing procedure largely removes the irregular component from the stationary series. If the time series is anticipated to remain stationary with the same mean level in future periods $t + 1$, $t + 2$, and so on, then the current estimate A_t is an appropriate forecast of the series for these future periods.

(25.19)

The **forecasting procedure for stationary time series** is as follows.

Forecast for period $t + k$: $F_{t+k} = A_t$ $k = 1, 2, \ldots$

where: F_{t+k} is the forecast for k periods ahead
  A_t is the smoothed estimate for current period t

FIGURE 25.4
**Plots of original and expo-
nentially smoothed series—
Rice orders example**

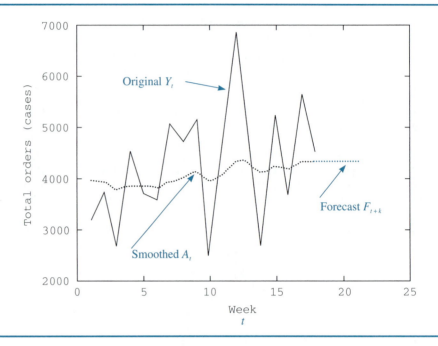

Note that the exponential smoothing model provides a single number for a forecast, and not an interval. A point forecast is often required, such as in the rice orders example.

□ EXAMPLE

For the rice orders example in Table 25.3a, forecasts are desired at the end of week 18 for weeks 19, 20, and 21. The current smoothed estimate is $A_{18} = 4393$. Hence, this value is the forecast of total orders for each of the following three weeks. In notation, we write $F_{19} = F_{20} = F_{21} = A_{18} = 4393$. These forecasts are shown in Table 25.3b and in Figure 25.4. Note from Figure 25.4 that the exponential smoothing forecast for a stationary time series is simply a horizontal straight-line projection from the current smoothed estimate level $A_{18} = 4393$.   □

Properties of Exponential Smoothing Model

The exponential smoothing procedure in (25.18) has a number of important properties.

Weighting of Observations.   Exponential smoothing is a special kind of *weighted moving average* procedure. Consider the smoothed estimate for time period t given in (25.18):

$$A_t = aY_t + (1 - a)A_{t-1}$$

Now, in turn, smoothed estimate A_{t-1} is given by:

$$A_{t-1} = aY_{t-1} + (1 - a)A_{t-2}$$

Substituting this expression for A_{t-1} in the preceding equation gives:

$$A_t = aY_t + (1 - a)[aY_{t-1} + (1 - a)A_{t-2}]$$
$$= aY_t + a(1 - a)Y_{t-1} + (1 - a)^2A_{t-2}$$

Repeating this process by substituting for A_{t-2}, then for A_{t-3}, and so on, we find that the current smoothed estimate A_t is the following weighted average of the observations for all earlier periods and the starting value A_0.

(25.20) $A_t = aY_t + a(1 - a)Y_{t-1} + a(1 - a)^2Y_{t-2} + \cdots + a(1 - a)^{t-1}Y_1$
$\qquad\qquad + (1 - a)^tA_0$

Note that the weight for observation Y_{t-r} is $a(1 - a)^r$. Since the smoothing constant a is a value between 0 and 1, it follows that the earlier is the observation Y_{t-r} (i.e., the larger is r), the smaller is its weight $a(1 - a)^r$. Indeed, the term *exponential smoothing* refers to the fact that the weight $a(1 - a)^r$ declines geometrically as r increases, that is, the further the period is in the past. For example, if $a = 0.10$, then the current observation Y_t has weight $a = 0.10$, while the observation that occurred 8 periods earlier, Y_{t-8}, has weight $0.10(1 - 0.10)^8 = 0.043$.

Thus, observations far in the past have little weight in determining the current smoothed estimate. In this sense, the smoothed estimate is a moving average because updating by means of (25.18) takes into account the most recent observation with weight a and reduces the weights of all earlier observations by a factor of $1 - a$. For instance, if $a = 0.10$, the weights of all earlier observations are reduced by 10 percent each time a new observation is obtained.

Adaptive Capability.   The smoothing procedure in (25.18), while strictly appropriate only for stationary time series, is capable of adapting to occasional step changes or to cyclical changes in the mean level of the series. This adaptive capability is a consequence of the moving average property of the exponential smoothing process. For instance, if the series shifts to a new mean level, the smoothed estimate will approach the new level as more observations from the new level are obtained and less weight is given to observations from the old level.

Data Storage Requirements.   An advantage of exponential smoothing over alternative computer-based forecasting systems is its economical use of data storage. With exponential smoothing procedure (25.18), the only information needed for calculating the current smoothed estimate A_t is the current observation Y_t and the preceding smoothed estimate A_{t-1}. Thus, A_{t-1} is the only number that must be stored and carried over to compute A_t in the next period. Although the cost of computer data storage is rapidly declining, the small storage requirement of the exponential smoothing procedure is still an attractive feature for systems with large numbers of time series.

Selection of Starting Value and Smoothing Constant

For a stationary time series, the starting value A_0 may be set equal to the mean of a few observations from the immediate past. If no past observations are available, an informed judgment can be used for the starting value A_0.

The choice of the smoothing constant a is largely governed by its effect on the responsiveness of the exponential forecasting system. The larger is the smoothing constant, the smaller is the weight given to past observations relative to more recent ones. For example, we noted earlier that when $a = 0.10$, Y_{t-8} is given weight $0.10(1 - 0.10)^8 = 0.043$ in the calculation of A_t. In contrast, when the smoothing constant is $a = 0.50$, then Y_{t-8} is only given weight $0.50(1 - 0.50)^8 = 0.0020$. Thus, if a is too large, the smoothed estimates can be influenced unduly by the irregular components of current and recent observations. On the other hand, if a is too small, the smoothed estimates lose their adaptability and respond too slowly to any changes in the underlying mean level of the series. Hence, the smoothing constant a must be chosen to strike an appropriate balance between these two extremes. Frequently, preliminary calculations are made with different values for the smoothing constant to see which value yields the smallest forecasting errors in aggregate for the time series under consideration. Values of a between 0.10 and 0.30 are commonly used in practice. It is always possible to change the value of a when the variability of the irregular component of the series changes or the frequency or magnitude of changes in the mean level departs from past experience.

Comments

1. Exponential smoothing calculations are usually done by computer. Some packages have an option that automatically selects the starting value A_0 and the smoothing constant a based on an input of some initial observations for the series and a specified criterion.
2. When an exponential smoothing system involves numerous time series, the series may be placed into relatively homogeneous groups, with all series in a group receiving the same smoothing constant.

25.4 EXPONENTIAL SMOOTHING FOR TIME SERIES WITH TREND COMPONENT

We have seen that exponential smoothing for a stationary time series involves one updating step for each new observation. When a time series contains a trend component, exponential smoothing involves two updating steps, one for the smoothed estimate and one for a trend estimate. The smoothed estimate A_t for period t still describes the mean level of the series at period t, including now the trend component for that time period. A *trend estimate* for period t, which we shall denote by B_t, is required for forecasting and subsequent smoothing. The trend estimate B_t may be considered as the estimated slope coefficient of a linear trend at period t.

Updating Procedure

To obtain the smoothed estimate A_t for a time series with a trend component, we again compute a weighted average of two separate estimates of the mean level of the series at period t. The first estimate is Y_t, the current observation for period t. The second estimate is $A_{t-1} + B_{t-1}$. This latter estimate of the current mean level in period t is the sum of (1) A_{t-1}, which reflects the mean level in the preceding period, and (2) B_{t-1}, which reflects the trend change in the mean level between periods $t - 1$ and t as anticipated in the preceding period.

The weights of the two estimates are a and $1 - a$, where, as before, a is the smoothing constant. The resulting formula for updating the smoothed estimate is as follows:

$$A_t = aY_t + (1 - a)(A_{t-1} + B_{t-1}) \qquad 0 < a < 1$$

The updating formula for the trend estimate B_t is also based on a weighted average of two estimates. The first estimate is $A_t - A_{t-1}$, which measures the change in the smoothed series between period $t - 1$ and period t and thus reflects the trend component of the most recent observation Y_t. The second estimate is B_{t-1}, the estimate of the trend change as of the preceding period. The weights of the two estimates are b and $1 - b$, where the weight b is called the *trend adjustment constant*. This constant, as with the smoothing constant, is generally a value between 0 and 1. The resulting formula for updating the trend estimate is as follows:

$$B_t = b(A_t - A_{t-1}) + (1 - b)B_{t-1} \qquad 0 < b < 1$$

In summary, exponential smoothing for a time series with a trend component involves a two-step updating procedure.

(25.21)
The **exponential smoothing procedure for time series with trend component** is as follows.

(25.21a) *Step 1.* Update the smoothed estimate: $A_t = aY_t + (1 - a)(A_{t-1} + B_{t-1})$
(25.21b) *Step 2.* Update the trend estimate: $B_t = b(A_t - A_{t-1}) + (1 - b)B_{t-1}$

where: $A_t, A_{t-1}, Y_t,$ and a are defined in (25.18)
   B_t is the trend estimate for the current period t
   B_{t-1} is the trend estimate for the preceding period $t - 1$
   b is the trend adjustment constant, $0 < b < 1$

Starting values A_0 and B_0 are required to apply the updating formulas in (25.21). These values correspond, respectively, to the mean level and the slope coefficient of the trend component for period $t = 0$. We shall discuss how these starting values and the weighting constants a and b are chosen in practice after the following example. In the example, the weighting constants and starting values are prespecified.

EXAMPLE

The annual time series for the lake level example is presented in Table 15.4. It represents the maximum level of a lake (in meters) each year during a 30-year period. The test of runs up and down in Chapter 15 established that the series has a trend. We now wish to forecast the maximum lake level for the next five years.

   A portion of the observations in Table 15.4 is reproduced in column 1 of Table 25.4. The corresponding smoothed estimates and trend estimates, calculated by using the two-step updating procedure in (25.21), appear in columns 2 and 3 of the table.

TABLE 25.4
Exponential smoothing with trend component—Lake level example

Year t	(1) Lake Level Y_t	(2) Smoothed Estimate A_t	(3) Trend Estimate B_t
1	6.63	6.6989	−0.0707
2	6.59	6.6244	−0.0715
3	6.46	6.5436	−0.0734
4	6.49	6.4722	−0.0730
⋮	⋮	⋮	⋮
27	4.85	4.9154	−0.0749
28	4.79	4.8354	−0.0759
29	4.73	4.7566	−0.0765
30	4.76	4.6881	−0.0749

Starting values of $A_0 = 6.7758$ and $B_0 = -0.0692$ were used to begin the procedure. The smoothing and trend adjustment constants were set at $a = 0.10$ and $b = 0.20$. The following calculations illustrate how the smoothed and trend estimates are obtained for the first two years:

Year 1

$$A_1 = aY_1 + (1 - a)(A_0 + B_0)$$
$$= 0.10(6.63) + 0.90(6.7758 - 0.0692) = 6.6989$$

$$B_1 = b(A_1 - A_0) + (1 - b)B_0$$
$$= 0.20(6.6989 - 6.7758) + 0.80(-0.0692) = -0.0707$$

Year 2

$$A_2 = aY_2 + (1 - a)(A_1 + B_1)$$
$$= 0.10(6.59) + 0.90(6.6989 - 0.0707) = 6.6244$$

$$B_2 = b(A_2 - A_1) + (1 - b)B_1$$
$$= 0.20(6.6244 - 6.6989) + 0.80(-0.0707) = -0.0715$$

Figure 25.5 contains a time series plot of the original and the smoothed series. Observe that the actual lake levels Y_t are fluctuating about the smoothed estimates A_t and that the smoothed series is tracking the original series quite well. ☐

Forecasting Procedure

The exponential smoothing model for a time series with trend uses the current smoothed estimate plus an allowance for the expected trend change as the forecast for a future period. Specifically, for a forecast k periods ahead to period $t + k$, the forecast will be based on the smoothed estimate A_t of the current mean level plus the amount kB_t, representing the current anticipated trend change in the series for k periods.

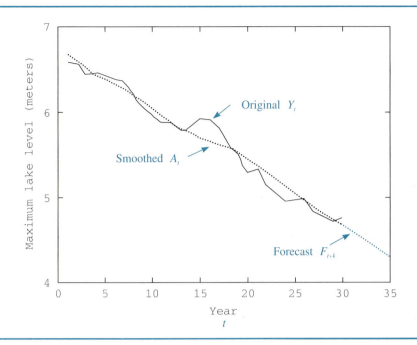

FIGURE 25.5
Plots of original and expo-
nentially smoothed series—
Lake level example

(25.22)
The **forecasting procedure for time series with trend component** is as follows.

$$\text{Forecast for period } t + k: \qquad F_{t+k} = A_t + kB_t$$

where: F_{t+k} is the forecast k periods ahead for future period $t + k$
A_t is the smoothed estimate for the current period t
B_t is the trend estimate for the current period t

EXAMPLE □

The forecasts for the lake level example for the next five-year period are shown in the MYSTAT output in Figure 25.6. These forecasts of the maximum lake level in years 31 through 35 were made at the end of year 30, using the forecasting procedure (25.22). Note that the MYSTAT output repeats the smoothing constant $a = 0.10$ and the trend adjustment constant $b = 0.20$, as well as the starting values $A_0 = 6.7758$ and $B_0 = -0.0692$. Also presented in the output are the smoothed estimate $A_{30} = 4.6881$ and the trend estimate $B_{30} = -0.0749$ for the most recent year ($t = 30$), as shown in Table 25.4. We illustrate the calculation of the forecasts for years 31 and 35:

$$F_{31} = 4.6881 + 1(-0.0749) = 4.613 \text{ meters}$$
$$F_{35} = 4.6881 + 5(-0.0749) = 4.314 \text{ meters}$$

The forecast projections for years 31 through 35 are also shown in Figure 25.5. Note that the forecasts are simply linear trend projections with slope $B_{30} = -0.0749$, starting from the most recent smoothed estimate level $A_{30} = 4.6881$. □

FIGURE 25.6
MYSTAT output for exponential smoothing with trend component—Lake level example

```
Smooth location parameter with coefficient= .100
Linear trend with smoothing coefficient= .200

Initial values

Initial smoothed value   =      6.7758
Initial trend parameter  =     -0.0692

Final values

Final smoothed value   =        4.6881
Final trend parameter  =       -0.0749

        OBS          Forecast

         31           4.6132
         32           4.5384
         33           4.4635
         34           4.3886
         35           4.3137
```

Selection of Starting Values and Weighting Constants

Starting values for the exponential smoothing procedure with trend can be set with the assistance of the time series methods discussed in Chapter 24 and earlier in this chapter. For example, a trend line can be fitted to past observations and the slope of this line used as the starting value B_0 for the trend estimate. The level of this trend line at period $t = 0$ or the estimated trend-cyclical component at period $t = 0$ can be taken as the starting value A_0 for the smoothed estimate.

The values of the weighting constants a and b should be chosen with care and need not be the same. In practical applications, these constants tend to range between 0.10 and 0.30. If the constants are too small, the estimates will react sluggishly to underlying changes. On the other hand, if the constants are too large, the estimates will respond excessively to random changes in the time series.

25.5 EXPONENTIAL SMOOTHING FOR TIME SERIES WITH TREND AND SEASONAL COMPONENTS

For time series with both trend and seasonal components, exponential smoothing requires three updates for each new observation. As before, updates of the smoothed estimate A_t and the trend estimate B_t must be obtained for the current period t. In addition, an update is required for the *seasonal component estimate* of the series. We shall denote this estimate by C_t. We also need a symbol for the number of periods in the seasonal cycle and shall use p for this purpose. Thus, for a quarterly series, $p = 4$; for a monthly series, $p = 12$.

Updating Procedure

The updating of the smoothed estimate A_t involves a weighted average of two estimates of the current mean level:

$$A_t = a(Y_t - C_{t-p}) + (1 - a)(A_{t-1} + B_{t-1}) \qquad 0 < a < 1$$

The first estimate, $Y_t - C_{t-p}$, is simply the seasonally adjusted value of the current observation Y_t. The term C_{t-p} is the most recent estimate of the seasonal component made in period $t - p$, one seasonal cycle before the current period t. To illustrate this last point, consider a quarterly series where the current quarter, $t = 11$, is the third quarter. Then, $C_{t-p} = C_{11-4} = C_7$ represents the seasonal component estimate for period 7, the third quarter of the preceding year. The seasonal component is subtracted from the observed value here because we are considering the additive exponential smoothing model. Because this first estimate is seasonally adjusted, it follows that the smoothed series A_t is deseasonalized.

The second estimate of the mean level, $A_{t-1} + B_{t-1}$, and the weighting of the two estimates are the same as in updating formula (25.21a) for time series with trend only.

The updating formula for the trend estimate B_t is identical to the one in (25.21b):

$$B_t = b(A_t - A_{t-1}) + (1 - b)B_{t-1} \qquad 0 < b < 1$$

The updating of the seasonal component estimate C_t is made by averaging two estimates of the seasonal component:

$$C_t = c(Y_t - A_t) + (1 - c)C_{t-p} \qquad 0 < c < 1$$

Here, c is a weighting constant that we shall call the *seasonal adjustment constant*. The first seasonal component estimate, $Y_t - A_t$, is based on the current time period. It is a seasonal component estimate because A_t is an estimate of the deseasonalized mean level of the series for period t. Hence, the difference $Y_t - A_t$ is the specific seasonal deviation for period t. The second seasonal component estimate, C_{t-p}, is the seasonal component estimate from p periods earlier—that is, from one seasonal cycle earlier.

In summary, exponential smoothing for a time series with trend and seasonal components involves a three-step updating procedure.

(25.23)
The **exponential smoothing procedure for time series with trend and seasonal components** is as follows.

(25.23a) *Step 1.* Update the smoothed estimate:
$$A_t = a(Y_t - C_{t-p}) + (1 - a)(A_{t-1} + B_{t-1})$$
(25.23b) *Step 2.* Update the trend estimate:
$$B_t = b(A_t - A_{t-1}) + (1 - b)B_{t-1}$$
(25.23c) *Step 3.* Update the seasonal component estimate:
$$C_t = c(Y_t - A_t) + (1 - c)C_{t-p}$$

where: $A_t, A_{t-1}, Y_t,$ and a are defined in (25.18)

$B_t, B_{t-1},$ and b are defined in (25.21)

C_t is the seasonal component estimate for current period t

C_{t-p} is the seasonal component estimate for period $t - p$, one seasonal cycle earlier

c is the seasonal adjustment constant, $0 < c < 1$

The updated trend and seasonal component estimates are used for forecasting and for subsequent smoothing. Also, note that only one seasonal component estimate is updated in each period t—namely, the one that corresponds to the same seasonal phase as period t. For instance, if the current period t is the third quarter in a quarterly series, then only the seasonal component for the third quarter is updated in that period.

To begin the updating procedure in (25.23), we again need starting values A_0 and B_0. We also need starting values for the seasonal components, denoted by $C_{-p+1}, C_{-p+2}, \ldots,$ C_{-1}, and C_0. We discuss how these starting values and the weighting constants a, b, and c are set in practice after the following example. In the example, the starting values and weighting constants are prespecified.

□ **EXAMPLE**

The quarterly time series of revenue passenger miles for the air taxi example is presented in Table 24.6. A portion of this series is repeated in Table 25.5a, column 1. We wish to forecast the revenue passenger miles for 1992 and 1993. We have applied the exponential smoothing procedure (25.23) to the time series, utilizing a computer package. The weighting constants were $a = 0.10$, $b = 0.20$, and $c = 0.40$. The starting values for the smoothed estimate and the trend estimate were $A_0 = 87.00$ and $B_0 = 1.000$. The starting values for the seasonal component estimates were:

Quarter:	1	2	3	4
Starting Value:	$C_{-3} = -50.000$	$C_{-2} = 5.000$	$C_{-1} = 70.000$	$C_0 = -25.000$

A portion of the updated smoothed, trend, and seasonal component estimates is shown in Table 25.5a, columns 2, 3, and 4.

We illustrate the calculations by updating A_t, B_t, and C_t for periods 1, 2, and 5:

Period 1

$A_1 = 0.10[43.20 - (-50.000)] + 0.90(87.00 + 1.000) = 88.52$

$B_1 = 0.20(88.52 - 87.00) + 0.80(1.000) = 1.104$

$C_1 = 0.40(43.20 - 88.52) + 0.60(-50.000) = -48.128$

Period 2

$A_2 = 0.10(90.00 - 5.000) + 0.90(88.52 + 1.104) = 89.16$

$B_2 = 0.20(89.16 - 88.52) + 0.80(1.104) = 1.012$

$C_2 = 0.40(90.00 - 89.16) + 0.60(5.000) = 3.335$

Period 5

$A_5 = 0.10[42.35 - (-48.128)] + 0.90(91.24 + 1.016) = 92.08$

$B_5 = 0.20(92.08 - 91.24) + 0.80(1.016) = 0.980$

$C_5 = 0.40(42.35 - 92.08) + 0.60(-48.128) = -48.769$ □

Forecasting Procedure

The exponential smoothing procedure for time series with trend and seasonal components uses the current smoothed estimate, adjusted for the expected trend change and seasonal

TABLE 25.5
Exponential smoothing with trend and seasonal components—Air taxi example

(a) Smoothing Procedure

Year and Quarter	Period t	(1) Miles (thousands) Y_t	(2) Smoothed Estimate A_t	(3) Trend Estimate B_t	(4) Seasonal Component Estimate C_t
1986 1	1	43.20	88.52	1.104	−48.128
2	2	90.00	89.16	1.012	3.335
3	3	162.00	90.36	1.048	70.658
4	4	64.80	91.24	1.016	−25.577
1987 1	5	42.35	92.08	0.980	−48.769
⋮	⋮	⋮	⋮	⋮	⋮
1991 1	21	47.64	95.59	0.165	−51.524
2	22	99.25	96.23	0.261	0.422
3	23	174.68	97.70	0.502	70.481
4	24	75.43	98.76	0.613	−26.337

(b) Forecasting Procedure

Year and Quarter	Period $t + k$	Forecast F_{t+k}
1992 1	25	$98.76 + 1(0.613) - 51.524 = 47.8$
2	26	$98.76 + 2(0.613) + 0.422 = 100.4$
3	27	$98.76 + 3(0.613) + 70.481 = 171.1$
4	28	$98.76 + 4(0.613) - 26.337 = 74.9$
1993 1	29	$98.76 + 5(0.613) - 51.524 = 50.3$
2	30	$98.76 + 6(0.613) + 0.422 = 102.9$
3	31	$98.76 + 7(0.613) + 70.481 = 173.5$
4	32	$98.76 + 8(0.613) - 26.337 = 77.3$

effects, as the forecast for a future period. First we project the deseasonalized series k periods ahead by adjusting for the expected trend. This yields the projection $A_t + kB_t$, just as in (25.22) for forecasting a series with only a trend component. We then add the appropriate seasonal component estimate C to the deseasonalized projection. The value of C depends on the period in the seasonal cycle that corresponds to the forecast period $t + k$.

(25.24)
The **forecasting procedure for time series with trend and seasonal components** is as follows.

Forecast for period $t + k$: $\qquad F_{t+k} = A_t + kB_t + C$

where: F_{t+k} is the forecast k periods ahead for future period $t + k$

A_t is the smoothed estimate for the current period t

B_t is the trend estimate for the current period t

C is the current seasonal component estimate corresponding to the seasonal phase of period $t + k$

☐ **EXAMPLE**

For the air taxi example, quarterly forecasts were needed at the end of 1991 for 1992 and 1993. The calculations are shown in Table 25.5b. Note that the smoothed estimate and the trend estimate used in the forecasts are those for the most recent quarter, namely, $A_{24} = 98.76$ and $B_{24} = 0.613$. Also note that the seasonal component estimates used in the forecasts are those for the most recent four quarters. For example, the forecasts for the first quarters of 1992 and 1993 use the first-quarter seasonal component estimate for 1991, $C_{21} = -51.524$.   ☐

Selection of Starting Values and Weighting Constants

The starting values A_0 and B_0 and the weighting constants a and b in updating procedure (25.23) are set on the basis of the same considerations as for a time series with only a trend component.

To obtain the starting values for the seasonal component estimates, we can use the seasonal deviations calculated from past observations by the method described in Section 24.7. The seasonal adjustment constant c may need to be relatively large compared with the weighting constants a and b because each individual seasonal component estimate is only updated once in each seasonal cycle. Thus, the seasonal component estimates have fewer opportunities to adjust to changes in the underlying seasonal pattern. Larger values of c compensate for this reduced opportunity.

> **Comment**
>
> Seasonal deviations produced by the method described in Section 24.7 are adjusted to give them an average value of 0. There is no mechanism in the exponential smoothing system that forces the seasonal component estimates to have an average value of 0. As a general rule, however, their average values will be fairly close to 0, as illustrated by the component values in column 4 of Table 25.5a.

PROBLEMS

* **25.1** Refer to **Electricity Sales** Problem 24.26. When regression model (25.3) is fitted to the quarterly sales revenues series (with $X_{t1} = 1$ in the first quarter of 1986), the estimated regression coefficients are: $b_0 = 172.34$, $b_1 = 0.10536$, $b_2 = 57.06$, $b_3 = 131.29$, $b_4 = 43.85$.

  a. Obtain the fitted value and the residual for the third quarter of 1990.
  b. Estimate the difference between expected revenues in any third quarter and expected revenues in the preceding first quarter that is attributable to (1) differential seasonal effects, (2) linear trend.
  c. Estimate the difference between expected revenues in any third quarter and expected revenues in the following fourth quarter that is attributable to (1) differential seasonal effects, (2) linear trend.
  d. Obtain a point estimate of expected revenues in the second quarter of 1992.
  e. Obtain a point estimate of the change in expected revenues between the second quarter of one year and the second quarter of the following year. Is the expected year-to-year change the same for the other quarters? Explain.

* **25.2** Refer to **Electricity Sales** Problems 24.26 and 25.1. Additional regression results are: $s\{b_0\} = 4.322$, $s\{b_1\} = 0.2464$, $s\{b_2\} = 4.767$, $s\{b_3\} = 4.786$, $s\{b_4\} = 4.818$, $SSR = 53{,}761.2$, $SSE = 1291.7$.

a. Construct a 90 percent prediction interval for revenues in the second quarter of 1992. Computer output shows that $s\{Y_{h(new)}\} = 9.551$ for this case. Interpret the prediction interval.

b. Test whether $\beta_4 = 0$, controlling the α risk at 0.05 when $\beta_4 = 0$. State the alternatives, the decision rule, the value of the test statistic, and the conclusion. What is the implication of the conclusion?

25.3 Refer to **Bergen Corporation** Problem 24.40. When regression model (25.3) is fitted to the quarterly sales series (with $X_{t1} = 1$ in the first quarter of 1986), the estimated regression coefficients are: $b_0 = 35.424$, $b_1 = 0.35536$, $b_2 = -1.189$, $b_3 = 3.123$, $b_4 = 13.767$.

a. Obtain the fitted value and the residual for the fourth quarter of 1991.

b. Estimate the difference between expected sales in any second quarter and expected sales in the preceding first quarter that is attributable to (1) differential seasonal effects, (2) linear trend.

c. Estimate the difference between expected sales in any second quarter and expected sales in the following third quarter that is attributable to (1) differential seasonal effects, (2) linear trend.

d. Obtain a point estimate of (1) expected sales in the first quarter of 1992, (2) total sales in 1992.

e. Give the trend function for sales in the third quarter.

25.4 Refer to **Bergen Corporation** Problems 24.40 and 25.3. Additional regression results are: $s\{b_0\} = 1.1475$, $s\{b_1\} = 0.065418$, $s\{b_2\} = 1.2657$, $s\{b_3\} = 1.2707$, $s\{b_4\} = 1.2791$, $SSR = 1076.890$, $SSE = 91.068$.

a. Construct a 99 percent prediction interval for sales in the first quarter of 1992. Computer output shows that $s\{Y_{h(new)}\} = 2.5359$ for this case.

b. Test whether $\beta_2 = 0$, controlling the α risk at 0.05 when $\beta_2 = 0$. State the alternatives, the decision rule, the value of the test statistic, and the conclusion. What is the implication of the conclusion?

c. Construct a 99 percent confidence interval for β_1. Interpret the confidence interval.

25.5 **Dishwashers.** The following version of regression model (25.1) was employed by a firm to forecast its annual sales of dishwashers:

$$Y_t = \beta_0 + \beta_1 X_{t1} + \beta_2 X_{t-1,2} + \varepsilon_t$$

Here Y_t denotes sales in year t (in thousand units), X_{t1} denotes a numerical code for year t, and $X_{t-1,2}$ denotes number of utility connections for new housing in year $t - 1$ (in thousands). Data for the period 1977–1992 follow:

Year t:	1977	1978	1979	1980	1981	1982	1983	1984
X_{t1}:	1	2	3	4	5	6	7	8
$X_{t-1,2}$:	21.4	21.6	19.1	19.3	18.9	20.8	22.3	28.1
Y_t:	3.4	3.8	3.8	4.2	4.5	5.3	6.0	7.5

Year t:	1985	1986	1987	1988	1989	1990	1991	1992
X_{t1}:	9	10	11	12	13	14	15	16
$X_{t-1,2}$:	19.6	24.7	26.9	27.3	19.9	20.4	26.7	20.7
Y_t:	6.4	7.7	8.5	9.0	8.0	8.4	10.1	9.3

The estimated regression coefficients are: $b_0 = -1.11428$, $b_1 = 0.40487$, $b_2 = 0.19196$.

a. Give a point estimate of the expected year-to-year increase in dishwasher sales when the number of utility connections for new housing in the preceding year remains constant.

b. Obtain the fitted value for 1990.

c. In 1992, there were 24.3 thousand utility connections for new housing. Obtain a point estimate of expected sales of dishwashers in 1993.

25.6 Refer to **Dishwashers** Problem 25.5. Additional regression results are: $s\{b_0\} = 0.07871$, $s\{b_1\} = 0.00253$, $s\{b_2\} = 0.00371$, $SSR = 74.2796$, $SSE = 0.024798$.

a. Construct a 99 percent prediction interval for dishwasher sales in 1993. Computer output shows that $s\{Y_{h(new)}\} = 0.04932$ for this case. Does it seem likely that dishwasher sales in 1993 will be higher than in 1992? Comment.

b. Construct a 99 percent confidence interval for β_1. Interpret the confidence interval.

25.7 Advertising Impact. An advertising executive is examining the relationship between monthly sales of a breakfast cereal product (Y_t, in thousand cases) and monthly advertising expenditures for the product (X_t, in \$ thousand). The executive conjectures that distributed-lag regression model (25.6) is appropriate. The available data follow:

Year:					1991					
Month t:	1	2	3	4	5	6	7	8	9	10
Y_t:	—	—	210	208	207	223	199	192	189	195
X_t:	66	74	78	75	80	65	66	57	60	59

Year:	1991					1992				
Month t:	11	12	13	14	15	16	17	18	19	20
Y_t:	190	233	211	216	228	207	215	227	186	219
X_t:	99	62	80	88	59	77	83	52	91	60

The estimated regression coefficients when regression model (25.6) is fitted to these data are: $b_0 = 102.810$, $b_1 = 1.09300$, $b_2 = 0.37348$.

a. Obtain the fitted value and the residual for June 1992.

b. What is the estimated change in expected monthly sales with an additional \$2 thousand of advertising expenditures in each of the preceding two months?

c. Obtain a point estimate of expected sales in September 1992.

25.8 Refer to **Advertising Impact** Problem 25.7. Additional regression results are: $s\{b_0\} = 12.0702$, $s\{b_1\} = 0.099356$, $s\{b_2\} = 0.105508$, $SSR = 3139.237$, $SSE = 389.041$.

a. Test whether a regression relation exists, controlling the α risk at 0.01. State the alternatives, the decision rule, the value of the test statistic, and the conclusion.

b. Construct a 99 percent confidence interval for β_2. Interpret the confidence interval.

c. Obtain $\sqrt{MSE}$. Does the standard deviation of the error terms appear to be relatively large for this model? Explain.

* 25.9 Refer to **Electricity Sales** Problems 24.26 and 25.1.

a. Figure 25.7 shows a time sequence plot of the residuals for the fitted regression model. Do the error terms appear to be serially correlated? Comment.

b. The Durbin–Watson test statistic is $d = 2.674$. Test whether the error terms are positively autocorrelated, controlling the α risk at 0.01 when $\rho = 0$. State the alternatives, the decision rule, and the conclusion.

25.10 Refer to **Bergen Corporation** Problems 24.40 and 25.3. The Durbin–Watson test statistic is $d = 2.241$. Test whether the error terms are positively autocorrelated, controlling the α risk at 0.05 when $\rho = 0$. State the alternatives, the decision rule, and the conclusion.

25.11 Refer to **Dishwashers** Problem 25.5. The Durbin–Watson test statistic is $d = 2.269$. Test whether the error terms are positively autocorrelated, controlling the α risk at 0.05 when $\rho = 0$. State the alternatives, the decision rule, and the conclusion.

25.12 Refer to **Advertising Impact** Problem 25.7.

a. Figure 25.8 shows a time sequence plot of the residuals for the fitted regression model. Do the error terms appear to be serially correlated? Comment.

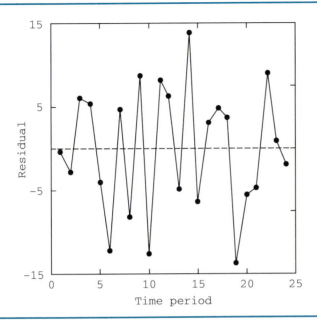

FIGURE 25.7
Residual time sequence plot
for Problem 25.9

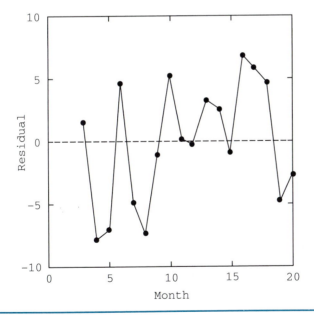

FIGURE 25.8
Residual time sequence plot
for Problem 25.12

b. The Durbin–Watson test statistic for the residuals is $d = 1.548$. Test whether the error terms are positively autocorrelated, controlling the α risk at 0.05 when $\rho = 0$. State the alternatives, the decision rule, and the conclusion.

* **25.13 Empire Data Processing.** Data on the numbers of payroll-processing accounts serviced by Empire Data Processing (X_t), the annual payroll-processing revenues received by Empire (Y_t, in \$ thousand), and the residuals (e_t) obtained in fitting regression model (19.1) follow:

Year:	1982	1983	1984	1985	1986	1987
t:	1	2	3	4	5	6
X_t:	20	25	30	40	35	45
Y_t:	402	526	603	803	756	972
e_t:	26.7	22.5	−28.8	−85.4	−4.1	−44.7

Year:	1988	1989	1990	1991	1992
t:	7	8	9	10	11
X_t:	50	50	45	50	55
Y_t:	1172	1172	1072	1165	1258
e_t:	27.0	27.0	55.3	20.0	−15.3

a. Plot the residuals in time sequence to assess whether the errors are time-dependent. State your findings.
b. Fit first-differences model (25.15) and state the estimated regression function.
c. Obtain the residuals for the first-differences model in part b and plot them in time sequence. Do the error terms of this model appear to be uncorrelated? Explain.
d. The firm has contracted to service 50 payroll-processing accounts in 1993. Obtain a point prediction of the payroll-processing revenues to be received in 1993.

* **25.14** Refer to **Empire Data Processing** Problem 25.13.

a. Obtain *MSE* for the first-differences model. How many degrees of freedom are associated with *MSE* here?
b. Obtain a 90 percent confidence interval for β_1. Interpret the confidence interval.

25.15 **University Applicants.** Data follow on the numbers of applicants accepted by a state university to a Master's program (X_t), the numbers of applicants who actually entered the program (Y_t), and the residuals (e_t) obtained in fitting regression model (19.1):

Year:	1982	1983	1984	1985	1986	1987
t:	1	2	3	4	5	6
X_t:	330	342	348	355	365	373
Y_t:	140	151	154	161	167	171
e_t:	1.76	1.57	−1.02	−0.55	−3.87	−7.33

Year:	1988	1989	1990	1991	1992	1993
t:	7	8	9	10	11	12
X_t:	375	367	376	381	375	385
Y_t:	170	176	184	192	185	192
e_t:	−10.19	3.26	2.87	6.21	4.81	2.48

a. Plot the residuals in time sequence. Do the error terms appear to be positively autocorrelated? Explain.
b. Fit first-differences model (25.15) and state the estimated regression function.
c. Obtain the residuals for the first-differences model in part b and plot them in time sequence. Do the error terms of this model appear to be time-independent as required by model (25.15)? Explain.
d. The number of applicants accepted for 1994 is 380. Obtain a point prediction of the number of applicants who actually will enter the program in 1994.

25.16 Refer to **University Applicants** Problem 25.15.

a. Obtain *MSE* for the fitted first-differences model. How many degrees of freedom are associated with *MSE* here?
b. Obtain a 95 percent confidence interval for β_1. Interpret the confidence interval.

25.17 Policy Loans. Data follow from an insurance company on the values of total assets (X_t, in $ million), the values of policy loans (Y_t, in $ million), and the residuals (e_t) obtained in fitting regression model (19.1):

Year:	1983	1984	1985	1986	1987
t:	1	2	3	4	5
X_t:	241	268	300	339	384
Y_t:	12.1	12.9	13.5	14.4	16.5
e_t:	0.432	0.240	−0.336	−0.869	−0.423

Year:	1988	1989	1990	1991	1992
t:	6	7	8	9	10
X_t:	439	489	541	602	669
Y_t:	19.5	21.4	22.9	24.7	27.2
e_t:	0.556	0.619	0.209	−0.233	−0.195

a. Plot the residuals in time sequence. Do the error terms appear to be positively autocorrelated? Explain.
b. Fit first-differences model (25.15) and state the estimated regression function. Interpret b_1.
c. Obtain the residuals for the first-differences model in part b and plot them in time sequence. Do the error terms of this model appear to be time-independent as required by model (25.15)? Explain.
d. The value of total assets of the company in 1993 is anticipated to be $683 million. Obtain a point prediction of the value of policy loans in 1993.

25.18 Refer to **Policy Loans** Problem 25.17.

a. Obtain $\sqrt{MSE}$. What does this statistic estimate here?
b. Obtain a 90 percent prediction interval for $Y'_{h(new)} = Y'_{11}$. Convert this interval into a 90 percent prediction interval for Y_{11}. Interpret the latter prediction interval.

* **25.19 Paper Mill.** The inventory manager of a paper mill uses exponential smoothing procedure (25.18) with smoothing constant $a = 0.25$ to smooth the weekly demand for a certain type of newsprint. The smoothed estimate from last week is 41.3 metric tons.

a. Give a forecast of demand for this week. What assumption is required for using exponential smoothing forecasting procedure (25.19) here?
b. Demand for this week is 46.5 metric tons. Give (1) the smoothed estimate for this week, (2) the forecast for next week, (3) the forecast for four weeks hence.

25.20 Data on the number of bed sheets washed by a hospital laundry in the past five weeks follow:

Week t:	1	2	3	4	5
Sheets Y_t:	1863	1961	1940	1876	1890

Assume that the number of sheets washed is a stationary time series.

a. Use exponential smoothing procedure (25.18) with smoothing constant $a = 0.15$ and starting value $A_0 = 1900$ to obtain the smoothed estimates for this series.
b. Use the results in part a to forecast the number of sheets to be washed in weeks 6, 7, and 8.
c. If the actual number of sheets washed in week 6 is 1936, what is the error in the forecast in part b? Revise the forecasts for weeks 7 and 8 based on the actual outcome for week 6.

25.21 Refer to **Communications Link** Problem 24.3. The use of exponential smoothing procedure (25.18) with smoothing constant $a = 0.10$ and starting value $A_0 = 15$ gives the following smoothed estimates for the series:

Hour t:	1	2	3	4	5	6	7	8
A_t:	14.0	15.0	17.1	15.6	14.1	13.0	13.6	17.8

Hour t:	9	10	11	12	13	14	15	16
A_t:	16.1	14.7	13.6	14.2	13.2	12.0	13.2	14.9

Hour t:	17	18	19	20	21	22	23	24
A_t:	13.4	12.2	11.7	10.8	12.4	—	—	—

Assume that the number of erroneous data bits transmitted each hour is a stationary time series.

a. Obtain the smoothed estimates for hours 22, 23, and 24.

b. Plot the original series and the smoothed estimates on the same graph. Does the exponential smoothing procedure appear to identify the smooth component of the series effectively? Comment.

c. Forecast the number of erroneous data bits for hours (1) 25, (2) 30, (3) 25 to 30 in total.

25.22 Refer to **Paper Mill** Problem 25.19b.

a. In the calculation of the smoothed estimate for this week, what weight is given to the observation for last week? To the observation for four weeks ago? Are these results consistent with the fact that exponential smoothing procedure (25.18) gives earlier observations less weight in determining the current smoothed estimate?

b. What would have been the weights in part a if the smoothing constant had been $a = 0.10$? Does the smaller smoothing constant give more or less weight to earlier observations in the calculation of the current smoothed estimate? Explain.

25.23 Consider the following statement: "The best indication of sales of a fashion good next month is given by sales in the most recent few months. Sales further in the past are relatively unimportant." To forecast sales of this fashion good one month ahead using exponential smoothing procedure (25.18), should the smoothing constant a be close to 0 or 1? Explain.

25.24 Consider the following series and smoothed estimates:

t:	1	2	3	4	5	6
Y_t:	200	200	300	300	300	300
A_t:	200	200	—	—	—	—

Obtain the missing smoothed estimates for periods 3 to 6 by using exponential smoothing procedure (25.18) with smoothing constant (1) $a = 0.10$, (2) $a = 0.40$. How does the magnitude of the smoothing constant affect the responsiveness of the smoothed estimate to the permanent shift in the level of the series?

* **25.25** Refer to Table 25.4 for the *Lake Level* example. The weighting constants used here are $a = 0.10$ and $b = 0.20$.

a. The lake level for year 31 is 4.84 meters. Obtain the smoothed and trend estimates for year 31.

b. Use the information for year 31 to obtain a new forecast of the lake level for year 35.

25.26 Refer to **Pulp and Paper Production** Problem 24.1. The smoothed and trend estimates for this series follow. They were obtained by using exponential smoothing procedure (25.21) with starting values $A_0 = 3.50$ and $B_0 = 0.050$ and weighting constants $a = 0.10$ and $b = 0.10$.

Year t:	1972	1973	1974	1975	1976	1977	1978
A_t:	3.53	3.59	3.66	3.69	3.66	3.64	3.66
B_t:	0.048	0.049	0.051	0.049	0.041	0.035	0.033

Year t:	1979	1980	1981	1982	1983	1984	1985
A_t:	3.68	3.66	3.67	3.72	3.84	3.90	4.02
B_t:	0.032	0.027	0.025	0.028	0.037	0.039	0.048

Year t:	1986	1987	1988	1989	1990	1991
A_t:	4.13	4.20	4.28	—	—	—
B_t:	0.054	0.055	0.058	—	—	—

a. Obtain the smoothed and trend estimates for 1989, 1990, and 1991.
b. Plot the original series and the smoothed estimates on the same graph. Do the smoothed estimates track the original series well? Comment.
c. Forecast production for 1992 and also for the year 2000. What precisely do these forecasts represent?

25.27 Refer to **Consulting Contracts** Problem 24.9.

a. Obtain smoothed and trend estimates for this series by means of exponential smoothing procedure (25.21). Use starting values $A_0 = 500$ and $B_0 = 100$ and weighting constants $a = 0.20$ and $b = 0.20$.
b. Forecast the total value of contracts for 1993 and also for 1994.
c. If the value of contracts series had an exponential trend component rather than a linear trend component, would exponential smoothing procedures (25.21) and (25.22) still be appropriate for smoothing and forecasting the series? Explain.

25.28 Refer to **Consulting Contracts** Problems 24.9 and 25.27. What starting values for the smoothed and trend estimates are suggested by the fitted linear trend function obtained in Problem 24.9a if exponential smoothing procedure (25.21) will be applied to the series beginning in 1993?

25.29 A business researcher wishes to apply exponential smoothing procedure (25.21) to an annual business indicator series with trend. Data for the series are available for years 1 to 4 inclusive. The linear trend function fitted to these data is $\hat{Y}_t = 103.2 + 2.264X_t$, where $X_t = 1$ in year 1. What starting values are suggested by this trend line for applying exponential smoothing procedure (25.21) to the series beginning with the observation for year 5?

* 25.30 Refer to **Electricity Sales** Problem 24.26. Exponential smoothing procedure (25.23) has been applied to this quarterly sales revenues series, using weighting constants $a = 0.10$, $b = 0.10$, and $c = 0.20$. The values of the smoothed, trend, and seasonal component estimates and the original observations for several recent quarters follow:

Year:	1990				1991			
Quarter:	1	2	3	4	1	2	3	4
Y_t:	179	235	292	213	170	241	307	217
A_t:	232.9	233.4	231.9	231.4	231.0	232.0	—	—
B_t:	0.208	0.239	0.066	0.007	−0.030	0.076	—	—
C_t:	−57.6	−0.6	72.5	−14.1	−58.3	1.3	—	—

a. Obtain the smoothed, trend, and seasonal component estimates for the third and fourth quarters of 1991.
b. At the end of 1991, forecast the company's sales revenues for each of the quarters of 1992 and for the first quarter of 1993.
c. Obtain the values of (1) $A_t - A_{t-1}$ and (2) $Y_t - C_{t-p}$ for the fourth quarter of 1991. Interpret each of these quantities.

25.31 Exponential smoothing procedure (25.23) has been applied to the quarterly number of travel bookings to Europe made by a travel agency. The weighting constants used in the procedure were

$a = 0.10$, $b = 0.20$, and $c = 0.30$. The values of the smoothed, trend, and seasonal component estimates and the original observations for several recent quarters follow:

Year:	1990				1991			
Quarter:	1	2	3	4	1	2	3	4
Y_t:	381	408	344	353	370	391	335	347
A_t:	367.4	364.8	363.2	361.8	359.4	356.3	—	—
B_t:	−2.92	−2.85	−2.61	−2.36	−2.37	−2.51	—	—
C_t:	11.1	41.0	−26.8	−16.6	10.9	39.1	—	—

a. Obtain the smoothed, trend, and seasonal component estimates for the third and fourth quarters of 1991.
b. At the end of 1991, forecast the number of bookings for each quarter of 1992. What is the total number of bookings forecast for 1992?
c. Does the seasonal pattern of this series appear to have been stable for years 1990 and 1991? Comment.

25.32 Refer to **Bus Traffic** Problem 24.2. Exponential smoothing procedure (25.23) has been used to smooth and project this series, with weighting constants $a = 0.10$, $b = 0.10$, and $c = 0.30$. At the end of Friday of week 4, the smoothed and trend estimates were 162.4 and 0.488, respectively, and the seasonal component estimates were:

Day of Week:	S	S	M	T	W	T	F
Seasonal Component Estimate C:	−25.0	−113.8	40.6	24.9	26.9	14.6	44.9

a. According to the set of seasonal component estimates at the end of Friday of week 4, which day of the week tends to have the most bus traffic? Does this set of seasonal component estimates sum to zero? Must the set sum to zero? Comment.
b. Obtain the smoothed, trend, and seasonal component estimates for Saturday and Sunday of week 4.
c. At the end of week 4, forecast the traffic for each day of week 5. What is the total traffic forecast for week 5?

25.33 An assistant asked a company's forecasting specialist, "Should the seasonal adjustment constant c be set equal to 0 when smoothing and forecasting a series that has a stable seasonal component?" Answer the question. Itemize the factors that should be taken into account in setting the value of the seasonal adjustment constant c in exponential smoothing procedure (25.23).

25.34 A monthly series currently has a pronounced seasonal component but no significant trend-cyclical component. Data are only available for the most recent year, as follows:

Month:	J	F	M	A	M	J	J	A	S	O	N	D
Y_t:	44	38	40	45	55	68	81	79	65	60	55	49

Suggest starting values for the smoothed, trend, and seasonal component estimates to be used with exponential smoothing procedure (25.23) when the observation for the coming January becomes available.

25.35 Refer to Tables 24.6 and 24.10 for the *Air Taxi* example. Exponential smoothing procedure (25.23) is to be applied to this series, commencing with the first quarter of 1992.

a. Use the seasonal deviations in Table 24.10b to obtain deseasonalized observations for the eight quarters of 1990 and 1991 and fit a linear trend function to these deseasonalized observations.
b. Using the seasonal deviations and the linear trend function, suggest starting values for the smoothed, trend, and seasonal component estimates.

25.36 Refer to regression model (25.3). What is the implication of each of the following cases: (1) $\beta_1 = 0$, (2) $\beta_4 = 0$, (3) $\beta_2 = \beta_3 = \beta_4 = 0$?

25.37 An analyst stated: "Quarterly sales of this product are primarily affected by three factors: the size of the current target market, advertising expenditures, and quarter of the year. The current size of the market has an immediate impact. Advertising expenditures in a given quarter affect sales in the following two quarters. Summer-related factors, which tend to boost sales in the second and third quarters, are absent in the first and fourth quarters. There are still other factors that are not as important, but together, they tend to have a linear trend effect on sales."

a. Construct a regression model that might be suitable here. What does each regression coefficient denote?
b. Does the model in part a contain any of the following: (1) a trend component, (2) a seasonal component, (3) a distributed time lag? If so, identify the relevant terms.
c. Will all of the independent variables in the model in part a have to be predicted in order to forecast sales? Explain.

25.38 Show that when first-differences model (25.15) is applied to a series with linear trend defined by:

$$Y_t = \beta_0 + \beta_1 X_t + \varepsilon_t \qquad \text{where } X_t = t$$

then $b_1 = (Y_n - Y_1)/(n - 1)$, where n is the number of observations in the original series.

25.39 If X_t is the same in two consecutive time periods, will Y_t also remain unchanged with first-differences model (25.15)? Explain.

25.40 When exponential smoothing procedure (25.18) is applied to a time series with linear trend, the smoothed estimate will lag behind the series. Let $Y_t = \beta_0 + \beta_1 t$ denote a series with linear trend and no irregular component. Use the expansion of A_t in (25.20) to show that the difference $Y_t - A_t$, when A_t is calculated from (25.18), approaches the value $\beta_1(1 - a)/a$ as t increases. Explain how this result demonstrates the lag of the smoothed estimate.

$$\left[\text{Hint: } \sum_{r=0}^{\infty} a(1 - a)^r = 1 \quad \text{and} \quad \sum_{r=1}^{\infty} ar(1 - a)^r = \frac{1 - a}{a} \right]$$

25.41 Refer to **University Enrollment** Problem 24.27.

a. Fit regression model (25.1) to the enrollment data, employing a linear trend variable X_1 (with $X_{t1} = 1$ in trimester 1 of 1984) and two indicator variables (X_2 and X_3) for differential seasonal effects for trimesters 2 and 3, respectively. State the estimated regression function.
b. Obtain point estimates of the expected enrollment in each trimester of 1992.
c. Obtain the Durbin–Watson test statistic and test whether the error terms of the regression model are positively autocorrelated. Control the α risk at 0.05 when $\rho = 0$. State the alternatives, the decision rule, the value of the test statistic, and the conclusion.
d. Examine the appropriateness of regression model (25.1) by means of (1) a plot of the residuals against the fitted values, (2) a time sequence plot of the residuals, (3) a symbolic scatter plot of the residuals against time order, using the trimester (1, 2, 3) as the plotting symbol. Discuss your findings.

25.42 Refer to the **Beer Sales** data set (Appendix D.4). Monthly beer sales are to be regressed against the monthly average of the maximum daily temperatures.

a. Using regression model (19.1), obtain the fitted regression function and the residuals.
b. Plot the residuals in time sequence. Does it appear that the error terms in the regression model are serially correlated? Explain.

c. Test the error terms for randomness using the lower-tail runs-up-and-down test in Section 15.4 as an approximate test. Control the α risk at 0.05 when the error terms are generated by a random process. State the alternatives, the decision rule, the value of the test statistic, and the conclusion.

d. Use the Durbin–Watson test to determine whether the error terms are positively autocorrelated. Control the α risk at 0.05 when $\rho = 0$. State the alternatives, the decision rule, the value of the test statistic, and the conclusion. Is the test conclusion here consistent with the one in part c? Comment.

25.43 Consider first-order autoregressive error model (25.8).

a. Generate a random sequence of 50 u_t observations ($t = 1, 2, \ldots, 50$) from the $N(0, 1)$ distribution using a standard normal random number generator.

b. Let $\varepsilon_0 = 0$ and compute ε_t ($t = 1, 2, \ldots, 50$) from the u_t sequence generated in part a for (1) $\rho = 0.8$, (2) $\rho = 0$, (3) $\rho = -0.8$.

c. Plot the three series of ε_t observations in time order in three separate graphs. Contrast the time series patterns in the graphs.

25.44 Refer to **Marriages** Problem 15.38.

a. Use exponential smoothing procedure (25.23) to obtain the smoothed, trend, and seasonal component estimates of the series for the period from July of year 3 to June of year 5, as follows:

1. Use weighting constants $a = 0.10$, $b = 0.10$, and $c = 0.30$.
2. Obtain the monthly seasonal deviations based on the first 24 months of data (July of year 1 to June of year 3) using the method described in Section 24.7. Use these seasonal deviations as starting values for the seasonal component estimates.
3. Deseasonalize the first 24 months of data by using the seasonal deviations obtained in step 2. Fit a linear trend function to the deseasonalized series of 24 observations (set $X_t = 1$ at July of year 1). Use the trend line to set the starting values for the smoothed and trend estimates.

b. Use exponential smoothing procedure (25.24) to forecast the numbers of marriages for each of the six months from July to December of year 5.

Price and Quantity Indexes

26

Price and quantity indexes are summary measures of relative price and quantity changes over time in a set of items. For example, the consumer price index summarizes relative price changes in goods and services purchased by urban households; the producer price indexes measure relative changes in prices received in primary markets by producers of commodities in all stages of processing; and the index of industrial production measures relative changes in output in manufacturing, mining, and utilities.

Price and quantity indexes not only serve as summary measures of price and quantity changes but also are employed in many statistical analyses. Regression models, for instance, frequently contain price or quantity indexes as independent variables. Another use of price indexes is to adjust time series expressed in monetary units, such as annual sales revenues during the past 20 years, so that changes in the series other than price changes can be studied.

In this chapter, we discuss price and quantity indexes in their several roles. First we take up the construction of price indexes and consider various uses of them. Then we discuss quantity indexes.

We begin with the measurement of price changes for a single item, such as a gallon of 89-octane unleaded gasoline or a 3½-inch high-density computer disk.

26.1
PERCENT RELATIVES AND LINK RELATIVES

Percent Relatives

Percent relatives are useful for studying the pattern of relative changes in the price of an item over time.

> **(26.1)**
> A **percent relatives series** expresses the value of the series in each period as a percent of the value in the base period.

EXAMPLE ☐ **Chemical Compound**

The average price of a chemical compound for each year in the period 1988–1991 is shown in Table 26.1, column 1. The corresponding percent relatives, calculated on the base period 1988, are shown in column 2. For instance, the percent relative for 1989 is $100(22/20) = 110$. This relative tells us that the price of the compound in 1989 was 110 percent as great as in 1988. ☐

		(1)	(2)	(3)	(4)
		Average	Percent Relatives		
		Price			Link
	Year	(dollars)	1988 = 100	1991 = 100	Relative
1988		20	100	40	—
1989		22	110	44	110
1990		36	180	72	164
1991		50	250	100	139

TABLE 26.1
Calculation of percent relatives and link relatives series—Chemical compound example

The base period for a percent relatives series is usually identified for reporting purposes in the manner shown in Table 26.1; for example, 1988 = 100. The base period may be any period that facilitates the comparisons of interest. For instance, the choice of 1988 in the chemical compound example facilitates the study of the price increases during a period when demand was increasing rapidly because of a new industrial use for the compound.

Interpretation of Percent Relatives.   Percent relatives must be interpreted with care. We discuss briefly three important considerations.

1. The absolute magnitudes of percent relatives are affected by the choice of the base period, but their proportional magnitudes are not affected by this choice. We illustrate this for the chemical compound example by presenting in column 3 of Table 26.1 the percent relatives calculated on the base period 1991. The percent relative for 1988 now is 100(20/50) = 40, not 100 as for the 1988 base period series. However, the proportional magnitudes of the percent relatives in columns 2 and 3 are the same. For instance, the percent relatives for 1989 and 1988, with base period 1988, have the ratio 110/100 = 1.1. For the percent relatives with the 1991 base period, this ratio is the same; that is, 44/40 = 1.1. Similarly, for both series of percent relatives, the ratio of the 1990 price relative to the 1989 price relative is 1.64.

2. A comparison of percent relatives for two different price series indicates nothing about the actual prices unless we know the actual magnitudes in the base period. For example, information that the percent relatives for large grade-A eggs last month were 120 for Montreal and 123 for Toronto tells us nothing about how the egg prices in the two cities compared last month unless we know the prices in the base period.

3. In analyzing percent relatives, we distinguish between *percent points of change* and *percent change*. To illustrate the difference, refer to the percent relatives with base period 1988 in column 2 of Table 26.1 for the chemical compound example. Between 1990 and 1991, the percent points of change were 250 − 180 = 70. On the other hand, the percent change was 100[(250 − 180)/180] = 100(70/180) = 38.9. Thus, percent points of change refers to the absolute change in the percent relatives and is dependent on the choice of the base period. Percent change, in contrast, refers to the relative change in percent relatives and, as we noted previously, is not affected by the choice of the base period.

Link Relatives

Link relatives are useful in studying relative period-to-period changes instead of changes from a fixed base period. For instance, link relatives enable us to study whether or not the price of an item is increasing at a constant rate, an increasing rate, or a decreasing rate.

(26.2)
A **link relatives series** expresses the value of the series in each period as a percent of the value in the immediately preceding period.

EXAMPLE ☐

The link relatives for the chemical compound example are given in column 4 of Table 26.1. For instance, the link relative for 1989 is $100(22/20) = 110$. This tells us that the price in 1989 was 110 percent as great as in 1988. The link relative for 1990 is $100(36/22) = 164$, indicating that the price in 1990 was 164 percent as great as in 1989. Note that the percent increase from 1990 to 1991 was smaller than that from 1989 to 1990 (column 4), even though the amount of price increase ($14) did not decline (column 1).

☐

Comment

We have illustrated percent relatives and link relatives for price series, but these relatives can be utilized for any time series (for example, annual sales or monthly production).

Price indexes are summary measures that combine the price changes for a group of items, using weights to give each item its appropriate importance. The consumer price index is such an index, measuring the combined effect of price changes in many goods and services purchased by urban households. Two basic methods are widely used for calculating price indexes: the method of weighted aggregates and the method of weighted average of relatives. We shall explain each in turn, using the following illustration.

**PRICE INDEXES 26.2
BY METHOD OF
WEIGHTED
AGGREGATES**

EXAMPLE ☐ **School Maintenance Supplies**

Officials of a large school district needed to develop a price index for maintenance supplies for the school buildings in the district. In compiling this index, the officials could not include all supply items in the index because the number of supply items used is very large. Instead, a sample of supply items was selected to represent all items used. The selected items are shown in Table 26.2, column 1. Also shown in this table, in columns 3, 4, and 5, are the unit prices of these items for 1988, 1989, and 1990, respectively. Finally, column 2 of Table 26.2 presents typical quantities consumed annually for each of the selected supply items.

☐

Method of Weighted Aggregates

The method of weighted aggregates for compiling a price index simply compares the cost of the typical quantities consumed at the prices of a given period with the corresponding

TABLE 26.2   **Calculation of price index series by method of weighted aggregates—School maintenance supplies example**

	(1)	(2)	(3)	(4)	(5)	(6)	(7)	(8)
				Unit Prices				
i	Schedule of Items	Quantity Q_{ia}	1988 P_{i0}	1989 P_{i1}	1990 P_{i2}	$P_{i0}Q_{ia}$ (3) × (2)	$P_{i1}Q_{ia}$ (4) × (2)	$P_{i2}Q_{ia}$ (5) × (2)
1	Window glass	15 sheets	15.55	15.86	16.21	233.25	237.90	243.15
2	Fluorescent tube	130 boxes	39.17	39.81	40.55	5,092.10	5,175.30	5,271.50
3	Floor detergent	290 cans	21.95	23.55	24.90	6,365.50	6,829.50	7,221.00
4	Floor finish	100 cans	49.39	52.27	52.53	4,939.00	5,227.00	5,253.00
5	Latex interior paint	175 cans	11.13	11.50	11.56	1,947.75	2,012.50	2,023.00
6	Mop head	200 pieces	2.86	2.93	2.99	572.00	586.00	598.00
					Total	19,149.60	20,068.20	20,609.65

$$I_{88} = 100\left(\frac{19{,}149.60}{19{,}149.60}\right) = 100.0$$

$$I_{89} = 100\left(\frac{20{,}068.20}{19{,}149.60}\right) = 104.8$$

$$I_{90} = 100\left(\frac{20{,}609.65}{19{,}149.60}\right) = 107.6$$

cost in the base period. We shall first explain the notation and terminology used and then illustrate the method of calculation by an example.

The *schedule of items* is the list of items included in the price index. Often, the schedule of items consists of a sample of all items of interest, as in the school maintenance supplies example. Sometimes, it is possible to include all items of interest in the schedule.

The price of the ith item in the schedule in any *given period t* is denoted by P_{it}, and the price of this item in the *base period* is denoted by P_{i0}. The typical quantity consumed for the ith item is denoted by Q_{ia}, where the subscript a stands for an average or typical period. These quantities are used as weights to reflect the importance of each item.

The method of weighted aggregates compares the cost of the typical quantities at period t prices with the cost of the same quantities at base period prices. The cost at period t prices is:

$$\text{Cost at period } t \text{ prices} = \sum_i P_{it}Q_{ia}$$

where the summation is over all the items in the schedule. The cost at base period prices is:

$$\text{Cost at period 0 prices} = \sum_i P_{i0}Q_{ia}$$

The price index for period t by the method of weighted aggregates is the ratio of these two costs, expressed as a percent. The price index for period t will be denoted by I_t.

(26.3)

The price index I_t for period t calculated by the **method of weighted aggregates** is:

$$I_t = 100 \, \frac{\sum\limits_i P_{it} Q_{ia}}{\sum\limits_i P_{i0} Q_{ia}}$$

where: P_{i0} is the unit price of the ith item in period 0 (base period)

P_{it} is the unit price of the ith item in period t (given period)

Q_{ia} is the quantity weight assigned to the ith item

EXAMPLE □

For the school maintenance supplies example, the price index series for 1988, 1989, and 1990 is calculated in Table 26.2 by the method of weighted aggregates. Columns 6, 7, and 8 contain the costs of the typical quantities at the prices of each year. For example, the cost of 15 sheets of window glass at 1988 prices is 15(15.55) = 233.25, while at 1989 prices the cost is 15(15.86) = 237.90. The aggregate costs of the schedule of items at the prices of the three periods are shown at the bottoms of columns 6, 7, and 8, respectively. The index numbers are calculated at the bottom of the table.

Since the same items and quantities are priced in each period, the differences in the index are attributable wholly to changes in the prices of the items. Thus, the price index $I_{89} = 104.8$ indicates that the prices of school maintenance supplies increased, in terms of their aggregate effect, by 4.8 percent between 1988 and 1989. Since the index for 1990 ($I_{90} = 107.6$) is above the index for 1989 ($I_{89} = 104.8$), the prices, in their aggregate effect, increased still more between 1989 and 1990. We calculate the percent price increase between 1989 and 1990 by expressing the percent point change, 107.6 − 104.8 = 2.8, as a percent of the 1989 index and obtain 100(2.8/104.8) = 2.7 percent as the relative price increase between 1989 and 1990. □

Comments

1. Just as with percent relatives, the proportional magnitudes of the index numbers in Table 26.2 are not affected by which period is chosen as the base period.
2. The period from which the quantity weights are derived is called the *weight period*. The base period for the price index and the weight period need not coincide.
3. When the typical quantity weights Q_{ia} are the quantities consumed in the base period—that is, Q_{i0}—the weighted aggregates price index is called a *Laspeyres* price index.

A second method of compiling a price index series is the method of weighted average of relatives. This method utilizes the percent relatives of the prices for the items in the schedule. The index for period t is simply a weighted average of the percent relatives for all schedule items. The weights required by this method are *value weights* rather than quantity weights. We shall denote the value weight for the ith item by V_{ia}.

PRICE INDEXES 26.3 BY METHOD OF WEIGHTED AVERAGE OF RELATIVES

(26.4)

The price index I_t for period t calculated by the **method of weighted average of relatives** is:

$$I_t = \frac{\sum_i \left(100\, \frac{P_{it}}{P_{i0}}\right) V_{ia}}{\sum_i V_{ia}}$$

where: P_{i0} is the unit price of the ith item in period 0 (base period)

P_{it} is the unit price of the ith item in period t (given period)

V_{ia} is the value weight assigned to the ith item

The value weights V_{ia} are dollar values that reflect the importance of each item in the schedule. For example, the value weights may be obtained from the typical quantities consumed, Q_{ia}, by multiplying these quantities by typical prices.

☐ **EXAMPLE**

For the school maintenance supplies example, we shall calculate a price index series by the method of weighted average of relatives with base period 1988. The value weights will be based on the typical quantities and the unit prices for the base period 1988 in Table 26.2; that is:

$$V_{ia} = P_{i0} Q_{ia}$$

Table 26.3 contains the necessary calculations. Column 1 repeats the schedule of items, and column 2 contains the value weights as obtained from Table 26.2. For example, the value weight for window glass is $V_{1a} = P_{10}Q_{1a} = 15.55(15) = 233.25$. The sum of the value weights is $\sum V_{ia} = 19,149.60$.

Columns 3, 4, and 5 contain the percent relatives for the prices of each schedule item in the three years based on the data in Table 26.2. Note that all percent relatives are expressed on a 1988 base because the price index is to have 1988 as the base period. For example, the 1989 percent relative for window glass is $100(15.86/15.55) = 101.99$.

We now take a weighted average of the percent relatives for each year. The weighting is done in columns 6, 7, and 8, and the indexes are obtained at the bottom of the table. The index $I_{89} = 104.8$ for 1989 indicates that prices of school maintenance supplies increased, in terms of their aggregate effect, by 4.8 percent between 1988 and 1989.   ☐

Comment

The price indexes in Tables 26.2 and 26.3 are identical. This happened because we used base year prices P_{i0} in obtaining the value weights. In that case, the method of weighted average of relatives index (26.4) reduces to the method of weighted aggregates index (26.3) as follows:

$$\frac{\sum\limits_{i}\left(100\frac{P_{it}}{P_{i0}}\right)V_{ia}}{\sum\limits_{i}V_{ia}} = \frac{\sum\limits_{i}\left(100\frac{P_{it}}{P_{i0}}\right)P_{i0}Q_{ia}}{\sum\limits_{i}P_{i0}Q_{ia}} = 100\frac{\sum\limits_{i}P_{it}Q_{ia}}{\sum\limits_{i}P_{i0}Q_{ia}}$$

When prices other than base period prices are used for obtaining the value weights V_{ia}, the two methods will not lead to identical results. Generally, though, the differences in the indexes calculated by the two methods will not be great.

We now take up several considerations that arise in compiling index numbers series.

Choice of Base Period

Any period within the coverage of the index series can be used as the base period. Of course, when a special-purpose index is designed to measure price changes occurring since a particular period, such as since the lifting of price controls, that period would be taken as

CONSIDERA- **26.4**
TIONS IN COMPILING
INDEX SERIES

Calculation of price index series by method of weighted average of relatives—School maintenance supplies example

TABLE 26.3

(1)		(2)	(3)	(4)	(5)	(6)	(7)	(8)
				Percent Relatives, 1988 = 100		Percent Relative × Value Weight		
			1988	1989	1990	1988	1989	1990
i	Schedule of Items	Value Weight $V_{ia} = P_{i0}Q_{ia}$	$100\left(\dfrac{P_{i0}}{P_{i0}}\right)$	$100\left(\dfrac{P_{i1}}{P_{i0}}\right)$	$100\left(\dfrac{P_{i2}}{P_{i0}}\right)$	$\left(100\dfrac{P_{i0}}{P_{i0}}\right)V_{ia}$ $(3)\times(2)$	$\left(100\dfrac{P_{i1}}{P_{i0}}\right)V_{ia}$ $(4)\times(2)$	$\left(100\dfrac{P_{i2}}{P_{i0}}\right)V_{ia}$ $(5)\times(2)$
1	Window glass	233.25	100.00	101.99	104.24	23,325	23,789	24,314
2	Fluorescent tube	5,092.10	100.00	101.63	103.52	509,210	517,510	527,134
3	Floor detergent	6,365.50	100.00	107.29	113.44	636,550	682,954	722,102
4	Floor finish	4,939.00	100.00	105.83	106.36	493,900	522,694	525,312
5	Latex interior paint	1,947.75	100.00	103.32	103.86	194,775	201,242	202,293
6	Mop head	572.00	100.00	102.45	104.55	57,200	58,601	59,803
	Total	19,149.60			Total	1,914,960	2,006,790	2,060,958

$$I_{88} = \frac{1,914,960}{19,149.60} = 100.0$$

$$I_{89} = \frac{2,006,790}{19,149.60} = 104.8$$

$$I_{90} = \frac{2,060,958}{19,149.60} = 107.6$$

the base period. Whenever possible, the base period should involve relatively normal or standard conditions, because many index users assume that the base period represents such conditions. Sometimes an interval of two or more years is selected as the base period. In this case, the index numbers are calculated so that the indexes for the base period years average to 100. The base period is then reported in a form such as 1990–1991 = 100.

Shifting of Base Period

Sometimes it is necessary to shift the base period of an index series that is already compiled, as when two index series on different base periods need to be placed on a common base period to facilitate comparison. Usually it is not possible to recalculate a published index on the new base period because the needed data are not available. A shortcut method can be employed, however, that does not entail recalculation of the index.

To illustrate the shortcut procedure, let us consider the following price index series with 1989 as the base period:

Year:	1986	1987	1988	1989	1990	1991
Price Index (1989 = 100):	75	88	92	100	110	122

Suppose the base period is to be shifted to 1986. We simply divide each index number by 0.75, the index number for the new base period in decimal form. The index number for 1986 becomes 75/0.75 = 100.0, that for 1987 becomes 88/0.75 = 117.3, and so on. The new price index series with base period 1986 is:

Year:	1986	1987	1988	1989	1990	1991
Price Index (1986 = 100):	100.0	117.3	122.7	133.3	146.7	162.7

Because the proportional magnitudes of the index numbers are not affected by this shifting, the new series conveys the same information about year-to-year relative price changes as the original series.

The shortcut procedure yields results identical to those obtained by recalculating the index series on the new base period when the index series is calculated by the method of weighted aggregates with fixed quantity weights. With most other formulas, however, the shortcut procedure only provides approximate results.

Splicing

When an index series is revised, the new series can be joined or spliced to the older series to yield a single continuous series by a procedure similar to that just described for shifting the base period of an index series. Suppose the following two price index series are to be joined:

Year:	1983	1984	1985	1986	1987	1988	1989	1990	1991
Old Series (1983 = 100):	100	103	106	113	130				
Revised Series (1987 = 100):					100	120	126	131	134

To splice the two series together, we simply shift the level of one of the series so that both have a common value for 1987. For instance, to obtain a combined series with 1983 as the base period, we multiply every index in the revised series by 1.30. The spliced series is:

Year:	1983	1984	1985	1986	1987	1988	1989	1990	1991
Spliced Series (1983 = 100):	100	103	106	113	130	156	164	170	174

Collection of Price and Quantity Data

All of the data-collection methods discussed in Chapter 1 are used in collecting data for price indexes. Price and quantity data are obtained by observation, interview, and self-enumeration. Often sampling is used, as in the consumer price index where probability samples of retail outlets and of items carried by the selected outlets are employed to obtain price data. Sampling is also used for the consumer price index in periodic consumer expenditures surveys to obtain quantity data for weights.

In collecting price data for an item over a period of years, it is important that the quality and other price-determining characteristics be held as constant as possible. This typically requires that detailed specifications be developed and adhered to in pricing the item. Here is an example of such a specification:

> Interior latex paint. Professional- or commercial-grade latex interior house paint, matte or flat finish, off white, first-line or quality.

Note that color, finish, and quality are specified because each of these characteristics can affect the price.

A different problem in collecting price data in ongoing index series is the adjustment for quality changes in the items covered by the index. In principle, a price change that is caused by a quality change should not be reflected in the price index. In practice, minor quality changes are ignored. On the other hand, attempts are made in better price indexes to adjust for important quality changes. However, the conceptual and measurement problems are difficult. For instance, if an oven door is modified by the manufacturer to exclude the glass window while the list price remains unchanged, should this be treated as a price increase and reflected in the index? If so, what should be the magnitude of the imputed increase? Extensive research has been undertaken on how problems connected with quality changes should be handled.

Another issue in compiling an index series concerns weights. In some indexes, the schedule includes all or almost all of the items used in the activity. For instance, in a price index for highway construction, where a relatively small number of major items account for most of the total cost, it suffices for practical purposes to limit the schedule to these major items. Here, each weight reflects the importance of the particular item.

Often, however, the schedule must be limited to a sample of items, since thousands of items may be involved in the activity and it is neither feasible nor necessary to include them all. Each item is then selected for the schedule to represent a class of related items. In such cases, the weight for an item normally reflects the importance of the entire class represented by the item. Consider, for instance, a particular type of insulation that was selected for a price index for residential building construction to represent all types of insulation and vapor barriers. The weight for the insulation in the schedule here needs to represent the importance of the entire class of insulation and vapor barriers.

Maintenance of Price Index Series

A key issue in the maintenance of price index series involves changes in the items included in the index and in the weights assigned to the items to keep them up to date. Should the schedule of items and the weights be changed frequently so that they are always up to date, or should they be held fixed over a relatively long sequence of years? If frequent changes are made, changes in the price index series over time will reflect both changes in prices and changes in the composition of the index. Another disadvantage of frequent revision is that it may be very expensive. On the other hand, an index can become badly outdated if the schedule of items and the weights are held fixed over too long a period.

The procedure generally followed in practice is to hold the schedule of items and the weights essentially fixed for 5 to 10 years and then to revise them. The schedule and weights need not be held exactly fixed, since procedures are available for making interim adjustments without disturbing the index. For instance, a new item can be substituted for an item that has been taken off the market, and seasonal items, such as fresh produce, can be included in the schedule in season.

In addition to periodic updating of the schedule of items and the weights, revisions often also include upgrading of data collection and data handling procedures and modernizing of definitions. A new series of index numbers is then initiated with each revision. The new series can be joined to the preceding series, if desired, by the splicing procedure already explained.

26.5 USES OF PRICE INDEXES

Price indexes play important roles in a variety of applications. We now consider two important uses of price indexes.

Measuring Real Earnings

As prices change, so does the quantity of goods and services that can be purchased by a fixed sum of money. In economic analysis, it is frequently important to measure changes in *real earnings* (that is, in the quantity of goods and services that can be purchased). Price indexes are a basic tool in making this measurement.

Weekly Earnings ☐ **EXAMPLE**

Table 26.4 shows, in column 1, average weekly dollar earnings of production or non-supervisory workers on nonagricultural payrolls in the United States during the period 1985–1989. These earnings data are expressed in *current dollars*. For example, 1985 earnings are expressed in terms of the buying power of workers' dollars in 1985, and earnings in 1986 are expressed in terms of the buying power of workers' dollars in 1986. We see that earnings in current dollars increased steadily during the period.

The extent to which the prices of goods and services purchased for daily living by the workers and their dependents changed during the period is shown in column 2 by the consumer price index for urban wage earners and clerical workers. This index is expressed on a 1985 base period in the table. Note that prices in 1986 stood at 101.6 percent of their 1985 level. Thus, on the average, families had to spend $1.016 in 1986 for every $1 spent in 1985 in purchasing goods and services for daily living. Consequently, the average weekly earnings of $304.85 in 1986 were equivalent in purchasing

	(1)	(2)	(3)	(4)	(5)
			Weekly Earnings in 1985 Dollars	Relatives for Column 3	
		Consumer			Percent
	Weekly	Price Index	$(1) \div \frac{(2)}{100}$	Link	Relative
Year	Earnings	(1985 = 100)		Relative	(1985 = 100)
1985	299.09	100.0	299.09	—	100.0
1986	304.85	101.6	300.05	100.3	100.3
1987	312.50	105.2	297.05	99.0	99.3
1988	322.36	109.4	294.66	99.2	98.5
1989	335.20	114.7	292.24	99.2	97.7

Source: Basic data from *Monthly Labor Review*

TABLE 26.4
Earnings in current and 1985 dollars—Weekly earnings example

power to 304.85/1.016 = $300.05 at 1985 prices. Since 1985 average weekly earnings were $299.09, we see that most of the increase in average weekly earnings between 1985 and 1986 (from $299.09 to $304.85) was offset by price increases and that real earnings increased only slightly.

The average weekly earnings in the other years at 1985 prices are obtained by the same procedure: Each current earnings figure is divided by the decimal value of the price index for that year. Column 3 of Table 26.4 contains the average weekly earnings data expressed at 1985 prices. These data are said to be in *constant dollars* or *1985 dollars* since they are expressed in terms of the purchasing power of the dollar in 1985.

Relative changes in constant-dollars earnings indicate the relative changes in purchasing power associated with the money earnings—that is, the relative changes in *real earnings*. To show these relative changes explicitly, we have expressed the constant-dollars data as link relatives in column 4 of Table 26.4 and as percent relatives in column 5 with 1985 = 100. The link relatives show that real earnings increased slightly between 1985 and 1986 and then declined a little each year between 1986 and 1989. The percent relatives show that real earnings in 1989 were only at 97.7 percent of their 1985 level. □

Comment

We noted earlier that proportional magnitudes in an index series are not affected by the choice of the base period but that the absolute magnitudes are affected. The same holds for constant-dollars series. Thus, in the weekly earnings example of Table 26.4, if the consumer price index had been expressed on a 1989 base, the constant-dollars earnings would have differed from column 3 of Table 26.4, but the percent relatives and the link relatives based on these 1989 constant-dollars data would have remained exactly the same.

Measuring Quantity Changes

Many important business and economic series on volume of activity are expressed in current dollars. Changes in the dollar volume reflect quantity changes, or price changes, or both. Often, there is interest in the quantity changes alone. For example, if retail sales this

TABLE 26.5
Manufacturer's sales in current and 1989 dollars—Appliance sales example

Year	(1) Annual Sales ($ thousand)	(2) Product Price Index (1989 = 100)	(3) Sales in 1989 Dollars ($ thousand) $(1) \div \dfrac{(2)}{100}$	(4) Link Relative	(5) Percent Relative (1989 = 100)
				Relatives for Column 3	
1989	38,500	100	38,500	—	100.0
1990	43,538	103	42,270	109.8	109.8
1991	49,050	105	46,714	110.5	121.3
1992	54,950	107	51,355	109.9	133.4

year are 5 percent higher than last year's sales, is this due to price increases only or did the physical volume of goods sold increase? We can measure relative changes in quantities by the use of price indexes. The procedure is similar to that for expressing current-dollars earnings as constant-dollars earnings.

Appliance Sales   □ **EXAMPLE**

Table 26.5, column 1, contains annual sales (in $ thousand) by a manufacturer of household appliances during 1989–1992. A relevant price index for the products of the manufacturer is shown in column 2. The constant-dollars sales in 1989 dollars are shown in column 3. The calculations parallel those for obtaining constant dollars earnings in Table 26.4. The link relatives and the percent relatives (1989 = 100) of the constant-dollars sales are shown in columns 4 and 5, respectively. We see that the quantity of household appliances sold increased by 33 percent between 1989 and 1992 and that the annual rate of increase was about 10 percent per year.   □

Comments

1. When current-dollars series are converted into constant dollars, the price index must be relevant to the series. For instance, we would not use a food price index to adjust a series on appliances sales, or the consumer price index to adjust the sales of a steel manufacturer.
2. When a current-dollars series is converted into a constant-dollars series, the latter is often referred to as a *deflated* series.

26.6   QUANTITY INDEXES

A quantity index is a summary measure of relative changes over time in the quantities of a set of items—for instance, in the quantities of automobiles, trucks, and other vehicles imported annually by the United States. In such index series, the quantities can change from one period to another, but the prices or other weights remain fixed. The index formulas are analogous to those for price indexes.

(26.5)
The quantity index I_t for period t calculated by the **method of weighted aggregates** is:

(26.5a) $\qquad I_t = 100 \dfrac{\sum\limits_i Q_{it} P_{ia}}{\sum\limits_i Q_{i0} P_{ia}}$

The quantity index I_t for period t calculated by the **method of weighted average of relatives** is:

(26.5b) $\qquad I_t = \dfrac{\sum\limits_i \left(100 \dfrac{Q_{it}}{Q_{i0}}\right) V_{ia}}{\sum\limits_i V_{ia}}$

A problem in constructing quantity indexes arises in the choice of weights. The usual price or value weights may not be appropriate here. Consider a company that wishes to measure quantity changes in it outputs of different electronic circuits assembled from purchased components. Here, price weights or value weights for the different circuits would reflect mainly the prices or values of the purchased components and not the quantities of assembly work performed by the company per se. Weights derived from the typical number of hours expended in assembling each circuit or from the *value added* to each circuit in the assembly would be more appropriate. Value added here would be the value of the circuit when shipped from the plant less the cost of the purchased components and other purchased goods and services (for example, fuel, electricity, containers) used in assembling the circuit and packaging it for shipment. The index of industrial production utilizes value-added weights in many of its segments.

PROBLEMS

* **26.1** An aquarium supply company sells two exotic strains of guppy (Monarch and Emperor). Prices (in dollars) for breeding pairs of these strains during 1987–1991 follow:

Year:	1987	1988	1989	1990	1991
Monarch:	6.00	6.25	7.00	7.50	8.25
Emperor:	8.00	8.50	9.00	10.00	10.75

a. For each of the two price series, obtain (1) percent relatives on a 1987 base, (2) link relatives.
b. For a Monarch pair, what is the percent point change in the percent relatives between 1989 and 1990? What is the corresponding percent change?
c. For which strain was the relative price increase between 1987 and 1991 greater? For which strain was the relative price increase between 1990 and 1991 greater?

26.2 Average prices (in cents per 100 grams) of peanut butter and canned tuna in a supermarket chain during 1988–1991 follow:

Year:	1988	1989	1990	1991
Peanut butter:	40	43	46	50
Canned tuna:	65	69	77	78

a. For each of the two price series, obtain (1) percent relatives on a 1988 base, (2) link relatives.

b. For peanut butter, what is the percent point change in the percent relatives between 1990 and 1991? What is the corresponding percent change?

c. For which food product was the absolute price increase between 1988 and 1991 greater? For which food product was the relative price increase greater? Is any contradiction involved here? Explain.

d. If 1991, instead of 1988, were used as the base period for the two series of percent relatives, would the magnitudes of the percent relatives differ from those in part a? Would the proportional magnitudes of the percent relatives differ from those in part a? Explain.

26.3 Annual Catch. The annual catches (in thousand metric tons) of three species of fish brought in at a port during 1988–1991 follow:

Year:	1988	1989	1990	1991
Flounder:	13.4	12.2	15.0	15.8
Haddock:	2.1	2.0	2.3	2.6
Cod:	4.0	4.3	4.3	4.5

a. For each of the three series, obtain (1) percent relatives on a 1988 base, (2) link relatives.

b. For which species was the relative increase in annual catch between 1988 and 1991 greatest? For which species was the relative increase greatest between 1990 and 1991?

c. Were the calculations in part a sufficient to answer part b or were additional calculations required? Explain.

* **26.4 Resort Complex.** Many types of items must be replaced periodically in routine maintenance of appliances and plumbing in a resort complex. A schedule of items that is representative of all these items follows, together with the quantity weights and the unit prices of the items in 1989, 1990, and 1991. A price index series for appliances and plumbing supplies is to be compiled.

Item	Quantity	Unit Price ($)		
		1989	1990	1991
Ice-making machine	1	909.00	920.00	940.00
Air conditioner	12	246.00	246.00	250.00
Color television	15	440.00	416.50	395.00
Sink faucet	80	22.16	22.45	23.00

a. Obtain the aggregate cost of the schedule of items in each year. Does this cost vary from year to year because of price differences only, quantity differences only, or both price and quantity differences?

b. Obtain the price index series by the method of weighted aggregates. Use 1989 as the base period. Interpret the index number for 1991.

26.5 Refer to **Annual Catch** Problem 26.3. Prices (in dollars per metric ton) received by fishermen at the port follow:

Year:	1988	1989	1990	1991
Flounder:	938	1270	1259	1198
Haddock:	859	872	1095	1299
Cod:	602	731	806	752

Obtain a price index series by the method of weighted aggregates. Use 1988 as the base period and the annual catches in 1988 as weights. Interpret the index number for 1991.

26.6 Guard Agency. The following data for personnel of the Marlowe Guard Agency show hourly wage rates and numbers of hours utilized in 1989, 1990, and 1991 for each job classification:

Job Classification	Wage Rate ($)			Hours Utilized (thousand)		
	1989	1990	1991	1989	1990	1991
Plant protection	9.32	9.51	9.70	50	58	55
Construction site patrol	8.49	8.81	9.16	40	44	42
Escort service	12.38	13.25	13.56	52	47	48

a. Obtain an index series of wage rates by the method of weighted aggregates. Employ 1989 as the base period and 1989 hours utilized as weights. Interpret the index number for 1991.
b. Would the index series in part a be different if 1991 were used as the weight period of the index? Explain.

* **26.7** Refer to **Resort Complex** Problem 26.4.

a. Obtain a price index series by the method of weighted average of relatives. Use 1989 as the base period and 1989 prices in the value weights.
b. Must your results in part a be the same as those in Problem 26.4b? Explain.
c. If the base period for the index series in part a had been specified as 1990, what percent relatives would have been utilized?

26.8 Refer to **Annual Catch** Problems 26.3 and 26.5.

a. Obtain a price index series by the method of weighted average of relatives. Use 1988 as the base period and derive the value weights from 1989 prices and annual catches. Interpret the index number for 1991.
b. Would the price index series in part a be different if the value weights were derived from 1988 prices and annual catches? Explain.

26.9 Refer to **Guard Agency** Problem 26.6.

a. Obtain an index series of wage rates by the method of weighted average of relatives. Use 1989 as the base period and derive the value weights from 1989 wage rates and hours utilized. Interpret the index number for 1990.
b. Must the results in part a agree with those in Problem 26.6a? Explain.

26.10 Why are value weights used when averaging percent relatives of prices while it suffices to use quantity weights when averaging unit prices? Explain.

26.11 In the compilation of a price index series for certain raw materials used principally by the textiles industry, why would it be inappropriate to select as a base period for the index a year when the textiles industry was economically depressed and demand for the raw materials was very low?

26.12 A price index series for products manufactured by the Zarthan Company during 1986–1991 has been compiled on a 1988 base and is as follows:

Year:	1986	1987	1988	1989	1990	1991
Index:	96.2	99.1	100.0	101.1	104.7	105.2

Shift the index series to a 1990 base. Are the proportional magnitudes of the index numbers maintained by this shift? Explain.

26.13 A price index series for exports of lumber products during 1987–1991 follows. The series has been compiled on a 1988 base.

Year:	1987	1988	1989	1990	1991
Index:	108.6	100.0	97.6	106.7	110.8

a. Shift the index series to a 1990 base.

b. Will the link relatives for the price index series on the 1990 base be the same as those for the price index series on the 1988 base? Explain.

26.14 Consider the following two price index series for certain industrial paints.

Year:	1986	1987	1988	1989	1990	1991
Old series (1986 = 100):	100	111	110			
New series (1990 = 100):			84	92	100	115

Splice the two index series into one continuous series with a 1986 base. Interpret the index number for 1990 in the spliced series.

26.15 The following two price index series for certain kinds of wood pulp are to be joined into one continuous series:

Year:	1985	1986	1987	1988	1989	1990	1991
Old series (1985 = 100):	100	109	113	110			
New series (1988 = 100):				100	97	109	114

a. Splice the two series into one continuous series with a 1988 base.
b. Obtain the link relatives for the spliced series. Would the link relatives be the same if a 1985 base had been used for splicing the two series? Explain.

26.16 One of the items to be included in a price index of school supplies is lead pencils. What are some price-affecting characteristics of lead pencils that you would consider in preparing specifications for obtaining price data for this item?

26.17 One of the items to be included in a food price index is canned mushrooms. What are some price-affecting characteristics of canned mushrooms that you would consider in preparing specifications for obtaining price data for this item?

26.18 Several years ago some manufacturers of clothes washers planned to eliminate the warm-rinse option because rinsing is a mechanical function and water temperature plays no part in the actual rinsing process. This change would yield an eight percent saving in energy use of clothes washers. However, many users appeared unwilling to accept the concept of cold rinsing. An increase in customer complaints and service calls could be anticipated if the warm-rinse option were eliminated. There would have been no price changes in the washers.

a. From the point of view of collecting price data for a price index, would this design change constitute a quality improvement (hence, a price decrease) or a quality deterioration (hence, a price increase)? Discuss.
b. What problem in the compilation of ongoing price index series is illustrated here?

26.19 An ammonia product has been selected to represent all cleaning products in the schedule of items used to compile a household maintenance price index. Should household expenditures for only this particular product or for all cleaning products be used as the value weight in calculating the price index based on this schedule of items? Explain.

* **26.20** Average weekly earnings (in current dollars) of production or nonsupervisory workers on private construction payrolls in the United States during 1985–1989 follow:

Year:	1985	1986	1987	1988	1989
Weekly earnings:	464.46	466.75	480.44	495.73	512.41

Source: Monthly Labor Review

a. Use the consumer price index in Table 26.4 to calculate each year's earnings in constant 1985 dollars. Interpret the constant-dollars earnings for 1989.
b. For the constant-dollars earnings series in part a, obtain (1) percent relatives on a 1985 base, (2) link relatives.

c. Make the following comparisons between the results in part b for construction workers and those in Table 26.4 for all nonagricultural workers: (1) the percent change in real earnings between 1985 and 1989, (2) the percent change in real earnings between 1988 and 1989.
d. Would the comparisons in part c be affected if each year's earnings were expressed in constant 1989 dollars? Explain.

26.21 The annual pensions (in current dollars) during 1986–1991 of former plant workers of the Penbrook Corporation who retired at the end of 1985 with a C-8 classification follow. An appropriate price index series, based on expenditures patterns of retirees, is also shown for the same period. The corporation's pension plan requires adjustments for cost-of-living changes every three years.

Year:	1986	1987	1988	1989	1990	1991
Annual Pension:	15,400	15,400	15,400	16,700	16,700	16,700
Index (1985 = 100):	111	115	120	124	127	131

a. Obtain each year's pension at 1985 prices. Interpret the constant-dollars pension for 1991.
b. For the constant-dollars pension series in part a, obtain (1) percent relatives on a 1986 base, (2) link relatives. Describe the changes in real pension earnings during 1986–1991.

26.22 Data on the value of shipments by a manufacturing firm follow, as well as an index of selling prices for the firm's products during 1986–1991:

Year:	1986	1987	1988	1989	1990	1991
Shipments ($ million):	7.1	9.1	10.3	12.7	14.8	18.4
Index (1988 = 100):	84	98	100	112	119	133

a. Obtain the value of shipments at 1988 prices for each year.
b. For the constant-dollars shipments series in part a, obtain (1) percent relatives on a 1986 base, (2) link relatives.
c. Has the increase in the value of shipments between 1986 and 1991 been due primarily to increases in the selling prices? Have the relative year-to-year changes in the constant-dollars series been fairly stable during the period? Discuss.

26.23 Data on annual imports and exports of merchandise by a developing country (in millions of national currency units) during 1986–1991 follow, as well as indexes of prices paid for imported merchandise and prices received for exported merchandise:

Year:	1986	1987	1988	1989	1990	1991
Imports:	380	383	435	420	525	706
Exports:	397	410	435	485	545	620
Imports index:	100	106	110	140	233	249
Exports index:	100	102	106	118	132	150

a. Express the imports and exports series in constant 1986 currency units. Interpret the constant-currency-units exports value for 1991.
b. Which constant-currency-units series in part a experienced the greater relative change between (1) 1986 and 1991, (2) 1990 and 1991?
c. Was the 86 percent increase in imports between 1986 and 1991 accounted for mainly by higher prices paid for imported merchandise or by a larger volume of imported merchandise? Explain.

26.24 Two alternative clauses concerning cost-of-living adjustments are under discussion in contract negotiations. One clause calls for a 15-cent change in the wage rate for each 1 percent change in the consumer price index. The other clause calls for a 15-cent change in the wage rate for each 1 percent point of change in the same index. Are the two clauses equivalent? Explain.

26.25 Refer to **Annual Catch** Problems 26.3 and 26.5.

a. Obtain a quantity index series for the annual catches by the method of weighted aggregates. Use 1988 as the base period and prices received in 1988 as weights.

b. By what percent did the quantity of annual catches change from 1988 to 1991 according to the index series?

26.26 Refer to **Guard Agency** Problem 26.6. Obtain a quantity index series for the utilization of personnel by the method of weighted aggregates. Employ 1989 as the base period and 1989 wage rates as weights. Interpret the index number for 1991.

26.27 Refer to **Annual Catch** Problems 26.3 and 26.5. Obtain a quantity index series for the annual catches by the method of weighted average of relatives. Use 1988 as the base period and derive the value weights from 1989 quantities and prices. Interpret the index number for 1991.

26.28 Refer to **Guard Agency** Problem 26.6.

a. Obtain a quantity index series for the utilization of personnel by the method of weighted average of relatives. Use 1989 as the base period and derive the value weights from 1989 wage rates and hours utilized.

b. By what percent did the quantity of personnel utilized change from 1989 to 1991 according to the index series?

EXERCISES

26.29 Refer to Table 26.1.

a. Obtain the geometric mean of the link relatives for 1989–1991 in column 4. Based on this mean value, what has been the average annual growth rate of the chemical compound's price during 1988–1991?

b. Show how the percent relative for 1991 (on a 1988 base) can be calculated directly from the geometric mean in part a.

26.30 Refer to Table 26.2. Obtain the price index series based on 1988–1989 = 100. Does the use of a two-year base period affect the year-to-year relative changes in the price index here? Explain.

26.31 Refer to **Resort Complex** Problem 26.4. You have been asked to examine whether the index should be updated. What are some of the questions you would investigate?

26.32 Explain why shifting the base of an index series by the shortcut method gives the same results as those obtained by recalculating the index series from the original data when the series is based on the method of weighted aggregates, but may not give the same results if the series is based on the method of weighted average of relatives.

STUDIES

26.33 Total annual construction costs of restaurants built by a fast-food chain during 1987–1990 follow, together with a construction cost index for restaurants of this type:

Year:	1987	1988	1989	1990
Cost ($ thousand):	1695	2501	4503	1938
Construction cost index:	100	108	126	135

Today's value of the construction cost index is 160. What is the approximate replacement cost today of the restaurants constructed during the period 1987–1990; that is, what would be the total construction cost if all of these restaurants were built today?

26.34 Explain how you would construct a daily index of stock prices for common stocks traded on a major exchange. In your explanation, describe (1) the types of stocks that would appear in your schedule of items, (2) the quantity or value weights to be used, (3) the choice of the base period, and (4) whether the method of weighted aggregates or the method of weighted average of relatives would be used in computing the index. Also comment on any special factors that would need to be considered in compiling and maintaining the index.

26.35

a. Consult the *Monthly Labor Review* to obtain the index values of the consumer price index for all urban consumers for each of the most recent six months reported. Also, obtain the index values for the following components: (1) food and beverages, (2) housing, (3) apparel and upkeep, (4) transportation.

b. Determine the percent changes in the all-items index and in each of the four component indexes for the six-month period covered by your data. Which components showed larger relative price changes than the all-items index? Which ones showed smaller relative changes?

Bayesian Decision Making

Bayesian Decision Making Without Sample Information

27

The methods of statistical inference presented in earlier chapters do not take explicit account of monetary or other economic consequences. For example, monetary factors are not considered directly in a statistical test for a population mean in choosing the test alternatives (one-sided versus two-sided) or in specifying the α and β risks, although these factors are considered implicitly. Furthermore, the earlier inference methods use only the information contained in the sample in drawing conclusions about the population parameters and make no use of other pertinent information that may be available. For example, a confidence interval for the population mean is based solely on information contained in the sample and takes no account of any prior information that might be available about the value of the mean.

In this chapter and the next one, we introduce the basic principles and methods of Bayesian decision making. This methodology takes explicit account of monetary and other economic consequences in making decisions and drawing statistical conclusions, and it utilizes information about the population that may be available prior to sampling. In this chapter, we discuss decision making when no sample information is available, and in the next chapter we take up decision making when sample information is available.

We begin by considering the common structure of decision problems. Then we take up decision making under certainty and decision making under uncertainty, in both of which the economic consequences of decisions are explicitly taken into account. We conclude this chapter by considering Bayesian decision making, where not only the consequences of decisions are taken into account but also prior information about the population is utilized.

Decision problems, whether in business, government, or any other sphere, have a common structure. To illustrate this common structure, we consider two examples.

DECISION PROBLEMS **27.1**

EXAMPLE **Engine Production**

Firm A has developed a new engine that will give greater fuel economy than conventional engines. Tooling up for mass production can be initiated now or can be delayed for another year in the hope that further research and development will yield additional major refinements. The engineering group assesses the probability of achieving additional major refinements in the next year to be 0.6. However, a rival—firm B—has also developed a new engine and has just begun to tool up for mass production. This engine is competitive with the present version of firm A's new engine. Thus, firm A will be at a long-term competitive disadvantage if it delays mass production of its new engine and then finds that no additional major refinements were achieved in the next year. However, if additional major refinements are achieved within the next year, firm

A will more than recoup the losses from the delay in getting into mass production. The executive committee of firm A must now decide whether to initiate immediate tooling up for mass production or undertake further research and development for another year.   □

Tanker   □   **EXAMPLE**

A small oil tanker has just discharged its cargo and is about to begin another round trip. The tanker's operator must decide whether to air-blow the ship's tanks to clear the gas residue or to fill the tanks with water ballast. The former procedure costs $200, while pumping the water ballast costs $2000. Air-blowing will suffice unless there is a collision affecting a tank. If this occurs, a tank that is merely air-blown will explode, with damage to the ship and possible risk to the crew—approximately a $10-million estimated loss altogether—whereas a tank filled with water ballast cannot explode. The tanker operator is not able to assess the probability of a tank collision but knows that the possibility cannot be ignored since two narrow and busy channels must be negotiated on the return trip for another cargo of oil.   □

Structural Elements of Decision Problems

The two examples just presented illustrate the common structural elements of decision problems.

(27.1)

The structural elements of decision problems are as follows.

1. **Decision maker:** The decision maker is either a single individual (the tanker operator) or a group of persons acting as a single individual (the executive committee).
2. **Acts or strategies A_j:** These are the alternative courses of action open to the decision maker (tool up now, delay; air-blow, use water ballast). They reflect the controllable variables in the decision problem—that is, the variables under the decision maker's control.
3. **Outcome states S_i:** These are the different situations that may prevail that affect the consequences of the acts (further refinements in engine will be achieved, no further refinements will be achieved; tank collision, no tank collision). The outcome states reflect the uncontrollable variables in the decision problem—that is, the variables not under the decision maker's control.
4. **Consequences or payoffs C_{ij}:** These are the different gains, rewards, or other consequences that may be experienced by the decision maker, measured in monetary or other relevant units. The notation C_{ij} denotes the gain or payoff experienced when the jth course of action is chosen and the ith outcome state prevails (when the operator uses air-blowing and a tank collision occurs).
5. **Outcome state probabilities $P(S_i)$:** These are probabilities that the decision maker may assign to the outcome states. Outcome state probabilities are not present in all decision problems (they are present in the engine production example but not in the tanker example).
6. **Decision criterion:** This is the basis for identifying the act that is "best" among the courses of action available to the decision maker (the examples did not illustrate any decision criterion).

Payoff Tables and Decision Trees

Two convenient devices for displaying the acts, outcome states, consequences, and probabilities in a decision problem are *payoff tables* and *decision trees.* They present the same information; hence, either one can be used depending on whether a tabular or graphic display is more effective in the given situation.

Payoff Table.   Table 27.1a shows the format and notation we shall employ for payoff tables. The number of outcome states is denoted by r and the number of available acts by c. The consequence of the jth act when the ith outcome state prevails is denoted by C_{ij}, as mentioned before. Table 27.1b presents the payoff table for the tanker example, where $r = c = 2$. Since the tanker operator was unable to assign probabilities $P(S_i)$ to the different possible outcome states S_i, "n.a." (not applicable) is entered in place of the $P(S_i)$. Ordinarily, we shall omit this column when probabilities are not used. The entries in Table 27.1b are negative because they entail losses to the operator. Thus, $C_{11} = -200$, since the cost of air-blowing is $200 when there is no tank collision.

Note how effectively Table 27.1b consolidates the information in the tanker decision problem. We see quickly that water ballast entails a loss of $2000 regardless of whether or not there is a collision. Air-blowing, on the other hand, entails a loss of only $200 if there is no tank collision but a loss of $10 million if a collision occurs.

Decision Tree.   Figure 27.1a shows the general format of a decision tree for a decision problem with two acts and two outcome states, and Figure 27.1b illustrates the decision tree for the tanker example. A decision tree consists of a sequence of *nodes* and *branches.* We enter from the left. The branches emanating from the first node represent the alternative courses of action A_j. Since this node entails a controllable variable, it is called a *decision node.* The next set of branches represents the possible outcome states S_i for each course of action. The nodes from which these branches emanate involve uncontrollable variables and are called *chance nodes.* The outcome state probabilities $P(S_i)$ are shown on the branches when applicable. Finally, the payoff for each sequence of act A_j and outcome state S_i is shown at the right. We follow the convention of indicating decision nodes by squares and chance nodes by circles.

TABLE 27.1
General format of a payoff table and payoff table for tanker example

(a) General Format

Outcome State	Probability	Act A_1	A_2	$\cdots$	A_c
S_1	$P(S_1)$	C_{11}	C_{12}	$\cdots$	C_{1c}
S_2	$P(S_2)$	C_{21}	C_{22}	$\cdots$	C_{2c}
$\vdots$	$\vdots$	$\vdots$	$\vdots$		$\vdots$
S_r	$P(S_r)$	C_{r1}	C_{r2}	$\cdots$	C_{rc}

(b) Tanker Example (payoffs in dollars)

Outcome State	Probability	Act Air-Blow	Water Ballast
No tank collision	n.a.	-200	-2000
Tank collision	n.a.	-10 million	-2000

FIGURE 27.1

General format of a decision tree and decision tree for tanker example

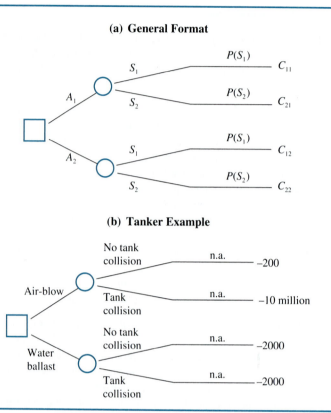

(a) General Format

(b) Tanker Example

Comments

1. The identification of feasible acts and relevant outcome states, and the measurement of payoffs, can be highly demanding in actual decision problems. For instance, in our tanker example, it is very difficult to assess the loss resulting from a tank explosion. In addition, the nature of the decision problem must be examined in great detail to identify all available acts so that no desirable courses of action are overlooked.

2. Only payoffs directly affected by a course of action need be considered in the payoff table. Gains or losses that run on unaffected by the action or that occurred in the past can be excluded. For instance, in the tanker example, there is no need to consider the crew's wages for the next trip because these will occur regardless of whether water ballast is used or the ship's tanks are air-blown.

27.2   DECISION MAKING UNDER CERTAINTY

The first type of decision problem we consider is one in which each act has a unique consequence. Decision problems of this type involve decision making under certainty.

(27.2)

In **decision making under certainty,** each act A_j has a unique consequence that is known in advance.

A key feature of decision making under certainty is that there are no uncontrollable variables present.

A municipality must select an investment firm to handle its forthcoming bond issue. Three highly reputable firms have made written bids guaranteeing the net interest rate the municipality would pay. The payoff table, indicating the guaranteed interest rate, is as follows:

	Firm A	Firm B	Firm C
Guaranteed interest rate	9.92%	9.80%	9.88%

Note there is only a single outcome state here, or equivalently, there are no uncontrollable variables present.

The municipality decided to employ the decision criterion: Minimize the total interest to be paid. Hence, firm B was selected.

☐

Decision problems involving certainty sometimes are referred to as *deterministic*. These decision problems are not necessarily simple. Sophisticated mathematical methods may be required to find the act that is best according to the decision criterion. But the characteristic feature of all such problems, whether simple or complex, is that for each act there is a unique payoff that is known in advance.

We turn next to decision problems where uncontrollable variables are present. Here several outcome states are possible, and the payoff for act A_j depends on the outcome state S_i that prevails. Since the outcome state S_i is not known with certainty, two different approaches may be taken.

DECISION MAKING UNDER UNCERTAINTY **27.3**

1. The decision maker is unwilling or unable to assign probabilities to the outcome states S_i and regards the outcome states simply as uncertain.
2. The decision maker assigns probabilities $P(S_i)$ to the outcome states.

We consider the first approach in this section. It is called decision making under uncertainty.

(27.3)
In **decision making under uncertainty**, there are two or more outcome states S_i that affect the payoff, and no assignment of probabilities $P(S_i)$ is made to these states.

The uncertainty approach tends to be appealing when there is no relevant past experience or other basis for assessing the outcome state probabilities $P(S_i)$.

A firm planning to produce a special camera and film for making photographs with a distinct three-dimensional effect is being sued for patent infringements in separate law-

TABLE 27.2
Payoff table ($ million)—
Camera firm example

	Act				
Outcome State	A_1 Tool Up for Both Camera and Film	A_2 Tool Up for Camera Only	A_3 Tool Up for Film Only	A_4 Do Not Tool Up for Either	Maximum Payoff
S_1: Win both suits	1450	1400	1100	1100	1450
S_2: Win camera suit only	1300	1425	1000	1000	1425
S_3: Win film suit only	700	825	850	850	850
S_4: Lose both suits	400	775	800	810	810
Minimum payoff	400	775	inadmissible	810	—

suits by a camera manufacturer and a film manufacturer. The alternative courses of action being considered by management are shown in Table 27.2. Note that they involve different degrees of tooling up to produce the camera and film. For example, act A_1 entails tooling up for both camera and film, while act A_2 entails tooling up for the camera but delaying tooling up for the film until the outcomes of the lawsuits are known. The firm has decided to oppose both suits and to obtain definitive legal decisions rather than seek any compromise settlements. Hence, the outcome states are the four possible combinations of legal decisions in the two suits shown in Table 27.2. The firm's legal counsel is unwilling to assess probabilities for the outcome states because legal issues without precedent are involved. Consequently, management has decided to regard the outcome states simply as uncertain. Thus, the problem is one of decision making under uncertainty.

The payoffs C_{ij} in Table 27.2 represent the estimated present value of future net income in millions of dollars. These payoffs are affected by two factors: (1) If the firm loses a lawsuit, it can still manufacture the item but at a higher production cost because of modifications to avoid patent infringement. (2) Competing firms will develop their own versions of the camera and film, so it is advantageous for the firm to place its products on the market quickly if the legal cases are won. □

Admissible Acts

Not all available acts need be considered in every case. An act may be so poor that it should be dropped immediately. To determine whether such poor acts are present, we compare the payoffs for each pair of acts.

☐ **EXAMPLE**

In the camera firm example, let us compare acts A_3 and A_4 in Table 27.2. We note that the payoffs are equally good under outcome states S_1, S_2, and S_3. However, under S_4, the payoff for act A_3 ($C_{43} = 800$) is less than that for act A_4 ($C_{44} = 810$). Hence, there is no need to consider act A_3 any further because the payoff with this act cannot be higher than with act A_4, but it can be lower. We say that act A_4 dominates act A_3 here, and hence act A_3 is inadmissible. □

A formal definition of admissible and inadmissible acts follows.

(27.4)
Act A_j **dominates** act $A_{j'}$ if A_j offers as good a payoff as $A_{j'}$ in every outcome state and a better payoff in at least one outcome state; that is, if $C_{ij} \geq C_{ij'}$ for all outcome states i, and $C_{ij} > C_{ij'}$ for at least one outcome state.

Act A_j is **admissible** if it is not dominated by any other available act, and it is **inadmissible** if it is dominated by another act.

EXAMPLE □

Table 27.2 for the camera firm example shows that acts A_1, A_2, and A_4 are not dominated by any other act and hence each is admissible. Thus, only acts A_1, A_2, and A_4 are candidates for best act here. □

Criteria for Selecting Best Act

To choose the "best" act, we require a decision criterion. We now discuss two criteria that have been proposed by decision theorists for decision making under uncertainty.

Maximin Criterion. A decision maker who tends to be cautious may emphasize the guaranteed payoff for each act—that is, the level below which the actual payoff cannot go—and select the act that offers the best guarantee. To put this another way, the decision maker would select the act that maximizes the minimum possible payoff.

(27.5)
The **maximin criterion** considers the minimum payoff that can be obtained with each act A_j—namely, $\min_i C_{ij}$—and selects that act for which the minimum payoff is maximized:

$$\max_j(\min_i C_{ij})$$

EXAMPLE □

In the camera firm example, the minimum payoffs for the three admissible acts A_1, A_2, and A_4 are the column minimums shown at the bottom in Table 27.2. For example, the minimum payoff for act A_1 is 400. We see from Table 27.2 that act A_4 maximizes the minimum possible payoff and would be chosen under the maximin criterion. Of course, the actual payoff with act A_4 can be higher than the minimum $810 million (if the outcome state is S_1, S_2, or S_3), but it cannot be lower. □

> **Comment**
>
> In some decision problems involving uncertainty, the outcome states are under the control of an adversary (for example, a business competitor) whose objective is to hold the decision maker to as poor a payoff as possible. The maximin criterion may have particular appeal to the decision maker here since it guarantees the best of the worst possible payoffs that can be forced on the decision maker by the adversary.

Minimax Regret Criterion.   A second criterion for decision making under uncertainty may appeal to a decision maker who reasons as follows: "I must commit myself now to a course of action before I know the outcome state. But once the outcome state is known, the 'Monday morning quarterbacks' will compare my result against the best that could have been obtained in light of the actual outcome state that prevailed. I would like to protect myself against falling too far short!"

   A decision maker taking this position is concerned about regrets or opportunity losses. We shall denote the regret for the ith outcome state and jth act by O_{ij}.

> **(27.6)**
>
> The **regret** or **opportunity loss** O_{ij} is the difference between the best payoff attainable under the ith outcome state—namely, $\max_j C_{ij}$—and the payoff attained with the jth act under that state:
>
> $$O_{ij} = (\max_j C_{ij}) - C_{ij}$$

The minimax regret criterion considers the maximum regret for each act and selects as best the act with the smallest maximum regret.

> **(27.7)**
>
> The **minimax regret criterion** considers the maximum regret that can be obtained with each act A_j—namely, $\max_i O_{ij}$—and selects that act for which the maximum regret is minimized:
>
> $$\min_j(\max_i O_{ij})$$

☐ **EXAMPLE**

The regrets for the camera firm example are shown in Table 27.3. They are obtained from Table 27.2, which contains both the payoffs and the maximum payoff for each outcome state (in the right column). Regret O_{11}, for instance, is obtained by noting that the maximum payoff attainable under outcome state S_1 is 1450. Hence, $O_{11} = 1450 - C_{11} = 1450 - 1450 = 0$. The regret of zero indicates that act A_1 is the best act when outcome state S_1 prevails. Similarly, $O_{12} = 1450 - 1400 = 50$, which indicates that the payoff under S_1 with act A_2 is \$50 million worse than with the best act for this outcome state. The other regrets are obtained in the same manner.

TABLE 27.3
Regret table ($ million)—
Camera firm example

Outcome	Act		
State	A_1	A_2	A_4
S_1	0	50	350
S_2	125	0	425
S_3	150	25	0
S_4	410	35	0
Maximum regret	410	50	425

The maximum regret for each act is shown at the bottom in Table 27.3. For example, the maximum regret for act A_1 is 410. We see that act A_2 has the smallest maximum regret. The decision maker, in selecting act A_2, is assured that the payoff actually achieved when the outcome state materializes will not fall short of the best attainable payoff by more than $50 million. □

Choice of Decision Criterion. We have considered two decision criteria for selecting a course of action in decision making under uncertainty. Still other criteria are available. Unfortunately, different criteria frequently select different acts as best. This occurred in the camera firm example, where the maximin criterion selected act A_4 while the minimax regret criterion selected act A_2. Since the theory of decision making under uncertainty does not identify any criterion as objectively best, the choice of a criterion remains subjective.

27.4 BAYESIAN DECISION MAKING WITH PRIOR INFORMATION ONLY

We now consider the second of the two approaches mentioned earlier for decision problems in which uncontrollable variables are present—namely, when the decision maker assigns probabilities $P(S_i)$ to the outcome states. This approach is called Bayesian decision making.

> **(27.8)**
> In **Bayesian decision making,** there are two or more outcome states S_i that affect the payoff, and probabilities $P(S_i)$ are assigned to these outcome states.

Bayesian decision making plays an important role in both the theory and practice of decision making, and we shall study it in the remainder of this chapter and in the following one. In this chapter, we take up Bayesian decision making using only prior information about the outcome states in the form of probabilities $P(S_i)$. In the next chapter, we consider Bayesian decision making when sample information is also available. Bayesian decision making with prior information only is also called *decision making under risk*.

EXAMPLE □ **Island Tour**

Mr. Adams operates a tourist service on an island noted for its historic sites. Tourists on packaged tours to a nearby mainland area may make optional side trips by air to this island. Mr. Adams handles the travel and lodging reservations for these side trips. He must make reservations for lodging and travel before he knows how many tourists will make the side trip. He gains $50 for each filled reservation up to four reservations but loses a $30 advance deposit for each unfilled reservation. Mr. Adams neither gains nor loses for any tourists in excess of the number of reservations made. Finally, if five or more persons in an incoming tour take the side trip, the travel agency handling the tour makes the side trip arrangements, reimbursing Mr. Adams for any deposits made and giving him a fixed fee of $150 for servicing the group's visit.

The alternative courses of action available to Mr. Adams are the number of reservations to be made for a tour, as shown in Table 27.4. No more than four reservations need be considered as available acts in this decision problem because any act for more than four reservations is dominated by act A_5 (four reservations). The outcome states are the actual numbers of tourists making the side trip, and these are also shown in Table 27.4.

We illustrate the entries in the payoff table by considering the payoffs for act A_3 (two reservations). If no tourists make the side trip, Mr. Adams forfeits two $30 deposits for a total loss of $60. If one tourist makes the side trip, he gains $50 for that one and loses $30 on the one unfilled reservation, for a net gain of $20. If two tourists make the side trip, he gains $50 for each, for a total gain of $100. If three or four tourists make the side trip, Mr. Adams' gain still is $100 since he neither gains nor loses on tourists in excess of the number of reservations. Finally, if five or more tourists make the side trip, the gain is the fixed fee of $150 paid to him by the travel agency.

Mr. Adams was able to assign probabilities to each of the outcome states from past experience. For example, in approximately 25 percent of past tours, no person elected the side trip; hence, he assessed the probability that no person will make the side trip on the next tour as $P(S_1) = 0.25$. The outcome state probabilities are also shown in Table 27.4. □

An objective interpretation can be given to Mr. Adams' outcome state probabilities since they are based on observed relative frequencies for past tours conducted under conditions similar to those prevailing currently. In many other cases, however, such relative frequencies are not available; instead, subjective or personal probability assessments may then be employed. We discussed such assessments in Chapter 4. Bayesian decision making proceeds in the same fashion whether the probabilities are objective or subjective.

TABLE 27.4
Payoff table (dollars)—Island tour example

Number of Tourists Making Side Trip	Probability $P(S_i)$	Number of Reservations				
		A_1 None	A_2 One	A_3 Two	A_4 Three	A_5 Four
S_1: None	0.25	0	−30	−60	−90	−120
S_2: One	0.10	0	50	20	−10	−40
S_3: Two	0.20	0	50	100	70	40
S_4: Three	0.20	0	50	100	150	120
S_5: Four	0.15	0	50	100	150	200
S_6: Five or more	0.10	150	150	150	150	150

Expected Payoff Criterion

When probabilities $P(S_i)$ are assigned to the outcome states, the payoff for an act is considered to be a random variable with an associated probability distribution. For example, we note from Table 27.4 that if Mr. Adams selects act A_5 (four reservations), the probability distribution of payoffs is as follows:

Payoff C_{i5}:	-120	-40	40	120	200	150
Probability $P(S_i)$:	0.25	0.10	0.20	0.20	0.15	0.10

For brevity, we use the term *lottery* to denote a probability distribution of payoffs. When a decision maker selects a given act in Bayesian decision making, in effect, he or she chooses a given lottery in preference to the other available lotteries.

Which of the available lotteries should Mr. Adams choose? Since his handling of the side trips is a recurrent activity, Mr. Adams may well prefer the lottery that yields the largest expected payoff per tour. This is equivalent to maximizing his total gain over the long run. The act with the largest expected payoff is said to satisfy the expected payoff criterion.

(27.9)

The **expected payoff criterion** selects that act for which the expected payoff is maximized.

We calculate the expected payoff in the manner of any other expected value. In adapting formula (5.5) to our present notation, we obtain the following.

(27.10) $$EP(A_j) = \sum_i C_{ij} P(S_i)$$

Here $EP(A_j)$ denotes the expected payoff for act A_j. The act with the largest expected payoff is called the Bayes act.

(27.11)

The act selected by the expected payoff criterion is called the **Bayes act.** The expected payoff for the Bayes act is denoted by BEP, where:

$$BEP = \max_j EP(A_j)$$

EXAMPLE □

To find the Bayes act for the island tour example, we need to obtain the expected payoff for each act. We illustrate the calculations for act A_5:

$$EP(A_5) = \sum_{i=1}^{6} C_{i5} P(S_i)$$
$$= (-120)(0.25) + (-40)(0.10) + \cdots + 150(0.10) = 43$$

Thus, if Mr. Adams makes four reservations, his expected gain per tour is $43. The other expected payoffs are obtained in similar fashion. They are as follows:

Act:	A_1	A_2	A_3	A_4	A_5
$EP(A_j)$:	15	40	57	58	43

We see that the expected payoff is largest for act A_4, so A_4 (three reservations) is the Bayes act, and the expected payoff for the Bayes act is $BEP = EP(A_4) = \$58$ per tour.

Note that the expected payoff $EP(A_j)$ becomes larger here and then declines as the number of reservations increases. The act of making three reservations strikes the best balance between losses arising from unused reservations and gains from reservations that are utilized.   □

Value of Information

Frequently a decision maker has the option of obtaining information about the outcome state before making a decision. Of course, this information normally entails a cost. We now consider how much a decision maker should be willing to pay to learn with certainty which outcome state prevails. We continue with the island tour example.

Expected Payoff with Perfect Information.   Suppose it were possible for Mr. Adams to ascertain in advance how many tourists will make the side trip (for example, by radiotelephone to the mainland). In that case, Mr. Adams would know the outcome state for each tour—that is, he would have *perfect information* about the outcome state. He would then be in the position of choosing the best act for the given outcome state. Thus, we see from Table 27.4 that if four tourists will make the side trip, he should select act A_5 (make four reservations), which will give him a payoff of $200. The best act for each outcome state and its payoff, and the probability of each outcome state, are as follows:

Outcome State:	S_1	S_2	S_3	S_4	S_5	S_6
Best act:	A_1	A_2	A_3	A_4	A_5	Any
$\max_j C_{ij}$:	0	50	100	150	200	150
$P(S_i)$:	0.25	0.10	0.20	0.20	0.15	0.10

Note that we have another probability distribution of payoffs here, this time for the case where there is perfect information about the outcome state and the best act is chosen for the outcome state. We can now obtain the expected payoff with perfect information, denoted by *EPPI:*

$$EPPI = 0(0.25) + 50(0.10) + 100(0.20) + 150(0.20) + 200(0.15)$$
$$+ 150(0.10) = 100$$

Thus, if Mr. Adams knows the outcome state in advance for each tour and selects the best act for that outcome state, his expected payoff is $EPPI = \$100$ per tour.

We now define the expected payoff with perfect information formally.

(27.12)

The **expected payoff with perfect information,** denoted by *EPPI*, is the expected payoff when the act is chosen based on information of the exact outcome state:

$$EPPI = \sum_i (\max_j C_{ij}) P(S_i)$$

Note that any cost incurred in obtaining the perfect information is not considered in the definition of *EPPI*.

Expected Value of Perfect Information. In the island tour example, we found earlier that the expected payoff of the Bayes act, which utilizes only the probabilistic information about the outcome states, is $BEP = 58$. Since the expected payoff with perfect information is $EPPI = 100$, the difference $EPPI - BEP = 100 - 58 = 42$ is the incremental gain due to perfect information about the outcome state. This incremental gain is called the expected value of perfect information. Assuming that the expected payoff criterion is relevant, Mr. Adams should be willing to pay up to \$42 per tour for perfect information about the number of tourists who will make the side trip.

We now define the expected value of perfect information formally.

(27.13)

The **expected value of perfect information,** denoted by *EVPI*, is the difference between the expected payoff with perfect information (*EPPI*) and the expected payoff of the Bayes act (*BEP*):

$$EVPI = EPPI - BEP$$

Comments

1. Assuming that the expected payoff criterion is relevant, *EVPI* is the maximum amount that should be spent for perfect information about the outcome state. Since sample information is imperfect, it is worth less than perfect information.
2. Regrets or opportunity losses can be used in Bayesian decision making. The Bayes act is then the one that minimizes the expected regret. Two results can be shown to hold:
   (a) The same act that maximizes expected payoff minimizes expected regret.
   (b) The magnitude of the minimum expected regret is identical to *EVPI*.
3. Frequently the data on outcome state probabilities and payoffs in decision problems are unreliable to some degree. In these cases, it is advisable to study whether the choice of the best act is sensitive to variations in the data. This type of analysis is called *sensitivity analysis*. Sensitivity analysis in large-scale decision problems is conducted by computerized mathematical or simulation methods. In small decision problems, new values for the probabilities or payoffs can simply be inserted to see whether the optimal act remains the same or, if not, whether the expected payoffs of the competing optimal acts differ greatly.

27.5   MULTISTAGE DECISION PROBLEMS

The decision problems discussed so far are called *single-stage* problems because a single decision is involved. Other decision problems involve a sequence of two or more decisions and are called *multistage* decision problems. Decision trees are useful for studying multistage problems.

Computer Manufacturer

 EXAMPLE

A computer manufacturer has determined that the firm's product line must be broadened to include a line of office machines utilizing the latest technology. Investigation shows that two alternative courses of action are worth considering:

A_1: Undertake research and development to achieve production of the desired line (R&D)

A_2: Acquire a company already producing a satisfactory line and undertake manufacturing integration (Acquisition)

These acts are represented in Figure 27.2 by the branches emanating from the first-stage decision node. (Ignore for now the blocking marks on some branches and the numerical entries in the nodes.) Note that if the research-and-development strategy is followed (A_1), the possible outcomes are failure and success and the respective probabilities are assessed at 0.4 and 0.6.

In turning to the path for acquisition and manufacturing integration (A_2), we note that the attempt to integrate manufacturing could lead to success (with probability 0.7) or failure (with probability 0.3). If success is achieved, the overall management struc-

FIGURE 27.2
Decision tree for a two-stage decision problem—Computer manufacturer example

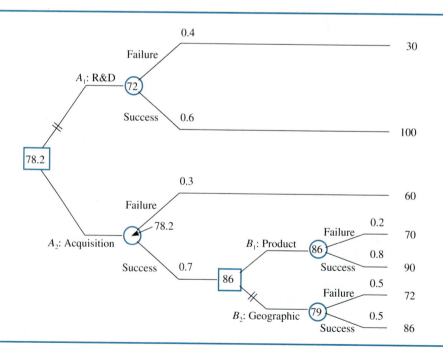

ture will need to be consolidated. The alternative courses of action in this second-stage decision are:

B_1: Consolidation on a product basis (Product)
B_2: Consolidation on a geographic basis (Geographic)

Thus, a second-stage decision node is indicated in the tree at this point. The branches emanating from this decision node each lead to a chance node and the possible outcomes failure and success, with probabilities as shown.

Finally, the payoffs are given for each of the branches. The payoffs are defined in terms of the estimated net-income position (in $ million) of the company five years hence.   □

Backward Induction

Bayesian decision making for multistage problems can be based on the principle of *backward induction*. This principle requires that the optimal acts at the last decision nodes be identified first, that the optimal acts at the preceding set of decision nodes then be identified by using the expected payoffs of the previously identified optimal acts, and so forth. In terms of a decision tree, the principle of backward induction requires us to work from right to left, using the expected payoffs of the optimal acts to the right in succeeding calculations.

EXAMPLE □

For the computer manufacturer example in Figure 27.2, the steps of backward induction are as follows:

1. There is only a single decision node at the second stage, which arises if first-stage act A_2 is successful. We therefore need to ascertain whether B_1 or B_2 is the Bayes act at this stage. The expected payoffs for the two acts are as follows:

$$EP(B_1) = 70(0.2) + 90(0.8) = 86$$
$$EP(B_2) = 72(0.5) + 86(0.5) = 79$$

These expected payoffs are recorded for convenience in the respective chance nodes. Since act B_1 maximizes the expected payoff in the second-stage decision, the expected payoff of the Bayes act (86) is entered in the second-stage decision node, and the B_2 path is blocked off.

2. The expected payoff in a decision node consolidates the data on probabilities and payoffs on all paths emanating rightward from the node. Thus, in Figure 27.2, act A_2 in the first-stage decision is now considered to lead to two chance outcomes: a payoff of 60 with probability 0.3 and a payoff of 86 with probability 0.7. Hence, we can calculate the expected payoffs for the acts emanating from the first-stage decision node as follows:

$$EP(A_1) = 30(0.4) + 100(0.6) = 72$$
$$EP(A_2) = 60(0.3) + 86(0.7) = 78.2$$

Thus, act A_2 maximizes the expected payoff and is selected under the expected payoff criterion. The A_1 path is blocked off.

3. The optimal strategy can now be seen by *forward induction*, that is, by tracing through the unblocked branches in Figure 27.2 from left to right. We see that the firm

should acquire a company already producing a satisfactory line and undertake manu-facturing integration (A_2). If this is successful, managerial consolidation on a product basis should be carried out (B_1). The possible payoffs for the optimal strategy, with attendant probabilities, are as follows:

Payoff	Probability
60	0.30
70	$0.7(0.2) = 0.14$
90	$0.7(0.8) = \underline{0.56}$
	Total 1.00

The expected payoff for the optimal strategy by forward induction therefore is $60(0.30) + 70(0.14) + 90(0.56) = 78.2$, the same as calculated before by backward induction.

$\square$

Comments

1. In multistage decision problems, second- or later-stage decisions are not necessarily acted on. If events intervene and new conditions arise, later-stage decisions must be determined anew in the light of the then-existing situation.
2. The probabilities in the branches emanating from any chance node always sum to 1. Thus, for any chance node after the first, these probabilities are conditional ones, the conditioning events being the relevant outcomes earlier in the tree.

27.6   OPTIONAL TOPIC—UTILITY

Problem with Monetary Payoffs

Our discussion of Bayesian decision making so far has utilized the expected payoff crite-rion where the payoffs are measured in monetary units. However, payoffs in monetary units may not always fully reflect a decision maker's attitude toward risk, and the expected pay-offs therefore may not indicate the decision maker's actual preferences. Consider the owner of a small construction company who must decide whether to be the general contractor on a large project (A_1) or simply a subcontractor (A_2). The payoff table is as follows:

		Act	
Outcome State	Probability	A_1	A_2
S_1: No delays	0.8	\$800,000	\$300,000
S_2: Delays	0.2	$-$\$400,000	$-$\$50,000

The expected payoffs of the two acts are $EP(A_1) = \$560,000$ and $EP(A_2) = \$230,000$ (calculations not shown). However, act A_1 is not the contractor's preferred act even though it leads to the larger expected payoff. The reason is the contractor knows that if delays occur, a loss of \$400,000 with act A_1 would place the firm in extreme financial difficulties, whereas a loss of \$50,000 with act A_2 could be absorbed. In other words, a \$400,000 loss would be much more than eight times as severe for the firm as a \$50,000 loss. Thus, the contractor finds act A_2 preferable.

When monetary consequences are not appropriate indicators of a decision maker's attitude toward risk, utility numbers should be used instead.

Basic Concepts of Utility Theory

Utility Function.   Utility theory takes into account the decision maker's attitude toward risk. By making some basic assumptions about the consistency of the decision maker's preferences among probability distributions of payoffs, utility theory shows that a utility function $u(C)$ exists that properly reflects the decision maker's attitude toward risk. Here C is the monetary amount, and $u(C)$ is the *utility number* associated with the monetary amount C. We call the function $u(C)$ the decision maker's *utility function for money*. Roughly interpreted, $u(C)$ is the subjective worth to the decision maker of a monetary payoff of amount C.

Expected Utility Criterion.   Utility theory shows that when utility numbers rather than monetary payoffs are used in Bayesian decision making, the preferred act will be the one with the largest expected value of the outcomes expressed in utility numbers. Consider a Bayesian decision problem where C_{ij} is the monetary payoff when the jth act is selected and the ith outcome state prevails. The expected payoff for the jth act, according to (27.10), is:

$$EP(A_j) = \sum_i C_{ij} P(S_i)$$

If we now express each monetary payoff C_{ij} in terms of its corresponding utility number $u(C_{ij})$, we can calculate the expected value of the outcomes for the jth act in utility numbers. We denote this expected value by $EU(A_j)$.

(27.14)
The **expected utility** for act A_j, denoted by $EU(A_j)$, is:

$$EU(A_j) = \sum_i u(C_{ij}) P(S_i)$$

As mentioned before, utility theory shows that the decision maker's preferred act is the one with the greatest expected utility.

(27.15)
The **expected utility criterion** selects that act for which the expected utility in (27.14) is maximized. The act selected by the expected utility criterion is called the **Bayes act.**

Assessing the Utility Function

A major task in employing utility theory is to assess the utility function of the decision maker. Once the utility function $u(C)$ has been ascertained, the finding of the Bayes act proceeds as before, except that utility numbers rather than monetary payoffs are used. We shall use an illustration to explain how the utility function of a decision maker might be assessed.

Investment Client   ☐   **EXAMPLE**

An investment counselor is offering a speculative venture to a client. If the venture is successful, the client will gain $10 thousand; but if it is unsuccessful, she will lose $6 thousand. Alternatively, the client can decline the venture, in which case her payoff is $0—she neither gains nor loses. Thus, there are three possible dollar outcomes—namely, $10 thousand, $0, and −$6 thousand. We wish to assign utility numbers to these dollar amounts that will reflect the client's attitude toward risk.   ☐

Specifying the Utility Scale.   Since utility is synonymous with subjective worth, the more desirable is an outcome, the larger is the utility number assigned to it. Thus, utility functions for money assign larger utility numbers $u(C)$ as the monetary amount C increases.

     There is no unique scale for utility, and two points on the utility function can be defined arbitrarily. It is often convenient to assign the utility number 0 to the smallest monetary outcome in the decision problem and the utility number 1 to the largest monetary outcome. For the investment client example, this practice leads us to define $u(-6) = 0$ and $u(10) = 1$, where the monetary outcomes are expressed in thousands of dollars.

Reference Lottery Method.   A variety of methods are available for assessing utility functions. We shall explain the *reference lottery method.*

☐   **EXAMPLE**

For the investment client example, we have defined $u(-6) = 0$ and $u(10) = 1$. We still require the utility of $0, $u(0)$, as well as a few other utility numbers so that we can assess the shape of the decision maker's utility function. We proceed through the following steps in utilizing the reference lottery method.

     1. First we wish to find $u(0)$. To begin, we give the investment client a choice between $0 for certain and the following reference lottery (made up of equal probabilities for the two extreme outcomes):

Outcome ($ thousand)	Probability
10	0.5
−6	0.5

The client states that she prefers $0 for certain to this reference lottery. Thus, $u(0)$ exceeds the expected utility of the reference lottery. Remembering that $u(10) = 1$ and $u(-6) = 0$, we compute the expected utility of the reference lottery as follows:

$$u(10)(0.5) + u(-6)(0.5) = 1(0.5) + 0(0.5) = 0.5$$

Hence, we know at this stage that $u(0) > 0.5$.

     2. To find the indifference point between $0 for certain and a reference lottery, we need to increase the probability of gaining $10 thousand in the reference lottery. After several trials, we reach the reference lottery:

Outcome ($ thousand)	Probability
10	0.7
−6	0.3

Now the client indicates that she is indifferent between $0 for certain and this reference lottery. Hence, $u(0)$ is equal to the expected utility of this reference lottery:

$$u(0) = u(10)(0.7) + u(-6)(0.3) = 1(0.7) + 0(0.3) = 0.7$$

Thus, we have found that $u(0) = 0.7$.

3. To ascertain some other points for the client's utility function, we wish next to find $u(2)$. We utilize the same approach, giving the client a choice each time between a $2 thousand gain for certain and a reference lottery. We eventually reach the indifference point for the following reference lottery:

Outcome ($ thousand)	Probability
10	0.8
−6	0.2

Hence, we have:

$$u(2) = u(10)(0.8) + u(-6)(0.2) = 1(0.8) + 0(0.2) = 0.8$$

4. By continuing in this manner, we obtain the following results for the investment client.

C:	−6	−3	−1	0	2	5	8	10
$u(C)$:	0	0.45	0.65	0.70	0.80	0.90	0.98	1.00

These points are plotted in Figure 27.3, and a curve has been drawn through them. This curve is the graph of the client's utility function for money. □

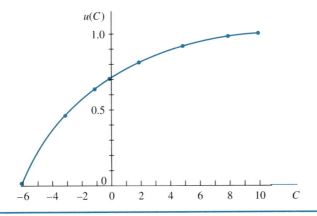

FIGURE 27.3
Utility function for money—
Investment client example

Comments

1. Often, a mathematical function is fitted to the utility assessment data. Two possible functions are as follows.

(27.16) **Quadratic utility function:** $u(C) = b_0 + b_1 C + b_2 C^2$

(27.17) **Exponential utility function:** $u(C) = b_0 + b_1 \exp(b_2 C)$

where b_0, b_1, and b_2 are constants.

2. Decision makers may give inconsistent responses when utility functions are assessed. Hence, it is good practice to check the responses for internal consistency with respect to the expected utility rule. In the investment client example, for instance, the client might be given an additional choice between $2 thousand for certain and a reference lottery offering a 50-50 chance at $0 and $5 thousand. Recall from earlier questioning that $u(2) = 0.8$ and note that the expected utility of this reference lottery is the same:

$$u(5)(0.5) + u(0)(0.5) = 0.9(0.5) + 0.7(0.5) = 0.8$$

If the client is not indifferent here, additional questioning and possible revisions of earlier responses will be required.

Application of Utility Function

Once a decision maker's utility function has been assessed, the finding of the Bayes act is routine, with the utility numbers simply replacing the monetary outcomes.

☐ EXAMPLES

1. For the investment client example, the speculative venture lottery is as follows:

Outcome ($ thousand)	Utility $u(C)$	Probability
10	1.00	0.5
−6	0	0.5

Let A_1 denote pursuing the speculative venture. We then have:

$$EU(A_1) = 1(0.5) + 0(0.5) = 0.5$$

Let A_2 denote the act not to pursue the speculative venture. Then, $EU(A_2) = u(0) = 0.7$. Hence, act A_2 has the higher expected utility and is the preferred act.

2. Management of a firm has the choice of conducting mineral exploration alone (A_1), as a joint venture with another firm (A_2), or not conducting the exploration (A_3).

The payoff table is as follows:

Outcome.State	Probability	A_1	A_2	A_3
Success	0.4	17	8.5	0
Failure	0.6	-10	-5.0	0

The utility function for the firm, within the range of payoffs for this problem, is:

$$u(C) = 0.2 + 0.03C + 0.001C^2$$

We obtain the needed utility numbers by substituting into the utility function. For example:

$$u(17) = 0.2 + 0.03(17) + 0.001(17)^2 = 1.0$$

Placing the utility numbers into the payoff table, we obtain:

Outcome State	Probability	A_1	A_2	A_3
Success	0.4	$u(17) = 1.0$	$u(8.5) = 0.53$	$u(0) = 0.2$
Failure	0.6	$u(-10) = 0$	$u(-5.0) = 0.075$	$u(0) = 0.2$

We now calculate the expected utility for each act:

$$EU(A_1) = 1.0(0.4) + 0(0.6) = 0.4$$
$$EU(A_2) = 0.53(0.4) + 0.075(0.6) = 0.257$$
$$EU(A_3) = 0.2(0.4) + 0.2(0.6) = 0.2$$

Hence, act A_1 is the preferred act.

Types of Utility Functions for Money

Researchers have investigated utility functions for money in some detail. Three prototype utility functions are graphed in Figure 27.4. The *concave utility function* in Figure 27.4a,

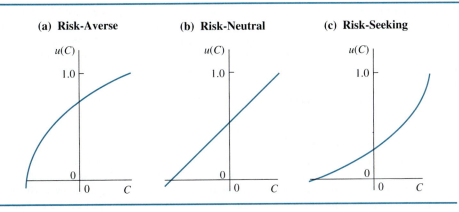

(a) Risk-Averse   **(b) Risk-Neutral**   **(c) Risk-Seeking**

FIGURE 27.4
Examples of utility functions for money showing different attitudes toward risk

labeled *risk-averse*, is the type found in the investment client example. It can be shown that a decision maker who is risk-averse will prefer a given dollar payoff for certain to a lottery offering an equal expected dollar payoff. For instance, a risk-averse person will prefer a sure gain of $5 thousand to a lottery offering a 50-50 chance at $0 or $10 thousand.

Figure 27.4b shows a *linear utility function*. A decision maker with this type of utility function for money is *risk-neutral*. It can be shown that the act with the largest expected payoff (in monetary terms) will be the preferred act for a decision maker with this type of utility function. Hence, the expected payoff criterion is the appropriate criterion for decision makers who are risk-neutral.

The *convex utility function* in Figure 27.4c pertains to decision makers who are *risk-seeking*. The quadratic utility function in example 2 is of this type. The attitude toward risk here is the opposite of risk aversion. For instance, a decision maker who is risk-seeking will prefer a lottery offering a 50-50 chance at $0 or $10 thousand to $5 thousand for certain.

Comments

1. Utility is subjective. We cannot say that an individual is correct or incorrect in his or her attitude toward risk. Moreover, a decision maker's utility function may change over time as new circumstances arise. Finally, because utility is a subjective measure and arbitrarily scaled, the utility numbers for different individuals cannot be compared.

2. In many practical situations, the decision maker's utility function for money is approximately linear within the range of the monetary outcomes encountered. Hence, as we have noted, in these situations utility assessments are not required because the expected payoff criterion (in monetary terms) will correctly identify the decision maker's preferred act.

PROBLEMS

27.1 A food company is launching a new product. The company's marketing group must choose one of three alternative package designs for the product. It is anticipated that sales of the new product will depend in part on the extent to which consumers find the chosen package design to be both functional and attractive. The marketing group is somewhat uncertain, however, about consumer reactions to each design. Identify each of the structural elements of a decision problem in (27.1) here.

27.2 An executive of a multinational corporation told a group of graduate students: "We employ a flexible credit policy in inflation-prone countries that is adjusted to the short-run outlook for inflation in the local currency. If inflation will be slight, we allow retailers a six-week credit limit, because experience shows this limit will generate the highest remitted profits in dollar terms. If inflation will be moderate, we cut this limit to four weeks, because a four-week limit then will result in the highest remitted dollar profits, and so on down to a limit of one day if inflation is severe. In one of the countries, the government has told us that we must announce a fixed credit policy for the next two years. We no longer can shift our limit back and forth, as in the other countries. In the next few days, I must decide what our two-year credit policy will be in that country." Identify each of the structural elements of a decision problem in (27.1) here.

* **27.3 Substitute Teachers.** Substitute teachers for a school can be hired either on contract or on a noncontract basis. Contract substitutes receive $40 each school day for the whole year simply for being available to teach and also receive an additional $200 on any day they actually teach. When the number of contract substitutes is inadequate on any school day, noncontract substitutes must be called in. Noncontract substitutes receive $300 for each day they teach but nothing when they do not teach. The school administrator must now decide how many substitutes should be given contracts for the coming school year. Records show that the following probability distribution describes the number of substitutes required on any school day:

S_i:	0	1	2
$P(S_i)$:	0.4	0.5	0.1

Construct the payoff table for the administrator's decision problem, showing the acts, outcome states, and payoffs for any given school day. Display the costs as negative payoffs.

27.4 Venture Capital. An investor is about to invest $100 thousand for a two-year period. The following four alternatives involving two ventures (V_1, V_2) are under consideration: (1) Invest the entire amount in V_1, (2) invest the entire amount in V_2, (3) invest half the amount in each of V_1 and V_2, (4) do not invest in either V_1 or V_2 and instead invest in riskless securities. The amount invested in V_1 will be doubled with probability 0.8 or lost completely with probability 0.2. The amount invested in V_2 will be tripled with probability 0.6 or lost completely with probability 0.4. The investment outcomes for V_1 and V_2 are independent. If the $100 thousand is invested in riskless securities, it will be worth $130 thousand at the end of the two-year period.

a. Construct the payoff table for the investor's decision problem, showing the acts, outcome states, and payoffs at the end of the two-year period.
b. Determine the probabilities associated with the outcome states identified in part a.

27.5 Odd–Even Game. Consider playing the following game with a friend. Your friend writes either "$0" or "$5" on a slip of paper. Concurrently, you write either "$2" or "$3" on your slip of paper. The two amounts are then added. If the sum is even, you pay your friend that amount; if the sum is odd, your friend pays you that amount.

a. Construct your payoff table for a single trial of this game.
b. If you knew the number your friend will write on the next play, would you have a decision problem on this play? If so, would it involve certainty or uncertainty?

27.6 Refer to **Substitute Teachers** Problem 27.3. Construct the decision tree corresponding to the payoff table. Show the probabilities associated with the branches of each chance node.

27.7 Refer to **Venture Capital** Problem 27.4. Construct the decision tree corresponding to the payoff table. Show the probabilities associated with the branches of each chance node.

27.8 Refer to **Odd–Even Game** Problem 27.5.

a. Construct the decision tree corresponding to the payoff table. Does the decision tree contain the same information as the payoff table? Explain.
b. Suppose you have already played this game twice and are $5 behind. In constructing a decision tree for the third play, should this loss be included in the payoffs? Explain.

27.9 Three workers (W_1, W_2, W_3) are to be assigned to carry out three tasks (T_1, T_2, T_3), one worker doing one task. Any worker can do any task, but they require different amounts of time (in hours), as shown by the following table:

Worker	Task T_1	T_2	T_3
W_1	43	35	30
W_2	32	32	29
W_3	36	37	33

a. Explain why this assignment problem is one of decision making under certainty.
b. Construct the payoff table for this decision problem when the consequence of interest is the total time required to do the three tasks. Which assignment yields the smallest total?
c. Construct the payoff table when the consequence of interest is the maximum of the three times required for the tasks. Which assignment yields the smallest maximum?

d. What would be the total number of available acts if 10 workers were to be assigned to 10 tasks, each worker being able to do each task?

27.10 Loose tea is to be packaged in a box with a total volume of 1000 cubic centimeters. It is desired to choose the height (h), width (w), and depth (d) for the box so that the surface area will be as small as possible to minimize boxing material requirements.

a. Explain why the choice of the dimensions of the box is a decision problem involving certainty.
b. Of the following alternative sets of dimensions under consideration, which is best?

	Dimensions (in centimeters)		
Alternative	h	w	d
1	5	20	10
2	10	10	10
3	5	25	8

* **27.11 Homecoming Brochure.** A student group at a university raises funds by undertaking small projects. The group's next project requires a decision concerning the number of copies of a brochure to be printed for homecoming weekend. The printing order must be a multiple of 500. The size of the homecoming crowd is uncertain because it depends on weather conditions. The payoff table follows (payoffs are net profits in dollars):

	Printing Order				
	A_1	A_2	A_3	A_4	A_5
Size of Crowd	1000	1500	2000	2500	3000
S_1: Above average	200	300	400	500	450
S_2: Average	200	300	250	200	150
S_3: Below average	200	150	100	50	0

a. Which acts are admissible? Does it appear that the group has missed an opportunity by not including printing sizes above 3000 among its possible courses of action? Comment.
b. Which act satisfies the maximin criterion? What is the guaranteed payoff with this act? Can you tell what will be the actual payoff with the maximin act? Can you always tell?
c. Convert the payoffs to regrets and ascertain which act satisfies the minimax regret criterion. Is this act the same as the one selected by the maximin criterion in part b? Is this necessarily the case?

27.12 Cooking Utensils. Union Appliance, Ltd., is a market leader in cast-iron cooking utensils. Last year, a firm specializing in advanced metal technology introduced a line of competitive lightweight utensils that also contained a new design. These utensils are made of a new alloy that exactly duplicates the properties of cast iron for cooking. The competitive line has cut sharply into sales of Union's line. Union can incorporate the new design into its own line but not the lighter weight. Changing its line to the new design will be expensive for Union, so management does not wish to undertake it unless the new design is the major factor in consumer acceptance of the competitive line. The payoff table follows (payoffs are discounted profits in $ million):

	Act	
	A_1	A_2
Importance of New Design	Incorporate New Design	Do not Incorporate New Design
S_1: Major factor	3.5	0.8
S_2: Not major factor	1.9	4.7

a. Which act satisfies the maximin criterion? What minimum payoff does this act guarantee? Can the actual payoff exceed the guaranteed minimum? Explain.

b. Convert the payoffs to opportunity losses and ascertain which act satisfies the minimax regret criterion. Is this act the same as the one selected by the maximin criterion in part a? Is this always the case?

27.13 Computer Graphics Manufacturer. A small manufacturer of computer graphics equipment has accumulated substantial cash reserves. The executive committee will now decide whether to expand vertically by buying another computer graphics company or to expand horizontally by buying a company that manufactures medical monitoring equipment. The outcome states correspond to different possible degrees of elimination of financially weaker companies in the computer graphics field in the next several years. The payoff table follows (payoffs are discounted profits in $ million):

	Act	
Degree of Elimination	A_1 Vertical	A_2 Horizontal
S_1: Moderate	8	15
S_2: Severe	16	10

a. Which act satisfies the maximin criterion? What minimum payoff does this act guarantee?

b. Convert the payoffs to regrets and ascertain which act satisfies the minimax regret criterion. For which outcome state would this act yield no opportunity loss?

* **27.14** Refer to **Homecoming Brochure** Problem 27.11. The group was able to assign the following probabilities to the outcome states: $P(S_1) = 0.2, P(S_2) = 0.5, P(S_3) = 0.3$.

a. Find the Bayes act. What is its expected payoff?

b. Could A_1 under some other set of probabilities ever be the Bayes act here? Could A_5 ever be the Bayes act? Explain.

27.15 Refer to **Cooking Utensils** Problem 27.12. Management has assessed the probabilities of the outcome states as follows: $P(S_1) = 0.6, P(S_2) = 0.4$.

a. Which act will give Union the larger expected payoff? What is the expected payoff with the Bayes act?

b. Give the probability distribution of payoffs that Union faces if it chooses the Bayes act.

27.16 Refer to **Computer Graphics Manufacturer** Problem 27.13. The executive committee has assessed the outcome state probabilities as follows: $P(S_1) = 0.5, P(S_2) = 0.5$.

a. Obtain the probability distributions of payoffs associated with acts A_1 and A_2, respectively.

b. Find the Bayes act. What is its expected payoff?

* **27.17** Refer to **Homecoming Brochure** Problems 27.11 and 27.14.

a. Obtain *EPPI* and *EVPI*. Interpret their meanings here.

b. By paying a $65 premium for overtime work, the group could delay the printing order until just prior to the weekend when the size of the crowd is known. Do you recommend that the group follow this option? Explain.

27.18 Refer to **Cooking Utensils** Problems 27.12 and 27.15. A consumer research study is under consideration to obtain information on whether the weight or the design of the competitive line is the major factor in consumer acceptance.

a. What is the maximum amount management should be willing to spend on a study providing perfect information? Explain.

b. Because the proposed study involves a sample survey, the survey findings will be subject to sampling error and perhaps other errors also. What effect does this have on the amount management should be willing to spend on the study?

27.19 Refer to **Computer Graphics Manufacturer** Problems 27.13 and 27.16. Obtain *EPPI* and *EVPI*. Interpret their meanings here.

FIGURE 27.5
Decision tree for Problem 27.22

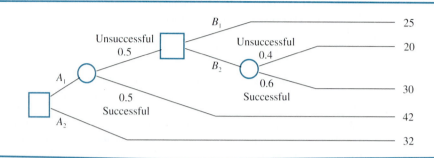

27.20 If *EPPI* = 160 and *EVPI* = 40 for a decision problem, what is the expected payoff of the Bayes act? What is the expected regret of the Bayes act?

* **27.21** Refer to Figure 27.2 for the *Computer manufacturer* example. A recent review disclosed that if management follows the R&D route and the outcome is failure, there is still an opportunity to broaden the product line by undertaking negotiations with another firm on either a merger or a license arrangement. (It would be extremely disadvantageous to attempt to negotiate in both areas, so one or the other must be selected for negotiation.) The acts, outcomes, payoffs, and probabilities of success for the negotiations are:

> Merger: success, 65; failure, 20; probability of success = 0.7
> License: success, 60; failure, 30; probability of success = 0.6

a. Draw a decision tree to reflect the modified situation.
b. Using backward induction, determine the optimal strategy. What is the expected payoff for the optimal strategy?
c. Confirm the expected payoff for the optimal strategy by forward induction.

27.22 Consumer By-Product. A firm making industrial products developed a marketable consumer item as a by-product of its research and development. A decision is to be made on whether to sell this item through the company's own sales and distribution channels (A_1) or to use independent sales agents (A_2). If the firm selects A_1 and is relatively unsuccessful, management must then decide between switching to sales agents immediately (B_1) or first attempting to dispose of the item by selling the key patent at favorable terms to another firm and switching only if the attempt is unsuccessful (B_2). The decision tree for this problem is shown in Figure 27.5 (payoffs are discounted profits in $ million).

a. Using backward induction, determine the optimal strategy. What is the expected payoff for the optimal strategy?
b. Confirm the expected payoff for the optimal strategy by forward induction.
c. What is the probability that the optimal strategy will yield a smaller payoff than act A_2?

27.23 A sophisticated robot welder in a manufacturing plant has just broken down. The plant manager must now decide whether the plant repair staff should attempt to repair it (A_1) or to bring in outside repair specialists immediately (A_2). The manager does not know if the repair problem is simple or complex but assesses the probabilities of these outcomes as 0.4 and 0.6, respectively. If the plant staff attempt to repair the robot welder and discover that the repair problem is simple, they will succeed in repairing the robot. On the other hand, if the repair problem turns out to be complex, the manager must decide either to have the staff attempt the repair (B_1) or to bring in the specialists at that stage (B_2). The probability is 0.7 that the staff will succeed in a complex repair problem. If the staff fail in a complex repair problem, the specialists would need to be brought in. Figure 27.6 shows the decision tree for this problem. The payoffs are costs (in $ thousand), which reflect the facts that (1) the specialists are more expensive than the plant staff, (2) a complex repair problem is more costly than a simple one, and (3) delays in repair are expensive.

a. Using backward induction, determine the optimal strategy. Remember that the consequences in Figure 27.6 are cost amounts.

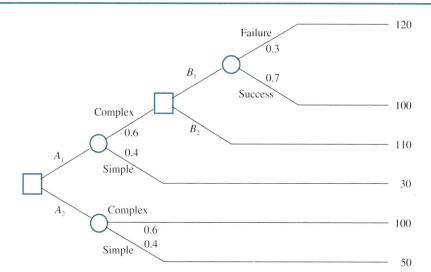

FIGURE 27.6
**Decision tree for Problem
27.23**

b. What is the expected cost for the optimal strategy identified in part a?

c. Give the probability distribution of the cost outcomes associated with the optimal strategy in part a.

∗ 27.24 A decision maker's utility function for money is $u(C) = C^2/90,000$, where $0 \le C \le 300$ and C is in $ thousand.

a. Plot $u(C)$. Is the decision maker risk-averse, risk-neutral, or risk-seeking?

b. Which will the decision maker prefer: (1) a lottery offering an even chance at $0 or $300 thousand or (2) $150 thousand for certain?

27.25 A decision maker's utility function for money is $u(C) = \sqrt{C/500}$, where $0 \le C \le 500$ and C is in $ thousand.

a. Plot $u(C)$. Is the decision maker risk-averse, risk-neutral, or risk-seeking?

b. Which will the decision maker prefer: (1) a lottery offering an even chance at $0 or $500 thousand or (2) $250 thousand for certain?

27.26 A decision maker's utility function for money is $u(C) = 1 - \exp(-0.01C)$, where $0 \le C < \infty$ and C is in $ thousand.

a. Plot $u(C)$. Is the decision maker risk-averse, risk-neutral, or risk-seeking?

b. Which will the decision maker prefer: (1) a lottery offering an even chance at $0 or $400 thousand or (2) $200 thousand for certain?

27.27 Refer to Figure 27.3 for the *Investment Client* example. The client, on being offered a lottery in which there is 1 chance in 3 of losing $6 thousand and 2 chances in 3 of winning $8 thousand, responded that she is indifferent between playing this lottery and not playing it. Is this response consistent with the client's utility function in Figure 27.3? Explain.

∗ 27.28 Refer to **Homecoming Brochure** Problems 27.11 and 27.14. The group's utility function for money is the one given in Problem 27.25.

a. Which of the five acts is optimal according to the expected utility criterion?

b. For each act, determine the amount of money offered with certainty that would be equally attractive to the group. Does the optimal act identified in part a correspond to the largest certain amount? Is this result to be expected? Explain.

c. Is the optimal act according to the expected utility criterion in part a the same as the optimal act according to the expected payoff criterion in Problem 27.14a? Is this result to be expected? Explain.

27.29 Refer to **Cooking Utensils** Problems 27.12 and 27.15. The utility function reflecting management's utility values for payoffs in this decision problem is $u(C) = \sqrt{(C - 0.8)/3.9}$, where C is in $ million.

a. Which act is optimal according to the expected utility criterion?

b. What monetary amount C, offered with certainty, would be equally attractive to management as the optimal act in part a?

27.30 Refer to **Computer Graphics Manufacturer** Problems 27.13 and 27.16. The utility function reflecting the executive committee's utility values for payoffs in this decision problem is $u(C) = 2 - (16/C)$, where C is in $ million.

a. Which act is optimal according to the expected utility criterion?

b. Would a different act be optimal if the executive committee were risk neutral? Explain.

EXERCISES

27.31 Show that the act selected by the maximin criterion is necessarily admissible.

27.32

a. Show that the act that maximizes expected payoff minimizes expected regret.

b. Show that the expected regret for the Bayes act equals *EVPI*.

27.33 Refer to **Cooking Utensils** Problems 27.12 and 27.15. For what value of $P(S_1)$ do the acts have equal expected payoffs? How is this information useful if management is unsure about the exact outcome state probabilities but believes that $P(S_1)$ is somewhere between 0.6 and 0.8? Between 0.2 and 0.4?

27.34 Refer to **Consumer By-Product** Problem 27.22. The firm is somewhat tentative about its assessed value for the probability of success of 0.5 if the company uses its own sales and distribution channels. How far off could this assessment be before it would affect the optimal strategy, assuming that all other assessments remain unchanged?

27.35 Refer to Figure 27.2 for the *Computer Manufacturer* example. The manufacturer conjectures that the payoff of 86 found on the bottom branch of the tree might be an understatement. How much larger could this payoff be before it would alter the manufacturer's optimal strategy, assuming that the other payoffs and the probabilities remained unchanged?

STUDIES

27.36 **Lottery Ticket.** Someone has given you a ticket in a state lottery. You now find that your ticket number is one of 10 in a final runoff. Three numbers will be drawn from the 10 at random without replacement. If your number is drawn first, second, or third, you will win $250,000, $100,000, or $50,000, respectively. If your number remains undrawn, you will win $1000. A consortium has offered you $25,000 for your ticket before the runoff. Assume that this offer is the only one you will receive.

a. Construct the decision tree associated with your decision problem. Also present this information in a payoff table. State the probability distribution of your winnings outcomes if you keep the lottery ticket.

b. Which act is optimal under (1) the maximin criterion, (2) the minimax regret criterion, (3) the expected payoff criterion?

c. Calculate the expected value of perfect information. Interpret its meaning here.

27.37 Refer to **Odd–Even Game** Problem 27.5. Assume that many trials of the game are to be played.

a. Suppose that you decide to employ a randomized strategy whereby in each trial you will write either "$2" or "$3" on your slip of paper with equal probability, the random choices being made independently from one trial to the next. Suppose your friend uses the same strategy to choose between "$0" and "$5," the random choices being made independently of yours. What is your expected payoff? Is the game played in this fashion a fair one in the sense that neither you nor your friend can expect to gain over many trials of the game?

b. Suppose that you employ the strategy whereby you will write either "$2" or "$3" on your slip of paper with probabilities p and $1 - p$, respectively, the random choices being made independently from one trial to the next. Is there a value of p that will make this an unfair game in your favor, irrespective of the strategy your friend employs? Explain, presenting any relevant calculations to support your answer. [*Hint:* What value of p gives you the same expected payoff irrespective of your friend's choice?]

27.38 Refer to **Lottery Ticket** Problem 27.36. Use the reference lottery method to assess your own utility function for the payoffs in this decision problem. Test your utility function for consistency, revising it if necessary, and then apply it to ascertain your optimal act. Is the result obtained from your derived utility function consistent with your actual preference? If not, what are the implications of this finding?

27.39 Refer to **Consumer By-Product** Problem 27.22 and the corresponding decision tree in Figure 27.5. Management's utility function is $u(C) = 1 - \exp(-0.05C)$, where $0 \leq C < \infty$ and C is in $ million.

a. What is management's optimal strategy in terms of maximizing expected utility? Is this the same strategy as the one that maximizes expected payoff? Discuss.

b. Suppose that the payoff associated with act A_2 (shown as 32 in Figure 27.5) were, in fact, somewhat uncertain. For what range of values for this payoff would A_1 be the optimal first-stage decision in terms of maximizing expected utility? For what range would A_2 be the optimal first-stage decision?

Bayesian Decision Making with Sample Information

28

In this chapter, we discuss the use of Bayesian decision making when (1) prior information about the outcome states is available in the form of probabilities $P(S_i)$, and (2) it is possible to obtain additional information about the prevailing outcome state by sampling. This type of decision situation is encountered frequently.

A decision must be made whether or not to accept a shipment from a supplier. The payoff is affected by the proportion of defective items in the shipment. The outcome states here are the different possible proportions defective in the shipment. From past experience with many shipments from the same supplier, probabilities can be assigned to the outcome states. In addition, a sample can be selected from the particular shipment under consideration to obtain information on the proportion defective in this shipment. ☐

When it is possible to obtain sample information about the prevailing outcome state to supplement prior information in the form of outcome state probabilities, two major questions arise.

NATURE OF   **28.1**
DECISION PROBLEM

1. For given sample size *n,* how do we select the best act based on both the prior probabilities and the sample information?
2. If the sample size is not predetermined, how do we ascertain the best sample size?

We shall discuss each question in turn, employing a detailed case illustration. Throughout this chapter, *we assume that the expected payoff criterion is the appropriate guide for finding the best act.*

In the context of decision making with sample information, the outcome state probabilities $P(S_i)$ are called *prior probabilities* because the sample information subsequently provides further information about the prevailing outcome state.

EXAMPLE ☐   **Pollution Detector**

A firm manufactures an air pollution detector. A key component is a synthetic membrane produced periodically by the firm in batches of several hundred and inventoried to meet production and replacement needs. A chemical purchased from a supplier and used in producing the membrane must be pure, or else some membranes will be inert and have to be discarded. From experience, it is known that if there are slight trace

TABLE 28.1
Payoff table (dollars)—Pollution detector example

Proportion of Inert Membranes	Prior Probability $P(S_i)$	Act A_1 Do Not Purify	Act A_2 Purify
S_1: 0	0.2	2200	500
S_2: 0.1	0.2	1600	500
S_3: 0.2	0.1	1000	500
S_4: 0.3	0.2	400	500
S_5: 0.4	0.3	-200	500

impurities, approximately 10 percent of membranes will be inert. With higher levels of impurity, the proportion of inert membranes will be higher.

To keep the illustration relatively simple, we assume that the proportion of inert membranes as a result of impurities in the chemical will be at one of five levels—these are the outcome states for the decision problem:

Outcome State:	S_1	S_2	S_3	S_4	S_5
Proportion of Inert Membranes:	0	0.1	0.2	0.3	0.4

When the firm receives a fresh shipment of the chemical from the supplier just before making a batch of membranes, the decision problem involves the following alternative acts:

A_1: Use the shipment as is.
A_2: Put the shipment through a purification process.

The purification process removes all impurities at a cost of $1700.

The outcome states S_i, prior probabilities $P(S_i)$, and payoffs C_{ij} for this decision problem are shown in Table 28.1. The prior probabilities are based on experience with previous shipments. The payoffs represent the anticipated dollar gain to the firm in producing a batch of the membranes. For act A_1, the payoff varies with the purity of the shipment. For act A_2, the payoff is the same ($500) for all outcome states—the gain when the chemical is pure ($2200) less the cost of purification ($1700).   □

Analysis Without Sample Information

Bayes Act.   If a decision in the pollution detector example had to be made without sample information, we would follow the procedures presented in Chapter 27. The expected payoffs for the two acts are, by (27.10):

$$EP(A_1) = 2200(0.2) + 1600(0.2) + 1000(0.1) + 400(0.2)$$
$$+ (-200)(0.3) = 880$$
$$EP(A_2) = 500(0.2) + 500(0.2) + 500(0.1) + 500(0.2)$$
$$+ 500(0.3) = 500$$

Hence, if no sample information about the purity of the shipment is available, the best act is A_1, to use the shipment as is. Act A_1 therefore is the Bayes act, and the Bayes expected payoff is $BEP(0) = 880$. We now use $BEP(0)$ to denote the Bayes expected payoff, indicating that a sample of size $n = 0$ is utilized.

Expected Payoffs	Expected Payoff with Perfect Information		TABLE 28.2
$EP(A_1) = 880$ $EP(A_2) = 500$	$EPPI = 1110$		Bayesian decision analysis without sample information—Pollution detector example
Bayes Act	Expected Value of Perfect Information		
A_1 $BEP(0) = 880$	$EVPI = 230$		

Value of Perfect Information.   If perfect information about the purity of each shipment were available, we see from the payoffs in Table 28.1 that the best act for outcome states S_1, S_2, and S_3 is A_1, and that act A_2 is the best act for outcome states S_4 and S_5. Hence, the expected payoff with perfect information is, by (27.12):

$$EPPI = 2200(0.2) + 1600(0.2) + 1000(0.1) + 500(0.2)$$
$$+ 500(0.3) = 1110$$

Finally, the expected value of perfect information is, by (27.13):

$$EVPI = EPPI - BEP(0) = 1110 - 880 = 230$$

Thus, the decision maker should pay no more than $230 per shipment for perfect information about the purity of the shipment.
   These results are summarized in Table 28.2.

Sample Information

In the pollution detector example, management need not make a decision without sample information about the purity of the shipment because an engineer has developed a process for making test membranes by hand. Although a test membrane is not sensitive enough to be used in the monitoring equipment, it can be checked for inertness. Thus, it is possible for the company to obtain sample information about the proportion of inert membranes for an incoming shipment by making a number of test membranes and ascertaining how many of these are inert.

28.2 OPTIMAL DECISION RULE FOR GIVEN SAMPLE SIZE

We now consider the first of the two questions posed earlier: For given sample size n, how do we select the best act based on both prior probabilities and sample information?
   In the pollution detector example, management has decided to make $n = 6$ test membranes for each shipment of chemical before choosing its course of action. We let f denote the sample frequency of test membranes that are inert.

Nature of Decision Rule

In Bayesian decision making with sample information, we use the same type of statistical decision rule as discussed in earlier chapters. Thus, since act A_1 yields a higher payoff than act A_2 when the chemical is relatively pure and a lower payoff when the chemical is relatively impure (Table 28.1), small values of the sample frequency f of inert membranes will lead to act A_1, while large values will lead to act A_2. Hence, the appropriate form of the decision rule is the following (where K denotes the action limit).

> **(28.1)**
>
>   If $f \le K$, select A_1 (do not purify).
>   If $f > K$, select A_2 (purify).
>
> Here f is the sample frequency in the sample of size n.

For compactness, decision rule (28.1) will be denoted by (n, K). For instance, $(6, 1)$ denotes the decision rule where the sample size is 6 and act A_1 is selected if $f \le 1$.

While decision rule (28.1) is of exactly the same form as earlier ones, in Bayesian decision analysis the action limit K is determined by explicitly utilizing information about the outcome state probabilities and payoffs.

Set of Decision Rules Under Consideration

The choice of the optimal action limit K when a sample of $n = 6$ test membranes is utilized must be made from among the eight possibilities shown in Table 28.3. Note that decision rule $(6, 6)$ always leads to act A_1 regardless of the sample outcome. The reason is that this rule leads to act A_1 whenever $f \le 6$, and in a sample of size $n = 6$, the sample frequency can never exceed 6. Similarly, rule $(6, -1)$ always leads to act A_2 since the sample frequency f can never be less than zero.

Expected Payoff for Decision Rule

We now seek the decision rule among the eight in Table 28.3 that has the greatest expected payoff. Hence, we need to calculate the expected payoff for each of the eight decision rules. Consider first decision rule $(6, 0)$. This rule requires that six test membranes be made and that act A_1 be selected if none of the six is inert; otherwise, act A_2 is to be selected. The procedure for calculating the expected payoff for this decision rule is shown in Table 28.4. The basic steps are as follows:

1. For each outcome state, determine the probabilities of the sample outcomes leading to acts A_1 and A_2.
2. For each outcome state, obtain the expected payoff.
3. Determine the expected payoff over all outcome states.

In columns 1 through 3 of Table 28.4, we repeat from Table 28.1 the outcome states and payoffs for the pollution detector example.

Act Probabilities.   Columns 4 and 5 of Table 28.4 show for each outcome state S_i the probabilities of selecting acts A_1 and A_2 for rule $(6, 0)$. We shall explain their calculation shortly. Note that these probabilities are conditional on the outcome state S_i. We call these probabilities act probabilities.

Decision Rule	Select A_1 If:	Select A_2 If:
(6, 0)	$f = 0$	$f > 0$
(6, 1)	$f \leq 1$	$f > 1$
(6, 2)	$f \leq 2$	$f > 2$
(6, 3)	$f \leq 3$	$f > 3$
(6, 4)	$f \leq 4$	$f > 4$
(6, 5)	$f \leq 5$	$f > 5$
(6, 6)	Always	Never
(6, −1)	Never	Always

TABLE 28.3
Set of possible decision rules based on sample frequency f when $n = 6$—Pollution detector example

(28.2)
An **act probability,** denoted by $P(A_j \mid S_i)$, is the conditional probability that act A_j will be selected under the given decision rule when the outcome state is S_i.

The act probabilities $P(A_j \mid S_i)$ are analogous to the acceptance and rejection probabilities in statistical tests; that is, the probabilities of selecting H_0 and H_1. These latter probabilities are conditional on the value of a parameter, such as the population mean μ.

Sampling Distribution of f. To ascertain the act probabilities for the pollution detector example, we must utilize for each outcome state the sampling distribution of the sample frequency f of inert membranes in the sample of six. It is reasonable to assume here that f is a binomial random variable, whose probability function is given by (13.3). The parameters of this probability function are $n = 6$ and the proportion of inert membranes for the batch, denoted by p. Recall that each outcome state is defined in terms of this latter parameter, as may be seen from column 1 of Table 28.4.

Calculation of Act Probabilities. Consider first outcome state S_1. Here the probability of an inert membrane is $p = 0$. Because no test membrane will be inert under this outcome state, rule (6, 0) will always lead to act A_1. The respective act probabilities therefore are

TABLE 28.4
Calculation of expected payoff for decision rule (6, 0)—Pollution detector example

(1) Proportion of Inert Membranes p	(2) Payoffs C_{i1}	(3) Payoffs C_{i2}	(4) Act Probabilities $P(A_1 \mid S_i)$	(5) Act Probabilities $P(A_2 \mid S_i)$	(6) Conditional Expected Payoff $(2) \times (4)$ + $(3) \times (5)$	(7) Prior Probability $P(S_i)$	(8) $(6) \times (7)$
S_1: 0	2200	500	1.0000	0	2200.00	0.2	440.000
S_2: 0.1	1600	500	0.5314	0.4686	1084.54	0.2	216.908
S_3: 0.2	1000	500	0.2621	0.7379	631.05	0.1	63.105
S_4: 0.3	400	500	0.1176	0.8824	488.24	0.2	97.648
S_5: 0.4	−200	500	0.0467	0.9533	467.31	0.3	140.193
							$EP(6, 0) = 957.85$

$P(A_1 \mid S_1) = 1.0$ and $P(A_2 \mid S_1) = 0$. These are entered in Table 28.4, columns 4 and 5, respectively, in the line for outcome state S_1.

When the outcome state is S_2, the probability of an inert membrane for the batch is $p = 0.1$. Since decision rule $(6, 0)$ leads to act A_1 if $f = 0$, we need to find $P(0)$ when $n = 6$ and $p = 0.1$. We use Table C.5 and find $P(0) = 0.5314$. Hence, $P(A_1 \mid S_2) = 0.5314$, and by complementation theorem (4.18), we obtain $P(A_2 \mid S_2) = 1 - 0.5314 = 0.4686$. These act probabilities are entered in Table 28.4 for outcome state S_2.

The act probabilities for the other outcome states are obtained in similar fashion.

Conditional Expected Payoffs.   We can now obtain for each outcome state the expected payoff for rule $(6, 0)$. Such an expected payoff is called a conditional expected payoff, because it is conditional on the particular outcome state.

(28.3)

The **conditional expected payoff** for decision rule (n, K) and outcome state S_i is denoted by $EP(n, K \mid S_i)$. It is the expected payoff for the decision rule conditional on outcome state S_i:

$$EP(n, K \mid S_i) = \sum_j C_{ij} P(A_j \mid S_i)$$

For outcome state S_1, we see from Table 28.4 that $C_{11} = 2200$, $C_{12} = 500$, $P(A_1 \mid S_1) = 1.0$, and $P(A_2 \mid S_1) = 0$. Hence, the conditional expected payoff for rule $(6, 0)$ when the outcome state is S_1 is:

$$EP(6, 0 \mid S_1) = 2200(1.0) + 500(0) = 2200$$

We interpret this to mean that if rule $(6, 0)$ is used repeatedly with many incoming shipments for which the proportion of inert membranes is $p = 0$, the average payoff is \$2200 per shipment.

The conditional expected payoffs under the other outcome states are calculated and interpreted in a similar manner. Column 6 of Table 28.4 shows all of the conditional expected payoffs for rule $(6, 0)$.

Expected Payoff.   We can now calculate the expected payoff for decision rule $(6, 0)$ over all outcome states. The prior probabilities $P(S_i)$ of the outcome states are repeated from Table 28.1 in column 7 of Table 28.4. The probability $P(S_1) = 0.2$ indicates that the probability of conditional expected payoff $EP(6, 0 \mid S_1) = 2200$ is 0.2 because S_1 will be the outcome state in 20 percent of the shipments. Thus, we can calculate the expected value over all outcome states in the usual fashion and obtain the expected payoff for decision rule $(6, 0)$.

(28.4)

The **expected payoff** for decision rule (n, K) is denoted by $EP(n, K)$. It is the expected value of the conditional expected payoffs under the prior probabilities $P(S_i)$:

$$EP(n, K) = \sum_i EP(n, K \mid S_i) P(S_i)$$

For the pollution detector example, we calculate (as shown in column 8 of Table 28.4):

$$EP(6, 0) = 2200(0.2) + 1084.54(0.2) + 631.05(0.1)$$
$$+ 488.24(0.2) + 467.31(0.3) = 957.85$$

We interpret this to mean that if decision rule $(6, 0)$ is used repeatedly with many incoming shipments, the average payoff per shipment in the long run will be $957.85.

Optimal Decision Rule

Now that we have obtained the expected payoff for decision rule $(6, 0)$, we need to find the expected payoffs for the other seven decision rules in Table 28.3. We obtain these in the same fashion as for rule $(6, 0)$ in Table 28.4. For decision rules $(6, 6)$ and $(6, -1)$, which do not utilize sample information, we require no further calculations. Their expected payoffs are given in Table 28.2 for decision making with prior information only. Decision rule $(6, 6)$ has expected payoff $EP(6, 6) = EP(A_1) = 880$, while rule $(6, -1)$ has expected payoff $EP(6, -1) = EP(A_2) = 500$.

The expected payoffs for all of the decision rules in Table 28.3 are as follows (further calculations are not shown):

K:	0	1	2	3	4	5	6	-1
$EP(6, K)$:	958	1010	972	918	889	881	880	500

Bayes Decision Rule. The expected payoff criterion leads to the selection of the decision rule with the largest expected payoff. In the pollution detector example, we see that this rule is $(6, 1)$. Rule $(6, 1)$ is called the Bayes decision rule for sample size $n = 6$.

> **(28.5)**
> The **Bayes decision rule** for a sample of size n is the one that maximizes the expected payoff among all possible decision rules for that sample size.
> The expected payoff for the Bayes decision rule for sample size n is denoted by $BEP(n)$, where:
>
> $$BEP(n) = \max_K EP(n, K)$$

In the pollution detector example, the expected payoff for the Bayes rule $(6, 1)$ is $BEP(6) = EP(6, 1) = 1010$.

Bayes Acts. The acts to be selected according to the Bayes decision rule for the different possible sample outcomes are called the *Bayes acts*. In the pollution detector example, where the Bayes rule for $n = 6$ is $(6, 1)$, the Bayes acts are as follows for the different sample outcomes f:

f:	0	1	2	3	4	5	6
Bayes Act:	A_1	A_1	A_2	A_2	A_2	A_2	A_2

In the most recent shipment sampled, $f = 2$ test membranes were found to be inert. Since the Bayes act for $f = 2$ is A_2, the shipment should be purified.

Summary of Procedure

We now summarize the procedure for finding the Bayes decision rule for a given sample size.

> **(28.6)**
>
> The steps for finding the Bayes decision rule for a given sample size n are as follows.
>
> *Step 1.* Identify all possible decision rules (n, K) for the given sample size n.
> *Step 2.* For each possible decision rule (n, K):
>   a. Calculate the act probabilities $P(A_j \mid S_i)$ for each outcome state S_i.
>   b. Obtain the conditional expected payoff $EP(n, K \mid S_i)$ for each outcome state S_i.
>   c. Use the prior probabilities $P(S_i)$ to obtain the expected payoff $EP(n, K)$ for the decision rule.
> *Step 3.* Select the decision rule for which the expected payoff $EP(n, K)$ is largest.

In actual applications, Bayesian decision rules are generally calculated by computer.

28.3 DETERMINATION OF OPTIMAL SAMPLE SIZE

We now come to the second question raised at the beginning of this chapter: If the sample size is not predetermined, how do we ascertain the best sample size? In practice, this case arises when no firm decision on sample size has been made as yet.

     The procedure for finding the optimal sample size is conceptually very simple, although extensive computations are generally required. We determine, for each possible sample size, the expected payoff of the Bayes rule net of sampling costs and select that sample size for which this net expected payoff is a maximum. We now consider this procedure in more detail.

Cost of Sampling

We denote the cost of a sample of size n by $s(n)$. We shall assume here that the total sample cost is a linear function of sample size.

> **(28.7)** $s(n) = \begin{cases} a + bn & n > 0 \\ 0 & n = 0 \end{cases}$
>
> where $a \geq 0$ and $b > 0$.

Here a is the overhead cost of setting up the sample and b is the variable cost of including one additional observation in the sample. Note that the sample cost is zero according to (28.7) when no sample is employed. This simple cost model is often an adequate representation.

Upper Bound on Optimal Sample Size

The expected value of perfect information $EVPI$ in (27.13) can be used to obtain a useful upper bound on the optimal sample size. Recall that $EVPI$ represents the maximum amount that a decision maker should spend for perfect information about the outcome state. Hence, the sample cost must not exceed $EVPI$; that is:

$$s(n) \le EVPI$$

For cost model (28.7), the inequality becomes:

$$a + bn \le EVPI$$

or:

$$n \le \frac{EVPI - a}{b}$$

(28.8)
For cost model (28.7), an **upper bound on the optimal sample size,** denoted by $\max(n)$, is:

$$\max(n) = \begin{cases} \dfrac{EVPI - a}{b} & \text{if } EVPI > a \\ 0 & \text{if } EVPI \le a \end{cases}$$

When the upper bound in (28.8) turns out to be zero, then the optimal sample size is zero and the decision should be made using prior information only.

Optimal Sample Size

We can now state more formally the procedure for finding the optimal sample size n.

(28.9)
The steps for finding the optimal size n are as follows.

Step 1. For each sample size in the interval $0 \le n \le \max(n)$, ascertain the Bayes decision rule.

Step 2. For each Bayes rule, determine the expected payoff $BEP(n)$ and the **net expected payoff,** denoted by $BNEP(n)$, where:

$$BNEP(n) = BEP(n) - s(n)$$

Step 3. Select that sample size and associated Bayes decision rule that maximizes the net expected payoff $BNEP(n)$.

In the pollution detector example, suppose that the sample size had not yet been determined and that an analyst has been asked to find the optimal sample size. Cost model (28.7) is considered to be applicable. There are no overhead costs and the variable cost per test membrane is $7.50, so the cost function is:

$$s(n) = 7.5n$$

Since $EVPI = 230$ (see Table 28.2), we obtain, by (28.8):

$$\max(n) = \frac{230}{7.5} = 30.7$$

The optimal sample size therefore cannot exceed 30 test membranes.

Consequently, the analyst needs to obtain the Bayes decision rule for each sample size from 0 to 30 and then determine the net expected payoff for each. Recall that we found earlier for $n = 6$ that the Bayes rule is (6, 1) and that its expected payoff is $BEP(6) = 1010$. Since the cost of six test membranes is $s(6) = 7.50(6) = 45$, the net expected payoff for the Bayes rule is:

$$BNEP(6) = BEP(6) - s(6) = 1010 - 45 = 965$$

The analyst used a computer routine to perform the needed calculations for the other sample sizes. Table 28.5 contains the results. It shows for each sample size the value of K in the Bayes decision rule and the net expected payoff of the Bayes rule. Note that for $n = 0$ there is no entry for K since no sampling is done. The Bayes net expected payoff in this case is simply the Bayes expected payoff with no sample information $BEP(0)$ in Table 28.2. Note also that the net expected payoff for the Bayes decision rule for $n = 6$ is shown to be $965.237, the same as our result except for a difference in rounding.

The analyst could readily see from the computer results in Table 28.5 that the optimal sample size is $n = 6$ and that the Bayes rule for this sample size is (6, 1).

Comment

The optimal sample size and decision rule depend on the interrelations between the payoffs, prior probabilities, and sample cost function. Computer-assisted calculations allow the decision maker to check whether the optimal sample size and decision rule are sensitive to variations in these parameters.

Value of Sample Information

We now present two measures that deal with the value of sample information in Bayesian decision making, and we shall illustrate them in terms of the pollution detector example.

Expected Value of Sample Information.   The first measure indicates the gain with use of sample information, neglecting the cost of sampling, thus producing a bound on how much should be paid for sample information.

n	K	$BNEP(n)$	n	K	$BNEP(n)$	n	K	$BNEP(n)$
0	—	880.000	11	2	957.439	21	5	910.427
1	0	930.500	12	2	950.946	22	5	906.189
2	0	949.800	13	3	948.537	23	5	900.552
3	0	951.260	14	3	946.992	24	5	893.740
4	1	946.628	15	3	942.743	25	6	887.155
5	1	960.117	16	3	936.259	26	6	882.091
6	1	965.237	17	4	931.306	27	6	875.977
7	1	963.502	18	4	928.187	28	6	868.974
8	1	956.604	19	4	923.158	29	7	862.244
9	2	959.824	20	4	916.534	30	7	856.560
10	2	960.661						

TABLE 28.5
Bayes decision rules and their expected payoffs, net of sampling costs, for sample sizes 0 to 30—Pollution detector example

(28.10)
For a sample of size n, the **expected value of sample information,** denoted by $EVSI(n)$, is the difference between the expected payoff of the Bayes decision rule for that n and the expected payoff for the Bayes act with prior information only:

$$EVSI(n) = BEP(n) - BEP(0)$$

In the pollution detector example, we calculate from earlier results for $n = 6$:

$$EVSI(6) = 1010 - 880 = 130$$

In a choice between making a decision using prior information only or taking a sample of size $n = 6$, the decision maker should not pay more than $130 for the sample information.

Expected Net Gain from Sampling. The second measure reflects the net gain with use of sample information, taking into account the cost of this information.

(28.11)
For a sample of size n, the **expected net gain from sampling,** denoted by $ENGS(n)$, is the difference between the net expected payoff for the Bayes rule for that sample size and the expected payoff of the Bayes act with prior information only:

$$ENGS(n) = BNEP(n) - BEP(0)$$

For the pollution detector example for $n = 6$, we have (see Table 28.5):

$$ENGS(6) = 965 - 880 = 85$$

Note that the expected net gain from sampling is equivalent to:

$$ENGS(6) = EVSI(6) - s(6) = 130 - 45 = 85$$

In using the optimal sample size, we automatically maximize the expected net gain from sampling.

28.4   DETERMINING BAYES ACT BY REVISION OF PRIOR PROBABILITIES

Once a sample has been selected and the result observed, it is possible to ascertain directly the Bayes act for the given sample result without developing the decision rule covering all possible sample results. The procedure employs Bayes' theorem. (A discussion of Bayes' theorem is presented in Section 6.1.)

We shall first restate Bayes' theorem (6.1) in terms of our present notation, using the pollution detector example for illustration. The variable of interest here is the outcome state S_i, denoting the proportion of inert membranes. The prior probabilities of the outcome states are denoted by $P(S_i)$. The additional information variable is the sample frequency f of inert membranes among the n test membranes in the sample. With this notation, Bayes' theorem is as follows.

$$(28.12) \qquad P(S_i \mid f) = \frac{P(S_i)P(f \mid S_i)}{\sum_i P(S_i)P(f \mid S_i)}$$

The posterior probability $P(S_i \mid f)$ is the probability that outcome state S_i is the prevailing one, given the sample frequency f.

To find the Bayes act for a given sample frequency f by means of Bayes' theorem, we employ a three-step procedure.

(28.13)
The steps for finding the Bayes act for a given sample frequency f by utilizing Bayes' theorem are as follows.

Step 1. For the prior probabilities $P(S_i)$ and the observed sample frequency f, obtain the posterior probabilities $P(S_i \mid f)$, using Bayes' theorem (28.12).

Step 2. For each act A_j, calculate the expected payoff under the posterior probabilities, denoted by $EP(A_j \mid f)$, as follows:

$$EP(A_j \mid f) = \sum_i C_{ij} P(S_i \mid f)$$

Step 3. Select the act A_j for which the expected payoff $EP(A_j \mid f)$ is largest.

We now illustrate this procedure for the pollution detector example where $n = 6$ test membranes were examined and $f = 2$ were found inert.

Obtaining Posterior Probabilities

The steps for revising the prior probabilities by Bayes' theorem (28.12) are shown in Table 28.6. Columns 1 and 2 repeat the outcome states and prior probabilities from Table 28.1.

TABLE 28.6

Revision of prior probabilities—Pollution detector example.   Here $n = 6$ membranes were tested and $f = 2$ membranes were found inert.

(1) Proportion of Inert Membranes p	(2) Prior Probability $P(S_i)$	(3) $P(f = 2 \mid S_i)$	(4) Joint Probability $P(S_i \cap f = 2)$ (2) × (3)	(5) Posterior Probability $P(S_i \mid f = 2)$ (4) ÷ 0.20238
S_1: 0	0.2	0	0	0
S_2: 0.1	0.2	0.0984	0.01968	0.0972
S_3: 0.2	0.1	0.2458	0.02458	0.1215
S_4: 0.3	0.2	0.3241	0.06482	0.3203
S_5: 0.4	0.3	0.3110	0.09330	0.4610
Total	1.0		0.20238	1.0000

Column 3 shows the conditional probabilities $P(f = 2 \mid S_i)$ of obtaining the sample frequency $f = 2$ for the different possible outcome states S_i. For instance, when the state is S_2—that is, when the probability of an inert membrane is $p = 0.1$—we find from Table C.5 for $n = 6$ and $p = 0.1$ that $P(2) = 0.0984$. Thus, $P(f = 2 \mid S_2) = 0.0984$. The other conditional probabilities are found in a similar manner.

In Table 28.6 the joint probabilities $P(S_i \cap f = 2)$ are calculated in column 4 by multiplying $P(f = 2 \mid S_i)$ by $P(S_i)$ for each outcome state. The sum of this column is the denominator in Bayes' theorem (28.12). The posterior probabilities $P(S_i \mid f = 2)$ are obtained in column 5 by dividing each entry in column 4 by the sum of column 4. We see, for instance, that $P(S_3 \mid f = 2) = 0.02458/0.20238 = 0.1215$. Thus, given that two membranes were inert among the six test membranes, the conditional probability that the shipment will produce proportion $p = 0.2$ of inert membranes (that is, that the outcome state is S_3) is 0.1215. The other posterior probabilities are similarly interpreted.

Note by comparing columns 2 and 5 in Table 28.6 how the prior probabilities have been revised in light of the sample information of two inert membranes among the six test membranes. The prior probabilities $P(S_1)$ and $P(S_2)$ have been revised downward substantially, and the prior probabilities $P(S_4)$ and $P(S_5)$ have been revised upward substantially. Thus, finding two inert test membranes makes it much more likely that the proportion p of inert membranes in the shipment is high.

Determining Expected Payoffs

Proceeding now to step 2 of (28.13), we need to find the expected payoff for each act under the posterior probabilities. We utilize the payoffs in Table 28.1 and the posterior probabilities in Table 28.6. The expected payoffs are calculated in the usual way and are denoted by $EP(A_1 \mid f = 2)$ and $EP(A_2 \mid f = 2)$ to indicate that they are conditional on the given sample outcome. We obtain:

$$EP(A_1 \mid f = 2) = 2200(0) + 1600(0.0972) + 1000(0.1215)$$
$$+ 400(0.3203) + (-200)(0.4610) = 312.94$$

$$EP(A_2 \mid f = 2) = 500(0) + 500(0.0972) + 500(0.1215)$$
$$+ 500(0.3203) + 500(0.4610) = 500$$

Determining Bayes Act

It is evident from the preceding calculations that A_2 is the Bayes act here since it has the larger expected payoff. Thus, when two test membranes among six are inert, the company should purify the shipment. Of course, this result is in accord with the Bayes decision rule (6, 1) obtained earlier.

Extensive and Normal Forms of Bayesian Analysis

The procedure just described for ascertaining the Bayes act for a given sample result is called the *extensive form* of Bayesian decision analysis. It contrasts with finding the Bayes rule first, called the *normal form.* As we just saw, the extensive form introduces the prior probabilities in the first stage of the analysis and revises them via Bayes' theorem. The normal form, on the other hand, introduces the prior probabilities as weights for the conditional expected payoffs in the final stage of the calculations and does not employ Bayes' theorem directly.

Each form of Bayesian analysis has its advantages. As noted earlier, if we wish to ascertain the Bayes act for a specific sample result, the extensive form will identify this act directly. Another advantage of the extensive form is its adaptability to shortcut mathematical procedures.

On the other hand, the normal form of Bayesian analysis has the advantage that the prior probabilities are introduced last. This makes it more suitable for sensitivity analysis because the more objective elements of the decision problem (the payoffs and the sample cost function) are introduced first. Also, many statisticians and management scientists find the normal form easier to explain to managers because it proceeds in straightforward steps without requiring Bayes' theorem.

PROBLEMS

* **28.1 Solar Heater.** A home products firm has developed a new solar energy heater and now must decide whether to market it. The success of the product, as measured by its profit contribution, depends on the proportion of households in the target market that will purchase it. Research by the marketing department has produced the following payoff table and prior probabilities (payoffs are in $ thousand):

Proportion Purchasing Product p	Prior Probability $P(S_i)$	Act	
		A_1 Don't Market Product	A_2 Market Product
S_1: 0.05	0.5	0	-500
S_2: 0.10	0.1	0	200
S_3: 0.15	0.4	0	900

a. Determine the Bayes act if a decision must be made without sample information. What is the expected payoff for the Bayes act? What is the probability of a negative payoff with the Bayes act?

b. Obtain *EPPI* and *EVPI.* Interpret the meaning of each of these measures.

28.2 Imported Fruit. A fruit importer is considering the purchase of a large shipment of grapefruit from a supplier abroad. The value of the shipment to the importer depends on the proportion of the shipment that is of premium grade. If the shipment is not purchased from the supplier abroad, the grapefruit will be bought from a local source. The following payoff table shows the profit contribution (in $ thousand) to the importer of each act for different proportions of premium-grade fruit in the shipment. The importer's prior probabilities for the shipment, based on extensive past experience with other shipments from the same supplier, are also shown.

| | | Act | |
Shipment Proportion of Premium-Grade Fruit p	Prior Probability $P(S_i)$	A_1 Buy From Local source	A_2 Buy From Supplier Abroad
S_1: 0.1	0.2	5	−2
S_2: 0.3	0.5	5	4
S_3: 0.5	0.3	5	10

a. Determine the Bayes act based on prior information alone. What is the expected payoff for the Bayes act?
b. Obtain *EPPI* and *EVPI*. Interpret their meanings here.
c. Suppose that a source of perfect information about the shipment proportion of premium-grade fruit were available. Prior to having this information, what is the probability that acting on the basis of this information will lead to a higher payoff than is offered by the Bayes act?

28.3 Control Boxes. A firm manufactures electronic control boxes for electric motors that are sold to industrial customers in large lots. The firm must choose between two policies for selling a lot of the boxes. The first policy, called the inspection policy, requires that every box be given a detailed inspection for nonconformance to quality requirements prior to shipment. The second policy, called the returns policy, entails shipment without detailed inspection for nonconformance, replacement of nonconforming boxes returned by customers, and reimbursement of customers for direct expenses and inconvenience. The payoff table showing the costs (expressed as negative payoffs in $ thousand) of these two policies, for different values of the proportion of nonconforming boxes in a lot, follows. Prior probabilities, based on the firm's extensive past experience with its production process for this type of box, are also given.

| | | Act | |
Lot Proportion Nonconforming p	Prior Probability $P(S_i)$	A_1 Returns Policy	A_2 Inspection Policy
S_1: 0.01	0.9	−1.5	−2.5
S_2: 0.05	0.1	−7.5	−4.5

a. Based on prior information alone, which policy should the firm adopt for any given lot? What is the expected cost per lot of this policy?
b. What would constitute perfect information about a lot in this decision problem? What is the expected value of perfect information to the firm?

* **28.4** Refer to **Solar Heater** Problem 28.1. A random sample of $n = 15$ households from the target market is being considered for obtaining information about the proportion of households that will purchase the heater. The decision rule to be employed is of the form: If $f \le K$, select A_1; if $f > K$, select A_2. Here f is the number of households in the sample that will purchase the heater and can be treated as a binomial random variable.

a. For outcome state S_1 and the decision rule $(n, K) = (15, 2)$, obtain (1) the act probabilities, (2) the conditional expected payoff. Interpret the latter value.
b. Obtain the expected payoff for decision rule (15, 2).
c. The expected payoffs for selected decision rules based on a sample of size 15 follow:

$(15, K)$:	$(15, 0)$	$(15, 1)$	$(15, 2)$	$(15, 3)$	$(15, 15)$	$(15, -1)$
$EP(15, K)$:	210.3	211.6	—	63.6	—	—

Obtain the missing expected payoffs.

d. The Bayes decision rule for $n = 15$ is $(15, 1)$. What is the value of $BEP(15)$? Interpret the meaning of this value.

e. What is the probability of a negative payoff with Bayes decision rule $(15, 1)$? How does this probability compare with the corresponding one for the Bayes decision rule based on no sampling?

28.5 Refer to **Imported Fruit** Problem 28.2. The importer is considering selecting a random sample of $n = 10$ grapefruit from the shipment prior to deciding whether to purchase the shipment from the supplier abroad or to buy from a local source. The importer will buy from a local source (A_1) if the number of premium-grade grapefruit in the sample f is less than or equal to a stipulated number K, and will purchase the shipment from the supplier abroad (A_2) if f exceeds K. Assume that f is a binomial random variable.

a. Suppose that the shipment proportion of premium-grade fruit is actually 0.5 (outcome state S_3). If $K = 1$ is used in the decision rule, what is the probability that the importer will purchase the shipment from the supplier abroad? What is the conditional expected payoff?

b. Obtain the expected payoff for decision rule $(n, K) = (10, 1)$.

c. The expected payoffs for selected decision rules based on a sample of size 10 follow:

$(10, K)$:	$(10, 0)$	$(10, 1)$	$(10, 2)$	$(10, 3)$	$(10, 4)$	$(10, 10)$	$(10, -1)$
$EP(10, K)$:	5.101	—	6.011	—	5.857	—	—

Obtain the missing expected payoffs.

d. The Bayes decision rule for $n = 10$ is $(10, 3)$. What is the value of $BEP(10)$? Interpret the meaning of this value.

28.6 Refer to **Control Boxes** Problem 28.3. It is proposed that a random sample of $n = 12$ control boxes be selected from each lot and inspected for nonconformance before deciding on the policy to be adopted for the lot. The decision rule to be employed is of the form: If $f \le K$, select A_1; if $f > K$, select A_2. Here f is the number of nonconforming boxes in the sample and can be treated as a binomial random variable.

a. Suppose that the lot proportion nonconforming is actually 0.01 (outcome state S_1). For the decision rule $(n, K) = (12, 1)$, obtain (1) the act probabilities, (2) the conditional expected payoff.

b. Obtain the expected payoffs for the following decision rules: (1) $(12, 12)$, (2) $(12, -1)$, (3) $(12, 0)$, (4) $(12, 1)$.

c. The Bayes decision rule for $n = 12$ is $(12, 0)$. If this rule is adopted here, what policy will be chosen for a lot with $f = 0$? With $f = 2$?

d. What proportion of all lots will be shipped under the returns policy (A_1) if the Bayes decision rule $(12, 0)$ is adopted? [*Hint*: $P(A_1) = P(A_1 \mid S_1)P(S_1) + P(A_1 \mid S_2)P(S_2)$.]

* 28.7 Refer to **Solar Heater** Problem 28.1. The setup cost of a survey is \$8 thousand and the cost per respondent is \$0.05 thousand.

a. State the sampling cost function.

b. Obtain the upper bound on the optimal sample size in (28.8). Interpret this bound.

28.8 Refer to **Imported Fruit** Problem 28.2. The sampling cost function (in \$ thousand) is $s(n) = 0.2 + 0.005n$. Obtain the upper bound on the optimal sample size in (28.8). Interpret this bound.

28.9 Refer to **Control Boxes** Problem 28.3. The firm has calculated that the setup cost for sampling a lot is \$0.1 thousand and the variable cost of inspecting each control box in the sample is \$0.003 thousand.

a. State the sampling cost function.

b. Obtain the upper bound on the optimal sample size in (28.8). Interpret this bound.

* 28.10 Refer to **Solar Heater** Problems 28.1, 28.4, and 28.7. The values of $BEP(n)$ for selected sample sizes follow:

n:	0	100	200	300
$BEP(n)$:	130.0	347.9	369.7	374.7

a. Obtain BNEP(n), EVSI(n), and ENGS(n) for each of the values of n considered here. Which of these sample sizes leads to the largest expected payoff net of sampling costs?

b. What is the meaning of (1) BNEP(100), (2) EVSI(100), (3) ENGS(100)?

c. Computer output for n > 0 shows that BNEP(n) attains a maximum value of 352.5 at n = 241 and that the Bayes decision rule for n = 241 is (241, 20). Should the marketing decision be made after sampling households or without any sampling? Explain. If sampling is to be done, describe the optimal sampling plan and give its net expected payoff. If no sampling is to be done, what is the optimal act and the expected payoff for this act?

28.11 Refer to **Imported Fruit** Problems 28.2, 28.5, and 28.8. The values of BEP(n) for selected sample sizes follow:

n:	0	20	40	60
BEP(n):	5.000	6.214	6.343	6.409

a. Obtain BNEP(n), EVSI(n), and ENGS(n) for each of the values of n considered here. Which of these sample sizes leads to the largest expected payoff net of sampling costs?

b. What is the meaning of (1) BNEP(20), (2) EVSI(20), (3) ENGS(20)?

c. Do the expected payoffs net of sampling costs differ greatly for sample sizes n = 20, 40, 60? What is the practical importance of this fact?

d. Computer output for n > 0 shows that BNEP(n) attains a maximum value of 5.944 at n = 37 and that the Bayes decision rule for n = 37 is (37, 13). Should the decision about the supplier of the grapefruit be made after sampling the imported shipment or without any sampling? Explain. If sampling is to be done, describe the optimal sampling plan and give its net expected payoff. If no sampling is to be done, what is the optimal act and the expected payoff for this act?

28.12 Refer to **Control Boxes** Problems 28.3, 28.6, and 28.9. The values of BEP(n) for selected sample sizes follow:

n:	0	20	40	60
BEP(n):	−2.100	−2.036	−1.974	−1.945

a. Obtain BNEP(n), EVSI(n), and ENGS(n) for each of the values of n considered here. Which of these sample sizes leads to the largest expected payoff net of sampling costs?

b. What is the meaning of (1) BNEP(20), (2) EVSI(20), (3) ENGS(20)?

c. Computer output for n > 0 shows that BNEP(n) attains a maximum value of −2.189 at n = 31 and that the Bayes decision rule for n = 31 is (31, 1). Should the firm do any sampling of a lot to decide on the policy for the lot? Explain. If sampling is to be done, describe the optimal sampling plan and give its net expected payoff. If no sampling is to be done, what is the optimal act and the expected payoff for this act?

* 28.13 Refer to **Solar Heater** Problems 28.1 and 28.4. A random sample of n = 15 households was selected; of these, f = 2 households indicated that they will purchase the heater.

a. Obtain the posterior probabilities for the outcome states based on this sample result.

b. Obtain the Bayes act and its expected payoff, given this sample result.

c. Explain how the procedure in parts a and b could be extended to determine the Bayes decision rule for the given sample size.

d. In advance of selecting the sample, what is the probability that the sample will contain exactly two households who indicate they will purchase the heater?

28.14 Refer to **Imported Fruit** Problems 28.2 and 28.5. The importer has selected a random sample of n = 10 grapefruit from the shipment and found f = 1 grapefruit to be of premium grade.

a. Obtain the importer's posterior probabilities for the outcome states based on this sample result.

b. Obtain the Bayes act and its expected payoff, given this sample result.

c. In advance of sampling, what is the probability that the sample will contain exactly one grapefruit of premium grade?

28.15   Refer to **Control Boxes** Problems 28.3 and 28.6. A random sample of $n = 12$ control boxes was selected from a lot and contained exactly $f = 2$ nonconforming boxes.

a. Obtain the posterior probabilities for the lot proportion nonconforming based on this sample result.

b. Obtain the Bayes act and its expected payoff, given this sample result.

c. The Bayes decision rule for a sample of this size is (12, 0). Is the Bayes act obtained in part b consistent with the rule? Explain.

EXERCISES

28.16   Prove that the expected opportunity loss associated with decision rule (n, K) equals $EPPI - EP(n, K)$.

28.17   Refer to Exercise 28.16. Show that, for a given sample size, maximizing the expected payoff and minimizing the expected opportunity loss yield the same optimal decision rule.

28.18   Is the upper bound for the optimal sample size in (28.8) applicable when the decision maker's utility function is not linear? Discuss.

28.19   Refer to **Imported Fruit** Problems 28.2 and 28.14.

a. Determine *EVPI* after the sample information has been obtained. Of what use is *EVPI* at this point to the importer?

b. Is the expected value of perfect information after the sample information has been obtained larger or smaller than it was before sampling? Must *EVPI* necessarily change in this direction after sample information is gathered? Discuss.

STUDIES

28.20   A quality control official for a government agency inspects large shipments of shirts that are received periodically from an overseas supplier. A shirt is classified as nonconforming if its workmanship, fabric, pattern, or certain other characteristics do not meet quality requirements. Rejecting a shipment costs $1500 because it entails 100 percent inspection and administrative costs of obtaining replacements for nonconforming shirts. The cost of accepting a shipment is $150,000p, where p is the proportion of shirts in the shipment that are nonconforming. Extensive past experience with the supplier indicates that shipment quality has the following probability distribution:

Proportion Nonconforming:	0	0.01	0.04	0.08
Probability:	0.30	0.50	0.15	0.05

a. Set up the payoff table for the official's decision problem. Show costs as negative payoffs. Denote the acts of accepting and rejecting a shipment by A_1 and A_2, respectively.

b. A random sample of $n = 10$ shirts from a shipment contains $f = 1$ that is nonconforming. Assume that the number of nonconforming shirts in the sample can be treated as a binomial random variable. Revise the prior probabilities and determine the Bayes act.

c. Obtain the expected payoffs for sample size $n = 10$ for the following values of K: $-1, 0, 1, 2$. What appears to be the Bayes decision rule for this sample size? Is the Bayes act identified in part b consistent with this rule?

d. Obtain the expected payoffs for sample size $n = 50$ for $K = -1, 0, 1, 2$. What appear to be the Bayes decision rule and the value of $BEP(50)$?

28.21   A food company imports nuts in the shell in large lots. Because consumer reaction to purchasing a package of nuts in the shell containing a high proportion of spoiled nuts is quite adverse, each lot is sampled to decide whether (1) to package the nuts in the shell for sale or (2) to shell all nuts first, removing the spoiled ones by sorters and packaging the remaining shelled nuts for sale.

The company currently follows a zero-defects inspection plan whereby a random sample of 20 nuts is selected from each lot and shelled. If no nuts in the sample are spoiled, the lot is judged acceptable and the nuts are packaged in the shell; otherwise, the lot is judged unacceptable and the nuts are shelled and sorted before being packaged.

The payoffs and prior probabilities for the quality of a lot (as measured by the proportion spoiled) follow. The payoffs (in dollars) represent profit contributions.

Proportion Spoiled p	Prior Probability $P(S_i)$	Act	
		A_1 Acceptable	A_2 Unacceptable
S_1: 0	0.6	700	600
S_2: 0.01	0.1	650	580
S_3: 0.10	0.3	−300	400

a. What is the expected payoff of the company's zero-defects inspection plan?

b. Construct the probability distribution of payoffs for a lot under the company's zero-defects inspection plan.

c. Does the company's zero-defects inspection plan correspond to the Bayes decision rule for sample size $n = 20$? [*Hint:* Consider the decision rules (20, 0), (20, 1), (20, 2), (20, 20), and (20, −1).]

28.22 Refer to **Solar Heater** Problems 28.1 and 28.13. Suppose that the utility function for money for the decision maker in this firm is $u(C) = 1 - \exp[-0.001(C + 500)]$, where $-500 \leq C < \infty$ and C is in $ thousand, and that the expected utility criterion is to be employed.

a. Based on the prior probabilities alone, which act is better?

b. What is the maximum amount that the firm should be willing to pay for perfect information about the proportion of households in the target market who will purchase the heater?

c. What is the optimal act, given that $f = 2$ households in the random sample of $n = 15$ indicated they will purchase the heater?

Mathematical Review

Single Summation

The symbol Σ (Greek capital sigma) is used in this book to designate summation. The expression:

$$\sum_{i=1}^{n} X_i$$

denotes the sum of the n values $X_1, X_2, \ldots, X_n$.

> **(A.1)** $\displaystyle\sum_{i=1}^{n} X_i = X_1 + X_2 + \cdots + X_n$

The subscript i on X is an index given to each value of X so that the n values may be distinguished from one another.

The following are interpretations of several expressions that employ the Σ symbol.

> **(A.2)** $\displaystyle\left(\sum_{i=1}^{n} X_i\right)^2 = (X_1 + X_2 + \cdots + X_n)^2$
>
> **(A.3)** $\displaystyle\sum_{i=1}^{n} X_i^2 = X_1^2 + X_2^2 + \cdots + X_n^2$
>
> **(A.4)** $\displaystyle\sum_{i=1}^{n} c = c + c + \cdots + c = nc$ (c is a constant.)
>
> **(A.5)** $\displaystyle\sum_{i=1}^{n} cX_i = cX_1 + cX_2 + \cdots + cX_n = c\left(\sum_{i=1}^{n} X_i\right)$
>
> **(A.6)** $\displaystyle\sum_{i=1}^{n} (X_i + Y_i) = (X_1 + Y_1) + (X_2 + Y_2) + \cdots + (X_n + Y_n)$
>
> $$= \sum_{i=1}^{n} X_i + \sum_{i=1}^{n} Y_i$$

EXAMPLE ☐

Suppose $n = 3$, $c = 11$, $X_1 = 3$, $X_2 = -1$, and $X_3 = 7$. Then:

$$\sum_{i=1}^{3} X_i = 3 - 1 + 7 = 9$$

$$\left(\sum_{i=1}^{3} X_i\right)^2 = (3 - 1 + 7)^2 = 9^2 = 81$$

$$\sum_{i=1}^{3} X_i^2 = 3^2 + (-1)^2 + 7^2 = 59$$

$$\sum_{i=1}^{3} 11 = 11 + 11 + 11 = 3(11) = 33$$

$$\sum_{i=1}^{3} 11X_i = 11 \sum_{i=1}^{3} X_i = 11(9) = 99$$

Double Summation

Some formulas use the double summation expression, such as $\sum_{i=1}^{r} \sum_{j=1}^{c} X_{ij}$. The double subscript ij on X is again an index given to each value of X so that the values may be distinguished from one another. In this case, the subscript i ranges from 1 to r and the subscript j ranges from 1 to c. The definition of $\sum_{i=1}^{r} \sum_{j=1}^{c} X_{ij}$ is as follows.

(A.7) $$\sum_{i=1}^{r} \sum_{j=1}^{c} X_{ij} = \sum_{i=1}^{r} \left(\sum_{j=1}^{c} X_{ij}\right) = \sum_{j=1}^{c} X_{1j} + \sum_{j=1}^{c} X_{2j} + \cdots + \sum_{j=1}^{c} X_{rj}$$

where, for instance:

$$\sum_{j=1}^{c} X_{1j} = X_{11} + X_{12} + \cdots + X_{1c} \qquad \sum_{j=1}^{c} X_{2j} = X_{21} + X_{22} + \cdots + X_{2c}$$

☐ EXAMPLE

Consider the following table of X_{ij} values:

	$j = 1$	$j = 2$	$j = 3$
$i = 1$	4	-3	1
$i = 2$	5	10	-6

The value X_{ij} is located at the intersection of the ith row and jth column in the table; for example, $X_{11} = 4$, $X_{12} = -3$, $X_{21} = 5$. Note that:

$$\sum_{j=1}^{3} X_{1j} = 4 - 3 + 1 = 2 \qquad \sum_{j=1}^{3} X_{2j} = 5 + 10 - 6 = 9$$

Hence:

$$\sum_{i=1}^{2} \sum_{j=1}^{3} X_{ij} = \sum_{j=1}^{3} X_{1j} + \sum_{j=1}^{3} X_{2j} = (4 - 3 + 1) + (5 + 10 - 6) = 11$$

Observe that the double summation is simply the sum of all six values in the table. ☐

The order of summation can be reversed if the ranges of the subscripts do not depend on one another, which is often the case, as in the following instance.

$$(A.8) \qquad \sum_{i=1}^{r}\sum_{j=1}^{c} X_{ij} = \sum_{j=1}^{c}\sum_{i=1}^{r} X_{ij}$$

Otherwise, the order of summation must be maintained.

EXAMPLE ☐

For the values in the table, we have:

$$\sum_{j=1}^{3}\sum_{i=1}^{2} X_{ij} = \sum_{i=1}^{2} X_{i1} + \sum_{i=1}^{2} X_{i2} + \sum_{i=1}^{2} X_{i3}$$
$$= (4 + 5) + (-3 + 10) + (1 - 6) = 11$$

which is the same as the value obtained earlier for $\sum_{i=1}^{2}\sum_{j=1}^{3} X_{ij}$.

☐

Abbreviated Notation

Where the context makes it clear which values are to be summed, it is common to omit the range of the summation index or even the index itself. In this case, we write:

$$\sum_{i} X_{i} \quad \text{or} \quad \sum X_{i} \quad \text{instead of} \quad \sum_{i=1}^{n} X_{i}$$

and

$$\sum_{i}\sum_{j} X_{ij} \quad \text{or} \quad \sum\sum X_{ij} \quad \text{instead of} \quad \sum_{i=1}^{r}\sum_{j=1}^{c} X_{ij}$$

Exponents

The following is a summary of algebraic rules for exponents. Here a, b, m, and n denote numbers. Each rule is illustrated.

ALGEBRAIC RULES FOR EXPONENTS AND LOGARITHMS

A.2

	Rule	Illustration
(A.9)	$a^0 = 1$	$3^0 = 1$
(A.10)	$a^m a^n = a^{m+n}$	$6^2(6^3) = 6^5 = 7776$
(A.11)	$(a^m)^n = a^{mn}$	$(3^2)^4 = 3^8 = 6561$
(A.12)	$a^{-n} = \dfrac{1}{a^n}$	$10^{-2} = \dfrac{1}{10^2} = 0.01$
(A.13)	$\dfrac{a^m}{a^n} = a^{m-n}$	$\dfrac{5^2}{5^4} = 5^{-2} = 0.04$
(A.14)	$(ab)^n = a^n b^n$	$[5(7)]^2 = 5^2(7^2) = 1225$

	Rule	Illustration
(A.15)	(i) $a^{1/n} = \sqrt[n]{a}$	$8^{1/3} = \sqrt[3]{8} = 2$
	(ii) $a^{1/2} = \sqrt[2]{a} = \sqrt{a}$	$9^{1/2} = \sqrt[2]{9} = \sqrt{9} = 3$
(A.16)	$a^{m/n} = \sqrt[n]{a^m}$	$5^{3/2} = \sqrt[2]{5^3} = 11.180$

Logarithms

Definitions. The symbol $\log_b A$, for any positive number A, denotes the exponent to which b must be raised to equal A. In other words, if $\log_b A = C$, then, by definition, $b^C = A$. The number b is called the *base* of the logarithm and may be any positive number except 1. Logarithms with base 10 are called *common logarithms*. Logarithms with base e (where $e = 2.71828\ldots$) are called *natural* or *Naperian logarithms*. The quantity $C = \log_b A$ is called the *logarithm* or *log* of A to base b, and $A = \text{antilog}_b C$ is called the *antilogarithm* or *antilog* of C to base b. Frequently, natural logarithms are denoted by ln (pronounced "lawn") rather than by $\log_e$. In this text, we use log for common logarithms to base 10 and ln or $\log_e$ for logarithms to base e. Most uses of logarithms in this text involve common logarithms.

☐ **EXAMPLES**

1. $\log 100 = 2$, $\log 5 = 0.6990$, $\log_e 7 = \ln 7 = 1.946$

2. antilog $(-3) = 10^{-3} = 0.001$, antilog $1.40 = 10^{1.40} = 25.12$, $\text{antilog}_e\, 4 = e^4 = 54.60$ ☐

Rules. The following is a summary of the algebraic rules for logarithms. Here A, B, b, and c denote numbers. Each rule is illustrated using common logarithms.

	Rule	Illustration
(A.17)	$\log_b AB = \log_b A + \log_b B$	$\log[10(100)] = \log 10 + \log 100$
		$= 1 + 2 = 3$
(A.18)	$\log_b \dfrac{A}{B} = \log_b A - \log_b B$	$\log\left(\dfrac{11}{5}\right) = \log 11 - \log 5$
		$= 1.0414 - 0.6990$
		$= 0.3424$
(A.19)	$\log_b A^c = c \log_b A$	$\log(100^3) = 3(\log 100)$
		$= 3(2) = 6$

Calculator Usage

Most calculators have one or more of the following operator keys for computing exponents and logarithms.

Key	Computes	Illustration
10^x	$\text{antilog}_{10}\, x$	$10^{0.5} = \text{antilog}_{10}\, 0.5 = 3.162$
$\log x$	$\log_{10}\, x$	$\log 11 = \log_{10} 11 = 1.0414$
e^x	$\text{antilog}_e\, x$	$e^{0.5} = \text{antilog}_e\, 0.5 = 1.649$
$\ln x$	$\log_e\, x$	$\ln 11 = \log_e 11 = 2.398$
y^x	y raised to power x	$10^{0.5} = 3.162, \;\; e^{0.5} = 1.649$
		$6.8^{2.3} = 82.18$

Permutations

**PERMUTATIONS   A.3
AND COMBINATIONS**

If r objects are selected from a set of n different objects and arranged in a particular order, the arrangement is called a *permutation of n objects taken r at a time*. The number of permutations of n different objects taken r at a time is denoted by $_nP_r$ and is as follows.

(A.20) $_nP_r = n(n - 1)(n - 2) \cdots (n - r + 1)$

The number of permutations of n different objects taken together is as follows.

(A.20a) $_nP_n = n! = n(n - 1)(n - 2) \cdots (1)$

The notation $x!$ (read "x factorial") stands for $x(x - 1)(x - 2) \cdots (1)$. Thus, $5! = 5(4)(3)(2)(1) = 120$. By definition, $0! = 1$.

EXAMPLES ☐

1. An experiment requires that two persons be selected from a group of four persons, denoted by A, B, C, and D. The first person selected is to be tested on day 1 and the second person on day 2. The number of different permutations of the $n = 4$ persons taken $r = 2$ at a time is given by (A.20) as $_4P_2 = 4(3) = 12$. The 12 permutations are:

$$
\begin{array}{cccccc}
\text{AB} & \text{AD} & \text{BC} & \text{CA} & \text{CD} & \text{DB} \\
\text{AC} & \text{BA} & \text{BD} & \text{CB} & \text{DA} & \text{DC}
\end{array}
$$

Note that the order of persons matters—AB and BA, for instance, are different testing sequences for persons A and B.

2. In Example 1, if all $n = 4$ persons are taken for the experiment and tested on different days, the total number of different permutations is given by (A.20a) as $4! = 4(3)(2)(1) = 24$. ☐

Permutations of Like Objects.   Consider n objects that are of c different types. There are r_1 objects of type 1, r_2 of type 2, and so on, where $r_1 + r_2 + \cdots + r_c = n$. The number of distinct permutations (that is, distinguishable ordered arrangements) when all n objects are taken together is as follows.

(A.21) $\dfrac{n!}{r_1!r_2!\cdots r_c!}$

where $r_1 + r_2 + \cdots + r_c = n$.

☐ **EXAMPLE**

Among $n = 5$ bottles coming off a filling line, $r_1 = 2$ have acceptable fills (A) and $r_2 = 3$ have nonconforming fills (N). One possible sequence or permutation for the five bottles as they are filled is ANNNA. The number of distinct permutations here is given by (A.21) as:

$$\frac{5!}{2!3!} = \frac{120}{2(6)} = 10$$

☐

Binomial Coefficient.   When there are only two different types of objects, r of one type and $n - r$ of the second type, the number of distinct permutations that can be formed from the n objects is sometimes denoted by the binomial coefficient $\binom{n}{r}$.

(A.22) $\dbinom{n}{r} = \dfrac{n!}{r!(n - r)!}$

Combinations

A set of r objects selected from n different objects, considered without regard to their order, is called a *combination of n objects taken r at a time.* The number of combinations of n different objects taken r at a time is denoted by $_nC_r$ and is as follows.

(A.23) $_nC_r = \dfrac{n!}{r!(n - r)!}$

Note that $_nC_r = {}_nC_{n-r}$.

☐ **EXAMPLE**

A litter contains five pups—A, B, C, D, and E—four of which are to be sold to a family. The number of different combinations of these $n = 5$ pups taken $r = 4$ at a time is given by (A.23) as $_5C_4 = 5!/(4!1!) = 5$. These five combinations are:

ABCD   ABCE   ABDE   ACDE   BCDE

Note that the order of the pups is immaterial here—ABCD and DCBA, for instance, are the same combination.   ☐

χ^2, t, and F Distributions

In this appendix, we discuss in more detail the χ^2, t, and F distributions, which are all related to the normal distribution and are widely used in statistical analysis. For each distribution, we describe its main characteristics and explain how to obtain probabilities and percentiles. Finally, we consider the theoretical relations between the χ^2, t, F, and normal distributions.

In Figure B.1, we have set out in summary form, for subsequent reference, several of the important characteristics of the χ^2, t, and F distributions, as well as those of the standard normal distribution.

Characteristics

The χ^2 (Greek chi-squared) distribution is a continuous probability distribution. It has one parameter, ν (Greek nu), which is called its *degrees of freedom* (abbreviated *df*). The parameter ν may be any positive whole number. Each different value of ν corresponds to a different member of the χ^2 family of distributions. The random variable that has a χ^2 distribution with ν degrees of freedom is denoted by $\chi^2(\nu)$ and may take on any positive value; that is, $0 < \chi^2(\nu) < \infty$. Graphs of the χ^2 probability density function for several values of ν are shown in Figure B.1. Figure B.1 also presents some of the characteristics of the χ^2 distribution.

Shape. The graphs of the χ^2 distribution in Figure B.1 show that the distributions are unimodal and right-skewed, although the skewness becomes smaller as ν increases. In fact, it can be shown that as ν increases, the shape of the χ^2 distribution approaches that of a normal distribution.

Mean and Variance. The mean and the variance of the $\chi^2(\nu)$ random variable are as follows.

$$(B.1) \qquad E\{\chi^2(\nu)\} = \nu$$

$$(B.2) \qquad \sigma^2\{\chi^2(\nu)\} = 2\nu$$

For example, the $\chi^2(5)$ random variable has a mean of 5 and a variance of $2(5) = 10$. Observe that the variance equals twice the mean and that both the mean and the variance increase as ν increases.

FIGURE B.1
Summary of characteristics of standard normal, χ^2, t, and F distributions

Random variable	Sample space	Mean	Variance	Density functions
Z	$(-\infty, \infty)$	0	1	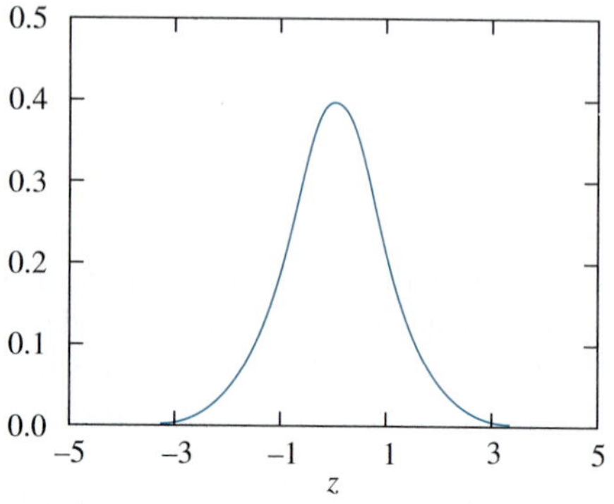
$\chi^2(v)$	$(0, \infty)$	v	$2v$	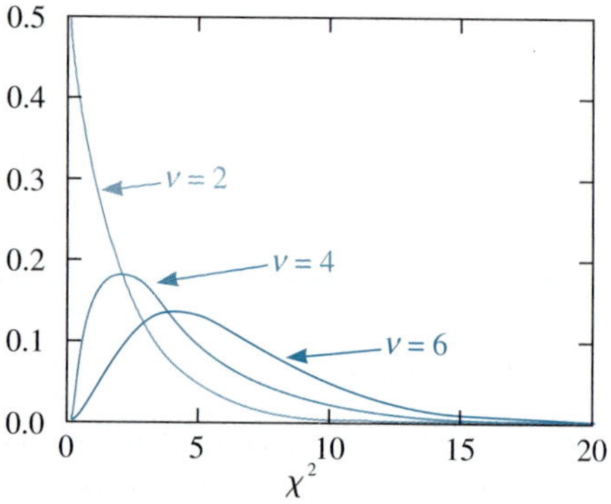
$t(v)$	$(-\infty, \infty)$	0	$\dfrac{v}{v-2}$ $v > 2$	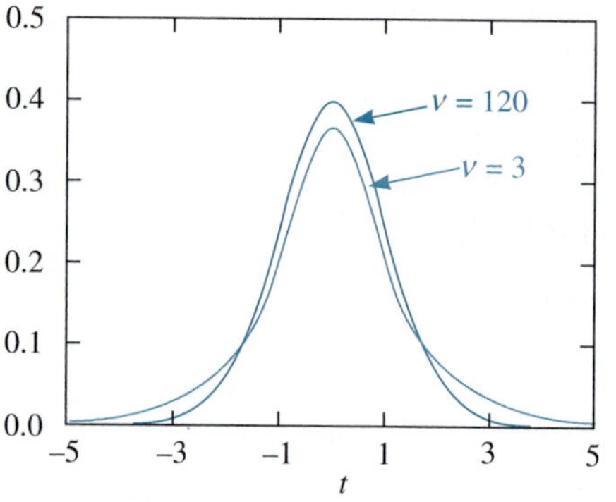
$F(v_1, v_2)$	$(0, \infty)$	$\dfrac{v_2}{v_2 - 2}$ $v_2 > 2$	$\dfrac{2v_2^2(v_1 + v_2 - 2)}{v_1(v_2 - 2)^2(v_2 - 4)}$ $v_2 > 4$	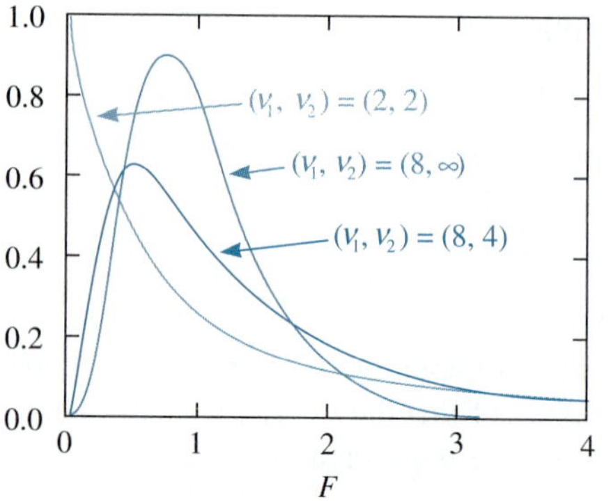

Determining Probabilities and Percentiles

We shall let $\chi^2(a; \nu)$ denote the $100a$ percentile of the χ^2 distribution with ν degrees of freedom.

(B.3) $P[\chi^2(\nu) \leq \chi^2(a; \nu)] = a$

Computer-Generated Probabilities and Percentiles.   Many statistical calculators and most statistical packages have routines for computing probabilities and percentiles for χ^2 distributions. Typically, an input of values for ν and k leads to an output of the probability $P[\chi^2(\nu) \leq k]$—that is, the probability that $\chi^2(\nu)$ is less than or equal to k. An input of values for ν and a leads to an output of the percentile $\chi^2(a; \nu)$.

EXAMPLES ☐

1. For $\nu = 6$ and $k = 8.4$, computer output (not shown) gives $P[\chi^2(6) \leq 8.4] = 0.7898$. Hence, the probability is 0.7898 that $\chi^2(6)$ is less than or equal to 8.4.

2. For $\nu = 14$ and $a = 0.30$, computer output (not shown) gives $\chi^2(0.30; 14) = 10.821$. Hence, the probability is 0.30 that $\chi^2(14)$ is less than or equal to 10.821. ☐

Table of Percentiles.   Table C.2 contains selected percentiles of χ^2 distributions for various degrees of freedom from 1 to 100.

EXAMPLES ☐

1. The 90th percentile of $\chi^2(5)$ is $\chi^2(0.90; 5) = 9.24$.

2. The 5th percentile of $\chi^2(60)$ is $\chi^2(0.05; 60) = 43.19$.   ☐

Normal Approximation.   The following transformation of a $\chi^2(\nu)$ random variable is approximately a standard normal random variable when ν is large.

(B.4) $\sqrt{2\chi^2(\nu)} - \sqrt{2\nu - 1} \simeq Z$ when ν is large

Thus, the following formulas may be used to obtain approximate probabilities and percentiles for $\chi^2(\nu)$ when ν is large.

(B.5) $P[\chi^2(\nu) \leq k] \simeq P[Z \leq \sqrt{2k} - \sqrt{2\nu - 1}]$

(B.6) $\chi^2(a; \nu) \simeq \frac{1}{2}[z(a) + \sqrt{2\nu - 1}]^2$

EXAMPLES ☐

1. To find the probability that $\chi^2(100)$ is less than or equal to $k = 82.36$—that is, $P[\chi^2(100) \leq 82.36]$—we use approximation (B.5). Since $k = 82.36$ and $\nu = 100$, we have:

$$\sqrt{2(82.36)} - \sqrt{2(100) - 1} = -1.27$$

Thus, the probability $P[\chi^2(100) \leq 82.36]$ is approximated by $P(Z \leq -1.27) = 0.1020$. Note in Table C.2 that 82.36 is shown as the 10th percentile of the $\chi^2(100)$ distribution, so the approximation is quite good.

2. To find the 95th percentile of the $\chi^2(100)$ distribution—that is, $\chi^2(0.95; 100)$—we use approximation (B.6). For $a = 0.95$, we know from Table C.1 that $z(0.95) = 1.645$. Since $\nu = 100$ and $z(0.95) = 1.645$, we obtain:

$$\chi^2(0.95; 100) \simeq \frac{1}{2}[1.645 + \sqrt{2(100) - 1}\,]^2 = 124.1$$

The exact percentile as shown in Table C.2 is 124.3, so again the approximation is quite good. $\square$

B.2 t DISTRIBUTIONS   Characteristics

The t distribution is a continuous probability distribution. As with the χ^2 distribution, it has only one parameter, ν, which is called its *degrees of freedom (df)*. The parameter ν may be any positive whole number. Each different value of ν corresponds to a different member of the family of t distributions. The random variable that has a t distribution with ν degrees of freedom is denoted by $t(\nu)$ and may take on any value; that is, $-\infty < t(\nu) < \infty$.

Graphs of the t probability density function for two values of ν are shown in Figure B.1. Figure B.1 also presents some of the characteristics of the t distribution.

Shape.   The graphs of the t distribution in Figure B.1 show that the t distribution is symmetrical and almost bell-shaped and that its appearance is like the standard normal distribution when ν is large. In fact, it can be shown mathematically that the t distribution approaches the standard normal distribution as ν becomes large.

Mean and Variance.   The mean and the variance of the $t(\nu)$ random variable are as follows.

(B.7) $E\{t(\nu)\} = 0$

(B.8) $\sigma^2\{t(\nu)\} = \dfrac{\nu}{\nu - 2}$ when $\nu > 2$

The variance is undefined for $\nu \leq 2$.

For example, the $t(6)$ random variable has mean 0 and variance $6/(6 - 2) = 1.50$. The $t(60)$ random variable still has mean 0, but the variance now is 1.03.

Note that the t distribution has a mean of 0 for all ν, and that the variance approaches 1 as ν gets larger. These facts are consistent with the earlier statement that the t distribution approaches the standard normal distribution as ν gets larger, since the latter distribution has mean 0 and variance 1.

Determining Probabilities and Percentiles

We shall let $t(a; \nu)$ denote the $100a$ percentile of the t distribution with ν degrees of freedom.

$$(B.9) \qquad P[t(\nu) \leq t(a; \nu)] = a$$

Computer-Generated Probabilities and Percentiles. Many statistical calculators and most statistical packages have routines for computing probabilities and percentiles for t distributions. Typically, an input of values for ν and k leads to an output of the probability $P[t(\nu) \leq k]$—that is, the probability that $t(\nu)$ is less than or equal to k. An input of values for ν and a leads to an output of the percentile $t(a; \nu)$.

EXAMPLES □

1. For $\nu = 6$ and $k = -2.30$, computer output (not shown) gives $P[t(6) \leq -2.30] = 0.0306$. Hence, the probability is 0.0306 that $t(6)$ is less than or equal to -2.30.
2. For $\nu = 14$ and $a = 0.70$, computer output (not shown) gives $t(0.70; 14) = 0.5366$. Hence, the probability is 0.70 that $t(14)$ is less than or equal to 0.5366. □

Table of Percentiles. Table C.3 contains selected percentiles of t distributions for various degrees of freedom ν.

EXAMPLES □

1. The 95th percentile of $t(10)$ is $t(0.95; 10) = 1.812$.
2. The 90th percentile of $t(30)$ is $t(0.90; 30) = 1.310$. □

To obtain percentiles less than 50 percent from Table C.3, we exploit the fact that the t distribution is symmetrical about its mean at 0.

$$(B.10) \qquad t(a; \nu) = -t(1 - a; \nu)$$

EXAMPLES □

1. The 5th percentile of $t(10)$ is $t(0.05; 10) = -t(0.95; 10) = -1.812$.
2. The 10th percentile of $t(30)$ is $t(0.10; 30) = -t(0.90; 30) = -1.310$. □

Normal Approximation. Since the t distribution approaches the standard normal distribution as ν increases, the percentiles of the standard normal distribution may be used to approximate those of the t distribution when ν is large. In fact, the last row in Table C.3, which corresponds to $\nu = \infty$, has the same percentiles as those for the standard normal distribution. For instance, we know from Table C.1 that $z(0.975) = 1.960$. In Table C.3, we see that $t(0.975; \infty) = 1.960$. As Table C.3 shows, once ν is larger than about 30, the standard normal distribution provides approximate percentiles for the t distribution that are adequate for most practical purposes. For example, $t(0.975; 217) \simeq z(0.975) = 1.960$.

B.3 F DISTRIBUTIONS

Characteristics

The F distribution is a continuous probability distribution. It has two parameters, ν_1 and ν_2, which are positive whole numbers and are called *degrees of freedom* (*df*). To each pair (ν_1, ν_2) corresponds a different member of the family of F distributions. The random variable associated with the F distribution is denoted by $F(\nu_1, \nu_2)$, where the parameter in the left position denotes the *numerator degrees of freedom* and the parameter in the right position denotes the *denominator degrees of freedom*. The $F(\nu_1, \nu_2)$ random variable may take on any positive value; that is, $0 < F(\nu_1, \nu_2) < \infty$.

Graphs of the F probability density function for several pairs of (ν_1, ν_2) values are shown in Figure B.1. Figure B.1 also presents some characteristics of the F distribution.

Shape.   The graphs of the F distribution in Figure B.1 show that the F distribution is unimodal and right-skewed. The mode of the F distribution (that is, the value of F for which the density is largest) approaches 1 as both degrees of freedom get large.

Mean and Variance.   The mean and the variance of the $F(\nu_1, \nu_2)$ random variable are as follows.

$$\textbf{(B.11)} \qquad E\{F(\nu_1, \nu_2)\} = \frac{\nu_2}{\nu_2 - 2} \qquad \text{when } \nu_2 > 2$$

$$\textbf{(B.12)} \qquad \sigma^2\{F(\nu_1, \nu_2)\} = \frac{2\nu_2^2(\nu_1 + \nu_2 - 2)}{\nu_1(\nu_2 - 2)^2(\nu_2 - 4)} \qquad \text{when } \nu_2 > 4$$

The mean is undefined for $\nu_2 \leq 2$, and the variance is undefined for $\nu_2 \leq 4$.

For example, the mean and the variance of the $F(3, 5)$ random variable are:

$$\frac{5}{5 - 2} = 1.67 \qquad \frac{2(5^2)(3 + 5 - 2)}{3(5 - 2)^2(5 - 4)} = 11.11$$

Hence, the standard deviation is $\sqrt{11.11} = 3.33$.

Note from (B.11) and (B.12) that the mean approaches 1 as ν_2 get large, while the variance approaches 0 as both degrees of freedom become large.

Determining Probabilities and Percentiles

We shall let $F(a; \nu_1, \nu_2)$ denote the $100a$ percentile of the F distribution with ν_1 and ν_2 degrees of freedom.

$$\textbf{(B.13)} \qquad P[F(\nu_1, \nu_2) \leq F(a; \nu_1, \nu_2)] = a$$

Computer-Generated Probabilities and Percentiles.   Many statistical calculators and most statistical packages have routines for computing probabilities and percentiles for F distributions. Typically, an input of values for ν_1, ν_2, and k leads to an output of the probability $P[F(\nu_1, \nu_2) \leq k]$—that is, the probability that $F(\nu_1, \nu_2)$ is less than or equal to k. An input of values for ν_1, ν_2, and a leads to an output of the percentile $F(a; \nu_1, \nu_2)$.

1. For values $\nu_1 = 2$, $\nu_2 = 35$, and $k = 2.46$, computer output (not shown) gives $P[F(2, 35) \leq 2.46] = 0.8999$. Hence, the probability is 0.8999 that $F(2, 35)$ is less than or equal to 2.46.

2. For values $\nu_1 = 18$, $\nu_2 = 10$, and $a = 0.90$, computer output (not shown) gives $F(0.90; 18, 10) = 2.215$. Hence, the probability is 0.90 that $F(18, 10)$ is less than or equal to 2.215. ☐

Table of Percentiles. Because the F distribution has two parameters, any extensive tabulation of it requires considerable space. Table C.4 contains the 95th and 99th percentiles of F distributions for selected numerator and denominator degrees of freedom.

1. The 95th percentile of $F(3, 5)$ is $F(0.95; 3, 5) = 5.41$.
2. The 99th percentile of $F(10, 6)$ is $F(0.99; 10, 6) = 7.87$. ☐

Percentiles below 50 percent can be obtained from percentiles above 50 percent by means of the following relationship.

$$\textbf{(B.14)} \qquad F(a; \nu_1, \nu_2) = \frac{1}{F(1 - a; \nu_2, \nu_1)}$$

Note the reversal of the degrees of freedom in the denominator on the right side of equation (B.14).

1. To find the 5th percentile of $F(3, 5)$, we first obtain $F(0.95; 5, 3) = 9.01$ from Table C.4. Then, by (B.14):

$$F(0.05; 3, 5) = \frac{1}{F(0.95; 5, 3)} = \frac{1}{9.01} = 0.111$$

Thus, the probability is 0.05 that $F(3, 5)$ is less than or equal to 0.111.

2. To find the 1st percentile of $F(6, 10)$, we first obtain $F(0.99; 10, 6) = 7.87$ from Table C.4. Then, by (B.14):

$$F(0.01; 6, 10) = \frac{1}{F(0.99; 10, 6)} = \frac{1}{7.87} = 0.127$$ ☐

B.4 OPTIONAL TOPIC—RELATIONS BETWEEN Z, t, χ^2, AND F

We first describe the relations of χ^2, t, and F to the standard normal random variable Z, and then we explain how Z, t, and χ^2 may be viewed as special cases of the random variable F.

χ^2 Random Variable

The χ^2 random variable is related directly to the standard normal variable. Let $Z_1, Z_2, \ldots, Z_\nu$ be ν independent standard normal variables. Then a $\chi^2(\nu)$ random variable can be defined in terms of these Z_i variables as follows.

> **(B.15)** $\chi^2(\nu) = Z_1^2 + Z_2^2 + \cdots + Z_\nu^2$
>
> where the Z_i are independent $N(0, 1)$.

In other words, the sum of ν independent squared standard normal variables is a χ^2 random variable with ν degrees of freedom.

We can find $E\{\chi^2(\nu)\}$ by applying theorem (6.9a) to definition (B.15). We obtain:

$$E\{\chi^2(\nu)\} = E\{Z_1^2\} + E\{Z_2^2\} + \cdots + E\{Z_\nu^2\}$$

For the standard normal variable Z, $E\{Z\} = 0$ and $\sigma^2\{Z\} = 1$. Hence, it follows from (5.7a) that $E\{Z^2\} = 1$. Consequently:

$$E\{\chi^2(\nu)\} = 1 + 1 + \cdots + 1 = \nu$$

which is the result given in (B.1).

t Random Variable

The $t(\nu)$ random variable is related to both the $\chi^2(\nu)$ and Z random variables as follows.

> **(B.16)** $t(\nu) = \dfrac{Z}{\left[\dfrac{\chi^2(\nu)}{\nu}\right]^{1/2}}$
>
> where Z and $\chi^2(\nu)$ are independent.

In other words, a t random variable with ν degrees of freedom is obtained by (1) taking a standard normal random variable Z and a χ^2 random variable with ν degrees of freedom that are independent of one another and (2) forming the ratio of Z and the positive square root of $\chi^2(\nu)$ divided by its degrees of freedom ν.

We can see intuitively why the t distribution approaches the standard normal distribution when ν is large by considering the random variable $\chi^2(\nu)/\nu$ in the denominator on the right of (B.16). The expected value of this random variable is:

$$E\left\{\frac{\chi^2(\nu)}{\nu}\right\} = \frac{1}{\nu}E\{\chi^2(\nu)\} = \frac{\nu}{\nu} = 1 \qquad \text{by (6.5b) and (B.1)}$$

while its variance is:

$$\sigma^2\left\{\frac{\chi^2(\nu)}{\nu}\right\} = \frac{1}{\nu^2}\sigma^2\{\chi^2(\nu)\} = \frac{2\nu}{\nu^2} = \frac{2}{\nu} \qquad \text{by (6.6b) and (B.2)}$$

Note that the variance approaches 0 as ν gets large. Hence, $\chi^2(\nu)/\nu$ becomes concentrated within an arbitrarily small neighborhood of the expected value 1 as ν approaches ∞. Thus, t approaches $Z/\sqrt{1} = Z$ as ν approaches ∞.

F Random Variable

The $F(\nu_1, \nu_2)$ random variable is related to two independent χ^2 random variables—$\chi^2(\nu_1)$ and $\chi^2(\nu_2)$—as follows.

$$(B.17) \qquad F(\nu_1, \nu_2) = \frac{\dfrac{\chi^2(\nu_1)}{\nu_1}}{\dfrac{\chi^2(\nu_2)}{\nu_2}}$$

where $\chi^2(\nu_1)$ and $\chi^2(\nu_2)$ are independent.

In other words, an F random variable is obtained by dividing two independent χ^2 random variables by their respective degrees of freedom and forming the ratio of these ratios.

The symmetrical relationship between the numerator and the denominator in (B.17) leads to the following result.

$$(B.18) \qquad F(\nu_1, \nu_2) = \frac{1}{F(\nu_2, \nu_1)}$$

This fact makes it possible to compute the $100a$ percentile of $F(\nu_1, \nu_2)$ from knowledge of the $100(1 - a)$ percentile of $F(\nu_2, \nu_1)$ by relationship (B.14).

Z, t, and χ^2 as Special Cases of F

Relations Between Random Variables. The random variables Z, t, and χ^2 may be considered as the following special cases of an F random variable.

(B.19)   The relations of Z, t, and χ^2 to F are as follows:

(B.19a) $Z^2 = F(1, \infty)$

(B.19b) $[t(\nu)]^2 = F(1, \nu)$

(B.19c) $\dfrac{\chi^2(\nu)}{\nu} = F(\nu, \infty)$

The relation between Z and F in (B.19a) follows because $v_2 = \infty$ implies, as we have seen, that $\chi^2(v_2)/v_2$ is essentially equivalent to 1. Thus:

$$F(1, \infty) = \frac{\chi^2(1)}{\dfrac{1}{1}} = Z^2$$

The other relations follow in the same manner.

Relations Between Percentiles. The percentiles of the Z, t, and χ^2 distributions are related to the percentiles of the F distribution in ways illustrated by the following examples.

☐ **EXAMPLES**

1. Consider $z(0.975) = 1.960$. We have $(1.960)^2 = 3.84 = F(0.95; 1, \infty)$.
2. Consider $t(0.975; 10) = 2.228$. We have $(2.228)^2 = 4.96 = F(0.95; 1, 10)$.
3. Consider $\chi^2(0.95; 8) = 15.51$. We have $15.51/8 = 1.94 = F(0.95; 8, \infty)$. ☐

In Examples 1 and 2, the reason why 97.5 percentiles for Z and t correspond to the 95th percentile of F is related to the fact that Z^2 and t^2 equal F in (B.19). Hence, both negative and positive outcomes of Z and t lead to the same F value. The following sequence of steps explains this for Example 1.

Step 1: $P[z(0.025) \le Z \le z(0.975)] = 0.95$
Step 2: $P[-z(0.975) \le Z \le z(0.975)] = 0.95$
Step 3: $P\{Z^2 \le [z(0.975)]^2\} = 0.95$

In step 3, we utilize the fact that an inequality of the form $-a \le x \le a$ is mathematically equivalent to $x^2 \le a$.

Step 4: $P[F(1, \infty) \le F(0.95; 1, \infty)] = 0.95$

In step 4, we utilize the fact that $Z^2 = F(1, \infty)$ by (B.19a). Hence, it follows that $[z(0.975)]^2$ must equal $F(0.95; 1, \infty)$.

PROBLEMS

* **B.1** Consider a $\chi^2(8)$ random variable.

a. Obtain the 10th and 90th percentiles of the probability distribution.
b. What are the mean and the standard deviation of the probability distribution?
c. Are the percentiles in part a equally spaced about the mean? What does this fact imply about the symmetry of the probability distribution?

B.2 Consider a $\chi^2(4)$ random variable.

a. Obtain the 5th and 95th percentiles of the probability distribution.
b. What are the mean and the standard deviation of the probability distribution?
c. Are the percentiles in part a equally spaced about the mean? What does this fact imply about the symmetry of the probability distribution?

* **B.3**

  a. Obtain $\chi^2(0.95; 2)$ and $\chi^2(0.10; 8)$. What does the latter value represent?
  b. Estimate $\chi^2(0.05; 160)$ by using the normal approximation.

B.4

  a. Obtain $\chi^2(0.01; 12)$ and $\chi^2(0.995; 7)$. What does the latter value represent?
  b. Estimate $\chi^2(0.90; 113)$ by using the normal approximation.

* **B.5** Consider a $t(11)$ random variable.

  a. Obtain the 5th and 95th percentiles of the probability distribution.
  b. What are the mean and the standard deviation of the probability distribution?
  c. Are the percentiles in part a equally spaced about the mean? What does this fact suggest about the symmetry of the probability distribution?

B.6 Consider a $t(7)$ random variable.

  a. Obtain the 10th and 90th percentiles of the probability distribution.
  b. What are the mean and the standard deviation of the probability distribution?
  c. Are the percentiles in part a equally spaced about the mean? What does this fact suggest about the symmetry of the probability distribution?

* **B.7**

  a. Obtain $t(0.90; 5)$ and $t(0.05; 21)$. What does the latter value represent?
  b. Estimate $t(0.01; 200)$ by using the normal approximation.

B.8

  a. Obtain $t(0.995; 11)$ and $t(0.10; 18)$. What does the latter value represent?
  b. Estimate $t(0.975; 410)$ by using the normal approximation.

B.9 On the same graph, plot the cumulative probability function of the $t(5)$ random variable and that of the standard normal random variable. Compare and contrast the two functions.

B.10 Compare the interquartile range of the $t(3)$ random variable and that of the standard normal random variable. Does the comparison reflect the greater variance of the t random variable? Explain.

* **B.11** Consider an $F(5, 8)$ random variable.

  a. Obtain the 1st and 99th percentiles of the probability distribution.
  b. What are the mean and the standard deviation of the probability distribution?
  c. Are the percentiles in part a equally spaced about the mean? What does this fact imply about the symmetry of the probability distribution?

B.12 Consider an $F(2, 6)$ random variable.

  a. Obtain the 5th and 95th percentiles of the probability distribution.
  b. What are the mean and the standard deviation of the probability distribution?
  c. Are the percentiles in part a equally spaced about the mean? What does this fact imply about the symmetry of the probability distribution?

* **B.13**

  a. Obtain $F(0.95; 2, 20)$. What does this value represent?
  b. Obtain $F(0.01; 3, 8)$.

B.14

  a. Obtain $F(0.99; 8, 7)$. What does this value represent?
  b. Obtain $F(0.05; 4, 4)$.

EXERCISES

B.15

  a. Given that $\sigma^2\{Z^2\} = 2$ for a standard normal random variable Z, show that $\sigma^2\{\chi^2(\nu)\} = 2\nu$.
  b. Explain why $\chi^2(\nu_1 + \nu_2) = \chi^2(\nu_1) + \chi^2(\nu_2)$ when $\chi^2(\nu_1)$ and $\chi^2(\nu_2)$ are independent.

B.16

a. Explain why $[t(\nu)]^2 = F(1, \nu)$.

b. Obtain $F(0.95; 1, 3)$ from Table C.3.

B.17

a. Explain why $\chi^2(\nu)/\nu = F(\nu, \infty)$.

b. Obtain $F(0.99; 8, \infty)$ from Table C.2.

B.18 Obtain the following probabilities and percentiles:

a. (1) $P[\chi^2(8) \leq 12.4]$, (2) $\chi^2(0.75; 8)$.

b. (1) $P[t(35) > 3.6]$, (2) $t(0.02; 35)$.

c. (1) $P[F(9, 17) > 4.9]$, (2) $F(0.90; 9, 17)$.

B.19 Obtain the following probabilities and percentiles:

a. (1) $P[\chi^2(13) > 23.7]$, (2) $\chi^2(0.20, 13)$.

b. (1) $P[t(18) \leq -1.5]$, (2) $t(0.80; 18)$.

c. (1) $P[F(4, 22) \leq 8.3]$, (2) $F(0.98; 4, 22)$.

Tables

Cumulative probabilities and percentiles of the standard normal distribution

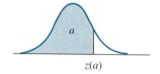

$z(a)$

(a) Cumulative probabilities

Entry is area a under the standard normal curve from $-\infty$ to $z(a)$.

z	0.00	0.01	0.02	0.03	0.04	0.05	0.06	0.07	0.08	0.09
0.0	0.5000	0.5040	0.5080	0.5120	0.5160	0.5199	0.5239	0.5279	0.5319	0.5359
0.1	0.5398	0.5438	0.5478	0.5517	0.5557	0.5596	0.5636	0.5675	0.5714	0.5753
0.2	0.5793	0.5832	0.5871	0.5910	0.5948	0.5987	0.6026	0.6064	0.6103	0.6141
0.3	0.6179	0.6217	0.6255	0.6293	0.6331	0.6368	0.6406	0.6443	0.6480	0.6517
0.4	0.6554	0.6591	0.6628	0.6664	0.6700	0.6736	0.6772	0.6808	0.6844	0.6879
0.5	0.6915	0.6950	0.6985	0.7019	0.7054	0.7088	0.7123	0.7157	0.7190	0.7224
0.6	0.7257	0.7291	0.7324	0.7357	0.7389	0.7422	0.7454	0.7486	0.7517	0.7549
0.7	0.7580	0.7611	0.7642	0.7673	0.7704	0.7734	0.7764	0.7794	0.7823	0.7852
0.8	0.7881	0.7910	0.7939	0.7967	0.7995	0.8023	0.8051	0.8078	0.8106	0.8133
0.9	0.8159	0.8186	0.8212	0.8238	0.8264	0.8289	0.8315	0.8340	0.8365	0.8389
1.0	0.8413	0.8438	0.8461	0.8485	0.8508	0.8531	0.8554	0.8577	0.8599	0.8621
1.1	0.8643	0.8665	0.8686	0.8708	0.8729	0.8749	0.8770	0.8790	0.8810	0.8830
1.2	0.8849	0.8869	0.8888	0.8907	0.8925	0.8944	0.8962	0.8980	0.8997	0.9015
1.3	0.9032	0.9049	0.9066	0.9082	0.9099	0.9115	0.9131	0.9147	0.9162	0.9177
1.4	0.9192	0.9207	0.9222	0.9236	0.9251	0.9265	0.9279	0.9292	0.9306	0.9319
1.5	0.9332	0.9345	0.9357	0.9370	0.9382	0.9394	0.9406	0.9418	0.9429	0.9441
1.6	0.9452	0.9463	0.9474	0.9484	0.9495	0.9505	0.9515	0.9525	0.9535	0.9545
1.7	0.9554	0.9564	0.9573	0.9582	0.9591	0.9599	0.9608	0.9616	0.9625	0.9633
1.8	0.9641	0.9649	0.9656	0.9664	0.9671	0.9678	0.9686	0.9693	0.9699	0.9706
1.9	0.9713	0.9719	0.9726	0.9732	0.9738	0.9744	0.9750	0.9756	0.9761	0.9767
2.0	0.9772	0.9778	0.9783	0.9788	0.9793	0.9798	0.9803	0.9808	0.9812	0.9817
2.1	0.9821	0.9826	0.9830	0.9834	0.9838	0.9842	0.9846	0.9850	0.9854	0.9857
2.2	0.9861	0.9864	0.9868	0.9871	0.9875	0.9878	0.9881	0.9884	0.9887	0.9890
2.3	0.9893	0.9896	0.9898	0.9901	0.9904	0.9906	0.9909	0.9911	0.9913	0.9916
2.4	0.9918	0.9920	0.9922	0.9925	0.9927	0.9929	0.9931	0.9932	0.9934	0.9936
2.5	0.9938	0.9940	0.9941	0.9943	0.9945	0.9946	0.9948	0.9949	0.9951	0.9952
2.6	0.9953	0.9955	0.9956	0.9957	0.9959	0.9960	0.9961	0.9962	0.9963	0.9964
2.7	0.9965	0.9966	0.9967	0.9968	0.9969	0.9970	0.9971	0.9972	0.9973	0.9974
2.8	0.9974	0.9975	0.9976	0.9977	0.9977	0.9978	0.9979	0.9979	0.9980	0.9981
2.9	0.9981	0.9982	0.9982	0.9983	0.9984	0.9984	0.9985	0.9985	0.9986	0.9986

TABLE C.I

(concluded)

z	0.00	0.01	0.02	0.03	0.04	0.05	0.06	0.07	0.08	0.09
3.0	0.99865	0.99869	0.99874	0.99878	0.99882	0.99886	0.99889	0.99893	0.99896	0.99900
3.1	0.99903	0.99906	0.99910	0.99913	0.99916	0.99918	0.99921	0.99924	0.99926	0.99929
3.2	0.99931	0.99934	0.99936	0.99938	0.99940	0.99942	0.99944	0.99946	0.99948	0.99950
3.3	0.99952	0.99953	0.99955	0.99957	0.99958	0.99960	0.99961	0.99962	0.99964	0.99965
3.4	0.99966	0.99968	0.99969	0.99970	0.99971	0.99972	0.99973	0.99974	0.99975	0.99976
3.5	0.99977	0.99978	0.99978	0.99979	0.99980	0.99981	0.99981	0.99982	0.99983	0.99983
3.6	0.99984	0.99985	0.99985	0.99986	0.99986	0.99987	0.99987	0.99988	0.99988	0.99989
3.7	0.99989	0.99990	0.99990	0.99990	0.99991	0.99991	0.99992	0.99992	0.99992	0.99992
3.8	0.99993	0.99993	0.99993	0.99994	0.99994	0.99994	0.99994	0.99995	0.99995	0.99995
3.9	0.99995	0.99995	0.99996	0.99996	0.99996	0.99996	0.99996	0.99996	0.99997	0.99997

(b) Selected percentiles

Entry is $z(a)$ where $P[Z \le z(a)] = a$.

a:	0.10	0.05	0.025	0.02	0.01	0.005	0.001
$z(a)$:	-1.282	-1.645	-1.960	-2.054	-2.326	-2.576	-3.090

a:	0.90	0.95	0.975	0.98	0.99	0.995	0.999
$z(a)$:	1.282	1.645	1.960	2.054	2.326	2.576	3.090

Example: $P(Z \le 1.96) = 0.9750$ so $z(0.9750) = 1.96$.
Text Reference: Use of this table is discussed on pp. 215–220.

TABLE C.2

Percentiles of the chi-square distribution
Entry is $\chi^2(a; \nu)$ where $P[\chi^2(\nu) \leq \chi^2(a; \nu)] = a$.

$\chi^2(a;\nu)$

| df ν | \multicolumn{10}{c}{a} |
	0.005	0.01	0.025	0.05	0.10	0.90	0.95	0.975	0.99	0.995
1	0.0⁴393	0.0³157	0.0³982	0.0²393	0.0158	2.71	3.84	5.02	6.63	7.88
2	0.0100	0.0201	0.0506	0.103	0.211	4.61	5.99	7.38	9.21	10.60
3	0.072	0.115	0.216	0.352	0.584	6.25	7.81	9.35	11.34	12.84
4	0.207	0.297	0.484	0.711	1.064	7.78	9.49	11.14	13.28	14.86
5	0.412	0.554	0.831	1.145	1.61	9.24	11.07	12.83	15.09	16.75
6	0.676	0.872	1.24	1.64	2.20	10.64	12.59	14.45	16.81	18.55
7	0.989	1.24	1.69	2.17	2.83	12.02	14.07	16.01	18.48	20.28
8	1.34	1.65	2.18	2.73	3.49	13.36	15.51	17.53	20.09	21.96
9	1.73	2.09	2.70	3.33	4.17	14.68	16.92	19.02	21.67	23.59
10	2.16	2.56	3.25	3.94	4.87	15.99	18.31	20.48	23.21	25.19
11	2.60	3.05	3.82	4.57	5.58	17.28	19.68	21.92	24.73	26.76
12	3.07	3.57	4.40	5.23	6.30	18.55	21.03	23.34	26.22	28.30
13	3.57	4.11	5.01	5.89	7.04	19.81	22.36	24.74	27.69	29.82
14	4.07	4.66	5.63	6.57	7.79	21.06	23.68	26.12	29.14	31.32
15	4.60	5.23	6.26	7.26	8.55	22.31	25.00	27.49	30.58	32.80
16	5.14	5.81	6.91	7.96	9.31	23.54	26.30	28.85	32.00	34.27
17	5.70	6.41	7.56	8.67	10.09	24.77	27.59	30.19	33.41	35.72
18	6.26	7.01	8.23	9.39	10.86	25.99	28.87	31.53	34.81	37.16
19	6.84	7.63	8.91	10.12	11.65	27.20	30.14	32.85	36.19	38.58
20	7.43	8.26	9.59	10.85	12.44	28.41	31.41	34.17	37.57	40.00
21	8.03	8.90	10.28	11.59	13.24	29.62	32.67	35.48	38.93	41.40
22	8.64	9.54	10.98	12.34	14.04	30.81	33.92	36.78	40.29	42.80
23	9.26	10.20	11.69	13.09	14.85	32.01	35.17	38.08	41.64	44.18
24	9.89	10.86	12.40	13.85	15.66	33.20	36.42	39.36	42.98	45.56
25	10.52	11.52	13.12	14.61	16.47	34.38	37.65	40.65	44.31	46.93
26	11.16	12.20	13.84	15.38	17.29	35.56	38.89	41.92	45.64	48.29
27	11.81	12.88	14.57	16.15	18.11	36.74	40.11	43.19	46.96	49.64
28	12.46	13.56	15.31	16.93	18.94	37.92	41.34	44.46	48.28	50.99
29	13.12	14.26	16.05	17.71	19.77	39.09	42.56	45.72	49.59	52.34
30	13.79	14.95	16.79	18.49	20.60	40.26	43.77	46.98	50.89	53.67
40	20.71	22.16	24.43	26.51	29.05	51.81	55.76	59.34	63.69	66.77
50	27.99	29.71	32.36	34.76	37.69	63.17	67.50	71.42	76.15	79.49
60	35.53	37.48	40.48	43.19	46.46	74.40	79.08	83.30	88.38	91.95
70	43.28	45.44	48.76	51.74	55.33	85.53	90.53	95.02	100.4	104.2
80	51.17	53.54	57.15	60.39	64.28	96.58	101.9	106.6	112.3	116.3
90	59.20	61.75	65.65	69.13	73.29	107.6	113.1	118.1	124.1	128.3
100	67.33	70.06	74.22	77.93	82.36	118.5	124.3	129.6	135.8	140.2

Source: Tabulated values adapted by permission from C. M. Thompson, "Table of Percentage Points of the Chi-Square Distribution," *Biometrika,* Vol. 32 (1941), pp. 188–189.

Example: $\chi^2(0.900; 4) = 7.78$ so $P[\chi^2(4) \leq 7.78] = 0.900$.   *Text Reference:* Use of this table is discussed on p. 911.

Percentiles of the *t* distribution   Entry is $t(a; \nu)$ where $P[t(\nu) \leq t(a; \nu)] = a$.

TABLE C.3

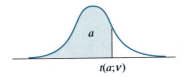

$t(a;\nu)$

df ν				a			
	0.75	0.90	0.95	0.975	0.99	0.995	0.9995
1	1.000	3.078	6.314	12.706	31.821	63.657	636.619
2	0.816	1.886	2.920	4.303	6.965	9.925	31.599
3	0.765	1.638	2.353	3.182	4.541	5.841	12.924
4	0.741	1.533	2.132	2.776	3.747	4.604	8.610
5	0.727	1.476	2.015	2.571	3.365	4.032	6.869
6	0.718	1.440	1.943	2.447	3.143	3.707	5.959
7	0.711	1.415	1.895	2.365	2.998	3.499	5.408
8	0.706	1.397	1.860	2.306	2.896	3.355	5.041
9	0.703	1.383	1.833	2.262	2.821	3.250	4.781
10	0.700	1.372	1.812	2.228	2.764	3.169	4.587
11	0.697	1.363	1.796	2.201	2.718	3.106	4.437
12	0.695	1.356	1.782	2.179	2.681	3.055	4.318
13	0.694	1.350	1.771	2.160	2.650	3.012	4.221
14	0.692	1.345	1.761	2.145	2.624	2.977	4.140
15	0.691	1.341	1.753	2.131	2.602	2.947	4.073
16	0.690	1.337	1.746	2.120	2.583	2.921	4.015
17	0.689	1.333	1.740	2.110	2.567	2.898	3.965
18	0.688	1.330	1.734	2.101	2.552	2.878	3.922
19	0.688	1.328	1.729	2.093	2.539	2.861	3.883
20	0.687	1.325	1.725	2.086	2.528	2.845	3.850
21	0.686	1.323	1.721	2.080	2.518	2.831	3.819
22	0.686	1.321	1.717	2.074	2.508	2.819	3.792
23	0.685	1.319	1.714	2.069	2.500	2.807	3.768
24	0.685	1.318	1.711	2.064	2.492	2.797	3.745
25	0.684	1.316	1.708	2.060	2.485	2.787	3.725
26	0.684	1.315	1.706	2.056	2.479	2.779	3.707
27	0.684	1.314	1.703	2.052	2.473	2.771	3.690
28	0.683	1.313	1.701	2.048	2.467	2.763	3.674
29	0.683	1.311	1.699	2.045	2.462	2.756	3.659
30	0.683	1.310	1.697	2.042	2.457	2.750	3.646
40	0.681	1.303	1.684	2.021	2.423	2.704	3.551
60	0.679	1.296	1.671	2.000	2.390	2.660	3.460
120	0.677	1.289	1.658	1.980	2.358	2.617	3.373
∞	0.674	1.282	1.645	1.960	2.326	2.576	3.291

Example: $t(0.95; 10) = 1.812$ so $P[t(10) \leq 1.812] = 0.95$.
Text Reference: Use of this table is discussed on p. 913.

TABLE C.4   **Percentiles of the F distribution**   Entry is $F(a; \nu_1, \nu_2)$ where $P[F(\nu_1, \nu_2) \leq F(a; \nu_1, \nu_2)] = a$.

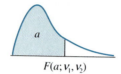

$F(a; \nu_1, \nu_2)$

$a = 0.95$

denominator df	numerator df								
	1	2	3	4	5	6	7	8	9
1	161.4	199.5	215.7	224.6	230.2	234.0	236.8	238.9	240.5
2	18.51	19.00	19.16	19.25	19.30	19.33	19.35	19.37	19.38
3	10.13	9.55	9.28	9.12	9.01	8.94	8.89	8.85	8.81
4	7.71	6.94	6.59	6.39	6.26	6.16	6.09	6.04	6.00
5	6.61	5.79	5.41	5.19	5.05	4.95	4.88	4.82	4.77
6	5.99	5.14	4.76	4.53	4.39	4.28	4.21	4.15	4.10
7	5.59	4.74	4.35	4.12	3.97	3.87	3.79	3.73	3.68
8	5.32	4.46	4.07	3.84	3.69	3.58	3.50	3.44	3.39
9	5.12	4.26	3.86	3.63	3.48	3.37	3.29	3.23	3.18
10	4.96	4.10	3.71	3.48	3.33	3.22	3.14	3.07	3.02
11	4.84	3.98	3.59	3.36	3.20	3.09	3.01	2.95	2.90
12	4.75	3.89	3.49	3.26	3.11	3.00	2.91	2.85	2.80
13	4.67	3.81	3.41	3.18	3.03	2.92	2.83	2.77	2.71
14	4.60	3.74	3.34	3.11	2.96	2.85	2.76	2.70	2.65
15	4.54	3.68	3.29	3.06	2.90	2.79	2.71	2.64	2.59
16	4.49	3.63	3.24	3.01	2.85	2.74	2.66	2.59	2.54
17	4.45	3.59	3.20	2.96	2.81	2.70	2.61	2.55	2.49
18	4.41	3.55	3.16	2.93	2.77	2.66	2.58	2.51	2.46
19	4.38	3.52	3.13	2.90	2.74	2.63	2.54	2.48	2.42
20	4.35	3.49	3.10	2.87	2.71	2.60	2.51	2.45	2.39
21	4.32	3.47	3.07	2.84	2.68	2.57	2.49	2.42	2.37
22	4.30	3.44	3.05	2.82	2.66	2.55	2.46	2.40	2.34
23	4.28	3.42	3.03	2.80	2.64	2.53	2.44	2.37	2.32
24	4.26	3.40	3.01	2.78	2.62	2.51	2.42	2.36	2.30
25	4.24	3.39	2.99	2.76	2.60	2.49	2.40	2.34	2.28
26	4.23	3.37	2.98	2.74	2.59	2.47	2.39	2.32	2.27
27	4.21	3.35	2.96	2.73	2.57	2.46	2.37	2.31	2.25
28	4.20	3.34	2.95	2.71	2.56	2.45	2.36	2.29	2.24
29	4.18	3.33	2.93	2.70	2.55	2.43	2.35	2.28	2.22
30	4.17	3.32	2.92	2.69	2.53	2.42	2.33	2.27	2.21
40	4.08	3.23	2.84	2.61	2.45	2.34	2.25	2.18	2.12
60	4.00	3.15	2.76	2.53	2.37	2.25	2.17	2.10	2.04
120	3.92	3.07	2.68	2.45	2.29	2.17	2.09	2.02	1.96
∞	3.84	3.00	2.60	2.37	2.21	2.10	2.01	1.94	1.88

(continued)

TABLE C.4

$a = 0.95$

				numerator *df*						denominator *df*
10	12	15	20	24	30	40	60	120	∞	
241.9	243.9	245.9	248.0	249.1	250.1	251.1	252.2	253.3	254.3	1
19.40	19.41	19.43	19.45	19.45	19.46	19.47	19.48	19.49	19.50	2
8.79	8.74	8.70	8.66	8.64	8.62	8.59	8.57	8.55	8.53	3
5.96	5.91	5.86	5.80	5.77	5.75	5.72	5.69	5.66	5.63	4
4.74	4.68	4.62	4.56	4.53	4.50	4.46	4.43	4.40	4.36	5
4.06	4.00	3.94	3.87	3.84	3.81	3.77	3.74	3.70	3.67	6
3.64	3.57	3.51	3.44	3.41	3.38	3.34	3.30	3.27	3.23	7
3.35	3.28	3.22	3.15	3.12	3.08	3.04	3.01	2.97	2.93	8
3.14	3.07	3.01	2.94	2.90	2.86	2.83	2.79	2.75	2.71	9
2.98	2.91	2.85	2.77	2.74	2.70	2.66	2.62	2.58	2.54	10
2.85	2.79	2.72	2.65	2.61	2.57	2.53	2.49	2.45	2.40	11
2.75	2.69	2.62	2.54	2.51	2.47	2.43	2.38	2.34	2.30	12
2.67	2.60	2.53	2.46	2.42	2.38	2.34	2.30	2.25	2.21	13
2.60	2.53	2.46	2.39	2.35	2.31	2.27	2.22	2.18	2.13	14
2.54	2.48	2.40	2.33	2.29	2.25	2.20	2.16	2.11	2.07	15
2.49	2.42	2.35	2.28	2.24	2.19	2.15	2.11	2.06	2.01	16
2.45	2.38	2.31	2.23	2.19	2.15	2.10	2.06	2.01	1.96	17
2.41	2.34	2.27	2.19	2.15	2.11	2.06	2.02	1.97	1.92	18
2.38	2.31	2.23	2.16	2.11	2.07	2.03	1.98	1.93	1.88	19
2.35	2.28	2.20	2.12	2.08	2.04	1.99	1.95	1.90	1.84	20
2.32	2.25	2.18	2.10	2.05	2.01	1.96	1.92	1.87	1.81	21
2.30	2.23	2.15	2.07	2.03	1.98	1.94	1.89	1.84	1.78	22
2.27	2.20	2.13	2.05	2.01	1.96	1.91	1.86	1.81	1.76	23
2.25	2.18	2.11	2.03	1.98	1.94	1.89	1.84	1.79	1.73	24
2.24	2.16	2.09	2.01	1.96	1.92	1.87	1.82	1.77	1.71	25
2.22	2.15	2.07	1.99	1.95	1.90	1.85	1.80	1.75	1.69	26
2.20	2.13	2.06	1.97	1.93	1.88	1.84	1.79	1.73	1.67	27
2.19	2.12	2.04	1.96	1.91	1.87	1.82	1.77	1.71	1.65	28
2.18	2.10	2.03	1.94	1.90	1.85	1.81	1.75	1.70	1.64	29
2.16	2.09	2.01	1.93	1.89	1.84	1.79	1.74	1.68	1.62	30
2.08	2.00	1.92	1.84	1.79	1.74	1.69	1.64	1.58	1.51	40
1.99	1.92	1.84	1.75	1.70	1.65	1.59	1.53	1.47	1.39	60
1.91	1.83	1.75	1.66	1.61	1.55	1.50	1.43	1.35	1.25	120
1.83	1.75	1.67	1.57	1.52	1.46	1.39	1.32	1.22	1.00	∞

(continues)

TABLE C.4   **Percentiles of the F distribution (continued)**

$$a = 0.99$$

denominator df	numerator df								
	1	2	3	4	5	6	7	8	9
1	4052	4999.5	5403	5625	5764	5859	5928	5981	6022
2	98.50	99.00	99.17	99.25	99.30	99.33	99.36	99.37	99.39
3	34.12	30.82	29.46	28.71	28.24	27.91	27.67	27.49	27.35
4	21.20	18.00	16.69	15.98	15.52	15.21	14.98	14.80	14.66
5	16.26	13.27	12.06	11.39	10.97	10.67	10.46	10.29	10.16
6	13.75	10.92	9.78	9.15	8.75	8.47	8.26	8.10	7.98
7	12.25	9.55	8.45	7.85	7.46	7.19	6.99	6.84	6.72
8	11.26	8.65	7.59	7.01	6.63	6.37	6.18	6.03	5.91
9	10.56	8.02	6.99	6.42	6.06	5.80	5.61	5.47	5.35
10	10.04	7.56	6.55	5.99	5.64	5.39	5.20	5.06	4.94
11	9.65	7.21	6.22	5.67	5.32	5.07	4.89	4.74	4.63
12	9.33	6.93	5.95	5.41	5.06	4.82	4.64	4.50	4.39
13	9.07	6.70	5.74	5.21	4.86	4.62	4.44	4.30	4.19
14	8.86	6.51	5.56	5.04	4.69	4.46	4.28	4.14	4.03
15	8.68	6.36	5.42	4.89	4.56	4.32	4.14	4.00	3.89
16	8.53	6.23	5.29	4.77	4.44	4.20	4.03	3.89	3.78
17	8.40	6.11	5.18	4.67	4.34	4.10	3.93	3.79	3.68
18	8.29	6.01	5.09	4.58	4.25	4.01	3.84	3.71	3.60
19	8.18	5.93	5.01	4.50	4.17	3.94	3.77	3.63	3.52
20	8.10	5.85	4.94	4.43	4.10	3.87	3.70	3.56	3.46
21	8.02	5.78	4.87	4.37	4.04	3.81	3.64	3.51	3.40
22	7.95	5.72	4.82	4.31	3.99	3.76	3.59	3.45	3.35
23	7.88	5.66	4.76	4.26	3.94	3.71	3.54	3.41	3.30
24	7.82	5.61	4.72	4.22	3.90	3.67	3.50	3.36	3.26
25	7.77	5.57	4.68	4.18	3.85	3.63	3.46	3.32	3.22
26	7.72	5.53	4.64	4.14	3.82	3.59	3.42	3.29	3.18
27	7.68	5.49	4.60	4.11	3.78	3.56	3.39	3.26	3.15
28	7.64	5.45	4.57	4.07	3.75	3.53	3.36	3.23	3.12
29	7.60	5.42	4.54	4.04	3.73	3.50	3.33	3.20	3.09
30	7.56	5.39	4.51	4.02	3.70	3.47	3.30	3.17	3.07
40	7.31	5.18	4.31	3.83	3.51	3.29	3.12	2.99	2.89
60	7.08	4.98	4.13	3.65	3.34	3.12	2.95	2.82	2.72
120	6.85	4.79	3.95	3.48	3.17	2.96	2.79	2.66	2.56
∞	6.63	4.61	3.78	3.32	3.02	2.80	2.64	2.51	2.41

(concluded)

TABLE C.4

$a = 0.99$

numerator df										denominator df
10	12	15	20	24	30	40	60	120	∞	
6056	6106	6157	6209	6235	6261	6287	6313	6339	6366	1
99.40	99.42	99.43	99.45	99.46	99.47	99.47	99.48	99.49	99.50	2
27.23	27.05	26.87	26.69	26.60	26.50	26.41	26.32	26.22	26.13	3
14.55	14.37	14.20	14.02	13.93	13.84	13.75	13.65	13.56	13.46	4
10.05	9.89	9.72	9.55	9.47	9.38	9.29	9.20	9.11	9.02	5
7.87	7.72	7.56	7.40	7.31	7.23	7.14	7.06	6.97	6.88	6
6.62	6.47	6.31	6.16	6.07	5.99	5.91	5.82	5.74	5.65	7
5.81	5.67	5.52	5.36	5.28	5.20	5.12	5.03	4.95	4.86	8
5.26	5.11	4.96	4.81	4.73	4.65	4.57	4.48	4.40	4.31	9
4.85	4.71	4.56	4.41	4.33	4.25	4.17	4.08	4.00	3.91	10
4.54	4.40	4.25	4.10	4.02	3.94	3.86	3.78	3.69	3.60	11
4.30	4.16	4.01	3.86	3.78	3.70	3.62	3.54	3.45	3.36	12
4.10	3.96	3.82	3.66	3.59	3.51	3.43	3.34	3.25	3.17	13
3.94	3.80	3.66	3.51	3.43	3.35	3.27	3.18	3.09	3.00	14
3.80	3.67	3.52	3.37	3.29	3.21	3.13	3.05	2.96	2.87	15
3.69	3.55	3.41	3.26	3.18	3.10	3.02	2.93	2.84	2.75	16
3.59	3.46	3.31	3.16	3.08	3.00	2.92	2.83	2.75	2.65	17
3.51	3.37	3.23	3.08	3.00	2.92	2.84	2.75	2.66	2.57	18
3.43	3.30	3.15	3.00	2.92	2.84	2.76	2.67	2.58	2.49	19
3.37	3.23	3.09	2.94	2.86	2.78	2.69	2.61	2.52	2.42	20
3.31	3.17	3.03	2.88	2.80	2.72	2.64	2.55	2.46	2.36	21
3.26	3.12	2.98	2.83	2.75	2.67	2.58	2.50	2.40	2.31	22
3.21	3.07	2.93	2.78	2.70	2.62	2.54	2.45	2.35	2.26	23
3.17	3.03	2.89	2.74	2.66	2.58	2.49	2.40	2.31	2.21	24
3.13	2.99	2.85	2.70	2.62	2.54	2.45	2.36	2.27	2.17	25
3.09	2.96	2.81	2.66	2.58	2.50	2.42	2.33	2.23	2.13	26
3.06	2.93	2.78	2.63	2.55	2.47	2.38	2.29	2.20	2.10	27
3.03	2.90	2.75	2.60	2.52	2.44	2.35	2.26	2.17	2.06	28
3.00	2.87	2.73	2.57	2.49	2.41	2.33	2.23	2.14	2.03	29
2.98	2.84	2.70	2.55	2.47	2.39	2.30	2.21	2.11	2.01	30
2.80	2.66	2.52	2.37	2.29	2.20	2.11	2.02	1.92	1.80	40
2.63	2.50	2.35	2.20	2.12	2.03	1.94	1.84	1.73	1.60	60
2.47	2.34	2.19	2.03	1.95	1.86	1.76	1.66	1.53	1.38	120
2.32	2.18	2.04	1.88	1.79	1.70	1.59	1.47	1.32	1.00	∞

Source: Tabulated values adapted from Table 5 of Pearson and Hartley, *Biometrika Tables for Statisticians,* Volume 2, 1972, published by the Cambridge University Press for the Biometrika Trustees, with the permission of the authors and publishers.

Example: $F(0.99; 8, 24) = 3.36$ so $P[F(8, 24) \leq 3.36] = 0.99$.

Text Reference: Use of this table is discussed on p. 915.

TABLE C.5　　　　**Binomial probabilities**　Entry is probability $P(x) = \binom{n}{x} p^x (1-p)^{n-x}$.

n	x	0.01	0.02	0.03	0.04	0.05	0.06	0.07	0.08	0.09		
2	0	0.9801	0.9604	0.9409	0.9216	0.9025	0.8836	0.8649	0.8464	0.8281	2	
	1	0.0198	0.0392	0.0582	0.0768	0.0950	0.1128	0.1302	0.1472	0.1638	1	
	2	0.0001	0.0004	0.0009	0.0016	0.0025	0.0036	0.0049	0.0064	0.0081	0	2
3	0	0.9703	0.9412	0.9127	0.8847	0.8574	0.8306	0.8044	0.7787	0.7536	3	
	1	0.0294	0.0576	0.0847	0.1106	0.1354	0.1590	0.1816	0.2031	0.2236	2	
	2	0.0003	0.0012	0.0026	0.0046	0.0071	0.0102	0.0137	0.0177	0.0221	1	
	3				0.0001	0.0001	0.0002	0.0003	0.0005	0.0007	0	3
4	0	0.9606	0.9224	0.8853	0.8493	0.8145	0.7807	0.7481	0.7164	0.6857	4	
	1	0.0388	0.0753	0.1095	0.1416	0.1715	0.1993	0.2252	0.2492	0.2713	3	
	2	0.0006	0.0023	0.0051	0.0088	0.0135	0.0191	0.0254	0.0325	0.0402	2	
	3			0.0001	0.0002	0.0005	0.0008	0.0013	0.0019	0.0027	1	
	4									0.0001	0	4
5	0	0.9510	0.9039	0.8587	0.8154	0.7738	0.7339	0.6957	0.6591	0.6240	5	
	1	0.0480	0.0922	0.1328	0.1699	0.2036	0.2342	0.2618	0.2866	0.3086	4	
	2	0.0010	0.0038	0.0082	0.0142	0.0214	0.0299	0.0394	0.0498	0.0610	3	
	3		0.0001	0.0003	0.0006	0.0011	0.0019	0.0030	0.0043	0.0060	2	
	4						0.0001	0.0001	0.0002	0.0003	1	
	5										0	5
6	0	0.9415	0.8858	0.8330	0.7828	0.7351	0.6899	0.6470	0.6064	0.5679	6	
	1	0.0571	0.1085	0.1546	0.1957	0.2321	0.2642	0.2922	0.3164	0.3370	5	
	2	0.0014	0.0055	0.0120	0.0204	0.0305	0.0422	0.0550	0.0688	0.0833	4	
	3		0.0002	0.0005	0.0011	0.0021	0.0036	0.0055	0.0080	0.0110	3	
	4					0.0001	0.0002	0.0003	0.0005	0.0008	2	
	5										1	
	6										0	6
7	0	0.9321	0.8681	0.8080	0.7514	0.6983	0.6485	0.6017	0.5578	0.5168	7	
	1	0.0659	0.1240	0.1749	0.2192	0.2573	0.2897	0.3170	0.3396	0.3578	6	
	2	0.0020	0.0076	0.0162	0.0274	0.0406	0.0555	0.0716	0.0886	0.1061	5	
	3		0.0003	0.0008	0.0019	0.0036	0.0059	0.0090	0.0128	0.0175	4	
	4				0.0001	0.0002	0.0004	0.0007	0.0011	0.0017	3	
	5								0.0001	0.0001	2	
	6										1	
	7										0	7
8	0	0.9227	0.8508	0.7837	0.7214	0.6634	0.6096	0.5596	0.5132	0.4703	8	
	1	0.0746	0.1389	0.1939	0.2405	0.2793	0.3113	0.3370	0.3570	0.3721	7	
	2	0.0026	0.0099	0.0210	0.0351	0.0515	0.0695	0.0888	0.1087	0.1288	6	
	3	0.0001	0.0004	0.0013	0.0029	0.0054	0.0089	0.0134	0.0189	0.0255	5	
	4			0.0001	0.0002	0.0004	0.0007	0.0013	0.0021	0.0031	4	
	5							0.0001	0.0001	0.0002	3	
	⋮										⋮	
	8										0	8
9	0	0.9135	0.8337	0.7602	0.6925	0.6302	0.5730	0.5204	0.4722	0.4279	9	
	1	0.0830	0.1531	0.2116	0.2597	0.2985	0.3292	0.3525	0.3695	0.3809	8	
	2	0.0034	0.0125	0.0262	0.0433	0.0629	0.0840	0.1061	0.1285	0.1507	7	
	3	0.0001	0.0006	0.0019	0.0042	0.0077	0.0125	0.0186	0.0261	0.0348	6	
	4			0.0001	0.0003	0.0006	0.0012	0.0021	0.0034	0.0052	5	
	5						0.0001	0.0002	0.0003	0.0005	4	
	⋮										⋮	
	9										0	9
		0.99	0.98	0.97	0.96	0.95	0.94	0.93	0.92	0.91	x	n

p

(continued)

TABLE C.5

n	x	0.01	0.02	0.03	0.04	0.05	0.06	0.07	0.08	0.09		
10	0	0.9044	0.8171	0.7374	0.6648	0.5987	0.5386	0.4840	0.4344	0.3894	10	
	1	0.0914	0.1667	0.2281	0.2770	0.3151	0.3438	0.3643	0.3777	0.3851	9	
	2	0.0042	0.0153	0.0317	0.0519	0.0746	0.0988	0.1234	0.1478	0.1714	8	
	3	0.0001	0.0008	0.0026	0.0058	0.0105	0.0168	0.0248	0.0343	0.0452	7	
	4			0.0001	0.0004	0.0010	0.0019	0.0033	0.0052	0.0078	6	
	5					0.0001	0.0001	0.0003	0.0005	0.0009	5	
	6									0.0001	4	
	⋮										⋮	
	10										0	10
12	0	0.8864	0.7847	0.6938	0.6127	0.5404	0.4759	0.4186	0.3677	0.3225	12	
	1	0.1074	0.1922	0.2575	0.3064	0.3413	0.3645	0.3781	0.3837	0.3827	11	
	2	0.0060	0.0216	0.0438	0.0702	0.0988	0.1280	0.1565	0.1835	0.2082	10	
	3	0.0002	0.0015	0.0045	0.0098	0.0173	0.0272	0.0393	0.0532	0.0686	9	
	4		0.0001	0.0003	0.0009	0.0021	0.0039	0.0067	0.0104	0.0153	8	
	5				0.0001	0.0002	0.0004	0.0008	0.0014	0.0024	7	
	6							0.0001	0.0001	0.0003	6	
	⋮										⋮	
	12										0	12
15	0	0.8601	0.7386	0.6333	0.5421	0.4633	0.3953	0.3367	0.2863	0.2430	15	
	1	0.1303	0.2261	0.2938	0.3388	0.3658	0.3785	0.3801	0.3734	0.3605	14	
	2	0.0092	0.0323	0.0636	0.0988	0.1348	0.1691	0.2003	0.2273	0.2496	13	
	3	0.0004	0.0029	0.0085	0.0178	0.0307	0.0468	0.0653	0.0857	0.1070	12	
	4		0.0002	0.0008	0.0022	0.0049	0.0090	0.0148	0.0223	0.0317	11	
	5			0.0001	0.0002	0.0006	0.0013	0.0024	0.0043	0.0069	10	
	6						0.0001	0.0003	0.0006	0.0011	9	
	7								0.0001	0.0001	8	
	⋮										⋮	
	15										0	15
20	0	0.8179	0.6676	0.5438	0.4420	0.3585	0.2901	0.2342	0.1887	0.1516	20	
	1	0.1652	0.2725	0.3364	0.3683	0.3774	0.3703	0.3526	0.3282	0.3000	19	
	2	0.0159	0.0528	0.0988	0.1458	0.1887	0.2246	0.2521	0.2711	0.2818	18	
	3	0.0010	0.0065	0.0183	0.0364	0.0596	0.0860	0.1139	0.1414	0.1672	17	
	4		0.0006	0.0024	0.0065	0.0133	0.0233	0.0364	0.0523	0.0703	16	
	5			0.0002	0.0009	0.0022	0.0048	0.0088	0.0145	0.0222	15	
	6				0.0001	0.0003	0.0008	0.0017	0.0032	0.0055	14	
	7						0.0001	0.0002	0.0005	0.0011	13	
	8								0.0001	0.0002	12	
	⋮										⋮	
	20										0	20
		0.99	0.98	0.97	0.96	0.95	0.94	0.93	0.92	0.91	x	n
						p						

(continues)

TABLE C.5   **Binomial probabilities (continued)**

n x	0.10	0.15	0.20	0.25	0.30	0.35	0.40	0.45	0.50		
2 0	0.8100	0.7225	0.6400	0.5625	0.4900	0.4225	0.3600	0.3025	0.2500	2	
1	0.1800	0.2550	0.3200	0.3750	0.4200	0.4550	0.4800	0.4950	0.5000	1	
2	0.0100	0.0225	0.0400	0.0625	0.0900	0.1225	0.1600	0.2025	0.2500	0	2
3 0	0.7290	0.6141	0.5120	0.4219	0.3430	0.2746	0.2160	0.1664	0.1250	3	
1	0.2430	0.3251	0.3840	0.4219	0.4410	0.4436	0.4320	0.4084	0.3750	2	
2	0.0270	0.0574	0.0960	0.1406	0.1890	0.2389	0.2880	0.3341	0.3750	1	
3	0.0010	0.0034	0.0080	0.0156	0.0270	0.0429	0.0640	0.0911	0.1250	0	3
4 0	0.6561	0.5220	0.4096	0.3164	0.2401	0.1785	0.1296	0.0915	0.0625	4	
1	0.2916	0.3685	0.4096	0.4219	0.4116	0.3845	0.3456	0.2995	0.2500	3	
2	0.0486	0.0975	0.1536	0.2109	0.2646	0.3105	0.3456	0.3675	0.3750	2	
3	0.0036	0.0115	0.0256	0.0469	0.0756	0.1115	0.1536	0.2005	0.2500	1	
4	0.0001	0.0005	0.0016	0.0039	0.0081	0.0150	0.0256	0.0410	0.0625	0	4
5 0	0.5905	0.4437	0.3277	0.2373	0.1681	0.1160	0.0778	0.0503	0.0312	5	
1	0.3280	0.3915	0.4096	0.3955	0.3601	0.3124	0.2592	0.2059	0.1562	4	
2	0.0729	0.1382	0.2048	0.2637	0.3087	0.3364	0.3456	0.3369	0.3125	3	
3	0.0081	0.0244	0.0512	0.0879	0.1323	0.1811	0.2304	0.2757	0.3125	2	
4	0.0004	0.0022	0.0064	0.0146	0.0283	0.0488	0.0768	0.1128	0.1562	1	
5		0.0001	0.0003	0.0010	0.0024	0.0053	0.0102	0.0185	0.0312	0	5
6 0	0.5314	0.3771	0.2621	0.1780	0.1176	0.0754	0.0467	0.0277	0.0156	6	
1	0.3543	0.3993	0.3932	0.3560	0.3025	0.2437	0.1866	0.1359	0.0938	5	
2	0.0984	0.1762	0.2458	0.2966	0.3241	0.3280	0.3110	0.2780	0.2344	4	
3	0.0146	0.0415	0.0819	0.1318	0.1852	0.2355	0.2765	0.3032	0.3125	3	
4	0.0012	0.0055	0.0154	0.0330	0.0595	0.0951	0.1382	0.1861	0.2344	2	
5	0.0001	0.0004	0.0015	0.0044	0.0102	0.0205	0.0369	0.0609	0.0938	1	
6			0.0001	0.0002	0.0007	0.0018	0.0041	0.0083	0.0156	0	6
7 0	0.4783	0.3206	0.2097	0.1335	0.0824	0.0490	0.0280	0.0152	0.0078	7	
1	0.3720	0.3960	0.3670	0.3115	0.2471	0.1848	0.1306	0.0872	0.0547	6	
2	0.1240	0.2097	0.2753	0.3115	0.3177	0.2985	0.2613	0.2140	0.1641	5	
3	0.0230	0.0617	0.1147	0.1730	0.2269	0.2679	0.2903	0.2918	0.2734	4	
4	0.0026	0.0109	0.0287	0.0577	0.0972	0.1442	0.1935	0.2388	0.2734	3	
5	0.0002	0.0012	0.0043	0.0115	0.0250	0.0466	0.0774	0.1172	0.1641	2	
6		0.0001	0.0004	0.0013	0.0036	0.0084	0.0172	0.0320	0.0547	1	
7				0.0001	0.0002	0.0006	0.0016	0.0037	0.0078	0	7
8 0	0.4305	0.2725	0.1678	0.1001	0.0576	0.0319	0.0168	0.0084	0.0039	8	
1	0.3826	0.3847	0.3355	0.2670	0.1977	0.1373	0.0896	0.0548	0.0312	7	
2	0.1488	0.2376	0.2936	0.3115	0.2965	0.2587	0.2090	0.1569	0.1094	6	
3	0.0331	0.0839	0.1468	0.2076	0.2541	0.2786	0.2787	0.2568	0.2188	5	
4	0.0046	0.0185	0.0459	0.0865	0.1361	0.1875	0.2322	0.2627	0.2734	4	
5	0.0004	0.0026	0.0092	0.0231	0.0467	0.0808	0.1239	0.1719	0.2188	3	
6		0.0002	0.0011	0.0038	0.0100	0.0217	0.0413	0.0703	0.1094	2	
7			0.0001	0.0004	0.0012	0.0033	0.0079	0.0164	0.0312	1	
8					0.0001	0.0002	0.0007	0.0017	0.0039	0	8
9 0	0.3874	0.2316	0.1342	0.0751	0.0404	0.0207	0.0101	0.0046	0.0020	9	
1	0.3874	0.3679	0.3020	0.2253	0.1556	0.1004	0.0605	0.0339	0.0176	8	
2	0.1722	0.2597	0.3020	0.3003	0.2668	0.2162	0.1612	0.1110	0.0703	7	
3	0.0446	0.1069	0.1762	0.2336	0.2668	0.2716	0.2508	0.2119	0.1641	6	
4	0.0074	0.0283	0.0661	0.1168	0.1715	0.2194	0.2508	0.2600	0.2461	5	
5	0.0008	0.0050	0.0165	0.0389	0.0735	0.1181	0.1672	0.2128	0.2461	4	
6	0.0001	0.0006	0.0028	0.0087	0.0210	0.0424	0.0743	0.1160	0.1641	3	
7			0.0003	0.0012	0.0039	0.0098	0.0212	0.0407	0.0703	2	
8				0.0001	0.0004	0.0013	0.0035	0.0083	0.0176	1	
9						0.0001	0.0003	0.0008	0.0020	0	9
	0.90	0.85	0.80	0.75	0.70	0.65	0.60	0.55	0.50	x	n

(concluded)

TABLE C.5

n	x	0.10	0.15	0.20	0.25	0.30	0.35	0.40	0.45	0.50		
10	0	0.3487	0.1969	0.1074	0.0563	0.0282	0.0135	0.0060	0.0025	0.0010	10	
	1	0.3874	0.3474	0.2684	0.1877	0.1211	0.0725	0.0403	0.0207	0.0098	9	
	2	0.1937	0.2759	0.3020	0.2816	0.2335	0.1757	0.1209	0.0763	0.0439	8	
	3	0.0574	0.1298	0.2013	0.2503	0.2668	0.2522	0.2150	0.1665	0.1172	7	
	4	0.0112	0.0401	0.0881	0.1460	0.2001	0.2377	0.2508	0.2384	0.2051	6	
	5	0.0015	0.0085	0.0264	0.0584	0.1029	0.1536	0.2007	0.2340	0.2461	5	
	6	0.0001	0.0012	0.0055	0.0162	0.0368	0.0689	0.1115	0.1596	0.2051	4	
	7		0.0001	0.0008	0.0031	0.0090	0.0212	0.0425	0.0746	0.1172	3	
	8			0.0001	0.0004	0.0014	0.0043	0.0106	0.0229	0.0439	2	
	9					0.0001	0.0005	0.0016	0.0042	0.0098	1	
	10							0.0001	0.0003	0.0010	0	10
12	0	0.2824	0.1422	0.0687	0.0317	0.0138	0.0057	0.0022	0.0008	0.0002	12	
	1	0.3766	0.3012	0.2062	0.1267	0.0712	0.0368	0.0174	0.0075	0.0029	11	
	2	0.2301	0.2924	0.2835	0.2323	0.1678	0.1088	0.0639	0.0339	0.0161	10	
	3	0.0852	0.1720	0.2362	0.2581	0.2397	0.1954	0.1419	0.0923	0.0537	9	
	4	0.0213	0.0683	0.1329	0.1936	0.2311	0.2367	0.2128	0.1700	0.1208	8	
	5	0.0038	0.0193	0.0532	0.1032	0.1585	0.2039	0.2270	0.2225	0.1934	7	
	6	0.0005	0.0040	0.0155	0.0401	0.0792	0.1281	0.1766	0.2124	0.2256	6	
	7		0.0006	0.0033	0.0115	0.0291	0.0591	0.1009	0.1489	0.1934	5	
	8		0.0001	0.0005	0.0024	0.0078	0.0199	0.0420	0.0762	0.1208	4	
	9			0.0001	0.0004	0.0015	0.0048	0.0125	0.0277	0.0537	3	
	10					0.0002	0.0008	0.0025	0.0068	0.0161	2	
	11						0.0001	0.0003	0.0010	0.0029	1	
	12								0.0001	0.0002	0	12
15	0	0.2059	0.0874	0.0352	0.0134	0.0047	0.0016	0.0005	0.0001		15	
	1	0.3432	0.2312	0.1319	0.0668	0.0305	0.0126	0.0047	0.0016	0.0005	14	
	2	0.2669	0.2856	0.2309	0.1559	0.0916	0.0476	0.0219	0.0090	0.0032	13	
	3	0.1285	0.2184	0.2501	0.2252	0.1700	0.1110	0.0634	0.0318	0.0139	12	
	4	0.0428	0.1156	0.1876	0.2252	0.2186	0.1792	0.1268	0.0780	0.0417	11	
	5	0.0105	0.0449	0.1032	0.1651	0.2061	0.2123	0.1859	0.1404	0.0916	10	
	6	0.0019	0.0132	0.0430	0.0917	0.1472	0.1906	0.2066	0.1914	0.1527	9	
	7	0.0003	0.0030	0.0138	0.0393	0.0811	0.1319	0.1771	0.2013	0.1964	8	
	8		0.0005	0.0035	0.0131	0.0348	0.0710	0.1181	0.1647	0.1964	7	
	9		0.0001	0.0007	0.0034	0.0116	0.0298	0.0612	0.1048	0.1527	6	
	10			0.0001	0.0007	0.0030	0.0096	0.0245	0.0515	0.0916	5	
	11				0.0001	0.0006	0.0024	0.0074	0.0191	0.0417	4	
	12					0.0001	0.0004	0.0016	0.0052	0.0139	3	
	13						0.0001	0.0003	0.0010	0.0032	2	
	14								0.0001	0.0005	1	
	15										0	15
20	0	0.1216	0.0388	0.0115	0.0032	0.0008	0.0002				20	
	1	0.2702	0.1368	0.0576	0.0211	0.0068	0.0020	0.0005	0.0001		19	
	2	0.2852	0.2293	0.1369	0.0669	0.0278	0.0100	0.0031	0.0008	0.0002	18	
	3	0.1901	0.2428	0.2054	0.1339	0.0716	0.0323	0.0123	0.0040	0.0011	17	
	4	0.0898	0.1821	0.2182	0.1897	0.1304	0.0738	0.0350	0.0139	0.0046	16	
	5	0.0319	0.1028	0.1746	0.2023	0.1789	0.1272	0.0746	0.0365	0.0148	15	
	6	0.0089	0.0454	0.1091	0.1686	0.1916	0.1712	0.1244	0.0746	0.0370	14	
	7	0.0020	0.0160	0.0545	0.1124	0.1643	0.1844	0.1659	0.1221	0.0739	13	
	8	0.0004	0.0046	0.0222	0.0609	0.1144	0.1614	0.1797	0.1623	0.1201	12	
	9	0.0001	0.0011	0.0074	0.0271	0.0654	0.1158	0.1597	0.1771	0.1602	11	
	10		0.0002	0.0020	0.0099	0.0308	0.0686	0.1171	0.1593	0.1762	10	
	11			0.0005	0.0030	0.0120	0.0336	0.0710	0.1185	0.1602	9	
	12			0.0001	0.0008	0.0039	0.0136	0.0355	0.0727	0.1201	8	
	13				0.0002	0.0010	0.0045	0.0146	0.0366	0.0739	7	
	14					0.0002	0.0012	0.0049	0.0150	0.0370	6	
	15						0.0003	0.0013	0.0049	0.0148	5	
	16							0.0003	0.0013	0.0046	4	
	17								0.0002	0.0011	3	
	18									0.0002	2	
	19										1	
	20										0	20
		0.90	0.85	0.80	0.75	0.70	0.65	0.60	0.55	0.50	x	n

p

Note: Blank entries in this table correspond to the probability value 0.0000.

Example: For $n = 12$, $p = 0.25$, and $x = 3$, $P(3) = 0.2581$. For $n = 15$, $p = 0.55$, and $x = 10$, $P(10) = 0.1404$.

Text Reference: Use of this table is discussed on pp. 194–195.

TABLE C.6

Poisson probabilities Entry is probability $P(x) = \dfrac{\lambda^x \exp(-\lambda)}{x!}$

					λ				
x	0.1	0.2	0.3	0.4	0.5	0.6	0.7	0.8	0.9
0	0.9048	0.8187	0.7408	0.6703	0.6065	0.5488	0.4966	0.4493	0.4066
1	0.0905	0.1637	0.2222	0.2681	0.3033	0.3293	0.3476	0.3595	0.3659
2	0.0045	0.0164	0.0333	0.0536	0.0758	0.0988	0.1217	0.1438	0.1647
3	0.0002	0.0011	0.0033	0.0072	0.0126	0.0198	0.0284	0.0383	0.0494
4		0.0001	0.0003	0.0007	0.0016	0.0030	0.0050	0.0077	0.0111
5				0.0001	0.0002	0.0004	0.0007	0.0012	0.0020
6							0.0001	0.0002	0.0003

					λ				
x	1.0	1.5	2.0	2.5	3.0	3.5	4.0	4.5	5.0
0	0.3679	0.2231	0.1353	0.0821	0.0498	0.0302	0.0183	0.0111	0.0067
1	0.3679	0.3347	0.2707	0.2052	0.1494	0.1057	0.0733	0.0500	0.0337
2	0.1839	0.2510	0.2707	0.2565	0.2240	0.1850	0.1465	0.1125	0.0842
3	0.0613	0.1255	0.1804	0.2138	0.2240	0.2158	0.1954	0.1687	0.1404
4	0.0153	0.0471	0.0902	0.1336	0.1680	0.1888	0.1954	0.1898	0.1755
5	0.0031	0.0141	0.0361	0.0668	0.1008	0.1322	0.1563	0.1708	0.1755
6	0.0005	0.0035	0.0120	0.0278	0.0504	0.0771	0.1042	0.1281	0.1462
7	0.0001	0.0008	0.0034	0.0099	0.0216	0.0385	0.0595	0.0824	0.1044
8		0.0001	0.0009	0.0031	0.0081	0.0169	0.0298	0.0463	0.0653
9			0.0002	0.0009	0.0027	0.0066	0.0132	0.0232	0.0363
10				0.0002	0.0008	0.0023	0.0053	0.0104	0.0181
11					0.0002	0.0007	0.0019	0.0043	0.0082
12					0.0001	0.0002	0.0006	0.0016	0.0034
13						0.0001	0.0002	0.0006	0.0013
14							0.0001	0.0002	0.0005
15								0.0001	0.0002

					λ				
x	5.5	6.0	6.5	7.0	7.5	8.0	9.0	10.0	11.0
0	0.0041	0.0025	0.0015	0.0009	0.0006	0.0003	0.0001		
1	0.0225	0.0149	0.0098	0.0064	0.0041	0.0027	0.0011	0.0005	0.0002
2	0.0618	0.0446	0.0318	0.0223	0.0156	0.0107	0.0050	0.0023	0.0010
3	0.1133	0.0892	0.0688	0.0521	0.0389	0.0286	0.0150	0.0076	0.0037
4	0.1558	0.1339	0.1118	0.0912	0.0729	0.0573	0.0337	0.0189	0.0102
5	0.1714	0.1606	0.1454	0.1277	0.1094	0.0916	0.0607	0.0378	0.0224
6	0.1571	0.1606	0.1575	0.1490	0.1367	0.1221	0.0911	0.0631	0.0411
7	0.1234	0.1377	0.1462	0.1490	0.1465	0.1396	0.1171	0.0901	0.0646
8	0.0849	0.1033	0.1188	0.1304	0.1373	0.1396	0.1318	0.1126	0.0888
9	0.0519	0.0688	0.0858	0.1014	0.1144	0.1241	0.1318	0.1251	0.1085
10	0.0285	0.0413	0.0558	0.0710	0.0858	0.0993	0.1186	0.1251	0.1194
11	0.0143	0.0225	0.0330	0.0452	0.0585	0.0722	0.0970	0.1137	0.1194
12	0.0065	0.0113	0.0179	0.0263	0.0366	0.0481	0.0728	0.0948	0.1094
13	0.0028	0.0052	0.0089	0.0142	0.0211	0.0296	0.0504	0.0729	0.0926
14	0.0011	0.0022	0.0041	0.0071	0.0113	0.0169	0.0324	0.0521	0.0728
15	0.0004	0.0009	0.0018	0.0033	0.0057	0.0090	0.0194	0.0347	0.0534
16	0.0001	0.0003	0.0007	0.0014	0.0026	0.0045	0.0109	0.0217	0.0367
17		0.0001	0.0003	0.0006	0.0012	0.0021	0.0058	0.0128	0.0237
18			0.0001	0.0002	0.0005	0.0009	0.0029	0.0071	0.0145
19				0.0001	0.0002	0.0004	0.0014	0.0037	0.0084
20					0.0001	0.0002	0.0006	0.0019	0.0046
21						0.0001	0.0003	0.0009	0.0024
22							0.0001	0.0004	0.0012
23								0.0002	0.0006
24								0.0001	0.0003
25									0.0001

(concluded)

TABLE C.6

					λ				
x	12	13	14	15	16	17	18	19	20
0									
1	0.0001								
2	0.0004	0.0002	0.0001						
3	0.0018	0.0008	0.0004	0.0002	0.0001				
4	0.0053	0.0027	0.0013	0.0006	0.0003	0.0001	0.0001		
5	0.0127	0.0070	0.0037	0.0019	0.0010	0.0005	0.0002	0.0001	0.0001
6	0.0255	0.0152	0.0087	0.0048	0.0026	0.0014	0.0007	0.0004	0.0002
7	0.0437	0.0281	0.0174	0.0104	0.0060	0.0034	0.0019	0.0010	0.0005
8	0.0655	0.0457	0.0304	0.0194	0.0120	0.0072	0.0042	0.0024	0.0013
9	0.0874	0.0661	0.0473	0.0324	0.0213	0.0135	0.0083	0.0050	0.0029
10	0.1048	0.0859	0.0663	0.0486	0.0341	0.0230	0.0150	0.0095	0.0058
11	0.1144	0.1015	0.0844	0.0663	0.0496	0.0355	0.0245	0.0164	0.0106
12	0.1144	0.1099	0.0984	0.0829	0.0661	0.0504	0.0368	0.0259	0.0176
13	0.1056	0.1099	0.1060	0.0956	0.0814	0.0658	0.0509	0.0378	0.0271
14	0.0905	0.1021	0.1060	0.1024	0.0930	0.0800	0.0655	0.0514	0.0387
15	0.0724	0.0885	0.0989	0.1024	0.0992	0.0906	0.0786	0.0650	0.0516
16	0.0543	0.0719	0.0866	0.0960	0.0992	0.0963	0.0884	0.0772	0.0646
17	0.0383	0.0550	0.0713	0.0847	0.0934	0.0963	0.0936	0.0863	0.0760
18	0.0255	0.0397	0.0554	0.0706	0.0830	0.0909	0.0936	0.0911	0.0844
19	0.0161	0.0272	0.0409	0.0557	0.0699	0.0814	0.0887	0.0911	0.0888
20	0.0097	0.0177	0.0286	0.0418	0.0559	0.0692	0.0798	0.0866	0.0888
21	0.0055	0.0109	0.0191	0.0299	0.0426	0.0560	0.0684	0.0783	0.0846
22	0.0030	0.0065	0.0121	0.0204	0.0310	0.0433	0.0560	0.0676	0.0769
23	0.0016	0.0037	0.0074	0.0133	0.0216	0.0320	0.0438	0.0559	0.0669
24	0.0008	0.0020	0.0043	0.0083	0.0144	0.0226	0.0328	0.0442	0.0557
25	0.0004	0.0010	0.0024	0.0050	0.0092	0.0154	0.0237	0.0336	0.0446
26	0.0002	0.0005	0.0013	0.0029	0.0057	0.0101	0.0164	0.0246	0.0343
27	0.0001	0.0002	0.0007	0.0016	0.0034	0.0063	0.0109	0.0173	0.0254
28		0.0001	0.0003	0.0009	0.0019	0.0038	0.0070	0.0117	0.0181
29		0.0001	0.0002	0.0004	0.0011	0.0023	0.0044	0.0077	0.0125
30			0.0001	0.0002	0.0006	0.0013	0.0026	0.0049	0.0083
31				0.0001	0.0003	0.0007	0.0015	0.0030	0.0054
32			0.0001		0.0001	0.0004	0.0009	0.0018	0.0034
33					0.0001	0.0002	0.0005	0.0010	0.0020
34						0.0001	0.0002	0.0006	0.0012
35							0.0001	0.0003	0.0007
36							0.0001	0.0002	0.0004
37								0.0001	0.0002
38									0.0001
39									0.0001

Note: Blank entries in this table correspond to the probability value 0.0000.

Example: For $\lambda = 14$ and $x = 8$, $P(8) = 0.0304$.

Text Reference: Use of this table is discussed on pp. 197–198.

TABLE C.7

Table of random digits

Row	1–5	6–10	11–15	16–20	21–25	26–30	31–35
				Column			
1	13284	16834	74151	92027	24670	36665	00770
2	21224	00370	30420	03883	94648	89428	41583
3	99052	47887	81085	64933	66279	80432	65793
4	00199	50993	98603	38452	87890	94624	69721
5	60578	06483	28733	37867	07936	98710	98539
6	91240	18312	17441	01929	18163	69201	31211
7	97458	14229	12063	59611	32249	90466	33216
8	35249	38646	34475	72417	60514	69257	12489
9	38980	46600	11759	11900	46743	27860	77940
10	10750	52745	38749	87365	58959	53731	89295
11	36247	27850	73958	20673	37800	63835	71051
12	70994	66986	99744	72438	01174	42159	11392
13	99638	94702	11463	18148	81386	80431	90628
14	72055	15774	43857	99805	10419	76939	25993
15	24038	65541	85788	55835	38835	59399	13790
16	74976	14631	35908	28221	39470	91548	12854
17	35553	71628	70189	26436	63407	91178	90348
18	35676	12797	51434	82976	42010	26344	92920
19	74815	67523	72985	23183	02446	63594	98924
20	45246	88048	65173	50989	91060	89894	36036
21	76509	47069	86378	41797	11910	49672	88575
22	19689	90332	04315	21358	97248	11188	39062
23	42751	35318	97513	61537	54955	08159	00337
24	11946	22681	45045	13964	57517	59419	58045
25	96518	48688	20996	11090	48396	57177	83867
26	35726	58643	76869	84622	39098	36083	72505
27	39737	42750	48968	70536	84864	64952	38404
28	97025	66492	56177	04049	80312	48028	26408
29	62814	08075	09788	56350	76787	51591	54509
30	25578	22950	15227	83291	41737	59599	96191
31	68763	69576	88991	49662	46704	63362	56625
32	17900	00813	64361	60725	88974	61005	99709
33	71944	60227	63551	71109	05624	43836	58254
34	54684	93691	85132	64399	29182	44324	14491
35	25946	27623	11258	65204	52832	50880	22273
36	01353	39318	44961	44972	91766	90262	56073
37	99083	88191	27662	99113	57174	35571	99884
38	52021	45406	37945	75234	24327	86978	22644
39	78755	47744	43776	83098	03225	14281	83637
40	25282	69106	59180	16257	22810	43609	12224
41	11959	94202	02743	86847	79725	51811	12998
42	11644	13792	98190	01424	30078	28197	55583
43	06307	97912	68110	59812	95448	43244	31262
44	76285	75714	89585	99296	52640	46518	55486
45	55322	07598	39600	60866	63007	20007	66819
46	78017	90928	90220	92503	83375	26986	74399
47	44768	43342	20696	26331	43140	69744	82928
48	25100	19336	14605	86603	51680	97678	24261
49	83612	46623	62876	85197	07824	91392	58317
50	41347	81666	82961	60413	71020	83658	02415

Source: Excerpt from *Table of 105,000 Random Decimal Digits.* Interstate Commerce Commission, Bureau of Transport Economics and Statistics, May 1949.
Text Reference: This table is discussed on pp. 243–244.

Confidence level associated with interval $L_r \le \eta \le U_r$ Entry is confidence level for given n and r.

TABLE C.8

				r			
n	1	2	3	4	5	6	7
2	0.500						
3	0.750						
4	0.875	0.375					
5	0.938	0.625					
6	0.969	0.781	0.312				
7	0.984	0.875	0.547				
8	0.992	0.930	0.711	0.273			
9	0.996	0.961	0.820	0.492			
10	0.998	0.979	0.891	0.656	0.246		
11	0.999	0.988	0.935	0.773	0.451		
12	1.000	0.994	0.961	0.854	0.612	0.226	
13	1.000	0.997	0.978	0.908	0.733	0.419	
14	1.000	0.998	0.987	0.943	0.820	0.576	0.209
15	1.000	0.999	0.993	0.965	0.882	0.698	0.393

Example: For $n = 7$ and $r = 1$, confidence interval $L_1 \le \eta \le U_1$ has confidence coefficient 0.984.
Text Reference: Use of this table is discussed on pp. 437–438.

Percentiles of the $D(n)$ **distribution**   Entry is $D(a; n)$ where $P[D(n) \le D(a; n)] = a$.

TABLE C.9

	a				a		
n	0.90	0.95	0.99	n	0.90	0.95	0.99
5	0.51	0.56	0.67	35	0.20	0.22	0.27
10	0.37	0.41	0.49	40	0.19	0.21	0.25
15	0.30	0.34	0.40	45	0.18	0.20	0.24
20	0.26	0.29	0.35	50	0.17	0.19	0.23
25	0.24	0.26	0.32	$n > 50$	$\dfrac{1.22}{\sqrt{n}}$	$\dfrac{1.36}{\sqrt{n}}$	$\dfrac{1.63}{\sqrt{n}}$
30	0.22	0.24	0.29				

Source: Tabulated values adapted by permission from Table 1 of L. H. Miller, "Table of Percentage Points of Kolmogorov Statistics," *Journal of the American Statistical Association,* Vol. 51 (1956), pp. 111–121.
Example: $D(0.95; 25) = 0.26$ so $P[D(25) \le 0.26] = 0.95$.
Text Reference: Use of this table is discussed on p. 493

TABLE C.10   **Durbin–Watson test bounds**

$$\alpha = 0.05$$

	Number of Independent Variables ($p - 1$)									
	1		2		3		4		5	
n	d_L	d_U	d_L	d_U	d_L	d_U	d_L	d_U	d_L	d_U
15	1.08	1.36	0.95	1.54	0.82	1.75	0.69	1.97	0.56	2.21
16	1.10	1.37	0.98	1.54	0.86	1.73	0.74	1.93	0.62	2.15
17	1.13	1.38	1.02	1.54	0.90	1.71	0.78	1.90	0.67	2.10
18	1.16	1.39	1.05	1.53	0.93	1.69	0.82	1.87	0.71	2.06
19	1.18	1.40	1.08	1.53	0.97	1.68	0.86	1.85	0.75	2.02
20	1.20	1.41	1.10	1.54	1.00	1.68	0.90	1.83	0.79	1.99
21	1.22	1.42	1.13	1.54	1.03	1.67	0.93	1.81	0.83	1.96
22	1.24	1.43	1.15	1.54	1.05	1.66	0.96	1.80	0.86	1.94
23	1.26	1.44	1.17	1.54	1.08	1.66	0.99	1.79	0.90	1.92
24	1.27	1.45	1.19	1.55	1.10	1.66	1.01	1.78	0.93	1.90
25	1.29	1.45	1.21	1.55	1.12	1.66	1.04	1.77	0.95	1.89
26	1.30	1.46	1.22	1.55	1.14	1.65	1.06	1.76	0.98	1.88
27	1.32	1.47	1.24	1.56	1.16	1.65	1.08	1.76	1.01	1.86
28	1.33	1.48	1.26	1.56	1.18	1.65	1.10	1.75	1.03	1.85
29	1.34	1.48	1.27	1.56	1.20	1.65	1.12	1.74	1.05	1.84
30	1.35	1.49	1.28	1.57	1.21	1.65	1.14	1.74	1.07	1.83
31	1.36	1.50	1.30	1.57	1.23	1.65	1.16	1.74	1.09	1.83
32	1.37	1.50	1.31	1.57	1.24	1.65	1.18	1.73	1.11	1.82
33	1.38	1.51	1.32	1.58	1.26	1.65	1.19	1.73	1.13	1.81
34	1.39	1.51	1.33	1.58	1.27	1.65	1.21	1.73	1.15	1.81
35	1.40	1.52	1.34	1.58	1.28	1.65	1.22	1.73	1.16	1.80
36	1.41	1.52	1.35	1.59	1.29	1.65	1.24	1.73	1.18	1.80
37	1.42	1.53	1.36	1.59	1.31	1.66	1.25	1.72	1.19	1.80
38	1.43	1.54	1.37	1.59	1.32	1.66	1.26	1.72	1.21	1.79
39	1.43	1.54	1.38	1.60	1.33	1.66	1.27	1.72	1.22	1.79
40	1.44	1.54	1.39	1.60	1.34	1.66	1.29	1.72	1.23	1.79
45	1.48	1.57	1.43	1.62	1.38	1.67	1.34	1.72	1.29	1.78
50	1.50	1.59	1.46	1.63	1.42	1.67	1.38	1.72	1.34	1.77
55	1.53	1.60	1.49	1.64	1.45	1.68	1.41	1.72	1.38	1.77
60	1.55	1.62	1.51	1.65	1.48	1.69	1.44	1.73	1.41	1.77
65	1.57	1.63	1.54	1.66	1.50	1.70	1.47	1.73	1.44	1.77
70	1.58	1.64	1.55	1.67	1.52	1.70	1.49	1.74	1.46	1.77
75	1.60	1.65	1.57	1.68	1.54	1.71	1.51	1.74	1.49	1.77
80	1.61	1.66	1.59	1.69	1.56	1.72	1.53	1.74	1.51	1.77
85	1.62	1.67	1.60	1.70	1.57	1.72	1.55	1.75	1.52	1.77
90	1.63	1.68	1.61	1.70	1.59	1.73	1.57	1.75	1.54	1.78
95	1.64	1.69	1.62	1.71	1.60	1.73	1.58	1.75	1.56	1.78
100	1.65	1.69	1.63	1.72	1.61	1.74	1.59	1.76	1.57	1.78

(concluded)

TABLE C.10

$$\alpha = 0.01$$

n	Number of Independent Variables $(p-1)$									
	1		2		3		4		5	
	d_L	d_U	d_L	d_U	d_L	d_U	d_L	d_U	d_L	d_U
15	0.81	1.07	0.70	1.25	0.59	1.46	0.49	1.70	0.39	1.96
16	0.84	1.09	0.74	1.25	0.63	1.44	0.53	1.66	0.44	1.90
17	0.87	1.10	0.77	1.25	0.67	1.43	0.57	1.63	0.48	1.85
18	0.90	1.12	0.80	1.26	0.71	1.42	0.61	1.60	0.52	1.80
19	0.93	1.13	0.83	1.26	0.74	1.41	0.65	1.58	0.56	1.77
20	0.95	1.15	0.86	1.27	0.77	1.41	0.68	1.57	0.60	1.74
21	0.97	1.16	0.89	1.27	0.80	1.41	0.72	1.55	0.63	1.71
22	1.00	1.17	0.91	1.28	0.83	1.40	0.75	1.54	0.66	1.69
23	1.02	1.19	0.94	1.29	0.86	1.40	0.77	1.53	0.70	1.67
24	1.04	1.20	0.96	1.30	0.88	1.41	0.80	1.53	0.72	1.66
25	1.05	1.21	0.98	1.30	0.90	1.41	0.83	1.52	0.75	1.65
26	1.07	1.22	1.00	1.31	0.93	1.41	0.85	1.52	0.78	1.64
27	1.09	1.23	1.02	1.32	0.95	1.41	0.88	1.51	0.81	1.63
28	1.10	1.24	1.04	1.32	0.97	1.41	0.90	1.51	0.83	1.62
29	1.12	1.25	1.05	1.33	0.99	1.42	0.92	1.51	0.85	1.61
30	1.13	1.26	1.07	1.34	1.01	1.42	0.94	1.51	0.88	1.61
31	1.15	1.27	1.08	1.34	1.02	1.42	0.96	1.51	0.90	1.60
32	1.16	1.28	1.10	1.35	1.04	1.43	0.98	1.51	0.92	1.60
33	1.17	1.29	1.11	1.36	1.05	1.43	1.00	1.51	0.94	1.59
34	1.18	1.30	1.13	1.36	1.07	1.43	1.01	1.51	0.95	1.59
35	1.19	1.31	1.14	1.37	1.08	1.44	1.03	1.51	0.97	1.59
36	1.21	1.32	1.15	1.38	1.10	1.44	1.04	1.51	0.99	1.59
37	1.22	1.32	1.16	1.38	1.11	1.45	1.06	1.51	1.00	1.59
38	1.23	1.33	1.18	1.39	1.12	1.45	1.07	1.52	1.02	1.58
39	1.24	1.34	1.19	1.39	1.14	1.45	1.09	1.52	1.03	1.58
40	1.25	1.34	1.20	1.40	1.15	1.46	1.10	1.52	1.05	1.58
45	1.29	1.38	1.24	1.42	1.20	1.48	1.16	1.53	1.11	1.58
50	1.32	1.40	1.28	1.45	1.24	1.49	1.20	1.54	1.16	1.59
55	1.36	1.43	1.32	1.47	1.28	1.51	1.25	1.55	1.21	1.59
60	1.38	1.45	1.35	1.48	1.32	1.52	1.28	1.56	1.25	1.60
65	1.41	1.47	1.38	1.50	1.35	1.53	1.31	1.57	1.28	1.61
70	1.43	1.49	1.40	1.52	1.37	1.55	1.34	1.58	1.31	1.61
75	1.45	1.50	1.42	1.53	1.39	1.56	1.37	1.59	1.34	1.62
80	1.47	1.52	1.44	1.54	1.42	1.57	1.39	1.60	1.36	1.62
85	1.48	1.53	1.46	1.55	1.43	1.58	1.41	1.60	1.39	1.63
90	1.50	1.54	1.47	1.56	1.45	1.59	1.43	1.61	1.41	1.64
95	1.51	1.55	1.49	1.57	1.47	1.60	1.45	1.62	1.42	1.64
100	1.52	1.56	1.50	1.58	1.48	1.60	1.46	1.63	1.44	1.65

Source: Reprinted by permission of the Biometrika Trustees from J. Durbin and G. S. Watson, "Testing for Serial Correlation in Least Squares Regression. II," *Biometrika,* Vol. 38 (1951), pp. 173 and 175.
Example: For $n = 20$, $\alpha = 0.01$, and two independent variables, $d_L = 0.86$ and $d_U = 1.27$.
Text Reference: Use of this table is discussed on pp. 801–803.

TABLE C.11   **Table of random standard normal numbers**

Row	Column									
	1	2	3	4	5	6	7	8	9	10
1	0.464	0.137	2.455	−0.323	−0.068	0.296	−0.288	1.298	0.241	−0.957
2	0.060	−2.526	−0.531	−0.194	0.543	−1.558	0.187	−1.190	0.022	0.525
3	1.486	−0.354	−0.634	0.697	0.926	1.375	0.785	−0.963	−0.853	−1.865
4	1.022	−0.472	1.279	3.521	0.571	−1.851	0.194	1.192	−0.501	−0.273
5	1.394	−0.555	0.046	0.321	2.945	1.974	−0.258	0.412	0.439	−0.035
6	0.906	−0.513	−0.525	0.595	0.881	−0.934	1.579	0.161	−1.885	0.371
7	1.179	−1.055	0.007	0.769	0.971	0.712	1.090	−0.631	−0.255	−0.702
8	−1.501	−0.488	−0.162	−0.136	1.033	0.203	0.448	0.748	−0.423	−0.432
9	−0.690	0.756	−1.618	−0.345	−0.511	−2.051	−0.457	−0.218	0.857	−0.465
10	1.372	0.225	0.378	0.761	0.181	−0.736	0.960	−1.530	−0.260	0.120
11	−0.482	1.678	−0.057	−1.229	−0.486	0.856	−0.491	−1.983	−2.830	−0.238
12	−1.376	−0.150	1.356	−0.561	−0.256	−0.212	0.219	0.779	0.953	−0.869
13	−1.010	0.598	−0.918	1.598	0.065	0.415	−0.169	0.313	−0.973	−1.016
14	−0.005	−0.899	0.012	−0.725	1.147	−0.121	1.096	0.481	−1.691	0.417
15	1.393	−1.163	−0.911	1.231	−0.199	−0.246	1.239	−2.574	−0.558	0.056
16	−1.787	−0.261	1.237	1.046	−0.508	−1.630	−0.146	−0.392	−0.627	0.561
17	−0.105	−0.357	−1.384	0.360	−0.992	−0.116	−1.698	−2.832	−1.108	−2.357
18	−1.339	1.827	−0.959	0.424	0.969	−1.141	−1.041	0.362	−1.726	1.956
19	1.041	0.535	0.731	1.377	0.983	−1.330	1.620	−1.040	0.524	−0.281
20	0.279	−2.056	0.717	−0.873	−1.096	−1.396	1.047	0.089	−0.573	0.932
21	−1.805	−2.008	−1.633	0.542	0.250	−0.166	0.032	0.079	0.471	−1.029
22	−1.186	1.180	1.114	0.882	1.265	−0.202	0.151	−0.376	−0.310	0.479
23	0.658	−1.141	1.151	−1.210	−0.927	0.425	0.290	−0.902	0.610	2.709
24	−0.439	0.358	−1.939	0.891	−0.227	0.602	0.873	−0.437	−0.220	−0.057
25	−1.399	−0.230	0.385	−0.649	−0.577	0.237	−0.289	0.513	0.738	−0.300
26	0.199	0.208	−1.083	−0.219	−0.291	1.221	1.119	0.004	−2.015	−0.594
27	0.159	0.272	−0.313	0.084	−2.828	−0.439	−0.792	−1.275	−0.623	−1.047
28	2.273	0.606	0.606	−0.747	0.247	1.291	0.063	−1.793	−0.699	−1.347
29	0.041	−0.307	0.121	0.790	−0.584	0.541	0.484	−0.986	0.481	0.996
30	−1.132	−2.098	0.921	0.145	0.446	−1.661	1.045	−1.363	−0.586	−1.023
31	0.768	0.079	−1.473	0.034	−2.127	0.665	0.084	−0.880	−0.579	0.551
32	0.375	−1.658	−0.851	0.234	−0.656	0.340	−0.086	−0.158	−0.120	0.418
33	−0.513	−0.344	0.210	−0.735	1.041	0.008	0.427	−0.831	0.191	0.074
34	0.292	−0.521	1.266	−1.206	−0.899	0.110	−0.528	−0.813	0.071	0.524
35	1.026	2.990	−0.574	−0.491	−1.114	1.297	−1.433	−1.345	−3.001	0.479
36	−1.334	1.278	−0.568	−0.109	−0.515	−0.566	2.923	0.500	0.359	0.326
37	−0.287	−0.144	−0.254	0.574	−0.451	−1.181	−1.190	−0.318	−0.094	1.114
38	0.161	−0.886	−0.921	−0.509	1.410	−0.518	0.192	−0.432	1.501	1.068
39	−1.346	0.193	−1.202	0.394	−1.045	0.843	0.942	1.045	0.031	0.772
40	1.250	−0.199	−0.288	1.810	1.378	0.584	1.216	0.733	0.402	0.226
41	0.630	−0.537	0.782	0.060	0.499	−0.431	1.705	1.164	0.884	−0.298
42	0.375	−1.941	0.247	−0.491	−0.665	−0.135	−0.145	−0.498	0.457	1.064
43	−1.420	0.489	−1.711	−1.186	0.754	−0.732	−0.066	1.006	−0.798	0.162
44	−0.151	−0.243	−0.430	−0.762	0.298	1.049	1.810	2.885	−0.768	−0.129
45	−0.309	0.531	0.416	−1.541	1.456	2.040	−0.124	0.196	0.023	−1.204
46	0.424	−0.444	0.593	0.993	−0.106	0.116	0.484	−1.272	1.066	1.097
47	0.593	0.658	−1.127	−1.407	−1.579	−1.616	1.458	1.262	0.736	−0.916
48	0.862	−0.885	−0.142	−0.504	0.532	1.381	0.022	−0.281	−0.342	1.222
49	0.235	−0.628	−0.023	−0.463	−0.899	−0.394	−0.538	1.707	−0.188	−1.153
50	−0.853	0.402	0.777	0.833	0.410	−0.349	−1.094	0.580	1.395	1.298

Source: Reprinted from W. H. Beyer, ed., *Handbook of Tables for Probability and Statistics,* 2nd ed., copyright The Chemical Rubber Company, 1968, p. 484, with the permission of CRC Press, Inc.

Data Sets

FINANCIAL CHARACTERISTICS **D.1**

This data set contains information on net assets, net income, and net sales for 383 firms classified into eight industry groups.

Each line of the data set contains an identification number and information on seven other variables for a single firm. The eight variables are as follows.

Variable Number	Variable Name	Description
1	FIRM	Firm identification number (1–383)
2	INDUSTRY	Industry number coded as follows:
		1 Crude oil producers
		2 Textile products
		3 Textile apparel manufacturers
		4 Paper
		5 Electronic computer equipment
		6 Electronics
		7 Electronic components
		8 Auto parts and accessories
3	ASSETS1	Net assets in year 1 (in $ million)
4	ASSETS2	Net assets in year 2 (in $ million)
5	INCOME1	Net income in year 1 (in $ million)
6	INCOME2	Net income in year 2 (in $ million)
7	SALES1	Net sales in year 1 (in $ million)
8	SALES2	Net sales in year 2 (in $ million)

1	2	3	4	5	6	7	8
1	1	33.037	35.197	2.281	2.908	8.498	11.537
2	1	320.156	355.848	17.661	24.378	370.733	384.491
3	1	307.332	321.408	23.718	26.656	31.049	49.118
4	1	41.395	45.792	2.489	3.641	10.226	11.559
5	1	144.090	180.534	10.934	12.407	84.864	101.029
6	1	46.611	60.950	2.585	4.389	12.297	16.143
7	1	9.767	12.350	0.923	0.350	4.705	6.124
8	1	3.252	3.141	−0.174	−0.155	1.098	1.017
9	1	98.651	114.403	2.564	3.825	17.522	26.788
10	1	19.605	23.518	0.730	1.027	6.761	6.827
11	1	21.404	20.982	0.572	0.809	4.720	4.474
12	1	24.046	26.714	0.508	0.614	3.730	4.001
13	1	173.007	198.311	13.208	8.936	54.328	60.160
14	1	98.261	93.740	0.563	−0.792	13.442	13.269
15	1	6.958	5.708	−0.359	−0.558	2.479	1.521
16	1	23.682	23.996	−0.907	0.502	4.312	5.940
17	1	273.146	308.475	13.774	14.895	56.038	71.194
18	1	24.983	27.505	1.990	2.723	11.279	13.755
19	1	43.478	52.257	3.444	4.490	17.569	18.148
20	1	8.996	14.643	−0.176	0.270	1.107	1.953
21	1	307.403	322.025	11.521	20.836	68.764	67.442
22	1	365.377	363.423	7.070	10.577	40.995	48.238
23	1	17.987	27.188	1.064	1.631	5.173	6.310
24	1	446.506	480.794	31.259	37.461	128.664	146.655
25	1	67.860	94.029	4.824	6.101	17.067	20.074
26	1	321.744	547.854	80.528	85.031	182.449	202.052
27	1	255.357	312.386	13.991	17.638	127.806	149.302
28	1	157.964	237.078	5.631	7.459	41.197	45.764
29	1	18.203	23.921	0.099	0.190	4.679	5.594
30	1	141.333	255.155	17.173	20.555	118.166	123.335
31	1	40.345	42.460	2.464	2.740	5.326	5.509
32	1	20.308	23.382	2.028	2.966	4.097	4.814
33	1	3483.039	3458.764	−64.824	26.561	3557.532	3673.144
34	1	32.736	41.457	2.368	3.339	15.155	17.123
35	1	2085.829	2340.733	63.693	79.184	986.517	1093.102
36	1	20.862	22.203	1.355	1.957	4.783	11.781
37	1	31.069	33.268	1.420	1.808	2.716	3.532
38	1	89.658	90.445	1.351	0.394	11.331	11.260
39	1	27.925	48.831	2.133	2.581	9.669	11.885
40	1	730.017	772.251	5.802	6.900	182.627	194.079
41	1	69.692	87.668	1.696	2.742	17.690	24.122
42	1	14.672	23.909	1.679	0.883	5.006	4.906
43	1	32.733	24.743	−4.766	−0.419	7.270	5.346
44	1	53.121	63.963	2.079	2.708	9.596	10.996
45	1	3.517	3.318	0.020	−0.169	2.294	2.045
46	1	17.077	24.944	1.264	1.354	2.849	4.289
47	1	17.203	22.230	−0.026	1.580	4.929	5.334
48	1	52.296	62.005	−1.500	−2.318	5.165	6.365
49	1	59.748	60.552	1.933	3.020	14.828	16.871
50	1	14.660	14.579	0.408	0.687	2.529	3.282
51	1	36.516	38.493	2.921	1.322	9.442	9.608
52	1	316.345	374.205	−2.618	1.041	34.545	40.330
53	1	4.915	3.895	0.248	0.579	0.949	1.309
54	1	43.039	47.474	−1.887	−4.212	0.909	1.291
55	1	44.989	34.052	−4.253	−0.277	34.668	24.866
56	1	7.587	8.077	0.375	0.158	2.267	2.279
57	1	67.094	65.312	3.470	3.742	20.651	20.831
58	1	17.042	16.720	0.169	0.801	6.398	5.216
59	1	16.315	17.071	0.678	0.689	6.141	8.510
60	1	49.406	49.457	−5.277	0.655	22.226	21.437

1	2	3	4	5	6	7	8
61	2	98.291	105.288	7.814	9.729	23.778	26.808
62	2	120.361	129.735	7.722	7.054	194.511	206.686
63	2	86.612	80.549	3.178	2.732	97.496	95.772
64	2	78.705	88.090	3.819	3.939	128.598	125.086
65	2	1876.768	1945.958	54.190	66.969	2331.511	2451.761
66	2	103.526	111.185	3.730	5.577	170.813	234.989
67	2	242.555	278.809	23.159	21.080	393.291	441.342
68	2	271.921	283.952	9.486	11.119	430.817	447.119
69	2	2227.966	2630.037	118.818	188.623	2208.021	2601.829
70	2	177.614	153.874	6.144	6.087	81.555	84.332
71	2	378.621	372.959	−3.456	5.374	421.747	494.932
72	2	168.253	177.291	7.424	2.470	190.932	184.217
73	2	35.087	45.125	3.366	2.701	43.570	48.507
74	2	33.296	35.537	3.984	1.828	52.414	46.413
75	2	3.325	3.484	0.467	0.616	5.114	6.287
76	2	17.075	21.426	2.749	2.294	32.083	30.235
77	2	216.766	228.358	10.673	10.130	306.821	329.509
78	2	34.459	26.707	−0.941	0.576	53.361	44.423
79	2	8.806	12.289	1.569	1.176	10.966	15.003
80	2	37.476	50.316	1.667	2.095	29.588	40.524
81	2	119.556	125.491	5.018	6.896	182.525	218.244
82	2	73.948	81.329	5.983	7.090	61.690	69.895
83	2	40.027	40.460	2.716	0.274	68.498	55.874
84	2	25.240	18.237	0.212	−1.393	29.230	24.659
85	2	421.222	466.554	12.058	10.720	599.505	634.442
86	2	63.540	70.280	0.477	2.561	76.270	94.361
87	2	48.743	54.090	2.615	3.059	103.068	111.544
88	2	42.856	44.796	1.044	1.072	50.015	54.776
89	2	146.614	153.471	7.079	7.162	236.523	250.703
90	2	180.981	181.323	3.816	6.205	231.001	273.785
91	2	70.844	82.104	2.534	3.800	81.564	101.269
92	2	468.790	482.852	11.371	19.201	439.613	538.514
93	2	55.277	54.014	2.275	1.929	71.302	77.232
94	2	22.546	32.187	3.213	2.238	44.455	49.048
95	2	833.932	869.886	−0.867	20.994	1162.439	1279.287
96	2	171.237	201.282	15.247	9.659	221.647	241.229
97	2	1100.631	1173.978	22.140	20.669	996.824	1062.485
98	2	30.260	31.820	5.281	3.974	40.137	38.814
99	2	7.016	11.883	0.828	0.441	10.774	11.043
100	2	345.825	367.921	8.413	13.765	479.578	551.020
101	2	38.295	38.497	3.827	1.457	61.619	56.399
102	2	19.868	16.748	−0.421	0.134	20.733	22.599
103	2	145.220	153.824	−7.868	0.540	152.852	156.510
104	2	19.741	23.351	2.003	1.675	34.386	32.993
105	2	359.381	369.360	22.725	22.073	433.403	477.104
106	2	55.337	58.428	4.389	5.198	89.018	101.382
107	2	12.768	15.499	1.098	1.962	17.998	26.777
108	2	7.871	8.037	1.127	1.000	14.630	16.386
109	2	9.796	12.007	0.995	0.740	26.989	23.657
110	2	23.693	28.103	−0.497	1.181	39.135	50.421
111	2	23.205	21.246	−0.959	−0.312	20.879	23.698
112	2	24.434	25.793	2.514	2.797	41.562	41.965
113	2	52.288	67.090	2.218	0.680	73.175	84.460
114	2	48.435	57.810	1.196	1.528	48.429	59.755
115	2	74.096	104.981	9.403	2.502	88.318	78.569
116	2	27.379	24.694	−1.439	−1.898	29.088	27.760
117	2	55.835	66.343	4.922	5.515	87.476	106.024
118	2	242.812	303.727	17.905	20.268	396.723	465.007
119	2	148.376	155.930	4.791	4.385	236.810	268.119
120	2	415.396	427.906	15.732	18.221	674.024	738.672

1	2	3	4	5	6	7	8
121	3	21.657	26.793	1.188	1.544	39.635	57.129
122	3	18.498	21.041	1.903	2.076	46.088	49.336
123	3	65.791	81.270	0.668	0.510	102.888	116.853
124	3	28.219	33.504	0.775	1.717	35.752	44.686
125	3	149.359	136.670	8.139	−11.507	222.170	210.068
126	3	30.541	32.527	3.245	4.520	66.321	76.073
127	3	22.911	24.662	0.852	1.563	43.243	46.969
128	3	832.097	832.483	34.706	17.298	1764.240	1883.795
129	3	30.256	31.564	2.086	2.824	32.864	36.671
130	3	27.655	32.011	2.129	1.925	50.676	56.214
131	3	216.942	229.044	4.570	11.120	237.709	330.275
132	3	337.690	362.123	13.973	19.156	502.230	571.204
133	3	15.891	19.193	2.666	2.622	30.633	32.883
134	3	10.361	17.169	0.864	1.458	17.812	27.200
135	3	29.480	38.321	0.671	0.899	44.515	51.431
136	3	63.427	66.513	3.087	3.761	102.076	112.988
137	3	305.329	331.933	22.288	24.886	407.090	448.536
138	3	256.412	284.656	8.012	9.007	381.683	414.912
139	3	81.581	112.385	2.854	3.025	124.636	157.575
140	3	81.716	97.902	4.925	6.113	139.016	169.551
141	3	327.182	414.532	26.162	33.781	546.893	680.541
142	3	12.766	13.032	0.543	0.748	13.666	15.683
143	3	16.697	24.152	0.895	1.705	48.217	61.023
144	3	132.795	156.450	5.630	6.822	245.480	290.168
145	3	54.738	61.331	0.612	1.035	105.897	118.966
146	3	55.727	50.776	0.258	0.618	70.245	81.732
147	3	23.634	27.333	1.305	1.301	43.990	46.182
148	3	70.405	77.150	5.416	6.133	108.545	123.516
149	3	8.087	10.656	1.841	2.612	11.514	14.926
150	3	8.512	10.629	1.034	0.085	18.546	21.148
151	3	13.585	16.143	0.999	1.754	22.880	25.259
152	3	5.678	6.433	0.190	0.679	15.668	19.428
153	3	131.981	159.208	7.638	10.760	256.439	274.423
154	3	56.938	66.077	0.397	2.029	78.640	102.272
155	3	227.935	232.482	9.580	11.943	342.660	402.085
156	3	12.685	19.903	2.021	2.144	32.874	36.647
157	3	47.874	52.075	−1.069	1.156	38.075	49.028
158	3	95.637	116.532	6.440	7.490	179.816	192.254
159	3	82.075	85.915	9.783	10.965	154.724	172.012
160	3	88.359	100.690	4.057	5.634	154.888	196.680
161	3	28.130	38.817	2.763	3.841	30.108	40.127
162	3	60.762	58.273	0.771	1.477	85.118	87.549
163	3	12.898	13.940	2.438	2.300	26.583	24.775
164	3	13.769	14.909	1.040	0.425	21.777	21.395
165	3	108.794	139.760	5.114	5.939	85.822	84.229
166	3	19.011	25.038	1.143	1.457	25.318	29.369
167	3	183.176	219.225	19.873	23.540	372.702	401.209
168	3	29.655	39.998	3.310	4.444	38.295	50.903
169	3	212.932	233.429	9.343	12.458	352.894	381.388
170	3	56.958	54.895	2.520	2.263	81.279	79.415
171	3	23.844	26.703	2.036	2.331	42.379	48.497
172	3	10.981	10.976	1.426	1.094	21.230	21.998
173	3	9.692	13.096	0.535	1.230	26.622	31.720
174	3	9.468	13.334	2.792	2.979	30.912	38.223
175	3	17.057	22.557	1.546	1.886	42.827	55.133
176	3	25.695	25.526	2.485	2.137	69.533	57.854
177	3	6.090	6.808	0.244	0.348	15.926	19.719
178	3	33.182	38.151	1.696	2.128	55.619	63.864
179	3	3.479	4.374	0.733	0.822	8.636	10.179
180	3	11.818	11.987	0.922	0.626	34.672	33.195

1	2	3	4	5	6	7	8
181	3	12.220	16.128	0.689	1.066	27.842	36.828
182	3	14.957	15.548	0.377	0.306	31.217	30.129
183	3	10.427	12.913	0.738	1.345	19.093	25.784
184	3	8.240	10.938	1.894	2.580	26.476	31.406
185	3	41.907	42.941	3.740	5.017	55.481	69.854
186	3	14.525	17.563	2.531	2.954	19.825	22.887
187	3	23.706	30.623	1.462	3.676	51.307	62.905
188	4	43.060	47.135	2.734	3.687	36.353	43.569
189	4	994.853	1562.930	11.333	32.293	879.184	1887.666
190	4	1391.506	1471.317	42.559	59.544	1330.982	1501.588
191	4	688.492	697.367	14.175	23.582	697.108	757.052
192	4	106.703	108.282	1.665	2.905	60.862	64.626
193	4	595.544	622.023	19.236	24.566	479.868	553.250
194	4	483.549	486.714	4.340	6.282	499.886	532.888
195	4	2751.196	2802.900	93.590	138.694	2658.893	2825.996
196	4	1266.933	1300.139	42.779	75.060	1266.331	1364.198
197	4	1173.023	1169.280	31.466	35.165	1425.591	1523.867
198	4	443.204	482.408	14.236	22.390	477.943	509.535
199	4	1292.187	1387.853	30.853	55.782	1226.328	1367.681
200	4	1159.126	1171.518	35.612	52.115	1007.854	1098.618
201	4	30.312	31.779	−1.411	0.009	36.129	41.804
202	4	29.018	31.579	0.855	0.894	46.539	52.793
203	4	774.129	812.662	35.259	52.402	699.699	812.175
204	4	715.082	736.833	6.776	17.685	581.754	637.162
205	4	71.040	67.924	0.905	0.049	68.168	77.653
206	4	34.637	61.263	0.529	1.253	43.365	58.621
207	4	188.390	200.197	6.534	9.129	188.456	204.339
208	4	140.033	138.510	1.328	2.707	121.527	128.775
209	4	32.912	32.955	1.220	1.523	43.876	45.754
210	4	4.529	20.428	0.510	0.865	7.367	34.267
211	4	196.256	221.863	15.668	16.562	107.785	120.837
212	5	25.632	43.231	1.949	1.473	32.442	42.881
213	5	23.537	25.615	1.002	1.787	36.713	43.357
214	5	1.561	7.428	−1.052	1.161	0.556	9.076
215	5	32.403	46.818	2.252	5.261	20.462	40.937
216	5	100.240	89.427	0.583	1.971	68.700	80.715
217	5	202.692	259.761	14.310	20.655	198.246	253.197
218	5	67.605	73.362	2.660	1.558	101.272	103.194
219	5	2947.183	3025.063	88.722	103.402	2627.272	2869.352
220	5	429.474	427.247	−18.076	1.611	148.771	196.320
221	5	245.226	318.560	−0.635	−0.417	161.723	193.342
222	5	18.563	20.268	−1.916	−5.549	13.323	18.988
223	5	19.358	20.193	2.641	2.664	28.844	30.895
224	5	2232.655	2484.858	82.053	121.577	2462.315	3009.492
225	5	22.095	72.295	−5.515	3.164	4.946	35.520
226	5	29.426	28.844	−11.480	1.085	17.244	21.221
227	5	127.468	183.368	4.134	−18.051	98.294	90.477
228	5	4.948	12.141	−2.367	0.432	2.993	17.326
229	5	6.774	6.965	0.207	−0.128	8.756	7.960
230	5	15.032	18.720	0.166	0.458	5.643	7.942
231	5	4.103	6.225	−2.654	0.398	3.421	5.210
232	5	1.675	5.018	−0.170	0.425	2.916	6.581
233	5	6.719	8.278	−0.089	0.150	11.108	8.751
234	5	38.596	54.448	−9.167	−2.866	11.995	39.982
235	5	25.140	45.206	−6.001	−7.097	5.243	17.661
236	5	10.176	9.651	0.015	0.158	15.275	17.472
237	5	10.504	13.862	−5.063	−2.997	4.001	7.304
238	5	2.608	13.271	−2.577	−3.406	0.244	4.585
239	5	6.305	5.696	−0.721	0.363	6.704	6.029
240	5	21.044	29.722	−3.411	0.397	18.021	29.925

1	2	3	4	5	6	7	8
241	5	12.481	19.111	0.004	1.239	14.348	21.639
242	5	21.854	39.536	−6.966	0.649	5.434	29.502
243	5	24.700	46.787	1.396	2.331	10.855	20.521
244	5	2.627	4.724	−1.922	0.485	3.133	8.428
245	5	84.004	81.315	−0.594	0.551	52.551	58.084
246	5	22.095	72.295	−5.515	3.164	4.946	35.520
247	5	8.431	13.859	−1.910	1.554	11.278	21.128
248	5	11.413	17.487	0.952	1.200	25.596	29.834
249	6	53.546	51.957	1.040	−0.262	64.782	63.373
250	6	184.036	187.940	4.031	5.932	181.903	177.285
251	6	277.861	329.313	32.481	44.808	323.525	407.816
252	6	53.838	62.830	2.600	3.218	66.057	78.744
253	6	107.916	74.338	0.377	0.486	96.629	88.244
254	6	37.926	40.133	0.432	0.363	49.600	52.207
255	6	45.367	50.225	0.136	1.021	54.903	53.225
256	6	204.968	256.404	−10.585	10.430	260.669	302.260
257	6	401.301	444.267	7.347	12.485	372.515	430.150
258	6	75.383	78.563	0.124	1.412	114.122	124.160
259	6	59.889	70.428	−1.107	0.921	44.851	81.252
260	6	34.603	40.125	−0.269	0.405	24.867	37.557
261	6	28.712	30.841	0.251	0.824	43.508	46.930
262	6	42.916	72.028	2.754	5.021	80.719	133.688
263	6	826.922	852.024	47.500	55.584	1765.533	1977.791
264	6	134.881	145.097	−33.251	2.596	197.293	200.795
265	6	6.286	7.568	0.510	0.447	8.934	9.742
266	6	30.609	33.084	0.833	2.480	41.858	61.578
267	6	24.926	27.066	−1.374	1.567	42.449	48.981
268	6	2.946	17.348	0.131	1.007	6.263	15.119
269	6	5.177	7.451	0.358	0.394	6.352	8.192
270	7	491.453	359.448	−121.041	1.530	383.297	346.415
271	7	9.783	13.555	2.141	3.042	13.716	19.243
272	7	258.823	308.337	16.783	21.618	399.929	482.695
273	7	103.572	116.888	5.219	6.386	119.848	133.649
274	7	76.125	91.839	8.100	13.427	114.130	145.126
275	7	20.064	21.133	0.304	0.436	25.379	28.065
276	7	3.056	3.395	−0.524	0.135	4.917	6.094
277	7	5.797	5.563	0.250	0.263	4.926	5.584
278	7	35.080	41.402	1.527	1.955	33.776	37.660
279	7	9.651	11.212	−0.023	0.662	11.004	14.495
280	7	1.773	1.530	−0.381	−0.347	1.743	1.746
281	7	8.006	7.277	−0.151	−0.450	5.378	5.463
282	7	9.466	10.025	−0.576	0.259	11.754	16.598
283	7	39.955	44.133	−0.473	2.816	46.724	58.257
284	7	17.887	20.594	−1.012	0.119	19.080	24.358
285	7	9.280	10.831	0.576	0.883	14.646	18.355
286	7	47.550	47.781	−3.671	0.736	46.603	59.519
287	7	516.657	492.824	8.370	15.976	691.172	751.541
288	7	143.136	168.244	7.641	9.998	215.780	256.349
289	7	32.473	36.204	2.124	2.799	41.907	54.490
290	7	5.827	5.909	0.209	0.447	8.509	8.334
291	7	76.873	82.094	1.912	3.776	112.952	131.262
292	7	1066.853	1061.011	41.017	40.784	853.079	945.749
293	7	14.997	15.819	0.076	0.107	14.318	16.286
294	7	25.923	20.142	−0.298	−0.078	24.091	25.411
295	7	6.323	7.938	0.713	1.332	8.705	11.734
296	7	14.228	15.215	0.544	0.913	20.119	22.640
297	7	104.990	120.159	5.530	7.270	132.893	163.156
298	7	72.548	70.039	0.036	1.215	27.895	33.645
299	7	173.356	179.659	−10.831	0.900	159.120	197.973
300	7	7.470	8.586	0.278	−0.063	8.725	6.373

1	2	3	4	5	6	7	8
301	7	782.829	856.462	45.526	64.841	1031.749	1273.987
302	7	79.974	70.361	1.783	2.803	95.707	106.215
303	7	60.445	64.317	3.497	4.081	41.531	46.513
304	7	115.657	97.123	−2.241	−1.172	89.658	86.283
305	7	20.165	22.599	1.678	2.363	18.487	26.527
306	7	46.189	45.390	0.869	1.763	40.148	47.731
307	7	19.156	18.650	0.544	1.220	29.685	36.424
308	7	14.436	15.593	−0.649	0.398	15.104	21.765
309	7	29.106	28.966	0.463	−2.460	42.485	31.948
310	7	41.592	42.399	0.344	0.117	35.806	38.969
311	7	11.767	13.177	0.872	1.269	16.624	21.109
312	7	38.790	41.032	2.979	4.119	52.419	66.423
313	7	5.176	8.205	−4.242	−4.928	0.942	1.434
314	7	47.106	48.975	−1.933	1.654	63.322	74.446
315	7	8.145	12.899	0.633	0.998	10.328	12.196
316	7	19.962	29.624	−0.556	2.673	12.374	31.011
317	7	12.108	10.303	−1.935	−2.020	6.030	7.509
318	7	13.831	15.193	0.041	0.867	21.518	28.862
319	7	5.531	22.394	0.166	3.043	4.906	23.941
320	7	7.756	10.240	0.703	1.195	8.489	13.842
321	7	12.112	13.072	−0.138	0.173	12.853	13.025
322	7	14.881	16.331	0.616	0.790	13.302	16.332
323	7	3.147	8.936	−1.080	0.554	2.326	6.132
324	7	29.650	30.896	0.996	1.823	26.993	30.811
325	7	4.872	5.319	0.054	0.327	5.705	6.101
326	7	4.698	5.503	0.189	0.493	5.544	6.942
327	8	26.590	24.798	1.233	1.520	42.840	47.912
328	8	135.421	151.732	4.886	5.196	166.394	201.787
329	8	28.351	36.032	2.479	3.740	55.229	78.438
330	8	163.767	177.817	7.584	10.103	245.157	277.402
331	8	55.625	64.745	6.309	6.515	125.118	137.080
332	8	1619.015	1667.520	56.835	79.920	2159.730	2380.186
333	8	1300.048	1378.921	63.967	80.020	1550.054	1732.303
334	8	550.678	554.082	7.669	19.981	729.176	905.594
335	8	19.386	21.273	0.339	1.813	23.566	36.507
336	8	370.446	407.549	47.122	53.730	439.957	495.959
337	8	665.819	757.080	38.073	60.519	858.155	1109.792
338	8	1129.579	1279.626	77.004	95.029	1398.144	1651.144
339	8	45.738	59.208	1.922	3.275	52.758	68.441
340	8	271.663	318.553	17.948	19.668	363.968	391.314
341	8	13.335	20.886	1.434	1.843	37.071	49.414
342	8	74.743	99.174	5.457	7.077	82.798	93.439
343	8	408.271	540.327	19.320	26.291	461.924	636.240
344	8	20.415	21.497	0.936	1.314	30.617	32.713
345	8	76.976	90.557	3.370	8.109	109.582	154.571
346	8	195.265	213.925	11.321	17.285	271.581	334.040
347	8	68.306	76.343	3.506	5.053	98.653	118.558
348	8	586.732	613.967	66.840	71.024	716.704	802.459
349	8	179.264	218.990	6.973	15.256	325.448	365.481
350	8	71.284	85.826	5.912	5.816	146.322	149.899
351	8	29.886	25.449	0.198	0.467	31.795	32.767
352	8	132.485	166.323	24.014	27.626	153.326	172.155
353	8	32.521	31.236	0.586	0.899	36.569	36.162
354	8	134.451	160.434	12.932	15.551	240.858	295.538
355	8	293.471	329.299	15.574	19.229	386.584	452.600
356	8	136.326	147.780	3.669	6.360	204.375	205.760
357	8	88.918	131.481	5.212	7.707	145.113	210.109
358	8	5.366	7.156	−1.079	0.016	8.114	11.256
359	8	44.238	49.625	4.294	4.618	83.835	89.282
360	8	93.654	94.370	3.217	2.408	94.851	99.816

1	2	3	4	5	6	7	8
361	8	75.159	100.290	5.798	9.242	99.889	133.685
362	8	166.925	184.460	7.289	8.555	275.438	309.597
363	8	382.097	407.795	17.291	13.424	616.742	665.240
364	8	37.384	45.877	3.980	3.784	74.313	83.476
365	8	40.821	54.390	3.195	5.249	88.393	106.061
366	8	44.951	48.009	1.839	2.600	54.470	63.508
367	8	12.157	14.886	1.281	1.700	18.538	22.333
368	8	1506.522	1668.217	90.891	102.747	2088.407	2278.139
369	8	62.097	66.840	2.229	3.008	146.312	175.388
370	8	567.894	594.967	51.458	57.116	554.328	635.523
371	8	167.801	191.273	10.939	13.426	289.663	319.323
372	8	69.428	75.824	3.241	4.955	111.572	126.815
373	8	10.842	15.981	1.933	2.643	26.641	31.759
374	8	25.402	27.937	0.707	1.717	32.731	34.171
375	8	23.277	25.307	1.408	2.403	43.447	47.335
376	8	46.671	53.533	3.430	4.732	59.015	70.728
377	8	49.480	52.981	4.429	4.807	103.741	117.489
378	8	13.272	14.407	0.292	−0.028	32.507	35.084
379	8	24.986	25.325	0.324	0.088	30.752	31.196
380	8	64.899	73.919	4.386	7.440	108.813	129.973
381	8	83.894	108.963	9.303	14.074	90.550	96.532
382	8	28.883	36.202	3.166	4.031	59.360	72.837
383	8	8.668	8.514	0.205	0.108	18.535	21.288

D.2   POWER CELLS

This data set contains the results of an experiment on 54 identical silver–zinc power cells that were run until failure. The experiment was conducted to measure two responses (number of discharge–charge cycles until failure and failure mode) for each experimental cell. Four experimental control variables were employed (charge rate, discharge rate, depth of discharge, and ambient temperature), with particular levels of the four factors specified for each cell.

Each line of the data set contains an identification number and information on the two response variables and the levels of the four experimental factors for a single power cell. The seven variables are as follows.

Variable Number	Variable Name	Description
1	CELL	Cell identification number (1–54)
2	CYCLES	Number of discharge–charge cycles a cell survives before failure
3	MODE	Mode of failure (0 denotes shorting failure, 1 denotes low-voltage failure)
4	CHARGE	Charge rate (0.4, 1.0, 1.6 amperes)
5	DISCHRG	Discharge rate (1, 3, 5 amperes)
6	DEPTH	Depth of discharge (40, 80 percent of rated ampere hours)
7	TEMP	Ambient temperature (10°, 20°, 30°C)

Source: The experimental setting and data are adapted from S.M. Sidik, H.F. Leibecki, and J.M. Bozek, *Cycles Till Failure of Silver–Zinc Cells with Competing Failure Modes—Preliminary Data Analysis,* NASA Technical Memorandum 81556, 1980.

1	2	3	4	5	6	7		1	2	3	4	5	6	7
1	138	1	0.4	1	40	10		28	16	1	1.0	3	80	10
2	325	1	0.4	1	40	20		29	96	1	1.0	3	80	20
3	246	0	0.4	1	40	30		30	141	0	1.0	3	80	30
4	101	0	0.4	1	80	10		31	3	1	1.0	5	40	10
5	281	1	0.4	1	80	20		32	383	0	1.0	5	40	20
6	150	0	0.4	1	80	30		33	469	1	1.0	5	40	30
7	142	1	0.4	3	40	10		34	2	1	1.0	5	80	10
8	179	1	0.4	3	40	20		35	308	1	1.0	5	80	20
9	398	0	0.4	3	40	30		36	385	0	1.0	5	80	30
10	44	1	0.4	3	80	10		37	55	1	1.6	1	40	10
11	170	1	0.4	3	80	20		38	73	1	1.6	1	40	20
12	314	1	0.4	3	80	30		39	216	1	1.6	1	40	30
13	12	1	0.4	5	40	10		40	3	1	1.6	1	80	10
14	103	1	0.4	5	40	20		41	69	1	1.6	1	80	20
15	503	0	0.4	5	40	30		42	62	1	1.6	1	80	30
16	10	1	0.4	5	80	10		43	35	1	1.6	3	40	10
17	145	1	0.4	5	80	20		44	15	1	1.6	3	40	20
18	159	1	0.4	5	80	30		45	305	0	1.6	3	40	30
19	78	1	1.0	1	40	10		46	2	1	1.6	3	80	10
20	122	1	1.0	1	40	20		47	13	1	1.6	3	80	20
21	283	1	1.0	1	40	30		48	159	0	1.6	3	80	30
22	8	1	1.0	1	80	10		49	3	1	1.6	5	40	10
23	143	1	1.0	1	80	20		50	6	1	1.6	5	40	20
24	130	0	1.0	1	80	30		51	233	0	1.6	5	40	30
25	33	1	1.0	3	40	10		52	2	1	1.6	5	80	10
26	80	1	1.0	3	40	20		53	7	1	1.6	5	80	20
27	217	0	1.0	3	40	30		54	2	1	1.6	5	80	30

This data set contains quarterly unit values of the equity and fixed-income funds of a retirement plan for an 11-year period. A quarterly unit value represents the value on the last day of the quarter of $1 invested at the inception of the fund. On January 1 of year 1, the unit values of the equity and fixed-income funds were 1.0780 and 1.0549, respectively.

Each line of the data set contains an identification number, the year and quarter, and information on the values of the two funds for the quarter. The five variables are as follows.

INVESTMENT FUNDS

Variable Number	Variable Name	Description
1	PERIOD	Period identification number (1–44)
2	YEAR	Year (1–11)
3	QUARTER	Quarter of year (1–4)
4	EQUITY	Unit value of equity fund at last day of quarter
5	FIXED	Unit value of fixed-income fund at last day of quarter

1	2	3	4	5	1	2	3	4	5
1	1	1	1.0620	1.0658	23	6	3	1.4399	1.6177
2	1	2	0.9699	1.0584	24	6	4	1.4668	1.6249
3	1	3	1.0322	1.0763	25	7	1	1.6055	1.6525
4	1	4	0.9702	1.0848	26	7	2	1.7342	1.7076
5	2	1	0.9944	1.0868	27	7	3	1.8593	1.7139
6	2	2	0.8842	1.0374	28	7	4	1.8624	1.6641
7	2	3	0.7435	1.0357	29	8	1	2.1086	1.6238
8	2	4	0.7794	1.0757	30	8	2	2.1378	1.7563
9	3	1	0.8943	1.1355	31	8	3	2.4167	1.7521
10	3	2	0.9251	1.1373	32	8	4	2.6049	1.7679
11	3	3	0.9438	1.1376	33	9	1	2.6610	1.8090
12	3	4	0.9425	1.1638	34	9	2	2.7680	1.7686
13	4	1	1.0418	1.2117	35	9	3	2.5949	1.7322
14	4	2	1.0502	1.2460	36	9	4	2.5433	1.9082
15	4	3	1.0394	1.2841	37	10	1	2.4322	1.9749
16	4	4	1.0038	1.3463	38	10	2	2.3985	2.0606
17	5	1	1.0418	1.4009	39	10	3	2.5211	2.1866
18	5	2	1.0656	1.4358	40	10	4	2.9551	2.4261
19	5	3	1.1246	1.4740	41	11	1	3.2605	2.5456
20	5	4	1.1550	1.5155	42	11	2	3.7334	2.6241
21	6	1	1.1852	1.5428	43	11	3	3.8861	2.5947
22	6	2	1.3033	1.5785	44	11	4	4.0496	2.6512

D.4

BEER SALES

This data set contains monthly time series for (1) total domestic beer sales in a region, (2) the average daily maximum temperature, and (3) the average daily minimum temperature for a five-year period. The temperatures are those recorded for a major city in the region.

Each line of the data set contains an identification number, the year and month, and information on sales and temperatures for the month. The six variables are as follows.

Variable Number	Variable Name	Description
1	PERIOD	Period identification number (1–60)
2	YEAR	Year (1–5)
3	MONTH	Month of year (1–12)
4	SALES	Monthly total domestic beer sales in region (in thousands of hectoliters)
5	MAX	Monthly average of maximum daily temperatures (in degrees Celsius)
6	MIN	Monthly average of minimum daily temperatures (in degrees Celsius)

D.4   Beer Sales

1	2	3	4	5	6	1	2	3	4	5	6
1	1	1	130.2	3.5	−3.2	31	3	7	258.5	20.6	13.5
2	1	2	137.0	6.9	0.8	32	3	8	241.3	23.1	14.4
3	1	3	171.4	11.5	2.9	33	3	9	202.2	19.0	10.8
4	1	4	178.5	13.0	4.7	34	3	10	154.9	12.6	6.1
5	1	5	211.9	16.9	8.6	35	3	11	149.3	10.8	4.4
6	1	6	227.4	19.9	10.3	36	3	12	194.4	6.4	1.3
7	1	7	253.8	22.7	13.0	37	4	1	128.9	4.3	−0.5
8	1	8	254.6	22.1	13.2	38	4	2	150.7	7.0	1.4
9	1	9	152.5	19.4	11.5	39	4	3	173.4	9.3	1.6
10	1	10	177.2	13.9	7.1	40	4	4	187.0	11.8	3.5
11	1	11	161.8	8.8	0.8	41	4	5	214.0	16.4	7.8
12	1	12	179.3	8.6	3.1	42	4	6	252.0	21.2	12.2
13	2	1	156.8	3.7	−2.8	43	4	7	239.2	20.8	13.2
14	2	2	130.4	8.4	2.9	44	4	8	231.0	20.7	12.9
15	2	3	166.1	8.8	2.5	45	4	9	158.5	18.7	10.5
16	2	4	205.4	14.2	5.7	46	4	10	161.9	13.7	6.7
17	2	5	209.1	16.0	8.5	47	4	11	155.2	7.7	0.7
18	2	6	235.4	17.8	10.6	48	4	12	175.2	6.9	1.3
19	2	7	227.3	20.7	12.4	49	5	1	139.2	9.0	3.5
20	2	8	203.9	20.3	10.2	50	5	2	141.8	9.5	3.2
21	2	9	182.5	17.5	10.3	51	5	3	195.9	11.8	4.5
22	2	10	200.4	15.0	6.5	52	5	4	196.4	13.4	5.0
23	2	11	161.0	9.8	4.1	53	5	5	192.6	17.9	9.4
24	2	12	195.9	7.1	2.7	54	5	6	220.9	19.1	11.4
25	3	1	142.8	8.2	2.7	55	5	7	200.8	20.5	12.7
26	3	2	161.8	8.4	1.9	56	5	8	242.7	21.8	13.6
27	3	3	173.3	11.8	3.8	57	5	9	165.2	17.5	9.4
28	3	4	185.6	12.4	5.0	58	5	10	174.4	13.0	5.9
29	3	5	192.1	16.0	8.9	59	5	11	159.5	9.9	5.2
30	3	6	223.8	17.4	10.3	60	5	12	167.9	3.8	−2.5

Business MYSTAT® Manual

APPENDIX E

CONTENTS

®Registered trademark of SYSTAT, Inc. This appendix reproduces the user's manual for Business MYSTAT, Version 1.2, an instructional business version of SYSTAT for IBM-PC/compatibles. Copyright © 1990 SYSTAT, Inc., Evanston, Ill.

BUSINESS MYSTAT DOCUMENTATION

This is a real statistics program—it is not just a demonstration. You can use Business MYSTAT to enter, transform, and save data, and to perform a wide range of statistical evaluations. Please use Business MYSTAT to solve *real* problems.

Business MYSTAT is a subset of SYSTAT, our premier statistics package. We've geared Business MYSTAT especially for teaching business statistics, with special forecasting and time series routines all in a single, easy-to-use package.

Business MYSTAT provides descriptive statistics, cross-tabulation, Pearson and Spearman correlation coefficients, multiple regression, time series forecasting, and much more.

Business MYSTAT is available in Macintosh, IBM-PC/compatible, and VAX/VMS versions. Copies are available at a nominal cost.

For more information about SYSTAT, FASTAT, and our other top-rated professional statistics and graphics packages, please call or write.

Installation

Business MYSTAT requires 512K of RAM and a floppy or hard disk drive. It can handle up to 50 variables and up to 32,000 cases.

Your Business MYSTAT disk contains three files: MYSTAT.EXE (the program), MYSTAT.HLP (a file with information for on-line help), and DEMO.CMD (a demonstration that creates a data file and demonstrates some of Business MYSTAT's features).

Hard disk

Set up the CONFIG.SYS file   The CONFIG.SYS file in your root directory must have a line that says FILES=20. You can modify your existing CONFIG.SYS file, or create one if you don't have one, by typing the following lines from the DOS prompt (>).

```
COPY CONFIG.SYS + CON: CONFIG.SYS
FILES=20
```

To finish, press the F6 key and then press Enter or Return.

Set up the AUTOEXEC.BAT file   You may find it convenient to install MYSTAT in its own directory and add that directory to the PATH statement in your AUTOEXEC.BAT file. This file should be located in the root directory (\) of your boot disk (the hard disk). *If you do not already have an AUTOEXEC.BAT file, or if your AUTOEXEC.BAT file does not have a PATH statement already, please do the following:*

```
COPY AUTOEXEC.BAT + CON: AUTOEXEC.BAT
PATH=C:\;C:\MYSTAT
[F6, ENTER]
```

If you aleady have an AUTOEXEC.BAT file that has a PATH statement, use a text editor to add the MYSTAT directory to the existing PATH statement.

Reboot your machine

Copy the files on the MYSTAT disk into a \SYSTAT directory   Now, make a \SYSTAT directory, insert the MYSTAT disk in drive A, and copy the Business MYSTAT files into the directory. (You must have the help file in the same directory as MYSTAT.EXE. If you put MYSTAT.EXE in a di-

rectory other than \SYSTAT, the help file MYSTATB.HLP must be either in that directory or the \SYSTAT directory.)

```
>MD \SYSTAT
>CD \SYSTAT
>COPY A:*.*
```

You are now ready to begin using Business MYSTAT. From now on, all you need to do to get ready to use Business MYSTAT is boot and move (CD) into the \SYSTAT directory. Save your MYSTAT master disk as a backup copy.

Floppy disk

Boot your machine   Insert a "boot disk" into drive A. Close the door of the disk drive and turn on the machine.

Set up a CONFIG.SYS file on your boot disk   The boot disk must contain a file named CONFIG.SYS with a line FILES=20. You can modify your existing CONFIG.SYS file, or create one if you don't have one, by typing the following lines from the DOS prompt (>).

```
COPY CONFIG.SYS + CON: CONFIG.SYS
FILES=20
```

To finish, press the F6 key and then press Enter or Return.

Set up an AUTOEXEC.BAT file on your boot disk   You must be sure that you have a file called AUTOEXEC.BAT containing the line PATH=A:\;B:\ on your boot disk. It you do not, type the following lines from the DOS prompt (>).

```
COPY CON AUTOEXEC.BAT + CON: AUTOEXEC.BAT
PATH=A:\;B:\
[F6, Enter]
```

Reboot your machine

Make a copy of the Business MYSTAT disk   When you get a DOS prompt (>), remove the boot disk. Put the Business MYSTAT disk in drive A and a blank, formatted disk in drive B and type the COPY command at the prompt:

```
>COPY A:*.* B:
```

Remove the master disk from drive A and store it. If anything happens to your working copy, use the master to make a new copy.

Use Business MYSTAT   Now, switch your working copy into drive A. If necessary, make drive A the "logged" drive (the drive your machine reads from) by issuing the command A: at the DOS prompt (>).

Business MYSTAT reads and writes its temporary work files to the currently logged drive. Since there is limited room on the Business MYSTAT disk, you should read and write all your data, output, and command files from a data disk in drive B.

From now on, all you need to do to use Business MYSTAT is boot, insert your working copy, and log the A drive.

Getting Started

To start, type MYSTAT and press Enter.

```
>MYSTAT
```

When you see the MYSTAT logo, press Enter. You'll see a command menu listing all the commands you can use in MYSTAT.

Command menu

The command menu shows a list of all the commands that are available for Business MYSTAT. The menu divides the commands into six groups: information, file handling miscellaneous, graphics, statistics, and forecasting.

Business MYSTAT --- A Personal Version of SYSTAT					
DEMO	EDIT	MENU	PLOT	STATS	LOG
HELP		NAMES	BOX	TABULATE	MEAN
SYSTAT	USE	LIST	HISTOGRAM	TTEST	SQUARE
	SAVE	FORMAT	STEM	CORRELATE	TREND
	PUT	NOTE			PCNTCHNG
	SUBMIT				DIFFRNCE
					INDEX
QUIT	OUTPUT	SORT	CHARSET	MODEL	SMOOTH
		RANK		CATEGORY	ADJSEAS
		WEIGHT		ANOVA	EXP
				ESTIMATE	TPLOT
					ACF, PACF
					CLEAR

As you become more experienced with MYSTAT, you might want to turn the menu off. Turn it on and off with the MENU command.

```
>MENU
```

Data Editor

MYSTAT has a built-in Data Editor with its own set of commands. To enter the Editor, use the EDIT command.

```
>EDIT
```

Inside the Editor, you can enter, view, edit, and transform data. When you are done with the Editor, type QUIT to get out of the Editor and back to MYSTAT, where you can do statistical and graphical analyses. To quit MYSTAT itself, type QUIT again.

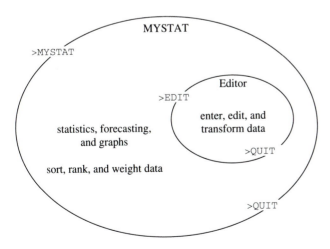

The Data Editor is an independent program inside MYSTAT and has its own commands. Five commands—USE, SAVE, HELP, QUIT, and FORMAT—appear both inside *and* outside the Editor.

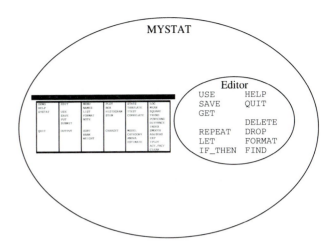

Demo

To see an on-line demonstration of MYSTAT, use the DEMO command. (Remember to press Enter after you type the command.)

```
>DEMO
```

When the demonstration is finished, Business MYSTAT returns you to the command menu. After you've seen the demo, you might want to remove the DEMO.CMD file and the CITIES.SYS file that it creates to save disk space.

Help

The HELP command provides instructions for any command—inside *or* outside the Editor. HELP lists and describes all the commands.

```
>HELP
```

You can get help for any specific command. For example:

```
>HELP EDIT
EDIT starts the MYSTAT full screen editor.

EDIT [filename]

EDIT (edit a new file)
EDIT CITIES (edit CITIES.SYS)

For further information, type EDIT [Enter],
[ESC], and then type HELP [Enter] inside the
data editor.
```

The second line shows a summary of the command. You see that any EDIT command must begin with the command word EDIT. The brackets indicate that specifying a file is optional. Anything in lowercase, like "filename," is just a placeholder—you should type a real file name (or a real variable name, or whatever).

Customizing DEMO and HELP All help information is stored in the text file MYSTATB.HLP. You can use a text editor to customize your help information. Teachers can design special demonstrations by editing the file EMO.CMD.

Information about SYSTAT For information about SYSTAT and how to order it, use the SYSTAT command.

Data Editor

The Business MYSTAT Data Editor lets you enter and edit data, view data, and transform variables. First, we enter data; later, we show you how to use commands.

To use the Editor to create a new file, type EDIT and press Enter. If you already have a MYSTAT data file that you want to edit, specify a filename with the EDIT command.

```
>EDIT [<filename>]
```

If you do not specify a filename, you get an empty Editor like the one above. MYSTAT stores data in a rectangular worksheet. *Variables* fill vertical columns and each horizontal row represents a case.

Entering data

First enter variable names in the top row. Variable names *must* be surrounded by single or double quotation marks, must begin with a letter, and can be no longer than 8 characters.

Character variables (those whose values are words and letters) must have names ending with a dollar sign ($). The quotation marks and dollar sign do not count toward the eight character limit.

Numeric variables (those whose values are numbers) can be named with subscripts; e.g., ITEM(3). Subscripts allow you to specify a range of variables for analyses. For example, STATS ITEM(1-3) does descriptive statistics on the first three ITEM(*i*) variables.

The cursor is already positioned in the first cell in the top row of the worksheet. Type 'CITY$' or "CITY$" and then press Enter.

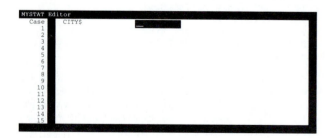

Business MYSTAT enters variable names in upper-case whether you enter them in lower- or upper-case.

The cursor automatically moves to the second column. You are now ready to name the rest of the variables, pressing Enter to store each name in the worksheet.

```
'STATE$'
"POP"
'RAINFALL'
```

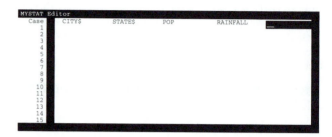

Now you can enter values. Move the cursor to the first blank cell under CITY$ by pressing home. (On most machines, Home is the 7 key on the numeric keypad. If pressing the 7 key types a 7 rather than moving the cursor, press the NumLock key and try again.)

When the cell under CITY$ is selected, enter the first data value:

```
'New York'
```

When you press Enter, Business MYSTAT accepts the value and moves the cursor to the right. You can also use the cursor arrow keys to move the cursor.

Character values can be no longer than twelve characters. Like variable names, character values must be surrounded by single or double quotation marks. Unlike variable names, character values are case sensitive—upper-case is not the same as lower-case (for example, 'TREE' is not the same as 'tree' or 'Tree'). Enter a blank space surrounded by quotation marks for missing character values. To use single or double quotation marks as part of a value, surround the whole value with the opposite marks.

Numeric values can be up to 10^{35} in absolute magnitude. Scientific notation is used for long numbers; e.g., .000000000015 is equivalent to 1.5E-11. Enter a decimal (.) for missing numeric values.

Enter the first few cases: type a value, press Enter, and type the next value. The cursor automatically moves to the beginning of the next case when a row is filled.

"New York"	"NY"	7164742	57.03
"Los Angeles"	"CA"	3096721	7.81
"Chicago"	"IL"	2992472	34

(etc.)

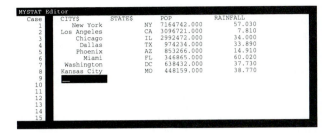

```
MYSTAT Editor
Case   CITY$         STATE$     POP          RAINFALL
  1    New York        NY    7164742.000       57.030
  2    Los Angeles     CA    3096721.000        7.810
  3    Chicago         IL    2992472.000       34.000
  4    Dallas          TX     974234.000       33.890
  5    Phoenix         AZ     853266.000       14.910
  6    Miami           FL     346865.000       60.020
  7    Washington      DC     638432.000       37.730
  8    Kansas City     MO     448159.000       38.770
  9
 10
 11
 12
 13
 14
 15
```

Moving around

Use the cursor keys on the numeric keypad to move around in the Editor.

Esc		toggle between Editor and command prompt(>)
Home	7	move to first cell in worksheet
↑	8	move upward one cell
PgUp	9	scroll screen up
←	4	move left one cell
→	6	move right one cell
End	1	move to last case in worksheet
↓	2	move down one cell
PgDn	3	scroll screen down

If these keys type numbers rather than move the cursor, press the Num-Lock key which toggles the keypad back and forth between typing numbers and performing the special functions. (If your computer does not have a NumLock key or something similar, consult the manual that came with your machine.)

Editing data

To change a value or variable name move to the cell you want, type the new value, and press Enter. Remember to enclose character values and variable names in quotation marks.

Data Editor Commands

When you have entered your data, press the Esc key to move the cursor to the prompt (>) below the worksheet. You can enter editor commands at this prompt. Commands can be typed in upper- or lower-case. Items in <angle brackets> are placeholders; for instance, you should type a specific filename in place of <filename>.

DELETE and DROP

DELETE lets you remove an entire case (row) from the dataset in the Editor. You can specify a range or list of cases to be deleted. The following are valid DELETE commands.

DELETE 3	Deletes third case from dataset.
DELETE 3-10	Deletes cases 3 through 10.
DELETE 3, 5-8, 10	Deletes cases 3, 5, 6, 7, 8, and 10.

DROP removes variables from the dataset in the Editor. You can specify several variables or a range of subscripted variables.

DROP RAINFALL	Drops RAINFALL from dataset.
DROP X(1-3)	Drops subscripted variables X(1-3).
DROP X(1-3), GROUP$	Drops X(1-3) and GROUP$.

Saving files

The SAVE command saves the data in the Editor to a Business MYSTAT data file. You must save data in a data file before you can analyze them with statistical and graphic commands.

SAVE <filename>	Saves data in a MYSTAT file.
/DOUBLE | SINGLE	Choose single or double precision.

Business MYSTAT filenames can be up to 8 characters long and must begin with a letter. Business MYSTAT adds a ".SYS" extension that labels the file as a Business MYSTAT data file. To specify a path name for a file, enclose the entire file name, including the file extension, in single or double quotation marks.

Business MYSTAT stores data in double-precision by default. You can choose single precision if you prefer: add /SINGLE to the end of the command. Always type a slash before command options.

SAVE CITIES/SINGLE	Saves CITIES.SYS in single precision.

Single precision requires approximately half as much disk space as double precision and is accurate to about 9 decimal places. The storage option (single or double precision) does not affect computations, which always use double precision arithmetic (accurate to about 15 places).

SAVE CITIES	Creates data file CITIES.SYS.
SAVE b:new	Creates file NEW.SYS on B drive.
SAVE 'C:\DATA\FIL.SYS'	Creates FIL.SYS in \DATA directory of C.

You can save data in text files for exporting to other programs with the PUT command. PUT is not an Editor command, though. First QUIT the Editor, USE the data file, and PUT the data to a text file.

Reading files

USE reads a MYSTAT data file into the Editor.

USE [<filename>]	Reads data from MYSTAT data file.

Starting new data files

Use the NEW command to clear the worksheet and start editing a new data file.

Importing data from other programs

You can import data from other programs through a plain text (ASCII) file. ASCII files contain only plain text and numbers—they have no special characters or formatting commands.

Start the program and save your data in a plain ASCII file according to the instructions given by that program's manual. The ASCII file must have a ".DAT" extension, data values must be separated by blanks or commas, and each case must begin on a new line.

Then, start Business MYSTAT. Use EDIT to get an empty worksheet, and enter variable names in the worksheet for each variable in the ASCII file. Next, use GET to read the text file. Finally, SAVE the data.

GET [<filename>]	Reads data from ASCII text file.

Finding a case

FIND searches through the Editor starting from the current cursor position, and moves the cursor to the first value that meets the condition you specify. Try this:

```
>FIND POP>1000000
```

This moves the cursor to case 4. After Business MYSTAT finds a value, use the FIND command without an argument (that is, "FIND" is the entire command) to find the next case meeting the same condition. All functions, relations, and operators listed below are available.

Some valid FIND commands:

```
>FIND AGE>45 AND SEX$='MALE'
>FIND INCOME<10000 AND STATE$='NY'
>FIND (TEST1+TEST2+TEST3)>90
```

Decimal places in the Editor

The FORMAT command specifies the number (0–9) of decimal places to be shown in the Editor. The default is 3. Numbers are stored the way you enter them regardless of the FORMAT setting; FORMAT affects the Editor display only.

FORMAT=<#>	Sets number of decimal places to <#>.
/UNDERFLOW	Displays tiny numbers in scientific notation.

For example, to set a two-place display with scientific notation for tiny numbers use the command:

```
>FORMAT=2/UNDERFLOW
```

Transforming variables

Use LET and IF. . . THEN to transform variables or create new ones.

LET <var>=<exprn>	Transforms <var> according to <exprn>.
IF <exprn> THEN LET <var>=<exprn>	Transforms <var> conditionally according to <exprn>.

For example, we can use LET to create LOGPOP from POP.

```
>LET LOGPOP=LOG(POP)
```

```
MYSTAT Editor
Case   CITY$          STATE$     POP          RAINFALL     LOGPOP
  1    New York         NY    7164742.000      57.030      15.785
  2    Los Angeles      CA    3096721.000       7.810      14.946
  3    Chicago          IL    2992472.000      34.000      14.912
  4    Dallas           TX     974234.000      33.890      13.789
  5    Phoenix          AZ     853266.000      14.910      13.657
  6    Miami            FL     346865.000      60.020      12.757
  7    Washington       DC     638432.000      37.730      13.367
  8    Kansas City      MO     448159.000      38.770      13.013
  9
 10
 11
 12
 13
 14
 15
```

LET labels the last column of the worksheet LOGPOP and sets the values to the natural logs of the POP values. (If LOGPOP had already existed, its values would have been replaced.)

Use IF. . . THEN for *conditional* transformations. For example:

```
>IF POP>1000000 THEN LET SIZE$='BIG'
```

creates a new character variable, SIZE$, and assigns the value BIG for every city that has population greater than one million.

For both LET and IF. . . THEN, character values must be enclosed in quotation marks and are case sensitive (i.e., "MALE" is not the same as "male"). Use a period to indicate missing values.

Some valid LET and IF-THEN commands:

```
>LET ALPHA$='abcdef'
>LET LOGIT1=1/(1+EXP(A+B*X)
>LET TRENDY=INCOME>40000 AND CAR$='BMW'
>IF SEX$='Male' THEN LET GROUP=1
>IF group>2 THEN LET NEWGROUP=2
>IF A=-9 AND B<10 OR B>20 THEN LET
C=LOG(D)*SQR(E)
```

Functions, relations, and operators for LET and IF . . . THEN

+	addition	CASE	current case number
-	subtraction	INT	integer truncation
*	multiplication	URN	uniform random number
/	division	ZRN	normal random number
^	exponentiation		
<	less than	AND	logical and
<=	less than or equal to	OR	logical or
=	equal to	SQR	square root
<>	not equal to	LOG	natural log
>=	greater than or equal	EXP	exponential function
>	greater than	ABS	absolute value

Logical expressions Logical expressions evaluate to one if true and to zero if false. For example, for LET CHILD=AGE<12, a variable CHILD would be filled with ones for those cases where AGE is less than 12 and zeros whenever AGE is 12 or greater.

Random data You can generate random numbers using the REPEAT, LET and SAVE commands in the Editor. First, enter variable names. Press Esc to move the cursor to the command line. Then, use REPEAT to fill cases with missing values and LET to redefine the values.

REPEAT 20	Fills 20 cases with missing values.
LET A=URN	Fills A with uniform random data.
LET B=ZRN	Fills B with normal random data.
SAVE RANDOM	Saves data in file RANDOM.SYS

Distribution functions MYSTAT can compute the following distribution functions.

Cumulative distribution functions compute the probability that a random value from the specified distribution falls below a given value; that is, it shows the proportion of the distribution below that value. *Inverse* distribution functions do the same thing backwards: you specify an alpha value (a probability or proportion) between zero and one, and MYSTAT shows the critical value below which lies that proportion.

Distribution	Cumulative	Inverse
Normal	**ZCF(z)**	**ZIF(alpha)**
Exponential	**ECF(x)**	**EIF(alpha)**
Chi-square	**XCF(chisq,df)**	**XIF(alpha,df)**
T	**TCF(t,df)**	**TIF(alpha,df)**
F	**FCF(F,df1,df2)**	**FIF(alpha,df1,df2)**
Poisson	**PCF(k,lambda)**	**PIP(alpha,k)**
Binomial	**NCF(p,k,n)**	**NIF(alpha,k,n)**
Logistic	**LCF(x)**	**LIF(alpha)**
Studentized range	**SCF(x,k,df)**	**SIF(alpha,k,df)**
Weibull	**WCF(x,p,q)**	**WIF(alpha,p,q)**

The uniform distribution is uniformly distributed from zero to one. The normal distribution is a standard normal distribution with mean zero and standard deviation one. The exponential distribution has parameter one. The cumulative Poisson function calculates the probability of the number of random events between zero and k when the expected value is *lambda*. The binomial cumulative function provides the probability of k or more occurrences in n trials with the binomial probability p.

Use the cumulative distribution functions to obtain probabilities associated with observed sample statistics. Use the inverse distributions to determine critical values and to construct confidence intervals. Finally, to generate pseudo-random data for the functions, apply the appropriate inverse cumulative distribution function to uniform random data.

Leaving the Data Editor

Use the QUIT command to leave the Editor and return to the main Business MYSTAT menu.

```
QUIT
```

General MYSTAT Commands

Once your data are in a Business MYSTAT data file, you can use Business MYSTAT's statistical and graphics routines to examine them.

Open a data file First, you must open the file containing the data you want to analyze:

```
USE <filename>        Reads the data in <filename>.
```

To analyze the data we entered earlier, type:

```
>USE CITIES
```

MYSTAT responds by listing the variables in the file.

```
VARIABLES IN MYSTAT FILE   ARE:
     CITY$      STATE$        POP   RAINFALL   LOGPOP
```

See variable names and data values The NAMES command shows the variable names in the current file.

```
>NAMES
VARIABLES IN MYSTAT FILE ARE:
     CITY$      STATE$        POP   RAINFALL   LOGPOP
```

The LIST command displays the values of variables you specify. If you specify no variables, all variables are shown.

```
>LIST CITY$
                          CITY$

CASE     1            New York
CASE     2         Los Angeles
CASE     3             Chicago
CASE     4              Dallas
CASE     5             Phoenix
CASE     6               Miami
CASE     7          Washington
CASE     8         Kansas City

   8 CASES AND    5 VARIABLES PROCESSED
```

Decimal places Use the FORMAT command to specify the number of digits to be displayed after the decimal in statistical output. This FORMAT command has the same syntax and works the same as in the Editor:

```
FORMAT=<#>           Sets number of decimal places to <#>.
  /UNDERFLOW         Uses scientific notation for tiny numbers.
```

Sorting and ranking data

SORT reorders the cases in a file in ascending order according to the variables you specify. You can specify up to ten numeric or character variables for nested sorts. Use a SAVE command after the USE command to save the sorted data into a MYSTAT file. Then, open the sorted file to do analysis.

```
>USE MYDATA
>SAVE SORTED
>SORT CITY$ POP
>USE SORTED
```

RANK replaces each value of a variable with its rank order within that variable. Specify an output file before ranking.

```
>USE MYDATA
>SAVE RANKED
>RANK RAINFALL
>USE RANKED
```

Weighting data

WEIGHT replicates cases according to the integer parts of the values of the weighting variable you specify.

```
WEIGHT <variable>        Weights according to variable
                         specified.
```

To turn weighting off, use WEIGHT without an argument.

```
>WEIGHT
```

Quitting

When you are done with your analyses, you can end your session with the QUIT command. Remember that the Data Editor also has a QUIT. To quit MYSTAT from the Editor, enter QUIT twice.

```
>QUIT
```

Notation used in command summaries

The box below describes the notation we use for command summaries in this manual.

Any item in angled brackets(<>) is representative—insert an actual value or variable in its place. Replace <var> with a variable name, replace <#> with a number, <var$> with a character variable,and <gvar> with a numeric grouping variable.

Some commands have *options* you can use to change the type of output you get. Place a slash / before listing any options for your command. You only need one slash, no matter how many options.

A vertical line (|) means "or." Items in brackets ([]) are optional. Commas and spaces are interchangeable, except that *you must use a comma at the end of the line when a command continues to a second line.* You can abbreviate commands and options to the first two characters and use upper-and lower-case interchangeably.

Most commands allow you to specify particular variables. If you don't specify variables, MYSTAT uses its defaults (usually the first numeric variable or all numeric variables, depending on the command).

Statistics

Descriptive statistics

STATS produces basic descriptive statistics.

```
>STATS
TOTAL OBSERVATIONS: 8
```

	POP	RAINFALL	LOGPOP
N OF CASES	8	8	8
MINIMUM	346865.000	7.810	12.757
MAXIMUM	7164742.000	60.020	15.785
MEAN	2064361.375	35.520	14.028
STANDARD DEV	2335788.226	18.032	1.068

Here is a summary of the STATS command.

STATS <var1> <var2>...	Statistics for the variables specified.

MEAN SD SKEWNESS,
    KURTOSIS MINIMUM,
    MAXIMUM RANGE SUM,
    SEM
	Choose which statistics you want.

/BY <gvar>	Statistics for each group defined by the grouping variable <gvar>. The data must first be SORTed on the grouping variable.

For example, you can get the mean, standard deviation, and range for RAINFALL with the following command.

```
>STATS RAINFALL / MEAN SD RANGE
 TOTAL OBSERVATIONS:   8
```

	RAINFALL
N OF CASES	8
MEAN	35.520
STANDARD DEV	18.032
RANGE	52.210

Tabulation

TABULATE provides one-way and multi-way frequency tables. For two-way tables, MYSTAT provides the Pearson chi-square statistic. You can produce a table of frequencies, percents, row percents, or column percents. You can tell MYSTAT to ignore missing data with the MISS option and suppress the chi-square statistic with NOSTAT.

TAB <var1>*<var2>...	Tabulates the variables you specify.

/LIST
  FREQUENCY PERCENT
    ROWPCT COLPCT MISS
    NOSTAT
	Special list format table.
	Different types of tables.

TABULATE	Frequency tables of all numeric variables.
TABULATE AGE/LIST	Frequency table of AGE in list format.
TAB AGE*SEX	Two-way table with chi-square
TAB AGE*SEX$*STATE$, /ROWPCT	Three-way row percent table.

TAB A,AGE*SEX/FREQ, PERC	Two two-way frequency and cell percent tables (A*SEX and AGE*SEX).
TAB AGE*SEX/MISS	Two-way table excluding missing values.
TAB AGE*SEX/NOSTAT	Two-way table excluding chi-square.

One-way frequency tables show the number of times a distinct value appears in a variable. Two-way and multi-way tables count the appearances of each unique combination of values. Multi-way tables count the appearances of a value in each subgroup. Percent tables convert the frequencies to percentages of the total count; row percent tables show percentages of the total for each row; and column percent tables show percentages of the total for each column.

T-tests

TTEST does dependent and independent *t*-tests. A dependent (paired samples) *t*-test tests whether the means of two continuous variables differ. An independent test tests whether the means of two groups of a single variable differ.

To request an independent (two-sample) *t*-test, specify one or more continuous variables and one grouping variable. Separate the continuous variable(s) from the grouping variable with an asterisk. The grouping variable must have only two values.

To request a dependent (paired) *t*-test, specify two or more continuous variables. MYSTAT does separate dependent *t*-tests for each possible pairing of the variables.

TTEST <var1>...[*<gvar>]	Does *t*-test of the variables you specify.
TTEST A B	Dependent (paired) *t*-test of A and B.
TTEST A B C	Paired tests of A and B, A and C, B and C.
TTEST A*SEX$	Independent test.
TTEST A B C*SEX	Three independent tests.

You can also do a one-sample test by adding a variable to your data file that has a constant value corresponding to the population mean of your null hypothesis. Then do a dependent *t*-test on this variable and your data variable.

Correlation

PEARSON computes Pearson product moment correlation coefficients for the variables you specify (or all numerical variables). You can select pairwise or listwise deletion of missing data; pairwise is the default. RANK the variables before correlating to compute Spearman rank-order correlations. CORRELATE is a synonym for PEARSON.

PEARSON <var1> ...	Pearson correlation matrix.
/PAIRWISE | LISTWISE	Pairwise or listwise deletion of missing values.
PEARSON	Correlation matrix of all numeric variables.
CORR HEIGHT IQ AGE	Matrix of three variables.
PEARSON /LISTWISE	Listwise deletion rather than pairwise.

Correlation measures the strength of linear association between two variables. A value of 1 or -1 indicates a perfect linear relationship; a value of 0 indicates that neither variable can be linearly predicted from the other.

Regression and ANOVA

Business MYSTAT computes simple and multiple regression and balanced or unbalanced ANOVA designs. For unbalanced designs, MYSTAT uses the method of weighted squares of means.

The MODEL and ESTIMATE commands provide linear regression. MODEL specifies the regression equation and ESTIMATE tells MYSTAT to start working. Your MODEL should almost always include a CONSTANT term.

```
>MODEL Y=CONSTANT+X       Simple linear regression.
>ESTIMATE

>MODEL Y=CONSTANT+X+Z     Multiple linear regression.
>ESTIMATE
```

Use CATEGORY and ANOVA commands for fully factorial ANOVA. CATEGORY specifies the number of categories (levels) for one or more variables used as categorical predictors (factors). ANOVA specifies the dependent variable and produces a fully-factorial design from the factors given by CATEGORY.

All CATEGORY variables must have integer values from 1 to *k*, where *k* is the number of categories.

```
>CATEGORY SEX=2       One-way design with independent
>ANOVA SALARY           variable SALARY and one factor
>ESTIMATE               (SEX) with 2 levels.

>CATEGORY A=2, B=3    Two-by-three ANOVA.
>ANOVA Y
>ESTIMATE
```

Saving residuals Use a SAVE command before MODEL or ANOVA to save residuals in a file. MYSTAT saves model variables, estimated values, residuals, and standard errors of prediction as the variables ESTIMATE, RESIDUAL, and SEPRED. When you use SAVE with a linear model, MYSTAT lists cases with extreme studentized residuals or leverage values and prints the Durbin-Watson and autocorrelation statistics.

```
>SAVE RESIDS          >SAVE RESID2
>MODEL Y=CONSTANT+X+Z >CATEGORY SEX=2
>ESTIMATE             >ANOVA SALARY
                      >ESTIMATE
```

You can USE the residuals file to analyze your residuals with MYSTAT's statistical and graphic routines.

Graphics

Use the CHARSET command to choose the type of graphic characters to be used for printing and screen display. If you have IBM screen or printer graphic characters, use GRAPHICS; if not, use GENERIC. The GENERIC setting uses characters like +, −, and |.

```
CHARSET GRAPHICS     For IBM graphic characters.
CHARSET GENERIC      For any screen or printer.
```

Scatterplots

PLOT draws a two-way scatterplot of one or more Y variables on a vertical scale against an X variable on a horizontal scale. Use different plotting symbols to distinguish Y variables.

```
PLOT <yvar1>...*<xvar>    Plots <yvar(s)> against
                            <xvar>.
  /SYMBOL=<var$> |        Use character variable values or
    '<char>'               character string as plotting
                            symbol.
  YMAX=<#> YMIN=<#>       Specify range of X and Y
    XMAX=<#> XMIN=<#>      values.
  LINES=<#>              Specify number of screen lines
                            for graph.
PLOT A*B/SYMBOL='*'      Uses asterisk as plotting
                            symbol.
PLOT A*B/SYMBOL=SEX$     Uses SEX$ values for plotting
                            symbol.
PL Y1 Y2*X/SY='1','2'    Plot Y1 points as 1 and Y2
                            points as 2.
PLOT A*B/LINES=40        Limits graph size to 40 lines on
                            screen.
```

For example, we can plot LOGPOP against RAINFALL using the first letter of the values of CITY$ for plotting symbols.

```
>PLOT LOGPOP*RAINFALL/SYMBOL=CITY$
```

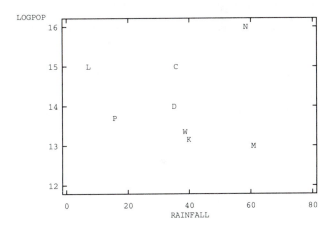

The SYMBOL option is powerful. If you are plotting several Y variables, you can label each variable by specifying its own plotting symbol.

```
>PLOT Y1 Y2*X/SYMBOL='1','2'
```

Or, you can name a character variable to plot each point with the first letter of the variable's value for the corresponding case:

```
>PLOT WEIGHT*AGE/SYMBOL=SEX$
```

Box-and-whisker plots

BOX produces box-and-whisker plots. Include an asterisk and a grouping variable for grouped box plots.

```
BOX <var1>... [*<gvar>]   [Grouped] box plots of the
                            variables.
  /GROUPS=<#>,            Show only the first <#>
                            groups.
  MIN=<#> MAX=<#>         Specify scale limits.
BOX                      Box plots of every numeric
                            variable.
BOX SALARY              Box plot of SALARY only.
```

```
BOX SALARY*RANK              Grouped box plots of
                               SALARY by RANK.
BOX INCOME*STATE$/GR=10      Box plots of first 10 groups
                               only.
```

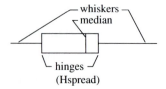

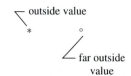

The center line of the box marks the *median*. The edges of the box show the upper and lower *hinges*. The median splits the ordered batch of numbers in half and the hinges split these halves in half again. The distance between the hinges is called the *Hspread*. The *whiskers* show the range of points within 1.5 Hspreads of the hinges. Points outside this range are marked by asterisks and those more than 3 Hspreads from the hinges are marked with circles.

Histograms

HISTOGRAM displays histograms for one or more variables.

```
HISTOGRAM <var1> ...        Histograms of variables specified.
  /BARS=<#>                 Limits the number of bars used.
  SCALE,                    Forces round cutpoints between
                              bars.
  MIN=<#> MAX=<#>           Specifies scale limits.
HISTOGRAM                    Histograms of every numeric
                              variable.
HISTOGRAM A B/BARS=18        Forces 18 bars for histograms of
                              A and B.
HIST A/MIN=0 MAX=10          Histogram of A with scale from 0
                              to 10.
```

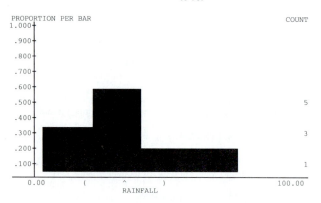

A histogram shows the *distribution* of a variable. The data are divided into equal-sized intervals along the horizontal axis. The number of values in each interval is represented by a vertical bar. The height of each bar can be measured in two ways: the axis on the right side shows the actual number of cases that have that value, and the left axis shows the proportion of the sample in each bar.

Stem-and-leaf diagrams

STEM plots a stem-and-leaf diagram.

```
STEM <var1>...              Diagrams of variables specified.
  /LINES=<#>                Specify number of lines in diagram.
```

```
STEM                        Stem-and-leaf of every numeric
                              variable.
STEM TAX                    Stem-and-leaf of TAX.
STEM TAX/LINES=20           Stem-and-leaf with 20 lines.
```

The numbers on the left side are *stems* (the most significant digits in which variation occurs). The *leaves* (the subsequent digits) are printed on the right. For example, in the following plot, the stems are 10's digits and the leaves are 1's digits.

```
STEM AND LEAF PLOT OF VARIABLE: RAINFALL, N = 48

MINIMUM IS:           7.000
LOWER HINGE IS:          26.500
MEDIAN IS:        36.000
UPPER HINGE IS:          43.000
MAXIMUM IS:       60.000

          0    78
          1    0114
          1    55556
          2
          2 H  588
          3    00113333
          3 M  5557999
          4 H  001123333
          4    55567999
          5    0
          5    9
          6    0
```

(We added 40 more cases to the dataset for this plot. You will get a different plot if you try this example with the CITIES.SYS data file.)

Hinges are defined under "Box-and-whisker plots," above.

Forecasting

MYSTAT's forecasting routines let you transform, smooth, and plot time series data. All of the forecasting routines operate in the computer's memory; that is, the results are saved in the working memory rather than being saved to a data file. When you execute a tranformation or a smooth, the only results you see are messages like "Series is smoothed" or "Series is transformed."

Your usual strategy will be to plot the series with TPLOT, ACF, or PACF; do a transformation or smoothing operation; and then replot the data to see your progress. You can save a modified series in a new data file by preceding the transformation or smoothing command with a SAVE command. You can use the CLEAR command at any time to restore the variable's original values.

Usually, you will do your work in memory, and when you find some satisfactory results, CLEAR the series, and then repeat the successful transformation using the SAVE command. Suppose, for example, that you wanted to save a demeaned and then smoothed series:

```
>MEAN MYSERIES
>SAVE A:NEW
>SMOOTH MYSERIES
```

This saves the smoothed, demeaned series in the file NEW.SYS. If you also want to save the intermediate (demeaned but not smoothed) series, use two SAVE commands:

```
>SAVE A:MEAN
>MEAN MYSERIES
>SAVE A:FINAL
>SMOOTH MYSERIES
```

Transforming

LOG replaces each value in a series with its natural logarithm. Logging removes nonstationary variability, such as increasing variances across time.

LOG <var>	Logs the variable specified.
LOG	Logs first numerical variable in file.
LOG SPEED	Logs SPEED.

MEAN subtracts the variable's mean from each value. Demeaning centers a series vertically on zero.

MEAN <var>	Demeans the variable specified.
MEAN	Demeans the first numeric variable.
MEAN SUNSPOTS	Demeans SPEED.

SQUARE squares each value. Squaring can help normalize variance in a series.

```
SQUARE <var>
SQUARE
SQUARE MAGNITUD
```

TREND removes linear trend to make a series "level."

```
TREND <var>
TREND
TREND WEATHER
```

PCNTCHNG replaces each value with its percentage difference from the immediately preceding value.

```
PCNTCHNG <var>
PCNTCHNG
PCNTCHNG SERIES
```

DIFFRNCE replaces each value with its difference from the immediately preceding value; you can specify a different lag to show differences from a greater number of positions previous. Differencing helps make some series "stationary" by removing the dependence of one point on the previous point.

DIFFRNCE /LAG=<#>]	Compute differences between values <#> cases apart.
DIFFRNCE	
DIFFRNCE STROKES/LAG=3	Difference interval of 3 values.

INDEX divides each value by the first value of the series, or by the base you specify.

INDEX <var>	
INDEX SERIES/BASE=12	Indexes series to its 12th value.

Smoothing

Smoothing removes local variations from a series and makes it easier to see the general shape of the series. Running smoothers replace each value

with the mean or median of that value and its neighbors. The "window" is the number of values considered at a time; 3 is the default. You can optionally specify weights for each value, separated by commas.

SMOOTH <var1>... /MEAN=<#>|MEDIAN=<#>	Smooths the variables specified. Mean or median smoothing with window size of <#>.
WT=<#,#,...>	Specifies smoothing weights for each value.
SMOOTH SERIES	Running mean smoother, window of 3.
SMOOTH SERIES/MEAN=5	Running mean smoother, window of 5.
SMOOTH SERIES/MEDIAN=4	Running median smoother, window of 4.
SMOOTH SERIES, /WT=.1,.2,.1	Weighted mean smoother, window of 3.

Seasonal decomposition removes additive or multiplicative seasonal effects according to the period you specify; the default is 12.

ADJSEAS <var> /SEASON=<#>,	Smooths specified variable. Number of observations per period.
ADDI=<#>|MULT=<#>	Removes additive or multiplicative seasonal effects with weight <#>.
ADJSEAS	Adjusts first variable with period 12.
ADJSEAS SALES/SEASON=4	Adjusts SALES with period 4.

Exponential smoothing forecasts future observations as a weighted average (running smooth) of previous observations. You can optionally specify a linear or percentage growth component and an additive or multiplicative seasonal effects component. Specify the weights for each component and the number or range of cases to be forecast.

EXP <var>	Smooths specified variable.
/LINEAR=<#>|PERCENT=<#>,	Remove linear or percentage growth trend.
SEASON=<#>,	Number of observations per period.
ADDI=<#>|MULT=<#>,	Removes additive or multiplicative seasonal effects with weight <#>.
FORECAST=<#>|<#>-<#>,	Forecasts number or range of cases.
SMOOTH=<#>	Specifies weight for level component.
EXP/FORE=4	Default model with 4 forecasted cases.
EXP SALES/LINEAR ADD	Holt-Winter's model: linear trend, additive seasonals with period 12.
EXP SERIES/MULT SEAS=4	Multiplicative seasonality with period 4.

Plotting

TPLOT produces time series plots, which plot the variable against CASE (time). TPLOT's STANDARDIZE option removes the series mean from each value and divides each by the standard deviation. MYSTAT uses the first fifteen cases unless you specify otherwise with LAG.

By default, TPLOT fills the area from the left axis to the plotted point. Include the NOFILL option to specify that the plot not be filled. Use the CENTER option to fill the plot from the observed value to the mean of the series.

```
TPLOT <var>              Case plot of variable specified.
  /LAG=<#>               Plots first <#> cases.
  STANDARDIZE,           Standardizes before plotting.
  CENTER|NOFILL,         Changes the way the plot is filled.
  MIN=<#>,MAX=<#>        Sets scale limits.
TPLOT                    Case plot of first numeric variable.
TPLOT PRICE/LAG=10       First 10 cases of PRICE.
TPLOT PRICE/STAN         Standardizes before plotting.
```

ACF plots show how closely the points in a series relate to the values immediately preceding them.

```
ACF <var>                Autocorrelation plot of specified
                           variable.
  /LAG=<#>               Plots first <#> cases.
ACF                      Autocorrelation plot of first variable.
ACF PRICE/LAG=10         First 10 cases of PRICE.
```

PACF plots show the relationship of values in a series to preceding points after partialing out the influence of intervening points.

```
PACF <var>               Partial autocorrelation plot of variable.
  /LAG=<#>43             Plots first <#> cases.
PACF                     Partial autocorrelation plot of first
                           variable.
PACF PRICE/LAG=10        First 10 cases of PRICE.
```

Clearing the series

CLEAR restores the initial values of the series, removing the effects of smoothings and transformations.

Submitting Files of Commands

You can operate MYSTAT in batch mode, where MYSTAT executes a series of commands from a file and you sit back and watch. (You've already seen a command file in action: the DEMO demonstration uses a file of commands, DEMO.CMD.)

The SUBMIT command reads commands from a file and executes the commands as though they were typed from the keyboard. Command files must have a ".CMD" file extension.

Use a word processor to create a file of commands—one command per line, with no extraneous characters. Save the file as a text (ASCII) file. (Use the command "COPY CON BATCH.CMD" and [F6] to type commands into a file if you have no word processor.)

```
SUBMIT <filename>        Submits file of commands.
SUBMIT COMMANDS          Reads commands from
                           COMMANDS.CMD.
SUBMIT B:NEWJOB          Reads commands from file on drive B.
```

Redirecting Output

Ordinarily, Business MYSTAT sends its results to the screen. OUTPUT routes *subsequent* output to an ASCII file or a printer.

```
OUTPUT *                 Sends output to the screen only.
OUTPUT @                 Sends output to the screen and the
                           printer.
OUTPUT <filename>        Sends output to the screen and a text
                           file.
```

MYSTAT adds a .DAT suffix to ASCII files produced by OUTPUT. You must use OUPUT * or QUIT the program to stop redirecting.

Printing and saving analysis results   To print analysis results, use the command OUTPUT @ before doing the analysis or analyses. Use OUTPUT <filename> to save analysis results in a file. Use OUTPUT * to turn saving or printing off when you are finished.

Printing data or variable names   You can print your data by using the command OUTPUT @ and then using the LIST command. Use LIST <var1> ... to print only certain variables. Don't forget to turn printing off when you are done.

You can print your variable names by using OUTPUT @ and then NAMES. Don't forget to turn printing off when you are done.

Putting Comments in Output

NOTE allows you to write comments in your output. Surround each line with quotation marks, and issue another NOTE command for additional lines:

```
>NOTE 'Following are decriptive statistics for
   the POP'
>NOTE "and RAINFALL variables of the CITIES
   dataset."
>STATS POP RAINFALL
TOTAL OBSERVATIONS:   8

                        POP      RAINFALL

N OF CASES               8         8
MINIMUM            346865.000     7.810
MAXIMUM           7164742.000    60.020
MEAN              2064361.375    35.520
STANDARD DEV      2335788.226    18.032
>NOTE "Note that the average annual rainfall for
   these."
>NOTE 'cities is 35.52 inches.'
```

Saving Data in Text Files

You can save datasets to ASCII text files with the PUT command. PUT saves the current dataset in a plain text file suitable for use with most other programs. Text files have a .DAT extension.

```
USE <filename>           Opens the dataset to be exported.
PUT <filename>           Saves the dataset as a text file.
```

Note that the PUT command is *not* an Editor command. To save a newly created dataset in a plain text file, you must QUIT from the Editor, USE the datafile, and finally PUT the data in an ASCII file.

```
>EDIT
 [editing session here]
>SAVE A:NEWSTUFF
>QUIT
```

```
>USE A:NEWSTUFF
>PUT A:NEWTEXT
```

The above commands would create a plain ASCII file called NEWTEXT.DAT on the A disk.

Index of Commands

This index lists all of MYSTAT's commands and the Editor commands, and it gives a brief description of each.

Editor commands

DELETE	delete a row (case)
DROP	drop a column (variable)
Esc key	toggle between command line and worksheet
FIND	find a particular data value
FORMAT	set number of decimal places in Editor
GET	read data from an ASCII file
HELP	get help for Editor commands
IF...THEN	conditionally transform or create a variable
LET	transform or create a variable
NEW	create a new data file
QUIT	quit the Editor and return to MYSTAT
REPEAT	fill cases with missing values
SAVE	save data in a data file
USE	read a data file into Editor

MYSTAT commands

ACF	autocorrelation plot
ADJSEAS	seasonal decomposition
ANOVA	analysis of variance
BOX	box-and-whisker plot
CATEGORY	specify factors for ANOVA
CHARSET	choose type of characters for graphs
CLEAR	restore original values of series
CORRELATE	Pearson correlaton matrix
DEMO	demonstration of Business MYSTAT
DIFFRNCE	difference transformation
EDIT	edit a new or existing data file
ESTIMATE	start computations for regression
EXP	exponential smoothing
FORMAT	set number of decimal places in output
HELP	get help for MYSTAT commands
HISTOGRAM	draw histogram
INDEX	index transformation
LIST	display data values
LOG	log transformation
MEAN	demean a series
MENU	turn the command menu on/off
MODEL	specify a regression model
NAMES	display variable names
NOTE	put comment in output
OUTPUT	redirect output to printer or text file
PACF	partial autocorrelation plot
PCNTCHNG	percent transformation
PEARSON	Pearson correlation matrix
PLOT	scatterplot (X-Y plot)
PUT	save data in text file
QUIT	quit the MYSTAT program
RANK	rank data
SAVE	save results in a file
SMOOTH	smooth a series
SORT	sort data
SQUARE	square transformation
STATS	descriptive statistics
STEM	stem-and-leaf diagram
SUBMIT	submit a batch file of commands
SYSTAT	get information about SYSTAT
TABULATE	one-way or multi-way tables
TPLOT	case plot
TREND	remove trend from a series
TTEST	independent and dependent t-tests
USE	read a data file for analysis
WEIGHT	weight data

SHORT QCTOOLS DOCUMENTATION

Herbert H. Stenson

QCTOOLS is a set of quality control programs. It is designed to be used in conjunction with data files created by MYSTAT. The results of all analyses that are performed will be printed on your screen in either graphical or tabular form (or both, if you request it). You also have the option of saving the screen output into an ASCII file and/or saving numerical output into a MYSTAT file. You can switch over to MYSTAT from within QCTOOLS.

Installation

QCTOOLS requires a minimum of 256K of RAM and a floppy or hard disk drive. One of the disks that you received contains QCTOOLS.EXE (the program), QCTOOLS.HLP (a file with information for on-line help), and two sample data files (files with a .SYS suffix in their names).

If you have not yet installed MYSTAT, please do so now by following the directions given for MYSTAT installation. Note that you cannot use QCTOOLS to analyze data unless the data files have been created with MYSTAT.

Once you have MYSTAT installed, the procedure for installing QCTOOLS is very similar. Just follow the directions given for MYSTAT that correspond to the type of drive that you have, but substitute the word QCTOOLS for MYSTAT in these procedures. However, you will not have to set up or alter the AUTOEXEC.BAT or CONFIG.SYS files again if you have already installed MYSTAT: just skip these steps. Be sure to make backup copies of both QCTOOLS and MYSTAT and their associated files.

If you are using a hard disk, the MYSTAT program and the MYSTATQ.HLP file must be in the same directory as the QCTOOLS program and QCTOOLS.HLP. If you are using a system with no hard disk and a 720K or more floppy disk with both QCTOOLS and MYSTAT on it, both programs must be in the same directory of that disk. If you are using a system with two 360K floppy disk drives then the QCTOOLS disk and the MYSTAT disk will both be used alternately in the A: drive. If these conditions are not met you will have trouble with certain aspects of the programs.

Getting Started

First, be sure that you are in the disk directory containing QCTOOLS and the associated files. Then, at your system prompt type:

```
QCTOOLS
```

after which you should press the Return or Enter key.

Command menu

The first thing that you will see on the screen is the QCTOOLS logo. Follow the instruction on the bottom of the screen and press your Enter or Return key. Next you will see the QCTOOLS command menu. It shows all of the commands available in QCTOOLS. Notice that they are split into two major groups. General system commands are in the left half of the menu and commands to perform specific quality control procedures are in the right half of the menu. Here is a list of the commands you will see on the left half of the menu, and a brief description of what each command does. Asterisks indicate the default settings, if any.

DEMO	Run a tutorial demonstration of QCTOOLS.
HELP	Obtain on-line help for any command.
MENU	Turn the menu off if it is on*, or on if it is off.
MYSTAT	Switch to the MYSTAT program from within QCTOOLS.
QUIT	Leave QCTOOLS and return to DOS.
CHARSET	Set graphics for output to IBM graphics* or generic.
FORMAT	Set number of decimals to be used in output. (3*)
NAMES	List the names of the variables in a file that is open.
NOTE	Write a note or title on the output.
OUTPUT	Route output to a file, a printer, or to the screen* only.
PRINT	Print long or short* results.
SAVE	Save the output into a MYSTAT data file.
SUBMIT	Run a batch job from commands listed in a command file.
USE	Open a particular MYSTAT data file.
WEIGHT	Designate a particular variable as a weighting variable.

Except for the PRINT and MYSTAT commands, all of the commands listed above function in the same way as their counterparts in the MYSTAT program. The PRINT command allows you to see extended output for many QCTOOLS commands. Most often this extra output is a tabular listing of the numerical values that are plotted on a control chart. To see this output enter the command:

```
PRINT LONG
```

prior to the command that produces the output that you wish to see. To switch back to the default output enter:

```
PRINT SHORT
```

The PRINT command stays in effect until you change it or QUIT.

The command MYSTAT allows you to switch directly to the MYSTAT program from within QCTOOLS. Any options that you have set within QCTOOLS (such as FORMAT or CHARSET) will be carried over to MYSTAT, as will any open files. Thus, if a data file is open (in use) in QCTOOLS it will still be open when you switch to MYSTAT. However, this command is sensitive to the installation of QCTOOLS and MYSTAT. Thus, you must follow the installation directions precisely for this command to work. The remainder of the commands in the list above are described in the text for the MYSTAT program and also on the HELP screens in QCTOOLS.

The WEIGHT command functions similarly to the WEIGHT command in MYSTAT, but needs some further clarification here. This command replicates cases according to a weighting variable that you specify, just as in MYSTAT. The values of the weighting variable are truncated to

integers, and the sample size used in the entire analysis is the sum of these integers. Cases weighted less than one are excluded from analysis. EXCEPTION: Truncation to integers is not performed in the case of a Shewhart U plot where the "sample size" for a case is the number of sample units, which may be fractional. (See the discussion of Poisson charts later for more on this). The command syntax is:

```
WEIGHT=<var>
```

The WEIGHT command must precede any analysis for which it is to be used. Type just WEIGHT with nothing after it to cancel the WEIGHT command. You will use this command most often in QCTOOLS to indicate individual sample sizes if you are analyzing data that are already aggregated into a single statistic, such as the mean, by samples or subgroups. (See the later section on Input Data for more on this.)

The commands on the right half of the menu are to provide specific quality control analyses and charts. They are discussed in the paragraphs that follow.

TIP: If, after you have entered any QCTOOLS command, you wish to use the same command again, simply press the F9 key and the command will reappear. Then press the ENTER key to activate that command. You also can edit the previous command before pressing ENTER. QCTOOLS stores the 6 most recent commands, so you can use the F9 key to scroll back through them.

Shewhart Control Charts

Eight types of Shewhart charts can be produced with QCTOOLS. The types are:

X	An X-bar chart of subgroup means
S2	A chart of subgroup variances
S	A chart of subgroup standard deviations
R	A chart of subgroup ranges
NP	Binomial count by subgroup
P	Binomial proportion by subgroup
C	Poisson count by subgroup
U	Poisson rate per sample unit

The generic form of the command to produce any of these Shewhart charts is:

```
SHEW (<type>) <yvar>*<xvar> [/ <option list> ]
```

Replace <type> with one of the eight types listed above, and enclose it in parentheses. Replace <yvar> with the name of a file variable that contains the quality control data for each subgroup. Replace <xvar> with the name of a file variable that identifies the subgroup to which each data entry belongs. You must enter the asterisk between <yvar> and <xvar>. The <yvar> must be a numeric variable, but the <xvar> can be either character (ending with $) or numeric (not ending with $).

Options

Each type of chart has a list of options that you may use to produce the chart. These options are very similar from type to type. Thus, they are described only once here. Any slight deviations in their meaning for a particular type of chart are described on the HELP screens and examples are given in the DEMO program. The options may be entered in any order after the / symbol. They must be separated from each other by either a comma or a space. The options are:

`CENTER=<#>`	Place the center line of the chart at the number specified. The default center line is the estimated expected value of the statistic being plotted.
`SLIMITS`	Enter this as an option if you wish to plot sigma limits instead of probability limits.
`LCL=<#>`	Place the lower control limit at the number specified. If SLIMITS is used, then the value for LCL is interpreted as so-called "sigma" units. Otherwise the number is taken to be the actual value for the lower control limit. The default LCL is a probability limit corresponding to the value of ALPHA used, or is −3.00 of SLIMITS is used.
`UCL=<#>`	Same as the LCL, but now applying to the upper control limit instead of the lower. Default value is +3.00 if SLIMITS is used.
`SIGMA=<#>`	Specify an a priori value for sigma, the population standard deviation. Note that this usually is not the standard deviation of the sampling distribution.
`ALPHA=<#>`	The proportion of the sampling distribution to be outside the LCL and/or UCL. If both control limits are to be used, then ALPHA/2 of the distribution is outside each control limit. If only one limit is used, then ALPHA of the distribution is outside that limit. Default value of ALPHA = .0027. This option is ignored if either LCL or UCL is specified in the option list.
`UPPER|LOWER`	Specify UPPER if you wish only an upper probability limit as the UCL. Specify LOWER to get only a lower probability limit. This option is ignored if either LCL or UCL is specified in the option list.
`AGG`	Indicates that the input data are already aggregated as the statistic that is to be plotted on the chart. You probably will want to use the WEIGHT command to specify individual sample (subgroup) sizes when you use the AGG option.
`YMAX=<#>`	Specify a maximum value for the Y-axis of the control chart.
`YMIN=<#>`	Specify a minimum value for the Y-axis of the control chart.

The last two options are useful for "zooming in" on a part of your data by rescaling the Y-axis of the control chart.

Input data

Each chart type will accept either Raw Data or Aggregated Data as input. By Raw Data we mean that a value of <yvar> is provided for each and every individual case in each subgroup of data. Aggregated data consist of a statistic that has already been computed for each subgroup. For example, if you already have a file that contains the mean of some variable for each subgroup, then those means constitute aggregated input data for an X chart.

The easiest way to think about aggregated data is to consider the statistic that is being plotted for each type of Shewhart chart. That statistic is the definition of aggregated input data for the chart type in question. Thus, subgroup means are the aggregated input data for an X chart, subgroup variances are the aggregated data for an S2 chart, etc. If you use the AGG

option, you also will have to use the WEIGHT command to inform QCTOOLS of the sample size for each subgroup whose aggregated data is entered. An example of this is given in the DEMO program.

Control limits

QCTOOLS computes probability limits for each chart by default. That means that the upper and lower control limits are points on a statistical distribution that have a specific proportion of the distribution outside those limits. The proportion of the distribution outside the limits is termed ALPHA, and its default value is 0.0027. That is the value of ALPHA for control limits set at the mean plus or minus 3 standard errors of the mean for a normal distribution of means. If only one control limit is being used, then all of alpha is one tail of the distribution.

By default, QCTOOLS computes probability limits by using the statistical distribution that is appropriate for the chart in question. Thus, a normal distribution is used for an X chart, a chi-square distribution is used for an S2 chart, the binomial distribution is used for an NP chart, etc. If, instead, you want to use the so-called "Sigma limits" that are described in many textbooks, you may do so by using the SLIMITS option.

Sample size vs. sample units for Poisson charts

The theoretical sample size for a Poisson distribution is infinitely large. Thus, the usual concept of a sample size does not apply. Instead, we define "sample units"; an arbitrary unit of time, weight, area, volume, etc. A sample unit is assumed to have a very large number of opportunities for an event of interest to occur, with a small probability of occurrence. For example, the sample unit might be 100 square yards of cloth being examined for defects. Logical subgroups could contain different numbers of sample units. Then we might want to plot the average number of defects per sample unit (e.g. defects per 100 square yards). For example, one subgroup might contain, say, two 100-square-yard bolts, while another contained three 50-square-yard bolts. Then the former subgroup would contain 2 sample units, and the latter would contain 1.5 sample units (with a sample unit defined as 100 square yards). To reflect the number of sample units in a subgroup, use the WEIGHT command (along with the AGG option). WEIGHT can be used this way for Shewhart C or U charts only. THIS IS THE ONLY PLACE IN QCTOOLS WHERE THE WEIGHT VALUES WILL NOT BE ROUNDED DOWN TO INTEGERS AUTOMATICALLY.

UWMA command

This command is very much like the SHEW (X) command, except that it plots the unweighted moving average by subgroup, instead of the simple average for each subgroup. Its syntax is:

```
UWMA <yvar>*<xvar>
[/WIDTH=<#>,CENTER=<#>,LCL=<#>,UCL=<#>,
   SIGMA=<#>,ALPHA=<#>,UPPER|LOWER,
   AGG,YMAX=<#>,YMIN=<#>]
```

Replace <yvar> with the name of a variable containing the data for which the moving average is desired. Replace <xvar> with the name of a variable identifying the subgroup to which each <yvar> observation belongs.

All of the options except for WIDTH are the same as for the X chart. The WIDTH option allows you to select the number of subgroups over which the moving average is to be computed. For example, if you choose WIDTH=4, then the plotted value for each successive subgroup will be the average of all the data for the current subgroup along with the previous

3 subgroups. Notice that the very first subgroup has no subgroups previous to it, so the point plotted for it will just be the mean of that subgroup. The second subgroup has only one previous subgroup, so the moving average for it can contain only the data from the first and second subgroups, etc. The default value of WIDTH is 1, so that the default UWMA chart is identical to a Shewhart X chart. The default CENTER value is the mean of all observations being analyzed. The default probability limits are calculated using a normal distribution model, the same as for the X chart. However, the sample size on which the limits are based is the total number of observations encompassed by the subgroups selected by the WIDTH option. See the DEMO program for an example of the UWMA chart.

Other QCTOOLS Commands

The last column of the QCTOOLS menu contains a number of utility commands that are useful in analyzing quality control data.

FUNCTION command

This command will replace most of your statistical tables. It calculates the cumulative probability for a statistical function or the inverse function of a probability, plus some statistics for the range of a standard normal distribution. Its general syntax is:

```
FUNCTION <type>
```

The following table shows what to substitute for <type> for each of the available distributions. Lower case letters represent numbers.

Distribution	Cumulative Function	Inverse Function
Normal	ZCF(x)	ZIF(prob)
T	TCF(x,df)	TIF(prob,df)
F	FCF(x,df1,df2)	FIF(prob,df1,df2)
Chi-Square	XCF(x,df)	XIF(prob,df)
Exponential	ECF(x)	EIF(prob)
Binomial	NCF(x,n,p)	NIF(prob,n,p)
Poisson	PCF(x,mean)	PIF(prob,mean)
Relative Range	RCF(x,n)	RIF(prob,n)
Expected Relative Range	ER(n)	
SD of the Relative Range	SDR(n)	

In this notation, x always refers to the value of a variable, prob refers to a probability, df is degrees of freedom, n is sample size, and p is a binomial proportion. The Relative Range refers to the range of a standard normal distribution.

Examples of this command are:

```
FUNCTION ZCF (1.96)
      (The program will respond with P=0.975)
FUNCTION NIF (.75,2,.5)
      (The program will respond with X=1.000)
```

PARETO command

This command creates a chart showing frequencies of occurrence of an attribute sorted in descending order. They are plotted as a function of a variable that identifies each subgroup. Its syntax is:

```
PARETO <yvar>*<xvar> [/CUM,P,AGG]
```

The <yvar> must consist of zeros and ones if individual instances of an event are in the input file. The <xvar> identifies subgroups and may be either numeric or character. If <yvar> contains the number of instances of an event already aggregated by subgroup, then the program must be informed by using the AGG option. Note that the WEIGHT command simply replicates cases in the file and has nothing to do with sample sizes in this case. All sample sizes are assumed to be equal.

The default output shows frequencies. The CUM option will change this to cumulative frequencies. The P option will produce proportions (relative frequencies) instead. CUM and P used together will produce cumulative proportions. Use the SAVE command prior to PARETO to save the chart data into a file. Use PRINT LONG prior to PARETO in order to print the chart values on the screen after the plot.

Here are some examples of the command:

PARETO DEFECTS*SAMPLE	Input data are zeros and ones.
PARETO ERRORS*DAY$ / AGG	Input data are frequencies by sample.
PARETO NUMBER*WEEK$ / CUM,P,AGG	Cumulative proportions will be shown.

See the DEMO program for an example of the output for this command.

PCA command

The PCA command produces Process Control Indices for a variable that you name. Its syntax is:

```
PCA (<lsl,usl>) <var>
```

The numerical lower and upper specification limits for a process are the numbers that you enter in the parentheses in place of <lsl,usl>. You must type these limits in parentheses. Replace <var> with the name of a MYSTAT variable that contains the empirical data for which you have stated the upper and lower specification limits. You will get output showing basic statistics on the data in <var> along with the process control indices Cp and Cpk. (These indices are described in many quality control textbooks.) No chart will appear.

The WEIGHT command may be used to indicate replications of input data for this analysis, but the use of PRINT=LONG and SAVE will have no effect.

As an example, USE the sample file BOXES.SYS from your QCTOOLS disk, and then issue the following command:

```
PCA (15,25) OHMS
```

This will produce the process control indices and other statistics whose meanings are self-evident. This example assumes that the specification limits for OHMS are 15 and 25 ohms.

PLOT command

This command creates a scatterplot of a numeric variable, <yvar>, against an <xvar> that may be either numeric or character. Its syntax is:

```
PLOT <yvar>*<xvar> / XMIN=<#>,XMAX=<#>,
   YMIN=<#>,YMAX=<#>
```

If the type of the <xvar> is character, each successive value of the variable as it exists in your file will be plotted along with its corresponding <yvar> value. In this case, the use of XMIN and XMAX will be ignored. If the x-variable is numeric, a normal scatter-plot will be produced. In this case XMIN, XMAX, YMIN, and YMAX can be used to set limits on the axes of the plot, thus allowing you to 'zoom in' on parts of the graph by enlarging it. Note that the Y axis of the plot is the horizontal dimension of your screen and the X axis is the vertical dimension.

The SAVE command is inoperative for the PLOT command, since the data that would be saved are already in a file.

PPLOT command

This command plots the values of a numeric variable against the corresponding percentage points of a normal distribution. Its syntax is:

```
PPLOT <var> / XMIN=<#>,XMAX=<#>,
    YMIN=<#>,YMAX=<#>
```

The program rank orders the variable given and then computes cumulative percentages of the data falling below each rank. These percentages are converted to expected standard normal deviates (Expected Z-values). The Expected Z-values are plotted against the ordered data values. If the data are approximately normally distributed a linear plot should be evident, as should points that do not conform to this pattern.

Your variable is the vertical axis of the plot, and Expected Z is the horizontal axis. The SAVE command is inoperative for this plot.

An example of the command is:

```
PPLOT EFFECTS / YMIN=-2,YMAX=2
```
Produces a PPLOT with Z values between -2 and 2.

References for QCTOOLS Sample Data Files

Data in the file BOXES.SYS represent ohms of resistance measured for 100 computer boxes and are from Messina, W. S., *Statistical Quality Control for Manufacturing Managers* (New York: Wiley, 1987).

Data in the file JUICE.SYS represents number of defects found in 1200 orange juice cans and are from Montgomery, D. C., *Introduction to Statistical Quality Control,* 2nd ed. (New York: Wiley, 1991).

Answers to Selected Problems

CHAPTER 2

2.1 a. 0 0 1 1 1 1 2 2 2 2 3 4 4 5 8
19 20 22 24 24 27 30 36 55

b.

0	001111222234458
1	9
2	02447
3	06
4	
5	5

2.6 b. (1) 0, 4; (2) 16

2.11 a. 5.0   **b.** 5.0

2.14 a.

Erroneous Bits	Frequency
0– 9	15
10–19	1
20–29	5
30–39	2
40–49	0
50–59	1
Total	24

2.19 a. The adjusted percent frequencies are:

Number of Shares Held	Percent of Stockholders per 100 Shares Width
0–under 50	50.000
50–under 100	38.000
100–under 500	10.250
500–under 1000	2.400
1000–under 5000	0.075

b. 0–under 50

2.24 a.

Less Than This Time (minutes)	Cumulative Percent of Alarms
2.5	0
7.5	10
12.5	46
17.5	76
22.5	94
27.5	100

b. 85

2.28 a.

Outcome	Number	Percent
A	38	63.3
I	7	11.7
C	5	8.3
S	3	5.0
G	2	3.3
O	5	8.3
Total	60	100.

c. (1) A; (2) 31.8

2.37 a. Tour; cost and duration

b. (1) 246; (2) 77; (3) 73; (4) A tour of 4 days duration costing between $601 and $1200.

2.40 a. 71.7

b.

Duration (days)	Number of Workers
1–7	57
8–30	35
More than 30	33
Total	125

c.

Duration (days)	Age (years)	
	Under 35	35 or more
1–7	45.0	46.7
8–30	37.5	11.1
More than 30	17.5	42.2
Total	100.0% (80)	100.0% (45)

2.46

Quarter	Residual
1	−2.1
2	2.2
3	2.7
4	−2.8
5	−2.9
6	2.6

Quarter	Residual
7	3.5
8	−3.1
9	−1.7
10	2.2
11	2.0
12	−2.7

CHAPTER 3

3.1 731.75, 878.10

3.8 a. 687.67

3.11 a. 746

3.21 985

3.25 a. 2063, 696

3.29 362,720.9, 602.26

3.33 a. 82.3

3.36 b. 155,315,596.6, 0.711

3.40 2063, 2.210

3.44 a.

Minimum	1st Quartile	Median	3rd Quartile	Maximum
0	275.5	746.0	971.5	2063

3.48 a. (1) 13.7, (2) 13.2, (3) 9.6, (4) 5.33

CHAPTER 4

4.4 a. $E_1 = \{B, C\}$, $E_1^* = \{A, D\}$ **b.** $E_2 = \{A, B, C\}$

4.14 (1) $\{o_2, o_3, o_4, o_6, o_7, o_8\}$; (2) $\{o_5\}$; (3) $\{o_1, o_3, o_4, o_5\}$; (4) Empty set

4.21 a. 0.857

4.25 a. Univariate **b.** 0, 0.10 **c.** 0.20

4.28 a. (1) $P(A_1)$; (2) $(P(A_1 \cap B_1)$; (3) $P(B_1 \mid A_1)$; (4) $P(A_1 \mid B_1)$
c. a: (1) 0.2; (2) 0.2; (3) 1.0; (4) 0.67
b: (1) 0; (2) 0.125; (3) 0.7

4.29 a.

B_i	$P(B_i)$
B_1	0.3
B_2	0.7
Total	1.0

b.

B_i	$P(B_i \mid A_2)$
B_1	0.125
B_2	0.875
Total	1.000

4.35 (1) 0.3; (2) 0.1; (3) 0.2

4.38 a. 0.01 **b.** 0.9596

4.47 a. 0.04, 0.008 **b.** 0.20

4.50 b.

B_i	Conditional On	
	A_1	A_2
B_1	1.000	0.125
B_2	0.000	0.875
Total	1.000	1.000

CHAPTER 5

5.3 b. (1) 0.60; (2) 0.80; (3) 0.35

5.7 b.

x:	0	1	2	3	4
$P(X \le x)$:	0.60	0.80	0.90	0.95	1.00

$P(X \le 3) = 0.95$

5.13 (1) 0.30; (2) 0.07; (3) 0.03; (4) 0.405; (5) 0.167

5.17 a.

x	$P(x)$
0	0.57
1	0.36
2	0.07
Total	1.00

b.

x	$P(X = x \mid Y = 0)$
0	0.541
1	0.405
2	0.054
Total	1.000

5.20 0.75

5.24 1.2875, 1.1347

5.28 a.

y:	−0.661	0.220	1.102	1.983	2.864
$P(y)$:	0.60	0.20	0.10	0.05	0.05

5.31 a. (1) 0.159; (2) 0.500; (3) 0.841

CHAPTER 6

6.1 a. $P(A_1) = 0.2$, $P(A_2) = 0.8$, $P(B_1 \mid A_1) = 1.0$, $P(B_1 \mid A_2) = 0.125$, $P(A_1 \mid B_1) = 0.67$, $P(A_2 \mid B_1) = 0.33$

6.5 a. $Y = 9.50X$
b. $E\{X\} = 3.5$, $\sigma\{X\} = 1.1619$, $E\{Y\} = 33.25$, $\sigma\{Y\} = 11.04$

6.9 a. $E\{T\} = 1.6$, $\sigma\{T\} = 1.140$

b.

t:	0	1	2	3	4
$P(t)$:	0.20	0.27	0.32	0.15	0.06

6.10 a. $E\{D\} = 0.2$, $\sigma\{D\} = 1.140$

b.

d:	−2	−1	0	1	2
$P(d)$:	0.08	0.18	0.35	0.24	0.15

0.35, 0.39

6.14 a. 100,000, 300,000 **b.** 1581, 2739

6.17 a. 6.8, 35 **b.** -4.0 **c.** -2.0

6.20 a. -0.408

6.23 a. 1.10, 0, 1.62, 0.60 **b.** 1.10, 0, 1.22, 1.22

CHAPTER 7
7.1 b. (1) 0.0714; (2) 0.2143; (3) 0.3571 **c.** 6.5, 4.031

7.7 b. 0.0048

7.12 a. {0, 1, 2, 3, 4}
b. (1) 0.8145; (2) 0.0135; (3) 0.9995

7.17 a. 0.2, 0.19, 0.436

7.24 a. (1) 0.0302; (2) 0.1850; (3) 0.3209
b. 3.5, 3.5

7.31 a. {0, 1, 2, 3} **b.** (1) 0.300; (2) 0.667
c. 1.8, 0.748

CHAPTER 8
8.1 b. 275, 208.33 **c.** (1) 0.4; (2) 255

8.9 a. (1) 0.5000; (2) 0.8554; (3) 0.9938
b. (1) 0.5000; (2) 0.0062; (3) 0.8492
c. (1) 0.00; (2) 1.32; (3) -1.32
d. (1) 2.326; (2) -1.645

8.12 a. (1) 0.8413; (2) 0.0918; (3) 0.8904
b. 2901 **c.** 2699 to 2901

8.20 a. 0.005 **b.** 200
c. (1) 0.3935; (2) 0.1353; (3) 460.5

d.

x:	0	100	400	1000
$F(x)$:	0	0.3935	0.8647	0.9933

8.27 b. 0.333 **c.** 0.5276

CHAPTER 9
9.3 0.80

9.16 a. *AB AC AD BC BD CD* **b.** 1/6

9.35 a. 37.00, 6.949 **b.** 36.67, 9.292, -0.33

CHAPTER 10
10.1 a. (1) 3540; (3) 40,741.67, 37,201.67
b. 1200, 60,512.50

10.6 (1) 1.740, 0.010; (2) 1.740, 0.00424

10.17 a. (1) 0.1190; (2) 0.7620
b. 1.733 to 1.747, 1.735 to 1.745

10.23 0.4972

10.36 a.

λ:	0	0.5	1.0	1.5	2.0
$L(\lambda)$:	0	0.0139	0.0249	0.0187	0.0099

CHAPTER 11
11.1 a. 0.185 **b.** 4.840, 5.580, 0.954
c. $4.906 \leq \mu \leq 5.514$ **d.** approximately 0.90

11.9 a. $107.7 \leq \mu \leq 117.9$

11.13 \$9.81 million $\leq \tau \leq$ \$11.03 million

11.16 a. 152

11.20 a. $\mu \geq 4.906$

11.24 a. $-2.019 \leq X_{new} \leq 3.486$

CHAPTER 12
12.10 a. $H_0: \mu \leq 60, H_1: \mu > 60$. If $z^* \leq 2.326$, conclude H_0; if $z^* > 2.326$, conclude H_1. $z^* = 2.16$. Conclude H_0.

12.14 a. $H_0: \mu = 8.50, H_1: \mu \neq 8.50$. If $|z^*| \leq 1.960$, conclude H_0; if $|z^*| > 1.960$, conclude H_1. $z^* = 2.54$. Conclude H_1.

12.20 a. 0.0154
b. If P-value ≥ 0.01, conclude H_0; if P-value < 0.01, conclude H_1. Conclude H_0.

12.24 a. 0.0110, two-sided
b. If P-value ≥ 0.05, conclude H_0; if P-value < 0.05, conclude H_1. Conclude H_1.

12.27 a. $H_0: \mu = 110, H_1: \mu \neq 110$. If $|t^*| \leq 2.131$, conclude H_0; if $|t^*| > 2.131$, conclude H_1. $t^* = 1.167$. Conclude H_0.
b. $0.20 < P$-value < 0.50 (exact P-value $= 0.262$), two-sided.

12.32 $107.7 \leq \mu \leq 117.9$. Conclude H_0. $\alpha = 0.05$.

12.36 b.

μ:	2.15	2.45
$P(H_1; \mu)$:	0.239	0.879

d. H_0 is incorrect. $P(H_0; \mu = 2.45) = 0.121$. β risk.

12.42 $P(H_0; \mu = 70) = 0.367$

12.46 $P(H_0; \mu = 10.0) = 0.005$

12.49 a. 108
b. If $z^* \leq 1.645$, conclude H_0; if $z^* > 1.645$, conclude H_1. $z^* = 2.07$. Conclude H_1.

CHAPTER 13
13.4 a. Partial results follow:

f	$\bar{p}$	$P(f) = P(\bar{p})$
0	0	0.0000
1	1/9	0.0000
⋮	⋮	⋮
7	7/9	0.1722
8	8/9	0.3874
9	1	0.3874

c. (1) 0.3874; (2) 0.3874; (3) 0.9470
d. $E\{f\} = 8.1$, $\sigma\{f\} = 0.9$, $E\{\bar{p}\} = 0.9$, $\sigma\{\bar{p}\} = 0.1$

13.10 a. 0.40, 0.03098
b. (1) 0.0537; (2) 0.8926 **c.** 0.9988

13.17 a. 0.80, 0.02673 **b.** $0.748 \leq p \leq 0.852$

13.21 a. 609 **b.** 714

13.27 **a.** H_0: $p \geq 0.70$, H_1: $p < 0.70$. If $z^* \geq -2.326$, conclude H_0; if $z^* < -2.326$, conclude H_1. $z^* = -1.09$. Conclude H_0.
  c. 0.1379, one-sided

13.31 **a.** If $\bar{p} \geq 0.5934$, conclude H_0; if $\bar{p} < 0.5934$, conclude H_1.
  b. (1) 0.9693; (2) 0.4483; (3) 0.0099
  c. $P(H_1; p = 0.60) = 0.4483$

13.35 **a.** 493
  b. H_0: $p \leq 0.01$, H_1: $p > 0.01$. If $z^* \leq 1.645$, conclude H_0; if $z^* > 1.645$, conclude H_1. $z^* = 4.02$. Conclude H_1.

13.39 $p \geq 0.756$

13.42 $0.00067 \leq p \leq 0.368$

CHAPTER 14

14.5 **b.** 4290.50   **c.** $-14.04 \leq \mu_2 - \mu_1 \leq 54.04$

14.8 **b.** $33 \leq \mu_2 - \mu_1 \leq 123$

14.11 **a.** H_0: $\mu_2 - \mu_1 = 0$, H_1: $\mu_2 - \mu_1 \neq 0$. If $|t^*| \leq 1.684$, conclude H_0; if $|t^*| > 1.684$, conclude H_1. $t^* = 0.989$. Conclude H_0.
  b. $0.20 < P\text{-value} < 0.50$ (exact P-value = 0.328)

14.14 **a.** H_0: $\mu_2 - \mu_1 \leq 0$, H_1: $\mu_2 - \mu_1 > 0$. If $z^* \leq 2.576$, conclude H_0; if $z^* > 2.576$, conclude H_1. $z^* = 4.47$. Conclude H_1.
  b. 0.0000

14.17 **b.** $1.821 \leq \mu_D \leq 2.119$

14.20 **b.** $-0.265 \leq \mu_D \leq -0.095$

14.23 **a.** H_0: $\mu_D \leq 0$, H_1: $\mu_D > 0$. If $t^* \leq 1.895$, conclude H_0; if $t^* > 1.895$, conclude H_1. $t^* = 25.08$. Conclude H_1.
  b. $P\text{-value} < 0.0005$ (exact P-value = 0.0000)

14.26 **a.** H_0: $\mu_D \geq 0$, H_1: $\mu_D < 0$. If $z^* \geq -2.326$, conclude H_0; if $z^* < -2.326$, conclude H_1. $z^* = -4.95$. Conclude H_1.
  b. 0.0000

14.30 **a.** 0.011333, 0.021273
  b. $-0.0237 \leq p_2 - p_1 \leq 0.0463$

14.34 **a.** 0.44318
  b. H_0: $p_2 - p_1 = 0$, H_1: $p_2 - p_1 \neq 0$. If $|z^*| \leq 1.645$, conclude H_0; if $|z^*| > 1.645$, conclude H_1. $z^* = 0.53$. Conclude H_0.
  c. 0.5962

14.38 **a.** $0.801 \leq \sigma^2 \leq 1.945$
  b. $0.895 \leq \sigma \leq 1.395$

14.42 If $X^2 \leq 59.34$, conclude H_0; if $X^2 > 59.34$, conclude H_1. $X^2 = 43.204$. Conclude H_0.

14.46 **a.** $0.3528 \leq \sigma_1^2/\sigma_2^2 \leq 1.7853$
  b. $0.594 \leq \sigma_1/\sigma_2 \leq 1.336$

14.49 H_0: $\sigma_1^2 \geq \sigma_2^2$, H_1: $\sigma_1^2 < \sigma_2^2$. If $F^* \geq 0.4545$, conclude H_0; if $F^* < 0.4545$, conclude H_1. $F^* = 0.8115$. Conclude H_0.

CHAPTER 15

15.3 **a.** 2.83   **b.** $2.31 \leq \eta \leq 3.00$

15.8 **a.** $1192 \leq \eta \leq 2040$

15.11 **a.** H_0: $\eta \leq 1800$, H_1: $\eta > 1800$. H_0: $p \leq 0.5$, H_1: $p > 0.5$.
  b. If $z^* \leq 1.645$, conclude H_0; if $z^* > 1.645$, conclude H_1. $z^* = -0.65$. Conclude H_0.
  c. 0.7422

15.16 **a.**

Brochure Only		Brochure and Inspection
4	2	
9 6 5 3 1 0	3	
7 5 5 4 3	4	8
5 4 3	5	1 2 9
	6	0 3 4 5 7 9
	7	1 5

  b. H_0: $\eta_2 - \eta_1 \leq 0$, H_1: $\eta_2 - \eta_1 > 0$. If $z^* \leq 1.645$, conclude H_0; if $z^* > 1.645$, conclude H_1. $z^* = 3.95$. Conclude H_1.
  c. 0.0000, one-sided

15.22 **a.** $0 \leq \eta_D \leq 13$

15.25 H_0: $\eta_D \leq 5$, H_1: $\eta_D > 5$. If $z^* \leq 1.960$, conclude H_0; if $z^* > 1.960$, conclude H_1. $z^* = 1.04$. Conclude H_0.

15.28 **b.** H_0: $\eta_D \leq 0$, H_1: $\eta_D > 0$. If $z^* \leq 1.282$, conclude H_0; if $z^* > 1.282$, conclude H_1. $z^* = 2.00$. Conclude H_1.
  c. 0.0228

15.36 **a.** 20, 23
  b. H_0: Sequence generated by a random process, H_1: Sequence generated by a process containing either persistence or frequent changes in direction. If $|z^*| \leq 1.645$, conclude H_0; if $|z^*| > 1.645$, conclude H_1. $z^* = -1.24$. Conclude H_0.
  c. 0.2150

15.39 $4.21 \leq \sigma \leq 16.05$

CHAPTER 16

16.1 **a.**

1	6 8
2	1 3 6 6 7 9
3	0 1 3 6 8
4	0 1

  b. 14.9, 31.6

16.8 **a.** H_0: The numbers of accompanying persons are Poisson distributed, H_1: The numbers of accompanying persons are not Poisson distributed. If $X^2 \leq 4.61$, conclude H_0; if $X^2 > 4.61$, conclude H_1. $X^2 = 1.98$. Conclude H_0.

16.13 **a.** H_0: Weight losses of persons assigned to regimen 2 are normally distributed, H_1: Weight losses of persons assigned to regimen 2 are not normally distributed. If $X^2 \leq 6.25$, conclude H_0; if $X^2 > 6.25$, conclude H_1. $X^2 = 2.40$. Conclude H_0.

16.19 **a.** $D(0.90; 15) = 0.30$. Selected values for the confidence band are:

x:	16	18	21	23	26
$L(x)$:	0	0	0	0	0.10
$U(x)$:	0.37	0.43	0.50	0.57	0.70
x:	27	29	30	$\cdots$	41
$L(x)$:	0.17	0.23	0.30	$\cdots$	0.70
$U(x)$:	0.77	0.83	0.90	$\cdots$	1.00

b. $L(20) = 0$

16.22 $F_0(x) = (x - 15)/30 \qquad 15 \le x \le 45$

$H_0: F(x) = F_0(x)$ for all x, $H_1: F(x) \ne F_0(x)$ for some x. Selected values of $F_0(x)$ are:

x	$F_0(x)$
16	0.033
18	0.100
21	0.200
23	0.267
26	0.367
27	0.400
29	0.467
30	0.500
$\vdots$	$\vdots$
41	0.867

$F_0(x)$ lies entirely within the confidence band at all values of x. Conclude H_0.

CHAPTER 17

17.1 **b.** 0.1372 **c.** 0.3087

17.4 **a.** $0.344 \le p_1 \le 0.489$
b. $H_0: p_2 = 0.20$, $H_1: p_2 \ne 0.20$. If $|z^*| \le 2.576$, conclude H_0; if $|z^*| > 2.576$, conclude H_1. $z^* = 3.35$. Conclude H_1.

17.8 **a.** $H_0: p_1 = 0.35$, $p_2 = 0.45$, $p_3 = 0.20$, H_1: The p_i values are not those stated in H_0.
b. If $X^2 \le 9.21$, conclude H_0; if $X^2 > 9.21$, conclude H_1. $X^2 = 17.54$. Conclude H_1.

c.

i:	1	2	3
$(f_i - F_i)^2/F_i$:	2.286	9.000	6.250

17.12 **a.** H_0: Disease duration and degree of self-reliance are independent, H_1: Disease duration and degree of self-reliance are not independent. If $X^2 \le 11.34$, conclude H_0; if $X^2 > 11.34$, conclude H_1. $X^2 = 12.395$. Conclude H_1.

b.

	Conditional upon Disease Duration			
Self-reliance	Less than 5	5–9	10–14	15 or more
Considerable	0.900	0.781	0.679	0.500
Little	0.100	0.219	0.321	0.500
Total	1.000	1.000	1.000	1.000

17.16 **a.** H_0: Preference patterns in the two periods are identical, H_1: Preference patterns in the two periods are not identical. If $X^2 \le 4.61$, conclude H_0; if $X^2 > 4.61$, conclude H_1. $X^2 = 6.68$. Conclude H_1.
b. $0.051 \le p_{12} - p_{11} \le 0.222$

CHAPTER 18

18.5 (1) 150; (2) -4

18.12 **b.** $\hat{Y} = 72.776 - 1.7726X$

18.16 **b.** 39.10

18.20 **a.**

i:	1	2	3	4	5	6	7	8
e_i:	-0.551	-3.233	-1.006	2.449	2.222	-0.869	-0.324	1.312

18.24 **a.**

Source	SS	df	MS
Regression	134.717	1	134.717
Error	25.283	6	4.214
Total	160.000	7	

18.28 **b.** 0.8420, -0.918

CHAPTER 19

19.1 $30.03 \le E\{Y_h\} \le 33.98$

19.8 **a.** $26.61 \le Y_{h(\text{new})} \le 37.40$

19.12 $-2.540 \le \beta_1 \le -1.005$

19.16 **a.** $H_0: \beta_1 = 0$, $H_1: \beta_1 \ne 0$. If $|t^*| \le 2.447$, conclude H_0; if $|t^*| > 2.447$, conclude H_1. $t^* = -5.654$. Conclude H_1.
b. 0.0013

19.20 **a.** $55.90 \le \beta_0 \le 89.65$

19.24 **a.** $H_0: \beta_1 = 0$, $H_1: \beta_1 \ne 0$. If $F^* \le 5.99$, conclude H_0; if $F^* > 5.99$, conclude H_1. $F^* = 31.97$. Conclude H_1.
b. 0.0013

19.28 **a.**

i	e_i	$\hat{Y}_i$
1	-0.551	35.551
2	-3.233	30.233
3	-1.006	32.006
4	2.449	35.551
5	2.222	33.778
6	-0.869	40.869
7	-0.324	37.324
8	1.312	26.688

CHAPTER 20

20.4 **a.** X_1　　**b.** (1) -0.5874; (2) 0.4977; (3) 0.7945

20.7 **a.** $\hat{Y} = 162.876 - 1.2103X_1 - 0.66591X_2 - 8.6130X_3$
b. 61.31

20.10 **a.**

Source	SS	df	MS
Regression	4133.633	3	1377.878
Error	2011.584	19	105.873
Total	6145.217	22	

b. $0.6727, 0.820$　　**c.** $0.6210, 0.788$

20.13 **b.** $0.02539, -0.159$

20.19 **a.** $H_0: \beta_1 = \beta_2 = \beta_3 = 0$, H_1: Not all $\beta_k = 0$, $k = 1, 2, 3$. If $F^* \le 5.01$, conclude H_0; if $F^* > 5.01$, conclude H_1. $F^* = 13.014$. Conclude H_1.
c. 0.00007

20.22 **a.** $-2.384 \le \beta_2 \le 1.052$
b. $H_0: \beta_3 = 0$, $H_1: \beta_3 \ne 0$. If $|t^*| \le 2.093$, conclude H_0; if $|t^*| > 2.093$, conclude H_1. $t^* = -0.704$. Conclude H_0.

20.25 **a.** $50.99 \le E\{Y_h\} \le 71.63$
b. $37.43 \le Y_{h(\text{new})} \le 85.19$

20.29 **b.** $0.244 \le \beta_2 \le 0.432$

20.32 **b.** $\hat{Y} = 4.084 + 5.8478X - 0.10736X^2$
c. $H_0: \beta_2 = 0$, $H_1: \beta_2 \ne 0$. If $|t^*| \le 2.262$, conclude H_0; if $|t^*| > 2.262$, conclude H_1. $t^* = -2.803$. Conclude H_1.

20.35 **b.** $H_0: \beta_3 = 0$, $H_1: \beta_3 \ne 0$. If P-value ≥ 0.05, conclude H_0; if P-value < 0.05, conclude H_1. Conclude H_1.

CHAPTER 21

21.7 **a.** (1) 15; (2) 5; (3) 88; (4) 88.0; (5) 81.0

c.

Source	SS	df	MS
Treatments	930.0	2	465.0000
Error	460.0	12	38.3333
Total	1390.0	14	

21.11 **a.** $H_0: \mu_1 = \mu_2 = \mu_3$, H_1: Not all μ_j are equal, $j = 1, 2, 3$. $F(2, 12)$
b. If $F^* \le 3.89$, conclude H_0; if $F^* > 3.89$, conclude H_1. $F^* = 12.130$. Conclude H_1.
d. 0.0013

21.15 **a.** $78.97 \le \mu_1 \le 91.03$
b. $-5.53 \le \mu_2 - \mu_1 \le 11.53$

21.19 $-7.50 \le \mu_2 - \mu_1 \le 13.50$, $7.50 \le \mu_2 - \mu_3 \le 28.50$, $4.50 \le \mu_1 - \mu_3 \le 25.50$

21.23 **a.** Residuals e_{ij}:

i	$j = 1$	$j = 2$	$j = 3$
1	1	2	8
2	-3	-9	0
3	9	0	-5
4	-8	-1	4
5	1	8	-7

21.29 **b.**

i	X_{i1}	X_{i2}	Y_i
1	0	0	86
2	0	0	82
3	0	0	94
4	0	0	77
5	0	0	86
6	1	0	90
7	1	0	79
8	1	0	88
9	1	0	87
10	1	0	96
11	0	1	78
12	0	1	70
13	0	1	65
14	0	1	74
15	0	1	63

c. (1) $78.97 \le \beta_0 \le 91.03$; (2) $-5.53 \le \beta_1 \le 11.53$

CHAPTER 22

22.10 **a.** 0.3077　　**b.** 0.010 percent

22.14 **a.** $335, 314, 356$

22.18 **a.** $21, 7.087, 37.534$

22.21 **a.** $0.25, 0.05, 0.50$

b.

Day:	1	2	3	4	5	6	7
$\bar{p}$:	0.25	0.20	0.40	0.10	0.20	0.25	0.30

Day:	8	9	10	11	12	13	14	15
$\bar{p}$:	0.35	0.20	0.25	0.25	0.20	0.30	0.30	0.35

c. (1) 0.0072; (2) 0.0534

22.29 **a.** Reject shipment. Accept shipment.
b. If $\bar{p} \le 0.05$, conclude H_0 (accept shipment); if $\bar{p} > 0.05$, conclude H_1 (reject shipment).

22.32 **a.**

p:	0.02	0.10	0.20
$P(H_0; p)$:	0.9401	0.3918	0.0691

c. 0.0599, supplier's risk

22.38 **a.** (1) 0.7734; (2) 0.3085; (3) 0.0401
b. 0.9896

22.41 **a.** 0.9698　　**b.** 0.9508

CHAPTER 23

23.3 $196.1 \le \mu \le 203.9$

23.6 **a.** 611 **b.** 1245

23.9 $0.455 \le p \le 0.645$

23.12 **a.** 546 **b.** 565

23.17 $452.05 \le \mu \le 486.52$

CHAPTER 24

24.1 **a.** 4.27, 4.03, 3.80

24.4 3.42, 3.43, 3.36

24.7 **a.** $\hat{Y}_t = 3.2900 + 0.051429X_t$, $X_t = 1$ at 1972, X in one-year units
c. 0.051429 or 51,429 metric tons
d. 4.473, 5.144

24.10 **a.** $\hat{Y}_t' = 3.30527 + 0.0156986X_t$, $X_t = 1$ at 1977, X in one-year units
c. 3.68 percent **d.** 4809, 29,307

24.16 **a.** Partial results follow:

Year	(1) $C \cdot I$	(2) C	(3) I
1972	98.8		
1973	109.1		
1974	113.2	100.6	112.6
1975	100.1	98.1	102.1
.	.	.	.
1988	108.1	101.5	106.4
1989	94.9	96.0	98.9
1990	84.4		
1991	88.0		

24.19 **a.** Partial results follow:

Year	(1) $C \cdot I$	(2) C	(3) I
1977	99.8		
1978	98.5	100.4	98.1
1979	102.9	100.1	102.9
.	.	.	.
1989	99.6	101.3	98.3
1990	100.5	99.6	100.8
1991	98.9		

24.26 **b.** Partial results follow:

Year	1986	
Quarter:	3	4
$S \cdot I$:	133.405	96.156

Year:	1987	$\cdots$	1991	
Quarter:	1	2 $\cdots$	1	2
$S \cdot I$:	73.598	95.666 $\cdots$	73.633	103.323

c.

Quarter:	1	2	3	4
Seasonal Index:	73.081	101.078	132.404	93.437

24.33 **a.**

Quarter	Deseasonalized Revenues
1	232.6
2	238.4
3	231.9
4	232.2

b. 929

c.

Quarter:	1	2	3	4
I:	105.29	100.76	96.46	99.87

24.37 **a.**

Quarter:	1	2	3	4
Forecast:	170	235	308	217

b. 230, 920

24.40 **a.** Partial results follow:

Year:	1986	
Quarter:	3	4
(1) $T + C$:	40.875	41.000
(2) $S + I$:	−0.875	11.000

Year:	1987		$\cdots$	1991	
Quarter:	1	2	$\cdots$	1	2
(1) $T + C$:	41.250	41.625	$\cdots$	46.750	47.500
(2) $S + I$:	−4.250	−5.625	$\cdots$	−3.750	−5.500

b.

Quarter:	1	2	3	4
Seasonal Deviation:	−4.344	−5.594	−0.969	10.906

c.

Quarter:	1	2	3	4
Deseasonalized Sales:	47.3	47.6	51.0	45.1

Seasonally adjusted annual rate is 180.4.

CHAPTER 25

25.1 **a.** 305.6, −13.6 **b.** (1) 131.29; (2) 0.211
c. (1) 87.44; (2) −0.10536 **d.** 232.14 **e.** 0.421

25.2 **a.** $215.6 \le Y_{h(\text{new})} \le 248.7$
b. $H_0: \beta_4 = 0$, $H_1: \beta_4 \ne 0$. If $|t^*| \le 2.093$, conclude H_0; if $|t^*| \ge 2.093$, conclude H_1. $t^* = 9.10$. Conclude H_1.

25.9 **b.** $H_0: \rho \le 0$, $H_1: \rho > 0$. If $d > 1.53$, conclude H_0; if $d < 0.80$, conclude H_1; if $0.80 \le d \le 1.53$, the test is inconclusive. Conclude H_0.

25.13 **b.** $\hat{Y}_t' = 20.88X_t'$

c.

t:	2	3	4	5	6
$Y_t' - \hat{Y}_t'$:	19.6	−27.4	−8.8	57.4	7.2

t:	7	8	9	10	11
$Y_t' - \hat{Y}_t'$:	95.6	0.0	4.4	−11.4	−11.4

d. 1154

25.14 a. 1553.07, 9 **b.** $17.15 \leq \beta_1 \leq 24.61$

25.19 a. 41.3
   b. (1) 42.6; (2) 42.6; (3) 42.6

25.25 a. 4.6359, −0.0704 **b.** 4.354

25.30

a.

Quarter	A_t	B_t	C_t
3	232.318	0.1002	72.936
4	232.287	0.0871	−14.337

b.

Year:	1992				1993
Quarter:	1	2	3	4	1
Forecast:	174.07	233.76	305.48	218.30	174.42

c. (1) −0.03, (2) 231.1

CHAPTER 26
26.1 a.

Year:	1987	1988	1989	1990	1991
(1) Monarch:	100.0	104.2	116.7	125.0	137.5
Emperor:	100.0	106.3	112.5	125.0	134.4
(2) Monarch:		104.2	112.0	107.1	110.0
Emperor:		106.3	105.9	111.1	107.5

b. 8.3, 7.1 **c.** Monarch, Monarch

26.4 a., b.

Year:	1989	1990	1991
Aggregate Cost:	12,233.80	11,915.50	11,705.00
Index:	100.0	97.4	95.7

26.7 a.

Year:	1989	1990	1991
Index:	100.0	97.4	95.7

26.20 a., b.

Year:	1985	1986	1987
Earnings (in 1985 dollars):	464.46	459.40	456.69
(1) Percent relative:	100.0	98.9	98.3
(2) Link relative:		98.9	99.4

Year:	1988	1989
Earnings (in 1985 dollars):	453.14	446.74
(1) Percent relative:	97.6	96.2
(2) Link relative:	99.2	98.6

c.

	Construction Workers	All Nonagricultural Workers
(1)	−3.8	−2.3
(2)	−1.4	−0.8

CHAPTER 27
27.3

		A_j		
S_i	$P(S_i)$	0	1	2
0	0.4	0	−40	−80
1	0.5	−300	−240	−280
2	0.1	−600	−540	−480

27.11 a. A_1, A_2, A_3, A_4
   b. A_1, 200
   c.

S_i	A_1	A_2	A_3	A_4
S_1	300	200	100	0
S_2	100	0	50	100
S_3	0	50	100	150

A_3 is the minimax regret act.

27.14 a. A_2, 255

27.17 a. 310, 55

27.21 b. Optimal strategy is to undertake research and development (A_1) and, if the outcome is failure, to negotiate a merger arrangement. 80.6

27.24 a. Risk-seeking
   b. (1) $EU = 0.50$ (2) $EU = 0.25$. Prefers (1).

27.28 a. A_2
   b.

Act:	A_1	A_2	A_3	A_4	A_5
Certainty Equivalent:	200	250	222	187	107

CHAPTER 28
28.1 a. A_2, 130, 0.5 **b.** 380, 250

28.4 a. (1) $P(A_1 \mid S_1) = 0.9639$, $P(A_2 \mid S_1) = 0.0361$;
   (2) −18.05
   b. 137.1
   c.

$(15, K)$:	(15, 2)	(15, 15)	(15, −1)
$EP(15, K)$:	137.1	0.0	130.0

   d. 211.6 **e.** 0.0855

28.7 a. $s(n) = \begin{cases} 8 + 0.05n & n > 0 \\ 0 & n = 0 \end{cases}$
   b. 4840

28.10 a.

n:	0	100	200	300
$BNEP(n)$:	130.0	334.9	351.7	351.7
$EVSI(n)$:	0.0	217.9	239.7	244.7
$ENGS(n)$:	0.0	204.9	221.7	221.7

$n = 200$ or $n = 300$.

28.13 a.

S_i:	S_1	S_2	S_3
$P(S_i \mid f = 2)$:	0.3235	0.1281	0.5484

b. A_2, 357.4   **d.** 0.20833

APPENDIX B

B.1 **a.** 3.49, 13.36   **b.** 8, 4

B.3 **a.** 5.99, 3.49   **b.** 131.5

B.5 **a.** -1.796, 1.796   **b.** 0, 1.106

B.7 **a.** 1.476, -1.721 **b.** -2.326

B.11 **a.** 0.0972, 6.63   **b.** 1.333, 1.398

B.13 **a.** 3.49   **b.** 0.0364

Index

TABLE C.2 Percentiles of the chi-square distribution

Entry is $\chi^2(a;\nu)$ where $P[\chi^2(\nu) \le \chi^2(a;\nu)] = a$.

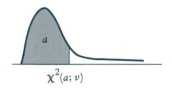

$$\chi^2(a;\nu)$$

df ν	.005	.010	.025	.050	.100	.900	.950	.975	.990	.995
1	0.0⁴393	0.0³157	0.0³982	0.0²393	0.0158	2.71	3.84	5.02	6.63	7.88
2	0.0100	0.0201	0.0506	0.103	0.211	4.61	5.99	7.38	9.21	10.60
3	0.072	0.115	0.216	0.352	0.584	6.25	7.81	9.35	11.34	12.84
4	0.207	0.297	0.484	0.711	1.064	7.78	9.49	11.14	13.28	14.86
5	0.412	0.554	0.831	1.145	1.61	9.24	11.07	12.83	15.09	16.75
6	0.676	0.872	1.24	1.64	2.20	10.64	12.59	14.45	16.81	18.55
7	0.989	1.24	1.69	2.17	2.83	12.02	14.07	16.01	18.48	20.28
8	1.34	1.65	2.18	2.73	3.49	13.36	15.51	17.53	20.09	21.96
9	1.73	2.09	2.70	3.33	4.17	14.68	16.92	19.02	21.67	23.59
10	2.16	2.56	3.25	3.94	4.87	15.99	18.31	20.48	23.21	25.19
11	2.60	3.05	3.82	4.57	5.58	17.28	19.68	21.92	24.73	26.76
12	3.07	3.57	4.40	5.23	6.30	18.55	21.03	23.34	26.22	28.30
13	3.57	4.11	5.01	5.89	7.04	19.81	22.36	24.74	27.69	29.82
14	4.07	4.66	5.63	6.57	7.79	21.06	23.68	26.12	29.14	31.32
15	4.60	5.23	6.26	7.26	8.55	22.31	25.00	27.49	30.58	32.80
16	5.14	5.81	6.91	7.96	9.31	23.54	26.30	28.85	32.00	34.27
17	5.70	6.41	7.56	8.67	10.09	24.77	27.59	30.19	33.41	35.72
18	6.26	7.01	8.23	9.39	10.86	25.99	28.87	31.53	34.81	37.16
19	6.84	7.63	8.91	10.12	11.65	27.20	30.14	32.85	36.19	38.58
20	7.43	8.26	9.59	10.85	12.44	28.41	31.41	34.17	37.57	40.00
21	8.03	8.90	10.28	11.59	13.24	29.62	32.67	35.48	38.93	41.40
22	8.64	9.54	10.98	12.34	14.04	30.81	33.92	36.78	40.29	42.80
23	9.26	10.20	11.69	13.09	14.85	32.01	35.17	38.08	41.64	44.18
24	9.89	10.86	12.40	13.85	15.66	33.20	36.42	39.36	42.98	45.56
25	10.52	11.52	13.12	14.61	16.47	34.38	37.65	40.65	44.31	46.93
26	11.16	12.20	13.84	15.38	17.29	35.56	38.89	41.92	45.64	48.29
27	11.81	12.88	14.57	16.15	18.11	36.74	40.11	43.19	46.96	49.64
28	12.46	13.56	15.31	16.93	18.94	37.92	41.34	44.46	48.28	50.99
29	13.12	14.26	16.05	17.71	19.77	39.09	42.56	45.72	49.59	52.34
30	13.79	14.95	16.79	18.49	20.60	40.26	43.77	46.98	50.89	53.67
40	20.71	22.16	24.43	26.51	29.05	51.81	55.76	59.34	63.69	66.77
50	27.99	29.71	32.36	34.76	37.69	63.17	67.50	71.42	76.15	79.49
60	35.53	37.48	40.48	43.19	46.46	74.40	79.08	83.30	88.38	91.95
70	43.28	45.44	48.76	51.74	55.33	85.53	90.53	95.02	100.4	104.2
80	51.17	53.54	57.15	60.39	64.28	96.58	101.9	106.6	112.3	116.3
90	59.20	61.75	65.65	69.13	73.29	107.6	113.1	118.1	124.1	128.3
100	67.33	70.06	74.22	77.93	82.36	118.5	124.3	129.6	135.8	140.2

SOURCE: Tabulated values adapted by permission from C. M. Thompson, "Table of Percentage Points of the Chi-Square Distribution," *Biometrika*, Vol. 32 (1941), pp. 188–189.
EXAMPLE: $\chi^2(0.900;4) = 7.78$ so $P[\chi^2(4) \le 7.78] = 0.900$.
TEXT REFERENCE: Use of this table is discussed on p. 911.

TABLE C.3 Percentiles of the *t* distribution

Entry is $t(a;\nu)$ where $P[t(\nu) \le t(a;\nu)] = a$.

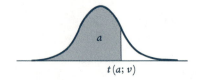

$t(a;\nu)$

df ν	.75	.90	.95	.975	.99	.995	.9995
1	1.000	3.078	6.314	12.706	31.821	63.657	636.619
2	0.816	1.886	2.920	4.303	6.965	9.925	31.599
3	0.765	1.638	2.353	3.182	4.541	5.841	12.924
4	0.741	1.533	2.132	2.776	3.747	4.604	8.610
5	0.727	1.476	2.015	2.571	3.365	4.032	6.869
6	0.718	1.440	1.943	2.447	3.143	3.707	5.959
7	0.711	1.415	1.895	2.365	2.998	3.499	5.408
8	0.706	1.397	1.860	2.306	2.896	3.355	5.041
9	0.703	1.383	1.833	2.262	2.821	3.250	4.781
10	0.700	1.372	1.812	2.228	2.764	3.169	4.587
11	0.697	1.363	1.796	2.201	2.718	3.106	4.437
12	0.695	1.356	1.782	2.179	2.681	3.055	4.318
13	0.694	1.350	1.771	2.160	2.650	3.012	4.221
14	0.692	1.345	1.761	2.145	2.624	2.977	4.140
15	0.691	1.341	1.753	2.131	2.602	2.947	4.073
16	0.690	1.337	1.746	2.120	2.583	2.921	4.015
17	0.689	1.333	1.740	2.110	2.567	2.898	3.965
18	0.688	1.330	1.734	2.101	2.552	2.878	3.922
19	0.688	1.328	1.729	2.093	2.539	2.861	3.883
20	0.687	1.325	1.725	2.086	2.528	2.845	3.850
21	0.686	1.323	1.721	2.080	2.518	2.831	3.819
22	0.686	1.321	1.717	2.074	2.508	2.819	3.792
23	0.685	1.319	1.714	2.069	2.500	2.807	3.768
24	0.685	1.318	1.711	2.064	2.492	2.797	3.745
25	0.684	1.316	1.708	2.060	2.485	2.787	3.725
26	0.684	1.315	1.706	2.056	2.479	2.779	3.707
27	0.684	1.314	1.703	2.052	2.473	2.771	3.690
28	0.683	1.313	1.701	2.048	2.467	2.763	3.674
29	0.683	1.311	1.699	2.045	2.462	2.756	3.659
30	0.683	1.310	1.697	2.042	2.457	2.750	3.646
40	0.681	1.303	1.684	2.021	2.423	2.704	3.551
60	0.679	1.296	1.671	2.000	2.390	2.660	3.460
120	0.677	1.289	1.658	1.980	2.358	2.617	3.373
∞	0.674	1.282	1.645	1.960	2.326	2.576	3.291

EXAMPLE: $t(0.95;10) = 1.812$ so $P[t(10) \le 1.812] = 0.95$.
TEXT REFERENCE: Use of this table is discussed on p. 913.